CONVERSION FACTORS

Length

1 in. = 2.54 cm
1 ft = 0.3048 m
1 mi = 5280 ft = 1.609 km
1 m = 3.281 ft
1 km = 0.6214 mi
1 angstrom ($\mathring{A}$) = 10^{-10} m

Mass

1 slug = 14.59 kg
1 kg = 1000 grams = 6.852×10^{-2} slug
1 atomic mass unit (u) = 1.6605×10^{-27} kg
(1 kg has a weight of 2.205 lb where the
   acceleration due to gravity is 32.174 ft/s^2)

Time

1 day = 24 h = 1.44×10^3 min = 8.64×10^4 s
1 yr = 365.24 days = 3.156×10^7 s

Speed

1 mi/h = 1.609 km/h = 1.467 ft/s = 0.4470 m/s
1 km/h = 0.6214 mi/h = 0.2778 m/s = 0.9113 ft/s

Force

1 lb = 4.448 N
1 N = 10^5 dynes = 0.2248 lb

Work and Energy

1 J = 0.7376 ft·lb = 10^7 ergs
1 kcal = 4186 J
1 Btu = 1055 J
1 kWh = 3.600×10^6 J
1 eV = 1.602×10^{-19} J

Power

1 hp = 550 ft·lb/s = 745.7 W
1 W = 0.7376 ft·lb/s

Pressure

1 Pa = 1 N/m^2 = 1.450×10^{-4} lb/in.^{2}
1 lb/in.^{2} = 6.895×10^3 Pa
1 atm = 1.013×10^5 Pa = 1.013 bar =
   14.70 lb/in.^{2} = 760 torr

Volume

1 liter = 10^{-3} m^3 = 1000 cm^3 = 0.03531 ft^3
1 ft^3 = 0.02832 m^3 = 7.481 U.S. gallons
1 U.S. gallon = 3.785×10^{-3} m^3 = 0.1337 ft^3

Angle

1 radian = 57.30°
1° = 0.01745 radian

STANDARD PREFIXES USED TO DENOTE MULTIPLES OF TEN

Prefix	Symbol	Factor	Prefix	Symbol	Factor
Tera	T	10^{12}	Centi	c	10^{-2}
Giga	G	10^9	Milli	m	10^{-3}
Mega	M	10^6	Micro	μ	10^{-6}
Kilo	k	10^3	Nano	n	10^{-9}
Hecto	h	10^2	Pico	p	10^{-12}
Deka	da	10^1	Femto	f	10^{-15}
Deci	d	10^{-1}			

P H Y S I C S

PHYSICS

JOHN D. CUTNELL/KENNETH W. JOHNSON

Southern Illinois University at Carbondale

JOHN WILEY & SONS
New York • Chichester • Brisbane • Toronto • Singapore

Library of Congress Cataloging in Publication Data:

Cutnell, John D.
  Physics.

  Bibliography: p.
  Includes index.
  1. Physics.   I. Johnson, Kenneth W.   II. Title.

QC23.C985   1989   530   88-27802
ISBN 0-471-89850-3

Printed in the United States of America

10 9 8 7 6 5 4 3

For my wife Joan Cutnell, whose patience, encouragement, and help have been beyond measure and generously given.

For my wife Anne Johnson and children, Lauri, Rick, and Rob, who have generously offered help and understanding in this special venture.

PREFACE

If you are a student, we welcome you to the world of physics and hope that this text will help you learn from and enjoy your studies. If you are a colleague, we greet you and hope that together we can help our students learn physics and even absorb some of our own enthusiasm for the topic.

It is easy for physicists to be enthusiastic about physics. After all, it is our profession. We are familiar with it, and we enjoy it. In contrast, students often wonder why they are taking a physics course. The best reason to study physics is that is provides a coherent and logical approach to understanding the world around us. And a person who understands his or her environment is able to deal with it rationally and effectively. Students, however, are often unaware of the ability of physics to explain the environment in terms that they can understand, and in our teaching we find that this fact comes into play constantly. Therefore, we decided to write this text to help convince students that (1) physics is a unified body of knowledge that can speak directly to them about the technological world in which they live, and (2) they can understand what physics has to say.

We have incorporated several features into the text that are intended to convey the power of physics as a window on the world. For one thing, there are a large number of explanations of how physics principles play a role in the operation of various devices and techniques. These devices and techniques have been chosen from a wide variety of areas and include medical applications (e.g., cavitron ultrasonic surgical aspirator and electroretinography), automobile features (e.g., inertial seat belt mechanism), home entertainment (e.g., compact disc player), home appliances (e.g., electric stove), information processing (e.g., laser printer), detection devices (e.g., metal detector and sonar), camera technology (e.g., autofocusing mechanism), satellite technology (e.g., geosynchronous satellite), and many more.

We have also included a number of discussions and examples that focus on human physiology. Among these, for instance, are muscle forces, blood pressure and blood flow, breathing, the detection of sound by the ear, the refraction of light by the eye, and the physiological effects of radioactivity. Topics such as these have been selected because of the straightforward connection they have to physics principles.

The text contains more than 268 calculational examples that are used to illustrate important material and are oriented toward realistic and familiar situations, as much as possible. We work these examples out thoroughly, so that they can serve as models for the students' own work. In this regard, we believe that students benefit from seeing calculations done in more than one fashion, so a number of multiple-method calculations are included, as in Examples 5 and 7 of Chapter 2.

The emphasis on physics as a window on the world is carried over into the 2160 problems and questions provided for homework assignment. The problems deal with the material quantitatively, whereas the questions deal with it qualitatively. Out of the total, 469 are questions. Wherever possible, the problems and questions are related to real-life situations. Moreover, we have taken special care to include problems and questions that combine current chapter material with previous chapter material. Such problems and questions emphasize the important point that various areas of physics usually come into play simultaneously in the real world. In addition, this multiple-topic approach provides a way for the students to review what they have studied and helps to improve problem-solving skills.

Homework questions (qualitative) are a traditional but underutilized part of physics texts. In hopes of stimulating the use of such questions, we have provided a collection that is distinctively different from the typical one. We have avoided the standard type of question that, in effect, asks the student to summarize text material. Instead, the student is asked to use the material in the text to interpret various situations. Many of the situations deal with familiar real-life experiences. Others relate explicitly to the material discussed in the text, but in a way that makes the student extend the reasoning, rather than merely repeat it. We hope that teachers and students will find the questions interesting and enjoy discussing the issues that they raise.

We believe students and teachers alike will find that the design of the book helps to make the learning process an efficient one. An important part of the design is the artwork, and to optimize its effectiveness, we have tried to focus only on one main idea per picture whenever possible. Therefore, the figures in the text often are divided into two or more parts, to avoid the confusion of superimposing several ideas on the same drawing. We have also tried to use the second color with judicious restraint, reserving it to call attention to the important features of a drawing.

Learning from a textbook is like a guided tour, in a way. One purpose of a guide is to help the people on the tour avoid dangerous mistakes, and textbook authors have this same responsibility. Thus, we have tried to anticipate the common mistakes that students make. Consequently, the book is liberally sprinkled with explanations and cautionary notes to clarify the meanings of difficult concepts and the conditions under which the concepts can be applied. For additional reinforcement we also include these conditions along with numbered equations as, for example, in the equation $v_T = r\omega$. This equation relates the angular speed ω to the tangential speed v_T and can be applied only if angles are measured in radians, and not in degrees.

Since reviewing is an essential step in the learning process, we have provided a detailed summary of the material at the end of each chapter. These summaries are condensed but thorough expositions of the material presented in the chapters and are not just compilations of equations. Equations are included, however, unless prohibited by complexity, in which case a reference to the text equation is given.

Reviewing is not something to be done only once, after a chapter is read. Reviewing should be an ongoing process that is done often. To encourage frequent reviewing, we have inserted references to the location of earlier material when we believe it is pertinent.

One of the tasks that face students as they read an introductory physics book is distinguishing between important basic concepts and other related, but less fundamental ideas. To identify the important basic concepts, we have enclosed them within a "box" formed by a prominant colored band. Since applying these concepts entails using correct units, we have included the appropriate SI units within the "box." The boxes are used sparingly so that they can serve effectively as a guide to the truly basic concepts.

Both learning and teaching are easier when the material is arranged in a logical order. Often, however, logic is in the eye of the beholder, and there is more than one sequence in which the material can be organized. Therefore, we do not claim that our ordering of the material is unique, only that we have tried to achieve maximum clarity.

Consistent with clarity, we have chosen section and subsection headings in a way that will lend itself to judicious editing. Material that is a likely candidate for omission is typically located in a subsection at the end of a main section or in a separate section near the end of a chapter. Sections marked with an asterisk can be omitted with little or no impact on the overall development of the material.

Several additional design elements of the book deserve special mention. These elements are summarized below.

Solved Problems. These are an innovative type of illustrative calculation that we have included at the end of certain chapters, between the chapter summary and the homework material. They differ from the standard numbered examples that occur within the text material of the chapter. The Solved Problems deal with applications that are more difficult and elaborate than those treated in the numbered examples. Furthermore, at the end of each Solved Problem there is a summary of the important points that have been illustrated.

The Solved Problems are intended for use in conjunction with the homework problems. At the beginning of each Solved Problem there is a convenient reference list that identifies three to five homework problems dealing with the same general issues as the Solved Problem. These associated homework problems have a high level of difficulty, and *we have avoided a simple repetition of the Solved Problem with only the data changed.* Each of the associated homework problems includes a phrase such as "See Solved Problem 2 for a related problem." Thus, once the basic material of the chapter has been discussed, teachers can focus on the applications in the solved problems by assigning the associated homework problems. Teachers wishing to skip these applications can do so without affecting the use of the text.

It is our experience that students need the most help with problem-solving techniques in the early part of their exposure to physics. The Solved Problems allow teachers to assign more challenging homework problems early in the text in a way that provides students with an added measure of help. As students gain experience, however, we believe that the need for the Solved Problems diminishes. Therefore, the number of Solved Problems is gradually reduced as the text develops.

Homework Problems. The homework problems are ranked according to difficulty. The most difficult problems are marked with a double asterisk (**), while problems of intermediate difficulty are marked with a single asterisk (*). The easiest problems are unmarked. Some of the problems are organized by section, and some are grouped without reference to any particular section under the heading "Additional Problems."

Free-Body Diagrams. Teachers are familiar with the importance of free-body diagrams when using Newton's second law of motion. And all students will learn about them as they study physics. We use free-body diagrams throughout this text, not just in the early chapters where Newton's second law is introduced and applied. For instance, in Chapter 12, when the relation between pressure and depth in a fluid is developed, a free-body diagram clarifies the discussion considerably. Similarly, in Chapter 20, when the expression for the speed of a transverse wave on a string is derived, a free-body diagram helps enormously.

Significant Figures. Standard procedures for significant figures are followed throughout this text. They are not just introduced at the beginning of the book and then ignored. A review of these procedures is given in Appendix B.

We have tried to produce an error-free book, but no doubt some errors still remain, all of which are solely our own responsibility. Please feel free to let us know of any errors that you find.

We hope that our efforts in this text are useful to both students and teachers and look forward to hearing from you concerning your experiences with it. We also hope that our efforts will make your lives a little easier and your work more enjoyable.

JOHN D. CUTNELL
KENNETH W. JOHNSON

Carbondale, Illinois
1988

ACKNOWLEDGMENTS

The preparation of a text such as this involves many people, directly and indirectly, and we would like to single out those to whom we are especially indebted. Foremost among them are our students. They have contributed much to this book through their influence on the development of our teaching philosophies and techniques. Without our students, we and this text would be much diminished.

We are especially grateful to George Black and Cathy Fahey, science librarians at Southern Illinois University, Carbondale. Their extraordinarily competent and enthusiastic help enabled us to overcome many obstacles.

For the first three quarters of this project we were fortunate to have Robert A. McConnin for our editor. He is an author's editor and a superb one. His patient guidance and encouragement and his expertise are appreciated as much now as they were at the time.

To Ed and Lorraine Burke of Hudson River Studio we extend our gratitude for their monumental efforts on our behalf in the production stages of the text. They are true masters of their craft and two of the finest professionals we have ever met.

At John Wiley & Sons, Inc., we extend our thanks to Catherine Faduska, our present editor, for her tireless efforts to bring this project to fruition. We are grateful to Maddy Lesure for the design of the cover and layout of the book. We also appreciate the help of John Balbalis in the development of our illustration program.

Many typists helped prepare the manuscript, but two deserve special thanks, Kat Chamberlin and Ann Dreyer. Thanks are also owed to James Bangs, who worked many of the homework problems. And to the many colleagues, friends, and acquaintances whom we have consulted, pestered, and brain-drained, we extend our appreciation.

One of us (J.D.C.) wishes to acknowledge Professor E. O. Stejskal (North Carolina State University) for support during a sabbatical year in his laboratory. The time-consuming production stages of the text occurred during that year, and his patience, cooperation, and advice will always be remembered and appreciated.

It has been a pleasure to work with the physicists who reviewed the manuscript and offered so many helpful suggestions for improving our writing style and removing ambiguities and inaccuracies. These individuals are an invaluable resource to the physics community, and we applaud them. They do what they do out of dedication to and love for their profession. Our thanks go to each of them:

Joseph Alward (University of the Pacific), Chi Kwan Au (University of South Carolina), William A. Barker (Santa Clara University), Edward E. Beasley (Gallaudet University), Robert Brehme (Wake Forest University), Michael E. Browne (University of Idaho), William S. Chow (University of Cincinnati), Albert C. Claus (Loyola University of Chicago), Lawrence Coleman (University of California at Davis), Henry L. Cote (Catonsville Community College), James E. Dixon (Iowa State University), Miles J. Dresser (Washington State University), Dewey Dykstra (Boise State University), Robert J. Friauf (University of Kansas), C. Sherman Frye, Jr. (Northern Virginia Community College), John Gagliardi (Rutgers University), Simon George (California State University at Long Beach), John Gieniec (Central Missouri State University), D. Wayne Green (Knox College), Lawrence A. Hitchingham (Jackson Community College), Paul R. Holody (Henry Ford Community College), David A. Jerde (St. Cloud State University), R. Lee Kernell (Old Dominion University), Gary Kessler (Illinois Wesleyan University), I. K. Kothari (Tuskegee Institute), Robert A. Kromhout (Florida State University), Theodore Kruse (Rutgers University), Rubin H. Landau (Oregon State University), Christopher P. Landee (Clark University), R. Wayne Major (University of Richmond), Kenneth Mucker (Bowling Green State University), David Newton (DeAnza College), R. Chris Olsen (University of Alabama), Harold Romero (University of Southern Mississippi), Larry Rowan (University of North Carolina), Michael Swift (Catonsville Community College), Howard G. Voss (Arizona State University), James M. Wallace (Jackson Community College), Walter G. Wesley (Moorhead State University), and Jerry H. Wilson (Metropolitan State College)

Professor Mario Iona (Emeritus, University of Denver) provided a line-by-line review of the manuscript that exceeded all of our expectations. It was a remarkable tour de force of careful attention to detail and unflagging insistance on correct physics. We came to respect his work greatly, and it had a marked influence on the manuscript.

J.D.C.
K.W.J.

SUPPLEMENTS

PHYSICS is accompanied by a complete supplementary package.

For the Student

STUDY GUIDE with Selected Solutions

Charles R. McKenzie, *Salisbury State College*
Andrew J. Pica, *Salisbury State College*
John D. Cutnell, *Southern Illinois University at Carbondale*
Kenneth W. Johnson, *Southern Illinois University at Carbondale*

In addition to selected solutions to the textbook problems; this study guide provides chapter objectives, key ideas and vocabulary, additional example problems and trial problems, chapter quizzes, and sample examinations. If this supplement is not in stock, you may ask the bookstore manager to order a copy for you.

WONDERING ABOUT PHYSICS . . . Using Spreadsheets to Find Out

Dewey I. Dykstra, Jr., *Boise State University*
Robert G. Fuller, *University of Nebraska at Lincoln*

Composed of 53 investigations, this specially developed supplement leads students to explore the real world of physical phenomena by using spreadsheet software on a personal computer.

For the Instructor

A complete supplementary package of teaching and learning materials is available for instructors. It includes a complete solutions manual, test bank (bound and for the IBM/MacIntosh personal computers), instructor's manual, overhead transparency acetates, and more. Contact your local Wiley representative for further information.

CONTENTS

PHYSICS

CHAPTER 1

Introduction and Mathematical Concepts

Mathematics is used in many fields, even art. This picture was produced by using a mathematical technique called fractal imagery.

1.1 THE NATURE OF PHYSICS

Physics is a science that has developed out of our efforts to describe how and why our physical environment behaves as it does. These efforts have been so successful that the physics of today encompasses a remarkably diverse set of phenomena. The planets orbiting the sun, a jetliner streaking through the sky, transistors working in stereo and computer systems, lasers being used in eye surgery — physics relates to all of these and more.

The laws of physics are also remarkable for their scope. They describe the behavior of particles many times smaller than an atom and objects many times larger than our sun. The same laws apply to the heat generated by a burning match and the heat generated by a rocket engine. The same laws allow an astronomer to use the light from a distant star to determine how fast the star is moving and a police officer to use radar to catch a speeder. Physics can be applied fruitfully to objects as diverse as subatomic particles, distant stars, or speeding automobiles because it focuses on issues that are truly basic to the way nature works.

The key to understanding the strength of physics is to recognize that its laws are based on experimental fact. This is not to say that intuition and educated guesses are unimportant in physics. The great creative geniuses in science, as in art, work in leaps and bounds that no one can fully understand. However, in physics a "flash of insight" never becomes accepted law unless its implications can be verified by experiment. Until such verification takes place, a "flash of insight" provides at best only a theory, often one among many. This insistence on experimental verification has enabled physicists to build a rational and coherent understanding of nature.

The exciting feature of physics is its capacity for predicting the way nature will behave in one situation on the basis of experimental data obtained in another situation. Such predictions can have a tremendous impact in our lives, for in this sense physics is at the heart of modern technology. Rocketry and the development of space travel have their roots firmly planted in the physical laws of Galileo Galilei (1564–1642) and Isaac Newton (1642–1727). The transportation industry relies heavily on physics in the development of engines and the design of aerodynamic vehicles. Entire electronics and computer industries owe their existence to the impetus provided by the invention of the transistor, which grew directly out of physical laws describing the electrical behavior of solids. The telecommunications industry depends extensively on electromagnetic waves, whose existence was predicted by James Clerk Maxwell (1831–1879) in his theory of electricity and magnetism. The medical profession uses x-ray, ultrasonic, and magnetic resonance methods for obtaining images of the interior of the human body, and physics lies at the core of all these. Perhaps the most widespread impact in modern technology is that due to the laser. Fields ranging from space exploration to medicine benefit from this incredible device, which is a direct result of the laws of atomic physics.

Because physics is so fundamental, it is a required course for students in a wide range of major areas. We welcome you to the study of this fascinating topic. You will learn how to see the world through the "eyes" of physics, and, in so doing, you will learn how to apply physics principles to a wide range of problems. We hope that you will come to recognize that physics has important things to say about your environment.

1.2 UNITS

SYSTEMS OF UNITS

Since physics is an experimental science, it involves the measurement of a variety of quantities. A great deal of effort goes into making these measurements as accurate and reproducible as possible. The first step toward ensuring accurate and reproducible measurements is defining the units in which the measurements are made.

In this text, we will stress the system of units known according to the French phrase "Le Système International d'Unites," referred to simply as *SI units.* This system, by international agreement, employs the *meter* (m) as the unit of length, the *kilogram* (kg) as the unit of mass, and the *second* (s) as the unit of time. Two other systems of units are worth mentioning. The CGS system utilizes the centimeter (cm), the gram (g), and the second for length, mass, and time, respectively, whereas the BE or British Engineering system (the gravitational version) uses the foot (ft), the slug (sl), and the second. Both the CGS and the BE systems will be used occasionally in this text, in recognition of their continued, although declining importance. Table 1.1 summarizes the units used for length, mass, and time in the three systems.

TABLE 1.1 Units for Measurement

Unit	System		
	SI	BE	CGS
Length	meter (m)	foot (ft)	centimeter (cm)
Mass	kilogram (kg)	slug (sl)	gram (g)
Time	second (s)	second (s)	second (s)

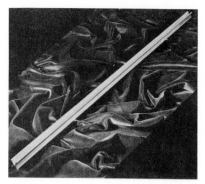

FIGURE 1.1 The standard platinum–iridium meter bar.

DEFINITION OF STANDARD UNITS

As physics has developed as a science, the definition of the meter has undergone changes that reflect this development. Originally, the meter was defined in terms of the distance measured along the earth's surface between the north pole and the equator. A more accurate measurement standard was eventually agreed upon internationally, and the meter was redefined as the distance between two marks on a bar of platinum–iridium alloy (see Figure 1.1) kept at a temperature of 0 °C. The present definition was adopted in 1983. Today, the meter is the distance that light travels in a vacuum in a time of 1/299 792 458 second. This definition was chosen because the speed at which light travels in a vacuum has been measured very accurately and has a value of 299 792 458 meters/second ± 1 meter/second.

The definition of a kilogram as a unit of mass has also undergone changes over the years. As Chapter 4 discusses, the mass of an object indicates the tendency of the object to continue in motion with a steady velocity. This tendency depends on the amount of material present in the object. Originally, the kilogram was defined in terms of a specific amount of water. Today, one kilogram is defined to be the mass of a standard cylinder of platinum–iridium alloy, like that in Figure 1.2.

As with the units for length and mass, the present definition of the second as a unit of time is different than the original definition. Originally, the second was defined according to the average time for the earth to rotate once about its axis, one day being set equal to 86 400 seconds. Now, however, the second is defined in terms of a cesium atomic clock like that in Figure 1.3.

FIGURE 1.2 The standard platinum–iridium kilogram is kept at the International Bureau of Weights and Measures in Sevres, France.

FIGURE 1.3 A cesium atomic clock at the National Bureau of Standards.

TABLE 1.2 Standard Prefixes Used to Denote Multiples of Ten

Prefix	Symbol	Factor[a]
Tera	T	10^{12}
Giga[b]	G	10^{9}
Mega	M	10^{6}
Kilo	k	10^{3}
Hecto	h	10^{2}
Deka	da	10^{1}
Deci	d	10^{-1}
Centi	c	10^{-2}
Milli	m	10^{-3}
Micro	μ	10^{-6}
Nano	n	10^{-9}
Pico	p	10^{-12}
Femto	f	10^{-15}

[a] Appendix A contains a discussion of powers of ten and scientific notation.

[b] Pronounced jig′a.

BASE UNITS AND DERIVED UNITS

The units for length, mass, and time, along with a few other units that will arise later in this text, are regarded as **base** SI units. The word "base" refers to the fact that these units will be used along with various laws to define additional units for other important physical quantities, such as force and energy. The units for these other physical quantities are referred to as **derived** units, since they are combinations of the base units. Derived units will be introduced as they arise naturally along with the related physical laws.

The value of a quantity in terms of base or derived units is sometimes a very large or very small number. In such cases, it is convenient to introduce larger or smaller units that are related to the normal units by multiples of ten. Table 1.2 summarizes the standard prefixes that are used to denote multiples of ten. For example, 1000 or 10^3 meters are referred to as 1 kilometer (km), and 0.001 or 10^{-3} meter is called 1 millimeter (mm). Similarly, 1000 grams and 0.001 gram are referred to as 1 kilogram (kg) and 1 milligram (mg), respectively.

1.3 THE ROLE OF UNITS IN PROBLEM SOLVING

THE CONVERSION OF UNITS

Since any quantity, such as length, can be measured in several different units, it is important to know how to convert from one unit to another. For instance, the foot can be used to express the distance between the two marks on the standard platinum–iridium meter bar. There are 3.28 feet in one meter, and this number can be used to convert from meters to feet, as the following example demonstrates.

EXAMPLE 1

How many feet are there in a length of 2.67 meters?

SOLUTION

Since 3.28 feet = 1 meter, it follows that (3.28 feet)/(1 meter) = 1. Multiplying by a factor of unity does not alter an equation. Therefore,

$$\text{Length} = (2.67 \text{ meters})(1) = (2.67 \text{ meters})\left(\frac{3.28 \text{ feet}}{1 \text{ meter}}\right)$$

$$= \boxed{8.76 \text{ feet}}$$

A calculator gives this answer as 8.7576 feet. Standard procedures for significant figures, however, indicate that the answer should be rounded off to three significant figures, since the value of 2.67 meters is accurate to only three significant figures. In this regard, the "1 meter" in the denominator does not limit the significant figures of the answer, because this number is precisely one meter by definition of the conversion factor. Appendix B contains a review of significant figures.

When converting between units, it is useful to write down the units explicitly. The units are treated like any algebraic quantity; in Example 1 the colored lines show how the "meter" cancels when the multiplication is performed, leaving only the desired unit of "feet" to describe the answer. In other words, **if the units do not combine algebraically to give the desired result, the conversion has not been carried out properly.** The next example also stresses the importance of writing down the units and illustrates a typical situation in which several conversions are required.

EXAMPLE 2

Express the speed limit of 55 miles/hour in terms of meters/second.

SOLUTION

To begin with, 5280 feet = 1 mile and 3600 seconds = 1 hour. As a result, (5280 feet)/(1 mile) = 1 and (3600 seconds)/(1 hour) = 1. Multiplying and dividing by these factors of unity does not alter an equation, so that

$$\text{Speed} = \left(55 \frac{\text{miles}}{\text{hour}}\right) \frac{(1)}{(1)} = \left(55 \frac{\text{miles}}{\text{hour}}\right) \frac{\left(\dfrac{5280 \text{ feet}}{1 \text{ mile}}\right)}{\left(\dfrac{3600 \text{ seconds}}{1 \text{ hour}}\right)}$$

$$= 81 \frac{\text{feet}}{\text{second}}$$

To convert feet into meters we use the fact that (3.28 feet)/(1 meter) = 1:

$$\text{Speed} = \frac{\left(81 \dfrac{\text{feet}}{\text{second}}\right)}{(1)} = \frac{\left(81 \dfrac{\text{feet}}{\text{second}}\right)}{\left(\dfrac{3.28 \text{ feet}}{1 \text{ meter}}\right)} = \boxed{25 \frac{\text{meters}}{\text{second}}}$$

In Example 2, it is necessary to know whether a conversion factor such as (5280 feet)/(1 mile) goes into the numerator or the denominator. In general, you can check to see if your choice is the correct one by verifying that the units combine algebraically to give the desired result for the answer. A collection of useful conversion factors is given on the page facing the inside of the front cover.

UNITS AS A PROBLEM SOLVING AID

In addition to their role in guiding the proper application of conversion factors, units serve a useful purpose in solving problems. They can provide an internal check to eliminate certain kinds of errors, if they are carried along during each step of a calculation and treated like any algebraic factor.

Suppose, for instance, that the tank of a car contains 2.0 gallons of gas to start with and that gas is added at a rate of 7.0 gallons/minute. The total amount of gas in the tank 96 seconds later can be obtained by adding the amount put into the tank to the amount present initially. The amount put in can be calculated by multiplying the filling rate by the time the gas pump is on. But if attention is not paid to the units used in the calculation, an erroneous result can be obtained:

$$\begin{aligned} \text{Total amount} \atop \text{of gas} &= \text{Gas initially} \atop \text{present} + \text{Gas} \atop \text{added} \\ &= 2.0 \text{ gallons} + \left(7.0 \frac{\text{gallons}}{\text{minute}}\right)(96 \text{ seconds}) \\ &= 2.0 \text{ gallons} + 672 \frac{\text{gallons} \cdot \text{seconds}}{\text{minute}} \end{aligned}$$

The answer cannot be 2.0 + 672 = 674, because the units for the two added terms are not the same. *Only quantities that have exactly the same units can be added (or subtracted).* With the filling rate expressed as 7.0 gallons/minute, the correct answer can be obtained only if the time of 96 seconds is converted into minutes:

$$\text{Time} = (96 \, \cancel{\text{seconds}}) \left(\frac{1 \text{ minute}}{60 \, \cancel{\text{seconds}}} \right) = 1.6 \text{ minutes}$$

$$\begin{matrix} \text{Total amount} \\ \text{of gas} \end{matrix} = 2.0 \text{ gallons} + \left(7.0 \, \frac{\text{gallons}}{\cancel{\text{minute}}} \right) (1.6 \, \cancel{\text{minutes}})$$

$$= 2.0 \text{ gallons} + 11 \text{ gallons} = 13 \text{ gallons}$$

As indicated by the colored lines, the units of time now cancel algebraically when the multiplication is carried out, leaving only the desired units of gallons. The procedure of "carrying along the units" prevents the inadvertent combination of numbers that do not have the same units. At the same time, the procedure serves as an automatic reminder to convert all data used in a calculation into a consistent set of units. For these important reasons, we will always carry along the units when working out examples in this text.

1.4 TRIGONOMETRY

BASIC TRIGONOMETRIC FUNCTIONS

Trigonometry is an important part of mathematics that is used in physics to describe how the physical universe works. There are three trigonometric functions that are applied throughout this text. They are the sine, the cosine, and the tangent of the angle θ (Greek theta), abbreviated as sin θ, cos θ, and tan θ, respectively. These functions are defined below in terms of the symbols given along with the right triangle in Figure 1.4.

DEFINITION OF SIN θ, COS θ, AND TAN θ

$$\sin \theta = \frac{h_o}{h} \tag{1.1}$$

$$\cos \theta = \frac{h_a}{h} \tag{1.2}$$

$$\tan \theta = \frac{h_o}{h_a} \tag{1.3}$$

h = length of the **hypotenuse** of a right triangle
h_o = length of the side **opposite** the angle θ
h_a = length of the side **adjacent** to the angle θ

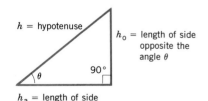

h = hypotenuse
h_o = length of side opposite the angle θ
90°
θ
h_a = length of side adjacent to the angle θ

FIGURE 1.4 A right triangle.

The sine, cosine, and tangent are numbers without units, because each is expressed as the ratio of the lengths of two sides of a right triangle. Example 3 illustrates a typical application of Equation 1.3.

EXAMPLE 3

On a sunny day, a tall building casts a shadow that is 67.2 m long. The angle between the sun's rays and the ground is $\theta = 50.0°$, as Figure 1.5 shows. Determine the height of the building.

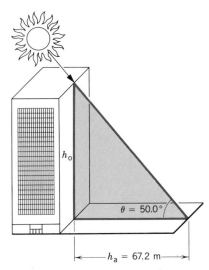

FIGURE 1.5 From a value for the angle θ and the length h_a of the shadow, the height h_o of the building can be found using trigonometry.

SOLUTION

In the right triangle in Figure 1.5, the height of the building is the side h_o opposite the angle θ, and the length of the shadow is the side h_a adjacent to the angle. These observations suggest using the tangent function to find the height of the building, since it is known that $\theta = 50.0°$ and $h_a = 67.2$ m:

$$\tan \theta = \frac{h_o}{h_a} \tag{1.3}$$

$$h_o = h_a \tan \theta = (67.2 \text{ m})(\tan 50.0°)$$

$$= (67.2 \text{ m})(1.19) = \boxed{80.1 \text{ m}}$$

The value of tan 50.0° is found by using a calculator or a table of values, such as that in Appendix F.

Either the sine, cosine, or tangent may be used in calculations such as that in Example 3, depending on which side of the triangle has a known value and which side is asked for. However, *the choice of which side of the triangle to label h_o (opposite) and which to label h_a (adjacent) can only be made after the angle θ is identified.*

Often the values for two sides of the right triangle in Figure 1.4 are available, and the value of the angle θ is unknown. The concept of *inverse* trigonometric functions plays an important role in such situations. Equations 1.4 – 1.6 below give the inverse sine, inverse cosine, and inverse tangent in terms of the symbols used in the drawing. These equations are just a convenient shorthand way of saying that the angle θ can be determined if the sides of the right triangle are known. For instance, Equation 1.6 is read as "θ equals the angle whose tangent is h_o/h_a."

$$\theta = \sin^{-1}\left(\frac{h_o}{h}\right) \tag{1.4}$$

$$\theta = \cos^{-1}\left(\frac{h_a}{h}\right) \tag{1.5}$$

$$\theta = \tan^{-1}\left(\frac{h_o}{h_a}\right) \tag{1.6}$$

The use of "-1" as an exponent in Equations 1.4 – 1.6 *does not mean* "take the reciprocal." For instance, $\tan^{-1}\left(\frac{h_o}{h_a}\right)$ does not equal $1/\tan\left(\frac{h_o}{h_a}\right)$. Another way to express the inverse trigonometric functions uses arc sin, arc cos, and arc tan instead of $\sin^{-1}$, $\cos^{-1}$, and $\tan^{-1}$. Example 4 illustrates the use of an inverse trigonometric function.

EXAMPLE 4

A lakefront beach drops off gradually at an angle θ, as Figure 1.6 indicates. For safety reasons, it is necessary to know how deep the lake is at various distances from the shore. To provide some information about the depth, a lifeguard rows straight out from the shore a distance of 14.0 m and drops a weighted fishing line. By measuring the length of the line, the lifeguard determines the depth to be 2.25 m. (a) What is the value of θ? (b) What is the depth d of the lake at a distance of 22.0 m from the shore?

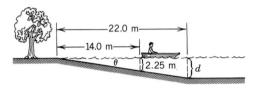

FIGURE 1.6 If the distance from the shore and the depth of the water at any one point is known, the angle θ can be found with the aid of trigonometry. Knowing the value of θ is useful, because then the depth d at another point can be determined.

SOLUTION

(a) Figure 1.6 identifies the sides of the right triangle that are opposite and adjacent to the angle θ; they are $h_o = 2.25$ m and $h_a = 14.0$ m, respectively. Equation 1.3 can be used to calculate $\tan \theta$:

$$\tan \theta = \frac{h_o}{h_a} = \frac{2.25 \text{ m}}{14.0 \text{ m}} = 0.161$$

Now that the value of $\tan \theta$ is known, the angle θ can be obtained by using the inverse tangent:

$$\theta = \tan^{-1}(0.161) = \boxed{9.15°}$$

(b) Farther from the shore, the sides of the right triangle opposite and adjacent to the angle θ are $h_o = d$ and $h_a = 22.0$ m. Since $\theta = 9.15°$, the tangent function can be used to find the unknown depth;

$$\tan \theta = \frac{h_o}{h_a}$$

$$h_o = h_a \tan \theta$$

$$d = (22.0 \text{ m})(\tan 9.15°) = \boxed{3.54 \text{ m}}$$

THE PYTHAGOREAN THEOREM

The right triangle in Figure 1.4 provides the basis for defining the various trigonometric functions according to Equations 1.1–1.3. These functions always involve an angle and two sides of the triangle. There is also a relationship among the three sides of a right triangle. This relationship is known as the *Pythagorean theorem* and is used often in this text.

PYTHAGOREAN THEOREM

The square of the hypotenuse of a right triangle is equal to the sum of the squares of the other two sides:

$$h^2 = h_o^2 + h_a^2 \qquad (1.7)$$

1.5 THE NATURE OF PHYSICAL QUANTITIES: SCALARS AND VECTORS

SCALARS

When measuring a quantity, it is sometimes sufficient to consider only the aspect of magnitude, as is the case with the volume of an object or the time of a race. The volume of water in a swimming pool may be 50 cubic meters, or the winning time of a race may be 11.3 seconds. The only important issue is the magnitude of these numbers, or *how much* volume or time there is. The "50" specifies the amount of water in units of cubic meters, while the "11.3" specifies the amount of time in seconds. Volume and time are examples of scalar quantities. A *scalar quantity,* or scalar for short, is one that can be described by a single number (including any units) giving its magnitude. Some other commonly found examples of scalars are temperature = 20 °C, heat = 25 calories, and mass = 85 kg.

VECTORS

While many quantities in physics are scalars, there are also many that are not scalars, quantities for which magnitude tells only part of the story. Consider Figure 1.7, in which a car has moved 2 km along a straight line from point *A* to point *B*. When describing how the car moved, it is incomplete to say that "the car moved a distance of 2 km," for this statement would indicate only that the car ends up somewhere on a circle whose center is point *A* and whose radius is 2 km. A complete description would include the direction, as well as the magnitude of the distance. For instance, a sufficient statement would be that "the car moved a distance of 2 km in a direction 30° north of east." A quantity that inherently deals with both magnitude and direction, is called a *vector quantity,* or vector for short. Because direction is an important characteristic of vectors, arrows are used to represent them; *the direction of the arrow gives the direction of the vector.* The colored arrow in Figure 1.7, for example, is called the displacement vector, because it shows how the car is displaced from point *A.* Chapter 2 discusses this particular vector.

It is natural to use the length of the arrow in Figure 1.7 to represent the magnitude of the displacement vector. If the car had moved 4 km instead of 2 km from the starting point, the arrow would have been drawn twice as long. *The length of a vector arrow is proportional to the magnitude of the vector.*

The procedure of using the length of an arrow to represent the magnitude of a vector applies to any kind of vector. And in physics there are many important vectors, in addition to the displacement vector. All forces, for instance, are vectors. A force is a push or a pull, and the direction in which a force acts is just as important as the strength or magnitude of the force. The magnitude of a force is measured in SI units called newtons (N) and BE units called pounds (lb). An arrow representing a force of 20 pounds is drawn twice as long as one representing a force of 10 pounds.

SYMBOLS USED FOR SCALARS AND VECTORS

Often, for the sake of convenience, quantities such as volume, time, displacement, and force are represented by symbols when they appear in sentences or equations. This text follows the usual practice of writing vectors in boldface symbols* **(this is boldface)** and writing scalars in italics symbols *(this is italics).* Thus, a displacement vector is written as "**s** = 750 m, due east," where the **s** is a boldface symbol. By itself, however, separated from the direction, the magnitude of this vector is a scalar quantity. Therefore, the magnitude is written as "*s* = 750 m," where the *s* is an italics symbol.

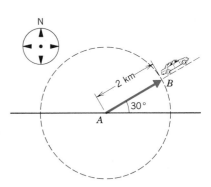

FIGURE 1.7 A vector quantity has a magnitude and a direction. Therefore, an arrow is used to represent a vector, the length of the arrow being proportional to the magnitude. The arrow in this drawing represents a displacement vector.

---

* A vector quantity can also be represented without boldface symbols, by including an arrow above the symbol, e.g., $\vec{s}$.

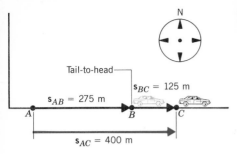

FIGURE 1.8 Two colinear displacement vectors, s_{AB} and s_{BC}, add to give the resultant displacement vector s_{AC}.

1.6 VECTOR ADDITION AND SUBTRACTION

ADDITION OF COLINEAR VECTORS

The need to add vector quantities arises repeatedly in physics, and the process of addition must take into account both the magnitude and the direction of the vectors. The simplest situation occurs when the vectors point along the same direction, that is, when they are colinear, as in Figure 1.8. This drawing shows a car that has moved due east along a straight line from point A to point B and then, after a pause, from point B to point C. The corresponding two displacement vectors are s_{AB} and s_{BC}. As usual, boldface symbols denote the vector quantities. These two separate vectors add to give the total displacement vector s_{AC}, which would apply if the car had moved directly from A to C. With the tail of the second arrow located at the head of the first arrow, the two lengths simply add to give the length of the total displacement. This kind of vector addition is identical to the addition of two scalar numbers ($2 + 3 = 5$), with which everyone is familiar, *and can be carried out here only because the vectors point along the same direction.* In such cases, then, we add the individual magnitudes to get the magnitude of the total, knowing in advance what the direction must be. Formally, the addition is written as follows for the data in Figure 1.8:

$$s_{AC} = s_{AB} + s_{BC}$$

$$s_{AC} = 275 \text{ m} + 125 \text{ m} = 400 \text{ m, pointing due east}$$

The total vector s_{AC} that results from the addition is referred to as the ***resultant vector.***

ADDITION OF PERPENDICULAR VECTORS

A tail-to-head arrangement characterizes the addition of vectors that have the same direction. A tail-to-head arrangement also characterizes the addition of vectors that do not have the same direction. However, the differences in directions must be taken into account.

For instance, perpendicular vectors are frequently encountered, and Figure 1.9 indicates how they are added. In this figure a car travels due east, from A to B, a distance of 275 m. The displacement vector is s_{AB}. The car then travels due north, from B to C, a distance of 125 m. The second displacement vector is s_{BC}. The resultant displacement vector of the car relative to its starting point is s_{AC}. Once again, the vectors to be added are arranged in a tail-to-head fashion, and the resultant vector points from the tail of the first to the head of the last vector added. The resultant displacement is given by the vector equation

$$s_{AC} = s_{AB} + s_{BC}$$

The addition in this equation cannot be carried out by writing $s_{AC} = 275$ m + 125 m, because the vectors are not colinear. Instead, we take advantage of the fact that the triangle in Figure 1.9 is a right triangle and use the Pythagorean theorem (Equation 1.7). According to this theorem, the magnitude of s_{AC} is

$$s_{AC} = \sqrt{(275 \text{ m})^2 + (125 \text{ m})^2} = 302 \text{ m}$$

The angle θ in Figure 1.9 gives the direction of the resultant vector. Since the lengths

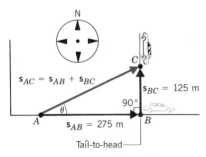

FIGURE 1.9 The addition of two perpendicular displacement vectors s_{AB} and s_{BC} gives the resultant vector s_{AC}.

of all three sides of the right triangle are now known, either $\sin\theta$, $\cos\theta$, or $\tan\theta$ can be used to determine θ:

$$\tan\theta = \frac{125\text{ m}}{275\text{ m}} = 0.455$$

$$\theta = \tan^{-1}(0.455) = 24.5°$$

Thus, relative to the starting point, the resultant displacement of the car has a magnitude of 302 m and points north of east at an angle of 24.5°.

ADDITION OF VECTORS THAT ARE NEITHER COLINEAR NOR PERPENDICULAR

When two vectors to be added are not perpendicular, the tail-to-head arrangement does not lead to a right triangle, and the Pythagorean theorem cannot be used. Figure 1.10a illustrates such a case. A car moves due east, from A to B, through a distance of 275 m. It then moves in a direction 55.0° north of west, from B to C, through a distance of 125 m. The corresponding displacement vectors are $\mathbf{s}_{AB}$ and $\mathbf{s}_{BC}$. As usual, the resultant displacement vector $\mathbf{s}_{AC}$ is directed from the tail of the first to the head of the last vector added. The vector addition is still given according to

$$\mathbf{s}_{AC} = \mathbf{s}_{AB} + \mathbf{s}_{BC}$$

However, since the triangle in the drawing is not a right triangle, some means other than the Pythagorean theorem must be used to carry out the vector addition.

One approach uses a graphical technique to carry out the addition. In this method, a diagram is constructed in which the lengths of the vector arrows are drawn to scale and the angles are drawn accurately (with a protractor, perhaps). Then, the length of the arrow representing the resultant vector is measured with a ruler. This length is converted into the magnitude of the resultant vector by using the scale factor with which the drawing is constructed. In Figure 1.10, for example, a scale of one centimeter of arrow length for each 10.0 m of displacement is used. It can be seen in part b of the drawing that the length of the arrow representing $\mathbf{s}_{AC}$ is 22.8 cm. Since each centimeter corresponds to 10.0 m of displacement, the magnitude of $\mathbf{s}_{AC}$ is 228 m. The angle θ, which gives the direction of $\mathbf{s}_{AC}$, can be measured with a protractor to be $\theta = 26.7°$. The disadvantage of the graphical technique is that it is not particularly convenient, so Section 1.8 discusses a better way to add vectors.

SUBTRACTION OF VECTORS

Suppose that two vectors, $\mathbf{A}$ and $\mathbf{B}$, add together to give a third vector $\mathbf{C}$, according to $\mathbf{C} = \mathbf{A} + \mathbf{B}$. Figure 1.11a shows these vectors. If values for the vectors $\mathbf{C}$ and $\mathbf{B}$ are available, we can calculate vector $\mathbf{A}$ as $\mathbf{A} = \mathbf{C} - \mathbf{B}$. This result is an example of vector subtraction, and it is important to know how to carry out such an operation. In practice, vector subtraction is performed exactly as vector addition, except that one of the vectors added ($\mathbf{B}$ in this case) is multiplied by a scalar factor of -1. To see why, rewrite the equation for vector $\mathbf{A}$ as follows: $\mathbf{A} = \mathbf{C} + (-\mathbf{B})$. When a vector is multiplied by -1, the magnitude of the vector remains the same, but the direction of the vector is reversed. With this in mind, Figure 1.11b shows how to calculate vector $\mathbf{A}$ from the vectors $\mathbf{C}$ and $-\mathbf{B}$. Notice that vectors $\mathbf{C}$ and $-\mathbf{B}$ are arranged tail-to-head and that any suitable method of vector addition can be employed to determine $\mathbf{A}$.

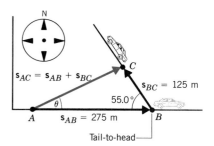
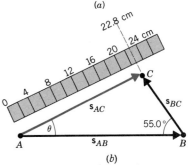

FIGURE 1.10 (a) The two displacement vectors $\mathbf{s}_{AB}$ and $\mathbf{s}_{BC}$ are neither colinear nor perpendicular. They add to give the resultant vector $\mathbf{s}_{AC}$. (b) In one method for carrying out the addition, an accurate scale drawing of the vectors (arranged tail-to-head) is constructed, using a ruler for measuring lengths and a protractor for measuring angles.

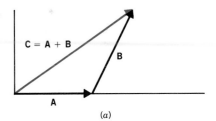
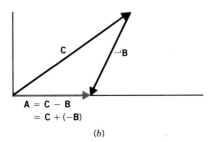

FIGURE 1.11 (a) Vector addition according to $\mathbf{C} = \mathbf{A} + \mathbf{B}$. (b) Vector subtraction according to $\mathbf{A} = \mathbf{C} - \mathbf{B} = \mathbf{C} + (-\mathbf{B})$.

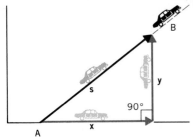

FIGURE 1.12 The displacement vector **s** and its vector components **x** and **y**.

1.7 VECTOR COMPONENTS

THE MEANING OF VECTOR COMPONENTS

The concept of vector components is very important in physics, and Figure 1.12 provides some insight into the meaning of this concept as applied in two dimensions. In this drawing, a car moves along a straight line from A to B. The magnitude and direction of the displacement vector **s** give the distance and direction traveled by the car along the line AB. However, the car could also have arrived at B by first moving horizontally, turning through 90°, and then moving vertically. This alternative path is shown in color in the drawing and is associated with the two displacement vectors **x** and **y**. The vectors **x** and **y** are called the x component and the y component of the vector **s**.

Two basic features of vector components are apparent in Figure 1.12. One feature is that the components add together to equal the original vector, as expressed by the following vector equation:

$$\mathbf{s} = \mathbf{x} + \mathbf{y}$$

In other words, the two components **x** and **y**, when added vectorially, convey exactly the same meaning as does the original vector **s**; that is, they indicate how point B is displaced relative to point A. Thus, ***the components of a vector can be used in place of the vector itself in any calculation where it is convenient to do so.*** It will be convenient to use the components of a vector, rather than the vector itself, many times in this text. The other feature of vector components that is apparent in Figure 1.12 is that the components **x** and **y** are not just any two vectors that add together to give the original vector **s**; they are perpendicular vectors.* This perpendicularity is a valuable characteristic of vector components, as we will soon see.

Any vector may be expressed in terms of its components, in a way similar to that illustrated for the displacement vector in Figure 1.12. Figure 1.13 shows an arbitrary vector **A** and its components $\mathbf{A}_x$ and $\mathbf{A}_y$. The components are drawn parallel to convenient x, y axes and are perpendicular. They add vectorially to equal the original vector **A**:

$$\mathbf{A} = \mathbf{A}_x + \mathbf{A}_y$$

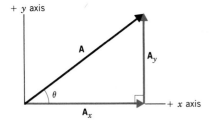

FIGURE 1.13 An arbitrary vector **A** and its vector components $\mathbf{A}_x$ and $\mathbf{A}_y$.

Boldface symbols are used for vector **A** and its components $\mathbf{A}_x$ and $\mathbf{A}_y$. However, the magnitudes of **A**, $\mathbf{A}_x$, and $\mathbf{A}_y$ are scalar quantities and, therefore, are represented by the italics symbols A, A_x, and A_y.

There are times when a drawing such as Figure 1.13 is not the most convenient way to represent vector components. Figure 1.14 presents an alternative method. The disadvantage of Figure 1.14 is that the tail-to-head arrangement of $\mathbf{A}_x$ and $\mathbf{A}_y$ is missing, an arrangement that is a nice reminder that $\mathbf{A}_x$ and $\mathbf{A}_y$ add together to equal **A**.

The definition given below summarizes the meaning of vector components:

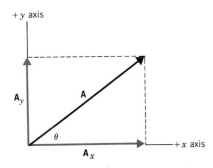

FIGURE 1.14 This alternative way of drawing the vector **A** and its components is completely equivalent to that shown in Figure 1.13.

> **DEFINITION OF VECTOR COMPONENTS**
>
> In two dimensions, the vector components of a vector **A** are two other perpendicular vectors $\mathbf{A}_x$ and $\mathbf{A}_y$ that are parallel to the axes of an x, y axes system and that add together vectorially so that $\mathbf{A} = \mathbf{A}_x + \mathbf{A}_y$.

* It is possible to introduce vector components that are not perpendicular, but, in general, they are not as useful as those introduced here.

This definition can be extended to include vectors in three dimensions, but such vectors are not needed in this text.

RESOLVING A VECTOR INTO ITS COMPONENTS

Starting with a known vector and finding its components is a process that is called "resolving the vector into its components." This process can be carried out with the aid of trigonometry, because the two perpendicular components and the original vector form a right triangle, as Figures 1.12–1.14 indicate. Example 5 shows how trigonometry is used to find the components of a vector.

EXAMPLE 5

A displacement vector **s** has a magnitude of $s = 175$ m and points at an angle of 50.0° relative to the x axis in Figure 1.15. Find the x and y components of this vector.

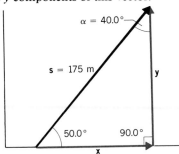

FIGURE 1.15 The x and y components of the displacement vector **s** can be found using trigonometry.

SOLUTION 1

The magnitude of the y component can be obtained using the 50.0° angle and the definition of sin θ from Equation 1.1:

$$\sin 50.0° = \frac{y}{s}$$

$$y = s \sin 50.0° = (175 \text{ m})(\sin 50.0°) = \boxed{134 \text{ m}}$$

In a similar fashion the magnitude of the x component can be obtained using the 50.0° angle and the definition of cos θ from Equation 1.2:

$$\cos 50.0° = \frac{x}{s}$$

$$x = s \cos 50.0° = (175 \text{ m})(\cos 50.0°) = \boxed{112 \text{ m}}$$

SOLUTION 2

It is not necessary to use the 50.0° angle to find the x and y components. The angle α in Figure 1.15 can also be employed. Since $\alpha + 50.0° = 90.0°$, it follows that $\alpha = 40.0°$. The solution using α proceeds differently than that using the 50.0° angle, but the same answers are obtained as in Solution 1.

$$\cos 40.0° = \frac{y}{s}$$

$$y = s \cos 40.0° = (175 \text{ m})(\cos 40.0°) = \boxed{134 \text{ m}}$$

$$\sin 40.0° = \frac{x}{s}$$

$$x = s \sin 40.0° = (175 \text{ m})(\sin 40.0°) = \boxed{112 \text{ m}}$$

Notice in Example 5 that it does not matter whether the 50.0° angle or the 40.0° angle is used in the trigonometric calculation. Both angles lead to the same answers, and the choice is a matter of convenience. In any event, it is possible to check the validity of the answers. Since the components and the original vector form a right triangle, the Pythagorean theorem can be applied to verify that the magnitude of the original vector is indeed 175 m, as given initially:

$$s = \sqrt{(112 \text{ m})^2 + (134 \text{ m})^2} = 175 \text{ m}$$

The values calculated for vector components depend on the orientation of the vector relative to the x, y axes used as a reference. Figure 1.16 illustrates this fact for a vector **A**, by showing two sets of axes, one set being rotated relative to the other. With respect to the x, y axes (black), vector **A** has perpendicular components labeled A_x

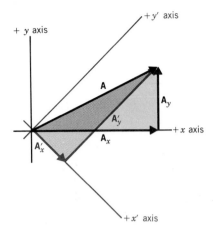

FIGURE 1.16 There is more than one set of components for the vector **A**. The components of the vector depend on the orientation of the axes used as a reference.

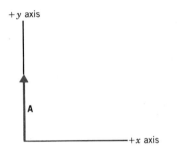

FIGURE 1.17 The x component of the vector **A** is zero, although the vector itself is not zero.

and A_y; with respect to the rotated x', y' axes (colored), vector **A** has different components labeled A_x' and A_y'. The choice of which set of components to use is purely a matter of convenience.

VECTORS THAT HAVE ZERO COMPONENTS

Depending on the orientation of the x, y axes used as a reference, it is possible that one of the components of a vector can be zero. Figure 1.17 shows an example of this situation to emphasize that a vector is not zero merely because one of its components is zero. In this drawing, the y component is itself the vector **A**, the x component being zero. Vector **A** would be expressed as the sum of its components according to the following vector equation: $\mathbf{A} = 0 + \mathbf{A}_y$.

For a vector to be zero, every component must individually be zero. Thus, in two dimensions, saying that $\mathbf{A} = 0$ is equivalent to saying that $A_x = 0$ and $A_y = 0$. This seemingly trivial fact plays an important role in physics. In particular, it will be used in Chapters 4 and 5 when we describe the equilibrium of an object by saying that the net force acting on the object is zero.

VECTORS THAT ARE EQUAL

Two vectors are equal if, and only if, they have the same magnitude and direction. Thus, if one displacement vector points east and another points north, they are *not* equal, even if each has the same magnitude of 480 m. In terms of components, two vectors, **A** and **B**, are equal if, and only if, each component of one is equal to the corresponding component of the other. In two dimensions, if $\mathbf{A} = \mathbf{B}$, then $A_x = B_x$ and $A_y = B_y$.

1.8 ADDITION OF VECTORS BY MEANS OF VECTOR COMPONENTS

Vector components provide the most convenient way of adding (or subtracting) any number of vectors. Suppose that the vectors to be added are **A**, **B**, and **C**, yielding a vector sum $\mathbf{D} = \mathbf{A} + \mathbf{B} + \mathbf{C}$. The essential step in the component method for addition is that **A**, **B**, and **C** are each replaced by their components, as indicated schematically in the following equations:

$$
\begin{aligned}
\mathbf{A} &= \mathbf{A}_x + \mathbf{A}_y \\
\mathbf{B} &= \mathbf{B}_x + \mathbf{B}_y \\
\mathbf{C} &= \mathbf{C}_x + \mathbf{C}_y \\
\downarrow \quad & \quad \downarrow \quad\quad \downarrow \\
\mathbf{D} &= \mathbf{D}_x + \mathbf{D}_y
\end{aligned}
$$

The value of this step is that the x components all point parallel to the x axis, and the y components all point parallel to the y axis. Therefore, the x components can be added as colinear vectors, and so can the y components. The addition of colinear vectors is the simplest kind of vector addition. Thus, $\mathbf{D}_x = \mathbf{A}_x + \mathbf{B}_x + \mathbf{C}_x$ and $\mathbf{D}_y = \mathbf{A}_y +$

$B_y + C_y$. In this fashion, D_x and D_y, the x and y components of the resultant vector, can be obtained. To see how this procedure is used, consider the resultant vector $\mathbf{D}$, which is the sum of the three vectors shown in Figure 1.18a.

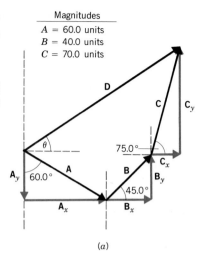

Magnitudes
A = 60.0 units
B = 40.0 units
C = 70.0 units

Step 1. Find the magnitude of the x and y components of each vector by using trigonometry. These magnitudes, being scalar quantities, are represented by italics symbols such as A_x, rather than boldface symbols such as $\mathbf{A}_x$.

$$A_x = (60.0 \text{ units})(\sin 60.0°) = 52.0 \text{ units}$$

$$A_y = (60.0 \text{ units})(\cos 60.0°) = 30.0 \text{ units}$$

$$B_x = (40.0 \text{ units})(\cos 45.0°) = 28.3 \text{ units}$$

$$B_y = (40.0 \text{ units})(\sin 45.0°) = 28.3 \text{ units}$$

$$C_x = (70.0 \text{ units})(\cos 75.0°) = 18.1 \text{ units}$$

$$C_y = (70.0 \text{ units})(\sin 75.0°) = 67.6 \text{ units}$$

(a)

Step 2. Assign a plus or minus sign to each component to account correctly for its direction, a plus sign to denote a component that points in the $+x$ or $+y$ direction and a minus sign to denote a component that points in the $-x$ or $-y$ direction.

Vector	x component	y component
A	$+52.0$ units	-30.0 units
B	$+28.3$ units	$+28.3$ units
C	$+18.1$ units	$+67.6$ units

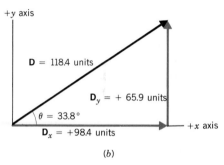

+y axis

D = 118.4 units

D_y = + 65.9 units

θ = 33.8°

D_x = +98.4 units

+x axis

(b)

Step 3. Find the x component of the resultant vector by adding the individual x components, and find the y component of the resultant vector by adding the individual y components. The individual components are added as colinear vectors.

$$D_x = +52.0 \text{ units} + 28.3 \text{ units} + 18.1 \text{ units} = +98.4 \text{ units}$$

$$D_y = -30.0 \text{ units} + 28.3 \text{ units} + 67.6 \text{ units} = +65.9 \text{ units}$$

FIGURE 1.18 (a) The three vectors **A**, **B**, and **C** add to give the resultant vector **D**. The components of **A**, **B**, and **C** are shown in color. (b) Once the components of the resultant vector **D** are known, the magnitude and direction of **D** can be determined.

The plus signs for the answers indicate that the x and y components of $\mathbf{D}$ point in the $+x$ and $+y$ directions, respectively.

Step 4. Knowing the x and y components, calculate the magnitude and direction of the resultant vector. Figure 1.18b shows the right triangle formed by the resultant vector $\mathbf{D}$ and its components. According to the Pythagorean theorem, the magnitude of $\mathbf{D}$ is given by

$$D = \sqrt{(98.4 \text{ units})^2 + (65.9 \text{ units})^2} = 118.4 \text{ units}$$

The angle θ in the drawing specifies the direction of the resultant vector and is given by

$$\tan \theta = \frac{+65.9 \text{ units}}{+98.4 \text{ units}} = +0.670$$

$$\theta = \tan^{-1}(0.670) = 33.8°$$

Calculations such as those above can always be checked by resorting to the graphical technique for adding vectors outlined in Section 1.6.

SUMMARY

Physics is an experimental science that uses precisely defined **units of measurement**. This text emphasizes SI (Système International) units, a system that includes the meter (m), the kilogram (kg), and the second (s) as base units for length, mass, and time, respectively. Other occasionally used systems of units are the CGS system (centimeter, gram, second) and the British Engineering or BE system (foot, slug, second). Units play an important role in solving problems, because the units on the left side of an equation must match the units on the right side. If the units on both sides do not match, either the equation is written incorrectly or the variables and constants in the equation are not expressed in a consistent set of units.

Trigonometry is used throughout physics. Particularly important are the sine, cosine, and tangent functions of an angle θ. These functions can be defined in terms of a right triangle that contains θ. The side of the triangle opposite θ is h_o, the side adjacent to θ is h_a, and the hypotenuse is h. In terms of these quantities $\sin \theta = h_o/h$, $\cos \theta = h_a/h$, and $\tan \theta = h_o/h_a$. Once the value of the sine, cosine, or tangent is known, the angle itself can be obtained using inverse trigonometric functions. The Pythagorean theorem, $h^2 = h_o^2 + h_a^2$, is useful when dealing with the sides of a right triangle.

Two kinds of physical quantities are important in this text, scalars and vectors. A **scalar quantity** can be described completely by giving its magnitude. For a **vector quantity**, however, both magnitude and direction must be specified. Vectors are often represented by arrows, with the length of the arrow being proportional to the magnitude of the vector and the direction of the arrow indicating the direction of the vector. The **addition of vectors** to give a resultant vector must correctly account for both magnitude and direction. When the vectors are all colinear, the addition proceeds in the same way as the simple addition of scalar quantities. When the vectors are not colinear, one procedure for addition utilizes a graphical technique, in which the vectors to be added are arranged in a tail-to-head fashion. The subtraction of a vector is treated as the addition of a vector that has been multiplied by a scalar factor of -1. Multiplying a vector by a scalar factor of -1 reverses the direction of the vector.

In two dimensions, the **vector components** of a vector **A** are two other perpendicular vectors $\mathbf{A}_x$ and $\mathbf{A}_y$ that are parallel to the axes of an x, y axes system and add together vectorially so that $\mathbf{A} = \mathbf{A}_x + \mathbf{A}_y$. Vector components provide the best way of adding any number of vectors. A vector is zero if, and only if, each of its components is zero. Two vectors are equal in two dimensions if, and only if, the x components of each are equal and the y components of each are equal.

SOLVED PROBLEMS

SOLVED PROBLEM 1
Related Problems: *34   *35   **36   **37

Vector **A** has a magnitude of 188 units and points 30.0° north of west. Vector **B** points 50.0° east of north. Vector **C** points 20.0° west of south. These three vectors add to give a resultant vector that is zero. Find the magnitudes of vectors **B** and **C** by using components.

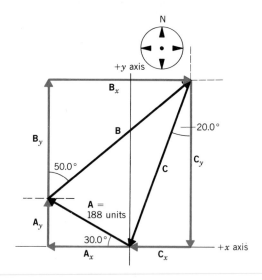

Solution In the drawing the vectors **A**, **B**, and **C** are arranged in a tail-to-head fashion. The fact that the resultant vector is zero ($A + B + C = 0$) means that the head of vector **C** is located at the tail of vector **A**, thus forming a closed triangle. Since the resultant vector is zero, each of its components must be zero. Therefore, the solution involves adding the three vectors by the component method and then setting the total of the x components and the total of the y components separately equal to zero. The components of each vector are listed below.

Vector	x component
A	$-188 \cos 30.0° = -163$
B	$+B \sin 50.0° = +B(0.766)$
C	$-C \sin 20.0° = -C(0.342)$
Total	$-163 + B(0.766) - C(0.342)$

Vector	y component
A	$+188 \sin 30.0° = +94.0$
B	$+B \cos 50.0° = +B(0.643)$
C	$-C \cos 20.0° = -C(0.940)$
Total	$+94.0 + B(0.643) - C(0.940)$

Notice that plus and minus signs have been used to denote whether the components point along the plus or minus axes, as is necessary in the component method. Setting the total x and y components separately equal to zero gives the following two equations:

$$-163 + B(0.766) - C(0.342) = 0$$
$$+94.0 + B(0.643) - C(0.940) = 0$$

These two equations can be solved simultaneously (see Appendix C for an algebra review) for the two unknown quantites B and C: $\boxed{B = 371 \text{ units}}$ and $\boxed{C = 354 \text{ units}}$.

Summary of Important Points Whenever a number of vectors add to give a resultant vector that is zero, the components of the resultant vector must separately be zero. Therefore, in two dimensions, the x components of the individual vectors may be added as colinear vectors and the resulting sum set equal to zero. The y components of the individual vectors may be treated in the same way. This approach leads to two separate equations, and provided enough initial information is available, these equations can be solved to yield useful results concerning the original vectors.

QUESTIONS

1. The table below lists four variables along with their units.

Variable	Units
x	meters (m)
v	meters per second (m/s)
t	seconds (s)
a	meters per second squared (m/s²)

These variables appear in the following six equations, along with a few numbers that have no units. In which of the equations are the units on the left side of the equal sign consistent with the units on the right side of the equal sign?

(a) $x = vt$
(b) $x = vt + \frac{1}{2}at^2$
(c) $v = at$
(d) $v = at + \frac{1}{2}at^3$
(e) $v^3 = 2ax^2$
(f) $t = \sqrt{\dfrac{2x}{a}}$

2. The variables x and v have the units shown in the table that accompanies question 1. Is it possible for x and v to be related to an angle θ according to $\tan \theta = x/v$? Account for your answer.

3. The variables x, v, and a have the units shown in the table that accompanies question 1. These variables are related by an equation that has the form $v^n = 2ax$, where n is an integer constant (1, 2, 3, etc.) without units. What must be the value of n, so that both sides of the equation have the same units? Explain your reasoning.

4. In the following equation the units of the variables x, v, and t are those shown in the table that accompanies question 1: $v = \frac{1}{3}zxt^2$. What must be the units of the variable z, such that both sides of the equation have the same units? Show how you determined your answer.

5. Using your calculator or a table of trigonometric values, verify that $\sin \theta$ divided by $\cos \theta$ is equal to $\tan \theta$, for any angle θ. Try 30°, for example. Prove that this result is true in general by using the definitions for $\sin \theta$, $\cos \theta$, and $\tan \theta$ given in Equations 1.1–1.3.

6. $\sin \theta$ and $\cos \theta$ are called sinusoidal functions of the angle θ. The way in which these functions change as θ changes leads to a characteristic pattern when they are graphed. This pattern arises many times in physics. (a) To familiarize yourself with the sinusoidal pattern, use a calculator or table of values and construct a graph, with $\sin \theta$ plotted on the vertical axis and θ on the horizontal axis. Use 15° increments for θ between 0° and 720°. (b) Repeat for $\cos \theta$.

7. Which of the following quantities (if any) can be considered a vector: (a) the number of people attending a football game, (b) the number of days in a month, and (c) the number of pages in a book? Explain your reasoning.

8. Which of the following displacement vectors (if any) are equal? Explain your reasoning.

Vector	Magnitude	Direction
A	100 m	30° north of east
B	100 m	30° south of west
C	50 m	30° south of west
D	100 m	60° east of north

9. Which two of the following displacement vectors would yield zero when added together? Account for your answer(s).

Vector	Magnitude	Direction
A	50 m	20° west of north
B	25 m	20° west of north
C	50 m	70° north of west
D	50 m	20° east of south

10. Can two vectors with unequal magnitudes be added together so their sum is zero? Justify your answer.

11. Can two nonzero perpendicular vectors be added together so their sum is zero? Explain.

12. Can three or more vectors with unequal magnitudes be added together so their sum is zero? If so, show by means of a tail-to-head arrangement of the vectors how this could occur.

13. Vectors **A** and **B** satisfy the vector equation **A** + **B** = 0. (a) How does the magnitude of **B** compare with the magnitude of **A**? (b) How does the direction of **B** compare with the direction of **A**? Give your reasoning.

14. Vectors **A**, **B**, and **C** satisfy the vector equation **A** + **B** = **C**, and their magnitudes are related by the scalar equation $A^2 + B^2 = C^2$. How is vector **A** oriented with respect to vector **B**? Account for your answer.

15. Vectors **A**, **B**, and **C** satisfy the vector equation **A** + **B** = **C**, and their magnitudes are related by the scalar equation $A + B = C$. How is vector **A** oriented with respect to vector **B**? Explain your reasoning.

16. A vector has a component of zero along the x axis of a certain x, y axes system. Does this vector necessarily have a component of zero along the x axis of another (rotated) x, y axes system? Use a drawing to justify your answer.

PROBLEMS

Problems that are not marked with a star are considered the easiest to solve. Problems that are marked with a single star (*) are more difficult, while those marked with a double star (**) are the most difficult.

Section 1.3 The Role of Units in Problem Solving

1. How many seconds are there in (a) one hour and thirty-five minutes and (b) one day?

2. Most running events in a track meet are now measured in meters, rather than in yards. How many meters are there in a one-hundred-yard race?

3. Sometimes, highway signs indicate distances to the upcoming exits in both miles and kilometers. One such distance is given as 17.0 miles. What is this distance in kilometers?

4. Express a speed of 110 km/h in mi/h.

5. One acre contains 43 560 ft². How many square meters (m²) are in one acre?

* 6. Acceleration is the rate of change of the velocity. The acceleration of a car can be expressed in units of feet/second/second (ft/s²). Another unit for acceleration is meter/hour/hour (m/h²). Convert an acceleration of 1.35 ft/s² to units of m/h².

Section 1.4 Trigonometry

7. A highway is to be built between two towns, one of which lies 35.0 km south and 72.0 km west of the other. What is the shortest length of highway that can be built between the two towns, and at what angle would this highway be directed with respect to due west?

8. A hiker sees a mountain in the distance, the peak of which is known to be 2900 m above sea level. The hiker estimates her line of sight with the top of the peak to be 11° above the horizontal. Assuming the hiker is at sea level and ignoring her height, find the horizontal distance to the point directly under the mountain top.

9. The corners of a square lie on a circle whose radius is 0.500 m. What is the length of a side of a square?

10. The silhouette of a Christmas tree is an isosceles triangle. The angle at the top of the triangle is 30.0°, and the base measures 2.00 m across. How tall is the tree?

* 11. Three buildings, A, B, and C, form the corners of a triangle. Building B is located 210 m from A at an angle of 41° east of north. Building C is located 320 m from A at an angle of 62° east of south. What is the distance between B and C? *(Hint: Consider the law of cosines given in Appendix E.)*

* 12. The drawing shows part of the fundamental unit of the crystal structure of sodium chloride (common table salt). In the drawing is a cube with sodium and chlorine ions positioned at the corners. The edge of the cube is 0.281 nm (1 nm = 1 nanometer = 10^{-9} m) in length. Calculate the distance (in nanometers) between the sodium ion located at one corner of the cube and the chlorine ion located on the diagonal at the opposite corner.

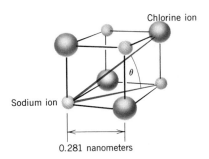

Chlorine ion

θ

Sodium ion

0.281 nanometers

* 13. What is the value of the angle θ in the drawing that accompanies problem 12?

**14. The height H of a regular tetrahedron is the perpendicular distance from one corner to the center of the opposite triangular base. The faces of such a tetrahedron are equilateral triangles, and all of its edges are equal in length. Show that the ratio between H and the length L of an edge of the tetrahedron is $H/L = \sqrt{2/3}$.

Section 1.6 Vector Addition and Subtraction

15. Two displacement vectors s_1 and s_2 each point 35.0° north of west. One has a magnitude of 4.60 km, while the other has a magnitude of 3.20 km. What are the magnitude and direction of the resultant vector $s_1 + s_2$?

16. Two ropes are attached to a heavy box to pull it along the floor. One rope applies a force of 475 pounds in a direction due west; the other applies a force of 315 pounds in a direction due south. Like all forces, these two forces are vector quantities. How much force should be applied by a single rope, and in what direction, if it is to accomplish the same effect as the two forces added together?

17. A jogger travels due south and in the process his displacement vector has a magnitude of 4.68 km. He then jogs due west. (a) What is the magnitude of his displacement vector in the due west direction, if the magnitude of his total displacement vector is 7.41 km? (b) What is the direction of his total displacement vector with respect to a north/south line?

18. One displacement vector s_A has a magnitude of 2.43 km and points due north. A second displacement vector s_B has a magnitude of 7.74 km and also points due north. (a) Find the magnitude and direction of $s_A - s_B$. (b) Find the magnitude and direction of $s_B - s_A$.

19. Vector **A** has a magnitude of 48.0 units and points due west, while vector **B** has the same magnitude but points due south.

Determine the magnitude and direction of (a) **A** + **B** and (b) **A** − **B**.

* 20. A boat travels due east for a distance of 5.74 km and then travels 25.0° south of east for a distance of 6.28 km. Use the graphical technique to find the magnitude and direction of the total displacement vector of the boat.

* 21. A car is being pulled out of the mud by two forces that are applied by the two ropes shown in the drawing. The dashed line in the drawing bisects the 30.0° angle. The magnitude of the force applied by each rope is 650 pounds. Arrange the force vectors tail-to-head and use the graphical technique to answer the following questions. (a) How much force would a single rope need to apply to accomplish the same effect as the two forces added together? (b) How would the single rope be directed relative to the dashed line?

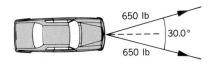

650 lb

30.0°

650 lb

* 22. Vector **A** has a magnitude of 8.00 units and points due west. Vector **B** points due north. (a) What is the magnitude of **B** if **A** + **B** has a magnitude of 10.00 units? (b) What is the direction of **A** + **B**? (c) What is the magnitude of **B** if **A** − **B** has a magnitude of 10.00 units? (d) What is the direction of **A** − **B**?

Section 1.7 Vector Components

23. Vector **A** points along the $+y$ axis and has a magnitude of 100.0 units. Vector **B** points at an angle of 60.0° above the $+x$ axis and has a magnitude of 200.0 units. Vector **C** points along the $+x$ axis and has a magnitude of 87.0 units. Which vector has (a) the largest x component and (b) the largest y component?

24. A displacement vector has a magnitude of 145 m and points at an angle of 28.0° above the $-x$ axis. What is the magnitude and direction of (a) the x component and (b) the y component of the vector?

25. The magnitude of the force vector **F** is 280 pounds. The x component of this vector is directed along the $+x$ axis and has a magnitude of 150 pounds. The y component points along the $+y$ axis. (a) Find the direction of **F** relative to the $+x$ axis. (b) Find the component of **F** along the $+y$ axis.

26. An ocean liner leaves New York City and travels 18.0° north of east for 155 km. How far east and how far north has it gone? In other words, what are the magnitudes of the components of the ship's displacement vector in the directions due east and due north?

27. A helicopter is traveling with a speed of 67.0 m/s toward a point that is located 38.0° south of east. The speed of the helicopter and the direction constitute a vector quantity known as the velocity. Obtain the magnitude of the velocity component that is directed (a) due south and (b) due east.

* 28. A bicyclist is headed due east. A 5.00-m/s wind is blowing partially into the rider's face and is coming from a direction that is 35.0° south of east. The speed of the wind and the direction constitute a vector quantity known as the velocity. In effect, then, the rider must "pump" against a component of the wind's velocity vector. What is the magnitude of this component?

* 29. Determine the magnitude and direction of the x and y components of the vector **A**, relative to (a) the black x, y axes and (b) the colored x', y' axes in the drawing.

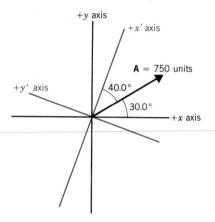

Section 1.8 Addition of Vectors by Means of Vector Components

30. The force vector $\mathbf{F}_A$ has a magnitude of 45.0 newtons and points 30.0° north of east. The force vector $\mathbf{F}_B$ has a magnitude of 75.0 newtons and points due north. Find the magnitude and direction of the resultant $\mathbf{F}_A + \mathbf{F}_B$ by using the component method.

31. A sailboat travels due east for a distance of 1.60 km and then heads 35.0° north of east for another 3.40 km. Using the method of vector components, calculate the vector sum (both magnitude and direction) of these two displacement vectors. Express the direction relative to due east.

32. A pilot flies from point A to point B to point C, in two straight line segments. The displacement vector $\mathbf{s}_{AB}$ for the first leg has a magnitude of 243 km and a direction 50.0° north of east. The displacement vector $\mathbf{s}_{BC}$ for the second leg has a magnitude of 57.0 km and a direction 20.0° south of east. The resultant displacement vector is $\mathbf{s}_{AC} = \mathbf{s}_{AB} + \mathbf{s}_{BC}$. What are the magnitude and direction of $\mathbf{s}_{AC}$? Use the component method.

33. Starting from point A, a football player runs the pattern given in the drawing by the three displacement vectors $\mathbf{s}_1$, $\mathbf{s}_2$, and $\mathbf{s}_3$. Using the component method, find the magnitude and direction θ of the resultant vector $\mathbf{s}_1 + \mathbf{s}_2 + \mathbf{s}_3$.

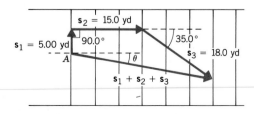

* 34. Three vectors, **A**, **B**, and **C**, add together so that the resultant is zero. Vector **A** points 30.0° north of east. Vector **B** points due north. Vector **C** points 50.0° south of west and has a magnitude of 225 units. By using components, find (a) the magnitude of **A** and (b) the magnitude of **B**. (*See Solved Problem 1 for a related problem.*)

* 35. Three vectors **A**, **B**, and **C** add together so that the resultant vector is zero. Vector **A** points 75.0° north of east. Vector **B** points due west. Vector **C** points due south and has a magnitude of 185 units. By using components find (a) the magnitude of **A** and (b) the magnitude of **B**. (*See Solved Problem 1 for a related problem.*)

**36. What are the x and y components of the vector that must be added to the following three vectors, so that the sum of the four vectors is zero? (*See Solved Problem 1 for a related problem.*)

$$\mathbf{A} = 113 \text{ units, } 60.0° \text{ south of west}$$
$$\mathbf{B} = 222 \text{ units, } 35.0° \text{ south of east}$$
$$\mathbf{C} = 177 \text{ units, } 23.0° \text{ north of east}$$

**37. A sailboat race course consists of four legs, defined by the displacement vectors $\mathbf{s}_A$, $\mathbf{s}_B$, $\mathbf{s}_C$, and $\mathbf{s}_D$, as the drawing indicates. The finish line of the course coincides with the starting line. Using the data in the drawing, find the distance of the fourth leg and the angle θ. (*See Solved Problem 1 for a related problem.*)

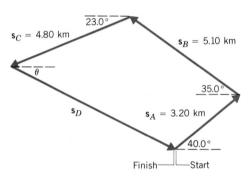

ADDITIONAL PROBLEMS

38. A displacement vector $\mathbf{s}_A$ has a magnitude of 1.62 km and points due north. Another displacement vector $\mathbf{s}_B$ has a magnitude of 2.48 km and points due east. Determine the magnitude and direction of (a) $\mathbf{s}_A + \mathbf{s}_B$ and (b) $\mathbf{s}_A - \mathbf{s}_B$.

39. An observer, whose eyes are 6.0 ft above the ground, is standing 105 feet away from a tree. The ground is level, and the tree is growing perpendicular to it. The observer's line of sight with the treetop makes an angle of 20.0° above the horizontal. How tall is the tree?

40. The x component of a displacement vector **s** has a magnitude of 125 m and points along the $-x$ axis. The y component has a magnitude of 184 m and points along the $-y$ axis. Find the magnitude and direction of **s**. Specify the direction with respect to the x axis.

41. A swimming pool has a volume of 4050 ft³. What is the volume in cubic meters (m³)?

42. A lead brick weighs 35.0 pounds and is resting on an incline, as the drawing shows. Weight is a force and therefore is a vector. It is represented by the arrow labeled **W**. The components of **W** are W_x and W_y, relative to the x, y axes in the drawing. This set of axes is parallel and perpendicular to the incline. Find the magnitudes of W_x and W_y.

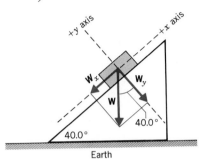

43. Find the resultant of the three displacement vectors in the drawing by means of the component method.

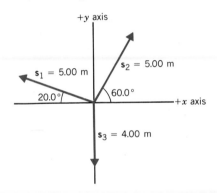

* 44. Consider the two vectors s_1 and s_2 in the drawing for problem 43. Determine the vector sum $s_1 + s_2$ and the vector difference $s_1 - s_2$ using the method of components. In each case, specify both magnitude and direction.

* 45. What is the value of each of the angles of a triangle whose sides are 95, 150, and 190 cm in length? *(Hint: Consider using the law of cosines given in Appendix E.)*

* 46. Vector **A** has a magnitude of 6.00 units and points due east. Vector **B** points due north. (a) What is the magnitude of **B**, if the vector **A** + **B** points 60.0° north of east? (b) Find the magnitude of **A** + **B**.

* 47. The force vector F_A has a magnitude of 90.0 newtons and points 30.0° north of east. The force vector F_B has a magnitude of 150 newtons and points due north. Use the graphical method and find the magnitude and direction of (a) $F_A - F_B$ and (b) $F_B - F_A$.

CHAPTER 2

Kinematics in One Dimension

Bicycle racers are aware of the importance of velocity and acceleration, concepts that are explored in the discussion that follows.

2.1 THE DESCRIPTION OF MOTION

Motion is one of the most noticeable features of the world around us. A moving object may be extremely large and traveling at high speed, like the earth in its journey around the sun. It may be a Saturn rocket, slowly acquiring enough lift-off speed for a successful launch. Or it may be an ordinary football, thrown by a quarterback to a receiver racing toward the end zone.

There are always two aspects to any example of motion. First, in a purely descriptive sense, there is the movement itself. A moving object does not remain in the same place; it travels from one location to another, perhaps quickly, perhaps slowly. Second, there is the important issue of what causes the motion or what changes it, once it has begun. In this second aspect of motion, forces that act on the object come into play. The Saturn rocket moves upward because of the enormous force generated by its engines, while the touchdown pass is sent on its way by the strong arm of the quarterback. And both the rocket and the football continually experience the force of gravity. In fact, the motion of an object can be affected significantly by the forces that act on the object.

Mechanics is the branch of physics that deals with the motion of objects and the forces that change it. Generally speaking, mechanics is divided into two general areas, called kinematics and dynamics. *Kinematics* describes the motion of objects without explicit reference to the forces that act on them. *Dynamics,* on the other hand, is the study of the explicit relationship between forces and their effect on motion.

This chapter discusses the fundamental concepts that form the basis of kinematics and applies them to the description of motion in one dimension. The concepts include displacement, velocity, and acceleration, and Chapter 3 applies them to the description of motion in two dimensions. The study of dynamics depends centrally on three laws of motion discovered by Isaac Newton in the seventeenth century, and Chapter 4 presents these laws.

2.2 DISPLACEMENT

To describe the motion of an object, we must be able to specify the location of the object at all times. Figure 2.1 shows one way of accomplishing this for one-dimensional motion, such as a car traveling along a straight road. Suppose that the initial position of the car is indicated by the vector labeled s_0. As the drawing shows, the length of s_0 is the distance of the car from an arbitrarily chosen origin. At a later time the car has moved to a new position that is indicated by the vector s. The *displacement* of the car Δs (read as "delta s") is a vector drawn from the initial position to the final position. Displacement is a vector quantity in the sense discussed in Section 1.5, for it conveys both a magnitude (the distance between the initial and final positions) and a direction. The displacement can be related to s_0 and s by noting from the drawing that

$$s_0 + \Delta s = s \quad \text{or} \quad \Delta s = s - s_0$$

Thus, the displacement Δs is the difference between s and s_0, and the Greek letter delta (Δ) is used to denote this difference.

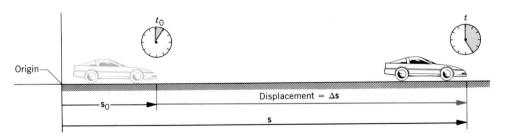

FIGURE 2.1 The displacement Δs is a vector that points from the initial position to the final position. The magnitude of the displacement is the shortest distance between the two positions.

> **DEFINITION OF DISPLACEMENT**
> The displacement of an object is the vector whose magnitude is the shortest distance between the initial and final positions of the motion and whose direction points from the initial position to the final position.
>
> **SI unit of displacement:** meter (m)

The SI unit for displacement is the meter (m), but there are other units as well, such as the inch and the centimeter. When converting between inches (in.) and centimeters (cm) remember that 1 inch = 2.54 centimeters.

We often deal with motion along a straight line. In such a case, a displacement that points in one direction along the line is assigned a positive value, and a displacement pointing in the opposite direction is assigned a negative value. For instance, assume that a car is moving along an east–west direction, and a positive (+) sign is used to denote a direction due east. Then $\Delta \mathbf{s} = +500$ m represents a displacement that points to the east and has a magnitude of 500 meters. Conversely, $\Delta \mathbf{s} = -500$ m is a displacement that has the same magnitude, but points in the opposite direction, due west.

2.3 SPEED AND VELOCITY

AVERAGE SPEED

One of the most obvious features of an object in motion is how fast it is moving. If a car travels 200 meters in 10 seconds, we say its average speed is 20 meters per second. In general, the *average speed* of an object is defined as the distance traveled divided by the amount of time required to cover this distance.

$$\text{Average speed} = \frac{\text{Distance}}{\text{Elapsed time}} \qquad (2.1)$$

Equation 2.1 indicates that the unit for average speed is distance divided by time, or meters per second (m/s) in SI units. Average speed is often expressed in other units as well, such as miles per hour (mi/h) or kilometers per hour (km/h).

EXAMPLE 1
How far does a jogger run in 1.5 hours (5400 s) if his average speed is 2.22 m/s?

SOLUTION
Since we want to find the distance run, we rewrite Equation 2.1 as Distance = (Average speed)(Elapsed time). Therefore,

$$\text{Distance} = (2.22 \text{ m/s})(5400 \text{ s}) = \boxed{12\ 000 \text{ m}}$$

AVERAGE VELOCITY

As useful as it is, the average speed of an object does not reveal anything about the direction of the motion. To take into account the direction of the motion, the vector concept of velocity is now introduced. Consider again the car in Figure 2.1. Suppose that the car's initial position is s_0 when the time is t_0. A little later the car arrives at the final position s at the time t. The time for the car to travel between these two positions is the *difference* between t and t_0, which is denoted by the symbol Δt:

$$\text{Elapsed time} = t - t_0 = \Delta t$$

Dividing the displacement Δs of the car by the elapsed time Δt gives the **average velocity** of the car. It is customary to denote the average value of a quantity by placing a horizontal "bar" above the symbol that represents it, so the average velocity is written as $\bar{v}$. The definition of average velocity is as follows:

DEFINITION OF AVERAGE VELOCITY

$$\text{Average velocity} = \frac{\text{Displacement}}{\text{Elapsed time}}$$

$$\bar{v} = \frac{s - s_0}{t - t_0} = \frac{\Delta s}{\Delta t} \qquad (2.2)$$

SI unit of average velocity: meter per second (m/s)

Equation 2.2 indicates that the unit for average velocity is that of a length divided by a time, or meters per second (m/s) in SI units. Velocity can also be expressed in other units, such as miles per hour (mi/h) or kilometers per hour (km/h).

Average velocity is a vector that points in the same direction as the displacement. As with displacement, we will use plus and minus signs to indicate the two possible directions of the velocity for motion along a straight line. Thus, if the displacement of the object points in the positive direction, the average velocity is also positive. Conversely, if the displacement points in the negative direction, the average velocity is also negative. Example 2 illustrates these features of average velocity.

EXAMPLE 2

The world record for the fastest jet-engined car is 274 m/s (613 mi/h) set by Craig Breedlove in the car *Spirit of America*. Such a measurement is made by calculating the average velocity of the car over a measured course. The driver must make two runs through the course, one in each direction, to nullify any effects of the wind. Figure 2.2a shows that during the first run the car travels from left to right. The displacement of the car for this run is $\Delta s = +604$ m, where the "+" sign denotes that the displacement points to the right. The elapsed time for the run is $\Delta t = 2.19$ s. Part b of the drawing shows the second run. The displacement is now $\Delta s = -604$ m, the minus sign indicating that the displacement points to the left, and the elapsed time is $\Delta t = 2.22$ s. From these data, calculate the average velocity for each run.

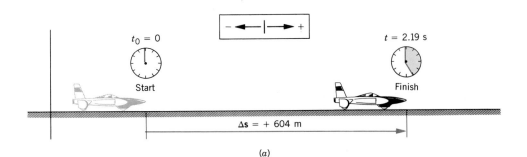

(a)

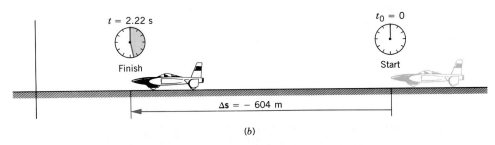

FIGURE 2.2 (a) During the first run, the displacement Δs and the average velocity of the car point to the right. (b) For the second run, the displacement and the average velocity point to the left. The arrows in the box at the top of the drawing indicate the positive and negative directions.

SOLUTION

According to Equation 2.2, the average velocities are

Run 1 $\bar{v} = \dfrac{\text{Displacement}}{\text{Elapsed time}} = \dfrac{\Delta s}{\Delta t} = \dfrac{+604 \text{ m}}{2.19 \text{ s}} = \boxed{+276 \text{ m/s}}$

Run 2 $\bar{v} = \dfrac{\text{Displacement}}{\text{Elapsed time}} = \dfrac{\Delta s}{\Delta t} = \dfrac{-604 \text{ m}}{2.22 \text{ s}} = \boxed{-272 \text{ m/s}}$

The "−" sign that appears in the average velocity for run 2 indicates that the average velocity, like the displacement, points to the left in Figure 2.2b. The magnitudes of the average velocities are 276 m/s and 272 m/s, and the average of these two numbers, 274 m/s, is recorded in the record book.

INSTANTANEOUS VELOCITY

Suppose you made a long trip and the magnitude of your average velocity was 20 m/s. This value, being an average, does not convey any information on how fast you were moving at any instant during the trip. Surely there were times when the car traveled faster than 20 m/s and times when it traveled slower. To have a complete knowledge of how the trip was made, we need to know the car's velocity at every instant. The ***instantaneous velocity*** v of the car indicates both how fast the car moves and the direction of the motion at each instant of time. The magnitude of the instantaneous velocity is called the ***instantaneous speed,*** and it is the number (with units) indicated by the speedometer.

To determine the instantaneous velocity at any point during a trip, we measure the time Δt it takes for the car to travel a *very small* displacement Δs centered on the point of interest. We then compute the average velocity of the car over this interval. If the time Δt is small enough, the instantaneous velocity of the car does not change much during the period over which the average velocity is calculated. Then, the instantaneous velocity v at the point of interest is approximately equal to the average velocity $\bar{v}$ computed over the interval, or $v \approx \bar{v} = \Delta s/\Delta t$ (for sufficiently small Δt). In fact, in the limit that Δt becomes infinitesimally small so it approaches zero, the instantaneous velocity and the average velocity become equal, and we can write

$$v = \lim_{\Delta t \to 0} \frac{\Delta s}{\Delta t} \tag{2.3}$$

The notation $\lim_{\Delta t \to 0} (\Delta s/\Delta t)$ means that the ratio $\Delta s/\Delta t$ is defined by a limiting process in which smaller and smaller values of Δt are used. As smaller values of Δt are used, Δs also becomes smaller. However, the ratio $\Delta s/\Delta t$ does *not* become zero but, rather, approaches the value of the instantaneous velocity.

For brevity, we will hereafter use the word *velocity* to mean "instantaneous velocity" and *speed* to mean "instantaneous speed." In a wide range of realistic motions, the velocity changes from moment to moment. To describe the manner in which the velocity changes, however, the concept of acceleration is needed, as the next section discusses.

2.4 ACCELERATION

Whenever the velocity of an object is changing, the object is said to be "accelerating." A car, temporarily stopped at a traffic signal, accelerates when the light turns green and the driver steps on the gas; or, if traveling along the highway, a car accelerates to a greater velocity to pass another car. While these examples involve an increase in velocity, there are also many examples of acceleration where the velocity decreases, such as a car slowing down.

The meaning of *average acceleration* can be illustrated by considering a plane during takeoff. Figure 2.3 focuses attention on how the velocity of the plane changes as the plane moves down the runway. During an elapsed time interval $\Delta t = t - t_0$, the velocity changes from an initial value of $\mathbf{v_0}$ to a final value of $\mathbf{v}$, the change in velocity being $\Delta \mathbf{v} = \mathbf{v} - \mathbf{v_0}$. The average acceleration is defined in the following manner, to provide a measure of how much the velocity changes per unit of elapsed time.

DEFINITION OF AVERAGE ACCELERATION

Average acceleration = $\dfrac{\text{Change in velocity}}{\text{Elapsed time}}$

$$\overline{\mathbf{a}} = \frac{\mathbf{v} - \mathbf{v_0}}{t - t_0} = \frac{\Delta \mathbf{v}}{\Delta t} \qquad (2.4)$$

SI unit of average acceleration: meter per second squared (m/s²)

The average acceleration $\overline{\mathbf{a}}$ is a vector that points in the same direction as $\Delta \mathbf{v}$, the change in the velocity. Following the usual custom, plus and minus signs will be used to indicate the two possible directions for the acceleration vector when the motion is along a straight line.

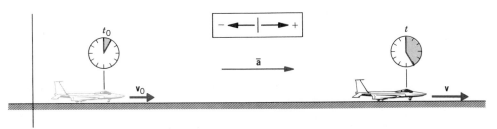

FIGURE 2.3 During takeoff, the plane accelerates from an initial velocity $\mathbf{v_0}$ to a final velocity $\mathbf{v}$ during the time interval $\Delta t = t - t_0$.

As with velocity, we are often interested in an object's acceleration at any instant of time. The **instantaneous acceleration a** can be defined by analogy with the procedure used for instantaneous velocity:

$$\mathbf{a} = \lim_{\Delta t \to 0} \frac{\Delta \mathbf{v}}{\Delta t} \qquad (2.5)$$

Equation 2.5 indicates that the instantaneous acceleration is a limiting case of the average acceleration. When the time interval Δt for measuring the average acceleration becomes extremely small (approaching zero in the limit), the average acceleration and the instantaneous acceleration become equal. Moreover, in many situations the acceleration is constant, so the acceleration has the same value at any instant of time. In the future, we will use the word *acceleration* to mean "instantaneous acceleration." Let us now consider some examples.

EXAMPLE 3

Suppose the plane in Figure 2.3 starts from rest ($v_0 = 0$) when $t_0 = 0$. The plane accelerates down the runway, and at $t = 29$ s attains a velocity of $\mathbf{v} = +260$ km/h, where the "+" sign indicates the velocity points to the right in the drawing. Determine the average acceleration of the plane.

SOLUTION

The average acceleration of the plane can be found from Equation 2.4 as

$$\bar{\mathbf{a}} = \frac{\mathbf{v} - \mathbf{v}_0}{t - t_0} = \frac{+260 \text{ km/h} - 0 \text{ km/h}}{29 \text{ s} - 0 \text{ s}} = \boxed{+9.0 \frac{\text{km/h}}{\text{s}}}$$

The average acceleration calculated in Example 3 is read as "nine kilometers per hour per second." Assuming the acceleration of the plane is constant, an acceleration of $9.0 \frac{(\text{km/h})}{\text{s}}$ means the plane changes its velocity by 9 km/h during each second of the motion. During the first second, the plane's velocity increases from 0 to 9 km/h; during the next second, the velocity increases by another 9 km/h to 18 km/h, and so on. By the end of the 29th second, the velocity is 260 km/h. Figure 2.4 illustrates how the velocity changes during the first three seconds.

It is customary to express the units for acceleration solely in terms of SI units. One way to obtain SI units for the acceleration in Example 3 is to convert the velocity units from km/h to m/s:

$$260 \frac{\text{km}}{\text{h}} \left(\frac{\frac{1000 \text{ m}}{1 \text{ km}}}{\frac{3600 \text{ s}}{\text{h}}} \right) = 72 \text{ m/s}$$

The average acceleration then becomes

$$\bar{\mathbf{a}} = \frac{+72 \text{ m/s} - 0 \text{ m/s}}{29 \text{ s} - 0 \text{ s}} = +2.5 \text{ m/s}^2$$

where we have used $2.5 \frac{\text{m/s}}{\text{s}} = 2.5 \frac{\text{m}}{\text{s} \cdot \text{s}} = 2.5 \frac{\text{m}}{\text{s}^2}$. An acceleration of $2.5 \frac{\text{m}}{\text{s}^2}$ is read as "2.5 meters per second per second" (or "2.5 meters per second squared"), and it means that the velocity is changing by 2.5 m/s during each second of the accelerated motion.

$$\bar{a} = \frac{+9.0 \text{ km/h}}{s}$$

$\Delta t = 0$
$v_0 = 0$

$\Delta t = 1.0$ s
$v = +9.0$ km/h

$\Delta t = 2.0$ s
$v = +18$ km/h

$\Delta t = 3.0$ s $v = +27$ km/h

FIGURE 2.4 An acceleration of $+9 \dfrac{\text{km/h}}{s}$ means that the velocity of the plane changes by $+9$ km/h during each second of the motion. The "+" direction for **a** and **v** is to the right.

EXAMPLE 4

A drag racer crosses the finish line, and the driver applies the brakes to slow down, as Figure 2.5 illustrates. The brakes are initially applied when $t_0 = 9.0$ s and the car's velocity is $v_0 = +28$ m/s. When $t = 12.0$ s, the velocity has been reduced to $v = +13$ m/s. What is the average acceleration of the dragster?

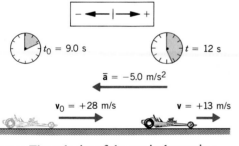

$- \longleftarrow \mid \longrightarrow +$

$t_0 = 9.0$ s $t = 12$ s

$\bar{a} = -5.0$ m/s²

$v_0 = +28$ m/s $v = +13$ m/s

FIGURE 2.5 The velocity of the car is decreasing, giving rise to an average acceleration that points opposite to the velocity.

SOLUTION

According to Equation 2.4, the average acceleration is

$$\bar{a} = \frac{v - v_0}{t - t_0} = \frac{+13 \text{ m/s} - 28 \text{ m/s}}{12.0 \text{ s} - 9.0 \text{ s}} = \boxed{-5.0 \text{ m/s}^2}$$

Figure 2.6 shows the velocity of the dragster during the braking, assuming the acceleration is constant throughout the motion. The average acceleration calculated in Example 4 is negative, indicating the acceleration points to the left in the drawing. As such, the average acceleration and the velocity point in *opposite* directions. Whenever the acceleration and velocity are in opposite directions, the object slows down and is said to be "decelerating." In contrast, the acceleration and velocity in Figure 2.4 point in the *same* direction, and the object speeds up.

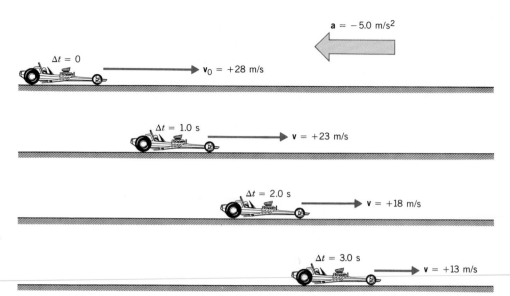

$a = -5.0 \text{ m/s}^2$

$\Delta t = 0$

$v_0 = +28 \text{ m/s}$

$\Delta t = 1.0 \text{ s}$

$v = +23 \text{ m/s}$

$\Delta t = 2.0 \text{ s}$

$v = +18 \text{ m/s}$

$\Delta t = 3.0 \text{ s}$

$v = +13 \text{ m/s}$

FIGURE 2.6 Here, an acceleration of -5.0 m/s^2 means the velocity decreases by 5.0 m/s during each second of elapsed time.

2.5 EQUATIONS OF KINEMATICS FOR CONSTANT ACCELERATION

We are now in a position to describe the motion of an object traveling with constant acceleration along a straight line. For convenience in what follows, it will be assumed that the object is located at the coordinate origin $s_0 = 0$ at the time $t_0 = 0$. With this assumption, the displacement $\Delta s = s - s_0$ becomes $\Delta s = s$. Furthermore, in what follows it is customary to dispense with the use of boldface symbols to denote the displacement, velocity, and acceleration vectors.

Suppose, for example, that a motorcycle has an initial velocity of $v_0 = +17 \text{ m/s}$ and moves for 4.0 s with a constant acceleration of $a = +3.0 \text{ m/s}^2$. For a complete description of the motion, it is also necessary to know the velocity and displacement of the motorcycle at the end of the 4.0-s interval. The final velocity v can be obtained from Equation 2.4:

$$\bar{a} = a = \frac{v - v_0}{t}$$

Solving for v yields

$$v = v_0 + at = +17 \text{ m/s} + (+3.0 \text{ m/s}^2)(4.0 \text{ s}) = +29 \text{ m/s}$$

The displacement of the motorcycle after 4.0 s can be obtained from Equation 2.2, if a value for the average velocity $\bar{v}$ can be determined.

$$\bar{v} = \frac{s - s_0}{t - t_0} = \frac{s}{t} \quad \text{or} \quad s = \bar{v}t$$

Because the acceleration is constant, the velocity increases at a constant rate. Thus, the average velocity $\bar{v}$ is midway between the initial and final velocities:

$$\bar{v} = \tfrac{1}{2}(v_0 + v) \qquad \text{(constant acceleration)} \qquad (2.6)$$

Equation 2.6 applies only if the acceleration is constant and cannot be used when the acceleration is changing. The average velocity of the motorcycle is, then,

$$\bar{v} = \tfrac{1}{2}(+17 \text{ m/s} + 29 \text{ m/s}) = +23 \text{ m/s}$$

The displacement of the motorcycle can now be determined by using $+23$ m/s for the average velocity:

$$s = \bar{v}t = (+23 \text{ m/s})(4.0 \text{ s}) = +92 \text{ m}$$

The preceding calculation for the displacement can be summarized algebraically as

$$s = \bar{v}t = \tfrac{1}{2}(v_0 + v)t \quad \text{(constant acceleration)} \quad\quad (2.7)$$

By using Equations 2.4 and 2.7, we were able to complete the description of the motorcycle's motion. These equations are repeated below for the sake of convenience:

$$v = v_0 + at \quad\quad (2.4) \quad\quad s = \tfrac{1}{2}(v_0 + v)t \quad\quad (2.7)$$

Notice that there are five kinematic variables that play a role in these equations:

1. $s =$ displacement
2. $a = \bar{a} =$ acceleration (constant)
3. $v =$ final velocity at time t
4. $v_0 =$ initial velocity at time $t_0 = 0$
5. $t =$ time elapsed since $t_0 = 0$

In addition, each equation contains four variables, so if three of them are known, the fourth variable can always be found. Example 5 further illustrates how Equations 2.4 and 2.7 are used to describe the motion of an object.

EXAMPLE 5

The speedboat in Figure 2.7 has a constant acceleration of $+2.0$ m/s². If the initial velocity of the boat is $+6.0$ m/s, calculate its displacement after 8.0 seconds.

SOLUTION

The three known variables are listed in the table:

Speedboat data

s	a	v	v_0	t
?	$+2.0$ m/s²		$+6.0$ m/s	8.0 s

We can use $s = \tfrac{1}{2}(v_0 + v)t$ to find the displacement of the boat if a value for the final velocity v can be found. The final velocity can be determined directly from Equation 2.4:

$$v = v_0 + at = +6.0 \text{ m/s} + (+2.0 \text{ m/s}^2)(8.0 \text{ s}) = +22 \text{ m/s}$$

The displacement of the boat is then

$$s = \tfrac{1}{2}(v_0 + v)t \quad\quad (2.7)$$
$$= \tfrac{1}{2}(+6.0 \text{ m/s} + 22 \text{ m/s})(8.0 \text{ s}) = \boxed{+110 \text{ m}}$$

A calculator would give the answer for the displacement as 112 m, but this number must be rounded to 110 m, since the data are accurate to only two significant figures.

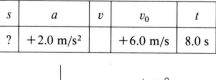

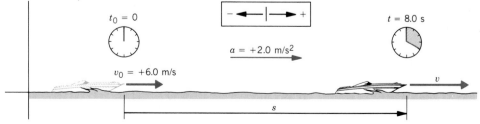

FIGURE 2.7 An accelerating speedboat.

The solution to Example 5 involved two steps: first finding v and then calculating s. For future work, it would be helpful if we could find an equation that would allow us to determine the displacement in a single step. Using Example 5 as a guide, we can obtain such an equation by substituting v from Equation 2.4 ($v = v_0 + at$) for the v that appears in Equation 2.7 [$s = \frac{1}{2}(v_0 + v)t$]:

$$s = \tfrac{1}{2}(v_0 + v)t = \tfrac{1}{2}(v_0 + \boxed{v_0 + at})t = \tfrac{1}{2}(2v_0 t + at^2)$$

$$s = v_0 t + \tfrac{1}{2}at^2 \qquad \text{(constant acceleration)} \tag{2.8}$$

The reader can verify that Equation 2.8 gives the displacement of the speedboat directly without the intermediate step of determining the final velocity. The first term ($v_0 t$) on the right side of Equation 2.8 represents the displacement that would result if the acceleration were zero, and the velocity remained constant at its initial value of v_0. The second term on the right ($\frac{1}{2}at^2$) gives the additional displacement that arises because the acceleration changes the velocity to values that are different from its initial value.

In Example 6, as in Example 5, two steps are required to complete the solution.

EXAMPLE 6

A jet is taking off from the deck of an aircraft carrier, as Figure 2.8 shows. Starting from rest, the jet is catapulted with a constant acceleration of $+31$ m/s² along a straight line and reaches a velocity of $+62$ m/s. Find the displacement of the jet.

SOLUTION

The data are as follows:

Jet data

s	a	v	v_0	t
?	$+31$ m/s²	$+62$ m/s	0	

The displacement s of the aircraft can be obtained from $s = \frac{1}{2}(v_0 + v)t$ if we can determine the time t during which the plane is being accelerated. For this purpose, Equation 2.4 is used:

$$t = \frac{v - v_0}{a} = \frac{+62 \text{ m/s} - 0}{+31 \text{ m/s}^2} = 2.0 \text{ s}$$

Since the time is known, the displacement can be found:

$$s = \tfrac{1}{2}(v_0 + v)t = \tfrac{1}{2}(0 + 62 \text{ m/s})(2.0 \text{ s}) = \boxed{+62 \text{ m}} \tag{2.7}$$

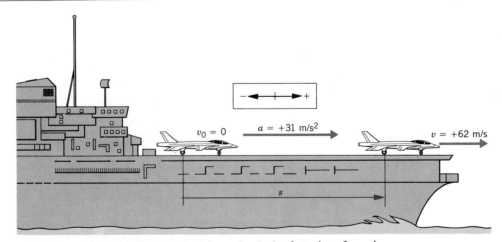

FIGURE 2.8 A plane is being launched from the deck of an aircraft carrier.

We can derive a single equation for finding the displacement s when a, v, and v_0 are known, but the time t is not known, as in Example 6. Solving Equation 2.4 for the time $[t = (v - v_0)/a]$, and then substituting t into Equation 2.7 reveals that

$$s = \tfrac{1}{2}(v_0 + v)t = \tfrac{1}{2}(v_0 + v)\boxed{\frac{v - v_0}{a}} = \frac{v^2 - v_0^2}{2a}$$

Solving for v^2, it follows that

$$v^2 = v_0^2 + 2as \qquad \text{(constant acceleration)} \tag{2.9}$$

It is a straightforward exercise to verify that Equation 2.9 can be used to find the displacement of the jet in Example 6 by using just the original data, without having to solve first for the time.

Table 2.1 presents a summary of the equations that we have been considering; these equations are called the ***equations of kinematics.*** Each equation contains four variables, as indicated by the check marks (✓) in the table. In the next section the equations of kinematics are applied to situations involving constant acceleration.

2.6 APPLICATIONS OF THE EQUATIONS OF KINEMATICS

In this section the equations of kinematics are applied to a variety of situations. We have chosen the examples to illustrate important points regarding the use of these equations.

Before attempting to solve a problem, verify that the given information contains values for at least three of the five kinematic variables (s, a, v, v_0, t). A glance back at the data boxes in Examples 5 and 6 will show which three variables were given for these problems. Only when values for at least three variables are known can the equations listed in Table 2.1 be used to find the values for the fourth and fifth variables.

Decide at the start which directions are to be called positive (+) and negative (−) relative to a conveniently chosen coordinate origin. While this decision is arbitrary, it is nonetheless an important one, for displacement, velocity, and acceleration are vectors, and their directions must always be taken into account. In the examples that follow, the positive and negative directions will be shown in the drawing that accompanies the problem. It does not matter which direction is chosen to be positive and which is chosen to be negative, because the equations of kinematics yield answers that are interpreted according to the choice for these directions. However, once the choices have been made, they should not be changed during the course of the calculation. Example 7 illustrates these important issues.

TABLE 2.1 Equations of Kinematics for Constant Acceleration

Equation Number	Equation	Variables				
		s	a	v	v_0	t
(2.4)	$v = v_0 + at$		✓	✓	✓	✓
(2.7)	$s = \tfrac{1}{2}(v_0 + v)t$	✓		✓	✓	✓
(2.8)	$s = v_0 t + \tfrac{1}{2}at^2$	✓	✓		✓	✓
(2.9)	$v^2 = v_0^2 + 2as$	✓	✓	✓	✓	

EXAMPLE 7

(a) Repeat Example 5 using the same data, but now assume the negative direction is to the right, rather than to the left (see Figure 2.9). (b) Calculate the final velocity of the boat.

SOLUTION

(a) The three known variables are as follows:

Speedboat data

s	a	v	v_0	t
?	-2.0 m/s^2		-6.0 m/s	8.0 s

Note that a and v_0 are now negative numbers, consistent with our new choice for the negative direction. The displace-

ment of the boat can be found from Equation 2.8 in Table 2.1. This equation is selected because it contains the four variables of interest, s, a, v_0, and t, with s being the only unknown.

$$s = v_0 t + \tfrac{1}{2}at^2 = (-6.0 \text{ m/s})(8.0 \text{ s}) + \tfrac{1}{2}(-2.0 \text{ m/s}^2)(8.0 \text{ s})^2$$
$$= \boxed{-110 \text{ m}}$$

(b) The final velocity of the boat can be determined from Equation 2.4:

$$v = v_0 + at = (-6.0 \text{ m/s}) + (-2.0 \text{ m/s}^2)(8.0 \text{ s})$$
$$= \boxed{-22 \text{ m/s}}$$

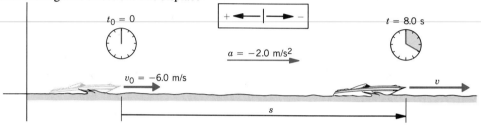

FIGURE 2.9 This drawing is the same as Figure 2.7, except the direction to the right is chosen to be negative.

The magnitude of the displacement in Example 7 is 110 m, the same as in Example 5. Now, however, the displacement is in the negative direction. Since the negative direction now points to the right, the meaning of this answer is exactly the same as it is in Example 5, that is, the boat moves to the right. Likewise, the magnitude of the final velocity in Example 7 is again 22 m/s, and the direction is to the right. The physical meaning of the answers in both examples is the same when interpreted in the context of the initial choice for the positive and negative directions.

Sometimes there are two possible answers to a kinematics problem, and the answers correspond to different situations. Example 8 illustrates this feature.

EXAMPLE 8

The spacecraft shown in Figure 2.10a is traveling with a velocity of $+3250$ m/s. Suddenly the retrorocket is fired, and the spacecraft begins to slow down with an acceleration whose magnitude is 10.0 m/s^2. What is the velocity of the spacecraft when the displacement of the craft is $+215$ km, relative to the point where the retrorocket began firing?

SOLUTION

Since the spacecraft is slowing down at this stage of the motion, the acceleration must be opposite to the velocity. The velocity points to the right in the drawing, so the acceleration must point to the left, which is the negative direction; thus, $a = -10.0$ m/s^2.

The three known variables are listed below.

Spacecraft data

s	a	v	v_0	t
$+215\,000$ m	-10.0 m/s^2	?	$+3250$ m/s	

The final velocity v of the spacecraft can be calculated using Equation 2.9, since it contains the four pertinent variables:

$$v^2 = v_0^2 + 2as = (+3250 \text{ m/s})^2 + 2(-10.0 \text{ m/s}^2)(+215\,000 \text{ m})$$
$$= 6.3 \times 10^6 \text{ m}^2/\text{s}^2$$

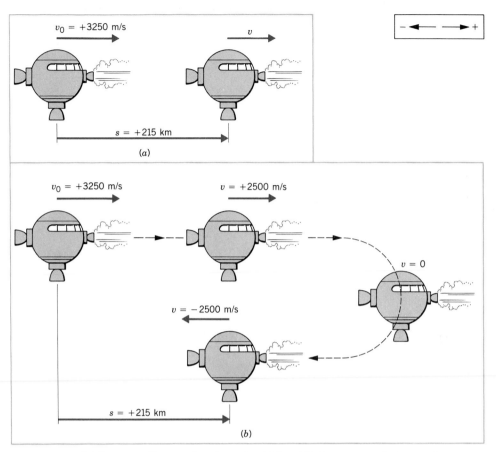

FIGURE 2.10 *(a)* Because of an acceleration of -10.0 m/s^2, the spacecraft changes its velocity from v_0 to v. *(b)* Continued firing of the retrorocket changes the direction of the craft's motion.

The two possible solutions are

$$v = +2.5 \times 10^3 \text{ m/s} \quad \text{or} \quad v = -2.5 \times 10^3 \text{ m/s}$$

Each of the two answers for the final velocity corresponds to the *same* displacement ($s = +215$ km) of the spacecraft, but each arises in a different part of the motion. The first answer, $v = +2500$ m/s, corresponds to the situation in Figure 2.10a. Here the spacecraft has slowed down to a speed of 2500 m/s, but is still traveling to the right. The second answer, $v = -2500$ m/s, arises because the retrorocket eventually brings the spacecraft to a momentary halt and causes it to reverse its direction of travel, after which it moves toward the left. As the craft moves toward the left, its speed increases due to the continually firing retrorocket. There comes a point in time when the velocity of the craft becomes $v = -2500$ m/s, giving rise to the situation shown in part *b* of the drawing. In both parts of the drawing the spacecraft has the same displacement, but a greater travel time is required in part *b* compared to part *a*. The reader can verify that it takes 75 s for the spacecraft to reach the final velocity of $v = +2500$ m/s, whereas it requires 575 s to reach $v = -2500$ m/s, since the spacecraft travels a greater distance in acquiring this velocity.

The motion of two objects may be interrelated, so they share a common variable. The fact that the motions are interrelated is an important piece of information. In such cases, data for only two variables need be specified for each object. Example 9 involves the interrelationship between two moving objects, a person and a bus. At first glance it appears that the problem cannot be solved, because only two kinematic variables are specified for each object. In this case, however, the third variable, the time t, is common between the person and the bus. Even though no value is given for the common variable, the problem can, in fact, be solved. The answer turns out to be quite interesting.

EXAMPLE 9

In Figure 2.11 a bus has stopped to pick up riders. A woman is running at a constant velocity of $+5.0$ m/s in an attempt to catch the bus. However, when she is 11 m from the bus, it pulls away with a constant acceleration of $+1.0$ m/s². From this point, how much time does it take her to reach the bus if she keeps running with the same velocity?

SOLUTION

As the drawing shows, when the woman catches the bus, her displacement s_{woman} is related to the displacement of the bus s_{bus} according to

$$s_{bus} + 11 \text{ m} = s_{woman}$$

Assuming the clock is zeroed at the instant when she is 11 m from the bus, her displacement is $s_{woman} = (5.0 \text{ m/s})t$, since she runs with a constant velocity. Therefore,

$$s_{bus} + 11 \text{ m} = (5.0 \text{ m/s})t$$

This result indicates that we must obtain a value for s_{bus} before t can be calculated. Thus, the motion of the bus is now considered, and the known variables for the bus are indicated below.

Bus data

s_{bus}	a	v	v_0	t
?	$+1.0$ m/s²		0	✓

A check mark has been placed in the box above for the time t. Although we do not have an explicit value for t, it is the same for both the bus and the woman. In other words, the time is a "common" variable, and the check mark reminds us of this fact.

Equation 2.8 can be used to relate the displacement of the bus and the time:

$$s_{bus} = v_0 t + \tfrac{1}{2}at^2 = \tfrac{1}{2}(1.0 \text{ m/s}^2)t^2$$

This result for the displacement of the bus can be substituted into the expression obtained earlier to give an equation containing only one unknown, the time t:

$$s_{bus} + 11 \text{ m} = (5.0 \text{ m/s})t$$

$$\tfrac{1}{2}(1.0 \text{ m/s}^2)t^2 + 11 \text{ m} = (5.0 \text{ m/s})t$$

or

$$(0.50 \text{ m/s}^2)t^2 - (5.0 \text{ m/s})t + 11 \text{ m} = 0$$

This last equation is a quadratic equation for t. Solving for t by using the quadratic formula (see Appendix C) reveals that there are two solutions:

$$\boxed{t = 3.3 \text{ s} \quad \text{or} \quad t = 6.7 \text{ s}}$$

Evidently there are two times when the woman can catch the bus, and they correspond to different situations. When she catches the bus for the first time at $t = 3.3$ s, she has a *different* velocity than the bus. In fact, the woman has a greater velocity, because the bus started from rest and in 3.3 s has not developed much velocity. Remember, the condition for catching the bus is that the woman and the bus must be at the same place at the same time; it does not matter if they have different velocities when they meet. If she fails to get the attention of the driver when she reaches the bus, she will actually run past it, since she has the greater velocity. Soon, however, the bus will catch up with her, because it is accelerating while she is maintaining a constant velocity. The two will meet for the second time when $t = 6.7$ s. If, after this time, the driver still does not notice her, the bus will pull ahead of the woman, and the two will never meet again.

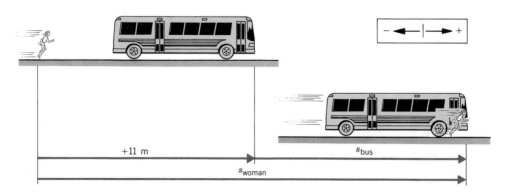

FIGURE 2.11 A person attempting to catch a bus.

2.7 FREELY FALLING BODIES

FREELY FALLING BODIES AND THE EQUATIONS OF KINEMATICS

Everyone has observed the effect of gravity as it causes objects to fall downward. A diver springing from a three-meter board falls downward toward the pool; an arrow shot upward eventually returns to earth. In the absence of air resistance, it is found that all bodies at the same location on the earth fall vertically with the same acceleration. Furthermore, if the distance of the fall is small compared to the radius of the earth, the acceleration remains constant throughout the fall. This idealized motion, in which air resistance is neglected and the acceleration is nearly constant, is known as *free fall.*

The acceleration of a freely falling body is called the *acceleration due to gravity,* and its magnitude is denoted by the symbol g. The acceleration due to gravity is directed downward, toward the center of the earth, and it decreases with increasing altitude; g is smaller on top of a mountain than at sea level. Furthermore, g varies slightly with latitude. However, near the earth's surface g is approximately

$$g = 9.80 \text{ m/s}^2 \quad \text{or} \quad 32.2 \text{ ft/s}^2$$

Unless circumstances warrant otherwise, we will use either of these values for g in subsequent calculations.

Figure 2.12a shows the well-known effect in which a rock falls faster than a sheet of paper. The effect of air resistance is responsible for the erroneous notion that gravity makes heavy objects fall faster than light objects. Figure 2.12b, in contrast, illustrates free fall motion. When the air is removed from the tube, the rock and paper fall with exactly the same acceleration. Free fall is closely approximated for objects falling near the surface of the moon, where there is no air to retard the motion. A nice demonstration of lunar free fall was performed in 1971 by astronaut David Scott who dropped a hammer and a feather simultaneously from the same height. Both experienced the same acceleration due to lunar gravity and consequently hit the ground at the same time. The acceleration due to gravity near the surface of the moon is approximately one-sixth as large as that on the earth.

We now turn our attention to several examples that illustrate how the equations of kinematics are applied to freely falling bodies.

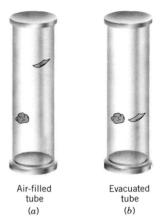

Air-filled tube
(a)

Evacuated tube
(b)

FIGURE 2.12 *(a)* In the presence of air resistance, the acceleration of the rock is greater than that of the paper. *(b)* In the absence of air resistance, both the rock and the paper have the same acceleration.

EXAMPLE 10

A stone is dropped from rest from the top of a tall building, as Figure 2.13 indicates. After 3.00 s of free fall, what is (a) the displacement and (b) the velocity of the stone?

SOLUTION

(a) For convenience, the downward direction is arbitrarily chosen as the positive direction. The three known variables are shown in the box. The initial velocity v_0 of the stone is zero, because the stone is dropped from rest.

Stone data

s	a	v	v_0	t
?	$+9.80 \text{ m/s}^2$		0	3.00 s

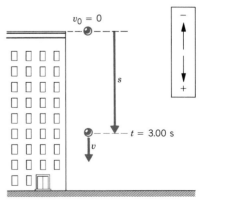

FIGURE 2.13 The stone, starting with zero velocity at the top of the building, is accelerated downward by gravity.

Equation 2.8 employs the appropriate variables and offers a direct solution to the problem:

$$s = v_0 t + \tfrac{1}{2}at^2 = \tfrac{1}{2}(9.80 \text{ m/s}^2)(3.00 \text{ s})^2 = \boxed{+44.1 \text{ m}}$$

(b) The final velocity is found most conveniently from Equation 2.4:

$$v = v_0 + at = (9.80 \text{ m/s}^2)(3.00 \text{ s}) = \boxed{+29.4 \text{ m/s}}$$

The "+" signs associated with the answers for s and v indicate these vectors point in the positive direction, toward the earth.

Example 10 reveals an interesting feature of uniformly accelerated motion when a body starts from rest. Equations 2.4 and 2.8 predict that the velocity and displacement at any time during the motion are

$$\left.\begin{array}{l} v = at \\ s = \tfrac{1}{2}at^2 \end{array}\right\} \text{ if } v_0 = 0$$

It is clear from these equations that v is proportional to t, while s is proportional to t^2. For example, if the time is tripled, the velocity is also tripled, but the displacement is increased by a factor of nine. Figure 2.14 shows the velocity and displacement for the first 5 seconds of free fall, assuming the body starts from rest. Notice, in particular, that the object is traveling at a speed of 29.4 m/s (65.8 mi/h) after only 3 s of free fall! There are very few cars that can accelerate from 0 to 29.4 m/s in 3 s, a fact that shows how large the acceleration due to gravity really is.

The acceleration due to gravity is always a downward-pointing vector and acts to increase the speed of an object falling downward. This same acceleration also acts to decrease the speed of an object moving upward, eventually bringing it to a halt and causing it to fall back to earth. Example 11 shows how the equations of kinematics are applied to the latter situation.

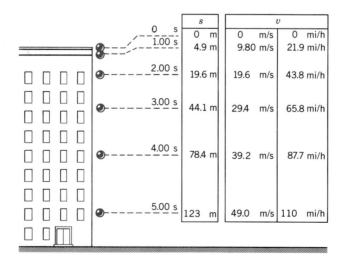

	s	v	
0 s	0 m	0 m/s	0 mi/h
1.00 s	4.9 m	9.80 m/s	21.9 mi/h
2.00 s	19.6 m	19.6 m/s	43.8 mi/h
3.00 s	44.1 m	29.4 m/s	65.8 mi/h
4.00 s	78.4 m	39.2 m/s	87.7 mi/h
5.00 s	123 m	49.0 m/s	110 mi/h

FIGURE 2.14 The displacement s and velocity v of a freely falling body are illustrated for the first five seconds of fall. The body is dropped from rest.

EXAMPLE 11

A ball is thrown vertically upward with an initial speed of 98.5 ft/s. In the absence of air resistance, (a) how high does the ball go, (b) how much time does it take for the ball to reach its maximum height, and (c) what is the total time the ball is in the air before striking the ground?

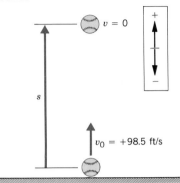

FIGURE 2.15 The ball is thrown upward with an initial velocity of $v_0 = +98.5$ ft/s. The velocity of the ball is momentarily zero when the ball reaches its maximum height.

SOLUTION

(a) The final velocity v is zero when the ball reaches its highest point, just before beginning its descent back to earth. The acceleration due to gravity is -32.2 ft/s^2, the minus sign appearing because downward has been selected as the negative direction in Figure 2.15. The four pertinent variables appear in the box, and Equation 2.9 ($v^2 = v_0^2 + 2as$) can be solved to find the maximum height s:

Ball data

s	a	v	v_0	t
?	-32.2 ft/s^2	0	$+98.5$ ft/s	

$$s = \frac{v^2 - v_0^2}{2a} = \frac{-(98.5 \text{ ft/s})^2}{2(-32.2 \text{ ft/s}^2)} = \boxed{+151 \text{ ft}}$$

(b) Equation 2.4 ($v = v_0 + at$) allows us to calculate the time required for the ball to reach the maximum height of 151 ft:

$$t = \frac{v - v_0}{a} = \frac{-(98.5 \text{ ft/s})}{-32.2 \text{ ft/s}^2} = \boxed{3.06 \text{ s}}$$

(c) The equations of kinematics can be used to show that the ball takes 3.06 s to fall from its maximum height back to the ground. In other words, the time for the ball to reach its maximum height is equal to the time for it to fall back to the ground. Hence, the total time the ball is in flight is $\boxed{t_{\text{total}} = 6.12 \text{ s.}}$ It is interesting to note that t_{total} can be determined by another method—a method that is not so obvious. When the ball is thrown into the air and returns, the displacement for the *entire trip* is $s = 0$, because the ball returns to its original position. Substituting the data below into Equation 2.8, we can find the time for the entire trip:

Ball data for entire trip

s	a	v	v_0	t
0	-32.2 ft/s^2		$+98.5$ ft/s	?

$$s = v_0 t + \tfrac{1}{2}at^2 = (v_0 + \tfrac{1}{2}at)(t)$$
$$0 = [98.5 \text{ ft/s} + \tfrac{1}{2}(-32.2 \text{ ft/s}^2)t](t)$$

There are two solutions to this equation: $t = 0$ s, and $t = 6.12$ s. As mentioned in the last section, the two solutions correspond to different parts of the motion. The first solution, $t = 0$ s, corresponds to the situation where the ball has not yet left the ground, so its displacement is zero. The second solution, $t = 6.12$ s, gives the time for the ball to make a complete up-and-down trip, the displacement also being zero.

Example 11 illustrates that the expression "freely falling" does not necessarily mean an object is falling down. A freely falling object is any object moving freely under the influence of gravity. Once the ball is released with an initial speed of 98.5 ft/s, its upward motion is free fall, just like its downward motion during the return trip. A freely falling object, whether moving upward or downward, always experiences the same *downward acceleration* due to gravity.

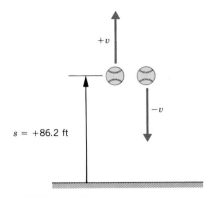

FIGURE 2.16 For a given displacement along the ball's path, the upward speed of the ball is equal to its downward speed. Note that the two velocities point in opposite directions.

SYMMETRY IN THE MOTION OF FREELY FALLING BODIES

The motion of an object that is thrown upward and eventually returns to earth, as in Example 11, contains a symmetry that is useful to keep in mind from the point of view of problem solving. The calculations just completed indicate that a time symmetry exists in free fall motion, in the sense that the time required for the object to reach maximum height equals the time for it to return to its starting point.

Another type of symmetry involving the speed also exists. Figure 2.16 shows the ball considered in Example 11. At any displacement s above the ground, the speed of the ball during the upward trip equals the speed at the same point during the downward trip. For example, when $s = +86.2$ ft, Equation 2.9 gives two possible values for v, as shown below:

$$v^2 = v_0{}^2 + 2as = (98.5 \text{ ft/s})^2 + 2(-32.2 \text{ ft/s}^2)(86.2 \text{ ft}) = 4150 \text{ ft}^2/\text{s}^2$$

$$v = \pm 64.4 \text{ ft/s}$$

The value $v = +64.4$ ft/s is the velocity of the ball on its upward trip, while $v = -64.4$ ft/s is its velocity on the downward trip. The speed in both cases is identical and equals 64.4 ft/s. Likewise, the speed of the ball just before it strikes the ground is equal to its initial speed of 98.5 ft/s. Another way to view this result is to note that the upward-moving ball loses 32.2 ft/s in speed each second on the way up toward the highest point. Then, the ball gains 32.2 ft/s for each second during the descent. So for each point along the path, the upward speed of the ball equals the downward speed.

*2.8 GRAPHICAL ANALYSIS OF VELOCITY AND ACCELERATION FOR LINEAR MOTION

Suppose a bicyclist is riding with a constant velocity of $v = +4$ m/s. The position s of the bicycle can be plotted as a function of time t on a graph in which s is measured along the vertical axis and t is measured along the horizontal axis. Since the position of the bike increases by 4 m every second, the graph of s vs. t is a straight line. Furthermore, if the bike is assumed to be at $s = 0$ when $t = 0$, the straight line passes through the origin of the axes, as Figure 2.17 shows. Each point on this line gives the position of the bike at a particular time. For instance, at $t = 1$ s its position is 4 m, while at $t = 3$ s its position is 12 m.

In constructing the graph in Figure 2.17, we used the fact that the velocity was +4 m/s. Suppose, however, that we were given this graph, but did not have any prior knowledge of the velocity. The velocity could be determined in the following manner. Consider what happens, for example, between the times of 1 and 3 s. During this time interval the position of the bike changes from +4 to +12 m. The change in time is $\Delta t = 2$ s, while the corresponding change in position is $\Delta s = +8$ m. The ratio $\Delta s / \Delta t$ is called the **slope** of the straight line:

$$\text{Slope} = \frac{\Delta s}{\Delta t} = \frac{+8 \text{ m}}{2 \text{ s}} = +4 \text{ m/s}$$

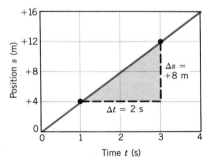

FIGURE 2.17 A graph of position vs. time for an object moving with a constant velocity of $v = \Delta s/\Delta t = +4$ m/s.

Notice that the slope is equal to the velocity of the bike. This result is no accident, because Equation 2.2 reveals that $\Delta s / \Delta t$ is, by definition, the average velocity. Thus, for an object moving with a constant velocity (which is also equal to its average velocity), the slope of the straight line in a position–time graph gives the velocity. Since the position–time graph is a straight line, any time interval Δt could have been chosen to calculate the velocity. Choosing a different Δt would yield a different Δs, but the velocity $\Delta s/\Delta t$ would not change.

In the real world, objects rarely move with a constant velocity at all times. On a hypothetical trip, a bicyclist might maintain a constant velocity on the outgoing leg of the journey, zero velocity while stopped for lunch, and another constant velocity on the return trip. Using a position–time graph, Figure 2.18 illustrates the motion of this trip. Each of the three straight line segments is associated with one part of the trip. The velocity of each segment can be determined from the corresponding straight line in the graph. Using the time and position intervals shown in the drawing, we obtain the following velocities:

Segment 1 $\qquad\qquad\qquad \bar{v} = \dfrac{\Delta s}{\Delta t} = \dfrac{+400 \text{ m}}{200 \text{ s}} = +2 \text{ m/s}$

Segment 2 $\qquad\qquad\qquad \bar{v} = \dfrac{\Delta s}{\Delta t} = \dfrac{0 \text{ m}}{400 \text{ s}} = 0 \text{ m/s}$

Segment 3 $\qquad\qquad\qquad \bar{v} = \dfrac{\Delta s}{\Delta t} = \dfrac{-400 \text{ m}}{400 \text{ s}} = -1 \text{ m/s}$

In the second segment of the journey the velocity is zero, reflecting the fact that the bike is stationary. Since the position of the bike does not change, segment 2 is a horizontal line that has a zero slope. In the third part of the motion the velocity is negative, because the position of the bike decreases from $s = +800$ m to $s = +400$ m during the 400-s interval shown in the graph. In this segment the bicyclist is returning home, since the position of the bicyclist is decreasing. As a result, segment 3 has a negative slope and the velocity is negative.

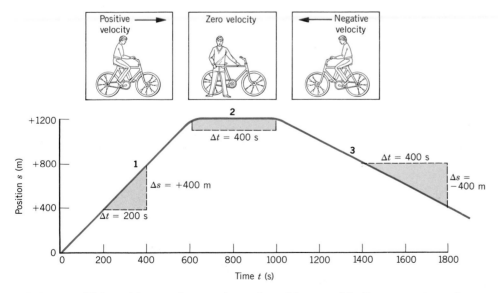

FIGURE 2.18 This position vs. time graph consists of three straight line segments, each corresponding to a different constant velocity.

If the object is accelerating, its velocity is changing. When the velocity is changing, the s vs. t graph is not a straight line, but is a curve like that in Figure 2.19. This curve was drawn using Equation 2.8, assuming an acceleration of $a = 0.26$ m/s² and an initial velocity of $v_0 = 0$:

$$s = \tfrac{1}{2}at^2 = \tfrac{1}{2}(0.26 \text{ m/s}^2)t^2$$

The velocity at any instant of time can be determined by measuring the slope of the curve at that instant. The slope at any point along the curve is defined to be the slope of the tangent to the curve at that point. In Figure 2.19 a tangent is drawn at $t = 20.0$ s. To determine the slope of the tangent, a triangle is constructed using an arbitrarily chosen time interval of $\Delta t = 5.0$ s. The change in Δs associated with this time interval can be read from the tangent line as $\Delta s = +26$ m. The slope of the tangent is

$$\text{Slope of tangent} = \frac{\Delta s}{\Delta t} = \frac{+26 \text{ m}}{5.0 \text{ s}} = +5.2 \text{ m/s}$$

The slope of the tangent is the instantaneous velocity, which in this case is $v = +5.2$ m/s. This graphical result can be verified by using Equation 2.4 with $v_0 = 0$: $v = at = (+0.26 \text{ m/s}^2)(20.0 \text{ s}) = +5.2$ m/s.

Insight into the meaning of acceleration can also be gained with the aid of a graphical representation. Consider an object moving with a constant acceleration of $a = +6$ m/s². If the object has an initial velocity of $v_0 = +5$ m/s, its velocity at any time is represented by Equation 2.4 as

$$v = v_0 + at = (5 \text{ m/s}) + (6 \text{ m/s}^2)t$$

This relation is plotted as the velocity vs. time graph in Figure 2.20. The graph of v vs. t is a straight line that intercepts the vertical axis at $v_0 = 5$ m/s. The slope of this straight line can be calculated from the data shown in the drawing:

$$\text{Slope} = \frac{\Delta v}{\Delta t} = \frac{+12 \text{ m/s}}{2 \text{ s}} = +6 \text{ m/s}^2$$

The ratio $\Delta v/\Delta t$ is, by definition, equal to the average acceleration (Equation 2.4), so we have the result that the slope of the straight line in a velocity–time graph is the average acceleration.

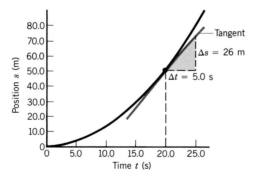

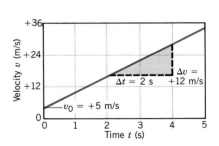

FIGURE 2.19 When the velocity is changing, the position vs. time graph is a curved line. The slope $\Delta s/\Delta t$ of the tangent drawn to the curve at a given time is the instantaneous velocity at that time.

FIGURE 2.20 A velocity vs. time graph that represents an object with an acceleration of $\Delta v/\Delta t = +6$ m/s². The initial velocity is $v_0 = +5$ m/s when $t = 0$.

SUMMARY

Displacement is a vector that points from an object's initial position to its final position. The magnitude of the displacement is the shortest distance between the two positions.

The **average speed** of an object is the distance traveled by the object divided by the time required to cover the distance: Average speed = (Distance)/(Elapsed time).

The **average velocity** $\bar{v}$ of an object is defined as the object's displacement Δs divided by the elapsed time Δt: $\bar{v} = \Delta s/\Delta t$. Average velocity is a vector that has the same direction as the displacement. When the elapsed time Δt is infinitesimally small, the average velocity becomes equal to the **instantaneous velocity v,** the velocity at an instant of time.

Average acceleration $\bar{a}$ is a vector that is equal to the change in the velocity Δv divided by the elapsed time Δt, the "change in velocity" being the difference between the final and initial velocities: $\bar{a} = \Delta v/\Delta t$. When Δt is infinitesimally small, the average acceleration becomes equal to the **instantaneous acceleration a.** Acceleration is the rate at which the velocity is changing.

When an object moves with a constant acceleration along a straight line, its displacement s, final velocity v, initial velocity v_0, acceleration a, and the elapsed time t are related by the following equations, assuming that $s = 0$ at $t = 0$:

$$s = \bar{v}t = \tfrac{1}{2}(v_0 + v)t$$
$$v = v_0 + at$$

These two equations can be combined algebraically to give two additional equations that are listed in Table 2.1. The equations in this table are known as the **equations of kinematics.**

In **free fall** motion, an object experiences a constant acceleration due to gravity and negligible air resistance. All objects at the same location on the earth have the same acceleration due to gravity. The acceleration due to gravity is directed toward the center of the earth and has a magnitude of approximately 9.80 m/s² or 32.2 ft/s² near the earth's surface.

SOLVED PROBLEMS

SOLVED PROBLEM 1

Related Problems: *5 *6 **7 *50 **54

During a trip, a bus travels 11 km with an average velocity of 75 km/h, but then travels in the same direction for the next 1.0 km at a smaller average velocity of 15 km/h, due to the presence of highway construction crews (see the drawing). Determine the average velocity of the bus for the entire trip.

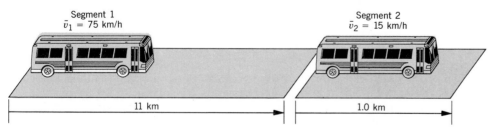

Segment 1
$\bar{v}_1 = 75$ km/h

11 km

Segment 2
$\bar{v}_2 = 15$ km/h

1.0 km

Solution The important point to realize is that the average velocity of the bus is *not* obtained by simply adding the two velocities and dividing the sum by 2:

$$\bar{v} \neq \frac{75 \text{ km/h} + 15 \text{ km/h}}{2}$$

This method is incorrect, because it does not take into account that most of the trip takes place at the greater velocity, while only a small part of the trip takes place at the smaller velocity. The correct procedure for calculating the average velocity is to use Equation 2.2:

$$\bar{v} = \frac{\text{Displacement}}{\text{Elapsed time}}$$

where the values for the displacement and elapsed time must be those for the entire trip. The displacement for the entire trip is 12 km. The elapsed time is the sum of two times, t_1 and t_2. The time t_1 for the bus to travel 11 km at an average velocity of 75 km/h (21 m/s) is

$$t_1 = \frac{11 \times 10^3 \text{ m}}{21 \text{ m/s}} = 520 \text{ s}$$

Likewise, the time t_2 for the bus to travel the remaining 1.0 km at a velocity of 15 km/h (4.2 m/s) is

$$t_2 = \frac{1.0 \times 10^3 \text{ m}}{4.2 \text{ m/s}} = 240 \text{ s}$$

Hence, the elapsed time for the trip is $t_1 + t_2 = 760$ s. The average velocity for the entire trip is, then,

$$\bar{v} = \frac{12 \times 10^3 \text{ m}}{760 \text{ s}} = \boxed{16 \text{ m/s (58 km/h)}}$$

Summary of Important Points The average velocity of a segmented trip is obtained by dividing the total displacement of the trip (which equals the vector sum of the displacements for each segment) by the total time of the trip (which equals the sum of the times for each segment). The average velocity *cannot* be calculated by adding together the velocities of the individual segments and dividing the sum by the number of segments.

SOLVED PROBLEM 2

Related Problems: *21 *22 *33

A motorcycle, starting from rest, has an acceleration of $+2.6$ m/s². When the motorcycle has traveled a distance of 120 m, it slows down with an acceleration of -1.5 m/s² until its velocity is $+12$ m/s (see the drawing). What is the total displacement of the motorcycle?

Segment 1 data

s	a	v	v_0	t
$+120$ m	$+2.6$ m/s²	?	0	

The final velocity v of the first segment can be calculated from Equation 2.9 ($v^2 = v_0^2 + 2as$):

$$v = \sqrt{v_0^2 + 2as} = \sqrt{2(2.6 \text{ m/s}^2)(120 \text{ m})} = +25 \text{ m/s}$$

With $v = +25$ m/s as the initial velocity for the second segment, the data for this phase of the motion are listed below:

Segment 2 data

s	a	v	v_0	t
?	-1.5 m/s²	$+12$ m/s	$+25$ m/s	

The displacement can be calculated for segment 2 by solving $v^2 = v_0^2 + 2as$ for s:

$$s = \frac{v^2 - v_0^2}{2a} = \frac{(12 \text{ m/s})^2 - (25 \text{ m/s})^2}{2(-1.5 \text{ m/s}^2)} = +160 \text{ m}$$

The total displacement of the motorcycle is 120 m + 160 m = $\boxed{280 \text{ m.}}$

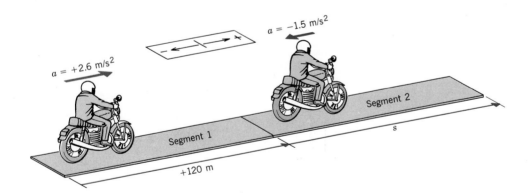

Solution The total displacement is the sum of the displacements for the first ("speeding up") and second ("slowing down") segments. The displacement for the first segment is $+120$ m. The displacement for the second segment can be found if the initial velocity for this segment can be determined, since values for two other variables are already known. The key point here is that the initial velocity for the second segment is the final velocity of the first segment. With this in mind, we can determine this velocity from the known data for the first segment:

Summary of Important Points Often the motion of an object is divided into "segments," each of which has a different acceleration. An important point to realize about solving such problems is that the final velocity for one segment becomes the initial velocity for the next segment.

SOLVED PROBLEM 3

Related Problems: *35 **39 *51 **56

An airplane is flying horizontally at an altitude of 1450 m and a constant speed of 75 m/s. A projectile is fired vertically from the ground with an initial speed of 375 m/s. For the projectile to strike the bottom of the plane, what must be the value of the angle θ at the moment of firing (see the drawing)?

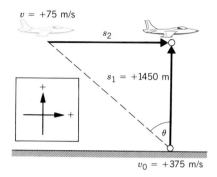

v = +75 m/s

s_2

$s_1 = +1450$ m

θ

$v_0 = +375$ m/s

Solution From the drawing it is evident that the angle θ is related to s_1 and s_2 by

$$\tan \theta = \frac{s_2}{s_1} = \frac{s_2}{1450 \text{ m}}$$

where s_2 and s_1 are the magnitudes of the displacements of the plane and the projectile, respectively. This equation indicates that we need to calculate s_2 to determine θ. The displacement of the plane is equal to its (constant) velocity times the time

$$s_2 = (75 \text{ m/s})t$$

The result above indicates that a value for t must be obtained before s_2 can be found. With this in mind, we note that the time for the plane to travel a displacement s_2 is identical to the time for the projectile to travel a displacement s_1. Since three variables are

known for the projectile, we can use the data below and Equation 2.8 to calculate the time.

Projectile data

s_1	a	v	v_0	t
$+1450$ m	-9.80 m/s^2		$+375$ m/s	?

$$s_1 = v_0 t + \tfrac{1}{2}at^2$$

$$1450 \text{ m} = (375 \text{ m/s})t + \tfrac{1}{2}(-9.80 \text{ m/s}^2)t^2$$

Solving this quadratic equation (see Appendix C) yields two values for the time:

$$t = 4.1 \text{ s} \quad \text{and} \quad t = 72 \text{ s}$$

We discard the 72-s answer, because it represents the time for the projectile to reach its maximum height above the ground and then fall downward to strike the plane from the top, rather than from the bottom. A value for s_2 can now be obtained and then used to calculate θ:

$$s_2 = (+75 \text{ m/s})t = (+75 \text{ m/s})(4.1 \text{ s}) = +310 \text{ m}$$

$$\tan \theta = \frac{s_2}{1450 \text{ m}} = \frac{310 \text{ m}}{1450 \text{ m}} = 0.21$$

$$\theta = \tan^{-1} 0.21 = \boxed{12°}$$

Summary of Important Points The important feature of this problem is that the time t is the same for both objects. There is enough information given about one of the objects (the projectile, in this case) to calculate t explicitly. This value of t can then be used to find one of the kinematic variables (e.g., the displacement s_2) of the other object.

QUESTIONS

1. A honeybee leaves the hive and travels 2 km before returning. Is the displacement for the trip the same as the distance traveled? If not, why not?

2. Two buses depart from Chicago, one going to New York and one to San Francisco. Each bus travels at a speed of 30 m/s. Do they have equal velocities? Explain.

3. Often, traffic lights are timed so that if you travel at a certain constant speed, you can avoid all red lights. Discuss how the timing of traffic lights is determined, considering that the distance between them varies from one light to the next.

4. Give an example from your own experience in which the velocity of an object is zero for just an instant of time, but its acceleration is not zero.

5. At a given instant of time, a car and a truck are traveling side by side in adjacent lanes of a highway. The car has a greater velocity than the truck. Does the car necessarily have a greater acceleration? Explain.

6. The average velocity for a trip has a positive value. Is it possible for the instantaneous velocity at any point during the trip to have a negative value? Justify your answer.

7. An experimental vehicle slows down and comes to a halt with an acceleration whose magnitude is 9.80 m/s². After reversing direction in a negligible amount of time, the vehicle speeds up with an acceleration of 9.80 m/s². Other than being horizontal, how is this motion different, if at all, from the motion of a ball that is thrown straight upward, comes to a halt, and falls back to earth?

8. A person standing on a bridge fires a rifle bullet straight up, and then fires another bullet straight down. Neglecting air resistance, which bullet, if either, strikes the water with a greater velocity? Provide a reason for your answer.

9. A ball thrown into the air reaches a maximum height and begins to fall down. Does the acceleration of gravity change, either in magnitude or in direction, when the velocity of the ball changes direction? Explain.

10. For each equation below, show that the units of the left side are the same as the units of each term on the right side:

$$s = v_0 t + \tfrac{1}{2}at^2$$

$$v^2 = v_0^2 + 2as$$

PROBLEMS

Section 2.3 Speed and Velocity

1. Sound travels at a constant speed of 343 m/s in air. Approximately how much time (in seconds) does it take for the sound of thunder to travel one mile (1609 m)?

2. A car is traveling at a constant speed of 96 km/h. The driver looks away from the road for 2.0 s to tune in a station on the radio. How far (in meters) does the car go during this time?

3. In 1980, Steve Ovett set the world record for the 1500-m race in a time of 3 minutes, 31.36 seconds. What was his average speed?

4. An 18-year-old runner can complete a 10.0-km course with an average speed of 4.38 m/s. A 50-year-old runner can cover the same distance with an average speed of 4.27 m/s. How much later should the younger runner start in order to finish the 10.0-km course *at the same time* as the older runner?

* 5. A sky diver, with parachute unopened, falls 625 m in 15.0 s. Then she opens her parachute and falls another 356 m in 142 s. What is her average velocity (both magnitude and direction) for the entire fall? *(See Solved Problem 1 for a related problem.)*

* 6. A bicyclist makes a trip that consists of three parts, each in the same direction (due north) along a straight road. During the first part, he rides for 22 minutes at an average speed of 7.2 m/s. During the second part, he rides for 36 minutes at an average speed of 5.1 m/s. Finally, during the third part, he rides for 8.0 minutes at an average speed of 13 m/s. (a) How far has the bicyclist traveled during the entire trip? (b) What is the average velocity of the bicyclist for the trip? *(See Solved Problem 1 for a related problem.)*

** 7. A car makes a 60.0-km trip with an average velocity of 40.0 km/h in a direction due north. The trip consists of three parts. The car moves with a constant velocity of 25 km/h due north for the first 15 km, and 62 km/h due north for the next 32 km. With what constant velocity does the car travel for the last 13-km segment of the trip? *(See Solved Problem 1 for a related problem.)*

Section 2.4 Acceleration

8. An airplane, starting from rest, moves south and attains a lift-off speed of 205 km/h in 18.0 s. What is the magnitude and direction of its average acceleration (in m/s²)?

9. If a sports car can go from rest to 60.0 mi/h in 9.00 s, what is the magnitude of its average acceleration (in ft/s²)?

10. The velocity of a train is +95.0 km/h. At an average acceleration of −1.50 m/s², how much time (in seconds) is required for the train to decrease its velocity to +35.0 km/h?

11. A runner accelerates to a velocity of 12.0 mi/h due west in 3.00 s. His average acceleration is 2.10 ft/s², also directed westward. What was his velocity (in ft/s) when he began accelerating?

* 12. A small particle is moving with an initial velocity of +110 m/s at $t = 0$. The particle has no acceleration for the first 3.0 s of its motion. Then, for the next 2.0 s, the particle has an acceleration of $a = −4.0$ m/s². Thereafter, the acceleration is again zero. What is the velocity of the particle at (a) $t = 5.0$ s, and (b) $t = 7.0$ s?

** 13. Two motorcycles are traveling due east with different velocities. However, four seconds later, they have the same velocity. During this four-second interval, one motorcycle has an average acceleration of 2.0 m/s² due east, while the other has an average acceleration of 4.0 m/s² due east. By how much did the speeds *differ* at the beginning of the four-second interval, and which motorcycle was moving faster?

Section 2.5 Equations of Kinematics for Constant Acceleration, Section 2.6 Applications of the Equations of Kinematics

14. A jetliner, traveling northward, is landing with a speed of 250 km/h. Once the jet touches down, it has 750 m of runway in which to reduce its speed to 22 km/h. (a) Compute the average acceleration (magnitude and direction) of the plane during landing. (b) How much time is required to reduce the speed of the plane?

15. A skier, starting from rest, accelerates down a slope at 1.6 m/s². (a) How far has she gone at the end of 5.0 seconds? (b) What is her velocity at this time?

16. A truck, traveling at a velocity of 33 m/s due east, comes to a halt by decelerating at 11 m/s². (a) How far does the truck travel in the process of stopping? (b) What is the velocity of the truck when the truck has traveled 23 m after beginning to slow down?

17. The length of the barrel of a primitive blowgun is 1.2 m. Upon leaving the barrel, a dart has a speed of 14 m/s. Assuming the dart is uniformly accelerated, how long does it take for the dart to travel the length of the barrel?

18. With the plane standing on the runway, the pilot brings the engines to full thrust before releasing the brakes. The aircraft accelerates at 9.6 ft/s² and reaches a takeoff speed of 190 ft/s. (a) Find the time from rest to takeoff. (b) Determine the displacement of the plane. (c) What is the average velocity of the plane?

19. Rederive the equations of kinematics presented in Table 2.1, assuming the initial position is s_0, instead of $s_0 = 0$ as was used in the text.

* 20. A drag racer, starting from rest, speeds up for 402 m with an acceleration of $+17.0$ m/s². A parachute then opens, slowing the car down with an acceleration of -6.10 m/s². (a) How fast is the racer moving 3.50×10^2 m after the parachute opens? (b) How much time has elapsed *from the start of the race* until the racer has traveled 3.50×10^2 m with the parachute opened?

* 21. Suppose a car is traveling at 12.0 m/s, and the driver sees a traffic light turn red. After 0.510 s has elapsed (the reaction time), the driver applies the brakes, and the car decelerates at 6.20 m/s². What is the stopping distance of the car, as measured from the point where the driver first notices the red light? *(See Solved Problem 2 for a related problem.)*

* 22. A speedboat starts from rest and accelerates at $+6.60$ ft/s² for 7.00 s. At the end of this time, the boat continues for an additional 6.00 s with an acceleration of $+1.70$ ft/s². Following this, the boat accelerates at -4.90 ft/s² for 8.00 s. (a) What is the velocity of the boat at $t = 21.0$ s? (b) Find the total displacement of the boat. (c) Find the average velocity of the boat during the trip. *(See Solved Problem 2 for a related problem.)*

* 23. A race driver has made a pit stop to refuel. After refueling, he leaves the pit area with an acceleration whose magnitude is 6.0 m/s², and after 4.0 s he enters the main speedway. At the same instant, another race car that is on the speedway and traveling at a constant speed of 70.0 m/s overtakes and passes the entering car. (a) If the entering car maintains its acceleration, how much time is required for it to catch the other car? (b) How far has the other car traveled during the time calculated in part (a)?

**24. A locomotive is accelerating at 1.6 m/s². It passes through a 20.0-m-wide crossing in a time of 2.4 s. After the locomotive leaves the crossing, how much time is required until its speed reaches 32 m/s?

**25. In the one-hundred-yard dash a sprinter accelerates from rest to a top speed with an acceleration whose magnitude is 9.0 ft/s². After achieving this speed, he runs the remainder of the race without speeding up or slowing down. If the total race is run in 11 s, how far does he run during the acceleration phase?

Section 2.7 Freely Falling Bodies

26. A penny is dropped from rest from the top of the Sears Tower in Chicago. Considering that the height of the building is 1400 ft and ignoring air resistance, find the speed of the penny when the penny strikes the ground.

27. A baseball is thrown upward with an initial speed of 35.0 m/s. (a) What is its speed at $t = 2.00$ s? (b) What is its speed at $t = 5.00$ s? (c) How high does the ball go?

28. Suppose you are visiting a planet in a distant part of the galaxy. To determine the acceleration due to gravity on this planet, you drop a small rock from a height of 55 m. The rock strikes the ground 1.9 s later. How many times greater is the acceleration due to gravity on this planet than that on earth?

29. A rifle bullet is shot vertically upward. Twenty-three seconds later the bullet has a velocity of 72.0 m/s, downward. (a) What is the velocity of the bullet when the bullet leaves the rifle? (b) What is the highest point reached by the bullet? (c) How much time is required for the bullet to reach its highest point?

30. With what initial speed must an arrow be fired straight upward to attain a height of 110 m in 5.4 s?

31. Suppose a ball is thrown vertically upward. Eight seconds later it returns to its point of release. (a) What is the initial velocity of the ball? (b) How high does the ball go? (c) What is the velocity of the ball just before the ball returns to its point of release?

32. A diver springs upward with an initial speed of 1.8 m/s from a 3.0-m board. (a) Find the velocity with which he strikes the water. [*Hint: Note that when the diver reaches the water, his displacement is $s = -3.0$ m (measured from the board), assuming the downward direction is chosen as the negative direction.*] (b) What is the highest point he reaches above the water? (c) What is his acceleration at the highest point?

* 33. A rocket is launched from rest with an acceleration of 20.0 m/s², upward. At an altitude of 415 m the engines shut off, but the rocket continues to coast upward. Find the total time that the rocket is in the air, from lift-off until it strikes the ground. *(See Solved Problem 2 for a related problem.)*

* 34. A roof tile falls from rest from the top of a building. An observer inside the building notices that it takes 0.20 s for the tile to pass her window, whose height is 1.6 m. (a) How far above the top of this window is the roof? (b) What is the velocity of the tile when the tile reaches the top of the window?

* 35. A log is floating on swiftly moving water. A stone is dropped from rest from a 75-m-high bridge and lands on the log as it passes under the bridge. If the log moves with a constant speed of

5.0 m/s, what is the horizontal distance between the log and the bridge when the stone is released? *(See Solved Problem 3 for a related problem.)*

* **36.** A spelunker (cave explorer) drops a stone from rest into a hole. The speed of sound is 343 m/s in air, and the sound of the stone striking the bottom is heard 1.50 s later. How deep is the hole?

* **37.** A gun fires a pellet straight upward at a glass target. The pellet has an initial speed of 31.0 m/s. After 2.51 s, the sound of breaking glass is heard. The speed of sound is 343 m/s in air. How high above the ground is the glass target?

** **38.** A ball is thrown straight upward with a speed of 45.0 m/s from the base of an apartment building. A renter notices that it takes 0.200 s for the ball to pass by her 1.00-m-high window on the way up. How far is the bottom of the window from the base of the building?

** **39.** A ball is thrown upward from the top of a 25.0-m-tall building. The ball's initial speed is 12.0 m/s. At the same instant, a person is running on the ground at a distance of 31.0 m from the building. What must be the average speed of the person if he is to catch the ball at the bottom of the building? *(See Solved Problem 3 for a related problem.)*

** **40.** A stone is hurled upward with an initial speed of 35 m/s from the top of an 84-m-tall building. At a later time t, another stone is thrown downward with an initial speed of 41 m/s. Both stones hit the ground at the same moment. Determine t.

Section 2.8 Graphical Analysis of Velocity and Acceleration for Linear Motion

41. A bus makes a trip according to the position–time graph shown in the illustration. What is the average velocity (magnitude and direction) of the bus during each of the three segments of its motion?

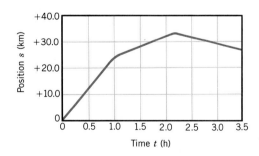

42. A snowmobile moves according to the velocity–time graph shown in the drawing. What is the snowmobile's average acceleration during each of the segments labeled A, B, and C?

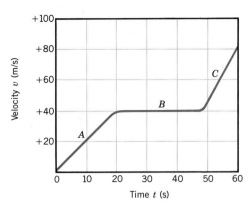

* **43.** Using the position–time graph that accompanies this problem, draw the corresponding velocity–time graph.

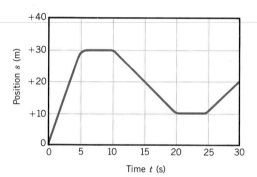

ADDITIONAL PROBLEMS

44. A jogger accelerates from rest to 3.0 m/s in 2.0 s. A car accelerates from 38 to 41 m/s also in 2.0 s. (a) Find the acceleration (magnitude only) of the jogger. (b) Determine the acceleration (magnitude only) of the car. (c) Does the car travel further than the jogger during the 2.0 s? If so, how much further?

45. Two runners in a one-mile (1609-m) race finish with times of 3:52.46 (3 minutes and 52.46 seconds) and 3:52.72. Assuming that both run at their average speeds during the entire time, what distance (in meters) separates them at the end of the race?

46. A cement block accidentally falls from rest from the ledge of a 53.0-m-high building. When the block is 14.0 m above the ground, a man, 2.00 m tall, looks up and notices that the block is directly above him. How much time, at most, does the man have to get out of the way?

47. A motorcycle has a constant acceleration of 2.5 m/s². Both the velocity and acceleration of the motorcycle point in the same direction. How much time is required for the motorcycle to change its speed (a) from 21 m/s to 31 m/s, and (b) from 51 to 61 m/s?

48. An arrow is shot straight up with an initial speed of 50.0 m/s. After reaching its maximum height, the arrow starts down. On the descent, a slight breeze blows the arrow laterally, and it strikes a tree limb that is 30.0 m above the ground. (a) Determine the velocity of the arrow just before the arrow strikes the limb. (b) Find the total time the arrow is in the air.

* 49. A sports car, picking up speed, passes between two markers in a time of 4.1 s. The markers are separated by 120 m. All the while, the car has an acceleration of 1.8 m/s². What is its speed at the second marker?

* 50. Suppose that the first one-fourth of the distance between two points is covered with an average velocity of +18 m/s. The average velocity for the remainder of the trip is +51 m/s. What is the average velocity for the entire trip? *(See Solved Problem 1 for a related problem.)*

* 51. A person on a bridge 90.0 m high sees a log floating at a constant speed on the river below. She drops a stone from rest in an attempt to hit the log. The stone is released when the log has 6.00 m more to travel before passing under the bridge. The stone hits the water 2.00 m in front of the log. Find the speed of the log. *(See Solved Problem 3 for a related problem.)*

* 52. A speed trap is set up with two pressure-activated strips placed 110 m apart across a highway. A car is speeding along at 120 km/h, while the speed limit is 75 km/h. At the instant the car activates the first strip, the driver begins slowing down. What deceleration is needed in order that the average speed of the car is within the speed limit by the time the car crosses the second marker?

**53. A ball is thrown downward with an initial speed of 25 m/s from the top of a 210-m-tall building. At the same time, another ball is thrown upward from ground level with a speed of 25 m/s. At what distance from the bottom do the two balls pass each other?

**54. In reaching her destination, a backpacker walked with an average velocity of 3.0 mi/h due west. This average velocity resulted, because she hiked for 4.0 mi with an average velocity of 6.0 mi/h due west, turned around, and hiked with an average velocity of 1.0 mi/h due east. How far did she walk while moving east? *(See Solved Problem 1 for a related problem.)*

**55. An express elevator goes from the 30th floor to the 1st floor, a distance of 150 m. The maximum acceleration and deceleration is limited to 3.7 m/s², and the maximum allowed vertical speed is 6.0 m/s. (a) What is the minimum time required to make the trip? (b) Find the average velocity for the trip.

**56. A book accidentally falls from a shelf 4.2 m high. A librarian is standing nearby and moves 0.80 m, starting from rest, to catch the book. What must be his average acceleration if he catches the book when it is 1.8 m above the floor? *(See Solved Problem 3 for a related problem.)*

**57. A football player, starting from rest at the line of scrimmage, accelerates along a straight line for a time of 3.0 s. Then, during a negligible amount of time, he changes the magnitude of his acceleration to a value of 1.1 m/s². With this acceleration, he continues in the same direction for another 2.0 s, until he reaches a speed of 6.4 m/s. What is the value of his acceleration (assumed to be constant) during the initial 3.0-s period?

CHAPTER 3

Kinematics in Two Dimensions

The motions of an airborne ballet dancer and a baseball have much in common, for both are influenced by the same acceleration due to gravity. The effect of gravity on objects moving through space is one of the topics of this chapter.

3.1 DISPLACEMENT, VELOCITY, AND ACCELERATION

DISPLACEMENT

In Chapter 2 the concepts of displacement, velocity, and acceleration are used to describe an object moving along a horizontal or a vertical straight line. There are also situations in which the motion is along a curved line, as in the famous Indianapolis 500 race where the cars move on an oval track. This type of "two-dimensional" motion can also be described using the same concepts. Figure 3.1 shows a race car at two different positions along the track. These positions are identified by the vectors s_0 and s that are drawn from an arbitrary coordinate origin. The ***displacement*** Δs of the car is the vector drawn from the initial position of the car at time t_0 to the final position at time t, the magnitude of Δs being the shortest distance between the two positions. From the drawing it is evident that s is the vector sum of s_0 and Δs, so $s = s_0 + \Delta s$, or

$$\text{Displacement} = \Delta s = s - s_0$$

The displacement here is defined in the same manner as that in Chapter 2, except now the displacement vector can lie anywhere in a plane, rather than just along a straight line.

VELOCITY

The average velocity $\bar{\mathbf{v}}$ of the car between two positions is defined in the usual way as the displacement, $\Delta\mathbf{s} = \mathbf{s} - \mathbf{s}_0$, divided by the elapsed time $\Delta t = t - t_0$:

$$\bar{\mathbf{v}} = \frac{\mathbf{s} - \mathbf{s}_0}{t - t_0} = \frac{\Delta\mathbf{s}}{\Delta t} \tag{3.1}$$

Since both sides of Equation 3.1 must agree in direction, the average velocity vector has the same direction as the displacement. The velocity of the car at an instant of time is its ***instantaneous velocity*** (or just "velocity") $\mathbf{v}$. The average velocity becomes equal to the instantaneous velocity $\mathbf{v}$ in the limit as Δt becomes infinitesimally small:

$$\mathbf{v} = \lim_{\Delta t \to 0} \frac{\Delta\mathbf{s}}{\Delta t}$$

Figure 3.2 shows the velocity components $\mathbf{v}_x$ and $\mathbf{v}_y$, which are parallel to the x and y axes, respectively. Using the components of a vector is advantageous when describing two-dimensional motion.

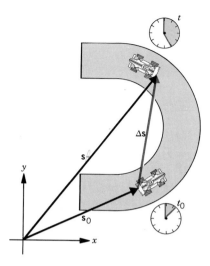

FIGURE 3.1 The displacement $\Delta\mathbf{s}$ of the car is a vector that points from the initial position of the car to the final position. The magnitude of $\Delta\mathbf{s}$ is the shortest distance between the two positions.

ACCELERATION

The ***average acceleration*** $\bar{\mathbf{a}}$ is defined in the same manner as that for one-dimensional motion, namely as the change in the velocity, $\Delta\mathbf{v} = \mathbf{v} - \mathbf{v}_0$, divided by the elapsed time Δt:

$$\bar{\mathbf{a}} = \frac{\mathbf{v} - \mathbf{v}_0}{t - t_0} = \frac{\Delta\mathbf{v}}{\Delta t} \tag{3.2}$$

The average acceleration vector has the same direction as the change in velocity. In the limit that the elapsed time becomes infinitesimally small, the average acceleration becomes equal to the ***instantaneous acceleration*** (or just "acceleration") $\mathbf{a}$:

$$\mathbf{a} = \lim_{\Delta t \to 0} \frac{\Delta\mathbf{v}}{\Delta t}$$

The acceleration has a component $\mathbf{a}_x$ along the x direction and a component $\mathbf{a}_y$ along the y direction.

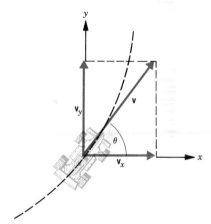

FIGURE 3.2 The instantaneous velocity $\mathbf{v}$ and its two components $\mathbf{v}_x$ and $\mathbf{v}_y$.

3.2 EQUATIONS OF KINEMATICS IN TWO DIMENSIONS

To illustrate how displacement, velocity, and acceleration are applied to two-dimensional motion, consider a spacecraft equipped with two engines that are mounted perpendicular to each other. The craft is far from other bodies, so the only forces acting on it are those produced by its own engines. As Figure 3.3 indicates, the spacecraft is assumed to be at the coordinate origin when $t_0 = 0$, so that $\mathbf{s}_0 = 0$. Therefore, the displacement of the spacecraft is $\Delta\mathbf{s} = \mathbf{s} - \mathbf{s}_0 = \mathbf{s}$. The components of the displacement vector $\mathbf{s}$ are $\mathbf{x}$ and $\mathbf{y}$, along the x and y axes, respectively.

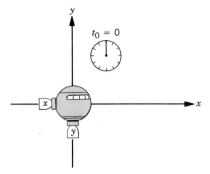

FIGURE 3.3 At $t_0 = 0$, the spacecraft is assumed to be at the coordinate origin, so $\mathbf{s}_0 = 0$.

In Figure 3.4 only the engine oriented along the x direction is firing, and the vehicle accelerates along this direction. Although the craft does have a second engine, the y engine, it contributes nothing to the motion at this time, since the engine is turned off. Further, it is assumed that the velocity in the y direction is also zero. The motion of the spacecraft along the x direction is described by the five kinematic variables x, a_x, v_x, v_{0x}, and t. Here the symbol "x" reminds us that we are dealing with the x components of the displacement, velocity, and acceleration vectors. If the spacecraft has a constant acceleration along the x direction, the motion is exactly like that described in Chapter 2, and the equations of kinematics can be used to relate the kinematic variables. For convenience, these equations are written in the left column of Table 3.1.

With only the y engine firing, the spacecraft accelerates along the y direction. Figure 3.5 indicates that such a motion can be described in terms of the kinematic variables y, a_y, v_y, v_{0y}, and t. And if the acceleration along the y direction is constant, these variables are related by the equations of kinematics written in the right column of Table 3.1.

If both engines of the spacecraft are firing *at the same time,* the resulting motion takes place in part along the x axis and in part along the y axis, as Figure 3.6 illustrates. The thrust of each engine gives the vehicle a corresponding acceleration component. The x engine accelerates the ship in the x direction and causes a change in the x component of the velocity. Likewise, the y engine causes a change in the y component

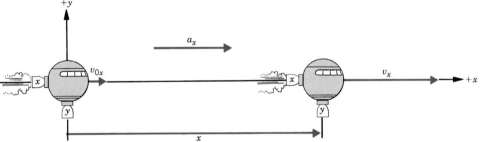

FIGURE 3.4 The spacecraft is moving with a constant acceleration a_x parallel to the x axis. There is no motion in the y direction, and the y engine is turned off.

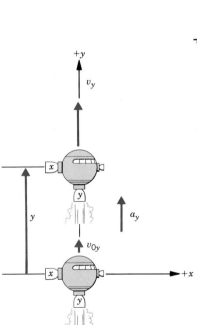

FIGURE 3.5 The spacecraft is moving with a constant acceleration a_y parallel to the y axis. There is no motion in the x direction, and the x engine is not firing.

TABLE 3.1 Equations of Kinematics for Constant Acceleration in Two-Dimensional Motion

x component	Variable	y component
x	Displacement	y
a_x	Acceleration	a_y
v_x	Final velocity	v_y
v_{0x}	Initial velocity	v_{0y}
t	Elapsed time	t

x component		y component	
$v_x = v_{0x} + a_x t$	(3.3a)	$v_y = v_{0y} + a_y t$	(3.3b)
$x = \frac{1}{2}(v_{0x} + v_x)t$	(3.4a)	$y = \frac{1}{2}(v_{0y} + v_y)t$	(3.4b)
$x = v_{0x}t + \frac{1}{2}a_x t^2$	(3.5a)	$y = v_{0y}t + \frac{1}{2}a_y t^2$	(3.5b)
$v_x^2 = v_{0x}^2 + 2a_x x$	(3.6a)	$v_y^2 = v_{0y}^2 + 2a_y y$	(3.6b)

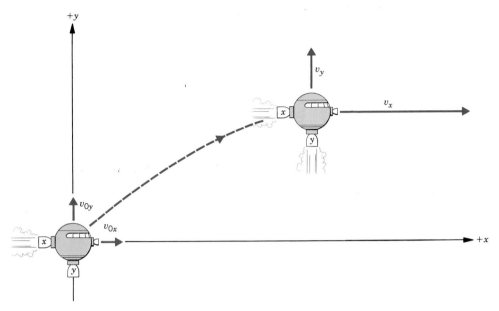

FIGURE 3.6 The two-dimensional motion of the spacecraft can be viewed as the combination of the separate x and y motions.

of the velocity. ***It is important to realize that the x part of the motion occurs exactly as it would if the y part did not occur at all. Similarly, the y part of the motion occurs exactly as it would if the x part of the motion did not exist.*** In other words, the x and y motions are independent of each other, and a problem dealing with two-dimensional motion can be considered as two one-dimensional problems. Example 1 illustrates this point.

EXAMPLE 1

In the x direction, the spacecraft in Figure 3.6 has an initial velocity of $v_{0x} = +25$ m/s and an acceleration of $a_x = +24$ m/s². In the y direction, the analogous quantities are $v_{0y} = +14$ m/s and $a_y = +12$ m/s². After a time of 7.0 s, find (a) x and v_x, (b) y and v_y, and (c) the final velocity (magnitude and direction) of the spacecraft.

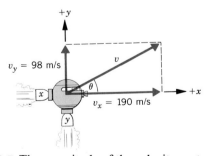

FIGURE 3.7 The magnitude of the velocity vector gives the speed of the spacecraft, and the angle θ gives the direction of travel relative to the positive x direction.

SOLUTION

(a) The x component of the spacecraft's displacement can be obtained from the data below and Equation 3.5a.

x-direction data

x	a_x	v_x	v_{0x}	t
?	$+24$ m/s²	?	$+25$ m/s	7.0 s

$$x = v_{0x}t + \tfrac{1}{2}a_x t^2 = (25 \text{ m/s})(7.0 \text{ s}) + \tfrac{1}{2}(24 \text{ m/s}^2)(7.0 \text{ s})^2$$
$$= \boxed{+760 \text{ m}}$$

The velocity component v_x can be calculated using Equation 3.3a:

$$v_x = v_{0x} + a_x t = (25 \text{ m/s}) + (24 \text{ m/s}^2)(7.0 \text{ s})$$
$$= \boxed{+190 \text{ m/s}}$$

(b) The data for the motion in the y direction are listed below.

y-direction data

y	a_y	v_y	v_{0y}	t
?	$+12$ m/s²	?	$+14$ m/s	7.0 s

Proceeding in the same manner as in part (a), we find that

$$y = +390 \text{ m} \qquad \text{and} \qquad v_y = +98 \text{ m/s}$$

(c) Figure 3.7 shows that the velocity of the vehicle is the vector sum of its x and y components. The magnitude v of the velocity can be found by using the Pythagorean theorem:

$$v = \sqrt{v_x^2 + v_y^2} = \sqrt{(190 \text{ m/s})^2 + (98 \text{ m/s})^2} = \boxed{210 \text{ m/s}}$$

The direction of the velocity vector is given by the angle θ in the drawing:

$$\tan \theta = \frac{v_y}{v_x} = \frac{98 \text{ m/s}}{190 \text{ m/s}} = 0.52$$

$$\theta = \tan^{-1} 0.52 = \boxed{27°}$$

After 7.0 s, the spacecraft has a speed of 210 m/s and a velocity vector that points 27° above the positive x axis.

3.3 PROJECTILE MOTION

A common type of two-dimensional motion is called "projectile motion," one example of which is a baseball being thrown from one person to another. A good description of projectile motion can often be obtained with the assumption that the moving object (the projectile) experiences only the acceleration due to gravity. This assumption implies that air resistance, which can slow down the projectile, is negligible. Specifically, in the absence of air resistance, there is no acceleration in the horizontal or x direction, so that $a_x = 0$. Consequently, the x component of the velocity always remains the same as its initial value, namely, $v_x = v_{0x}$. For our purposes, then, the phrase "projectile motion" means that $a_x = 0$ and $a_y =$ acceleration due to gravity. If the trajectory of the projectile is near the surface of the earth, then $a_y = 9.80$ m/s². Example 2 illustrates how projectile motion can be described by the equations of kinematics.

EXAMPLE 2

Figure 3.8 shows an airplane moving horizontally with a constant velocity of $+115$ m/s at an altitude of 1050 m. In the drawing, the directions to the right and downward have been chosen as the positive directions. The plane releases a "care package" that falls to the ground along the trajectory indicated in the drawing. Ignoring air resistance, determine the time required for the package to hit the ground.

SOLUTION

The initial velocity of the package in the y direction is zero at the moment the package is released from the plane, so $v_{0y} = 0$. The package is moving at the instant of release, but only in the x direction, not in the y direction. Furthermore, when the package hits the ground, the y component of the package's displacement is $y = +1050$ m, as the drawing shows. The acceleration is that due to gravity, so $a_y = +9.80$ m/s². These data are summarized below.

y-direction data

y	a_y	v_y	v_{0y}	t
$+1050$ m	$+9.80$ m/s²		0	?

With this data, Equation 3.5b ($y = v_{0y}t + \frac{1}{2}a_y t^2$) can be employed to calculate the time for the package to strike the ground. Since $v_{0y} = 0$,

$$t = \sqrt{\frac{2y}{a_y}} = \sqrt{\frac{2(1050 \text{ m})}{9.80 \text{ m/s}^2}} = \boxed{14.6 \text{ s}}$$

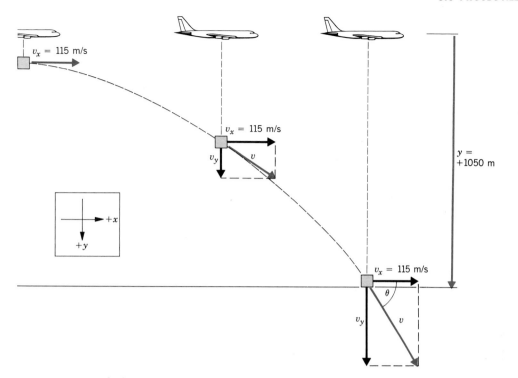

FIGURE 3.8 The falling package is an example of projectile motion. As the package descends, its horizontal velocity component v_x remains constant, while the vertical component v_y increases linearly with time.

In Example 2, the initial velocity of $v_{0x} = +115$ m/s in the horizontal direction plays no role in determining the time required for the package to reach the ground. In fact, for all types of projectile motion, the time of flight depends only on the y variables, and the x variables do not come into play. To emphasize this point, Figure 3.9 illustrates what happens to two packages, labeled A and B, that are released simultaneously from the same height. Package A is given an initial velocity of $v_{0x} = +115$ m/s in the horizontal direction, as in Example 2, and the package follows the curved path shown in the figure. Package B, on the other hand, is dropped from a stationary balloon and falls directly toward the ground, since $v_{0x} = 0$. What is surprising is that both packages hit the ground at the same time. The packages hit simultaneously, because the time of fall depends only on the y variables (y, a_y, and v_{0y}), and these variables are the same for both packages.

Although v_y increases during the 14.6 s of free fall in Example 2, the horizontal component of the velocity retains its initial value of $v_{0x} = +115$ m/s throughout the entire descent. Since the plane also travels at a constant horizontal velocity of $+115$ m/s, the plane remains directly above the falling package. Stated another way, the pilot always sees the falling package directly beneath the plane, as the dashed vertical lines in Figure 3.8 show. This result is a direct consequence of the fact that the box has no acceleration in the horizontal direction. In reality, air resistance would slow down the package, and a_x would no longer be zero, with the result that the package would not remain directly beneath the plane during the descent.

Not only do the packages reach the ground at the same time, but the y components

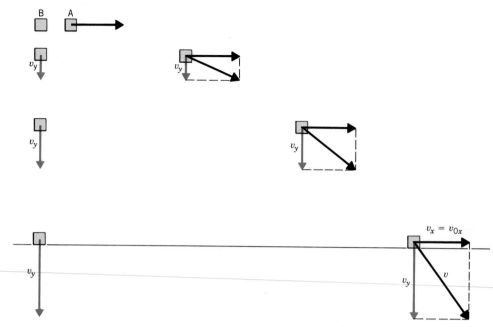

FIGURE 3.9 Both packages strike the ground at the same time, because their y variables (y, a_y, and v_{0y}) are the same.

of their velocities are also equal at all points on the way down, as Figure 3.9 indicates. However, package A does hit the ground with a greater speed than package B. Remember, speed is the magnitude of the velocity vector, and the velocity of A has an x component, whereas the velocity of B does not. The magnitude and direction of the velocity vector for package A at the instant the package hits the ground is computed in Example 3.

EXAMPLE 3

For the situation shown in Figure 3.8, find the speed of the package and the direction of the velocity vector just before the package hits the ground.

SOLUTION

Since the speed v of the package is given by $v = \sqrt{v_x^2 + v_y^2}$, it is necessary to know values for v_x and v_y at the instant of impact. The component v_x is constant with a magnitude of 115 m/s. The component v_y can be determined by using Equation 3.3b and the data from Example 2 ($a_y = +9.80$ m/s^2, $v_{0y} = 0$, $t = 14.6$ s):

$$v_y = v_{0y} + a_y t = (9.80 \text{ m/s}^2)(14.6 \text{ s}) = 143 \text{ m/s}$$

The speed of the package at the instant of impact is

$$v = \sqrt{(115 \text{ m/s})^2 + (143 \text{ m/s})^2} = \boxed{184 \text{ m/s}}$$

The velocity vector makes an angle θ with the horizontal, as Figure 3.8 indicates:

$$\tan \theta = \frac{v_y}{v_x} = \frac{143 \text{ m/s}}{115 \text{ m/s}} = 1.24$$

$$\theta = \tan^{-1} 1.24 = \boxed{51.1°}$$

Often projectiles, like footballs and baseballs, are thrown into the air at an angle with respect to the ground. From a knowledge of the projectile's initial velocity, a wealth of information can be obtained about the motion. For instance, Example 4 demonstrates how to calculate the maximum height reached by the projectile.

EXAMPLE 4

A projectile is fired from ground level at an angle of $\theta = 40.0°$ above the horizontal axis, as Figure 3.10 shows. The initial speed of the projectile is $v_0 = 25$ m/s. From these data, find the maximum height H that the projectile attains.

SOLUTION

The maximum height reached by the projectile is a characteristic of the vertical part of the motion. Therefore, we begin by calculating the vertical component of the initial velocity:

$$v_{0y} = v_0 \sin 40.0° = (25 \text{ m/s}) \sin 40.0° = +16 \text{ m/s}$$

The maximum height H can be determined by noting that the y component of the velocity v_y decreases as the projectile moves upward. Eventually, $v_y = 0$ at the top of the trajectory. The data

below can be used in Equation 3.6b ($v_y^2 = v_{0y}^2 + 2a_y y$) to find the maximum height:

y-direction data

y	a_y	v_y	v_{0y}	t
$H = ?$	-9.80 m/s^2	0	$+16$ m/s	

$$y = H = \frac{v_y^2 - v_{0y}^2}{2a_y} = \frac{-(16 \text{ m/s})^2}{2(-9.80 \text{ m/s}^2)} = \boxed{+13 \text{ m}}$$

The height H depends only on the y variables; the same height would have been reached had the projectile been thrown *straight up* with an initial velocity of $v_{0y} = +16$ m/s.

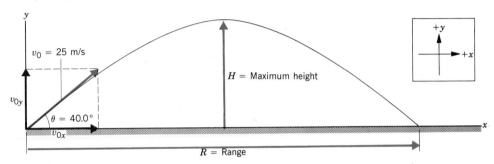

FIGURE 3.10 The trajectory of a projectile showing its maximum height H and range R. The projectile is fired with an initial speed of $v_0 = 25$ m/s at an angle of $\theta = 40.0°$ above the ground.

It is also possible to find the total time during which the projectile in Figure 3.10 is in the air before returning to the ground. Example 5 shows how to determine this time.

EXAMPLE 5

For the motion illustrated in Figure 3.10, determine the total time of flight between launching and landing, assuming that the projectile returns to ground level.

SOLUTION

When the projectile starts at ground level and returns to ground level, the displacement in the y direction is zero. The initial velocity in the y direction is the same as that in Example 4, i.e., $v_{0y} = +16$ m/s. Therefore, we have

y-direction data

y	a_y	v_y	v_{0y}	t
0	-9.80 m/s^2		$+16$ m/s	$?$

The time that the projectile spends in the air can be determined from Equation 3.5b:

$$y = v_{0y}t + \tfrac{1}{2}a_y t^2$$

$$0 = [(16 \text{ m/s}) + \tfrac{1}{2}(-9.80 \text{ m/s}^2)t]t$$

$$\boxed{t = 3.3 \text{ s}}$$

The second solution, $t = 0$, is discarded, because it represents the situation where the projectile has not yet begun its trip. As in Example 2, the time of flight depends only on the y variables, and the same result would have been obtained had the projectile been thrown *straight up* with an initial velocity of $v_{0y} = +16$ m/s.

An alternative way to compute the total time of flight in a range problem is to calculate the time for the projectile to reach its maximum height and then multiply this time by two.

Another important feature of projectile motion is called the "range." The range, as Figure 3.10 shows, is the horizontal distance traveled between launching and landing, assuming the projectile returns to the *same vertical level* at which it was fired. Example 6 shows how to obtain the range.

EXAMPLE 6

Using the data shown in Figure 3.10, calculate the range R of the projectile.

SOLUTION

The range of the projectile is a characteristic of the horizontal part of the motion. Thus, our starting point is to determine the horizontal component of the initial velocity:

$$v_{0x} = v_0 \cos \theta = (25 \text{ m/s}) \cos 40.0° = +19 \text{ m/s}$$

Recall from Example 5 that the time spent by the projectile in the air is $t = 3.3$ s. Since there is no acceleration in the x direction, v_{0x} remains constant, and the range is simply the product of v_{0x} and the time:

$$x = R = v_{0x}t = (19 \text{ m/s})(3.3 \text{ s}) = \boxed{+63 \text{ m}}$$

The range in the previous example depends on the angle θ at which the projectile is fired above the horizontal. It can be shown that the maximum range results when $\theta = 45°$, ignoring the effects of air resistance.

In all the examples in this section, the projectiles follow a curved trajectory. In general, if the only acceleration is that due to gravity, the shape of the path can be shown to be a *parabola*.

Section 2.7 points out that certain types of symmetry with respect to time and velocity are present for freely falling bodies. These symmetries are also found in projectile motion, since projectiles are falling freely in the vertical direction. In particular, the time required for a projectile to reach its maximum height H is equal to the time spent returning to the ground. In addition, Figure 3.11 shows that the speed of the object at any height above the ground on the upward part of the trajectory is equal to the speed at the same height on the downward part. Although the two speeds are the same, the velocities are different, because they point in different directions.

A third type of symmetry also exists in projectile motion. If you have ever used a hose, you might be aware that there are two possible angles at which to point the nozzle such that the range of the water is the same. Figure 3.12 illustrates three trajectories that have initial angles of 20°, 45°, and 70°. As mentioned earlier, the 45° angle gives rise to the maximum range. The 20° and 70° angles give rise to *identical ranges* that are less than the maximum range. In general, there are always two angles that give rise to the same range; one angle is less than 45° and the other angle is greater than 45°. The drawing indicates that the two angles are symmetrically placed about the 45° line. In other words, the 20° angle is 25° below the 45° angle, while the 70° angle is 25° above the 45° angle, as the insert in the picture emphasizes. Or equivalently, the two angles that produce the same range always add up to 90°.

FIGURE 3.11 The speed of a projectile at a given height above the ground is the same on the upward and downward parts of the trajectory. The velocities are different, however, since they point in different directions.

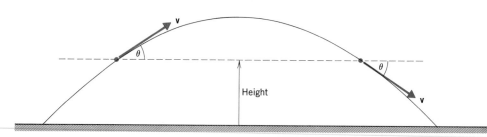

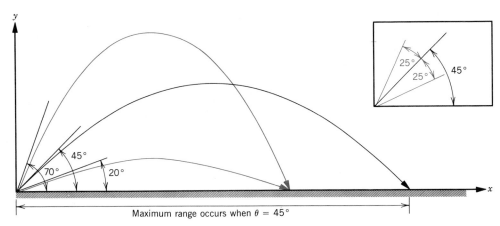

FIGURE 3.12 The 20° trajectory and the 70° trajectory have identical ranges. Note from the insert that the two angles are 25° on either side of the 45° line.

3.4 RELATIVE VELOCITY

RELATIVE VELOCITY IN ONE DIMENSION

The velocity of an object is relative to the observer who is making the measurement. Figure 3.13 illustrates the concept of relative velocity by showing a passenger walking toward the front of a moving train. The people sitting on the train see the passenger walking forward with a velocity of $+2.0$ m/s relative to them, where the plus sign denotes a direction to the right. In addition, suppose the train is moving with a velocity of $+9.0$ m/s relative to an observer standing on the ground. The ground-based observer sees the passenger moving with a velocity of $+11$ m/s, due in part to the walking motion and in part to the train's motion. As an aid in describing relative velocity, let us define the following symbols:

$\mathbf{v}_{\boxed{PT}}$ = velocity of the $\boxed{\text{Passenger}}$ relative to the $\boxed{\text{Train}}$ = $+2.0$ m/s

$\mathbf{v}_{\boxed{TG}}$ = velocity of the $\boxed{\text{Train}}$ relative to the $\boxed{\text{Ground}}$ = $+9.0$ m/s

$\mathbf{v}_{\boxed{PG}}$ = velocity of the $\boxed{\text{Passenger}}$ relative to the $\boxed{\text{Ground}}$ = $+11$ m/s

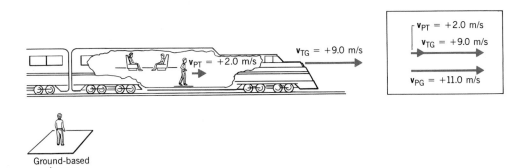

FIGURE 3.13 The velocity $\mathbf{v}_{PG}$ of the passenger relative to the ground-based observer is the vector sum of the velocity $\mathbf{v}_{PT}$ of the passenger relative to the train and the velocity $\mathbf{v}_{TG}$ of the train relative to the ground: $\mathbf{v}_{PG} = \mathbf{v}_{PT} + \mathbf{v}_{TG}$.

In terms of these symbols, the situation in Figure 3.13 is summarized as follows:

$$\mathbf{v_{PG}} = \mathbf{v_{PT}} + \mathbf{v_{TG}} \tag{3.7}$$

or

$$\mathbf{v_{PG}} = (2.0 \text{ m/s}) + (9.0 \text{ m/s}) = +11 \text{ m/s}$$

According to Equation 3.7, $\mathbf{v_{PG}}$ is the vector sum of $\mathbf{v_{PT}}$ and $\mathbf{v_{TG}}$, and this sum is indicated in the drawing. Had the passenger been walking toward the rear of the train, rather than toward the front, the velocity of the passenger relative to the ground-based observer would have been $\mathbf{v_{PG}} = (-2.0 \text{ m/s}) + (9.0 \text{ m/s}) = +7.0 \text{ m/s}$.

Each velocity symbol in Equation 3.7 contains a two-letter subscript. The first letter in the subscript refers to the body that is moving, while the second letter indicates the object relative to which the velocity is measured. For example, $\mathbf{v_{TG}}$ and $\mathbf{v_{PG}}$ are the velocities of the **Train** and **Passenger** measured relative to the **Ground**. Similarly, $\mathbf{v_{PT}}$ is the velocity of the **Passenger** measured by an observer who is sitting on the **Train**.

The ordering of the subscript symbols in Equation 3.7 follows a definite pattern. The first subscript (P) on the left side of the equation is also the first subscript on the right side of the equation. Likewise, the last subscript (G) on the left side of the equation is also the last subscript on the right side of the equation. The third subscript (T) appears only on the right side of the equation as the two "inner" subscripts. The colored boxes below emphasize the pattern of the symbols in the subscripts:

$$\mathbf{v}\boxed{\text{PG}} = \mathbf{v}\boxed{\text{P}}\text{T} + \mathbf{v_T}\boxed{\text{G}}$$

In other situations, the subscripts will not necessarily be P, G, and T, but will be chosen to be compatible with the names of the objects involved in the motion. In any event, the pattern of the subscript symbols will be the same as that illustrated above.

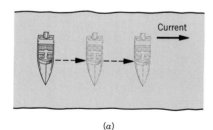

(a)

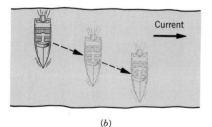

(b)

FIGURE 3.14 (a) A boat with its engine turned off is carried along by the current, so the velocity of the boat relative to the water is zero: $\mathbf{v_{BW}} = 0$. (b) With the engine turned on, the boat moves across the river in a diagonal fashion.

RELATIVE VELOCITY IN TWO DIMENSIONS

Figure 3.14 depicts a common situation that deals with relative velocity in two dimensions. Part *a* of the drawing shows a boat being carried downstream by a river. The engine of the boat is turned off. In part *b*, the engine has been turned on, and now the boat moves across the river in a diagonal fashion because of the combined motion produced by the current and the engine. The list below gives the velocities for this type of motion and the objects relative to which they are measured:

$\mathbf{v}\boxed{\text{BW}}$ = velocity of the $\boxed{\text{Boat}}$ relative to the $\boxed{\text{Water}}$

$\mathbf{v}\boxed{\text{WS}}$ = velocity of the $\boxed{\text{Water}}$ relative to the $\boxed{\text{Shore}}$

$\mathbf{v}\boxed{\text{BS}}$ = velocity of the $\boxed{\text{Boat}}$ relative to the $\boxed{\text{Shore}}$

The velocity $\mathbf{v_{BW}}$ of the boat relative to the water is the velocity measured by an observer who, for instance, is floating on an inner tube and drifting downstream with the current. When the engine is turned off, the boat also drifts downstream with the current and $\mathbf{v_{BW}}$ is zero. When the engine is turned on, however, the boat can move relative to the water, and $\mathbf{v_{BW}}$ is no longer zero. The velocity $\mathbf{v_{WS}}$ of the water relative to the shore is the velocity of the current measured by an observer on the shore. The velocity $\mathbf{v_{BS}}$ of the boat relative to the shore is due to the combined motion of the boat relative to the water and the motion of the water relative to the shore. In symbols,

$$\mathbf{v}\boxed{\text{BS}} = \mathbf{v}\boxed{\text{B}}\text{W} + \mathbf{v_W}\boxed{\text{S}}$$

The ordering of the subscripts in this equation is identical to that in Equation 3.7, although the letters have been changed to reflect a different physical situation. Example 7 illustrates the concept of relative velocity in two dimensions.

EXAMPLE 7

The engine of a boat drives it relative to the water at a velocity of $v_{BW} = 4.0$ m/s, directed perpendicular to the current (see Figure 3.15). (a) If the velocity of the water relative to the shore is $v_{WS} = 2.0$ m/s, what is the velocity v_{BS} of the boat relative to the shore? (b) If the river is 1800 m wide, how long does it take for the boat to cross it?

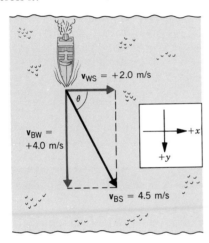

FIGURE 3.15 The velocity v_{BS} of the boat relative to the shore is the vector sum of the velocity v_{BW} of the boat relative to the water and the velocity v_{WS} of the water relative to the shore: $v_{BS} = v_{BW} + v_{WS}$.

SOLUTION

(a) The three velocity vectors are related by $v_{BS} = v_{BW} + v_{WS}$. Since the vectors v_{BW} and v_{WS} are perpendicular to each other, the magnitude of v_{BS} can be determined from the Pythagorean theorem:

$$v_{BS} = \sqrt{v_{BW}^2 + v_{WS}^2} = \sqrt{(4.0 \text{ m/s})^2 + (2.0 \text{ m/s})^2} = \boxed{4.5 \text{ m/s}}$$

Thus, the boat moves at a speed of 4.5 m/s with respect to an observer on the shore. The direction of the boat relative to the shoreline is given by the angle θ in the drawing:

$$\tan \theta = \frac{v_{BW}}{v_{WS}} = \frac{4.0 \text{ m/s}}{2.0 \text{ m/s}} = 2.0$$

$$\theta = \tan^{-1} 2.0 = \boxed{63°}$$

(b) The component of v_{BS} that is parallel to the width of the river (see Figure 3.15) indicates how fast the boat is moving across the river; the magnitude of this parallel component is $v_{BS} \sin \theta = v_{BW} = 4.0$ m/s. The time for the boat to cross the river is equal to the width of the river divided by the magnitude of the boat's velocity component parallel to the width:

$$t = \frac{\text{Width}}{v_{BS} \sin \theta} = \frac{1800 \text{ m}}{4.0 \text{ m/s}} = \boxed{450 \text{ s}}$$

Occasionally, situations arise when two vehicles are in relative motion, and it is useful to know the relative velocity of one with respect to the other. Example 8 considers this type of relative motion.

EXAMPLE 8

Figure 3.16a shows two cars, labeled A and B, that are approaching an intersection along two perpendicular roads. The cars have the following velocities:

$v_{\boxed{AG}}$ = velocity of $\boxed{\text{car A}}$ relative to the $\boxed{\text{Ground}}$

= 82 ft/s, eastward

$v_{\boxed{BG}}$ = velocity of $\boxed{\text{car B}}$ relative to the $\boxed{\text{Ground}}$

= 52 ft/s, northward

Find the magnitude and direction of v_{AB}, where

$v_{\boxed{AB}}$ = velocity of $\boxed{\text{car A}}$ as measured by a passenger
in $\boxed{\text{car B}}$

SOLUTION

To find v_{AB}, we would write down an equation whose subscripts followed the order outlined earlier. Thus,

$$v_{\boxed{AB}} = v_{\boxed{A}G} + v_{G\boxed{B}}$$

The second term on the right side of this equation involves v_{GB}, the velocity of the ground relative to a passenger in car B, rather than v_{BG}, which was given as 82 ft/s, northward. In other words, the subscripts are reversed. However, v_{GB} is related to v_{BG} according to

$$v_{GB} = -v_{BG}$$

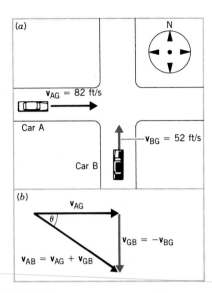

FIGURE 3.16 Two cars are approaching an intersection along two perpendicular roads.

This relationship reflects the fact that a passenger in car B, moving northward relative to the ground, looks out the car window and sees objects on the ground moving southward, that is, in the opposite direction. Therefore, the equation $v_{AB} = v_{AG} + v_{GB}$ may be used to find v_{AB}, provided we recognize v_{GB} as a vector that points opposite to the given velocity v_{BG}. With this in mind, Figure 3.15b illustrates how v_{AG} and v_{GB} are added vectorially to give v_{AB}. From the vector triangle shown in the drawing, the magnitude and direction of v_{AB} can be calculated as

$$v_{AB} = \sqrt{v_{AG}^2 + v_{GB}^2} = \sqrt{(82 \text{ ft/s})^2 + (52 \text{ ft/s})^2} = \boxed{97 \text{ ft/s}}$$

and

$$\tan \theta = \frac{v_{GB}}{v_{AG}} = \frac{52 \text{ ft/s}}{82 \text{ ft/s}} = 0.63$$

$$\theta = \tan^{-1} 0.63 = \boxed{32°}$$

SUMMARY

Motion that occurs in two dimensions can be described in terms of the time t and the x and y components of four vectors: the displacement, the acceleration, the final velocity, and the initial velocity. When the acceleration vector is constant, the x components of these vectors (x, a_x, v_x, and v_{0x}) are related by the equations of kinematics, as are the y components (y, a_y, v_y, and v_{0y}). Table 3.1 summarizes the equations of kinematics for two-dimensional motion.

Projectile motion—that of an object (the projectile) moving through the air—is one particular kind of motion in two dimensions. If air resistance can be neglected, the projectile experiences only the acceleration due to gravity. In projectile motion, the horizontal component of the projectile's velocity stays constant at all times, while the vertical component changes because of the acceleration due to gravity.

SOLVED PROBLEMS

SOLVED PROBLEM 1
Related Problems: *19 *20 *21

A meteorite is being tracked by radar as it falls through the earth's atmosphere. When its altitude is 3.00×10^4 m, the radar screen shows that the meteorite is traveling with a velocity of 583 m/s at an angle of 28.3° below the horizontal (see the drawing). (a) In the absence of air resistance, how much time elapses before the meteorite strikes the earth? (b) What is the velocity (magnitude and direction) of the meteorite just before impact with the earth?

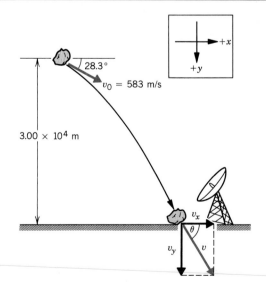

Solution (a) As Section 3.3 discusses, the time that a projectile is in the air is governed by the values of the y variables of the motion. The vertical component of the initial velocity of the meteorite is $v_{0y} = (583 \text{ m/s}) \sin 28.3° = 276 \text{ m/s}$, where the sign convention is shown in the drawing. These variables are summarized in the box below.

y-direction data

y	a_y	v_y	v_{0y}	t
$+3.00 \times 10^4$ m	$+9.80$ m/s²		$+276$ m/s	?

From these data and Equation 3.5b, the time can be found:

$$y = v_{0y}t + \tfrac{1}{2}a_y t^2$$

$$3.00 \times 10^4 \text{ m} = (276 \text{ m/s})t + \tfrac{1}{2}(9.80 \text{ m/s}^2)t^2$$

Solving this quadratic equation for the time t yields $\boxed{t = 55.0 \text{ s}}$. The second solution to the quadratic equation, $t = -111$ s, is ignored, because a negative value for the time is not physically meaningful here. (b) The speed v of the meteorite just before impact is given by the Pythagorean theorem as $v = \sqrt{v_x^2 + v_y^2}$. Since $a_x = 0$ in projectile motion, v_x remains constant at all times, so

$$v_x = v_{0x} = (583 \text{ m/s}) \cos 28.3° = 513 \text{ m/s}$$

The value of v_y can be obtained with the aid of Equation 3.3b:

$$v_y = v_{0y} + a_y t = (276 \text{ m/s}) + (9.80 \text{ m/s}^2)(55.0 \text{ s}) = 815 \text{ m/s}$$

The speed just before impact is

$$v = \sqrt{v_x^2 + v_y^2} = \sqrt{(513 \text{ m/s})^2 + (815 \text{ m/s})^2} = \boxed{963 \text{ m/s}}$$

From the drawing, it can be seen that the angle θ is given by

$$\tan \theta = \frac{v_y}{v_x} = \frac{815 \text{ m/s}}{513 \text{ m/s}} = 1.59$$

$$\theta = \tan^{-1} 1.59 = \boxed{57.8°}$$

Summary of Important Points The time that a projectile spends in the air is governed by the y variables of the motion; the time does not depend on the x variables. However, once the time has been determined, it can be used to find kinematic variables in either the y or x direction.

SOLVED PROBLEM 2
Related Problems: *23 *24 *25

A baseball player hits a home run, and the ball lands in the left-field seats, 25 ft above the point at which the ball was hit. The ball lands with a velocity of 160 ft/s at an angle of 32° to the horizontal (see the drawing). What is the initial velocity of the ball when the ball leaves the bat? See illustration below.

Solution The initial speed v_0 and angle θ of the baseball are related to the magnitudes of the horizontal and vertical components of the initial velocity (v_{0x} and v_{0y}) by the relations

$$v_0 = \sqrt{v_{0x}^2 + v_{0y}^2} \quad \text{and} \quad \tan \theta = \frac{v_{0y}}{v_{0x}}$$

where θ is shown in the drawing. The equations of kinematics can be employed to find v_{0x} and v_{0y}. Since $a_x = 0$ in projectile motion, v_x remains constant throughout the motion, so

$$v_{0x} = v_x = (160 \text{ ft/s}) \cos 32° = 140 \text{ ft/s}$$

The value for v_{0y} can be obtained from Equation 3.6b ($v_y^2 = v_{0y}^2 + 2a_y y$) and the data displayed below (see drawing for sign convention):

y-component data

y	a_y	v_y	v_{0y}	t
-25 ft	$+32.2$ ft/s²	$(+160 \sin 32°)$ ft/s	?	

$$v_{0y}^2 = (160 \sin 32° \text{ ft/s})^2 - 2(32.2 \text{ ft/s}^2)(-25 \text{ ft})$$

$$v_{0y} = 94 \text{ ft/s}$$

The initial speed v_0 and angle θ of the baseball are

$$v_0 = \sqrt{v_{0x}^2 + v_{0y}^2} = \sqrt{(140 \text{ ft/s})^2 + (94 \text{ ft/s})^2} = \boxed{170 \text{ ft/s}}$$

$$\tan \theta = \frac{v_{0y}}{v_{0x}} = \frac{94 \text{ ft/s}}{140 \text{ ft/s}} = 0.67$$

$$\theta = \tan^{-1} 0.67 = \boxed{34°}$$

Summary of Important Points If the final parameters of the motion of a projectile are known (y, v_x, and v_y in this problem), the equations of kinematics allow the initial parameters to be calculated (v_0 and θ in this problem).

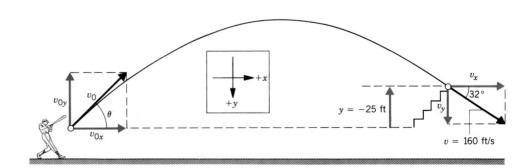

SOLVED PROBLEM 3
Related Problems: *40 *41 *42 **44

Aircraft A has a velocity relative to the ground of 410 m/s while traveling due north. Aircraft B, traveling at a slightly lower altitude, has a velocity relative to the ground of 640 m/s and is headed in a direction 40.0° west of south. Find the relative velocity (magnitude and direction) of plane B as observed by the passengers in plane A. Ignore any difference in altitude.

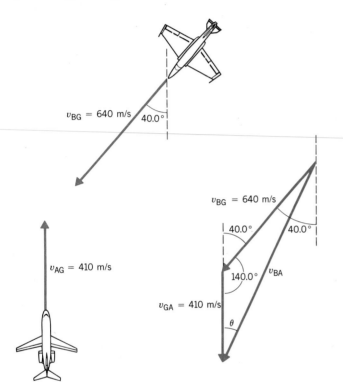

Solution The velocities in this problem are

$\mathbf{v}_{AG}$ = velocity of plane A relative to the ground = 410 m/s, north

$\mathbf{v}_{BG}$ = velocity of plane B relative to the ground = 640 m/s, at 40.0° west of south

$\mathbf{v}_{BA}$ = velocity of plane B as seen by the passengers in plane A

The equation that relates these velocities is of the form

$$\mathbf{v}_{BA} = \mathbf{v}_{BG} + \mathbf{v}_{GA} = \mathbf{v}_{BG} + (-\mathbf{v}_{AG})$$

The drawing shows how $\mathbf{v}_{BG}$ and $\mathbf{v}_{GA}$ are added vectorially to give $\mathbf{v}_{BA}$.

To find the magnitude of $\mathbf{v}_{BA}$, the law of cosines from trigonometry (see Appendix E) is applied to the triangle in the figure:

$$v_{BA}^2 = v_{BG}^2 + v_{GA}^2 - 2v_{BG}v_{GA} \cos 140.0°$$
$$= (640 \text{ m/s})^2 + (410 \text{ m/s})^2$$
$$- 2(640 \text{ m/s})(410 \text{ m/s}) \cos 140.0°$$

$$\boxed{v_{BA} = 990 \text{ m/s}}$$

The direction of $\mathbf{v}_{BA}$ can be specified by calculating the angle θ in the drawing through the use of the law of sines (see Appendix E):

$$\frac{\sin \theta}{v_{BG}} = \frac{\sin 140.0°}{v_{BA}}$$

$$\frac{\sin \theta}{640 \text{ m/s}} = \frac{\sin 140.0°}{990 \text{ m/s}}, \quad \boxed{\theta = \sin^{-1} 0.42 = 25°}$$

Summary of Important Points   Sometimes in relative velocity problems the three velocities do not form a right triangle when they are added according to an equation of the form, $\mathbf{v}_{BA} = \mathbf{v}_{BG} + \mathbf{v}_{GA}$. In these cases, one way to find the magnitude and direction of the unknown velocity is to use the law of cosines and the law of sines.

QUESTIONS

1. Suppose an object could move in three dimensions. What additions to the equations of kinematics in Table 3.1 would be necessary to describe three-dimensional motion?

2. At a given time, can an object be moving horizontally and yet have a vertical acceleration? If so, give an example.

3. If an object has a negative acceleration, does it necessarily mean that the object is slowing down? If not, why not?

4. A little league baseball player and a major league baseball player each hit a fly ball to center field. Once in flight, which ball, if either, has the greater acceleration? Provide a reason for your answer.

5. A tennis ball is hit into the air and moves along an arc. Neglecting air resistance, where along the arc is the speed of the ball a minimum? Where along the arc is the speed a maximum? Justify your answers.

6. A projectile is shot straight upward with an initial speed of v_0. Another projectile, fired at the same time from the same place, has an initial speed of v and is directed at an angle θ above the horizontal. The projectile shot at an angle has an initial velocity whose vertical component is $v \sin \theta = v_0$. In the absence of air resistance, which projectile, if either, strikes the ground first? Explain.

7. A rifle, at a height H above the ground, fires a bullet parallel to the ground. At the same instant and at the same height, a second bullet is dropped from rest. In the absence of air resistance, which bullet strikes the ground first? Explain.

8. A projectile is fired upward from the surface of the earth. Another projectile is fired upward from the surface of the moon and has the same initial velocity as the projectile on earth. Which projectile has the greater range, and which attains the greater height? Why?

9. A railroad flatcar is equipped with a cannon that points straight up. The train is moving due east at a constant velocity, and the cannon is fired. Neglecting air resistance, discuss where the shell will strike when it returns to earth.

10. On a riverboat cruise, a plastic bottle is accidentally dropped overboard. A passenger on the boat estimates that the boat pulls ahead of the bottle by 5 meters each second. Is it possible to conclude that the boat is moving at 5 m/s with respect to the shore? Give a reason.

11. Three swimmers can swim equally fast. They have a race to see who can swim across a river in the least time. Swimmer A swims perpendicular to the current and lands on the far shore downstream, because the current has swept him in that direction. Swimmer B swims upstream at an angle to the current and lands on the far shore directly opposite the starting pont. Swimmer C swims downstream at an angle to the current in an attempt to take advantage of the current. Who crosses the river in the least time? Justify your answer.

PROBLEMS

Section 3.1 Displacement, Velocity, and Acceleration

1. A radar antenna is tracking a satellite orbiting the earth. At a certain time, the radar screen shows the satellite to be 162 km away. The radar antenna is pointing upward at an angle of 62.3° from the ground. Find the x and y components (in km) of the position of the satellite.

2. A person stands 20.0 m in front of a tall building and looks up toward a window. The person's line of sight makes an angle of 55.0° with the horizontal. How far above the person's eyes is this window?

3. A jetliner is moving at a speed of 883 km/h. The vertical component of the plane's velocity is 146 km/h and points downward. Determine (a) the magnitude of the horizontal component of the plane's velocity, and (b) the angle that the velocity vector of the plane makes with the horizontal.

4. A dart is thrown upward at an angle of 25° from the horizontal. A vertical component of the dart's velocity is $v_y = +7.1$ ft/s. Determine (a) the x component of the velocity, and (b) the speed of the dart.

Section 3.2 Equations of Kinematics in Two Dimensions, Section 3.3 Projectile Motion

5. The punter on a football team tries to kick a football so that it stays in the air for a long "hang time" while traveling as far down the field as possible. If the ball is kicked with an initial velocity of 25.0 m/s at an angle of 60.0° above the ground, what is (a) the "hang time," and (b) the distance between the punter and the point where the ball lands?

6. With a particular club, the maximum speed that a golfer can impart to a ball is 109 km/h. (a) What is the longest hole in one (in

kilometers) that the golfer can make, if the ball does not roll when it hits the green? (b) How much time does the ball spend in the air? (c) What is the maximum height reached by the ball?

7. A bullet is fired from a rifle that is held 1.6 m above the ground in a horizontal position. The initial speed of the bullet is 1100 m/s. Find (a) the time it takes for the bullet to strike the ground and (b) the horizontal distance traveled by the bullet.

8. (a) If a projectile has a launching angle of 52.0° above the horizontal and an initial speed of 18.0 m/s, what is the highest barrier that the projectile can clear? (b) What is the horizontal distance of this barrier from the launching point?

9. A major league pitcher can often throw a baseball in excess of 92 mi/h. If a ball is thrown horizontally at this speed, how much can it be expected to drop (in inches) due to gravity by the time it reaches a catcher who is 56 ft away from the point of release?

10. A car drives straight off the edge of a cliff that is 54 m high. The police at the scene of the accident note that the point of impact is 130 m from the base of the cliff. How fast was the car traveling when it went over the cliff?

11. A jet fighter is traveling horizontally with a speed of 4.00×10^2 km/h at an altitude of 3.00×10^2 m, when the pilot accidentally releases an outboard fuel tank. (a) How much time elapses before the tank hits the ground? (b) What is the speed of the tank just before it hits the ground? (c) What is the horizontal distance traveled by the tank?

12. An archer is standing inside a building whose ceiling is 11 m high. (a) An arrow is shot from ground level at an initial speed of 62 m/s. Calculate the angle of firing (above the horizontal) that gives the greatest possible range inside the building. Using the result from part (a), what is (b) the range of the arrow and (c) its time of flight?

13. A motorcycle daredevil is attempting to jump across as many buses as possible (see the drawing). The takeoff ramp makes an angle of 18° to the horizontal, and the landing ramp is identical to the takeoff ramp. The buses are parked side by side, and each bus is 9.0 ft wide. The cyclist leaves the ramp with a speed of 75 mi/h. What is the maximum number of buses over which the cyclist can jump?

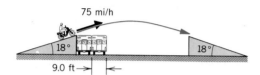

75 mi/h
18°   18°
9.0 ft

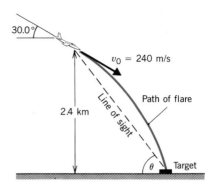

30.0°
v_0 = 240 m/s
Path of flare
2.4 km
Line of sight
θ   Target

14. A rifle is held horizontally and fired at a bull's-eye. The muzzle speed of the bullet is 2200 ft/s. The bullet strikes the target one inch below the bull's-eye. What is the horizontal distance between the end of the rifle and the bull's-eye?

15. A criminal is escaping across a rooftop and runs off the roof horizontally, landing on the roof of an adjacent building. The horizontal distance between the two buildings is 3.4 m, and the roof of the adjacent building is 2.0 m below the jumping off point. What would be the minimum speed needed by the criminal?

16. A rock, thrown horizontally from the top of a lighthouse, strikes the water 2.6 s later. An imaginary straight line is drawn from the top of the lighthouse to the point where the rock strikes the water. This line makes an angle of 35° with respect to the lighthouse. Calculate the initial speed (in m/s) of the rock.

17. During a fireworks display, a rocket is launched with an initial velocity of 35 m/s at an angle of 75° above the ground. The rocket explodes 3.7 s later. (a) What is the height of the rocket when it explodes? (b) What is the velocity (magnitude and direction) of the rocket just before it explodes?

* 18. From the edge of a 60.0-m cliff, a projectile is launched upward with an initial velocity of 23.0 m/s at an angle of 50.0° with respect to the horizontal. At what point above the ground does the rocket strike the wall of a vertical cliff located 20.0 m away?

* 19. A golfer is standing on a fairway and hits a shot to a green that is elevated 6.0 m above the point where she is standing. If the ball leaves her club with a velocity of 43 m/s at an angle of 40.0° to the ground, find (a) the time for the ball to come down on the green, (b) the speed of the ball just before impact, and (c) the angle of the velocity vector when the ball hits the green. *(See Solved Problem 1 for a related problem.)*

* 20. An airplane is flying with a speed of 240 m/s at an angle of 30.0° with the horizontal, as the drawing shows. When the altitude of the plane is 2.4 km, a flare is released from the plane. The flare hits the target on the ground. What is the angle θ? *(See Solved Problem 1 for a related problem.)*

* 21. An airplane, with a speed of 218 mi/h, is climbing upward at an angle of 50.0° with respect to the horizontal. When the plane's altitude is 2.40×10^3 ft, the pilot releases a package. (a) Calculate the distance along the ground, measured from a point directly beneath the plane, to the point where the package hits the earth. (b) Find the speed of the package just before impact. (c) Relative to the ground, determine the angle of the velocity vector just before impact. *(See Solved Problem 1 for a related problem.)*

* 22. If the *maximum* horizontal distance that a ball can be thrown is 47.0 m, how high can it be thrown straight upward, assuming the same throwing speed in each case?

* 23. A diver springs upward from a three-meter board. At the instant she contacts the water her speed is 8.90 m/s and her body makes an angle of 75.0° with respect to the surface of the water. (a) Determine her initial velocity, both magnitude and direction. *(See Solved Problem 2 for a related problem.)* (b) How much time does she spend in the air?

* 24. After leaving the end of a ski ramp, a ski jumper lands downhill at a point that is displaced 180 ft horizontally from the end of the ramp. His velocity, just before landing, is 81 ft/s and points in a direction 38° below the horizontal. Neglecting air resistance and any lift that he experiences while airborne, find his initial velocity (magnitude and direction) when he left the end of the ramp. *(See Solved Problem 2 for a related problem.)*

* 25. A golf ball is driven from a level fairway. At a time of 5.10 s later, the ball is traveling downward with velocity of 48.6 m/s at an angle of 22.2° below the horizontal. (a) Calculate the initial velocity (magnitude and direction) of the golf ball. *(See Solved Problem 2 for a related problem.)* (b) Find the x and y components of the ball's displacement at t = 5.10 s.

* 26. The drawing shows an exaggerated view of a rifle that has been "sighted in" for a 1.00×10^2-yard target. If the muzzle speed of the bullet is $v_0 = 1.40 \times 10^3$ ft/s, what are the two possible angles θ_1 and θ_2 between the rifle barrel and the horizontal such that the bullet will hit the target? One of these angles is so large that it is never used in target shooting. (*Hint: The following trigonometric identity may be useful:* $2 \sin \theta \cos \theta = \sin 2\theta$.)

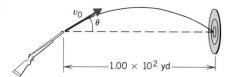

* 27. Stones are thrown horizontally with the same velocity from the tops of two different buildings. One stone lands twice as far from the base of the building from which it was thrown as does the other stone. How do the heights of the buildings compare?

* 28. A projectile is fired at an angle θ above the horizontal. Prove that the time for the projectile to travel from the ground to its maximum height is equal to the time to travel from its maximum height back to the ground.

* 29. An arrow is shot with an initial speed of $v_0 = 71.0$ m/s at an angle of $\theta = 25.0°$ from the horizontal. The horizontal distance from the arrow to the wall is $x = 260.0$ m, as the drawing illustrates. If the arrow is initially aimed at the point P, find the distance y below P where the arrow strikes the wall.

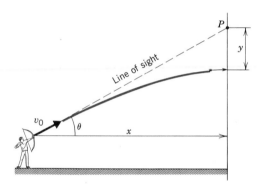

**30. Suppose the arrow in problem 29 is shot with an initial speed of $v_0 = 58.0$ m/s at an angle of $\theta = 35.0°$ above the horizontal. The archer notices that the arrow strikes a vertical wall at a distance of $y = 26.0$ m below the line of sight. What is the horizontal displacement x of the arrow?

**31. A garden hose, pointed at an angle of 25° above the horizontal, splashes water on a sunbather lying on the ground 4.4 m away in the horizontal direction. If the hose is held 1.4 m above the ground, at what speed does the water leave the nozzle?

**32. A ball is thrown downward at an angle from a cliff whose height is 107 m. A few moments later, a 1.70-m-tall observer, standing below and at some distance from the base of the cliff, looks up and notices that the ball is 85.0 m directly above her, as the drawing shows. The ball strikes the ground 2.38 s after passing over the observer and lands at a point that is 57.0 m from her. Determine (a) the horizontal velocity component with which the ball was thrown, (b) the vertical velocity component with which the ball was thrown, (c) the initial velocity (magnitude and direc-

tion) of the ball, and (d) the horizontal distance between the base of the cliff and the point where the ball strikes the ground.

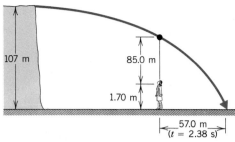

**33. A rocket launcher is sitting on a surface that is inclined at 25°, as the drawing shows. The launcher is inclined at an angle of 15° with respect to the surface. The initial speed of the projectile is 81 m/s. Find the distance D up the incline at which the rocket lands.

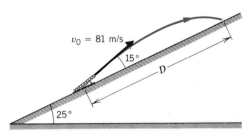

**34. A placekicker is about to a kick a field goal. The ball is 30.0 yards from the goal post. The ball is kicked with an initial velocity of 65 ft/s at an angle θ above the ground. Between what two angles, θ_1 and θ_2, will the ball clear the 3.0-yard-high crossbar? (*Hint: The following trigonometric identities may be useful:* $\sec \theta = 1/\cos \theta$ *and* $\sec^2 \theta = 1 + \tan^2 \theta$.)

**35. A small can is hanging from the ceiling by an electromagnet. A rifle is aimed directly at the can, as the figure illustrates. At the instant the gun is fired, the can is released. Ignore air resistance and show that the bullet will always strike the can, regardless of the initial speed of the bullet. Assume that the bullet strikes the can before the can reaches the ground.

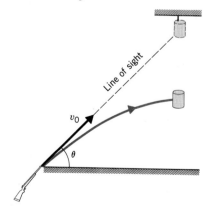

Section 3.4 Relative Velocity

36. Two cars, A and B, are traveling in the same direction, although car B is 186 m behind car A. The speed of B is 88.0 km/h and the speed of A is 67.1 km/h. How much time (in seconds) does it take for B to catch A?

37. Two passenger trains are passing each other on adjacent tracks. Train A is moving east with a speed of 42 ft/s, and train B is traveling west with a speed of 93 ft/s. (a) What is the velocity (magnitude and direction) of train A as seen by the passengers in train B? (b) What is the velocity (magnitude and direction) of train B as seen by the passengers in train A?

38. A swimmer, capable of swimming at a speed of 1.4 m/s in still water (i.e., the swimmer can swim with a speed of 1.4 m/s relative to the water), starts to swim directly across a 2.8-km-wide river. However, the current is 0.91 m/s, and it carries the swimmer downstream. (a) How long does it take the swimmer to cross the river? (b) How far downstream will the swimmer be upon reaching the other side of the river? (c) What is the swimmer's speed relative to the shore?

39. A remote-controlled model airplane is flying due east in still air. The airplane travels with a speed of 22.6 m/s relative to the air. A wind suddenly begins to blow from the north toward the south with a speed of 8.70 m/s. Find the velocity (magnitude and direction) of the airplane as seen by the controller who is standing on the ground.

* 40. A ferry boat is traveling in a direction 35.1° north of east with a speed of 5.12 m/s relative to the water. A passenger is walking with a velocity of 2.71 m/s due east relative to the boat. What is the velocity (magnitude and direction) of the passenger with respect to the water? *(See Solved Problem 3 for a related problem.)*

* 41. A bicyclist is riding on a windless day and moves due west at a speed of 11.4 m/s. (a) Relative to the bicycle, what is the velocity (magnitude and direction) of the air that he feels blowing against him? (b) If a wind starts to blow from the northeast toward the southwest (along a line that is 45° north of east) at 6.10 m/s relative to the earth, what is the velocity (magnitude and direction) of the air that the cyclist feels blowing against him? *(See Solved Problem 3 for a related problem.)*

* 42. An oceanliner is heading due north with a speed of 28 ft/s relative to the water. A small sailboat is heading 45° east of north with a speed of 3.3 ft/s relative to the water. Find the relative velocity (magnitude and direction) of the sailboat as observed by the passengers on the oceanliner. *(See Solved Problem 3 for a related problem.)*

* 43. A yacht can travel at a speed of 12.0 km/h relative to the water. The captain wishes to reach a marina that is 3.00 km *directly across* the river from his location. The water is flowing at a speed of 4.00 km/h. (a) At what angle (measured *upstream* relative to the perpendicular line crossing the river) must the captain steer the boat to reach his destination? (b) How much time is required for the yacht to reach the marina?

**44. A sailboat is moving with a speed of 10.9 km/h relative to the (still) water at an angle of 62.0° east of north. A wind is blowing from the north to the south. A flag, attached to the mast, indicates that the velocity of the wind relative to the boat is directed at an angle of 28.0° from the centerline of the boat (see the drawing). What is the speed of the wind relative to the still water? *(See Solved Problem 3 for a related problem.)*

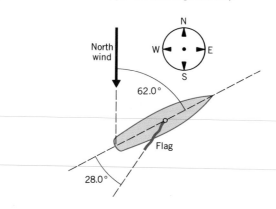

ADDITIONAL PROBLEMS

45. What is the smallest muzzle velocity that a rifle can have, if a horizontally fired bullet is to hit a 1.00-inch-diameter target located 86.0 ft away? Assume that the center of the target is on the same horizontal line as is the barrel of the rifle.

46. Two cars with different velocities are approaching an intersection. Car A, traveling due east, has a speed of 15 m/s. Car B, traveling due north, has a speed of 21 m/s. What is the velocity (magnitude and direction) of B as seen by the passengers in A?

47. At a certain point along its trajectory, a golf ball has the following velocity components: $v_x = +46$ ft/s and $v_y = +62$ ft/s, with upward and to the right being positive. (a) What is the speed of the golf ball at this point? (b) At what angle does the velocity vector point relative to the horizontal direction?

48. A box of .22-caliber bullets has the following message written on it: "Warning! Range, $1\frac{1}{4}$ miles." Assume the muzzle speed of a bullet is 1100 ft/s and calculate the maximum range in miles. Compare your answer with the "warning." (Of course, air resistance will slow down the bullet, so its actual range will be substantially less than that calculated.)

* 49. A remote-controlled "target" plane is flying horizontally at an altitude of 4.2 km with a speed of 810 km/h. When the plane is directly overhead, a projectile is fired at an angle θ with the ground and has an initial speed of 1400 km/h, as the diagram shows. The projectile hits the plane. (a) Find the angle θ. (b) Find the time (in seconds) required for the projectile to hit the plane. (The calculation yields two values. Interpret each one.)

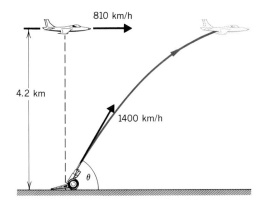

810 km/h

4.2 km

1400 km/h

θ

* 50.   A small aircraft is headed due south with a speed of 208 km/h with respect to the still air. Then, for 15.0 minutes a wind blows the plane so that it moves in a direction 45.0° west of south, even though the plane continues to point due south. The plane travels 81.0 km with respect to the ground in this time. Determine the velocity (magnitude and direction) of the wind with respect to the ground.

* 51.   An archer shoots an arrow at a target located 30.0 m away. Initially, the target and the arrow are at the same height. If the arrow leaves the bow at a speed of 48.0 m/s, how high (in meters) above the bull's-eye must the archer aim to hit the target? There are two answers to this problem. What does each answer represent physically? (*Hint: The following trigonometric identity may be useful:* $2 \sin \theta \cos \theta = \sin 2\theta$.)

* 52.   A soccer player kicks the ball toward a goal that is 29.0 m in front of him. The ball leaves his foot at a speed of 19.0 m/s and an angle of 32.0° above the ground. Find the speed of the ball when the goalie catches the ball in front of the net.

**53.   Two identical dormitory buildings are situated 10.0 m apart. Standing on top of one building is a college prankster, throwing pebbles at the window of a friend's room in the other building. If the window is 4.00 m below the point of release, and if the throwing speed is 15.0 m/s, at what angle below the horizontal should the initial velocity be directed? (*Hint: The following trigonometric identities may be useful:* $\sec \theta = 1/\cos \theta$, and $\sec^2 \theta = 1 + \tan^2 \theta$.)

**54.   A jetliner can fly 6.00 hours on a full load of fuel. Without any wind it flies at a speed of 865 km/h. The plane is to make a round-trip by heading due west for a certain distance, turning around, and then heading due east for the return trip. During the entire flight, however, the plane encounters a 208 km/h wind from the jet stream, which blows from west to east. What is the maximum distance that the plane can travel due west and just be able to return home?

**55.   A baseball is hit into the air at an initial speed of 1.20×10^3 ft/s and an angle of 50.0° above the horizontal. At the same time, the centerfielder starts running away from the batter and catches the ball 3.00 ft above the level at which it was hit. If the centerfielder is initially 3.60×10^2 ft from home plate, what must be his average speed?

CHAPTER 4

Forces and Newton's Laws of Motion

Romania's Mihaela Loghin exerts a considerable force in putting the shot. According to Isaac Newton, the acceleration of the shot is determined by the vector sum of this force, and all other forces that act on the shot, including its weight and the retarding force of the air.

4.1 THE CONCEPTS OF FORCE AND MASS

The previous two chapters discuss how to describe the motion of an accelerating object but do not explain what causes acceleration. The present chapter explores the subject of what causes acceleration, and in the process, two new ideas will come into play, force and mass. *Force* is a common word in our vocabulary, and it usually means a pull or a push. A tow truck pulls a stalled car by applying a force to the car. The force of gravity continually pulls all objects downward, so it causes a fly ball in a baseball game to return to the earth. And a bulldozer pushes over the wall of a dilapidated building by exerting a force on the wall. Figure 4.1 illustrates each of these examples and uses arrows to represent the forces. It is appropriate to use arrows, because a force is a vector quantity and has both a magnitude and a direction. The direction of the arrow gives the direction of the force, and the length is proportional to the strength, or magnitude, of the force.

The word *mass* also occurs in our vocabulary, although more recognizably as part of the word "massive." A massive supertanker, for instance, is one that contains an

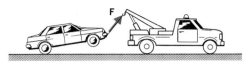
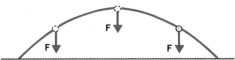
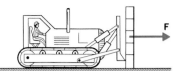

FIGURE 4.1 In each case, the arrow labeled **F** represents the force acting on each object.

enormous amount of matter or mass. In comparison, a penny does not contain much mass. The emphasis here is on the amount of matter, and the idea of direction is of no concern. Therefore, mass is a scalar quantity.

During the seventeenth century, Isaac Newton, starting with the work of Galileo, developed three important laws that deal with force and mass. They are now collectively called "Newton's laws of motion," and they provide an accurate basis for understanding the effect that any force has on an object. Because of the importance of these laws, separate sections will be devoted to each of them.

4.2 NEWTON'S FIRST LAW OF MOTION

THE FIRST LAW

To gain some insight into Newton's first law, consider the game of ice hockey (Figure 4.2). Obviously, if a player does not hit a stationary puck, it will remain at rest on the ice. After the puck is struck, however, it coasts on its own across the ice, slowing down only slightly because of the friction present when two surfaces slide against one another. Since ice is very slippery, there is only a relatively small amount of friction to slow down the puck. In fact, if it were possible to remove all friction and wind resistance, and if the rink were infinitely large, the puck would coast forever in a straight line at a constant speed. Left on its own, the puck would lose none of the speed imparted to it at the time it was struck. Here we have the essence of Newton's first law of motion:

> **NEWTON'S FIRST LAW OF MOTION**
> An object continues in a state of rest or in a state of motion at a constant speed along a straight line, unless compelled to change that state by a net force.

FIGURE 4.2 Ice hockey.

Notice that the first law uses the phrase "net force." Often, several forces act simultaneously on a body, and *the net force is the vector sum of all of them.* Individual forces matter only to the extent that they contribute to the total. For instance, if friction and other opposing forces were absent, a car could travel forever at 55 mi/h in a straight line, without using any gas after it has come up to speed. In reality, of course, gas is needed, but only so that the engine can produce the necessary driving force to cancel opposing forces such as friction. This cancellation ensures that there is no net force to change the state of motion of the car.

When an object moves at a constant speed along a straight line, its velocity vector is constant. Newton's first law indicates that a state of rest (zero velocity) and a state of constant velocity are completely equivalent, in the sense that neither one requires the application of a net force to sustain it. The purpose served when a net force acts on an object is not to sustain the velocity of the object, but, rather, to *change the velocity.*

INERTIA AND MASS

Some objects have a greater tendency than others to remain at rest or in motion at a constant speed along a straight line. For instance, a freight train has more of a tendency to keep moving than a bicycle does. This tendency is referred to as *inertia*, and, in fact, Newton's first law is sometimes called the law of inertia. The more matter an object contains, the greater is its inertia. Quantitatively, the inertia of an object is measured by its *mass*, a large mass, such as that of a freight train, indicating a large inertia:

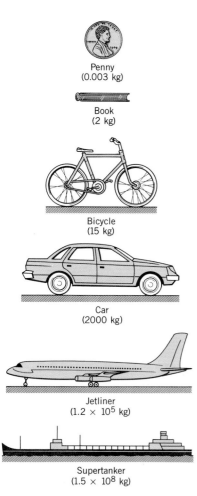

Penny
(0.003 kg)

Book
(2 kg)

Bicycle
(15 kg)

Car
(2000 kg)

Jetliner
(1.2×10^5 kg)

Supertanker
(1.5×10^8 kg)

FIGURE 4.3 The masses of various objects, expressed in kilograms.

DEFINITION OF INERTIA AND MASS

Inertia is the natural tendency of an object to remain at rest or in motion at a constant speed along a straight line. The mass of an object is a quantitative measure of inertia.

   SI unit of inertia and mass: kilogram (kg)

As Section 1.2 discusses, the SI unit for mass is the kilogram (kg), whereas the units employed in the CGS system and the BE system are the gram (g) and the slug (sl), respectively. Conversion factors between these units are given on the page facing the inside of the front cover. Figure 4.3 gives the masses of various objects, ranging from a penny to a supertanker. The larger the mass of the object, the greater is the inertia. Often the words "mass" and "weight" are used interchangeably, but they should not be. Mass and weight are different concepts, and Section 4.6 will discuss the distinction between them.

Figure 4.4 shows a useful application of inertia. Seat belts unwind freely when pulled gently, so they can be buckled. But in an accident, they do not unwind. They hold you safely in place. One mechanism used in these seat belts consists of a ratchet wheel, a locking bar, and a pendulum. The seat belt is wound around a spool mounted on the ratchet wheel. While the car is at rest or moving at a constant velocity, the pendulum hangs straight down, and the locking bar rests horizontally, as the black-lined part of the drawing shows. Consequently, there is nothing to prevent the ratchet wheel from turning, and the seat belt can be pulled out easily. When the car suddenly slows down in an accident, however, the relatively massive, lower part of the pendulum keeps moving forward because of its inertia. The pendulum swings on its pivot into the position shown in color and causes the locking bar to block the rotation of the ratchet wheel, thus preventing the seat belt from unwinding.

AN INERTIAL REFERENCE FRAME

Sometimes Newton's first law (and also the second law) can appear to be invalid to certain observers. Suppose, for instance, that you are a passenger riding in a friend's car. While the car moves at a constant speed along a straight line, you do not feel the seat pushing against your back to any unusual extent. This experience is consistent with the first law, which indicates that in the absence of a net force you should move with a constant velocity. Suddenly the driver floors the gas pedal. Immediately you feel the seat pressing against your back as the car accelerates, and, therefore, you sense that a force is being applied to you. The first law leads you to believe that your motion should change, and, relative to the ground outside, your motion does change. But *relative to the car,* you can see that your motion does *not* change, because you remain stationary with respect to the car. Clearly, Newton's first law does not hold for observers who use the accelerating car as a frame of reference. As a result, such a reference frame is said to be noninertial. All accelerating reference frames are noninertial. In contrast, observers for whom the law of inertia is valid are said to be using *inertial reference frames* for their observations, as defined below:

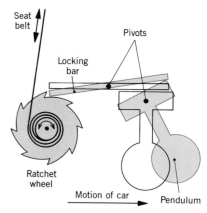

FIGURE 4.4 Inertia plays the central role in one mechanism that allows a seat belt to unwind in response to a gentle pull, but prevents it from unwinding in an accident. The colored parts of the drawing show what happens when the car suddenly slows down, as in an accident.

DEFINITION OF AN INERTIAL REFERENCE FRAME
An inertial reference frame is one in which Newton's law of inertia is valid.

All of Newton's laws of motion are valid in inertial reference frames, and when we apply these laws, we will be assuming such a reference frame unless otherwise stated. In particular, the earth itself is a good approximation of an inertial reference frame.

4.3 NEWTON'S SECOND LAW OF MOTION

THE SECOND LAW

Newton's first law indicates that if the net force applied to an object is zero, the velocity of the object remains unchanged. The second law deals with what happens when the net force is not zero. To see the basic ideas expressed by the second law, consider a hockey puck once again. When a hockey player strikes a stationary puck, he causes the velocity of the puck to change. In other words, he makes the puck accelerate. The cause of the acceleration is the force that the hockey stick applies. As long as this force acts, the velocity increases, and the puck accelerates.

Now, suppose that another player strikes the puck and applies twice as much force as the first player does. The greater force acting for the same amount of time produces a greater acceleration. In fact, if the friction between the puck and the ice is negligible, and if there is no wind resistance, the acceleration of the puck is directly proportional to the force. Twice the force produces twice the acceleration. Moreover, the acceleration is a vector quantity, just as the force is, and the acceleration of the puck points in the same direction as the force.

So far, only a single force has been considered, namely, the force with which a player strikes the puck. Often, however, several forces act on an object simultaneously. Friction and wind resistance, for instance, do have some effect on the puck. In such cases, it is the net force, or the vector sum of all the forces acting, that is important. Mathematically, the net force is written as $\Sigma \mathbf{F}$, where the Greek capital letter Σ (sigma) denotes the vector sum. Newton's second law states that the acceleration is proportional to the net force acting on the object.

In Newton's second law, the net force is only one of two factors that determine the acceleration of an object. The other factor is the inertia or mass of the object. After all, the same net force that imparts an appreciable acceleration to the hockey puck (small mass) will impart very little acceleration to a semitrailer (large mass). Newton's second law states that for a given net force, the magnitude of the acceleration is inversely proportional to the mass. Twice the mass means one-half the acceleration, if the same net force acts on both objects. Thus, the second law specifies the manner in which the acceleration depends on both the net force and the mass, as summarized below in Equation 4.1.

NEWTON'S SECOND LAW OF MOTION

When a net force ΣF acts on an object of mass m, the acceleration **a** that results is directly proportional to the net force and has a magnitude that is inversely proportional to the mass. The direction of the acceleration is the same as the direction of the net force.

$$\mathbf{a} = \frac{\Sigma \mathbf{F}}{m} \quad \text{or} \quad \Sigma \mathbf{F} = m\mathbf{a} \tag{4.1}$$

SI unit of force: $\text{kg} \cdot \text{m/s}^2 = \text{newton (N)}$

UNITS AND THE SECOND LAW

According to Equation 4.1, the SI unit for force is the unit for mass (kg) times the unit for acceleration (m/s²), or

$$\text{SI unit for force} = (\text{kg}) \left(\frac{\text{m}}{\text{s}^2} \right) = \frac{\text{kg} \cdot \text{m}}{\text{s}^2}$$

The combination of kg·m/s² is called a *newton* (N) and is a derived SI unit, not a base unit; 1 newton = 1 N = 1 kg·m/s².

In the CGS system, the procedure for establishing the units is the same as with SI units, except that mass is expressed in grams (g) and acceleration in cm/s². The resulting unit for force is the *dyne;* 1 dyne = 1 g·cm/s².

In the BE system, the unit for force is defined to be the pound (lb),* while the unit for acceleration is ft/s². With this procedure, Newton's second law can then be used to derive the unit for mass:

$$\text{BE unit for force} = \text{lb} = m \left(\frac{\text{ft}}{\text{s}^2} \right)$$

$$m = \frac{\text{lb} \cdot \text{s}^2}{\text{ft}}$$

The combination of lb·s²/ft is the unit for mass in the BE system and is called the *slug* (sl); 1 slug = 1 sl = 1 lb·s²/ft.

---

* We refer here to the gravitational version of the BE system, in which a force of one pound is defined to be the pull of the earth on a certain standard body at a location where the acceleration due to gravity is 32.174 ft/s².

TABLE 4.1 Units for Mass, Acceleration, and Force

System	Mass	Acceleration	Force
SI	kilogram (kg)	meter/second² (m/s²)	newton (N)
CGS	gram (g)	centimeter/second² (cm/s²)	dyne (dyn)
BE	slug (sl)	foot/second² (ft/s²)	pound (lb)

Table 4.1 summarizes the units for force, mass, and acceleration in the various systems. Conversion factors between force units from different systems are provided on the page facing the inside of the front cover. When using Newton's second law to solve problems, do not mix the force units from one system with the mass and acceleration units from another system. Always use a consistent set of units for $\Sigma\mathbf{F}$, m, and $\mathbf{a}$.

FREE-BODY DIAGRAMS AND THE SECOND LAW

When using the second law to calculate the acceleration, it is necessary to determine correctly the net force that acts on the object. In this determination a *free-body diagram* helps enormously. A free-body diagram is a diagram that represents the object and *all* of the forces that act on it. Only the forces that *act on the object* appear in a free-body diagram. Forces that the object exerts on its environment are not included. Example 1 illustrates the use of a free-body diagram.

EXAMPLE 1

Two people are pushing a stalled car, as Figure 4.5*a* indicates. The mass of the car is 1850 kg. One person applies a force of 275 N to the car, while the other applies a force of 395 N. Both of these forces act in the same direction. A third force of 560 N also acts on the car, but in a direction exactly opposite to that in which the people are pushing. This force arises because of friction and the extent to which the gravel on the road opposes the motion of the tires. Find the acceleration of the car.

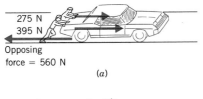

275 N
395 N
Opposing force = 560 N

(a)

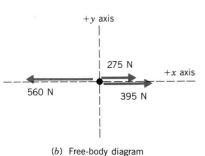

+*y* axis

275 N

+*x* axis

560 N

395 N

(b) Free-body diagram

SOLUTION

Before Newton's second law can be used to obtain the acceleration, the net force acting on the car must be determined. Figure 4.5*b* shows the free-body diagram for the car, and this diagram is helpful in determining the net force. In this drawing the car is represented as a "dot," and the motion of the car is chosen to be along the *x* axis. The diagram makes it clear that the forces all act along one direction. Therefore, they can be added as colinear vectors to give a net force of

$$\Sigma F = +275 \text{ N} + 395 \text{ N} - 560 \text{ N} = +110 \text{ N}$$

The acceleration can now be obtained:

$$a = \frac{\Sigma F}{m} = \frac{+110 \text{ N}}{1850 \text{ kg}} = \boxed{+0.059 \text{ m/s}^2} \qquad (4.1)$$

The plus sign indicates that the acceleration vector points along the +*x* axis, in the same direction as the net force.

FIGURE 4.5 *(a)* Two people push a stalled car, in opposition to a force created by friction and the gravel on the road. *(b)* A free-body diagram shows all the forces acting on the car.

NEWTON'S FIRST LAW AS A SPECIAL CASE OF THE SECOND LAW

We have been discussing Newton's first and second laws as if they were independent. In fact, the first law can be considered as a special case of the more general second law. To see this, consider the case when the net force acting on an object is zero, so that $\Sigma \mathbf{F} = 0$. According to Newton's second law, $\Sigma \mathbf{F} = m\mathbf{a}$, the acceleration must also be zero. A zero acceleration means that the object travels with a constant velocity, that is, with a constant speed along a straight line, which is exactly the kind of motion that the first law deals with.

4.4 THE VECTOR NATURE OF NEWTON'S SECOND LAW OF MOTION

Newton's second law, as expressed in Equation 4.1, deals with vectors. In two dimensions, this equation can be written in an equivalent form as two equations, one for the x components of the net force and acceleration vectors (ΣF_x and a_x) and one for the y components (ΣF_y and a_y):

$$\Sigma F_x = ma_x \tag{4.2a}$$

$$\Sigma F_y = ma_y \tag{4.2b}$$

The procedure here is similar to that employed for writing the equations of two-dimensional kinematics in Chapter 3. There, one set of equations applies to the x components of the displacement, velocity, and acceleration vectors, and another similar set applies to the y components (see Table 3.1). The vector components themselves in Equations 4.2a and 4.2b will be either positive or negative numbers, depending on whether they point along the positive or negative x or y axis. The remainder of this section deals with an example that shows how these equations are used.

EXAMPLE 2

In Figure 4.6 a 15.0-kg block moves on a flat, friction-free, horizontal surface. At the instant shown in part a, the block has a velocity of 20.0 m/s directed along the $+x$ axis, and at this instant two forces, $\mathbf{F}_1$ and $\mathbf{F}_2$, begin acting. $\mathbf{F}_1$ has a magnitude of 35.0 N and acts at an angle of 60.0° with respect to the $+x$ axis, while $\mathbf{F}_2$ has a magnitude of 10.0 N and acts along the $+x$ axis. (a) Find the x and y components of the resulting acceleration. (b) Where is the block located relative to the x and y axes after 8.00 s?

SOLUTION

(a) In preparation for using Newton's second law, we begin by finding ΣF_x and ΣF_y from the x and y components of $\mathbf{F}_1$ and $\mathbf{F}_2$ (see part a of the figure):

Force	x component
$\mathbf{F}_1$	$+(35.0 \text{ N}) \cos 60.0° = +17.5 \text{ N}$
$\mathbf{F}_2$	$+10.0 \text{ N}$

$$\Sigma F_x = +17.5 \text{ N} + 10.0 \text{ N} = +27.5 \text{ N}$$

Force	y component
$\mathbf{F}_1$	$+(35.0 \text{ N}) \sin 60.0° = +30.3 \text{ N}$
$\mathbf{F}_2$	0

$$\Sigma F_y = +30.3 \text{ N}$$

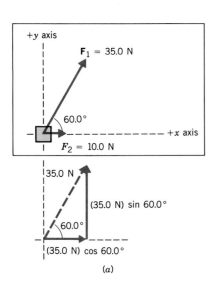

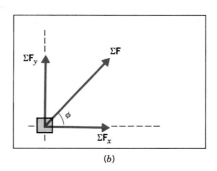

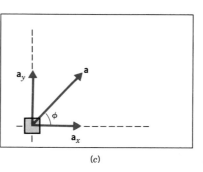

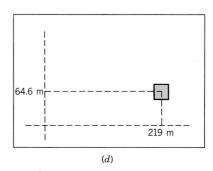

(b)

(d)

(a)

(c)

FIGURE 4.6 *(a)* Two forces, $\mathbf{F}_1$ and $\mathbf{F}_2$, act on a block and accelerate it. *(b)* The x and y components of the net force can be used with Newton's second law to find *(c)* the x and y components of the acceleration. *(d)* The equations of kinematics can be used to determine the displacements $x = +219$ m and $y = +64.6$ m.

The plus signs indicate that these components point in the direction of the $+x$ and $+y$ axes, as part b of the drawing shows. The x and y components of the acceleration can now be obtained:

$$a_x = \frac{\Sigma F_x}{m} = \frac{+27.5 \text{ N}}{15.0 \text{ kg}} = \boxed{+1.83 \text{ m/s}^2} \quad (4.2a)$$

$$a_y = \frac{\Sigma F_y}{m} = \frac{+30.3 \text{ N}}{15.0 \text{ kg}} = \boxed{+2.02 \text{ m/s}^2} \quad (4.2b)$$

Figure 4.6*c* shows these components of the acceleration.

(b) To find where the block is located after 8.00 s, the equations of kinematics can now be used. Since $a_x = +1.83$ m/s^2 and

the initial value of the x component of the velocity is $v_{0x} = +20.0$ m/s, the x component of the displacement can be obtained from Equation 3.5a:

$$x = v_{0x}t + \tfrac{1}{2}a_x t^2$$

$$= (+20.0 \text{ m/s})(8.00 \text{ s}) + \tfrac{1}{2}(+1.83 \text{ m/s}^2)(8.00 \text{ s})^2$$

$$= \boxed{+219 \text{ m}}$$

In a similar fashion the y component of the displacement can be obtained from Equation 3.5b, since $a_y = = +2.02$ m/s^2 and the initial value of the y component of the velocity is $v_{0y} = 0$: $\boxed{y = +64.6 \text{ m}}$. Figure 4.6*d* shows the final location of the block.

4.5 NEWTON'S THIRD LAW OF MOTION

Newton's first and second laws of motion specify how bodies move under the influence of forces. Newton's third law deals with a characteristic of all forces.

NEWTON'S THIRD LAW OF MOTION
Whenever one body exerts a force on a second body, the second body exerts an oppositely directed force of equal magnitude on the first body.

The third law is often called the "action–reaction" law, for it is sometimes quoted as follows: "for every action (force) there is an equal, but opposite, reaction." In other words, forces always occur in pairs.

FIGURE 4.7 The astronaut pushes on the spacecraft with a force $+\mathbf{F}$, causing the spacecraft to accelerate to the right. According to Newton's third law, the spacecraft simultaneously pushes back on the astronaut with a force $-\mathbf{F}$, causing the astronaut to accelerate to the left.

Figure 4.7 illustrates how the third law applies to an astronaut drifting just outside a spacecraft. Suppose the astronaut pushes on the spacecraft with a force $\mathbf{F}$. According to the third law, the spacecraft pushes back on the astronaut with a force $-\mathbf{F}$ that is equal in magnitude, but opposite in direction. Notice especially that the two forces are exerted on *different* objects; one is exerted by the astronaut on the spacecraft, while the other is exerted by the spacecraft on the astronaut. Let us now examine the accelerations produced by each of these forces.

Suppose the mass of the spacecraft is $m_S = 11\,000$ kg, and the mass of the astronaut is $m_A = 92$ kg. In addition, assume the astronaut exerts a force of $\mathbf{F} = +36$ N on the spacecraft. The acceleration of the spacecraft $\mathbf{a}_S$ depends on the force and the mass m_S according to Newton's second law:

$$\mathbf{a}_S = \frac{\mathbf{F}}{m_S} = \frac{+36 \text{ N}}{11\,000 \text{ kg}} = +0.0033 \text{ m/s}^2$$

As long as the astronaut pushes on the spacecraft, it will accelerate in the positive direction at 0.0033 m/s². On the other hand, the astronaut must also accelerate because of the force $-\mathbf{F}$ exerted on him by the spacecraft. The acceleration $\mathbf{a}_A$ of the astronaut depends on the force $-\mathbf{F}$ and the mass m_A:

$$\mathbf{a}_A = \frac{-\mathbf{F}}{m_A} = \frac{-36 \text{ N}}{92 \text{ kg}} = -0.39 \text{ m/s}^2$$

The astronaut, having a smaller mass, experiences a much larger acceleration than the spacecraft and accelerates in the opposite direction. Even though the two forces have the same magnitude, they produce different accelerations, because they act on different objects that have different masses.

Action–reaction situations occur all the time. Figure 4.8 illustrates the process of walking, for example. When a person walks, his shoe exerts a force on the earth due to friction. The earth accelerates under the application of this force, but the acceleration is imperceptibly small, since the earth is so massive (approximately 6×10^{24} kg). According to Newton's third law, the earth exerts a reaction force of equal magnitude on the shoe. It is the presence of this reaction force that causes a person to accelerate forward and, hence, walk. It is difficult to walk on ice, because ice, with its slippery surface, does not permit a walker to exert a large frictional force on the earth. Correspondingly, the reaction force exerted by the earth is small, and the walker does not accelerate forward very much.

Newton's three laws of motion make it very clear that forces play a central role in determining the motion of an object. To predict the motion accurately, we must know all the forces that act on the object. In nature there are four fundamental types of forces, and these will be encountered at various places in this text. They are "fundamental" in the sense that all forces are manifestations of one or more of these four. One of these fundamental forces is the gravitational force, which is the topic of the next section.

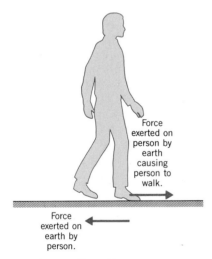

Force exerted on person by earth causing person to walk.

Force exerted on earth by person.

FIGURE 4.8 To walk, a person exerts a force on the earth. Consistent with Newton's third law, the earth exerts an oppositely directed force of equal magnitude on the person's foot, thus causing the foot to accelerate forward.

4.6 THE GRAVITATIONAL FORCE

NEWTON'S LAW OF UNIVERSAL GRAVITATION

Everyone knows that an object falls downward because of gravity. Chapters 2 and 3 discuss how to describe the effects of gravity by using a value of $g = 9.80$ m/s² for the downward acceleration it causes. However, nothing has been said about why g is 9.80 m/s². The present section sheds some light on this matter.

The acceleration due to gravity is like any other acceleration, and Newton's second law indicates that it must be caused by a net force. In addition to his famous three laws of motion, Newton also provided a coherent understanding of the **gravitational force.** His "law of universal gravitation" is stated below.

NEWTON'S LAW OF UNIVERSAL GRAVITATION

Every particle in the universe exerts an attractive force on every other particle. The force between two point particles, which have masses m_1 and m_2 and are separated by a distance r, is directed along the line joining the particles and has a magnitude given by

$$F = G \frac{m_1 m_2}{r^2} \qquad (4.3)$$

A point particle is one that is so small that it can be regarded as a mathematical point. G is the universal gravitational constant, whose value is found experimentally to be

$$G = 6.673 \times 10^{-11} \text{ N} \cdot \text{m}^2/\text{kg}^2$$

The constant G that appears in Equation 4.3 is called the **universal gravitational constant,** because it has the same value for all pairs of point particles anywhere in the universe, no matter what their separation.

To see the main features of Newton's law of universal gravitation, consider the two point particles in Figure 4.9. They have masses m_1 and m_2 and are separated by a distance r. The picture assumes that a force pointing to the right is positive. The gravitational forces point along the line joining the particles and are

$+\mathbf{F}$, the gravitational force exerted on m_1 by m_2
$-\mathbf{F}$, the gravitational force exerted on m_2 by m_1

These two forces have equal magnitudes and opposite directions. They act on different bodies, causing them to be mutually attracted. In fact, these forces are the action–reaction pair required by Newton's third law.

Example 3 shows that the magnitude of the gravitational force is extremely small when the masses are of ordinary size.

FIGURE 4.9 The two point particles, whose masses are m_1 and m_2, are attracted by gravitational forces. Assuming that a force pointing to the right is positive, the force $+\mathbf{F}$ acts on m_1 and $-\mathbf{F}$ acts on m_2. The particles are drawn large for the sake of clarity.

EXAMPLE 3

What is the magnitude of the gravitational force that acts on each particle in Figure 4.9, assuming $m_1 = 12.0$ kg (approximately the mass of a bicycle), $m_2 = 25.0$ kg, and $r = 1.20$ m?

SOLUTION

The magnitude of the gravitational force can be found using Equation 4.3:

$$F = G \frac{m_1 m_2}{r^2} \qquad (4.3)$$

$$= (6.67 \times 10^{-11} \text{ N} \cdot \text{m}^2/\text{kg}^2) \frac{(12.0 \text{ kg})(25.0 \text{ kg})}{(1.20 \text{ m})^2}$$

$$= \boxed{1.39 \times 10^{-8} \text{ N}}$$

For comparison, you exert a force of about 1 N when pushing a doorbell. It is apparent, then, that the gravitational force is exceedingly small in circumstances such as those here. This result is due to the fact that G itself is so small. However, if one of the bodies has a large mass, like that of the earth (5.98×10^{24} kg), the gravitational force can be large.

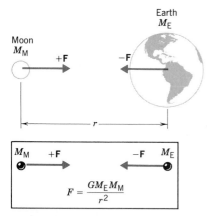

FIGURE 4.10 The gravitational force between two uniform spheres of matter is the same as if each sphere were a point particle with its mass concentrated at its center. The earth (mass M_E) and the moon (mass M_M) approximate such uniform spheres.

As expressed by Equation 4.3, Newton's law of gravitation applies only to point particles. Most familiar objects are certainly not point particles because of their relatively large physical sizes. With the aid of calculus, however, the law of universal gravitation can still be applied to such objects. Moreover, Newton was able to prove that an object of finite size can be considered to be a point particle for purposes of using the gravitation law, provided the mass of the object is distributed with spherical symmetry about its center. For example, Equation 4.3 can be applied when each object is a sphere whose mass is spread uniformly over its entire volume. Figure 4.10 shows this kind of application, assuming that the earth and the moon are such uniform spheres of matter. In this case, r is the distance *between the centers of the spheres* and not the distance between the outer surfaces. The gravitational forces that the spheres exert on each other are the same as if the entire mass of each was concentrated at its center. Even if the objects are not uniform spheres, Equation 4.3 can be used to a good degree of approximation if the sizes of the objects are small relative to the distance of separation r.

MEASUREMENT OF THE UNIVERSAL GRAVITATIONAL CONSTANT

One way to determine G is to measure the gravitational force between two objects of known masses that are separated by a known distance. With such experimental data, Equation 4.3 can be used to determine the value of G $[G = Fr^2/(m_1 m_2)]$. Because the gravitational force between ordinary objects is exceedingly small, this experiment is difficult, and it was not carried out until more than a century after Newton proposed his law of universal gravitation. The English scientist Henry Cavendish (1731–1810) was the first to determine the universal gravitational constant.

The Cavendish experiment utilized an instrument called a ***torsion balance.*** Figure 4.11 illustrates the type of torsion balance used by Cavendish. It consists of a horizontal rod that is suspended by a thin fiber. To each end of the rod is attached a small uniform lead ball of known mass, labeled m_1 and m_2 in the drawing. When large uniform lead spheres of known masses M_1 and M_2 are positioned near the small spheres, the gravitational force of attraction causes the rod to rotate and twist the supporting fiber. A mirror attached to the fiber causes a light beam to deflect on a measurement scale. From the deflection, the gravitational force can be determined. These experiments were performed so carefully by Cavendish that he obtained a value for G that differs by only about one percent from the presently accepted value.

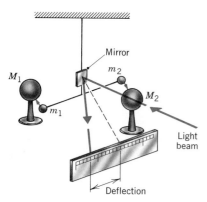

FIGURE 4.11 The torsion balance used by Henry Cavendish to measure the value of the universal gravitational constant G. The small uniform lead balls (masses m_1 and m_2) attached to the thin rod are attracted toward the large uniform lead spheres (masses M_1 and M_2) because of the gravitational force.

THE MASS OF THE EARTH

Once the value of the gravitational constant G is known, it becomes possible to determine the mass of the earth M_E. To see how this feat is accomplished, consider an object of mass m falling freely near the surface of the earth. Such an object falls with an acceleration $g = 9.80$ m/s², if the effects of air resistance are negligible. According to Newton's second law, this accelerated motion is produced by a force whose magnitude is given by the product of the mass and the acceleration, $F = mg$. However, this force is provided by the earth's gravitational attraction, and Equation 4.3 can also be used to express the force as

$$F = mg = G \frac{M_E m}{R_E^2}$$

The radius R_E of the earth has been substituted for the distance r in Equation 4.3, since the value $g = 9.80$ m/s² applies near the earth's surface. Canceling the mass m of the object from both sides of this equation and solving for M_E shows that

$$M_E = \frac{gR_E^2}{G}$$

What is remarkable about this result is that it gives the mass of the earth in terms of parameters (g, R_E, and G) that can be experimentally measured. If we consider the earth to be a perfect sphere of radius $R_E = 6.38 \times 10^6$ m, the mass of the earth is

$$M_E = \frac{gR_E^2}{G} = \frac{(9.80 \text{ m/s}^2)(6.38 \times 10^6 \text{ m})^2}{6.67 \times 10^{-11} \text{ N} \cdot \text{m}^2/\text{kg}^2} = 5.98 \times 10^{24} \text{ kg}$$

THE WEIGHT OF AN OBJECT

A meaning can now be given to the concept of weight.

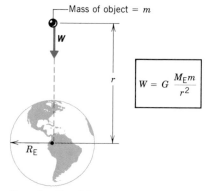

FIGURE 4.12 On Earth, the weight **W** of an object is the gravitational force exerted on the object by the earth.

DEFINITION OF WEIGHT

The weight of an object on the earth is the gravitational force that the earth exerts on the object. The weight always acts downward, toward the center of the earth. On another astronomical body, the weight is the gravitational force exerted on the object by that body.

   SI unit of weight: newton (N)

Newton's law of universal gravitation can be used to express how much weight an object has due to the gravitational pull of the earth. Using W for the magnitude of the weight,* m for the mass of the object, and M_E for the mass of the earth, it follows that

$$W = G\frac{M_E m}{r^2} \tag{4.4}$$

Equation 4.4 and Figure 4.12 both emphasize that an object has weight whether or not it is resting on the earth's surface. In other words, the gravitational force is acting even when the distance r is not equal to the radius R_E of the earth. Certainly, the gravitational force becomes weaker as r increases, since r is in the denominator of Equation 4.4. But no matter how great r is, the weight is never truly zero. Example 4 demonstrates how the weight changes, depending on where an object is located with respect to the center of the earth.

---

* Often, the word "weight" and the phrase "magnitude of the weight" are used interchangeably, even though weight is a vector. Generally, the context makes it clear when the direction of the weight vector must be taken into account.

EXAMPLE 4

The mass of a satellite is 1500 kg. Determine the weight of the satellite when the satellite is (a) resting on the earth and (b) 910 km above the earth's surface.

SOLUTION

(a) The weight of the satellite on the surface of the earth can be obtained from Equation 4.4 by letting $r = 6.38 \times 10^6$ m, which is the radius of the earth:

$$W = G\frac{M_E m}{r^2} = (6.67 \times 10^{-11}\ \text{N} \cdot \text{m}^2/\text{kg}^2)$$

$$\times \frac{(5.98 \times 10^{24}\ \text{kg})(1500\ \text{kg})}{(6.38 \times 10^6\ \text{m})^2}$$

$$\boxed{W = 1.5 \times 10^4\ \text{N (3300 lb)}}$$

(b) When the satellite is 910 km above the surface, its distance from the center of the earth is

$$r = 6.38 \times 10^6\ \text{m} + 910 \times 10^3\ \text{m} = 7.29 \times 10^6\ \text{m}$$

The weight now can be calculated as in part (a), except the new value of r must be used: $\boxed{W = 1.1 \times 10^4\ \text{N (2500 lb)}}$. As expected, the weight has decreased.

The space age has forced us to broaden our ideas about weight. For instance, you may be aware that an astronaut weighs only about one-sixth as much on the moon as on the earth. To obtain the weight of the astronaut on the moon from Equation 4.4, it is only necessary to replace M_E by M_M (the mass of the moon) and let $r = R_M$ (the radius of the moon). The ratio of the weight W_M of the astronaut on the moon to the weight W_E on the earth is

$$\frac{W_M}{W_E} = \frac{G\dfrac{M_M m}{R_M^2}}{G\dfrac{M_E m}{R_E^2}} = \frac{M_M R_E^2}{M_E R_M^2}$$

Substituting the known values for the masses and radii of the earth and moon reveals that

$$\frac{W_M}{W_E} = \frac{(7.35 \times 10^{22}\ \text{kg})(6.38 \times 10^6\ \text{m})^2}{(5.98 \times 10^{24}\ \text{kg})(1.74 \times 10^6\ \text{m})^2} = 0.165 \approx \frac{1}{6}$$

In spite of the moon's smaller radius, the moon's smaller mass leads to its reduced gravitational force.

RELATION BETWEEN MASS AND WEIGHT

Although massive objects weigh a lot when they are near the earth, mass and weight are not the same quantity. The two words should not be used interchangeably. As Section 4.2 discusses, mass is a quantitative measure of inertia and is related to the quantity of matter present. As such, mass is an intrinsic property of matter and does not change as an object is moved from one location to another. Weight, on the other hand, is the gravitational force that is exerted on the object and can vary, depending on how far the object is above the earth's surface or whether the object is located near another body such as the moon.

The correct relation between weight W and mass m can be written in either one of two ways, as shown below:

$$W = \boxed{G\frac{M_E}{r^2}}m \tag{4.4}$$

$$W = m\boxed{g} \tag{4.5}$$

The first of these is Newton's law of universal gravitation, while the second is Newton's second law incorporating the acceleration g due to gravity. These expressions make the distinction between mass and weight stand out. The weight of an object whose mass is m depends on the values for the universal gravitation constant G, the mass M_E of the earth, and the distance r. These three parameters together determine the acceleration g due to gravity. The specific value of $g = 9.80$ m/s^2 applies only when r equals the radius R_E of the earth. For larger values of r, as would be the case on top of a mountain, the effective value of g decreases below 9.80 m/s^2. The fact that g decreases as the distance r increases means that the weight likewise decreases. The mass of the object, however, does not depend on these effects.

4.7 THE NORMAL FORCE

In many situations, an object is resting or moving on a surface, such as a tabletop. Because of the contact with the surface, there is a force acting on the object. The present section discusses only one component of this force, the component that acts perpendicular to the surface. The next section discusses the component that acts parallel to the surface. The perpendicular component is called the ***normal force,*** where the word "normal" is used as a synonym for the word "perpendicular."

> **DEFINITION OF THE NORMAL FORCE**
> The normal force $\mathbf{F_N}$ is one component of the force that a surface exerts on an object with which it is in contact, namely, the component that is perpendicular to the surface.

Figure 4.13 shows a block resting on a horizontal table and identifies the two forces that act on the block, the gravitational force or weight $\mathbf{W}$ and the normal force $\mathbf{F_N}$. For some, it is difficult to understand how an inanimate object, such as a tabletop, can exert the normal force. To understand the role played by the tabletop, think about what happens when you sit on a mattress. Your weight causes the springs in the mattress to compress. As a result, the compressed springs exert an upward force (the normal force) on you. In a similar manner, the weight of the block causes invisible "atomic springs" in the surface of the table to compress, thus producing a normal force on the block. The normal force is not one of nature's fundamental forces, in the sense that the gravitational force is. Instead, the normal force arises at the atomic level as a manifestation of the fundamental electric force between the electrically charged particles within the atoms of the block and the table.

Newton's third law always comes into play in connection with the normal force. In Figure 4.13, for instance, the block exerts a force on the table. Consistent with the third law, the table exerts an oppositely directed force of equal magnitude on the block. This reaction force is the normal force. In this sense, the magnitude of the normal force indicates how hard the two objects are pressing against each other, and we will make use of this idea in the next section when friction is introduced.

If an object is resting on a horizontal surface, and there are no vertically acting forces except the object's weight and the normal force, the magnitudes of these two forces are equal, i.e., $F_N = W$. This is the situation in Figure 4.13. The weight must be balanced by the normal force for the object to remain at rest on the table. If the

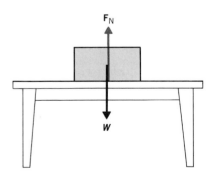

FIGURE 4.13 Two forces act on the block, its weight $\mathbf{W}$ and the normal force $\mathbf{F_N}$. The normal force $\mathbf{F_N}$ is exerted by the surface of the table, and the direction of $\mathbf{F_N}$ is perpendicular to the surface.

magnitudes of these forces were not equal, there would be a net force acting on the block, and the block would accelerate either upward or downward, in accord with the second law.

If other forces in addition to **W** and **F**$_N$ act in the vertical direction, the magnitudes of the normal force and the weight are no longer equal. In Figure 4.14a, for instance, a box whose weight is 15 N is being pushed downward against a table. The pushing force has a magnitude of 11 N. Thus, the total downward force exerted on the box is 26 N, and this must be balanced by the upward-acting normal force if the box is to remain at rest. In this situation, then, the normal force is 26 N, which is considerably larger than the weight of the box.

Figure 4.14b illustrates a somewhat different situation. Here, the box is being pulled upward by a rope that applies a force of 11 N. The net force acting on the box due to its weight and the rope is only 4 N, downward. To balance this force, the normal force needs to be only 4 N. It is not hard to imagine what would happen if the force applied by the rope were increased to 15 N — exactly equal to the weight of the box. In this situation, the normal force would become zero. In fact, the table could be removed completely, since the block would be supported entirely by the rope. The situations depicted in Figure 4.14 are consistent with the idea that the magnitude of the normal force indicates how hard two objects are pressing against each other. Clearly, the box and the table are pressing against each other harder in part a of the picture than in part b.

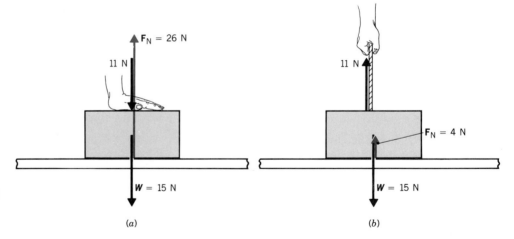

FIGURE 4.14 (a) The normal force has a greater magnitude than the weight of the box, because the box is being pressed downward with an 11-N force. (b) The normal force has a smaller magnitude than the weight, because the rope supplies an upward force of 11 N that partially supports the box.

EXAMPLE 5

In a circus balancing act, a woman performs a headstand on top of her husband's head, as Figure 4.15a illustrates. The woman weighs 490 N, and her husband's head and neck weigh 50 N. It is primarily the seventh cervical vertebra in the spine that supports all the weight above the shoulders. What is the normal force that this vertebra exerts on the neck and head of the man (a) before the act and (b) during the act?

SOLUTION

(a) Figure 4.15b shows the free-body diagram for the man's body above the shoulders. The only forces acting are the

normal force **F**$_N$ and the weight **W** = 50 N. These two forces must balance for the man's head and neck to remain at rest. Therefore, the seventh cervical vertebra exerts a normal force of $\boxed{F_N = 50 \text{ N}}$.

(b) Figure 4.15c shows the free-body diagram that applies during the act. Now, the total downward force exerted on the man's head and neck is 50 N + 490 N = 540 N, which must be balanced by the upward normal force provided by the vertebra: $\boxed{F_N = 540 \text{ N}}$.

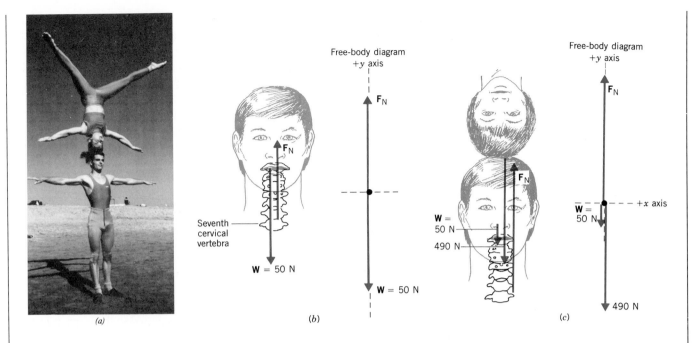

FIGURE 4.15 *(a)* A husband and wife balancing act and free-body diagrams for the man's body above the shoulders *(b)* before the act and *(c)* during the act. For convenience, the scales used for the vectors in parts *b* and *c* are different.

Like the box and the table in Figure 4.14, various parts of the human body press against one another and exert normal forces. Example 5 illustrates the remarkable ability of the human skeleton to withstand a wide range of normal forces.

In summary, the normal force does not necessarily have the same magnitude as the weight of the object. The value of the normal force depends on what other forces are present. The next chapter will show how to calculate the normal force under a variety of situations, including inclined surfaces and accelerating objects.

4.8 FRICTIONAL FORCES

When an object is in contact with a surface, there is a force acting on the object. The previous section discusses the component of this force that is perpendicular to the surface. When the object moves or attempts to move along the surface, there is also a component that is parallel to the surface. This parallel force component is called *friction.* Like the normal force, frictional forces are not fundamental forces in the sense that the gravitational force is.

In many situations considerable engineering effort is expended trying to reduce friction. For example, oil is used to reduce the friction that causes wear and tear in the pistons and cylinder walls of an automobile engine. Sometimes, however, friction is absolutely necessary. Without friction we could not walk. The difficulty in walking on an icy surface, which is almost frictionless, is well known. Thus, friction can be either detrimental or absolutely essential.

When the surface of one object slides over the surface of another, *each object* exerts

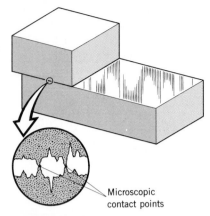

FIGURE 4.16 Even when two highly polished surfaces are in contact, they touch only at a relatively few points.

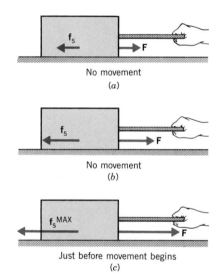

No movement
(a)

No movement
(b)

Just before movement begins
(c)

FIGURE 4.17 Applying a small force **F** to the box, as in parts *a* and *b*, produces no movement, because the static frictional force **f**$_s$ exactly balances the applied force. (*c*) The box just begins to move when the applied force is slightly greater than the maximum static frictional force **f**$_s^{MAX}$.

a frictional force on the other. This frictional force is called the **_kinetic_*** or **_sliding frictional force,_** and, for example, it comes into play when a baseball player slides into home plate. Another frictional force, called the **static frictional force,** can act even when there is no relative motion between the surfaces of two objects. The static frictional force is what makes it so difficult to start a heavy box moving across a rough floor.

Surfaces that appear to be highly polished can actually look quite rough when examined under a microscope. Such an examination reveals that two surfaces in contact touch only at relatively few spots, as Figure 4.16 illustrates. The microscopic area of contact for these spots is substantially less than the apparent macroscopic area of contact between the two objects—perhaps thousands of times less. At these contact points the molecules of the different bodies are close enough together to exert strong attractive intermolecular forces on one another, leading to what are known as "cold welds." Frictional forces are associated with these welded spots, but the exact details of how frictional forces arise are not well understood. However, some empirical relations have been developed, and they will enable us to account for the effects of friction in many situations.

Figure 4.17 helps to explain the main features of static friction. The block in this drawing is initially at rest, and as long as no attempt is made to move the block, there is no static frictional force. Then, a horizontal force **F** is applied to the block by means of a rope. If **F** is small, as in part *a*, experience tells us that the block still does not move. Why? It does not move because the static frictional force **f**$_s$ exactly cancels the effect of the applied force. The direction of **f**$_s$ is opposite to that of **F**, and the magnitude of **f**$_s$ equals the magnitude of the applied force, $f_s = F$.

Increasing the applied force in Figure 4.17 by a small amount still does not produce any movement of the block. There is no movement because the static frictional force also increases, by an amount that cancels out the increase in the applied force (see part *b* of the drawing). If the applied force continues to increase, however, there comes a point when the block finally "breaks away" and begins to slide. This breakaway force represents the *maximum static frictional force* **f**$_s^{MAX}$ that the table can exert on the block (see part *c* of the drawing). Any applied force that is greater than **f**$_s^{MAX}$ cannot be balanced by static friction, and the resulting net force accelerates the block to the right. Thus, the magnitude f_s of the static frictional force can assume any value from zero up to a maximum of f_s^{MAX}, depending on the magnitude of the applied force:

$$f_s \leq f_s^{MAX}$$

The symbol "$\leq$" is read as "less than or equal to," and the equality holds only when f_s attains its maximum value.

Experimental evidence shows that, to a good degree of approximation, the maximum static frictional force between a pair of dry, unlubricated surfaces has two main characteristics. In the first place, **f**$_s^{MAX}$ is independent of the apparent macroscopic area of contact between the objects. For instance, in Figure 4.18 the maximum static frictional force that the surface of the table can exert on the block is the same, whether the block is resting on its largest side or its smallest side. The other main characteristic of **f**$_s^{MAX}$ is that its magnitude is proportional to the magnitude of the normal force **F**$_N$. As Section 4.7 points out, the magnitude of the normal force indicates how hard the two surfaces are being pressed together. The harder the surfaces are pressed together, the larger is f_s^{MAX}, presumably because the number of "cold-welded" microscopic

---

* The word "kinetic" is derived from the Greek word "kinetikos," meaning "of motion."

points of contact is increased. Equation 4.6 expresses this proportionality with the aid of a proportionality constant μ_s, which is called the **coefficient of static friction.**

FIGURE 4.18 The maximum static frictional force $\mathbf{f_s}^{MAX}$ is the same, no matter which side of the block is in contact with the table.

STATIC FRICTIONAL FORCE

The magnitude f_s of the static frictional force can have any value from zero up to a maximum value of f_s^{MAX}, depending on the applied force. In other words, $f_s \leq f_s^{MAX}$, where

$$f_s^{MAX} = \mu_s F_N \qquad (4.6)$$

In Equation 4.6, μ_s is the coefficient of static friction, and F_N is the magnitude of the normal force.

It should be emphasized that Equation 4.6 relates only the magnitudes of $\mathbf{f_s}^{MAX}$ and $\mathbf{F_N}$, *not the vectors themselves.* Equation 4.6 does not imply that these two forces have the same directions. In fact, $\mathbf{f_s}^{MAX}$ is parallel to the surface, while $\mathbf{F_N}$ is perpendicular to the surface.

The coefficient of static friction, being the ratio of the magnitudes of two forces $(\mu_s = f_s^{MAX}/F_N)$, is a unitless number. It depends on the type of material from which each surface is made (steel on wood, rubber on concrete, etc.), the condition of the surfaces (polished, rough, lubricated, etc.), and other variables such as temperature. Typical values for μ_s range from around 0.01 for smooth surfaces to around 1.5 for rough surfaces.

Once two surfaces begin sliding over one another, the static frictional force is no longer of any concern. Instead, the kinetic frictional force comes into play and opposes the relative sliding motion. If you have ever pushed an object across a floor, you may have noticed that it takes less force to keep the object sliding than it takes to get it going in the first place. In other words, the kinetic frictional force is usually less than the static frictional force.

Experimental evidence indicates that the kinetic frictional force $\mathbf{f_k}$ has three main characteristics, to a good degree of approximation. It is independent of the apparent area of contact between the surfaces (see Figure 4.18). It is independent of the speed at which the sliding motion occurs, if the speed is small. And lastly, the magnitude of the kinetic frictional force is proportional to the magnitude of the normal force. Equation 4.7 expresses this proportionality with the aid of a proportionality constant μ_k, which is called the **coefficient of kinetic friction.**

KINETIC FRICTIONAL FORCE

The magnitude f_k of the kinetic frictional force is given by the relation

$$f_k = \mu_k F_N \qquad (4.7)$$

In Equation 4.7, μ_k is the coefficient of kinetic friction, and F_N is the magnitude of the normal force.

Equation 4.7, like Equation 4.6, is a relationship between only the magnitudes of the frictional and normal forces. The directions of these forces are perpendicular. Moreover, like the coefficient of static friction, the coefficient of kinetic friction is a unitless number and depends on the type and condition of the two surfaces that are in contact. Values for μ_k are typically less than those for μ_s, reflecting the fact that kinetic friction is generally less than static friction.

The next example illustrates the effects of kinetic and static friction.

EXAMPLE 6

A sled, traveling at 13.0 ft/s, enters a horizontal stretch of snow, as Figure 4.19a illustrates. The sled, together with its rider, weighs 80.0 lb. The coefficients of kinetic and static friction are $\mu_k = 0.0500$ and $\mu_s = 0.350$, respectively. Determine (a) the kinetic frictional force, (b) the distance that the sled slides before stopping, and (c) the force needed to get the sled just barely moving again.

SOLUTION

(a) To determine the kinetic frictional force, it is necessary to know the magnitude of the normal force. Part b of the drawing shows the forces acting on the sled, and part c gives the free-body diagram. Since the sled does not accelerate in the vertical direction, there can be no net force acting vertically on the sled, so the normal force and the weight must balance. Consequently, the magnitude of the normal force is $F_N = 80.0$ lb, and the magnitude of the kinetic frictional force is

$$f_k = \mu_k F_N = (0.0500)(80.0 \text{ lb}) = \boxed{4.00 \text{ lb}} \qquad (4.7)$$

The kinetic frictional force opposes the sliding of the sled and is directed to the left in Figure 4.19.

(b) The sled slows down because of the kinetic frictional force. Equation 3.6a from the equations of kinematics ($v_x^2 =$

$v_{0x}^2 + 2a_x x$) can be used to determine the stopping distance x from the sled's initial velocity ($v_{0x} = 13.0$ ft/s), final velocity ($v = 0$), and acceleration a_x. The acceleration can be obtained from the kinetic frictional force and Newton's second law, once the mass of the sled is known. The mass of the sled is related to its weight and the acceleration due to gravity:

$$m = \frac{W}{g} = \frac{80.0 \text{ lb}}{32.2 \text{ ft/s}^2} = 2.48 \text{ slugs} \qquad (4.5)$$

$$a_x = \frac{\Sigma F_x}{m} = \frac{-4.00 \text{ lb}}{2.48 \text{ slugs}} = -1.61 \text{ ft/s}^2 \qquad (4.2a)$$

The minus sign indicates that the frictional force points along the $-x$ axis in the free-body diagram, and, therefore, so does the acceleration. The stopping distance can now be obtained by solving the expression $v_x^2 = v_{0x}^2 + 2a_x x$ for x:

$$x = \frac{v_x^2 - v_{0x}^2}{2a_x} = \frac{-(13.0 \text{ ft/s})^2}{2(-1.61 \text{ ft/s}^2)} = \boxed{52.5 \text{ ft}}$$

(c) To set the sled into motion again, sufficient force is needed to overcome the maximum force of static friction. The magnitude of the necessary force must be greater than

$$f_s^{\text{MAX}} = \mu_s F_N = (0.350)(80.0 \text{ lb}) = \boxed{28.0 \text{ lb}} \qquad (4.6)$$

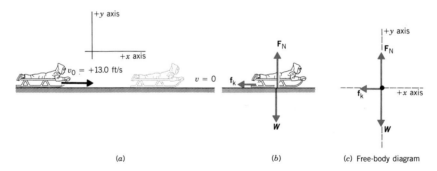

(a)   (b)   (c) Free-body diagram

FIGURE 4.19 (a) The moving sled decelerates because of the kinetic frictional force. (b) Three forces act on the moving sled, its weight **W**, the normal force **F**$_N$, and the kinetic frictional force **f**$_k$. (c) The free-body diagram for the sled.

Static friction opposes the impending relative motion between two objects, while kinetic friction opposes the relative sliding motion that actually does occur. In either case, *relative motion* is opposed. However, this opposition to relative motion does not mean that friction prevents or works against the motion of *all* objects. In Figure 4.8, for instance, the foot of a person walking exerts a force on the earth, and the earth exerts a reaction force on the foot. This reaction force is, in fact, a static frictional force, and it opposes the impending backward motion of the foot, propelling the person forward in the process. Kinetic friction, too, can cause an object to move, all the while opposing relative motion, as it does in Example 6. In this example the kinetic frictional force acts on the sled and opposes the relative motion of the sled and the earth. Newton's third law indicates, however, that if the earth exerts the kinetic frictional force on the sled, the sled must exert a reaction force on the earth. In response, the earth accelerates, but because of the earth's huge mass, the motion is too slight to be noticed.

4.9 THE TENSION FORCE

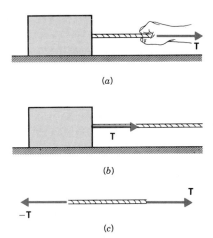

FIGURE 4.20 *(a)* A force **T** is being applied to the right end of the rope. *(b)* If the rope is massless or is not being accelerated, the force is transmitted undiminished to the box. *(c)* Forces are applied to both ends of the rope. These forces have equal magnitudes and opposite directions and lead to the common usage of the word "tension" as meaning "the tendency of the rope to be pulled apart."

Forces are often applied by means of cables or ropes that are used to pull on an object, and it is important to understand the way in which such pulling forces come about. Figure 4.20a shows a force **T** being applied to the right end of a rope that is attached to a box. Each particle in the rope, in turn, applies a force to its neighbor, and in the process the force is transmitted along the rope. Eventually, the force is applied to the box at the other end of the rope, as part *b* of the drawing shows.

In situations such as that in Figure 4.20, it is often said that "the tension force in the rope is responsible for applying the force **T** to the box." This statement means that both the tension in the rope and the force applied to the box have the same magnitude. However, the common usage of the word "tension" is in the following context: "the tension is the tendency of the rope to be pulled apart." To see the relationship between these two uses of the word "tension," consider Figure 4.20c. The left end of the rope applies the force **T** to the box, and, in accordance with Newton's third law, the box applies a reaction force to the rope. The reaction force has the same magnitude as **T** but is oppositely directed. In other words, a force $-\mathbf{T}$ acts on the left end of the rope. Thus, forces of equal magnitude act on both ends of the rope. Since these forces point in opposite directions, they tend to pull the rope apart. In any event, the tension force, like the normal and the frictional forces, is not one of nature's fundamental forces. Tension is merely a manifestation of the fundamental forces that exist at the atomic level.

In the discussion above we have used the concept of a "massless" rope without saying so. A massless rope has zero mass, and, of course, such ropes do not exist. The advantage of introducing such an idealization can be understood with the aid of Newton's second law, which indicates that whenever an object has mass, there must be a net force if the object is to be accelerated. By assuming a massless rope, we are able to set the mass m equal to zero in the second law, with the result that $\Sigma\mathbf{F} = ma = 0$. Therefore, no net force is needed to accelerate the rope. To put it simply, we can ignore the rope and assume that a force **T** applied to one end is transmitted undiminished to the object attached at the other end.* In contrast, if the rope in Figure 4.20 had mass, part of the force **T** applied on the right in part *a* would have to be used to accelerate the rope. In part *b*, then, the force applied to the box would not be **T**, but some diminished value. In this text we will assume that a rope connecting one object to another is a massless rope, unless stated otherwise. The ability of a massless rope to transmit a force undiminished from one end to the other is not affected when the rope passes around objects such as the pulley (assumed to be massless and frictionless) in Figure 4.21.

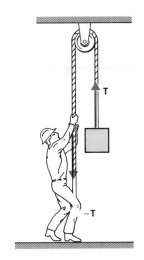

FIGURE 4.21 The force **T** applied at one end of a massless rope is transmitted undiminished to the other end, even when the rope bends around a pulley, provided the pulley is also massless and there is no friction.

* If a rope is not accelerating, **a** is zero in the second law and $\Sigma\mathbf{F} = ma = 0$, regardless of the mass of the rope. Then, the rope can be ignored, no matter what mass it has.

SUMMARY

Newton's first law of motion or **law of inertia** states that an object continues in a state of rest or in a state of motion at a constant speed along a straight line unless compelled to change that state by a net force. **Inertia** is the natural tendency of an object to remain at rest or in motion at a constant speed along a straight line. The **mass** of a body is a quantitative measure of inertia and is measured in an SI unit called the **kilogram** (kg). An **inertial reference frame** is one in which Newton's law of inertia is valid.

Newton's second law of motion states that the acceleration **a** of an object is directly proportional to the net force ΣF acting on the object and inversely proportional to the mass m of the object; $\mathbf{a} = \Sigma\mathbf{F}/m$ or $\Sigma\mathbf{F} = m\mathbf{a}$. The SI unit of force is the **newton** (N). When determining the net force, a **free-body diagram** is helpful. A free-body diagram is a diagram that represents the object and all the forces acting on it.

Newton's third law of motion, often called the "action–reaction law," states that whenever one object exerts a force on a second object, the second object exerts an oppositely directed force of equal magnitude on the first object. Newton's third law indicates that forces always exist in pairs and that each member of a pair acts on a different body.

Newton's law of universal gravitation states that every particle in the universe exerts an attractive force on every other particle. According to the law, the magnitude F of the force between two point particles with masses m_1 and m_2 is directly proportional to the product of the masses and inversely proportional to the square of the distance r between them; $F = Gm_1m_2/r^2$. The force is directed along the line between the two particles. The constant G has a value of $G = 6.673 \times 10^{-11} \text{ N} \cdot \text{m}^2/\text{kg}^2$ and is called the universal gravitational constant. The gravitational force is one of nature's four fundamental forces. The **weight** of an object on earth is the gravitational force that the earth exerts on the object. Weight and mass are different quantities.

A surface exerts a force on an object with which it is in contact. The component of the force that is perpendicular to the surface is called the **normal force**. The component that is parallel to the surface is called friction. The **force of static friction** between any two surfaces opposes any impending relative motion of the surfaces. The magnitude of the force of static friction depends on the magnitude of the applied force and can assume any value up to a maximum of $f_s^{\text{MAX}} = \mu_s F_N$, where μ_s is the **coefficient of static friction** and F_N is the magnitude of the normal force. The **force of kinetic friction** between any two surfaces that are sliding against one another opposes the relative motion of the surfaces. This force has a magnitude given by $f_k = \mu_k F_N$, where μ_k is the **coefficient of kinetic friction**. Both the static and kinetic frictional forces are independent of the apparent area of contact between the surfaces and are very sensitive to the nature and condition of the surfaces.

The word **tension** is commonly used to mean the tendency of a rope to be pulled apart due to forces that are applied at each end. Neither the tension force, the normal force, nor the friction force is one of nature's fundamental forces. Because of tension, a rope transmits a force from one end to the other. When a rope is accelerating, the force is transmitted undiminished only if the rope is massless.

QUESTIONS

1. The instructions for mounting a phonocartridge on the tone arm of a stereo turntable say to "adjust the tracking force of the cartridge so that it is less than three grams." From the point of view of correct physics, is there anything wrong with this statement? Explain.

2. Why do you lunge forward when your car suddenly comes to a halt? Why are you "thrown backward" when your car rapidly accelerates? In your explanation, refer to the most appropriate one of Newton's three laws of motion.

3. Is a net force being applied to an object when the object is moving downward (a) with a constant acceleration of 9.80 m/s² and (b) with a constant velocity of 9.80 m/s? Give your reasoning.

4. Newton's second law indicates that when a body is accelerating, a net force must be acting on the body. Does this mean that when two or more forces are applied to an object simultaneously, the object must always accelerate? Explain.

5. A father and his seven-year-old daughter are facing each other on ice skates. With their hands, they push off against one another. (a) Compare the magnitudes of the forces that they experience. (b) Compare the magnitudes of the accelerations that they experience. Account for your answers.

6. According to Newton's third law, when you push on an object, the object pushes back on you with an oppositely directed force of equal magnitude. If the object is an enormously massive crate resting on the floor, it will probably not move. Some people think the reason the crate does not move is that the two oppositely directed pushing forces cancel. Explain why this logic is faulty and why the crate does not move.

7. When a body is moved from sea level to the top of a tall mountain, what changes—the body's mass, its weight, or both? Explain.

8. Suppose you wish to calculate the acceleration of gravity on the surface of a distant planet. What properties of the planet must you know to make this calculation? Justify your answers.

9. Does the acceleration of a freely falling object depend to any extent on the location, i.e., whether the object is on top of Mt. Everest or in Death Valley, California? Explain.

10. A person has a choice of either pushing or pulling a sled at a constant velocity, as the drawing illustrates. If the angle θ is the same in both cases, does it require less force to push or to pull? Account for your answer.

11. Assume the surfaces of a box and a horizontal platform are identical on the moon and on the earth. In an experiment, an astronaut is to push horizontally on the box to start it sliding across the platform. He practices this maneuver on earth and performs it on the moon. In which case does he exert a greater pushing force? Justify your answer.

12. A rope is used in a standoff tug-of-war between two teams of five people each. An identical rope is tied to a tree and the same ten people pull just as hard on the loose end as they did in the tug-of-war. In both cases, the people pull steadily with no jerking. Which rope, if any, is more likely to break? Explain.

13. Three point particles have identical masses. Each particle experiences only the gravitational forces due to the other two particles. How should the particles be arranged so each one experiences a total gravitational force that has the same magnitude? Give your reasoning.

PROBLEMS

Section 4.3 Newton's Second Law of Motion

1. An empty airplane, whose mass is 30 400 kg has a maximum takeoff acceleration of 1.20 m/s². What is its maximum acceleration when it is carrying a load of 8200 kg? Ignore friction.

2. A bicycle (0.90 slug) and its rider (5.60 slug) are traveling at a speed of 10.0 ft/s. The rider exerts himself so that a horizontal net force of 2.20 lb accelerates them. What is the speed after 8.00 s?

3. A 1580-kg car is traveling with a speed of 15.0 m/s. What is the magnitude of the horizontal net force that is required to bring the car to a halt in a distance of 50.0 m?

4. A catapult on an aircraft carrier is capable of launching a plane from 0 to 56.0 m/s in a distance of 80.0 m. Find the average net force that the catapult exerts on a 13 300-kg jet.

* 5. A net force acts on mass m_1 and creates an acceleration. A mass m_2 is added to mass m_1. The same net force acting on the two masses together creates one-third the acceleration. Determine the ratio m_2/m_1.

Section 4.4 The Vector Nature of Newton's Second Law of Motion

6. A force vector has a magnitude of 720 N and a direction of 38° north of east. Determine the magnitude and direction of the components of the force that point along the north–south line and along the east–west line.

7. Find the magnitude and direction of the net force acting on each of the three objects shown in the drawing. See below.

8. (a) If the masses of the objects in problem 7 are each 5.00 kg, find their accelerations (both magnitude and direction). (b) Is

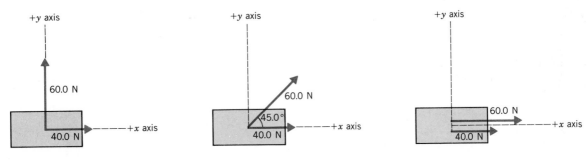

there any orientation of the two force vectors that will give the object a zero acceleration? Explain.

* 9. A 2.50-slug sphere is acted upon by a 30.0-lb force that is directed along the $+x$ axis. What must be the magnitude and direction of a second force, such that the sphere experiences an acceleration of 13.0 ft/s² directed 21.0° above the $+x$ axis?

** 10. A 4160-kg space probe is traveling with a speed of 2170 m/s in the direction shown in the drawing. The probe has four engines, A, B, C, and D, as indicated. Each engine delivers a thrust of 68 300 N when turned on. Which engines should be fired and for how long, to change the velocity of the probe so the velocity has twice the original magnitude and points in a direction 90° clockwise relative to its original direction?

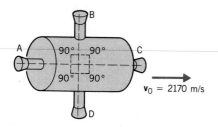

Section 4.6 The Gravitational Force

11. In one model of the hydrogen atom, an electron is in a circular orbit around the proton. The radius of the orbit is 5.29×10^{-11} m. (a) The mass of the electron is 9.11×10^{-31} kg, and the mass of the proton is 1.67×10^{-27} kg. Determine the gravitational force exerted on the electron by the proton. (b) Explain how the gravitational force exerted on the proton differs, either in magnitude or in direction, from the gravitational force exerted on the electron.

12. Three masses are positioned along a straight line. From left to right their masses are 181, 70.0, and 405 kg. The 70.0-kg mass is 0.310 m from the 181-kg mass and 0.460 m from the 405-kg mass. What is the net gravitational force acting on the 181-kg mass?

13. Using the following data, find the ratio F_{ES}/F_{EM}, where F_{ES} and F_{EM} denote, respectively, the magnitudes of the gravitational forces exerted on the earth by the sun and by the moon: mass of sun $= 1.99 \times 10^{30}$ kg, mass of moon $= 7.35 \times 10^{22}$ kg, distance from Earth to sun $= 1.50 \times 10^{11}$ m, and distance from Earth to moon $= 3.85 \times 10^{8}$ m. The distances given are the distances between the centers of the objects.

14. Two identical space probes, each of mass 2.00×10^{4} kg, are "parked" close to each other in outer space. If they are separated initially by 1.00 km, how long (in hours) will it take for the distance between them to decrease by 1.00 m, due to gravitational attraction? Assume that the probes behave as point particles and that the gravitational force remains constant as the probes move closer to each other.

15. A person has a mass of 42.0 kg. What is the person's weight on (a) Earth and (b) the moon?

16. A can of coffee has a mass of 2.27 kg. What is its weight in (a) newtons and (b) pounds?

17. A space traveler weighs 130 lb on Earth. What will the traveler weigh on another planet whose radius is three times that of the earth and whose mass is twice that of the earth?

18. Mount Everest has an altitude of 8850 m above sea level. Approximately, what is the ratio between the acceleration due to gravity on the mountaintop and that at sea level?

* 19. Jupiter has a mass that is approximately 318 times greater than the mass of Earth. The acceleration due to gravity on Jupiter's surface is 2.64 times larger than that on Earth. From these data, obtain the radius of Jupiter expressed as a multiple of the earth's radius.

* 20. A spacecraft is on a journey to the moon. The masses of Earth and the moon are, respectively, 5.98×10^{24} kg and 7.35×10^{22} kg. The distance between the centers of Earth and the moon is 3.85×10^{8} m. At what point, as measured from the center of Earth, does the gravitational force exerted on the craft by Earth precisely balance the gravitational force exerted by the moon?

* 21. Two identical masses m_1 are fixed to opposite corners of a square and exert a gravitational force on one another. Identical masses m_2 are added to each of the remaining corners. The total gravitational force acting on either mass m_1 is observed to be twice what it was originally. Determine the mass ratio m_1/m_2.

Section 4.7 The Normal Force, Section 4.8 Frictional Forces

22. A 60.0-kg crate rests on a level floor at a shipping dock. The coefficients of static and kinetic friction are 0.760 and 0.410, respectively. What horizontal pushing force is required to (a) just start the crate moving and (b) slide the crate across the dock at a constant speed?

23. A block whose weight is 45.0 N rests on a horizontal table. A horizontal force of 36.0 N is applied to the block. The coefficients of static and kinetic friction are 0.650 and 0.420, respectively. Will the block move under the influence of the force, and, if so, what will be the block's acceleration? Explain your reasoning.

24. A freezer weighs 206 lb. A person is attempting to push the freezer across the room with a horizontal force of 60.0 lb, but the freezer does not move. (a) What is the static frictional force that the floor exerts on the freezer? (b) What can be said about the coefficient of static friction? Explain your reasoning.

25. A 65.0-kg man is about to run on ice. The coefficient of static friction between his shoes and the ice is 0.160. What is his maximum possible acceleration?

* 26. A skater with an initial speed of 7.60 m/s is gliding across the ice. Air resistance is negligible. (a) The skater's weight is 615 N and the coefficient of kinetic friction between the ice and the skate blades is 0.100. Find the deceleration caused by kinetic friction.

(b) How far will the skater travel before coming to rest? (c) Would a lighter skater, with the same initial speed and coefficient of kinetic friction, travel farther before coming to rest? Explain.

* 27. A block rests on a horizontal surface and weighs 425 N. A force is applied to the block and has a magnitude of 142 N. The force is directed upward at an angle θ relative to the horizontal. The block begins to move horizontally when $\theta = 60.0°$. Determine the coefficient of static friction between the block and the surface.

**28. A 661-N force is being applied to a 121-kg block, as in the drawing. What is the minimal amount of additional mass that must be added on top of the block, so as to prevent the block from moving?

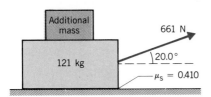

ADDITIONAL PROBLEMS

29. Saturn has an equatorial radius of 6.00×10^7 m and a mass of 5.67×10^{26} kg. (a) Compute the acceleration of gravity at the equator of Saturn. (b) What would be the weight of an 82.0-kg person at the equator of Saturn? (c) How many times greater is a person's weight on Saturn compared to that on Earth?

30. A net force of 525 N gives an object an acceleration of 4.20 m/s². (a) What net force is needed to give the object an acceleration of 13.7 m/s²? (b) What net force is required to keep the object moving at a constant velocity?

31. The mass of one of the small spheres of a Cavendish torsion balance is 0.00150 kg, and the mass of one of the larger spheres is 0.870 kg. If the center-to-center distance between these two spheres is 0.100 m, find the magnitude of the gravitational force that each exerts on the other.

32. Three forces are acting on a moving object. One force has a magnitude of 80.0 N and is directed due north. Another has a magnitude of 60.0 N and is directed due west. (a) What must be the magnitude and direction of the third force, such that the object continues to move in the same direction at a constant speed? (b) If the object in part (a) is initially at rest, what must be the magnitude and direction of the third force, such that the object continues to remain at rest?

33. At what altitude above the earth's surface would the value of g be one-half that at the surface?

34. A spacecraft has a mass of 3.50×10^4 kg and is drifting along a straight line through deep space. Its speed is 1820 m/s. An engine is suddenly turned on and provides a thrust of 2240 N in the direction of the motion. (a) Find the acceleration of the spacecraft. (b) Determine the time needed for the spacecraft to increase its speed to 2310 m/s. (c) What is the distance (in km) traveled during this time?

* 35. Traveling at a speed of 58.0 km/h, the driver of an automobile suddenly locks the wheels by slamming on the brakes. The coefficient of kinetic friction between the tires and the road is 0.720. How far does the car skid before coming to a halt? Ignore the effects of air resistance.

* 36. Three uniform spheres, whose masses are 5.00, 8.00, and 9.00 kg, are placed at the corners of a right triangle, as the drawing indicates. The distances given in the drawing are center-to-center values. Determine the resultant gravitational force (both magnitude and direction) acting on the 8.00-kg mass.

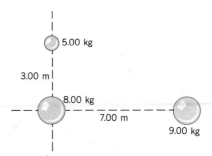

** 37. A 225-kg crate rests on a surface that is inclined above the horizontal at an angle of 20.0°. A horizontal force (parallel to the ground, not the incline) whose magnitude is 535 N is required to start the crate moving down the incline. What is the coefficient of static friction between the crate and the surface of the incline?

CHAPTER 5

Applications of Newton's Laws of Motion

A jet being launched from the deck of an aircraft carrier is simultaneously subjected to many forces, such as its weight, the thrust from its engines, the lift force on its wings, and the force from the catapult. The influence of these forces on the flight of the jet is described by Newton's laws of motion.

5.1 THE SIGNIFICANCE OF NEWTON'S LAWS OF MOTION

Newton's laws of motion may be applied to obtain various types of information. In this chapter, the focus of the applications is on the second law, although the first and third laws will also be involved. In a sense, the kinds of information that can be obtained from Newton's laws are obvious. After all, the second law is $\Sigma\mathbf{F} = m\mathbf{a}$, and if the mass is known, the information obtained must relate either to the net force or the acceleration. However, the significance of Newton's laws goes far beyond the simple use of $\Sigma\mathbf{F} = m\mathbf{a}$ to calculate one unknown quantity from values for two known quantities. The real significance of the laws of motion lies in the basis they provide for building a coherent understanding of motion. The quantities of force, mass, and acceleration are so fundamental that understanding the relationship between them leads to additional insights about the physical world. This ability of the laws of motion to point the way to further knowledge is the mark of a great scientific achievement.

Two kinds of applications are presented in this chapter. One kind deals with objects that move with a constant velocity and thus have zero acceleration. In this case, Newton's second law indicates that the net force acting on such an object must be zero. We will see that this condition of zero net force can be used to predict the magnitude and direction of some of the individual forces that act. When designing any structure, such as a bridge or an airplane, it is essential to know what forces will come to bear on the various elements of the structure. Predictions of these forces make it possible to incorporate adequate safety margins into the design.

A second kind of application deals with objects that are accelerating. If the mass and the forces are known, it is possible to calculate the acceleration. The accurate control of a spacecraft on the way toward the rings of Saturn depends, for example, on knowing the acceleration produced by the gravitational forces that the craft experiences. Conversely, if the mass of the craft is known, and the acceleration can be measured, then information about gravitational forces can be obtained. In either the first or second kind of application, we ignore rotation, deferring this topic until Chapters 9 and 10.

5.2 EQUILIBRIUM APPLICATIONS OF NEWTON'S LAWS OF MOTION

An object that is at rest or travels at a constant speed along a straight line has zero acceleration and is said to be in "equilibrium," according to the following definition.

DEFINITION OF EQUILIBRIUM
An object is in equilibrium when the object has zero acceleration.

According to Newton's second law, this definition implies that the net force acting on an object in equilibrium is zero. To put it another way, the forces acting on an object in equilibrium must balance. If the net force is zero, it follows in two dimensions that the x component and the y component of the net force must each be zero. Consequently, the equilibrium condition can be expressed conveniently by two equations:

$$\Sigma F_x = 0 \qquad\qquad (5.1a)$$

$$\Sigma F_y = 0 \qquad\qquad (5.1b)$$

This section deals with the application of Equations 5.1a and 5.1b to various equilibrium situations, and there are five steps that are followed in each situation.

Step 1. Select the object (often called the "system") to which Equations 5.1a and 5.1b are to be applied. Generally, this will be the object about which the most information is known. It may be that two or more objects are connected together by means of a rope or a cable. In this case, it may be necessary to treat each object separately according to the following steps.

Step 2. Draw a "free-body" diagram for each object chosen above. As Section 4.3 discusses, a free-body diagram is a drawing that represents the object and shows *all* the forces that act on it, each force with its proper direction. Be sure to include only forces that act on the object. Do not include forces that the object exerts on its environment.

Step 3. Choose a convenient set of x, y axes for each object and resolve all forces in the free-body diagram into components that point along these axes. The emphasis here is on the word "convenient," because the axes are typically selected so that as many forces as possible point directly along the x axis or the y axis. Such a choice minimizes the number of calculations needed to determine the components.

Step 4. Apply Equations 5.1a and 5.1b by setting the sum of the x components of the forces equal to zero and the sum of the y components of the forces to zero.

Step 5. Solve the two equations obtained in Step 4 for the desired unknown quantities, remembering that two equations can yield answers for only two unknowns at most.

Example 1 is a particularly straightforward illustration of how these steps are followed, because only two forces act together to establish the equilibrium.

EXAMPLE 1

During recuperation from a neck injury, the cervical vertebrae are kept under tension by means of a traction device, as Figure 5.1a illustrates. The device creates tension in the vertebrae by pulling to the left on the head with a force **T**, which, in effect, is applied to the first vertebra at the top of the spine. This vertebra remains in equilibrium, because it is simultaneously pulled to the right by a force **F** that is supplied by the next vertebra in line. The force **F** comes about in reaction to the pulling effect of force **T**, in accord with Newton's third law. If it is desired that **F** have a magnitude of 34 N, how much mass m should be suspended from the rope?

SOLUTION

Since the forces **T** and **F** act on the first cervical vertebra, we choose it as the object for analysis. In Figure 5.1b, the free-body diagram for this vertebra shows only the two forces **T** and **F**. Friction between the head and the table is assumed to be negligible, since the head rests on a small rolling platform. At equilibrium the net force must be zero, so that $F = T = mg$. The necessary mass, then, is

$$m = \frac{F}{g} = \frac{34 \text{ N}}{9.80 \text{ m/s}^2} = \boxed{3.5 \text{ kg}}$$

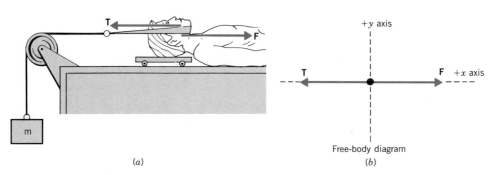

FIGURE 5.1 (a) A traction device for the neck. (b) The free-body diagram for the first vertebra.

The next example also deals with a traction device, but now three forces act together to bring about the equilibrium.

EXAMPLE 2

Figure 5.2a shows a traction device used with a foot injury. The weight of the 2.2-kg mass creates a tension in the rope that passes around the pulleys. Therefore, tension forces T_1 and T_2 are applied to the pulley on the foot, and they have the same magnitude T. It may seem surprising that the rope applies a force to either side of the foot pulley. A similar effect occurs when you place a finger inside a rubber band and pull downward. You can feel each side of the rubber band pulling upward on the finger. The foot pulley is kept in equilibrium, because the foot also applies a force F to it. This force arises in reaction (Newton's third law) to the pulling effect of the forces T_1 and T_2. Find the magnitude of F.

SOLUTION

Figure 5.2b shows the free-body diagram of the pulley on the foot. The x axis is chosen to be along the direction of force F, and the components of the tension forces are indicated in the drawing. (See Section 1.7 for a review of vector components.) Since the pulley is in equilibrium, there can be no net force acting on it. Consequently, the sum of the x components and the sum of the y components must separately be zero. For the x components, it follows that

$$\Sigma F_x = T_1 \cos 35° + T_2 \cos 35° - F = 0$$

This equation can be used to calculate F, once the magnitudes of the tension forces are known. The tension in the rope is determined by the weight of the 2.2-kg mass; $T = mg = (2.2 \text{ kg})(9.80 \text{ m/s}^2) = 22 \text{ N}$. Thus, $T_1 = T_2 = 22 \text{ N}$ and

$$F = 2(22 \text{ N}) \cos 35° = \boxed{36 \text{ N}}$$

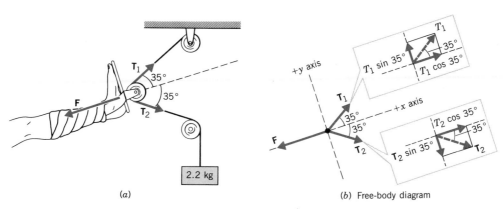

(a)   (b) Free-body diagram

FIGURE 5.2 (a) A traction device for the foot. (b) The free-body diagram for the pulley on the foot.

Example 3 presents another situation in which three forces are responsible for the equilibrium of an object. However, in this example all the forces have different magnitudes.

EXAMPLE 3

An automobile engine (weight $W = 3150 \text{ N}$) is being positioned above an engine compartment, as Figure 5.3a illustrates. To position the engine, a worker is using a rope. Find the tension T_1 in the supporting cable and the tension T_2 in the positioning rope.

SOLUTION

Under the influence of the forces W, T_1, and T_2 the ring is at rest and therefore at equilibrium. Consequently, the sum of the x components and the sum of the y components of these forces

must each be zero. Figure 5.3b shows the free-body diagram of the ring and the force components along the axes of a suitable x, y axes system. Using the angles given in the drawing, the table below lists the components for each of the three forces:

Force	x component	y component
T_1	$-T_1 \sin 10.0°$	$+T_1 \cos 10.0°$
T_2	$+T_2 \sin 80.0°$	$-T_2 \cos 80.0°$
W	0	-3150 N

The plus signs in the table denote components that point along the positive axes, while the minus signs denote components that point along the negative axes. Setting the sum of the x components and the sum of the y components equal to zero leads to the following two equations:

$$\Sigma F_x = -T_1 \sin 10.0° + T_2 \sin 80.0° = 0$$

$$\Sigma F_y = +T_1 \cos 10.0° - T_2 \cos 80.0° - 3150 \text{ N} = 0$$

Solving the first of these equations for T_1 shows that

$$T_1 = \left(\frac{\sin 80.0°}{\sin 10.0°}\right) T_2 = 5.67 T_2$$

Substituting this expression for T_1 into the second equation gives

$$(5.67 T_2) \cos 10.0° - T_2 \cos 80.0° - 3150 \text{ N} = 0$$

which can be solved to show that $\boxed{T_2 = 582 \text{ N}}$. Since $T_1 = 5.67 T_2$, it follows that $\boxed{T_1 = 3.30 \times 10^3 \text{ N}}$. The equations solved above are simultaneous equations with two unknown quantities. Appendix C contains a review of the procedure used to solve such equations.

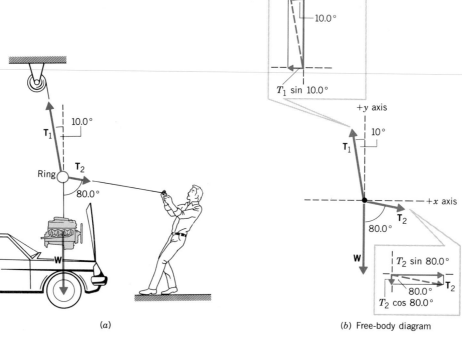

(a)   (b) Free-body diagram

FIGURE 5.3 (a) The engine is in equilibrium because of the three forces $\mathbf{T}_1$ (the tension force in the supporting cable), $\mathbf{T}_2$ (the tension force in the positioning rope), and $\mathbf{W}$ (the weight of the engine). (b) The free-body diagram for the ring.

An object can be moving and still be in equilibrium, provided there is no acceleration. Example 4 illustrates such a case, and the solution is again obtained using the five steps summarized at the beginning of the section.

EXAMPLE 4

A jet plane is flying with a constant speed along a straight line, at an angle of 30.0° above the horizontal, as Figure 5.4a indicates. The plane has a weight $\mathbf{W}$ of 86 500 N and its engines provide a forward thrust $\mathbf{T}$ of 103 000 N. In addition, the lift force $\mathbf{L}$ (directed perpendicular to the wings) and the force $\mathbf{R}$ of air resistance (directed opposite to the motion) act on the plane. Find $\mathbf{L}$ and $\mathbf{R}$.

SOLUTION

Figure 5.4b shows the free-body diagram of the plane, including the forces $\mathbf{W}$, $\mathbf{L}$, $\mathbf{T}$, and $\mathbf{R}$. Since the plane is not accelerating, it is in equilibrium, and the sum of the x components and the sum of the y components of these forces must be zero. To calculate the components, we have chosen axes in the free-body diagram that are rotated by 30.0° from their usual horizontal–vertical positions. This has been done purely for convenience, since the weight **W** is then the only force that does not lie along either axis. The components of the forces are as follows:

Force	x component	y component
W	$-(86\ 500\ \text{N}) \sin 30.0°$	$-(86\ 500\ \text{N}) \cos 30.0°$
L	0	$+L$
T	$+103\ 000\ \text{N}$	0
R	$-R$	0

Setting the sum of the x components and the sum of the y components of the forces equal to zero yields

$$\Sigma F_x = -(86\ 500\ \text{N}) \sin 30.0° + 103\ 000\ \text{N} - R = 0$$

$$\Sigma F_y = -(86\ 500\ \text{N}) \cos 30.0° + L = 0$$

These equations can be solved to show that $\boxed{R = 59\ 800\ \text{N}}$ and $\boxed{L = 74\ 900\ \text{N}}$.

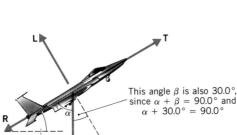

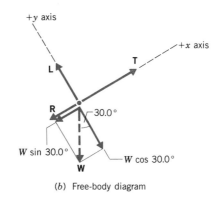

This angle β is also 30.0°, since $\alpha + \beta = 90.0°$ and $\alpha + 30.0° = 90.0°$

(a)

(b) Free-body diagram

FIGURE 5.4 (*a*) A plane moves with a constant velocity at an angle of 30.0° above the horizontal due to the action of four forces, the weight **W**, the lift **L**, the engine thrust **T**, and the air resistance **R**. (*b*) The free-body diagram for the plane.

Static friction is a force that sometimes plays a role in keeping an object at rest, that is, in equilibrium. The next example illustrates one way in which information about the static friction force can be obtained.

EXAMPLE 5

A block of mass m rests on a hinged board whose angle of elevation is adjustable, as in Figure 5.5*a*. When the right end of the board is raised, the block remains at rest until a maximum angle θ is reached. If the angle is increased beyond θ, the block breaks loose and slides down the board. Obtain an equation that relates the coefficient of static friction μ_s to the angle θ.

SOLUTION

There are three forces that act on the block, the weight **W** ($W = mg$), the normal force $\mathbf{F_N}$, and the maximum force of static friction $\mathbf{f_s}^{\text{MAX}}$. Recall that the magnitude of the maximum static frictional force is given by Equation 4.6 as $f_s^{\text{MAX}} = \mu_s F_N$. These forces keep the block in equilibrium, and, therefore, they must balance. To express this balancing of forces, we use the x, y axes shown in Figure 5.5*b*, along with the free-body diagram for the block. The geometry in this diagram is the same as that in

Figure 5.4*b*. The x and y components of the forces are given below:

Force	x component	y component
W	$-mg \sin \theta$	$-mg \cos \theta$
$\mathbf{F_N}$	0	$+F_N$
$\mathbf{f_s}^{\text{MAX}}$	$+\mu_s F_N$	0

The forces must balance, so the sum of the x components and the sum of the y components must each be zero:

$$\Sigma F_x = -mg \sin \theta + \mu_s F_N = 0$$

$$\Sigma F_y = -mg \cos \theta + F_N = 0$$

The second equation reveals that $F_N = mg \cos \theta$. Substituting this relation for F_N into the first equation yields

$$-mg \sin \theta + \mu_s (mg \cos \theta) = 0$$

Solving for μ_s gives

$$\mu_s = \frac{\sin \theta}{\cos \theta} = \tan \theta$$

Thus, from a measured value for the maximum angle θ, the coefficient of static friction can be obtained.

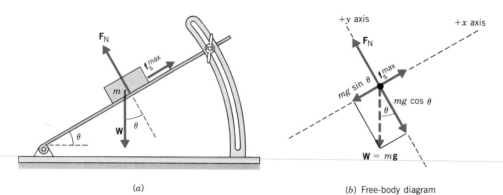

(a)

(b) Free-body diagram

FIGURE 5.5 (a) A block resting on an inclined board. The maximum angle of the incline, just before the block begins to slip, is given by θ. The forces acting on the block are its weight **W**, the normal force $\mathbf{F_N}$, and the maximum force of static friction $\mathbf{f_s}^{\text{MAX}}$. (b) The free-body diagram for the block.

5.3 NONEQUILIBRIUM APPLICATIONS OF NEWTON'S LAWS OF MOTION

When an object is accelerating, it is not in equilibrium, and the forces acting on it are not balanced. The net force is not zero in Newton's second law. However, with one exception, the steps followed in solving nonequilibrium problems are identical to those used in equilibrium situations. The exception is in Step 4 of the five steps outlined in Section 5.2. Now, since the object is accelerating, the two-dimensional representation of Newton's second law in Equations 4.2a and 4.2b applies. These equations are repeated here for convenience:

$$\Sigma F_x = ma_x \quad \text{(4.2a)} \qquad \text{and} \qquad \Sigma F_y = ma_y \quad \text{(4.2b)}$$

Example 6 uses Equations 4.2a and 4.2b in a situation where the forces are applied in directions similar to those in Example 2, except that now an acceleration is present.

EXAMPLE 6

A supertanker (mass $= 1.50 \times 10^8$ kg) is being towed by two tugboats, as in Figure 5.6a. The tensions in the towing cables apply the forces $\mathbf{T_1}$ and $\mathbf{T_2}$ at equal angles (30.0°) with respect to the tanker's axis. In addition, the tanker's engines produce a forward drive force **D**, whose magnitude is 75.0×10^3 N. Moreover, the water applies an opposing force **R**, whose magnitude is 40.0×10^3 N. The tanker moves forward with an acceleration **a** that points along the tanker's axis and has a magnitude of 2.00×10^{-3} m/s². Find the magnitudes of the tensions $\mathbf{T_1}$ and $\mathbf{T_2}$ in the towing cables.

SOLUTION

Figure 5.6b shows the free-body diagram for the tanker, with the axis of the ship chosen to be the x axis. (The geometry here is similar to that in Figure 5.2b.) The acceleration of the tanker is not zero, so the tanker must be experiencing a net force. The individual force components are summarized below:

Force	x component	y component
$\mathbf{T_1}$	$+T_1 \cos 30.0°$	$+T_1 \sin 30.0°$
$\mathbf{T_2}$	$+T_2 \cos 30.0°$	$-T_2 \sin 30.0°$
$\mathbf{D}$	$+D$	0
$\mathbf{R}$	$-R$	0

Since the acceleration points along the x axis, there is no y component of the acceleration. Consequently, the sum of the y components of the forces must be zero:

$$\Sigma F_y = +T_1 \sin 30.0° - T_2 \sin 30.0° = 0$$

This result shows that the magnitudes of the tensions in the cables are equal, $T_1 = T_2$. Since the ship accelerates along the x direction, the sum of the x components of the forces is not zero. The second law indicates that

$$\Sigma F_x = T_1 \cos 30.0° + T_2 \cos 30.0° + D - R = ma_x$$

Using the fact that $T_1 = T_2 = T$ and the given values for D, R, m, and a_x, it can be seen that

$$2T \cos 30.0° + 75.0 \times 10^3 \text{ N} - 40.0 \times 10^3 \text{ N}$$
$$= (1.50 \times 10^8 \text{ kg})(2.00 \times 10^{-3} \text{ m/s}^2)$$

$$\boxed{T = 1.53 \times 10^5 \text{ N}}$$

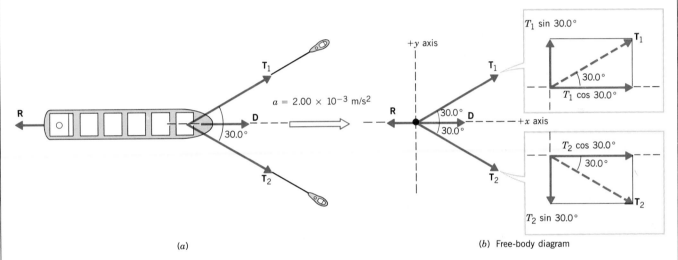

(a)

(b) Free-body diagram

FIGURE 5.6 (a) Four forces act on a supertanker: $\mathbf{T_1}$ and $\mathbf{T_2}$ are the tension forces due to the towing cables, $\mathbf{D}$ is the forward drive force produced by the tanker's engines, and $\mathbf{R}$ is the force with which the water opposes the tanker's motion. (b) The free-body diagram for the tanker.

It often happens that two objects are connected by some device, such as the drawbar that is used when a truck pulls a trailer. If the tension in the connecting device is of no interest, the two objects can be treated as a single composite object when applying Newton's second law. However, if it is necessary to find the tension, as in the next example, then the second law must be applied separately to each object.

EXAMPLE 7

An 8500-kg truck is hauling a 27 000-kg trailer along a level road, as Figure 5.7a illustrates. The acceleration is 0.78 m/s². Ignoring the retarding forces of friction and air resistance, determine (a) the magnitude of the tension in the horizontal drawbar between the trailer and the truck and (b) the tractive force $\mathbf{D}$ that acts on the drive wheels to propel the truck forward.

SOLUTION

(a) Since the truck and the trailer accelerate along the horizontal direction and friction is being ignored, only forces that have components in the horizontal direction are of interest here. Therefore, the weight and the normal force are omitted in Figure 5.7, since they act vertically. Note, however, that these vertical forces balance, since there is no vertical component to the acceleration.

We begin with the trailer, whose free-body diagram is shown in Figure 5.7b. There is only one horizontal force acting on the trailer, the tension force **T** due to the drawbar. Therefore, it is straightforward to obtain the tension from $\Sigma F_x = ma_x$, since the mass of the trailer and the acceleration are known:

$$T = (27\ 000\ \text{kg})(+0.78\ \text{m/s}^2) = \boxed{21\ 000\ \text{N}}$$

(b) Two horizontal forces act on the truck, as the free-body diagram in Figure 5.7b shows. One is the desired tractive force **D**. The other is the force **T'**. According to Newton's third law, **T'** is the force with which the trailer pulls back on the truck, in reaction to the truck pulling forward. If the drawbar has negligible mass, the magnitude of **T'** is equal to the magnitude of **T**, namely, 21 000 N. In other words, we are assuming that the drawbar behaves as a massless rope does (see Section 4.9). Since the magnitude of **T'**, the mass of the truck, and the acceleration are known, $\Sigma F_x = ma_x$ can be used to determine the drive force:

$$D - (21\ 000\ \text{N}) = (8500\ \text{kg})(+0.78\ \text{m/s}^2)$$

$$\boxed{D = 28\ 000\ \text{N}}$$

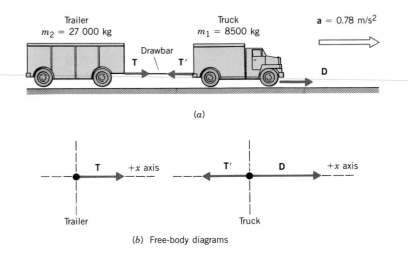

FIGURE 5.7 (a) The horizontal forces that act when the truck pulls the trailer are the tractive force **D** exerted on the drive wheels of the truck by the road, the tension force **T'** exerted on the truck by the drawbar, and the tension force **T** exerted on the trailer by the drawbar. (b) The free-body diagrams for the truck and the trailer, ignoring the vertical forces.

The force of gravity is often present among the forces that affect the acceleration of an object. Example 8 deals with the force of gravity in a familiar situation.

EXAMPLE 8
San Francisco is famous for its hills. A car (mass = 1350 kg) is being propelled up one of these hills by a drive force that is parallel to the hill and has a magnitude of 7200 N. As Figure 5.8a illustrates, the hill makes a 25° angle with the horizontal. Assuming that the retarding forces of friction and air resistance are negligible, find (a) the acceleration of the car and (b) the normal force exerted on the car by the hill.

SOLUTION
(a) Figure 5.8b gives the free-body diagram for the car. The 7200-N drive force acts along the +x axis, chosen to be parallel to the hill. The normal force $\mathbf{F_N}$ acts along the +y axis, perpendicular to the hill. The weight of the car [$W = mg = (1350\ \text{kg})(9.80\ \text{m/s}^2) = 13\ 200\ \text{N}$] has the x and y components shown in the diagram. Since the car accelerates up the hill, Newton's second law can be used to determine the acceleration from the sum of the x components of the forces:

$$\Sigma F_x = 7200\ \text{N} - W \sin 25° = ma_x$$

$$7200\ \text{N} - (13\ 200\ \text{N}) \sin 25° = (1350\ \text{kg})a_x$$

$$\boxed{a_x = 1.2\ \text{m/s}^2}$$

(b) Since the acceleration of the car has no component perpendicular to the hill, the sum of the y components of the forces must be zero:

$$\Sigma F_y = F_N - (13\ 200\ \text{N}) \cos 25° = 0$$

$$\boxed{F_N = 12\ 000\ \text{N}}$$

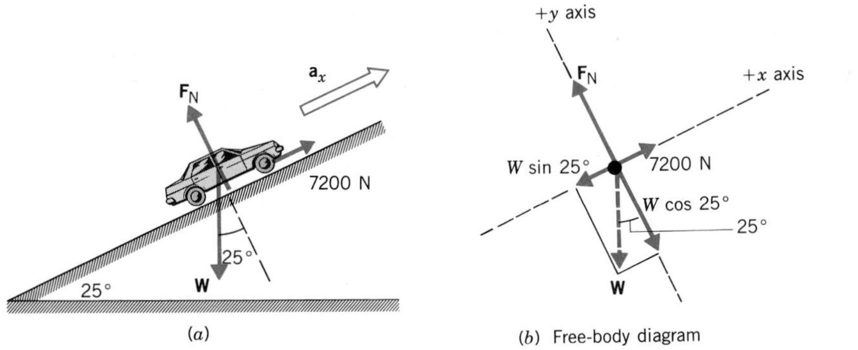

FIGURE 5.8 (*a*) The car accelerates up the hill under the influence of the 7200 N drive force, the normal force $\mathbf{F}_N$, and the weight $\mathbf{W}$ of the car. (*b*) The free-body diagram for the car.

(*a*)

(*b*) Free-body diagram

This section concludes with another example in which the force of gravity plays an important role.

EXAMPLE 9

A block (mass $m_1 = 8.00$ kg) is moving on a frictionless incline plane whose angle is 30.0°. This block is connected to a second block (mass $m_2 = 22.0$ kg) by a cord that passes over a small, frictionless pulley (see Figure 5.9*a*). Find the acceleration of each mass and the tension in the cord.

SOLUTION

It is important to realize that both blocks have accelerations of the same magnitude *a*, since they move together as a unit. In setting up the problem, we assume arbitrarily that m_1 accelerates up the incline and choose this direction to be the $+x$ axis for m_1. If m_1 in reality accelerates down the incline, then the value obtained for the acceleration will turn out to be a negative number. There are three forces that act on this block: (1) $\mathbf{W}_1$ is the weight

$[W_1 = m_1 g = (8.00$ kg$)(9.80$ m/s²$) = 78.4$ N], (2) $\mathbf{T}$ is the force applied because of the tension in the cord, and (3) $\mathbf{F}_N$ is the normal force that the surface of the incline exerts. Figure 5.9*b* shows the free-body diagram for m_1. The components of the weight are given in this diagram, and it is the only force that does not point along the *x*, *y* axes. Applying the second law to the accelerated motion of m_1 along the *x* axis shows that

$$\Sigma F_x = -W_1 \sin 30.0° + T = m_1 a_x$$

$$-(78.4 \text{ N}) \sin 30.0° + T = (8.00 \text{ kg})a$$

where we have set $a_x = a$. This equation cannot be solved as it stands, since both *T* and *a* are unknown quantities. To complete the solution, we must consider block m_2.

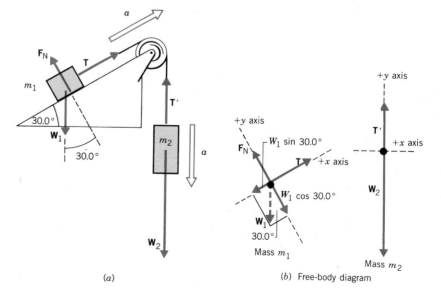

FIGURE 5.9 (*a*) The three forces that act on m_1 are its weight $\mathbf{W}_1$, the normal force $\mathbf{F}_N$, and the force $\mathbf{T}$ due to the tension in the cord. The two forces that act on m_2 are its weight $\mathbf{W}_2$ and the force $\mathbf{T}'$ due to the tension in the cord. The acceleration is labeled $\mathbf{a}$. (*b*) The free-body diagrams for the two masses.

(*a*)

(*b*) Free-body diagram

There are two forces that act on m_2, as the free-body diagram in Figure 5.9b indicates: (1) $\mathbf{W}_2$ is the weight [$W_2 = m_2 g = (22.0 \text{ kg})(9.80 \text{ m/s}^2) = 216$ N] and (2) $\mathbf{T}'$ is exerted as a result of m_1 pulling back on the connecting cord. If the mass of the cord is negligible, and if the frictionless pulley also has negligible mass, the magnitudes of $\mathbf{T}'$ and $\mathbf{T}$ are the same: $T' = T$. Applying the second law to m_2 reveals that

$$\Sigma F_y = T' - W_2 = m_2 a_y$$

$$+T - 216 \text{ N} = (22.0 \text{ kg})(-a)$$

The acceleration a_y has been set equal to $-a$ since block m_2 moves downward along the $-y$ axis in the free-body diagram,

which is consistent with the original assumption that block m_1 moves up the incline. Now there are two equations in two unknowns, and following the procedure discussed in Appendix C, they may be solved simultaneously to give the values of the tension T and the acceleration a:

$$\boxed{T = 86.4 \text{ N}} \quad \text{and} \quad \boxed{a = 5.89 \text{ m/s}^2}$$

The positive value for a confirms the initial arbitrary assumption that m_1 accelerates up the incline. The acceleration is less than the acceleration due to gravity, because block m_1 prevents block m_2 from falling freely.

5.4 APPARENT WEIGHT

The weight of an object on the earth is the downward-acting gravitational force that the earth exerts on the object. Usually, the weight can be determined with the aid of a scale. However, even though a scale is working properly, there are situations in which it does not give the correct weight. In such situations, the reading on the scale is called the "apparent" weight to distinguish it from the gravitational force or "true" weight.

To see the kinds of discrepancies that can arise between true weight and apparent weight, consider a scale that has been placed inside an elevator, as in Figure 5.10. The reasons for these discrepancies will be explained shortly. A person whose true weight is 700 N steps on the scale. If the elevator is at rest or moving with a constant velocity (either upward or downward), the apparent weight equals the true weight, as Figure 5.10a illustrates. A constant velocity represents a condition of equilibrium, so if the scale with the person on it is in equilibrium, the apparent weight equals the true weight.

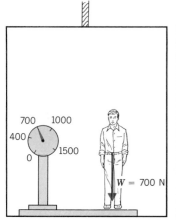

(a) No acceleration (**v** = constant)

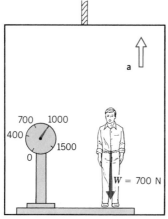

(b) Upward acceleration

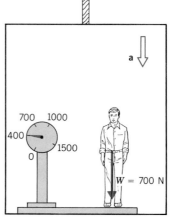

(c) Downward acceleration

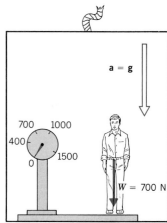

(d) Free-fall

FIGURE 5.10 (a) When the elevator is in equilibrium, the apparent weight (the reading on the scale) equals the true weight ($W = 700$ N) of the person. (b) When the elevator accelerates upward, the apparent weight exceeds the true weight. (c) When the elevator accelerates downward, the apparent weight is less than the true weight. (d) The apparent weight is zero if the elevator falls freely, that is, if it falls with the acceleration of gravity.

If the elevator is not in equilibrium, that is, if it is accelerating, the apparent weight and the true weight are not equal. When the elevator accelerates upward, the apparent weight is greater than the true weight, as Figure 5.10b shows. Conversely, if the elevator accelerates downward, as in part c, the apparent weight is less than the true weight. In fact, if the elevator falls freely, so its acceleration is equal to the acceleration due to gravity, the apparent weight becomes zero, as part d indicates. In a situation such as this, where the apparent weight is zero, the person is said to be "weightless." The apparent weight, then, does not equal the true weight if the scale and the person on it are accelerating.

The discrepancies between true weight and apparent weight can be understood with the aid of Newton's second law. Figure 5.11 shows a diagram of the person riding in the elevator and the two forces that act on him: (1) $\mathbf{W}$ = true weight = mg and (2) $\mathbf{F_N}$ = the normal force exerted by the platform of the scale. Applying Newton's second law in the vertical direction gives

$$\Sigma F_y = +F_N - mg = m(\pm a)$$

where $+a$ is used if the person's acceleration points upward and $-a$ is used if the acceleration points downward. Solving for F_N shows that

$$\underbrace{F_N}_{\substack{\text{Apparent} \\ \text{weight}}} = \underbrace{mg}_{\substack{\text{True} \\ \text{weight}}} + m(\pm a) \tag{5.2}$$

In Equation 5.2, F_N is the magnitude of the normal force exerted on the person by the platform of the scale. But in accord with Newton's third law, F_N is also the magnitude of the downward force that the person exerts on the scale, namely, the apparent weight.

Equation 5.2 contains all the features shown in Figure 5.10. If the elevator is in equilibrium, $a = 0$, and the apparent weight equals the true weight. If the elevator accelerates upward (acceleration $= +a$), the equation shows that the apparent weight is greater than the true weight. If the elevator accelerates downward (acceleration $= -a$), the apparent weight is less than the true weight. If the elevator falls freely (acceleration $= -a = -g$), the apparent weight is zero. The scale registers an apparent weight of zero, because when both the person and the scale fall freely, they cannot push against one another. In this text, when the weight of an object is given, it is assumed to be the true weight, unless otherwise indicated.

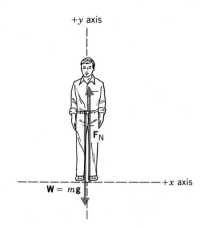

FIGURE 5.11 A diagram showing the forces acting on the person riding in the elevator of Figure 5.10. W is the true weight and $\mathbf{F_N}$ is the normal force exerted on the person by the platform of the scale.

SUMMARY

An object is in **equilibrium** when the object moves at a constant velocity (which may be zero), or, in other words, when it is not accelerating. The sum of the forces that act on an object in equilibrium is zero. Under equilibrium conditions in two dimensions, the separate sums of the force components in the x direction and in the y direction must each be zero.

There are five steps that facilitate the application of Newton's laws of motion. **Step 1.** Select the object or objects to which the laws of motion are to be applied. **Step 2.** Draw a free-body diagram for each object chosen in Step 1.

A free-body diagram is a diagram that shows only the forces acting on the object of interest. **Step 3.** Choose a convenient set of axes for each object and resolve all forces in the free-body diagram into components along these axes. **Step 4.** Apply Newton's second law ($\Sigma F_x = ma_x$ and $\Sigma F_y = ma_y$). **Step 5.** Solve the equations obtained in Step 4.

The **apparent weight** is the force that an object exerts on the platform of a scale and may be larger or smaller than the true weight, depending on the acceleration of the object and the scale.

SOLVED PROBLEMS

SOLVED PROBLEM 1
Related Problems: *22 *23 **27 **28

A person on a scaffold is hoisting the scaffold by pulling downward on a rope, as part *a* of the drawing illustrates. The magnitude of the pulling force is 540 N, and the combined mass of the person and the scaffold is 155 kg. (a) Find the upward acceleration of the unit. (b) Find the tension in the rope when the scaffold is hanging motionless.

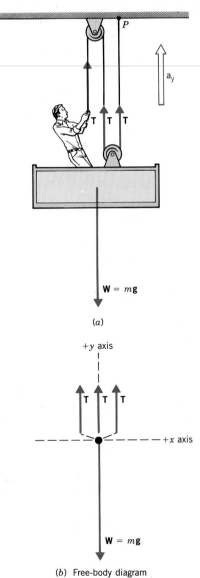

(a)

(b) Free-body diagram

Solution (a) In this problem, the person and the scaffold are considered together as a single unit or object, since the rope and

pulley arrangement is attached to both of them. The rope exerts a force on the unit in three places. The left end of the rope exerts an upward force **T** on the person's hands. This force arises because the person pulls downward on the rope with a 540-N force, and the rope exerts an oppositely directed force of equal magnitude on the person, in accord with Newton's third law. Thus, the magnitude of the upward force **T** is 540 N, which is equal to the magnitude of the tension in the rope, as Section 4.9 discusses. If the masses of the rope and each pulley are negligible and if the pulleys are friction-free, the tension is transmitted undiminished along the rope. Then, a 540-N tension force **T** acts upward on the left side of the scaffold pulley (see part *a* of the drawing). The tension is also transmitted to the point P, where the rope attaches to the roof. The roof pulls back on the rope in accord with the third law, and this pull leads to the force $T = 540$ N that acts on the right side of the scaffold pulley.

In addition to the three upward forces, the weight of the entire unit must be taken into account $[W = mg = (155 \text{ kg})(9.80 \text{ m/s}^2) = 1520 \text{ N}]$. Part *b* of the drawing shows the free-body diagram and a convenient set of x, y axes. Only the y axis is of interest here, and Newton's second law can be applied to calculate the acceleration a_y ($\Sigma F_y = ma_y$):

$$+T + T + T - 1520 \text{ N} = (155 \text{ kg})a_y$$

$$3(540 \text{ N}) - 1520 \text{ N} = (155 \text{ kg})a_y$$

$$\boxed{a_y = 0.65 \text{ m/s}^2}$$

(b) To find the tension in the rope when the unit is motionless, we need only set a_y equal to zero in the previous analysis:

$$3T - 1520 \text{ N} = 0 \qquad \boxed{T = 507 \text{ N}}$$

Thus, the tension in the rope is less when the scaffold is at rest than when it is accelerating upward ($T = 540$ N). When the scaffold is at rest, the tension serves only to support the weight of the unit, by virtue of the three upward forces **T**. But additional tension is needed to accelerate the unit upward.

Summary of Important Points The essential point here is that the tension is the same everywhere in a massless rope, even when the rope is wrapped around pulleys (which are assumed to be massless and frictionless). Because of the tension, the rope on each side of a pulley exerts a force on the pulley. In this problem, for instance, a net upward force of 2**T** acts on the scaffold pulley.

SOLVED PROBLEM 2
Related Problems: *24 *25 **29 **30

A flatbed truck is carrying a crate up a 10.0° hill, as part *a* of the drawing shows. The coefficient of static friction between the truck bed and the crate is $\mu_s = 0.350$. Find the maximum acceleration that the truck can attain before the crate begins to slip backward relative to the truck.

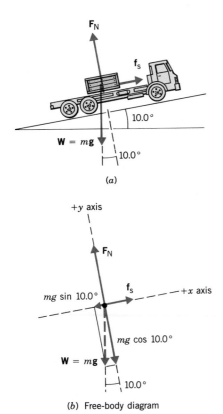

(a)

(b) Free-body diagram

Solution As the truck accelerates up the hill, the crate will not slip as long as it has the same acceleration as the truck. Therefore, a net force must act on the crate to accelerate it, and the static frictional force $\mathbf{f}_s$ contributes in a major way to this net force. As the acceleration of the truck increases, $\mathbf{f}_s$ must also increase, to produce a corresponding increase in the acceleration of the crate. However, the static frictional force can increase only until its maximum magnitude $f_s^{MAX} = \mu_s F_N$ is reached, at which point the crate experiences its maximum acceleration a^{MAX}. If the acceleration of the truck increases even more, the crate will not be able to "keep up" and slipping will occur.

To find a^{MAX}, we focus our attention on the crate, and part b of the drawing shows the free-body diagram, along with a convenient set of x, y axes. The three forces acting on the crate at the instant slipping begins are (1) its weight $\mathbf{W} = m\mathbf{g}$, (2) the normal force $\mathbf{F}_N$ exerted by the bed of the truck, and (3) the maximum static frictional force $\mathbf{f}_s^{MAX}$. The x and y components of these forces are listed below:

Force	x component	y component
$\mathbf{W}$	$-mg \sin 10.0°$	$-mg \cos 10.0°$
$\mathbf{F}_N$	0	$+F_N$
$\mathbf{f}_s$	$+\mu_s F_N$	0

Using the sum of the x components of the forces and the fact that $a_x = a^{MAX}$, Newton's second law ($\Sigma F_x = ma_x$) can be applied to show that

$$-mg \sin 10.0° + \mu_s F_N = ma^{MAX}$$

Before this equation can be solved for a^{MAX}, however, a value is needed for F_N, the magnitude of the normal force. This value can be obtained by considering the force components along the y axis. Since the crate does not accelerate along this axis, the sum of the y components of the forces must be zero according to Newton's second law ($\Sigma F_y = ma_y = 0$):

$$-mg \cos 10.0° + F_N = 0$$

It can be seen that $F_N = mg \cos 10.0°$. Substituting this value into the previous equation reveals that

$$-g \sin 10.0° + \mu_s g \cos 10.0° = a^{MAX}$$

The mass m does not appear here, because it occurs in each term on either side of the equation and, thus, is eliminated algebraically. Letting $g = 9.80$ m/s^2 and $\mu_s = 0.350$, it can be seen that $\boxed{a^{MAX} = 1.68 \text{ m/s}^2}$.

Summary of Important Points The main point of this problem is that the static frictional force (not the kinetic frictional force) keeps two objects that are in contact from slipping relative to each other. However, there is a limit beyond which slipping cannot be prevented. This limit occurs when the static frictional force attains its maximum magnitude f_s^{MAX}.

QUESTIONS

1. A stone is thrown from the top of a cliff. As the stone falls, is it in equilibrium? Explain, ignoring air resistance.

2. Can an object ever be in equilibrium if the object is acted on by (a) a single nonzero force, (b) two forces that point in mutually perpendicular directions, and (c) two forces that point in directions that are not perpendicular? Account for your answers.

3. During the final stages of descent, a parachutist approaches the ground with a constant velocity. The wind does not blow him from side to side. Is the parachutist in equilibrium and, if so, what forces are responsible for the equilibrium?

4. A weight hangs from a ring at the middle of a rope, as the drawing illustrates. Can the person who is pulling on the right end of the rope ever make the rope perfectly horizontal? Explain your answer in terms of the forces that act on the ring.

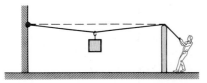

5. Two identical picture frames are hung from the midpoints of wires of the same type, as the drawing shows. In which case does the wire sustain a greater tension and why?

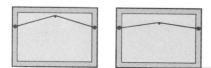

6. A 10-kg mass is placed on a scale that is in an elevator. Is the elevator accelerating up or down when the scale reads (a) 75 N and (b) 120 N? Justify your answers.

7. An object whose true weight is 165 N is placed on a scale in an elevator. The scale reads 165 N. Can you tell from this information whether the elevator is moving with a constant velocity of 2 m/s upward or 2 m/s downward or whether the elevator is at rest? Explain.

8. Suppose that you are riding in an elevator that is moving upward with a constant velocity. A scale inside the elevator shows your weight to be 600 N. (a) Does the scale register a value that is greater than, less than, or equal to 600 N during the time when the elevator slows down as it comes to a stop? (b) What is the reading when the elevator is stopped? (c) How does the value registered on the scale compare to 600 N during the time when the elevator picks up speed again on its way back down? Give your reasoning in each case.

PROBLEMS

Section 5.2 Equilibrium Applications of Newton's Laws of Motion

1. A 12.0-kg lantern is suspended from the ceiling by two vertical wires. What is the magnitude of the tension in each wire?

2. A supertanker (mass $= 1.70 \times 10^8$ kg) is moving with a constant velocity. Its engines generate a forward thrust of 7.40×10^5 N. Determine (a) the resistive force exerted on the tanker by the water and (b) the upward buoyant force exerted on the tanker by the water.

3. A wire is stretched between the tops of two identical buildings. When a tightrope walker is at the middle of the wire, the tension in the wire is 2220 N. Each half of the wire makes an angle of 8.00° with respect to the horizontal. Find the weight of the performer.

4. An 82.0-kg block is resting on a frictionless surface that is inclined at an angle of 25.0° above the horizontal. The block is held in place by a rope that is parallel to the incline. What is the tension in the rope?

5. A 20.0-kg box is being pulled across a horizontal surface at a constant velocity. The pulling force has a magnitude of 80.0 N and is directed at an angle of 30.0° above the horizontal. Determine the coefficient of kinetic friction.

* 6. A 1.60×10^6-kg submarine is submerged and heading upward along a straight line that makes an angle of 15.0° with respect to the horizontal. The submarine has a constant speed. Its engines are producing a forward thrust of 2.10×10^5 N. The water also applies forces to the submarine: a vertically upward buoyant force and a resistive force that opposes the motion of the submarine. Ignoring any other forces applied to the submarine by the water, determine the magnitudes of (a) the buoyant force and (b) the resistive force.

* 7. A bicyclist coasts at a constant velocity along a road that slopes downward at an angle of 20.0° with respect to the horizontal. The combined mass of the bicycle and rider is 75.0 kg. Find the resistive force that opposes the motion.

* 8. A 425-kg crate is hanging motionless from the end of a massless horizontal strut, as the drawing indicates. (a) Find the tension in the cable that supports the strut. *(Hint: Consider the forces that act on the stationary ring, and assume that the strut exerts a force on it that is directed horizontally to the left.)* (b) If the angle of the cable is increased beyond 35.0°, does the tension increase or decrease? Why?

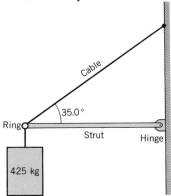

9. A skier is being pulled up a slope at a constant velocity by a tow bar. The slope is inclined at 25.0° with respect to the horizontal. The force applied to the skier by the tow bar is parallel to the slope. The skier's mass is 55.0 kg and the coefficient of kinetic friction between the skis and the snow is 0.120. Find the magnitude of the force that the tow bar exerts on the skier.

10. The weight of the block in the drawing is 20.0 lb. The coefficient of static friction between the block and the vertical wall is 0.560. What minimum force **F** is required to (a) prevent the block from sliding down the wall and (b) start the block moving up the wall?

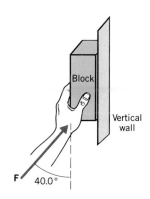

11. A 0.600-kg kite is being flown at the end of a string. Assume that the string is straight and makes an angle of 55.0° above the horizontal. The kite is stationary and the tension in the string is 35.0 N. Determine the force (both magnitude and direction) that the wind exerts on the kite.

Section 5.3 Nonequilibrium Applications of Newton's Laws of Motion

12. A 350-kg sailboat has an acceleration of 0.62 m/s² at an angle of 64° north of east. Find the magnitude and direction of the net force that acts on the sailboat.

13. A car is moving due east with an initial speed of 27.0 m/s. After 8.00 s the car has slowed down to 17.0 m/s. The mass of the car is 1380 kg. Find the magnitude and direction of the net force that produces the deceleration.

14. In the drawing, the weight of the block on the table is 25.0 lb and that of the hanging block is 58.0 lb. Ignoring all frictional effects and assuming the pulley to be massless, find (a) the acceleration of the two blocks and (b) the tension in the cord.

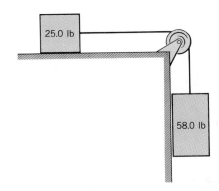

15. A rescue helicopter is lifting a man (weight = 185 lb) from a capsized boat by means of a cable and harness. (a) What is the tension in the cable when the man is given an initial upward acceleration of 3.60 ft/s²? (b) What is the tension during the remainder of the rescue when he is pulled upward at a constant velocity?

16. A lunar landing craft (mass = 11 400 kg) is about to touch down on the surface of the moon, where the acceleration due to gravity is 1.60 m/s². At an altitude of 165 m the craft's downward velocity is 18.0 m/s. To slow down the craft, a retrorocket is firing to provide an upward thrust. Assuming the descent is vertical, find the magnitude of the thrust needed to reduce the velocity to zero at the instant when the craft touches the lunar surface.

17. A 165-kg astronaut, equipped with a portable propulsion unit, is about to travel along a straight line from one spacecraft to another. He accelerates for 15.0 s under the influence of a 142-N propulsion force. Upon reaching point A, he shuts off the propulsion unit and continues to move with a constant velocity until point B is reached. At point B, which is 425 m from the second spacecraft, he turns on the propulsion unit to produce a reverse thrust, so that his velocity will be reduced to zero at the instant he reaches the second spacecraft. Find (a) the acceleration of the astronaut from the first spacecraft to point A, (b) the speed from A to B, and (c) the amount of reverse thrust necessary to reduce the velocity to zero at the second spacecraft.

18. Solve problem 14, assuming a coefficient of kinetic friction of 0.300 between the block and the table.

19. A locomotive is pulling two freight cars with an acceleration of 0.520 m/s². The mass of the first car is 51 300 kg, while that of the second car is 18 400 kg. Find the tension in the coupling mechanism between the engine and the first car and between the first car and the second car.

* 20. A 205-kg log is being pulled up a ramp by means of a rope that is parallel to the ramp. The ramp is inclined at 30.0° with respect to the horizontal. The coefficient of kinetic friction between the log and the ramp is 0.900, and the log has an acceleration of 0.800 m/s². Find the tension in the rope.

* 21. A person is sledding down a slope that is inclined at 30.0° with respect to the horizontal. A moderate wind is aiding the motion by providing a steady force of 105 N that is directed parallel to the slope in the direction the sled is moving. The combined mass of the person and sled is 65.0 kg, and the coefficient of kinetic friction between the runners of the sled and the snow is 0.150. How much time is required for the sled to travel down a 175-m slope, starting from rest?

* 22. As the drawing indicates, an electric motor is lowering a 452-kg crate with an acceleration of 1.60 m/s². Determine the tension in the cable. *(See Solved Problem 1 for a related problem.)*

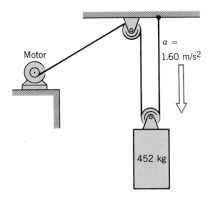

* 23. A 185-kg cart is pulled up a 25.0° incline by means of the cable and pulley arrangement shown in the drawing. (a) The tension in the cable is 304 N. Assuming there is no friction, find the acceleration of the cart. (b) What is the tension in the cable if the cart is either at rest or is being pulled up the incline at a constant speed? *(See Solved Problem 1 for a related problem.)*

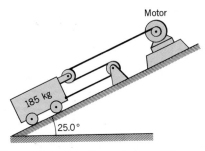

* 24. A book is resting on a piece of paper. The paper is resting on a flat table. The coefficient of static friction between the book and the paper is 0.72. If a person pulls on the paper, what is the

maximum acceleration that the book can have, before the book begins to slip relative to the paper? *(See Solved Problem 2 for a related problem.)*

* 25. A crate is resting on the bed of a moving truck. The coefficient of static friction between the crate and the truck bed is 0.40. The driver hits the brakes. Assuming the truck is traveling on level ground, determine the maximum deceleration that the truck can have without the crate slipping forward relative to the truck. *(See Solved Problem 2 for a related problem.)*

**26. As part *a* of the drawing shows, two blocks are connected by a rope that passes over a set of pulleys. One block has a weight of 412 N, and the other has a weight of 908 N. The rope and the pulleys are massless and there is no friction. (a) What is the acceleration of the lighter block? (b) Suppose that the heavier block is removed, and a downward force of 908 N is provided by someone pulling on the rope, as part *b* of the drawing shows. Find the acceleration of the remaining block. (c) Explain why the answers in (a) and (b) are different.

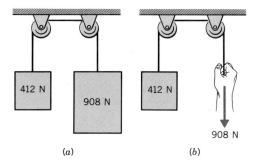

(a)　　　　　　　(b)

**27. In the drawing, the rope and the pulleys are massless, and there is no friction. Find (a) the tension in the rope and (b) the acceleration of the 10.0-kg block. *(Hint: The larger mass moves twice as far as the smaller mass.) (See Solved Problem 1 for a related problem.)*

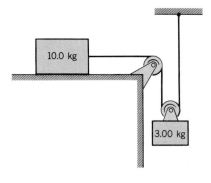

**28. Two masses are connected together by a rope and pulley system, as in the drawing. Ignore the masses of the rope and the pulleys, and also ignore friction. (a) Find the acceleration of each mass. *(Hint: The larger mass moves only half as far as the smaller*

mass.) (b) Determine the tension in the rope. *(See Solved Problem 1 for a related problem.)*

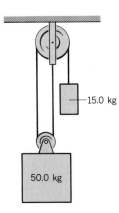

15.0 kg

50.0 kg

29. A 5.00-kg block is placed on top of a 12.0-kg block that rests on a frictionless table. The coefficient of static friction between the two blocks is 0.600. What is the maximum horizontal force that can be applied before the 5.00-kg block begins to slip relative to the 12.0-kg block, if the force is applied to (a) the more massive block and (b) the less massive block? *(See Solved Problem 2 for a related problem.)*

30. A truck is traveling at a speed of 25.0 m/s. A crate is resting on the bed of the truck, and the coefficient of static friction between the crate and the truck bed is 0.650. Determine the shortest distance in which the truck can come to a halt without causing the crate to slip forward relative to the truck. *(See Solved Problem 2 for a related problem.)*

Section 5.4 Apparent Weight

31. A 95.0-kg person stands on a scale in an elevator. What is the apparent weight when the elevator is (a) accelerating upward with an acceleration of 1.80 m/s^2, (b) moving upward at a constant speed, (c) accelerating downward with an acceleration of 1.30 m/s^2, and (d) moving downward at a constant speed?

32. A woman stands on a scale in a moving elevator. Her mass is 60.0 kg, and the combined mass of the elevator and scale is an additional 815 kg. Starting from rest, the elevator accelerates upward. During the acceleration, there is a tension of 9410 N in the hoisting cable. (a) What is the reading on the scale during the acceleration? (b) Assuming the elevator starts from rest, how fast is it traveling at the end of 3.00 s?

33. A 55.0-kg person is riding in a hot air balloon, and a scale shows this person's weight to be 549 N. The acceleration of gravity at the location of the balloon is known to be 9.79 m/s^2. Determine the magnitude and direction of the vertical component of the balloon's acceleration.

34. A person has an apparent weight of 750 N when accelerating upward and 650 N when accelerating downward. In each case, the magnitude of the acceleration is the same. What is the person's true weight?

ADDITIONAL PROBLEMS

35. A 15.0-g bullet is fired from a rifle. It takes 2.50×10^{-3} s for the bullet to travel the length of the barrel, and it exits the barrel with a speed of 715 m/s. Assuming that the acceleration of the bullet is constant, find the average net force exerted on the bullet.

36. A stuntman is being pulled along a rough road at a constant velocity, by a cable attached to a moving truck. The cable is parallel to the ground. The mass of the stuntman is 7.50 slugs and the coefficient of kinetic friction between the road and him is 0.870. Find (a) the normal force acting on the stuntman and (b) the tension in the cable.

37. A fisherman is fishing from a bridge and is using a "10-lb test line." In other words, the line will sustain a maximum force of 10.0 lb without breaking. (a) What is the heaviest fish that can be pulled up vertically, when the line is reeled in at a constant speed? (b) Repeat part (a), assuming that the line is given an upward acceleration of 6.50 ft/s^2.

38. A 750-kg sailboat is being towed at a constant velocity by a power boat, as the drawing illustrates. The tension in the towing cable is 3200 N. Calculate (a) the buoyant force (vertically upward) and (b) the horizontal resisting force that the water exerts on the sailboat.

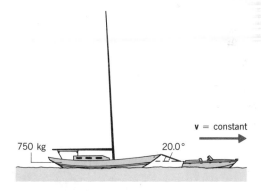

750 kg   20.0°   v = constant

39. Two masses (45.0 and 21.0 kg) are connected by a massless string that passes over a massless, frictionless pulley. The pulley hangs from the ceiling. Find (a) the acceleration of the masses and (b) the tension in the string.

40. A block slides down an incline and has a constant speed. The angle of the incline is 25.0° with respect to the horizontal. What is the coefficient of kinetic friction between the surface of the incline and the block?

* 41. A 3.00-slug sign is suspended by two ropes, as the drawing shows. Find the magnitude of the tension in each of the ropes.

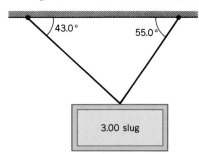

* 42. A person whose weight is 117 lb is being pulled up vertically by a rope from the bottom of a cave that is 115 ft deep. The maximum tension that the rope can withstand without breaking is 128 lb. What is the shortest time, starting from rest, in which the person can be brought out of the cave?

* 43. A 45.0-kg box is sliding up an incline that makes an angle of 15.0° with respect to the horizontal. The coefficient of kinetic friction between the box and the surface of the incline is 0.180. The initial speed of the box at the bottom of the incline is 1.50 m/s. How far does the box travel along the incline before coming to rest?

* 44. A person is trying to judge if a picture (mass = 1.10 kg) is properly positioned by temporarily pressing it against a wall. The pressing force is perpendicular to the wall. The coefficient of static friction between the picture and the wall is 0.660. (a) What is the minimum amount of pressing force that must be used? (b) In general, must the pressing force exceed the weight of the picture if the coefficient of static friction is less than 1.00? Explain.

**45. A small sphere is hung by a string from the ceiling of a van. When the van is stationary, the sphere hangs vertically. However, when the van accelerates, the sphere swings backward so that the string makes an angle of θ with respect to the vertical. (a) Derive an expression for the magnitude a of the acceleration of the van in terms of the angle θ and the magnitude g of the acceleration due to gravity. (b) Find the acceleration of the van when $\theta = 10.0°$. (c) What is the angle θ when the van moves with a constant velocity? Can this device be used to measure the velocity? Explain.

**46. The drawing shows three objects. They are connected by strings that pass over massless and friction-free pulleys. The objects move, and the coefficient of kinetic friction between the middle object and the surface of the table is 0.100. (a) What is the acceleration of the three masses? (b) Find the tension in each of the two strings.

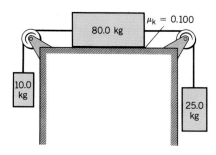

CHAPTER 6

Dynamics of Uniform Circular Motion

Many rides at amusement parks go 'round and round.' Those that rotate at a constant speed are examples of uniform circular motion.

Newton's laws of motion provide powerful tools for analyzing motion, and they apply equally well to motion along a straight line and motion along a curved line. This chapter concentrates on motion along a specific kind of curved line, the arc of a circle. The goal is to understand the nature of the acceleration and forces that are present when an object moves at a constant speed on a circular path. To reach this goal we will search for any acceleration that may exist and then use Newton's second law to tell us how much force is required to maintain the acceleration. In the process, we will develop a deeper insight into the motion in many familiar scenes, scenes ranging from race cars negotiating banked turns to satellites circling the earth.

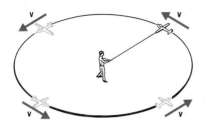

FIGURE 6.1 The motion of an airplane flying at a constant speed on a horizontal circular path is an example of uniform circular motion. The direction of the velocity vector changes as the plane moves, although the magnitude of the vector (the speed) does not change.

6.1 UNIFORM CIRCULAR MOTION

The definition of uniform circular motion is given below:

DEFINITION OF UNIFORM CIRCULAR MOTION
Uniform circular motion is the motion of an object traveling at a constant (uniform) speed on a circular path.

As an example of this kind of motion, Figure 6.1 shows a model airplane attached to a guideline. The speed of the plane is the magnitude of the velocity vector **v**, and since the speed is constant, the vectors in the drawing have the same magnitude at various points along the circular path.

Sometimes it is more convenient to describe uniform circular motion by specifying the period of the motion, rather than the speed. The *period T* is the time required for the object to travel once around the circle, that is, to make one complete revolution. There is a relationship between period and speed, since speed v is the distance traveled (circumference $= 2\pi r$) divided by the time T:

$$v = \frac{2\pi r}{T} \tag{6.1}$$

Therefore, if the radius is known, as in Example 1, the speed can be calculated from the period or vice versa.

EXAMPLE 1

The wheel of a car has a radius of 0.29 m and is being rotated at 830 revolutions per minute (rpm) on a tire-balancing machine. Determine the speed at which the outer edge of the wheel is moving.

SOLUTION

The speed v can be obtained directly from $v = 2\pi r/T$, but first the period T is needed. Since the tire makes 830 revolutions in one minute, the number of minutes required for a single revolution is

$$\frac{1}{830 \text{ revolutions/min}} = 1.2 \times 10^{-3} \text{ min/revolution}$$

Therefore, the period is $T = 1.2 \times 10^{-3}$ min, which corresponds to 0.072 s. Equation 6.1 can now be used to find the speed:

$$v = \frac{2\pi r}{T} = \frac{2\pi (0.29 \text{ m})}{0.072 \text{ s}} = \boxed{25 \text{ m/s}}$$

The definition of uniform circular motion emphasizes that the magnitude of the velocity vector is constant. It is equally significant that the direction of the vector is *not constant*. In Figure 6.1, for instance, the velocity vector changes direction as the plane moves around the circle, and any change in the velocity vector, even if it is only a change in direction, means that an acceleration is occurring. This particular acceleration is called "centripetal acceleration," because it points toward the center of the circle, as will become clear in the next section.

6.2 CENTRIPETAL ACCELERATION

In Figure 6.2a an object (symbolized by a dot ●) is in uniform circular motion, and the velocity vector is drawn at two different times. At time t_0 the velocity is tangent to

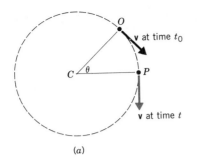

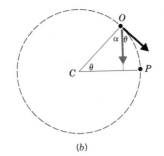

FIGURE 6.2 (*a*) For an object in uniform circular motion, the velocity vector **v** has different directions at different points on the circle. (*b*) The velocity vector has been removed from point *P*, shifted parallel to itself, and redrawn with its tail at the starting point *O*. The angles between the two velocity vectors and between the two radii have the same value θ.

(*a*)

(*b*)

the circle at point *O*, while at a later time *t* the velocity is tangent at point *P*. As the object moves from *O* to *P*, the radius traces out the angle θ, and the velocity vector changes direction. To emphasize the change in direction, part *b* of the picture shows the velocity vector removed from point *P*, shifted parallel to itself, and redrawn with its tail at the starting point *O*. The change in direction is indicated by the angle θ between the two velocity vectors, the same angle that is between the two radii. The two angles are equal, because the radii *CO* and *CP* are perpendicular to the tangents at points *O* and *P*, with the result that each angle θ is complementary to the angle α.

Figure 6.3 reveals the features of Figure 6.2 that are important in determining the acceleration that uniform circular motion entails. As always, the acceleration is the change $\Delta \mathbf{v}$ in velocity divided by the elapsed time Δt, or $\mathbf{a} = \Delta \mathbf{v}/\Delta t$. Part *a* of the drawing shows the two velocity vectors oriented at the angle θ with respect to one another, together with the vector $\Delta \mathbf{v}$ that represents the change in velocity. The change $\Delta \mathbf{v}$ is the increment that must be added to the velocity at time t_0, so that the resultant velocity has the new direction after an elapsed time $\Delta t = t - t_0$. Figure 6.3*b* shows the sector of the circle *COP*, the geometry of which is needed to determine $\Delta \mathbf{v}/\Delta t$. (For convenience, this sector has been rotated a quarter of a turn clockwise, relative to its orientation in Figure 6.2.) The arc length *OP* can be approximated as a straight line in the limit of a very small elapsed time Δt, making *COP* a triangle. In this limit, the two triangles in the drawing are isosceles and have equal apex angles θ. Thus, they are similar triangles, so that

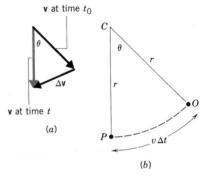

FIGURE 6.3 (*a*) The direction of the velocity vector at time *t* differs from the direction at time t_0 by an amount given by the angle θ. (*b*) When the object moves along the circumference of the circle from *O* to *P*, the radius *r* traces out the same angle θ. In the limit of a very short elapsed time Δt, the two triangles are similar.

$$\frac{\Delta v}{v} = \frac{v \, \Delta t}{r}$$

We have used the fact that side *OP* equals $v \, \Delta t$, the distance traveled by the object. This equation can be solved for $\Delta v/\Delta t$, to show that the magnitude a_c of the centripetal acceleration is given by $a_c = v^2/r$.

Centripetal acceleration is a vector quantity and has a direction as well as a magnitude. The direction is toward the center of the circle, a fact that can be understood with the aid of Figure 6.4. This drawing illustrates that an object in uniform circular motion would fly off on a tangent if suddenly released from the circular path at point *O*. The object would move on a straight line (Newton's first law) to point *A* in the time it would have taken to travel on the circle to point *P*. It is as if the object drops through the distance *AP* in the process of remaining on the circle, and *AP* is directed toward the center of the circle in the limit that the angle θ is small. Thus, the object accelerates toward the center of the circle at every moment. The acceleration is called **centripetal acceleration,** because the word "centripetal" means "center-seeking."

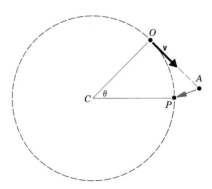

FIGURE 6.4 The object must accelerate in a direction that points toward the center of the circle, if the object is to remain on the circular path.

The concept of centripetal acceleration is summarized below.

CENTRIPETAL ACCELERATION

Magnitude: The centripetal acceleration of an object moving with a speed v on a circular path of radius r has a magnitude a_c given by

$$a_c = \frac{v^2}{r} \tag{6.2}$$

Direction: The centripetal acceleration vector always points toward the center of the circle and continually changes direction as the object moves.

The following example illustrates the effect of the radius r on the centripetal acceleration.

EXAMPLE 2

A car is driven around one curve whose radius is 320 m and then another whose radius is 960 m. Figure 6.5 shows the two curves. In both cases the speed is 28 m/s. Compare the centripetal accelerations for both turns.

SOLUTION

In each case, the magnitude of the acceleration can be obtained from $a_c = v^2/r$:

$$[\text{Radius} = 320 \text{ m}] \quad a_c = \frac{(28 \text{ m/s})^2}{320 \text{ m}} = \boxed{2.5 \text{ m/s}^2}$$

$$[\text{Radius} = 960 \text{ m}] \quad a_c = \frac{(28 \text{ m/s})^2}{960 \text{ m}} = \boxed{0.82 \text{ m/s}^2}$$

When the radius is larger, the centripetal acceleration is smaller. In fact, when r becomes very large, the centripetal acceleration approaches zero. Motion along the arc of an infinitely large circle is just like motion at a constant speed along a straight line, in which case there is no acceleration.

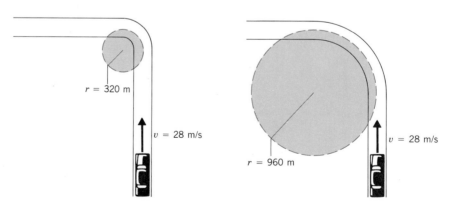

FIGURE 6.5 A car travels at the same speed around two curves with different radii. For the turn with the smaller radius, the car has the larger centripetal acceleration.

Because of the different centripetal accelerations, driving around each turn in Figure 6.5 would "feel" different, as most drivers know from their experiences with tight turns (smaller r) and gentle turns (larger r). The "feeling" is associated with the force that must be present in uniform circular motion, and we now turn to this topic.

6.3 CENTRIPETAL FORCE

Newton's second law indicates that whenever an object accelerates, there must be a net force to create the acceleration. Thus, in uniform circular motion there must be a net force to produce the centripetal acceleration. Moreover, the magnitude of the net force can be calculated in the usual fashion, by multiplying the mass of the object by the magnitude of the acceleration: $F_c = ma_c = mv^2/r$. This force points in the same direction as the centripetal acceleration, that is, toward the center of the circle. Because of its direction, the force is called the **centripetal force.**

CENTRIPETAL FORCE

   Magnitude: The centripetal force is the net force required to keep an object of mass m moving at a speed v on a circular path of radius r and has a magnitude given by

$$F_c = \frac{mv^2}{r} \qquad (6.3)$$

   Direction: The centripetal force always points toward the center of the circle and continually changes direction as the object moves.

   In some cases it is easy to identify the source of the centripetal force. In Figure 6.1, for instance, the only force pulling the airplane inward is the tension force of the guideline, so this force is the centripetal force. Figure 6.6 and Example 3 illustrate the fact that higher speeds and smaller circles require greater tension in the guideline.

EXAMPLE 3

Suppose that the guideline used in Figure 6.6a has a length of 14 m and can sustain a maximum tension of 85 N without breaking. What is the maximum speed that a 0.90-kg model airplane can have?

SOLUTION

Equation 6.3 can be used to calculate the maximum speed:

$$F_c = \frac{mv^2}{r}$$

$$v = \sqrt{\frac{rF_c}{m}} = \sqrt{\frac{(14\text{ m})(85\text{ N})}{0.90\text{ kg}}} = \boxed{36\text{ m/s}}$$

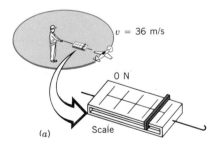

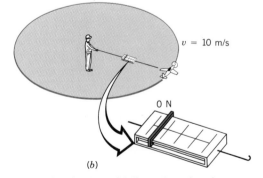

FIGURE 6.6 The scale shows that a greater tension exists in the guideline when the plane has (*a*) a greater speed and smaller turning radius than when it has (*b*) a smaller speed and greater turning radius.

FIGURE 6.7 (*a*) When the car moves without skidding around a circular curve, the force of static friction between the road and the tires provides the centripetal force to keep the car on the road. (*b*) If the upholstery on the seat cannot provide enough static friction, the side of the car ultimately provides the centripetal force to keep the passenger on the circular path.

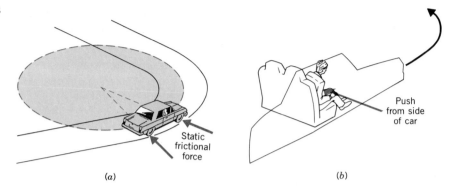

(*a*)   (*b*)

When a car moves at a steady speed around an unbanked curve, the centripetal force keeping the car on the curve comes from the static friction between the road and the tires, as Figure 6.7*a* indicates. It is static, rather than kinetic friction, because the tires are not slipping with respect to the radial direction. If the static friction is insufficient, given the speed and the radius of the turn, the car will skid off the road, away from the center of the circle. Conditions of limited friction occur when the road becomes wet or icy. Example 4 shows how such conditions can limit safe driving.

EXAMPLE 4

Compare the maximum speeds at which a car can safely negotiate an unbanked turn (radius = 50.0 m) under normal condition (coefficient of static friction = 0.900) and icy condition (coefficient of static friction = 0.100).

SOLUTION

At the maximum speed, the maximum centripetal force acts on the tires of the car. The magnitude of the maximum force of static friction is specified by Equation 4.6 as $f_s^{MAX} = \mu_s F_N$, where μ_s is the coefficient of static friction and F_N is the magnitude of the normal force. Since the car does not accelerate in the vertical direction, the weight mg of the car is balanced by the normal force, so $F_N = mg$. From Equation 6.3 it follows, then, that

$$F_c = \mu_s F_N = \mu_s mg = \frac{mv^2}{r}$$

Consequently, $\mu_s g = v^2/r$, and

$$v = \sqrt{\mu_s g r}$$

The mass m of the car has been eliminated algebraically from this result. The maximum speeds can now be calculated:

Normal conditions ($\mu_s = 0.900$)

$$v = \sqrt{(0.900)(9.80 \text{ m/s}^2)(50.0 \text{ m})}$$

$$\boxed{v = 21.0 \text{ m/s}}$$

Icy conditions ($\mu_s = 0.100$)

$$v = \sqrt{(0.100)(9.80 \text{ m/s}^2)(50.0 \text{ m})}$$

$$\boxed{v = 7.00 \text{ m/s}}$$

The icy turn requires a slower speed for safe driving, as most people know.

The passenger in Figure 6.7 must also experience a centripetal force to remain on the circular path. However, if the upholstery is very slippery, there may not be enough static friction to keep the passenger in place as the driver negotiates a tight turn at high speed. Then, when viewed from inside the car, the passenger appears to be thrown toward the outside of the turn. What really happens is that he slides off on a tangent to the circle, until he encounters a source of centripetal force to keep him in place while the car turns. In Figure 6.7 this occurs when the passenger bumps into the side of the car, which pushes on him with the necessary force.

Sometimes the source of the centripetal force is not obvious. A pilot making a turn, for instance, banks or tilts his plane at an angle to create the centripetal force. When a plane flies, the air pushes upward on the wing surfaces with a lifting force **L** that is perpendicular to the wing surfaces, as Figure 6.8*a* shows. When the plane is banked into the turn at an angle θ above the horizontal, a component L sin θ of the lifting force is directed toward the center of the turn, as part *b* of the drawing indicates. It is this component of the lift that provides the centripetal force. Greater speeds and/or tighter turns require greater centripetal forces. In such situations, the pilot must bank the plane at a larger angle, so that a larger component of the lift points toward the center of the turn. The technique of banking into a turn also has an application in the construction of high speed roadways, where the road itself is banked to achieve a similar effect, as the next section discusses.

There must always be a centripetal force if there is uniform circular motion, and it is possible that more than one source contributes to the centripetal force at the same time. Thus, the phrase "centripetal force" does not denote a new and separate force created by nature. The phrase merely labels the net force pointing toward the center of the circular path, and this net force is the vector sum of *all* the forces that point along the radial direction.

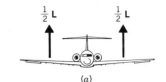

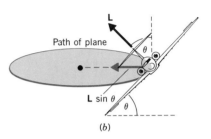

FIGURE 6.8 (*a*) The air exerts an upward lifting force $\frac{1}{2}$L on each wing. (*b*) When a plane executes a circular turn, the plane banks at an angle θ. The component L sin θ of the lifting force is directed toward the center of the circle and provides the required centripetal force.

6.4 BANKED CURVES

When a car travels without skidding around an unbanked curve, the static frictional force between the tires and the road provides the centripetal force. Depending on the condition of the pavement (ice, oil slicks, etc.), however, the friction may vary from day to day. The reliance on friction can be eliminated completely for a given speed if the curve is banked at an angle relative to the horizontal, much in the same way that a plane is banked while making a turn.

Figure 6.9*a* shows a car going around a friction-free banked curve. The radius of the curve is *r*, where *r* is measured parallel to the horizontal and not to the slanted road surface. Part *b* shows the normal force $\mathbf{F_N}$ that the road applies to the car, the normal force being directed perpendicular to the road surface. Because the roadbed makes an angle θ with respect to the horizontal, $\mathbf{F_N}$ has a component $F_N \sin \theta$ pointing toward the center *C* of the circle, and this component provides the centripetal force:

$$F_c = F_N \sin \theta = \frac{mv^2}{r}$$

The force $\mathbf{F_N}$ can be evaluated by recognizing that the car does not accelerate in the vertical direction. Therefore, the vertical component $F_N \cos \theta$ of the normal force must balance the weight $m\mathbf{g}$ of the car, so $F_N \cos \theta = mg$. Dividing this equation into the previous one shows that

$$\frac{F_N \sin \theta}{F_N \cos \theta} = \frac{mv^2/r}{mg}$$

$$\tan \theta = \frac{v^2}{rg} \qquad (6.4)$$

Equation 6.4 indicates that, for a given speed *v*, all the centripetal force needed for a turn of radius *r* can be obtained from the normal force by banking the turn at an angle θ, independent of the mass of the vehicle. Greater speeds and smaller radii require more steeply banked curves, that is, larger values of θ. However, at a speed that is too small for a given θ, a car would slide down a frictionless banked curve; at a

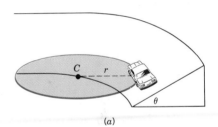

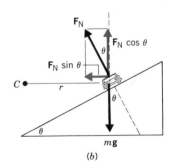

FIGURE 6.9 (*a*) A car travels in a circular path of radius *r* on a frictionless banked road. The banking angle is θ, and the center of the circle is at *C*. (*b*) The forces that act on the car are its weight $m\mathbf{g}$ and the normal force $\mathbf{F_N}$. The component of the normal force pointing toward the center of the circle ($F_N \sin \theta$) provides the centripetal force.

speed that is too large, a car would slide off the top. Typical banking conditions are calculated in the next example.

EXAMPLE 5

Determine the angle at which a frictionless curve should be banked for a speed of 35 m/s and a turning radius of 550 m.

SOLUTION

A straightforward application of Equation 6.4 gives the answer:

$$\tan \theta = \frac{v^2}{rg} = \frac{(35 \text{ m/s})^2}{(550 \text{ m})(9.80 \text{ m/s}^2)} = 0.23$$

$$\theta = \tan^{-1} 0.23 = \boxed{13°}$$

6.5 SATELLITES IN CIRCULAR ORBITS

THE RELATION BETWEEN ORBITAL RADIUS AND ORBITAL SPEED

Satellites in circular orbits about the earth are examples of uniform circular motion, and their motional characteristics can be understood in that context. For instance, given the orbital radius, there is only one speed that the satellite can have, if it is to remain in orbit. To see how this fundamental characteristic arises, consider the gravitational force acting on the satellite of mass m in Figure 6.10. In accord with Newton's law of gravitation (Equation 4.3), the gravitational force is directed toward the center of the earth and has a magnitude given by $F = GmM_E/r^2$. Here G is the universal gravitational constant (6.67×10^{-11} N $\cdot$ m²/kg²), M_E is the mass of the earth (5.98×10^{24} kg), and r is the distance measured from the center of the earth. Since the gravitational force is the only force acting on the satellite in the radial direction, it alone provides the centripetal force. Therefore,

$$\frac{mv^2}{r} = G\frac{mM_E}{r^2}$$

Solving for v gives

$$v = \sqrt{\frac{GM_E}{r}} \qquad (6.5)$$

If the satellite is to remain in the orbit of radius r, the speed must have precisely the value indicated by this result. The closer the satellite is to the earth, the greater the orbital speed must be. Once in orbit at the correct speed, the satellite continues in uniform circular motion forever, assuming that effects such as friction due to residual atmosphere do not reduce the speed.

Equation 6.5 applies to man-made earth satellites or to the moon, which is a natural satellite of the earth. This equation also holds for satellites in circular orbits about the sun or another planet, provided M_E is replaced by the mass of the object about which the satellite moves.

The mass m of the satellite does not appear in Equation 6.5, having been eliminated algebraically from the result. Consequently, for a given orbit, a satellite with a large mass has exactly the same orbital speed as a satellite with a small mass. However, more effort is certainly required to lift the larger mass into orbit. The orbital speed of a typical artificial satellite is determined in the following example.

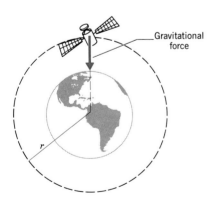

FIGURE 6.10 For a satellite in circular orbit around the earth, the gravitational force provides the required centripetal force. The symbol r denotes the distance between the center of the earth and the satellite.

Gravitational force

EXAMPLE 6

Determine the speed of a satellite orbiting at a height of 340 000 m (210 mi) above the earth's surface.

SOLUTION

Before Equation 6.5 can be applied, the orbital radius r must first be determined *relative to the center of the earth*. Since the radius of the earth is approximately 6.38×10^6 m, and the height of the satellite above the earth's surface is 0.34×10^6 m, the orbital radius is $r = 6.72 \times 10^6$ m. The orbital speed is

$$v = \sqrt{\frac{GM_E}{r}} = \sqrt{\frac{(6.67 \times 10^{-11} \text{ N} \cdot \text{m}^2/\text{kg}^2)(5.98 \times 10^{24} \text{ kg})}{6.72 \times 10^6 \text{ m}}}$$

$$\boxed{v = 7.70 \times 10^3 \text{ m/s } (17\ 200 \text{ mi/h})}$$

THE PERIOD OF THE SATELLITE

The period T of a satellite is the time required for one complete orbital revolution. As in any uniform circular motion, the period is related to the speed of the motion by $v = 2\pi r/T$. Substituting v from Equation 6.5 shows that

$$\sqrt{\frac{GM_E}{r}} = \frac{2\pi r}{T}$$

Solving this expression for T gives

$$T = \frac{2\pi r^{3/2}}{\sqrt{GM_E}} \qquad (6.6)$$

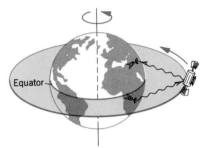

FIGURE 6.11 A synchronous satellite orbits the earth in a circular path that is in the plane of the equator. The period of the satellite is chosen to be one day, the same as the period of the earth's rotation. A synchronous satellite appears stationary to observers on the earth.

Although derived for earth orbits, Equation 6.6 can also be used for calculating the periods of those planets that have nearly circular orbits about the sun, if M_E is replaced by the mass M_S of the sun and r is interpreted as the distance between the center of the planet and the center of the sun. The fact that the period is proportional to the three-halves power of the orbital radius is known as Kepler's third law, for it was one of the laws discovered by Johannes Kepler (1571–1630) during his studies of the planets. Kepler's third law also holds for elliptical orbits, which will be discussed in Chapter 10.

An important application of Equation 6.6 occurs in the field of communications, where so-called "synchronous satellites" are put into a circular orbit that is in the plane of the equator, as Figure 6.11 shows. The period of such a satellite is chosen to be one day, which is also the time it takes for the earth to turn once about its axis. Therefore, these satellites move around their orbits in a way that is synchronized with the rotation of the earth, and, for earth-based observers, they have the useful characteristic of appearing in fixed positions in the sky. Thus, they can serve as stationary targets or relay stations for communication signals sent up from the earth's surface. According to Equation 6.6, the orbit of a synchronous satellite can have only one radius, and this value is determined in the next example.

EXAMPLE 7

What is the height above the earth's surface at which all synchronous satellites (regardless of mass) must be placed in orbit?

SOLUTION

The period of the satellite is one day* or 8.64×10^4 s. This value for T can be used in Equation 6.6 to determine the orbital radius:

$$T = \frac{2\pi r^{3/2}}{\sqrt{GM_E}}$$

* Successive appearances of the sun define the solar day of 24 h or 8.64×10^4 s. The sun moves against the background of the stars, however, and the time required for the earth to turn once on its axis relative to the fixed stars is 23 h 56 min, which is called the sidereal day. The sidereal day should be used in Example 7, but the neglect of this effect introduces an error of less than 0.4 percent in the answer.

$$8.64 \times 10^4 \text{ s} = \frac{2\pi r^{3/2}}{\sqrt{(6.67 \times 10^{-11} \text{ N} \cdot \text{m}^2/\text{kg}^2)(5.98 \times 10^{24} \text{ kg})}}$$

By squaring and then taking the cube root, we can solve this equation to show that $r = 4.23 \times 10^7$ m. Since the radius of the earth is approximately 6.38×10^6 m, the height H of the satellite above the earth's surface is

$$H = 4.23 \times 10^7 \text{ m} - 0.64 \times 10^7 \text{ m}$$
$$= \boxed{3.59 \times 10^7 \text{ m (22 300 mi)}}$$

6.6 APPARENT WEIGHTLESSNESS AND ARTIFICIAL GRAVITY

The idea of living on board a satellite captures the imagination of many because of the popularized concept of "weightlessness." Actually, the concept should be called "apparent weightlessness," because it is similar to the condition of zero apparent weight that occurs in an elevator during free-fall. As a reminder of our earlier discussion concerning apparent weight, Figure 6.12a essentially reproduces Figure 5.10d. The apparent weight is the force that an object exerts on the platform of a scale. In free-fall, the apparent weight becomes zero, because both the scale and a person on it fall together and hence cannot push against one another. In contrast, the true weight is the gravitational force ($F = GmM_E/r^2$) that the earth exerts on an object, and this force is not zero in a freely falling elevator or aboard an orbiting satellite.

Within an orbiting satellite, apparent weightlessness has a meaning similar to that illustrated by the elevator in Figure 6.12a. In uniform circular motion, an object constantly accelerates or falls toward the center of the circle, in order to remain on the circular path. Therefore, the astronaut in part b of the drawing and the scale beneath his feet both fall toward the center of the orbit at all times and cannot push against one another. As a result, the apparent weight is zero, and the astronaut experiences apparent weightlessness. The only difference between the satellite and the elevator is that the satellite moves on a circle, so that its "falling" does not bring it closer to the surface of the earth.

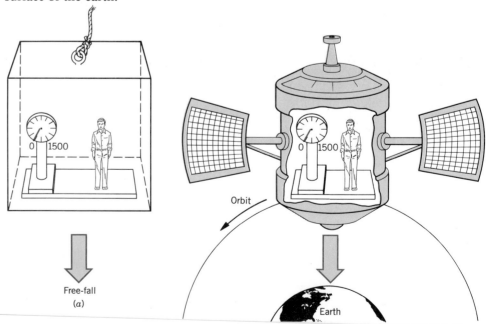

FIGURE 6.12 (a) During free-fall, the elevator accelerates downward with the acceleration of gravity. The person is said to be in a state of apparent weightlessness, because his apparent weight (the reading on the scale) is zero. (b) The astronaut is also in a state of apparent weightlessness, because the orbiting space station is in free-fall toward the center of the earth.

FIGURE 6.13 The surface of the rotating space station pushes on an object with which it is in contact and thereby provides the centripetal force needed to keep it moving on a circular path.

The physiological effects of prolonged apparent weightlessness are only partially known. To minimize such effects, it is likely that artificial gravity will be provided in large space stations of the future. Artificial gravity can be understood with the aid of Figure 6.13, which shows a space station rotating about an axis. Because of this rotational motion, any object located at a point P on the surface of the station experiences a centripetal force directed toward the axis. The surface of the structure provides this force, pushing on the feet of an astronaut much in the same way the surface of the earth pushes up on our feet. The centripetal force can be adjusted to match the astronaut's earth weight by properly selecting the rotational speed of the space station, as Example 8 illustrates.

EXAMPLE 8

At what speed must the surface of the space station ($r = 1700$ m) move in Figure 6.13, so that the astronaut at point P experiences a push on his feet that equals his earth weight?

SOLUTION

The earth weight of the astronaut (mass $= m$) is mg, and with this substitution, Equation 6.3 can be used to determine the required speed:

$$F_c = \frac{mv^2}{r} = mg$$

The speed is

$$v = \sqrt{rg} = \sqrt{(1700 \text{ m})(9.80 \text{ m/s}^2)} = \boxed{130 \text{ m/s (290 mi/h)}}$$

---

*6.7 VERTICAL CIRCULAR MOTION

In uniform circular motion, the speed of the object is constant. If the speed varies, the motion is said to be "nonuniform." Using the concepts we have developed so far, it is possible to gain considerable insight into nonuniform circular motion.

As an example, consider the motion of a motorcycle being driven around the vertical circular track shown in Figure 6.14a. Usually, the speed varies in this stunt, decreasing as the cycle moves upward and increasing as the cycle comes downward. When the speed changes, the magnitude of the centripetal force also changes from point to point on the circle. There are four points on the circle where the centripetal force can be identified easily, as part b of the drawing indicates. At each point the centripetal force is the net sum of all the forces that are oriented along the radial direction. The drawing shows only the weight of the cycle plus rider (magnitude $= mg$) and the normal force pushing on the cycle (magnitude $= F_N$). The propulsion and braking forces are omitted for the sake of simplicity and, in any event, do

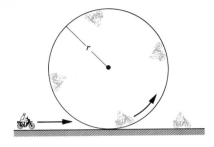

(a)

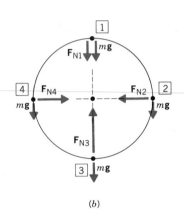

(b)

FIGURE 6.14 (a) A vertical loop-the-loop motorcycle stunt. (b) The normal force $\mathbf{F_N}$ and the weight mg of the cycle and the rider are shown here at four locations on the loop.

not act in the radial direction. The magnitude of the centripetal force at each of the four points is expressed below in terms of mg and F_N:

$$(1) \quad F_{c1} = \frac{mv_1^2}{r} = F_{N1} + mg$$

$$(2) \quad F_{c2} = \frac{mv_2^2}{r} = F_{N2}$$

$$(3) \quad F_{c3} = \frac{mv_3^2}{r} = F_{N3} - mg$$

$$(4) \quad F_{c4} = \frac{mv_4^2}{r} = F_{N4}$$

As the cycle goes around, the magnitude of the normal force changes. It changes because the speed changes and because the weight does not have the same effect at every point on the circle. At the top, the normal force and the weight reinforce each other to provide a centripetal force whose magnitude is $F_{N1} + mg$. At the bottom, in contrast, the normal force and the weight oppose one another, giving a centripetal force of magnitude $F_{N3} - mg$. At points 2 and 4 on either side, only F_{N2} and F_{N4} provide the centripetal force, the weight acting tangent to the circle and, thus, having no component pointing toward the center. If the speed at each of the four points is known, along with the mass and the radius, the normal force at each point can be determined from the equations above.

Riders who perform the loop-the-loop trick know that they must have at least a minimum speed at the top of the circle to remain on the track. This speed can be determined by considering the centripetal force at point 1 in Figure 6.14b:

$$\frac{mv_1^2}{r} = F_{N1} + mg$$

The speed v_1 calculated from this equation is a minimum when F_{N1} is zero, in which case the speed is given by $v_1 = \sqrt{rg}$. At this speed, the track does not exert a normal force to keep the cycle on the circle at point 1, because the weight mg provides all the centripetal force. For a radius of 9.0 m, for example, this expression predicts a minimum speed of 9.4 m/s (21 mi/h). Under these conditions, the rider experiences an apparent weightlessness like that discussed in Section 6.6, because for an instant the rider and the cycle are falling freely toward the center of the circle.

SUMMARY

In **uniform circular motion,** an object of mass m travels at a constant speed v on a circular path of radius r. The **period** T of the motion is the time required to make one complete revolution. The speed, the period, and the radius are related according to $v = 2\pi r/T$. The velocity vector in such motion is always changing direction, and, therefore, an acceleration exists. This acceleration is called **centripetal acceleration,** and its magnitude is $a_c = v^2/r$, while its direction is toward the center of the circle. To create this acceleration, a net force pointing toward the center of the circle is needed. This net force is called the **centripetal force,** and its magnitude is $F_c = mv^2/r$.

There are many examples of uniform circular motion, including the **banking of curves** and the **orbiting of satellites**. The angle θ at which a friction-free curve is banked depends on the radius of the curve and the speed at which the curve is to be negotiated, according to $\tan \theta = v^2/(rg)$. The speed and the period of a satellite in a circular orbit about the earth depend on the radius of the orbit, according

to $v = \sqrt{GM_E/r}$ and $T = 2\pi r^{3/2}/\sqrt{GM_E}$, where G is the universal gravitational constant and M_E is the mass of the earth. **Apparent weightlessness** and **artificial gravity** in satellites can be explained in terms of uniform circular motion. **Motion in a vertical circle** is usually nonuniform, and the centripetal force varies from point to point.

SOLVED PROBLEMS

SOLVED PROBLEM 1
Related Problems: **7 **14 **25

A space station is rotating to create artificial gravity, as the drawing indicates. The rate of rotation is chosen so the outer ring A ($r_A = 2150$ m) simulates the acceleration of gravity on the surface of the planet Venus (8.62 m/s²). (a) How long does it take the space station to turn once around its axis; in other words, what is its period T? (b) What should be the radius r_B, so the inner ring B simulates the acceleration of gravity on the surface of the planet Mercury (3.63 m/s²)?

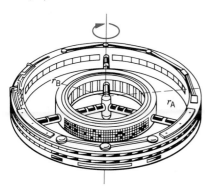

Solution (a) Equation 6.1 ($v_A = 2\pi r/T$) can be used with the given radius r_A to calculate the period T, once the speed v_A is known. This speed can be obtained from Equation 6.2 for centripetal acceleration, using the given acceleration and radius:

$$a_c = \frac{v_A^2}{r_A}$$

$$v_A = \sqrt{a_c r_A} = \sqrt{(8.62 \text{ m/s}^2)(2150 \text{ m})} = 136 \text{ m/s}$$

$$T = \frac{2\pi r_A}{v_A} = \frac{2\pi(2150 \text{ m})}{136 \text{ m/s}} = \boxed{99.3 \text{ s}}$$

(b) The key to solving this part of the problem is to realize that the period is characteristic of the entire space station, since the various sections of the station are rigidly connected and rotate as a unit. Thus, the inner ring B also has a period $T = 99.3$ s. It follows,

then, that

$$T = \frac{2\pi r_A}{v_A} = \frac{2\pi r_B}{v_B}$$

$$r_B = \frac{r_A v_B}{v_A}$$

Considering that r_A and v_A are known, a value for v_B is needed to calculate the radius of the inner ring. The speed v_B is such that the centripetal acceleration of ring B is 3.63 m/s². Therefore,

$$3.63 \text{ m/s}^2 = \frac{v_B^2}{r_B}$$

Solving this expression for v_B in terms of r_B and substituting the result into $r_B = r_A v_B/v_A$ reveals that

$$r_B = \frac{r_A \sqrt{(3.63 \text{ m/s}^2)r_B}}{v_A}$$

Squaring both sides of this expression, we find

$$r_B = \frac{r_A^2(3.63 \text{ m/s}^2)}{v_A^2} = \frac{(2150 \text{ m})^2(3.63 \text{ m/s}^2)}{(136 \text{ m/s})^2} = \boxed{907 \text{ m}}$$

Summary of Important Points There are two important ideas in this problem. One is that the speed in uniform circular motion can always be represented as the circumference of the circle divided by the period. The other is that when a rigid object rotates about an axis, all points on the object have the same period.

SOLVED PROBLEM 2
Related Problems: *30 **31 **32

A 1350-kg sports car is sprinting along the hilly road shown in the drawing. The hill and the dip both have a radius $r = 74.4$ m. The car has a speed of 27.0 m/s at the top of the hill and 30.0 m/s at the bottom of the dip. Find the normal force that the road applies to the car at (a) the top of the hill and (b) the bottom of the dip. (c)

Compare these normal forces and interpret them in the light of common driving experiences.

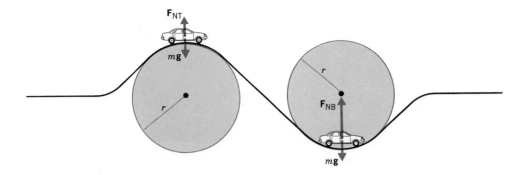

Solution (a) Two sources contribute to the centripetal force at the moment the car crests the hill. One source is the weight mg, which points downward. The other is the normal force $\mathbf{F}_{NT}$, which points upward, exactly opposite to the weight and away from the center of the circle. Since the centripetal force is the net force pointing toward the center of the circle, the magnitude of mg must be greater than the magnitude of $\mathbf{F}_{NT}$. Therefore, the magnitude of the centripetal force at the top of the hill is $mg - F_{NT}$, and Equation 6.3 becomes

$$mg - F_{NT} = \frac{mv^2}{r}$$

The normal force, then, is

$$F_{NT} = mg - \frac{mv^2}{r}$$

$$F_{NT} = (1350 \text{ kg})(9.80 \text{ m/s}^2) - \frac{(1350 \text{ kg})(27.0 \text{ m/s})^2}{74.4 \text{ m}} = \boxed{0}$$

A value of zero for F_{NT} means that the car has lost contact with the ground, and a condition of apparent weightlessness exists. Usually, F_{NT} is not zero at the top of a hill. However, F_{NT} may be zero if the speed and the radius are properly matched ($v^2/r = g$), as is the case here.

(b) At the bottom of the dip, there are again two sources contributing to the centripetal force. The weight mg acts downward, this direction now being away from the center of the circle, and the normal force $\mathbf{F}_{NB}$ points upward. The normal force must exceed the weight, if these two forces acting jointly are to create a net force pointing toward the center of the circle. Therefore,

$$F_{NB} - mg = \frac{mv^2}{r}$$

$$F_{NB} = mg + \frac{mv^2}{r}$$

$$F_{NB} = (1350 \text{ kg})(9.80 \text{ m/s}^2) + \frac{(1350 \text{ kg})(30.0 \text{ m/s})^2}{74.4 \text{ m}}$$

$$= \boxed{29\ 600 \text{ N}}$$

(c) The normal force at the bottom of the dip is over twice the weight of the car ($mg = 13\ 200$ N), while there is no normal force at the top of the hill. These results reflect the common driving experiences of being "pressed" more tightly against the seat when speeding through the bottom of a dip and feeling a "falling-elevator-sensation" when zooming over the top of a hill.

Summary of Important Points The important step in this problem is recognizing that two sources contribute to the centripetal force. It is necessary to identify all such sources and to take into account the direction of each individual force, using vector components if necessary. The centripetal force is the net force pointing toward the center of the circle. A second important point is that the condition known as apparent weightlessness exists when the normal force is zero.

QUESTIONS

1. A car is moving with a constant speed along a circular path. According to the definition of equilibrium given in Section 5.2, is the car in equilibrium? If not, why not?

2. For uniform circular motion, complete the following table by answering "yes" or "no" in the appropriate spaces. Provide a reason for each of your answers.

	Velocity vector	*Acceleration vector*
Constant magnitude?		
Constant direction?		

3. The equations of kinematics (Equations 3.3–3.6) describe the motion of an object that has a constant acceleration. These equations cannot be applied to uniform circular motion. Why not?

4. A car is traveling at 55 mi/h along the road *ABCDE* shown in the drawing. Sections *AB* and *DE* are straight. Rank the accelerations in each of the four sections according to magnitude, listing the smallest first.

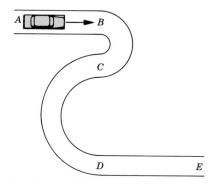

5. At an amusement park there is a rotating circular platform that resembles a large turntable. For the price of a ticket, anyone can try to remain in place without falling off, as the platform picks up speed. Initially each person stands at the same distance from the center. Explain why people with rubber soles on their shoes have an easier time on this ride than people whose shoes have hard leather soles.

6. What are the chances of a light car safely rounding an unbanked curve on an icy road as compared to the chances for a heavy car: worse, the same as, or better? Assume that both cars have the same speed and are equipped with identical tires. Explain.

7. A propeller, operating under test conditions, is being made to rotate at ever faster speeds. (a) What provides the centripetal force that acts on any given atom in the rotating propeller? (b) Explain what is likely to happen, and why, when the maximum rated speed of the propeller is exceeded.

8. The platter on a stereo turntable rotates at $33\frac{1}{3}$ rpm. Where should a penny be placed on the platter so as to experience the largest centripetal force? Give your reasoning.

9. Explain why a real airplane must bank as it flies in a circle, while a model airplane on a guideline can fly in a circle without banking.

10. Would a change in the earth's mass affect (a) the banking of airplanes as they turn, (b) the banking of roadbeds, (c) the speeds with which satellites are put into circular orbits, and (d) the performance of the loop-the-loop motorcycle stunt? In each case, give your reasoning.

11. A space station is in circular orbit about the earth and has no artificial gravity. A book is on a table in this space station. Is any kinetic frictional force encountered when the book slides across the table? Explain.

PROBLEMS

Section 6.2 Centripetal Acceleration

1. Magnetic tape is being spooled from a supply reel to a takeup reel. The tape is moving with a speed of 7.50 in./s, and the radius of the takeup reel is 3.00 in. What is the magnitude and direction of the centripetal acceleration of the tape as it is wound on the outer layer of the takeup reel?

2. A bicycle chain is wrapped around a rear sprocket ($r = 3.9$ cm) and a front sprocket ($r = 10.0$ cm). The chain moves with a speed of 140 cm/s around the sprockets, while the bike moves at a constant velocity. Find the magnitude of the acceleration of a chain link that is in contact with (a) the rear sprocket, (b) neither sprocket, and (c) the front sprocket.

3. Two cars are going around curves, one car traveling at 60.0 mi/h and the other traveling at 30.0 mi/h. Each car experiences a centripetal acceleration of the same magnitude. How do the radii of the two curves compare?

4. (a) How long does it take a plane, traveling at a steady speed of 110 m/s, to fly once around a circle whose radius is 2850 m? (b) What is the magnitude of the centripetal acceleration of the plane?

* 5. The earth rotates once per day about an axis passing through the north and south poles, an axis that is perpendicular to the plane of the equator. Assuming the earth is a sphere with a radius of 6.38×10^6 m, determine the speed and centripetal acceleration of a person situated (a) at the equator and (b) at a latitude of 45.0°.

* 6. A centrifuge is a device in which a small container of material is rotated at a high speed on a circular path. Such a device is used in medical laboratories, for instance, to cause the more dense red blood cells to settle through the less dense blood serum and collect at the bottom of the container. Suppose the centripetal acceleration of the sample is 6.25×10^3 times larger than the acceleration due to gravity. How many revolutions per minute is the sample making, if it is located at a radius of 5.00 cm from the axis of rotation?

**7.   A person is riding a merry-go-round at a distance of 7.00 m from its center. This person experiences a centripetal acceleration of 7.50 m/s². What centripetal acceleration is experienced by another person who is riding at a distance of 3.00 m from the center? *(See Solved Problem 1 for a related problem.)*

Section 6.3 Centripetal Force

8.   A 0.015-kg ball is shot from the plunger of a pinball machine. Because of a centripetal force of 0.028 N, the ball follows a circular arc whose radius is 0.25 m. What is the speed of the ball?

9.   A child is twirling a 12.0-g ball on a string in a horizontal circle whose radius is 10.0 cm. The ball travels once around the circle in 0.500 s. (a) Determine the tension in the string. (b) If the speed is doubled, does the tension double? If not, by what factor does the tension increase?

10.   A model airplane is being flown on a guideline that can sustain at most 180 N of tension. At a speed of 28 m/s, what is the radius of the smallest horizontal circle in which a 0.75-kg plane can be flown?

11.   A car rounds an unbanked curve (radius = 92 m) without skidding at a speed of 26 m/s. What is the smallest possible coefficient of static friction between the tires and the road?

* 12.   A "swing" ride at a carnival consists of chairs that are swung in a circle by 12.0-m cables attached to a vertical rotating pole, as the drawing shows. Suppose the total mass of a chair and its occupants is 220 kg. (a) Determine the tension in the cable attached to the chair. (b) Find the speed of the chair.

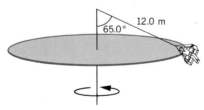

65.0°   12.0 m

* 13.   What is the *minimum* coefficient of static friction necessary to allow a penny to rotate along with a 33⅓-rpm record (diameter = 0.300 m), when the penny is placed anywhere on the record?

**14.   A rigid massless rod is rotated about one end in a horizontal circle. There is a mass m_1 attached to the center of the rod and a mass m_2 attached to the outer end of the rod. The inner section of the rod sustains twice as much tension as the outer section. Find the ratio m_2/m_1. *(See Solved Problem 1 for a related problem.)*

Section 6.4 Banked Curves

15.   At what angle should a curve of radius 150 m be banked, so cars may travel safely at 25 m/s under icy conditions?

16.   A curve of radius 120 m is banked at an angle of 18°. At what speed can it be safely negotiated under icy conditions?

* 17.   There is a similarity between a plane banking into a turn and a car going around a banked curve. The lifting force **L** in Figure 6.8 plays the same role as the normal force $\mathbf{F_N}$ in Figure 6.9. (a) Derive an expression that relates the banking angle to the speed of the plane, the radius of the turn, and the acceleration due to gravity. (b) At what angle with respect to the horizontal should a plane be banked when traveling at 195 m/s around a turn whose radius is 8250 m?

**18.   Refer to problem 17 before attempting to solve this problem. A jet ($m = 2.00 \times 10^5$ kg), flying at 123 m/s, banks to make a horizontal circular turn. The radius of the turn is 3810 m. Calculate the necessary lifting force.

Section 6.5 Satellites in Circular Orbits, Section 6.6 Apparent Weightlessness and Artificial Gravity

19.   A satellite is placed in orbit 6.00×10^5 m above the surface of Jupiter. Jupiter has a mass of 1.90×10^{27} kg and a radius of 7.14×10^7 m. Find the orbital speed of the satellite.

20.   Suppose the surface (radius = r) of the space station in Figure 6.13 is rotating at 80.0 mi/h. What must be the value of r for the astronauts to weigh one-half of their earth weight?

21.   Venus rotates very slowly about its axis, the period being 243 days. The mass of Venus is 4.87×10^{24} kg. Determine the radius for a synchronous satellite in orbit about Venus.

22.   The moon orbits the earth at a distance of 3.85×10^8 m. Assume that this distance is between the centers of the earth and the moon and that the mass of the earth is 5.98×10^{24} kg. Find the period for the moon's motion around the earth. Express the answers in days and compare it to the length of a month.

* 23.   If the earth had no atmosphere and were a perfectly smooth sphere of radius 6.38×10^6 m and mass 5.98×10^{24} kg, what would be the greatest orbital speed that an artificial satellite could have in a circular orbit about the earth?

* 24.   The earth orbits the sun once a year at a distance of 1.50×10^{11} m. Venus orbits the sun at a distance of 1.08×10^{11} m. These distances are between the centers of the planets and the sun. How long does it take for Venus to complete one orbit around the sun?

**25.   To create artificial gravity, the space station shown in the drawing is rotating at a rate of 1.00 rpm. The radii of the cylindrically shaped chambers have the ratio $r_A/r_B = 4.00$. Each chamber A simulates an acceleration due to gravity of 10.0 m/s². Find values for (a) r_A, (b) r_B, and (c) the acceleration due to gravity that is simulated in chamber B. *(See Solved Problem 1 for a related problem.)*

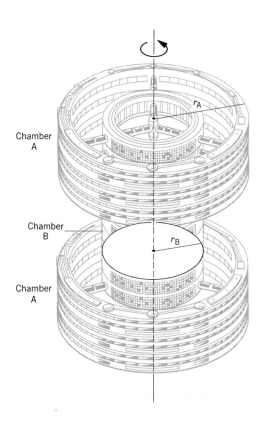

Chamber
A

r_A

Chamber
B

r_B

Chamber
A

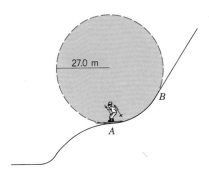

27.0 m

B

A

* 30. A motorcycle is traveling up one side of a hill and down the other side. The crest is a circular arc with a radius of 45.0 m. Determine the maximum speed that the cycle can have while moving over the crest without losing contact with the road. *(See Solved Problem 2 for a related problem.)*

**31. A car is traveling at a constant speed along a hilly road, as the drawing shows. Point A is at the bottom of the dip, while point B is at the top of the hill. The radius of curvature is 35.0 m at both points. The apparent weight of the car at point A exceeds the apparent weight at point B by a factor of four. What is the speed of the car? (Recall from Section 5.4 that the apparent weight of an object is the reading on a scale, that reading being equal to the magnitude of the normal force). *(See Solved Problem 2 for a related problem.)*

Section 6.7 Vertical Circular Motion

26. The condition of apparent weightlessness can be created for a brief instant when a plane flies over the top of a vertical circle. At a speed of 480 mi/h, what is the radius of the vertical circle that the pilot must use?

27. A motorcycle has a constant speed of 25.0 m/s as it passes over the top of a hill whose radius of curvature is 126 m. The mass of the motorcycle and driver is 342 kg. Find the magnitude of (a) the centripetal force and (b) the normal force that acts on the cycle.

28. A roller coaster at an amusement park has a dip that bottoms out in a vertical circle of radius r. Passengers on this ride feel the seat of the car pushing on them with a force equal to twice their weight as they go through the dip. If $r = 20.0$ m, how fast is the roller coaster car traveling at the bottom of the dip?

29. A downhill skier, whose mass is 50.0 kg, attains a speed of 21.0 m/s just as she reaches the point where a jump is necessary (see point A in the drawing). When she leaves the ground, her velocity is horizontal. In other words, point A is at the bottom of the circular arc AB (radius = 27.0 m). Determine the normal force acting on the skis at point A.

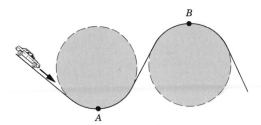

B

A

**32. A car (weight = 4130 lb) is moving at a constant speed over a hill and dip that both have the same radius of curvature (see the drawing for problem 31). The speed and radius are unknown. In passing over the crest of the hill, the car experiences a normal force equal to one-half its weight. Determine the normal force that the car experiences when passing through the bottom of the dip. *(See Solved Problem 2 for a related problem.)*

ADDITIONAL PROBLEMS

33. The earth travels around the sun once a year in an approximately circular orbit whose radius is 1.50×10^{11} m. From these data determine (a) the orbital speed of the earth and (b) the mass of the sun.

34. Speedboat A negotiates a curve whose radius is 120 m. Speedboat B negotiates a curve whose radius is 240 m. Each boat experiences the same centripetal acceleration. What is the ratio v_A/v_B of the speeds of the boats?

35. A satellite circles the earth in an orbit whose radius is twice the earth's radius. The earth's mass is 5.98×10^{24} kg and its radius is 6.38×10^6 m. What is the period of the satellite?

36. A dust particle of mass 1.5×10^{-7} kg is located on the outer edge of a compact disc recording (diameter = 0.12 m). The disc is rotating at 3.5 rev/s. Find the magnitude of the centripetal force that acts on the dust particle.

* 37. A racetrack has the shape of an inverted cone, as the drawing indicates. On this surface the cars race in circles that are parallel to the ground. For a steady speed of 27.0 m/s, at what value of the distance d should a driver locate his car, if he wishes to stay on a circular path without depending on friction?

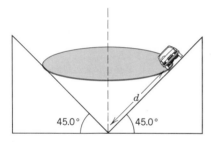

* 38. A demolition ball (weight = 4550 lb) swings at the end of a 50.0-ft cable on the arc of a vertical circle. At the lowest point of the swing, the ball is moving at a speed of 25.0 ft/s. Determine the tension in the cable.

* 39. An unbanked curve in the road has a radius of 75.0 m. The greatest speed that a motorcycle can have without skidding around this curve is 25.0 m/s. Another curve has an identical surface but has a radius of 125 m. What is the maximum speed that the motorcycle can have around the second curve without skidding?

* 40. A helicopter rotor turns at a rate of 315 rpm. The tip of one of the blades moves on a circle whose radius is 7.50 m. (a) What is the speed of the tip of the blade? (b) What is the centripetal acceleration of the tip of the blade? Express the answer in terms of multiples of $g = 9.80$ m/s^2.

**41. At amusement parks, there is a popular ride where the floor of a rotating cylindrical room falls away, leaving the riders "plastered" against the wall (see the drawing). Suppose the radius of the room is 3.30 m and the speed of the wall is 10.0 m/s when the floor falls away. (a) What is the source of the centripetal force acting on the riders? (b) How much centripetal force acts on a 55.0-kg rider? (c) What is the minimum coefficient of static friction that must exist between a rider's back and the wall, if the rider is to remain in place when the floor drops away?

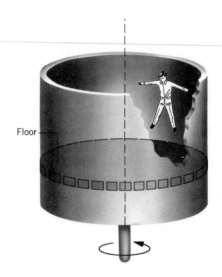

**42. A 125-kg crate rests on the flatbed of a truck that moves at a speed of 15.0 m/s around a curve whose radius is 66.0 m. A second identical crate rests on top of the first crate. The crates do not slip relative to the truck or each other. Obtain the magnitude of the static frictional force that (a) the bottom crate exerts on the top crate and (b) the truck bed exerts on the bottom crate.

CHAPTER 7

Work and Energy

The canoeist is expending a considerable amount of energy to do the work of moving the craft through the water. The relation between work and energy is an important one in physics, as will be seen in the following pages.

7.1 WORK

WORK DONE BY A FORCE THAT POINTS IN THE DIRECTION OF THE MOTION

Work is a familiar idea. For example, it takes work to push a heavy object such as a stalled car. In fact, the more pushing force that is used and the greater the displacement over which the car is moved, the greater the amount of work that is done. Force and displacement, then, are the two essential elements of work, and they are emphasized in Figure 7.1. The drawing shows a situation in which a constant force **F** points in the same direction as the displacement **s** of the car. In such a case, the work W is defined as the magnitude F of the force times the magnitude s of the displacement:

$$\text{Work} = W = Fs \qquad \text{(Force parallel to displacement)}$$

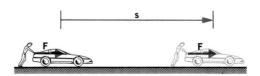

FIGURE 7.1 Work is done in pushing a stalled car along a straight line. The work done equals the product of F, the magnitude of the constant pushing force, and s, the magnitude of the displacement of the car.

TABLE 7.1 Units of Measurement for Work

System	Force	×	Distance	=	Work
SI	newton (N)		meter (m)		joule (J)
BE	pound (lb)		foot (ft)		foot · pound (ft · lb)
CGS	dyne (dyn)		centimeter (cm)		erg

The equation indicates that the unit of work is that of force times distance, or newton · meter in SI units. One newton · meter is referred to as a *joule* (J) (pronounced "jewel"), in honor of James Joule (1818–1889) and his research into the nature of work, energy, and heat. Table 7.1 summarizes the units for work in several systems of measurement.

Work is a scalar quantity, for work is determined by the magnitude of the force and the magnitude of the displacement (the distance) over which the force is applied. The work done, for example, to push a car down a straight road is the same when the car is moved north to south or east to west, provided that the amount of force used and the distance moved are the same. Work does not convey directional information, and, therefore, is not a vector.

Defining work as $W = Fs$ does have one surprising feature. If the distance s is zero, the work is zero, even if a force is applied. Thus, pushing on an immovable object, such as a brick wall, may tire your muscles, but there is no work done of the type we are discussing. In physics, the idea of work is intimately tied up with the idea of motion. If there is no movement of the object, the work done by the force acting on the object is zero.

WORK DONE BY A FORCE THAT POINTS AT AN ANGLE RELATIVE TO THE DIRECTION OF THE MOTION

When the force does not point in the same direction as the displacement of the object, the relation $W = Fs$ must be modified. Figure 7.2a shows a situation where a force is directed at an angle θ relative to the displacement. In such a case, only the component of the force along the displacement is used in defining work. As part *b* of the drawing illustrates, this component is $F \cos \theta$. The more general definition of work is presented below.

DEFINITION OF WORK

The work done on an object by a constant force **F** is

$$W = (F \cos \theta)s \qquad (7.1)$$

where $F \cos \theta$ is the force component along the displacement of the object, s is the magnitude of the displacement, and θ is the angle between the force and the displacement.

SI unit of work: newton · meter = joule (J)

When the force points in the same direction as the displacement, $\theta = 0°$, and Equation 7.1 reduces to $W = Fs$. The following example illustrates how Equation 7.1 is used to calculate work.

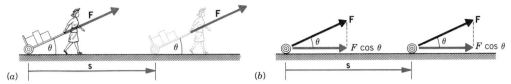

FIGURE 7.2 (*a*) Work can be done by a force **F** that points at an angle θ relative to the displacement **s**. (*b*) If the force is constant, the work equals the product of $F \cos \theta$ and s. The term $F \cos \theta$ is the force component that points along the displacement, and s is the magnitude of the displacement.

EXAMPLE 1

Find the work done by a 45.0-N force in pulling the luggage carrier shown in Figure 7.2a at an angle $\theta = 50.0°$ for a distance $s = 75.0$ m.

$W = (F \cos \theta)s = [(45.0 \text{ N}) \cos 50.0°](75.0 \text{ m}) = \boxed{2170 \text{ J}}$

The answer is expressed in newton · meters or joules (J).

SOLUTION

According to Equation 7.1, the work done on the luggage carrier by the 45.0-N force is

---

The definition of work presented in Equation 7.1 takes into account only the component of the force in the direction of the displacement. It is natural to ask if anything should be done about the force component perpendicular to the displacement. The answer is no. To do work, there must be a force *and* a displacement, and since there is no displacement in the perpendicular direction, there is no work produced by the perpendicular component of the force.

POSITIVE AND NEGATIVE WORK

Work can be either positive or negative, depending on whether a component of the force points in the same direction as the displacement or in the opposite direction. Example 2 illustrates how positive and negative work arise.

EXAMPLE 2

The weight lifter in Figure 7.3 is bench-pressing a barbell whose weight is 710 N. He raises the barbell a distance of 0.65 m above his chest and then lowers the barbell back down. The weight is raised and lowered at a constant velocity. Determine the work done on the barbell by the weight lifter during (a) the lifting phase and (b) the lowering phase.

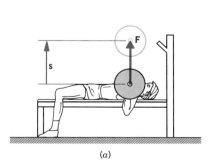

(*a*)

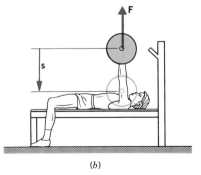

(*b*)

FIGURE 7.3 (*a*) During the lifting phase of a bench press, the force **F** does positive work on the barbell. (*b*) During the lowering phase, the force does negative work.

SOLUTION

(a) The barbell is being lifted at a constant velocity and, therefore, is in equilibrium. Consequently, the force **F** exerted on the barbell by the weight lifter must balance the weight of the barbell, so $F = 710$ N. As part *a* of the drawing shows, the force and the displacement are in the same direction. The angle between them is $\theta = 0°$, and the work done by the force **F** is

$$W = (F \cos \theta)s = [(710 \text{ N}) \cos 0°](0.65 \text{ m}) \quad (7.1)$$
$$= \boxed{460 \text{ J}}$$

(b) When the barbell is lowered, the force and the displacement are in opposite directions, as part *b* of the drawing indicates. The angle between the force and displacement is now $\theta = 180°$, and the work is

$$W = (F \cos \theta)s = [(710 \text{ N}) \cos 180°](0.65 \text{ m}) = \boxed{-460 \text{ J}}$$

since $\cos 180° = -1$. The work is negative, because the force is opposite to the displacement. Weight lifters call each complete up-and-down movement of the barbell a repetition, or "rep." The lifting of the weight is referred to as the positive part of the rep, and the lowering is known as the negative part of the rep.

As Example 2 illustrates, work can be positive or negative. If the force has a component in the *same* direction as the displacement of the object, the work done by the force is *positive*. On the other hand, if a force component points in a direction *opposite* to the displacement, the work is *negative*. If the force is perpendicular to the displacement, the force has no component in the direction of the displacement, and the work is zero. This latter situation corresponds to $\theta = 90°$ in Equation 7.1, which yields $W = 0$. The next example deals with the work done by a static frictional force when it accelerates an object.

EXAMPLE 3

Figure 7.4*a* shows a 120-kg crate sitting on the flatbed of a truck that is moving with an acceleration of $a = +1.5 \text{ m/s}^2$ along the positive *x* axis. The crate does not slip with respect to the truck. What is the work done on the crate when the truck has moved a distance of $s = 65$ m?

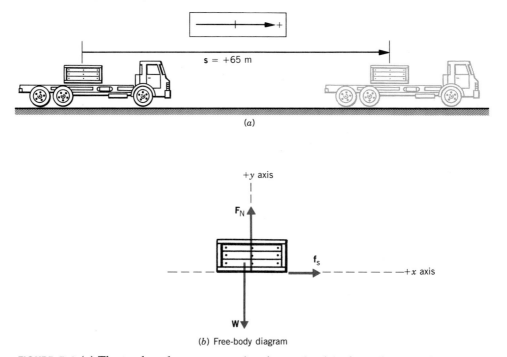

(a)

(b) Free-body diagram

FIGURE 7.4 (*a*) The truck and crate are accelerating to the right for a distance of $s = 65$ m. (*b*) The free-body diagram for the crate. Only the static frictional force $\mathbf{f_s}$ does work on the crate.

SOLUTION

To determine the work done on the crate, it is necessary to find the net force exerted on the crate in the direction of the displacement. Part *b* of the drawing shows the free-body diagram for the crate. The weight **W** of the crate and the normal force $\mathbf{F_N}$ are directed perpendicular to the displacement, so they do no work on the crate. The static frictional force $\mathbf{f_s}$ between the truck bed and the crate is responsible for accelerating the crate, since it is the only force acting on the crate in the *x* direction. Since the crate does not slip, it has the same acceleration as the truck, and the magnitude of the static frictional force can be obtained from Newton's second law:

$$f_s = ma = (120 \text{ kg})(1.5 \text{ m/s}^2) = 180 \text{ N}$$

The work done on the crate by the static frictional force is

$$W = (f_s \cos \theta)s = (180 \text{ N})(\cos 0°)(65 \text{ m}) \qquad (7.1)$$
$$= \boxed{1.2 \times 10^4 \text{ J}}$$

The work is positive, because the frictional force is in the same direction as the displacement. If the truck and crate were slowing down (decelerating), the static frictional force would point *opposite* to the displacement, and the work done by $\mathbf{f_s}$ would be *negative*.

7.2 THE WORK–ENERGY THEOREM AND KINETIC ENERGY

Most people expect that if you do work, you should get something as a result. In physics, when a net force performs work on an object, there is always a result from the effort. The result is a change in the *kinetic energy* of the object. The important relationship that relates work to the change in kinetic energy is known as the *work–energy theorem.*

To gain some insight into the idea of kinetic energy and the work–energy theorem, look at Figure 7.5. This picture shows a moving object of mass *m* being acted upon by a constant net force **F**. The net force is the vector sum of all the forces acting on the object. For simplicity, the direction of the net force is assumed to be parallel to the displacement **s**. According to Newton's second law, the net force produces an acceleration whose magnitude *a* is given by $a = F/m$. Consequently, the speed of the object changes from an initial value of v_0 to a final value of v_f. Multiplying both sides of $F = ma$ by the distance *s* gives

$$Fs = mas$$

The term *as* on the right side can be related to v_0 and v_f by using Equation 2.9 $(v_f^2 = v_0^2 + 2as)$, which is one of the equations of kinematics. Solving this equation for *as* and substituting the result into $Fs = mas$ gives

$$\underbrace{Fs}_{\text{Work}} = \underbrace{\tfrac{1}{2}mv_f^2}_{\substack{\text{Final} \\ \text{kinetic} \\ \text{energy}}} - \underbrace{\tfrac{1}{2}mv_0^2}_{\substack{\text{Initial} \\ \text{kinetic} \\ \text{energy}}}$$

This expression is the work–energy theorem. Its left side is the work *W* done by the net force, while its right side involves the difference between two terms, each of which has the form $\tfrac{1}{2}(\text{mass})(\text{speed})^2$. The quantity $\tfrac{1}{2}(\text{mass})(\text{speed})^2$ is called "kinetic energy" and plays a significant role in physics, as we will soon see.

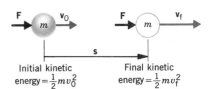

DEFINITION OF KINETIC ENERGY

The kinetic energy KE of an object with mass *m* and speed *v* is given by

$$KE = \tfrac{1}{2}mv^2 \qquad (7.2)$$

SI unit of kinetic energy: joule (J)

FIGURE 7.5 When a constant net force **F** acts over a displacement **s**, the force does work on the object. As a result of the work done, the kinetic energy of the object increases from an initial value of $\tfrac{1}{2}mv_0^2$ to a final value of $\tfrac{1}{2}mv_f^2$.

The SI unit of kinetic energy is the same as the unit for work, namely, the joule. Kinetic energy, like work, is a scalar quantity. These observations are not surprising, for work and kinetic energy are closely related, as is clear from the following statement of the work–energy theorem.

THE WORK–ENERGY THEOREM

When a net force does work W on an object, the kinetic energy of the object changes from its initial value of KE_0 to a final value of KE_f, the difference between the two values being equal to the work:

$$W = KE_f - KE_0 = \tfrac{1}{2}mv_f^2 - \tfrac{1}{2}mv_0^2 \qquad (7.3)$$

The work–energy theorem may be derived for any direction of the force relative to the displacement, not just the situation depicted in Figure 7.5. In fact, the force may even vary from point to point along a path that is curved rather than straight, and the theorem remains valid. According to the work–energy theorem, a moving object has kinetic energy, because work was done to accelerate the object from rest to a speed v. Conversely, an object with kinetic energy can perform work, if it is allowed to push or pull on another object.

Example 4 illustrates the work–energy theorem and considers a single force that does work to change the kinetic energy of a space probe.

EXAMPLE 4

A 5.00×10^4-kg space probe is traveling at a speed of $v_0 = 1.10 \times 10^4$ m/s through deep space. No forces act on the probe except that generated by its own engine. The engine exerts a constant force of 4.00×10^5 N, directed parallel to the displacement (Figure 7.6). The engine fires continually while the probe moves in a straight line for a distance of 2.50×10^6 m. Determine the final speed of the probe.

SOLUTION

The final speed of the probe can be obtained from the work–energy theorem, but first the work done on the probe by the engine must be calculated:

$$W = (F \cos \theta)s \qquad (7.1)$$

$$= [(4.00 \times 10^5 \text{ N}) \cos 0°](2.50 \times 10^6 \text{ m})$$

$$= 1.00 \times 10^{12} \text{ J}$$

The work is positive, because the force and displacement are in the same direction (see the drawing). Since $W = KE_f - KE_0$ according to the work–energy theorem, the final kinetic energy of the probe is

$$KE_f = W + KE_0$$

$$= (1.00 \times 10^{12} \text{ J}) + \tfrac{1}{2}(5.00 \times 10^4 \text{ kg})(1.10 \times 10^4 \text{ m/s})^2$$

$$= 4.03 \times 10^{12} \text{ J}$$

The final kinetic energy is $KE_f = \tfrac{1}{2}mv_f^2$, so the final speed is

$$v_f = \sqrt{\frac{2(KE_f)}{m}} = \sqrt{\frac{2(4.03 \times 10^{12} \text{ J})}{(5.00 \times 10^4 \text{ kg})}} = \boxed{1.27 \times 10^4 \text{ m/s}}$$

Since the force of the engine does positive work, the final speed of the probe is greater than its initial speed, in accord with the work–energy theorem.

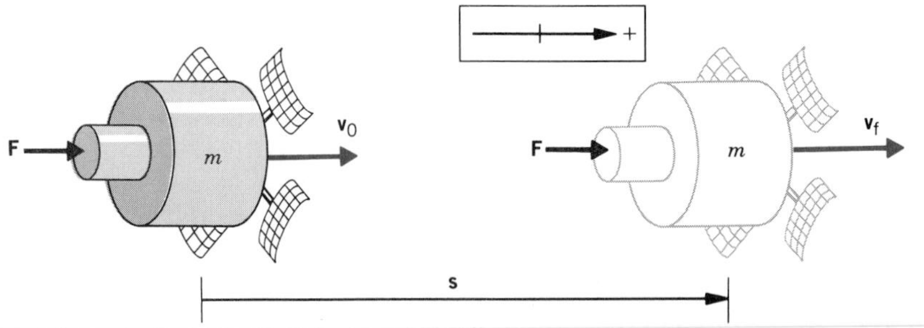

FIGURE 7.6 A space probe has a mass m. The engine of the probe generates a force **F** that points in the same direction as the displacement **s**. The force performs positive work, causing the craft to gain kinetic energy.

In Example 4 only one force, that of the engine, does work on the space probe. If several forces act on an object, they must be added together vectorially to give the net force. The work done by the net force can then be related to the change in the object's kinetic energy by using the work–energy theorem, as in the next example.

EXAMPLE 5

A 58-kg skier is coasting down a 25° slope, as Figure 7.7a shows. A kinetic frictional force of magnitude $f_k = 70$ N opposes her motion. Near the top of the slope, the skier's speed is $v_0 = 3.6$ m/s. Ignoring air resistance, determine the speed v_f at a point that is displaced 57 m downhill.

SOLUTION

The final speed of the skier can be calculated from the work–energy theorem, provided the work done by the net force acting on the skier is known. The free-body diagram in part b of the drawing shows all the forces acting on the skier. The normal force $\mathbf{F_N}$ and the component of the skier's weight perpendicular to the slope, $mg \cos 25°$, do no work, because they are perpendicular to the displacement $\mathbf{s}$. Only the forces along the displacement do work. The net force along the displacement is

$$\mathbf{F} = +mg \sin 25° - f_k = (58 \text{ kg})(9.80 \text{ m/s}^2) \sin 25° - 70 \text{ N}$$
$$= +170 \text{ N}$$

The work done by the net force is

$$W = (F \cos \theta)s = [(170 \text{ N}) \cos 0°](57 \text{ m}) = 9700 \text{ J} \quad (7.1)$$

The work is positive, because the net force and the displacement point in the same direction. From the work–energy theorem $(W = \text{KE}_f - \text{KE}_0)$, it follows that the final kinetic energy of the skier is

$$\text{KE}_f = W + \text{KE}_0 = 9700 \text{ J} + \tfrac{1}{2}(58 \text{ kg})(3.6 \text{ m/s})^2 = 10\ 100 \text{ J}$$

Since the final kinetic energy is $\text{KE}_f = \tfrac{1}{2}mv_f^2$, the final speed of the skier is

$$v_f = \sqrt{\frac{2(\text{KE}_f)}{m}} = \sqrt{\frac{2(10\ 100 \text{ J})}{58 \text{ kg}}} = \boxed{19 \text{ m/s}}$$

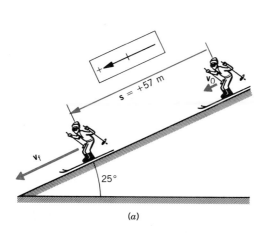

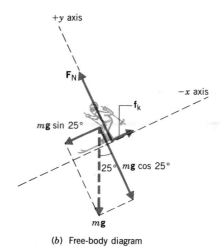

(a) (b) Free-body diagram

FIGURE 7.7 (a) A skier, coasting downhill, has a displacement of $\mathbf{s} = +57$ m. (b) The free-body diagram for the skier.

Example 5 emphasizes that the work–energy theorem deals with the work done by the *net force* when a number of forces act on an object. The work–energy theorem does *not* apply to the work done by an individual force, unless that force happens to be the only one present, in which case it is the net force. If the work done by the net force is *positive,* as in Example 5, the kinetic energy of the object *increases.* If the work done by the net force is *negative,* the kinetic energy *decreases.*

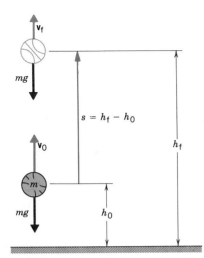

FIGURE 7.8 Gravity exerts a force of magnitude mg on a basketball of mass m. When the ball is moving upward, the force of gravity performs negative work on it, since the force and displacement are in opposite directions.

7.3 GRAVITATIONAL POTENTIAL ENERGY

WORK DONE BY THE FORCE OF GRAVITY

The gravitational force is a well-known force that can do positive or negative work. Figure 7.8 helps to show how the work done by this force can be determined. This drawing depicts a basketball of mass m moving vertically upward, the force of gravity mg being the only force acting on the ball. The initial height of the ball is h_0, and the final height is h_f, both distances measured from the earth's surface. The displacement **s** is directed upward and has a magnitude of $s = h_f - h_0$. When calculating the work W_{gravity} done on the ball by the force of gravity, we use Equation 7.1 with $\theta = 180°$, since the force and displacement are in opposite directions:

$$W = (F \cos \theta)s \qquad (7.1)$$

$$W_{\text{gravity}} = (mg \cos 180°)(h_f - h_0) = -mg(h_f - h_0) \qquad (7.4)$$

Equation 7.4 is valid for *any path* taken between the initial and final positions, and not just for the straight-up path shown in Figure 7.8. For example, the same expression can be derived for the three additional paths shown in Figure 7.9. Thus, only the *change in vertical distance* $(h_f - h_0)$ need be considered when calculating the work done by gravity. Since the change in the vertical distance is the same for each path in the drawing, the work done by gravity is the same in each case.

In Equation 7.4 only the difference between h_f and h_0 appears. Therefore, the vertical distances themselves need not be measured from the earth. For instance, they could be measured relative to a zero level that is one meter above the ground, and $h_f - h_0$ would still have the same value. It is assumed here that the object remains close to the surface of the earth, so we can ignore the dependence of g on height and use the value of $g = 9.80 \text{ m/s}^2$. Example 6 illustrates how the work done by gravity is used in conjunction with the work–energy theorem.

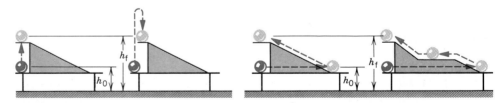

FIGURE 7.9 An object can be moved along any number of different paths in going from an initial height of h_0 to a final height of h_f. In each case, the work done by the gravitational force is the same $[W_{\text{gravity}} = -mg(h_f - h_0)]$, since the change in vertical height $(h_f - h_0)$ is the same.

EXAMPLE 6

An arrow ($m = 0.0500$ kg) is fired vertically from a height of $h_0 = 5.0$ m above the ground and reaches a height of $h_f = 51.0$ m. Ignoring air resistance, determine (a) the speed v_0 with which the arrow was fired and (b) the speed the arrow has after falling back to a height of 20.0 m above the ground.

SOLUTION

(a) Part *a* of Figure 7.10 shows the arrow on its upward flight, when only the force of gravity acts on it. According to Equation 7.4, the work done by gravity is

$$W_{\text{gravity}} = -mg(h_f - h_0)$$

$$= -(0.0500 \text{ kg})(9.80 \text{ m/s}^2)(51.0 \text{ m} - 5.0 \text{ m})$$

$$= -22.5 \text{ J}$$

The initial speed v_0 of the arrow follows from an application of the work–energy theorem, $W = KE_f - KE_0$. This theorem indicates that $W = -KE_0$, since the final speed and kinetic energy are zero at the arrow's highest point. As a result, $W = -\frac{1}{2}mv_0^2$ and

$$v_0 = \sqrt{\frac{-2W}{m}} = \sqrt{\frac{-2(-22.5 \text{ J})}{0.0500 \text{ kg}}} = \boxed{30.0 \text{ m/s}}$$

(b) Part b of the drawing shows the arrow on its downward flight. The calculation here proceeds in the same fashion as in part (a), except now the initial position is at the top of the flight path, so $h_0 = 51.0$ m and $h_f = 20.0$ m:

$$W_{\text{gravity}} = -mg(h_f - h_0)$$

$$= -(0.0500 \text{ kg})(9.80 \text{ m/s}^2)(20.0 \text{ m} - 51.0 \text{ m})$$

$$= +15.2 \text{ J}$$

The difference in heights, $h_f - h_0$, is now a negative number (-31.0 m), because h_f is smaller than h_0, as part b of the picture shows. The fact that the work is now positive means that the force of gravity and the displacement are in the *same* direction (downward) as the arrow falls back to earth. Since the initial speed and kinetic energy are zero at the top of the flight path, the work–energy theorem indicates that $W = KE_f - KE_0 = KE_f = \frac{1}{2}mv_f^2$. The final speed at a height of 20.0 m is

$$v_f = \sqrt{\frac{2W}{m}} = \sqrt{\frac{2(15.2 \text{ J})}{0.0500 \text{ kg}}} = \boxed{24.7 \text{ m/s}}$$

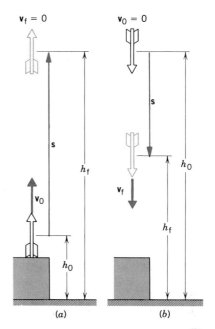

FIGURE 7.10 (*a*) An arrow, fired upward with an initial speed v_0, reaches maximum height with a final speed of zero. The displacement **s** of the arrow points upward. (*b*) Starting at the top of its flight path, where its initial speed is zero, the arrow falls back toward its point of origin. Here the displacement **s** points downward. Notice in the two parts of the drawing that h_0 and h_f have different values and are not drawn to scale.

GRAVITATIONAL POTENTIAL ENERGY

Kinetic energy is not the only kind of energy an object can have. An object can possess energy by virtue of position, even if the object has no kinetic energy. To see why such a "positional" energy can exist, let us reconsider the work–energy theorem in the light of Equation 7.4, which gives the work done by the force of gravity. Along with the force of gravity, suppose there are other forces acting. Such other forces could include frictional, tension, and thrust forces, for instance. Let the work done by the forces *other than gravity* be labeled as W_{other}. The work W done by all the forces (including gravity) can then be written as the sum of two terms,

$$W = W_{\text{gravity}} + W_{\text{other}} = -mg(h_f - h_0) + W_{\text{other}}$$

and the work–energy theorem takes the following form:

$$W = -mg(h_f - h_0) + W_{\text{other}} = \frac{1}{2}mv_f^2 - \frac{1}{2}mv_0^2$$

The work done by the gravitational force can be moved to the right side of the equation, with the result that

$$W_{\text{other}} = (\tfrac{1}{2}mv_f^2 - \tfrac{1}{2}mv_0^2) + (mgh_f - mgh_0) \qquad (7.5)$$

Look closely at Equation 7.5; it states that W_{other}, the work done by forces other than gravity, can produce two kinds of changes. First, W_{other} can change the kinetic energy from its initial to its final value. Second, W_{other} can change the quantity mgh from its initial value of mgh_0 to its final value of mgh_f. Thus, the quantity mgh, like $\frac{1}{2}mv^2$, is a kind of energy. The quantity mgh is called the ***gravitational potential energy.***

DEFINITION OF GRAVITATIONAL POTENTIAL ENERGY

The gravitational potential energy PE is the energy that an object of mass m has by virtue of position above the surface of the earth, that position being measured by the height h of the object relative to an arbitrary zero level:

$$PE = mgh \qquad (7.6)$$

SI unit of gravitational potential energy: joule (J)

Gravitational potential energy, like work and kinetic energy, is a scalar quantity, and has the same SI unit as they do, namely, the joule. The work–energy theorem, as expressed in Equation 7.5, shows that only the *difference* between two potential energies is significant. Therefore, the zero level for the heights can be taken anywhere, as long as both h_0 and h_f are measured relative to the same zero level. The gravitational potential energy depends on both the object and the earth (m and g, respectively), as well as the height h of the object above the surface of the earth. Therefore, the gravitational potential energy is an energy that is possessed by the object–earth system, although one often speaks of the object alone as possessing the gravitational potential energy.

Using the definition above of gravitational potential energy, we can write Equation 7.5 in a form that will be useful later on:

$$W_{\text{other}} = \underbrace{(KE_f - KE_0)}_{\substack{\text{Change in} \\ \text{kinetic energy}}} + \underbrace{(PE_f - PE_0)}_{\substack{\text{Change in} \\ \text{gravitational} \\ \text{potential energy}}} \qquad (7.7a)$$

This equation contains the difference in the kinetic energies and the difference in the potential energies. It is customary to use the delta symbol (Δ) to denote such differences; thus, $\Delta KE = (KE_f - KE_0)$ and $\Delta PE = (PE_f - PE_0)$. With the delta notation, the work–energy theorem takes the form

$$W_{\text{other}} = \Delta KE + \Delta PE \qquad (7.7b)$$

Equation 7.7b emphasizes that W_{other}, the work done by forces other than gravity, can change either the kinetic energy, or the potential energy, or both, depending on how the work is done. Example 7 illustrates the effect of W_{other} on changing the kinetic and potential energies.

EXAMPLE 7

A Fourth-of-July rocket (0.200 kg) is launched from rest and follows an erratic flight path to reach the point P, as Figure 7.11 shows. Point P is 30.0 m above the starting point. In the process, $+425$ J of work is done on the rocket by the burning chemical propellant. Ignoring air resistance and the amount of mass lost due to the burning propellant, find the speed v_f of the rocket at the point P.

SOLUTION

The only force acting on the rocket other than gravity is the force generated by the burning propellant, and the work done by this force is $W_{\text{other}} = +425$ J. According to Equation 7.7b, $W_{\text{other}} = \Delta\text{KE} + \Delta\text{PE}$, so that $\Delta\text{KE} = W_{\text{other}} - \Delta\text{PE}$. The change in the gravitational potential energy is

$$\Delta\text{PE} = mg(h_f - h_0) = (0.200 \text{ kg})(9.80 \text{ m/s}^2)(30.0 \text{ m}) = 58.8 \text{ J}$$

Note that we did not use separate values for h_f and h_0. All that is necessary is $(h_f - h_0)$, and this difference is given as 30.0 m in Figure 7.11. The change in kinetic energy can now be obtained:

$$\Delta\text{KE} = W_{\text{other}} - \Delta\text{PE} = 425 \text{ J} - 58.8 \text{ J} = 366 \text{ J}$$

Because the initial speed of the rocket is zero, the initial kinetic energy of the rocket is also zero, and it follows that $\Delta\text{KE} = \frac{1}{2}mv_f^2 - \frac{1}{2}mv_0^2 = \frac{1}{2}mv_f^2$. The final speed of the rocket is

$$v_f = \sqrt{\frac{2(\Delta\text{KE})}{m}} = \sqrt{\frac{2(366 \text{ J})}{0.200 \text{ kg}}} = \boxed{60.5 \text{ m/s}}$$

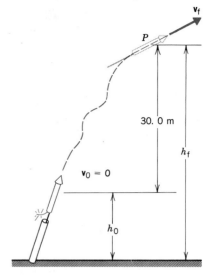

FIGURE 7.11 A Fourth-of-July rocket, moving along an erratic flight path, reaches a point P that is located 30.0 m above the launch point.

7.4 THE CONSERVATION OF MECHANICAL ENERGY

The work–energy theorem has introduced us to kinetic energy and potential energy. The sum of these two kinds of energy is called the ***total mechanical energy*** E, so that $E = \text{KE} + \text{PE}$. The work–energy theorem of Equation 7.7a can be expressed in terms of the total mechanical energy as

$$W_{\text{other}} = (\text{KE}_f - \text{KE}_0) + (\text{PE}_f - \text{PE}_0) \tag{7.7a}$$
$$= \underbrace{(\text{KE}_f + \text{PE}_f)}_{= E_f} - \underbrace{(\text{KE}_0 + \text{PE}_0)}_{= E_0}$$

or

$$W_{\text{other}} = E_f - E_0 \tag{7.8}$$

Equation 7.8 states that W_{other}, the work done by forces other than gravity, changes the total mechanical energy from an initial value of E_0 to a final value of E_f.

The reason we put the work–energy theorem into the form $W_{\text{other}} = E_f - E_0$ is that its conciseness allows an important basic principle of physics to stand out. To see how this principle arises, suppose that only the gravitational force does work on the object. Thus, the net work done by all other forces is zero, and $W_{\text{other}} = 0$. Then, Equation 7.8 reduces to

$$E_f = E_0 \tag{7.9}$$

This result indicates that the final mechanical energy is equal to the initial mechanical energy. Consequently, the total mechanical energy *remains constant at all points along the path between the initial and final points,* never varying from the initial value of E_0. A quantity, such as the total mechanical energy, that stays constant throughout the motion is said to be "conserved." The fact that the total mechanical energy is conserved when $W_{other} = 0$ is called the *principle of conservation of mechanical energy.*

THE PRINCIPLE OF CONSERVATION OF MECHANICAL ENERGY
The total mechanical energy ($E = KE + PE$) of an object remains constant as the object moves, provided that no net work is done by forces other than gravity.

The principle of conservation of mechanical energy offers penetrating insight into the way in which the physical universe operates. While the sum of the kinetic and potential energies at any point is conserved, the two forms may be interconverted or transformed into one another. Kinetic energy of motion is converted into potential energy of position, for instance, when a moving object coasts up a hill. Conversely, potential energy of position is converted into kinetic energy of motion when an object above the earth's surface is allowed to fall. Figure 7.12 illustrates such transformations of energy for a bobsled run, assuming that frictional forces and wind resistance can be ignored. The normal force, being directed perpendicular to the path, does no work. Only the force of gravity does work, so the total mechanical energy E remains constant at all points along the run. The conservation principle is well known for the ease with which it can be applied, as in the next example.

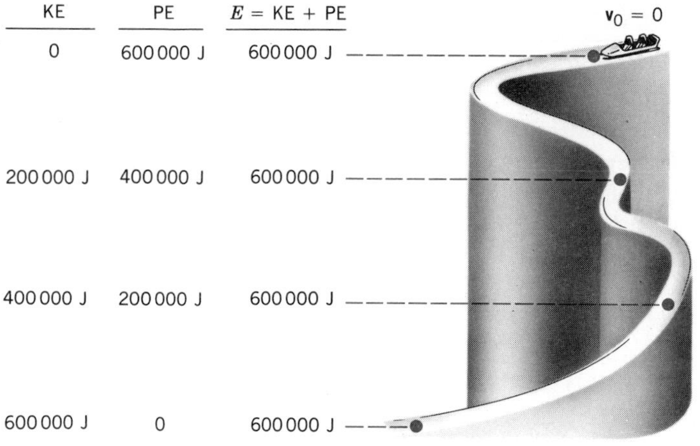

KE	PE	$E = KE + PE$	
0	600 000 J	600 000 J	$v_0 = 0$
200 000 J	400 000 J	600 000 J	
400 000 J	200 000 J	600 000 J	
600 000 J	0	600 000 J	

FIGURE 7.12 If friction and wind resistance are ignored, a bobsled run provides an example of how kinetic energy and potential energy can be interconverted, while the total mechanical energy remains constant at each point along the run. The sled starts from rest at the top with 600 000 J of mechanical energy, all potential energy. The 600 000 J of mechanical energy remains constant as the sled moves down the run.

EXAMPLE 8

A motorcycle rider is trying to leap across the canyon shown in Figure 7.13 by driving horizontally off the cliff. When it leaves the cliff, the cycle has a speed of 55.0 m/s. Ignoring air resistance, find the speed with which the cycle strikes the ground on the other side.

SOLUTION

Once the cycle leaves the cliff, there are no forces acting on the cycle other than gravity, since air resistance is being ignored. Thus, $W_{other} = 0$, and the principle of conservation of mechanical energy applies:

$$\underbrace{\tfrac{1}{2}mv_f^2 + mgh_f}_{E_f} = \underbrace{\tfrac{1}{2}mv_0^2 + mgh_0}_{E_0}$$

The mass m of the rider and cycle can be eliminated algebraically from this equation, since m appears as a factor in every term:

$$\tfrac{1}{2}v_f^2 + (9.80 \text{ m/s}^2)(35.0 \text{ m})$$
$$= \tfrac{1}{2}(55.0 \text{ m/s})^2 + (9.80 \text{ m/s}^2)(70.0 \text{ m})$$

Solving for v_f gives $\boxed{v_f = 60.9 \text{ m/s}}$.

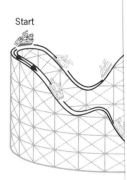

thrown into the air and returns to its starting point with the same k
originally. *It is a property of any conservative force that the work it*
zero when the object moves around any closed path, starting and fi
point.

Not all forces are conservative forces. A force is nonconservativ
on a moving object depends on the path of the motion. The kinetic
good example of a nonconservative force. When an object slides
kinetic frictional force points opposite to the sliding motion and
work equal to the magnitude of the kinetic frictional force multipli
the path. Between any two points, greater amounts of work are
paths between the points. The work, thus, depends on the choice
kinetic frictional force is a nonconservative force.

For a closed path, the total work done by a nonconservative forc
for a conservative force. In Figure 7.15, for instance, a frictional
reality. The frictional force is always directed opposite to the motio
the car. Unlike gravity, friction does negative work on the car thr
trip, on *both* the up and down parts of the motion. Assuming the ca
the starting point, the car will have *less* kinetic energy than it had o
mechanical energy is *not* conserved when a nonconservative forc

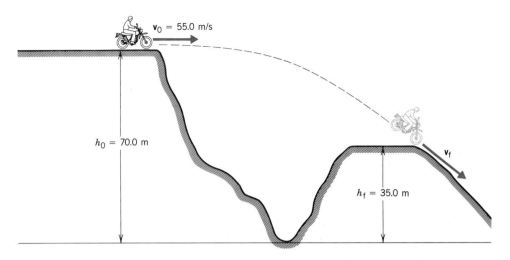

FIGURE 7.13 A motorcycle rider jumping a canyon. With the aid of the principle of conservation of mechanical energy, the speed v_f on impact can be determined from the data given.

Example 9 illustrates that the principle of conservation of mechanical energy can be applied even when forces act perpendicular to the path of a moving object.

7.6 POWER

In many situations of practical interest, the time it takes to do work
the amount of work that is done. When a construction compan
moving equipment and cranes, for instance, both matters are co
machines are needed that will do a large amount of work in the
time. As an aid in considering the relationship between work and ti
power is useful, because power is the work done per unit of time.

DEFINITION OF AVERAGE POWER
Average power $\overline{P}$ is the average rate at which work W is done, and
dividing W by the time t required to perform the work:

$$\overline{P} = \frac{\text{Work}}{\text{Time}} = \frac{W}{t}$$

SI unit of power: joule/s = watt (W)

EXAMPLE 9
A 6.00-m rope is tied to a tree limb and used as a swing. A person starts from rest with the rope held in a horizontal orientation, as in Figure 7.14. Ignoring friction and air resistance, determine how fast the person is moving at the lowest point on the circular arc of the swing.

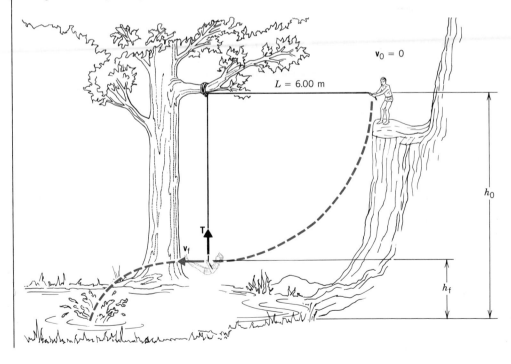

FIGURE 7.14 A person swinging on a rope. The tension **T** in the rope acts perpendicular to the circular arc and, hence, does no work on the person. Therefore, the principle of conservation of mechanical energy applies, and the speed v_f can be determined.

SOLUTION

First let us decide whether the conservation of mechanical energy applies to this problem. After all, there is a force other than gravity that acts on the person, namely, the tension $\mathbf{T}$ in the rope. This force, however, does not do any work, because it points perpendicular to the circular path of the motion. Thus, $W_{other} = 0$, and the conservation principle is applicable:

$$\underbrace{\tfrac{1}{2}mv_f^2 + mgh_f}_{E_f} = \underbrace{\tfrac{1}{2}mv_0^2 + mgh_0}_{E_0}$$

As in
Since

v_f

where
ing a

7.5 CONSERVATIVE

The gravitational force has th
from one place to another, the
of path. In Figure 7.9, for insta
height h_f along several differe
gravity depends only on the
between these heights. Moreo
with each height, PE = mgh, s
the difference between the ini

$$W_{gravity} = -mg$$

If the work done by a force
dent of the path of the motion, t
gravitational force is a conserv
examples of conservative force
force between electrically char
tial energy can be associated
depends only on the position, a
an object from one position to
the final potential energy.

A conservative force has an
nature of the work done by the
the influence of gravity. The we
starting and ending points of th
are the same, so that PE$_0$ = PE
for $W_{gravity} = $ PE$_0 -$ PE$_f = 0$.
coaster car racing through dip
point. Assume that there is no
Of course, the track exerts a no
ular to the motion and, hence
force that does work on the ca:
force does positive work, inc
upward parts of the motion, t
the car's kinetic energy. Over
positive work as negative wor
net work is zero. The car retu
had to begin with. This is the

Since both work and time are scalar quantities, power is also a scalar quantity. The unit in which power is expressed is that of work divided by time, or a joule per second in SI units. One joule per second is called a "watt" (W), in honor of James Watt (1736–1819), the developer of the steam engine. The unit of power in the BE system is the foot-pound per second (ft · lb/s), although the familiar horsepower (hp) unit is frequently used for specifying the power generated by electric motors and internal combustion engines:

$$1 \text{ horsepower} = 550 \text{ foot} \cdot \text{pounds/second} = 746 \text{ watts}$$

Table 7.2 summarizes the units for power in the various systems of measurement.

Frequently, an alternative expression for power is used. This expression can be obtained from Equation 7.1 for the work W done when a constant net force of magnitude F points in the same direction as the displacement, $W = (F \cos 0°)s = Fs$. Dividing both sides of this expression by t, the time it takes the force to move the object through the distance s, gives

$$\frac{W}{t} = \frac{Fs}{t}$$

Recognizing W/t as the average power $\overline{P}$, and s/t as the average speed $\overline{v}$, we have

$$\overline{P} = F\overline{v} \tag{7.11}$$

TABLE 7.2 Units of Measurement for Power

System	Work	÷ Time	= Power
SI	joule (J)	second (s)	watt (W)
BE	foot · pound (ft · lb)	second (s)	foot · pound per second (ft · lb/s)
CGS	erg	second (s)	erg per second (erg/s)

The next example illustrates the use of Equation 7.11.

EXAMPLE 10

A 1.10×10^3-kg car, starting from rest, accelerates for 5.00 s. The magnitude of the acceleration is $a = 4.60$ m/s^2. Determine the average power generated by the net force that accelerates the vehicle.

SOLUTION

To calculate the average power using $\overline{P} = F\overline{v}$, we need to find the magnitude F of the net force and the average speed $\overline{v}$ of the car. The magnitude of the net force can be obtained from Newton's second law:

$$F = ma = (1.10 \times 10^3 \text{ kg})(4.60 \text{ m/s}^2) = 5060 \text{ N}$$

Since the car starts from rest ($v_0 = 0$) and has a constant acceleration, the average speed $\overline{v}$ of the car is one-half of its final speed v:

$$\overline{v} = \tfrac{1}{2}(v_0 + v) = \tfrac{1}{2}v \tag{2.6}$$

Because the initial speed of the car is zero, the final speed of the car after 5.00 s is the product of its acceleration and time:

$$v = v_0 + at = (4.60 \text{ m/s}^2)(5.00 \text{ s}) = 23.0 \text{ m/s} \tag{2.4}$$

Thus, the average speed is $\overline{v} = 11.5$ m/s, and the average power is

$$\overline{P} = F\overline{v} = (5060 \text{ N})(11.5 \text{ m/s}) = \boxed{5.82 \times 10^4 \text{ W (78.0 hp)}}$$

The reader can verify that the average power can also be calculated by using Equation 7.10, $\overline{P} = W/t$. In this case, the work W done by the net force must be determined and then divided by the time of 5.00 s.

7.7 OTHER FORMS OF ENERGY AND THE CONSERVATION OF ENERGY

Up to now, we have considered only two kinds of energy, kinetic energy and gravitational potential energy. There are many other types of energy, however. Electrical energy, for instance, is used to run electrical appliances. Thermal energy (heat) is utilized in cooking food. Moreover, the work done by the kinetic frictional force appears as thermal energy, as anyone can experience by rubbing his hands back and forth. Another form of energy is chemical energy. Chemical energy is the energy stored in the molecules of fuels and food. When gasoline is burned, some of the stored chemical energy is released, and this energy does the work of moving cars, airplanes, and boats. Likewise, the chemical energy stored in food provides the energy needed for metabolic processes.

One of the most controversial forms of energy is nuclear energy. The research of many scientists, most notably Albert Einstein, led to the discovery that mass itself is one manifestation of energy. Einstein's famous equation, $E = mc^2$, describes how mass m and energy E are related, where c is the speed of light and has a value of 3.00×10^8 m/s. Because the speed of light is so large, this equation implies that very small masses are equivalent to large amounts of energy. The relationship between mass and energy will be discussed further in Chapter 35.

We have seen that kinetic energy can be converted into gravitational potential energy and vice versa. In general, energy of all types can be converted from one form to another. Part of the chemical energy stored in food is transformed into the kinetic energy of walking and into the thermal energy needed to keep our bodies near 98.6 °F. Similarly, in a moving car the chemical energy of gasoline is converted into kinetic energy, as well as electrical energy (to power the radio, headlights, and air conditioner), and thermal energy (to heat the car during the winter). Whenever energy is transformed from one form to another, it is found that no energy is gained or lost in the process; the sum total of all the energies before the process is equal to the sum total of the energies after the process. This observation leads to the following important principle:

> **THE PRINCIPLE OF CONSERVATION OF ENERGY**
> Energy can neither be created nor destroyed, but can only be converted from one form to another.

Learning how to convert energy from one form to another more efficiently, while obeying the principle of conservation of energy, is one of the main goals of modern science and technology.

SUMMARY

The **work** W done by a constant force acting on an object is $W = (F \cos \theta)s$, where $F \cos \theta$ is the magnitude of the force component along the displacement of the object, s is the magnitude of the displacement, and θ is the angle between the force and the displacement. Work can be positive or negative, depending on whether the force component along the displacement points in the same direction as the displacement or opposite to the displacement.

The **kinetic energy** KE of an object of mass m and speed v is $KE = \frac{1}{2}mv^2$. The **work-energy theorem** states that the work W done by the net force acting on an object equals the difference between the final kinetic energy KE_f and the initial kinetic energy KE_0 of the object: $W = KE_f - KE_0$. If the net force does positive work, the kinetic energy increases; if the net force does negative work, the kinetic energy decreases.

Gravitational potential energy PE is the energy that an object has by virtue of its position. For an object near the surface of the earth, the gravitational potential energy is given by $PE = mgh$, where h is the height of the object relative to an arbitrary zero level. The work–energy theorem can be expressed in an alternative form as $W_{other} = \Delta KE + \Delta PE$, where W_{other} is the work done by all forces other than gravity, and ΔKE and ΔPE are the changes in the kinetic and potential energies, respectively.

The **total mechanical energy** E is the sum of the kinetic energy and the gravitational potential energy: $E = KE + PE$. The **principle of conservation of mechanical energy** states that the total mechanical energy remains constant along the path of an object, provided that no net work is done by forces other than gravity. While E is constant, however, KE and PE may be transformed into one another.

A **conservative force** is one that, in moving an object between two points, does work that is independent of the path taken between the points. Alternatively, a force is conservative if the work it does in moving an object around any closed path is zero.

Average power $\overline{P}$ is the work done per unit time, $\overline{P} = $ Work/Time, or the rate at which work is done.

The **principle of conservation of energy** states that energy can neither be created nor destroyed, but can only be transformed from one form to another.

SOLVED PROBLEMS

SOLVED PROBLEM 1
Related Problems: **7   *44   *46

A 6.00×10^3-kg plane is diving at an angle of $10.0°$ for a distance of 1.70×10^3 m, as in the drawing. Four forces act on the plane: its weight $\mathbf{W} = m\mathbf{g}$, the lift force $\mathbf{L}$ that acts perpendicular to the surfaces of the wings, the forward thrust $\mathbf{T}$ (magnitude $= 1.80 \times 10^4$ N) generated by the plane's engine, and the force of air resistance $\mathbf{R}$ that opposes the plane's motion. The total work done by these four forces is $+2.90 \times 10^7$ J. (a) Find the work done by $\mathbf{R}$ alone and (b) the magnitude of $\mathbf{R}$.

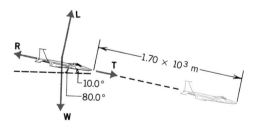

Solution   (a) We begin by determining the work that each force contributes to the total, with a view toward setting the sum of these contributions equal to the given value of $+2.90 \times 10^7$ J. Equation 7.1 is used four times:

Force	Work = (F cos θ)s
W	$[(6.00 \times 10^3 \text{ kg})(9.80 \text{ m/s}^2) \cos 80.0°]$ $\times (1.70 \times 10^3 \text{ m}) = +1.74 \times 10^7$ J
L	$[L \cos 90.0°](1.70 \times 10^3 \text{ m}) = 0$
T	$[(1.80 \times 10^4 \text{ N}) \cos 0°](1.70 \times 10^3 \text{ m}) = +3.06 \times 10^7$ J
R	Work $= W_R$

The work done by gravity is positive, reflecting the fact that the component of the plane's weight ($mg \cos 80.0°$) is in the same direction as the displacement. The lift force $\mathbf{L}$ does no work, since the force is perpendicular to the displacement. The individual contributions to the work add together to equal $+2.90 \times 10^7$ J:

$$+1.74 \times 10^7 \text{ J} + 3.06 \times 10^7 \text{ J} + W_R = +2.90 \times 10^7 \text{ J}$$

Solving for W_R gives $\boxed{W_R = -1.90 \times 10^7 \text{ J}}$. The work done by the resistive force $\mathbf{R}$ is negative, since $\mathbf{R}$ points opposite to the displacement.

(b) Now that W_R is known, the magnitude of $\mathbf{R}$ can be found from Equation 7.1 $[W_R = (R \cos \theta)s]$:

$$R = \frac{W_R}{s \cos \theta} = \frac{-1.90 \times 10^7 \text{ J}}{(1.70 \times 10^3 \text{ m}) \cos 180°} = \boxed{1.12 \times 10^4 \text{ N}}$$

Summary of Important Points   Part (a) shows that the total work W is equal to the sum of the individual work contributions, one contribution for each of the forces that act on the object. The work done by an individual force may be positive, negative, or zero. When all but one of these contributions are known, the unknown contribution can be determined. Part (b) shows that knowing the work done by a single force can lead to a value for the magnitude of the force, if the distance and the angle θ are known.

SOLVED PROBLEM 2

Related Problems: *31 **32 **49

A motorcycle stunt driver is doing the loop-the-loop trick. As he passes the top of the circular track ($r = 10.0$ m), he has a speed of 10.0 m/s, as part a of the drawing indicates. The mass of the cycle and rider is 405 kg. Ignoring friction and air resistance, and assuming the driver disengages the engine at the top of the track, determine the normal force that the track exerts when the cycle reaches the bottom.

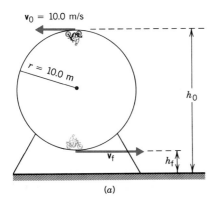

$v_0 = 10.0$ m/s

$r = 10.0$ m

h_0

v_f

h_f

(a)

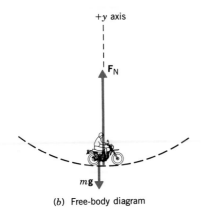

+y axis

F_N

mg

(b) Free-body diagram

Solution At the bottom of the loop two forces act on the cycle, the normal force $\mathbf{F_N}$ and the weight mg of the rider and cycle. These two forces acting together provide the centripetal force ($F_c = mv_f^2/r$) needed to keep the cycle on the circular path, as Section 6.7 discusses. From the free-body diagram in part b of the drawing, it can be seen that

$$\underbrace{+F_N - mg}_{\substack{\text{Centripetal}\\\text{force}}} = \frac{mv_f^2}{r}$$

To calculate F_N from this result, it is necessary to have a value for v_f, the speed at the bottom of the loop. The principle of conservation of mechanical energy (Equation 7.9) can be applied to obtain this value. This principle applies, because the only force acting other than gravity is the normal force $\mathbf{F_N}$, which is perpendicular to the motion and does no work:

$$\underbrace{\tfrac{1}{2}mv_f^2 + mgh_f}_{E_f} = \underbrace{\tfrac{1}{2}mv_0^2 + mgh_0}_{E_0}$$

Eliminating m algebraically from this result, using $v_0 = 10.0$ m/s, and noting that $h_0 - h_f = 2r = 20.0$ m (see the drawing and note that h_0 is greater than h_f), we find that

$$\tfrac{1}{2}v_f^2 = \tfrac{1}{2}v_0^2 + g(h_0 - h_f)$$
$$= \tfrac{1}{2}(10.0 \text{ m/s})^2 + (9.80 \text{ m/s}^2)(20.0 \text{ m})$$

Solving for v_f gives $v_f = 22.2$ m/s. This result for v_f can now be used to determine the magnitude F_N of the normal force:

$$F_N = \frac{mv_f^2}{r} + mg$$

$$= \frac{(405 \text{ kg})(22.2 \text{ m/s})^2}{10.0 \text{ m}} + (405 \text{ kg})(9.80 \text{ m/s}^2) = \boxed{23\ 900 \text{ N}}$$

Summary of Important Points The main point in this problem is that the principle of conservation of mechanical energy can be used profitably as one step in a multistep problem. In the present case, the principle is used to calculate intermediate information (v_f) needed so that the required answer (F_N) can be obtained.

QUESTIONS

1. Can work be done on an object that remains at rest? Explain.

2. Two forces $\mathbf{F_1}$ and $\mathbf{F_2}$ are acting on the box shown in the drawing, causing the box to move across the floor. The two force vectors are drawn to scale. Which force does more work? Justify your answer.

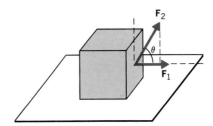

F_2

θ

F_1

3. A train, traveling at a steady speed, makes a 180° turn on a semicircular section of track and heads in a direction that is exactly opposite to its initial direction. Even though a centripetal force acts on the train during the turn, this force does no work on the train. Why?

4. The drawing shows a box with a velocity **v** being moved by a force **P** along a level horizontal floor. The normal force is F_N, the kinetic frictional force is f_k, and the weight of the box is mg. Complete the table below by marking a +, 0, or −, to indicate whether the corresponding force does positive, zero, or negative work on the box. Provide a reason for each of your answers.

P	F_N	f_k	mg

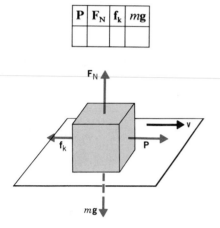

5. A sailboat is moving at a constant velocity. (a) Is any work being done by the net force that acts on the boat? Explain. (b) Recognizing that the wind propels the boat forward and the water resists the boat's motion, what does your answer in part (a) imply about the work done by the force of the wind compared to the work done by the resistive forces?

6. A ball has a speed of 15 m/s. Only one force acts on the ball. After this force acts, the speed of the ball is reduced to 7 m/s. Has the force done positive or negative work? Give a reason for your answer.

7. Three forces, F_A, F_B, and F_C, act on a bicycle and do positive work, zero work, and negative work, respectively. The bicycle moves on a horizontal surface. (a) Describe what each force would do to the kinetic energy of the bicycle, if that force alone acted. (b) With respect to the displacement of the bicycle, what can be said about the direction of these forces?

8. By measuring the speed of a boat at one moment and comparing it to the speed at another moment, one can tell whether the net force has done work on the boat. This statement is possible because of the work–energy theorem. Suppose the two speeds are measured to be exactly the same, and the boat moves horizontally. Is it safe to conclude that no force has acted on the boat? Explain, making sure to distinguish between the phrases "no force" and "no net force."

9. A motorcycle is being driven at a steady speed up a hill. (a) Is the term W_{other} in Equation 7.7a positive, negative, or zero? Why? (b) The motorcycle is acted on by a drive force that moves it forward and a frictional force that retards its motion. What does your answer to part (a) imply about the relative amounts of work done by each of these forces?

10. Suppose the total mechanical energy of an object is conserved. (a) If the kinetic energy decreases, what must be true about the gravitational potential energy? (b) If the potential energy decreases, what must be true about the kinetic energy? (c) If the kinetic energy does not change, what must be true about the potential energy?

11. Consider the following two situations in which the retarding effects of friction and air resistance are negligible. Car A approaches a hill. The driver turns off the engine at the bottom of the hill, and the car coasts up the hill. Car B, its engine running, is driven up the hill at a constant speed. Which situation is an example of the principle of conservation of mechanical energy? Provide a reason for your answer.

12. A trapeze artist, starting from rest, swings downward, lets go of the bar at the bottom of the swing, and falls freely to the net (see the drawing). An assistant, standing on a platform, jumps from rest straight downward. The platform is at the same height as that of the trapeze artist at the bottom of the swing. Friction and air resistance are negligible. (a) On which person, if either, does gravity do the greatest amount of work? Explain. (b) Who strikes the net with a greater speed? Why?

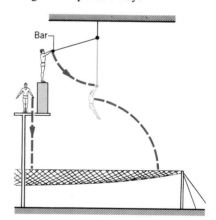

13. Is it correct to conclude that one engine is doing twice the work of another just because it is generating twice the power? Explain, neglecting friction and taking into account the time of operation of the engines.

PROBLEMS

Section 7.1 Work

1. The cable of a large crane applies a force of 2.2×10^4 N to a demolition ball as it lifts the ball vertically upward a distance of 7.6 m. (a) How much work does this force do on the ball? (b) Is the work positive or negative? Explain.

2. Suppose in Figure 7.2 that $+1.10 \times 10^3$ J of work are done by the force **F** (magnitude = 30.0 N) in moving the luggage carrier a distance of 50.0 m. At what angle θ is the force oriented with respect to the ground?

3. The drawing shows a boat being pulled by two locomotives through a canal of length 2.00 km. The tension in each cable is 5.00×10^3 N, and $\theta = 20.0°$. What is the total work done on the boat by the two locomotives?

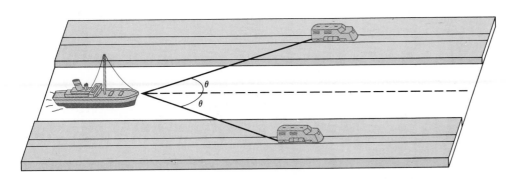

4. To help lower a piano gently down a flight of stairs, two ropes are attached, as in the drawing. The tension in each rope is 1500 N, and the ropes are oriented parallel to the surface of the incline. The piano moves a distance of 4.00 m along the incline. How much work is done on the piano by *each rope?*

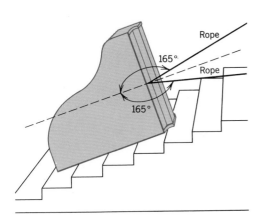

* 5. A 2.40×10^2-N force is pulling an 85.0-kg block across a horizontal surface. The force acts at an angle of 20.0° above the surface. The coefficient of kinetic friction is 0.200, and the block moves a distance of 8.00 m. Find (a) the work done by the pulling force, (b) the work done by the kinetic frictional force, and (c) the total work done by all the forces. In each case, be sure to include the proper plus or minus sign for the work. (d) Why can the work done by the normal force and the weight be ignored when calculating the total work in part (c)?

* 6. A 1.00×10^2-kg crate is being pulled across a horizontal floor by a force **P** that makes an angle of 30.0° above the horizontal. The coefficient of kinetic friction is 0.200. What should be the magnitude of **P**, so that the total work done by it and the kinetic frictional force is zero?

7. A 1200-kg car is being driven up a 5.0° hill, as the drawing illustrates. The frictional force is directed opposite to the motion of the car and has a magnitude of $f = 5.0 \times 10^2$ N. The tractive force **F is applied to the car by the road and propels the car forward. In addition to these two forces, two other forces act on the car: its weight **W**, and the normal force $\mathbf{F_N}$ directed perpendicular to the road surface. The length of the road up the hill is 3.00×10^2 m. What should be the magnitude of **F**, so that the total work done by all the forces acting on the car is $+150\ 000$ J? *(See Solved Problem 1 for a related problem.)*

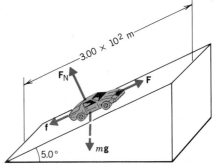

Section 7.2 The Work–Energy Theorem and Kinetic Energy

8. A 0.075-kg arrow is fired horizontally. The bow string exerts an average force of 65 N on the arrow over a distance of 0.90 m. What speed does the arrow have upon leaving the bow?

9. A water-skier whose weight is 155 lb has an initial speed of 20.0 ft/s. Later on, the speed of a skier is 37.0 ft/s. Determine the work done by the net force acting on the skier.

10. A 1.20×10^3-kg automobile coasts through a 50.0-m-long snowdrift that has been blown onto the road. The automobile has a speed of 20.0 m/s as it approaches the drift and emerges with a speed of 8.00 m/s. Find the average net force acting on the car in the drift. Relative to the motion of the car, what is the direction of this net force?

11. A 5.0×10^4-kg space probe is traveling at a speed of 11 000 m/s through deep space. The retrorockets are fired along the line of motion to reduce the probe's speed. The retrorockets generate a force of 4.0×10^5 N over a distance of 2500 km. What is the final speed of the probe?

12. When a 45-g golf ball takes off after being hit, its speed is 41 m/s. (a) How much work was done on the ball by the club? (b) Assume that the force of the golf club acts parallel to the motion of the ball and that the club is in contact with the ball for a distance of 1.0 cm. Ignoring the weight of the ball, determine the average force applied to the ball by the club.

* 13. The speed of a hockey puck decreases from 45.00 to 44.67 m/s in coasting 16 m across the ice. Find the coefficient of kinetic friction between the puck and the ice.

* 14. A wind-driven iceboat has a mass of 4.00×10^2 kg. The boat starts from rest and reaches a speed of 16.0 m/s after traveling a distance of 60.0 m. The coefficient of kinetic friction between the ice and the runners of the boat is 0.100. Determine the work done on the boat by the wind.

* 15. The head of a sledge hammer weighs 5.0 lb and is moving at a speed of 25 ft/s when it strikes a stake. The stake moves one inch into the ground in response to the blow from the hammer. Assume that forty percent of the hammer's kinetic energy is converted into the initial kinetic energy of the stake. Apply the work–energy theorem to the stake, and obtain the average resistive force applied to the stake by the ground.

Section 7.3 Gravitational Potential Energy

16. A 0.15-kg ball is thrown straight upward for a distance of 9.0 m. (a) Find the work done by the gravitational force. Be sure to include the correct plus or minus sign. (b) What is the change ($\Delta PE = PE_f - PE_0$) in the gravitational potential energy?

17. A shot-putter puts a shot (16 lb) that leaves his hand at a distance of 5.0 ft above the ground. (a) Find the work done by the gravitational force when the shot has risen to a height of 7.0 ft. Include the correct plus or minus sign for the work. (b) Determine the change ($\Delta PE = PE_f - PE_0$) in the gravitational potential energy of the shot.

18. A roller coaster (375 kg) moves from A to B, as the drawing indicates. In the process, friction does -2.00×10^4 J of work on the car. Also, a chain mechanism does $+3.00 \times 10^4$ J of work on the car to help it up the long climb. What is the change in the car's kinetic energy, $\Delta KE = KE_f - KE_0$, when the car moves from A to B?

19. A 55.0-kg skateboarder starts out with a speed of 1.80 m/s. After doing $+80.0$ J of work on himself and after friction has done -265 J of work on him, he has a speed of 6.00 m/s. (a) Calculate the change ($\Delta PE = PE_f - PE_0$) in the gravitational potential energy. (b) How much has the vertical height of the skater changed, and is the skater above or below the starting point?

20. A 5.0×10^2-kg hot-air balloon takes off from rest at the surface of the earth. The wind and lift forces take the balloon up, doing $+9.7 \times 10^4$ J of work on the balloon in the process. At what height above the surface of the earth does the balloon have a speed of 8.0 m/s?

* 21. A gymnast is bouncing on a trampoline. On the upward part of the motion, the mat of the trampoline pushes on the gymnast over a distance of 0.300 m. After leaving the mat, the 55.0-kg gymnast rises into the air for an additional 2.00 m before falling back down. What average force does the mat exert on the gymnast?

* 22. At a carnival, the hammer-and-bell concession uses a 9.00-kg hammer and a 0.400-kg sliding metal piece to strike a bell that is located 5.00 m above the ground, as drawing 22 shows. Suppose that 25.0% of the hammer's kinetic energy is used to do the work of sending the metal piece upward, and that friction does -50.0 J of work on the metal piece as the piece rises to strike the bell. How fast must the hammer be moving when it strikes the target, so that the bell just barely rings?

**23. A 3.00-kg model rocket is launched vertically straight upward with sufficient initial speed to reach a height of 1.00×10^2 m, even though air resistance performs -8.00×10^2 J of work on the rocket. How high would the rocket have gone if there were no air resistance?

Section 7.4 The Conservation of Mechanical Energy, Section 7.5 Conservative Forces and Potential Energy

24. A pole-vaulter approaches the takeoff point at a speed of

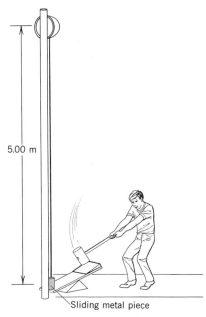

Drawing 22

9.00 m/s. Assuming that only this speed determines the height to which he can rise, find the maximum height at which the vaulter can clear the bar.

25. A 2.00-kg rock is released from rest at a height of 20.0 m. Ignore air resistance and determine the kinetic energy, gravitational potential energy, and total mechanical energy at each of the heights indicated in the table below:

Height	KE	PE	E = KE + PE
20.0 m			
15.0 m			
10.0 m			
5.00 m			
0 m			

26. A person is sled-riding down a hill that is 15.0 ft high. Starting at the top with a speed of 10.0 ft/s, the sled reaches the bottom with a speed of 25.0 ft/s. (a) Determine whether mechanical energy has been conserved. (b) Why might mechanical energy not be conserved?

27. A cyclist approaches the bottom of a gradual hill at a speed of 11 m/s. The hill is 5.0 m high, and the cyclist estimates that he is going fast enough to coast up and over it without peddling. Ignoring air resistance and friction, find the speed at which the cyclist crests the hill.

28. A water-skier lets go of the boat tow upon leaving the end of a jump ramp at a speed of 14.0 m/s. As the drawing indicates, the skier has a speed of 13.0 m/s at the highest point of the jump.

Ignoring air resistance, determine the skier's height H above the *top of the ramp* at the highest point of the jump.

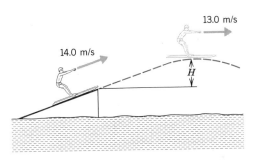

★ 29. A grappling hook, attached to a 1.5-m rope, is whirled in a circle that lies in the vertical plane. The hook is whirled at a constant rate of three revolutions per second. In the absence of air resistance, to what maximum height can the hook be cast?

★ 30. A metal ball of negligible mass, starting from point A in the drawing, is projected down the curved runway. Upon leaving the runway at point B, the ball is traveling straight upward and reaches a height of 4.00 m above the floor before falling back down. Ignoring friction and air resistance, find the speed of the ball at point A.

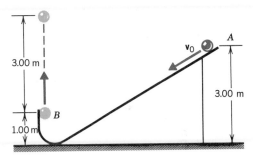

★ 31. A water slide is constructed so that swimmers, starting from rest at the top of the slide, leave the end of the slide traveling horizontally. As the drawing shows, one person is observed to hit the water 5.00 m from the end of the slide in 0.500 s after leaving the slide. Ignoring friction and air resistance, find the height H in the drawing. *(See Solved Problem 2 for a related problem.)*

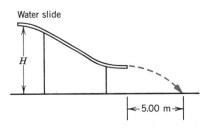

★★32. A swing is made from a rope that will tolerate a maximum tension of 8.00×10^2 N without breaking. Initially, the swing hangs vertically. The swing is then pulled back through an angle of 60.0° with respect to the vertical and is released from rest. What is the mass of the heaviest person who can ride on the swing? *(See Solved Problem 2 for a related problem.)*

Section 7.6 Power

33. One kilowatt · hour (kWh) is the amount of work or energy generated when one kilowatt of power is supplied for a time of one hour. A kilowatt · hour is the unit of energy used by power companies when figuring your electric bill. Determine the number of joules of energy in one kilowatt · hour.

34. The floors in a typical house are separated by a vertical distance of approximately 8.0 ft. A teenager (1.0×10^2 lb) climbs the stairs between floors at a steady speed. Find the power necessary to accomplish this, if the stairs are climbed in (a) 10.0 s and (b) 2.0 s. Express the answers in units of horsepower.

35. A 3.00×10^2-kg piano is being lifted at a steady speed from ground level straight upward to an apartment 10.0 m above the ground. The crane that is doing the lifting produces a steady power of 4.00×10^2 W. How much time does it take to lift the piano?

* 36. A car accelerates uniformly from rest to 65 mi/h in 12 s along a level stretch of road. Ignoring friction, determine the average power required to accelerate the car if (a) the weight of the car is 2600 lb, and (b) if the weight of the car is 3600 lb. Express your answers in units of horsepower.

* 37. A 73-kg sprinter, starting from rest, reaches a speed of 7.0 m/s in 1.8 s, with a negligible effect due to air resistance. The sprinter then runs the remainer of the race at a steady speed of 7.0 m/s under the influence of a 35-N force due to air resistance. What is the average power needed (a) to accelerate the runner and (b) to sustain the steady speed at which most of the race is run? Express the answers in watts and horsepower.

**38. A motorcycle (mass of cycle plus rider = 2.50×10^2 kg) is traveling at a steady speed of 20.0 m/s over a 1.00-km stretch of road. The force of air resistance acting on the cycle and rider is 2.00×10^2 N. Find the power necessary to sustain this speed if (a) the 1.00-km stretch of road is level and (b) if the road is sloped upward at 37.0° with respect to the horizontal.

**39. A 1.20×10^3-kg car has a speed of 11.0 m/s at the bottom and a speed of 23.0 m/s at the top of a hill that makes an angle of 5.00° with respect to the horizontal. The length of the hill is 1.50 km, and the force of friction opposing the car's motion has a magnitude of 6.00×10^2 N. Determine the power required to accelerate the car up the hill. Express the answer in watts and horsepower.

ADDITIONAL PROBLEMS

40. A slingshot fires a pebble from the top of a building at a speed of 10.0 m/s. The building is 20.0 m tall. Ignoring air resistance, find the speed with which the pebble strikes the ground when the pebble is fired (a) horizontally, (b) vertically straight up, and (c) vertically straight down.

41. Two cars, A and B, are traveling with the same speed of 40.0 m/s, each having started from rest. Car A has a mass of 1.20×10^3 kg, and car B has a mass of 2.00×10^3 kg. Compared to the work required to bring car A up to speed, how much *additional* work is required to bring car B up to speed?

42. The brakes of a runaway truck cause a retarding force of 3.0×10^3 N to be applied to the truck over a distance of 850 m. How much work does this force perform on the truck? Is the work positive or negative? Why?

43. A pitcher throws a 0.140-kg baseball, and it approaches the bat at a speed of 40.0 m/s. The bat does $+1.40 \times 10^2$ J of work on the ball in hitting it. Ignoring air resistance, determine the speed of the ball after the ball leaves the bat and is 25.0 m above the point of impact.

* 44. The force propelling a 1.50×10^3-kg car up a mountain road does $+4.70 \times 10^6$ J of work on the car. The car starts from rest at sea level and has a speed of 27.0 m/s at an altitude of 2.00×10^2 m above sea level. Obtain the work done on the car by friction and air resistance. *(See Solved Problem 1 for a related problem.)*

* 45. A 95-kg refrigerator is resting on a frictionless horizontal surface and is attached by a rope to a motor-driven winch. The winch is turned on and for 4.0 s supplies an average power of 110 W while the refrigerator is pulled across the surface. What is the speed of the refrigerator at the end of the time interval?

* 46. A 55-kg box is being pushed a distance of 7.0 m across the floor by a force **P** whose magnitude is 150 N. The force **P** is parallel to the displacement of the box. The coefficient of kinetic friction is 0.25. Determine the work done on the box by each of the *four* forces that act on the box. Be sure to include the proper plus or minus sign for the work done by each force. *(See Solved Problem 1 for a related problem.)*

* 47. A wrecking ball swings at the end of a 10.0-m cable on a vertical circular arc. In getting the ball going, the crane operator manages to give the ball a speed of 6.00 m/s as it passes through the lowest point of its swing, moving *toward* the crane (see drawing 47). The operator gives the ball no further assistance once the ball has been given its initial speed of 6.00 m/s. Of course, the ball ultimately swings back and moves *away* from the crane. Friction and air resistance are negligible. What speed v_f does the ball have as the ball moves away from the crane, when the cable makes an angle of 30.0° with respect to the vertical?

* 48. A car is approaching an intersection on a level road, and the traffic light turns from green to red. The driver slams on the brakes, and the car skids to a stop in a distance of 50.0 m. The coefficient of kinetic friction between the tires and the road is 0.850. Was the driver exceeding the 24.6-m/s speed limit when the brakes were first applied? If so, by how much (in m/s) was the driver exceeding the speed limit?

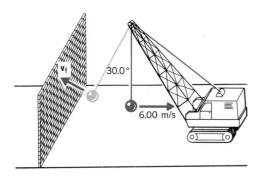

Drawing 47

**49. The drawing shows a version of the loop-the-loop trick for a small windup car. If the car is given an initial speed of 4.0 m/s, what is the largest value that the radius r can have if the car is to remain in contact with the circular track at all times? *(See Solved Problem 2 for a related problem.)*

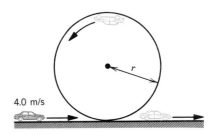

**50. A truck is traveling at 40.0 km/h down a hill when the brakes on all four wheels lock. The hill makes an angle of 15.0° with respect to the horizontal. The coefficient of kinetic friction between the tires and the road is 0.750. How far does the truck skid before coming to a stop?

CHAPTER 8

Impulse and Momentum

This "rocket" man is flying around the stadium at the opening ceremony of the 1984 Olympics in Los Angeles. The concepts of impulse and momentum are important for describing the operation of the propulsion unit, which is a kind of rocket engine.

8.1 INTRODUCTION

The way in which an object responds to forces can often be predicted with the help of the equations of kinematics, Newton's laws of motion, and the concepts of work and energy. However, the necessary calculations are not always easy to perform. For example, Newton's second law can always be applied in principle, but in practice it is only when the net force is constant or known to be changing in a predictable manner that the application is straightforward. In many situations in the real world, the net force does not behave in such a convenient way.

Consider what happens when a baseball is hit. The force that the bat applies to the ball rises from zero at the instant the bat touches the ball, reaches a maximum value as the bat makes full contact, and then falls back to zero as the ball leaves the bat. In general, the exact behavior of the force is rather complicated, and few experimental data are available to indicate what happens from moment to moment. In the absence of detailed information about the force, it is not possible to use Newton's second law to determine the acceleration of the ball at each instant.

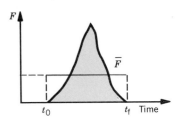

There are many situations, like that of the bat and the ball, where the motion of one object strongly influences the motion of another. Similar effects occur any time two objects collide, such as a hammer and a nail or two billiard balls. In addition, there are a host of other examples involving two objects, although they are not regarded as collisions in the usual sense. When an arrow is shot from a bow, for instance, the taut string of the bow propels the arrow. Or when a rifle is fired, the burning of the powder drives the bullet forward, while giving the rifle a backward "kick."

All such examples have in common the fact that one object interacts with another. This chapter examines a useful method by which information can be obtained concerning the motion of interacting objects. The method has its roots in Newton's second and third laws, which deal with the acceleration caused by a net force and the action–reaction forces that occur when two objects interact.

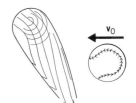

8.2 THE IMPULSE–MOMENTUM THEOREM

Figure 8.1 shows a baseball being hit by a bat. The ball has an initial velocity $\mathbf{v}_0$ just before contact is made and a final velocity $\mathbf{v}_f$ just after leaving the bat. In general, the final velocity does not equal the initial velocity, either in magnitude or in direction. During the time interval $\Delta t = t_f - t_0$, the bat and the ball are in contact, and the force $\mathbf{F}$ exerted on the ball changes. As the graph in the drawing indicates, the magnitude of the net force rises from zero to a maximum value and then returns to zero when the ball leaves the bat. The graph also shows the magnitude $\overline{F}$ of the average force, for the sake of comparison.

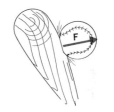

If a baseball is to be hit well, both the size of the force and the time of contact are important. When a sufficiently large average force acts on the ball for a long enough time, the ball is hit solidly. Therefore, we are motivated to bring together the average force and the time of contact, calling the product of the two the *impulse* of the force.

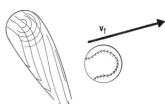

FIGURE 8.1 The initial velocity of a baseball is $\mathbf{v}_0$, just before contact is made with the bat. Just after the ball leaves the bat, the ball has a final velocity $\mathbf{v}_f$. The force exerted on the baseball at any instant during contact with the bat is $\mathbf{F}$. The average force acting on the ball during the time of contact is $\overline{\mathbf{F}}$. The graph suggests how the magnitude of the force changes during contact for the time interval $\Delta t = t_f - t_0$.

DEFINITION OF IMPULSE

The impulse of a force is the product of the average force $\overline{\mathbf{F}}$ and the time interval Δt during which the force acts:

$$\text{Impulse} = \overline{\mathbf{F}} \, \Delta t \qquad (8.1)$$

The impulse is a vector quantity and has the same direction as the average force.

   SI unit of impulse: newton · second (N · s)

When a ball is hit, it responds to the value of the impulse. A large impulse produces a large response, that is, a well-hit ball. Of course, the phrase "large response" also means that the ball departs from the bat with a large velocity, although the more massive the ball, the less velocity it picks up in a given interaction with the bat. Therefore, both mass and velocity play a role in how an object responds to a given impulse. The effect of mass and velocity is conveyed by the concept of *momentum,* as defined on page 158.

> **DEFINITION OF LINEAR MOMENTUM**
>
> The linear momentum **p** of an object is the product of the object's mass m and velocity **v**:
>
> $$\mathbf{p} = m\mathbf{v} \tag{8.2}$$
>
> Linear momentum is a vector quantity that points in the same direction as the velocity.
>
> **SI unit of momentum:** kilogram · meter/second (kg·m/s)

In the definition above, the word "linear" emphasizes that the object is moving along a straight line and distinguishes linear momentum from angular momentum. Angular momentum is associated with rotational motion and will be discussed in Chapter 10.

Newton's second law can be used to establish the relationship between impulse and momentum. When an object changes its velocity from $\mathbf{v}_0$ to $\mathbf{v}_f$ during a time interval Δt, the average acceleration $\bar{\mathbf{a}}$ is given by Equation 2.4 as

$$\bar{\mathbf{a}} = \frac{\Delta \mathbf{v}}{\Delta t} = \frac{\mathbf{v}_f - \mathbf{v}_0}{\Delta t}$$

According to Newton's second law, the cause of the acceleration is an average net force $\bar{\mathbf{F}} = m\bar{\mathbf{a}}$:

$$\bar{\mathbf{F}} = m\left(\frac{\mathbf{v}_f - \mathbf{v}_0}{\Delta t}\right) = \frac{m\mathbf{v}_f - m\mathbf{v}_0}{\Delta t} \tag{8.3}$$

In this result, the numerator on the right is the final momentum minus the initial momentum, so the average net force is given by the change in momentum per second. The equality between force and the time rate of change of momentum is, in fact, the version of the second law of motion presented originally by Newton. Multiplying both sides of Equation 8.3 by Δt yields Equation 8.4, which is known as the *impulse–momentum theorem* and gives the relationship between impulse and momentum.

> **IMPULSE–MOMENTUM THEOREM**
>
> When a net force **F** acts on an object, the impulse of the net force is equal to the change in momentum of the object:
>
> $$\bar{\mathbf{F}}\,\Delta t = \underbrace{m\mathbf{v}_f}_{\substack{\text{Final} \\ \text{momentum}}} - \underbrace{m\mathbf{v}_0}_{\substack{\text{Initial} \\ \text{momentum}}} \tag{8.4}$$
>
> Impulse = Change in momentum

Example 1 illustrates how this important theorem is used.

EXAMPLE 1

A baseball ($m = 0.14$ kg) has an initial velocity of $\mathbf{v}_0 = -38$ m/s as it approaches the bat. We have chosen the direction of approach as the negative direction. The bat applies a force that is much larger than the weight of the ball, and the ball departs from the bat with a final velocity of $\mathbf{v}_f = +58$ m/s. The contact time between the bat and the ball is $\Delta t = 4.0 \times 10^{-3}$ s. Find (a) the initial and final momenta of the ball, (b) the impulse produced by the bat, and (c) the average force exerted on the ball.

SOLUTION

(a) Initial momentum: $\mathbf{p}_0 = m\mathbf{v}_0 = (0.14 \text{ kg})(-38 \text{ m/s})$
$$= \boxed{-5.3 \text{ kg}\cdot\text{m/s}}$$

Final momentum: $\mathbf{p}_f = m\mathbf{v}_f = (0.14 \text{ kg})(+58 \text{ m/s})$
$$= \boxed{+8.1 \text{ kg}\cdot\text{m/s}}$$

(b) Equation 8.1 cannot be used directly to calculate the impulse, since the average force that the bat applies to the ball is not known. However, since the weight of the ball is negligible, the force applied by the bat is the net force, and the impulse of the net force equals the change in momentum, according to the impulse–momentum theorem:

$$\text{Impulse} = m\mathbf{v}_f - m\mathbf{v}_0$$
$$= (+8.1 \text{ kg}\cdot\text{m/s}) - (-5.3 \text{ kg}\cdot\text{m/s})$$
$$= \boxed{+13.4 \text{ kg}\cdot\text{m/s}}$$

(c) Now that the impulse is known, the time of contact can be used in Equation 8.1 to find the average force:

$$\overline{\mathbf{F}} = \frac{\text{Impulse}}{\Delta t} = \frac{+13.4 \text{ kg}\cdot\text{m/s}}{4.0 \times 10^{-3} \text{ s}}$$
$$= \boxed{+3400 \text{ N}}$$

The force is positive, reflecting the fact that it points opposite to the velocity of the approaching ball. A force of 3400 N corresponds to 760 lb, such a large value being necessary to produce the change in momentum during the brief contact time.

The impulse–momentum theorem is similar in form to the work–energy theorem discussed in Chapter 7. The impulse–momentum theorem states that the impulse produced by a net force is equal to the change in the object's momentum, momentum being a vector quantity that is proportional to the velocity. The work–energy theorem states that the work done by a net force is equal to the change in the object's kinetic energy, kinetic energy being a scalar quantity that is proportional to the square of the speed. The work–energy theorem leads directly to the principle of conservation of mechanical energy when only the gravitational force does work. The impulse–momentum theorem also leads to a very powerful principle called the **principle of conservation of linear momentum,** which the next section considers.

8.3 THE PRINCIPLE OF CONSERVATION OF LINEAR MOMENTUM

To explain the conservation of linear momentum, we apply the impulse–momentum theorem to a midair collision between two objects. In such applications, the word "system" is often used to refer to the collection of objects being studied. In the present case, the system consists of the two objects (masses m_1 and m_2), which are approaching each other with initial velocities $\mathbf{v}_{01}$ and $\mathbf{v}_{02}$, as Figure 8.2a shows. The objects interact during the collision in part b of the drawing and then depart with the final velocities $\mathbf{v}_{f1}$ and $\mathbf{v}_{f2}$ shown in part c. Because of the collision, the initial and final velocities are not the same.

To apply the impulse–momentum theorem, it is necessary to identify the forces acting on the system. There are two types of forces:

1. **Internal forces**—Forces that the objects within the system exert on each other.
2. **External forces**—Forces exerted on the objects by agents that are external to the system.

The forces $\mathbf{F}_{12}$ and $\mathbf{F}_{21}$ in Figure 8.2b are action–reaction forces that arise during the collision. They are internal forces: $\mathbf{F}_{12}$ is the force exerted on object 1 by object 2, while $\mathbf{F}_{21}$ is the force exerted on object 2 by object 1. The force of gravity also acts on each object, their weights being $\mathbf{W}_1$ and $\mathbf{W}_2$. These forces are external forces, because

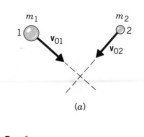

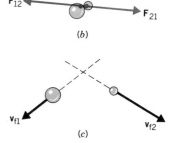

FIGURE 8.2 (a) The velocities of two objects (masses m_1 and m_2) are $\mathbf{v}_{01}$ and $\mathbf{v}_{02}$, just before a midair collision. (b) During the collision, each object exerts a force on the other object. These forces are labeled $\mathbf{F}_{12}$ and $\mathbf{F}_{21}$. (c) The velocities are $\mathbf{v}_{f1}$ and $\mathbf{v}_{f2}$, just after the collision.

they are applied by the earth, which is outside the system. Any friction and air resistance would also be considered as external forces, although these forces are being ignored for the sake of simplicity. The impulse–momentum theorem, then, gives the following results for each of the objects:

[Object 1]
$$(\underbrace{\mathbf{W}_1}_{\substack{\text{External} \\ \text{force}}} + \underbrace{\overline{\mathbf{F}}_{12}}_{\substack{\text{Internal} \\ \text{force}}})\,\Delta t = m_1\mathbf{v}_{f1} - m_1\mathbf{v}_{01}$$

[Object 2]
$$(\underbrace{\mathbf{W}_2}_{\substack{\text{External} \\ \text{force}}} + \underbrace{\overline{\mathbf{F}}_{21}}_{\substack{\text{Internal} \\ \text{force}}})\,\Delta t = m_2\mathbf{v}_{f2} - m_2\mathbf{v}_{02}$$

Adding these equations produces a single result for the system as a whole:

$$(\underbrace{\mathbf{W}_1 + \mathbf{W}_2}_{\substack{\text{External} \\ \text{forces}}} + \underbrace{\overline{\mathbf{F}}_{12} + \overline{\mathbf{F}}_{21}}_{\substack{\text{Internal} \\ \text{forces}}})\,\Delta t = \underbrace{(m_1\mathbf{v}_{f1} + m_2\mathbf{v}_{f2})}_{\substack{\text{Total final} \\ \text{momentum } \mathbf{P}_f}} - \underbrace{(m_1\mathbf{v}_{01} + m_2\mathbf{v}_{02})}_{\substack{\text{Total initial} \\ \text{momentum } \mathbf{P}_0}}$$

On the right side of this equation, $m_1\mathbf{v}_{f1} + m_2\mathbf{v}_{f2}$ is the vector sum of the individual final momenta for each object, or the total final momentum $\mathbf{P}_f$ of the system. Likewise, $m_1\mathbf{v}_{01} + m_2\mathbf{v}_{02}$ is the total initial momentum $\mathbf{P}_0$. Therefore, the above result can be rewritten in the following more convenient form:

$$\left(\begin{array}{c}\textbf{Sum of} \\ \textbf{external forces}\end{array} + \begin{array}{c}\textbf{Sum of average} \\ \textbf{internal forces}\end{array}\right)\Delta t = \mathbf{P}_f - \mathbf{P}_0 \tag{8.5}$$

The advantage of the internal/external force classification is that the internal forces always add together to give zero, as a consequence of Newton's third law of action–reaction. The action–reaction law indicates that $\mathbf{F}_{12} = -\mathbf{F}_{21}$, so that $\mathbf{F}_{12} + \mathbf{F}_{21} = 0$. Such cancellation of the internal forces occurs no matter how many parts there are to the system and allows us to ignore the internal forces from this point on, as Equation 8.6 indicates:

$$\textbf{(Sum of external forces) } \Delta t = \mathbf{P}_f - \mathbf{P}_0 \tag{8.6}$$

We developed this result with gravity as the only external force. But, in general, the sum of the external forces on the left side of the equation includes *all* the external forces.

With the aid of Equation 8.6, it is possible to see how the conservation of linear momentum arises. Suppose that the sum of the external forces is zero. A system for which this is true is called an **isolated system.** Consider, for example, a system composed of two billiard balls colliding on a frictionless pool table. The weight of each ball and the normal forces provided by the table are the external forces. Since the weights and the normal forces balance, the sum of the external forces is zero, and the balls constitute an isolated system. In such a case, Equation 8.6 indicates that

$$0 = \mathbf{P}_f - \mathbf{P}_0 \quad \text{or} \quad \mathbf{P}_0 = \mathbf{P}_f \tag{8.7}$$

In other words, the final total momentum of the isolated system after the collision or interaction is the same as the initial total momentum. This result is known as the ***principle of conservation of linear momentum.***

> **PRINCIPLE OF CONSERVATION OF LINEAR MOMENTUM**
> The total linear momentum of an isolated system remains constant (is conserved). An isolated system is one for which the vector sum of the external forces acting on the system is zero.

This principle applies to a system containing any number of objects, regardless of the internal forces, provided the system is isolated. Whether a force is considered to be internal depends on what objects are included as members of the system. In the case of two billiard balls, the collision forces are considered to be internal, if both balls are included. However, if *only one ball* is included, the collision force exerted on it by the other ball is an external force, since the other ball is then outside the system. Clearly, the total linear momentum of a one-ball system is *not* conserved in the presence of this external collision force; the momentum (and, hence, velocity) of a single billiard ball always changes during a collision. Example 2 illustrates an application of momentum conservation.

EXAMPLE 2

A freight train is being assembled in a switching yard, and Figure 8.3 illustrates two boxcars in the process of being coupled together. Car 1 has a mass of $m_1 = 65 \times 10^3$ kg and moves at a velocity of $v_{01} = +0.80$ m/s. Car 2, with a mass of $m_2 = 92 \times 10^3$ kg and a velocity of $v_{02} = +1.2$ m/s, overtakes car 1 and couples to it. Neglecting friction, find the common velocity v_f of the two cars after they become coupled.

SOLUTION

The two boxcars constitute the system. The sum of the external forces acting on the system is zero, because the weight of each car is balanced by a corresponding normal force, and external frictional forces are being neglected. Thus, the system is isolated, and the principle of conservation of linear momentum applies. The coupling forces that each car exerts on the other are internal forces and, therefore, do not affect the applicability of this principle. Momentum conservation, then, indicates that

$$\underbrace{m_1 v_{01} + m_2 v_{02}}_{\substack{\text{Total momentum} \\ \text{before collision}}} = \underbrace{(m_1 + m_2)v_f}_{\substack{\text{Total momentum} \\ \text{after collision}}}$$

This equation can be solved for v_f, the common velocity of the two cars after the collision:

$$\begin{aligned} v_f &= \frac{m_1 v_{01} + m_2 v_{02}}{m_1 + m_2} \\ &= \frac{(65 \times 10^3 \text{ kg})(+0.80 \text{ m/s}) + (92 \times 10^3 \text{ kg})(+1.2 \text{ m/s})}{(65 \times 10^3 \text{ kg} + 92 \times 10^3 \text{ kg})} \\ &= \boxed{+1.0 \text{ m/s}} \end{aligned}$$

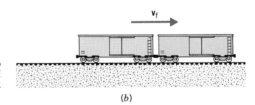

$m_2 = 92 \times 10^3$ kg
$v_{02} = +1.2$ m/s

$m_1 = 65 \times 10^3$ kg
$v_{01} = +0.80$ m/s

v_f

(a)   (b)

FIGURE 8.3 (*a*) One boxcar eventually catches up with the other and couples to it. (*b*) The coupled cars move together with a common velocity $\mathbf{v}_f$ after the collision.

From the result in Example 2, it can be seen that car 1 accelerates (increases its velocity), while car 2 decelerates as a result of the collision. The acceleration and deceleration arise at the moment the cars become coupled, because the cars exert internal forces on each other. These forces are equal in magnitude and opposite in direction, in accord with Newton's third law. The powerful feature of the momentum conservation principle is that it allows us to determine the changes in velocity without having to know explicitly what the internal forces are. Example 3 further illustrates this feature.

EXAMPLE 3

Two skaters, a woman ($m_1 = 54$ kg) and a man ($m_2 = 88$ kg), are initially facing each other on smooth level ice, where friction is negligible. Starting from rest, they "push off" against each other. As Figure 8.4 shows, the woman moves away with a velocity of $v_{f1} = +2.5$ m/s. Find the "recoil" velocity of the man.

SOLUTION

For a system consisting of the two skaters, the sum of the external forces is zero, because the weight of each skater is balanced by a corresponding normal force and the ice is assumed to be frictionless. The two skaters, then, constitute an isolated system, and the principle of conservation of linear momentum applies.

The total momentum of the two skaters before they push on each other is zero, since they are standing at rest. Momentum conservation requires that the total momentum remains zero at

all times, even after the skaters have separated, as in part *b* of the drawing:

$$\underbrace{0}_{\substack{\text{Total momentum} \\ \text{before pushing}}} = \underbrace{m_1 v_{f1} + m_2 v_{f2}}_{\substack{\text{Total momentum} \\ \text{after pushing}}}$$

Solving for the recoil velocity of the man gives

$$v_{f2} = \frac{-m_1 v_{f1}}{m_2} = \frac{-(54 \text{ kg})(+2.5 \text{ m/s})}{88 \text{ kg}}$$
$$= \boxed{-1.5 \text{ m/s}}$$

The minus sign indicates that the man moves to the left in the drawing. After the skaters separate, the total momentum of the system remains zero, because momentum is a vector quantity, and the momenta of the man and the woman have equal magnitudes but opposite directions.

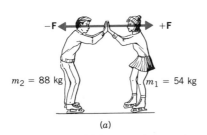

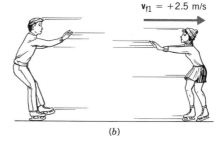

FIGURE 8.4 (*a*) Two stationary skaters push on each other with oppositely directed forces of equal magnitude. The two skaters constitute the system. (*b*) As the skaters move away from each other, the total linear momentum of the system remains zero, which is what it was initially.

It is important to realize that the total linear momentum may be conserved even when the total kinetic energy is not constant. In Example 3, for instance, the initial kinetic energy is zero since the skaters are stationary. But after they push off, the skaters are moving and each has some kinetic energy. The kinetic energy changes, because work is done by the internal forces that each skater exerts on the other. This work causes the kinetic energy to increase, as required by the work–energy theorem (see Section 7.2). However, internal forces cannot change the total linear momentum of a system, since the total linear momentum of an isolated system is conserved in the presence of such forces.

8.4 COLLISIONS IN ONE DIMENSION

As discussed in the last section, linear momentum is conserved when two objects collide, provided they constitute an isolated system. When the objects are atoms or subatomic particles, it is often found, in addition, that the total kinetic energy of the particles before the collision equals the total kinetic energy of the particles after the collision. In such a case, whatever kinetic energy is gained by one particle is lost by the other.

In general, however, when two macroscopic objects collide, such as two cars, the total kinetic energy after the collision is less than that before the collision. Thus, there is a loss of kinetic energy. During a collision, kinetic energy is lost predominantly in two ways. First, it can be converted into heat because of friction. Second, kinetic energy is lost whenever an object suffers permanent distortion and does not return to its original shape after the collision. In this case, energy is spent in creating the permanent damage, as in an automobile collision. With very hard objects, such as a solid steel ball and a marble floor, the permanent distortions suffered upon collision are much smaller and, consequently, less kinetic energy is spent.

Collisions are often classified according to whether kinetic energy changes during the collision:

1. *Elastic collision*—One in which the total kinetic energy after the collision is equal to the total kinetic energy before the collision.

2. *Inelastic collision*—One in which the total kinetic energy is *not* the same before and after the collision; if the objects stick together after colliding, the collision is said to be completely inelastic.

The boxcars coupling together in Figure 8.3 is an example of a completely inelastic collision. When a collision is completely inelastic, the greatest amount of kinetic energy is lost. Example 4 shows how one particular elastic collision is described using the conservation of linear momentum and the fact that no kinetic energy is lost during the collision.

EXAMPLE 4

As Figure 8.5 illustrates, a ball of mass $m_1 = 0.250$ kg and velocity $v_{01} = +5.00$ m/s collides head-on with a ball of mass $m_2 = 0.800$ kg that is initially at rest. No external forces act on the balls. If the collision is elastic, what are the velocities of the balls after the collision?

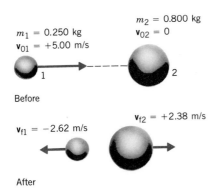

$m_1 = 0.250$ kg
$v_{01} = +5.00$ m/s

$m_2 = 0.800$ kg
$v_{02} = 0$

1   2

Before

$v_{f2} = +2.38$ m/s

$v_{f1} = -2.62$ m/s

After

FIGURE 8.5 A 0.250-kg ball, traveling with an initial velocity of $v_{01} = +5.00$ m/s, undergoes an elastic collision with a 0.800-kg ball initially at rest.

SOLUTION

The total linear momentum of the two-ball system is conserved, because no external forces act on the system. Momentum conservation applies whether or not the collision is elastic:

$$\underbrace{m_1 v_{01} + 0}_{\substack{\text{Total momentum} \\ \text{before collision}}} = \underbrace{m_1 v_{f1} + m_2 v_{f2}}_{\substack{\text{Total momentum} \\ \text{after collision}}}$$

For an elastic collision, the total kinetic energy is the same before and after the collision:

$$\underbrace{\tfrac{1}{2} m_1 v_{01}^2 + 0}_{\substack{\text{Total kinetic energy} \\ \text{before collision}}} = \underbrace{\tfrac{1}{2} m_1 v_{f1}^2 + \tfrac{1}{2} m_2 v_{f2}^2}_{\substack{\text{Total kinetic energy} \\ \text{after collision}}}$$

There are now two equations containing the two unknown quantities v_{f1} and v_{f2}. These equations can be solved simultaneously to give

$$v_{f1} = \left(\frac{m_1 - m_2}{m_1 + m_2} \right) v_{01} \tag{8.8a}$$

$$v_{f2} = \left(\frac{2m_1}{m_1 + m_2} \right) v_{01} \tag{8.8b}$$

With the given values for m_1, m_2, and v_{01}, Equations 8.8 yield the following values for v_{f1} and v_{f2}:

$$\boxed{v_{f1} = -2.62 \text{ m/s}} \quad \text{and} \quad \boxed{v_{f2} = +2.38 \text{ m/s}}$$

The negative value for v_{f1} indicates that m_1 rebounds to the left after the collision in Figure 8.5, while the positive value for v_{f2} indicates that m_2 moves to the right.

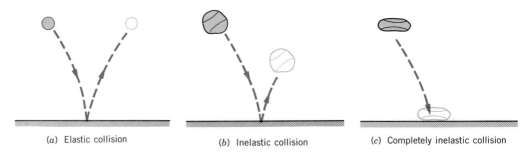

(a) Elastic collision   (b) Inelastic collision   (c) Completely inelastic collision

FIGURE 8.6 (a) A hard steel ball would rebound to its original height after striking a hard marble surface, if the collision were elastic. (b) A partially deflated basketball has little bounce on a soft asphalt surface. (c) A deflated basketball has no bounce at all.

We can get a "feel" for an elastic collision by dropping a steel ball onto a hard surface, such as a marble slab resting on the ground. If the collision is elastic, the ball will rebound to its original height, as Figure 8.6a illustrates. This observation can be seen to be consistent with the results of Example 4, if it is recognized that the colliding objects are the steel ball (mass = m_1) and the earth itself ($m_2 = 6 \times 10^{24}$ kg). With v_{01} being the velocity of the ball just before impact, Equation 8.8a gives the velocity v_{f1} of the ball just after impact. Since m_1 is negligible compared with m_2, this equation shows that

$$v_{f1} = -\left(\frac{m_2}{m_2}\right) v_{01} = -v_{01}$$

The velocity v_{f1} is directed opposite to v_{01} but has the same magnitude. Thus, the kinetic energy of the steel ball is the same before and after the elastic collision, and the ball rebounds to its original height as kinetic energy is converted into gravitational potential energy. In contrast, a partially deflated basketball exhibits little rebound from a relatively soft asphalt surface, as in part b, indicating that a large fraction of the ball's kinetic energy is dissipated during the collision. The completely deflated basketball in part c has no bounce at all, since a maximum amount of kinetic energy is lost during the collision.

The next example illustrates a completely inelastic collision in a device called a "ballistic pendulum." This device can be used to measure the speed of a bullet.

EXAMPLE 5

A ballistic pendulum consists of a 2.50-kg block of wood suspended by a wire of negligible mass. A 10.0-g bullet is fired into the block, as Figure 8.7a illustrates. The block with the bullet in it then swings to a maximum height of 0.650 m above the initial position, as part b of the drawing indicates. Find the speed of the bullet.

SOLUTION

To find the speed of the bullet, it is convenient to deal first with the completely inelastic collision between the bullet and the block and then with the subsequent motion of the pair as they swing upward after the impact. The total linear momentum of the system consisting of the bullet and the block is conserved during the collision, because the sum of the external forces acting on the system is very nearly zero.* Using the symbols in the drawing, we find that

$$\underbrace{m_1 v_{01}}_{\substack{\text{Total linear momentum} \\ \text{before collision}}} = \underbrace{(m_1 + m_2)v_f}_{\substack{\text{Total linear momentum} \\ \text{after collision}}}$$

This equation can be solved for the initial speed v_{01} of the bullet:

$$v_{01} = \frac{m_1 + m_2}{m_1} v_f$$

Before this result can be used to determine the speed of the bullet, a value is needed for the speed v_f immediately after the collision. This value can be obtained from the maximum height to which the system swings. As the system swings upward after the collision, kinetic energy is converted into gravitational potential energy. During the swing, the tension force in the wire acts perpendicular to the motion and does no work. Therefore, the principle of conservation of mechanical energy is applicable:

$$\underbrace{\tfrac{1}{2}(m_1 + m_2)v_f^2}_{\substack{\text{Total mechanical energy} \\ \text{at the bottom of the} \\ \text{swing, all kinetic}}} = \underbrace{(m_1 + m_2)gh_f}_{\substack{\text{Total mechanical energy} \\ \text{at the top of the swing,} \\ \text{all potential}}}$$

It follows that

$$v_f = \sqrt{2gh_f} = \sqrt{2(9.80 \text{ m/s}^2)(0.650 \text{ m})} = 3.57 \text{ m/s}$$

With this value for v_f, it is now possible to determine the speed of the bullet:

$$v_{01} = \frac{m_1 + m_2}{m_1} v_f = \frac{0.0100 \text{ kg} + 2.50 \text{ kg}}{0.0100 \text{ kg}} (+3.57 \text{ m/s})$$
$$= \boxed{+896 \text{ m/s}}$$

Now that the speed of the bullet is known, it can be verified that the total kinetic energy is not the same before and after the completely inelastic collision between the bullet and the block. The kinetic energy of the bullet initially is $\tfrac{1}{2}m_1 v_{01}^2 = 4010$ J, and the kinetic energy of the system (bullet and block) just after the collision is $\tfrac{1}{2}(m_1 + m_2)v_f^2 = 16.0$ J. Thus, a scant 0.4% of the original kinetic energy remains after the collision.

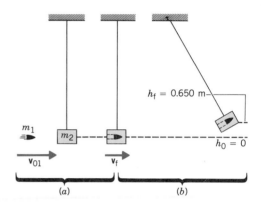

FIGURE 8.7 (a) The bullet, whose velocity just before impact is $\mathbf{v}_{01}$, makes a completely inelastic collision with the block of a ballistic pendulum. Just after the collision, the velocity of the bullet and the block is $\mathbf{v}_f$. (b) The bullet/block combination swings upward to a maximum height of 0.650 m.

* The sum of the external forces acting on the system before the collision is not exactly zero, because the force of gravity acts on the incoming bullet and is not balanced by another force. However, the force of gravity changes the momentum of the bullet by a negligibly small amount during the collision, since the collision occurs so quickly. Therefore, momentum conservation is a very good approximation.

8.5 COLLISIONS IN TWO DIMENSIONS

The collisions discussed so far have been "head-on" or one-dimensional collisions, in the sense that the velocities of the objects point along a single line before and after contact is made. In the real world, however, collisions often occur in two or three dimensions. Figure 8.8 shows a two-dimensional case in which two balls collide on a horizontal frictionless table.

For the system consisting of the two balls, the external forces include the weights of the balls and the corresponding normal forces produced by the table. Since each weight is balanced by a normal force, the sum of the external forces is zero. Consequently, the total linear momentum of the system is conserved, as Equation 8.7

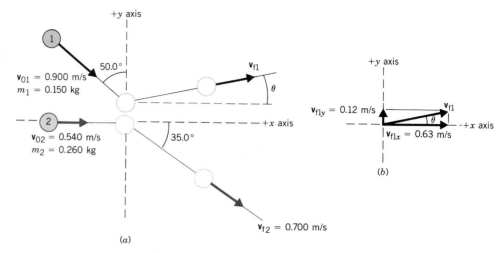

FIGURE 8.8 (*a*) The principle of conservation of linear momentum applies to two balls as they collide on a horizontal frictionless table. (*b*) By applying the principle in two dimensions, the *x* and *y* components of the velocity of ball 1 after the collision can be determined.

indicates: $P_0 = P_f$. Momentum is a vector quantity, however, and in two dimensions the *x* and *y* components of the total linear momentum are conserved separately. In other words, Equation 8.7 is equivalent to the following two equations:

[*x* component]	$P_{0x} = P_{fx}$	(8.9a)
[*y* component]	$P_{0y} = P_{fy}$	(8.9b)

Example 6 shows how to deal with a two-dimensional collision when the total linear momentum of the system is conserved.

EXAMPLE 6

For the data given in Figure 8.8, use momentum conservation to determine the magnitude and direction of the velocity of ball 1 after the collision.

SOLUTION

To determine the velocity of ball 1 after the collision, we calculate the components of this velocity, which are v_{f1x} and v_{f1y}. Momentum conservation applied to the *x* component of the total linear momentum (Equation 8.9a) shows that

x component

$$\underbrace{(0.150 \text{ kg})(0.900 \text{ m/s})(\sin 50.0°)}_{\text{Ball 1, before}} + \underbrace{(0.260 \text{ kg})(0.540 \text{ m/s})}_{\text{Ball 2, before}}$$

$$= \underbrace{(0.150 \text{ kg})(v_{f1x})}_{\text{Ball 1, after}} + \underbrace{(0.260 \text{ kg})(0.700 \text{ m/s})(\cos 35.0°)}_{\text{Ball 2, after}}$$

This equation can be solved to show that $v_{f1x} = +0.63$ m/s.

Momentum conservation applied to the *y* component of the total linear momentum (Equation 8.9b) shows that

y component

$$\underbrace{(0.150 \text{ kg})[-(0.900 \text{ m/s})(\cos 50.0°)]}_{\text{Ball 1, before}} + \underbrace{0}_{\text{Ball 2, before}}$$

$$= \underbrace{(0.150 \text{ kg})(v_{f1y})}_{\text{Ball 1, after}} + \underbrace{(0.260 \text{ kg})[-(0.700 \text{ m/s})(\sin 35.0°)]}_{\text{Ball 2, after}}$$

The solution to this equation reveals that $v_{f1y} = +0.12$ m/s.

Figure 8.8*b* shows the *x* and *y* components of the final velocity of ball 1. The magnitude of the velocity is

$$v_{f1} = \sqrt{(0.63 \text{ m/s})^2 + (0.12 \text{ m/s})^2} = \boxed{0.64 \text{ m/s}}$$

The direction of the velocity is given by the angle θ:

$$\tan \theta = \frac{0.12 \text{ m/s}}{0.63 \text{ m/s}} = 0.19 \qquad \theta = \tan^{-1} 0.19 = \boxed{11°}$$

By using the masses and the speeds of the balls in Example 6, it can be verified that the final total kinetic energy is 0.094 J, slightly less than the initial value of 0.099 J. Therefore, the collision is inelastic, and once again we see that the total linear momentum of a system can be conserved, even though the kinetic energy of the system changes.

*8.6 ROCKET PROPULSION

Space Shuttle launch.

Space exploration and the widespread use of artificial satellites have been made possible by advances in rocket propulsion. Rocket engines are based on the familiar notion of "recoil," the same idea that is involved when the two skaters in Example 3 push off against each other. Recoil is also involved when a bullet is fired from a rifle. The rifle and the bullet push off against each other with the aid of the burning gunpowder, and the gun recoils as the bullet is propelled forward. In all examples of recoil, internal forces act on the two objects that constitute the system, accelerating them in opposite directions.

In a rocket, fuel is burned to create fast moving hot gases that are forced out the rear of the rocket. According to Newton's third law, the hot gases, in turn, apply an oppositely directed force of equal magnitude on the rocket, making it recoil. In this sense, the rocket behaves like a large rifle aimed downward. When the bullet (the hot gases) is fired, the rifle (the rocket) recoils upward. The internal force applied to the rocket by the ejected gases is called the "thrust," and the impulse–momentum theorem can be applied to identify the factors that determine the thrust.

Consider the rocket in Figure 8.9. This rocket moves away from the earth at a velocity v_0. The drawing focuses on the momentum of a small mass Δm of propellant before and after a burn. The gravitational force acting on Δm is negligible. This mass is changed into hot gases in a time interval Δt. According to the impulse–momentum theorem, the impulse that causes the momentum of this mass to change is given by

$$\overline{F}\,\Delta t = (\Delta m)v_f - (\Delta m)v_0 \tag{8.4}$$

$\overline{F}$ is the average net force exerted on Δm during the time interval Δt, v_f is the final velocity of the propellant (in the form of the ejected gases), and v_0 is the initial velocity of the propellant (and rocket) before the burn. Both velocities are measured with respect to the earth.

Since the gravitational force acting on Δm is negligible, the average net force is just the downward force T' exerted by the engine on the propellant. Consequently, the impulse–momentum theorem can be rewritten as

$$\overline{F} = T' = \left(\frac{\Delta m}{\Delta t}\right)(v_f - v_0)$$

According to Newton's third law, $-T'$ is the thrust T that the ejected gases exert *on* the rocket. Therefore, multiplying both sides of the above result by -1, we find that

$$\text{Thrust} = T = -\left(\frac{\Delta m}{\Delta t}\right)(v_f - v_0) \tag{8.10}$$

One factor that determines the thrust is $\Delta m/\Delta t$, which is the mass of propellant burned per second. A larger burn rate creates a larger thrust. Another factor is $v_f - v_0$,

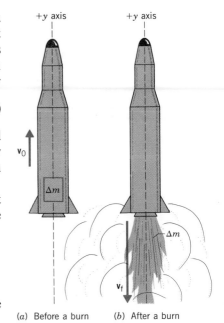

(a) Before a burn (b) After a burn

FIGURE 8.9 (a) Before a burn, the mass Δm of propellant has an upward velocity of v_0 relative to the earth. (b) As a result of a burn, hot gases of mass Δm are ejected with a velocity of v_f relative to the earth.

which is the velocity of the ejected gases *relative to the moving rocket*. Example 7 illustrates the use of Equation 8.10.

EXAMPLE 7

The rocket in Figure 8.9 is moving away from the earth at a velocity of $v_0 = +4.0 \times 10^3$ m/s and is burning propellant at a rate of $\Delta m/\Delta t = 550$ kg/s. The hot gases are ejected at a velocity of $v_f = -8.0 \times 10^3$ m/s with respect to the earth. Find the thrust exerted on the rocket by the ejected gases.

$$T = -\left(\frac{\Delta m}{\Delta t}\right)(v_f - v_0)$$
$$T = -(550 \text{ kg/s})(-8.0 \times 10^3 \text{ m/s} - 4.0 \times 10^3 \text{ m/s})$$
$$= \boxed{6.6 \times 10^6 \text{ N}}$$

SOLUTION

The thrust is given directly by Equation 8.10:

SUMMARY

The **impulse** of a force is the product of the average force $\overline{\mathbf{F}}$ and the time interval Δt during which the force acts: **Impulse** $= \overline{\mathbf{F}} \Delta t$. Impulse is a vector quantity.

The **linear momentum p** of an object is the product of the object's mass m and velocity $\mathbf{v}$: $\mathbf{p} = m\mathbf{v}$. Linear momentum is a vector quantity. The total linear momentum of a system of objects is the vector sum of the linear momenta of the individual objects.

The **impulse–momentum theorem** states that an impulse produces a change in an object's momentum, according to $\overline{\mathbf{F}} \Delta t = m\mathbf{v}_f - m\mathbf{v}_0$, where $m\mathbf{v}_f$ is the final momentum and $m\mathbf{v}_0$ is the initial momentum.

The **principle of conservation of linear momentum** states that the total linear momentum of an isolated system remains constant. An isolated system is one for which the sum of the external forces acting on the system is zero. This principle is independent of the principle of conservation of mechanical energy.

An **elastic collision** is one in which the total kinetic energy is the same before and after the collision. An **inelastic collision** is one in which the total kinetic energy is not the same before and after the collision. If the objects stick together after colliding, the collision is said to be **completely inelastic.**

The **thrust T** developed by a rocket engine can be determined with the aid of the impulse–momentum theorem: $\mathbf{T} = -(\Delta m/\Delta t)(\mathbf{v}_f - \mathbf{v}_0)$, where $\Delta m/\Delta t$ is the mass of propellant burned per second, and $\mathbf{v}_f - \mathbf{v}_0$ is the velocity of the ejected gases relative to the moving rocket. The terms $\mathbf{v}_f$ and $\mathbf{v}_0$ are, respectively, the velocities of the ejected gases and the rocket relative to the earth.

QUESTIONS

1. Two identical automobiles have the same speed, one traveling east and one traveling west. Do these cars have the same momentum? Explain.

2. If two different objects have the same momentum, do they necessarily have the same kinetic energy? Give a reason for your answer.

3. Can a single object have kinetic energy but no momentum? Can a system of two or more objects have a total kinetic energy that is not zero but a total momentum that is zero? Account for your answers.

4. An airplane is flying horizontally with a constant momentum during a time interval Δt. (a) With the aid of Equation 8.4, decide whether a net impulse is acting on the plane during this time interval. (b) In the horizontal direction, the thrust and air resistance both act on the plane. What does the answer in part (a) imply about the impulse of the resistive force and the impulse of the thrust?

5. An object slides along the surface of the earth and slows down because of kinetic friction. If the object itself is considered as the system, the kinetic frictional force must be identified as an external force that, according to Equation 8.4, decreases the momentum of the system. If *both* the object and the earth are considered to be part of the system, is the force of kinetic friction still an external force? Can the friction force change the total linear momentum of the two-body system? Give your reasoning for both answers.

6. In movies, Superman hovers stationary in midair, grabs a villain by the neck, and throws him forward. Superman, however, remains stationary. Using the conservation of linear momentum, explain what is wrong with this sequence of events.

7. A satellite explodes in outer space, far from any other body, sending thousands of pieces in all directions. How does the linear momentum of the satellite before the explosion compare with the total linear momentum of all the pieces after the explosion? Account for your answer.

8. Can the total linear momentum of a system be conserved if the total mechanical energy of the system is not conserved? If so, give an example.

9. A blank cartridge is one in which the lead bullet is replaced by a thin paper cap. When a gun fires a blank, is the recoil greater than, the same as, or less than when the gun fires a standard bullet? Give your reasoning in terms of the law of conservation of linear momentum.

10. A collision occurs between three moving billiard balls such that no net external force acts on the three-ball system. Is the momentum of *each* ball conserved during the collision? If so, explain why. If not, what quantity is conserved?

11. In an elastic collision, is the kinetic energy of *each* object the same before and after the collision? Explain.

12. In a completely inelastic collision, is the total kinetic energy of the system always zero after the collision? Support your answer with two examples.

13. The drawing shows a garden sprinkler. Using the impulse–momentum theorem, explain how the sprinkler works.

14. On a distant asteroid, a large catapult is used to "throw" large chunks of stone into space. Could such a device be used as a propulsion system to move the asteroid closer to the earth? Explain.

15. Review Example 4. Now, suppose both objects have the same mass, $m_1 = m_2$. Describe what happens to the velocities of both objects as a result of the collision, using Equations 8.8a and 8.8b to justify your answers.

PROBLEMS

Section 8.2 The Impulse–Momentum Theorem

1. (a) What is the momentum and the kinetic energy of a car (mass $= 2.00 \times 10^3$ kg) that is traveling due north at a speed of 15.0 m/s? If the speed is tripled, by what factor does (b) the momentum increase and (c) the kinetic energy increase?

2. (a) The earth travels with an approximate speed of 29.9 km/s (66 900 mi/h) in its journey around the sun. The mass of the earth is 5.98×10^{24} kg. Find the magnitude of its linear momentum. (b) Is the direction of the earth's linear momentum constant? If not, describe how it changes and specify the force that causes it to change.

3. A freight train moves due north with a speed of 17 m/s. The mass of the train is 6.0×10^6 kg. (a) Find the linear momentum of the train. (b) How fast would a 1500-kg automobile have to be moving due north to have the same momentum as the train? (c) By determining the kinetic energy of each, show whether the car or the train has the larger kinetic energy.

4. Two arrows are fired horizontally with the same speed of 30.0 m/s. Each arrow has a mass of 0.100 kg. One is fired due east and the other due south. Find the magnitude and direction of the total momentum of this two-arrow system.

5. A woman, driving a golf ball off the tee, gives the ball a speed of 28 m/s. The mass of the ball is 0.045 kg, and the duration of the impact with the golf club is 6.0×10^{-3} s. (a) What is the change in momentum of the ball? (b) Determine the average force applied by the club.

* 6. A 1220-kg car is traveling due south at a speed of 20.0 m/s. A 1540-kg car is moving due east at a speed of 30.0 m/s. A third car has a mass of 935 kg and is heading 45.0° south of east at a speed of 15.0 m/s. (a) Compute the total linear momentum (magnitude and direction) of the three-car system. (b) Find the total kinetic energy of the system.

* 7. A 0.060-kg ball is dropped from a distance of 4.0 m above a concrete floor. (a) Neglecting air resistance, use the conservation of mechanical energy to compute the speed of the ball just before impact. (b) Find the momentum (magnitude and direction) of the ball just before impact. (c) Assuming the ball loses one-tenth of its speed during impact, find its momentum (magnitude and direction) just as it rebounds from the floor. (d) What is the change in momentum during the collision? (e) If the ball remains in contact with the floor for 0.050 s, find the average net force (magnitude and direction) exerted on the ball by the floor.

**8.   A 1080-kg car moves with a speed of 28.0 m/s and is headed 30.0° north of east. A second car has a mass of 1630 kg and is headed due south. A third car has a mass of 1350 kg and is headed due west. What must be the speeds of the second and third cars, so that the total linear momentum of the three-car system is zero?

**9.   Two vehicles A and B are moving in the same direction along straight lines. Vehicle A has four times the momentum and twice the kinetic energy as B. Determine (a) the ratio of the masses of the vehicles and (b) the ratio of their speeds.

Section 8.3 The Principle of Conservation of Linear Momentum

10.   For tests using a device called a *ballistocardiograph,* a patient lies on a horizontal platform that is supported from beneath by jets of air. Because of the air jets, the friction impeding the horizontal motion of the platform is negligible. Each time the heart beats, blood is pushed out of the heart in a direction that is approximately parallel to the platform. Since momentum must be conserved, the body and the platform recoil, and this recoil can be detected to provide information about the heart. For each beat, suppose that 0.050 kg of blood is pushed out of the heart with a velocity of 0.25 m/s and that the mass of the patient and platform is 85 kg. Assuming the patient does not slip with respect to the platform, determine the recoil speed following a heartbeat.

11.   A 55-kg swimmer is standing on a stationary 210-kg floating raft. The swimmer then runs off the raft horizontally with a speed of 4.6 m/s relative to the shore. (a) Find the recoil speed that the raft would have, if there were no friction and resistance due to the water. (b) What is the impulse that acts on the raft?

12.   An astronaut, whose mass is 9.00 slugs, is motionless in outer space. Upon command, the portable propulsion unit strapped to his back ejects 0.300 slugs of gas with a speed of 50.0 ft/s. Find the recoil speed of the astronaut.

13.   Show that the kinetic energy KE_2 of the recoiling man in Example 3 is related to the kinetic energy KE_1 of the recoiling woman according to $KE_2 = (m_1/m_2)KE_1$.

* 14.   A two-stage rocket moves in space at a constant velocity of 4900 m/s. The two stages are then separated by a small explosive charge placed between them. Immediately after the explosion the velocity of the 1200-kg upper stage is 5700 m/s in the same direction as before the explosion. What is the velocity of the 2400-kg lower stage after the explosion?

**15.   A wagon is coasting at a speed v_A along a straight and level road. When ten percent of the wagon's mass is thrown off the wagon, parallel to the ground and in the forward direction, the wagon is brought to a halt. If the direction in which this mass is thrown is exactly reversed, everything else remaining the same, the wagon accelerates to a new speed v_B. Calculate v_B/v_A.

Section 8.4 and Section 8.5 Collisions in One and Two Dimensions

16.   A 31-kg swimmer runs with a horizontal velocity of 4.0 m/s off a boat dock into a stationary 8.0-kg rubber raft. Find the velocity that the swimmer and raft would have after the impact, if there were no friction and resistance due to the water.

17.   A 2.50×10^{-3}-kg bullet, traveling at a speed of 425 m/s, strikes the wooden block of a ballistic pendulum. The block has a mass of 0.200 kg. (a) Find the speed of the bullet/block combination immediately after the collision. (b) How much kinetic energy is lost during the collision? (c) Ignoring friction and air resistance, is any mechanical energy lost *after* the collision has occurred and the combination swings upward? Why? (d) How high does the combination rise above its initial position?

18.   A golf ball, starting with a vertical velocity component of zero, bounces down a flight of steel stairs, striking each stair once on the way down. If all the collisions with the stairs are elastic, and if the vertical height of the staircase is 3.00 m, determine the bounce height when the ball reaches the bottom of the stairs. Neglect air resistance.

19.   A 0.150-kg projectile is fired with a velocity of $+715$ m/s at a 2.00-kg wooden block that rests on a frictionless stand, as the drawing shows. The velocity of the block, immediately after the projectile passes through it, is $+40.0$ m/s. Find the velocity with which the projectile exits from the block.

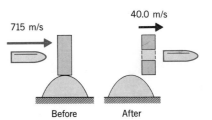

715 m/s   40.0 m/s

Before   After

20.   A 5.00-kg ball, moving to the right at a speed of 2.00 m/s on a frictionless table, collides head-on with a stationary 7.50-kg ball. Find the final velocities of the balls if (a) the collision is elastic and (b) the collision is completely inelastic.

* 21.   A 60.0-kg person, running horizontally with a speed of 3.80 m/s, jumps onto a 12.0-kg sled that is initially at rest. (a) Ignoring the effects of friction, find the velocity of the sled and person as they move away. (b) The sled and the person coast 30.0 m on the level snow before coming to rest. What is the coefficient of kinetic friction between the sled and the snow?

* 22.   By accident, a large ceramic plate is dropped vertically onto the floor and breaks into three pieces. The pieces fly apart parallel to the floor. As the plate falls, its momentum has only a vertical component, and no component parallel to the floor. After the collision, the component of the total momentum parallel to the floor must remain zero, since the external force acting on the

plate has no component parallel to the floor. Using the data shown in the drawing, find the masses of pieces 1 and 2.

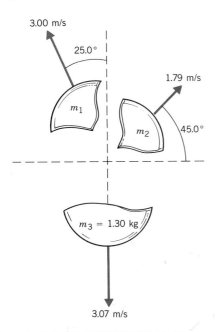

3.00 m/s

25.0°

m_1

1.79 m/s

m_2

45.0°

m_3 = 1.30 kg

3.07 m/s

* 23. A mine car, whose mass is 440 kg, rolls at a speed of 0.50 m/s on a horizontal track, as the drawing shows. A 150-kg chunk of coal has a speed of 0.80 m/s when it leaves the chute. Determine the velocity of the car/coal system after the coal has come to rest in the car.

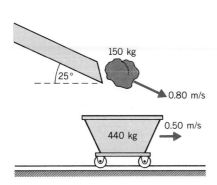

150 kg

25°

0.80 m/s

0.50 m/s

440 kg

* 24. A 50.0-kg skater is traveling due east at a speed of 3.00 m/s. A 70.0-kg skater is moving due south at a speed of 7.00 m/s. They collide and hold on to each other after the collision, managing to move off at an angle θ south of east, with a speed of v_f. Find (a) the angle θ and (b) the speed v_f, assuming that friction can be ignored.

* 25. Two identical balls are traveling toward each other with speeds of 4.0 and 7.0 m/s, and they experience an elastic head-on collision. Obtain the velocities (magnitude and direction) of each ball after the collision.

**26. A ball is dropped from rest at the top of a 20.0-ft-tall building, falls straight downward, collides inelastically with the ground, and bounces back. The ball loses 10.0% of its kinetic energy every time the ball collides with the ground. How many bounces can the ball make before it fails to reach a window sill that is 8.00 ft above the ground?

Section 8.6 Rocket Propulsion

27. A rocket burns propellant at a rate of 1.5 kg/s, ejecting gases with a speed of 7800 m/s relative to the rocket. Find the thrust exerted on the rocket.

28. During a launch, a rocket (weight = 30.0 tons) lifts off vertically from rest. The engine burns propellant at a rate of 40.0 slugs/s and ejects gases with a speed of 5500 ft/s relative to the rocket. (a) Determine the thrust developed by the engine. (b) Find the initial upward acceleration of the rocket. (c) Noting that the mass of the rocket is smaller because fuel has been consumed, determine the acceleration of the rocket at the end of 10.0 s.

* 29. During a severe storm, rain comes straight down with a speed of 15.0 m/s. A person holds out his hand, palm up, to feel how hard it is raining. The rain strikes his palm at the rate of 6.00×10^{-2} kg/s. Assuming that the rain comes to rest upon striking the hand, find the average force exerted by the rain on the hand.

* 30. A stream of water strikes a stationary turbine blade, as the drawing illustrates. The incident water stream has a velocity of +18.0 m/s, while the exiting water stream has a velocity of −18.0 m/s. The rate of flow of the water is 25.0 kg/s. Find the magnitude of the net force exerted on the water by the blade.

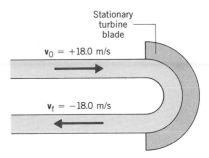

Stationary turbine blade

v_0 = +18.0 m/s

v_f = −18.0 m/s

**31. A dump truck is being filled with sand, as in the drawing. The sand falls from rest from a height of 2.00 m above the truck bed and fills the truck at a rate of 50.0 kg/s. The truck is parked on the platform of a weight scale. By how much does the scale reading exceed the weight of the truck and sand?

ADDITIONAL PROBLEMS

32. A 1550-kg car, traveling with a speed of 12.0 m/s, plows into the rear of a 1220-kg stationary car. During the collision, the two cars lock bumpers and then move together as a unit. (a) What is their common speed just after the impact? (b) What fraction of the initial kinetic energy remains after the collision?

33. In problem 32, suppose that both cars have special bumpers, so the collision is elastic. (a) What is the total linear momentum of the two-car system before and after the collision? (b) What is the total kinetic energy of the system before and after the collision? (c) What is the velocity of each car after the collision?

34. Two joggers are running with a common velocity of 4.00 m/s due south. Their masses are 90.0 and 55.0 kg. (a) Find the magnitude of the total linear momentum and the total kinetic energy of the two-jogger system. (b) Repeat part (a), assuming that one of the joggers is running due north at a speed of 4.00 m/s.

35. A person stands in a stationary canoe and throws a 5.00-kg stone with a velocity of 8.00 m/s at an angle of 30.0° above the horizontal. The person and canoe have a combined mass of 105 kg. Ignoring air resistance and effects of the water, find the horizontal recoil velocity of the canoe.

36. A car weighing 12 000 N has an initial velocity of 13 m/s, eastward. The driver applies the brakes and brings the car to rest in 6.0 s. Determine the average net force (magnitude and direction) exerted on the car.

* 37. A 3.00-kg block of wood rests on the muzzle opening of a vertically oriented rifle, the stock of the rifle being firmly planted on the ground. When the rifle is fired, an 8.00-g bullet (speed = 8.00×10^2 m/s) is completely embedded in the block. (a) Using the conservation of linear momentum, find the velocity of the block/bullet system immediately after the collison. (b) How high does the block/bullet system rise above the muzzle opening of the rifle?

* 38. A machine gun is mounted on a cart that is free to roll on a flat, frictionless surface. The gun fires parallel to the surface at a rate of ten bullets per second, each bullet having a mass of 9.7 g and a speed of 790 m/s relative to the cart. (a) Determine the average thrust exerted on the machine gun by the bullets. (b) The total mass of the gun/cart unit is 45 kg; find its acceleration.

* 39. A .22-caliber bullet is fired from a rifle that has a 61-cm barrel. The bullet has a mass of 2.6 g, and it exits the barrel with a velocity of 410 m/s, eastward. (a) Find the impulse of the force that acts on the bullet. (b) Assuming the acceleration of the bullet is constant, use the appropriate equation of kinematics from Chapter 2 and find the time that the bullet spends in the barrel. (c) What is the average net force (magnitude and direction) that acts on the bullet?

* 40. A Fourth-of-July rocket is moving as in part *a* of the drawing. The rocket suddenly breaks into two pieces of equal mass, and

they fly off with velocities $\mathbf{v}_1$ and $\mathbf{v}_2$, as in part *b* of the drawing. What are the magnitudes of $\mathbf{v}_1$ and $\mathbf{v}_2$?

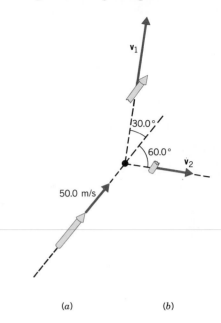

(a)   (b)

**41. A 75-kg skater is standing on ice (assumed to be frictionless) and is holding two rocks, each with a mass of 3.0 kg. (a) What is the recoil velocity of the skater, when the first rock is thrown horizontally with a speed of 14 m/s relative to the earth? (b) What is the velocity of the skater after the second rock is thrown in the same direction as the first, assuming that its speed is also 14 m/s relative to the earth? (c) Would the final velocity of the skater be different, if both rocks were thrown simultaneously, at a speed of 14 m/s relative to the earth in the same direction as in (a) and (b), rather than one at a time? Explain.

**42. Starting with an initial speed of 5.00 m/s at a height of 0.300 m, a 1.50-kg ball swings downward and strikes a 4.60-kg ball that is at rest, as the drawing shows. (a) Using the principle of conservation of mechanical energy, find the speed of the 1.50-kg ball just before impact. (b) Assuming the collision is elastic, find the velocities (magnitude and direction) of both balls just after the collision. (c) How high does each ball swing after the collision, ignoring air resistance?

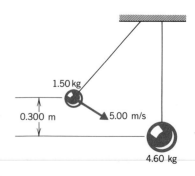

CHAPTER 9

Rotational Kinematics

Roulette is one of the favorite games at the casinos of Monte Carlo. If the angular acceleration is constant, the motion of the roulette wheel can be described by using the concepts of rotational kinematics.

Examples of rotational motion are widespread: a rotating compact disc recording, the tires on a moving automobile, and the propeller on an airplane in flight, to mention only a few. Our analysis of such motion begins with a discussion of kinematics. As you will recall from the discussion of linear motion in Chapter 2, kinematics is the area of mechanics in which the motion of an object is described without reference to the forces that act on the object. That chapter introduces the concepts of displacement, velocity, and acceleration and then develops the equations of kinematics for constant acceleration. The present chapter proceeds in a similar fashion, beginning with the analogous concepts of angular displacement, angular velocity, and angular acceleration and then turning to the equations of rotational kinematics for constant angular acceleration.

9.1 ROTATIONAL MOTION AND ANGULAR DISPLACEMENT

When a rigid object exhibits only rotational motion, points on the object move on circular paths. Figure 9.1 shows, for example, the circles traversed by points A, B, and C on a spinning skater. The centers of all such circular paths define a line, and this line

Axis of rotation

A

B

C

FIGURE 9.1 When an object exhibits only rotational motion, points on the object, such as *A*, *B*, or *C*, move on circular paths. The centers of the circles form a line that is called the axis of the rotation.

is called the **axis of rotation.** When the axis remains fixed at all times, the motion is said to be "pure rotation."

The angle through which a rigid object rotates about a fixed axis is called the **angular displacement.** Figure 9.2 shows how the angular displacement is measured for the rotating take-up reel on a tape deck. First, we identify the axis of rotation and then draw a radial line through any point on the object. A radial line is one that intersects the axis of rotation perpendicularly. As the reel turns, we observe the angle through which this line moves relative to a convenient reference orientation. The axis of rotation is determined by the spindle on which the reel is mounted, and because of the rotation, the radial line marked r moves from its initial orientation at angle θ_0 to a final orientation at angle θ. In the process, the line sweeps out the angle $\Delta\theta = \theta - \theta_0$. The angle $\Delta\theta$ is the angular displacement. Clearly, a rotating object may rotate either counterclockwise or clockwise. Standard convention distinguishes between these alternatives by calling the angular displacement positive when it is counterclockwise and negative when it is clockwise.

DEFINITION OF ANGULAR DISPLACEMENT

When a rigid body rotates about a fixed axis, the angular displacement is the angle $\Delta\theta$ swept out by a line passing through any point on the body and intersecting the axis of rotation perpendicularly. By convention, the angular displacement is positive if it is counterclockwise and negative if it is clockwise.

SI unit of angular displacement: radian (rad) *

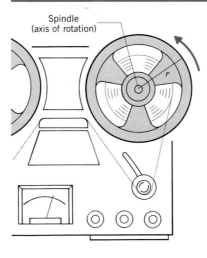

Spindle
(axis of rotation)

r

FIGURE 9.2 The take-up reel of a tape deck is rotating in the counter-clockwise direction. The angular displacement of the reel is the angle $\Delta\theta$ swept out by the radial line r as the reel turns about its axis of rotation.

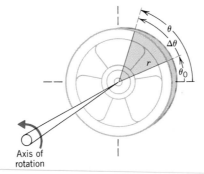

θ

$\Delta\theta$

r

θ_0

Axis of
rotation

---

* The radian is neither a base nor a derived SI unit. It is regarded as a supplementary SI unit.

The angular displacement can be expressed in three units. The first of these is the familiar **degree,** and it is well known that there are 360 degrees in one circle. The second unit is the **revolution (rev),** one revolution representing one complete turn of 360°. The most useful unit from a scientific viewpoint is the SI unit called the **radian (rad).** Figure 9.3 shows how the radian is defined, again using the take-up reel of a tape deck as an example. The picture focuses attention on a point P on the rotating reel. This point starts out on the horizontal axis, so that $\theta_0 = 0$, and the angular displacement is $\Delta\theta = \theta - \theta_0 = \theta$. As the reel rotates, the point traces out an arc of length s, which is measured along a circle of radius r. Equation 9.1 defines the angle θ in radians:

$$\theta \text{ (in radians)} = \frac{\text{arc length}}{\text{radius}} = \frac{s}{r} \qquad (9.1)$$

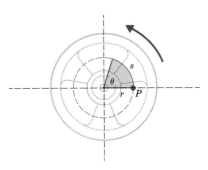

FIGURE 9.3 In radian measure, the angle θ is defined to be the arc length s divided by the radius r: $\theta = s/r$.

According to this definition, an angle in radians is the ratio of two lengths, for example, meters/meters. In calculations, therefore, the radian is treated as a dimensionless number and has no effect on other units that it multiplies or divides.

To convert between degrees and radians, it is only necessary to remember that the arc length of an entire circle of radius r is the circumference $2\pi r$ of the circle. Therefore, according to Equation 9.1, **the number of radians that corresponds to one revolution or 360° is**

$$\theta = \frac{s}{r} = \frac{2\pi r}{r} = 2\pi \text{ rad}$$

It is useful to express an angle θ in radians, because then the arc length s subtended at any radius r can be calculated conveniently by multiplying θ by r. Example 1 illustrates this point and shows how to convert between degrees and radians.

EXAMPLE 1

Synchronous or "stationary" communications satellites are put into an orbit whose radius is $r = 4.23 \times 10^7$ m, as Figure 9.4 illustrates. The orbit is in the plane of the equator. It has been proposed that two adjacent satellites should have an angular separation of no less than $\theta = 2.00°$. Find the arc length s (see drawing) that would separate adjacent satellites if this proposal were adopted.

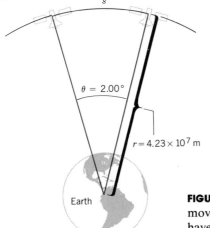

SOLUTION

Once the angle θ is converted into radians, the relation $\theta = s/r$ can be used to calculate the arc length:

$$2.00° = (2.00 \text{ degrees})\left(\frac{2\pi \text{ radians}}{360 \text{ degrees}}\right) = 0.0349 \text{ radians}$$

From Equation 9.1, it follows that

$$s = r\theta = (4.23 \times 10^7 \text{ m})(0.0349 \text{ rad})$$

$$= \boxed{1.48 \times 10^6 \text{ m (917 miles)}}$$

The radian unit, being a dimensionless quantity, is dropped from the final result, leaving the answer expressed in meters.

FIGURE 9.4 Two adjacent synchronous satellites, moving in a circular orbit of radius $r = 4.23 \times 10^7$ m, have an angular separation of $\theta = 2.00°$. The distances and angles have been exaggerated for the sake of clarity.

9.2 ANGULAR VELOCITY AND ANGULAR ACCELERATION

ANGULAR VELOCITY

According to Equation 2.2 ($\bar{\mathbf{v}} = \Delta\mathbf{s}/\Delta t$), the average linear velocity can be calculated by dividing the linear displacement of the object by the time required for the displacement to occur. For rotational motion about a fixed axis, the *average angular velocity* $\bar{\omega}$ (Greek letter "omega") can be obtained in a similar fashion, by dividing the angular displacement by the elapsed time during which the displacement occurs.

DEFINITION OF AVERAGE ANGULAR VELOCITY

$$\text{Average angular velocity} = \frac{\text{Angular displacement}}{\text{Elapsed time}}$$

$$\bar{\omega} = \frac{\theta - \theta_0}{t - t_0} = \frac{\Delta\theta}{\Delta t} \tag{9.2}$$

SI unit of angular velocity: radian per second (rad/s)

The SI unit for angular velocity is the radian per second (rad/s), although other units such as revolutions per minute (rev/min) are often encountered. In agreement with the sign convention adopted for angular displacement, angular velocity is considered to be positive when the rotation is counterclockwise and negative when the rotation is clockwise.

On a moment-to-moment basis, the angular velocity of a rotating object can be greater or less than the average value. It is natural, then, to speak of the *instantaneous angular velocity* ω as the angular velocity that exists at any given instant. To measure the instantaneous angular velocity, we follow the same procedure used in Chapter 2 for instantaneous linear velocity. In this procedure, a small angular displacement $\Delta\theta$ occurs during a small time interval Δt. The time interval is chosen to be so small that it approaches zero ($\Delta t \to 0$), and in this limit, the measured average angular velocity, $\bar{\omega} = \Delta\theta/\Delta t$, becomes the instantaneous velocity ω:

$$\omega = \lim_{\Delta t \to 0} \bar{\omega} = \lim_{\Delta t \to 0} \frac{\Delta\theta}{\Delta t} \tag{9.3}$$

The magnitude of the instantaneous angular velocity, without reference to whether it is a positive or negative quantity, is called the *instantaneous angular speed.* If the instantaneous angular velocity has the same value at every moment, the rotating object has a constant angular velocity ω, which is identical to the average value $\bar{\omega}$.

ANGULAR ACCELERATION

In linear motion, a changing velocity means that an acceleration is occurring. Such is also the case in rotational motion; a changing angular velocity means that an *angular acceleration* α (Greek letter "alpha") is occurring. There are many examples of angular acceleration. For instance, as a compact disc recording is played, the disc turns with an angular velocity that is continually increasing. And when the push buttons of an electric blender are changed from a high setting to a lower setting, the angular velocity of the blades decreases.

For a changing linear velocity, Equation 2.4 ($\bar{\mathbf{a}} = \Delta\mathbf{v}/\Delta t$) defines the average linear acceleration as the change in velocity per unit time. When the angular velocity changes from an initial value of ω_0 at time t_0 to a final value of ω at time t, the average angular acceleration is defined in an analogous fashion:

DEFINITION OF AVERAGE ANGULAR ACCELERATION

$$\text{Average angular acceleration} = \frac{\text{Change in angular velocity}}{\text{Elapsed time}}$$

$$\bar{\alpha} = \frac{\omega - \omega_0}{t - t_0} = \frac{\Delta\omega}{\Delta t} \qquad (9.4)$$

SI unit of average angular acceleration: radian per second squared (rad/s^2)

The SI unit for average angular acceleration is the unit for angular velocity divided by the unit for time, or $(\text{rad/s})/\text{s} = \text{rad/s}^2$. An angular acceleration of $+5$ rad/s^2, for example, is read as "five radians per second per second," or "five radians per second squared," and means that the angular velocity of the rotating object changes by $+5$ radians per second during each second of acceleration.

The *instantaneous angular acceleration* is the angular acceleration at a given instant. Previously, in discussing linear motion, a condition of constant acceleration was assumed, so that the average and instantaneous accelerations were identical ($\bar{\mathbf{a}} = \mathbf{a}$). Similarly, in discussing rotational motion, it is assumed that the angular acceleration is constant. Consequently, the instantaneous angular acceleration α and the average angular acceleration $\bar{\alpha}$ are the same ($\bar{\alpha} = \alpha$), and the horizontal "bar" denoting an average quantity can be omitted. The next example illustrates the concept of angular acceleration.

EXAMPLE 2

A jet, awaiting clearance for takeoff, is momentarily stopped on the runway. As the engines idle, the fan blades are rotating counterclockwise (see Figure 9.5) with an angular velocity of 110 rad/s. As the plane takes off, the angular velocity of the blades reaches 330 rad/s in a time of 14 s. Find the angular acceleration, assuming it to be constant.

SOLUTION

In applying Equation 9.4 to determine the angular acceleration, we use positive values for the angular velocities, since the rotation is counterclockwise:

$$\alpha = \bar{\alpha} = \frac{\omega - \omega_0}{t - t_0} = \frac{(+330 \text{ rad/s}) - (+110 \text{ rad/s})}{14 \text{ s}}$$

$$= \boxed{+16 \text{ rad/s}^2}$$

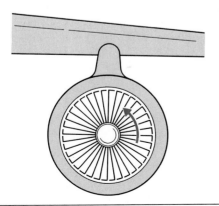

FIGURE 9.5 The fan blades of a jet engine accelerate from $\omega_0 = +110$ rad/s to $\omega = +330$ rad/s.

9.3 THE EQUATIONS OF ROTATIONAL KINEMATICS

To describe rotational motion completely requires values for the angular displacement $\Delta\theta$, the angular acceleration α, the final angular velocity ω, the initial angular velocity ω_0, and the elapsed time Δt. In Example 2, for instance, only the angular displacement of the fan blades during the 14-s interval is missing. Such missing information can be calculated, however. For convenience in the calculations, we assume that the orientation of the rotating object is given by $\theta_0 = 0$ at time $t_0 = 0$. Then, the angular displacement becomes $\Delta\theta = \theta - \theta_0 = \theta$, and the time interval becomes $\Delta t = t - t_0 = t$.

In Example 2, the angular velocity of the fan blades increases at a constant rate from an initial value of $\omega_0 = +110$ rad/s to a final value of $\omega = +330$ rad/s. Therefore, the average angular velocity is midway between the initial and final values:

$$\overline{\omega} = \tfrac{1}{2}[(+110\ \text{rad/s}) + (+330\ \text{rad/s})] = +220\ \text{rad/s}$$

In other words, when the angular acceleration is constant, the average angular velocity is given by

$$\overline{\omega} = \tfrac{1}{2}(\omega_0 + \omega) \tag{9.5}$$

With a value for the average angular velocity, Equation 9.2 can be used to obtain the angular displacement of the fan blades:

$$\theta = \overline{\omega}t = (+220\ \text{rad/s})(14\ \text{s}) = +3100\ \text{rad} \quad (+490\ \text{rev})$$

In general, when the angular acceleration is constant, the angular displacement can be obtained from

$$\theta = \overline{\omega}t = \tfrac{1}{2}(\omega_0 + \omega)t \tag{9.6}$$

In Example 2, then, known values for the initial and final angular velocities and the time interval can be used with Equation 9.4 to determine the angular acceleration and Equation 9.6 to determine the angular displacement. These two equations are all that is needed to provide a complete description of rotational motion under the condition of constant angular acceleration.

In linear kinematics, there are two analogous equations, and they are compared below with Equations 9.4 and 9.6:

Linear motion (a = constant)		Rotational motion (α = constant)	
$v = v_0 + at$	(2.4)	$\omega = \omega_0 + \alpha t$	(9.4)
$s = \tfrac{1}{2}(v_0 + v)t$	(2.7)	$\theta = \tfrac{1}{2}(\omega_0 + \omega)t$	(9.6)

The purpose of this comparison is to emphasize that the mathematical forms of Equations 2.4 and 9.4 are identical, as are the forms of Equations 2.7 and 9.6. Of course, the symbols used for the rotational variables are different than those used for the linear variables, as Table 9.1 indicates.

In Chapter 2, Equations 2.4 and 2.7 are used to derive the remaining two equations of kinematics (Equations 2.8 and 2.9). These additional equations convey no new information but are convenient to have when solving problems. Similar derivations can be carried out, starting with Equations 9.4 and 9.6. The results are listed as

TABLE 9.1 Symbols Used in Linear
and Rotational Kinematics

Quantity	Linear Motion	Rotational Motion
Displacement	s ⟶	θ
Initial velocity	v_0 ⟶	ω_0
Final velocity	v ⟶	ω
Acceleration	a ⟶	α
Time	t ⟶	t

Equations 9.7 and 9.8 and can be inferred directly from their counterparts in linear motion by making the substitution of symbols indicated in Table 9.1:

Linear motion ($a = $ constant)		Rotational motion ($\alpha = $ constant)	
$s = v_0 t + \frac{1}{2}at^2$	(2.8)	$\theta = \omega_0 t + \frac{1}{2}\alpha t^2$	(9.7)
$v^2 = v_0^2 + 2as$	(2.9)	$\omega^2 = \omega_0^2 + 2\alpha\theta$	(9.8)

Equations 9.4, 9.6, 9.7, and 9.8 are called the ***equations of rotational kinematics for constant angular acceleration*** and are used in the same fashion as the equations of linear kinematics, as the following example illustrates.

EXAMPLE 3

The blades of an electric blender are whirling with an angular velocity of 375 rad/s while the "puree" button is pushed in, as Figure 9.6 shows. When the "blend" button is depressed, the blades accelerate and reach a greater angular velocity in 44.0 rad (seven revolutions). The angular acceleration has a constant value of 1740 rad/s². Find the new angular velocity of the blades.

$\omega_0 = +375$ rad/s

FIGURE 9.6 The blades of an electric blender spin with an angular velocity of $\omega_0 = +375$ rad/s while the "puree" button is pushed in. This value changes when other push-button settings are chosen.

SOLUTION

The three known variables are listed in the table below, along with a question mark indicating that a value for the final angular velocity ω is being sought. The positive values indicate counterclockwise rotation, in accord with standard convention.

θ	α	ω	ω_0	t
$+44.0$ rad	$+1740$ rad/s²	?	$+375$ rad/s	

Equation 9.8 provides a solution to this problem, since it relates the angular variables θ, α, ω, and ω_0:

$$\omega^2 = \omega_0^2 + 2\alpha\theta = (+375 \text{ rad/s})^2$$
$$+ 2(+1740 \text{ rad/s}^2)(+44.0 \text{ rad})$$
$$= 2.94 \times 10^5 \text{ rad}^2/\text{s}^2$$
$$\boxed{\omega = +542 \text{ rad/s}}$$

The negative root is discarded, since the blades do not reverse their direction of rotation.

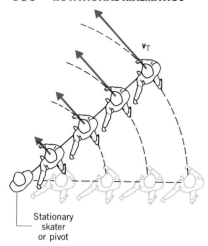

FIGURE 9.7 Each skater along the radial line moves on a circular arc. The tangential velocity $\mathbf{v}_T$ of each skater is represented by an arrow that is tangent to each arc. The magnitude of the tangential velocity is different for each skater, being smallest for the one closest to the stationary skater (pivot) and greatest for the one farthest away.

The equations of rotational kinematics can be used with any self-consistent set of units for $\theta, \alpha, \omega, \omega_0$, and t. Radians are used in Example 3 only because data are given in terms of radians. The equations of kinematics are valid, however, whether or not radians are used. Had the data for θ, α, and ω_0 been provided in rev, rev/s², and rev/s, respectively, then Equation 9.8 could have been used to determine the answer for ω directly in rev/s.

9.4 ANGULAR VARIABLES AND TANGENTIAL VARIABLES

In the familiar ice-skating stunt known as "crack-the-whip," a number of skaters attempt to maintain a straight line as they skate around the one person (the pivot) who remains in place. Figure 9.7 shows each skater moving on a circular arc and includes the corresponding velocity vector at the instant portrayed in the picture. For every individual skater, the vector is drawn tangent to the appropriate circle and, therefore, is called the **tangential velocity** $\mathbf{v}_T$. As usual, the magnitude of the velocity is measured in SI units of meters/second and is referred to as the **tangential speed.**

Of all the skaters involved in the stunt, the one farthest from the pivot has the hardest job. Why? Because, in keeping the line straight, this skater covers more distance than anyone else. To accomplish this, he must skate faster than anyone else or, in other words, must have the largest tangential speed. In fact, the line remains straight only if each person skates with exactly the correct tangential speed. Those skaters closer to the pivot must move with smaller tangential speeds than those farther out, as indicated by the magnitudes of the vectors drawn in Figure 9.7.

With the aid of Figure 9.8, it is possible to show that the tangential speed of any skater is directly proportional to his distance r from the pivot, assuming a given angular speed for the rotating line. When the line rotates as a rigid unit for a time t, it sweeps out the angle θ. The distance s through which a skater moves along a circular arc can be calculated from the relation $s = r\theta$, provided θ is measured in radians. Dividing both sides of this equation by t gives $s/t = r(\theta/t)$. The term s/t is the tangential speed v_T (e.g., in meters/second) of the skater, while the term θ/t is the angular speed ω (in radians/second) at which the line rotates:

$$v_T = r\omega \qquad (\omega \text{ in rad/s}) \tag{9.9}$$

Thus, for a given angular speed ω, the tangential speed v_T is directly proportional to the radius r. In this expression, the terms v_T and ω refer to the magnitudes of the tangential and angular velocities, respectively, and are numbers without algebraic signs.

It is important to emphasize that the angular speed ω in $v_T = r\omega$ must be expressed in radian measure (e.g., in rad/s); no other units, such as revolutions per second, are acceptable. This restriction arises because the equation was derived by using the definition of radian measure, $s = r\theta$.

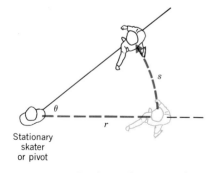

FIGURE 9.8 During a time t, the line of skaters sweeps through an angle θ. Any individual skater, located at a distance r from the stationary skater (pivot), moves through a distance s on a circular arc.

The relationship $v_T = r\omega$ is valid for any rigid object rotating about a fixed axis, and it reveals an important advantage of using the angular speed ω when describing the motion. The advantage is that ω describes the motion of the *entire rotating object.* In principle, the motion could also be described by specifying the tangential speeds v_T for all of the particles within the object. But to do so would be very awkward. As Equation 9.9 indicates, it would be necessary to specify a different value of v_T for each point on the object, according to its distance r from the axis.

The real challenge for the "crack-the-whip" skaters is to keep the line straight, while making it pick up angular speed, that is, while giving it an angular acceleration. To make the angular speed ω of the line increase, each skater must increase his tangential speed v_T, since the two quantities are related according to $v_T = r\omega$. Of course, the fact that a skater must skate faster and faster means that he must accelerate in the tangential direction, and his tangential acceleration a_T can be related to the angular acceleration α of the straight line. If time is measured relative to $t_0 = 0$, the definition of linear acceleration is given by $a_T = (v_T - v_{T0})/t$ (Equation 2.4). Substituting $v_T = r\omega$ for the tangential speed shows that

$$a_T = \frac{v_T - v_{T0}}{t} = \frac{(r\omega) - (r\omega_0)}{t} = r\left(\frac{\omega - \omega_0}{t}\right)$$

Since $\alpha = (\omega - \omega_0)/t$ according to Equation 9.4, it follows that

$$a_T = r\alpha \qquad (\alpha \text{ in rad/s}^2) \tag{9.10}$$

Equation 9.10 shows that for a given value of α the tangential acceleration a_T is proportional to the radius r, so the skater farthest from the pivot must generate the largest tangential acceleration. In this expression, the terms a_T, r, and α refer to the magnitudes of the numbers involved, without reference to any algebraic sign. Moreover, as is the case for ω in $v_T = r\omega$, only radian measure can be used for α in $a_T = r\alpha$.

As with the angular speed ω, there is an advantage to using the angular acceleration α to describe the motion of a rigid object rotating about a fixed axis. The angular acceleration describes the motion of the *entire object*. In contrast, the tangential acceleration describes only the motion of a single point on the object, and Equation 9.10 indicates that different points located at different distances r have different tangential accelerations. Example 4 stresses this advantage.

EXAMPLE 4

A helicopter blade starts from rest and, with a constant angular acceleration, reaches an operational angular speed of 6.50 rev/s in 5.00 s. For points 1 and 2 on the blade in Figure 9.9, find (a) the operational tangential speeds and (b) the magnitudes of the tangential accelerations.

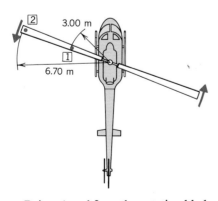

FIGURE 9.9 Points 1 and 2 on the rotating blade of the helicopter have the same angular speed and acceleration, but they have *different* tangential speeds and accelerations.

SOLUTION

(a) Using the radii shown in the drawing, we can compute the tangential speeds of each point with the aid of $v_T = r\omega$. However, since this equation can only be used with radian measure, the given angular speed must first be converted from rev/s to rad/s:

$$\omega = \left(6.50 \frac{\text{rev}}{\text{s}}\right)\left(\frac{2\pi \text{ rad}}{1 \text{ rev}}\right) = 40.8 \frac{\text{rad}}{\text{s}}$$

[Point 1] $v_T = r\omega = (3.00 \text{ m})(40.8 \text{ rad/s})$

$$= \boxed{122 \text{ m/s (273 mph)}} \tag{9.9}$$

[Point 2] $v_T = r\omega = (6.70 \text{ m})(40.8 \text{ rad/s})$

$$= \boxed{273 \text{ m/s (611 mph)}} \tag{9.9}$$

The "rad" unit, being dimensionless, does not appear in the final answers.

(b) The tangential accelerations of points 1 and 2 can be determined using $a_T = r\alpha$. First, however, it is necessary to determine the angular acceleration α, which is the same for both points since the blade is rigid. Since the blade starts from rest and attains an angular velocity of $+40.8$ rad/s in 5.00 s, α

can be obtained directly from the definition of angular acceleration:

$$\alpha = \frac{\omega - \omega_0}{t} = \frac{+40.8 \text{ rad/s} - 0}{5.00 \text{ s}} = +8.16 \text{ rad/s}^2 \quad (9.4)$$

The tangential accelerations can now be determined:

[Point 1] $a_T = r\alpha = (3.00 \text{ m})(8.16 \text{ rad/s}^2)$

$$= \boxed{24.5 \text{ m/s}^2} \quad (9.10)$$

[Point 2] $a_T = r\alpha = (6.70 \text{ m})(8.16 \text{ rad/s}^2)$

$$= \boxed{54.7 \text{ m/s}^2} \quad (9.10)$$

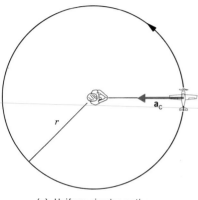

(a) Uniform circular motion

(b) Nonuniform circular motion

FIGURE 9.10 (a) A model airplane on a guide wire is flying with a constant tangential speed in a horizontal circle and, thus, is an example of uniform circular motion. The airplane has a centripetal acceleration a_c but no tangential acceleration a_T. (b) Nonuniform circular motion occurs when the tangential speed changes, in which case there is a tangential acceleration in addition to the centripetal acceleration. These two acceleration components combine vectorially to give the total acceleration a that the airplane experiences.

9.5 CENTRIPETAL ACCELERATION AND TANGENTIAL ACCELERATION

The tangential acceleration discussed in the last section should not be confused with the centripetal acceleration discussed in Chapter 6; the purpose of this section is to distinguish between them. Chapter 6 deals with **uniform circular motion,** in which a point particle moves at a constant (uniform) tangential speed on a circular path. The tangential speed v_T is the magnitude of the tangential velocity vector. Even though the magnitude of the tangential velocity is constant, an acceleration is present, since the direction of the velocity changes continually. Because the resulting acceleration points toward the center of the circle, it is called the centripetal acceleration. Figure 9.10a shows the centripetal acceleration a_c for a model airplane flying on a guide wire, the magnitude of a_c being given by

$$a_c = \frac{v_T^2}{r} \quad (6.2)$$

The subscript "T" has been included in this equation as a reminder that it is the tangential speed that appears in the numerator.

The centripetal acceleration can be expressed in terms of the angular speed ω by using the substitution $v_T = r\omega$:

$$a_c = \frac{v_T^2}{r} = \frac{(r\omega)^2}{r} = r\omega^2 \quad (9.11)$$

Only radian measure, such as rad/s, can be used for ω in this expression, since the derivation depends on the relation $v_T = r\omega$, which presumes radian measure.

While considering uniform circular motion in Chapter 6, we ignored the details of how the motion is established in the first place. For instance, how did the plane, starting from rest, attain the tangential speed that it has in Figure 9.10a? The answer, of course, is that the engine of the plane produced a thrust in the tangential direction and this force led to a tangential acceleration. In response, the tangential speed of the plane increased from moment to moment, until the situation shown in the drawing is reached. While the tangential speed is changing, the motion is called **nonuniform circular motion.**

Figure 9.10b illustrates an important feature of nonuniform circular motion. Since both the direction and the magnitude of the tangential velocity are changing, the airplane experiences two acceleration components simultaneously. The changing direction means that there is a centripetal acceleration a_c. The magnitude of a_c at any moment can be calculated using the value of the instantaneous angular speed and the radius: $a_c = r\omega^2$. The fact that the magnitude of the tangential velocity is changing means that there is also a tangential acceleration a_T. The magnitude of a_T can be determined from the angular acceleration α according to $a_T = r\alpha$, as the previous section explains. Alternatively, a_T can be calculated using Newton's second law, $F_T = ma_T$, if the magnitude F_T of the net tangential force and the mass m are known.

Figure 9.10b shows the two acceleration components. The total acceleration is given by the vector sum of $\mathbf{a}_c$ and $\mathbf{a}_T$, an addition that is particularly easy, since $\mathbf{a}_c$ and $\mathbf{a}_T$ are perpendicular. Thus, the magnitude of the total acceleration $\mathbf{a}$ can be obtained from the Pythagorean theorem as $a = \sqrt{a_c{}^2 + a_T{}^2}$, while the angle ϕ can be obtained from $\tan \phi = a_T / a_c$.

9.6 ROLLING MOTION

Rotation plays an important role in many situations, and one situation, that of rolling motion, is so familiar that it deserves special attention. Figure 9.11 shows the rolling motion of an automobile tire. The essence of rolling motion is that there is *no slipping* at the point of contact where the tire touches the ground. To a good approximation, the tires on a normally moving automobile roll and do not slip. On the other hand, the squealing tires that accompany the start of a drag race are certainly rotating, but they are not rolling as they rapidly spin and slip against the ground.

When the tires in Figure 9.11 roll, there is a relationship between the angular speed at which the tires rotate and the linear speed at which the car moves forward. With the help of part b of the drawing and the assumption that the car has a constant linear speed, it is possible to determine this relationship. As a tire rolls along the ground, the axle moves through the linear distance d. Provided that the tire does not slip, the distance d must be equal to the circular arc length s, measured along the outer edge of the tire: $d = s$. Dividing both sides of this equation by the elapsed time t shows that $d/t = s/t$. The term d/t is the speed at which the axle moves parallel to the ground, namely, the linear speed v of the car. The term s/t is the tangential speed v_T at which a

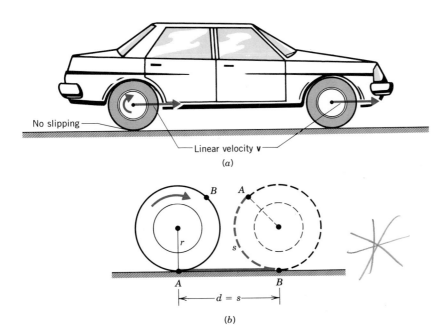

No slipping

Linear velocity **v**

(a)

$d = s$

(b)

FIGURE 9.11 (a) As an automobile moves with a linear speed v, its tires roll along the ground and have an angular speed ω with respect to the axles. (b) If the tires roll and do not slip, the distance d, through which an axle moves, equals the circular arc length s measured along the outer edge of a tire.

point on the outer edge of the tire moves relative to the axle. In addition, v_T is related to the angular speed ω about the axle according to $v_T = r\omega$. Therefore, it follows that

$$v = v_T = r\omega \qquad (\omega \text{ in rad/s}) \qquad (9.12)$$

If the car in Figure 9.11 has a linear acceleration **a** (parallel to the ground), a point on the tire's outer edge experiences a tangential acceleration $\mathbf{a}_T$ relative to the axle. The same kind of reasoning used in the last paragraph can be applied to show that the magnitudes of these accelerations are the same and that they are related to the angular acceleration α of the wheel relative to the axle:

$$a = a_T = r\alpha \qquad (\alpha \text{ in rad/s}^2) \qquad (9.13)$$

Equations 9.12 and 9.13 may be applied to any rolling motion, as long as the object does not slip against the surface on which it is rolling. Example 5 illustrates the basic features of rolling motion.

EXAMPLE 5

An automobile starts from rest and for 20.0 s has a constant linear acceleration of 0.800 m/s². During this period, the tires do not slip. The radius of the tires is 0.330 m. At the end of the 20.0-s interval what is (a) the linear velocity of the car, (b) the angular velocity of each wheel, and (c) the angle through which each wheel has rotated?

SOLUTION

(a) This part of the example is a problem in one-dimensional kinematics. In the data summary below, the forward direction of the car has been chosen as the positive direction. The linear velocity can be determined from Equation 2.4:

s	a	v	v_0	t
	+0.800 m/s²	?	0	20.0 s

$$v = v_0 + at = (0.800 \text{ m/s}^2)(20.0 \text{ s}) = \boxed{+16.0 \text{ m/s}}$$

(b) The tires do not slip. Therefore, relative to the axle, the tangential speed of a point on the edge of a tire is equal to the linear speed of the car, as in Equation 9.12; $v = v_T = r\omega$:

$$\omega = \frac{v}{r} = \frac{16.0 \text{ m/s}}{0.330 \text{ m}} = 48.5 \text{ rad/s}$$

This result gives only the magnitude of the angular velocity. In Figure 9.11, the wheels rotate in the clockwise sense as the car moves forward. According to standard convention, then, the angular velocity is $\boxed{\omega = -48.5 \text{ rad/s}}$.

(c) The angle through which each wheel turns can be obtained from Equation 9.6 of the equations of rotational kinematics by using the data summarized below:

θ	α	ω	ω_0	t
?		−48.5 rad/s	0	20.0 s

$$\theta = \tfrac{1}{2}(\omega_0 + \omega)t = \tfrac{1}{2}(-48.5 \text{ rad/s})(20.0 \text{ s}) = \boxed{-485 \text{ rad}}$$

This angle corresponds to 77.2 revolutions in the negative, or clockwise, direction.

---

*9.7 THE VECTOR NATURE OF ANGULAR VARIABLES

Angular velocity and angular acceleration have been presented with the aid of the analogy between angular variables and linear variables. Like the linear velocity and the linear acceleration, the angular quantities are also vectors and have a direction as well as a magnitude. As yet, however, the directions of these vectors have not been specified.

When a rigid object rotates about a fixed axis, it is the axis that identifies the motion, and the angular velocity vector points along this axis. Figure 9.12 shows how the direction is determined using a *right-hand rule:*

Right-Hand Rule. Grasp the axis of rotation with your right hand, so that your fingers circle the axis in the same sense as the rotation. The thumb points along the axis in the direction of the angular velocity vector.

Note that no part of the rotating object moves in the direction of the angular velocity vector.

Angular acceleration arises when the angular velocity changes, and the acceleration vector also points along the axis of rotation. The acceleration vector has the same direction as the *change* in the angular velocity, as Figure 9.13 illustrates.

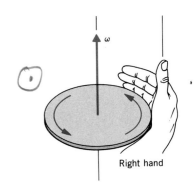

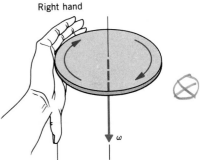

FIGURE 9.12 The angular velocity vector ω of a rotating object points along the axis of rotation. The direction along the axis depends on the sense of the rotation and can be determined with the aid of a right-hand rule (see text), as this drawing illustrates.

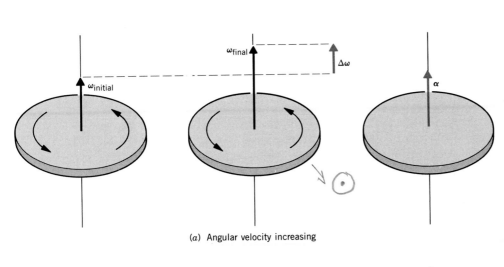

(a) Angular velocity increasing

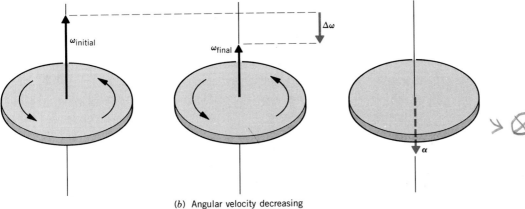

(b) Angular velocity decreasing

FIGURE 9.13 The angular acceleration vector α of a rotating object points along the axis of rotation. The direction of the acceleration is the same as the direction of the change $\Delta\omega$ in the angular velocity. Thus, α has the direction shown in part *a* when the angular velocity increases and the direction shown in part *b* when the angular velocity decreases.

SUMMARY

When a rigid body rotates about a fixed axis, the **angular displacement** is the angle swept out by a line passing through any point on the body and intersecting the axis of rotation perpendicularly. The **radian (rad)** is the SI unit of angular displacement. In radians, the angle θ is defined as the circular arc length s traveled by a point on the rotating body divided by the radial distance r of the point from the axis: $\theta = s/r$.

The **average angular velocity** $\bar{\omega}$ is the angular displacement $\Delta\theta$ divided by the elapsed time Δt: $\bar{\omega} = \Delta\theta/\Delta t$. When Δt is infinitesimally short, the average angular velocity becomes equal to the **instantaneous angular velocity** ω. The magnitude of the instantaneous angular velocity is called the **instantaneous angular speed.**

The **average angular acceleration** $\bar{\alpha}$ is the change $\Delta\omega$ in the angular velocity divided by the elapsed time Δt: $\bar{\alpha} = \Delta\omega/\Delta t$. When Δt is infinitesimally short, the average angular acceleration becomes equal to the **instantaneous angular acceleration** α.

When a rigid body rotates with constant angular acceleration about a fixed axis, the angular displacement θ, final angular velocity ω, initial angular velocity ω_0, angular acceleration α, and the elapsed time t are related by the following equations, assuming that $\theta = 0$ at $t = 0$:

$$\theta = \tfrac{1}{2}(\omega_0 + \omega)t$$

$$\omega = \omega_0 + \alpha t$$

These two equations can be combined algebraically to give two additional equations, Equations 9.7 and 9.8. Together, the four equations are known as the **equations of rotational kinematics for constant angular acceleration.** These equations may be used with any self-consistent set of units and are not restricted to radian measure.

When a rigid body rotates about a fixed axis, any single point on the body moves on a circular path of radius r. Such a point moves through an arc length s and has a tangential velocity $\mathbf{v_T}$ and, possibly, a tangential acceleration $\mathbf{a_T}$. The **angular and tangential variables are related** by the equations $s = r\theta$, $v_T = r\omega$, and $a_T = r\alpha$. These equations refer to the magnitudes of the variables involved, without reference to positive or negative signs, and only radian measure can be used when applying them.

A point on an object rotating with **nonuniform circular motion** experiences a total acceleration that is the vector sum of two perpendicular acceleration components, the tangential acceleration $\mathbf{a_T}$ and the centripetal acceleration $\mathbf{a_c}$.

The essence of **rolling motion** is that there is no slipping at the point where the object touches the surface upon which it is rolling. As a result, the tangential speed v_T (relative to the axis through the center of the object) of a point on the outer edge of a rolling object (radius r) is equal to the linear speed v with which the object moves parallel to the surface. The relation $v = v_T = r\omega$ expresses this conclusion. The magnitudes of the tangential acceleration a_T and the linear acceleration a of a rolling object are similarly related: $a = a_T = r\alpha$.

QUESTIONS

1. Explain the difference in the shape of the path traveled by a point in a rigid object and that traveled by a point in a nonrigid object, as each rotates about a fixed axis.

2. In the drawing, the flat triangular sheet ABC is lying in the plane of the paper. This sheet is going to rotate about an axis that also lies in the plane of the paper and passes through point A. Draw all such axes that are orientated so that points B and C will move on circular paths having the same radii. There are more than one.

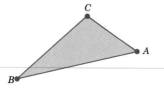

3. Two rigid objects start from rest and achieve the same angular velocity after ten revolutions about the same axis. Is it correct to say that each has (a) the same average angular acceleration and (b) the same instantaneous angular acceleration at all times? Account for your answer in both cases.

4. Are the equations of rotational kinematics valid when the angular displacement, angular velocity, and angular acceleration are expressed in terms of degrees, rather than radians? Explain why or why not.

5. Starting with Equations 9.4 and 9.6, derive Equations 9.7 and 9.8.

6. Does the *tangential* acceleration depend on the angular velocity or on the *change* in the angular velocity? Justify your choice.

7. A thin rod rotates at a constant angular speed. Consider the tangential speed of each point on the rod for the case when the axis of rotation is perpendicular to the rod (a) at its center and (b) at one end. Explain for each case whether there are any points on the rod that have the same tangential speeds.

8. A car is up on a hydraulic lift at a garage. The drive wheels are rotating with a constant angular velocity. Does a point on the rim of a wheel have (a) any tangential acceleration and (b) any centripetal acceleration? In each case, give your reasoning.

9. Two points are located on a rigid wheel that is rotating with an increasing angular velocity about a fixed axis. The axis is perpendicular to the wheel at its center. Point 1 is located on the rim and point 2 is halfway between the rim and the axis. (a) Which point turns through the greater angle in a given amount of time? At any given instant, which point has the greater (b) angular velocity, (c) angular acceleration, (d) tangential speed, (e) tangential acceleration, and (f) centripetal acceleration? Provide a reason for each of your answers.

10. Section 6.6 discusses how the uniform circular motion of a space station can be used to create "artificial" gravity for the astronauts. This can be done by adjusting the angular speed at which the station rotates, so that the centripetal acceleration at the astronaut's feet equals *g*, the acceleration due to gravity (see Figure 6.13). If such an adjustment is made, will the acceleration due to the "artificial" gravity also equal *g* at the astronaut's head? Account for your answer.

11. Explain why a given point on the rim of a tire has an acceleration when the tire is on a car that is moving at a constant linear velocity.

12. Suppose that the speedometer of a truck is set to read the linear speed of the truck, but uses a device that actually measures the angular speed of the tires. If larger diameter tires are mounted on the truck, will the reading on the speedometer be correct? If not, will the reading be greater than or less than the true linear speed of the truck? Why?

13. The automobile tire in the drawing rolls and does not slip. The axle of the wheel at point *C* moves with a constant linear velocity **v** parallel to the ground. For any point on the wheel, two velocity vectors contribute to the total velocity of that point with respect to the ground: (1) the vector representing the forward (linear) motion of the wheel and (2) the vector representing the tangential velocity caused by the rotation of the wheel. (a) Using vectors, explain why the *total* velocity of point *A* with respect to the ground has a magnitude of 2*v* at the instant shown. (b) Also using vectors, explain why the total velocity of point *B* with respect to the ground is zero.

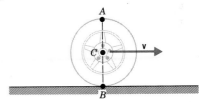

PROBLEMS

Section 9.1 Rotational Motion and Angular Displacement, Section 9.2 Angular Velocity and Angular Acceleration

1. For each of the following angles, give its equivalent in revolutions and radians: (a) 45°, (b) 180°, (c) 360°, and (d) 470°.

2. The moon has a diameter of 3.48×10^6 m and is 3.85×10^8 m from the earth. The sun has a diameter of 1.39×10^9 m and is 1.50×10^{11} m from the earth. Show that the angle θ subtended by the moon, as measured by a person standing on the earth, is approximately equal to the angle subtended by the sun.

3. On a wristwatch, what is the angular velocity (magnitude and direction) of (a) the second hand, (b) the minute hand, and (c) the hour hand? Express your answers in rev/min and rad/s.

4. There is a useful grass and weed cutting tool that utilizes a 0.21-m length of nylon "string," rotating at 6200 rev/min about an axis perpendicular to one end of the string. (a) What is the time needed for the string to sweep out an angle of 35°? (b) Assuming that the length of the string does not change, find the distance

through which the tip of the string moves during this interval.

5. After being turned on, a turntable reaches its rated angular speed of 45.0 rev/min in a time of 4.10 s. What is the average angular acceleration in rad/s²?

6. An automatic drier spins wet clothes at an angular speed of 65 rev/min. Starting from rest, the drier reaches its operating speed with an average angular acceleration of 7.0 rad/s². How long (in seconds) does it take the drier to come up to speed?

7. An electric fan is set on its HIGH setting. After the LOW push button is depressed, the angular speed of the fan blades decreases to a value of 8.00×10^2 rev/min in 1.75 s. The deceleration is 42.0 rad/s². Determine the initial angular speed of the blades in rev/min.

* 8. A space station consists of two donut-shaped living chambers, A and B, that have the radii shown in the drawing. As the station rotates to create artificial gravity, an astronaut in chamber A is moved 240 m along a circular arc. How far along a circular arc is an astronaut in chamber B moved during the same time period?

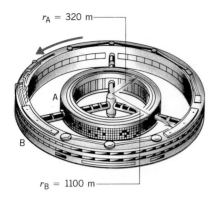

$r_A = 320$ m

$r_B = 1100$ m

* 9. Suppose that wedge-shaped pieces are cut from a pie, so that the arc length along the outer crust of each piece exactly equals the radius of the piece. (a) What is the largest number of such pieces that can be cut from the pie? (b) What is the apex angle (in degrees) of any remaining portion of the pie?

* 10. A stroboscope is a light that flashes on and off at a constant rate. It can be used to illuminate a rotating object, and if the flashing rate is adjusted properly, the object can be made to appear stationary. What is the shortest time between flashes of light that will make a three-bladed propeller appear stationary, even though it is rotating with an angular speed of 1.00×10^3 rev/min?

**11. A baton twirler throws a spinning baton directly upward. As it goes up and returns to the twirler's hand, the baton turns through four revolutions. Ignoring air resistance and assuming that the average angular speed of the baton is 108 rev/min, determine the height to which the center of the baton travels above the point of release.

Section 9.3 The Equations of Rotational Kinematics

12. The drill bit of a variable-speed electric drill has a constant angular acceleration of 2.50 rad/s². The angular speed of the bit is increasing from an initial value of 5.00 rad/s. After 4.00 s, (a) what angle has the bit turned through and (b) what is the angular speed of the bit?

13. The angular speed of the rotor in a centrifuge increases from 420 to 1420 rad/s in a time of 5.00 s. (a) Obtain the angle through which the rotor turns during this time. (b) What is the angular acceleration (assumed constant)?

14. A wheel accelerates so that its angular speed increases steadily from 150 to 580 rad/s in 16 revolutions. What is the angular acceleration (assumed constant) in rad/s²?

15. A flywheel has a constant angular deceleration of 2.0 rad/s². (a) Find the number of revolutions through which the flywheel turns as it comes to rest from an angular speed of 2100 rev/min. (b) Find the time required for the flywheel to come to rest.

* 16. After 10.0 s, a spinning roulette wheel has slowed down to an angular speed of 0.300 rev/s. During this time, the wheel makes 7.00 revolutions. Determine the angular acceleration (assumed constant) of the wheel in rev/s².

* 17. A fan blade, whose angular acceleration is a constant 2.00 rad/s², rotates through an angle of 285 radians in 11.0 s. How long did it take the blade, starting from rest, to reach the *beginning* of the 11.0-s interval?

* 18. The drive propeller of a ship starts from rest and accelerates at 3.50×10^{-3} rad/s² for 30.0 min. For the next 60.0 min, the propeller rotates at a constant angular speed. Then it decelerates at 2.00×10^{-3} rad/s², until it slows (without reversing direction) to an angular speed of 5.00 rad/s. Find the total number of revolutions made by the propeller.

**19. A child, hunting for his favorite wooden horse, is running on the ground around the edge of a stationary merry-go-round. The angular speed of the child has a constant value of 0.250 rad/s. At the instant the child spots the horse, one quarter of a turn away, the merry-go-round begins to move (in the direction the child is running) with a constant angular acceleration of 0.0100 rad/s². What is the shortest time it takes for the child to catch up with the horse?

Section 9.4 Angular Variables and Tangential Variables

20. Two fans, whose blade diameters are 6.00 and 10.0 in., are set to run at the same constant angular speed of 125 rad/s. (a) By how much does the tangential speed at the tip of the 10.0-in. blade exceed that at the end of the 6.00-in. blade? (b) Is there any point along the 10.0-in. blade that has the same tangential speed as the tip of the 6.00-in. blade? If so, where?

21. A small disk (radius = 2.00 mm) is attached to a high-speed drill at a dentist's office and is turning at 750 000 rev/min. Determine the tangential speed of a point on the outer edge of this disk. Express your answer in m/s and mi/h.

22. A meter stick is rotating with a constant angular acceleration of 12.0 rad/s² about an axis that passes through one end of the stick and is perpendicular to it. What point on the stick has a tangential acceleration whose magnitude equals that of the acceleration due to gravity?

* 23. A compact disc (CD) recording, like an LP record, contains a spiral track, along which the music is found. During playback, an LP record is a rotated at a constant angular speed. However, music is put onto the spiral track of a CD with the assumption that, during playback, the music will be detected at a *constant tangential speed* at any point. Consequently, since $v_T = r\omega$, a CD rotates at a smaller angular speed for music near the outer edge and a larger angular speed for music near the inner part of the disc. A CD has a radius of about 0.060 m and rotates at 3.5 rev/s for music at the outer edge. Find (a) the constant tangential speed at which music is detected and (b) the angular speed (in rev/s) for music at a distance of 0.025 m from the center of a CD.

* 24. A thin straight rod is rotating with an angular speed of 30.0 rev/min about a vertical axis, as the drawing shows. In one second, the end of the rod moves through a circular arc whose length equals the length of the rod. Calculate the value of the angle θ.

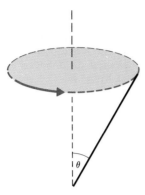

**25. One particular type of slingshot can be made from a length of rope and a leather pocket for holding the stone. The stone can be thrown by whirling it rapidly in a horizontal circle and releasing it at the right moment. Such a slingshot is used to throw a stone from the edge of a cliff, the point of release being 20.0 m above the base of the cliff. The stone lands on the ground below the cliff at a point X. The horizontal distance of point X from the base of the cliff (directly beneath the point of release) is thirty times the radius of the circle on which the stone is whirled. Determine the angular speed of the stone at the moment of release.

Section 9.5 Centripetal Acceleration and Tangential Acceleration

26. A circular disk, whose radius is 0.100 m, is rotating at a constant angular speed about an axis that is perpendicular to the disk at its center. Determine the ratio of the centripetal acceleration at point A on the circumference to that at point B located 0.0700 m from the center.

27. A $5\frac{1}{4}$-in. floppy disk for a personal computer rotates with a constant angular speed of 3.00×10^2 rev/min about an axis perpendicular to the disk at its center. (a) Find the angular speed in rad/s. (b) Find the tangential speed (in ft/s) of a point that is 2.00 in. from the center of the disk. (c) What is the magnitude of the centripetal acceleration (in ft/s^2) of the point mentioned in part (b)?

28. A 220-kg speedboat is negotiating a circular turn (radius = 32 m) around a buoy. During the turn, the engine causes a constant net tangential force of 550 N to be applied to the boat. The initial tangential speed of the boat going into the turn is 5.0 m/s. (a) Find the tangential acceleration. After the boat is 2.0 s into the turn, find (b) the centripetal acceleration and (c) the total acceleration (magnitude and direction with respect to the radius).

* 29. A disk has a constant angular acceleration of 4.00 rad/s^2 about an axis perpendicular to the disk at its center. Find the radius of a point on the disk where, 0.500 s after the disk begins to rotate, the magnitude of the total acceleration (centripetal plus tangential) equals that of the acceleration due to gravity.

* 30. A thin rigid rod is rotating with a constant angular acceleration about an axis that passes perpendicularly through one of its ends. At one instant, the total acceleration vector (centripetal plus tangential) at the other end of the rod makes a 60.0° angle with respect to the rod and has a magnitude of 15.0 m/s^2. The rod has an angular speed of 2.00 rad/s at this instant. What is (a) the rod's length and (b) its angular speed one second later?

**31. The blades of a windmill start from rest and rotate with an angular acceleration of 22.0 rad/s^2. At any point on a blade, how much time passes before the magnitude of the tangential acceleration equals the magnitude of the centripetal acceleration?

Section 9.6 Rolling Motion

Note: All problems in this section assume that there is no slipping of the surfaces in contact during the rolling motion.

32. (a) A bicycle travels a linear distance of 515 m. How many revolutions does one of the tires (diameter = 0.711 m) make? (b) What does this angle correspond to in radians?

33. An automobile tire has a radius of 0.330 m, and its center moves forward with a linear speed of $v = 15.0$ m/s. (a) Determine the angular speed of the wheel. Relative to the axle, what is the tangential speed of (b) a point on the circumference of the tire and (c) a point located 0.175 m from the axle?

34. On an open-reel tape deck, the tape is being pulled past the playback head at a constant linear speed of 15.0 in./s. (a) Using the data in part a of the drawing, find the angular speed of the take-up reel (in rad/s). (b) After forty minutes of playing time the take-up reel is almost full, as part b of the drawing indicates. Find the average angular acceleration (in rad/s^2) of the reel and specify whether the acceleration indicates an increasing or decreasing angular velocity.

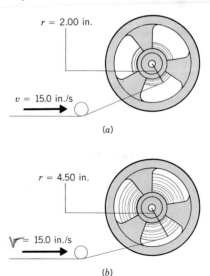

$r = 2.00$ in.

$v = 15.0$ in./s

(a)

$r = 4.50$ in.

$v = 15.0$ in./s

(b)

* 35. A person lowers a bucket into a well by turning the hand crank, as the drawing illustrates. The crank handle moves with a constant tangential speed of 1.20 m/s on its circular path. Find the linear speed (in m/s) with which the bucket moves down the well.

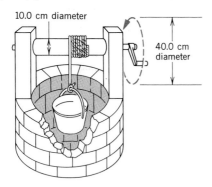

36. The two-gear combination shown in the drawing is being used to hoist the load L with a constant upward speed of 2.50 m/s. The rope attached to the load is being wound onto a cylinder (radius = 0.300 m) behind the big gear. Assuming that the depth of the teeth of the gears is negligible compared to the radii, determine the angular velocity (magnitude and direction) of the smaller gear.

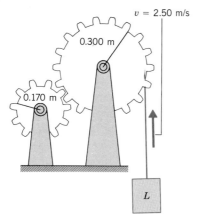

37. A bicycle, moving along the ground, has a constant linear acceleration. What must be the angular acceleration of the bike tires, so that in 5.00 s the bike moves through a distance equal to ten times the radius of the tires and, in the process, doubles its forward linear speed?

ADDITIONAL PROBLEMS

38. A circular disk rotates with a constant angular speed about an axis perpendicular to the disk at its center. Point A on the disk is located 0.100 m from the center and has a tangential speed of 1.50 m/s. (a) What is the angular speed of the disk? (b) Find the tangential speed of point B, which is located 0.170 m from the center.

39. Assuming the earth rotates once about its axis in 23.9 hours and orbits the sun once in 365 days, find the average angular

speed (in rad/s) of the earth's (a) rotational motion and (b) orbital motion.

40. A train is rounding a circular curve whose radius is 2.00×10^2 m. At one instant, the train has an angular acceleration of 1.50×10^{-3} rad/s^2 and an angular speed of 0.0500 rad/s. (a) Find the magnitude of the total acceleration (centripetal plus tangential) of the train. (b) Determine the direction of the total acceleration relative to the radial direction.

41. The shaft of a pump starts from rest and has an angular acceleration of 3.00 rad/s^2 for 18.0 s. At the end of this interval, what is (a) the shaft's angular speed and (b) the angle through which the shaft has turned?

42. An electric circular saw is designed to reach its operating angular speed, starting from rest, in 1.50 s. Its average angular acceleration is 328 rad/s^2. Obtain its operating speed (a) in rad/s and (b) in rev/min.

43. The drawing shows a view of the platter (from beneath) on a belt-drive turntable. The platter has an angular speed of $33\frac{1}{3}$ rev/min. The pulley on the motor shaft has a radius of 0.500 inches. Assuming that the belt does not slip, determine the angular speed of the motor shaft in rev/min.

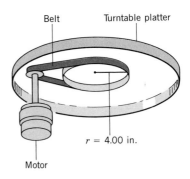

44. A dentist causes the bit of a high-speed drill to accelerate from an angular speed of 1.00×10^5 rev/min to an angular speed of 3.00×10^5 rev/min. In the process, the bit turns through 3.00×10^3 revolutions. Assuming a constant angular acceleration, how long would it take the bit to reach its maximum speed of 7.50×10^5 rev/min, starting from rest?

45. A thin rigid wooden rectangle is rotating with a constant angular acceleration about an axis that passes perpendicularly through one corner, as the drawing shows. The tangential acceleration measured at corner A has twice the magnitude of that measured at corner B. What is the ratio L_1/L_2 of the lengths of the sides of the rectangle? See Drawing 45.

46. At the local swimming hole, a favorite trick is to run horizontally off a cliff that is 9.0 ft above the water. One show-off claims to have done this trick in a new way, which involves tucking into a "ball" and (he claims) rotating through three revolutions before hitting the water. Ignoring air resistance and assuming the "ball" (diameter = 3.0 ft) is already rotating at the instant it begins falling vertically, evaluate the likelihood of doing such a trick by calculating the constant angular speed (in rev/s) that would be necessary.

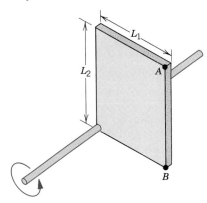

Drawing 45

**47. Two disks start from rest and rotate as the drawing shows. The angular accelerations of these disks are constant but have different magnitudes. The solid (not dashed) colored reference lines on the disks were directly above each other at the start. The drawing shows the disks at a later time, when the lines are separated by 30.0°, the angular speed of disk A exceeds that of disk B by 1.50 rad/s, and each disk has made less than one revolution. The angular acceleration of disk A is 2.50 rad/s². Find the angle through which each disk has turned.

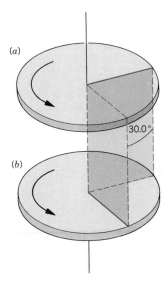

CHAPTER 10

Rotational Dynamics

Paddle wheelers were once a frequent sight on the Mississippi River, and the *Delta Queen* shown here is a fine example of this type of boat. In order to accelerate the boat, the paddle wheel must have an angular acceleration. As the text discusses, an angular acceleration occurs when a net torque acts on a rotating object. The engines of the boat contribute to the net torque that acts on the paddle wheel.

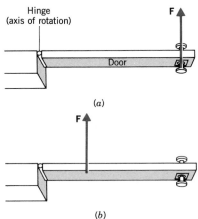

FIGURE 10.1 It is noticeably easier to open the door with a force of a given magnitude by (*a*) pushing at the door's outer edge than by (*b*) pushing closer to the axis of rotation (the hinge).

10.1 TORQUE

When there is a change in the linear velocity of a moving object, such as a car, one or more forces are responsible for the linear acceleration. But when there is a change in the angular acceleration of a rotating object, such as a turntable platter, what is responsible for the angular acceleration? The quantity that can produce an angular acceleration is called a *torque,* and is represented by the symbol τ (Greek letter "tau"). In fact, it is found experimentally that the angular acceleration of a rotating object is proportional to the net torque.*

To provide insight into the concept of torque, Figure 10.1 shows a door. If you push on the door with a force **F**, as in part *a* of the drawing, you will find that the door opens more quickly when the force is larger; that is, a larger force causes a larger angular acceleration of the door. Thus, larger forces lead to larger torques, other

* This relation between the angular acceleration and the net torque is analogous to Newton's second law for linear motion and, like it, is valid only in an inertial reference frame.

things being equal. However, the door does not open so quickly if you apply the same force at a point closer to the hinge, as in part *b* of the drawing. The force now produces less angular acceleration, so the torque of the force is less. Furthermore, if the direction of the force is not perpendicular to the door, the angular acceleration, and hence torque, is less than when the force is perpendicular to the door. Thus, the torque depends on the magnitude of the force, on the point where the force is applied relative to the axis of rotation, and on the direction of the force.

For convenience, we deal with situations in which the force lies in a plane that is perpendicular to the axis of rotation. In Figure 10.2, for instance, the axis is perpendicular to the page and the force lies in the plane of the paper. In the definition of torque, the line of action and the lever arm of a force play an important role. The *line of action* is an extended line drawn colinear with the force, as in Figure 10.2. The *lever arm* is the distance between the line of action and the axis of rotation, measured on a line that is perpendicular to both. The lever arm is denoted by the symbol ℓ, and it is also shown in the drawing. The magnitude of the torque is defined as the magnitude of the force times the lever arm, as expressed below:

DEFINITION OF TORQUE

Magnitude: Torque = (Force) × (Lever arm)

$$\tau = F\ell \qquad\qquad (10.1)$$

Direction: The torque is a positive quantity if the force tends to produce a counterclockwise angular acceleration about the axis, and negative if the force tends to produce a clockwise angular acceleration.

SI unit of torque: newton · meter (N · m)

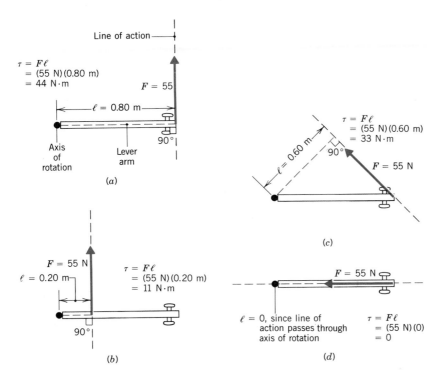

FIGURE 10.2 Four applications of the definition of torque, $\tau = F\ell$. The view is the top view of a door, and the hinges defining the axis of rotation appear as a dot. While the magitude of the applied force is 55 N in each case, the resulting torques are different, because the lever arms ℓ (see the dashed, colored lines) are different.

Figure 10.2 shows a 55-N force being applied to a door in four different ways. In each case the torque is calculated according to $\tau = F\ell$, and the results show that the *same* force can produce *different* torques, depending on the value of the lever arm ℓ. Notice in part d of the drawing that the line of action of **F** passes through the axis of rotation (the hinge) and, hence, the lever arm is zero. Consequently, the torque is zero in this case, and there is no angular acceleration.

Viewed along the axis of rotation, an object can rotate either counterclockwise or clockwise. Following the sign convention used for angular acceleration in Chapter 9, we assign a positive value to a torque that tends to produce a counterclockwise angular acceleration and a negative value to a torque that tends to produce a clockwise angular acceleration.

Example 1 illustrates how the Achilles tendon produces a torque about the ankle joint.

EXAMPLE 1

Figure 10.3 shows the Achilles tendon exerting a force of magnitude $F = 720$ N on the heel at the point P. Determine the torque (magnitude and direction) of this force about the ankle joint.

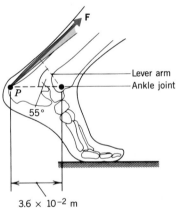

FIGURE 10.3 The force **F** generated by the Achilles tendon produces a clockwise (negative) torque about the ankle joint.

SOLUTION

To calculate the magnitude of the torque, it is necessary to have a value for the lever arm ℓ. From the drawing, it can be seen that the lever arm is $\ell = (3.6 \times 10^{-2} \text{ m}) \cos 55° = 2.1 \times 10^{-2}$ m. The magnitude of the torque is

$$\tau = F\ell = (720 \text{ N})(2.1 \times 10^{-2} \text{ m}) = 15 \text{ N} \cdot \text{m} \quad (10.1)$$

The force **F** tends to produce a clockwise rotation about the ankle joint, so the torque is negative: $\boxed{\tau = -15 \text{ N} \cdot \text{m}}$.

10.2 RIGID OBJECTS IN EQUILIBRIUM

TRANSLATIONAL AND ROTATIONAL MOTION

Figure 10.4 illustrates the two kinds of motion that are possible for a rigid object. Part a of the drawing depicts an example of **translational motion**. Translational motion is any motion where all points on the body travel in parallel paths (not necessarily straight lines). In pure translation there is no rotation of any line in the body. Because translational motion can occur along a curved line, it is often referred to as curvilinear motion or, more simply, linear motion. All points within an object have the same motion in pure translation, so the motion can be completely specified by giving values for the displacement, velocity, and acceleration of any one point.

In addition to translational motion, *rotational motion* is also possible. As discussed in Chapter 9, rotational motion can be described by giving values for the angular displacement, the angular velocity, and the angular acceleration. Frequently, rotational motion occurs in conjunction with translational motion, as indicated by the somersaulting gymnast in Figure 10.4b. An object can have either a translational acceleration, an angular acceleration, or both. As will now be discussed, both types of acceleration are zero when the object is in equilibrium.

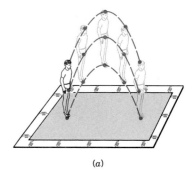

(a)

EQUILIBRIUM

If a rigid body is in equilibrium, its translational acceleration is zero, and the net force applied to the body is also zero, since $\Sigma F = ma = 0$. For two-dimensional motion, the condition $\Sigma F = 0$ means that the x and y components of the net force are separately zero: $\Sigma F_x = 0$ and $\Sigma F_y = 0$. These two equations are Equations 5.1a and 5.1b. When calculating the net force, it is sufficient to include only *external forces*, that is, those forces applied to the object by external agents. The internal forces that occur between the internal parts of an object need not be included, since they always occur in action–reaction pairs. As far as the motion of the entire object is concerned, the action–reaction pairs of internal forces do not have any effect, since each pair always consists of oppositely directed forces of equal magnitude, and the effect of one force cancels the effect of the other.

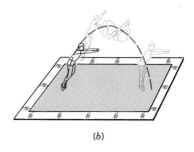

(b)

FIGURE 10.4 An example of (a) translational motion and (b) combined translational and rotational motion.

A rigid body can rotate as well as translate, and if it is in equilibrium, its angular acceleration α is zero. As mentioned in Section 10.1, the angular acceleration of an object is proportional to the net torque that acts on the object. Each positive torque tends to generate a counterclockwise angular acceleration, while each negative torque tends to generate a clockwise angular acceleration. If the angular acceleration of an object is zero, the sum of the positive torques is balanced by the sum of the negative torques, the net torque being zero. Using the symbol $\Sigma\tau$ to represent the net torque (the sum of all positive and negative torques), we write this condition as

$$\Sigma\tau = 0 \qquad (10.2)$$

The conditions that must be met if a rigid body is to be in equilibrium are summarized below.

EQUILIBRIUM OF A RIGID BODY

A rigid body is in equilibrium if it has zero translational acceleration and zero angular acceleration. In equilibrium, the sum of the externally applied forces is zero, and the sum of the externally applied torques is zero:

$$\Sigma F_x = 0 \qquad \Sigma F_y = 0 \qquad \text{(5.1a and 5.1b)}$$

$$\Sigma\tau = 0 \qquad \text{(10.2)}$$

The procedure used to analyze the forces and torques acting on a body in equilibrium is very similar to the procedure outlined in Section 5.2. For use here, the first four steps of that procedure are summarized below. There are two additional steps that are needed to account for any torques that may be present:

Step 1. Select the object to which the conditions for equilibrium are to be applied.

Step 2. Draw a free-body diagram that shows all the external forces that act on the object, each force with its proper direction.

Step 3. Choose a convenient set of x, y axes and resolve all forces into components that lie along these axes.

Step 4. Apply the conditions that specify the balance of forces at equilibrium: $\Sigma F_x = 0$ and $\Sigma F_y = 0$.

Step 5. Select a convenient axis of rotation. Clearly identify the point where each force acts on the object, and calculate the torque produced by each force about the axis of rotation. Set the sum of the torques about this axis equal to zero: $\Sigma \tau = 0$.

Step 6. Solve the equations in Steps 4 and 5 for the desired unknown quantities.

Example 2 illustrates how the conditions for equilibrium are applied to a diving board.

EXAMPLE 2

A woman whose weight is 535 N is poised at the right end of a diving board. The board has negligible weight and is bolted down at the left end, while being supported near its middle by a fulcrum, as Figure 10.5 shows. Find the forces $\mathbf{F}_1$ and $\mathbf{F}_2$ that the bolt and the fulcrum, respectively, exert on the board.

SOLUTION

The diving board is selected as the object for analysis, and part b of the figure shows the free-body diagram. The three forces that act on the board are $\mathbf{F}_1$, $\mathbf{F}_2$, and the force due to the diver's weight $\mathbf{W}$. In choosing the directions of $\mathbf{F}_1$ and $\mathbf{F}_2$ we have used our intuition: $\mathbf{F}_1$ points downward, because the bolt must pull in that direction to counteract the tendency of the board to rotate clockwise about the fulcrum; $\mathbf{F}_2$ points upward because the board pushes downward against the fulcrum and, in reaction, the fulcrum pushes upward on the board.

Since the board is in equilibrium, the sum of the vertical forces must be zero:

$$\Sigma F_y = -F_1 + F_2 - 535\ \text{N} = 0 \qquad (5.1\text{b})$$

Similarly, the sum of the torques must be zero, $\Sigma \tau = 0$. For calculating torques, we select an axis that passes through the left end of the board and is perpendicular to the page. (We will see shortly that this choice is arbitrary.) Note that $\mathbf{F}_1$ produces no torque since it passes through the axis and has a zero lever arm, $\mathbf{F}_2$ creates a counterclockwise (positive) torque, and $\mathbf{W}$ produces a clockwise (negative) torque. The free-body diagram shows the lever arms for the torques.

$$\Sigma \tau = +F_2 \ell_2 - W\ell_w$$
$$= +F_2(1.60\ \text{m}) - (535\ \text{N})(4.50\ \text{m}) = 0 \qquad (10.2)$$
$$\boxed{F_2 = 1500\ \text{N}}$$

This value for F_2 can be substituted into Equation 5.1b above to show that $\boxed{F_1 = 970\ \text{N}}$.

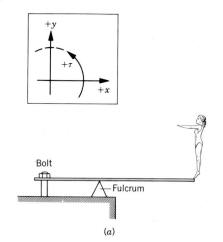

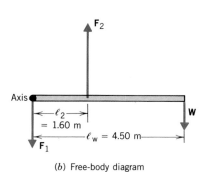

(b) Free-body diagram

(a)

FIGURE 10.5 (a) A diver stands at the end of a diving board that is in equilibrium. (b) The free-body diagram shows the three forces that act on the board. The insert indicates the positive x and y directions for the forces, as well as the positive (counterclockwise) direction for the torques.

THE AXIS USED FOR CALCULATING TORQUES IS ARBITRARY

In Example 2 the sum of the external torques is calculated using an axis that passes through the left end of the diving board. *However, the location of the axis is completely arbitrary, for an object in equilibrium is in equilibrium with respect to any axis whatsoever.* Thus, the sum of the external torques is zero, no matter where the axis is placed. Example 3 illustrates this important point.

EXAMPLE 3

Repeat Example 2, choosing another rotational axis for computing the torques.

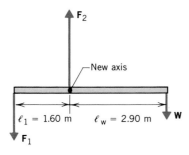

FIGURE 10.6 The free-body diagram for the diving board shown in Figure 10.5a. The rotational axis is now at the fulcrum.

SOLUTION

A new axis is selected so that it passes through the fulcrum and is perpendicular to the page, as Figure 10.6 indicates. The equation representing the balance of the vertical forces is the same as it is in Example 2:

$$\Sigma F_y = -F_1 + F_2 - 535 \text{ N} = 0$$

The equation representing the balance of torques with respect to the new axis is

$$\Sigma \tau = +F_1 \ell_1 - W\ell_w = +F_1(1.60 \text{ m}) - (535 \text{ N})(2.90 \text{ m}) = 0$$

Solving for F_1 yields $\boxed{F_1 = 970 \text{ N}}$. This value for F_1 can be substituted into the equation for the balance of the vertical forces to show that $\boxed{F_2 = 1500 \text{ N}}$. These answers are identical to those obtained in Example 2 and illustrate the fact that the axis for computing torques may be chosen arbitrarily.

While Example 3 illustrates that the location of the rotational axis is arbitrary, *in practice* one usually chooses its location so the lines of action of one or more of the unknown forces pass through the axis. Such a choice simplifies the torque equation, because the torques produced by these forces are zero.

DETERMINATION OF THE LEVER ARMS

In a calculation of torque, the lever arm of the force must be determined relative to the axis of rotation. In Examples 2 and 3 the lever arms are rather obvious, but sometimes this is not the case. Example 4 illustrates a situation in which a little care must be exercised in determining lever arms.

EXAMPLE 4

In Figure 10.7a an 8.00-m ladder of weight $W_L = 355$ N leans against a smooth vertical wall. (The term "smooth" means that the wall can exert only a normal force directed perpendicular to the surface and cannot exert a friction force directed parallel to the surface.) A firefighter, whose weight is $W_F = 875$ N, stands on the ladder. The weight of the ladder and the force due to the weight of the firefighter are assumed to act on the ladder at the points identified in parts b and c of the figure. (See part c for distances.) Find the forces that the wall and ground exert on the ladder.

SOLUTION

Focusing attention on the ladder itself, part b of the figure shows the corresponding free-body diagram. The forces exerted on the ladder are (1) its weight W_L acting at point A, (2) the force W_F due to the firefighter standing at point B, (3) the force P applied to the top of the ladder by the wall and directed perpendicular to the wall, and (4) the forces G_x and G_y, which are the horizontal and vertical components of the force exerted by the ground on the bottom of the ladder. The force G_x is produced by static friction and must be present to prevent the ladder from slipping away from the wall. The force G_y is the normal force applied to the ladder by the ground.

We begin by noting that the net force acting on the ladder must be zero, since the ladder is in equilibrium:

$$\Sigma F_x = +G_x - P = 0 \qquad (5.1a)$$

$$\Sigma F_y = +G_y - 355\text{ N} - 875\text{ N} = 0 \qquad (5.1b)$$

$$\boxed{G_y = 1230\text{ N}}$$

Equation 5.1a cannot be solved as it stands, because it contains two unknown variables. However, another equation can be obtained from the fact that the net torque acting on an object in equilibrium is zero. In calculating torques, it is convenient to use an axis at the left end of the ladder, directed perpendicular to the page, as Figure 10.7c indicates. This axis is convenient, because G_x and G_y produce no torques about it, their lever arms being zero. Consequently, these forces will not appear in the equation representing the balance of torques. The lever arms for the re-

maining forces are shown as colored, dashed lines in the drawing. The following list summarizes these forces, the lever arms, and the torques:

Force	Lever arm	Torque
$W_L = 355$ N	$\ell_L = (4.00\text{ m})\cos 50.0°$	$-W_L\ell_L$
$W_F = 875$ N	$\ell_F = (6.30\text{ m})\cos 50.0°$	$-W_F\ell_F$
P	$\ell_P = (8.00\text{ m})\sin 50.0°$	$+P\ell_P$

Setting the sum of the torques equal to zero gives

$$\begin{aligned}\Sigma\tau &= -W_L\ell_L - W_F\ell_F + P\ell_P \qquad (10.2)\\ &= -(355\text{ N})(4.00\cos 50.0°\text{ m})\\ &\quad -(875\text{ N})(6.30\cos 50.0°\text{ m})\\ &\quad +P(8.00\sin 50.0°\text{ m}) = 0\end{aligned}$$

This equation can be solved directly to give $\boxed{P = 727\text{ N}}$. Equation 5.1a indicates that $G_x = P$, so $\boxed{G_x = 727\text{ N}}$.

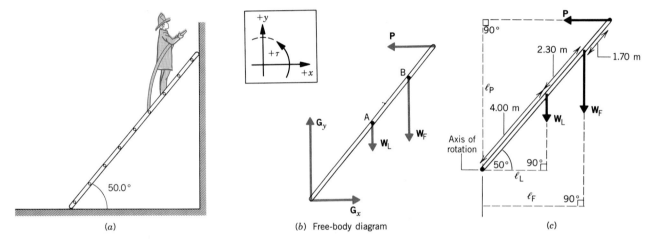

FIGURE 10.7 (a) A ladder leaning against a smooth (frictionless) wall. (b) The free-body diagram for the ladder. (c) The forces and their lever arms (drawn in color). The axis of rotation is at the lower end of the ladder and is perpendicular to the page.

SELECTING THE DIRECTIONS OF THE FORCES IN THE FREE-BODY DIAGRAM

To a large extent the directions of the forces acting on an object in equilibrium can be deduced using intuition. Sometimes, however, the direction of an unknown force is not obvious, and it is inadvertently drawn reversed in the free-body diagram. This kind of mistake causes no difficulty. *Choosing the direction of an unknown force backward in the free-body diagram simply means that the value determined for the force will be a negative number,* as the next example illustrates.

EXAMPLE 5

A bodybuilder, strengthening his shoulder muscles, holds a dumbbell of weight W_d as in Figure 10.8a. His arm is fully

extended, horizontal, and weighs $W_a = 7.00$ lb. The weights W_d and W_a act on the arm at the points identified in parts b and c of the drawing. (See part c for distances.) The deltoid muscle is assumed to be the only muscle acting and is attached to the arm as shown. The maximum force that the deltoid muscle can supply to keep the arm horizontal is $T = 415$ lb. What is the weight of the largest dumbbell that can be held? Also, determine the horizontal and vertical force components, $\mathbf{S}_x$ and $\mathbf{S}_y$, that the shoulder joint applies to the arm.

SOLUTION

Figure 10.8b shows the free-body diagram of the arm. Note that $\mathbf{S}_x$ is directed to the right, because the deltoid muscle pulls the arm in toward the shoulder joint, and the joint pushes back in accordance with Newton's third law. The direction of the force $\mathbf{S}_y$, however, is less obvious, and we are alert for the possibility that the direction chosen in the free-body diagram might be backward. If so, the value for $\mathbf{S}_y$ will turn out negative in the following analysis.

The arm is in equilibrium, so the forces must balance:

$$\Sigma F_x = S_x - (415 \text{ lb}) \cos 13.0° = 0 \qquad (5.1a)$$

$$\boxed{S_x = 404 \text{ lb}}$$

$$\Sigma F_y = +S_y + (415 \text{ lb}) \sin 13.0° - 7.00 \text{ lb} - W_d = 0 \qquad (5.1b)$$

Equation 5.1b cannot be solved at this point, because it contains two unknowns. However, since the arm is in equilibrium, the torques acting on the arm must also balance, and this fact provides another equation. To calculate torques, we choose an axis that passes through the left end of the arm and is perpendicular to the page. With this axis, the torques due to $\mathbf{S}_x$ and $\mathbf{S}_y$ are zero. According to the data in part c of the drawing, the condition specifying a zero net torque can be written as

$$\Sigma \tau = -W_a \ell_a - W_d \ell_d + T\ell_t \qquad (10.2)$$
$$= -(7.00 \text{ lb})(0.900 \text{ ft}) - W_d(2.00 \text{ ft})$$
$$+ (415 \text{ lb})(0.500 \sin 13.0° \text{ ft}) = 0$$

Equation 10.2 can be solved directly to show that the maximum dumbbell weight is only $\boxed{W_d = 20.2 \text{ lb}}$. Finally, this value for W_d can be substituted in Equation 5.1b to show that the value for S_y is $\boxed{S_y = -66.2 \text{ lb}}$. The minus sign indicates that the choice of direction for S_y in the free-body diagram is wrong. In reality, S_y has a magnitude of 66.2 lb but is directed downward, not upward.

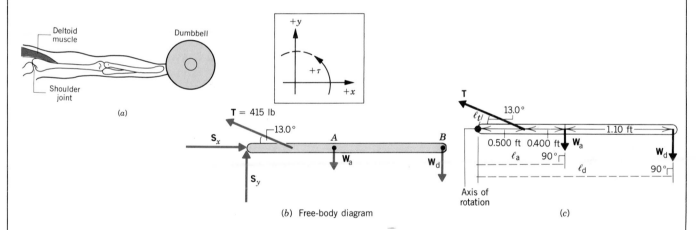

(a)

(b) Free-body diagram

(c)

FIGURE 10.8 (a) The fully extended, horizontal arm of a bodybuilder supports a dumbbell. (b) The free-body diagram for the arm. (c) The forces that act on the arm and their lever arms (drawn in color). The axis of rotation is at the left end of the arm and is perpendicular to the page.

10.3 CENTER OF GRAVITY

Often, it is important to know the torque produced by the weight of an *extended* body. In Examples 4 and 5, for instance, it is necessary to determine the torques caused by the weight of the ladder and arm, respectively. In both cases the weight of the object is considered to act at a definite point for the purpose of calculating the torque. This point is called the ***center of gravity*** (abbreviated "cg").

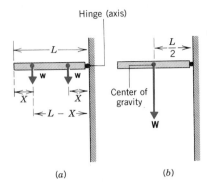

FIGURE 10.9 A thin uniform rod of length L is attached to a vertical wall by a hinge. (a) Two identical particles, each of weight w, are located symmetrically with respect to the center of the rod. The weights have lever arms of X and $L - X$. (b) The center of gravity of the rod is at its geometrical center.

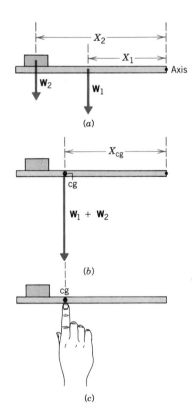

FIGURE 10.10 (a) A board is horizontal, and a box rests near the left end of the board. (b) The total weight acts at the center of gravity of the group. (c) The group can be balanced by applying an external force at the center of gravity.

DEFINITION OF CENTER OF GRAVITY

The center of gravity of a rigid body is the point at which its weight can be considered to act when calculating the torque due to the weight.

When an object has a symmetric shape and its weight is distributed uniformly, the center of gravity lies at its geometrical center. To see why this is so, consider a thin, uniform, horizontal rod of length L that is attached to a vertical wall by a hinge (Figure 10.9). Imagine the rod as being subdivided into a large number of identical particles, each having a weight w. Part a of the figure singles out two particles that are located symmetrically on either side of the rod's center. One particle has a lever arm of X, while the other has a lever arm of $L - X$. The net torque produced by these two weights is $\Sigma\tau = wX + w(L - X) = (2w)(L/2)$. This same value for the net torque can be obtained by treating the combined weight $2w$ of both particles as if it acted at the location $L/2$, that is, at the center of the rod. Since the weight of the rod is distributed uniformly, the entire rod can be treated in terms of such pairs of particles. Thus, the total weight W of the rod can be considered to act at the center of the rod when calculating torque, as in Figure 10.9b. The center of gravity of the rod, then, is located at the geometrical center, and the torque due to the weight W of the rod is $\tau = W(L/2)$. In a similar fashion, it can be demonstrated that the center of gravity of any symmetrically shaped and uniform object, such as a sphere, disk, cube, cylinder, etc., is located at its geometrical center. The center of gravity need not necessarily lie within the object itself. The center of gravity of a stereo record, for instance, lies at the center of the spindle hole and, hence, is located "outside" the record.

Suppose we have a group of extended objects, each of whose weight and center of gravity are known, and it is necessary to know the center of gravity for the group as a whole. As an example, Figure 10.10a shows a group that is composed of two parts: a horizontal uniform board (weight $\mathbf{W}_1$) and a uniform box (weight $\mathbf{W}_2$) resting near the left end of the board. The center of gravity of the group can be determined by calculating the net torque generated by the board and box about an axis that is arbitrarily picked to lie at the right end of the board. For this purpose, part a of the figure shows the weights $\mathbf{W}_1$ and $\mathbf{W}_2$ and the corresponding lever arms X_1 and X_2. The net torque can also be calculated by treating the total weight, $\mathbf{W}_1 + \mathbf{W}_2$, as if it were located at the center of gravity, which has the lever arm X_{cg}, as part b of the drawing indicates. Since the two values for the net torque must be the same, they can be set equal to obtain the following result:

$$W_1 X_1 + W_2 X_2 = (W_1 + W_2)X_{cg}$$

This expression can be solved for X_{cg}, which locates the center of gravity relative to the axis:

$$\begin{bmatrix} \text{Center} \\ \text{of} \\ \text{gravity} \end{bmatrix} \qquad X_{cg} = \frac{W_1 X_1 + W_2 X_2 + \cdots}{W_1 + W_2 + \cdots} \qquad (10.3)$$

where the notation "$+ \cdots$" indicates that the reasoning above can be extended to account for any number of weights distributed along a horizontal line. Figure 10.10c illustrates that the group can be balanced by a single external force, if the line of action of that force passes through the center of gravity, and if the force is equal in magnitude, but opposite in direction, to the weight of the group.

Example 6 demonstrates how to calculate the center of gravity for the human arm.

EXAMPLE 6

The horizontal arm in Figure 10.11 is composed of three parts: the upper arm (weight $W_1 = 3.7$ lb), the lower arm ($W_2 = 2.4$ lb), and the hand ($W_3 = 0.95$ lb). The drawing shows the center of gravity of each part, measured with respect to the shoulder joint. Find the center of gravity of the entire arm, relative to the shoulder joint.

SOLUTION

The coordinate X_{cg} of the center of gravity is given by

$$X_{cg} = \frac{W_1 X_1 + W_2 X_2 + W_3 X_3}{W_1 + W_2 + W_3} \qquad (10.3)$$

$$= \frac{(3.7 \text{ lb})(5.2 \text{ in.}) + (2.4 \text{ lb})(15.0 \text{ in.}) + (0.95 \text{ lb})(24 \text{ in.})}{3.7 \text{ lb} + 2.4 \text{ lb} + 0.95 \text{ lb}}$$

$$= \boxed{11 \text{ in.}}$$

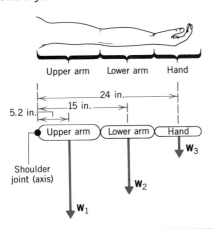

FIGURE 10.11 The three parts of the human arm, and the weight and center of gravity for each.

The center of gravity of a two-dimensional object with an irregular shape and a nonuniform weight distribution can be found by suspending the object from two different points P_1 and P_2, one at a time. Figure 10.12a shows the object at the moment of release, when its weight **W**, acting at the center of gravity, has a nonzero lever arm ℓ. At this instant the weight produces a torque about the axis. The tension force **T**, applied to the object by the suspension cord produces no torque, because its line of action passes through the axis and, thus, has a zero lever arm. Hence, in part a there is a net torque applied to the object, and the object experiences an angular acceleration. Friction will eventually bring the object to rest as in part b, where the center of gravity lies directly below the point of suspension, on a vertical line that

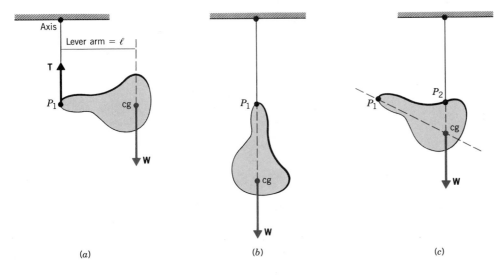

FIGURE 10.12 The center of gravity (cg) of a two-dimensional object can be located experimentally by suspending the object from two different points, one at a time.

passes through the axis. In such an orientation, the line of action of the weight also passes through the axis, so there is no longer any net torque. In the absence of a net torque there can be no angular acceleration and the object remains at rest.

By suspending the object from a second point (see Figure 10.12c), a second line through the object can be established, along which the center of gravity must also lie. The center of gravity, then, must be at the intersection of the two lines, as the drawing indicates. If the object is three-dimensional, three points of suspension must be used in this experimental method. These three points and the center of gravity must not all lie in the same plane.

The center of gravity is closely related to another concept that is known as the **center of mass.** If the weight $W = mg$ of each object in Equation 10.3 is replaced by its mass m, the resulting equation gives the x coordinate of the center of mass. In general, the center of mass is not identical to the center of gravity. However, the two points can be shown to be identical if the acceleration g due to gravity does not vary over the physical extent of the objects. For most ordinary-sized objects this assumption is quite good, and the two centers can be considered to coincide.

10.4 NEWTON'S SECOND LAW FOR ROTATIONAL MOTION

The goal of this section is to put Newton's second law into a form that is suitable for describing the rotational motion of a rigid object. We begin by considering a particle moving on a circular path about an axis of rotation. Figure 10.13 presents a good approximation of this situation by using a small model plane flying on a guideline of negligible mass. Suppose a net tangential force F_T acts on the plane and gives it a tangential acceleration a_T. In accord with Newton's second law, $F_T = ma_T$. The tangential acceleration, however, is related to the angular acceleration α by the relation $a_T = r\alpha$ (Equation 9.10), where r is the radius of the circular path and α must be expressed in rad/s². The net torque τ produced by the net tangential force is $\tau = F_T r$, since r is also the lever arm of the force. It follows that $F_T = \tau/r$, and substitution of this result into $F_T = mr\alpha$ reveals that

$$\tau = \underbrace{(mr^2)}_{\substack{\text{Moment} \\ \text{of} \\ \text{inertia } I}}\alpha \qquad (10.4)$$

Equation 10.4 is the form of Newton's second law we have been seeking, for it indicates that the net torque τ is directly proportional to the angular acceleration α. The constant of proportionality is $I = mr^2$, a quantity that is called the **moment of inertia of the particle.** The SI unit for moment of inertia is kg·m².

If all objects were particles, it would be just as convenient to use the second law in its original form $F_T = ma_T$, as in the form $\tau = I\alpha$. However, the advantage in using $\tau = I\alpha$ lies in the fact that it can be applied to the motion of any rigid body rotating about a fixed axis, and not just to a particle. To illustrate how this advantage arises, Figure 10.14 shows a flat sheet of material that rotates about an axis perpendicular to the sheet. The sheet is composed of a number of mass particles, $m_1, m_2, \ldots, m_N$, where N is very large, although only a few are shown for the sake of clarity. Each

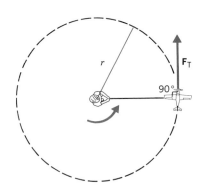

FIGURE 10.13 A model airplane of mass m is flying on a guideline of length r. A net tangential force F_T acts on the plane and produces a torque $\tau = F_T r$ about the axis of rotation.

particle behaves in much the same way as the model airplane does in Figure 10.13, so the relation $\tau = I\alpha$ applies to each particle:

$$\tau_1 = (m_1 r_1^2)\alpha$$
$$\tau_2 = (m_2 r_2^2)\alpha$$
$$\cdot$$
$$\cdot$$
$$\cdot$$
$$\tau_N = (m_N r_N^2)\alpha$$

In these equations each particle has the same value for the angular acceleration α, since the rotating object is assumed to be rigid. Adding together the N equations produces the following result:

$$\underbrace{\Sigma\tau}_{\substack{\text{Net} \\ \text{torque}}} = \underbrace{(\Sigma m r^2)}_{\substack{\text{Moment of} \\ \text{inertia}}} \alpha \qquad (10.5)$$

where $\Sigma\tau = \tau_1 + \tau_2 + \cdots + \tau_N$ is the sum of the torques, and $\Sigma m r^2 = m_1 r_1^2 + m_2 r_2^2 + \cdots + m_N r_N^2$ represents the sum of the individual moments of inertia. This latter quantity is the ***moment of inertia of the body*** and is represented by the symbol I:

$$\begin{bmatrix} \text{Moment of} \\ \text{inertia of} \\ \text{a body} \end{bmatrix} \qquad I = \Sigma m r^2 \qquad (10.6)$$

In Equation 10.6, r is the perpendicular radial distance of each particle from the axis of rotation. Combining Equation 10.6 with Equation 10.5 gives the following result:

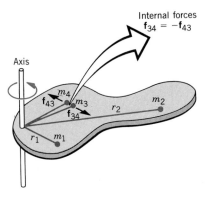

FIGURE 10.14 A rigid body consists of a large number of particles. Four such particles are shown, along with one example of the internal forces that two particles exert on each other.

ROTATIONAL ANALOG OF NEWTON'S SECOND LAW FOR RIGID BODIES

$$\text{Net torque} = \left(\begin{array}{c} \text{Moment of} \\ \text{inertia} \end{array}\right) \times \left(\begin{array}{c} \text{Angular} \\ \text{acceleration} \end{array}\right)$$

$$\Sigma\tau = I\alpha \qquad (10.7)$$

Requirement: α must be expressed in rad/s^2.

The form of the rotational analog of the second law, $\Sigma\tau = I\alpha$, is similar to that for translational (linear) motion, $\Sigma F = ma$. The moment of inertia I is analogous to the mass m, and plays the same role for rotational motion that mass does for translational motion. Thus, I is a measure of the rotational inertia of a body. When using Equation 10.7, α must be expressed in rad/s^2, because the relation $a_T = r\alpha$ (which requires radian measure) was used in the derivation.

When calculating the sum of torques in Equation 10.7, it is necessary to include only the *external torques,* those applied by agents outside the body. The torques produced by internal forces need not be considered for the following reason. Internal forces are those that one particle within the body exerts on another particle. The internal forces between two particles always occur in pairs of oppositely directed forces of equal magnitude, in accord with Newton's third law (see m_3 and m_4 in Figure 10.14). The forces in such a pair have the same line of action, so the forces have identical lever arms and produce torques of equal magnitudes. One member of the pair produces a counterclockwise torque, while the other produces a clockwise torque, the net torque from the pair being zero.

It can be seen from Equation 10.6 that the moment of inertia depends on both the mass of each particle and its distance from the axis of rotation. The farther a particle is from the axis, the greater is its contribution to the moment of inertia. Therefore, although a rigid object possesses a unique total mass, it does not have a unique moment of inertia, for **the moment of inertia depends on the location and orientation of the axis relative to the particles that make up the object.** Example 7 shows how the moment of inertia can change when the axis of rotation changes.

EXAMPLE 7

Three identical particles are fixed to the corners of an isosceles right triangle by means of massless connecting rods, as Figure 10.15 indicates. Each of the two equal sides has a length d. Find the moment of inertia of this rigid object (a) when the axis of rotation coincides with the hypotenuse of the triangle, and (b) when the axis is parallel to the hypotenuse, but passes through the opposite corner of the triangle.

SOLUTION

(a) Particles 1 and 2 lie on the axis, as part a of the drawing shows, so each has a zero radial distance: $r_1 = r_2 = 0$. Particle 3, however, has a perpendicular distance from the axis of $r_3 = \sqrt{2}\, d/2$, which can be calculated by using the Pythagorean theorem and the geometry illustrated in the drawing. Applying the definition of the moment of inertia to this situation yields

$$I = \Sigma mr^2 = m_3 r_3^2 = m(\sqrt{2}\, d/2)^2 = \boxed{md^2/2} \quad (10.6)$$

(b) Figure 10.15b shows that particles 1 and 2 no longer lie on the rotation axis. Each has the same perpendicular radial distance, $r_1 = r_2 = \sqrt{2}\, d/2$, the value being the same as that calculated for r_3 in part (a). Since particle 3 now lies on the axis, it has a zero radial distance and $r_3 = 0$. Thus,

$$I = \Sigma mr^2 = m_1 r_1^2 + m_2 r_2^2$$
$$= m(\sqrt{2}\, d/2)^2 + m(\sqrt{2}\, d/2)^2 = \boxed{md^2}$$

This value is different from that obtained in part (a), because the location of the axis is different, and the moment of inertia depends on the location of the axis.

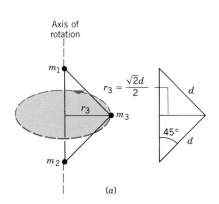

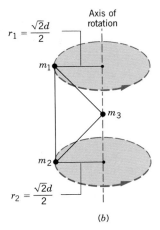

(a)   (b)

FIGURE 10.15 Three identical particles are attached together by massless rods to form a rigid isosceles right triangle, whose two equal sides have a length d. (a) The axis of rotation is the hypotenuse. (b) The axis of rotation is parallel to the hypotenuse, but passes through the opposite corner of the triangle.

The procedure illustrated in Example 7 can be extended using integral calculus to evaluate the moment of inertia of a rigid object with a continuous mass distribution, and Table 10.1 gives some typical results. Notice that these results depend on the total mass of the object, its shape, and the location and orientation of the axis. The only object in this table for which the moment of inertia is obvious is the thin hoop, with the axis perpendicular to the plane of the hoop and passing through its center. All the mass of the hoop is concentrated at the same distance $r = R$ from the axis. Thus, its moment of inertia is $I = \Sigma mR^2 = (\Sigma m)R^2 = MR^2$, where $M = \Sigma m$ is the total mass of the hoop.

In Example 8, the relation $\Sigma\tau = I\alpha$ is applied to the platter of a stereo turntable.

TABLE 10.1 Moments of Inertia for Various Rigid Objects of Mass M

Thin-walled hollow cylinder or hoop		$I = MR^2$
Solid cylinder or disk		$I = \frac{1}{2}MR^2$
Thin rod, axis perpendicular to rod and passing through center		$I = \frac{1}{12}ML^2$
Thin rod, axis perpendicular to rod and passing through one end		$I = \frac{1}{3}ML^2$
Solid sphere, axis through center		$I = \frac{2}{5}MR^2$
Solid sphere, axis tangent to surface		$I = \frac{7}{5}MR^2$
Thin-walled spherical shell, axis through center		$I = \frac{2}{3}MR^2$
Thin rectangular sheet, axis parallel to one edge and passing through center of other edge		$I = \frac{1}{12}ML^2$
Thin rectangular sheet, axis along one edge		$I = \frac{1}{3}ML^2$

EXAMPLE 8

Most turntables can bring a record from rest up to the rated angular speed of $33\frac{1}{3}$ rev/min in one-half a revolution. The platter of one turntable has a moment of inertia of 0.0500 kg·m² (including the effect of the record). Neglecting frictional effects, what torque (assumed constant) must the turntable motor apply to the platter to achieve this performance level?

SOLUTION

Newton's second law for rotational motion can be used to find the desired torque, once the angular acceleration is determined. The angular acceleration can be calculated by using the data listed in the table and the appropriate equation of rotational kinematics. The rotational variables ω and θ are negative, because the rotation of a turntable platter is clockwise when viewed from above. Also, the data for ω and θ have been converted to radian measure, since $\Sigma\tau = I\alpha$ requires that α be expressed in radian measure.

θ	α	ω	ω_0	t
$-\pi$ rad ($-\frac{1}{2}$ rev)	?	-3.49 rad/s ($-33\frac{1}{3}$ rev/min)	0	

Employing $\omega^2 = \omega_0^2 + 2\alpha\theta$ (Equation 9.8) gives

$$\alpha = \frac{\omega^2 - \omega_0^2}{2\theta} = \frac{(-3.49 \text{ rad/s})^2}{2(-\pi \text{ rad})} = -1.94 \text{ rad/s}^2$$

The second law for rotational motion can now be used to obtain the torque:

$$\Sigma\tau = I\alpha = (0.0500 \text{ kg·m}^2)(-1.94 \text{ rad/s}^2) \quad (10.7)$$
$$= \boxed{-0.0970 \text{ N·m}}$$

The next example deals with an interesting situation in which both an angular acceleration and a translational acceleration are present.

EXAMPLE 9

A 425-kg crate is being lifted by the hoisting mechanism shown in Figure 10.16a. The two cables are securely wrapped around their respective pulleys, which have radii of 0.600 and 0.200 m. The pulleys are fastened together to form a "dual" pulley and

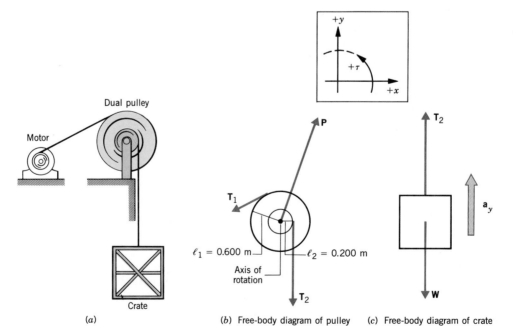

FIGURE 10.16 (a) The crate is lifted upward by the motor and pulley arrangement. The free-body diagram for (b) the dual pulley and (c) the crate.

turn as a single unit about the center axle, relative to which the combined moment of inertia is $I = 50.0$ kg·m². If a tension of magnitude $T_1 = 2150$ N is maintained in the cable attached to the motor, find the angular acceleration of the "dual" pulley and the tension in the cable connected to the crate.

SOLUTION

Three external forces act on the dual pulley, as its free-body diagram shows (Figure 10.16b). These forces are (1) the tension $\mathbf{T}_1$ in the cable connected to the motor, (2) the tension $\mathbf{T}_2$ in the cable attached to the crate, and (3) the reaction force $\mathbf{P}$ exerted on the dual pulley by the axle. The force $\mathbf{P}$ arises because the two cables pull the pulley into the axle and the axle pushes back, thus keeping the pulley in place. Notice that $\mathbf{P}$ has a zero lever arm with respect to the axle, since the line of action of $\mathbf{P}$ passes directly through the axle. Using the lever arms shown in part b of the figure, we can apply the second law to the rotational motion of the pulley:

$$\Sigma\tau = I\alpha \qquad (10.7)$$

$$+(2150 \text{ N})(0.600 \text{ m}) - T_2(0.200 \text{ m}) = (50.0 \text{ kg·m}^2)\alpha$$

This equation contains two unknown quantities, so a second equation is needed. To obtain this additional equation, consider the crate. The crate accelerates upward under the action of its weight $[W = (425 \text{ kg})(9.80 \text{ m/s}^2) = 4170 \text{ N}]$ and the cable tension $\mathbf{T}_2$, as the free-body diagram in part c of the drawing indicates. Applying Newton's second law to the translational motion of the crate gives

$$\Sigma F_y = ma_y \qquad (4.2b)$$

$$+T_2 - (4170 \text{ N}) = (425 \text{ kg})(a_y)$$

Because the cable attached to the crate rolls on the pulley without slipping, it is possible to relate the linear acceleration a_y of the crate to the angular acceleration α of the pulley via Equation 9.13:

$$a_y = r\alpha = (0.200 \text{ m})\alpha$$

With the aid of this relation for a_y, Equations 10.7 and 4.2b can be solved simultaneously for T_2 and α to yield

$$\boxed{T_2 = 4750 \text{ N}} \quad \text{and} \quad \boxed{\alpha = 6.81 \text{ rad/s}^2}$$

We have seen that Newton's second law for translational motion, $\Sigma F = ma$, has the same form as that for rotational motion, $\Sigma\tau = I\alpha$, so each translational variable has a rotational analog: force F and torque τ are analogous quantities, as are mass m and moment of inertia I, and linear acceleration a and angular acceleration α. The other physical concepts developed for studying translational motion, such as kinetic energy and momentum, also have rotational analogs. For future reference, Table 10.2 itemizes these concepts and their rotational analogs.

TABLE 10.2 Analogies between Translational and Rotational Concepts

Physical Concept	Translational	Rotational
Displacement	s	θ
Velocity	v	ω
Acceleration	a	α
The cause of acceleration	Force F	Torque τ
Inertia	Mass m	Moment of inertia I
Newton's second law	$\Sigma F = ma$	$\Sigma\tau = I\alpha$
Work	Fs	$\tau\theta$
Kinetic energy	$\frac{1}{2}mv^2$	$\frac{1}{2}I\omega^2$
Momentum	$p = mv$	$L = I\omega$

10.5 ROTATIONAL WORK AND ENERGY

ROTATIONAL WORK

Work and energy are among the most fundamental and useful concepts in physics. Chapter 7 discusses their application to translational motion. These concepts are

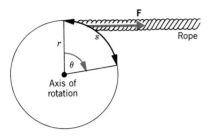

FIGURE 10.17 The force **F** does work in rotating the wheel through the angle θ.

equally useful for understanding rotational motion, provided they are recast into forms that involve angular variables.

The work W done by a constant force that points in the same direction as the displacement is $W = Fs$ (Equation 7.1). In this expression, F and s are the magnitudes of the force and displacement, respectively. Now focus on the situation illustrated in Figure 10.17, in which a rope is wrapped around a wheel. The rope is under a constant tension F and is being pulled so as to give the wheel an angular acceleration. If the rope is pulled out a distance s, the wheel rotates through an angle $\theta = s/r$ (Equation 9.1), where r is the radius of the wheel and θ is in radians. The work done by the tension force in turning the wheel is $W = Fs = Fr\theta$. But Fr is the torque τ applied to the wheel by the tension, so the rotational work can be written in angular variables as follows:

DEFINITION OF ROTATIONAL WORK

The work W_R done by a constant torque τ in turning an object through an angle θ is

$$W_R = \tau\theta \qquad (10.8)$$

Requirement: θ must be expressed in radians.

SI unit of rotational work: joule (J)

Example 10 considers the rotational work done by a common power tool.

EXAMPLE 10

An electric drill is turned on (Figure 10.18), and the chuck exerts a constant torque of 4.0×10^{-4} N·m to turn the drill bit through 9.0 revolutions. Find the rotational work done by the chuck.

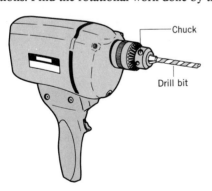

FIGURE 10.18 An electric drill.

SOLUTION

Before we can use $W_R = \tau\theta$ for calculating work, it is necessary to express the angular displacement θ in radians: $\theta = (9.0 \text{ rev})$ $[2\pi \text{ rad}/(1 \text{ rev})] = 57$ rad. The rotational work done by the chuck is

$$W_R = \tau\theta = (4.0 \times 10^{-4} \text{ N·m})(57 \text{ rad}) = \boxed{2.3 \times 10^{-2} \text{ J}}$$

ROTATIONAL KINETIC ENERGY

A rotating body possesses kinetic energy, because its constituent particles are in motion. If the body is rotating with an angular speed ω, the tangential speed v_T of a particle at a distance r from the axis is $v_T = r\omega$. Figure 10.19 shows two such particles in a rotating object. If the particle's mass is m, its kinetic energy is $\frac{1}{2}mv_T^2 = \frac{1}{2}mr^2\omega^2$.

The kinetic energy of the entire rotating body is the sum of the kinetic energies of the particles:

$$\text{Rotational KE} = \Sigma \tfrac{1}{2}mr^2\omega^2 = \tfrac{1}{2}(\Sigma mr^2)\omega^2$$

Since all particles in a rigid body have the same angular velocity ω, this variable has been factored outside the summation in the equation above. The term in parentheses is the moment of inertia of the body, $I = \Sigma mr^2$, so the rotational kinetic energy takes the form given below:

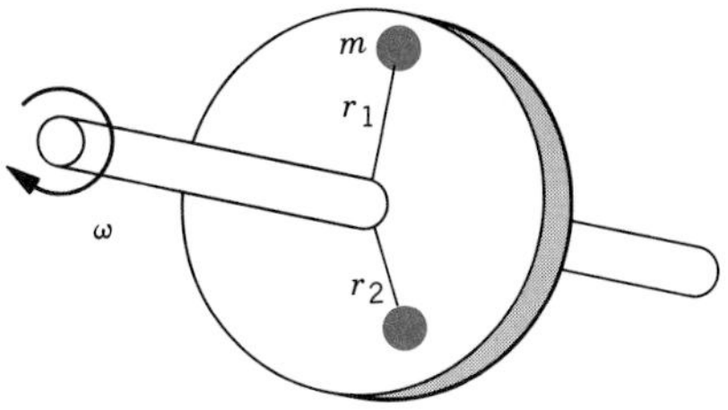

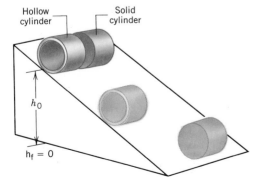

FIGURE 10.19 The rotating wheel is composed of many particles, two of which are shown. The tangential speeds of the two particles are $r_1\omega$ and $r_2\omega$, where ω is the angular speed of the wheel.

DEFINITION OF ROTATIONAL KINETIC ENERGY

The rotational kinetic energy of a rigid object rotating with an angular speed ω about a fixed axis and having a moment of inertia I is

$$\text{Rotational KE} = \tfrac{1}{2}I\omega^2 \qquad (10.9)$$

Requirement: ω must be expressed in rad/s.

SI unit of rotational kinetic energy: joule (J)

When a bicycle or a car is in motion, its tires are both translating and rotating. The total kinetic energy KE of an object that is simultaneously translating and rotating is the sum of its translational and rotational kinetic energies:

$$\text{Total KE} = \tfrac{1}{2}mv^2 + \tfrac{1}{2}I\omega^2$$

where v is the translational speed of the object's center of mass, m is the total mass, I is the moment of inertia about an axis through the center of mass, and ω is the angular speed. The next example deals with combined translational and rotational motion.

EXAMPLE 11

A hollow cylinder (radius $= r_h$) and a solid cylinder (radius $= r_s$) start from rest at the top of an inclined plane (Figure 10.20). Both cylinders have the same mass m and start at the same vertical height h_0. Ignoring energy losses due to retarding forces, determine which cylinder has the greatest translational speed upon reaching the bottom of the incline.

SOLUTION

The principle of conservation of mechanical energy can be used here, provided the rotational kinetic energy is included in the total mechanical energy of each cylinder. The total mechanical energy E at any height h above the base of the incline is the sum of the translational and rotational kinetic energies and the gravitational potential energy:

$$E = \tfrac{1}{2}mv^2 + \tfrac{1}{2}I\omega^2 + mgh$$

FIGURE 10.20 A hollow cylinder and a solid cylinder each have the same mass and start together from rest at the top of an incline. The conservation of mechanical energy can be used to show that the solid cylinder, having the greatest translational speed, reaches the bottom first.

Since it is assumed that only the force of gravity does work on the cylinders, the total mechanical energy is conserved as they roll down the incline. Thus, the total mechanical energy E_0 at the top of the incline ($h = h_0$, $v_0 = 0$, $\omega_0 = 0$) is the same as the total mechanical energy E_f at the bottom ($h_f = 0$):

$$\tfrac{1}{2}mv_0^2 + \tfrac{1}{2}I\omega_0^2 + mgh_0 = \tfrac{1}{2}mv_f^2 + \tfrac{1}{2}I\omega_f^2 + mgh_f$$

$$mgh_0 = \tfrac{1}{2}mv_f^2 + \tfrac{1}{2}I\omega_f^2$$

Since each cylinder rolls without slipping, the final rotational speed ω_f and the final translational speed v_f of its center of mass are related according to Equation 9.12, $\omega_f = v_f/r$, where r is the radius of the cylinder. Substituting this expression for ω_f into the equation above and solving for v_f yields

$$v_f = \sqrt{\frac{2mgh_0}{m + \dfrac{I}{r^2}}}$$

Setting $r = r_h$ and $I = mr_h^2$ for the hollow cylinder (see Table 10.1), and then setting $r = r_s$ and $I = \tfrac{1}{2}mr_s^2$ for the solid cylinder, we find that the two cylinders have the following translational speeds at the bottom of the incline:

Hollow cylinder $v_f = \sqrt{gh_0}$

Solid cylinder $v_f = \sqrt{\dfrac{4gh_0}{3}} = 1.15\sqrt{gh_0}$

Thus, the solid cylinder, having the greatest translational speed, arrives at the bottom first, a result that is independent of the masses and radii of the cylinders. It is of interest to note that in the absence of retarding forces an object like a box, which slides down the incline without rolling, reaches the bottom with an even greater translational speed of $v_f = 1.41\sqrt{gh_0}$.

10.6 ANGULAR MOMENTUM

Chapter 8 deals with the concept of linear momentum and shows how it can be used profitably in situations where Newton's second law is difficult to apply directly. The linear momentum p of an object is defined as the product of its mass m and linear velocity v, $p = mv$. In rotational motion the analogous concept is called the *angular momentum L*. The mathematical form of angular momentum is identical to that of linear momentum, with the mass m and the linear velocity v being replaced with their rotational counterparts, the moment of inertia I and the angular velocity ω.

DEFINITION OF ANGULAR MOMENTUM

The angular momentum L of a body rotating about a fixed axis is the product of the body's moment of inertia I and its angular velocity ω:

$$L = I\omega \tag{10.10}$$

Requirement: ω must be expressed in rad/s.

SI unit of angular momentum: $\text{kg} \cdot \text{m}^2/\text{s}$

Linear momentum is an important concept in physics, because the total momentum of a system is a conserved quantity when the net external force acting on the system is zero. Similarly, the total angular momentum of a system is conserved when the net external torque acting on the system is zero. The starting point for developing the conservation of angular momentum is the impulse–momentum theorem in the form given by Equation 8.6. In this form, the theorem states that when a net external force ΣF acts for a time Δt, the linear momentum of the system changes from an initial value of P_0 to a final value of P_f: $(\Sigma F)\,\Delta t = P_f - P_0$. The rotational analog of this important theorem, using torque in place of force and angular momentum in place of linear momentum, is

$$(\Sigma\tau)\,\Delta t = L_f - L_0 = I_f\omega_f - I_0\omega_0$$

This result indicates that a net external torque, acting for a time Δt, changes the angular momentum of a system from an initial value of $L_0 = I_0\omega_0$ to a final value of $L_f = I_f\omega_f$. When the net external torque is zero, $\Sigma\tau = 0$, the final angular momentum equals the initial angular momentum, so $L_f = L_0$. Since the final and initial times are arbitrary, the statement $L_f = L_0$ means that the angular momentum remains constant at all times; this is the ***principle of conservation of angular momentum.***

PRINCIPLE OF CONSERVATION OF ANGULAR MOMENTUM
The total angular momentum of a system remains constant (is conserved) if the net external torque acting on the system is zero.

FIGURE 10.21 (*a*) A skater spins slowly on one skate, with both arms and one leg outstretched. (*b*) As she pulls her arms and leg in toward the rotational axis, her moment of inertia I decreases, and the angular speed ω increases.

Since the angular momentum is the product of the moment of inertia I and the angular velocity ω, one interesting consequence of the conservation principle is that any change in one of these variables must be accompanied by a corresponding change in the other so as to keep the product constant. Consider, for example, the skater in Figure 10.21*a*, who is spinning with both arms and a leg outstretched. If the external torques produced by air resistance and friction are so small that they can be ignored, the skater would spin forever at the same angular velocity, because her angular momentum would be conserved. It is also because of the conservation of angular momentum that the skater can spin at a higher angular velocity by pulling in her arms and leg, as part *b* of the drawing shows. When the mass of each arm and the leg is moved closer to the rotational axis, the skater's moment of inertia decreases. The angular velocity, therefore, must increase to keep the product $L = I\omega$ constant. The last example in this chapter illustrates a quantitative application of the conservation of angular momentum.

EXAMPLE 12
An artificial satellite is placed into an elliptical orbit about the earth, as in Figure 10.22. Radar has determined that its point of closest approach (called the *perigee*) is $r_P = 8.37 \times 10^6$ m from the center of the earth, while its point of greatest distance (called the *apogee*) is $r_A = 25.1 \times 10^6$ m from the center of the earth. The speed of the satellite at the perigee is $v_P = 8450$ m/s. Find its speed v_A at the apogee.

SOLUTION
The only force of any significance that acts on the satellite is the gravitational force exerted by the earth. However, at any instant, this force is directed toward the center of the earth and passes through the axis about which the satellite instantaneously rotates. Therefore, the gravitational force exerts *no torque* on the satellite (the lever arm is zero). Consequently, the angular momentum of the satellite remains constant at all times. In particular, the angular momentum is the same at the apogee (A) and the perigee (P): $I_A\omega_A = I_P\omega_P$. Since the orbiting satellite can be considered a point mass, its moment of inertia is $I = mr^2$. In addition, the angular speed ω of the satellite is related to its tangential speed v_T by $\omega = v_T/r$. If these relations are used at the apogee and

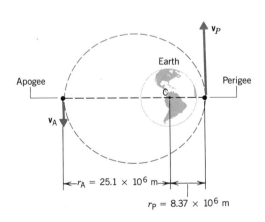

FIGURE 10.22 A satellite moving in an elliptical orbit about the earth. The gravitational force **F** exerts no torque on the satellite and, as a result, the angular momentum of the satellite is conserved at all points of the orbit.

perigee, the conservation of angular momentum gives the following result:

$$I_A \omega_A = I_P \omega_P$$

$$(mr_A{}^2)\left(\frac{v_A}{r_A}\right) = (mr_P{}^2)\left(\frac{v_P}{r_P}\right)$$

$$v_A = \frac{r_P v_P}{r_A} = \frac{(8.37 \times 10^6 \text{ m})(8450 \text{ m/s})}{25.1 \times 10^6 \text{ m}} = \boxed{2820 \text{ m/s}}$$

The answer is independent of the mass of the satellite. The speed at the apogee is less than the speed at the perigee. The satellite behaves just like the skater in Figure 10.21, because its speed is greater when the moment of inertia is smaller.

The result in Example 12 indicates that a satellite does not have a constant speed in an elliptical orbit. The speed changes from a maximum at the perigee to a minimum at the apogee, and, in general, the closer the satellite comes to the earth, the faster it travels. Planets moving around the sun in elliptical orbits exhibit the same kind of behavior, and Johannes Kepler (1571–1630) formulated his famous second law based on observations of such characteristics of planetary motion. Kepler's second law states that in a given amount of time a line joining any planet to the sun sweeps out the same amount of area no matter where the planet is on its elliptical orbit, as Figure 10.23 illustrates. The conservation of angular momentum can be used to show why the law is valid, by means of a calculation similar to that in Example 12.

FIGURE 10.23 Kepler's second law of planetary motion states that a line joining a planet to the sun sweeps out equal areas in equal time intervals, regardless of the position of the planet in its orbit. This law is a direct consequence of the principle of conservation of angular momentum.

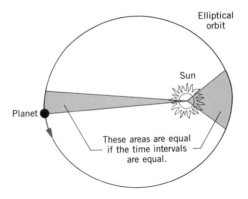

Elliptical orbit

Sun

Planet

These areas are equal if the time intervals are equal.

SUMMARY

The **torque** τ of a force is the magnitude F of the force times the lever arm ℓ: $\tau = F\ell$. The lever arm is the perpendicular distance between the line of action of the force and the axis of rotation.

A rigid body is in **equilibrium** if it has zero translational acceleration and zero angular acceleration, in which case the net external force and the net external torque acting on the body are zero. For forces acting only in the x, y plane, the conditions for equilibrium are $\Sigma F_x = 0$, $\Sigma F_y = 0$, and $\Sigma \tau = 0$. When an object is in equilibrium, the axis used for calculating torques can be chosen arbitrarily.

The **center of gravity** of a rigid object is the point where its entire weight can be considered to act when calculat-

ing the torque due to the weight of the object. For a symmetrical body with uniformly distributed weight, the center of gravity is located at the geometrical center of the body. When a number of objects whose weights W_1, W_2, . . . are distributed along the x axis at locations X_1, X_2, . . . , the center of gravity is located at $X_{cg} = (W_1 X_1 + W_2 X_2 + \cdots)/(W_1 + W_2 + \cdots)$.

Newton's second law for rotational motion is $\Sigma \tau = I\alpha$, where $\Sigma \tau$ is the net torque applied to a body, I is the moment of inertia of the body, and α is the angular acceleration (in rad/s^2). The **moment of inertia** I of an object composed of N particles is $I = (m_1 r_1^2 + m_2 r_2^2 + \cdots + m_N r_N^2)$, where m_1, m_2, . . . , m_N are the masses of the particles and $r_1, r_2, \ldots, r_N$ are the perpendicular distances of the particles from the axis of rotation.

The **rotational work** W_R done by a constant torque τ in turning a rigid body through an angular displacement θ (in radians) is expressed by $W_R = \tau\theta$. The **rotational kinetic energy** of an object with angular speed ω and moment of inertia I is $KE = \frac{1}{2}I\omega^2$. The **total mechanical energy** E of a rigid object is the sum of its translational kinetic energy, its rotational kinetic energy, and its gravitational potential energy.

The **angular momentum** L of a body rotating with angular velocity ω (in rad/s) about a fixed axis and having a moment of inertia I is $L = I\omega$. The **principle of conservation of angular momentum** states that the total angular momentum of a system remains constant if the net external torque acting on the system is zero.

Table 10.2 summarizes the analogies between the physical concepts that characterize translational and rotational motion.

SOLVED PROBLEMS

SOLVED PROBLEM 1
Related Problems: **17 **18 **56

In gymnastics a number of beautiful routines are performed on the still rings, and one particular routine involves the so-called "iron cross." To perform this feat, a gymnast who weighs 655 N pushes with each hand against the rings to support himself as shown in part a of the drawing. The gymnast's arms each weigh 33.0 N, have a length of 0.600 m, and are held parallel to the ground. Find (a) the tension in each supporting rope and (b) the forces that are applied to each arm by the shoulder joint and the latissimus dorsi muscle (assumed, as an approximation, to be the only muscle acting).

Solution (a) Part a of the figure shows the free-body diagram for the gymnast and indicates that the pertinent forces are the tension **T** in each rope and the 655-N weight acting at the gymnast's center of gravity. The tension in each rope is the same, assuming that both sides of the gymnast's body are identical, so that his center of gravity is located midway between the two rings. Since the gymnast is in equilibrium, the magnitude T of the tension can be obtained by setting the sum of the vertical force components equal to zero:

$$\Sigma F_y = +T \sin 78.0° + T \sin 78.0° - 655 \text{ N} = 0 \quad (5.1b)$$

$$T = \frac{655 \text{ N}}{2 \sin 78.0°} = \boxed{335 \text{ N}}$$

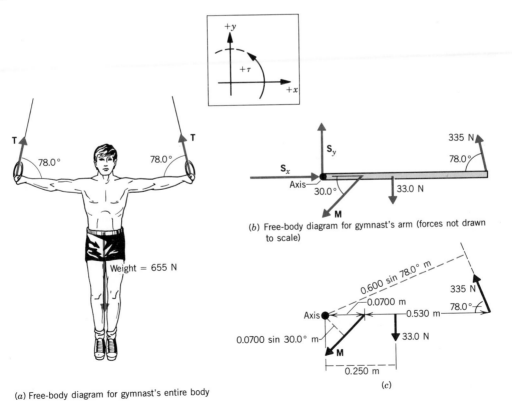

(b) Free-body diagram for gymnast's arm (forces not drawn to scale)

(c)

(a) Free-body diagram for gymnast's entire body

(b) For this part of the problem we select one arm as the object of our analysis, and part b of the figure shows its free-body diagram. The forces acting on the arm are as follows: the 335-N force applied to the arm by the ring, as calculated in part (a), the 33.0-N force of gravity acting at the arm's center of gravity, the muscle force **M** that is assumed to act on the arm at a 30.0° angle, and the horizontal and vertical force components S_x and S_y that are exerted on the arm by the shoulder joint. The directions of S_x and S_y are drawn using our intuition, and if they should be wrong, the values for these forces will turn out to be negative numbers. Since the arm is in equilibrium, the net horizontal force, the net vertical force, and the net torque are zero. When calculating torques, we select an axis parallel to the ground and perpendicular to the arm at the shoulder joint and use the lever arms (colored, dashed lines) given in part c of the drawing:

$$\Sigma F_x = +S_x - M\cos 30.0° \\ - (335\text{ N})\cos 78.0° = 0 \qquad (5.1a)$$

$$\Sigma F_y = +S_y - M\sin 30.0° - 33.0\text{ N} \\ + (335\text{ N})\sin 78.0° = 0 \qquad (5.1b)$$

$$\Sigma\tau = -M(0.0700\sin 30.0°\text{ m}) - (33.0\text{ N})(0.250\text{ m}) \\ +(335\text{ N})(0.600\sin 78.0°\text{ m}) = 0 \qquad (10.2)$$

Equation 10.2 can be solved directly to show that $\boxed{M = 5380\text{ N}}$. With this value for M, Equations 5.1a and 5.1b can be used to show that

$$\boxed{S_x = 4730\text{ N}} \quad\text{and}\quad \boxed{S_y = 2400\text{ N}}$$

The answers for S_x and S_y are positive, so the directions shown for these vector components in part b of the drawing are correct.

Summary of Important Points Two points are emphasized in this problem. One is that the selection of the object to be analyzed is a very important step in problem solving. Sometimes, as is the

Gymnast Scott Johnson

case here, it is necessary to choose different objects for analysis at different stages of the same problem. A second important point is that large forces may be necessary to keep a rigid object in equilibrium, if the lever arms of the forces are short. The large force (more than eight times the body weight) exerted by the latissimus dorsi muscle in this problem is needed because the muscle attaches to the upper arm so close to the shoulder joint (0.0700 m).

QUESTIONS

1. Explain (a) how it is possible for a large force to produce only a small, or even zero, torque, and (b) how it is possible for a small force to produce a large torque.

2. A magnetic tape is being played on a cassette deck. The tension in the tape applies a torque to the supply reel. Assuming the tension remains constant during playback, discuss how this torque varies as the reel becomes empty.

3. A flat rectangular sheet of plywood is fixed so that it can rotate about an axis perpendicular to the sheet through one corner. How should a force (acting in the plane of the sheet) be applied to the plywood so as to create the largest possible torque? Give your reasoning.

4. A torque is the product of a force and a distance (lever arm). Work is also the product of a force and a distance. Yet, torque and work *are different*. What is it about the distances that makes torque and work different?

5. Starting in the spring, fruit begins to grow on the outer end of a branch on a pear tree. Explain how the center of gravity of the pear-growing branch shifts during the course of the summer.

6. The free-body diagram shown in the drawing has been made by a student to show the forces that act on a thin rod in equilibrium. According to the student, the three forces are drawn to scale and lie in the plane of the paper. Are these forces sufficient to keep the rod in equilibrium, or are additional forces necessary? Explain.

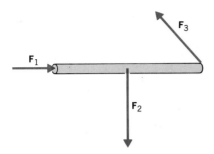

7. An A-shaped step ladder is standing on frictionless ground. The ladder consists of two sections joined at the top and kept from spreading apart by a horizontal crossbar. Draw a free-body diagram showing the forces that keep *one* section of the ladder in equilibrium.

8. For each of the two examples shown in the drawing, which rotating system, (a) or (b), has the *larger* moment of inertia? Give the reason for your answers.

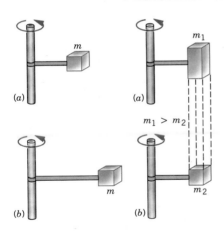

9. Two wheels have the same shape and radii, and they rotate about axes through their centers. The wheels are made from different substances and have different masses. In each case the material is distributed uniformly throughout the wheel. Which wheel, if any, has the larger moment of inertia? Account for your answer.

10. In the drawing two flat sheets of uniform material are shown rotating about an axis. Each sheet is formed from two identical right triangles, one shaded in color and the other in gray. Furthermore, the two right triangles in the top sheet are identical to those in the bottom sheet. The masses of the two sheets are the same. Which sheet (if either) has the larger moment of inertia? Justify your answer.

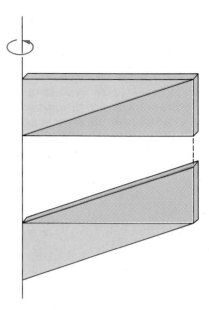

11. A flat triangular sheet of uniform material is shown in the drawing. There are three possible axes of rotation, each perpendicular to the sheet and passing through one corner, *A*, *B*, or *C*. For which axis is the greatest torque required to bring the triangle up to an angular speed of 100 rev/min in 10.0 s, starting from rest? Explain, assuming that the torque is kept constant while it is being applied.

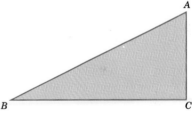

12. An object has an angular acceleration due to torques that are present. Therefore, the angular velocity of the object is changing. What happens to the angular velocity if (a) additional torques are applied so as to make the net torque suddenly equal to zero and (b) all the torques are suddenly removed?

13. The satellite shown in the drawing is initially moving with a constant translational velocity and zero angular velocity through outer space. (a) When the two engines are fired, each generating a thrust of magnitude *T*, will the translational velocity increase, decrease, or remain the same? Why? (b) Explain what will happen to the angular velocity.

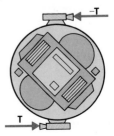

14. Can the mass of a rigid body be considered as concentrated at its center of mass for purposes of computing (a) the body's translational kinetic energy, and (b) the body's moment of inertia? If not, discuss the reason why it cannot be so treated.

15. A contest is held between identical twins. Each uses an identical automobile tire. The goal is to roll down a hill in the least amount of time, starting from rest at the top. One twin fashions a crude unicycle, using a negligible mass of material to build the seat and axle system. The other twin simply curls up inside the tire. Ignoring friction and air resistance, who wins? Give your reasoning.

16. A woman is sitting on the spinning seat of a piano stool with her arms folded. What happens to her angular velocity and her angular momentum when she extends her arms outward? Justify your answers.

17. Suppose the ice cap at the South Pole melted and the water

was distributed uniformly over the earth's oceans. Would the earth's angular velocity increase, decrease, or remain the same? Explain.

18. Many rivers, like the Mississippi river, flow from north to south toward the equator. These rivers often carry a large amount of sediment that they deposit when entering the ocean. What effect does this redistribution of the earth's soil have on the angular velocity of the earth? Why?

19. A person is sitting in a chair and swinging his leg back and forth. Is the angular momentum of the leg itself conserved? Explain.

20. A person is hanging motionless from a vertical rope over a swimming pool. He lets go of the rope and drops straight down. After letting go, is it possible for him to curl into a ball and start spinning? Justify your answer.

PROBLEMS

Section 10.1 Torque

1. A force of 110 N is applied perpendicularly to the left edge of the rectangle shown in the drawing. (a) Find the torque (magnitude and direction) produced by this force with respect to an axis perpendicular to the plane of the rectangle at corner A and (b) with respect to a similar axis at corner B.

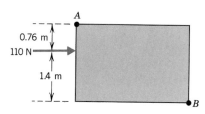

2. Find the net torque (magnitude and direction) produced by the forces F_1 and F_2 about the rotation axis shown in the drawing. The forces are acting on a thin rigid rod.

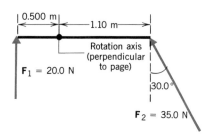

3. One end of a meter stick is pinned to a table, so the stick can rotate freely in a plane parallel to the tabletop. Two forces, both parallel to the tabletop, are applied to the stick in such a way that

the net torque is zero. One force has a magnitude of 2.00 N and is applied perpendicular to the free end of the stick. The other force has a magnitude of 6.00 N and acts at a 30.0° angle with respect to the stick. Where along the stick should the 6.00-N force be applied? Express this distance with respect to the end that is pinned.

* 4. A pair of forces with equal magnitudes, opposite directions, and different lines of action is called a "couple." When a couple acts on a rigid object, it produces a torque that does *not* depend on the location of the axis. The drawing shows an example of a couple acting on a rigid rod. The axis is located at the left end of the rod and is normal to the plane containing the two force vectors. Determine an expression for the torque produced by the couple, thereby showing that the torque depends only on F and d.

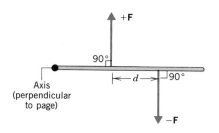

**5. A rotational axis is directed perpendicular to the plane of a square and is located somewhere within the square. Two forces, F_1 and F_2, are applied to diagonally opposite corners, and act along the sides of the square, first as shown in part *a* and then as shown in part *b* of the drawing. In each case the net torque produced by the forces is zero. The square is one meter on a side, and the magnitude of F_2 is three times that of F_1. Locate the point where the axis intersects the plane of the square; measure this point relative to the lower right-hand corner of the square.

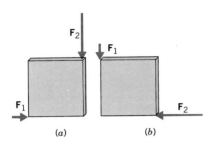

(a)   (b)

Section 10.2 Rigid Objects in Equilibrium, Section 10.3 Center of Gravity

6. The tonearm on a stereo turntable is positioned in a horizontal manner with the stylus resting in the record groove and a fulcrum providing support near the other end. The tracking force is defined to be the normal force F_N that acts at the point where the stylus touches the record. Usually F_N is adjusted by properly positioning a sliding counterweight on the other end of the tonearm, as the drawing shows. Suppose the tonearm has a total mass of 60.0 g (excluding the counterweight), with a center of gravity as indicated in the drawing. Find the location X for a 1.00×10^2-g counterweight, so that the tracking force has a value of 9.80×10^{-3} N. (An audio equipment manufacturer would specify this as a one "gram" tracking force.)

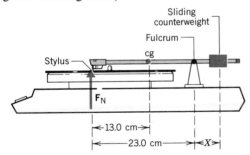

7. In an isometric exercise a man places his hand on a scale and pushes vertically downward, keeping the forearm horizontal. He can do this because the triceps muscle applies an upward force M perpendicular to the arm, as the drawing indicates. The forearm weighs 5.00 lb and has a center of gravity as indicated. The scale registers 25.0 lb. Determine (a) the magnitude of M and (b) the magnitude and direction of the force applied by the upper arm bone to the forearm at the elbow joint.

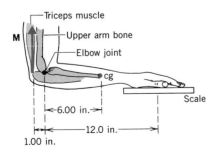

8. An empty wheelbarrow weighs 95.0 N and can carry a 525-N load. It is well known that the wheelbarrow is much easier to use if the center of gravity of the load is placed directly over the axle. Verify this fact by calculating the vertical lifting force F required to support each of the two wheelbarrows illustrated in the drawing.

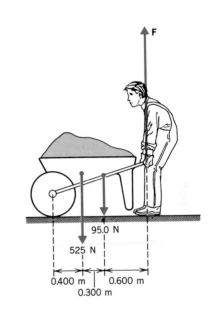

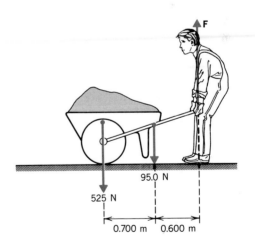

9. A lunch tray is being held in one hand, as the drawing illustrates. The mass of the tray itself is 0.200 kg, and its cg is located at its geometrical center. On the tray is a 1.00-kg plate of food and a 0.250-kg cup of coffee. Obtain the force T exerted by the thumb and the force F exerted by the four fingers. Both forces act perpendicular to the tray, which is being held parallel to the ground.

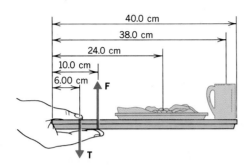

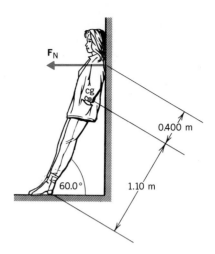

10. A man holds a 40.0-lb weight in his hand, with the forearm horizontal (see the drawing). He can support the weight in this position because of the flexor muscle force **M**, which is applied perpendicular to the forearm. The forearm weighs 5.00 lb and has a center of gravity as indicated. Find (a) the magnitude of **M** and (b) the magnitude and direction of the force applied by the upper arm bone to the forearm at the elbow joint.

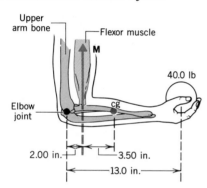

11. Three objects are situated on the x axis. Their masses and positions are as follows: (a) 4.00 kg at $x = +1.00$ m, (b) 2.00 kg at $x = -0.500$ m, and (c) 2.50 kg at $x = -1.50$ m. Where on the x axis is the center of gravity of this collection of objects located?

12. A jet transport is fully loaded and has a weight of 2.25×10^5 lb. The plane is at rest on the runway. The two rear wheels are 15.0 m behind the front wheel, and the plane's center of gravity is 12.6 m behind the front wheel. Determine the normal force exerted on the front wheel and on each of the two rear wheels.

* 13. A 5.00×10^2-N woman is leaning against a smooth vertical wall, as the drawing shows. Find the force $\mathbf{F_N}$ (directed perpendicular to the wall) exerted on her shoulder by the wall and the horizontal and vertical components of the force exerted on her shoes by the ground.

* 14. A person is sitting with one leg outstretched, so that it makes an angle of 30.0° with the horizontal, as the drawing indicates. The weight of the leg below the knee is 10.0 lb, with a center of gravity located 10.0 inches below the knee joint. The leg is being held in this position because of the force **M** applied by the quadriceps muscle. Obtain (a) the magnitude of **M** and (b) the magnitude and direction of the reaction force applied to the leg by the knee joint. Specify the direction of the reaction force with respect to the leg.

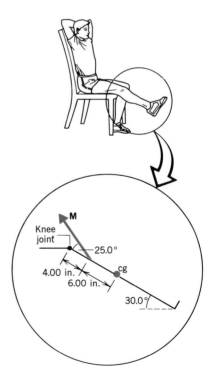

* **15.** A 125-kg uniform beam is attached to a vertical wall and is supported by a wire. The beam is 3.00 m long, and a 2.00×10^2-kg crate hangs from it. Using the data shown in the drawing, find (a) the tension T in the wire and (b) the horizontal and vertical components of the force that the wall exerts on the left end of the beam.

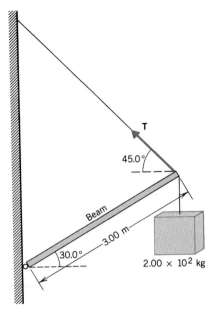

* **16.** A massless, rigid board is placed across two bathroom scales that are separated by a distance of 2.00 m. A person lies on the board. The scale under his head reads 425 N and the scale under his feet reads 315 N. (a) Find the weight of the person. (b) Locate the center of gravity of the person relative to the scale beneath his head.

** **17.** The drawing shows a ladder that has the shape of an "A." Both sides of the ladder are equal in length. This ladder is standing on a frictionless horizontal surface and only the crossbar (which has a negligible mass) of the "A" keeps the ladder from collapsing. The ladder is uniform and has a mass of 20.0 kg, while the man standing on it has a mass of 80.0 kg. Determine the tension in the crossbar of the ladder. (See Solved Problem 1 for a related problem.)

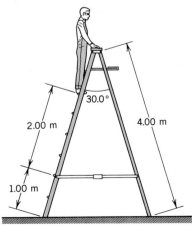

** **18.** Two uniform boards of very nearly the same length are leaning against one another, each board making a 30.0° angle with the vertical, as the drawing shows. One board weighs 20.0 lb, and the other weighs 40.0 lb. Find the force of friction that must act on the lower end of each board. (See Solved Problem 1 for a related problem.)

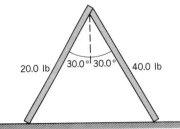

Section 10.4 Newton's Second Law for Rotational Motion

19. A $33\frac{1}{3}$-rev/min record has a diameter of 0.305 m and a mass of 0.122 kg. Assuming the record to be a thin uniform disk, determine the moment of inertia relative to the axis about which the record rotates.

20. A uniform solid disk of mass 30.0 kg is free to rotate about a frictionless axle. Forces of 90.0 N and 125 N are applied to the disk, as the drawing illustrates. (a) What is the net torque produced by the two forces and (b) the angular acceleration of the disk?

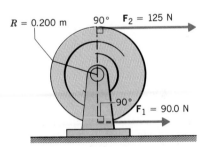

21. A bicycle wheel has a radius of 0.330 m and a rim whose mass is 1.20 kg. The wheel has 50 spokes, each with a mass of 0.010 kg. (a) Calculate the moment of inertia of the rim about the axle. (b) Determine the moment of inertia of *any one spoke*, assuming it to be a long thin rod that can rotate about one end. (c) Find the *total* moment of inertia of the wheel, including the rim and all 50 spokes.

22. The circular blade on a radial arm saw is turning at 2500 rev/min at the instant the motor is turned off. In 18 s the speed of the blade is reduced to 810 rev/min. Assume the blade to be a uniform disk of radius 0.13 m and mass 0.40 kg. Find (a) the angular deceleration in rad/s² (assumed constant) of the blade, (b) the number of revolutions the blade makes during this phase, and (c) the net torque applied to the blade.

23. A turntable platter has a radius of 0.150 m and is rotating at $33\frac{1}{3}$ rev/min. When the power is shut off, the platter slows down and comes to rest in 15 s, due to a net retarding torque of 6.2 $\times$

10^{-3} N·m. Assume the platter to be a uniform solid disk. Determine the mass of the platter.

24. A clay vase on a potter's wheel experiences an angular acceleration of 8.00 rad/s² due to the application of a 10.0-N·m net torque. Find the total moment of inertia of the vase and potter's wheel.

* 25. The drawing shows a model for the motion of the human forearm in throwing a dart. Because of the force **M** applied by the triceps muscle, the forearm can rotate about an axis at the elbow joint. Assume that the forearm has the dimensions shown in the drawing and a moment of inertia of 0.065 kg·m² (including the effect of the dart) relative to the axis at the elbow. Assume also that the force **M** acts perpendicular to the forearm. Ignoring the effect of gravity and any frictional forces, determine the magnitude of the force **M** needed to give the fingertips a tangential speed of 5.0 m/s in 0.10 s, starting from rest.

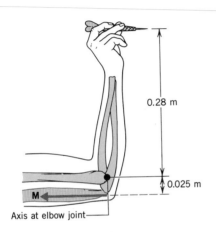

0.28 m

0.025 m

M

Axis at elbow joint

* 26. A thin, rigid, uniform rod has a mass of 2.00 kg and a length of 2.00 m. (a) Find the moment of inertia of the rod relative to an axis that is perpendicular to the rod at one end. (b) Suppose all the mass of the rod were located at a single point. Determine the perpendicular distance of this point from the axis in part (a), such that this point particle has the same moment of inertia as the rod. This distance is called the **radius of gyration** of the rod. The radius of gyration may be determined in a similar fashion for any rigid object and rotational axis. (c) Explain why the radius of gyration of the rod does not equal the distance between the axis in part (a) and the center of gravity.

* 27. One part of a fireworks display uses a 10.0-cm-square platform, on which tubes of gun powder are mounted along each outer edge (see the drawing). As the powder burns, the square spins about an axis perpendicular to its center. The unit reaches an angular speed of 5.0 rev/s in one-half second, starting from rest, because each of three tubes generates a force of 0.30 N and the fourth tube generates an unknown force. The moment of inertia of the unit has a constant value of 1.0×10^{-3} kg·m², and the forces are constant. What is the magnitude of the force generated by the fourth tube?

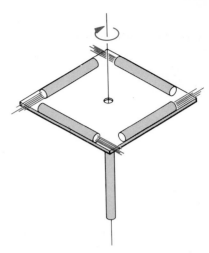

* 28. The **parallel axis theorem** provides a useful way to calculate the moment of inertia I about an arbitrary axis. The theorem states that $I = I_{cm} + Mh^2$, where I_{cm} is the moment of inertia of the object relative to an axis that passes through the center of mass and is parallel to the axis of interest, M is the total mass of the object, and h is the perpendicular distance between the two axes. Use this theorem to determine an expression for the moment of inertia of a solid cylinder of radius R relative to an axis that lies on the surface of the cylinder and is perpendicular to the circular ends.

**29. A thin uniform rod has a length of 3.00 m and is cut into two pieces. The moment of inertia of each piece is measured relative to an axis that is perpendicular to one end of the piece. These moments of inertia are found to have a ratio of 2 : 1. Find the lengths of the two pieces.

**30. By means of a rope whose mass is negligible, two blocks are suspended over a pulley, as the drawing shows. The pulley is a uniform cylindrical disk. The downward acceleration of the 3.00-slug mass is observed to be exactly one-half the acceleration due to gravity. Noting that the tension in the rope is not the same on each side of the pulley, determine the mass of the pulley.

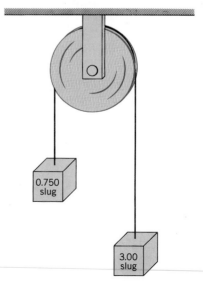

0.750 slug

3.00 slug

Section 10.5 Rotational Work and Energy

31. The mass of the earth is 6.0×10^{24} kg, its radius is 6.4×10^6 m, and it has an angular speed of 1 rev/day. (a) Assuming the earth to be a uniform sphere, what is its rotational kinetic energy (in joules)?

32. A four-bladed ceiling fan rotates with an angular speed of 30.0 rad/s. The length of each blade is 0.600 m, and the mass of each blade is 2.00 kg. Each blade can be approximated as a uniform rod. Determine the total rotational kinetic energy of the four blades.

33. A car starts from rest and accelerates to a speed of 20.0 m/s. Each wheel has a radius of 0.30 m and a moment of inertia of 0.70 kg·m². The car has a total mass (including the wheels) of 1900 kg. Find (a) the translational kinetic energy of the entire car, (b) the total rotational kinetic energy of the four wheels, and (c) the total kinetic energy of the car.

* 34. A thin-walled spherical shell is rolling on a surface. What is the fraction of its total kinetic energy that is in the form of rotational kinetic energy about the center of mass?

* 35. A marble and a cube have the same mass. Starting from rest, the marble rolls and the cube slides (no kinetic friction) down a ramp. Determine the ratio of the center of mass speed of the cube to the center of mass speed of the marble at the bottom of the ramp.

* 36. A solid cylinder and a thin-walled hollow cylinder (see Table 10.1) have the same mass and radius. They are rolling horizontally toward the bottom of an inclined plane. The center of mass of each has the same translational speed. The cylinders roll up the incline and reach their highest points. Calculate the ratio of the distances along the incline through which each center of mass moves.

**37. A bowling ball encounters a 2.50-ft vertical rise on the way back to the ball rack, as the drawing illustrates. Neglect frictional losses and assume the mass of the ball is distributed uniformly. If the linear speed of the ball is 11.5 ft/s at the bottom of the rise, find the linear speed at the top.

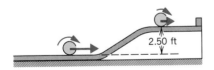

**38. A yo-yo consists of two uniform solid cylinders, each of which has a diameter of 7.00 cm. The two cylinders are joined at their centers by a short rod (negligible mass) that is perpendicular to each cylinder. The diameter of this rod is 1.00 cm. As usual, a string (negligible thickness) is wound around the center rod, and the yo-yo is allowed to roll vertically down the string starting from rest. (a) Assuming the yo-yo does not slip on the string, how fast is the center of mass of the yo-yo moving after it has descended through a vertical distance of 1.00 m? (b) What is the angular speed of the yo-yo at this point?

Section 10.6 Angular Momentum

39. For a certain satellite with an apogee distance of $r_A = 1.30 \times 10^7$ m, the ratio of the orbital speed at perigee to the orbital speed at apogee is 1.20. Find the perigee distance r_P.

40. A woman stands at the center of a platform. The woman and the platform rotate with an angular speed of 5.00 rad/s. Friction is negligible. Her arms are outstretched, and she is holding a dumbbell in each hand. In this position the total moment of inertia of the rotating system (platform, woman, and dumbbells) is 5.40 kg·m². By pulling in her arms, the moment of inertia is reduced to 3.80 kg·m². Find her new angular speed.

41. A baggage carousel at an airport is rotating with an angular speed of 0.20 rad/s when the baggage begins to be loaded onto it. The moment of inertia of the carousel is 1500 kg·m². Ten pieces of baggage with an average mass of 15 kg each are dropped vertically onto the carousel and come to rest at a perpendicular distance of 2.0 m from the axis of rotation. (a) Assuming that no external torques act on the system of carousel and baggage, find the final angular speed. (b) In reality, the angular speed of a baggage carousel does not change. Therefore, speaking qualitatively, what can you say about the external torque acting on this kind of system?

* 42. A cylindrically shaped space station is rotating about the axis of the cylinder to create artificial gravity. The diameter of the cylinder is 165 m. The moment of inertia of the station without people is 3.00×10^9 kg·m². Suppose 500 people, with an average mass of 70.0 kg each, live on this station. As they move radially from the outer surface of the cylinder toward the axis, the angular speed of the station changes. What is the maximum possible percentage change in the station's angular speed due to the radial movement of the inhabitants?

**43. A small 0.500-kg object moves on a frictionless horizontal table in a circular path of radius 1.00 m. The angular speed is 1.00 rev/s. The object is attached to a cord that has a negligible mass and passes through a small hole in the table at the center of the circle. Someone under the table begins to pull the string downward to make the circle smaller. If the string will tolerate a tension of no more than 105 N, what is the radius of the smallest possible circle on which the object can move?

ADDITIONAL PROBLEMS

44. A cylinder is rotating about an axis that passes through the centers of each circular end piece. The cylinder has a radius of 0.0750 m, an angular speed of 840 rev/min, and a moment of inertia of 0.850 kg·m². A brake shoe presses against the surface of the cylinder and applies a tangential frictional force to it. The frictional force reduces the angular speed of the cylinder by a factor of two during a time of 5.00 s. (a) Find the angular deceleration (assumed constant) of the cylinder. Express your answer in rad/s². (b) Find the force of friction applied by the brake shoe.

45. An automobile weighs 2550 lb and the horizontal distance between its front and rear axles is 8.33 ft. The center of gravity of the car is between the front and rear tires, and the horizontal distance between the center of gravity and the front axle is 3.33 ft. Determine the normal force that the ground applies to *each* of the two front wheels and to *each* of the two rear wheels.

46. A playground carousel is free to rotate about its center on frictionless bearings. The carousel has an angular speed of 0.500 rev/s, a moment of inertia of 125 kg·m², and a radius of 1.50 m. A 40.0-kg person, standing still next to the carousel, jumps onto it very close to the outer edge. Find the resulting angular speed of the carousel and person.

47. Three point masses are located on the x axis as follows: (1) 4.00 kg at $x = 0.200$ m, (2) 10.0 kg at $x = 0.500$ m, and (3) 1.5 kg at $x = 0.900$ m. (a) Calculate the moment of inertia of *each* mass with respect to the y axis. (b) Find the total moment of inertia. (c) Based on the results in parts (a) and (b), decide whether it is true that the smallest mass necessarily contributes the smallest amount to the total moment of inertia. Explain.

48. Three objects lie in the x, y plane. Each rotates about the z axis with an angular speed of 6.00 rad/s. The mass m of each object and its perpendicular distance r from the z axis are as follows: (1) $m_1 = 6.00$ kg and $r_1 = 2.00$ m, (2) $m_2 = 4.00$ kg and $r_2 = 1.50$ m, (3) $m_3 = 3.00$ kg and $r_3 = 3.00$ m. (a) Find the tangential speed of each object. (b) Determine the total kinetic energy of this system using the expression $\text{KE} = \frac{1}{2}m_1v_1^2 + \frac{1}{2}m_2v_2^2 + \frac{1}{2}m_3v_3^2$. (c) Obtain the total moment of inertia of the system. (d) Find the rotational kinetic energy of the system using the relation $\frac{1}{2}I\omega^2$ to verify that the answer is the same as that in (b).

* 49. A uniform steel beam of length 5.00 m has a weight of 4.5×10^3 N. One end of the beam is bolted to a vertical wall. The beam is held in a horizontal position by a cable attached between the other end of the beam and a point on the wall. The cable makes an angle of 25.0° above the horizontal. A load whose weight is 12.0×10^3 N is hung from the beam at a point that is 3.50 m from the wall. Find (a) the magnitude of the tension in the supporting cable and (b) the magnitude of the force exerted on the end of the beam by the bolt that attaches the beam to the wall.

* 50. A thin uniform stick is initially positioned in the vertical direction, with its lower end attached to a frictionless axis that is mounted on the floor. The stick has a length of 2.00 m and is allowed to fall, starting from rest. Find the linear speed of the free end of the stick, just before the stick hits the floor after rotating through 90°.

* 51. In raising water from a well, a force **P** must be applied to rotate the cylindrical shaft shown in the drawing. The shaft is uniform and has a mass of 15 kg, while the mass of the handle attached to it is negligible. The force **P** is applied so that it has a perpendicular distance of 0.40 m from the shaft's rotational axis

at all times. Ignoring friction, how much force must be exerted to raise 8.0 kg of water with an acceleration of 0.70 m/s²?

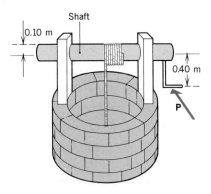

* 52. An equilateral triangle has three corners, labeled A, B, and C. A force $\mathbf{F}_A$ is applied to corner A, and a force $\mathbf{F}_B$ is applied to corner B. Both forces point in the same direction and are parallel to side BC of the triangle. Determine the ratio of the magnitudes of the forces, F_A/F_B, so that the net torque is zero relative to an axis perpendicular to the plane of the triangle and passing through its geometrical center.

* 53. A flat uniform circular disk (radius $= 2.00$ m, mass $= 1.00 \times 10^2$ kg) is initially stationary and fixed so that it can rotate in the horizontal plane about a frictionless axis perpendicular to the center of the disk. A 40.0-kg person, standing 1.25 m from the axis, begins to run on the disk in a circular path and has a tangential speed of 2.00 m/s relative to the ground. (a) Find the resulting angular speed of the disk (in rev/s) and describe the direction of the rotation. (b) Determine the time it takes for a spot marking the starting point to pass again beneath the runner's feet.

**54. A solid uniform cylinder is attached to a cord, which is wound around an identical cylinder fixed so as to rotate without friction about an axis parallel to the ground. Two identical uniform solid spheres are set up in a similar fashion, as the drawing indicates. The spheres and cylinders have different masses and radii. The cords have negligible masses. The hanging objects are released from rest with their centers at the same height above ground level. How far apart will the centers be after 2.00 s have elapsed, and which object will have fallen farther?

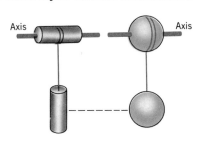

55. The drawing shows the top view of two doors. The doors are uniform and have identical widths. However, door A rotates about an axis along one edge, while door B rotates about an axis through its center. The same force **F is applied perpendicular to each door at its edge, and the force remains perpendicular to each door as the door turns. Starting from rest, door A rotates through 90.0° in 3.00 s. How long does it take door B to rotate through the same angle?

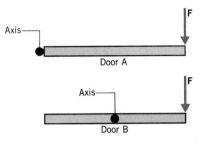

**56. The drawing shows an inverted "A" that is suspended from the ceiling by two vertical ropes. Each leg of the "A" has a length of 2L and a weight of 120 N. The horizontal crossbar also has a weight of 120 N. Find the force that the crossbar applies to each of the two legs. *(See Solved Problem 1 for a related problem.)*

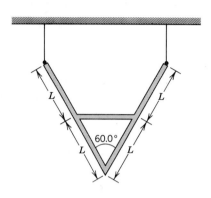

CHAPTER 11

Elasticity and Simple Harmonic Motion

Being thrown into the air by the elastic surface of a trampoline is an exhilarating experience. Many other systems also exhibit elastic behavior, as we will explore in this chapter.

11.1 INTRODUCTION

Loosely speaking, an "elastic" material is one that has the ability to return to its original shape after being stretched, compressed, or otherwise distorted. A rubber band, for instance, snaps back into shape when the stretching force is removed, provided the force is not too large. Similarly, billiard balls retain their "roundness" under repeated use, because the ivory from which they are made is sufficiently elastic to return to its original shape after each collision.

From an atomic viewpoint, the elastic behavior of any material has its origin in the forces between the atoms that comprise the material. Figure 11.1 symbolizes these forces with the aid of springs. In this picture, the springs are drawn between adjacent atoms, but it should be realized that forces can also exist between atoms that are not nearest neighbors. It is well known that a spring exerts a force when its coils are pulled apart or pushed together. Thus, because of atomic-level "springs," a material exhibits a tendency to return to its initial shape once the forces that cause a deformation are removed. The exact details of a material's elastic behavior depend on the kind of interatomic forces that are present.

In this chapter, our study of elasticity begins with a description of the kinds of deformations that elastic materials can experience. Then, attention will be focused on one particularly important elastic object, the spring.

11.2 ELASTIC DEFORMATION

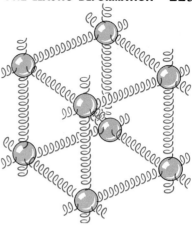

FIGURE 11.1 The forces between atoms are symbolized by springs. The atoms are represented by black spheres, and the springs between some adjacent atoms have been omitted for the sake of clarity.

STRETCHING, COMPRESSION, AND YOUNG'S MODULUS

The forces that hold the atoms of a solid together are particularly strong, so considerable force must be applied to stretch a solid object. The force needed depends on several factors, as Figure 11.2 illustrates. Naturally, the force depends on the extent of the stretching. For two identical rods, part *a* of the drawing indicates that more force is required to produce a greater amount of stretch than to produce a smaller amount. Part *b* indicates another obvious fact: To stretch two equally long rods by the same amount, more force is required for the rod with the larger cross-sectional area, assuming the rods are made from the same material. Finally, part *c* shows that, for a given amount of stretch, more force is required for a shorter rod than for a longer rod, provided the rods are made from the same material and have the same cross-sectional area. Simply take a long rubber band, for example, and stretch it an arbitrary amount. Then take a much shorter rubber band with the same cross-sectional area and stretch it by the same amount. You will be easily convinced that more force is required to stretch the shorter rubber band.

If the amount of stretching is small compared to the original length of the object, the elastic behavior discussed above can be described with the aid of the following relation:

$$F = Y\left(\frac{\Delta L}{L_0}\right) A \tag{11.1}$$

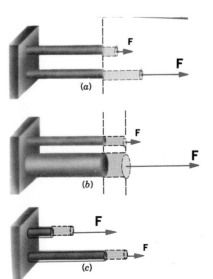

FIGURE 11.2 The amount of force F needed to stretch a solid rod depends on (*a*) the amount by which the rod is stretched, (*b*) the cross-sectional area of the rod, and (*c*) the unstretched length of the rod.

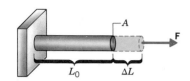

FIGURE 11.3 In this diagram, **F** denotes the stretching force, A the cross-sectional area, L_0 the original length of the rod, and ΔL the amount of stretch.

As Figure 11.3 shows, F denotes the magnitude of the stretching force that is applied perpendicular to the cross-sectional area A, ΔL is the increase in length, and L_0 is the original length. The term Y is a proportionality constant called **Young's modulus,** after Thomas Young (1773–1829). Solving Equation 11.1 for Y shows that Young's modulus has units of force per unit area (N/m^2), the exact value depending on the nature of the material, as Table 11.1 reveals. Equation 11.1 is found experimentally to hold for a variety of solid materials and indicates that the required stretching force is directly proportional to the fractional increase in length $\Delta L/L_0$ and to the cross-sectional area A.

It is important to realize that the magnitude of the force in Equation 11.1 is proportional to the *fractional increase* in length $\Delta L/L_0$, rather than the absolute increase ΔL. In addition, the cross-sectional area need not be circular, as in Figure 11.3, but can have any shape (e.g., rectangular).

Forces that are applied as in Figure 11.3 and cause stretching are called "tensile" forces because they create a tension in the material, much like the tension in a rope. The proportionality expressed in Equation 11.1 also applies when the force compresses the material along its length. In this situation, the force is applied in a direction opposite to that shown in Figure 11.3, and ΔL stands for the amount by which the original length L_0 decreases.

Most solids have Young's moduli that are rather large, reflecting the fact that a large force is needed to change the length of a solid object by even a small amount, as Example 1 illustrates.

EXAMPLE 1

In the "living pyramid" act, a circus performer carries six colleagues on his shoulders. The colleagues have a combined weight of 4200 N. Each thighbone (femur) of this performer has a length of 0.55 m and an effective cross-sectional area of 7.7×10^{-4} m^2. Determine the amount by which each thighbone compresses under the extra weight.

SOLUTION

The additional weight supported by each bone is 2100 N, and Table 11.1 indicates that Young's modulus for compression is

9.4×10^9 N/m^2. The amount of compression can be obtained from Equation 11.1:

$$\Delta L = \frac{FL_0}{YA} = \frac{(2100 \text{ N})(0.55 \text{ m})}{(9.4 \times 10^9 \text{ N/m}^2)(7.7 \times 10^{-4} \text{ m}^2)}$$

$$= \boxed{1.6 \times 10^{-4} \text{ m}}$$

This is a very small change in length, the fractional decrease being $\Delta L/L_0 = 0.00029$.

Example 2 illustrates how structural design must take into account the large forces that can come into play whenever the length of a solid object changes.

EXAMPLE 2

A steel beam, 9.8 m in length and 0.10 m^2 in cross-sectional area, is used in the roadbed of a bridge. The beam is mounted between two concrete supports, as Figure 11.4 shows, and is fitted perfectly into place, with no room provided for expansion. In response to a rise in temperature of 19 Celsius degrees, the beam would expand by 2.2 mm if it were free to do so. To prevent this small expansion from occurring, what compressive force must be supplied by the concrete supports?

FIGURE 11.4 The steel beam is fitted between concrete supports with no room provided for expansion of the beam. As Example 2 shows, disastrous results can occur.

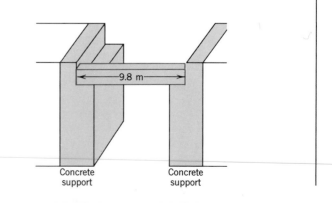

SOLUTION

The force required to compress the beam by 2.2 mm can be calculated from Equation 11.1, using the given information and the value of 2.0×10^{11} N/m² for Young's modulus of steel from Table 11.1:

$$F = Y \left(\frac{\Delta L}{L_0} \right) A$$

$$= (2.0 \times 10^{11} \text{ N/m}^2) \left(\frac{2.2 \times 10^{-3} \text{ m}}{9.8 \text{ m}} \right) (0.10 \text{ m}^2)$$

$$= \boxed{4.5 \times 10^6 \text{ N}}$$

If the concrete supports cannot supply this large force (about one million pounds), they will have to move or crack to provide the needed expansion space. Notice how large the force is, even though 2.2 mm is only 0.022% of the original length of the beam.

SHEAR DEFORMATION AND THE SHEAR MODULUS

It is possible to deform a solid object in a way that is not a simple stretch or compression. For instance, place a book on a rough table and push on the book as in Figure 11.5a. The resulting deformation is called **shear deformation.** Part b of the drawing indicates that the deformation occurs because of the combined effect of the force **F** applied to the top of the book and the reaction force $-$**F** applied to the bottom of the book by the table. Equation 11.2 gives the magnitude F of the force needed to produce an amount of shear ΔX for an object with cross-sectional area A and thickness L_0:

$$F = S \left(\frac{\Delta X}{L_0} \right) A \qquad (11.2)$$

This equation is very similar to Equation 11.1. The constant of proportionality S is called the **shear modulus** and, like Young's modulus, has units of force per unit area (N/m²). The value of S depends on the nature of the material, and Table 11.2 gives some representative values.

Although Equations 11.1 and 11.2 are similar, they refer to different kinds of deformations. The shearing force in Figure 11.5 is parallel to the area A, whereas the tensile force in Figure 11.3 is perpendicular to the area A. Furthermore, the ratio $\Delta X/L_0$ in Equation 11.2 is different than the ratio $\Delta L/L_0$ in Equation 11.1: the distances ΔX and L_0 are perpendicular, whereas ΔL and L_0 are parallel. Clearly, Young's modulus and the shear modulus reflect different elastic deformations. Young's modulus refers to a *change in length* of one dimension of a solid object as a result of tensile or compressive forces. In contrast, the shear modulus refers to a *change in shape* of a solid object as a result of shearing forces.

VOLUME DEFORMATION AND THE BULK MODULUS

When a compressive force is applied along one dimension of a solid, the length of that dimension decreases. It is also possible to apply compressive forces so that the size of

TABLE 11.1 Values for the Young's Modulus of Solid Materials

Material	Young's Modulus Y (N/m²)
Aluminum	6.9×10^{10}
Bone (compression)	9.4×10^9
Bone (tension)	1.6×10^{10}
Brass	9.0×10^{10}
Brick	1.4×10^{10}
Copper	1.1×10^{11}
Mohair	2.9×10^9
Nylon	3.7×10^9
Pyrex glass	6.2×10^{10}
Steel	2.0×10^{11}
Teflon	3.7×10^8
Tungsten	3.6×10^{11}

TABLE 11.2 Values for the Shear Modulus of Solid Materials

Material	Shear Modulus S (N/m²)
Aluminum	2.4×10^{10}
Bone	8.0×10^{10}
Brass	3.5×10^{10}
Copper	4.2×10^{10}
Lead	5.4×10^9
Nickel	7.3×10^{10}
Steel	8.1×10^{10}
Tungsten	1.5×10^{11}

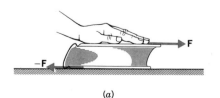

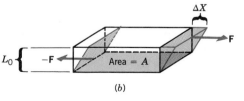

FIGURE 11.5 (*a*) An example of a shear deformation. (*b*) The shearing forces **F** and $-$**F**, applied parallel to the cross-sectional area A and perpendicular to the thickness L_0, cause a solid object to change shape. The shear deformation ΔX is the distance that the top surface is shifted relative to the fixed bottom surface.

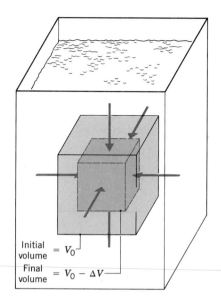

FIGURE 11.6 The colored arrows denote the forces that push perpendicularly on every available surface of an object that is immersed beneath the surface of a liquid. The force per unit area is called the pressure. When the pressure increases, the volume of the object decreases from an original value of V_0 to a final value of $V_0 - \Delta V$.

every dimension (length, width, and depth) decreases, leading to a decrease in volume, as Figure 11.6 illustrates. This kind of overall compression occurs, for example, when an object is submerged in a liquid, and the liquid presses inward everywhere on the object. The forces acting in such situations are applied perpendicular to every available surface. For this reason it is more convenient to speak of the perpendicular force per unit area, rather than the amount of any one force in particular. The perpendicular force per unit area is called the *pressure P*.

In general, the result of increasing the pressure on an object by an amount ΔP is that the volume of the object decreases by an amount ΔV (see Figure 11.6). Such a pressure increase occurs, for example, when a swimmer dives deeper and deeper into the water. Experiment reveals that the amount of pressure increase needed to decrease the volume is directly proportional to the fractional volume change $\Delta V/V_0$, where V_0 is the initial volume:

$$\Delta P = -B\left(\frac{\Delta V}{V_0}\right) \tag{11.3}$$

Equation 11.3 is analogous to Equations 11.1 and 11.2, except that the area A in the latter equations does not appear here explicitly; the area is already taken into account on the left side of Equation 11.3 by the concept of pressure (force per unit area). The proportionality constant B is known as the *bulk modulus.* The minus sign occurs because an increase in pressure (ΔP positive) always creates a decrease in volume (ΔV negative), and B is given as a positive quantity. Like Young's modulus and the shear modulus, the bulk modulus has units of force per unit area (N/m²), and its value depends on the nature of the material.

Table 11.3 gives representative values of the bulk modulus and includes liquids as well as solids. The bulk modulus concept can also be applied to gases, although we will not do so here. In contrast, Young's modulus and the shear modulus do not pertain to liquids and gases, because the molecular structures of these materials do not allow tensile forces and compressive forces to be applied along only a single dimension, nor do they allow shear forces to be applied. Whenever an attempt is made to apply such forces, liquids and gases respond by flowing around the point of application. The bulk modulus, however, refers to a change in volume under conditions that prevent liquids and gases from flowing.

TABLE 11.3 Values for the Bulk Modulus of Solid and Liquid Materials

Material	Bulk Modulus B (N/m²)
Solids	
Aluminum	7.1×10^{10}
Brass	6.7×10^{10}
Copper	1.3×10^{11}
Lead	4.2×10^{10}
Nylon	6.1×10^{9}
Pyrex glass	2.6×10^{10}
Steel	1.4×10^{11}
Liquids	
Ethanol	8.9×10^{8}
Oil	1.7×10^{9}
Water	2.2×10^{9}

11.3 STRESS, STRAIN, AND HOOKE'S LAW

Equations 11.1–11.3 specify the amount of force needed for a given amount of elastic deformation, and they are repeated below to emphasize their common features:

$$\boxed{\frac{F}{A}} = Y \boxed{\left(\frac{\Delta L}{L_0}\right)} \tag{11.1}$$

$$\boxed{\frac{F}{A}} = S \boxed{\left(\frac{\Delta X}{L_0}\right)} \tag{11.2}$$

$$\boxed{\Delta P} = -B \boxed{\left(\frac{\Delta V}{V_0}\right)} \tag{11.3}$$

$$\boxed{\text{Stress}} \text{ proportional } \boxed{\text{Strain}}$$
$$\text{is} \qquad \text{to}$$

The left side of each equation is the magnitude of the force per unit area required to cause an elastic deformation. In general, the ratio of the force to the area is called the **stress**. The right side of each equation involves the change in a quantity (ΔL, ΔX, or ΔV) divided by a quantity (L_0 or V_0) that serves as a reference, relative to which the change is compared. The terms $\Delta L/L_0$, $\Delta X/L_0$, and $\Delta V/V_0$ are unitless ratios, and each is referred to as the **strain** that results from the applied stress. In the case of stretch and compression, the strain is the fractional change in length, whereas in volume deformation it is the fractional change in volume. In shear deformation the strain refers to a change in shape of the object. Experiments show that these three equations, with constant values for Young's modulus, the shear modulus, and the bulk modulus, apply to a wide range of materials. Therefore, stress and strain are directly proportional to one another. This simple relationship between elastic stress and strain was first discovered by Robert Hooke (1635–1703) and is referred to as **Hooke's law.**

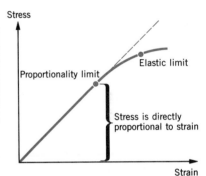

FIGURE 11.7 Hooke's law (stress is directly proportional to strain) is valid only up to the proportionality limit of a material. Between the proportionality limit and the elastic limit, stress is no longer proportional to strain, although the object will return to its original size and shape when the stress is removed. Beyond the elastic limit, the material remains deformed even when the stress is removed.

HOOKE'S LAW FOR STRESS AND STRAIN
Stress is directly proportional to strain.

SI unit of stress: newton per square meter = pascal (Pa)

SI unit of strain: Strain is a unitless quantity.

The SI unit of stress is the *pascal (Pa)*, named for the French scientist Blaise Pascal (1623–1662); $1 \text{ Pa} = 1 \text{ N/m}^2$.

In reality, materials obey Hooke's law only up to a certain limit. Figure 11.7 displays the region of applicability of Hooke's law in a graphical fashion. As long as stress remains proportional to strain, a plot of stress versus strain produces a straight line. The point on the graph where the material deviates from straight line behavior is called the "proportionality limit." Beyond the proportionality limit stress and strain are no longer directly proportional. However, if the stress does not exceed the "elastic limit" of the material, the object will return to its original size and shape once the stress is removed. The "elastic limit" is the point beyond which the object no longer returns to its original size and shape when the stress is removed; the object remains permanently deformed. Figure 11.8 shows objects that have been stretched to complete failure.

FIGURE 11.8 These objects have been stretched so severely that they have broken into several pieces.

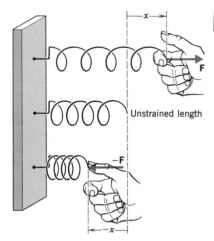

FIGURE 11.9 An ideal spring is one that obeys a Hooke's law type of equation, $F = kx$, where **F** denotes the force applied to the spring, x is the amount of stretch or compression, and k is the spring constant.

11.4 THE IDEAL SPRING AND SIMPLE HARMONIC MOTION

The spring in Figure 11.9 will return to its original length after being stretched or compressed, provided the applied force is not large enough to cause a permanent deformation. For relatively small deformations, the force required to stretch or compress a spring obeys the following equation:

$$F = kx \qquad (11.4)$$

In this expression x denotes the amount by which the spring is stretched or compressed from its unstrained length. The term k is a proportionality constant called the **spring constant** and has dimensions of force per unit length (N/m). Sometimes k is referred to as the **stiffness** of the spring, because a large value for k means the spring is "stiff," in the sense that a large force is required to stretch or compress it.

Equation 11.4 is a Hooke's law type of relationship, as can be seen by comparing it to Equation 11.1 for the stretching or compressing of a solid rod:

$$F = \underbrace{\frac{YA}{L_0}}_{}\underbrace{\Delta L}_{}$$
$$F = \quad k \quad x \qquad (11.1)$$

This comparison shows that x is analogous to ΔL, whereas k is analogous to the term YA/L_0, which is a constant for a given object. A spring that behaves according to the Hooke's law relationship $F = kx$ is said to be an **ideal spring**. Springs are put to good use in a wide variety of applications, and Example 3 illustrates one of them.

EXAMPLE 3

In a tire pressure gauge, the air in the tire pushes against a spring when the gauge is attached to the tire valve. Figure 11.10 shows such a gauge. Suppose the spring constant of the spring is 320 N/m and the bar indicator of the gauge extends 2.0 cm when the gauge is pressed against the air valve of the tire. What force does the air in the tire apply to the spring?

SOLUTION

Since the spring constant is known, the force applied to the spring can be obtained from Equation 11.4:

$$F = kx = (320 \text{ N/m})(0.020 \text{ m}) = \boxed{6.4 \text{ N}}$$

Thus, the exposed length of the bar indicator gives a measure of the force that the air pressure in the tire exerts on the spring. Since pressure is force per unit area and the area of the plunger surface (see drawing) is fixed, the bar indicator can be marked in units of pressure.

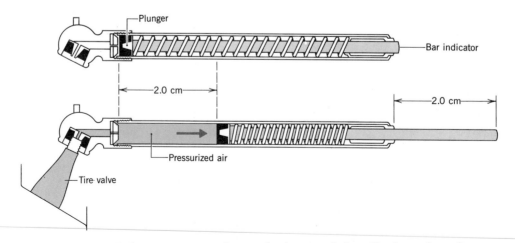

FIGURE 11.10 A tire pressure gauge is one of many practical applications of a spring.

To stretch or compress a spring, a force must be applied to it. In accord with Newton's third law, the spring exerts an oppositely directed force of equal magnitude. This reaction force is applied by the spring to the agent that does the pulling or pushing. In other words, the reaction force is applied to the object attached to the spring. The reaction force is also called a "restoring force," for a reason that will be clarified shortly. The restoring force of an ideal spring is obtained from the relation $F = kx$ by including the minus sign required by Newton's action/reaction law, as indicated below.

HOOKE'S LAW RESTORING FORCE OF AN IDEAL SPRING
The restoring force of an ideal spring is

$$F = -kx \qquad (11.5)$$

The minus sign indicates that the restoring force always points in a direction opposite to that in which the spring is deformed.

To understand the phrase "restoring force," consider Figure 11.11, in which a mass m is attached to a spring and is lying on a frictionless table. In part A the mass is at rest. The spring has its undeformed length and, hence, applies no force to the object. The object is in equilibrium. In part B, the spring has been stretched to the right, so it applies the leftward-pointing force $-\mathbf{F}$ to the mass m. When the object is released, this force pulls the object to the left, restoring it toward its equilibrium position. However, consistent with Newton's first law, the moving mass has inertia and coasts to the left beyond the equilibrium position, compressing the spring as in part C. The force exerted by the spring now points to the right and, after bringing the object to a momentary halt, acts to restore the object to its equilibrium position. But the moving mass again coasts beyond the equilibrium position, this time stretching the spring and leading to the restoring force $-\mathbf{F}$ shown in part D. The back-and-forth motion illustrated in the drawing then repeats itself, continuing forever, since there is no friction.

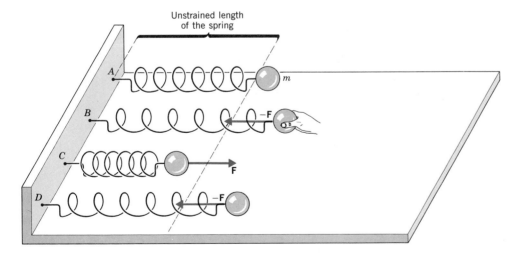

FIGURE 11.11 The restoring force (see colored arrows) produced by an ideal spring always points opposite to the direction in which the spring is deformed and leads to a back-and-forth motion of the mass m. If opposing forces such as friction are absent, the motion continues forever and is known as *simple harmonic motion*.

When the restoring force has the mathematical form given by $F = -kx$, the type of motion illustrated in Figure 11.11 is designated as **simple harmonic motion.** By attaching a pen to the mass m and moving a strip of paper past it at a steady rate, we can record the position of the vibrating object as time passes. Figure 11.12 illustrates the resulting graphical record of simple harmonic motion. The maximum excursion from equilibrium is the **amplitude A** of the motion. The shape of this graph is characteristic of simple harmonic motion and is called "sinusoidal," because it has the shape of a trigonometric sine or cosine function.

When an object attached to a horizontal spring is moved from its equilibrium position and released, the restoring force $F = -kx$ leads to simple harmonic motion. The restoring force also leads to simple harmonic motion when the object is attached to a vertical spring. When the spring is vertical, however, the weight of the object stretches the spring by a fixed amount to begin with. Therefore, simple harmonic motion occurs with respect to the equilibrium position of the object on the stretched spring, as Figure 11.13 indicates. The amount of initial stretching d_0 caused by the weight of the object can be calculated by equating the weight to the magnitude of the restoring force that supports it; thus, $mg = kd_0$, which gives $d_0 = mg/k$.

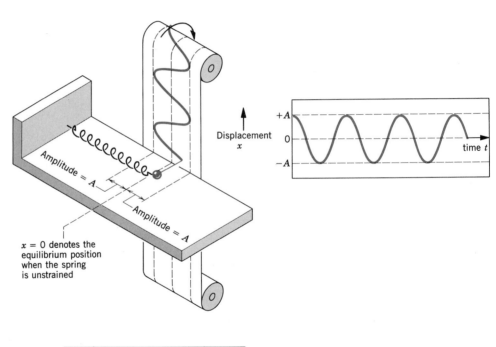

FIGURE 11.12 When an object moves in simple harmonic motion, a record of its position as a function of time is a graph that has a sinusoidal shape with an amplitude A.

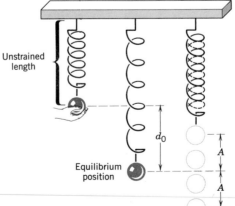

FIGURE 11.13 The weight of an object on a vertical spring initially stretches the spring by an amount d_0. Vertical simple harmonic motion of amplitude A occurs with respect to this equilibrium position of the object.

11.5 SIMPLE HARMONIC MOTION AND THE REFERENCE CIRCLE

Simple harmonic motion, like any motion, can be described by specifying the associated displacement, velocity, and acceleration, and the model in Figure 11.14 is helpful in explaining these characteristics. This model consists of a small ball moving in uniform circular motion (see Chapter 6 to review uniform circular motion) on a circle known as the **reference circle.** As the ball moves, its shadow falls on a strip of film, which is moving vertically upward at a steady rate and records the position of the shadow as time passes. A comparison of the film with the paper in Figure 11.12 reveals that they have recorded identical patterns, indicating that the shadow of the moving ball is a good model for simple harmonic motion.

DISPLACEMENT

Figure 11.15 takes a closer look at the reference circle (radius $= A$) and indicates how to determine the displacement of the shadow on the film. The ball starts out on the x axis and moves through the angle θ in a time t. Since the circular motion is uniform, the ball moves with a constant angular speed ω (in rad/s), and, therefore, the angle has a value (in rad) of $\theta = \omega t$. The position of the shadow on the film is just x, the projection of the radius A onto the horizontal axis:

$$x = A \cos \theta = A \cos \omega t \qquad (11.6)$$

Figure 11.16 shows a graph of this equation. As time passes, the shadow of the ball oscillates between the maximum and minimum values of $x = +A$ and $x = -A$, respectively, corresponding to the limiting values of $+1$ and -1 for the cosine of an angle. The radius A of the reference circle, then, determines the amplitude of the simple harmonic motion.

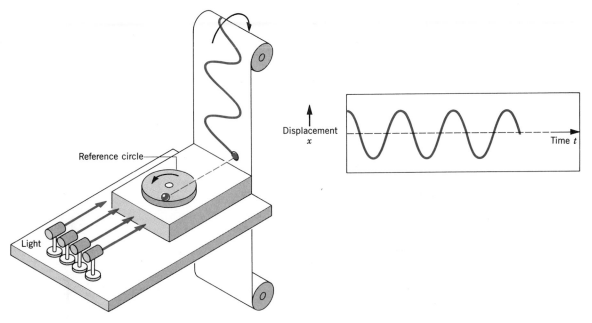

FIGURE 11.14 The ball mounted on the turntable moves in uniform circular motion, and its shadow, projected on a moving strip of film, executes simple harmonic motion.

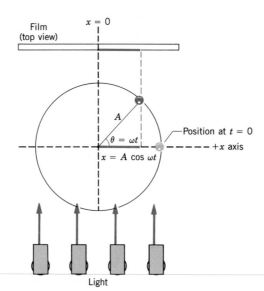

FIGURE 11.15 The ball's shadow on the film has a displacement x that depends on the angle θ through which the ball has moved on the reference circle.

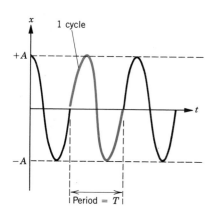

FIGURE 11.16 The graph of displacement x versus time t for an object undergoing simple harmonic motion is a sinusoidal curve. The period T is the time required for the object to complete one cycle of the motion.

As the ball moves one revolution or cycle around the reference circle, its shadow executes one cycle of back-and-forth motion. For the shadow, or for any object in simple harmonic motion, the time required for one of these cycles to occur is the **period** T, as Figure 11.16 indicates. The value of T depends on the angular speed ω of the ball. The relationship between ω and T can be obtained by using $\omega = \theta/t$ and noting that 2π radians of angular displacement correspond to the one cycle that occurs in T seconds:

$$\omega = \frac{2\pi}{T} \tag{11.7}$$

Often, instead of the period, it is more convenient to speak of the **frequency** f of the simple harmonic motion, the frequency being just the number of cycles of the motion per second. For example, if an object on a spring executes 10 cycles in one second, the frequency is $f = 10$ cycles/s. The period T, or the time to complete one cycle, would be $\frac{1}{10}$ s. Thus, frequency and period are related according to

$$f = \frac{1}{T} \tag{11.8}$$

Usually one cycle per second is referred to as one hertz (Hz), the unit being named after Heinrich Hertz (1857–1894). One thousand cycles per second is denoted as one kilohertz (kHz), with the result that five thousand cycles per second can be written as 5 kHz, for instance.

Using the relationships $\omega = 2\pi/T$ and $f = 1/T$, we can relate the angular speed ω (in rad/s) to the frequency f (in cycles/s or Hz):

$$\omega = \frac{2\pi}{T} = 2\pi f \tag{11.9}$$

Because ω is directly related to the frequency f, ω is sometimes called the **angular frequency** of the simple harmonic motion.

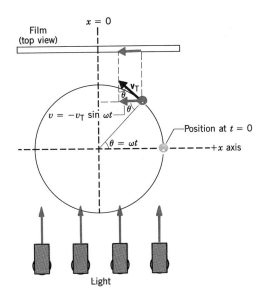

FIGURE 11.17 The velocity of the ball's shadow on the film is **v**, which is the horizontal component of the tangential velocity v_T of the ball on the reference circle.

VELOCITY

The reference circle can also be used to calculate the velocity of an object in simple harmonic motion. Figure 11.17 shows the tangential velocity v_T of the ball on the reference circle. The drawing indicates that the velocity **v** of the shadow is just the x component of the vector v_T, that is, $v = -v_T \sin \omega t$. The minus sign is necessary, since **v** points to the left, in the direction of the negative x axis. Since the tangential speed v_T is related to the angular speed ω by $v_T = r\omega$ (Equation 9.9) and since $r = A$, it follows that $v_T = A\omega$. Therefore, the velocity in simple harmonic motion is given by

$$v = -A\omega \sin \omega t \tag{11.10}$$

Note that this velocity is *not* constant but varies between maximum and minimum values as time passes. The velocity is zero when the shadow changes direction at either end of the oscillatory motion. The velocity has a maximum magnitude when the shadow passes through the $x = 0$ position. The largest magnitude that v can have is $A\omega$, since the sine of an angle is between $+1$ and -1:

$$v_{max} = A\omega \tag{11.11}$$

The maximum velocity depends, then, on both the amplitude A and the angular frequency ω, a fact that Example 4 emphasizes.

EXAMPLE 4

In response to a record groove, a phono stylus (monaural) generates electrical signals by vibrating back and forth in simple harmonic motion, as Figure 11.18 shows. If the frequency of the motion is 1.0 kHz and the amplitude is 8.0×10^{-6} m, what is the maximum speed of the stylus?

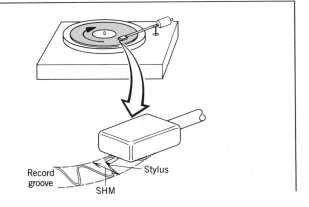

FIGURE 11.18 A phono stylus (monaural) generates electrical signals by vibrating back and forth in simple harmonic motion (SHM) within a record groove.

SOLUTION

The solution can be obtained from Equation 11.11 and the fact that $\omega = 2\pi f$:

$$v_{max} = A\omega = (8.0 \times 10^{-6} \text{ m})[2\pi(1.0 \times 10^3 \text{ Hz})] = \boxed{0.050 \text{ m/s}}$$

ACCELERATION

In simple harmonic motion, the velocity is not constant; consequently, there must be an acceleration. This acceleration can also be determined with the aid of the reference circle. As Figure 11.19 shows, the ball on the reference circle has a centripetal acceleration a_c that points toward the center of the circle. The acceleration of the shadow is **a** and is just the "shadow" of the vector a_c, in other words, its horizontal component; $a = -a_c \cos \omega t$. The minis sign is needed because the acceleration of the shadow points to the left. Recalling that the centripetal acceleration is related to the angular speed ω by $a_c = r\omega^2$ (Equation 9.11) and using $r = A$, we find that $a_c = A\omega^2$. With this substitution, the acceleration in simple harmonic motion becomes

$$a = -A\omega^2 \cos \omega t \tag{11.12}$$

The acceleration, like the velocity, does *not* have a constant value as time passes. The maximum magnitude of the acceleration is

$$a_{max} = A\omega^2 \tag{11.13}$$

Although both the amplitude A and the angular frequency ω determine the maximum value, the frequency has a particularly strong effect, for it enters as the frequency squared. Example 5 shows that the acceleration can be remarkably large in a practical situation.

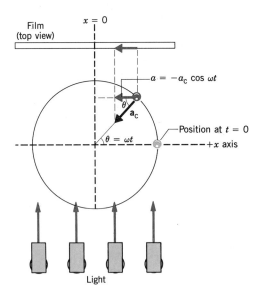

FIGURE 11.19 The acceleration of the ball's shadow on the film is **a** and is the horizontal component of the centripetal acceleration a_c of the ball on the reference circle.

EXAMPLE 5

The diaphragm of a loudspeaker moves back and forth in simple harmonic motion to create sound, as in Figure 11.20. The frequency of the vibratory motion is 1.0 kHz and the amplitude is 2.0×10^{-4} m. Find the maximum acceleration that the diaphragm experiences.

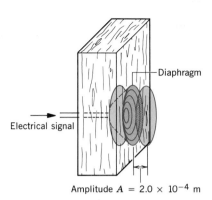

Diaphragm

Electrical signal

Amplitude $A = 2.0 \times 10^{-4}$ m

FIGURE 11.20 In response to electrical signals, the diaphragm of a loudspeaker generates a 1.0-kHz sound by moving back and forth in simple harmonic motion.

SOLUTION

Using Equation 11.13 and the fact that $\omega = 2\pi f$, we find that

$$a_{max} = A\omega^2 = (2.0 \times 10^{-4} \text{ m})[2\pi(1.0 \times 10^3 \text{ Hz})]^2$$
$$= \boxed{7.9 \times 10^3 \text{ m/s}^2}$$

This result is more than 800 times the acceleration due to gravity (9.8 m/s^2) and is significantly larger than the acceleration experienced by astronauts during a rocket launch.

FREQUENCY OF VIBRATION

With the aid of Newton's second law ($F = ma$), it is possible to determine the frequency at which an object of mass m vibrates on a spring. The mass of the spring itself is assumed to be negligible. The force that the spring applies to the object is the Hooke's law restoring force $F = -kx$, so that Newton's second law becomes $-kx = ma$. Using Equation 11.6 for x and Equation 11.12 for a, we find that

$$-k(A \cos \omega t) = m(-A\omega^2 \cos \omega t)$$

which yields

$$\omega = \sqrt{\frac{k}{m}} = 2\pi f \qquad (11.14)$$

In this expression, the angular frequency ω must be in radian measure, and larger values for k and smaller masses result in larger frequencies, as Example 6 illustrates.

EXAMPLE 6

Determine the frequency (in Hz) at which each of the masses will vibrate in Figure 11.21.

SOLUTION

This example asks for the vibrational frequency f in Hz (cycles/s), while the equation $\omega = \sqrt{k/m}$ gives the angular frequency ω in rad/s. The relationship between the two kinds of frequencies is given by $\omega = 2\pi f$, so that

$$f = \frac{\omega}{2\pi} = \frac{1}{2\pi}\sqrt{\frac{k}{m}}$$

[Spring 1] $\quad f = \dfrac{1}{2\pi}\sqrt{\dfrac{225 \text{ N/m}}{0.400 \text{ kg}}} = \boxed{3.77 \text{ Hz}}$

[Spring 2] $\quad f = \dfrac{1}{2\pi}\sqrt{\dfrac{225 \text{ N/m}}{0.100 \text{ kg}}} = \boxed{7.55 \text{ Hz}}$

[Spring 3] $\quad f = \dfrac{1}{2\pi}\sqrt{\dfrac{825 \text{ N/m}}{0.100 \text{ kg}}} = \boxed{14.5 \text{ Hz}}$

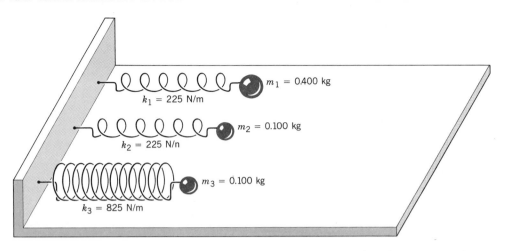

FIGURE 11.21 The angular frequency ω at which an object on a spring vibrates depends on the mass m of the object and the spring constant k, according to $\omega = \sqrt{k/m}$, where ω is in rad/s.

11.6 ENERGY AND SIMPLE HARMONIC MOTION

ELASTIC POTENTIAL ENERGY

Chapter 7 discusses the work–energy theorem, which states that when a net force does work on an object, the kinetic energy of the object changes by an amount that is equal to the work done. As we saw in that chapter, the work done by the gravitational force leads to the concept of gravitational potential energy. In terms of this concept, the work–energy theorem has the following form:

$$W_{\text{other}} = \Delta\text{KE} + \Delta\text{PE}_{\text{gravitational}} \qquad (7.7)$$

Equation 7.7 indicates that when work W_{other} is done by *forces other than gravity*, the return for the effort comes in the form of a change in kinetic energy (ΔKE), a change in gravitational potential energy ($\Delta\text{PE}_{\text{gravitational}}$), or both, depending on how the work is done. When an object is attached to a spring, the spring force can do work W_{elastic} on the object. Therefore, Equation 7.7 can be rewritten as

$$W_{\text{other}} + W_{\text{elastic}} = \Delta\text{KE} + \Delta\text{PE}_{\text{gravitational}} \qquad (11.15)$$

if it is understood that the term W_{other} no longer includes the work done by the spring force.

When the amount by which a spring is stretched changes from an initial value of x_0 to a final value of x_f (see Figure 11.22), the work done by the spring force can be calculated from the definition of work as $W = (F\cos\theta)s$ (Equation 7.1). In this definition, F stands for the magnitude of the spring force and s for the magnitude of the displacement $x_f - x_0$ through which the force moves. The spring force equals $-kx$, with the magnitude being just kx. Thus, as the spring stretches from $x = x_0$ to $x = x_f$, the magnitude of the spring force changes from kx_0 to kx_f. To account for the changing magnitude of the force, we can use the average of kx_0 and kx_f, because the dependence on x is linear: $F_{\text{ave}} = \frac{1}{2}(kx_0 + kx_f)$. The work W_{elastic} is, then,

$$W = (F\cos\theta)s \qquad (7.1)$$

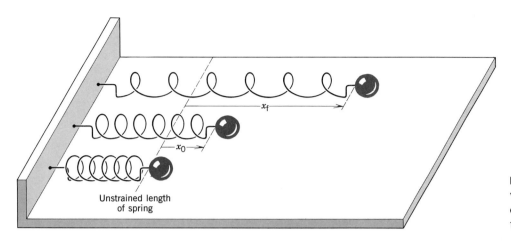

FIGURE 11.22 The amount by which the spring is stretched changes from an initial value of x_0 to a final value of x_f.

$$W_{\text{elastic}} = [\tfrac{1}{2}(kx_0 + kx_f) \cos 180°](x_f - x_0)$$
$$W_{\text{elastic}} = \tfrac{1}{2}kx_0^2 - \tfrac{1}{2}kx_f^2 \qquad (11.16)$$

In this calculation, θ has been set equal to $180°$, since the spring force points opposite to the direction of the displacement in Figure 11.22. After substitution of this expression for W_{elastic}, Equation 11.15 becomes

$$W_{\text{other}} = \Delta\text{KE} + \Delta\text{PE}_{\text{gravitational}} + (\tfrac{1}{2}kx_f^2 - \tfrac{1}{2}kx_0^2) \qquad (11.17)$$

Equation 11.17 shows that the work done by forces other than gravity and the spring force can cause (1) a change ΔKE in the kinetic energy, (2) a change $\Delta\text{PE}_{\text{gravitational}}$ in the gravitational potential energy, and (3) a change in the quantity $\tfrac{1}{2}kx^2$. This quantity changes from an initial value of $\tfrac{1}{2}kx_0^2$ to a final value of $\tfrac{1}{2}kx_f^2$ and is called the ***elastic potential energy.*** Note that the elastic potential energy has a different mathematical form ($\tfrac{1}{2}kx^2$) than the gravitational potential energy (mgh). Both, however, represent energy of position. It is possible to determine an elastic potential energy, because the force exerted by a spring, like the gravitational force, is conservative. (See Section 7.5 for a discussion of conservative forces.)

ELASTIC POTENTIAL ENERGY

The elastic potential energy $\text{PE}_{\text{elastic}}$ is the energy that an object has by virtue of its position on a spring. For an ideal spring that has a spring constant k and is stretched or compressed by an amount x relative to its unstrained length, the elastic potential energy is

$$\text{PE}_{\text{elastic}} = \tfrac{1}{2}kx^2 \qquad (11.18)$$

SI unit of elastic potential energy: joule (J)

THE CONSERVATION OF MECHANICAL ENERGY

As Section 7.4 discusses, the principle of conservation of mechanical energy arises from the work–energy theorem and specifies that the total mechanical energy E (kinetic plus potential) remains constant when only the gravitational force does work. The fact that the gravitational force is a conservative force leads to the conservation principle. But the force of a spring is also conservative. Therefore, the principle can be extended to include elastic potential energy. When only the spring force and the gravitational force are present, $W_{other} = 0$, and Equation 11.17 reduces to

$$W_{other} = 0 = \Delta KE + \Delta PE_{gravitational} + \Delta PE_{elastic}$$

In other words, $0 = E_f - E_0 = \Delta E$, where

$$E = KE + PE_{gravitational} + PE_{elastic}$$

Thus, the total mechanical energy, including kinetic and *both* forms of potential energy, does not change: $E_f = E_0$. As an illustration of the conservation principle, Example 7 deals with an object of mass m executing simple harmonic motion on a horizontal spring.

EXAMPLE 7

In Figure 11.12 an object of mass m is vibrating horizontally on a frictionless surface. The simple harmonic motion has an amplitude A, because the spring was stretched initially to $x = A$ and then released from rest. Determine the kinetic energy at $x = A$, $x = A/2$, and $x = 0$ and compare these results to the potential energy at the same points.

SOLUTION

The conservation principle indicates that, in the absence of friction (a nonconservative force), the initial and final total mechanical energies are the same:

$$E_f = E_0$$

$$KE_f + \tfrac{1}{2}kx_f^2 + mgh_f = KE_0 + \tfrac{1}{2}kx_0^2 + mgh_0$$

Since the spring is horizontal, gravitational potential energy plays no role. Algebraically, $h_f = h_0$, and the above equation becomes

$$KE_f + \tfrac{1}{2}kx_f^2 = KE_0 + \tfrac{1}{2}kx_0^2$$

Initially at $x = A$, the object is stationary, so $\tfrac{1}{2}kx_0^2 = \tfrac{1}{2}kA^2$ and $KE_0 = 0$. Consequently, the conservation principle indicates that

$$KE_f + \tfrac{1}{2}kx_f^2 = \tfrac{1}{2}kA^2$$

This expression can be used to calculate the kinetic energy KE_f at $x_f = A$, $x_f = A/2$, and $x_f = 0$. The results for the kinetic energy are compared to the potential energy in Table 11.4. All of the energy is in the form of elastic potential energy at $x_f = A$ and in the form of kinetic energy at $x_f = 0$. At intermediate points the energy is part kinetic and part potential. In the absence of friction, the total of the two kinds of energy is constant at all times; the simple harmonic motion merely converts the energy between one form and the other.

TABLE 11.4 Energy of a Mass m in Simple Harmonic Motion on a Horizontal Spring

Spring	KE	PE	$E = KE + PE$
$x_f = A$	0	$\tfrac{1}{2}kA^2$	$\tfrac{1}{2}kA^2$
$x_f = \tfrac{A}{2}$	$\tfrac{3}{8}kA^2$	$\tfrac{1}{8}kA^2$	$\tfrac{1}{2}kA^2$
$x_f = 0$	$\tfrac{1}{2}kA^2$	0	$\tfrac{1}{2}kA^2$

In the previous example, gravitational potential energy played no role because the spring was horizontal. The next example illustrates that gravitational potential energy must be taken into account when a spring is oriented vertically.

EXAMPLE 8

A 0.20-kg ball is attached to a vertical spring, as in Figure 11.23. The spring constant of the spring is 28 N/m. The ball, supported initially so that the spring is neither stretched nor compressed, is released from rest. In the absence of air resistance, how far does the ball fall before being brought to a momentary stop by the spring?

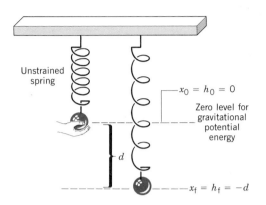

FIGURE 11.23 The ball is supported initially so that the spring is unstrained, being neither stretched nor compressed. After being released from rest, the ball falls through the distance d before being momentarily stopped by the spring.

SOLUTION

Since air resistance is absent, only the conservative forces of gravity and the spring act on the ball. Therefore, the principle of conservation of mechanical energy applies:

$$E_f = E_0$$

$$\tfrac{1}{2}mv_f^2 + \tfrac{1}{2}kx_f^2 + mgh_f = \tfrac{1}{2}mv_0^2 + \tfrac{1}{2}kx_0^2 + mgh_0$$

As Figure 11.23 indicates, the spring is unstrained to begin with, and the ball falls through a distance d, so that $x_0 = 0$ and $x_f = -d$. In addition, the drawing shows that the highest position of the ball is arbitrarily taken as the zero level for measuring the gravitational potential energy; thus, $h_0 = 0$ and $h_f = -d$. Furthermore, the initial and final speeds are zero, because the ball is released from rest and comes to a halt after falling through the distance d. Therefore, the conservation principle reduces to

$$\tfrac{1}{2}kd^2 - mgd = 0$$

reflecting the fact that gravitational potential energy (mgd) has been completely converted into elastic potential energy ($\tfrac{1}{2}kd^2$) when the ball comes to a halt. Solving this equation for d gives

$$d = \frac{2mg}{k} = \frac{2(0.20 \text{ kg})(9.80 \text{ m/s}^2)}{(28 \text{ N/m})} = \boxed{0.14 \text{ m}}$$

The distance d calculated in this example is not the same as the distance d_0 at which the ball would hang stationary on the spring. Determine d_0 to see whether it is greater or less than d and explain why.

11.7 THE PENDULUM

As Figure 11.24 shows, a ***simple pendulum*** consists of a particle of mass m, attached to a frictionless pivot P by a thin support of length L, whose mass is negligible. When the particle is pulled away from its equilibrium position by an angle θ and released, the particle swings back and forth. By attaching a pen to the bottom of the swinging mass and moving a strip of paper beneath it at a steady rate, we can record the position of the particle as time passes. The graphical record reveals a pattern that is similar (not identical) to the sinusoidal pattern that characterizes simple harmonic motion.

The force of gravity is responsible for the back-and-forth rotation about the axis at P. The rotation speeds up as the particle approaches the lowest point on the arc going clockwise and slows down on the upward part of the swing. Eventually the rotational speed is reduced to zero, and the particle returns in the counterclockwise direction. As Section 10.4 discusses, a net torque must be present when the rotational speed changes. The gravitational force $m\mathbf{g}$ produces this torque. (The tension $\mathbf{T}$ in the support creates no torque, because it points directly at the pivot and, therefore, has a zero lever arm.) Since Torque = Force × Lever arm, the torque τ is the gravitational force multiplied by the perpendicular distance d to the axis: $\tau = -(mg)d$. The minus sign is included since the torque is a restoring torque; that is, it acts in such a way as to reduce the angle θ. The perpendicular distance d is very nearly equal to the arc length

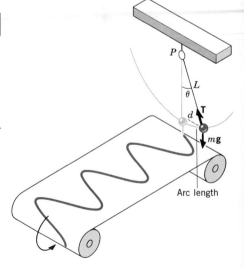

FIGURE 11.24 A simple pendulum swinging back and forth about the pivot P. If the angle θ is small, the swinging motion is approximately simple harmonic motion.

An elegant "grandfather" clock.

on the circular path if only small values (10° or less) of θ are considered. Furthermore, if θ is expressed in radians, the arc length and the radius L of the circular path are related, according to Arc length $= L\theta$. Under these conditions, it follows that $d \approx$ Arc length $= L\theta$, and the torque created by gravity can be written as

$$\tau \approx -\underbrace{mgL}_{k'}\,\theta \qquad (11.19)$$

In Equation 11.19, the term mgL has a constant value k', independent of θ. *For small angles,* then, the torque that restores the pendulum to its vertical equilibrium position is proportional to the angular displacement θ. The expression $\tau = -k'\theta$ has the same form as the Hooke's law restoring force for a particle on a spring, $F = -kx$. Therefore, by analogy with the simple harmonic motion of a particle on a spring, we expect the frequency of the back-and-forth movement of the pendulum to be given by an equation analogous to Equation 11.14 ($\omega = 2\pi f = \sqrt{k/m}$). In place of the spring constant k, the constant $k' = mgL$ will appear, and, as usual in rotational motion, in place of the mass m, the moment of inertia I will appear:

$$\omega = 2\pi f = \sqrt{\frac{mgL}{I}} \qquad (11.20)$$

The moment of inertia of a particle of mass m, rotating at a radius $R = L$ about an axis, is given by $I = mL^2$ (Equation 10.6). Substituting this expression for I into Equation 11.20 reveals for a simple pendulum that

$$\omega = 2\pi f = \sqrt{\frac{g}{L}} \qquad (11.21)$$

The mass of the particle has been eliminated algebraically from this expression, and we see that only the length L and the acceleration g due to gravity are important in determining the frequency of a simple pendulum. Equation 11.21 does not apply if the angle of oscillation is large, for then the pendulum does not exhibit simple harmonic motion. For small-angle motion, Equation 11.21 provides the basis for using a pendulum to keep time, as Example 9 demonstrates.

EXAMPLE 9

Determine the length of a simple pendulum that will swing back and forth in simple harmonic motion with a period of 1.00 s.

$$2\pi f = \frac{2\pi}{T} = \sqrt{\frac{g}{L}}$$

SOLUTION

Equation 11.21 can be applied directly. However, it is necessary to remember that the relationship between frequency f in cycles/s and period T in seconds is $f = 1/T$:

$$L = \frac{T^2 g}{4\pi^2} = \frac{(1.00\ \text{s})^2(9.80\ \text{m/s}^2)}{4\pi^2} = \boxed{0.248\ \text{m}}$$

It is not necessary that the mass in Figure 11.24 be a simple particle. It may be an extended object, in which case the pendulum is designed as a **physical pendulum.** For small oscillations, Equation 11.20 still applies, but the moment of inertia I is no longer mL^2. The proper value for the rigid object must be used. (See Section 10.4 for a discussion of moment of inertia.) In addition, the length L for a physical pendulum is the distance between the axis at P and the center of gravity of the object.

11.8 DAMPED HARMONIC MOTION

In simple harmonic motion, an object oscillates with a constant amplitude, because there is no mechanism for dissipating energy. In reality, however, friction or some other energy-dissipating mechanism is always present to some extent. In the presence of energy dissipation, the amplitude of oscillation decreases as time passes, and the motion is no longer simple harmonic motion. Instead, it is referred to as *damped harmonic motion,* the decrease in amplitude being called "damping."

One widely used application of damped harmonic motion is in the suspension system of an automobile. Figure 11.25*a* shows a shock absorber attached to a main suspension spring of a car. A shock absorber is designed so that it introduces damping forces. As part *b* of the drawing shows, a shock absorber consists of a piston that moves in a reservoir of oil. When the piston moves in response to a bump in the road, holes in the piston head permit the piston to move through the oil. Viscous forces that arise during this movement cause the damping.

Figure 11.26 illustrates the different degrees of damping that can exist. As applied to the example of a car's suspension system, these graphs show the vertical position of the chassis after it has been pulled upward by the amount A_0 at time $t = 0$ and then released. Part *a* of the figure compares undamped or simple harmonic motion in curve 1 to slightly damped motion in curve 2. In damped harmonic motion, the chassis oscillates with decreasing amplitude until it eventually comes to rest. As the degree of damping is increased from curve 2 to curve 3, the car makes fewer oscillations before coming to a halt. Part *b* of the drawing repeats curve 3 and shows that as the degree of damping is increased still further, there comes a point when the car does not oscillate at all after it is released but, rather, settles directly back to its equilibrium position, as in curve 4. The smallest degree of damping that completely eliminates the oscillations is termed "critical damping," and the motion is said to be *critically damped.*

Figure 11.26*b* also shows that the car takes the longest time to return to its equilibrium position in curve 5, where the degree of damping is increased above the critical value depicted in curve 4. When the damping exceeds the critical value, the motion is said to be *overdamped.* In contrast, when the damping is less than the critical level, the motion is said to be *underdamped* (curves 2 and 3). Typical automobile shock absorbers are designed to produce underdamped motion somewhat like that in curve 3.

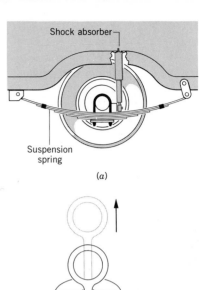

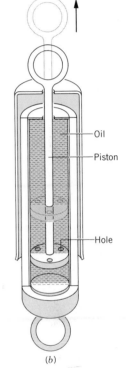

FIGURE 11.25 (*a*) A shock absorber mounted in the suspension system of an automobile and (*b*) a simplified, cutaway view of the shock absorber.

11.9 DRIVEN HARMONIC MOTION AND RESONANCE

In damped harmonic motion, a mechanism such as friction dissipates the energy of an oscillating system, with the result that the amplitude of the motion decreases. This section discusses the opposite effect, namely the increase in amplitude that results when energy is continually added to an oscillating system.

To set an object on an ideal spring into simple harmonic motion, some agent must apply a force that stretches or compresses the spring initially. Suppose that this force is applied at all times, not just for a brief initial moment. The force could be provided, for example, by a person who simply pushes and pulls the object back and forth. The resulting motion is known as *driven harmonic motion,* because the additional force

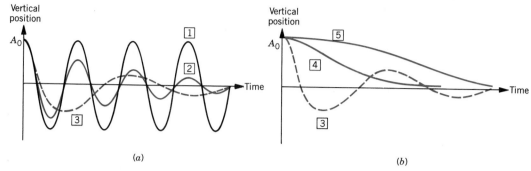

FIGURE 11.26 Damped harmonic motion. The degree of damping increases from curve 1 to curve 5. Curve 1 represents undamped or simple harmonic motion, while curve 4 represents critically damped harmonic motion. Curves 2 and 3 show underdamped motion, while curve 5 illustrates overdamped motion.

drives or controls the behavior of the object to a large extent. The additional force is identified as the **driving force.**

Figure 11.27 illustrates one particularly important example of driven harmonic motion. Here the frequency of the driving force is the same as the frequency of the spring system, and the driving force always points in the direction of the object's velocity. Since the driving force and the velocity always have the same direction, positive work is done on the oscillating object at all times, and the total mechanical energy of the system increases. As a result, the amplitude of the vibration becomes larger, which is exactly opposite to the effect of damping. In fact, the amplitude will increase without limit if there is no damping force to dissipate the energy being added by the driving force. The situation depicted in Figure 11.27 is known as **resonance.**

RESONANCE
Resonance is the condition under which an oscillating force can transmit large amounts of energy to an oscillating object, leading to a large amplitude motion. In the absence of damping, resonance occurs when the frequency of the force matches a natural frequency at which the object will oscillate.

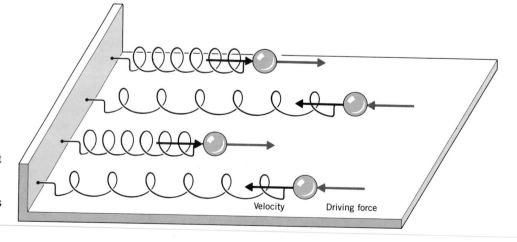

FIGURE 11.27 Resonance occurs when the frequency of the driving force exactly matches a frequency at which the object naturally vibrates. The colored arrows represent the driving force, while the black arrows represent the velocity of the object.

The role played by the frequency of a driving force is a critical one. The matching of this frequency with a natural frequency of vibration allows even a relatively weak force to produce a large amplitude vibration, because the effect of each push–pull cycle is cumulative.

Resonance can occur with any object that can oscillate, and springs need not be involved. For example, the diaphragm of a loudspeaker has a natural frequency of vibration. When designing loudspeakers, audio engineers must take this natural frequency into account, or else resonance will cause the diaphragm to vibrate with a much greater amplitude at one sound frequency than at others, leading to severe distortion in the reproduction of music.

SUMMARY

The possible kinds of **elastic deformation** include stretch and compression, shear deformation, and volume deformation. The forces required to create them are given by Equations 11.1–11.3 and are characterized with the aid of proportionality constants called, respectively, **Young's modulus,** the **shear modulus,** and the **bulk modulus.**

Stress is the force per unit area applied to an object and causes **strain.** For stretch/compression and volume deformation, strain is the resulting fractional change in length or in volume. For shear deformation, strain reflects the change in shape of an object. **Hooke's law** expresses the fact that stress is directly proportional to strain, up to a limit called the **proportionality limit.** The **elastic limit** of a material is the point beyond which stress causes permanent deformation.

The force F that must be applied to stretch or compress an **ideal spring** is $F = kx$, where k is the spring constant and x is the distance by which the spring is stretched or compressed from its unstrained length. A spring exerts a **restoring force** on an object attached to the spring. The restoring force produced by an ideal spring is $F = -kx$, where the minus sign indicates that the restoring force points opposite to the direction of the stretch or compression.

Simple harmonic motion is the oscillatory motion that occurs when a restoring force of the form $F = -kx$ acts on an object. A graphical record of position versus time for an object in simple harmonic motion is sinusoidal. The **amplitude** A of the motion is the maximum distance that the object moves away from its equilibrium position. The **period** T is the time required to complete one cycle of the motion, while the **frequency** f is the number of cycles per second that occur. Frequency and period are related according to $f = 1/T$. The frequency f (in Hz) is related to the angular frequency ω (in rad/s) according to $\omega = 2\pi f$. For an object of mass m on a spring with spring constant k, the frequency is determined by $2\pi f = \sqrt{k/m}$. The velocity and the acceleration in simple harmonic motion are continually changing with time, the maximum speed being $v_{\max} = A\omega$ and the maximum acceleration being $a_{\max} = A\omega^2$.

The **elastic potential energy** of an object attached to an ideal spring is PE $= \frac{1}{2}kx^2$. This energy must be taken into account along with gravitational potential energy when applying the work–energy theorem and the principle of conservation of mechanical energy.

A **simple pendulum** consists of a particle of mass m attached to a frictionless pivot by a thin support whose length is L and whose mass is negligible. The small-angle ($\leq 10°$) back-and-forth motion of a simple pendulum is simple harmonic motion, while large-angle motion is not. The frequency of small-angle motion is given by $2\pi f = \sqrt{g/L}$. A **physical pendulum** consists of a rigid object, with moment of inertia I and mass m, suspended from a frictionless pivot. For small-angle displacements, the frequency of simple harmonic motion for a physical pendulum is determined by $2\pi f = \sqrt{mgL/I}$, where L is the distance between the axis of rotation and the center of gravity of the rigid object.

Damped harmonic motion is motion in which the amplitude of oscillation decreases as time passes. **Critical damping** is the minimum degree of damping that eliminates any oscillations in the motion as the object returns to its equilibrium position.

Driven harmonic motion occurs when an additional driving force is applied to an object along with the restoring force. **Resonance** is the condition under which a driving force can transmit large amounts of energy to an oscillating object, leading to large amplitude motion. In the absence of damping, resonance occurs when the frequency of the driving force matches a natural frequency at which the object oscillates.

QUESTIONS

1. Three rods, made of the same material, have equal lengths and have square, circular, and triangular cross sections, as the drawing shows. Rank the rods according to the amount of force (smallest first) required to stretch each rod by the same amount. Give your reasoning.

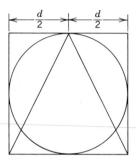

2. A trash compactor smashes a number of empty aluminum cans, thereby reducing the total volume of the cans by 75%. Can the value given in Table 11.3 for the bulk modulus of aluminum be used to calculate the pressure generated in the trash compactor? Explain.

3. Both sides of the relation $F = S(\Delta X/L_0)A$ (Equation 11.2) can be divided by the area A to give F/A on the left side. Why can this force per unit area *not* be called a pressure, such as the pressure that appears in $\Delta P = -B(\Delta V/V_0)$ (Equation 11.3)?

4. The block in the drawing rests on the ground. Which face, A, B, or C, experiences the largest and which experiences the smallest stress when the block is resting on it? Justify your answer.

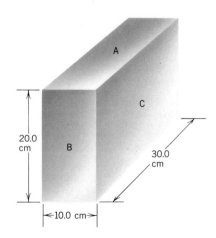

5. Where on the path followed by an object in simple harmonic motion is (a) the velocity equal to zero and (b) the acceleration equal to zero? Refer to Figures 11.17 and 11.19.

6. Ignoring the damping introduced by the shock absorbers, explain why the number of passengers in a car affects the vibration frequency of the car's suspension system.

7. Explain how an astronaut in an orbiting satellite can use a spring with a known spring constant to measure his mass. Since his apparent weight is zero, he *cannot* simply hang motionless from the spring, determine his weight mg from the spring constant and the amount by which the spring stretches, and calculate his mass by dividing the weight by the acceleration due to gravity g.

8. A block is attached to a horizontal spring and slides back and forth in simple harmonic motion on a frictionless horizontal surface. A second identical block is suddenly attached to the first block. In one case, the attachment is accomplished by joining the blocks at one extreme end of the oscillation cycle. In another case, the attachment is accomplished by joining the blocks at the point when the spring is not deformed. In either case, the velocities of the blocks are exactly matched at the instant of joining. For each method of attachment, explain how the amplitude, the frequency, and the maximum speed of the oscillation change.

9. Over a nightclub entrance is mounted a horizontal straight strip of equally spaced light bulbs, each one of which turns on in sequence for one half second. Thus, the lighted bulb appears to move from left to right to left to right, etc. Is the apparent motion of the lighted bulb simple harmonic motion? Give your reasoning.

10. An electric saber saw consists of a blade that is driven back and forth by a pin mounted on the circumference of a rotating circular disk. As the disk rotates at a constant angular speed, the pin engages a slot and forces the blade back and forth, as the drawing illustrates. Is the motion of the blade simple harmonic motion? Explain, remembering that simple harmonic motion results only when the force applied to the moving object is given by $F = -kx$, which predicts a maximum force at the extreme ends of the motion.

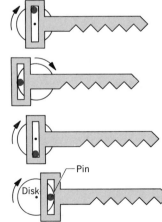

11. Is more elastic potential energy stored in a spring when the spring is compressed by one centimeter than when it is stretched by the same amount? Explain.

12. In principle, the motion of a simple pendulum and an object on an ideal spring can both be used to provide the basic time interval or period used in a clock. Which of the two kinds of clocks is likely to become more inaccurate when carried to the top of a high mountain? Justify your answer.

13. Suppose you were kidnapped and held prisoner by space invaders in a completely isolated room, with nothing but a digital watch and a pair of shoes (including two 27-inch shoelaces). Explain how you might determine whether this room is on earth or on the moon.

14. Damped harmonic motion occurs when frictional forces come into play and dissipate energy. Suppose in Figure 11.14 that friction arises from a bad bearing in the turntable motor, and thus the circular platter rotates slower and slower after the motor is shut off. Does the shadow of the ball exhibit damped harmonic motion? Account for your answer.

15. A car travels over a road that contains a series of equally spaced bumps. Explain why a particularly "jarring" ride can result if the horizontal velocity of the car, the bump spacing, and the oscillation frequency of the car's suspension system are properly "matched."

PROBLEMS

Note: Unless otherwise indicated, the necessary values for Young's modulus Y, the shear modulus S, and the bulk modulus B are given, respectively, in Table 11.1, Table 11.2, and Table 11.3.

Section 11.2 Elastic Deformation, Section 11.3 Stress, Strain, and Hooke's Law

1. An 82-kg mountain climber hangs freely on an 8.0-mm-diameter nylon rope. If the rope stretches by 0.10 m, what is the unstretched length of the rope?

2. A rectangular solid block has the dimensions shown in the drawing for question 4 and a shear modulus of 7.00×10^9 N/m². A force of 4.2×10^4 N is applied as in Figure 11.5, with one face of the block fixed to an immovable horizontal surface. Determine the resulting shear deformation ΔX when the face fixed to the horizontal surface is (a) face A, (b) face B, and (c) face C.

3. A copper cube, 0.30 m on a side, is subjected to a shearing force of $F = 6.0 \times 10^6$ N, as the drawing shows. Find the angle θ, which is one measure of how the shape of the block has been altered by the resulting shear deformation.

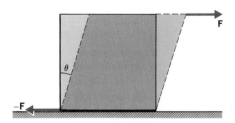

4. How much change in pressure is required to decrease by 0.050% the volume of a block of (a) steel and (b) water?

5. An 1800-kg car is being lifted at a steady speed by a crane and hangs at the end of a cable whose diameter is 1.2 cm. The cable is 15 m in length and stretches by 8.0 mm because of the weight of the car. Determine (a) the stress, (b) the strain, and (c) Young's modulus for the cable.

6. At the elastic limit for stretching a copper wire, the stress is 1.5×10^8 Pa. (a) Use this value in Hooke's law to calculate the corresponding strain. (b) Strictly speaking, Hooke's law does not apply at the elastic limit. Nevertheless, if strain values are kept below the value calculated in part (a), a copper wire will not be permanently deformed by stretching. Referring to Figure 11.7, explain why.

7. Two metal beams are joined together by four steel rivets, as the drawing indicates. Each rivet has a diameter of 1.0 cm and is to be exposed to a shearing stress of no more than 5.0×10^8 Pa. What is the maximum tension T that can be applied to each beam, assuming that each rivet carries one-fourth of the total load?

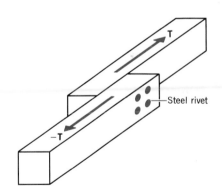

Steel rivet

8. The shovel of a backhoe is controlled by hydraulic cylinders that are moved by oil under pressure. (a) Determine the volume strain experienced by the oil, when the pressure increases from 1.0×10^5 Pa to 7.2×10^5 Pa while the shovel is digging a trench. (b) Express the answer as a percentage volume change.

* 9. Suppose that F newtons of force are required to stretch a wire (circular cross section) by an amount ΔL. This wire is melted down, and a new wire (circular cross section) is formed that has half the length of the first. How much force is needed to stretch this second wire by the same amount ΔL?

* 10. A helicopter is lifting a 2100-kg jeep. The steel suspension cable is 48 m long and has a diameter of 1.0 cm. (a) Find the amount that the cable is stretched when the jeep is suspended motionless in the air. (b) What is the amount of cable stretch when the jeep is hoisted upward with an acceleration of 1.5 m/s²?

* **11.** A gymnast does a one-arm handstand. The humerus, which is the upper arm bone between the elbow and the shoulder joint, may be approximated as a 0.30-m-long cylinder with an outer diameter of 2.00 cm and a hollow inner core with a diameter of 0.80 cm. Excluding the arm, the mass of the gymnast is 63 kg. (a) What is the compressional strain of the humerus? (b) By how much is the humerus compressed?

* **12.** An 8.0-kg stone at the end of a steel wire is being whirled at a constant speed of 12 m/s in a horizontal circle. The wire is 4.0 m long and has a diameter of 2.0 mm. Find the strain in the wire.

* **13.** A 1.0-g spider is hanging vertically by a thread that has a Young's modulus of 4.5×10^9 N/m² and a diameter of 25×10^{-6} m. Suppose that a 95-kg person is hanging vertically on an aluminum wire. What is the diameter of the wire that would exhibit the same strain as the spider's thread, when the thread is stressed by the full weight of the spider?

* **14.** A square plate is 1.0 cm thick, measures 0.30 cm on a side, and has a mass of 120 kg. The shear modulus of the material is 2.0×10^{10} N/m². One of the square faces rests on a flat horizontal surface, and the coefficient of static friction between the plate and the surface is 0.90. A force is applied as in Figure 11.5. Determine (a) the maximum possible amount of shear stress, (b) the maximum possible amount of shear strain, and (c) the maximum possible amount of shear deformation ΔX (see Figure 11.5) that can be created by the applied force.

** **15.** A thin rod is made of two sections joined together end to end. The sections are identical, except that one is steel and the other is tungsten. One end of this composite rod is attached to an immovable wall, while the other is pulled until the total length changes by one millimeter. By how much does the length of each section increase?

** **16.** A solid brass sphere is subjected to a pressure of 1.0×10^5 N/m² due to the Earth's atmosphere. On Venus the pressure due to the atmosphere is 9.0×10^6 N/m². By what percentage does the radius of the sphere decrease when it is exposed to the Venusian atmosphere? Assume that the change in radius is very small relative to the initial radius.

Section 11.4 The Ideal Spring and Simple Harmonic Motion

17. A spring has a spring constant of 248 N/m. Find the magnitude of the force needed (a) to stretch the spring by 3.00 cm from its unstrained length and (b) to compress the spring by the same amount.

18. In a room that is 8.00 ft high, a spring (unstrained length = 1.00 ft) hangs from the ceiling. From this spring a board hangs, so that its 6.50-ft length is perpendicular to the floor, the lower end just reaching to, but not touching the floor. The board weighs 23 lb. What is the spring constant of the spring?

19. A hand exerciser utilizes a coiled spring. A force of 20.0 lb is required to compress the spring 0.750 in. Determine the amount of force needed to compress the spring by 2.00 in.

* **20.** To measure the static friction coefficient between a 2.3-kg block and a vertical wall, the setup shown in the drawing is used. This setup employs a spring (spring constant = 480 N/m) attached to the block. Someone simply pushes on the end of the spring in a direction perpendicular to the wall until the block does not slip downward. If the spring in such a setup is compressed by 5.1 cm, what is the coefficient of static friction?

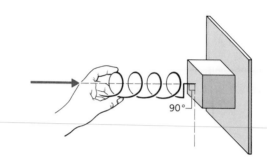

* **21.** An 11.5-kg uniform board is wedged into a corner and held by a spring at a 45.0° angle, as the drawing shows. The spring has a spring constant of 152 N/m and is parallel to the floor. Find the amount by which the spring is stretched from its unstrained length.

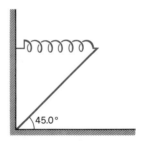

** **22.** A 15.0-kg block rests on a horizontal table and is attached to one end of a massless, horizontal spring. By pulling horizontally on the other end of the spring, someone causes the block to accelerate uniformly and reach a speed of 5.00 m/s in 0.500 s. In the process, the spring is stretched by 20.0 cm. The block is then pulled at a *constant speed* of 5.00 m/s, during which time the spring is stretched by only 5.00 cm. Find (a) the spring constant of the spring and (b) the coefficient of kinetic friction between the block and the table.

** **23.** A 30.0-kg block is resting on a flat horizontal table. On top of this block is resting a 15.0-kg block, to which a horizontal spring is attached, as the drawing illustrates. The spring constant of the spring is 325 N/m. The coefficient of kinetic friction between the lower block and the table is 0.600, while the coefficient of static friction between the two blocks is 0.900. A horizontal force **F** is applied to the lower block as shown. This force is increasing in such a way as to keep the blocks moving at a *constant speed*. At the point where the upper block begins to slip on the lower block determine (a) the amount by which the spring is compressed and (b) the magnitude of the force **F**.

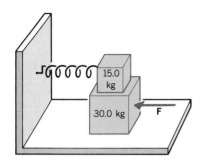

Section 11.5 Simple Harmonic Motion and the Reference Circle

24. A small ball is attached to the outer edge of a 33⅓-rev/min stereo record that is revolving on a turntable. The record has a diameter of 12.0 inches. In a fashion similar to that in Figure 11.14, the shadow of the ball is projected onto a screen. For the simple harmonic motion of the shadow, what is (a) the amplitude in inches, (b) the frequency in hertz, and (c) the maximum acceleration in in./s^2?

25. The shock absorbers in the suspension system of a car are in such bad shape that they have no effect on the behavior of the springs attached to the axles. Each of the identical springs attached to the front axle supports 320 kg. A person pushes down on the middle of the front end of the car and notices that it vibrates through five cycles in 3.0 s. Find the spring constant that characterizes either spring.

26. The diaphragm of a loudspeaker (see Figure 11.20) is moving back and forth in simple harmonic motion with a frequency of 128 Hz and an amplitude of 0.100 inch. Determine the maximum speed (in in./s) of the diaphragm.

27. A computer to be used in a satellite must be able to withstand accelerations of up to 25g. In a test to see if it meets this specification, the computer is bolted to a frame that is vibrated back and forth in simple harmonic motion at a frequency of 9.5 Hz. What is the minimum amplitude of vibration that must be used in this test?

* 28. In Figure 11.15, the radius of the reference circle is 0.500 m. Suppose the frequency of the simple harmonic motion of the shadow is 2.00 Hz. At time $t = 0.0500$ s calculate (a) the position x, (b) the magnitude of the velocity, and (c) the magnitude of the acceleration of the shadow.

* 29. Suppose that an object on a vertical spring oscillates up and down at a frequency of 5.00 Hz. By how much will this object, hanging at rest, stretch the spring?

* 30. When mass m_1 is hung on a vertical spring and set into vertical simple harmonic motion, the observed frequency is

10.0 Hz. When mass m_2 is hung on the spring along with m_1, the frequency of the motion is 5.00 Hz. Find the ratio m_2/m_1 of the masses.

**31. A spring (spring constant = 80.0 N/m) is mounted on the floor and is oriented vertically. A 0.600-kg mass is placed on top of this spring and pushed down to start it oscillating in simple harmonic motion. The mass is not attached permanently to the spring. (a) Obtain the frequency of the motion in Hz. (b) Determine the amplitude of the motion at which the mass will lose contact with the spring.

Section 11.6 Energy and Simple Harmonic Motion

32. A 0.22-kg mass is oscillating on a horizontal spring. The period and amplitude of the oscillation are 0.50 s and 0.15 m, respectively. How much work was done to start the simple harmonic motion?

33. A 10.0-g block is resting on a horizontal frictionless surface and is attached to a horizontal unstrained spring whose spring constant is 124 N/m. The block is shoved parallel to the spring axis and in the process is given an initial speed of 8.00 m/s. What is the amplitude of the resulting simple harmonic motion?

34. A 2.00-kg object is hanging from the end of a vertical spring. The spring constant is 50.0 N/m. The object is pulled 0.200 m downward and released from rest. Complete the table below by calculating the kinetic energy, the gravitational potential energy, the elastic potential energy, and the total mechanical energy E for each of the vertical positions indicated. The vertical positions h indicate points above the point of release, where $h = 0$.

h (meters)	KE	PE (gravity)	PE (spring)	E
0				
0.100				
0.200				
0.300				
0.400				

35. A heavy-duty stapling gun uses a 0.150-kg metal rod that rams against the staple to eject it. The rod is pushed by a stiff spring called a "ram spring" ($k = 34\ 000$ N/m). The mass of this spring may be ignored. Squeezing the handle of the gun first compresses the ram spring by 3.5 cm from its unstrained length and then releases it. Assuming that the ram spring is oriented vertically and is still compressed by 1.0 cm when the ram hits the staple, find the speed of the ram at the instant of contact. Express the answer in m/s and mi/h.

* 36. A 1.00-slug block and a 2.00-slug block are resting on a horizontal frictionless surface. Between the two is squeezed a spring (spring constant = 80.0 lb/ft). The spring is compressed by 0.500 ft from its unstrained length. With what speed does each block move away when the mechanism keeping the spring squeezed is released? *(Hint: Remember the principle of conservation of linear momentum, as well as the conservation of mechanical energy.)*

* 37. A 1.1-kg mass is suspended from a vertical spring whose spring constant is 120 N/m. (a) Find the amount by which the spring is stretched from its unstrained length. (b) The object is pulled straight down by an additional distance of 0.20 m and released from rest. Find the speed with which the mass passes through its original position on the way up.

* 38. A 70.0-kg circus performer is fired from a cannon that is elevated at an angle of 45.0° above the horizontal. The cannon uses strong elastic bands to propel the performer, much in the same way that a slingshot fires a stone. Setting up for this stunt involves stretching the bands by 3.00 m from their unstrained length. At the point where the performer flies free of the bands, his height above the floor is the same as that of the net into which he is shot. He takes 4.00 s to travel the horizontal distance of 50.0 m between this point and the net. Ignore friction and air resistance and determine the effective spring constant of the firing mechanism.

**39. A spring is mounted vertically on the floor. The mass of the spring is negligible. A certain object is placed on the spring to compress it. When the object is pushed down further by just a bit and then released, one up/down oscillation cycle occurs in 0.250 s. However, when the object is pushed down by 5.00 cm to point P and then released, the object flies entirely off the spring. To what height above point P does the object rise in the absence of air resistance?

**40. A 10.0-g bullet is fired horizontally into a 2.50-kg wooden block attached to one end of a massless, horizontal spring ($k = $ 845 N/m). The other end of the spring is fixed in place, and the spring is unstrained initially. The block rests on a horizontal, frictionless surface. The bullet strikes the block perpendicularly and quickly comes to a halt within it. As a result of this completely inelastic collision, the spring is compressed along its axis and causes the block/bullet to oscillate with an amplitude of 0.200 m. What is the speed of the bullet?

Section 11.7 The Pendulum

41. A grandfather clock can be approximated as a simple pendulum of length 1.00 m and keeps accurate time at a location where $g = 9.83$ m/s². In a location where $g = 9.78$ m/s², what must be the new length of the pendulum, such that the clock continues to keep accurate time?

42. A wrecking ball is hanging at the end of a long cable on a crane. A bright physics student wants to estimate the length of the cable and, therefore, improvises by using a simple pendulum made from a 0.500-m length of string and a stone. The student observes that, in swinging back and forth over a small amplitude, the wrecking ball makes one complete oscillation cycle in the time it takes the stone to complete five cycles. What is the length of the cable?

* 43. The period of a simple pendulum is 0.200% longer at location A on the Earth's surface than it is at location B. Find the ratio g_A/g_B of the acceleration due to gravity at these two locations.

* 44. Pendulum A is a physical pendulum made from a very thin rigid and uniform rod whose length is 1.00 m. One end of this rod is attached to the ceiling by a frictionless hinge, so the rod is free to swing back and forth. Pendulum B is a simple pendulum whose length is also 1.00 m. Obtain the ratio T_A/T_B of their periods for small-angle oscillations.

**45. A solid sphere (mass $= M$, radius $= R$) is suspended from a point on its surface, as the drawing shows. It swings back and forth as a physical pendulum with a small amplitude. What is the length of a simple pendulum that has the same period as this physical pendulum?

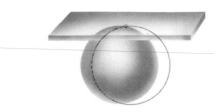

ADDITIONAL PROBLEMS

46. The drawing shows how a piston in an automobile engine is attached to the crankshaft, which is rotating with an angular speed of $\omega = 1200$ rev/min. The shadow of point P, if projected onto a screen, moves nearly in simple harmonic motion. Find (a) the amplitude in inches, (b) the period in seconds, and (c) the maximum speed in in./s for the simple harmonic motion.

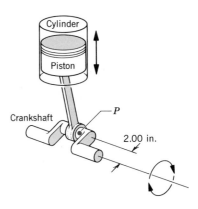

47. A die is designed to punch 2.00-cm-diameter holes in a metal sheet that is 3.0 mm thick, as the drawing illustrates. To punch through the sheet, the die must exert a shearing stress of

3.5 × 10⁸ Pa. What force **F** must be applied to the die? *(Hint: Consider carefully which area is used in your calculation.)*

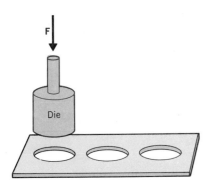

48. A rifle fires a 10.0-g pellet straight upward, because the pellet rests on a compressed spring that is released when the trigger is pulled. The spring has a negligible mass and is compressed by 7.50 cm from its unstrained length. The pellet rises to a height of 5.00 m above its position on the compressed spring. Ignoring air resistance, determine the spring constant of the spring.

49. A 2.5-m length of aluminum cable is to be used in an attempt to pull a car out of a ditch. The maximum tension that the cable will have to sustain is 15 000 N, and, to be on the safe side, only 1.0 mm of cable stretch will be tolerated. What is the diameter of the thinnest cable that should be used?

50. A 9.0-kg object hangs on a copper rod (6.0-mm diameter) from a support that is located 2.0 m directly above. The rod acts as a "spring," and the object oscillates vertically with a small amplitude. Find the frequency of the simple harmonic motion in cycles/s (Hz).

51. When used in an exercise apparatus, a spring is stretched 0.24 m when a bodybuilder exerts a force of 410 N. When used vertically to support a 12-kg mass, by how much does this spring compress?

52. The pressure increases by 9800 N/m² for every meter of depth beneath the surface of the ocean. At what depth does the volume of a Pyrex glass cube, 1.0 cm on an edge at the ocean's surface, decrease by 1.0 × 10⁻⁴ cm³?

* 53. A block is attached to a horizontal spring and oscillates back and forth on a frictionless horizontal surface at a frequency of 3.00 Hz. The amplitude of the motion is 2.00 in. At the point where the block has its maximum speed, it suddenly splits into two identical parts, only one part remaining attached to the spring. (a) What is the amplitude and the frequency of the simple harmonic motion that exists after the block splits? (b) Repeat part (a), assuming the block splits when it is at one of its extreme positions.

* 54. A piece of mohair from an Angora goat has a diameter of

62 × 10⁻⁶ m. What is the least number of identical pieces of mohair that should be used to suspend a 75-kg person, so the strain experienced by each piece is less than one percent? Assume that the tension is the same in all the pieces.

* 55. In 0.750 s, a 7.00-kg block is pulled through a distance of 4.00 m on a frictionless horizontal surface, starting from rest. The block has a constant acceleration and is pulled by means of a horizontal spring that is attached to the block. The spring constant of the spring is 415 N/m. By how much does the spring stretch?

* 56. Two solid cubes of the same size, one tungsten and one lead, are subjected to the same shearing force, as in the drawing that accompanies problem 3. Determine the ratio $\theta_{lead}/\theta_{tungsten}$ for the angles that characterize the resulting shear deformation, assuming the angles are small.

**57. A 0.200-m uniform bar has a mass of 0.750 kg and is released from rest in the vertical position, as the drawing indicates. The spring is initially unstrained and has a stiffness of $k = 25.0$ N/m. Find the speed with which end A strikes the horizontal surface.

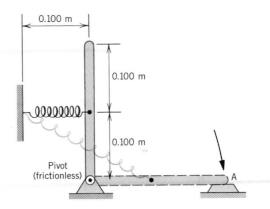

**58. A tray is moved horizontally back and forth in simple harmonic motion at a frequency of 2.00 Hz. On this tray is an empty cup. (a) Obtain the coefficient of static friction between the tray and the cup, given that the cup is observed to begin slipping when the amplitude of the motion is 5.00 cm. (b) At what amplitude would the cup begin to slip if the frequency of the simple harmonic motion were 3.00 Hz? (c) Repeat part (a) for a cup with twice the mass, the frequency remaining at 2.00 Hz.

**59. A 1.25-kg target is attached to one end of a horizontal spring and rests on a horizontal frictionless surface. The other end of the spring is attached to a wall. The spring is unstrained initially and has a spring constant of 425 N/m. A 0.250-kg ball is thrown at the target with a speed of 8.00 m/s and strikes it perpendicularly in an elastic collision lasting only a very brief moment. As a result, the spring is compressed along its axis. What is the amplitude of the resulting simple harmonic motion of the target?

CHAPTER 12

Fluid Statics

Hot air balloons rise in air just like a bubble rises in a can of soda. The buoyant force that propels the balloon and bubble upward is described by Archimedes' principle.

12.1 THE NATURE OF FLUIDS

Fluids are materials that flow. Since all liquids have the ability to flow, they are classified as fluids. Gases are also fluids, for they too can flow. For instance, the natural gas used in home heating flows through buried pipes, and the wind itself is just air, a mixture of gases (primarily nitrogen and oxygen), flowing from one place to another.

The study of fluids has two parts, statics and dynamics. Fluid statics concentrates on the properties of fluids at rest, while fluid dynamics focuses on properties related to the motion of fluids. In either case, the principles of statics and dynamics that have been discussed earlier in this text are applicable.

Fluids at rest display remarkable characteristics. For example, static fluids in the hydraulic systems of cars, trucks, and all heavy construction equipment transmit the large forces that such equipment utilizes so cleverly for pushing, pulling, lifting, and digging. And the property of static fluids that enables water in the ocean to keep a supertanker afloat also enables the air to keep a hot-air balloon suspended high above the ground. This chapter focuses on these and other characteristics of fluids at rest. The next chapter deals with fluids in motion.

12.2 MASS DENSITY

The *mass density* of a liquid or gas is one factor that determines its behavior as a fluid. As indicated below, the mass density of a substance is denoted by the Greek letter rho (ρ) and is the mass per unit volume.

DEFINITION OF MASS DENSITY

The mass density ρ is the mass m of a substance divided by its volume V:

$$\rho = \frac{m}{V} \qquad (12.1)$$

SI unit of mass density: kg/m^3

In general, equal volumes of different substances have different masses, so the density depends on the nature of the material. Table 12.1 lists the densities of some common solids, liquids, and gases and indicates that gases have the smallest densities. The densities of gases are small because gas molecules are relatively far apart, and a given volume of gas contains a relatively large fraction of empty space. In contrast, the molecules are much more tightly packed together in solids and liquids. The tighter packing leads to larger densities, as Table 12.1 reveals.

The density of a substance also depends on the temperature and pressure at which it is measured. However, for the range of temperatures and pressures encountered in this text, the densities of solids and liquids do not differ greatly from the values given in Table 12.1. On the other hand, the densities of gases are particularly sensitive to changes in temperature and pressure.

It is the mass of a substance, not its weight, that enters into the definition of density. In situations where weight is needed, it can be calculated from the mass density, the volume, and the acceleration of gravity, as Example 1 illustrates.

TABLE 12.1 Mass Densities[a] of Common Substances

Substance	Mass Density ρ (kg/m³)	Substance	Mass Density ρ (kg/m³)
Solids		*Liquids*	
Aluminum	2 700	Blood (whole, 37 °C)	1 060
Brass	8 470	Ethyl alcohol	806
Concrete	2 200	Mercury	13 600
Copper	8 890	Oil (hydraulic)	800
Diamond	3 520	Water (4 °C)	1.000×10^3
Gold	19 300		
Ice	917	*Gases*	
Iron (steel)	7 860	Air	1.29
Lead	11 300	Carbon dioxide	1.98
Quartz	2 660	Helium	0.179
Silver	10 500	Hydrogen	0.0899
Wood (yellow pine)	550	Nitrogen	1.25
		Oxygen	1.43

[a] Unless otherwise noted, densities are given at 0 °C and 1 atm pressure.

EXAMPLE 1

The body of a man whose weight is about 690 N typically contains about 5.2×10^{-3} m³ (5.5 qt) of blood. (a) Calculate the weight of the blood and (b) express this weight as a percentage of the body weight.

SOLUTION

(a) The weight W of the blood can be obtained with the aid of the acceleration due to gravity g ($W = mg$), provided the mass of the blood is known. The mass can be determined by using the given volume and the density of blood from Table 12.1:

$$m = \rho V = (1060 \text{ kg/m}^3)(5.2 \times 10^{-3} \text{ m}^3) = 5.5 \text{ kg} \quad (12.1)$$

$$W = mg = (5.5 \text{ kg})(9.80 \text{ m/s}^2) = \boxed{54 \text{ N}}$$

(b) The percentage of body weight contributed by the blood is

$$\text{Percentage} = \frac{54 \text{ N}}{690 \text{ N}} \times 100 = \boxed{7.8\%}$$

A convenient way to compare the densities of various materials is to use the concept of **specific gravity.** The specific gravity of a substance is its density divided by the density of a standard reference material, usually chosen to be water at 4 °C:

$$\text{Specific gravity} = \frac{\text{Density of substance}}{\text{Density of water at 4 °C}} = \frac{\text{Density of substance}}{1.000 \times 10^3 \text{ kg/m}^3} \quad (12.2)$$

Thus, specific gravity is just the ratio between the density of a material and the density of water. Being the ratio of two densities, specific gravity has no units. For example, Table 12.1 reveals that diamond has a specific gravity of 3.52, since diamond has a density that is 3.52 times greater than the density of water at 4 °C.

The next two sections deal with the important concept of pressure. We will see that the density of a fluid is one factor determining the pressure that the fluid creates.

12.3 PRESSURE

Most people who have fixed a flat tire know something about pressure. The final step in the repair process is to reinflate the tire to the proper pressure. The underinflated

FIGURE 12.1 (*a*) Because of collisions with other molecules and the containing surface, each molecule of air inside a tire moves along a path that is composed of randomly directed segments. (*b*) Because of molecular collisions with the surface, the air exerts a force on every part of the surface.

(*a*)

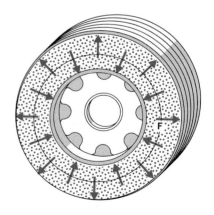
(*b*)

tire is soft and pliable, because there is an insufficient number of molecules of air to push outward against the rubber and give the tire that solid feel. When air is added from a pump, the number of molecules inside the tire and the collective force they exert are increased. When the tire is inflated to the proper pressure, the air pushes outward with enough force to give the tire the shape it needs to roll properly.

The air molecules inside a tire are free to wander throughout its entire volume. However, collisions between one molecule and another and between the molecules and the walls of the containing surface are continually occurring. Thus, the path followed by any one molecule is composed of short, randomly directed segments, as Figure 12.1*a* suggests. The collisions of the molecules with the walls allow the air to exert a force against every part of the wall area, as part *b* of the figure shows. The idea of *pressure* takes into account the force, as well as the area over which the force acts.

DEFINITION OF PRESSURE

The pressure P is the magnitude F of the force acting perpendicular to a surface divided by the area A over which the force acts:

$$P = \frac{F}{A} \tag{12.3}$$

SI unit of pressure: $N/m^2 =$ pascal (Pa)

Equation 12.3 indicates that the unit for pressure is the unit of force divided by the unit of area. The SI unit for pressure, then, is newton/meter^2 (N/m^2), a combination that is referred to as the *pascal (Pa)*. The pascal is the same unit that is used for measuring stress (see Section 11.3). A pressure of 1 Pa is an extremely small amount of pressure. Many common situations involve pressures that are approximately 10^5 Pa, an amount that is referred to as one *bar* of pressure. Alternatively, force can be measured in pounds and area in square inches, so another unit for pressure is the familiar pounds per square inch (psi = lb/inch^2).

Because of its pressure, the air in a tire applies a force to any surface with which it is in contact. Suppose, for instance, that a small cube is inserted inside the tire. As Figure 12.2 shows, the air pressure causes a force to be applied to each of the six faces of the cube, the force in each case being perpendicular to the corresponding face. In a similar fashion, a liquid also exerts pressure. Water pressure, for example, causes a force to be applied perpendicularly to the side of a swimming pool or the body of a swimmer. A swimmer feels the water pushing inward everywhere on her body, as Figure 12.3 illustrates. In general, a static fluid cannot produce a force parallel to a surface, for if it did, the surface would apply a reaction force to the fluid, consistent with Newton's action–reaction law. In response, the fluid would flow and would not be static, as assumed.

While fluid pressure can generate a force, the pressure itself is not a vector quantity, as is the force. In the definition of pressure, $P = F/A$, the symbol F refers only to the magnitude of the force, so that pressure has no directional characteristic. The direction of a force arising from the pressure of a fluid is known only if the orientation of the surface on which the fluid acts is known, as Example 2 illustrates.

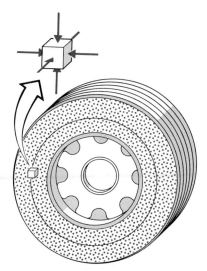

FIGURE 12.2 If a small cube were inserted inside a tire, the cube would experience forces acting perpendicular to each of its six faces. The air pressure in the tire would generate these forces.

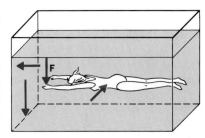

FIGURE 12.3 Water applies a force perpendicular to each surface within the water, including the walls and bottom of the swimming pool and all parts of the swimmer's body.

EXAMPLE 2

Suppose the pressure acting on the back of a swimmer's hand is 18 lb/in.², a realistic value near the bottom of the diving end of a pool. The surface area of the back of the hand is 13 in.². (a) Determine the magnitude of the force that acts on it. (b) Discuss the direction of the force.

SOLUTION

(a) A straightforward application of the definition of pressure shows that the force is surprisingly large:

$$F = PA = (18 \text{ lb/in.}^2)(13 \text{ in.}^2) = \boxed{230 \text{ lb}} \qquad (12.3)$$

(b) In Figure 12.3, the hand (palm downward) is oriented parallel to the bottom of the pool, and since the water pushes perpendicularly against the back of the hand, the force **F** is directed downward in the drawing. This downward-acting force is balanced by an upward-acting force on the palm, thus keeping the hand in equilibrium. If the hand were rotated by 90°, the direction of these forces would also be rotated by 90°, always being perpendicular to the hand.

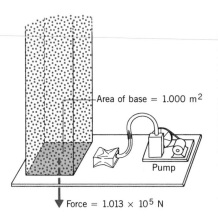

FIGURE 12.4 Atmospheric pressure at sea level is 1.013×10^5 Pa, which is sufficient to cause structural damage to a can if the inside air is pumped out.

A person need not be under water to experience the effects of pressure. Walking about on land, we are at the bottom of the earth's atmosphere, which, being a fluid, pushes inward on our bodies just like the water in a swimming pool. As Figure 12.4 indicates, there is enough air above the surface of the earth to create the following pressure at sea level:

$$\left[\begin{array}{c} \text{Atmospheric} \\ \text{pressure at} \\ \text{sea level} \end{array} \right] \qquad 1.013 \times 10^5 \text{ Pa} \quad \text{or} \quad 14.70 \text{ lb/in.}^2$$

This amount of pressure is referred to as one *atmosphere (atm)*. Just how significant one atmosphere of pressure really is can be appreciated by looking in Figure 12.4 at the results of pumping out all of the air from the inside of a gasoline can.

12.4 THE RELATION BETWEEN PRESSURE AND DEPTH IN A STATIC FLUID

The pressure that an underwater swimmer experiences depends on how far beneath the surface he is. The deeper he goes, the more strongly the water pushes inward on his body. To help determine the exact relation between pressure and depth, Figure 12.5 shows a container of fluid and focuses attention on one particular column of fluid. Since the fluid in the container is at rest, the column can be analyzed just like any object in equilibrium. In other words, the net force acting on the column must be zero. The free-body diagram in the figure shows all the vertical forces acting on the column.

On the top face (area = A), the fluid pressure P_1 generates a downward force whose magnitude is P_1A. Similarly, on the bottom face, the pressure P_2 generates an upward force of magnitude P_2A. Notice that the pressure P_2 is greater than the pressure P_1, because the lower of the two colored faces of the column supports the weight of more fluid than the upper one does. In fact, the excess weight supported by the lower face is exactly the weight of the molecules within the column. As the free-body diagram indicates, this excess weight is mg, where m is the mass of the fluid in the column and g is the acceleration due to gravity. Setting the sum of the vertical forces equal to zero, we find that

$$\Sigma F_y = 0 = P_2A - P_1A - mg \quad \text{or} \quad P_2A = P_1A + mg$$

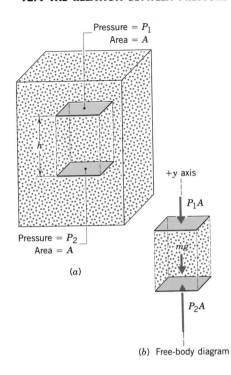

FIGURE 12.5 (*a*) A container of fluid molecules in which one particular column of molecules is outlined. The fluid is at rest. (*b*) The free-body diagram, showing the vertical forces acting on the column. One of these forces is the weight *mg* of the column.

The mass *m* is related to the density ρ and the volume *V* of the column by $m = \rho V = \rho A h$, where we have expressed the volume as the cross-sectional area *A* times the vertical dimension *h*. With this substitution, the condition for equilibrium becomes $P_2 A = P_1 A + \rho A h g$. The area *A* can be eliminated algebraically from this expression, leading to the result that

$$P_2 = P_1 + \rho g h \qquad (12.4)$$

This equation indicates that if the pressure P_1 is known at a higher level, the larger pressure P_2 at a deeper level can be calculated by adding the increment $\rho g h$. Example 3 illustrates how to apply this useful relation between pressure and depth.

EXAMPLE 3

Figure 12.6 shows the cross section of a swimming hole. Find the pressure beneath the surface at (a) point *A* and (b) point *B*.

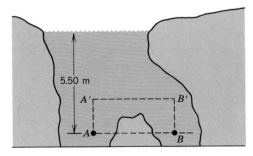

FIGURE 12.6 In the water shown here, the pressures at points *A* and *B* are the same, since both points are located at the same vertical distance of 5.50 m beneath the surface of the water.

SOLUTION

(a) The pressure acting on the surface of the water is the atmospheric pressure of 1.01×10^5 Pa. Using this value as P_1 in Equation 12.4, we can determine a value for the pressure P_2 at point *A*, which is located 5.50 m under the water:

$$P_2 = P_1 + \rho g h = 1.01 \times 10^5 \text{ Pa}$$
$$+ (1.000 \times 10^3 \text{ kg/m}^3)(9.80 \text{ m/s}^2)(5.50 \text{ m})$$
$$= \boxed{1.55 \times 10^5 \text{ Pa}}$$

(b) The pressure at point *B* is the same as that at point *A*, since both are located at the *same vertical distance* of 5.50 m beneath the surface. The fact that point *B* is displaced horizontally to the right of point *A* is of no concern, because only the vertical distance *h* affects the pressure increment $\rho g h$ in Equation 12.4. To understand this important feature more clearly, consider the path $AA'B'B$ in Figure 12.6. The pressure decreases on the way up along the vertical segment AA' and increases by the same amount on the way back down along vertical segment $B'B$. Since no change in pressure occurs along the horizontal segment $A'B'$, the pressure is the same at A and B.

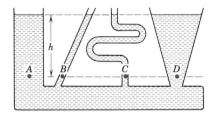

FIGURE 12.7 Since the points A, B, C, and D are located at the same distance h beneath the liquid surface, the pressure at each of these points is the same.

Figure 12.7 shows an irregularly shaped container of liquid. Reasoning similar to that used in Example 3 leads to the conclusion that the pressure is exactly the same at points A, B, C, and D since each of these points is at the same vertical distance h beneath the surface. In effect, the arteries in our bodies constitute an irregularly shaped "container" for the blood. The next example examines the blood pressure at different places in this "container."

EXAMPLE 4

Blood in the arteries is flowing, but as a first approximation, the effects of this flow can be ignored. Treating the blood as a static fluid, estimate the amount by which the blood pressure P_2 in the anterior tibial artery at the foot exceeds the blood pressure P_1 in the aorta at the heart when the body is (a) reclining horizontally as in Figure 12.8a and (b) standing as in part b of the drawing.

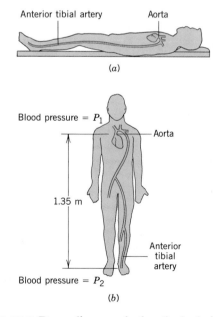

SOLUTION

(a) When the body is horizontal, there is little or no vertical separation between the feet and the heart. Since $h = 0$,

$$P_2 - P_1 = \rho g h = \boxed{0} \qquad (12.4)$$

(b) When an adult is erect, the vertical separation between the feet and the heart is about 1.35 m, as Figure 12.8b indicates. Table 12.1 gives the density of blood as 1060 kg/m³, so that

$$P_2 - P_1 = \rho g h = (1060 \text{ kg/m}^3)(9.80 \text{ m/s}^2)(1.35 \text{ m})$$
$$= \boxed{1.40 \times 10^4 \text{ Pa}}$$

FIGURE 12.8 Depending on whether the body is reclining horizontally, as in part a, or standing erect, as in part b, the blood pressure in the feet can exceed the blood pressure in the heart.

Example 5 applies the relation $P_2 = P_1 + \rho g h$ to obtain some useful information about the job of pumping water from a well.

EXAMPLE 5

Determine the maximum height to which water can be pumped from a well with the arrangement shown in Figure 12.9.

SOLUTION

The job of the pump is to draw the air out of the pipe. With the pipe empty of air, the atmospheric pressure in the well shaft pushes the water upward. The best the pump can do is to remove all the air, in which case the pressure at the top of the water in the pipe is $P_1 = 0$. The pressure P_2 at the bottom of the pipe at point A is the same as that at point B, namely, the atmospheric pressure of 1.01×10^5 Pa, because the two points are at the same elevation. Equation 12.4 can now be applied to obtain the maximum pumping height:

$$P_2 = P_1 + \rho g h = \rho g h$$

$$h = \frac{P_2}{\rho g} = \frac{1.01 \times 10^5 \text{ Pa}}{(1.000 \times 10^3 \text{ kg/m}^3)(9.80 \text{ m/s}^2)} = \boxed{10.3 \text{ m}}$$

If water is located more than 10.3 m underground, schemes other than that shown in the drawing must be used to obtain it.

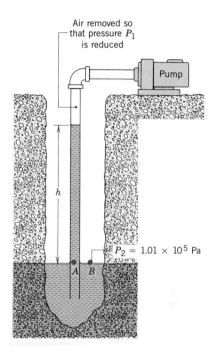

FIGURE 12.9 A pumping arrangement for drawing water out of a well.

It is tempting to apply the equation $P_2 = P_1 + \rho g h$ to all fluids, but this cannot be done. In determining the pressure increment $\rho g h$, we assumed that the density ρ is the same at any vertical distance h; that is, the fluid is incompressible. To a good degree of approximation this assumption is valid for liquids, since the bottom layers of a liquid can support the upper layers with little compression. In a gas, however, the lower layers may be compressed markedly by the weight of the upper layers, with the result that the density of a gas varies with vertical distance. For example, the density of our atmosphere is larger near the earth's surface than it is at higher altitudes. When applied to gases, the relation $P_2 = P_1 + \rho g h$ can be used only when h is small enough that any variation in ρ can be neglected.

12.5 PRESSURE GAUGES

One of the simplest pressure gauges is the mercury barometer used for measuring atmospheric pressure. As Figure 12.10 shows, this device is a tube sealed at one end, filled completely with mercury, and then inverted, so that the open end is under the surface of a pool of mercury. The relation $P_2 = P_1 + \rho g h$ can be used to explain how such a device measures atmospheric pressure. Except for a negligible amount of mercury vapor, the space above the mercury in the tube is empty, and the pressure P_1 is zero there. The pressure P_2 at point A at the bottom of the mercury column is the same as that at point B, namely, atmospheric pressure, for these two points are at the same level. With these values for P_1 and P_2, it follows that $P_{\text{atm}} = 0 + \rho g h$, and the atmospheric pressure can be determined from the height h of the mercury in the tube,

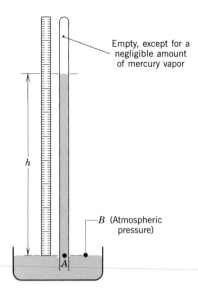

FIGURE 12.10 A mercury barometer. Points A and B are at the same pressure, because they are at the same elevation.

the density ρ of mercury, and the acceleration due to gravity. Usually, however, weather forecasters report the pressure in terms of the height h, expressing it in millimeters or inches of mercury. For instance, using $P_{atm} = 1.013 \times 10^5$ Pa and $\rho = 13.6 \times 10^3$ kg/m³ for the density of mercury, we find that $h = P_{atm}/(\rho g) = 760$ mm (29.9 inches). Slight variations from this value occur, depending on weather conditions and altitude. One millimeter of mercury is sometimes referred to as one *torr* of pressure, named after the inventor of the barometer, Evangelista Torricelli (1608–1647). Thus, one atmosphere of pressure is 760 torr.

Compare Figures 12.9 and 12.10 and notice how similar they are. The pump in Figure 12.9 reduces the pressure above the water column nearly to zero, just as it is above the mercury column in Figure 12.10. Could water or some other liquid, therefore, be used in a barometer instead of mercury? In principle, yes. However, the density of water is small compared to that of mercury, and a correspondingly longer vertical column would be needed to achieve the pressure increment $\rho g h$. In fact, for a pressure of one atmosphere, the column of water in a water barometer would be 10.3 m in length (see Example 5), 13.6 times longer than the mercury column. Such a length is inconvenient, and water barometers are rare.

Figure 12.11 shows another kind of pressure gauge, the open-tube manometer. The phrase "open-tube" refers to the fact that one side of the U-tube is open to atmospheric pressure. The tube contains a liquid, often mercury, and its other side is connected to the container whose pressure P_2 is to be measured. When the pressure in the container is equal to the atmospheric pressure, the liquid levels in both sides of the U-tube are the same, as Figure 12.11*a* indicates. However, when the pressure in the container is greater than atmospheric pressure, as in Figure 12.11*b*, the liquid in the tube falls on the left side and rises on the right side. The relation $P_2 = P_1 + \rho g h$ can be used to determine the container pressure. Atmospheric pressure exists at the top of the right column, so that $P_1 = P_{atm}$. Moreover, the pressure P_2 is the same at points A and B. Therefore, we find that $P_2 = P_{atm} + \rho g h$. Upon rearrangement this result becomes

$$P_2 - P_{atm} = \rho g h$$

The height h is proportional to $P_2 - P_{atm}$, which is called the **gauge pressure**. The gauge pressure is the amount by which the container pressure exceeds atmospheric pressure. The actual value for P_2 is called the **absolute pressure**.

The sphygmomanometer is a familiar pressure measuring device that is used for determining blood pressure. As Figure 12.12 illustrates, a squeeze bulb can be used to

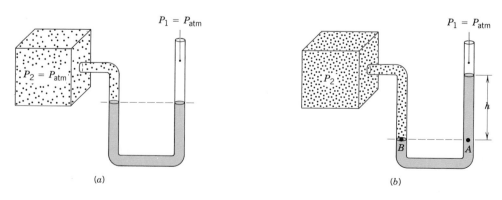

FIGURE 12.11 An open-tube manometer is used to measure the pressure P_2 in a container. In *a* the pressure P_2 equals P_{atm}, while in *b* the pressure P_2 exceeds P_{atm}.

inflate the cuff with air, so that the pressure applied to the arm cuts off the flow of blood through the artery below the cuff. When the release valve is opened, the cuff pressure starts to drop. Blood begins to flow again when the pressure created by the heart at the peak of its beating cycle exceeds the cuff pressure. Using a stethoscope to listen for the initial flow, the operator can measure the corresponding cuff gauge pressure with an open-tube manometer. This cuff gauge pressure is called the *systolic* pressure. As the amount of air in the cuff continues to drop, there comes a point when even the pressure created by the heart at the low point of its beating cycle is sufficient to cause blood to flow. Identifying this point with the stethoscope, the operator can again measure the corresponding cuff gauge pressure, which is referred to as the *diastolic* pressure. The systolic and diastolic pressures are reported in millimeters of mercury, and values of 125 and 85, respectively, are typical of a young healthy heart.

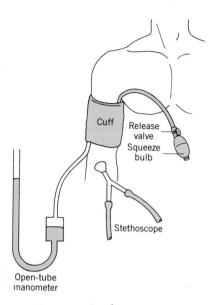

FIGURE 12.12 A sphygmomanometer is used to determine blood pressure.

12.6 PASCAL'S PRINCIPLE

As we have seen, the pressure in a fluid increases with depth, due to the weight of the fluid above the point of interest. In addition, a confined fluid may be subjected to an additional pressure by the application of an external force. Figure 12.13 illustrates this important aspect of fluid behavior.

Part *a* of the drawing shows two interconnected cylindrical chambers. The chambers have different diameters and, together with the connecting tube, are completely filled with a liquid. Suppose, for the moment, that the larger chamber is sealed at the top with a cap, while the smaller one is fitted with a movable piston. What determines the pressure P_1 at a point immediately beneath the piston? According to the definition of pressure, it is the magnitude F_1 of the external force divided by the area A_1 of the piston: $P_1 = F_1/A_1$. If it is necessary to know the pressure *at any other place in the liquid,* all we do, according to $P_2 = P_1 + \rho g h$, is add the value of P_1 to the increment $\rho g h$ that takes into account the vertical distance below the level of the piston. The important feature here is this—the pressure P_1 adds to the pressure $\rho g h$ due to the depth of the liquid at any point, whether that point is in the smaller chamber, the connecting tube, or the larger chamber. It is in the very nature of a liquid, therefore, that if the applied pressure P_1 is increased or decreased, the pressure at any other point within the confined liquid changes correspondingly. This behavior is in accord with **Pascal's principle.**

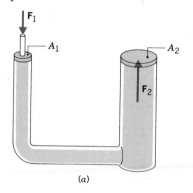

> **PASCAL'S PRINCIPLE**
> Any change in the pressure applied to a completely enclosed fluid is transmitted undiminished to all parts of the fluid and the enclosing walls.

The real usefulness of the arrangement in Figure 12.13*a* becomes apparent when the force F_2 applied by the liquid to the cap on the right side is calculated. The area of the cap is A_2 and the pressure there is P_2. As long as the tops of the left and right chambers are at the same level, the pressure increment $\rho g h$ is zero, and $P_2 = P_1 + \rho g h = P_1$. Consequently, $F_2/A_2 = F_1/A_1$, so that

$$F_2 = F_1 \left(\frac{A_2}{A_1} \right) \tag{12.5}$$

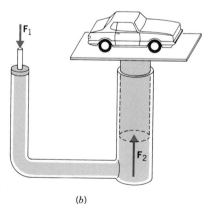

FIGURE 12.13 (*a*) An external force $\mathbf{F}_1$ is applied to the piston on the left. As a result, a force $\mathbf{F}_2$ is exerted on the cap on the chamber on the right. (*b*) The familiar hydraulic car lift.

FIGURE 12.14 Three examples of devices that employ a hydraulic fluid to generate a large output force, starting with a small input force.

This result shows that with area A_2 larger than area A_1, a large force $\mathbf{F}_2$ can be applied to the cap on the right chamber in the figure, starting with a smaller force $\mathbf{F}_1$ on the left. Depending on the ratio of the areas A_2/A_1, the force $\mathbf{F}_2$ can be large indeed, as in the familiar hydraulic car lift shown in part b of the drawing. In this device the force $\mathbf{F}_2$ is not applied to a cap that seals the larger chamber, but, rather, to a movable plunger that lifts a car.

In a device such as a hydraulic car lift, the same amount of work is done by both the input and output forces in the absence of friction. The larger output force $\mathbf{F}_2$ moves through a smaller distance, while the smaller input force $\mathbf{F}_1$ moves through a larger distance. The work, being the product of force and distance, is the same in either case, as it must be, since mechanical energy is conserved.

An enormous variety of clever devices use hydraulic fluids to create large forces, starting with small ones. In addition to the hydraulic car lift, such devices include a jack used when replacing tires on an airplane, and a backhoe used for digging. Figure 12.14 illustrates these applications.

12.7 ARCHIMEDES' PRINCIPLE

Anyone who has tried to push a beach ball under the surface of a swimming pool has felt how the water pushes back with a strong upward force. This upward force is called the ***buoyant force,*** and all fluids apply such a force to objects that are immersed in them. The buoyant force exists because fluid pressure is larger at greater depths.

Consider the cylinder of height h that is being held under the surface of the liquid in Figure 12.15a. The pressure P_1, acting on the top face, generates the downward force P_1A, where A is the area of the face. Similarly, the pressure P_2, acting on the bottom face, generates the upward force P_2A. Since the pressure is greater at greater depths, the upward force exceeds the downward force. Consequently, the liquid applies to the cylinder a net upward force, or buoyant force, whose magnitude is

$$\text{Buoyant force} = P_2A - P_1A = (P_2 - P_1)A$$

Substituting $P_2 - P_1 = \rho gh$ from Equation 12.4, we find that the buoyant force equals ρghA. Notice two things in this result: (1) hA is the volume of liquid that the cylinder moves aside or displaces in being submerged, and (2) ρ denotes the density of the liquid, not the density of the material from which the cylinder is made. Therefore, the term ρhA gives the mass m of the displaced fluid, so that the buoyant force equals mg, the weight of the displaced fluid. Part b of the drawing helps to clarify the meaning of the phrase "weight of the displaced fluid." This phrase refers to the weight of the fluid (colored in the drawing) that would spill out, if the container were filled to the brim before the cylinder is inserted into the liquid.

The cylindrical shape of the object in Figure 12.15 is not important. No matter what the shape, the buoyant force arises in a similar fashion, in accord with ***Archimedes' principle.*** It was an impressive accomplishment that the Greek scientist Archimedes (ca. 287–212 B.C.) discovered the essence of this principle so long ago.

> **ARCHIMEDES' PRINCIPLE**
> Any fluid applies a buoyant force to an object that is partially or completely immersed in it; the magnitude of the buoyant force equals the weight of the fluid that the object displaces.

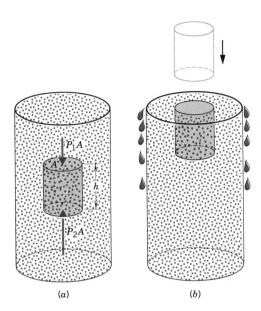

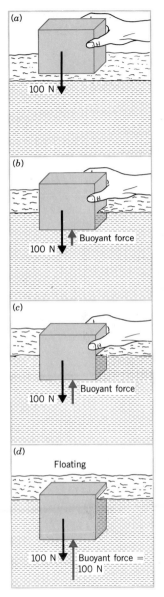

FIGURE 12.15 (*a*) A cylinder being held beneath the surface of a liquid (fluid). The pressure is larger at greater depths in a fluid, that is, $P_2 > P_1$. The fluid applies an upward force P_2A to the bottom face of the cylinder and a downward force P_1A to the top face. Since the upward force exceeds the downward force, there is a net upward force acting on the cylinder, and this net force is called the buoyant force. (*b*) The weight of the displaced fluid is the weight of the fluid (colored drops) that would spill out if the container were filled exactly to the brim before the cylinder is inserted into the liquid.

The effect that the buoyant force has depends on how strong it is compared with the other forces that are acting. For example, if the buoyant force is strong enough to balance the force of gravity, an object will float in a fluid. Figure 12.16 explores this possibility. In part *a*, a block that weighs 100 N is held above a liquid and displaces none of it, so the liquid exerts no buoyant force. In part *b*, the block displaces a small amount of liquid. Thus, the liquid applies a small buoyant force to the block, according to Archimedes' principle. Nevertheless, if the block were released, it would fall, because the buoyant force is not sufficiently strong to balance the weight. Part *c* is similar to part *b*, except that more of the block is immersed, leading to a larger buoyant force. Finally, in part *d* the buoyant force is strong enough to balance the 100-N weight, so the block is in equilibrium and does not fall when released. The block floats. If the buoyant force were not sufficiently large to balance the block's weight, even with all of the block submerged, the block would sink when released. Realize, however, that even if an object sinks, there is still a buoyant force acting on it; it's just that the buoyant force is not large enough to balance the weight. Example 6 provides additional insight into what determines whether an object will float or sink in a fluid.

FIGURE 12.16 Parts *a–d* show an object whose weight is 100 N being inserted deeper and deeper into a liquid. The deeper the object is, the more liquid it displaces, with the result that the buoyant force increases from *b* to *d*. In part *d*, the buoyant force just matches the 100-N weight of the object, so the object floats.

EXAMPLE 6

A solid, square, pinewood raft measures 4.0 m on a side and is 0.30 m thick. (a) Determine whether the raft floats in water, and (b) if so, how much of the raft is beneath the surface (see Figure 12.17).

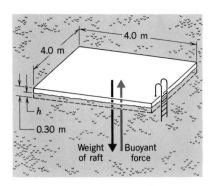

FIGURE 12.17 The pinewood raft floats with the distance h beneath the surface of the water. The buoyant force is equal in magnitude to the weight of the raft.

SOLUTION

(a) To determine whether the raft floats, we compare its weight to the maximum possible buoyant force. The weight of the raft can be calculated from the density $\rho_{pine} = 550$ kg/m^3 (Table 12.1), the volume of the wood, and the acceleration due to gravity. The volume of the wood is $V_{pine} = 4.0$ m $\times$ 4.0 m $\times$ 0.30 m $= 4.8$ m^3, so that

$$\frac{\text{Weight}}{\text{of raft}} = (\rho_{pine}V_{pine})g = (550 \text{ kg/m}^3)(4.8 \text{ m}^3)(9.80 \text{ m/s}^2)$$
$$= 26\ 000 \text{ N}$$

The maximum possible buoyant force occurs when the entire raft is under the surface, displacing a volume of water

equal to the 4.8-m^3 volume of the raft. The weight of this volume of water is the maximum buoyant force and can be obtained using the density of water:

$$\begin{aligned}\text{Maximum} \\ \text{buoyant} \\ \text{force}\end{aligned} = \rho_{water}(4.8 \text{ m}^3)g$$
$$= (1.000 \times 10^3 \text{ kg/m}^3)(4.8 \text{ m}^3)(9.80 \text{ m/s}^2)$$
$$= 47\ 000 \text{ N}$$

Since the maximum buoyant force exceeds the weight of the raft, the raft will float only partially submerged.

(b) The value of the distance h in Figure 12.17 can be obtained by utilizing the fact that the floating raft is in equilibrium. The buoyant force balances the 26 000-N weight of the raft. According to Archimedes' principle, the buoyant force is the weight of a displaced amount of water whose volume is 4.0 m $\times$ 4.0 m $\times$ h. The weight of this volume of water is ρ_{water}(4.0 m $\times$ 4.0 m $\times$ h)g. It follows that

$$\rho_{water}(4.0 \text{ m} \times 4.0 \text{ m} \times h)g = 26\ 000 \text{ N}$$
$$h = \frac{26\ 000 \text{ N}}{\rho_{water}(4.0 \text{ m} \times 4.0 \text{ m})g}$$
$$= \frac{26\ 000 \text{ N}}{(1.000 \times 10^3 \text{ kg/m}^3)(4.0 \text{ m} \times 4.0 \text{ m})(9.80 \text{ m/s}^2)}$$
$$= \boxed{0.17 \text{ m}}$$

In deciding whether the raft floats in part (a) of Example 6, we compared the raft's weight [$(\rho_{pine}V_{pine})g$] to the maximum possible buoyant force [$(\rho_{water}V_{pine})g$]. This comparison depends only on the densities ρ_{pine} and ρ_{water}. The take-home message is that any object that is *solid throughout,* no matter what its shape, will float in a liquid if and only if the object has a density equal to or smaller than the liquid. For instance, at 0 °C ice has a density of 917 kg/m^3, while water has a density of 1000 kg/m^3. Therefore, ice floats in water.

Although a solid piece of a high-density material like steel will sink in water, such materials can, nonetheless, be used to make floating objects. A supertanker, for example, floats because it is *not* solid metal. Such a ship contains enormous amounts of empty space and, because of its shape, displaces enough water to balance its own large weight. On the other hand, a submarine can float or travel submerged, rising or diving as needed. A sub can behave in this fashion because it contains ballast tanks and can admit seawater into or expel seawater from them, thus changing its own weight. When the weight of the sub exceeds the buoyant force pushing upward on the hull, the sub sinks; when its weight is less than the buoyant force, the sub rises. (We assume here that only the gravitational force and the buoyant force act on a submarine. In reality, there are also lift forces that arise from the motion of the sub through the water. The next chapter discusses such forces.)

Archimedes' principle has allowed us to determine how an object can float in a liquid. This principle also applies to gases, as the next example illustrates.

EXAMPLE 7

Normally, a Goodyear airship, such as that in Figure 12.18, contains about 5400 m³ of helium whose density is 0.179 kg/m³. Find the weight of the load W_L that the airship can carry in equilibrium at an altitude of 0.50 km, where the density of air is approximately 1.2 kg/m³.

SOLUTION

The airship and its load are in equilibrium. Thus, the buoyant force applied to the airship by the surrounding air balances the weight W_{helium} of the helium and the weight W_L of the load. Note that W_L includes the weight of the solid parts of the airship. The free-body diagram in Figure 12.18b shows these forces:

$$\underbrace{\rho_{helium}(5400 \text{ m}^3)g}_{\text{Weight of helium}} + \underbrace{W_L}_{\substack{\text{Weight}\\\text{of load}}} = \underbrace{\rho_{air}(5400 \text{ m}^3)g}_{\substack{\text{Buoyant force}\\\text{(weight of}\\\text{displaced air)}}}$$

$$(0.179 \text{ kg/m}^3)(5400 \text{ m}^3)(9.80 \text{ m/s}^2) + W_L$$
$$= (1.2 \text{ kg/m}^3)(5400 \text{ m}^3)(9.80 \text{ m/s}^2)$$

$$\boxed{W_L = 54\,000 \text{ N (12 000 lb)}}$$

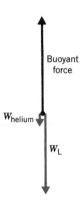

(a)

(b) Free-body diagram

FIGURE 12.18 (a) A helium-filled Goodyear airship carries a load weight W_L. (b) The free-body diagram.

SUMMARY

Fluids are materials that can flow. The **mass density** ρ of any substance is its mass m divided by its volume V: $\rho = m/V$. The **specific gravity** of a substance is its density divided by 1.000×10^3 kg/m³, the density of water at 4 °C.

Pressure P is the magnitude F of the force acting perpendicular to a surface divided by the area A over which the force acts: $P = F/A$. The SI unit of pressure is the pascal (Pa); 1 Pa = 1 N/m². In the presence of gravity, the upper layers of a fluid push downward on the layers beneath, with the result that **fluid pressure is related to depth**. In an incompressible static fluid whose density is ρ, the relation is $P_2 = P_1 + \rho gh$, where P_1 is the pressure at one level and P_2 is the pressure at a level that is h meters deeper. The **gauge pressure** is the amount by which a pressure P exceeds atmospheric pressure. The **absolute pressure** is the actual value for P.

The **buoyant force** is the net upward force that a fluid applies to any object that is immersed partially or completely in the fluid. **Archimedes' principle** states that the magnitude of the buoyant force equals the weight of the fluid that the immersed object displaces. A solid object will float in a fluid if the density of the material from which the object is made is equal to or smaller than the density of the fluid.

SOLVED PROBLEMS

SOLVED PROBLEM 1
Related Problems: *30   *32   **33

To verify his suspicion that a rock specimen is hollow, a geologist weighs the specimen in air and in water. He finds that the specimen weighs twice as much in air as it does in water. The solid part of the specimen has a density of 5.0×10^3 kg/m³. What is the fraction of the specimen's apparent volume that is solid?

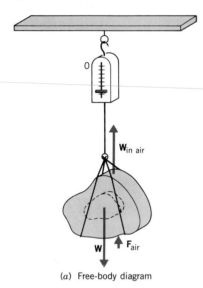

(a) Free-body diagram

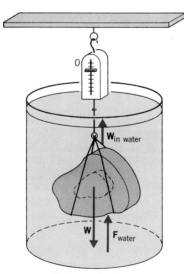

(b) Free-body diagram

Solution   Parts a and b of the drawing show the free-body diagrams that correspond to the rock being weighed in air and in

water. These two situations can be analyzed in exactly the same fashion. In each, the measured weight is the true weight W of the specimen diminished by the buoyant force of the surrounding fluid, air in one case and water in the other. The buoyant force F_{air} of the air can be ignored (the weight of the displaced air is small compared to the weight of the specimen), but the buoyant force F_{water} of the water must be taken into account (the weight of the displaced water is not small). According to Archimedes' principle, the buoyant force of the water is the weight of the water displaced by the total volume of the rock $(V_R + V_H)$, where V_R denotes the volume of the solid part of the rock and V_H the hollow part:

$$F_{water} = \rho_{water}(V_R + V_H)g$$

As usual, we have used Equation 12.1 to express the mass as density times volume and then multiplied the mass by the acceleration due to gravity to obtain the weight.

In air	*In water*
$W_{in\ air} = W - F_{air}$	$W_{in\ water} = W - F_{water}$
$W_{in\ air} = W$	$W_{in\ water} = W - \rho_{water}(V_R + V_H)g$

The weight in air is twice the weight in water, so that

$$W = 2[W - \rho_{water}(V_R + V_H)g]$$
$$W = 2\rho_{water}(V_R + V_H)g$$

The next step is to write the true weight W of the specimen as the sum of the weights of its parts, the solid rock (density $= \rho_R$) and the gas in the hollow space (density $= \rho_H$): $W = \rho_R V_R g + \rho_H V_H g$. Since the density of any gas in the hollow space is negligibly small, the true weight of the rock is simply $W = \rho_R V_R g$. With this value for W, the previous result can be used to find the fraction $V_R/(V_R + V_H)$ of the specimen that is solid:

$$\rho_R V_R g = 2\rho_{water}(V_R + V_H)g$$

$$\frac{V_R}{V_R + V_H} = \frac{2\rho_{water}}{\rho_R} = \frac{2(1.000 \times 10^3 \text{ kg/m}^3)}{5.0 \times 10^3 \text{ kg/m}^3} = \boxed{0.40}$$

Summary of Important Points   When an object is composed of two parts that have different densities, information about the individual parts can be obtained from the apparent weight of the object in a fluid. The apparent weight will be less than the true weight because of the buoyant force, which can be taken into account with the aid of Archimedes' principle. An object that floats in a fluid has a zero apparent weight, since the buoyant force balances the true weight. An expression for the apparent weight can often be used along with other known information to obtain the solution of a problem that deals with the individual parts of the object.

QUESTIONS

1. A pile of empty aluminum beer cans has a volume of 1.0 m³. The density of aluminum is 2700 kg/m³. Explain why the mass of the pile of cans is *not* equal to $\rho_{Al}V = (2700 \text{ kg/m}^3)(1.0 \text{ m}^3) = 2700$ kg.

2. A lead brick weighs less on the moon than on the earth, but the density of lead is the same in both places. Why?

3. The part of a magnetic phonograph cartridge that rests on a rotating stereo record is called the stylus and is usually a small polished diamond. The tip (radius ≈ 0.0007 inch) of this diamond touches the record. Explain why the pressure that the stylus applies to the record is large, even though the force the stylus applies to the record is quite small.

4. As you climb a mountain, your ears "pop" because of the changes in atmospheric pressure. In which direction does your eardrum move (a) as you climb up and (b) as you climb down? Give your reasoning.

5. A bottle of fruit juice is sealed with a lid on which a red dot or "button" is painted. Around the button the following phrase is printed: "Button pops up when seal is broken." Explain why the button remains pushed in when the seal is intact.

6. A sealed five-gallon jug of water is full, except for a tall thin tube that is attached to it, as the drawing shows. Water is poured into this tube, and before the tube is full the jug bursts. Explain why.

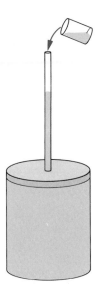

7. The drawing shows two tanks at a marine exhibit, each with identical observation windows. Each tank is filled to the top with seawater and exposed to the air, but one contains a much greater volume of water than the other, as shown. Compare the total force that each window must sustain. Account for your answer.

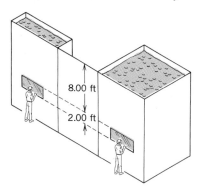

8. Why does the cork fly out of the bottle with a loud "pop" when a bottle of champagne is opened? To answer this question on an exam a student says "because the gas pressure in the bottle is about 0.3×10^5 Pa." Explain whether the student is referring to absolute or gauge pressure.

9. A scuba diver is some distance below the surface of the water when a storm approaches, dropping the air pressure above the water. Would a sufficiently sensitive pressure gauge attached to his wrist register this drop in air pressure? Give your reasoning.

10. Is either one (or both) of the following statements true? (1) Any object that floats in mercury also floats in water. (2) Any object that floats in water also floats in mercury. Justify your answers.

11. A glass beaker, filled to the brim with water, is resting on a scale. A block of wood is carefully placed in the water and floats. The water that spills over the beaker is wiped away, and the beaker is still filled to the brim. How do the initial and final readings on the scale compare? Explain.

12. Suppose that a solid metal cube has perfectly flat and smooth faces and that the bottom surface of a swimming pool is also perfectly flat and smooth. Under these conditions the cube could be placed on the bottom with no water whatsoever under its lower face. In such a situation, would there be a buoyant force acting on the cube? Account for your answer.

13. A glass of water has an ice cube floating in it. The glass is filled to the brim. When the ice cube melts, will the water level drop, remain the same, or rise, causing water to spill out? Give your reasoning.

14. A solid 10.0-kg sphere and a solid 10.0-kg cube are completely immersed in the same liquid. Which of the buoyant forces that act on these objects is larger, if (a) each object is made from lead and (b) the sphere is made from lead and the cube is made from aluminum? Justify your answers.

PROBLEMS

Section 12.2 Mass Density

1. Numerous jewelry items of solid silver are melted down and cast into a solid circular disk that is 2.00 cm thick. The total mass of the jewelry is 10.0 kg. Find the radius of the disk.

2. A pirate in a movie is carrying a chest (0.30 m × 0.30 m × 0.20 m) that is supposed to be filled with gold. To see how ridiculous this is, determine the weight of the gold that this fellow is carrying. Express your answer in (a) newtons and (b) pounds.

3. The *karat* is a dimensionless unit that is used to indicate the proportion of gold in a gold-containing alloy. An alloy that is one karat gold contains a weight of pure gold that is one part in twenty-four. What is the volume of gold in a 14.0-karat gold necklace whose weight is 1.00 N?

* 4. An irregularly shaped chunk of concrete has a hollow spherical cavity inside. The mass of the chunk is 33 kg, and the volume enclosed by the outside surface of the chunk is 0.025 m³. What is the radius of the spherical cavity?

* 5. Planners of an experiment are evaluating the design of a helium-filled (0 °C, 1 atm pressure) sphere of radius R. Ultrathin silver foil of thickness T is used to make the sphere, and the designers claim that the mass of helium in the sphere equals the mass of silver used. Assuming that T is much less than R, calculate the ratio T/R for such a sphere.

**6. An antifreeze solution is made by mixing ethylene glycol ($\rho = 1116$ kg/m³) with water. Suppose the specific gravity of such a solution is 1.0730. Assuming that the total volume of the solution is the sum of its parts, determine the volume percentage of ethylene glycol in the solution.

Section 12.3 Pressure

7. The circular top of a can of soda has a radius of 3.20 cm. The pull-tab has an area of 3.80 cm². The absolute pressure of the carbon dioxide in the can is 1.40×10^5 Pa. Find the force that this gas pressure generates (a) on the top of the can (including the pull-tab) and (b) on the pull-tab itself.

8. High-heeled shoes can cause tremendous pressure to be applied to a floor. Suppose the radius of a heel is 6.00 mm. At times during a normal walking motion, nearly the entire body weight acts perpendicularly against the area of such a heel. (a) Find the pressure that is applied to the floor under the heel because of the weight of a 50.0-kg woman. (b) What is the ratio of this pressure to atmospheric pressure?

9. A glass bottle of soda is sealed with a screw cap. The absolute pressure of the carbon dioxide in the bottle is 27.0 lb/inch². Assuming that the top and bottom surfaces of the cap each have an area of 0.800 inch², obtain the force that the screw threads must exert to keep the cap on the bottle.

* 10. A 75.0-kg person is sitting on a beach ball of negligible weight, as the drawing shows. Suppose the absolute pressure in the ball is 1.20×10^5 Pa. Determine the radius of the circular area where the beach ball is in contact with the ground.

* 11. A brick weighs 4.00 lb and is resting on the ground. The dimensions of the brick are 8.00 inches × 3.51 inches × 2.25 inches. A number of these bricks are stacked one on top of the other. What is the *smallest number* that could be used, so that the weight of the bricks creates a pressure of one atmosphere on the ground beneath the lowest brick?

* 12. The piston chamber shown in the drawing is oriented vertically. The piston itself has a mass of 10.0 kg, fits into the chamber without friction, and is in equilibrium in the position shown. The piston surface in contact with the fluid in the chamber is a rectangle whose dimensions are given in the drawing. There is no air above the piston. Find the pressure of the fluid, constructing a free-body diagram of the piston as the first step in your solution.

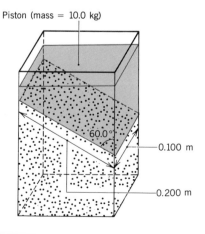

Piston (mass = 10.0 kg)

60.0°

0.100 m

0.200 m

Section 12.4 The Relation between Pressure and Depth in a Static Fluid, Section 12.5 Pressure Gauges

13. Express a pressure of 450 mm of mercury in (a) Pa, (b) pounds per square inch, and (c) atmospheres.

14. The Mariana trench is located in the Pacific Ocean and has a depth of approximately 11 000 m. The density of seawater is 1025 kg/m^3. (a) If a diving chamber were to explore such depths, what force would the water exert on the chamber's observation window (radius = 0.10 m)? (b) For comparison, determine the weight of a jetliner whose mass is 1.2×10^5 kg.

15. The drawing shows a typical setup for intravenous feeding. With the distance shown, nutrient solution ($\rho = 1030$ kg/m^3) can just barely enter the blood in the vein. What is the gauge pressure of the venous blood? Express your answer in millimeters of mercury.

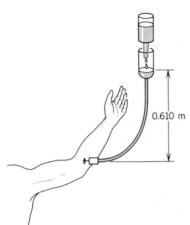

0.610 m

16. A meat baster consists of a squeeze bulb attached to a plastic tube. When the bulb is squeezed and released, with the open end of the tube under the surface of the basting sauce, the sauce rises in the tube and can then be squirted over the meat. Suppose water rises 0.15 m in the tube when this device is tested (see drawing). (a) Find the absolute pressure in the bulb, assuming that atmospheric pressure has the value of 1.013×10^5 Pa. (b) Do you expect this device to work better or worse on top of a mountain? Explain.

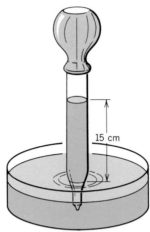

15 cm

* 17. A water tower is a familiar sight in many towns. The purpose of such a tower is to provide storage capacity and to provide sufficient pressure in the pipes that deliver the water to customers. The drawing shows a spherical reservoir that contains 5.25×10^5 kg of water when full. The reservoir is vented to the atmosphere at the top. For a full reservoir, find the gauge pressure that the water has at the basement faucet in (a) house A and (b) house B. Ignore the diameter of the delivery pipes.

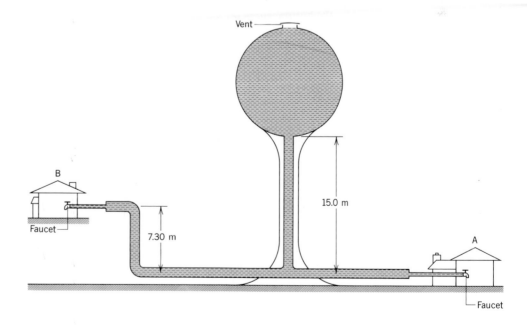

Vent

B

Faucet

7.30 m

15.0 m

A

Faucet

* **18.** A mercury barometer reads 750.0 mm on the roof of a building and 760.0 mm on the ground. Assuming a constant value of 1.29 kg/m³ for the density of air, determine the height of the building.

* **19.** A 1.00-m-tall container is filled to the brim, part way with mercury and the rest of the way with water. The container is open to the atmosphere. What must be the depth of each layer, so the absolute pressure on the bottom of the container is twice the atmospheric pressure?

**20. Two identical containers are open at the top and are connected at the bottom via a tube of negligible volume and a valve (which is initially closed). Both containers are filled initially to the same height of 1.00 m, one with water, the other with mercury, as the drawing indicates. The valve is then opened. Water and mercury are immiscible. Determine the fluid level in each container when equilibrium is reestablished.

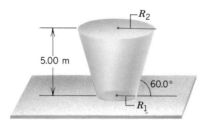

Water   Mercury

**21. As the drawing illustrates, a pond has the shape of an inverted cone with the tip sliced off and has a depth of 5.00 m. The atmospheric pressure above the pond is 1.01 × 10⁵ Pa. The circular top surface (radius = R_2) and circular bottom surface (radius = R_1) of the pond are both parallel to the ground. The magnitude of the force acting on the top surface is the same as the magnitude of the force acting on the bottom surface. Obtain R_2 and R_1.

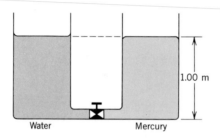

Section 12.6 Pascal's Principle

22. The drawing shows a hydraulic press used in a trash compactor, where the radii of the input piston and the output plunger are 0.25 in. and 2.0 in., respectively. If the height difference between the input piston and the output plunger can be neglected, what force is applied to the trash when the input force is 75 lb?

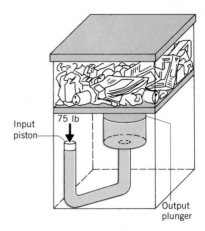

23. The atmospheric pressure above a swimming pool changes from 755 mm to 765 mm of mercury. The bottom of the pool is a 12 m × 24 m rectangle. By how much does the force on the bottom of the pool increase?

24. A dump truck uses a hydraulic cylinder, as the drawing illustrates. When activated by the operator, a pump injects hydraulic oil into the cylinder at an absolute pressure of 3.54 × 10⁶ Pa and drives the output plunger, which has a radius of 0.150 m. Assuming the plunger remains perpendicular to the floor of the load bed, find the torque that the plunger creates about the axis identified in the drawing.

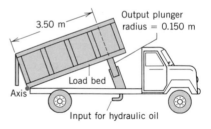

* **25.** In a hydraulic car lift, the radius of the input piston (negligible mass) is 1.20 cm, while the radius of the output plunger (mass = 350 kg) is 15.0 cm. The lift utilizes hydraulic oil. What input force **F** is needed to support a 1720-kg car (a) when the bottom surface of the input piston and the bottom surface of the output plunger are at the same level and (b) when the input piston and output plunger are in the positions shown in the drawing?

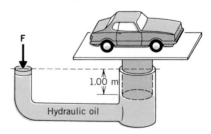

Section 12.7 Archimedes' Principle

26. Only a small part of an iceberg protrudes above the water, while the bulk lies below the surface. The density of ice is 917 kg/m³ and that of seawater is 1025 kg/m³. Find the percentage of the iceberg's volume that lies below the surface.

27. What is the total mass of swimmers that the raft in Example 6 can carry and float with its top surface at water level?

28. The 80.0-kg plastic hemispherical shell in the drawing is to carry a load of 2.00×10^3 kg in its interior space without sinking. What is the smallest possible outer radius of the shell?

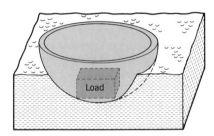

29. A hydrometer is a device used to measure the density of a liquid. The hydrometer is a cylindrical tube that is weighted at one end, so that it floats with the heavier end downward. As the drawing illustrates, the hydrometer is often contained inside a large "medicine dropper," into which the liquid is drawn using a squeeze bulb. The weighted hydrometer tube has a mass of 6.00 g and a radius of 5.00 mm. How far from the bottom of the tube should the mark be put that denotes (a) battery acid whose density is 1280 kg/m³ and (b) antifreeze solution whose density is 1073 kg/m³?

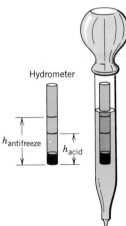

Hydrometer

$h_{\text{antifreeze}}$ h_{acid}

* 30. A 1967 Kennedy half-dollar has a mass of 11.50 g. The coin is a mixture of silver and copper, and in water the coin weighs 0.1011 N. Determine the number of grams of silver and the number of grams of copper in the coin. *(See Solved Problem 1 for a related problem.)*

* 31. A person can change the volume of his body by taking air into his lungs. The amount of change can be determined by weighing the person under water. Suppose that under water a person weighs 20.0 N with partially full lungs, while he weighs 40.0 N with empty lungs. Find the change in body volume.

* 32. A paperweight is made of glass ($\rho = 2.60 \times 10^3$ kg/m³) and has a hollow space within it. In air, the paperweight weighs 12.0 N. Moreover, in ethyl alcohol, the paperweight weighs 4.00 N more than it does in water. What percentage of the total volume of the paperweight is empty? *(See Solved Problem 1 for a related problem.)*

**33. One kilogram of glass ($\rho = 2.60 \times 10^3$ kg/m³) is shaped into a hollow spherical shell that just barely floats in water. What are the inner and outer radii of the shell? Do not assume the shell is thin. *(See Solved Problem 1 for a related problem.)*

**34. A cylinder (radius = 15.0 cm, height = 12.0 cm) has a mass of 7.00 kg. This cylinder is floating in water. Then oil ($\rho = 725$ kg/m³) is poured on top of the water until the situation shown in the drawing results. How much of the height of the cylinder is in the water and how much is in the oil?

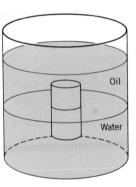

Oil

Water

ADDITIONAL PROBLEMS

35. What is the radius of a hydrogen-filled balloon that would carry a load of 5750 N when the density of air is 1.29 kg/m³?

36. To remove water from a deep well, the pumping arrangement discussed in Example 5 cannot be used. Instead a submersible pump is put under the water at the bottom of the well and is used to push the water up through a pipe. (a) What minimum output gauge pressure must the pump generate to make the water reach the nozzle at ground level, 71 m above the pump? (b) Describe what happens if the pump generates a greater output pressure than that calculated in (a).

37. Accomplished silver workers in India can pound silver into incredibly thin sheets, as thin as 3.00×10^{-4} mm (about one hundredth of the thickness of this sheet of paper). Find the number of square meters of such a sheet that can be formed from 1.00 kg of silver.

38. Determine the buoyant force that the air applies (a) to a solid 16-kg piece of aluminum and (b) to a hollow cube, 0.80 m on a side, made from this piece of aluminum.

39. (a) A mercury barometer is used at a location where the atmospheric pressure is 1.013×10^5 Pa and the acceleration due to gravity is 9.78 m/s². How high (in mm) does the mercury rise in the barometer? (b) At another location the atmospheric pressure is also 1.013×10^5 Pa, but the acceleration due to gravity is 9.83 m/s². How high (in mm) does the mercury rise here?

* 40. A cylinder (with circular ends) and a hemisphere are solid throughout and made from the same material. They are resting on the ground, the cylinder on one of its ends and the hemisphere on its flat side. The weight of each causes the same pressure to act on the ground. The cylinder is 0.500 m high. What is the radius of the hemisphere?

* 41. What is the smallest integral number of logs ($\rho = 725$ kg/m³, radius = 8.00 cm, length = 3.00 m) that can be used to build a raft that will carry four people, each of whom has a mass of 80.0 kg?

* 42. The vertical face of a reservoir dam is 120 m wide and 11 m high. (a) Find the total force (in newtons) that the water in a completely full reservoir exerts on this vertical face. *(Hint: The pressure varies linearly with depth, so you must use an average pressure.)* (b) Express your answer in pounds.

* 43. A geologist finds a solid rock that is composed solely of quartz and gold. The rock has a mass of 12.0 kg and a volume of 4.00×10^{-3} m³. What mass of gold is contained in the rock?

**44. A spring is attached permanently to the bottom of an empty swimming pool, with the axis of the spring perpendicular to the surface of the earth. An 8.00-kg block of wood ($\rho = 840$ kg/m³) is fixed to the top of the spring and compresses it. Then the pool is filled with water, completely covering the block. The spring is now observed to be stretched twice as much as it had been compressed. Determine the percentage of the block's total volume that is hollow. Ignore any air in the hollow space.

**45. A house has a roof with the dimensions shown in the drawing. The atmospheric pressure in the attic and outside the house is 758 mm of mercury. Determine the magnitude and direction of the net force that the atmosphere applies (a) to the outside surface of the roof and (b) to the inside surface of the roof. (c) What is the magnitude of the net force applied to the entire roof (both outside and inside surfaces) by the atmosphere? (d) Assuming the outside pressure rises suddenly by 10.0 mm of mercury before the pressure in the attic can adjust, repeat part (c).

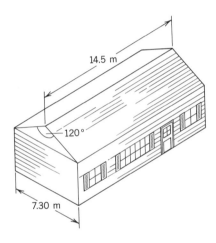

CHAPTER 13

Fluid Dynamics

Hang gliders, like birds and planes, can remain in the air because of the lift force that acts on their wings. The lift force occurs when air flows by the wings.

13.1 FLUIDS IN MOTION AND STREAMLINES

This chapter discusses moving or flowing fluids. There are many types of fluid flow and it is a formidable task to study them all. Therefore, some simplifying assumptions are necessary so we can focus on the essential features. To appreciate the nature of these simplifications, consider the following four types of fluid flow.

1. **Fluid flow can be steady or unsteady.** In steady flow the velocity of the fluid particles at any point is constant as time passes. For instance, Figure 13.1 depicts a fluid particle flowing with a velocity of $v_1 = +2$ m/s past point 1. In steady flow every particle that passes through this point has this same velocity, regardless of the arrival time. At another location the velocity may be different, as in a river, which usually flows fastest near its center and slowest near its banks. Thus, at point 2, the fluid velocity is $v_2 = +0.5$ m/s, and if the flow is steady, all particles passing through this point have a velocity of $+0.5$ m/s.

   Unsteady flow exists whenever the velocity at a point in the fluid changes as time passes. A particularly extreme kind of unsteady flow is *turbulent flow.*

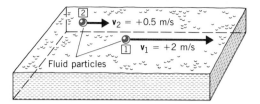

FIGURE 13.1 Two fluid particles in a stream. At different locations in the stream the particle velocities may be different, as indicated by $\mathbf{v}_1$ and $\mathbf{v}_2$.

Turbulence occurs when there are sharp obstacles or bends in the path of a fast-moving fluid, as in the rapids in Figure 13.2. Turbulence may also occur in situations where there are no sharp obstructions, if the flow velocity is high enough. In turbulent flow, the velocity at any particular point changes erratically from moment to moment, both in magnitude and direction.

2. **Fluid flow can be compressible or incompressible.** Most liquids are highly incompressible; that is, the density of a liquid remains nearly constant as the pressure changes. To a good approximation, liquids flow in an incompressible manner. In contrast, gases are highly compressible. However, there are situations in which the density of a flowing gas remains nearly constant, and in such cases the flow can be considered as incompressible.

3. **Fluid flow can be viscous or nonviscous.** A viscous fluid is one that does not flow readily, such as honey, and is said to have a large viscosity. In contrast, water is less viscous and flows more readily; water has a smaller viscosity than honey. The flow of a viscous fluid is an energy-dissipating process and is analogous to the energy-dissipating motion that occurs in the presence of kinetic friction. In this sense, the viscosity of a fluid is like an internal friction that hinders neighboring layers of fluid from sliding freely past one another. A fluid with zero viscosity flows in an unhindered manner with no dissipation of energy. Although no real fluid has zero viscosity at normal temperatures, many fluids possess negligibly small viscosities. An incompressible, nonviscous fluid is sometimes called an *ideal fluid.*

4. **Fluid flow can be rotational or irrotational.** The flow is said to be rotational when a part of the fluid has rotational as well as translational motion. To test for

FIGURE 13.2 The flow of water in a rapids is an example of turbulent flow.

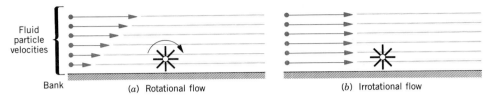

FIGURE 13.3 (*a*) In a river, water flowing near the bank usually has a smaller velocity than water flowing near the center. This top view of a river shows that a small paddle wheel placed near the bank would be turned by the flowing water, indicating rotational flow. (*b*) In an ideal (nonviscous, incompressible) fluid in which all portions of the fluid flow with the same velocity, a paddle wheel would not rotate, indicating irrotational flow.

rotational flow, a small paddle wheel can be immersed in the fluid. If the wheel rotates, as in Figure 13.3*a*, the flow is rotational. If the paddle wheel does not rotate, as in part *b* of the figure, the flow is irrotational, and the fluid exhibits only translational motion.

When the fluid flow is steady, *streamlines* are often used to represent the trajectories of the fluid particles. A streamline is a line drawn in the fluid such that a tangent to the streamline at any point is parallel to the fluid velocity at that point. Figure 13.4 shows the velocity vectors at three points along a streamline. The fluid velocity can vary (in both magnitude and direction) from point to point along a streamline, but at any given point, the velocity is constant in time, as required by the condition of steady flow. In fact, steady flow is often called *streamline flow.*

Figure 13.5*a* illustrates a method for making streamlines visible by using very small tubes to release a colored dye into the flowing fluid. The dye does not immediately mix with the fluid and is carried along a streamline. In the case of a flowing gas, such as that in a wind tunnel, streamlines are often revealed by smoke streamers, as part *b* of the figure shows.

In steady flow, the pattern of streamlines is steady in time, and, as Figure 13.5 indicates, no two streamlines cross one another. If they did, every particle arriving at the crossing point could go one way or another. This would mean that the velocity at the crossing point would change from moment to moment, a condition that does not exist in steady flow.

Streamlines provide a useful way of looking at fluid flow. Imagine, for instance, a tubular region of fluid whose sidewalls consist of streamlines, as in Figure 13.6. This region is known as a *tube of flow.* In accord with the meaning of streamlines, the tube of flow has the property that no fluid can flow through its sidewalls, for the fluid velocity is parallel to the sidewalls. In this sense, a tube of flow is analogous to a pipe of the same shape.

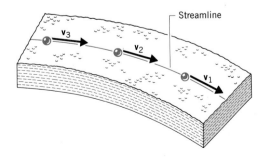

FIGURE 13.4 At any point along a streamline, the velocity vector of the fluid particle at that point is tangent to the streamline.

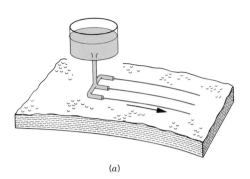
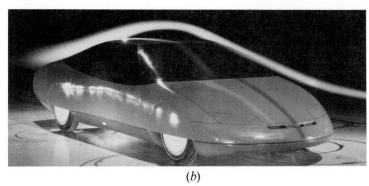

(a)
(b)

FIGURE 13.5 (a) In the steady flow of a liquid, the streamlines can be made visible with the aid of a colored dye. (b) A smoke streamer is used to reveal the streamline pattern for the air flowing around a car.

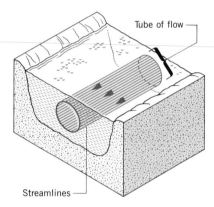

FIGURE 13.6 A tubular region of a moving fluid, whose sidewalls consist of streamlines, is known as a tube of flow.

13.2 THE EQUATION OF CONTINUITY

Have you ever used your thumb to control the water flowing from the end of a hose, as in Figure 13.7? If so, you probably have observed that the water velocity increases noticeably when your thumb reduces the cross-sectional area of the hose opening. This kind of fluid behavior is described quantitatively by a relation known as the *equation of continuity.* The equation of continuity expresses the simple idea that the mass of fluid that enters one end of a pipe must leave at the other end. For example, if fluid enters a pipe at a *mass flow rate* of 5 kilograms per second, then fluid must also leave at the same rate, assuming that there are no "sources" or "sinks" between the entry and exit points to add or remove fluid.

Figure 13.8 depicts a small mass of fluid or fluid element moving along a tube. Upstream at position 2, where the tube has a cross-sectional area A_2, the fluid has a speed v_2 and a density ρ_2. Downstream at location 1, the corresponding quantities are $v_1, \rho_1,$ and A_1. During a small time interval Δt, the fluid at point 2 moves a distance of $s_2 = v_2 \Delta t$. The volume of fluid that has flowed past this point is $A_2 s_2 = A_2 v_2 \Delta t$. The mass Δm of this fluid element is the product of the density and volume: $\Delta m = \rho_2 A_2 v_2 \Delta t$. Dividing Δm by Δt gives the mass flow rate (the mass per second):

$$\text{Mass flow rate at position 2} = \frac{\Delta m}{\Delta t} = \rho_2 A_2 v_2 \qquad (13.1a)$$

Identical reasoning leads to the mass flow rate at position 1:

$$\text{Mass flow rate at position 1} = \rho_1 A_1 v_1 \qquad (13.1b)$$

Since no fluid can cross the sidewalls of the tube, the mass flow rates at points 1 and 2 must be equal:

$$\text{Mass flow rate at position 1} = \text{Mass flow rate at position 2}$$

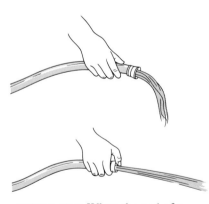

FIGURE 13.7 When the end of a hose is partially closed off, thus reducing its cross-sectional area, the fluid velocity increases.

Moreover, locations 1 and 2 were selected arbitrarily, so the mass flow rate has the same value at all positions in the tube, an important result that is known as the equation of continuity.

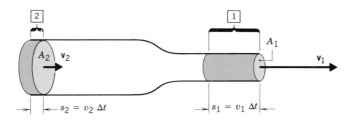

FIGURE 13.8 A fluid flowing in a tube that has different cross-sectional areas at positions 1 and 2.

EQUATION OF CONTINUITY
The mass flow rate (ρAv) has the same value at every position along a tube that has a single entry and a single exit point for fluid flow. For two positions along such a tube

$$\rho_1 A_1 v_1 = \rho_2 A_2 v_2 \qquad (13.2)$$

where $\rho =$ fluid density (kg/m^3)
$A =$ cross-sectional area of tube (m^2)
$v =$ fluid speed (m/s)

SI unit of mass flow rate: kg/s

The density of an incompressible fluid does not change during flow, so that $\rho_1 = \rho_2$. In such a case, the equation of continuity reduces to

$$A_1 v_1 = A_2 v_2 \qquad (13.3)$$

The quantity Av represents the volume of fluid per second that passes through the tube and is referred to as the **volume flow rate Q**:

$$Q = \text{volume flow rate} = Av \qquad (13.4)$$

Equation 13.3 shows that where the tube area is large, the fluid speed is small, and, conversely, where the tube area is small, the speed is large. Example 1 explores this behavior in more detail.

EXAMPLE 1

In the condition known as atherosclerosis, a deposit or atheroma forms on the arterial wall and reduces the opening through which blood can flow. In the carotid artery in the neck, the radius of the opening is $r_U = 3.5$ mm at an unobstructed point but is only $r_A = 2.0$ mm where an atheroma has formed. Determine the ratio of the speeds of the blood at these two locations.

SOLUTION

Since blood, like most liquids, is incompressible, the equation of continuity in the form of Equation 13.3 can be applied. Since the

area of a circle is πr^2, it follows that

$$\underbrace{(\pi r_U^2)v_U}_{\text{Unobstructed}} = \underbrace{(\pi r_A^2)v_A}_{\substack{\text{Obstructed by} \\ \text{atheroma}}}$$

The ratio of the speeds is

$$\frac{v_A}{v_U} = \frac{r_U^2}{r_A^2} = \frac{(3.5 \text{ mm})^2}{(2.0 \text{ mm})^2} = \boxed{3.1}$$

The next example shows that the equation of continuity predicts the response of the water flowing from a hose, when one end of the hose is partially closed off, as in Figure 13.7.

EXAMPLE 2

The water flowing from a garden hose fills a two-gallon bucket (1 gal = 231 in.³) in 30.0 s. Find the speed of the water that leaves the hose when the hose has (a) an unobstructed opening whose radius is $r = 0.375$ in. and (b) an obstructed opening whose area is reduced by a factor of two.

SOLUTION

(a) Once the volume flow rate Q is known, the desired speed can be obtained from $Q = Av$:

$$Q = \frac{(2.00 \text{ gal}) \left(\dfrac{231 \text{ in.}^3}{1 \text{ gal}} \right)}{30.0 \text{ s}} = 15.4 \text{ in.}^3/\text{s}$$

and the speed is

$$v = \frac{Q}{A} = \frac{Q}{\pi r^2} = \frac{15.4 \text{ in.}^3/\text{s}}{\pi (0.375 \text{ in.})^2} = \boxed{34.9 \text{ in./s}}$$

(b) Water can be considered incompressible. Therefore, the equation of continuity can be applied in the form $A_1 v_1 = A_2 v_2$, and since $A_2 = \frac{1}{2} A_1$, we find that

$$v_2 = \left(\frac{A_1}{A_2} \right) v_1 = \left(\frac{A_1}{\frac{1}{2} A_1} \right) (34.9 \text{ in./s}) = \boxed{69.8 \text{ in./s}}$$

While the fluid speed here is twice that in part (a), the volume flow rate is exactly the same, that is, 15.4 in.³/s.

13.3 BERNOULLI'S EQUATION

For the *steady, irrotational* flow of an *incompressible, nonviscous* fluid, it is possible to characterize the fluid at any point along its motional path by specifying its speed, pressure, and elevation. These three variables are related by **Bernoulli's equation,** named after its discoverer, Daniel Bernoulli (1700–1782). Before considering Bernoulli's equation, let us make two observations concerning the speed, pressure, and elevation of a moving fluid.

First, whenever a fluid flowing in a horizontal pipe encounters a region of reduced cross-sectional area, the pressure of the fluid drops, as Figure 13.9a indicates. The reason is as follows. When moving from region 2 downstream to region 1 in the picture, the fluid speeds up (accelerates), as required by the conservation of mass and

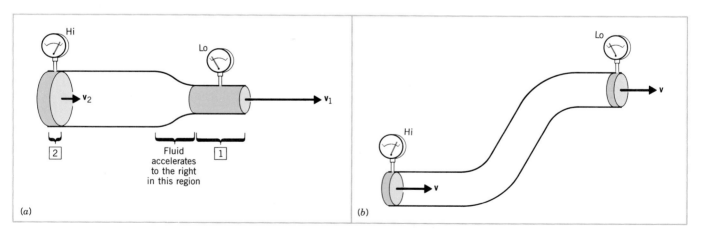

FIGURE 13.9 (a) In this horizontal pipe, the pressure in region 2 is greater than the pressure in region 1. The difference in pressure accelerates the fluid to a higher speed as the fluid moves from region 2 to region 1. (b) When the fluid changes elevation, the pressure at the bottom is greater than the pressure at the top, assuming the cross-sectional area of the pipe remains constant.

expressed by the equation of continuity. According to Newton's second law, the accelerating fluid must be subjected to an unbalanced force. Such an unbalanced force can exist only if the pressure in region 2 is greater than the pressure in region 1. Thus, in a horizontal pipe, the pressure drops wherever the fluid velocity increases (smaller cross-sectional area) and, conversely, rises wherever the fluid velocity decreases (larger cross-sectional area). The exact change in pressure is given by Bernoulli's equation.

Second, if the fluid experiences a rise in elevation, as in part *b* of the figure, the pressure at the bottom is greater than the pressure at the top. The basis for this statement is our previous study of static fluids. Bernoulli's equation indeed confirms the statement, provided the cross-sectional area of the pipe does not change.

To derive Bernoulli's equation, consider a fluid that is flowing in a pipe whose cross-sectional area and elevation both change. Figure 13.10*a* shows a small portion of fluid (a fluid element) of mass m, upstream in region 2 of such a pipe. The speed, pressure, and elevation in this region are v_2, P_2, and y_2, respectively. Downstream in region 1 these variables have the values v_1, P_1, and y_1. As Chapter 7 discusses, an object moving under the influence of gravity possesses a total mechanical energy E that is the sum of the kinetic energy KE and the gravitational potential energy PE: $E = \text{KE} + \text{PE} = \frac{1}{2}mv^2 + mgy$. If work W_{other} is done on the fluid element by *forces other than the force of gravity,* the total mechanical energy changes. According to the work–energy theorem (see Section 7.4), the work done equals the change in the total mechanical energy:

$$W_{\text{other}} = E_1 - E_2 = \underbrace{(\tfrac{1}{2}mv_1{}^2 + mgy_1)}_{\substack{\text{Total mechanical} \\ \text{energy in region 1}}} - \underbrace{(\tfrac{1}{2}mv_2{}^2 + mgy_2)}_{\substack{\text{Total mechanical} \\ \text{energy in region 2}}} \qquad (7.8)$$

To evaluate the term W_{other}, it is necessary to examine what forces, *other than gravity,* do work on the fluid. Since the flow is assumed to be nonviscous, there are no viscous forces. However, there are forces that come from the fluid surrounding the element of mass m. To see why, refer to Figure 13.10*b*. On the top surface of the mass

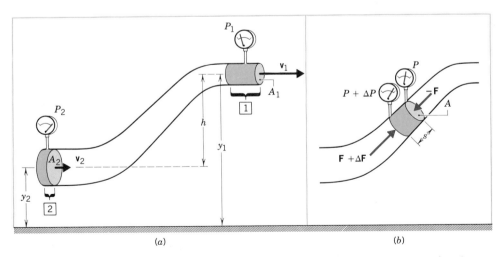

(*a*)   (*b*)

FIGURE 13.10 (*a*) A mass element of fluid moving through a pipe whose cross-sectional area and elevation both change. (*b*) The mass element experiences a force $-\mathbf{F}$ on its top surface due to the pressure of the fluid above it, and a force $\mathbf{F} + \Delta\mathbf{F}$ on its bottom surface due to the fluid below it. The net force on the element is $\Delta\mathbf{F}$.

element, the surrounding fluid exerts a pressure P. This pressure gives rise to a force of magnitude $F = PA$, where A is the cross-sectional area. On the bottom surface, the surrounding fluid exerts a slightly greater pressure, $P + \Delta P$, where ΔP is the pressure difference between the ends of the element. As a result, the force on the bottom surface is $F + \Delta F = (P + \Delta P)A$. The *net* force pushing the mass m up the tube is $\Delta F = (\Delta P)A$. When the mass element moves through its own length s, the work done is the product of the net force and the distance, according to Equation 7.1: Work $= (\Delta F)s = (\Delta P)As$. The quantity As is the volume V of the fluid element, so the work done is $(\Delta P)V$.

The total work done on the mass element in moving it from region 2 to region 1 is the sum of the small increments of work $(\Delta P)V$ done as the mass element moves along the tube. This sum amounts to $(P_2 - P_1)V$, where $P_2 - P_1$ is the pressure difference between the two regions. Thus, the work done on the mass element by forces other than gravity is $W_{other} = (P_2 - P_1)V$. With this expression for W_{other}, the work–energy theorem becomes

$$W_{other} = (P_2 - P_1)V = (\tfrac{1}{2}mv_1{}^2 + mgy_1) - (\tfrac{1}{2}mv_2{}^2 + mgy_2)$$

By dividing both sides of this result by the volume V, recognizing that m/V is the density ρ of the fluid, and rearranging the terms, we obtain Bernoulli's equation.

BERNOULLI'S EQUATION

For any two points (1 and 2) in the steady, irrotational flow of a nonviscous, incompressible fluid, the pressure P, the fluid speed v, and elevation y are related by

$$P_1 + \tfrac{1}{2}\rho v_1{}^2 + \rho g y_1 = P_2 + \tfrac{1}{2}\rho v_2{}^2 + \rho g y_2 \qquad (13.5)$$

Bernoulli's equation is a direct consequence of the work–energy theorem. Furthermore, the derivation makes it clear that the equation is applicable only if the flow is nonviscous, so that viscous losses are absent. Since the points 1 and 2 were selected arbitrarily, the term $P + \tfrac{1}{2}\rho v^2 + \rho g y$ has a constant value at all positions in the pipe. For this reason, Bernoulli's equation is sometimes expressed as $P + \tfrac{1}{2}\rho v^2 + \rho g y =$ constant.

Bernoulli's equation can be regarded as an extension of the earlier result $P_2 = P_1 + \rho g h$, which specifies how the pressure varies with depth in a static fluid. The terms $\tfrac{1}{2}\rho v_1{}^2$ and $\tfrac{1}{2}\rho v_2{}^2$ in Bernoulli's equation account for the effects that the different fluid speeds v_1 and v_2 have on the pressure at different points. Note, however, that Bernoulli's equation reduces to the result for static fluids when the speed of the fluid is the same everywhere ($v_1 = v_2$), as it is when the cross-sectional area remains constant ($A_1 = A_2$). Under such conditions, Bernoulli's equation is

$$P_1 + \rho g y_1 = P_2 + \rho g y_2$$

After rearrangement, this result becomes

$$P_2 = P_1 + \rho g(y_1 - y_2) = P_1 + \rho g h$$

which is the familiar result for static fluids.

13.4 APPLICATIONS OF BERNOULLI'S EQUATION

When a moving fluid is contained in a pipe, it is a straightforward matter to apply Bernoulli's equation. For example, when the pipe is horizontal, all parts of it have the same elevation ($y_1 = y_2$), and Bernoulli's equation simplifies to

$$P_1 + \tfrac{1}{2}\rho v_1{}^2 = P_2 + \tfrac{1}{2}\rho v_2{}^2 \tag{13.6}$$

This equation indicates that the quantity $P + \tfrac{1}{2}\rho v^2$ remains constant throughout a horizontal pipe; if v decreases, P increases and vice versa. This is exactly the result that we deduced qualitatively from Newton's second law at the beginning of Section 13.3. Example 3 illustrates the use of Equation 13.6.

EXAMPLE 3

An aneurysm is an abnormal enlargement of a blood vessel such as the aorta. Suppose that, because of an aneurysm, the cross-sectional area A_1 of the aorta increases to a value $A_2 = 1.7A_1$. The speed of the blood ($\rho = 1060$ kg/m³) through a normal portion of the aorta is $v_1 = 0.40$ m/s. Assuming the aorta is horizontal (the person is lying down), determine the amount by which the pressure in the enlarged region exceeds that in the normal region.

SOLUTION

The pressure is greater in the enlarged region, since the speed v_2 of the blood is smaller there. According to the equation of continuity ($A_1 v_1 = A_2 v_2$),

$$v_2 = \left(\frac{A_1}{A_2}\right) v_1 = \left(\frac{A_1}{1.7A_1}\right)(0.40 \text{ m/s}) = 0.24 \text{ m/s}$$

That this smaller speed leads to a higher pressure P_2 can be seen from Bernoulli's equation for horizontal flow (Equation 13.6):

$$P_1 + \tfrac{1}{2}\rho v_1{}^2 = P_2 + \tfrac{1}{2}\rho v_2{}^2$$
$$P_2 - P_1 = \tfrac{1}{2}\rho(v_1{}^2 - v_2{}^2)$$
$$= \tfrac{1}{2}(1060 \text{ kg/m}^3)[(0.40 \text{ m/s})^2 - (0.24 \text{ m/s})^2] = \boxed{54 \text{ Pa}}$$

The excess pressure puts added stress on the already weakened tissue of the arterial wall at the aneurysm.

A Venturi meter is a device for measuring the speed of a fluid. Suppose, for instance, that it is necessary to determine the speed of a fluid in a horizontal pipe. The speed can be obtained with a Venturi meter that is substituted for a section of the pipe, as Figure 13.11 illustrates. As the fluid moves into the narrow part of the

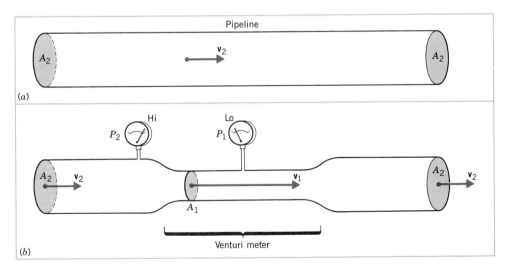

Pipeline

(a)

(b)

FIGURE 13.11 A Venturi meter can be used to measure the fluid speed v_2 in a gas pipeline.

Venturi meter, the speed of the fluid increases from v_2 to v_1, and the pressure of the fluid decreases from P_2 to P_1. From a measurement of these pressures and a knowledge of the cross-sectional areas of the pipe and the Venturi tube, the fluid speed (and volume flow rate) can be determined with the aid of Bernoulli's equation, as the next example shows.

EXAMPLE 4

In Figure 13.11a, gas is flowing through a horizontal pipe whose cross-sectional area is $A_2 = 0.0700$ m². The gas has a density of $\rho = 1.30$ kg/m³. In part b, a Venturi meter is substituted for a section of the pipe. The pressure difference measured with the Venturi tube (cross-sectional area $A_1 = 0.0500$ m²) is $P_2 - P_1 = 120$ Pa. Find (a) the speed v_2 of the gas in the pipe and (b) the volume flow rate Q of the gas.

$$v_1 = \left(\frac{A_2}{A_1}\right) v_2 = \left(\frac{0.0700 \text{ m}^2}{0.0500 \text{ m}^2}\right) v_2 = 1.40 v_2 \quad (13.3)$$

$$P_1 + \tfrac{1}{2}\rho v_1^2 = P_2 + \tfrac{1}{2}\rho v_2^2 \quad (13.6)$$

$$P_2 - P_1 = \tfrac{1}{2}\rho(v_1^2 - v_2^2) = \tfrac{1}{2}\rho[(1.40 v_2)^2 - v_2^2] = \tfrac{1}{2}\rho(0.96 v_2^2)$$

$$v_2 = \sqrt{\frac{2(P_2 - P_1)}{\rho(0.96)}} = \sqrt{\frac{2(120 \text{ Pa})}{(1.30 \text{ kg/m}^3)(0.96)}} = \boxed{14 \text{ m/s}}$$

SOLUTION

(a) The known pressure difference can be used with Bernoulli's equation for horizontal flow to determine the speed v_2 of the gas in the pipe. This speed and the speed v_1 of the gas in the Venturi tube are related by the equation of continuity. Assuming the gas in the Venturi meter is not compressed to any significant extent, we have

(b) The volume flow rate is

$$Q = A_2 v_2 = (0.0700 \text{ m}^2)(14 \text{ m/s}) = \boxed{0.98 \text{ m}^3/\text{s}} \quad (13.4)$$

In dealing with fluid flow in pipes that have both a change in elevation and a change in cross-sectional area, it is important to use the complete form of Bernoulli's equation (Equation 13.5). The complete form is required, because the effects on the pressure of flow speed and elevation can either offset or reinforce one another, as Figure 13.12 suggests. The effects depend on how the wide and narrow regions of the pipe are arranged with respect to elevation.

Figure 13.13 illustrates some instances where Bernoulli's equation can be used, even though the fluids are not flowing through actual pipes. Each of these instances occurs because higher fluid speeds are associated with lower pressures. Try blowing across the upper surface of a piece of paper, as in part a of the drawing. According to Bernoulli's equation, the pressure is lower in the rapidly flowing air above the paper than it is in the stationary air beneath. Therefore, the paper is pushed upward. In a similar vein, have you ever watched a tarpaulin-covered semitrailer moving down the highway and noticed that the canvas bulges upward, as in part b of the figure? The bulging is caused by the reduced pressure in the air that streams by the top surface of the tarpaulin, compared to the higher pressure of the stationary air inside the semi.

One of the most spectacular examples of the Bernoulli effect is the dynamic lift on airplane wings. Figure 13.14a shows a wing (in cross section) moving to the right, with the air flowing past the wing to the left. Because of the shape of the wing, the air travels faster over the curved upper surface than it does over the flatter lower surface. According to Bernoulli's equation, the pressure above the wing is low (faster moving air), while the pressure below the wing is high (slower moving air). Thus, the wing is lifted upward. Part b of the figure shows the wing of a jet airplane. Part c shows a race car that uses an inverted "wing," so that the inverted "lift" force pushes the car into the ground, thereby helping to improve stability at high speeds.

The curve ball, one of the main weapons in the arsenal of a baseball pitcher, is another illustration of the effects of fluid flow. Figure 13.15a shows a baseball moving to the right with no spin. *The view is from above*, looking down toward the ground. In

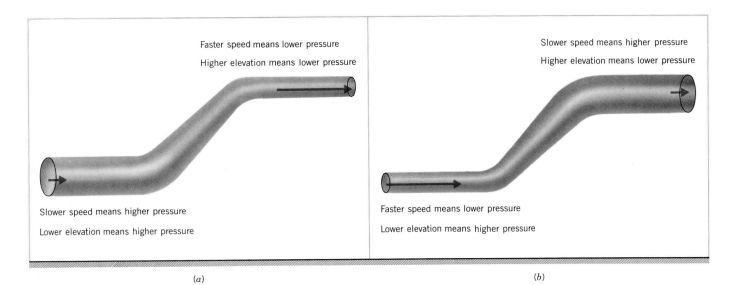

Faster speed means lower pressure

Higher elevation means lower pressure

Slower speed means higher pressure

Lower elevation means higher pressure

(a)

Slower speed means higher pressure

Higher elevation means lower pressure

Faster speed means lower pressure

Lower elevation means higher pressure

(b)

FIGURE 13.12 The complete form of Bernoulli's equation (Equation 13.5) must be used when a fluid flows in a pipe that has both a change in elevation and a change in cross-sectional area. As parts *a* and *b* of this figure summarize, the effects on the pressure of fluid speed and elevation can either offset or reinforce one another. The colored arrows indicate the fluid velocity vectors.

this situation, air flows with the same speed around both sides of the ball, and the pressure is the same on both sides. No net force exists to make the ball curve. However, when the ball is given a spin, the air close to its surface is dragged around with it, with the result that the air on one half of the ball is speeded up (lower pressure), and that on the other half is slowed down (higher pressure). Part *b* of the picture illustrates the effects of a counterclockwise spin. The baseball experiences a net deflection force (like the lift force on an airplane wing) and curves on its way from the pitcher's mound to the plate, as part *c* shows.

As a final application of Bernoulli's equation, Figure 13.16*a* shows a large tank from which water is emerging through a small pipe near the bottom. Bernoulli's

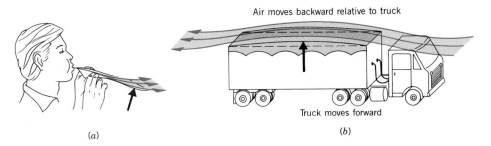

Air moves backward relative to truck

Truck moves forward

(a)   (b)

FIGURE 13.13 (*a*) When air is blown across the top surface of the paper, the paper rises, because the air pressure on the top surface is lower than that on the bottom surface. (*b*) When a semitrailer speeds down the highway, the air streaming over the top of the tarpaulin cover has a lower pressure than that inside the cargo area, so the tarpaulin bulges upward.

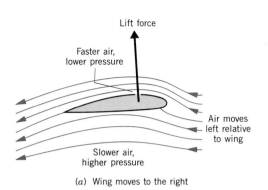

Lift force

Faster air,
lower pressure

Air moves
left relative
to wing

Slower air,
higher pressure

(a) Wing moves to the right

FIGURE 13.14 (a) Streamlines of air flow around an airplane wing. (b) An airplane. (c) A race car.

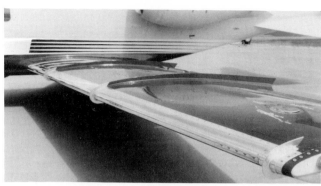

(b)

(c)

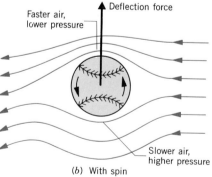

Air moves
left relative
to ball

(a) Ball moves to the right without spin

Deflection force

Faster air,
lower pressure

Slower air,
higher pressure

(b) With spin

FIGURE 13.15 These views of a baseball are from above, looking down toward the ground. (a) With no spin, the moving baseball does not curve to either side. (b) A spinning baseball, however, experiences a net force due to Bernoulli's effect and curves to one side or the other, depending on the direction of the spin. (c) The spin in part (b) causes the ball to follow the curved path shown here.

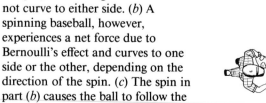

Spinning ball

(c)

equation can be used to determine the speed (called the efflux speed) at which the water leaves the pipe. In the drawing, point 2 is at one end of a tube of flow that begins at the top surface of the water. Point 1 is just outside the efflux tube, at the other end of the tube of flow. The pressure at both points is atmospheric pressure, so that $P_1 = P_2$, and Bernoulli's equation reduces to

$$\tfrac{1}{2}\rho v_1{}^2 + \rho g y_1 = \tfrac{1}{2}\rho v_2{}^2 + \rho g y_2$$

The density ρ can be eliminated algebraically from this result, which can then be solved for the square of the efflux speed v_1:

$$v_1{}^2 = v_2{}^2 + 2g(y_2 - y_1) = v_2{}^2 + 2gh$$

where we have substituted $h = y_2 - y_1$ for the height of the liquid above the efflux tube. If the tank is very large, the water level changes only slowly and the speed at point 2 can be set equal to zero.

The speed of the liquid coming out of the tank is exactly the same as if the liquid had freely fallen through a height h (see Equation 2.9 with $s = h$ and $a = g$), a result that is known as ***Torricelli's theorem.*** If the outlet pipe were pointed directly upward, as in part b of the drawing, the liquid would rise to a height h equal to the fluid level above the pipe. However, if the liquid is not an ideal fluid, its viscosity cannot be neglected. Then, the efflux speed would be less than that given by Bernoulli's equation, and the liquid would rise to a height less than h.

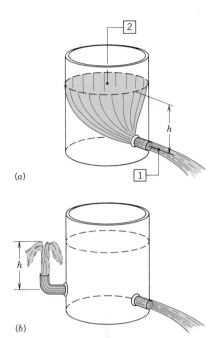

(a)

(b)

FIGURE 13.16 (a) Bernoulli's equation can be used to determine how fast the water is moving as the water leaves the small pipe. (b) If the liquid behaves as an ideal fluid (no viscosity), the liquid will rise a distance h after leaving the vertical nozzle.

*13.5 VISCOUS FLOW

VISCOSITY

In an ideal fluid there is no viscosity to hinder the fluid layers as they slide past one another. Thus, in a pipe of uniform cross section, every layer of an ideal fluid moves with the same velocity, even the layer next to the wall, as Figure 13.17a shows. In a real fluid, where viscosity is present, the fluid layers do not all have the same velocity, as part b of the drawing illustrates. The fluid closest to the wall does not move at all, while the fluid at the center of the pipe has a maximum velocity. The fluid layer next to the wall surface does not move, because it is held tightly by intermolecular forces. So strong are these intermolecular forces, in fact, that if a solid surface moves, the adjacent fluid layer moves right along with it and remains at rest *relative* to the moving surface. That is why a fine layer of dust remains on a car even at high driving speeds. The layer of air in immediate contact with the car has no velocity relative to the car, and, thus, does not blow off the dust.

To help introduce viscosity in a quantitative fashion, Figure 13.18a shows the flow of a viscous fluid between two parallel plates. The top plate is free to move while the bottom one is stationary. If the top plate is to move with a velocity **v** relative to the bottom plate, a force **F** is required. For a highly viscous fluid, like thick honey, a large force is needed; for a less viscous fluid, like water, a correspondingly smaller force is necessary. As part b of the drawing suggests, we may imagine the fluid to be composed of many thin horizontal layers. When the top plate moves, the intermediate fluid layers slide over each other. The velocity of each layer is different, changing uniformly from one layer to another, from **v** at the top plate to zero at the bottom plate.

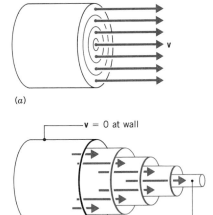

(a)

(b)

FIGURE 13.17 (*a*) In ideal (non-viscous) fluid flow, all fluid particles across the pipe have the same velocity. (*b*) In viscous flow, the speed of the fluid is zero at the surface of the pipe and gradually increases until it is a maximum along the center axis.

The resulting flow is called *laminar flow,* since a thin layer is often referred to as a lamina. As each layer moves, it is subjected to viscous forces from its neighbors, and the purpose of the force **F** is to compensate for the effect of these forces, so that any layer can move with a constant velocity.

The exact amount of force required in Figure 13.18 depends on several factors. The force is proportional to the area A of contact between the moving plate and the fluid ($F \propto A$); larger areas are in contact with more fluid and require larger forces. For a given area, the force is proportional to the speed ($F \propto v$), with the result that a larger force is required to achieve a greater speed. The force is also inversely proportional to the perpendicular distance y between the top and bottom plates ($F \propto 1/y$). The larger the distance y, the smaller is the force required to achieve a given speed with a given contact area. These three proportionalities can be expressed together in the following manner: $F \propto Av/y$. Equation 13.7 expresses this relationship with the aid of a proportionality constant η (Greek letter "eta"), which is called the *coefficient of viscosity* or simply the *viscosity.*

FORCE NEEDED TO MOVE A LAYER OF VISCOUS FLUID WITH A CONSTANT VELOCITY

The tangential force **F** required to move a fluid layer at a constant speed v, when the layer has an area A and is located a perpendicular distance y from an immobile surface, is given by

$$F = \frac{\eta A v}{y} \qquad (13.7)$$

where η is the coefficient of viscosity.

SI unit of viscosity: Pa·s

Common unit of viscosity: poise (P)

By solving this equation for the viscosity, $\eta = Fy/(vA)$, it can be seen that the SI unit for viscosity is $\text{N} \cdot \text{m}/[(\text{m/s}) \cdot \text{m}^2] = \text{Pa} \cdot \text{s}$. Another common unit for viscosity is the *poise* (P), which is used in the cgs system of units and is named after the French

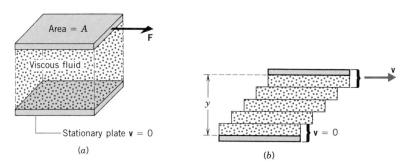

(a)

(b)

FIGURE 13.18 (*a*) A force **F** is applied to the top plate. The volume between the two plates is filled with a viscous fluid. (*b*) Under the application of the force **F**, the top plate and the adjacent layer of fluid move with a constant velocity **v**. The layers of fluid slide over one another, with the bottom layer having zero velocity.

TABLE 13.1 Coefficients of Viscosity of Common Fluids

Fluid	Temperature (°C)	Viscosity η (Pa·s)
Gases		
Air	0	0.0171×10^{-3}
	20	0.0182×10^{-3}
	40	0.0193×10^{-3}
Carbon dioxide	20	0.0147×10^{-3}
Helium	20	0.0196×10^{-3}
Liquids		
Whole blood	37	4×10^{-3}
Glycerine	20	1500×10^{-3}
Methanol	20	0.584×10^{-3}
Water	0	1.78×10^{-3}
	20	1.00×10^{-3}
	40	0.651×10^{-3}

physician Jean Poiseuille (1797–1869, pronounced, approximately, as Pwah-zoy'). The following relation exists between the two units:

$$1 \text{ poise (P)} = 0.1 \text{ Pa·s}$$

Values of viscosity depend on the nature of the fluid, and Table 13.1 lists data for some liquids and gases. Under ordinary conditions, the viscosities of gases are significantly *smaller* than those of liquids. Moreover, the viscosities of either liquids or gases depend markedly on temperature. Usually, the viscosities of liquids decrease as the temperature is increased. Anyone who has heated honey or oil, for example, knows that it flows much more freely at an elevated temperature. In contrast, the viscosities of gases increase as the temperature is raised. In general, fluids having smaller values of η are more nearly ideal fluids, because they flow more readily with only relatively weak viscous forces impeding their movement; an ideal fluid has $\eta = 0$.

POISEUILLE'S LAW

Viscous flow occurs in a wide variety of situations, such as oil moving through a pipeline, a liquid being forced through the needle of a hypodermic syringe, or blood moving in the human circulatory system. In cases like these, it is important to know what determines the volume flow rate Q (m³/s) of the fluid. Figure 13.19 identifies the significant factors.

First, a difference in pressure $P_2 - P_1$ must be maintained between any two locations along the pipe in order for the fluid to flow. In fact, Q is proportional to $P_2 - P_1$, a greater pressure difference leading to a larger flow rate. Second, a long pipe offers greater resistance to the flow than a short pipe does, and Q is inversely proportional to the length L. Because of this fact, long pipelines, such as the Alaskan pipeline, have pumping stations at various places along the line to compensate for a drop in pressure. Third, high-viscosity fluids flow less readily than low viscosity fluids do, and Q is inversely proportional to the viscosity η. Finally, the volume flow rate is larger in a pipe of larger radius, other things being equal. The dependence on the radius R is a surprising one, Q being proportional to R^4. If, for instance, the pipe radius is reduced to one half of its original value, the volume flow rate is reduced to one-sixteenth of its original value, assuming the other variables remain constant. The mathematical

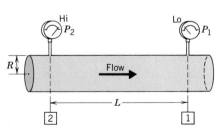

FIGURE 13.19 For viscous flow, the factors that influence the volume flow rate Q in the tube are the difference in pressure $P_2 - P_1$, the radius R and length L of the tube, as well as the viscosity η of the fluid.

relation for Q in terms of $P_2 - P_1$, L, η, and R was discovered by Poiseuille, and it is known as **Poiseuille's law:**

$$Q = \frac{\pi R^4 (P_2 - P_1)}{8 \eta L} \tag{13.8}$$

Example 5 illustrates the use of Poiseuille's law.

EXAMPLE 5

A hypodermic syringe is filled with a solution whose viscosity is 1.5×10^{-3} Pa·s. As Figure 13.20 shows, the plunger area of the syringe is 8.0×10^{-5} m², and the length of the needle is 0.025 m. The internal radius of the needle is 4.0×10^{-4} m. The gauge pressure in a vein is 1900 Pa (14 mm of mercury). What force must be applied to the plunger, so that 1.0×10^{-6} m³ of solution can be injected in 3.0 s?

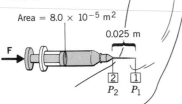

Area = 8.0×10^{-5} m²

0.025 m

F

2   1
P_2 P_1

FIGURE 13.20 The difference in pressure $P_2 - P_1$ required to sustain the fluid flow through a hypodermic needle can be found with the aid of Poiseuille's law.

SOLUTION

According to Poiseuille's law, the volume flow rate is proportional to the fourth power of the radius. Because the needle has a much smaller radius than the syringe barrel, the time needed to

empty the syringe is determined primarily by the volume flow rate through the needle. Applying Poiseuille's equation to the needle, we can find the difference in pressure $P_2 - P_1$ (see drawing) required to sustain a flow rate of $Q = (1.0 \times 10^{-6}$ m³)/$(3.0$ s$) = 3.3 \times 10^{-7}$ m³/s:

$$Q = \frac{\pi R^4 (P_2 - P_1)}{8 \eta L} \tag{13.8}$$

$$P_2 - P_1 = \frac{8 \eta L Q}{\pi R^4}$$

$$= \frac{8(1.5 \times 10^{-3}\ \text{Pa·s})(0.025\ \text{m})(3.3 \times 10^{-7}\ \text{m}^3/\text{s})}{\pi (4.0 \times 10^{-4}\ \text{m})^4} = 1200\ \text{Pa}$$

Since $P_1 = 1900$ Pa, the pressure P_2 must be $P_2 = 1200$ Pa + 1900 Pa = 3100 Pa. This pressure is nearly equal to the pressure at the plunger, because the barrel of the syringe is so large that relatively little pressure difference is required to sustain the flow up to point 2, where the fluid encounters the narrow needle. Since pressure is force per unit area, the force that must be applied to the plunger is the pressure times the plunger area:

$$F = (3100\ \text{Pa})(8.0 \times 10^{-5}\ \text{m}^2) = \boxed{0.25\ \text{N}}$$

SUMMARY

The **mass flow rate** (kg/s) of a fluid with a density ρ, flowing with a speed v in a pipe of cross-sectional area A, is the mass per second flowing past a point and is given by $\rho A v$. The **equation of continuity** expresses the fact that mass is conserved; what flows into one end of a section of pipe must flow out the other end, assuming there are no additional entry or exit points in between. In terms of the mass flow rate, the equation of continuity is expressed as $\rho_1 A_1 v_1 = \rho_2 A_2 v_2$, where the subscripts 1 and 2 denote two points along the pipe. If the fluid is incompressible, $\rho_1 = \rho_2$. The equation of continuity then becomes $A_1 v_1 = A_2 v_2$, where the product of the cross-sectional area and speed is called the **volume flow rate** Q; $Q = Av$.

Bernoulli's equation is a direct consequence of the work–energy theorem and describes the steady irrotational flow of an ideal fluid whose density is ρ. For any two points (1 and 2) in the fluid, this equation relates the pressure P, the speed v, and the elevation y: $P_1 + \frac{1}{2}\rho v_1^2 + \rho g y_1 = P_2 + \frac{1}{2}\rho v_2^2 + \rho g y_2$. When the flow is horizontal, Bernoulli's equation indicates that high-speed regions are associated with low fluid pressure, while low-speed regions are associated with high fluid pressure.

The **coefficient of viscosity** η is the proportionality constant that determines how much tangential force $\mathbf{F}$ is required to move a fluid layer at a constant speed v, when the layer has an area A and is located a perpendicular distance y

from an immobile surface. The magnitude of the force is $F = \eta A v / y$. The SI unit of viscosity is Pa·s. To make a viscous fluid flow from location 2 to location 1 along a pipe of radius R and length L, the pressure at location 2 must exceed that at location 1. **Poiseuille's law** gives the volume flow rate Q that results from such a pressure difference $P_2 - P_1$: $Q = \pi R^4 (P_2 - P_1)/(8\eta L)$. Bernoulli's equation does not apply to viscous flow.

QUESTIONS

1. In steady flow, the velocity **v** of a fluid particle at any point is constant in time. On the other hand, a fluid accelerates when it moves into a region of smaller cross-sectional area. (a) Explain what causes the acceleration. (b) Explain why the condition of steady flow does not preclude such acceleration.

2. The cross-sectional area of a stream of water becomes smaller as the water falls from a faucet. Account for this phenomenon in terms of the equation of continuity. What would you expect to happen to the cross-sectional area when the water is shot upward, as it is in a fountain?

3. Suppose you are driving your car along side a moving truck. Using your knowledge of the relation between air speed and pressure, determine whether the truck and car will tend to pull together or push apart. Give your reasoning.

4. Can Bernoulli's equation, $P + \frac{1}{2}\rho v^2 + \rho g y = $ constant, be used to describe the flow of water that is cascading down a rock-strewn spillway? Explain.

5. Hold two sheets of paper loosely together, one on top of the other, and blow air between them. Instead of being blown further apart, the two sheets will come together. Discuss this phenomenon in terms of Bernoulli's equation.

6. Which way would you have to spin a baseball, so that it curves upward on its way to the plate? In describing the spin, state how you are viewing the ball. Justify your answer.

7. A passenger is smoking in the backseat of a car that is traveling at 60 mph. To remove the smoke, the driver opens a window just a bit. Explain why the smoke is drawn to and out of the driver's window.

8. The same liquid is flowing through each of the pipes in the drawing. The pressure difference between the ends of each pipe is the same. Using Poiseuille's law, rank the pipes in descending order, according to the volume flow rate of the fluid (largest rate first).

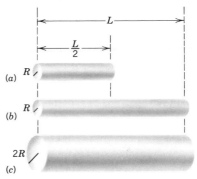

9. To change the oil in a car, you remove a plug beneath the engine and let the old oil run out. Your car has been sitting in the garage on a cold day. Before changing the oil, it is advisable to run the engine for a while. What is the reason for this?

PROBLEMS

Section 13.2 The Equation of Continuity

1. Suppose that blood flows through the aorta with a speed of 0.35 m/s. The cross-sectional area of the aorta is 2.0×10^{-4} m². (a) Find the volume flow rate and the mass flow rate of the blood. (b) The aorta branches into tens of thousands of capillaries whose total cross-sectional area is about 0.28 m². What is the average blood speed through them?

2. Three fire hoses are connected to a fire hydrant. Each hose has a radius of 2.0 cm. Water enters the hydrant through an underground pipe of radius 8.0 cm. In this pipe the water has a speed of 3.0 m/s. (a) Find the water speed in each hose. (b) How many kilograms of water are poured on a fire in one hour?

3. Oil is flowing with a speed of 4.00 ft/s through a 2.00-ft-diameter pipeline. How many gallons of oil (1 gal = 0.134 ft³) flow through this pipeline in one day?

4. Water flows with a volume flow rate of 1.50 m³/s in a pipe.

Find the speed of the water at a point where the pipe diameter is (a) 0.800 m and (b) 1.00 m.

Section 13.3 Bernoulli's Equation, Section 13.4 Applications of Bernoulli's Equation

5. Suppose that a 15-m/s wind is blowing across the roof of your house. The density of air is 1.29 kg/m³. (a) Determine the reduction in pressure (below atmospheric pressure of stationary air) that accompanies this wind. (b) Explain why some roofs are literally "blown outward" when the wind speed is high.

6. A small crack occurs at the base of a 50.0-ft-high dam. The effective crack area through which water leaves is 1.00×10^{-2} ft². (a) Ignoring viscous losses, what is the speed of the water flowing through the crack? (b) How many cubic feet of water leave the dam per second?

7. Water is flowing at a speed of 0.500 m/s through a 4.00-cm-diameter hose. The hose is horizontal. (a) What is the mass flow rate of the water? (b) At what speed does the water pass through a nozzle whose effective diameter is 0.600 cm? (c) What must be the absolute pressure of the water entering the hose if the pressure at the nozzle is atmospheric pressure?

8. A fountain sends a stream of water 5.00 m into the air. (a) Neglecting air resistance and any viscous effects, what must be the speed of the water at the point where the water leaves the pipe feeding the fountain? (b) The effective cross-sectional area of the pipe is 5.00×10^{-4} m². How many gallons per minute are being used by the fountain? (*Note:* 1 gal $= 3.79 \times 10^{-3}$ m³)

9. Water flows downward within a pipe of constant cross-sectional area. At a point somewhere down the pipe, the speed of the water is 7.00 m/s and the absolute pressure is 1.50×10^5 Pa. At a height of 3.00 m above this point, find (a) the speed of the water and (b) the absolute pressure.

* 10. An airplane wing is designed so that the speed of the air across the top of the wing is 248 m/s when the speed of the air below the wing is 225 m/s. The density of the air is 1.29 kg/m³. What is the lifting force on a wing of area 20.0 m²?

* 11. A water rocket is partially filled with water, as the drawing illustrates. The region above the water is filled with compressed air from a hand pump. The gauge pressure of the compressed air is 3.0×10^5 Pa. Neglecting the variation in pressure with water height and assuming that the area A_2 is much larger than A_1, find the speed v_1 at which the water leaves the nozzle. See drawing 11.

* 12. In a closed tank, the gauge pressure of the air above the water is 5.00×10^5 Pa. The water leaves the bottom of the tank through a nozzle that is directed straight upward. The opening of the nozzle is 4.00 m below the surface of the water. (a) Find the speed at which the water leaves the nozzle. (b) Ignoring air resistance and viscous effects, determine the height to which the water rises.

**13. Two circular holes, one larger than the other, are cut in the side of a large water tank whose top is open to the atmosphere. The center of one of these holes is located twice as far beneath the

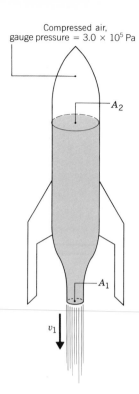

Compressed air,
gauge pressure $= 3.0 \times 10^5$ Pa

A_2

A_1

v_1

Drawing 11

surface of the water as the other. The volume flow rate of the water coming out of the holes is the same. (a) Decide which hole is located where. (b) Calculate the ratio of the radii of the holes.

**14. A liquid is flowing through a horizontal pipe whose radius is 2.00 cm. The pipe bends straight upward through a height of 10.0 m and joins another horizontal pipe whose radius is 4.00 cm. What volume flow rate will keep the pressures in the two horizontal pipes the same.

**15. A siphon tube is useful for removing liquid from a tank. The siphon tube is first filled with liquid, and then one end is inserted into the tank. Liquid then drains out the other end, as the drawing illustrates. (a) Using reasoning similar to that employed in obtaining Torricelli's theorem, derive an expression for the speed v of the fluid emerging from the tube. This expression gives v in terms of the vertical height y and the acceleration due to gravity g. (Note that this speed does not depend on the depth d of the tube below the surface of the liquid.) (b) At what value of the vertical distance y will the siphon stop working? (c) Derive an expression for the absolute pressure at the highest point in the siphon (point A) in terms of the atmospheric pressure P_0, the fluid density ρ, g, and the heights h and y. (Note that the fluid speed at point A is the same as the speed of the fluid emerging from the tube, because the cross-sectional area of the tube is the same everywhere.) See drawing 15.

Section 13.5 Viscous Flow

16. A blood vessel is 0.10 m in length and has a radius of 1.5×10^{-3} m. Blood flows at a rate of 1.0×10^{-7} m³/s through

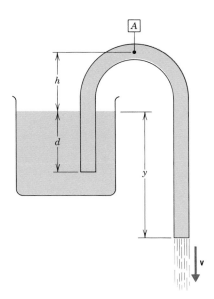

Drawing 15

this vessel. Determine the difference in pressure that must be maintained between the two ends of the vessel.

17. Poiseuille's law remains valid so long as the fluid flow is laminar. For sufficiently high speed, however, the flow becomes turbulent, even if the fluid is moving through a smooth pipe with no restrictions. It is found experimentally that the flow is laminar as long as the Reynolds number Re is less than about 2000:

$$\text{Re} = \frac{2\bar{v}\rho R}{\eta}$$

Here $\bar{v}$, ρ, and η are, respectively, the average speed, density, and viscosity of the fluid, and R is the radius of the pipe. (a) Calculate the highest average speed that blood ($\rho = 1060$ kg/m^3) could have and still remain in laminar flow when it flows through the aorta ($R = 8.0 \times 10^{-3}$ m). (b) When compared to the flow of water, why can the flow of blood attain a higher average speed and still remain laminar?

18. When an object moves through a fluid, as when a ball falls through air or a glass sphere falls through water, the fluid exerts a viscous force **F** on the object that tends to slow it down. For a small sphere of radius R, moving slowly with a speed v, the magnitude of the viscous force is given by Stoke's law, $F = 6\pi\eta Rv$, where η is the viscosity of the fluid. (a) What is the viscous force on a glass sphere of radius $R = 1.0$ mm falling through water ($\eta = 1.00 \times 10^{-3}$ Pa·s) when the sphere has a speed of 3.0 m/s? (b) The speed of the falling sphere increases until the viscous force balances the weight of the sphere. Thereafter, no net force acts on the sphere, and it falls with a constant speed called the terminal speed. If the sphere has a mass of 1.0×10^{-5} kg, what is its terminal speed?

* 19. Two hoses are connected to the same outlet by means of a

Y-connector. One hose has a diameter that is 1.50 times larger than the other, but each has the same length. Find the ratio of the average speed of the water in the larger hose to that in the smaller hose.

* 20. A certain volume of water ($\eta = 1.00 \times 10^{-3}$ Pa·s) is observed to flow through a tube in 115 s. The same volume of another liquid flows through the same tube, under the same conditions, in 175 s. What is the viscosity of the liquid?

ADDITIONAL PROBLEMS

21. The large water tower in the drawing is drained by a pipe that extends to the ground. Assume that the flow is nonviscous. (a) What is the absolute pressure at point 1 if the valve is *closed,* assuming that the top surface of the water at point 2 is at atmospheric pressure? (b) What is the absolute pressure at point 1 when the valve is opened and the water is flowing? (c) Assuming the effective cross-sectional area of the valve opening is 2.00×10^{-2} m^2, find the volume flow rate at point 1.

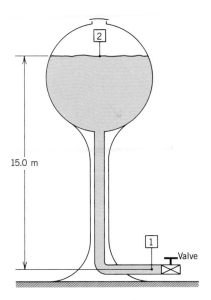

15.0 m

22. A water line with an internal diameter of 1.3 cm is connected to a shower head that has 12 holes. The speed of the water in the line is 1.2 m/s. (a) What is the volume flow rate (in m^3/s) in the line? (b) What is the mass flow rate? (c) At what speed does the water leave one of the holes (effective hole diameter = 0.091 cm) in the head?

23. During a heavy rain, a 3.0 m $\times$ 4.6 m family room is flooded to a depth of 15 cm. To remove the water ($\eta = 1.00 \times 10^{-3}$ Pa·s), a pump is used that does the job in two hours. The water flows through a horizontal pipe of radius 0.64 cm and length 6.7 m. What gauge pressure does the pump produce?

* **24.** A pump draws water into a horizontal pipe located 12 m beneath the surface of a reservoir. The speed of the water in the pipe causes the pressure in the pipe to decrease, in accord with Bernoulli's principle. Assuming nonviscous flow, what is the maximum speed with which water can flow through the intake pipe?

**25. The air speed of a plane can be measured with a Pitot-static tube, an example of which is illustrated in the drawing. The Pitot-static tube consists of two concentric tubes: the inner one is the static tube, and the outer one is the Pitot tube. The difference in air pressure between the two is measured by the U-tube manometer in the drawing. The air speed inside the static tube is zero, because the closed tube presents an immovable obstacle to the flow of air. In contrast, the air rushing past the holes in the Pitot tube has a high speed. (a) By applying Bernoulli's equation, show that the speed of the air is $v = \sqrt{2(P_2 - P_1)/\rho}$, where $P_2 - P_1$ is the difference in pressure and ρ is the air density at the altitude of the plane. (b) By expressing $P_2 - P_1$ in terms of the height h and density ρ_0 of the fluid in the U-tube, show that the air speed can be expressed as $v = \sqrt{2gh\rho_0/\rho}$.

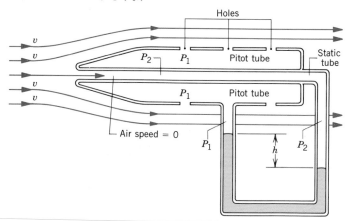

CHAPTER 14

Temperature and Thermal Expansion

Molten lava flows from Mount Helka in Reykjavík, Iceland during an eruption. The temperature of the lava is about 1000°C, considerably more than the 37°C of a normal human body.

14.1 TEMPERATURE AND THE COMMON TEMPERATURE SCALES

To measure temperature we use a thermometer. Many thermometers make use of the fact that materials usually expand when their temperatures increase. Figure 14.1 shows the familiar mercury-in-glass thermometer, which utilizes the expansion of liquid mercury to indicate the temperature. The thermometer consists of a mercury-filled glass bulb connected to a capillary tube. When the mercury is heated, it expands into the capillary, the amount of expansion being proportional to the change in temperature. The outside of the glass is marked with an appropriate scale, and the temperature is read by noting the distance that the mercury expands into the tube.

A number of different temperature scales have been devised, two popular choices being the *Celsius* (formerly, Centigrade) and *Fahrenheit scales.* Figure 14.1 illustrates these scales. Historically,* both scales were defined by assigning two temperature

* Today, the Celsius and Fahrenheit scales are defined in terms of the Kelvin temperature scale; the Kelvin scale is discussed in Section 14.2.

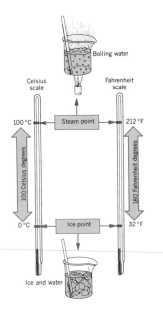

FIGURE 14.1 The Celsius and Fahrenheit temperature scales.

points on the scale arbitrarily and then dividing the distance between them into a number of equally spaced intervals. One point was chosen to be that at which ice melts under one atmosphere of pressure (the "ice point"), and the other was that at which water boils under one atmosphere of pressure (the "steam point"). On the Celsius scale, the ice point was selected to be 0 °C (0 degrees Celsius), and the steam point was chosen to be 100 °C. On the Fahrenheit scale, the ice point was defined to be 32 °F (32 degrees Fahrenheit), and the steam point was specified to be 212 °F. The Celsius scale is used worldwide, while the Fahrenheit scale is used mostly in the United States, often in home medical thermometers.

There is a subtle difference in the way the temperature of an object is reported, as compared to a *change* in its temperature. For example, the temperature of the human body is about 37 °C, where the combined symbol °C stands for "degrees Celsius." However, the *change* between two temperatures is specified in "Celsius degrees" (C°)—not in "degrees Celsius." Thus, if the body temperature rises to 39.0 °C, the change in temperature is 2 Celsius degrees or 2 C°, not 2 °C.

As Figure 14.1 indicates, the separation between the ice and steam points on the Celsius scale is divided into 100 Celsius degrees, while on the Fahrenheit scale the separation is divided into 180 Fahrenheit degrees. Therefore, the size of the Celsius degree is larger than that of the Fahrenheit degree by a factor of $\frac{180}{100}$, or $\frac{9}{5}$. Example 1 illustrates how to convert between the Celsius and Fahrenheit scales using this factor.

EXAMPLE 1

(a) A healthy person has an oral temperature of 98.6 °F. What would this reading be on the Celsius scale? (b) A "time and temperature" sign on a bank indicates the outdoor temperature is −20.0 °C. Find the corresponding temperature on the Fahrenheit scale.

SOLUTION

(a) To begin with, note that 98.6 °F is 66.6 Fahrenheit degrees above the ice point of 32.0 °F. The difference of 66.6 Fahrenheit degrees is equivalent to $\frac{5}{9}$(66.6) = 37.0 Celsius degrees.

Adding 37.0 Celsius degrees to the ice point on the Celsius scale (0 °C) gives a Celsius temperature of $\boxed{37.0\ °C}$.

(b) The temperature of −20.0 °C is 20.0 Celsius degrees *below* the ice point, and this difference corresponds to $\frac{9}{5}$(20.0) = 36.0 Fahrenheit degrees. Subtracting 36.0 Fahrenheit degrees from the ice point on the Fahrenheit scale (32.0 °F), gives a Fahrenheit temperature of $\boxed{−4.0\ °F}$.

The procedure used in Example 1 for converting between the Celsius and Fahrenheit temperature scales can be summarized as follows:

1. Determine the difference between the given temperature and the ice point on the given scale.

2. Convert this temperature difference from one scale to the other scale by using the fact that one Celsius degree is $\frac{9}{5}$ larger than one Fahrenheit degree.

3. Add or subtract the temperature difference on the new scale to or from the ice point on the new scale.

Although the Celsius and Fahrenheit scales are used in many applications, the Kelvin temperature scale has greater scientific significance, as Section 14.2 discusses.

14.2 THE KELVIN TEMPERATURE SCALE

THE SCALE ITSELF

The *Kelvin temperature scale* was introduced by the Scottish physicist William Thompson (Lord Kelvin, 1824–1907). In his honor each degree on the scale is called a kelvin (K). By international agreement, the symbol K is not written with a degree sign (°), nor is the word "degrees" used when quoting temperatures. For example, a temperature of 300 K (not 300 °K) is read as "three hundred kelvins," not "three hundred degrees kelvin." Because temperature cannot be expressed in terms of the three SI base units for mass (kilogram), length (meter), and time (second), it is necessary to define a fourth base unit for temperature measurement. The kelvin is the SI base unit for temperature.

The size of a kelvin is identical to that of a Celsius degree, for there are one hundred divisions between the ice and steam points on both scales. As we will discuss shortly, experiments with gases have shown that there exists a lowest possible temperature, below which nothing can be cooled. This lowest temperature is defined to be the zero point on the Kelvin scale, and is referred to as absolute zero. Moreover, the ice point (0 °C) occurs at 273.15 K on the Kelvin scale. Thus, the Kelvin temperature T and the Celsius temperature T_c are related by

$$T = T_c + 273.15 \qquad (14.1)$$

Figure 14.2 compares the Kelvin and Celsius scales.

THE CONSTANT-VOLUME GAS THERMOMETER

When a gas is heated, its pressure increases, and when the gas is cooled, its pressure decreases, assuming the gas is confined to a fixed volume. For example, the air pressure in automobile tires can rise by as much as 6 lb/in.² after the car has been driven a few miles and the tires become warm. The change in gas pressure with temperature is the basis of the *constant-volume gas thermometer.*

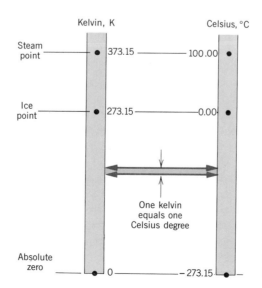

FIGURE 14.2 A comparison of the Kelvin and Celsius temperature scales. One kelvin is equal to one Celsius degree.

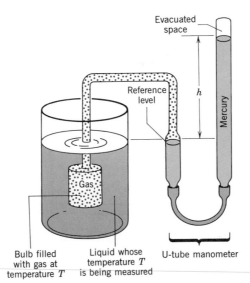

FIGURE 14.3 A constant-volume gas thermometer.

Figure 14.3 illustrates that a constant-volume gas thermometer consists of a gas-filled bulb to which a pressure gauge is attached. The gas is often hydrogen or helium at a low density, and the pressure gauge can be a U-tube manometer filled with mercury. The bulb is placed in thermal contact with the substance whose temperature is being measured. The volume of the gas is kept constant by raising or lowering the *right* column of the U-tube, so that the mercury level in the *left* column is always kept at the same reference level. A change in gas temperature causes a proportional change in gas pressure, so the height h of the mercury on the right must be changed proportionally to keep the gas volume constant; this height is used to indicate the temperature. Once the constant-volume gas thermometer has been calibrated, the thermometer can be used to measure the temperature of various substances.

ABSOLUTE ZERO

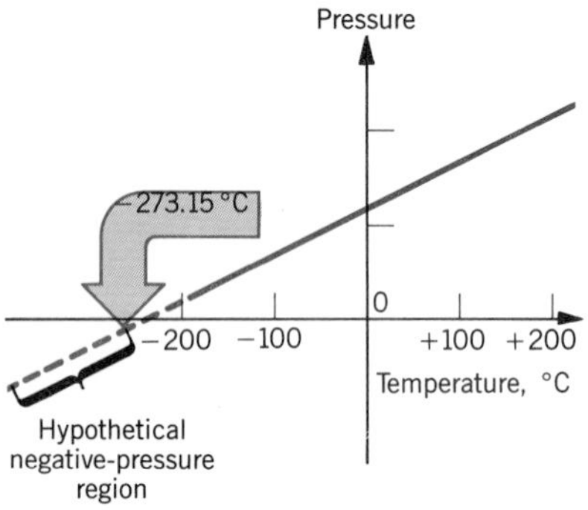

Suppose the pressure of the gas in Figure 14.3 is measured at different temperatures. If the results are plotted on a pressure versus temperature graph, a straight line is obtained, as in Figure 14.4. If the straight line is extrapolated to lower and lower temperatures, the line crosses the temperature axis at -273.15 °C. In reality, no gas can be cooled to this temperature, because all gases liquify before reaching it. However, helium and hydrogen liquify at such low temperatures that they are often used in the thermometer. This kind of graph can be obtained for different amounts of low-density gas and for different types of gas. In all cases, it is found that the straight line extrapolates back to the *same point* on the temperature axis.

The temperature of -273.15 °C is of fundamental significance, because it does not depend on the type of gas, as long as the density of the gas is sufficiently low. The value of -273.15 °C is the ***absolute zero point*** for temperature measurement, the phrase "absolute zero" meaning that temperatures lower than -273.15 °C cannot be reached by the continual cooling of a gas or any other substance.* If lower temperatures could be reached, then further extrapolation of the straight line in the graph

FIGURE 14.4 A plot of pressure versus temperature for a low-density gas at constant volume. The graph is a straight line and, when extrapolated (dashed line) to zero pressure, crosses the temperature axis at -273.15 °C.

---

* The unattainability of a temperature of absolute zero forms the basis for the third law of thermodynamics and will be discussed further in Chapter 19.

TABLE 14.1 Temperatures of Various Phenomena

Temperature (K)	Phenomenon
4.2	Helium liquifies
20	Hydrogen liquifies
77	Nitrogen liquifies
273	Water freezes
310	Human body temperature
373	Water boils
600	Lead melts
1 336	Gold melts
6 000	Surface temperature of Sun
16 000	Core temperature of Earth
10^7	Core temperature of Sun
10^9	Core temperature of hottest stars

would suggest that negative absolute gas pressures could exist. Such a situation would be impossible, because a negative absolute gas pressure has no meaning. Thus, the Kelvin temperature scale is chosen so that its point of zero temperature is the lowest temperature attainable, with negative temperatures not existing on the scale.

While the Kelvin scale establishes absolute zero as the lower limit for temperature, there is no upper limit to temperature. Table 14.1 indicates the Kelvin temperatures associated with various phenomena.

14.3 THERMOMETERS

All thermometers make use of the change in some physical property with temperature. The general name for such a property is a ***thermometric property.*** For example, the thermometric property of the mercury-in-glass thermometer is the length of the mercury column, while in the constant-volume gas thermometer the thermometric property is the pressure of the gas. Useful thermometric properties other than length and pressure also exist.

The *thermocouple* is a thermometer used extensively in scientific laboratories. The thermocouple consists of thin wires of different metals, welded together at the ends to form two junctions, as Figure 14.5 illustrates. Often the metals are copper and constantan (a copper–nickel alloy). One of the junctions, called the "hot" junction, is placed in thermal contact with the object whose temperature is being measured. The second junction, termed the "reference" junction, is kept at a known constant temperature (usually an ice–water mixture at 0 °C). The thermocouple generates a "voltage" that depends on the *difference in temperature* between the two junctions. This voltage is the thermometric property and is measured by a voltmeter, as the drawing indicates. With the aid of calibration tables or curves found in most physics handbooks, the temperature of the hot junction can be obtained from the voltage. Thermocouples are used to measure temperatures as high as 2300 °C or as low as −270 °C.

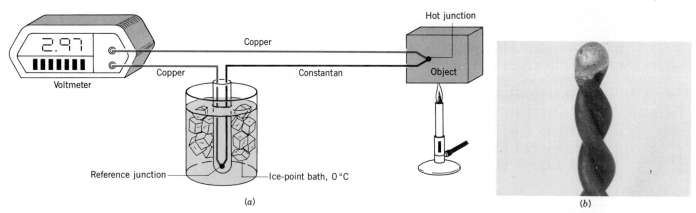

Hot junction

Copper

Copper

Constantan

Object

Voltmeter

Reference junction

Ice-point bath, 0 °C

(a)

(b)

FIGURE 14.5 (*a*) A thermocouple is made from two different types of wires, copper and constantan in this case. The "reference" junction is kept at a known temperature, such as that of an ice-point bath. The "hot" junction is placed in contact with the object whose temperature is being measured. The voltage produced by the thermocouple can be converted into a temperature with the aid of a calibration curve. (*b*) A thermocouple junction.

Most substances offer resistance to the flow of electricity. Because this electrical resistance changes with temperature, electrical resistance is another thermometric property. *Electrical resistance thermometers* are often made from platinum wire, because platinum has excellent mechanical and electrical properties in the temperature range from −270 °C to +700 °C. Since the electrical resistance of platinum wire is known as a function of temperature, the temperature of a substance can be determined by placing the resistance thermometer in thermal contact with the substance and measuring the resistance of the platinum wire.

The amount, or intensity, of radiation emitted by an object can also be used to indicate the temperature of the object. At low to moderate temperatures, the predominant type of emitted radiation is infrared radiation, sometimes called heat waves. As the temperature is raised, the intensity of the radiation increases substantially. In one interesting application, an infrared camera scans the human body and registers the intensity of the infrared radiation produced at different locations. The camera is connected to a color monitor that displays the different infrared intensities as different colors. This "thermal painting" of the body is called a *thermograph* and gives a picture of the hot areas of the body. Thermography is an important diagnostic tool in medicine. For example, breast cancer may be indicated in the thermograph by the elevated temperatures often generated by malignant tissue. The color insert between pages 300 and 301 shows typical thermographs used in medicine. The insert also illustrates how infrared aerial photography can detect heat loss from badly insulated houses and provide temperature maps for use in assessing agriculture crop damage.

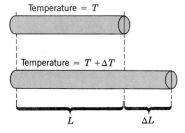

Temperature = T

Temperature = $T + \Delta T$

L ΔL

FIGURE 14.6 When the temperature of a rod is raised from T to $T + \Delta T$, the length of the rod expands from L to $L + \Delta L$.

14.4 LINEAR THERMAL EXPANSION

NORMAL SOLIDS

Most materials expand when heated and contract when cooled. The increase in any one dimension of such a solid is called a *linear expansion*, linear in the sense that the expansion occurs along a line. Figure 14.6 illustrates the linear expansion of a rod whose length is L when the temperature is T. When the temperature increases to $T + \Delta T$, the length becomes $L + \Delta L$, where ΔT and ΔL are the magnitudes of the changes in temperature and length, respectively. Conversely, when the temperature decreases to $T - \Delta T$, the length decreases to $L - \Delta L$.

The linear expansion of a solid depends on the amount of the temperature change. The greater the change in temperature, the greater the change in length. For modest temperature changes, experiments show that the change in length is directly proportional to the change in temperature ($\Delta L \propto \Delta T$). In addition, the change in length is proportional to the initial length of the rod, a fact that can be understood by considering Figure 14.7. Part a of the drawing shows two identical rods. Each rod has a length L and expands by ΔL when the temperature increases by ΔT. Part b shows the two heated rods combined into a single rod. The total expansion of this "combined" rod is the sum of the expansions of each part, namely, $\Delta L + \Delta L = 2 \Delta L$. Clearly, the amount of expansion doubles if the rod is twice as long to begin with. In other words, the change in length ΔL is directly proportional to the original length L ($\Delta L \propto L$). Equation 14.2 expresses the fact that ΔL is proportional to both L and ΔT ($\Delta L \propto L \Delta T$) by using a proportionality constant α, which is called the **coefficient of linear expansion.**

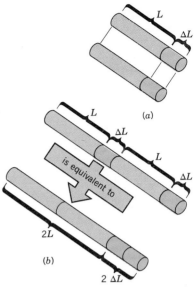

FIGURE 14.7 (*a*) Each of the two rods expands by an amount ΔL when heated. (*b*) When the two rods are combined into a single rod of length $2L$, the "combined" rod expands by $2 \Delta L$. In general, the amount of linear expansion is directly proportional to the initial length of the rod, for a given change in temperature.

LINEAR THERMAL EXPANSION OF A SOLID

The change in length ΔL that an object of initial length L experiences when its temperature is changed by an amount ΔT is given by

$$\Delta L = \alpha L \, \Delta T \tag{14.2}$$

where α is the coefficient of linear expansion.

Common unit for the coefficient of linear expansion: $\dfrac{1}{C^\circ} = (C^\circ)^{-1}$

Solving Equation 14.2 for α shows that $\alpha = \Delta L/(L \, \Delta T)$. Thus, the coefficient of linear expansion α has the unit of $(C^\circ)^{-1}$ when the temperature difference ΔT is expressed in Celsius degrees (C°), for the length units of ΔL and L cancel. Different materials with the same initial length expand and contract by different amounts as the temperature changes, so the value of α depends on the nature of the material, as Table 14.2 shows. A bar of lead expands 58 times more than a similar bar of fused quartz. Thus, lead has a value of α that is 58 times greater than that for fused quartz. Coefficients of linear expansion vary somewhat depending on the range of temperatures involved, but the values in Table 14.2 are adequate approximations for our purposes. Example 2 deals with a situation where the effect of thermal expansion can be observed, even though the change in temperature is small.

EXAMPLE 2

A concrete sidewalk is constructed between two buildings on a day when the temperature is 25 °C. The sidewalk consists of two slabs, each three meters in length and of negligible thickness (Figure 14.8a). No space is provided for thermal expansion of the slabs. When the temperature rises to 38 °C, the slabs expand. The buildings do not move, so the expanded slabs buckle upward. Determine the vertical buckling distance y (see part b of the drawing).

SOLUTION

The change in length of each concrete slab can be determined by using the coefficient of linear expansion for concrete given in Table 14.2 and noting that the change in temperature is 13 C°:

$$\Delta L = \alpha L \, \Delta T = [12 \times 10^{-6} \, (C^\circ)^{-1}](3.0 \text{ m})(13 \text{ C}^\circ)$$
$$= 0.000 \, 47 \text{ m} \tag{14.2}$$

The expanded length of each slab is 3.000 47 m. The vertical buckling distance y can be obtained by applying the Pythagorean theorem to the right triangle in part b of the drawing:

$$y = \sqrt{(3.000\ 47\ \text{m})^2 - (3.000\ 00\ \text{m})^2} = \boxed{0.053\ \text{m}}$$

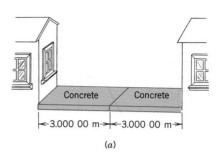

(a)

FIGURE 14.8 (*a*) Two concrete slabs completely fill the space between the buildings when the outdoor temperature is 25 °C. (*b*) When the temperature increases to 38 °C, each slab expands in length, causing the sidewalk to buckle.

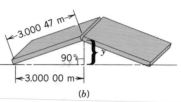

(b)

TABLE 14.2 Coefficients of Thermal Expansion for Solids and Liquids[a]

Substance	Coefficient of Thermal Expansion, $(\text{C}°)^{-1}$	
	Linear (α)	Volumetric (β)
Solids		
Aluminum	23×10^{-6}	69×10^{-6}
Brass	19×10^{-6}	57×10^{-6}
Concrete	12×10^{-6}	36×10^{-6}
Copper	17×10^{-6}	51×10^{-6}
Glass (common)	8.5×10^{-6}	26×10^{-6}
Glass (Pyrex)	3.3×10^{-6}	9.9×10^{-6}
Gold	14×10^{-6}	42×10^{-6}
Iron or steel	12×10^{-6}	36×10^{-6}
Lead	29×10^{-6}	87×10^{-6}
Nickel	13×10^{-6}	39×10^{-6}
Quartz (fused)	0.50×10^{-6}	1.5×10^{-6}
Silver	19×10^{-6}	57×10^{-6}
Liquids[b]		
Benzene	—	1240×10^{-6}
Carbon tetrachloride	—	1240×10^{-6}
Ethyl alcohol	—	1120×10^{-6}
Gasoline	—	950×10^{-6}
Mercury	—	182×10^{-6}
Methyl alcohol	—	1200×10^{-6}
Water	—	207×10^{-6}

[a] The values for α and β pertain to a temperature near 20 °C.

[b] Since liquids do not have fixed shapes, the coefficient of linear expansion is not defined for them.

The buckling of a sidewalk is one consequence of not providing sufficient room for thermal expansion, and Figure 14.9a shows another. It is common for builders to incorporate expansion joints or spaces at intervals along railroad tracks and bridge

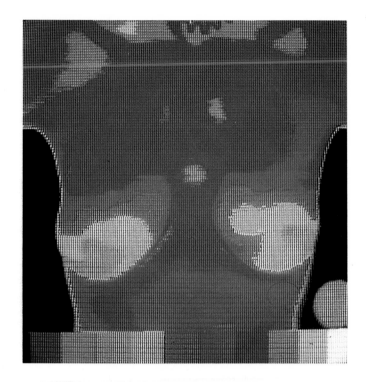

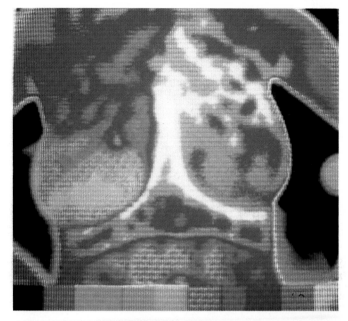

Thermography is used to map the temperature distribution over the surface of the body, giving a color-coded image of the temperature variations. The scale at the bottom of the thermograms is calibrated so that blue represents the coolest temperature and yellow/white the hottest temperature. (Above, left) A thermogram of a woman's normal, healthy breasts. (Above, right) A thermogram showing invasive carcinoma (cancer) of a woman's left breast, indicated by a marked temperature increase over the breast and surrounding areas.

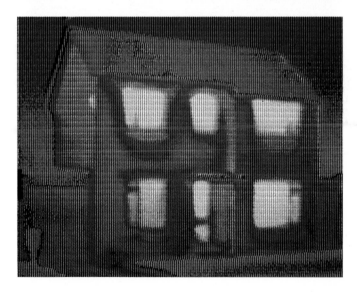

The thermogram shows the temperature distribution over the external surface of a house. The color coding ranges from white and orange for the warmest areas to green and blue for the coolest. The picture shows the greatest heat loss is through the windows. The roof is relatively well insulated, but there is heat loss through the red patches.

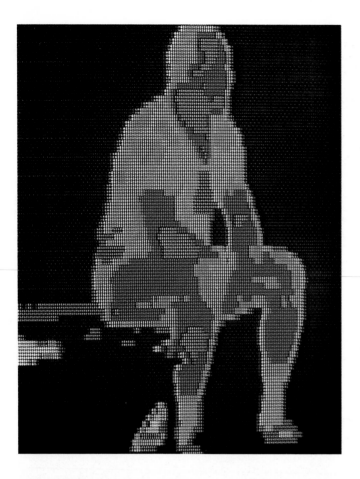

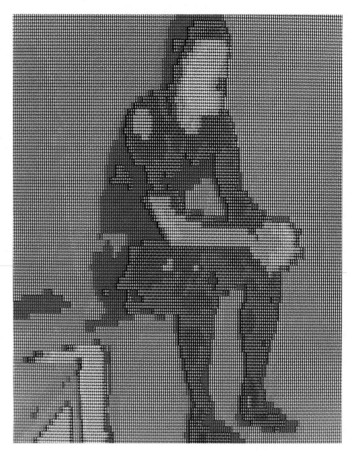

(Above, left) A thermogram of a man immediately before playing a game of squash, indicating normal body temperatures. The color code runs from white/yellow at the hottest extreme, through red, blue, green, purple, to black, the coolest color. (Above, right) A thermogram of a squash player cooling down just after a game. The face and arms are still hot (yellow) compared with the cooler regions (red and blue) of the rest of his body.

An aerial mapping of groves using thermography shows the healthy trees in red. The gaps in the tree rows indicate degrees of stress.

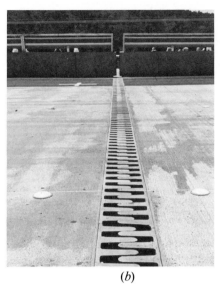

(*a*)   (*b*)

FIGURE 14.9 (*a*) The rails buckled because inadequate allowance was made for thermal expansion. (*b*) An expansion joint in a bridge.

roadbeds to alleviate such problems. Part *b* of the figure shows such an expansion joint in a bridge.

THERMAL STRESS

If the concrete slabs in Figure 14.8 had not buckled upward, they would have been subjected to immense forces from the buildings. The forces needed to keep a solid object from expanding must be strong enough to counteract any change in length that would occur due to a change in temperature. Although the change in temperature may be relatively small, the forces — and hence the stresses — that are needed can be enormous, as Example 3 illustrates.

EXAMPLE 3

A steel beam is used in the roadbed of a bridge. The beam is mounted between two concrete supports when the temperature is 23 °C, with no room provided for thermal expansion (Figure 14.10). What compressional stress must the concrete supports apply to each end of the beam, if the supports are to keep the beam from expanding when the temperature rises to 42 °C?

SOLUTION

Recall from Section 11.3 that the stress (force per unit cross-sectional area) required to change the length L of an object by an amount ΔL is

$$\text{Stress} = Y \frac{\Delta L}{L} \qquad (11.1)$$

where Y is Young's modulus. If the steel [$\alpha = 12 \times 10^{-6}\,(\text{C}°)^{-1}$] were free to expand because of the 19-C° change in temperature, the length of the beam would change by $\Delta L = \alpha L \, \Delta T$. However, the fractional change in the length of the beam would

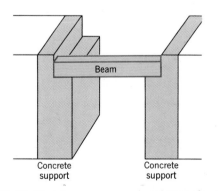

FIGURE 14.10 A large compressional stress is required to prevent a steel beam from expanding by even the small amount that occurs when the temperature rises.

be very small: $\Delta L/L = \alpha\,\Delta T = [12 \times 10^{-6}\ (\text{C}°)^{-1}](19\ \text{C}°) = 2.3 \times 10^{-4}$. Because the concrete supports do not permit any expansion, they must supply a stress to the beam that would, in effect, compress the beam by an amount ΔL. Thus,

$$\text{Stress} = Y\frac{\Delta L}{L} = Y\frac{\alpha L\,\Delta T}{L} = Y\alpha\,\Delta T$$

Young's modulus for steel is $Y = 2.0 \times 10^{11}\ \text{N/m}^2$ (Table 11.1), so the thermal stress is

$$\text{Stress} = Y\alpha\,\Delta T = (2.0 \times 10^{11}\ \text{N/m}^2)[12 \times 10^{-6}\ (\text{C}°)^{-1}](19\ \text{C}°)$$
$$= \boxed{4.6 \times 10^7\ \text{N/m}^2}$$

This stress is enormous, for if the I-beam has a cross-sectional area of $A = 0.10\ \text{m}^2$, the force applied to each end of the beam by a concrete support is $F = (\text{Stress})A = 4.6 \times 10^6\ \text{N}$ (over one million pounds).

THE BIMETALLIC STRIP

As an application of thermal expansion, the ***bimetallic strip*** has many uses in the home. A bimetallic strip is made from two thin strips of metal that have *different* coefficients of linear expansion, as Figure 14.11a shows. Often brass [$\alpha = 19 \times 10^{-6}$ (C°)$^{-1}$] and steel [$\alpha = 12 \times 10^{-6}$ (C°)$^{-1}$] are selected. The two pieces are welded or riveted together. When the bimetallic strip is heated, the brass, having the larger value of α, expands more than the steel. Since the two metals are bonded together, the bimetallic strip bends into an arc as in part b, with the longer brass piece having a larger radius than the steel piece. When the strip is cooled, the bimetallic strip bends in the opposite direction, as in part c.

Bimetallic strips are frequently used as adjustable automatic switches in electrical appliances. Consider a coffee maker that turns off automatically after the coffee is brewed to the selected strength. Figure 14.12a reveals that, while the brewing cycle is on, electricity passes through the heating coil that warms the water. The electricity can flow, because the contact mounted on the bimetallic strip touches the contact

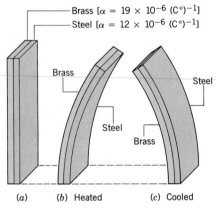

Brass [$\alpha = 19 \times 10^{-6}$ (C°)$^{-1}$]
Steel [$\alpha = 12 \times 10^{-6}$ (C°)$^{-1}$]

(a) (b) Heated (c) Cooled

FIGURE 14.11 (*a*) A bimetallic strip and its behavior when (*b*) heated and (*c*) cooled.

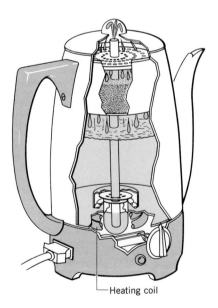

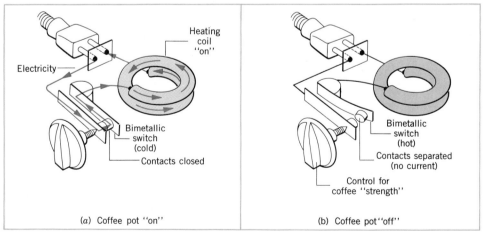

FIGURE 14.12 A bimetallic strip can be used for controlling the brewing time on an automatic coffee maker. (*a*) When the bimetallic strip is cold, it makes contact with the rest of the electrical circuit, allowing electricity to flow through the heater. (*b*) When sufficiently hot, the bimetallic strip bends away from the control knob contact, thus "breaking" the electrical circuit and turning off the electricity.

mounted on the "strength" adjustment knob, thus providing a continuous path for the electricity. When the bimetallic strip gets hot enough to bend away, as in part *b* of the drawing, the contacts separate. The electricity stops, because it no longer has a continuous path along which to flow, and the brewing cycle is shut off. Turning the "strength" control adjusts the distance through which the bimetallic strip must bend for the contact points to separate, thus adjusting the brewing time.

THE EXPANSION OF HOLES

An interesting example of linear expansion arises when there is a hole in a piece of solid material. For example, when a circular ring is heated, does the hole itself expand or contract? The surprising answer is that the hole expands. One way to visualize how the expansion arises is to imagine that the unheated ring is cut and straightened out to form a linear strip (Figure 14.13*a*). The linear strip is then heated, and the strip expands. When the expanded strip is rolled up again, it forms a larger ring with a larger hole (part *b* of the drawing). Thus, *a hole in a piece of solid material expands when heated and contracts when cooled.* The thermal expansion of the ring and its hole is analogous to a photographic enlargement: in both situations everything is enlarged, including holes. The hole expands and contracts just as if it were filled with the material that surrounds it. If the hole is circular, the equation $\Delta L = \alpha L \, \Delta T$ can be used to find the change in any linear dimension of the hole, such as its radius or diameter. For instance, if the ring were made of gold [$\alpha = 14 \times 10^{-6} \, (\text{C}°)^{-1}$] and the hole had a diameter of 1.5 cm, an increase in the temperature of the ring by 35 C° would produce an increase in the diameter of the hole of $\Delta L = [14 \times 10^{-6} \, (\text{C}°)^{-1}]$ (1.5 cm)(35 C°) $= 7.4 \times 10^{-4}$ cm.

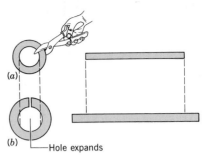
(a)
(b) └─Hole expands

FIGURE 14.13 (*a*) A circular ring is "unrolled" hypothetically to form a straight strip. (*b*) When heated, the strip expands. The heated strip is "rolled" back into a ring that has a larger outer diameter, as well as a larger hole in the center.

14.5 VOLUME THERMAL EXPANSION

NORMAL MATERIALS

A normal material is one whose volume increases as the temperature increases and decreases as the temperature decreases. Most solids and liquids behave in this fashion. By analogy with linear expansion, the change in volume ΔV over modest temperature changes is proportional to the change in temperature ΔT ($\Delta V \propto \Delta T$) and to the initial volume V ($\Delta V \propto V$). These two proportionalities can be combined ($\Delta V \propto V \, \Delta T$) and converted into Equation 14.3 below with the aid of a proportionality constant β, known as the *coefficient of volume expansion.* The algebraic form of this equation is similar to that for linear expansion, $\Delta L = \alpha L \, \Delta T$.

> **VOLUME THERMAL EXPANSION**
> The change in volume ΔV that an object of initial volume V experiences when its temperature changes by an amount ΔT is given by
>
> $$\Delta V = \beta V \, \Delta T \qquad (14.3)$$
>
> where β is the coefficient of volume expansion.
>
> **Common unit for the coefficient of volume expansion: $(\text{C}°)^{-1}$**

Solving Equation 14.3 for β shows that $\beta = \Delta V/(V \Delta T)$, and the unit of β, like the unit of α, is $(C°)^{-1}$. Values for β depend on the nature of the material, reflecting the fact that different materials with the same initial volume expand by different amounts when they experience the same change in temperature. Table 14.2 lists some representative values of β, measured near 20 °C. Notice that the values of β for liquids are substantially larger than those for solids, because liquids typically expand more than solids, given the same initial volumes and temperature changes. Table 14.2 also shows that, for most solids, the coefficient of volume expansion is three times greater than the coefficient of linear expansion: $\beta = 3\alpha$. (See problem 23.)

If a cavity exists within a solid object, the volume of the cavity increases when the object expands, just as if the cavity were filled with the surrounding material. The expansion of the cavity is analogous to the expansion of a hole in a circular ring. Accordingly, the change in volume of a cavity can be found using the relation $\Delta V = \beta V \Delta T$, where β is the coefficient of volume expansion of the material that surrounds the cavity. Example 4 illustrates this important point.

EXAMPLE 4

A small plastic container, called the coolant reservoir, catches the radiator fluid that overflows when an automobile engine becomes hot (see Figure 14.14). The radiator is made of copper, and the coolant has a coefficient of volume expansion of $\beta = 410 \times 10^{-6} \ (C°)^{-1}$. If the radiator is filled to its 15-quart capacity when the engine is "cold" (6.0 °C) and the coolant reservoir is initially empty, how much overflow from the radiator will be in the reservoir when the coolant reaches its operating temperature of 92 °C?

SOLUTION

When the temperature increases by 86 C°, the coolant expands by an amount

$$\Delta V = \beta V \Delta T = [410 \times 10^{-6} \ (C°)^{-1}](15 \text{ quarts})(86 \ C°)$$
$$= 0.53 \text{ quarts} \qquad (14.3)$$

This is not the answer we seek, however, because the volume of the radiator cavity also expands. If the coolant and the radiator cavity each expand by the same amount, there would be no overflow. The volume of the radiator cavity expands as if it were filled with the surrounding material, namely, copper [$\beta = 51 \times 10^{-6} \ (C°)^{-1}$]. Therefore, the expansion of the radiator cavity is

$$\Delta V = \beta V \Delta T = [51 \times 10^{-6} \ (C°)^{-1}](15 \text{ quarts})(86 \ C°)$$
$$= 0.066 \text{ quarts}$$

The amount of coolant overflow is the amount of coolant expansion *minus* the expansion of the radiator cavity, so the overflow volume is $\boxed{0.46 \text{ quarts}}$.

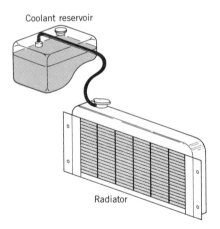

Coolant reservoir

Radiator

FIGURE 14.14 An automobile radiator and its plastic coolant reservoir for catching the overflow from the radiator.

THE ANOMALOUS BEHAVIOR OF WATER NEAR 4 °C

While most substances expand when heated, a few do not. Water is one such substance. If water at 0 °C is heated, its volume *decreases* until the temperature reaches 4 °C. Above 4 °C water behaves normally, and its volume increases as the temperature increases. Because a given mass of water has a minimum volume at 4 °C, the density (mass per unit volume) of water is greatest at 4 °C, as Figure 14.15 shows.

The fact that water has its greatest density at 4 °C, rather than at 0 °C, has important consequences for the way in which a lake freezes. When the air temperature drops, the surface layer of water is chilled. When the temperature of the water reaches 4 °C, the surface water becomes more dense than the warmer water below it. The denser water sinks, thereby pushing up warmer water from below, which in turn is chilled at the surface. This process continues until the temperature of the entire lake reaches 4 °C. Thereafter, further cooling of the surface water below 4 °C makes the water *less dense* than the lower layers; consequently, the surface layer does not sink, but stays on top of the lake. Continued cooling of the surface layer to 0 °C leads to the formation of ice that floats on the water, because ice has a smaller density than water at any temperature. Below the sheet of ice, however, the water temperature remains above 0 °C. The sheet of ice acts as an insulator and inhibits the loss of heat from the lake, especially if the ice is covered with a blanket of snow. Furthermore, heat from the ground beneath the lake helps to keep the water under the ice sheet from freezing. As a result, lakes usually do not freeze solid, even during prolonged cold spells, so fish and other aquatic life can exist below the ice during the winter.

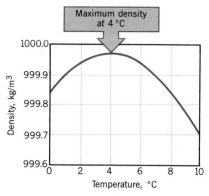

FIGURE 14.15 The density of water in the temperature range from 0 °C to 10 °C. Water has a maximum density of 999.973 kg/m³ at 4 °C. (This value for the maximum density of water is equivalent to the often-quoted density of 1.000 00 grams per milliliter.)

SUMMARY

On the **Celsius temperature scale,** there are 100 equal divisions between the ice point (0 °C) and the steam point (100 °C). On the **Fahrenheit temperature scale,** the interval between the ice point (32 °F) and the steam point (212 °F) contains 180 equal divisions. For most scientific work, the **Kelvin temperature scale** is the scale of choice. One kelvin is equal in size to one Celsius degree; however, the temperature T on the Kelvin scale differs from the temperature T_c on the Celsius scale by an additive constant of 273.15: $T = T_c + 273.15$. The lower limit of temperature is called **absolute zero** and is designated as 0 K on the Kelvin scale.

The operation of any thermometer is based on the change in some physical property with temperature; this physical property is called a **thermometric property.** Some types of thermometers are the mercury-in-glass thermometer, the constant-volume gas thermometer, the thermocouple, and the electrical resistance thermometer.

Most substances expand when heated and contract when cooled. For **linear expansion,** an object of length L experiences a change in length ΔL when the temperature changes by ΔT: $\Delta L = \alpha L \Delta T$, where α is the **coefficient of linear expansion.** For **volume expansion,** the change in volume ΔV of an object of volume V is given by $\Delta V = \beta V \Delta T$, where β is the **coefficient of volume expansion.** When the temperature changes, a hole in a plate or a cavity in a piece of solid material expands or contracts just as if the hole or cavity were filled with the surrounding material.

For an object held rigidly in place, a **thermal stress** can occur when the object attempts to expand or contract. The thermal stress can be extremely large, even when the temperature change is relatively small.

In the temperature range from 0 °C to 4 °C water is unlike most substances, because its volume decreases as the temperature increases. Above 4 °C, water behaves normally, because its volume increases as the temperature increases. Water has a minimum volume and a maximum density at 4 °C.

QUESTIONS

1. For the highest accuracy, would you choose an aluminum tape rule or a steel tape rule for year-round outdoor use. Why?

2. The first international standard of length was a metal bar kept at the International Bureau of Weights and Measures. One meter of length was defined to be the distance between two fine lines engraved near the ends of the bar. Why was it important that the bar be kept at a constant temperature?

3. A circular hole is cut through a flat aluminum plate. A spherical brass ball has a diameter that is slightly *smaller* than the diameter of the hole. The plate and the ball have the same temperature at all times. Should the plate and ball both be heated or both be cooled so the ball *cannot* fall through the hole? Give your reasoning.

4. For added strength, many highways and buildings are constructed with reinforced concrete (concrete that is reinforced with embedded steel rods). Table 14.2 shows that the coefficient of linear expansion for concrete is nearly the same as that for steel. Why is it important that these two coefficients be nearly the same?

5. At a certain temperature, an aluminum rod is hung from an aluminum frame, as the drawing shows. A small gap exists between the rod and the floor. The frame and rod are heated uniformly. (a) Is it possible that the rod will ever touch the floor? Explain. (b) Repeat part (a), assuming the frame is aluminum and the rod is lead.

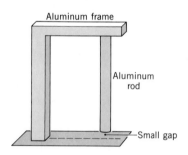

6. A simple pendulum is made using a long thin metal wire. When the temperature drops, does the period of the pendulum increase, decrease, or remain the same? Account for your answer.

7. Why is it often possible to loosen the metal lid of a glass jar by placing the lid in hot water?

8. The metal sheet shown in part *a* of the drawing has three holes cut through it. Each hole has a different shape. Someone claims that the sheet is heated uniformly to produce the sheet shown in part *b*. Does part *b* of the drawing correctly represent the heated sheet? Justify your answer.

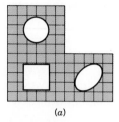

(a)

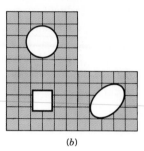

(b)

9. A heated steel ring fits snugly over a cold brass cylinder. The temperatures of the ring and cylinder are, respectively, above and below room temperature. Account for the fact that it is nearly impossible to pull the ring off the cylinder once the assembly has equilibrated at room temperature.

10. For glass baking dishes, Pyrex glass is used instead of common glass. A cold Pyrex dish taken from the refrigerator can be put directly into a hot oven without cracking from thermal stress. A dish made from common glass would crack. With the aid of Table 14.2, explain why Pyrex is better in this respect than common glass.

11. Suppose liquid mercury and glass both had the same coefficient of volume expansion. Explain why such a mercury-in-glass thermometer would not work.

12. When the bulb of a mercury-in-glass thermometer is inserted into boiling water, the mercury column first drops slightly before it begins to rise. Account for this phenomenon. *(Hint: Consider what happens initially to the glass envelope.)*

PROBLEMS

Note: For problems in this set, use the values of α and β given in Table 14.2, unless stated otherwise.

Section 14.1 Temperature and the Common Temperature Scales, Section 14.2 The Kelvin Temperature Scale

1. A personal computer is designed to operate over the temperature range from 10.0 °F to 105 °F. To what do these temperatures correspond (a) on the Celsius scale and (b) on the Kelvin scale?

2. A comfortable temperature for most people is around 24 °C. What is this temperature (a) on the Fahrenheit scale and (b) on the Kelvin scale?

3. A hamburger at 25 °C is heated to 175 °C. What is the *change* in its temperature in (a) C°, (b) F°, and (c) kelvins?

* 4. At what temperature will the reading on the Fahrenheit scale be numerically equal to that on the Celsius scale?

**5. Martians land on Earth. On the Martian temperature scale, the ice point is at 15 °M (M = Martian), and the steam point is at 165 °M. The Martian thermometer shows the temperature on Earth to be 42 °M. Using logic similar to that in Example 1 in the text, what would this temperature be on the Celsius scale?

Section 14.4 Linear Thermal Expansion

6. A steel aircraft carrier is 1200 ft long when moving through the icy North Atlantic at a temperature of 2.0 °C. By how much does the carrier lengthen when it is traveling in the warm Mediterranean Sea at a temperature of 21 °C?

7. Find the approximate length of the Golden Gate bridge if it is known that the steel in the roadbed expands by 0.53 m when the temperature changes from −2.0 °C to +32 °C.

8. An aluminum baseball bat has a length of 0.86 m at a temperature of 25 °C. When the temperature of the bat is raised, the bat expands by 0.000 16 m. Determine the final temperature of the bat.

9. A steel beam is used in the construction of a skyscraper. By what percentage does the length of the beam increase when the temperature changes from that on a cold winter day (−15 °F) to that on a hot summer day (+105 °F)?

10. A 45-m sidewalk is made from a special type of concrete. The sidewalk is a single piece and is observed to expand by 0.014 m when the temperature changes from 35 °F to 95 °F. What is the coefficient of linear expansion of this concrete in units of $(C°)^{-1}$?

11. A commonly used method of securely fastening one part to another part is called "shrink fitting." A steel rod has a diameter of 2.0026 cm, and a flat plate contains a hole whose diameter is 2.0000 cm. The rod is then cooled so that it just fits into the hole. When the rod warms up, the enormous thermal stress exerted by the plate holds the rod securely to the plate. By how many Celsius degrees should the rod be cooled?

* 12. The brass bar and the aluminum bar in the drawing are each attached to an immovable wall. At 22 °C the air gap between the rods is 1.0×10^{-3} m. At what temperature will the gap be closed?

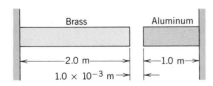

* 13. A lead sphere has a diameter that is 0.050% larger than the inner diameter of a steel ring when each has a temperature of

70.0 °C. Thus, the ring will not slip over the sphere. At what common temperature will the ring just slip over the sphere?

* 14. A rigid steel ruler is accurate when the temperature is 25 °C. When the temperature drops to −15 °C, the ruler no longer reads correctly, but it can be made to read correctly if a stress is applied to each end of the ruler. (a) Should the stress be a compression or a tension? Why? (b) What is the magnitude of the necessary stress?

**15. A 0.30-mm-diameter aluminum wire is stretched between the ends of a concrete block, as the drawing illustrates. When the system (wire and concrete) is at 35 °C, the tension in the wire is 50.0 N. What is the tension in the wire when the system is heated to 185 °C?

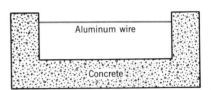

**16. A thin uniform aluminum rod is rotating freely at a constant angular speed about an axis perpendicular to its center. The temperature of the rod increases by 195 C°. (a) Does the angular speed increase or decrease? Why? (b) By what percentage does the angular speed change?

Section 14.5 Volume Thermal Expansion

17. A container holds 3.0 gallons of carbon tetrachloride at 22 °C. What is the change in the volume of the carbon tetrachloride when the liquid is heated to 72 °C?

18. Suppose that the gas tank in your car is completely filled when the temperature is 17 °C. How many gallons will spill out of the twenty-gallon steel tank when the temperature rises to 35 °C?

19. Many hot-water heating systems have a reservoir tank connected directly to the pipeline, so as to allow for expansion when the water becomes hot. The heating system of a house has 250 ft of copper pipe whose inside diameter is 0.75 inch. When the water and pipe are heated from 24 °C to 78 °C, what must be the minimum volume (in ft³) of the reservoir tank to hold the overflow of water?

20. Suppose you were selling apple cider for two dollars a gallon when the temperature is 4.0 °C. The coefficient of volume expansion of the cider is $280 \times 10^{-6} (C°)^{-1}$. If the expansion of the container is neglected, how much more money (in pennies) would you make per gallon by refilling the container on a day when the temperature is 26 °C?

21. A thin spherical shell of silver has an inner radius of 1.5 cm when the temperature is 25 °C. The shell is heated to 135 °C. Find (a) the change in the radius, and (b) the change in the interior volume of the shell. (c) Would the results of (a) and (b) be different if the shell were thick? Explain.

* 22. A mercury-in-glass thermometer, designed to measure body temperature orally, contains 1.0 gram of mercury. The density of mercury is 13 600 kg/m³. The capillary tube attached to the bulb reservoir has a radius of 2.3 × 10⁻⁵ m. Neglect the expansion of the glass, and calculate the distance (in millimeters) between the 98 °F and the 99 °F marks.

* 23. Each side of a cube has a length L. When the temperature of the cube is increased by ΔT, the enlarged volume of the cube becomes $(L + \Delta L)^3$. Expand this expression for the enlarged volume and show that the change in volume ΔV is given approximately by $\Delta V = (3\alpha)V \Delta T$, where $V = L^3$ is the initial volume and α is the coefficient of linear expansion. *(Hint: The terms involving $(\alpha \Delta T)^2$ and $(\alpha \Delta T)^3$ are much smaller than the term involving $\alpha \Delta T$ and, therefore, can be neglected; try substituting any reasonable numbers for α and ΔT to convince yourself of this fact.)* Comparing this expression with $\Delta V = \beta V \Delta T$ (Equation 14.3), we see that $\beta = 3\alpha$ for solids.

* 24. The bulk modulus of water is $B = 2.2 \times 10^9$ N/m². How many atmospheres of pressure are required to keep water from expanding when it is heated from 15 °C to 25 °C?

* 25. A solid aluminum sphere has a radius of 0.50 m and a temperature of 75 °C. The sphere is then completely immersed in a pool of water whose temperature is 25 °C. The sphere cools down, while the water temperature remains nearly at 25 °C, because the pool is very large. The sphere is weighed in the water immediately after being submerged and then again after cooling to 25 °C. (a) Which weight is larger? Why? (b) Use Archimedes' principle to find the magnitude of the *difference* between the values obtained for the weight.

**26. Two identical thermometers are made of Pyrex glass and they contain, respectively, identical volumes of mercury and methyl alcohol. If the expansion of the glass is taken into account, how many times greater is the distance between the degree marks on the methyl alcohol thermometer than that on the mercury thermometer?

**27. The column of mercury in a barometer has a height of 76.0 cm when the pressure is one atmosphere and the temperature is 0.0 °C. What will be the height of the mercury column for the same one atmosphere of pressure when the temperature rises to 38.0 °C on a hot day? *(Hint: The pressure in the barometer is given by Pressure = ρgh, and the density ρ of the mercury changes when the temperature changes.)*

ADDITIONAL PROBLEMS

28. A hole of radius 0.50 cm is drilled through a copper plate whose temperature is 11 °C. (a) When the temperature of the plate is increased, will the radius of the hole be larger or smaller than the

radius at 11 °C? Why? (b) When the plate is heated to 110 °C, by what percentage will the radius of the hole change?

29. On the Rankine temperature scale, which is sometimes used in engineering applications, the ice point is at 491.67 °R, while the steam point is at 671.67 °R. Determine a relationship (analogous to Equation 14.1) between the Rankine temperature and the Fahrenheit temperature.

30. A 1.0 × 10⁻³-m³ container made of common glass is filled with mercury at a temperature of 0.0 °C. The density of mercury is 13 600 kg/m³. How many grams of mercury overflow when the temperature is raised to 25 °C?

31. What is the length of a fused quartz rod whose length changes by the same amount as that of a 0.10-m lead rod when both experience the same temperature change?

* 32. A can is filled with a liquid to 97.0% of its capacity. The temperature of the can and the liquid is 0.0 °C. The material from which the can is made has a coefficient of volume expansion of 85 × 10⁻⁶ (C°)⁻¹. At a temperature of 100.0 °C, the can is observed to be filled exactly to the brim. Determine the coefficient of volume expansion of the liquid.

* 33. A bar of aluminum and a bar of copper each have the same length when the temperature is 25 °C. The aluminum bar is heated to 65 °C. To what temperature must the copper bar be heated so that it has the same length as the heated aluminum bar?

**34. A steel ruler is calibrated to read true at 20.0 °C. A draftsman uses the ruler at 40.0 °C to draw a line on a 40.0 °C copper plate. As indicated on the warm ruler, the length of the line is 0.50 m. To what temperature should the plate be cooled, such that the length of the line truly becomes 0.50 m?

**35. The drawing shows an automatic electrical "switch" made from a copper rod and a glass cylinder that contains mercury. When the copper rod is in contact with the mercury, the switch is "on," and electricity passes through the copper rod, the mercury, and the bulb (see the drawing). The switch shuts off the electricity when the copper rod no longer remains in contact with the mercury. The expansion of the glass can be ignored. When the temperature rises or drops sufficiently, the switch shuts off the light. (a) Does a temperature rise or drop turn the light off? Explain. (b) By how many Celsius degrees would the temperature have to change for the switch to shut off the light?

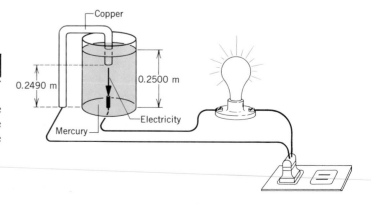

CHAPTER 15

Heat Energy and Phase Changes

This iceberg originated from the Weddel Ice Shelf in Anarctica. In general, matter can exist in three phases, solid, liquid, vapor. All three phases are present in this picture, although the water vapor is not visible. When matter changes from one phase to another, energy is involved, as this chapter discusses.

15.1 INTERNAL ENERGY AND HEAT

Individual molecules in a liquid or a gas can move from place to place, rotating and vibrating as they go. In a solid, molecules are relatively immobile, but they can vibrate. As a result of such motions, which have nothing to do with the overall motion of a macroscopic sample, individual molecules have kinetic energy. In addition, they have potential energy, which is related to the forces that act between the atoms making up the molecules. The sum of the various kinds of molecular energy is called the *internal energy* of the substance. The internal energy of a material increases when the temperature of the material goes up and decreases when the temperature goes down.

This chapter deals with one manifestation of internal energy, namely, **heat.** Heat flows from a hotter object to a cooler object when the two are placed in contact. It is for this reason that a pot of boiling water feels hot to the touch, while a glass of ice water feels cold. What happens in this situation is that a person's body maintains its own temperature at about 37 °C, which is lower than the temperature of boiling

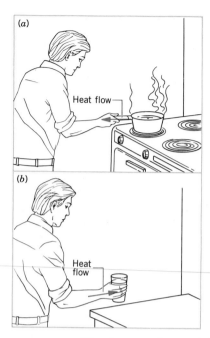

FIGURE 15.1 Heat is energy in transit from hot to cold. In part *a*, energy in the form of heat flows from the hotter pot of boiling water to the colder hand. In part *b*, the flow is from the warmer hand to the colder glass of ice water.

water, but higher than the temperature of ice water. When the person in Figure 15.1*a* touches the pot, heat flows from the hotter pot into the cooler hand. The response of the nerves in the hand to the arrival of heat prompts the brain to conclude that the pot is hot. A similar thing happens when the person touches the glass in part *b* of the drawing. Heat flows from hot to cold, in this case from the warmer hand to the colder glass. The response of the nerves in the hand to the departure of heat prompts the brain to conclude that the glass is cold.

But just what is heat? As the definition below indicates, heat is a form of energy, energy in transit from hot to cold.

DEFINITION OF HEAT
Heat is energy that flows from a high-temperature object to a low-temperature object because of the difference in temperatures.

   SI unit of heat: joule (J)

Being a kind of energy, heat is measured in the same units used for work, kinetic energy, and potential energy. Thus, the SI unit for heat is the joule.

The heat that flows from hot to cold in Figure 15.1 originates in the internal energy of the hot substance. When heat flows in such a circumstance, where the work done is negligible, the internal energy of the hot substance decreases and the internal energy of the cold substance increases. While heat may originate in the internal energy supply of a hot substance, *it is not correct to say that substances contain heat*. They contain internal energy, not heat. The word "heat" is used only when referring to the energy actually in transit from hot to cold.

Heat was not always recognized as a form of energy. Until the early nineteenth century, the generally accepted theory was that heat was a type of invisible fluid, called *caloric,* and that all materials contained this fluid. According to the theory, caloric flowed from hotter objects to cooler objects. Moreover, caloric could be released when substances were broken into small pieces, due to burning in a fire or the friction that occurs when two objects are rubbed together. The caloric theory was discarded, however, largely because of the famous experiments of Count Rumford (Benjamin Thompson, 1753–1814). While boring cannon barrels, he was struck by the fact that the water being used to cool the cannon continued to boil, even when only small amounts of metal were being ground into pieces by a blunt cutting tool. According to the caloric theory, the ever-boiling water meant that caloric was always flowing from the cannon and the boring tool into the water. In the absence of an apparent source, such an inexhaustible supply of caloric fluid made little sense to Rumford. He proposed instead that the rotating motion of the boring tool was somehow being transmitted (through friction) to the individual "particles" (atoms or molecules) of the cannon, boring tool, and water. Rumford's idea that heat is somehow related to motion is now accepted, for heat is related to the internal energy of a substance, and this supply of energy includes the motional or kinetic energy that all molecules have.

15.2 SPECIFIC HEAT CAPACITY

SOLIDS AND LIQUIDS

More and more heat is needed to raise the temperature of a solid or liquid to higher and higher values. In addition, a greater amount of heat is required to raise the temperature of a greater mass of material by a given amount. Similar comments apply when the temperature is lowered, except that heat must be removed. In fact, for limited ranges in temperature, experiment shows that the amount of heat Q is directly proportional to the change in temperature ΔT ($Q \propto \Delta T$) and to the mass m ($Q \propto m$). These two proportions are expressed below in Equation 15.1, with the help of a proportionality constant c that is referred to as the *specific heat capacity* of the material or simply the *specific heat.*

HEAT SUPPLIED OR REMOVED IN CHANGING THE TEMPERATURE OF A SUBSTANCE
The heat Q that must be supplied or removed to change the temperature of a substance of mass m by an amount ΔT is

$$Q = cm\,\Delta T \qquad (15.1)$$

where c is the specific heat capacity of the substance.

Common unit for specific heat capacity: $J/(kg \cdot C°)$

Solving Equation 15.1 for the specific heat capacity shows that $c = Q/(m\,\Delta T)$. If heat is expressed in joules (J), mass in kilograms (kg), and temperature change in

TABLE 15.1 Specific Heat Capacities[a] of Some Solids and Liquids

Substance	Specific Heat Capacity c	
	J/(kg·C°)	kcal/(kg·C°)[b]
Solids		
Aluminum	9.00×10^2	0.215
Copper	387	0.0924
Glass	840	0.20
Human body (37 °C, average)	3500	0.83
Ice (−15 °C)	2.00×10^3	0.478
Iron or steel	452	0.108
Lead	128	0.0305
Silver	235	0.0562
Liquids		
Benzene	1740	0.415
Ethyl alcohol	2450	0.586
Glycerin	2410	0.576
Mercury	139	0.0333
Water (15 °C)	4186	1.000

[a] Except as noted, the values are for 25 °C and 1 atm of pressure.
[b] The values given are the same in units of cal/(g·C°).

Celsius degrees (C°), then the unit of specific heat capacity is J/(kg·C°). Table 15.1 reveals that the value of the specific heat capacity depends on the nature of the material. Example 1 demonstrates the use of Equation 15.1.

EXAMPLE 1

In a half hour, a 65-kg jogger can generate 8.0×10^5 J of heat, which is removed from the jogger's body by a variety of means, including the body's own temperature regulating mechanisms. If the heat were not removed, how much would the body temperature increase?

SOLUTION

Table 15.1 gives the average specific heat capacity of the human body as 3500 J/(kg·C°). With this value, Equation 15.1 can be used to calculate the temperature increase as

$$\Delta T = \frac{Q}{cm} = \frac{8.0 \times 10^5 \text{ J}}{[3500 \text{ J/(kg·C°)}](65 \text{ kg})} = \boxed{3.5 \text{ C°}}$$

The phrase *heat capacity,* without the word "specific," is sometimes used to refer to the term cm in the relation $Q = cm \, \Delta T$. Thus, for example, the copper in a penny and in a long pipe has the same specific heat capacity of $c = 387$ J/(kg·C°), but the heat capacity cm of the more massive pipe is larger.

HEAT UNITS OTHER THAN THE JOULE

Equation 15.1 provides the basis for introducing units other than the joule for heat measurement. There are three other heat units in common use. One kilocalorie (1 kcal) was defined historically as the amount of heat needed to raise the temperature of one kilogram of water by one Celsius degree.* With $Q = 1.00$ kcal, $m = 1.00$ kg,

---

* From 14.5 °C to 15.5 °C.

and $\Delta T = 1.00$ C°, the equation $Q = cm\,\Delta T$ shows that such a definition is tantamount to stating that the specific heat capacity of water is $c = 1.00$ kcal/(kg·C°). Similarly, one calorie (1 cal) was defined as the amount of heat needed to raise the temperature of one gram of water by one Celsius degree, which is consistent with assigning a value of $c = 1.00$ cal/(g · C°) for the specific heat capacity of water. (Nutritionists use the word "Calorie," with a capital C, to specify the energy content of foods; this use is unfortunate, since 1 Calorie = 1000 calories = 1 kcal.) The British thermal unit (Btu) is the other commonly used heat unit and was defined historically as the amount of heat needed to raise the temperature of one pound of water by one Fahrenheit degree.

It was not until the time of James Joule (1818–1889) that the relationship between energy in the form of work (joules) and energy in the form of heat (kilocalories) was firmly established. Joule's experiments revealed that the performance of mechanical work can make the temperature of a substance rise, just as the absorption of heat can. His experiments and those of later workers have shown that

$$1 \text{ kcal} = 4186 \text{ joules} \quad \text{or} \quad 1 \text{ cal} = 4.186 \text{ joules}$$

Because of its historical significance, this conversion factor is known as the **mechanical equivalent of heat.** Example 2 illustrates the use of various heat units.

EXAMPLE 2

Cold water at a temperature of 15 °C enters a heater, and the resulting hot water has a temperature of 61 °C. Suppose a person uses 120 kg of hot water in taking a long shower. Find the number of (a) joules and (b) kilocalories needed to heat the water. (c) Assuming that the power utility charges $0.10 per kilowatt·hour for electrical energy, determine the cost of heating the water.

SOLUTION

(a) The number of joules of heat can be determined from Equation 15.1, since the specific heat capacity of water is known:

$$Q = cm\,\Delta T = [4186 \text{ J/(kg·C°)}](120 \text{ kg})(61 \text{ °C} - 15 \text{ °C})$$
$$= \boxed{2.3 \times 10^7 \text{ J}}$$

(b) The corresponding number of kilocalories can be obtained by using the mechanical equivalent of heat:

$$Q = (2.3 \times 10^7 \text{ J})\left(\frac{1 \text{ kcal}}{4186 \text{ J}}\right) = \boxed{5.5 \times 10^3 \text{ kcal}}$$

Alternatively, the same answer can be obtained directly from $Q = cm\,\Delta T$, if $c = 1.000$ kcal/(kg·C°) is used for the specific heat capacity of water.

(c) The kilowatt·hour (kWh) is the unit of energy used by utility companies to compute monthly electric bills. To calculate the cost, it is first necessary to determine the number of joules in one kilowatt·hour. Recalling that Energy = Power × Time (Equation 7.10), and that 1 watt = 1 joule/second, we find that 1 kWh = (1000 J/s)(3600 s) = 3.60 × 10^6 J. The number of kilowatt·hours of energy used to heat the water is

$$(2.3 \times 10^7 \text{ J})\left(\frac{1 \text{ kWh}}{3.60 \times 10^6 \text{ J}}\right) = 6.4 \text{ kWh}$$

At a charge of $0.10 per kWh, the cost of the heat is $\boxed{\$0.64}$.

GASES

As we will see in Chapter 18, the value of the specific heat capacity depends on whether the pressure or volume is held constant while energy in the form of heat is being added to or removed from a substance. The distinction between constant pressure and constant volume is usually not important for solids and liquids but is significant for gases. Different values are obtained when the specific heat capacity for a gas is measured under conditions of *constant pressure* and under conditions of

TABLE 15.2 Specific Heat Capacities[a] of Gases

| | Specific Heat Capacity | |
Gas	Constant Pressure c_P [J/(kg·C°)]	Constant Volume c_V [J/(kg·C°)]
Ammonia	2190	1670
Carbon dioxide	833	638
Nitrogen	1040	739
Oxygen	912	651
Water vapor (100 °C)	2020	1520

[a] Except as noted, the values are for 15 °C and 1 atm of pressure.

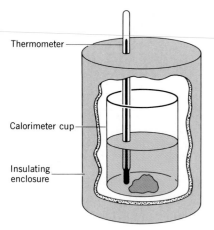

Thermometer

Calorimeter cup

Insulating enclosure

FIGURE 15.2 A calorimeter. The insulating enclosure minimizes the flow of heat energy to or from the calorimeter cup.

constant volume. Table 15.2 illustrates the difference for several gases and indicates that the value c_P at constant pressure is always greater than the value c_V at constant volume.

CALORIMETRY

Specific heat capacity can be measured using the technique of *calorimetry.* Figure 15.2 shows one kind of experimental apparatus, called a calorimeter. Essentially, a calorimeter is an insulated container like a thermos for hot coffee or iced tea. A perfect thermos would prevent any heat from leaking out or in. However, energy in the form of heat can flow *between* materials inside the thermos to the extent that they have different temperatures, for example, between ice cubes and warm tea. Such heat flow satisfies the conservation of energy, for the colder materials gain the energy that the hotter materials lose. The colder materials warm up and the hotter materials cool down, until eventually a common temperature is reached at thermal equilibrium. The next example shows how the specific heat capacity of a substance can be determined from the temperature changes that occur inside a calorimeter as thermal equilibrium is established.

EXAMPLE 3

The calorimeter cup in Figure 15.2 is made from 0.15 kg of aluminum and contains 0.20 kg of water. Initially, the water and the cup have a common temperature of 18.0 °C. Forty grams of an unknown material are heated to a temperature of 97.0 °C and then added to the water. The temperature of the water, the cup, and the unknown material is 22.0 °C after thermal equilibrium is reestablished. Ignoring the small amount of heat gained by the thermometer, find the specific heat capacity of the unknown material.

SOLUTION

Since energy is conserved and there is negligible heat flow between the calorimeter and the outside surroundings, the heat energy gained by the cold water and the aluminum cup as they warm up is equal to the heat energy lost by the unknown material as it cools down. Each quantity of heat can be calculated using $Q = cm \, \Delta T$:

$$\text{Heat gained} = \text{Heat lost}$$

$$(cm \, \Delta T)_{\text{aluminum}} + (cm \, \Delta T)_{\text{water}} = (cm \, \Delta T)_{\text{unknown}}$$

$$[9.00 \times 10^2 \text{ J/(kg·C°)}](0.15 \text{ kg})(22.0 °C - 18.0 °C)$$
$$+ [4186 \text{ J/(kg·C°)}](0.20 \text{ kg})(22.0 °C - 18.0 °C)$$
$$= c_{\text{unknown}}(0.040 \text{ kg})(97.0 °C - 22.0 °C)$$

$$\boxed{c_{\text{unknown}} = 1300 \text{ J/(kg·C°)}}$$

An important feature of the calculation in Example 3 is the way in which the temperature changes ΔT are written. Each is written as a positive number, that is, the higher temperature minus the lower temperature, so that the heat contribution is a positive number. In this fashion, we ensure that both sides of the equation, Heat gained = Heat lost, have the same algebraic sign, as they must.

15.3 THE LATENT HEAT OF PHASE CHANGE

It may come as a surprise to hear that there are important situations in which the addition or removal of heat energy does not cause a temperature change. To get a feel for how such circumstances arise, consider a well-stirred glass of iced tea that has come to thermal equilibrium. Even though heat energy enters the glass from the warmer room, the temperature of the iced tea does not rise above 0 °C as long as ice cubes are present. Apparently the heat is being used for some purpose other than raising the temperature. In fact, the heat is being used to melt the ice, and only when all of the ice is melted will the temperature of the liquid begin to rise.

One important point illustrated by the above example is that there is more than one type or phase of matter. For instance, some of the water in the glass is in the solid phase (ice) and some in the liquid phase. The gas or vapor* phase is the third familiar phase of matter. In the gas phase, water is referred to as water vapor or steam.

A second important point of the iced tea example is that matter can change from one phase to another, and heat plays a role in the change. Figure 15.3 summarizes the various possibilities for phase changes between solids, liquids, and gases. A solid can *melt* or *fuse* into a liquid if heat energy is added, while the liquid can *freeze* into a solid if the energy is removed. Similarly, a liquid can *evaporate* into a gas if heat energy is supplied, while the gas can *condense* into a liquid if the energy is taken away. Rapid evaporation, with the formation of vapor bubbles within the liquid, is called boiling. Finally, a solid can change directly into a gas if heat energy is provided. We say that the solid *sublimes* into a gas. Examples of sublimation are (1) solid carbon dioxide CO_2 (dry ice) turning into gaseous CO_2 and (2) solid naphthalene (moth balls) turning into naphthalene fumes. Conversely, if heat energy is removed under the right conditions, the gas will condense directly into a solid.

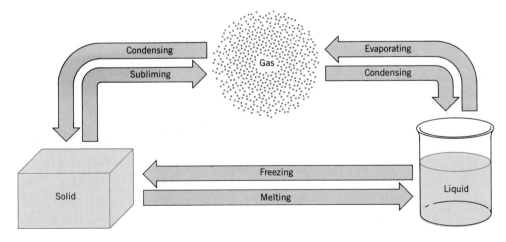

FIGURE 15.3 Three familiar phases of matter—solid, liquid, and gas—and the phase changes that can occur between any two of them.

* The words "gas" and "vapor" are used interchangeably in this text.

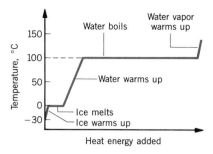

FIGURE 15.4 The graph shows the way the temperature of water changes as energy in the form of heat is added, starting with ice at −30 °C. The pressure is atmospheric pressure.

Figure 15.4 displays a graph that indicates what happens when energy in the form of heat is added to a material that changes phases. The graph records temperature versus energy and refers to water at the normal atmospheric pressure of 1.01×10^5 Pa. The water starts off as ice at the subfreezing temperature of −30 °C. As heat energy is added, the temperature of the ice increases, in accord with the specific heat capacity of ice [2000 J/(kg·C°)]. Not until the temperature reaches the normal melting/freezing point of 0 °C does the water begin to change phase. As long as energy is added, the solid changes into the liquid, the temperature staying at 0 °C until *all the ice has melted.* Once all of the material is in the liquid phase, the added heat energy causes the temperature to increase again, now in accord with the specific heat capacity of liquid water [4186 J/(kg·C°)]. When the temperature reaches the normal boiling/condensing point of 100 °C, the water begins to change from the liquid to the gas and continues to do so as long as energy is added. The temperature remains 100 °C *until all liquid is gone.* When all of the material is in the gas phase, the added heat energy once again causes the temperature to rise, this time according to the specific heat capacity of water vapor at constant atmospheric pressure [2020 J/(kg·C°)]. Materials other than water behave in a fashion similar to that in Figure 15.4, except that they have their own melting/freezing and boiling/condensing temperatures, as Table 15.3 indicates.

When a substance changes from one phase to another, the amount of energy that must be added or removed in the form of heat depends on the type of material and the nature of the phase change. The heat per kilogram associated with a phase change is referred to as *latent heat,* according to the following definition:

DEFINITION OF LATENT HEAT

The latent heat L is the amount of heat energy per kilogram that must be added or removed when a substance changes from one phase to another.

SI unit of latent heat: J/kg

The *latent heat of fusion* L_f refers to the change between solid and liquid phases, the *latent heat of vaporization* L_v applies to the change between liquid and gas phases,

TABLE 15.3 Latent Heats[a] of Fusion and Vaporization

Substance	Melting Point (°C)	Latent Heat of Fusion L_f (J/kg)	Boiling Point (°C)	Latent Heat of Vaporization L_v (J/kg)
Ammonia	−77.8	33.2×10^4	−33.4	13.7×10^5
Benzene	5.5	12.6×10^4	80.1	3.94×10^5
Copper	1083	20.7×10^4	2566	47.3×10^5
Ethyl alcohol	−114.4	10.8×10^4	78.3	8.55×10^5
Gold	1063	6.28×10^4	2808	17.2×10^5
Lead	327.3	2.32×10^4	1750	8.59×10^5
Mercury	−38.9	1.14×10^4	356.6	2.96×10^5
Nitrogen	−210.0	2.57×10^4	−195.8	2.00×10^5
Oxygen	−218.8	1.39×10^4	−183.0	2.13×10^5
Water	0.0	33.5×10^4	100.0	22.6×10^5

[a] The values pertain to 1 atm pressure.

and the *latent heat of sublimation* L_s refers to the change between solid and gas phases.

Table 15.3 gives some typical values of latent heats. For instance, the latent heat of fusion for water is 3.35×10^5 J/kg. Thus, 3.35×10^5 J of heat must be supplied to melt one kilogram of ice at 0 °C into liquid water at 0 °C; conversely, this amount of energy must be removed from one kilogram of liquid water at 0 °C to freeze the liquid into ice at 0 °C. By comparison, the latent heat of vaporization for water has the much larger value of 22.6×10^5 J/kg. When water boils at 100 °C, 22.6×10^5 J of heat energy must be supplied for each kilogram of liquid turned into vapor. And when water vapor condenses at 100 °C, this same amount of energy is released for each kilogram of vapor that changes back into liquid. Liquid water at 100 °C is hot enough by itself to cause a bad burn, and the additional effect of the large latent heat can cause severe tissue damage if condensation occurs on the skin. Example 4 illustrates how to take into account the effect of latent heat when using the conservation of energy principle.

EXAMPLE 4

A 7.00-kg glass bowl [$c = 840$ J/(kg·C°)] contains 16.0 kg of punch at 25.0 °C. Two and a half kilograms of ice [$c = 2.00 \times 10^3$ J/(kg·C°)] are added to the punch. The ice has an initial temperature of -20.0 °C, having been kept in a very cold freezer. The punch may be treated as if it were water [$c = 4186$ J/(kg·C°)], and it may be assumed that there is no heat flow between the punch bowl and the external environment. What is the temperature of the punch, ice, and bowl when they reach thermal equilibrium?

SOLUTION

First, it is necessary to check whether any ice is left at equilibrium. If so, the final temperature will be 0.0 °C, the melting/freezing point of water. In making the check, we apply the equation $Q = cm \, \Delta T$ to calculate the amount of heat that would be available if the punch and the bowl were cooled from 25.0 °C to 0.0 °C:

$$Q = \underbrace{[4186 \text{ J/(kg·C°)}](16.0 \text{ kg})(25.0 \text{ °C} - 0.0 \text{ °C})}_{\text{Punch}}$$
$$+ \underbrace{[840 \text{ J/(kg·C°)}](7.00 \text{ kg})(25.0 \text{ °C} - 0.0 \text{ °C})}_{\text{Glass bowl}} = 1.82 \times 10^6 \text{ J}$$

Is this enough heat to melt all the ice? The heat needed to melt the ice is

Heat needed
$$= \underbrace{[2.00 \times 10^3 \text{ J/(kg·C°)}](2.50 \text{ kg})[0.0 \text{ °C} - (-20.0 \text{ °C})]}_{\text{Heat to warm the ice from } -20.0 \text{ °C to } 0.0 \text{ °C}}$$
$$+ \underbrace{(2.50 \text{ kg})(3.35 \times 10^5 \text{ J/kg})}_{\text{Heat to melt ice at } 0.0 \text{ °C}} = 0.938 \times 10^6 \text{ J}$$

where the heat to melt the ice at 0.0 °C is the mass of the ice (2.50 kg) times the latent heat of fusion (3.35×10^5 J/kg). Thus, there is more than enough heat to melt all the ice and raise the final temperature above 0.0 °C. The final temperature T can be determined by using the conservation of energy to equate the total heat energy gained to the total heat energy lost:

(a) Heat gained when ice warms to 0.0 °C = $[2.00 \times 10^3 \text{ J/(kg·C°)}](2.50 \text{ kg}) \times [0.0 \text{ °C} - (-20.0 \text{ °C})]$

(b) Heat gained when ice melts at 0.0 °C = $(2.50 \text{ kg})(3.35 \times 10^5 \text{ J/kg})$

(c) Heat gained when melted ice (liquid) warms to temperature T = $[4186 \text{ J/(kg·C°)}](2.50 \text{ kg})(T - 0.0 \text{ °C})$

(d) Heat lost when punch cools to temperature T = $[4186 \text{ J/(kg·C°)}](16.0 \text{ kg}) \times (25.0 \text{ °C} - T)$

(e) Heat lost when bowl cools to temperature T = $[840 \text{ J/(kg·C°)}](7.00 \text{ kg})(25.0 \text{ °C} - T)$

$$\underbrace{(a) + (b) + (c)}_{\text{Heat gained}} = \underbrace{(d) + (e)}_{\text{Heat lost}}$$

This equation contains a single unknown variable, the equilibrium temperature T and can be solved to show that $\boxed{T = 11 \text{ °C}}$.

*15.4 THE PHASE DIAGRAM

THE EQUILIBRIUM LINES

Under specific conditions of temperature and pressure, a substance can exist in equilibrium in more than one phase at the same time. Consider Figure 15.5, which shows a container kept at a constant temperature by a large reservoir of heated sand. Initially the container is evacuated, and part *a* shows it just after it has been partially filled with a liquid. A few fast-moving molecules can escape the liquid and form a vapor phase, as part *b* suggests. These molecules pick up the required energy (the latent heat of vaporization) during collisions with neighboring molecules in the liquid. However, the reservoir of heated sand immediately replenishes the energy carried away, thus maintaining the constant temperature. At first, the movement of molecules is predominantly from liquid to vapor, although some molecules in the vapor phase do reenter the liquid. As the concentration of molecules builds up in the vapor, the number reentering the liquid eventually equals the number jumping into the vapor, and equilibrium becomes established, as in part *c*. From this point on, the concentration of molecules in the vapor phase does not change, and the vapor pressure remains constant. The pressure of the vapor that coexists in equilibrium with the liquid is called the ***equilibrium vapor pressure*** of the liquid.

The equilibrium vapor pressure does not depend on the volume of space above the liquid. If more space were provided, more liquid would vaporize, until equilibrium was reestablished at the same vapor pressure, assuming the same temperature is maintained. In fact, the equilibrium vapor pressure for a liquid depends only on the temperature of the liquid; a higher temperature causes a higher pressure, as the graph in Figure 15.6 indicates for the specific case of water. Only when the temperature and vapor pressure correspond to a point on the curved line, which is called the ***vapor pressure curve*** or the ***vaporization curve,*** can liquid and vapor phases coexist in equilibrium.

To illustrate the use of a vaporization curve, let us see what happens when water boils in a pot that is *open to the air.* Assume the air pressure acting on the water is 1.01×10^5 Pa (one atmosphere). When boiling occurs, bubbles of water vapor form throughout the liquid, rise to the surface, and break. For these bubbles to form and rise, the pressure of the vapor inside them must at least equal the air pressure acting on the surface of the water. According to Figure 15.6, a value of 1.01×10^5 Pa corresponds to a temperature of 100 °C. Consequently, water boils at 100 °C at 1 atmosphere of pressure. In general, a ***liquid boils at the temperature at which its vapor pressure equals the external pressure above the surface of the liquid.*** Water will not

FIGURE 15.5 (*a*) Initially, only a liquid is in the evacuated container. (*b*) Quickly, some of the molecules begin entering the vapor phase. (*c*) Eventually, equilibrium is reached when the number of molecules entering the vapor phase equals the number returning to the liquid. The pressure of the gas at equilibrium is called the equilibrium vapor pressure of the liquid.

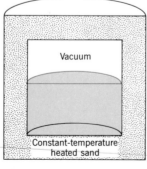
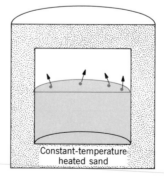
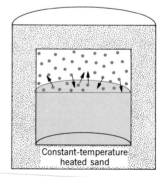

(*a*)   (*b*)   (*c*)

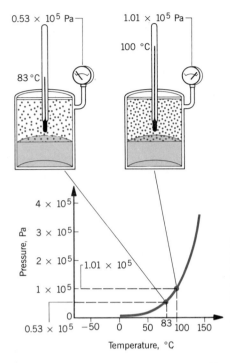

FIGURE 15.6 As the temperature increases, the equilibrium vapor pressure of the liquid increases. A plot of vapor pressure versus temperature is called the vapor pressure curve or the vaporization curve, the example shown being that for water.

boil, then, at sea level if the temperature is only 83 °C, because at this temperature the vapor pressure of water is only 0.53×10^5 Pa (see Figure 15.6), a value that is less than the external pressure of 1.01×10^5 Pa. However, water does boil at 83 °C on a mountain at an altitude of just under five kilometers, because the atmospheric pressure there is 0.53×10^5 Pa.

A solid and its vapor can also coexist in equilibrium; for instance, solid dry ice and gaseous CO_2. The vapor pressure of a solid arises in the same fashion as that of a liquid, and the vapor pressure curve, also called the **sublimation curve,** gives those particular combinations of temperature and pressure at which vapor and solid are in equilibrium. Figure 15.7 presents the vapor pressure curve for ice (in color) as an example. For comparison, the drawing also includes the vaporization curve of water.

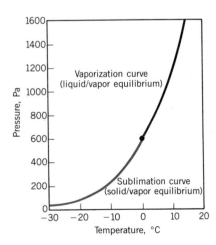

FIGURE 15.7 A plot of the vapor pressure of a solid versus temperature is known as the sublimation curve (in color). The curve given here is that for ice. The vaporization curve of water is included for comparison. Note that this drawing displays a much smaller range of pressures than in Figure 15.6.

Notice that the temperatures and pressures for the solid/vapor equilibrium are lower than those for the liquid/vapor equilibrium.

As is the case for liquid/vapor and solid/vapor equilibria, the solid/liquid equilibrium occurs only at specific conditions of temperature and pressure. For each temperature there is a single pressure for which the two phases will coexist in equilibrium. A plot of the equilibrium pressure versus equilibrium temperature is referred to as the *fusion curve,* and Figure 15.8a shows a typical curve for a normal substance. A normal substance is one that expands when it melts (e.g., carbon dioxide and sulfur). Since higher pressures make it more difficult for such materials to expand, a higher melting temperature is needed for a higher pressure, and the fusion curve slopes upward to the right. Part *b* of the picture illustrates the fusion curve for water, one of the few substances that contract when they melt. Higher pressures make it easier for such substances to melt. Consequently, a lower melting temperature is associated with a higher pressure, and the fusion curve slopes downward to the right in the drawing.

The unusual slope of the fusion curve of water explains in part why ice skating is possible; the large pressure generated by the skater's weight acting on the sharp blade causes the melting/freezing temperature to be lowered below the normal value of 0 °C (see Figure 15.8b). The lower melting temperature allows the heat arising from kinetic friction to melt a small amount of ice beneath the blade. The resulting thin layer of water acts as a lubricant, permitting the skater to glide on the "ice."

Figure 15.9 combines the vaporization, sublimation, and fusion curves into a single plot. Parts *a* and *b* show such plots for carbon dioxide (a normal substance) and water,* respectively. The composite plot is called a *phase diagram,* and several of its features merit discussion.

THE REGIONS BETWEEN THE EQUILIBRIUM LINES

The regions between the equilibrium lines in the phase diagram are labeled SOLID, LIQUID, and VAPOR. For temperature and pressure conditions corresponding to a point within one of these regions, the material can exist only as the appropriate single phase. If, for example, carbon dioxide vapor is subjected to the temperature and pressure conditions of point *A* in Figure 15.9a, the vapor would change into a liquid.

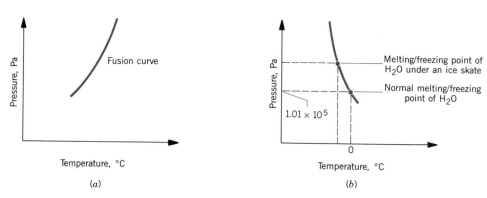

FIGURE 15.8 A plot of the pressure versus the temperature for which a solid coexists in equilibrium with a liquid is called the fusion curve. Part *a* illustrates the fusion curve for a normal substance that expands on melting. Part *b* shows the fusion curve for water, one of the few substances that contract on melting. Neither curve is drawn to scale.

* The plot for water is more complicated than that shown, involving at least eight different solid forms of ice at high pressures.

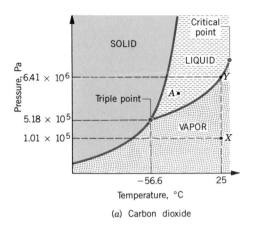

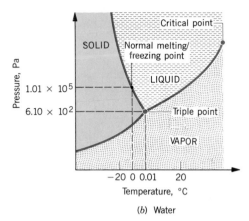

(a) Carbon dioxide

(b) Water

FIGURE 15.9 (a) Phase diagram for carbon dioxide. (b) Phase diagram for water. Neither diagram is drawn to scale.

THE TRIPLE POINT

The point where the three equilibrium lines meet on the phase diagram is called the *triple point.* Only at the conditions of temperature and pressure at the triple point can solid, liquid, and vapor phases of a pure substance coexist in equilibrium. Table 15.4 gives the triple point temperatures and pressures of various substances. For carbon dioxide the triple point occurs at a temperature of -56.6 °C and a pressure of 5.18×10^5 Pa. This temperature is so low and the pressure is so high that only carbon dioxide gas exists at normal conditions of pressure and temperature (see point X in Figure 15.9a, for example). At a temperature of 25 °C the pressure would have to be at least 6.41×10^6 Pa (see point Y in the figure) for liquid carbon dioxide to form. Such high pressures exist inside the familiar large metal cylinders, where CO_2 is stored as a liquid. For water, the triple point occurs at a temperature of $+0.01$ °C and a pressure of 6.10×10^2 Pa. Since the triple point pressure is low, a special apparatus is required to observe the three-phase equilibrium of water.

A cylinder of CO_2.

TABLE 15.4 Triple Point and Critical Point Conditions for Some Common Materials

Substance	Triple Point		Critical Point	
	Temperature (°C)	Pressure (Pa)	Temperature (°C)	Pressure (Pa)
Ammonia	−77.8	6.05×10^3	132.4	11.3×10^6
Carbon dioxide	−56.6	5.18×10^5	31.1	7.38×10^6
Hydrogen	−259.3	7.04×10^3	−239.9	1.30×10^6
Nitrogen	−210.0	1.25×10^4	−146.9	3.40×10^6
Oxygen	−218.8	1.52×10^2	−118.4	5.08×10^6
Sulfur dioxide	−75.5	1.67×10^3	157.6	7.88×10^6
Water	0.01	6.10×10^2	374.3	22.1×10^6

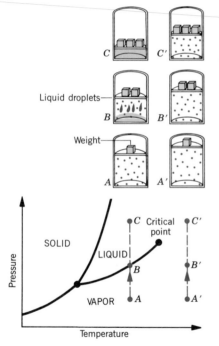

FIGURE 15.10 At a temperature below the critical point temperature, a gas can be liquified by increasing the pressure along line *ABC* in the phase diagram. Condensation occurs at point *B*. Condensation cannot be made to occur by increasing the pressure along line *A′B′C′*, because the temperature exceeds the critical point temperature.

THE CRITICAL POINT

The **critical point** specifies the conditions of temperature and pressure beyond which it is no longer possible to distinguish a liquid from a gas. On the phase diagram, the vaporization curve stops at the critical point (see Figure 15.9). If the temperature of a gas exceeds the critical temperature, the gas cannot be liquified, no matter how large a pressure is applied to it. To illustrate this fact, Figure 15.10 shows the same mass of a gaseous substance in each of two cylinders fitted with movable pistons. In one cylinder the temperature is held constant at a value that is below the critical temperature, while in the other the temperature is kept above the critical temperature. In either case, the starting vapor pressure is the same, corresponding to the points *A* and *A′* on the phase diagram. The pressure is then increased in both cylinders by adding weights to the top of each piston. In case *A*, the pressure increases along line *ABC*, pausing and remaining constant at point *B* on the vaporization curve, while all the vapor condenses into liquid droplets. Upon further increase in pressure from *B* to *C*, the substance exists only in its liquid phase. In case *A′*, however, the pressure never pauses during its rise from *A′* to *C′*, and no liquid droplets form, because the temperature exceeds the critical point value. Table 15.4 gives the critical point temperatures and pressures for some familiar materials.

EQUILIBRIUM VERSUS NONEQUILIBRIUM

We conclude this section with the following observation: the fact that multiple phases can coexist in equilibrium does not necessarily mean that they will. Other factors may prevent it. For example, water in an *open* bowl may never come into equilibrium with water vapor if air currents are present. What happens is that the liquid, perhaps at a temperature of 25 °C, attempts to establish the corresponding equilibrium vapor pressure of 3.2×10^3 Pa. If air currents continually blow the water vapor away, however, equilibrium will never be established, and eventually the water will evaporate completely. Each kilogram of water that goes into the vapor phase takes along energy in the form of the latent heat of vaporization. Because of this loss of energy, the remaining liquid would become cooler, except for the fact that the surroundings replenish the loss and prevent the cooling from taking place. In the case of the human body, however, evaporative cooling does occur. Water is exuded by the sweat glands and evaporates from a much larger area than the surface of a typical bowl of water. The removal of energy along with the water vapor is one mechanism that the body uses to maintain its constant temperature.

*15.5 HUMIDITY

The **absolute humidity** is the part of the total atmospheric pressure that is due to water vapor in the air. Out of the total pressure of 1.01×10^5 Pa, suppose, for example, that the part due to water vapor is 2.30×10^3 Pa. The absolute humidity, then, is 2.30×10^3 Pa.

On the evening weather report, however, it is the **relative humidity,** not the absolute humidity, that is quoted. The relative humidity depends on the temperature and the vaporization curve of water. For each temperature, the vaporization curve specifies the equilibrium vapor pressure, that is, the pressure of the water vapor in equilibrium with a pool of liquid. At a given temperature, the pressure of the water vapor in the air cannot exceed this value; if it did, the vapor would not be in equilibrium with liquid and would condense in the form of dew or rain to reestablish equilibrium. The relative humidity expresses the actual pressure of the water vapor in the air (the absolute humidity) as a percentage of the maximum pressure that the water vapor could have (the equilibrium vapor pressure of water):

$$\text{Percent relative humidity} = \frac{\text{Absolute humidity at the existing temperature}}{\text{Equilibrium vapor pressure of water at the existing temperature}} \times 100 \qquad (15.2)$$

When the absolute humidity equals the equilibrium vapor pressure of water at a given temperature, the relative humidity is 100%. In such a situation, the vapor is said to be *saturated,* because it is present in the maximum amount, as it would be above a pool of liquid at equilibrium in a closed container. If the relative humidity is less than 100%, the water vapor is said to be *unsaturated.* Example 5 demonstrates how to find the relative humidity.

EXAMPLE 5

On a certain day, the absolute humidity is 2.0×10^3 Pa. Using the vaporization curve for water given in Figure 15.11, determine the relative humidity if the temperature is (a) 32 °C and (b) 21 °C.

SOLUTION

(a) According to Figure 15.11, the equilibrium vapor pressure of water at 32 °C is 4.8×10^3 Pa. Equation 15.2 reveals that the relative humidity is

Relative humidity at 32 °C $= \dfrac{2.0 \times 10^3 \text{ Pa}}{4.8 \times 10^3 \text{ Pa}} \times 100 = \boxed{42\%}$

(b) A similar calculation shows that

Relative humidity at 21 °C $= \dfrac{2.0 \times 10^3 \text{ Pa}}{2.5 \times 10^3 \text{ Pa}} \times 100 = \boxed{80\%}$

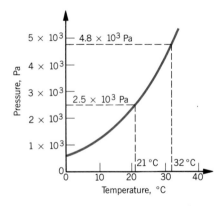

FIGURE 15.11 The vaporization curve of water.

Whatever the actual pressure of water vapor in the air, it is possible to locate this value on the vertical axis of the vaporization curve and identify a corresponding

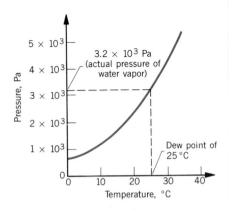

FIGURE 15.12 On the vaporization curve of water, the dew point is the temperature that corresponds to the actual partial pressure of water vapor in the air.

FIGURE 15.13 For fog to form, the air temperature must drop below the dew point.

temperature. This temperature is known as the **_dew point._** Figure 15.12 shows that if the actual pressure of water vapor is 3.2×10^3 Pa, the dew point is 25 °C. The actual partial pressure would correspond to a relative humidity of 100%, if the ambient temperature were equal to the temperature at the dew point. Hence, the dew point is the temperature at which water vapor in the air would condense in the form of liquid drops (dew or fog). The closer the actual temperature is to the dew point, the closer the relative humidity is to 100%. Thus, for fog to form around the bridge in Figure 15.13, the air temperature must drop below the dew point. Similarly, water condenses on the outside of a cold glass when the temperature of the air next to the glass falls below the dew point. And the cold coils in a home dehumidifier (see Figure 15.14) function very much in the same way that the cold glass does. The coils are kept cold by a circulating refrigerant. When the air blown across them by the fan cools below the dew point, water vapor condenses in the form of droplets, which collect in a receptacle.

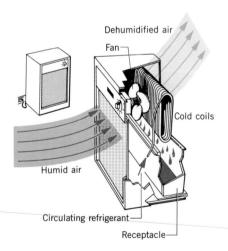

FIGURE 15.14 The cold coils of a dehumidifier cool the air blowing across them below the dew point, and water vapor condenses out of the air.

SUMMARY

The **internal energy** of a substance is the sum of the internal kinetic, potential, and other kinds of energy that the molecules of the substance have. **Heat** is energy that flows from a high-temperature substance to a low-temperature substance because of the difference in temperatures. The SI unit for heat is the joule. Other units for heat are the kilocalorie and the British thermal unit. The conversion factor between the joule and the kilocalorie is referred to as the **mechanical equivalent of heat;** 1 kilocalorie = 4186 joules.

The **specific heat capacity** c of a substance of mass m determines how much heat Q must be supplied to or removed from the substance to change its temperature by an amount ΔT: $Q = cm \Delta T$. When materials are placed in thermal contact within a perfectly insulated container, the **principle of energy conservation** requires that the energy lost in the form of heat by the warmer materials equals the energy gained by the cooler materials.

Heat energy must be supplied to make a material change from the solid to the liquid phase, the liquid to the vapor phase, or the solid to the vapor phase. Heat energy must be removed to make the reverse changes occur. The amount of heat per kilogram of material is called the latent heat of the phase change. The **latent heats of fusion, vaporization,** and **sublimation** refer, respectively, to the solid/liquid, liquid/vapor, and solid/vapor changes.

The **equilibrium vapor pressure** of a substance is the pressure of the vapor phase that is in equilibrium with the liquid or solid phase. Vapor pressure depends only on temperature. For a liquid, a plot of the equilibrium vapor pressure versus temperature is called the **vapor pressure curve** or the **vaporization curve.** For a solid, the plot is known as the **sublimation curve.** The vaporization and sublimation curves give those combinations of temperature and pressure at which the corresponding two phases can coexist in equilibrium. The **fusion curve** gives the combinations of temperature and pressure for equilibrium between solid and liquid phases. A **phase diagram** is a composite plot that includes the vaporization, sublimation, and fusion curves. The **triple point** is the one combination of temperature and pressure at which solid, liquid, and vapor coexist in equilibrium. The **critical point** defines the conditions of temperature and pressure beyond which it is no longer possible to distinguish a liquid from a vapor.

Absolute humidity is the actual pressure of water vapor in the air. The **relative humidity** expresses the absolute humidity as a percentage of the maximum pressure that the water vapor could have at the existing temperature (the equilibrium vapor pressure of water). The **dew point** is the temperature below which the water vapor in the air condenses. On the vaporization curve of water, the dew point is the temperature that corresponds to the actual pressure of water vapor in the air.

QUESTIONS

1. Two different objects are supplied with equal amounts of heat energy. Give the reason(s) why their temperature changes would not necessarily be the same.

2. Two objects are made from the same material. The more massive object has a higher temperature than the less massive object. If the two are placed in contact, which object will experience the greater temperature change? Justify your answer.

3. Near a large body of water, the fluctuations in air temperature are usually less extreme than they are far away from the water. Explain why.

4. To help lower the high temperature of a sick patient, an alcohol rub is sometimes used. In this procedure, isopropyl alcohol is rubbed over the patient's back, arms, legs, etc., and allowed to evaporate. Why does the procedure work?

5. Suppose the latent heat of vaporization of H_2O were one-tenth its actual value. (a) Other things being equal, would it take the same time, a shorter time, or a longer time for a pot of water on a stove to boil away? (b) Would the evaporative cooling mechanism of the human body be as effective? Account for both answers.

6. Fruit blossoms are permanently damaged when the temperature drops below about −4 °C (a "hard freeze"). Orchard owners sometimes spray a film of water over the blossoms to protect them when a hard freeze is expected in the spring. From the point of phase changes, give a reason for the protection.

7. If a beaker containing liquid water is placed in a closed container and water vapor is pumped away rapidly enough, the remaining liquid will turn to ice. Explain why the ice appears.

8. Freeze-drying is a process for preparing dried food. The food is frozen quickly and then put into a vacuum chamber where pumps remove water vapor. Using the phase diagram of H_2O, account for the fact that the process produces food with no water in it.

9. Wet clothes quickly freeze when hung outside during the winter. Nonetheless, they dry out eventually. Why?

10. A bowl of water is covered tightly and allowed to sit at a constant temperature of 23 °C for a long time. What is the relative humidity in the space between the surface of the water and the cover? Justify your answer.

11. When the relative humidity is high, perspiration evaporates

more slowly. Consequently, the body's evaporative cooling mechanism doesn't work as well. Explain.

12. Is it possible for dew to form on Tuesday night and not on Monday night, even though Monday night is the cooler night? Incorporate the idea of the dew point into your answer.

13. A jar is half filled with boiling water. The lid is then screwed on the jar. After the jar has cooled to room temperature, the lid is difficult to remove. Why?

PROBLEMS

Note: For problems in this set, use the values of c, L_f, and L_v given in Tables 15.1 and 15.3, unless stated otherwise.

Section 15.2 Specific Heat Capacity

1. Blood can carry excess energy from the interior to the surface of the body, where the energy is dispersed in a number of ways. While a person is exercising, 0.6 kg of blood flows to the surface of the body and releases 2000 J of energy. The blood arriving at the surface has the temperature of the body interior, namely, 37.0 °C. Assuming that blood has the same specific heat capacity as water, determine the temperature of the blood that leaves the surface and returns to the interior.

2. Find the heat capacity of (a) 3.00 kg of ethyl alcohol and of (b) 20.0 kg of ethyl alcohol.

3. If the price of electrical energy is $0.10 per kilowatt · hour, what is the cost of using electrical energy to heat the water in a swimming pool (12.0 m × 9.0 m × 1.5 m) from 15 °C to 27 °C?

4. Calculate (a) the number of joules and (b) the number of kilocalories that correspond to one Btu.

5. Into a 0.200-kg copper cup (20.0 °C) is put 0.100 kg of aluminum at 50.0 °C and 0.250 kg of water at 85.0 °C. Assuming there is no heat flow between the cup and its outside environment, find the final equilibrium temperature.

6. When you take a bath in 185 kg of water, how many kilograms of hot water (60.0 °C) and cold water (25.0 °C) must you mix, so that the temperature of the bath is 40.0 °C? Ignore any heat flow between the water and its surroundings.

* 7. Lead shot (0.600 kg, 90.0 °C) and steel shot (0.100 kg, 60.0 °C) are put into a can. How many kilograms of water at 74.0 °C must be added, so that in reaching thermal equilibrium the lead and the steel experience a temperature change of the *same magnitude*? Ignore the specific heat capacity of the can and any heat exchanged with the environment.

* 8. The box of a well-known breakfast cereal states that one ounce of the cereal contains 110 Calories (1 food Calorie = 1 kcal). If all this energy could be converted by a weight lifter's body

into work done in lifting a barbell, what is the heaviest barbell that could be lifted through a distance of 2.0 m? Express your answer in newtons and in pounds.

* 9. An electric hot water heater takes in cold water at 25.0 °C and delivers hot water. The hot water has a constant temperature of 30.0 °C, when the "hot" faucet is left wide open all the time and the flow rate is 9.0 liter/min (1 liter = 1000 cm³). What is the minimum power rating (in watts) of the hot water heater?

**10. A steel rod (ρ = 7860 kg/m³) has a length of 2.0 m. It is bolted at both ends between immobile supports. Initially there is no tension in the rod, because the rod just fits between the supports. Find the tension that develops when the rod loses 3300 J of heat energy.

Section 15.3 The Latent Heat of Phase Change

11. A 10.0-kg block of ice has a temperature of −10.0 °C. The pressure is one atmosphere. The block absorbs 982 kcal of heat energy. What is the final temperature of the water?

12. Assume the pressure is one atmosphere and determine the heat required to produce 2.00 kg of water vapor at 100.0 °C, starting with (a) 2.00 kg of water at 100.0 °C, (b) 2.00 kg of water at 0.0 °C, (c) 2.00 kg of ice at 0.0 °C, and (d) 2.00 kg of ice at −30.0 °C.

13. Suppose the amount of heat energy removed when 3.0 kg of water freezes at 0 °C were removed from ethyl alcohol at its freezing/melting point of −114 °C. How many kilograms of ethyl alcohol would freeze?

14. The latent heat of vaporization of H_2O at body temperature (37.0 °C) is 577 kcal/kg. To cool the body of a 75-kg jogger by 1.0 C°, how many kilograms of water in the form of sweat have to be evaporated?

15. Ice at −10.0 °C and water vapor at 130 °C are brought together at atmospheric pressure in a perfectly insulated container whose heat capacity can be neglected. After thermal equilibrium is reached, the liquid phase at 50.0 °C is present. Ignoring the equilibrium vapor pressure of the liquid at 50.0 °C, find the ratio of the mass of steam to the mass of ice.

16. In solar-assisted greenhouses some provision is usually made for storing solar energy in order to minimize heating costs. One way of storing energy is to use large containers of water. Another way is to use hydrated sodium sulfate (Glauber salt), which has a normal melting/freezing point of 32.4 °C, a latent heat of fusion of 239 000 J/kg, and a specific heat capacity of 2850 J/(kg·C°) in the liquid phase and 1900 J/(kg·C°) in the solid phase. On an equal volume basis, the sodium sulfate may have an advantage over water. To show this advantage, calculate the heat energy released when (a) 1.0 m³ of water (1.0 × 10³ kg) and (b) 1.0 m³ of sodium sulfate (1.6 × 10³ kg) cool from 35.0 °C to 18.0 °C. (c) Repeat parts (a) and (b) for a temperature interval of 30.0 °C to 18.0 °C and comment on how the two storage media compare under these conditions.

* 17. It is claimed that if a lead bullet goes fast enough, it can melt completely when it comes to a halt suddenly and all of its kinetic energy is converted into heat via friction. Find the minimum speed of a lead bullet (30.0 °C) for such an event to happen.

* 18. An unknown material has a normal melting/freezing point of −25.0 °C, and the liquid phase has a specific heat capacity of 160 J/(kg·C°). One hundred grams of the solid at −25.0 °C is put into a one-hundred-and-fifty-gram aluminum calorimeter cup that contains one hundred grams of glycerin. The temperature of the cup and the glycerin is initially 27.0 °C. All the unknown material melts, and the final temperature at equilibrium is 20.0 °C. The calorimeter loses no energy to the external environment. What is the latent heat of fusion of the unknown material?

* 19. (a) Express the latent heat of fusion of H_2O in Btu/lb. (b) Suppose an air conditioner can remove heat at a rate of 12 000 Btu/h. Determine the number of pounds of water at 0 °C that such an air conditioner could freeze into ice at 0 °C in 24 h. (c) Express the answer to part (b) in tons. This is the number of "tons of air conditioning" that the unit can provide, a phrase that is sometimes used to rate the cooling capacity of air conditioners.

* 20. Water is moving with a speed of 5.00 m/s just before it passes over the top of a waterfall. At the bottom, 5.00 m below, the water flows away with a speed of 3.00 m/s. What is the largest amount by which the temperature of the water at the bottom could exceed the temperature of the water at the top?

**21. A locomotive wheel is 1.00 m in diameter. A 25.0-kg steel band has a temperature of 20.0 °C and a diameter that is 0.600 mm less than that of the wheel. What is the smallest number of kilograms of water vapor at 100 °C that can be condensed on the steel band to heat it, so that it will fit onto the wheel? Do not ignore the water that results from the condensation.

Section 15.4 The Phase Diagram

22. The pressure is 5.0 × 10⁶ Pa. The phase diagram of carbon dioxide that accompanies this problem is to scale. Use the diagram to determine the temperature at which carbon dioxide will exist as an equilibrium of (a) solid and liquid phases and (b) liquid and vapor phases.

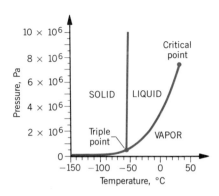

23. The equilibrium vapor pressure of liquid sulfur dioxide is 5.07 × 10⁵ Pa at 32.1 °C. What phase is present at equilibrium at a temperature of 32.1 °C when the pressure is one atmosphere? Explain.

24. Use the phase diagram for water in Figure 15.9*b* to answer this question, recognizing that the complete phase diagram is more complicated than that shown. (a) Assuming the pressure can be varied while the temperature remains at −20 °C, what are the possibilities (single phases and equilibrium combinations of phases) for the ways that water can exist in thermal equilibrium? (b) Repeat part (a) for a temperature of +20 °C.

* 25. A container is fitted with a movable piston of negligible mass and radius $r = 0.050$ m. Inside the container is liquid water in equilibrium with its vapor, as the drawing shows. The piston remains stationary with a 120-kg mass on top of it. The air pressure acting on the top of the piston is one atmosphere. By using the vaporization curve for water in Figure 15.6, find the temperature of the water.

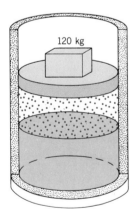

Section 15.5 Humidity

26. Using the vaporization curve for water that accompanies this problem, find the absolute humidity on a day when the weather forecast gives the relative humidity as 70.0% and the temperature as 38.0 °C.

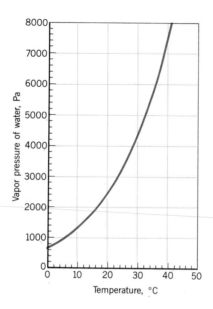

Temperature, °C

27. The relative humidity is 35% when the temperature is 27 °C. Using the vaporization curve for water that accompanies problem 26, determine the dew point.

* 28. A woman has been outdoors where the temperature is 10 °C. She walks into a 25 °C house, and her glasses "steam up." Using the vaporization curve for water that accompanies problem 26, find the smallest possible value for the relative humidity of the room.

**29. At a picnic, a glass contains 0.300 kg of tea at 30.0 °C, which is the air temperature. To make iced tea, someone adds 0.0670 kg of ice at 0.0 °C and stirs the mixture continually. When all the ice melts and the final temperature is reached, the glass begins to fog up, because water vapor condenses on the outer glass surface. Using the vaporization curve for water that accompanies problem 26, ignoring the specific heat capacity of the glass, and treating the tea as if it were water, estimate the relative humidity.

ADDITIONAL PROBLEMS

30. The critical point of ammonia is at a temperature of 132.4 °C. The normal boiling point of ammonia is −33.4 °C. What phase of ammonia is present at equilibrium at room temperature when the pressure is one atmosphere? Explain.

31. At a fabrication plant, a hot metal forging has a mass of 75 kg and a specific heat capacity of 430 J/(kg·C°). To harden it, the forging is quenched by immersion in 710 kg of oil that has a temperature of 32 °C and a specific heat capacity of 2700 J/(kg·C°). The final temperature of the oil and forging at thermal equilibrium is 47 °C. Assuming that heat energy flows only between the forging and the oil, determine the initial temperature of the metal forging.

32. In preparation for a party, a number of glass mugs and bottles of beer are put into a plastic ice chest (assumed to be perfectly insulated). The mass of the glass is 9.0 kg, while the mass of the beer is 8.0 kg. The initial temperature of the beer and the mugs is 27 °C. Ignoring the specific heat capacity of the chest and taking the specific heat capacity of beer to be the same as that of water, determine the smallest amount of ice (at 0 °C) that can be added to the chest to cool the contents to 0 °C.

33. The outdoor temperature is 15 °C and the relative humidity is 45%. The absolute humidity inside a house is the same as outdoors, but the temperature in the house is 25 °C. Using the vaporization curve for water that accompanies problem 26, determine the relative humidity in the house.

34. A precious-stone dealer wishes to find the specific heat capacity of a 0.030-kg gemstone. The specimen is heated to 95.0 °C and then placed in a 0.15-kg copper vessel that contains 0.080 kg of water at equilibrium at 25.0 °C. The heat flow with the external environment is negligible. When equilibrium is reestablished, the water temperature is 28.5 °C. What is the specific heat capacity of the specimen?

* 35. To help keep his barn warm on cold days, a farmer stores 840 kg of solar-heated water in barrels. For how many hours would a 2.0-kW electric space heater have to operate to provide the same amount of heat as the water does, when it cools from 10.0 °C to 0.0 °C and completely freezes?

* 36. A 0.25-kg coffee mug is made from a material that has a specific heat capacity of 950 J/(kg·C°) and contains 0.30 kg of water. The cup and the water are at 25 °C. To make a cup of coffee, a small electric heater is immersed in the water and brings it to a boil in two minutes. Assume that the cup and the water always have the same temperature and determine the minimum number of watts at which the heater is rated.

**37. A 1.0-kg steel sphere will not fit through a circular hole in a 1.0-kg aluminum plate, because the radius of the sphere is 0.10% larger than the radius of the hole. If both the sphere and the plate are always kept at the same temperature, how much heat energy must be put into the two so the ball just passes through the hole?

**38. Two rigid walls are located 0.50 m apart. On a cold day, a steel ($\rho = 7860$ kg/m³) beam just fits between these walls. Therefore, on a hot day, the beam is jammed tightly between the walls, each wall pushing on the beam with a force of 2.0×10^5 N. To free the beam, ice at 0 °C is applied and, in melting, cools and shrinks the beam. What is the smallest number of kilograms of ice that can be used to free the beam? Consider only the ice and the beam and ignore the water that results from the melted ice.

CHAPTER 16

The Transfer of Heat Energy

The Calgary Prehistoric Park in Canada contains life-size models of dinosaurs, such as the stegosaurus shown here. One popular theory holds that the triangular-like plates on the back of the stegosaurus helped to keep the animal cool by removing excess heat, much like the fins of an automobile radiator. In general, there are three ways to transfer heat, as we will soon discover.

16.1 INTRODUCTION

The transfer of heat energy from one object to another is important in many aspects of our lives. For instance, most of our energy (except for small amounts of nuclear energy) originates in the sun and is transferred to us over a distance of 93 million miles through the void of space. The sunlight of today provides the energy to drive photosynthesis in plants that provide food and, hence, metabolic energy, while the sunlight of eons ago nurtured the organic matter that eventually produced the important fossil fuels of oil, natural gas, and coal. Within the home, energy transfer occurs routinely. A heating unit transfers heat energy throughout a house on a cold day, while an air conditioner removes heat energy on a hot day. And without the transfer of heat energy, the automobile engine would be useless.

There are three fundamental processes by which energy is transferred: convection, conduction, and radiation. This chapter considers each of them.

16.2 CONVECTION

When a portion of a fluid is warmed, such as the air above a fire, the volume of the fluid expands, and the density decreases. According to Archimedes' principle (see Section 12.7), the surrounding cooler and denser fluid exerts a buoyant force on the warmer fluid, which pushes the warmer fluid upward. As warm fluid is pushed upward, the surrounding cooler fluid replaces it. This cooler fluid, in turn, is warmed and pushed upward. Thus, a continuous flow is established. The fluid flow carries along heat energy and is called a ***convection current.*** In general, whenever heat energy is transferred by the bulk movement of a gas or a liquid, the energy is said to be transferred by ***convection.***

CONVECTION

Convection is the process in which heat energy is carried from place to place by the bulk movement of a fluid.

The smoke rising above a campfire or a chimney is a visible result of convection. Figure 16.1 shows an example of convection currents in a pan of water that is being heated on a gas burner. The currents distribute the heat energy from the burning gas to all parts of the water.

Certain kinds of home heating systems also take advantage of convection to distribute heat energy throughout a room. Figure 16.2a illustrates the air convection current originating from a baseboard heating unit located near the floor. Had the heating unit been located near the ceiling, the warm air would remain there, and very little convection current would be generated to distribute the energy. Part *b* of the figure shows an analogous situation in a refrigerator, where the convection current is set up by the cooling coils. The coils are located near the *top* of the refrigerator, in

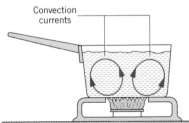

FIGURE 16.1 Convection currents are set up when a pan of water is heated.

The smoke from a campfire rises because of convection.

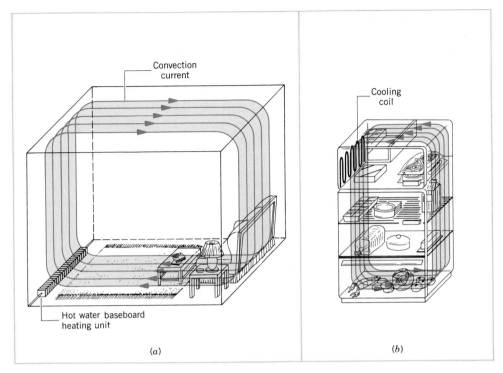

FIGURE 16.2 (*a*) The warm air, heated by the baseboard heating unit, is pushed to the top of the room by the surrounding cooler and denser air. A convection current is established in the room. (*b*) The air that is cooled by the cooling coils sinks to the bottom of the refrigerator and pushes up warmer air, thereby creating a convection current.

direct contrast to the placement of the heating unit in part *a*. As the temperature of the air in contact with the coils decreases, the volume decreases, and the density increases. This cooler and denser air settles to the bottom and forces warmer, less dense air upward to the cooling coils. The resulting convection current keeps all parts of the refrigerator uniformly cool. Placing the cooling coils at the bottom of the refrigerator, rather than at the top, would produce stagnant, cool air at the bottom and lead to inefficient cooling at the top.

Another example of convection occurs when the ground, heated by the sun's rays, warms the neighboring air. Surrounding cooler and denser air pushes the heated air upward. The resulting updraft or "thermal," as it is sometimes called, can be quite strong, depending on the amount of heat energy that the ground can supply. As Figure 16.3 illustrates, these thermals can be used by glider pilots to gain considerable altitude. Birds such as hawks utilize thermals in a similar fashion.

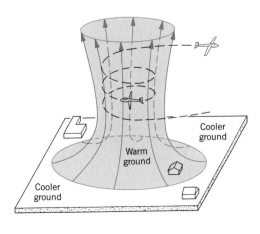

FIGURE 16.3 Updrafts, or thermals, are produced by the convective movement of air that has been warmed by the ground.

The smog rises about 100 meters above the ground. The upper portion of the Los Angeles City Hall is visible in the clear air above the smog.

Strong thermals are found where the temperature near ground level is markedly warmer than it is at higher altitudes. Although strong thermals are not found everywhere, it is usual for air temperature to decrease with increasing altitude, and the resulting upward-moving convection currents are an important mechanism for dispersing pollutants from industrial sources and automobile exhaust systems. Sometimes, however, meteorological conditions cause a layer to form in the atmosphere where the temperature increases with increasing altitude. Such a layer is called an *inversion layer,* because its temperature profile is inverted compared to the usual situation. An inversion layer arrests the normal upward-moving convection currents, causing a stagnant-air condition in which the concentration of pollutants increases.

The situations discussed above are examples of ***natural convection.*** Natural convection occurs because of buoyant forces that arise when a temperature difference causes the density at one place in the fluid to be different than at another. Sometimes, natural convection is inadequate to transfer sufficient amounts of heat energy. In such cases ***forced convection*** is often used, and an external device such as a fan mixes the warmer and cooler portions of the fluid. Figure 16.4 shows two examples of forced convection. In one, a fan mounted on a computer creates the forced convection that removes heat energy produced by the electrical components. In the other, a pump circulates radiator fluid through an automobile engine to remove the excessive heat energy due to the combustion process.

FIGURE 16.4 The forced convection created by a fan removes heat energy produced by the electrical components in a computer. The forced convection generated by a pump circulates radiator fluid through an automobile engine and removes excessive heat energy produced by combustion.

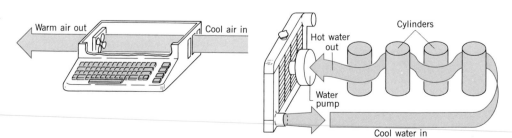

16.3 CONDUCTION

To anyone who has fried a hamburger in an all-metal skillet, one fact is clear: the metal handle becomes warm, and sometimes even hot, as heat energy is transferred from the heating element, through the skillet, to the handle. Clearly, the energy is not being transferred by the bulk movement of the metal or the surrounding air, so convection can be ruled out. Instead, energy is transferred directly through the metal itself. This situation is one example of the transfer of heat energy by *conduction.*

CONDUCTION
Conduction is the process whereby heat energy is transferred directly through a material, any bulk motion of the material playing no role in the transfer.

One mechanism for conduction occurs when the atoms or molecules in a hotter part of the material vibrate or move with greater energy than those in a cooler part. By means of collisions, the more energetic molecules pass on a portion of their energy to their less energetic neighbors. Figure 16.5 illustrates such a conduction mechanism in a gas. Molecules that strike the hotter wall absorb energy from it and rebound with a greater kinetic energy than when they arrived. As these more energetic molecules collide with their less energetic neighbors, they transfer some of their energy to the neighbors. Through such molecular collisions, heat energy is conducted from the hotter wall to the cooler wall. This collision mechanism does not depend on a bulk flow of material.

Another mechanism for the conduction of heat energy occurs in metals. Metals are different from most substances in having a pool of electrons that are more or less free to wander throughout the volume of the metal. These free electrons are capable of transporting energy and allow metals to transfer heat energy very well. The free electrons are also responsible for the excellent electrical conductivity that metals have.

Materials that conduct heat energy well are called *thermal conductors.* Most metals, such as aluminum, copper, gold, and silver, are excellent thermal conductors. Compared to thermal conductors, substances that conduct heat poorly are called *thermal insulators.* Some common thermal insulators are wood, glass, and most plastics. Thermal insulators have many important applications. For instance, vir-

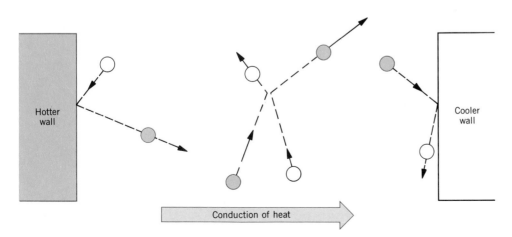

FIGURE 16.5 Heat conduction in a gas occurs when energetic molecules (colored circles) transfer some of their energy to less energetic molecules (open circles) through collisions. By this process, energy is conducted from the hotter wall to the cooler wall.

FIGURE 16.6 Energy in the form of heat is conducted along the bar when the two ends of the bar are maintained at different temperatures; the energy flows from the warmer end to the cooler end. Although not shown for the sake of clarity, the sides of the bar are insulated, so energy lost through the sides is negligible. The length of the rod is L, and the cross-sectional area is A.

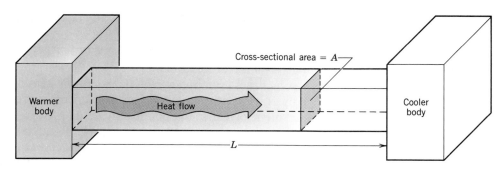

tually all new housing construction incorporates thermal insulation in attics and walls to reduce heating and cooling costs. And the wooden or plastic handles on many pots and pans reduce the flow of heat energy to the cook's hand.

Let us examine the conduction process a little more closely in order to determine the factors that influence it. Figure 16.6 displays a rectangular bar, the ends of which are in thermal contact with two bodies whose temperatures are constant. One of the bodies is maintained at a higher temperature, while the other is maintained at a lower temperature. Although not shown for the sake of clarity, the sides of the bar are insulated, so the energy lost through them is negligible. In this situation, an amount of heat energy Q is conducted along the bar from the warmer end to the cooler end, an amount that depends on a number of factors:

1. Q is proportional to the length of time t during which conduction has been taking place ($Q \propto t$). More heat energy flows in longer time periods.

2. Q is proportional to the temperature difference ΔT between the two ends of the bar ($Q \propto \Delta T$). A larger temperature difference causes more heat energy to flow. No energy flows when both ends of the bar have the same temperature and $\Delta T = 0$.

3. Q is proportional to the cross-sectional area A of the bar ($Q \propto A$). Figure 16.7 helps to explain this fact by showing two identical bars (insulated sides not shown) placed between the warmer and cooler bodies. Clearly, twice as much heat energy flows through two bars as through one. Since two bars are equivalent to one bar with twice the cross-sectional area, doubling the area doubles the energy flow. In other words, Q is proportional to A.

4. Q is inversely proportional to the length L of the bar ($Q \propto 1/L$). Greater lengths of material conduct less heat energy. To experience this effect, put two insulated

FIGURE 16.7 The amount of heat energy flowing through two identical bars is twice that flowing through one bar. The sides of the bars are insulated to ensure that the energy lost through the sides is negligible, but the insulation is omitted for the sake of clarity.

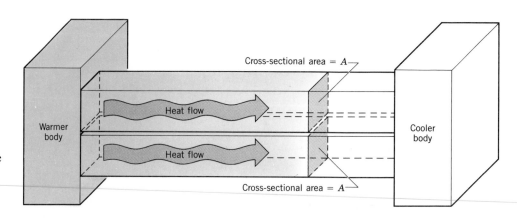

mittens (the kind that cooks keep around the stove) on the *same hand.* Then, touch a hot pot and notice that it feels cooler than when you wear only one mitten, signifying that less energy flows through the greater thickness ("length") of material.

The proportionalities above can be stated together as $Q \propto A\,\Delta Tt/L$. Equation 16.1 expresses this result with the aid of a proportionality constant k, which is called the *thermal conductivity.*

CONDUCTION OF HEAT ENERGY THROUGH A MATERIAL

The heat energy Q conducted during a time t through a bar of length L and cross-sectional area A is

$$Q = \frac{kA\,\Delta Tt}{L} \qquad (16.1)$$

where ΔT is the temperature difference between the ends of the bar and k is the thermal conductivity of the material.

SI unit of thermal conductivity: $J/(s \cdot m \cdot C°)$

The units for thermal conductivity can be obtained by solving the equation above to show that $k = QL/(tA\,\Delta T)$ and substituting the units for each term on the right. The SI unit for k is $J \cdot m/(s \cdot m^2 \cdot C°)$ or $J/(s \cdot m \cdot C°)$. Equation 16.1 can also be written as $Q/t = kA\,\Delta T/L$, where Q/t is the heat energy per unit time, or power. The SI unit of power is the joule/second (J/s) or watt (W), so the thermal conductivity is sometimes given in units of $J/(s \cdot m \cdot C°) = W/(m \cdot C°)$.

Different materials have different thermal conductivities, and Table 16.1 gives some representative values. Because metals are such good thermal conductors, they have the largest thermal conductivities. In comparison, liquids and gases generally have small thermal conductivities. In fact, in most fluids the heat transferred by conduction is negligible compared to that transferred by convection when there are

TABLE 16.1 Thermal Conductivities[a] of Selected Materials

Substance	Thermal Conductivity k [J/(s·m·C°)]	Substance	Thermal Conductivity k [J/(s·m·C°)]
Metals		**Other Materials**	
Aluminum	240	Asbestos	0.090
Brass	110	Body fat	0.20
Copper	390	Concrete	1.1
Iron	79	Glass	0.80
Lead	35	Goose down	0.025
Silver	420	Ice (0 °C)	2.2
Steel (stainless)	14	Styrofoam	0.010
Gases		Water	0.60
Air	0.0256	Wood (oak)	0.15
Hydrogen (H_2)	0.180	Wool	0.040
Nitrogen (N_2)	0.0258		
Oxygen (O_2)	0.0265		

[a] Except as noted, the values pertain to temperatures near 20 °C.

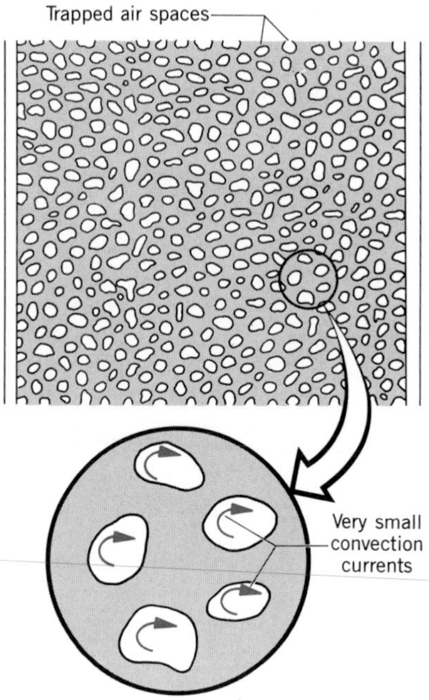

FIGURE 16.8 Styrofoam contains many small, dead-air spaces within it. These small spaces inhibit the formation of large convection currents. Since heat transfer by convection is small, and since air has a low thermal conductivity, Styrofoam is an excellent thermal insulator.

strong convection currents. Air, for instance, with its small thermal conductivity, is an excellent thermal insulator when confined to small spaces where no appreciable convection currents can be established, as Figure 16.8 illustrates. Goose down, Styrofoam, and wool derive their fine insulating properties in part from the small "dead-air" spaces within them.

Example 1 deals with the role that conduction through body fat plays in regulating body temperature.

EXAMPLE 1

When excessive heat energy is produced within the body, the energy must be transferred to the skin and dispersed if the temperature at the body interior is to be maintained at the normal value of 37.0 °C. One possible mechanism for transfer is conduction through the body fat. Suppose that heat energy travels through 3.0 cm of fat in reaching the skin, which has a total surface area of 1.7 m² and a temperature of 34.0 °C. Find the amount of energy that reaches the skin in half an hour.

SOLUTION

In Table 16.1, the thermal conductivity of body fat is given as $k = 0.20$ J/(s·m·C°). According to Equation 16.1,

$$Q = \frac{kA\,\Delta T t}{L}$$

$$= \frac{[0.20 \text{ J/(s·m·C°)}](1.7 \text{ m}^2)(37.0 \text{ °C} - 34.0 \text{ °C})(1800 \text{ s})}{0.030 \text{ m}}$$

$$= \boxed{6.1 \times 10^4 \text{ J}}$$

For comparison, a jogger can generate over ten times this amount of heat energy in a half hour. Thus, conduction through body fat is not a particularly effective way of removing excess heat energy. Heat transfer via blood flow to the skin is more effective and has the added advantage that the body can vary the blood flow as needed.

Example 2 examines the heat conduction through two common thermal insulators, Styrofoam and wood.

EXAMPLE 2
A portable ice chest has walls 2.0 cm thick. The area of the walls is 0.66 m². On a day when the temperature is 35 °C, the chest is partially filled with 3.0 kg of ice at 0 °C. Find the time required to melt the ice when the chest is made from (a) Styrofoam and (b) wood.

SOLUTION
(a) Recall from Section 15.3 that the latent heat of fusion L_f is the heat energy per kilogram needed to melt a solid. Table 15.3 gives $L_f = 3.35 \times 10^5$ J/kg for ice. Thus, the energy necessary to melt 3.0 kg of ice is

$$Q = mL_f = (3.0 \text{ kg})(3.35 \times 10^5 \text{ J/kg}) = 1.0 \times 10^6 \text{ J}$$

The time it takes for this amount of heat energy to penetrate the chest can be determined by using Equation 16.1 with the value of $k = 0.010$ J/(s·m·C°) for the thermal conductivity of Styrofoam (see Table 16.1):

$$t = \frac{QL}{kA\,\Delta T}$$

$$= \frac{(1.0 \times 10^6 \text{ J})(0.020 \text{ m})}{[0.010 \text{ J/(s·m·C°)}](0.66 \text{ m}^2)(35 \text{ °C} - 0 \text{ °C})}$$

$$= \boxed{87\,000 \text{ s or 24 h}}$$

(b) Table 16.1 indicates that the thermal conductivity of wood is 15 times greater than that of Styrofoam. Therefore, other factors being the same, heat energy will penetrate a wooden chest 15 times more quickly than it penetrates a Styrofoam chest: $t = (24 \text{ h})/15 = \boxed{1.6 \text{ h}}$. Styrofoam, with its smaller thermal conductivity, is clearly a better thermal insulator than wood.

16.4 RADIATION

Energy from the sun is brought to earth by large amounts of visible light waves, as well as substantial amounts of infrared and ultraviolet waves, all of which belong to a class of waves known as *electromagnetic waves.* This class also includes the microwaves used for cooking and the radio waves used for AM and FM broadcasts. Many common experiences are related to the fact that electromagnetic waves carry energy. Sunbathers feel hot because their bodies absorb energy from the sun's electromagnetic waves. And anyone who has stood by a roaring fire or put his hand near an incandescent light bulb has experienced a similar effect. Thus, fires and light bulbs also emit electromagnetic waves, and when the energy of such waves is absorbed, it can have the same effect as heat energy.

The process of transferring energy via electromagnetic waves is called *radiation* and, unlike the process of convection or conduction, it does not require a material medium. Electromagnetic waves from the sun, for example, travel through the void of space during their journey to earth.

RADIATION
Radiation is the process in which energy is transferred by means of electromagnetic waves.

All bodies, whether hot or cold, continuously radiate energy in the form of electromagnetic waves. Even an ice cube radiates energy, although so little of it is in the form of visible light that an ice cube cannot be seen in the dark. Likewise, the human body emits insufficient visible light to be seen in the dark. However, the human body can be detected in the dark by electronic cameras that record the infrared waves radiating

Microwaves provide a fast, efficient way to heat food.

Suntans are produced by ultraviolet rays.

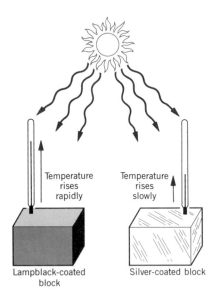

FIGURE 16.9 When two blocks are placed in direct sunlight, the temperature of the block coated with lampblack rises faster than the temperature of the block coated with silver. This result indicates that radiant energy from the sun is being absorbed at a greater rate by the black block than by the silver block.

from the body. Generally, an object does not emit appreciable amounts of visible light until the temperature of the object becomes greater than about 1000 K, and then a characteristic red glow appears, like that of a heating coil on an electric stove. It is not until its temperature reaches about 1700 K that an object glows white-hot, like the tungsten filament in an incandescent light bulb.

In the transfer of energy by radiation, the absorption of electromagnetic waves is just as important as the emission. The surface of an object plays a significant role in determining how much radiant energy the object will absorb. Consider, for example, the two blocks in Figure 16.9. They are identical, except that one has a rough surface coated with lampblack (a fine black soot), while the other has a highly polished silver surface. If a thermometer is inserted into each block and the blocks are placed in direct sunlight, it is found that the temperature of the black block rises at a much faster rate than that of the silvery block. The rapid temperature rise of the black block occurs because lampblack absorbs about 97% of the incident radiant energy, while the silvery surface absorbs only about 10%. As Figure 16.10 indicates, the remaining part of the incident energy is reflected in each case. In fact, we see the lampblack as black in color because so little of the light that falls on it is reflected. In contrast, the silvery surface looks like a mirror because it reflects so much light. Since the color black is associated with nearly complete absorption of visible light, the term *perfect blackbody* or, simply, *blackbody* is used when referring to an object that absorbs *all* the electromagnetic waves falling on it.

We have seen that all objects can emit and absorb electromagnetic waves. In fact, objects do both simultaneously. Moreover, when a body has the same constant temperature as its surroundings,* the amount of radiant energy that the body absorbs must balance the amount that it emits in a given interval of time, as Figure 16.11 illustrates. The block coated with lampblack absorbs and emits the same amount of

---

* The assumption is that *all* of the surroundings has the same temperature. Such would not be the case, for instance, when an object is exposed to sunlight, for the sun is part of the surroundings and has a different temperature than does the air.

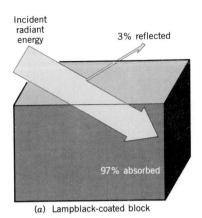

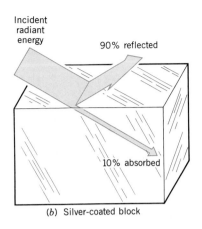

Incident radiant energy

3% reflected

97% absorbed

(a) Lampblack-coated block

Incident radiant energy

90% reflected

10% absorbed

(b) Silver-coated block

FIGURE 16.10 (*a*) The object coated with lampblack absorbs about 97% of the incident radiant energy, while reflecting only 3%. (*b*) The polished silver surface absorbs only about 10% of the incident radiant energy and reflects the remaining 90%.

radiant energy, and the silvery block does too. In either case, if absorption were greater than emission, the block would experience a net gain in energy, and the temperature of the block would rise. Conversely, if the emission were greater than the absorption, the temperature would fall. Consequently, *a material that is a good absorber, like lampblack, is also a good emitter, and a material that is a poor absorber, like polished silver, is also a poor emitter.* A perfect blackbody, being a perfect absorber, is also a perfect emitter.

The fact that a black surface is both a good absorber and a good emitter is the reason that people are uncomfortable wearing dark clothes during the summer. Dark clothes absorb a large fraction of the sun's radiation and then reemit it in all directions. About one-half of the emitted radiation is directed inward toward the body and creates the sensation of warmth. Light-colored clothes, on the other hand, are cooler to wear, since they absorb relatively little of the incident radiation.

The amount of radiant energy Q emitted by a perfect blackbody depends on several factors. First of all, Q is proportional to the radiation time interval t ($Q \propto t$). The longer the time, the greater the amount of energy radiated. Second, experiment shows that Q is proportional to the surface area A ($Q \propto A$). An object with a large surface area radiates more energy than one with a small surface area, other things being equal. Finally, experiment reveals that Q is proportional to the *fourth power of the Kelvin temperature T* ($Q \propto T^4$). This temperature dependence is a strong one and indicates that the emitted energy increases markedly with increasing temperature. If, for example, the Kelvin temperature of an object doubles, the object emits 2^4 or 16 times more energy. Combining these factors into a single proportionality, we see that

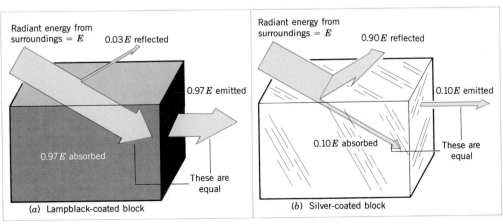

Radiant energy from surroundings = E

0.03E reflected

0.97E emitted

0.97E absorbed

These are equal

(a) Lampblack-coated block

Radiant energy from surroundings = E

0.90E reflected

0.10E emitted

0.10E absorbed

These are equal

(b) Silver-coated block

FIGURE 16.11 When a block and its surroundings have the same constant temperature, the block emits the same amount of radiant energy that it absorbs in a given time interval. Although the block emits radiation in all directions, the emission is represented by a single arrow in these drawings. (*a*) For an incident radiant energy E, a block coated with lampblack absorbs and emits radiant energy in the amount of 0.97E, while (*b*) a block coated with silver absorbs and emits radiant energy in the amount 0.10E.

$Q \propto T^4At$. This proportionality is converted into an equation by inserting a proportionality constant σ, known as the *Stefan-Boltzmann constant*. It has been found experimentally that $\sigma = 5.67 \times 10^{-8}$ J/(s·m²·K⁴):

$$Q = \sigma T^4 A t$$

The relationship above holds only for a perfect emitter. Most objects are not perfect emitters, however. For instance, a dark-colored human body radiates only about 80% of the visible light energy that a perfect emitter would radiate, so Q (for dark skin) $= (0.80)\sigma T^4 At$. A factor such as the 0.80 in this equation is called the *emissivity e* and is a dimensionless number between zero and one. The emissivity is the ratio of the energy an object actually radiates to the energy the object would radiate if it were a perfect emitter. For visible light, the value of e for the human body varies between about 0.65 and 0.80, the smaller values pertaining to lighter skin colors. For infrared radiation, e is nearly one for all skin colors. For a perfect black-body, $e = 1$. Including the factor e on the right side of the expression $Q = \sigma T^4 At$ leads to the *Stefan-Boltzmann law of radiation.*

THE STEFAN-BOLTZMANN LAW OF RADIATION
The radiant energy Q, emitted in a time t by an object that has a Kelvin temperature T, a surface area A, and an emissivity e, is given by

$$Q = e\sigma T^4 At \qquad (16.2)$$

where σ has a value of 5.67×10^{-8} J/(s·m²·K⁴).

In Equation 16.2, the Stefan-Boltzmann constant σ is a universal constant in the sense that its value is the same for all bodies, regardless of the nature of their surfaces. The emissivity e, however, depends on the condition of the surface. Example 3 shows how to apply the Stefan-Boltzmann law.

EXAMPLE 3

A wood-burning stove stands unused in a room where the temperature is 18 °C (291 K). A fire is started inside the stove. Eventually, the temperature of the stove surface reaches a constant 198 °C (471 K), and the room warms to a constant 29 °C (302 K). The stove has an emissivity of 0.900 and a surface area of 3.50 m². Determine the *net* radiant power (emitted power minus absorbed power) generated by the stove when the stove (a) is unheated and has a temperature equal to room temperature and (b) has a temperature of 198 °C.

SOLUTION

(a) The Stefan-Boltzmann law can be used to determine the power [in watts (W)] emitted by the unheated stove; power is energy per unit time or Q/t. Furthermore, in the Stefan-Boltzmann law the temperature is expressed in kelvins:

Power emitted
by stove at 18 °C $= \dfrac{Q}{t} = e\sigma T^4 A$ (16.2)

$= (0.900)[5.67 \times 10^{-8}$ J/(s·m²·K⁴)]
$\times (291 \text{ K})^4 (3.50 \text{ m}^2)$
$= 1280 \text{ W}$

The fact that the unheated stove emits 1280 W of power and yet maintains a constant temperature means that the stove must also absorb 1280 W of radiant power. The power that the stove absorbs originates from the walls, ceiling, and floor of the room, all of which emit radiation. Thus, the *net* power generated by the unheated stove is zero:

Net power
generated by $=$ 1280 W $-$ 1280 W $= \boxed{0}$
stove at 18 °C Power emitted Power emitted by
 by stove at 18 °C room at 18 °C and
  absorbed by stove

(b) The hot stove (198 °C) emits more radiant power than it absorbs from the cooler room (29 °C). The radiant power the stove emits is

Power emitted
by stove at 198 °C $= \dfrac{Q}{t} = e\sigma T^4 A$

$= (0.900)[5.67 \times 10^{-8}$ J/(s·m²·K⁴)]$(471 \text{ K})^4 (3.50 \text{ m}^2)$
$= 8790 \text{ W}$

The radiant power the stove absorbs from the room is identical to the power that the stove would emit at the constant room temperature of 29 °C (302 K). The reasoning here is exactly like that in part (a):

Power emitted by
room at 29 °C and $= \dfrac{Q}{t} = e\sigma T^4 A$
absorbed by stove

$= (0.900)[5.67 \times 10^{-8} \text{ J/(s} \cdot \text{m}^2 \cdot \text{K}^4)](302 \text{ K})^4(3.50 \text{ m}^2)$

$= 1490 \text{ W}$

The *net* radiant power produced by the stove is the difference between the power the stove emits and the power the stove absorbs:

$$\begin{matrix} \text{Net power} \\ \text{generated by} \\ \text{stove at 198 °C} \end{matrix} = \underbrace{8790 \text{ W}}_{\substack{\text{Power emitted} \\ \text{by stove at} \\ \text{198 °C}}} - \underbrace{1490 \text{ W}}_{\substack{\text{Power emitted by} \\ \text{room at 29 °C and} \\ \text{absorbed by stove}}}$$

$$= \boxed{7300 \text{ W}}$$

16.5 APPLICATIONS

When it is in the earth's shadow, an orbiting satellite is shielded from the intense electromagnetic waves emitted by the sun. But when a satellite moves out of the earth's shadow, the satellite experiences the full effect of these waves. As a result, the temperature within a satellite would increase and decrease sharply during an orbital period, to the detriment of sensitive electronic circuitry, unless precautions are taken. To minimize temperature fluctuations, satellites are often coated with a highly reflecting and, hence, poorly absorbing metal foil, as Figure 16.12 shows. By reflecting much of the sunlight, the foil minimizes temperature rises. Being a poor absorber, the foil is also a poor emitter and reduces radiant energy losses. Reducing these losses keeps the temperature from falling excessively when the satellite is in the earth's shadow.

Metal foil

FIGURE 16.12 The highly reflecting metal foil that covers the satellite is a poor absorber and, as such, it protects sensitive electronics against intense electromagnetic waves from the sun. The foil also functions as a poor emitter and helps to minimize radiant energy losses from the satellite when the satellite is on the dark side of the earth.

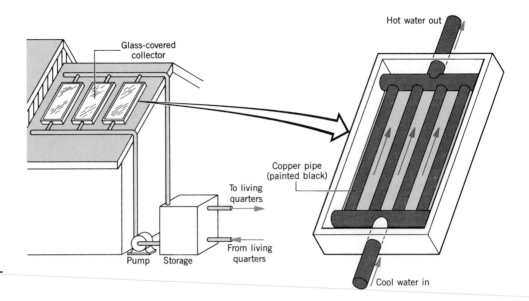

FIGURE 16.13 The design of a hot water solar collector takes into account energy transfer via convection, conduction, and radiation.

The design of solar collectors takes into account all three methods of energy transfer in order to capture radiant energy from the sun. As Figure 16.13 illustrates, cool water is pumped into the collector, heated by solar energy, and then sent into the living quarters. The entire inside of the collector, including the water pipes, is coated with a highly absorptive black paint, to capture as much radiant energy as possible. The copper from which the pipes are made has a large thermal conductivity, and, therefore, readily conducts the absorbed energy to the water. The glass cover minimizes the loss of heat energy due to air convection.

A thermos bottle, sometimes referred to as a Dewar flask, reduces the rate at which hot liquids cool down or cold liquids warm up. A thermos accomplishes its job by minimizing energy transfer via convection, conduction, and radiation. A thermos usually consists of a double-walled glass vessel with silvered inner walls (see Figure 16.14). The space between the walls is evacuated to minimize energy losses due to

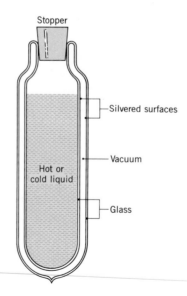

FIGURE 16.14 A thermos bottle or Dewar flask. A thermos bottle minimizes energy transfer due to convection, conduction, and radiation. Therefore, a thermos bottle minimizes the rate of change in the temperature of any hot or cold liquid that it contains.

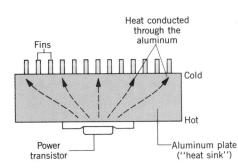

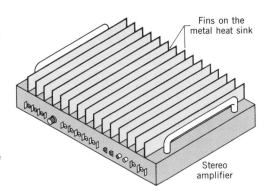

FIGURE 16.15 In an amplifier, the transistors that send electrical current to the speakers are mounted on a metal "heat sink." Heat energy generated by the transistor travels through the metal by conduction and is dissipated by convection, as the air in contact with the fins is heated.

conduction and convection. The silvered surfaces reflect most of the radiant energy that would otherwise enter or leave the liquid in the thermos. Finally, little energy is lost through the glass or the rubberlike stopper, since these materials have relatively small thermal conductivities.

The transfer of heat energy also plays an important role in the design of a stereo amplifier. Transistors in the amplifier send electrical current to the speakers and in so doing produce heat energy. The heat must be removed, since excessively hot transistors are prone to failure. Figure 16.15 illustrates that transistors are often mounted on an aluminum plate, so that heat can be conducted away readily; this plate is known as a "heat sink." Frequently, heat sinks are provided with fins, so as to increase the surface area in contact with the surrounding air. The large area allows the air to be warmed more efficiently, thereby facilitating convection.

SUMMARY

Convection is the process in which heat energy is carried by the bulk movement of a fluid. During natural convection, the warmer, less dense part of a fluid is pushed upward by the buoyant force exerted on it by the surrounding cooler and denser part. Forced convection arises whenever an external device, such as a fan or a pump, causes the fluid to move.

Conduction is the process whereby heat energy is transferred directly through a material, any bulk motion of the material playing no role in the transfer. The heat energy Q conducted during a time t through a bar of length L and cross-sectional area A is expressed as $Q = kA \, \Delta Tt/L$, where ΔT is the difference in temperature between the ends of the bar and k is the **thermal conductivity** of the material from which the bar is made. Materials that have large values of k, such as most metals, are known as thermal conductors. Materials that have small values of k, such as Styrofoam and wood, are referred to as thermal insulators.

Radiation is the process in which energy is transferred by means of electromagnetic waves. All objects, regardless of their temperature, simultaneously absorb and emit electromagnetic waves. A body that is at the same constant temperature as its surroundings absorbs and emits equal amounts of radiant energy per unit time. **Objects that are good absorbers of radiant energy are also good emitters, and objects that are poor absorbers are also poor emitters.** An object that absorbs all the radiation incident upon it is called a **perfect blackbody**. A perfect blackbody, being a perfect absorber, is also a perfect emitter.

The amount of radiant energy Q emitted during a time t by an object whose surface area is A and whose Kelvin temperature is T is given by the **Stefan-Boltzmann law**, $Q = e\sigma T^4 At$. In this equation, σ is the Stefan-Boltzmann constant $[\sigma = 5.67 \times 10^{-8} \text{ J/(s} \cdot \text{m}^2 \cdot \text{K}^4)]$ and e is the emissivity, a dimensionless number characterizing the surface of the object. The emissivity lies between 0 and 1, being zero for a (hypothetical) nonemitting surface and one for a perfect blackbody.

SOLVED PROBLEMS

SOLVED PROBLEM 1
Related Problems: *9 *10 *26

One wall of a house being remodeled consists of $\frac{3}{4}$-inch-thick (0.019 m) plywood backed by 3.0-inch-thick (0.076 m) insulation, as the drawing shows. The temperature at the inside surface is 25.0 °C, while the temperature at the outside surface is 4.0 °C, both being constant. The thermal conductivities of the insulation and plywood are, respectively, 0.030 J/(s·m·C°) and 0.080 J/(s·m·C°), and the area of the wall is 35 m². Find the heat conducted through the wall in one hour.

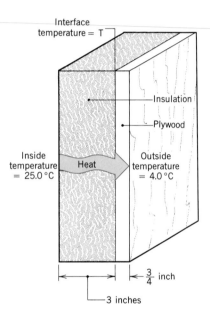

Interface temperature = T

Insulation

Plywood

Inside temperature = 25.0 °C

Heat

Outside temperature = 4.0 °C

$\frac{3}{4}$ inch

3 inches

Solution The temperature at the insulation–plywood interface (see drawing) must be determined before the heat conducted through the wall can be obtained. In calculating this temperature T, we observe that no energy is accumulating in the wall, for the inner and outer temperatures are constant. Therefore, the energy conducted through the insulation must equal the energy conducted through the plywood during the same time. Moreover, the energy conducted through either material can be expressed by the relation $Q = kA \, \Delta Tt/L$. As a result, it follows that

$$Q_{\text{insulation}} = Q_{\text{plywood}}$$

$$\frac{[0.030 \text{ J/(s·m·C°)}](35 \text{ m}^2)(25.0 \text{ °C} - T)(3600 \text{ s})}{0.076 \text{ m}}$$
$$= \frac{[0.080 \text{ J/(s·m·C°)}](35 \text{ m}^2)(T - 4.0 \text{ °C})(3600 \text{ s})}{0.019 \text{ m}}$$

Solving this equation for T reveals that the temperature at the insulation–plywood interface is $T = 5.8$ °C.

The heat energy conducted through the wall can now be found by using $T = 5.8$ °C in the expression for either $Q_{\text{insulation}}$ or Q_{plywood}, since the two quantities are equal. Choosing $Q_{\text{insulation}}$, we find that

$$\frac{\text{Energy conducted}}{\text{through wall}} = Q_{\text{insulation}}$$

$$= \frac{[0.030 \text{ J/(s·m·C°)}](35 \text{ m}^2)(25.0 \text{ °C} - 5.8 \text{ °C})(3600 \text{ s})}{0.076 \text{ m}}$$

$$= \boxed{9.5 \times 10^5 \text{ J}}$$

It is straightforward to calculate that the amount of energy flowing through the plywood in one hour would be $Q_{\text{plywood}} = 110 \times 10^5$ J, if the insulation were absent. Clearly, the insulation helps to reduce the loss of heat energy through the wall substantially.

Summary of Important Points The same amount of energy is conducted through each layer in a multiple-layer material when the temperatures at the layer interfaces are constant. Using this fact and applying the conduction equation ($Q = kA \, \Delta Tt/L$) to each layer makes it possible to determine the temperature at the interface between any two layers. With a knowledge of the interface temperature, the heat conducted through any one layer can be obtained.

QUESTIONS

1. One often hears about heat transfer by convection in gases and liquids, but not in solids. Why?

2. A heavy drape in front of a cold window reduces heat loss through the window considerably, by interfering primarily with one of the three processes of energy transfer. Explain which one.

3. The *windchill factor* is a term used by weather forecasters.

Roughly speaking, it refers to the fact that you feel colder when the wind is blowing than when it is not, even though the air temperature is the same in either case. Which of the three processes for energy transfer plays the principal role in the windchill factor? Explain your reasoning.

4. Often the following warning sign is seen on bridges:

"Caution—Bridge surface freezes before road surface." Account for the warning, singling out the most appropriate of the three transfer processes for heat energy, and remembering that, unlike the road, a bridge has both its top and bottom surfaces exposed to the air.

5. One way that heat energy is transferred from place to place inside the human body is by the flow of blood. Which one of the three energy transfer processes best describes this action of the blood? Justify your answer.

6. A poker used in a fireplace is held at one end, while the other end is in the fire. Why are pokers made of iron rather than copper? Ignore the fact that iron may be cheaper and stronger.

7. For heat conducted along a bar, the term ΔT in the equation $Q = kA\,\Delta Tt/L$ represents the temperature difference between the two ends of the bar. Does the term ΔT in the equation for the linear expansion of a bar, $\Delta L = \alpha L\,\Delta T$, also represent the temperature difference between the two ends of the bar? If not, what does ΔT represent?

8. Grandma says that it is quicker to bake a potato if you put a nail into it. In fact, she is right. Justify her baking technique in terms of one of the three processes of heat transfer.

9. Several days after a snowstorm, the roof on a house is uniformly covered with snow. On a neighboring house, however, the snow on the roof has completely melted. Which house is probably better insulated? Give your reasoning.

10. The metal cooling coils in a freezer have become coated with ice. Explain whether the ice helps or hinders the transfer of energy from warm food to the coolant in the coils.

11. Many high-quality pots have copper bases and polished stainless steel sides. A high-quality pot is designed so heat energy can enter readily and be distributed evenly, while the rate of energy loss from the pot is kept to a minimum. Based on conduction and radiation principles, explain why this design is better than all-copper or all-steel units.

12. Two objects have exactly the same shape. Object A has an emissivity of 0.3, and object B has an emissivity of 0.6. Each radiates the same power. Is the Kelvin temperature of A twice that of B? Give your reasoning.

13. A concave mirror can be used to start a fire by directing sunlight onto a small spot on a piece of paper. Explain why the mirror does not get as hot as the paper.

14. Two strips of material, A and B, are identical, except they have emissivities of 0.4 and 0.7, respectively. The strips are heated to the same temperature and have a red glow. A brighter glow signifies that more energy per second is being radiated. Which strip has the brighter glow? Justify your choice.

15. You have to leave a hot cup of coffee sitting for ten minutes. To have the warmest coffee to drink, should you add cold cream at the beginning or the end of the ten-minute period? Defend your choice.

16. What is the principal means by which heat energy is transferred from a hot body to a cold body when they are separated by (a) a vacuum and (b) a silver bar? Account for each of your answers.

17. Two identical hot cups of cocoa are sitting on a kitchen table. One has a metal spoon in it and one does not. After five minutes, which cup is cooler? Explain which of the three energy transfer processes plays a role here.

18. (a) Would a hot solid cube cool more rapidly if it were left intact or cut in half? Explain your answer in terms of one or more of the three energy transfer processes. (b) Using reasoning similar to that used in answering part (a), decide which cools faster, one pound of wide and flat lasagna noodles or one pound of spaghetti noodles. Assume that both kinds of noodles are made from the same pasta and start out with the same temperature.

PROBLEMS

Note: For problems in this set, use the values for thermal conductivities given in Table 16.1 unless stated otherwise.

Section 16.3 Conduction

1. A glass windowpane is 1.5 m high, 1.2 m wide, and 4.0 mm thick. In one hour, 5.5×10^6 J of heat energy are conducted through the glass. The temperature at the inside surface of the glass is 21 °C. What is the temperature at the outside surface? There are two answers, depending on the direction of the energy flow.

2. The temperature in an electric oven is 160 °C. The temperature at the outer surface in the kitchen is 50 °C. The oven (surface area = 1.6 m²) is insulated with 2.0-cm-thick material whose thermal conductivity is 0.045 J/(s·m·C°). (a) How much energy is used to operate the oven for six hours? (b) At a price of $0.10 per kilowatt·hour for electrical energy, what is the cost of operating the oven?

3. One end of an iron poker is placed in a fire where the temperature is 502 °C, and the other end is kept at a temperature of 26 °C. The poker is 1.2 m long and is 1.0 cm in diameter. Ignoring the heat energy lost along the length of the poker, find the amount of energy conducted from one end of the poker to the other in 5.0 s.

4. A skier wears a jacket that is filled with a 1.5-cm thickness of goose down. Another skier wears a wool sweater that is 5.0 mm thick. Both have the same surface area. Assuming the temperature difference between the inner and outer surfaces of each garment is the same, calculate the ratio (wool/goose down) of the heat energies lost due to conduction during the same time interval.

5. Ignoring air convection, what thickness of body fat is required to give the same insulating value as a 1.0 cm thickness of air?

6. Heat energy is conducted by two bars, one made from asbestos and the other from copper. One end of each bar is at 125 °C and the other end is at 25 °C. The bars have the same cross-sectional area of 4.00 cm² and the same length of 0.200 m. Ignore any energy lost through the sides of the bars. (a) Find the *total* power or energy per second conducted by the two bars. (b) Identify the bar through which the greater percentage of the total power is transmitted. For this bar, which single parameter (length, cross-sectional area, or thermal conductivity) is responsible for the large heat conduction?

7. A refrigerator has a surface area of 5.0 m². It is lined with 8.0-cm-thick insulation whose thermal conductivity is 0.040 J/(s·m·C°). The interior temperature is kept at 5 °C, while the outside temperature is 25 °C. At what rate (in J/s or watts) must heat energy be extracted from the refrigerator?

* 8. In the conduction equation $Q = kA\,\Delta Tt/L$, the combination of factors kA/L is called the *conductance*. The human body has the ability to vary the conductance of the tissue beneath the skin by means of vasoconstriction and vasodilation, in which the flow of blood to the veins and capillaries underlying the skin is decreased and increased, respectively. The conductance can be adjusted over a range such that the tissue beneath the skin is equivalent to a thickness of 0.080 mm of Styrofoam or 3.5 mm of air. By what factor can the body adjust the conductance?

* 9. Two rods, one of aluminum and the other of copper, are joined end to end. The cross-sectional area of each is 4.0 cm², and the length of each is 4.0 cm. The free end of the aluminum rod is kept at 302 °C, while the free end of the copper rod is kept at 25 °C. The loss of energy through the sides of the rods may be ignored. (a) What is the temperature at the aluminum–copper interface? (b) How much energy is conducted through the unit in 2.0 s? (c) What is the temperature in the aluminum rod at a distance of 1.5 cm from the hot end? *(See Solved Problem 1 for a related problem.)*

* 10. In an aluminum pot, 0.20 kg of water boils away in five minutes. The bottom of the pot is 2.5 mm thick and has a surface area of 0.020 m². To prevent the water from boiling too rapidly, a stainless steel plate is placed between the pot and the heating element. The plate is 1.2 mm thick and its area matches that of the pot. Assuming that heat energy is conducted into the water only through the bottom of the pot, find the temperature at (a) the aluminum–steel interface and (b) the steel surface in contact with the heating element. *(See Solved Problem 1 for a related problem.)*

* 11. One end of a brass bar is maintained at 295 °C, while the other end is kept at a constant but lower temperature. The cross-sectional area of the bar is 4.0 cm². Because of insulation, there is negligible energy loss through the sides of the bar. Heat energy

flows through the bar, however, at the rate of 2.7 J/s. What is the temperature of the bar at a point 0.20 m from the hot end?

**12. A 0.30-m-thick sheet of ice floats on the water in a lake. The air temperature at the ice surface is −15 °C. In five minutes, the ice thickens by a small amount. Assuming no heat energy flows from the ground below into the water, find the number of millimeters by which the ice thickens.

**13. The drawing shows a solid cylindrical rod made from a center cylinder of lead and an outer concentric jacket of copper. Except for its ends, the rod is insulated (not shown), so that the loss of energy from the curved surface of the rod is negligible. When a temperature difference is maintained between its ends, this rod conducts one-half the amount of heat energy that it would conduct if it were solid copper. Determine the ratio of the radii r_1/r_2.

Section 16.4 Radiation

14. An object emits 30 W of radiant power. If it were a perfect blackbody, other things being equal, it would emit 90 W of radiant power. What is the emissivity of the object?

15. The amount of radiant power produced by the sun is approximately 3.9×10^{26} W. Assuming the sun to be a perfect blackbody sphere with a radius of 6.96×10^8 m, find its surface temperature.

16. A perfect blackbody has a temperature of 605 °C. An identically shaped object whose emissivity is 0.400 emits the same radiant power as the blackbody. What is the Celsius temperature of this second object?

17. By what factor should the Kelvin temperature of an object be increased to (a) double and (b) triple the radiant energy per second emitted by the object?

18. How many days does it take a perfect blackbody cube (1.00 cm on a side, 30.0 °C) to radiate the amount of energy that a one-hundred-watt light bulb uses in one hour?

* 19. Suppose the skin temperature of a naked person is 34 °C when the person is standing inside a room whose temperature is 25 °C. The skin area of the individual is 1.5 m². (a) Assuming the emissivity is 0.80, find the net loss of radiant power from the body. (b) Determine the number of food Calories of energy (1 food Calorie = 1 kcal) that is lost in one hour due to the net loss rate obtained in part (a). Metabolic conversion of food into energy replaces this loss.

* 20. A solid aluminum sphere is coated with lampblack (emissivity = 0.97) and hung inside an evacuated container. The sphere has a radius of 2.0 cm and is initially at 20.0 °C. The container is maintained at a temperature of 70.0 °C. (a) Assuming that the temperature of the sphere does not change very much, what is the *net energy* gained by the sphere in ten seconds? (b) Estimate the change in temperature of the sphere.

**21. A small sphere (emissivity = 0.90, radius = r_1) is located at the center of a spherical asbestos shell (thickness = 1.0 cm, outer radius = r_2). The thickness of the shell is small compared to the inner and outer radii of the shell. The temperature of the small sphere is 800.0 °C, while the temperature of the inner surface of the shell is 600.0 °C, both temperatures remaining constant. Assuming that $r_2/r_1 = 10.0$ and ignoring any air inside the shell, find the temperature of the outer surface of the shell.

ADDITIONAL PROBLEMS

22. In a light bulb, the tungsten filament has a temperature of 3.0×10^3 °C and radiates sixty watts of power. Assume the emissivity of the filament is 0.36 and estimate the surface area of the filament.

23. A rod, made from an unknown metal, is 1.2 m long and has a cross-sectional area of 3.0×10^{-4} m². Except for its ends, the rod is insulated, so that no heat energy escapes through the sides of the rod. The rod conducts heat energy at a rate equal to or greater than 220 J per minute, when a temperature difference of 75 C° is maintained between its ends. From which one or more of the metals listed in Table 16.1 could the rod be made?

24. In an electrically heated home, the temperature of the ground in contact with a concrete basement wall is 12.8 °C. The temperature at the inside surface of the wall is 20.0 °C. The wall is 0.10 m thick and has an area of 9.0 m². Assume one kilowatt · hour of electrical energy costs $0.10. How many hours are required for one dollar's worth of energy to be conducted through the wall?

* 25. One day, the temperature of a radiator has to be 66 °C to

keep the surrounding walls of a room at 27 °C. The next day is warmer outside, so the temperature of the radiator needs to be only 52 °C to keep the walls at 27 °C. Assuming the room is heated only by radiation, determine the ratio of the *net* power radiated by the unit on the colder day to that radiated on the warmer day.

* 26. Three building materials, plasterboard [$k = 0.30$ J/(s · m · C°)], brick [$k = 0.60$ J/(s · m · C°)], and wood [$k = 0.10$ J/(s · m · C°)], are sandwiched together as the drawing illustrates. The temperatures at the indoor and outdoor surfaces are 27 °C and 0 °C, respectively. Each material has the same thickness and cross-sectional area. Find the temperature at the plasterboard–brick interface and at the brick–wood interface. *(See Solved Problem 1 for a related problem.)*

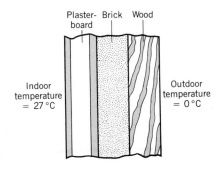

**27. A solid sphere has a temperature of 500.0 °C. The sphere is melted down and recast into a cube that has the same emissivity as the sphere. What should be the Celsius temperature of the cube, if the cube is to emit the same radiant power as the sphere?

**28. Two cylindrical rods have the same mass. One is made of silver (density = 10 500 kg/m³) and one is made of iron (density = 7860 kg/m³). Both rods conduct the same amount of heat energy per second when the same temperature difference is maintained across their ends. What is the ratio of (a) the lengths and (b) the radii of these rods?

CHAPTER 17

The Ideal Gas Law and Kinetic Theory

Scuba divers rely on a supply of compressed air to remain under water for extended periods. The duration of the dive is governed by the volume of the scuba tank, and the temperature and pressure of the compressed air. The kinetic theory of gases describes the relation between these variables.

17.1 INTRODUCTION

Earlier chapters have discussed some of the effects that heat and temperature have on materials. Since all materials are composed of large numbers of molecules, it is natural to search for an understanding of heat and temperature at the molecular level. The internal energy of the material plays a key role in the search, for, as we have seen in Section 15.1, heat is related to this energy supply.

For the sake of simplicity, the present chapter deals with gases and interprets their internal energy and temperature in molecular terms with the aid of a model that is referred to as an *ideal gas.* An application of Newton's second and third laws to the motion of the particles of an ideal gas leads to a relation between temperature and internal energy. Because this approach focuses on the motion of the particles, it is known as the *kinetic theory of gases,* where the word "kinetic" has the same connotation it has in the phrase "kinetic energy." The next section begins with some preliminary concepts that will allow us to deal in a convenient way with large numbers of particles.

17.2 MOLECULAR MASS, THE MOLE, AND AVOGADRO'S NUMBER

ATOMIC AND MOLECULAR MASSES

Often it is necessary to know the masses of the atoms or molecules from which a material is made. The relative masses of the atoms of different elements can be expressed in terms of their *atomic masses,** which indicate how massive an atom of one element is compared to an atom of another. The atomic masses of all the elements are listed in the periodic table, part of which is shown in Figure 17.1. The complete periodic table is given in Appendix H. In general, the masses listed are average values and take into account the various types or isotopes of an element that exist naturally. The periodic table indicates, for example, that the helium atom (He) has an atomic mass of 4.00260, while the corresponding value for the fluorine atom (F) is 18.9984; thus, atomic fluorine is 18.9984/4.00260 = 4.74651 times more massive than atomic helium.

To set up the atomic mass scale, a value must be chosen for one of the elements. The reference element has been chosen to be the most abundant isotope of carbon, called carbon-12, and its atomic mass is defined to be exactly twelve. The units on this scale are called *atomic mass units* (u). Thus, one atomic mass unit is exactly one-twelfth the mass of a carbon-12 atom. In Figure 17.1 the atomic mass of carbon is given as 12.011 u, rather than exactly 12 u, because a small amount (about 1%) of the naturally occurring material is an isotope other than carbon-12, called carbon-13. The value 12.011 u is an average that reflects the small contribution of carbon-13.

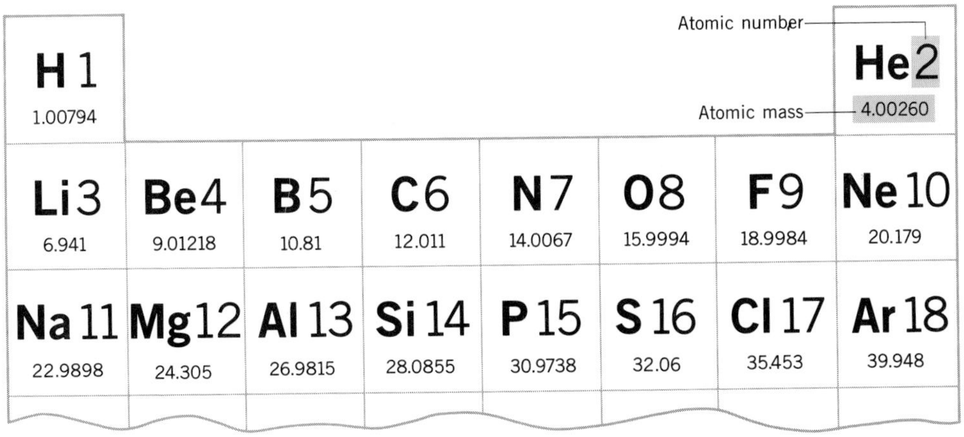

Atomic number								
H 1 1.00794								**He** 2 4.00260 Atomic mass
Li 3 6.941	**Be** 4 9.01218	**B** 5 10.81	**C** 6 12.011	**N** 7 14.0067	**O** 8 15.9994	**F** 9 18.9984	**Ne** 10 20.179
Na 11 22.9898	**Mg** 12 24.305	**Al** 13 26.9815	**Si** 14 28.0855	**P** 15 30.9738	**S** 16 32.06	**Cl** 17 35.453	**Ar** 18 39.948

FIGURE 17.1 A portion of the periodic table showing the atomic number and atomic mass of each element. Note that in the periodic table it is customary to omit the symbol "u" denoting atomic mass units.

* In chemistry the expression "atomic weight" is frequently used in place of "atomic mass."

The *molecular mass* of a molecule is the sum of the atomic masses of its constituent atoms. For instance, the molecular mass of the H_2O molecule is the sum of the atomic masses of the two hydrogen atoms and the oxygen atom; molecular mass of water $= 2(1.00794\ u) + 15.9994\ u = 18.0153\ u$. In the same manner, the molecular mass of carbon dioxide (CO_2) can be shown to be $44.010\ u$.

THE MOLE AND AVOGADRO'S NUMBER

Macroscopic amounts of materials contain large numbers of atoms or molecules. Even in a relatively small volume of gas, $1\ cm^3$ for example, the number is enormous. It is convenient to express such large numbers in terms of a single unit, the *gram-mole,* or simply the *mole* (symbol: *mol*). *One gram-mole of any substance contains as many particles (atoms or molecules) as there are atoms in 12 grams of the isotope carbon-12.* Experiment shows that 12 grams of carbon-12 contain 6.022×10^{23} atoms. This number is called *Avogadro's number* N_A, after the Italian scientist Amedeo Avogadro (1776–1856). Although defined in terms of numbers of carbon atoms, the concept of a mole can be applied to any collection of objects by noting that one mole is Avogadro's number of objects. Thus, one mole of atomic sulfur contains 6.022×10^{23} sulfur atoms, one mole of water contains 6.022×10^{23} H_2O molecules, and one mole of golf balls contains 6.022×10^{23} golf balls. Just as the meter is a unit for expressing length, the mole is a unit for expressing "the amount of substance." In fact, the mole is an SI base unit, along with the meter, the second, the kilogram, and the kelvin.

Although one mole of any substance contains Avogadro's number of particles, it is *not true* in general that a mole of one substance has the same mass as a mole of another substance. For example, one mole of carbon-12 is defined to have a mass of 12 grams. However, one mole of aluminum has a mass of 26.9815 grams for the following reason. An aluminum atom is more massive than a carbon-12 atom by the ratio of their atomic masses, $(26.9815\ u)/(12\ u)$. Since Avogadro's number (1 mole) of carbon-12 atoms has a mass of 12 grams, then Avogadro's number (1 mole) of aluminum atoms must have a mass of

$$\left(\frac{26.9815\ u}{12\ u}\right)(12\ grams) = 26.9815\ grams$$

Thus, one mole of aluminum has a mass that is exactly equal to its atomic mass, expressed in grams. This reasoning can be extended to any other atom or molecule, with the result that *one mole of any substance has a mass in grams that is equal to the atomic or molecular mass of its constituent particles.*

Example 1 illustrates how to determine the number of atoms or molecules present in a known mass of material.

EXAMPLE 1

Figure 17.2*a* shows the Hope diamond (44.5 carats), which is almost pure carbon. Figure 17.2*b* shows the Rosser Reeves ruby (138 carats), which is primarily aluminum oxide (Al_2O_3). Recognizing that one carat is equal to 0.200 g, determine (a) the number of carbon atoms in the diamond and (b) the number of Al_2O_3 molecules in the ruby.

SOLUTION

(a) The mass of the diamond is (44.5 carats)[(0.200 g)/(1 carat)] = 8.90 g. Since the average atomic mass of naturally occurring carbon is 12.011 u, one mole of it has a mass of 12.011 g. Thus, the number of moles of carbon in the diamond is

$$(8.90 \text{ g})\left(\frac{1 \text{ mol}}{12.011 \text{ g}}\right) = 0.741 \text{ mol}$$

Because one mole contains Avogadro's number N_A of carbon atoms, the number of atoms in the diamond is $(0.741)N_A$, or

$$(0.741 \text{ mol})\left(\frac{6.022 \times 10^{23} \text{ atoms}}{1 \text{ mol}}\right) = \boxed{4.46 \times 10^{23} \text{ atoms}}$$

(b) The mass of the ruby is (138 carats)[(0.200 g)/(1 carat)] = 27.6 g. The molecular mass of aluminum oxide (Al_2O_3) is the sum of the atomic masses of its atoms:

$$\text{Molecular mass} = 2(26.9815 \text{ u}) + 3(15.9994 \text{ u})$$
$$= 101.9612 \text{ u}$$

Thus, one mole of Al_2O_3 has a mass of 101.9612 g. Calculations like those in part (a) reveal that in the ruby there is 0.271 mol or $\boxed{1.63 \times 10^{23} \text{ molecules of } Al_2O_3}$.

(a) (b)

FIGURE 17.2 (a) The oval-shaped Hope diamond surrounded by 16 smaller diamonds. (b) The Rosser Reeves ruby. Both gems are on display at the Smithsonian Institution in Washington, D.C.

17.3 THE IDEAL GAS LAW AND THE BEHAVIOR OF GASES

THE IDEAL GAS LAW

The ideal gas law expresses the relationship between the pressure, the Kelvin temperature, the volume, and the number of moles of an ideal gas. An ideal gas is an idealized model for real gases. Real gases behave according to this model if their densities are sufficiently low. The condition of low density means that the molecules of the gas are so far apart that they do not interact (except during collisions that are effectively elastic).

In discussing the constant volume gas thermometer, Section 14.2 has already explained the relationship between the absolute pressure and Kelvin temperature of a low-density gas. This thermometer utilizes a small amount of gas (e.g., hydrogen or helium) placed inside a bulb of constant volume. Since the density, or mass of gas per unit volume, is kept low, the gas behaves as an ideal gas. Experiment reveals that a plot of the pressure of the gas versus the temperature is a straight line, as in Figure 14.4. This plot is redrawn in Figure 17.3, with the change that the temperature axis is now labeled in kelvins rather than in degrees Celsius. The graph indicates that the absolute pressure P is directly proportional to the Kelvin temperature T ($P \propto T$) for a fixed volume and a fixed number of molecules of an ideal gas.

The relation between absolute pressure and the number of molecules of an ideal gas is simple. Experience indicates that it is possible to increase the pressure of a gas by adding more molecules; this is exactly what happens when a tire is pumped up. When the volume and temperature of a low-density gas are kept constant, doubling the number of molecules doubles the pressure, and tripling the number triples the pres-

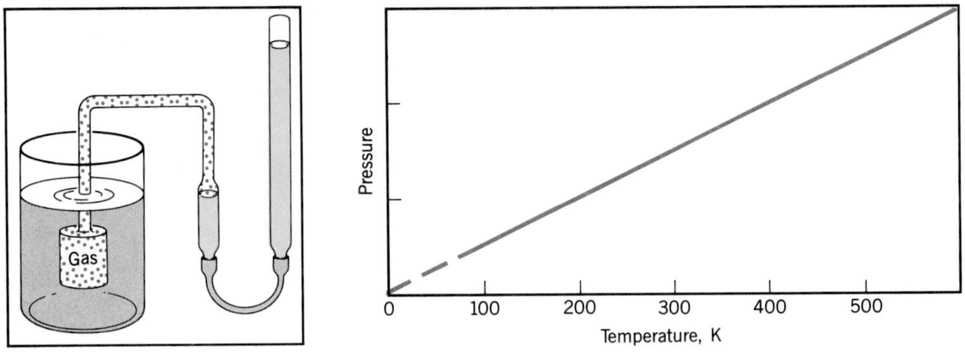

FIGURE 17.3 The pressure inside a constant-volume gas thermometer is directly proportional to the Kelvin temperature, because the gas behaves as an ideal gas.

sure. Thus, under conditions of constant volume and constant temperature, the absolute pressure of an ideal gas is proportional to the number of molecules or the number of moles n of the gas ($P \propto n$).

To see how the absolute pressure of a gas depends on the volume of the gas, consider the partially filled balloon in Figure 17.4a. This balloon is "soft," because the pressure of the air is too low to expand the balloon to its fullest. However, if all of the air in the balloon is squeezed into a small "bubble," as in part b of the figure, the "bubble" has a very tight feel, indicating that the pressure in the smaller space is high enough to stretch the rubber to the limit. Thus, it is possible to increase the pressure of a gas by reducing its volume. As a matter of fact, if the number of molecules and the temperature are kept constant, the pressure of an ideal gas is inversely proportional to its volume V ($P \propto 1/V$).

The three relations discussed above for the absolute pressure of an ideal gas can be expressed as a single proportionality, namely, $P \propto nT/V$. This proportionality can be written as an equation by inserting a proportionality constant R, called the **_universal gas constant._** The value of R has been determined experimentally to be 8.31 J/(mol·K) for any real gas whose density is sufficiently low to ensure ideal gas behavior. The resulting equation is known as the **_ideal gas law._**

IDEAL GAS LAW

The absolute pressure P of an ideal gas is directly proportional to the Kelvin temperature T and the number of moles n of the gas and is inversely proportional to the volume V of the gas: $P = R(nT/V)$. In other words,

$$PV = nRT \qquad (17.1)$$

where $R = 8.31$ J/(mol·K) is the universal gas constant.

Sometimes, it is convenient to express the ideal gas law in terms of the total number of particles N, instead of the number of moles n. In such a situation, we multiply and divide the right side of the ideal gas law by Avogadro's number $N_A = 6.022 \times 10^{23}$ particles/mol,* and recognize that the product nN_A is equal to the total number N of particles:

$$PV = nRT = nN_A \left(\frac{R}{N_A}\right) T = N \left(\frac{R}{N_A}\right) T$$

---

* Since "particles" is not an SI unit, it is often omitted. Then, particles/mol = 1/mol = mol^{-1}.

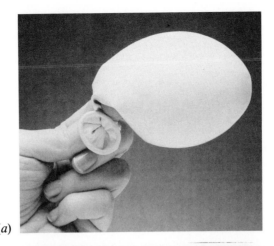

(a)

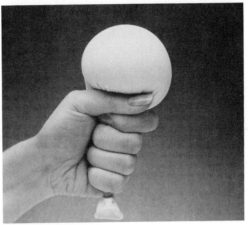

(b)

FIGURE 17.4 (a) The air pressure in the partially filled balloon can be increased by decreasing the volume of the balloon, as illustrated in (b).

The constant term R/N_A is referred to as **Boltzmann's constant,** in honor of the Austrian physicist Ludwig Boltzmann (1844–1906), and is represented by the symbol k:

$$k = \frac{R}{N_A} = \frac{8.31 \text{ J/(mol·K)}}{6.022 \times 10^{23} \text{ mol}^{-1}} = 1.38 \times 10^{-23} \text{ J/K}$$

With this substitution, the ideal gas law becomes

$$PV = NkT \qquad\qquad (17.2)$$

Example 2 presents an application of the ideal gas law.

EXAMPLE 2

In the lungs, the respiratory membrane separates tiny sacs of air (absolute pressure = 1.00×10^5 Pa) from the blood in the capillaries. These sacs are called alveoli, and it is from them that oxygen enters the blood. The average radius of the alveoli is 0.125 mm, and the air inside contains 14% oxygen, which is a somewhat smaller amount than in fresh air. Assuming that the air behaves as an ideal gas at body temperature (310 K), find the number of oxygen molecules in one of the sacs.

SOLUTION
The volume of a sac is $\frac{4}{3}\pi r^3 = 8.18 \times 10^{-12}$ m³. The form of the ideal gas law given in Equation 17.2 is convenient to use here, because it contains the total number N of molecules explicitly:

$$N = \frac{PV}{kT} = \frac{(1.00 \times 10^5 \text{ Pa})(8.18 \times 10^{-12} \text{ m}^3)}{(1.38 \times 10^{-23} \text{ J/K})(310 \text{ K})} = 1.9 \times 10^{14}$$

The number of oxygen molecules is 14% of this value or $0.14N$ = $\boxed{2.7 \times 10^{13}}$.

The next example shows that one mole of an ideal gas occupies a volume of 22.4 liters at a temperature of 0 °C and a pressure of one atmosphere (1.013×10^5 Pa). These conditions of temperature and pressure are known as *standard temperature and pressure (STP)*.

EXAMPLE 3
Find the volume occupied by one mole of an ideal gas at STP conditions.

SOLUTION
Before the ideal gas law can be used, the standard temperature of 0 °C must be converted to kelvins: $T = 0 + 273 = 273$ K. At this Kelvin temperature and at a pressure of 1.013×10^5 Pa, the volume of one mole is

$$V = \frac{nRT}{P} = \frac{(1.00 \text{ mol})[8.31 \text{ J/(mol} \cdot \text{K)}](273 \text{ K})}{1.013 \times 10^5 \text{ Pa}}$$

$$= 22.4 \times 10^{-3} \text{ m}^3 \qquad (17.1)$$

Since 1 liter = 1000 cm³ = 10^{-3} m³, the volume occupied by one mole of an ideal gas at STP is $\boxed{22.4 \text{ liters}}$.

BOYLE'S LAW

Historically the work of several investigators led to the formulation of the ideal gas law. The Irish scientist Robert Boyle (1627–1691) was one of these. He discovered that the absolute pressure of a fixed mass (fixed number of moles) of a low-density gas at constant temperature is inversely proportional to its volume ($P \propto 1/V$). This fact is often called Boyle's law and can be derived from the ideal gas law by noting that $P = nRT/V = \text{constant}/V$ when n and T are constants. Alternatively, if an ideal gas changes from an initial pressure and volume (P_i, V_i) to a final pressure and volume (P_f, V_f), it is possible to write $P_iV_i = nRT$ and $P_fV_f = nRT$, or

$$\begin{bmatrix} \text{Constant } T, \\ \text{constant } n \end{bmatrix} \qquad P_iV_i = P_fV_f \qquad (17.3)$$

This equation is a concise way of expressing Boyle's law.

Figure 17.5 illustrates graphically how pressure and volume change according to

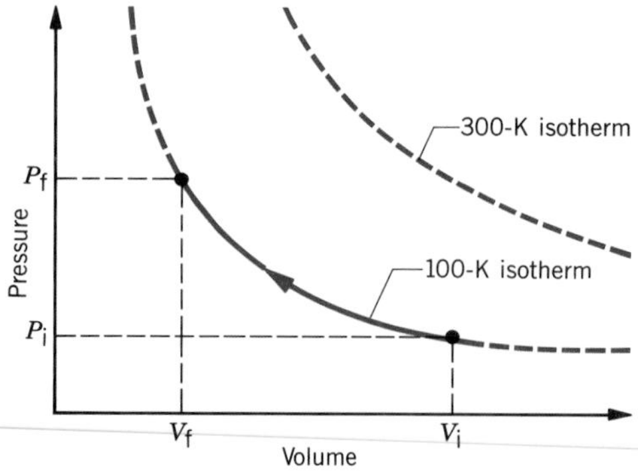

FIGURE 17.5 A pressure-versus-volume plot for a compression of an ideal gas at a constant temperature. Each isotherm is a plot of the equation $P = nRT/V = \text{constant}/V$, one with $T = 100$ K and the other with $T = 300$ K.

Boyle's law. The figure shows a plot of pressure versus volume for the compression of a fixed number of moles of an ideal gas at a constant temperature of 100 K. The gas begins with an initial pressure and volume of P_i and V_i, respectively. The pressure increases as the volume decreases, according to $P = nRT/V$, until the final pressure and volume are reached. This curve is called an **isotherm,** meaning "same temperature." If the temperature of the gas had been 300 K, rather than 100 K, the compression would have occurred along the 300 K isotherm shown in the figure. Different isotherms do not intersect. Example 4 deals with an application of Boyle's law.

EXAMPLE 4

In scuba* diving, a greater water pressure acts on the diver's body at greater depths. The air pressure inside the body cavities (e.g., lungs, sinuses) must be maintained at the same pressure as that of the surrounding water, otherwise they would collapse. For this reason, a special valve automatically adjusts the pressure of the air breathed from a scuba tank to ensure that the air pressure equals the water pressure at all times. The scuba gear in Figure 17.6a consists of a 0.0150-m³ tank that is filled with compressed air at an absolute pressure of 2.02×10^7 Pa. Assuming that air is consumed at a rate of 0.0300 m³ per minute and that the temperature is the same at all depths, determine how long the diver can stay under at a depth of (a) 10.0 m and (b) 30.0 m.

SOLUTION

(a) Since the temperature is constant, the ideal gas law applies in the form of Boyle's law. However, to use Boyle's law, it is necessary to know the pressure P_2 at a depth of $h = 10.0$ m below the surface, as Figure 17.6b indicates. According to Equation 12.4, this pressure is the sum of the atmospheric pressure $P_1 = 1.01 \times 10^5$ Pa that acts on the surface of the water and the pressure $\rho g h$ due to the weight of the water above the diver, where $\rho = 1025$ kg/m³ is the density of seawater and g is the acceleration due to gravity:

$$P_2 = P_1 + \rho g h \qquad (12.4)$$
$$= 1.01 \times 10^5 \text{ Pa} + (1025 \text{ kg/m}^3)(9.80 \text{ m/s}^2)$$
$$\times (10.0 \text{ m})$$
$$= 2.01 \times 10^5 \text{ Pa}$$

Boyle's law can now be used, together with the initial pressure and volume of the tank, to determine the volume of air V_f available at a pressure of $P_f = 2.01 \times 10^5$ Pa:

$$V_f = \frac{P_i V_i}{P_f} = \frac{(2.02 \times 10^7 \text{ Pa})(0.0150 \text{ m}^3)}{2.01 \times 10^5 \text{ Pa}} \qquad (17.3)$$
$$= 1.51 \text{ m}^3$$

Of this volume, only 1.51 m³ − 0.0150 m³ = 1.50 m³ is available for breathing, because 0.0150 m³ of air always remains in the tank. At a consumption rate of 0.0300 m³/min, the tank of compressed air will last for

$$t = \frac{1.50 \text{ m}^3}{0.0300 \text{ m}^3/\text{min}} = \boxed{50.0 \text{ min}}$$

(a)

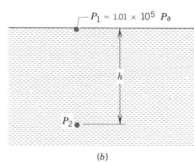

(b)

FIGURE 17.6 (a) A scuba diver. (b) The pressure P_2 at a depth h below the surface of the ocean is given by $P_2 = P_1 + \rho g h$, where $P_1 = 1.01 \times 10^5$ Pa is the air pressure at the surface and ρ is the density of seawater.

(b) The calculation here is analogous to that in part (a). Equation 12.4 indicates that at a depth of 30.0 m, the pressure is 4.02×10^5 Pa. Because this pressure is twice that at the 10.0-m depth, Boyle's law reveals that the volume of air provided by the tank is now only $V_f = 0.754$ m³. The air available for use by the diver is 0.754 m³ − 0.0150 m³ = 0.739 m³. At a consumption rate of 0.0300 m³/min, the air will last for $\boxed{t = 24.6 \text{ min}}$. Thus, the deeper dive must have a shorter duration.

---

* The word is an acronym for self-contained underwater breathing apparatus.

CHARLES' LAW

Another investigator whose work contributed to the formulation of the ideal gas law was the Frenchman, Jacques Charles (1746–1823). He discovered that the volume of a fixed mass (fixed number of moles) of a low-density gas at constant pressure is directly proportional to its Kelvin temperature ($V \propto T$). This relationship is known as Charles' law and can be obtained from the ideal gas law by noting that $V = nRT/P = (\text{constant})T$, if n and P are constant. Equivalently, when an ideal gas changes from an initial volume and temperature (V_i, T_i) to a final volume and temperature (V_f, T_f), it is possible to write $V_i/T_i = nR/P$ and $V_f/T_f = nR/P$, or

$$\begin{bmatrix} \text{Constant } P, \\ \text{constant } n \end{bmatrix} \qquad \frac{V_i}{T_i} = \frac{V_f}{T_f} \qquad\qquad (17.4)$$

Equation 17.4 is one way of stating Charles' law.

DALTON'S LAW OF PARTIAL PRESSURES

In a mixture of gases, two or more different gases are present at the same time. For example, fresh air is a mixture in which approximately 78.1% of the molecules are nitrogen (N_2), 21.0% oxygen (O_2), and 0.9% argon (Ar), with minute amounts of other species. It is a straightforward matter to deal with such mixtures, if the law first proposed by the Englishman John Dalton (1766–1844) is applicable. ***Dalton's law of partial pressures*** is stated as follows: ***The total pressure of a gas mixture is equal to the sum of the partial pressures of the component gases.*** The partial pressure of a gas is the pressure it would exert if it were present by itself in the container at the same temperature as the mixture.

Like Boyle's law and Charles' law, Dalton's law is a natural consequence of ideal gas behavior. Consider, for example, a container of volume V filled with air, and assume that the air is composed only of N_2, O_2, and Ar. If each substance behaves as an ideal gas, the partial pressures are

$$P_N = \frac{n_N RT}{V} \qquad P_O = \frac{n_O RT}{V} \qquad P_A = \frac{n_A RT}{V}$$

where n_N, n_O, and n_A are the number of moles of nitrogen, oxygen, and argon, respectively, in the mixture. The total number of moles is $n = n_N + n_O + n_A$, and the total pressure P of the mixture is

$$P = \frac{nRT}{V} = \frac{(n_N + n_O + n_A)RT}{V} = \underbrace{\frac{n_N RT}{V}}_{P_N} + \underbrace{\frac{n_O RT}{V}}_{P_O} + \underbrace{\frac{n_A RT}{V}}_{P_A}$$

Therefore, $P = P_N + P_O + P_A$, as stated in Dalton's law of partial pressures.

17.4 KINETIC THEORY OF GASES

As useful as it is, the ideal gas law provides little insight into the way that pressure and temperature are related to properties of the molecules themselves, such as their

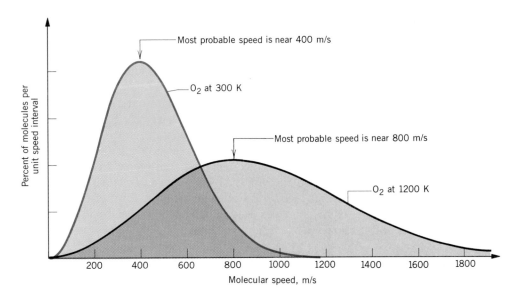

FIGURE 17.7 The Maxwell distribution curves for particle speeds in oxygen gas at temperatures of 300 and 1200 K.

speeds. To show how such microscopic properties are related to the pressure and temperature of an ideal gas, this section examines the dynamics of molecular motion with the aid of Newton's second and third laws.

THE DISTRIBUTION OF MOLECULAR SPEEDS

A macroscopic container filled with a gas at standard temperature and pressure contains a large number of particles (atoms or molecules). These particles are in constant, random motion, colliding with each other and with the walls of the container. In the course of one second, a particle typically undergoes many collisions, and each collision changes the particle's speed and direction of motion. As a result, the atoms or molecules, in general, have different speeds. It is possible, however, to speak about an average particle speed. At any given instant, some particles have speeds less than the average, some near the average, and some greater than the average. For conditions of low gas density, the distribution of speeds within a large collection of molecules at a constant temperature was calculated by the Scottish physicist James Clerk Maxwell (1831–1879). Figure 17.7 displays the speed distribution curves for O_2 gas at two different temperatures. Each curve is called a Maxwell distribution curve. When the gas temperature is 300 K, the maximum in the curve indicates that the most probable speed is about 400 m/s. At a temperature of 1200 K, the distribution curve is shifted to the right, and the most probable speed increases to about 800 m/s.

KINETIC THEORY

If a ball is thrown against a wall, it exerts a force on the wall. As Figure 17.8 illustrates, gas particles do the same thing, except that their mass is smaller, their speed is greater, and there are billions of them. In fact, the number of particles is so great that they strike the walls often enough for the effect of their individual impacts to appear as a

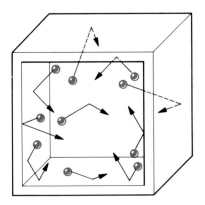

FIGURE 17.8 The pressure that a gas exerts is caused by the impact of its molecules on the walls of the container.

continuous force. Dividing the magnitude of this force by the area of the wall gives the pressure exerted by the gas.

Consider an ideal gas composed of N identical, infinitesimally small particles (point particles) contained in a cubical container whose sides have length L. Except for elastic* collisions, these particles do not interact. Figure 17.9 focuses attention on one particle of mass m as it strikes the right wall perpendicularly and rebounds elastically. While approaching the wall, the particle has a velocity $+v$ and linear momentum $+mv$ (see Section 8.2 for a review of linear momentum). The particle rebounds with a velocity $-v$ and momentum $-mv$, travels to the left wall, rebounds from it, and heads back toward the right wall. The time t between collisions with the right wall is the round-trip distance $2L$ divided by the speed of the particle, that is, $t = 2L/v$. According to Newton's second law of motion, in the form of the impulse-momentum theorem, the average force exerted on the particle by the wall is given by the change in the particle's momentum per unit time:

$$\text{Average force} = \frac{\text{Final momentum} - \text{Initial momentum}}{\text{Time between successive collisions}} \tag{8.4}$$

$$= \frac{(-mv) - (+mv)}{2L/v} = \frac{-mv^2}{L}$$

According to Newton's action-reaction law, the force applied to the wall by the particle is equal in magnitude to this value, but oppositely directed (i.e., $+mv^2/L$). The magnitude F of the *total* force exerted on the right wall is equal to the number of particles that collide with the wall during the time t multiplied by the average force produced by one particle. Since the N particles are moving randomly in three dimensions, one-third of them on the average strike the right wall during the time t. Therefore, the total force is

$$F = \left(\frac{N}{3}\right)\left(\frac{m\overline{v^2}}{L}\right)$$

In the above result, v^2 has been replaced by $\overline{v^2}$, the *average* value of the squared speed. The collection of particles possesses a Maxwell distribution of speeds, so an average value for v^2 must be used, rather than a value for any individual particle. Furthermore, it should be noted that the average value of the squared speed does not equal the square of the average speed: $\overline{v^2} \neq (\overline{v})^2$ (see problem 23).

Often, it is necessary to know the square root of the quantity $\overline{v^2}$. The result is called the ***root-mean-square speed,*** or, for short, the *rms-speed;* $v_{\text{rms}} = \sqrt{\overline{v^2}}$. With this substitution, the result for the total force becomes

$$F = \left(\frac{N}{3}\right)\left(\frac{mv_{\text{rms}}^2}{L}\right)$$

Pressure is force per unit area. Therefore, the pressure P acting on a wall of area L^2 is

$$P = \frac{F}{L^2} = \left(\frac{1}{L^2}\right)\left(\frac{N}{3}\right)\left(\frac{mv_{\text{rms}}^2}{L}\right)$$

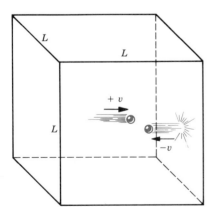

FIGURE 17.9 A gas is contained in a cubical box whose sides have length L. One gas particle is shown colliding elastically with the right wall and rebounding from it.

---

* The term "elastic" is used here to mean that *on the average,* in a large number of particles, there is no gain or loss of translational kinetic energy because of collisions.

Since the volume of the box is $V = L^3$, this equation can be rewritten as

$$PV = \tfrac{2}{3}N(\tfrac{1}{2}mv_{rms}^2) \qquad (17.5)$$

Equation 17.5 relates the macroscopic properties of the gas, its pressure and volume, to the microscopic properties of the constituent particles, their mass and speed. Moreover, the term $\tfrac{1}{2}mv_{rms}^2$ is the average translational kinetic energy $\overline{KE}$ of an individual particle. Thus, it follows that

$$PV = \tfrac{2}{3}N(\overline{KE})$$

This result is similar to the ideal gas law, $PV = NkT$, both equations having identical terms on the left. Consequently, $\tfrac{2}{3}N(\overline{KE}) = NkT$, and we find that

$$\overline{KE} = \tfrac{1}{2}mv_{rms}^2 = \tfrac{3}{2}kT \qquad (17.6)$$

This result is significant, for it provides an interpretation of temperature in terms of the motion of gas particles. Equation 17.6 indicates that the Kelvin temperature is directly proportional to the average translational kinetic energy of an individual particle in an ideal gas, no matter what the pressure and volume are. On the average, the particles in an ideal gas have greater speeds when the gas is hotter than when it is cooler.

If two ideal gases have the same temperature, the relation $\tfrac{1}{2}mv_{rms}^2 = \tfrac{3}{2}kT$ indicates that the average kinetic energy of each kind of gas particle is the same. In general, however, the rms-speed of each kind of particle is not the same, for the masses of the particles may be different. The next example illustrates this fact and shows how rapidly gas particles move at normal temperatures.

EXAMPLE 5

Fresh air is primarily a mixture of nitrogen N_2 (molecular mass = 28.0 u), oxygen O_2 (molecular mass = 32.0 u), and argon Ar (atomic mass = 39.9 u). Assume that each gas behaves as an ideal gas. At a temperature of 293 K, find (a) the average translational kinetic energy and (b) the rms-speed for each type of molecule.

SOLUTION

(a) According to kinetic theory, the particles in each gas have the *same* average translational kinetic energy, since each gas has the same temperature:

$$\overline{KE} = \tfrac{3}{2}kT = \tfrac{3}{2}(1.38 \times 10^{-23} \text{ J/K})(293 \text{ K}) \qquad (17.6)$$

$$= \boxed{6.07 \times 10^{-21} \text{ J}}$$

(b) The average translational speed can be obtained directly from Equation 17.6 as $v_{rms} = \sqrt{2(\overline{KE})/m}$, once the mass of each type of gas particle is known. To calculate the particle mass, recall that one mole, or Avogadro's number of particles, has a mass in grams equal to the atomic or molecular mass of the substance. For nitrogen, the particle mass is

$$m = \frac{28.0 \text{ g/mol}}{6.022 \times 10^{23} \text{ mol}^{-1}}$$

$$= 4.65 \times 10^{-23} \text{ g} = 4.65 \times 10^{-26} \text{ kg}$$

Similarly, we find the particle mass of oxygen to be 5.31×10^{-26} kg and that of argon to be 6.63×10^{-26} kg. The rms-speed for each particle can now be determined. The calculation for nitrogen is shown below, the results for oxygen and argon being obtained in a similar fashion:

[Nitrogen] $\quad v_{rms} = \sqrt{\dfrac{2(\overline{KE})}{m}} = \sqrt{\dfrac{2(6.07 \times 10^{-21} \text{ J})}{4.65 \times 10^{-26} \text{ kg}}}$

$$= \boxed{511 \text{ m/s}}$$

[Oxygen] $\quad v_{rms} = \boxed{478 \text{ m/s}}$

[Argon] $\quad v_{rms} = \boxed{428 \text{ m/s}}$

For comparison, the speed of sound at a temperature of 293 K is 343 m/s (767 mi/h).

The equation $\overline{KE} = \frac{3}{2}kT$ has also been applied to particles that are much larger than single atoms or molecules. The English botanist, Robert Brown (1773–1858) observed through a microscope that pollen grains suspended in water move on very irregular, zigzag paths. This Brownian motion can also be observed with other particle suspensions, such as fine smoke particles in air. In 1905, Albert Einstein (1879–1955) showed that Brownian motion could be explained as a response of the large suspended particles to impacts from the moving molecules of the fluid medium (e.g., water or air). As a result of the impacts, the suspended particles have the same average translational kinetic energy as the fluid molecules, namely, $\overline{KE} = \frac{3}{2}kT$. But unlike the fluid molecules, the particles are large enough to be seen through a microscope, and their average velocity is comparatively slow because of their relatively large mass. Einstein's explanation provided a means by which Avogadro's number could be measured.

THE INTERNAL ENERGY OF A MONATOMIC IDEAL GAS

Chapters 18 and 19 deal with the science of thermodynamics, in which the concept of internal energy plays an important role. Using the results just developed for the average translational kinetic energy, we conclude this section by expressing the internal energy of a monatomic ideal gas in a form that is suitable for use in these chapters.

The internal energy of a substance is the sum of the various kinds of energy that the atoms or molecules of the substance possess. A monatomic ideal gas is composed of single atoms. These atoms are assumed to be so small that the mass is concentrated at a point, with the result that the moment of inertia I about the center of mass is negligible. Thus, the rotational kinetic energy $\frac{1}{2}I\omega^2$ is also negligible. Moreover, vibrational kinetic and potential energies are absent, because the atoms are not connected by chemical bonds and, except for elastic collisions, do not interact. As a result, the internal energy U is simply the total kinetic energy of the N atoms that constitute the gas: $U = N(\frac{1}{2}mv_{rms}^2)$. Since $\frac{1}{2}mv_{rms}^2 = \frac{3}{2}kT$, the internal energy can be written in terms of the Kelvin temperature as

$$U = N(\tfrac{1}{2}mv_{rms}^2) = N(\tfrac{3}{2}kT)$$

Usually, U is expressed in terms of the number of moles n, rather than the number of atoms N. Using the fact that Boltzmann's constant is $k = R/N_A$, where R is the universal gas constant and N_A is Avogadro's number, and realizing that $N/N_A = n$, we find that

$$\begin{bmatrix} \textbf{Monatomic} \\ \textbf{ideal gas} \end{bmatrix} \qquad\qquad U = \tfrac{3}{2}nRT \qquad\qquad (17.7)$$

The internal energy of a monatomic ideal gas is proportional to the number of moles of the gas and the Kelvin temperature of the gas.

*17.5 DIFFUSION

The fragrance of a perfume can often be detected at a distance from an open bottle, because molecules in the perfume evaporate from the liquid, where they are relatively concentrated, and gradually spread out into the surrounding air, where they are less concentrated. During their journey, they collide with other molecules, so their paths resemble the zigzag paths characteristic of Brownian motion. The process in which molecules move from a region of higher concentration to one of lower concentration is called **diffusion.** Diffusion also occurs in liquids (see Figure 17.10) and solids. However, compared to the rate of diffusion in gases, the rate is generally smaller in liquids and even smaller in solids. The host medium, such as the air or water in the above examples, is referred to as the **solvent,** while the diffusing substance, like the perfume molecules or the ink in Figure 17.10, is known as the **solute.** In general terms, then, the solute diffuses through the solvent.

The diffusion process can be described in terms of the arrangement shown in Figure 17.11a. A hollow channel of length L and cross-sectional area A is filled with a fluid. The left end of the channel is connected to a container in which the solute concentration C_2 is relatively high, while the right end is connected to a container in which the solute concentration C_1 is lower. These concentrations are defined as the total mass of the solute molecules divided by the volume of the solute/solvent mixture (e.g., 0.1 kg/m³). Because of the difference in concentration between the ends of the channel, $\Delta C = C_2 - C_1$, there is a net diffusion of the solute from the left end to the right end.

The arrangement in Figure 17.11a is similar to that shown in Figure 16.6 for the conduction of heat along a bar; for convenience, this drawing is reproduced in Figure 17.11b. It has been explained that when the ends of the bar are maintained at different temperatures, T_2 and T_1, the heat energy Q conducted along the bar in a time t is

$$Q = \frac{kA \, \Delta T t}{L} \qquad (16.1)$$

where $\Delta T = T_2 - T_1$, and k is the thermal conductivity. Whereas conduction is the flow of heat energy from a region of higher temperature to a region of lower tempera-

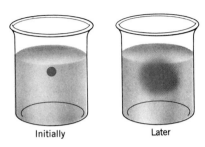

Initially   Later

FIGURE 17.10 When a small drop of ink is placed in water, the ink diffuses into the surrounding water, eventually becoming completely dispersed.

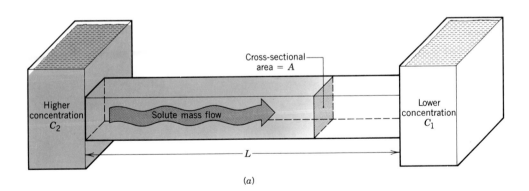

(a)

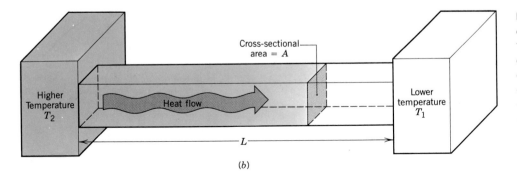

(b)

FIGURE 17.11 (a) Solute mass diffuses through the channel when the two ends are maintained at different concentrations; the solute diffuses from the region of higher concentration to the region of lower concentration. (b) Heat energy is conducted along a bar whose ends are maintained at different temperatures. This arrangement is analogous to that in part a.

ture, diffusion is the mass flow of solute from a region of higher concentration to a region of lower concentration. By direct analogy with Equation 16.1, it is possible to write an equation for diffusion: (1) replace Q by the mass m of solute that is diffusing through the channel, (2) replace $\Delta T = T_2 - T_1$ by the difference in concentrations $\Delta C = C_2 - C_1$, and (3) replace k by a constant known as the diffusion constant D. The resulting equation, first formulated by the German physiologist Adolf Fick (1829–1901), is referred to as *Fick's law of diffusion.*

FICK'S LAW OF DIFFUSION

The mass m of solute that diffuses in a time t through a solvent contained in a channel of length L and cross-sectional area A is

$$m = \frac{DA\,\Delta Ct}{L} \qquad (17.8)$$

where ΔC is the concentration difference between the ends of the channel and D is the diffusion constant.

SI unit for the diffusion constant: m²/s

It can be seen from Equation 17.8 that the diffusion constant has units of m²/s, the exact value depending on both the nature of the solute and the solvent. For example, the diffusion constant for ink in water is different than that for ink in benzene. Example 6 illustrates an important application of Fick's law.

EXAMPLE 6

Large amounts of water can be given off by plants. It has been estimated, for instance, that a single sunflower plant can lose up to a pint of water a day during the growing season. Figure 17.12 shows a cross-sectional view of a leaf. Inside the leaf, water passes

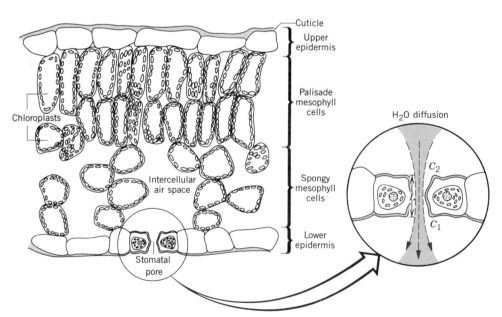

FIGURE 17.12 A cross-sectional view of a leaf, indicating the various types of cells, the intercellular air space, and a stomatal pore. Water vapor diffuses out of the leaf through the stomatal pores.

from the liquid phase to the vapor phase at the walls of the mesophyll cells. The water vapor then diffuses through the inter-cellular air spaces and eventually exits the leaf through small openings, called stomatal pores. The diffusion constant for water vapor in air is $D = 2.4 \times 10^{-5}$ m²/s. A stomatal pore has an area of about $A = 8.0 \times 10^{-11}$ m². The concentration of water vapor on the interior side of a pore is roughly $C_2 = 0.022$ kg/m³, while that on the outside is approximately $C_1 = 0.011$ kg/m³. The pore length is about $L = 2.5 \times 10^{-5}$ m. Determine the mass of water vapor that passes through a stomatal pore in one hour.

SOLUTION

Fick's law of diffusion shows that

$$m = \frac{DA\,\Delta Ct}{L} \tag{17.8}$$

$$= \frac{(2.4 \times 10^{-5} \text{ m}^2/\text{s})(8.0 \times 10^{-11} \text{ m}^2)}{2.5 \times 10^{-5} \text{ m}}$$
$$\times (0.022 \text{ kg/m}^3 - 0.011 \text{ kg/m}^3)(3600 \text{ s})$$

$$= \boxed{3.0 \times 10^{-9} \text{ kg}}$$

This mass of water may not seem like a significant amount. However, a single leaf may have a million or so stomatal pores, so the water lost by an entire plant can be substantial.

SUMMARY

Each element in the periodic table is assigned an **atomic mass.** One **atomic mass unit** (u) is exactly one-twelfth the mass of an atom of carbon-12. The **molecular mass** of a molecule is the sum of the atomic masses of its constituent atoms. One **gram-mole,** or one **mole,** of any substance contains **Avogadro's number** N_A of particles, where $N_A = 6.022 \times 10^{23}$ particles/mole. The mass in grams of one mole of a substance is equal to the atomic or molecular mass of its particles.

The **ideal gas law** relates the absolute pressure P, the volume V, the number of moles n, and the Kelvin temperature T of an ideal gas according to $PV = nRT$, where $R = 8.31$ J/(mol·K) is the universal gas constant. An alternative form of the ideal gas law is $PV = NkT$, where N is the number of particles and $k = R/N_A$ is Boltzmann's constant. A real gas behaves as an ideal gas when the density of the real gas is low enough so its particles do not interact, except via elastic collisions.

Dalton's law of partial pressures states that the total pressure of a gas mixture is equal to the sum of the partial pressures of the component gases. The partial pressure of a gas is the pressure the gas would exert if it were present by itself in the container at the same temperature as the mixture. Dalton's law is a natural consequence of the behavior of ideal gases.

The distribution of particle speeds in an ideal gas at constant temperature is the **Maxwell speed distribution.** According to the **kinetic theory of gases,** an ideal gas consists of a large number of tiny particles (atoms or molecules) that are in constant random motion. The particles are far apart compared to their dimensions, so they do not interact with one another except when elastic collisions occur. The pressure on the walls of a container is produced by the impact of the particles with the walls. One of the most important results of the kinetic theory is that the Kelvin temperature T of an ideal gas is a measure of the average translational kinetic energy $\overline{KE}$ per particle through the relation $\overline{KE} = \frac{3}{2}kT$, where k is Boltzmann's constant. The **internal energy** U of n moles of a monatomic ideal gas is $U = \frac{3}{2}nRT$.

Diffusion is the process whereby solute molecules move from a region of higher concentration to a region of lower concentration. **Fick's law of diffusion** states that the mass m of solute that diffuses in a time t through a solvent contained in a channel of length L and cross-sectional area A is given by $m = (DA\,\Delta Ct)/L$, where ΔC is the concentration difference between the ends of the channel and D is the **diffusion constant.**

QUESTIONS

1. (a) Which, if either, contains a greater number of molecules, a mole of hydrogen (H_2) or a mole of oxygen (O_2)? (b) Which one has more mass? Give reasons for your answers.

2. Why is it possible for two chemically different molecules to have the same molecular mass?

3. Suppose two different substances, A and B, have the same mass densities. (a) In general, does one mole of substance A have the same mass as one mole of substance B? (b) Does 1 m³ of substance A have the same mass as 1 m³ of substance B? Justify each answer.

4. Assuming that air behaves as an ideal gas, explain what happens to the pressure in a tightly sealed house when an electric furnace turns on for a while.

5. A slippery cork is being pressed into a very full (but not 100% full) bottle of wine. When released, the cork slowly slides back out. However, if some wine is removed from the bottle before the cork is inserted, the cork does not slide out. Account for these observations in terms of the ideal gas law.

6. The bubbles emitted by the breathing apparatus of a scuba diver expand as they rise to the surface of the water. Why? In giving your explanation, state any assumptions you are making.

7. A commonly used packing material consists of "bubbles" of air trapped between bonded layers of plastic, as the photograph shows. Using the ideal gas law, explain why this packing material offers less protection on cold days than on warm days.

8. If the translational speed of every molecule in an ideal gas were tripled, would the Kelvin temperature also triple? If not, by what factor would the Kelvin temperature increase? Account for your answer.

9. Suppose that the atoms in a container of helium (He) have the same translational rms-speed as the molecules in a container of oxygen (O₂). Treating each gas as an ideal gas, explain which of the two, if either, has the greater temperature.

10. The molecules in a tank of hydrogen (H₂) have the same translational rms-speed as the molecules in a tank of argon (Ar). What will happen to the rms-speeds of each type of gas (increase, decrease, or remain the same) if the gases are allowed to mix and reach thermal equilibrium? Justify your answer.

11. Example 5 in the text shows that, near room temperature, a molecule of a gas has a translational rms-speed on the order of hundreds of meters per second. At such a speed, a molecule could travel across an ordinary room in just a fraction of a second. Yet, it often takes several seconds, and sometimes minutes, for the smell of a perfume to travel across a room. Why does it take so long?

12. In the lungs, oxygen in very small sacs called alveoli diffuses into the blood. The diffusion occurs directly through the walls of the sacs. The walls are very thin, so the oxygen diffuses over a distance L that is quite small. Because there are so many alveoli, the effective area A across which diffusion occurs is very large. Use this information, together with Fick's law of diffusion, and explain why the mass of oxygen per second that diffuses into the blood is large.

PROBLEMS

Section 17.2 Molecular Mass, the Mole, and Avogadro's Number

1. Hemoglobin has a molecular mass of 64 500 u. Find the mass (in kg) of one molecule of hemoglobin.

2. Glucose is a sugar whose chemical formula is $C_6H_{12}O_6$. (a) Determine the molecular mass of glucose. (b) How many molecules are contained in 350 g of glucose?

3. The hydrogen in a certain amount of hydrochloric acid (HCl) is equivalent to four grams of hydrogen gas (H₂). To how many grams of chlorine gas (Cl₂) is the chlorine in the acid equivalent?

* 4. A mass of 0.135 kg of an element is known to contain 30.1×10^{23} atoms. What is the element?

* 5. Ethyl alcohol (C_2H_5OH) has a density of 806 kg/m³. (a) Determine the mass (in kg) of a molecule of ethyl alcohol, and (b) find the number of molecules in two liters of the liquid.

**6. Estimate the spacing between the centers of neighboring atoms in a piece of solid aluminum, based on a knowledge of the density (2700 kg/m³) and atomic mass of aluminum. *(Hint: Assume the volume of the solid is filled with many small cubes, with one atom at the center of each.)*

Section 17.3 The Ideal Gas Law and the Behavior of Gases

7. In a portable oxygen system, the oxygen (O₂) is contained in a cylinder whose volume is 0.0028 m³. A full cylinder has an absolute pressure of 1.5×10^7 Pa when the temperature is 296 K. Find the mass of oxygen in the cylinder.

8. At the start of a trip, a driver adjusts the gauge pressure in her tires to be 1.80×10^5 Pa when the outdoor temperature is 11 °C. At the end of the trip she measures the gauge pressure to be 2.00×10^5 Pa. Neglecting the expansion of the tires, find the air temperature (in degrees Celsius) inside the tires at the end of the trip.

9. A ten-liter container is initially evacuated, and then 2.0 g of water is placed in it. After a time, all the water evaporates. The temperature of the vapor is 75 °C. Find the absolute pressure.

10. The relative humidity is 67% on a day when the temperature is 34 °C. Using the graph that accompanies problem 26 in Chapter 15, determine the number of moles of water vapor per cubic meter of air.

11. In certain regions of outer space the temperature is about 3 K, and there are approximately 5 molecules per cubic centimeter. (a) How many moles are there per cubic meter? (b) What is the pressure exerted by this gas?

12. A young male adult takes in about 5.0×10^{-4} m³ of fresh air during a normal breath. Fresh air contains approximately 21% oxygen. Assuming that the pressure in the lungs is 1.0×10^5 Pa and that air is an ideal gas at a temperature of 310 K, find the number of oxygen molecules in a normal breath.

13. The gas in the space above the liquid in a bottle of soda pop contains primarily carbon dioxide, oxygen, water vapor, and nitrogen. The small amounts of other gases may be ignored. Of the total number of molecules in the gas, 19.0% is oxygen, 3.0% is water vapor, and 76.0% is nitrogen. The total pressure is 1.6×10^5 Pa. What is the partial pressure of the carbon dioxide in (a) pascals and (b) millimeters of mercury?

14. A frictionless air-filled cylinder is fitted with a movable piston, as the drawing shows. The height h is 12.0 cm when the temperature is 0 °C and increases as the temperature increases. What is the value of h (in cm) when the temperature reaches 45 °C?

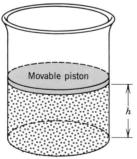

Movable piston

h

* 15. An air bubble is located at a depth h beneath the surface of a pond. The bubble rises and reaches the surface, where the volume of the bubble has expanded to twice its initial value. Assuming the temperature of the bubble remains constant and atmospheric pressure is 1.01×10^5 Pa, find the depth h.

* 16. An ideal gas exerts a pressure of 5.00×10^5 Pa when its temperature is 127 °C. What is the mass density if the gas is (a) helium (He) and (b) nitrogen (N₂)?

* 17. A primitive diving bell consists of a cylindrical tank with one end open and one end closed. The tank is lowered into a freshwater lake, open end downward. Water rises into the tank, compressing the trapped air, whose temperature remains constant during the descent. The tank is brought to a halt when the distance between the surface of the water in the tank and the surface of the lake is 40.0 m. Atmospheric pressure above the lake is 1.01×10^5 Pa. Find the fraction of the tank's volume that is filled with water.

* 18. One assumption of the ideal gas law is that the atoms or molecules themselves occupy a negligible volume. Verify that this assumption is reasonable by considering gaseous argon (Ar). Argon has an atomic radius of 0.70×10^{-10} m. For STP conditions, calculate the percentage of the total volume occupied by the atoms.

**19. A spherical balloon is made from a material whose mass is 2.00 kg. The thickness of the material is negligible compared to the 1.20-m radius of the balloon. The balloon is filled with helium gas at a temperature of 23.0 °C and just floats in air, neither rising nor falling. The density of the surrounding air is 1.19 kg/m³. Find the absolute pressure of the helium gas.

**20. A gas fills one part of a horizontal cylinder whose radius is 5.00 cm. The initial absolute pressure of the gas is 1.01×10^5 Pa. A frictionless movable piston separates the gas from another part of the cylinder that is evacuated and contains an ideal spring, as the drawing shows. The piston is initially held in place by a pin. The spring is initially unstrained and the length of the gas-filled chamber is 20.0 cm. When the pin is removed and the gas is allowed to expand at constant temperature, the length of the gas-filled chamber doubles. Determine the spring constant of the spring.

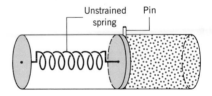

Unstrained   Pin
spring

Section 17.4 Kinetic Theory of Gases

21. Suppose that very fine smoke particles are suspended in air. The translational rms-speed of a smoke particle is 4.9×10^{-3} m/s and the temperature is 295 K. Find the mass of the particle.

22. The temperature of a gas is raised from 27.0 °C to 100.0 °C. (a) By what fraction is the translational rms-speed of the gas molecules increased? (b) By what fraction is the translational kinetic energy increased?

23. The average value of the squared speed $\overline{v^2}$ does not equal the square of the average speed $(\overline{v})^2$. To verify this fact, consider three particles with the following speeds: $v_1 = 2.0$ m/s, $v_2 = 5.0$ m/s, and $v_3 = 8.0$ m/s. Calculate (a) $\overline{v^2} = \frac{1}{3}(v_1^2 + v_2^2 + v_3^2)$ and (b) $(\overline{v})^2 = [\frac{1}{3}(v_1 + v_2 + v_3)]^2$.

24. (a) At what temperature would the translational rms-speed of hydrogen molecules (H_2) be equal to that of oxygen molecules (O_2) at 27 °C? (b) At what temperature would the average translational kinetic energy of the hydrogen molecules be equal to that of oxygen molecules at 27 °C?

25. Neon (Ne) is a monatomic gas. At a temperature of 50.0 °C, what is the internal energy of two grams of neon?

＊26. Helium (He), a monatomic gas, fills a ten-liter container. The absolute pressure of the gas is 6.2×10^5 Pa. How long would a 0.25-hp engine have to run (1 hp = 746 W) to produce an amount of energy equal to the internal energy of this gas?

＊27. In 10.0 s, 200 bullets strike and embed themselves in a wall. The bullets strike the wall perpendicularly. Each bullet has a mass of 5.0 g and a speed of 1200 m/s. (a) What is the average change in momentum per second for the bullets? (b) Determine the average force exerted on the wall. (c) Assuming the bullets are spread out over an area of 3.0 cm², obtain the average pressure they exert on this region of the wall.

＊28. The partial pressure of oxygen (O_2) in a room is 2.12×10^4 Pa. The room has a volume of 50.0 m³ and contains 421 moles of oxygen. Find the translational rms-speed of the oxygen molecules.

Section 17.5 Diffusion

29. The diffusion constant of ethanol in water is 12.4×10^{-10} m²/s. A cylinder has a cross-sectional area of 4.00 cm² and a length of 2.00 cm. A difference in ethanol concentration of 1.50 kg/m³ is maintained between the ends of the cylinder. In one hour, what is the mass (in kg) of ethanol that diffuses through the cylinder?

30. It is found that the amino acid glycine diffuses through water at a rate of 6.00×10^{-14} kg/s. The diffusion constant is 10.6×10^{-10} m²/s. A tube of water has a radius of 1.40 cm. What difference in concentration per unit length of the tube must be maintained to give this flow?

31. Carbon tetrachloride (CCl_4) is diffusing through benzene (C_6H_6), as the drawing illustrates. The concentration of CCl_4 at the left end of the tube is maintained at 1.00×10^{-2} kg/m³, and the diffusion constant is 20.0×10^{-10} m²/s. The CCl_4 enters the tube at the constant rate of 5.00×10^{-13} kg/s. Using this data, and that shown in the drawing, find (a) the mass of CCl_4 per second that passes point A, (b) the concentration at point A, and (c) the mass of CCl_4 per second that passes point B.

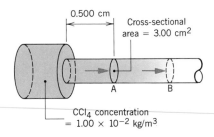

0.500 cm

Cross-sectional area = 3.00 cm²

A   B

CCl_4 concentration = 1.00×10^{-2} kg/m³

＊32. It is possible to convert Fick's law into a form that is useful in situations where the concentration is essentially zero at one end of the diffusion channel ($C_1 = 0$ in Figure 17.11a). In such circumstances, Fick's law becomes $m = DAC_2t/L$. (a) Noting that AL is the volume V of the channel and that m/V is the average concentration of solute in the channel, show that Fick's law becomes $t = L^2/(2D)$. This form of Fick's law can be used to estimate the time required for the first solute molecules to traverse the channel. (b) A bottle of perfume is opened in a room where convection currents are absent. Assuming the diffusion constant for perfume in air is 1.0×10^{-5} m²/s, estimate the minimum time required for the perfume to be smelled 2.5 cm away.

ADDITIONAL PROBLEMS

33. An ultrahigh vacuum pump can reduce the absolute pressure to 1.2×10^{-7} Pa. (a) In a volume of 2.0 m³ and at a temperature of 27 °C, how many molecules of gas are present at this pressure? (b) Suppose you had a penny for every molecule in part (a) and spent money day and night at the rate of one million dollars per hour. How many years would it take to spend this money?

34. A three-liter container holds a mixture of 4.00 g of carbon dioxide gas (CO_2) and 5.00 g of carbon monoxide gas (CO) at 0 °C. (a) What is the total absolute pressure inside the container? For each type of molecule, find (b) the average translational kinetic energy and (c) the translational rms-speed.

35. The chlorophyll-a molecule, $C_{55}H_{72}MgN_4O_5$, is important in photosynthesis. (a) Determine its molecular mass (in atomic mass units). (b) What is the mass (in grams) of 3.00 moles of chlorophyll-a molecules?

36. A tube has a length of 0.015 m and a cross-sectional area of 7.0×10^{-4} m². The tube is filled with a solution of sucrose in water. The diffusion constant of sucrose in water is 5.0×10^{-10} m²/s. A difference in concentration of 3.0×10^{-3} kg/m³ is maintained between the ends of the tube. How much time is required for 8.0×10^{-13} kg of sucrose to be transported through the tube?

37. Divers working at great depths must deal with the problem of nitrogen narcosis, in which nitrogen dissolves into the blood at toxic levels. One way to avoid the problem is to breathe a mixture containing only helium and oxygen. In a diving chamber, the 7.0-m³ work space contains 6600 moles of helium and 210 moles of oxygen and has a temperature of 21 °C. Find (a) the partial pressure of helium, (b) the partial pressure of oxygen, and (c) the total pressure.

＊38. If the translational rms-speed of the water vapor molecules in air is 648 m/s, what is the translational rms-speed of the carbon dioxide molecules?

＊39. A closed cylindrical tank has a height of 0.80 m. The tank is initially filled with air at an absolute pressure of 2.0 atm. Water is pumped into the tank until the absolute air pressure reaches 6.0 atm. Assuming the temperature of the air remains constant, determine the height of the water.

＊ 40. A drop of water has a radius of 1.20 mm. How many water molecules are in the drop?

＊ 41. Assume the pressure in a room remains constant at 1.01×10^5 Pa and the air is composed only of nitrogen (N_2). The volume of the room is 60.0 m³. When the temperature increases from 16 °C to 29 °C, what mass of air escapes from the room?

＊＊42. At the normal boiling point of a material, the liquid phase has a density of 958 kg/m³ and the vapor phase has a density of 0.598 kg/m³. Determine the ratio of the distance between neighboring molecules in the gas phase to that in the liquid phase. *(Hint: Assume the volume of each phase is filled with many small cubes, with one molecule at the center of each cube.)*

＊＊43. A frictionless movable piston has negligible mass and thickness. This piston traps a gas in a cylinder, as the drawing indicates. On top of the piston is a layer of mercury that extends to the top of the cylinder and is exposed to the atmosphere. Initially the temperature is 0 °C, and the heights of the mercury and the gas are both 76.0 cm. The temperature is then increased, and in the process, half of the mercury spills out. Ignoring the thermal expansion of the mercury itself, find the final temperature.

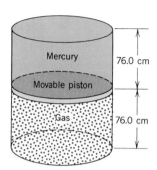

CHAPTER 18

Thermodynamics: The Zeroth Law and the First Law

The "Gumper" at Yellowstone National Park is a pool of mud at its boiling temperature. Scientists define the concept of temperature with the help of the zeroth law of thermodynamics and describe the process of heating the mud in terms of the first law of thermodynamics.

18.1 INTRODUCTION

As Chapter 15 explains, heat is energy that flows from a high-temperature body to a low-temperature body because of the difference in temperature. As Chapter 7 discusses, work is the energy needed to move an object through a distance along the direction of an applied force. Energy in the form of heat and work obeys certain fundamental laws of nature, and *thermodynamics* is the branch of physics that is built upon these laws. There are four laws of thermodynamics. This chapter considers the first two, and the next chapter deals with the remaining two.

It is necessary to clarify some terminology before we begin our study of thermodynamics. The *system* is the collection of objects on which attention is being focused, while the *surroundings* are everything else in the environment. The system and its surroundings are usually separated by walls of some kind. Walls that allow heat to flow through them are referred to as *diathermal walls.* Walls that do not permit heat to flow are called *adiabatic walls.* The *state of the system* refers to the physical condition of the system and is described by specifying the values of the physical parameters of the system, usually pressure, volume, and temperature.

Figure 18.1 uses a weather balloon to illustrate these ideas. The system of interest, for example, might be the helium gas within the balloon. In this event, the surroundings would be the skin of the balloon, the outside air, and the instrument payload carried by the balloon. Assuming the mass of helium in the balloon is known, the state of the system would be described by giving values for the pressure, volume, and temperature of the gas.

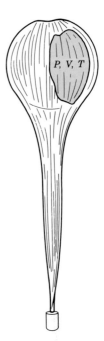

FIGURE 18.1 The helium gas in a weather balloon is one example of a thermodynamic system. Everything else in the drawing is considered as the surroundings, including the outside air. The state of the system is described by values for the pressure P, volume V, and temperature T of the helium.

18.2 THE ZEROTH LAW OF THERMODYNAMICS

The zeroth law of thermodynamics deals with the concept of *thermal equilibrium.* Two systems are said to be in thermal equilibrium if there is no net flow of heat energy between them when they are brought into thermal contact. For example, Figure 18.2a shows two systems, labeled X and Z, each within containers whose walls are made from thick slabs of insulation that prevents the flow of heat. Each system has the same temperature, as indicated by the thermometer Y. In part b, one wall of each container is replaced by a thin silver sheet, and the two silver sheets are touched together. Silver has a large thermal conductivity, so heat flows through it readily and the silver sheets function as diathermal walls. Even though the diathermal walls would permit it, no net flow of heat occurs in part b, indicating that system X and system Z are in thermal equilibrium. There is no net flow of heat because the two systems have the same temperature. Thus, *temperature is the indicator of thermal equilibrium in the sense that there is no net flow of heat between two systems in thermal contact that have the same temperature.*

In Figure 18.2, the role of system Y, the thermometer, is an important one. When the thermometer is in thermal equilibrium with system X, the thermometer is in the same state as when it is in equilibrium with system Z. In other words, the thermometer registers the same temperature in both cases. In effect, systems X and Z are both in thermal equilibrium with system Y; consequently, systems X and Z are found to be in thermal equilibrium with each other. This finding is an example of the *zeroth law of thermodynamics.*

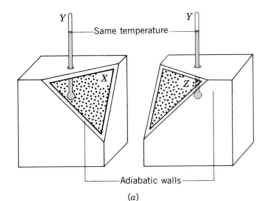
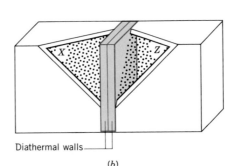

FIGURE 18.2 (*a*) Systems X and Z are surrounded by adiabatic walls and have the same temperature, as recorded on thermometer Y. (*b*) When system X is put into thermal contact with system Z through diathermal walls, there is no net flow of heat energy between the two systems.

THE ZEROTH LAW OF THERMODYNAMICS
Two systems individually in thermal equilibrium with a third system are in thermal equilibrium with each other.

The zeroth law establishes temperature as the indicator of thermal equilibrium. Moreover, the law implies that all parts of a system must be in thermal equilibrium if the system is to have a definable single temperature. In other words, there can be no net flow of heat within the system.

18.3 THE FIRST LAW OF THERMODYNAMICS

Heat flow is one way that a system can gain energy from or lose energy to its surroundings. A system also gains energy if the surroundings do work on the system. Conversely, a system loses energy by doing work on the surroundings. When a system gains energy as a result of heat flow and work, the energy gained becomes part of the internal energy of the system. The internal energy increases. When a system loses energy, its internal energy decreases. In either case, the change in the internal energy is governed by the principle of conservation of energy. The *first law of thermodynamics* is just an expression of the conservation principle, as Equation 18.1 indicates.

THE FIRST LAW OF THERMODYNAMICS
When a system absorbs an amount of heat Q and performs an amount of work W, the internal energy of the system changes by an amount ΔU, from an initial value of U_i to a final value U_f:

$$\Delta U = U_f - U_i = Q - W \qquad (18.1)$$

Notice that the work W enters with a minus sign on the right side of Equation 18.1. The minus sign is present as a matter of convention, because work done *by* the system is assigned a positive value and work done *on* the system is assigned a negative value. Likewise by convention, heat *absorbed* by the system is positive and heat *given off* by the system is negative. Example 1 illustrates the use of these sign conventions.

EXAMPLE 1

A system absorbs 1500 J of heat energy from its surroundings. Determine the change in the internal energy of the system when (a) the system performs 2200 J of work on the surroundings and (b) the surroundings perform 2200 J of work on the system.

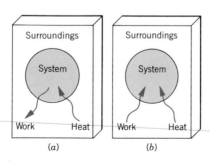

FIGURE 18.3 (*a*) The system gains energy in the form of heat from the surroundings. However, the system performs work on the surroundings and, therefore, also loses some energy. (*b*) The system gains energy in the form of heat from the surroundings and also in the form of work performed on the system by the surroundings.

SOLUTION

(a) Figure 18.3a illustrates this situation. According to convention, heat absorbed and work done by the system are both positive. The first law of thermodynamics gives the change in the internal energy of the system:

$$\Delta U = Q - W = (+1500 \text{ J}) - (+2200 \text{ J}) \quad (18.1)$$
$$= \boxed{-700 \text{ J}}$$

The negative value for ΔU means that the internal energy of the system *decreases* by 700 J. The system consumes 700 J of energy from its internal supply in performing an amount of work that exceeds the amount of heat energy absorbed.

(b) Figure 18.3b shows this case. The calculation is similar to that above, except that the work done on the system is negative, by convention:

$$\Delta U = Q - W = (+1500 \text{ J}) - (-2200 \text{ J}) = \boxed{+3700 \text{ J}}$$

The positive value for ΔU means that the internal energy of the system *increases* by 3700 J. The increase occurs because the system receives energy from the surroundings in the form of both heat and work.

In the first law of thermodynamics, the internal energy U, heat Q, and work W are energy quantities, and each is expressed in energy units such as joules. However, there is a fundamental difference between U, on the one hand, and Q and W on the other. The next example sets the stage for explaining this difference.

EXAMPLE 2

Suppose the temperature of three moles of a monatomic ideal gas is raised from $T_i = 350$ K to $T_f = 540$ K by two different methods. In the first method, 5500 J of heat energy are added to the gas, while in the second method, 1500 J are added. In each case determine (a) the change in internal energy and (b) the work done.

SOLUTION

(a) In both methods the change in internal energy of the gas is the same, for the internal energy of an ideal gas depends only on the Kelvin temperature ($U = \frac{3}{2}nRT$) and the change in temperature is the same. As the temperature increases, the internal energy increases:

$$\Delta U = \tfrac{3}{2}nR(T_f - T_i)$$
$$= \tfrac{3}{2}(3.0 \text{ mol})[8.31 \text{ J/(mol} \cdot \text{K})](540 \text{ K} - 350 \text{ K})$$
$$= \boxed{7100 \text{ J}}$$

(b) Since ΔU is now known and the heat is given in each case, the first law of thermodynamics can be used to determine the work:

[1st method] $W = Q - \Delta U = 5500 \text{ J} - 7100 \text{ J}$
$$= \boxed{-1600 \text{ J}}$$

[2nd method] $W = Q - \Delta U = 1500 \text{ J} - 7100 \text{ J}$
$$= \boxed{-5600 \text{ J}}$$

In the first method, more heat energy is supplied, so the magnitude of the work needed to produce the 7100-J increase in the internal energy is less.

To understand the difference between U, on the one hand, and Q and W on the other, consider the value for ΔU in Example 2. For both methods, ΔU is the same. The value of ΔU is determined once the initial and final temperatures of the system are specified, because the internal energy of an ideal gas depends only on the Kelvin temperature. Temperature is one of the variables that, along with pressure and volume, define the state of a system. Thus, in general,* the internal energy depends only on the state of the system and not on the method by which the system arrives at a given state. In recognition of this characteristic, internal energy is referred to as a *function of state*. In contrast, heat and work are not functions of state, because they

---

* The fact that an ideal gas is used in Example 2 does not restrict our conclusion here. Had a real (nonideal) gas or other material been used, the only difference would have been that the expression for the internal energy would have been more complicated. It might have involved the volume V, as well as the temperature T, for instance.

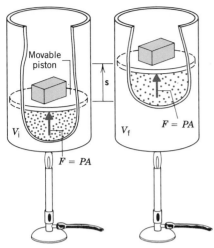

FIGURE 18.4 The substance in the chamber is expanding isobarically, because the pressure is held constant. The pressure is determined by the external atmospheric pressure and by the weight of the movable piston and the block resting on the piston.

have different values for each different method used to make the system change from one state to another, as in Example 2.

18.4 THERMAL PROCESSES INVOLVING PRESSURE, VOLUME, AND TEMPERATURE

A system can interact with its surroundings in many ways, and the heat and work that come into play always obey the first law of thermodynamics. This section introduces four common thermal processes that any system can undergo. In each case, it is assumed that the process is *quasi-static;* that is, the process occurs slowly enough that a uniform pressure and temperature exist throughout all regions of the system at all times.

To begin with, *an isobaric process is one that occurs at constant pressure.* For instance, Figure 18.4 shows a substance (solid, liquid, or gas) contained in a chamber fitted with a movable frictionless piston. The substance itself is the system, while everything else (the container, piston, block, and burner) is regarded as the surroundings. The pressure P experienced by the system is always the same and is determined by the external atmospheric pressure and by the weight of the piston and the block on the piston. Heating the substance makes it expand, and the substance does work W in lifting the piston and block through the displacement **s**. The work can be calculated from the definition of work as the magnitude F of the force times the magnitude s of the displacement, $W = Fs$ (Equation 7.1). In this case, the force is generated by the pressure P acting on the cross-sectional area A of the piston, according to $F = PA$. With this substitution for F, the work done by the expanding substance is $W = (PA)s$. But the product $A \cdot s$ is the change in volume of the material, $\Delta V = V_f - V_i$, where V_i and V_f are the initial and final volumes, respectively. Thus, the expression for the work is

$$\begin{bmatrix} \textbf{Isobaric} \\ \textbf{process} \end{bmatrix} \qquad W = P\,\Delta V = P(V_f - V_i) \qquad (18.2)$$

Consistent with our sign convention for specifying work, this result predicts a positive value for the work done *by a system* when the system expands isobarically (V_f exceeds V_i). Equation 18.2 also applies to an isobaric compression (V_f less than V_i). In this situation, the work is negative, again consistent with our sign convention, since work must be done *on the system* to compress it. Example 3 emphasizes that the expression $W = P\,\Delta V$ applies to any kind of system, solid, liquid, or gas, as long as the pressure remains constant while the volume changes.

EXAMPLE 3

One gram of water is placed in the cylinder in Figure 18.4, and the pressure is maintained at 2.0×10^5 Pa. The temperature of the water is raised by 31 C°. In one case, the water is in the liquid phase and expands by the small amount of 1.0×10^{-8} m³. In another case, the water is in the vapor phase and expands by the much greater amount of 7.1×10^{-5} m³. For each case, find (a) the work done by the water and (b) the change in the internal energy of the water.

SOLUTION

(a) In both cases, the process is isobaric, so the work done is given by Equation 18.2:

$$W_{\text{liquid}} = P\,\Delta V = (2.0 \times 10^5 \text{ Pa})(+1.0 \times 10^{-8} \text{ m}^3)$$

$$= \boxed{+0.0020 \text{ J}}$$

$$W_{\text{gas}} = P\,\Delta V = (2.0 \times 10^5 \text{ Pa})(+7.1 \times 10^{-5} \text{ m}^3)$$

$$= \boxed{+14 \text{ J}}$$

The work is positive, since it is done *by the system* on the surroundings. When a liquid or solid is heated (or cooled) under conditions of constant pressure, it expands (or contracts) by only a small amount compared to a gas. Hence, over comparable temperature ranges, the work of expansion (or compression) for liquids and solids is negligible compared to that for gases.

(b) The first law of thermodynamics can be used to obtain the change in internal energy, provided a value for the heat Q is available along with the work determined in part (a). In each case, the heat can be obtained by using the specific heat capacity in the expression $Q = cm\,\Delta T$ (Equation 15.1). For liquid water $c = 4186$ J/(kg·C°), while for water vapor the value at constant pressure is $c_P = 2020$ J/(kg·C°) (see Section 15.2 for a discussion of specific heat capacity):

$$\Delta U = Q - W \qquad (18.1)$$

$$\Delta U_{\text{liquid}} = [4186 \text{ J/(kg·C°)}](0.0010 \text{ kg})(31 \text{ C°}) - 0.0020 \text{ J}$$
$$= 130 \text{ J} - 0.0020 \text{ J} = \boxed{+130 \text{ J}}$$

$$\Delta U_{\text{gas}} = [2020 \text{ J/(kg·C°)}](0.0010 \text{ kg})(31 \text{ C°}) - 14 \text{ J}$$
$$= 63 \text{ J} - 14 \text{ J} = \boxed{+49 \text{ J}}$$

Virtually all the 130 J of heat energy added to the liquid serves to change the internal energy, since the volume change and the corresponding work of expansion are so small. In contrast, a significant fraction of the 63 J of heat energy added to the vapor causes work of expansion to be done, so that noticeably less than 63 J is left for the internal energy change.

It is often convenient to display thermal processes graphically. For instance, Figure 18.5 shows a plot of pressure versus volume for an isobaric expansion. Since the pressure is constant, the graph is a horizontal straight line, beginning at the initial volume V_i and ending at the final volume V_f. In terms of such a plot, the work $W = P(V_f - V_i)$ is the area under the graph, which is the shaded rectangle of height P and width $V_f - V_i$ in the figure.

Another common thermal process is an **isochoric process, one that occurs at constant volume.** Figure 18.6a illustrates an isochoric process in which a substance (solid, liquid, or gas) is heated. The substance would expand if it could, but the rigid walls of the container keep the volume constant. The expansion of the container itself is negligible. Because the volume is constant, the pressure inside rises, and the substance exerts more and more force on the walls. While enormous forces can be generated in the closed container, no work is done, since the system does not move. The pressure-volume plot shown in part *b* of the drawing also indicates that no work is done. The graph is a vertical straight line, for the volume is constant. The work is the area under the graph, as it is for an isobaric process, and because the area under the vertical line is zero, the work is zero. Since no work is done, the first law of thermodynamics indicates that the heat in an isochoric process serves only to change the internal energy of the system: $\Delta U = Q - W = Q$.

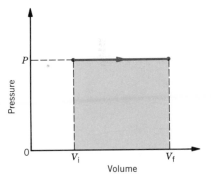

FIGURE 18.5 For an isobaric process, a pressure-versus-volume plot is a horizontal straight line, since the pressure is constant. The work done [$W = P(V_f - V_i)$] is the shaded rectangular area under the line.

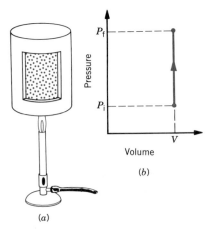

FIGURE 18.6 (*a*) The substance in the chamber is being heated isochorically, because the rigid walls of the chamber keep the volume constant. (*b*) The presure-volume plot for an isochoric process is a vertical straight line. The area under the line is zero, indicating that no work is done.

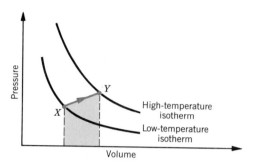

FIGURE 18.7 The thermal process that occurs from X to Y cannot be classified according to one of the four types discussed in the text. However, the work done by the gas is given by the shaded area.

A third important thermal process is an ***isothermal process, one that takes place at constant temperature.*** The next section illustrates the details of an isothermal process when the system is an ideal gas.

Last, there is the ***adiabatic process, one that occurs without the transfer of any heat.*** Since there is no heat transfer, Q equals zero, and the first law indicates that $\Delta U = Q - W = -W$. Thus, when a system does work adiabatically, the internal energy of the system decreases by exactly the amount of the work done. When work is done on a system adiabatically, the internal energy increases correspondingly. The next section discusses an adiabatic process in more detail.

The four thermal processes discussed above are not the only ones that a system can undergo. A process may be complex enough that no part of it is recognizable as one of the four. For instance, Figure 18.7 shows a process for an ideal gas in which the pressure, volume, and temperature are changed along the straight line from X to Y. No part of this process is strictly isobaric, isochoric, isothermal, or adiabatic. It has been pointed out that the area under a pressure-volume graph is the work in an isobaric process. With the aid of integral calculus, it can be shown that ***the area under a pressure-volume graph is the work for any kind of process.*** Therefore, the area representing the work has been shaded in Figure 18.7. The volume increases, and the gas does work. This work is positive by convention, as is the area. In contrast, if a process reduces the volume, work must be done on the gas; this work is negative by convention, and the area under the corresponding pressure-volume graph is also negative.

18.5 THERMAL PROCESSES THAT UTILIZE AN IDEAL GAS

ISOTHERMAL EXPANSION OR COMPRESSION

Figure 18.8*a* shows an example of a system that performs work isothermally, that is, at a constant temperature. The metal cylinder contains n moles of an ideal gas, and the large mass of hot sand maintains the cylinder and gas at a constant Kelvin temperature T. The piston is held in place initially so the volume of the gas is V_i. As the external force applied to the piston is reduced quasi-statically, the gas expands quasi-statically to the final volume V_f. Figure 18.8*b* gives a plot of pressure ($P = nRT/V$) versus volume for the process. The work W done by the gas is *not* given by $W = P\,\Delta V = P(V_f - V_i)$, because the pressure is not constant. Nevertheless, the work is equal to the area under the curve. The techniques of integral calculus lead to the following result* for W:

* In this result, "ln" denotes the natural logarithm to the base $e = 2.71828$. The natural logarithm is related to the common logarithm to the base ten by $\ln(V_f/V_i) = 2.303 \log(V_f/V_i)$.

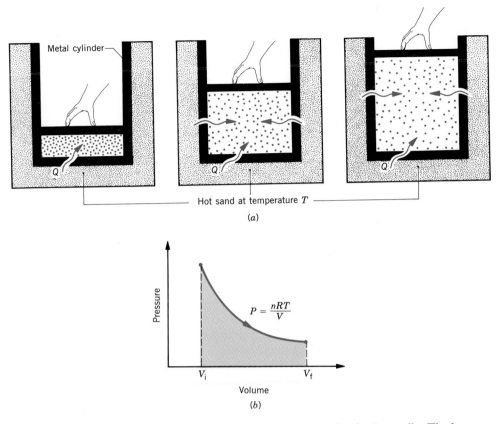

FIGURE 18.8 (*a*) The ideal gas in the metal cylinder is expanding isothermally. The large mass of hot sand keeps the temperature constant. The force holding the movable piston in place is reduced slowly, so the expansion occurs quasi-statically. (*b*) The work done by the expanding gas is given by the shaded area under the pressure-volume graph.

$$
\begin{bmatrix}
\textbf{Isothermal} \\
\textbf{expansion or} \\
\textbf{compression of} \\
\textbf{an ideal gas}
\end{bmatrix}
\qquad
W = nRT \ln\left(\frac{V_f}{V_i}\right)
\qquad (18.3)
$$

Where does the energy for this work originate? Since the internal energy of an ideal gas is proportional to the Kelvin temperature ($U = \frac{3}{2}nRT$ for a monatomic ideal gas), the internal energy remains constant throughout an isothermal process. Thus, the change in internal energy is zero, and the first law of thermodynamics becomes $\Delta U = 0 = Q - W$. In other words, $Q = W$, and the energy to do the work originates in the hot sand. Heat energy flows into the gas from the sand, as part *a* of the drawing illustrates. If the gas is compressed isothermally, Equation 18.3 still applies, and heat energy flows out of the gas into the sand. The following example deals with the isothermal expansion of an ideal gas.

EXAMPLE 4

Two moles of argon gas at 25 °C expand isothermally from an initial volume $V_i = 0.025 \text{ m}^3$ to a final volume of $V_f = 0.050 \text{ m}^3$. Assuming argon behaves as an ideal gas, find (a) the work done by the expanding gas, (b) the change in the internal energy of the gas, and (c) the amount of heat energy supplied to the gas.

SOLUTION

(a) The work done by the gas can be found from Equation 18.3:

$$W = nRT \ln\left(\frac{V_f}{V_i}\right)$$

$$= (2.0 \text{ mol})[8.31 \text{ J/(mol·K)}](298 \text{ K}) \ln\left(\frac{0.050 \text{ m}^3}{0.025 \text{ m}^3}\right)$$

$$= \boxed{+3400 \text{ J}}$$

(b) There is no change in the internal energy of an ideal gas during an isothermal process: $\boxed{\Delta U = 0}$.

(c) The heat energy Q added to the gas can be determined from the first law of thermodynamics:

$$Q = \Delta U + W = 0 + 3400 \text{ J} = \boxed{+3400 \text{ J}} \quad (18.1)$$

All the heat energy supplied is used for doing work, since the internal energy of the ideal gas does not change.

The behavior of a real gas under isothermal conditions is more complicated than that represented by Equation 18.3. A real gas does not, in general, obey the ideal gas law, and the internal energy of a real gas depends on factors in addition to the Kelvin temperature.

ADIABATIC EXPANSION OR COMPRESSION

Figure 18.9a shows an arrangement in which n moles of an ideal gas do work under adiabatic conditions, expanding quasi-statically from an initial volume V_i to a final volume V_f. The arrangement is similar to that used in Figure 18.8 for isothermal expansion. However, a different amount of work is done here, because the gas in the cylinder is now surrounded by insulating material that prevents the flow of heat, so

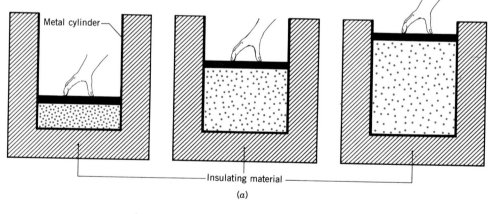

(a)

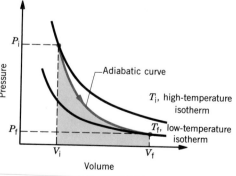

(b)

FIGURE 18.9 (a) The ideal gas in the metal cylinder is expanding adiabatically, because the insulating material prevents the flow of heat energy between the gas and its surroundings. The force holding the movable piston in place is reduced slowly, so the expansion occurs quasi-statically. (b) A plot of pressure versus volume is given by the curve shown in color. This adiabatic curve intersects the isotherms given by the ideal gas law at the initial temperature T_i and the final temperature T_f. The work done by the gas is given by the shaded area.

$Q = 0$. According to the first law of thermodynamics, the change in internal energy is $\Delta U = Q - W = -W$. Since the internal energy of an ideal monatomic gas is $U = \frac{3}{2}nRT$, it follows that $\Delta U = U_f - U_i = \frac{3}{2}nR(T_f - T_i)$, where T_i and T_f are the initial and final Kelvin temperatures. With this substitution, the relation $\Delta U = -W$ becomes

$$\left[\begin{array}{l}\textbf{Adiabatic} \\ \textbf{expansion or} \\ \textbf{compression of} \\ \textbf{a monatomic} \\ \textbf{ideal gas}\end{array}\right] \qquad W = \tfrac{3}{2}nR(T_i - T_f) \qquad (18.4)$$

Equation 18.4 indicates that if the ideal gas expands adiabatically and does work (W is positive), the final temperature of the gas is less than the initial temperature. The internal energy of the gas is reduced to provide the necessary energy, and because the internal energy is proportional to the Kelvin temperature, the temperature decreases. Figure 18.9b shows a plot of pressure versus volume for this process. The adiabatic curve (in color) intersects the isotherms given by the ideal gas law at the higher initial temperature [$T_i = P_iV_i/(nR)$] and the lower final temperature [$T_f = P_fV_f/(nR)$]. These two isotherms are the solid black curves in the drawing. The area under the adiabatic curve represents the work done.

The reverse of an adiabatic expansion is an adiabatic compression (W is negative), and Equation 18.4 indicates that the final temperature exceeds the initial temperature. The energy provided by the agent doing the work increases the internal energy of the gas. As a result, the gas becomes hotter.

The equation that gives the adiabatic curve between the initial pressure and volume (P_i, V_i) and the final pressure and volume (P_f, V_f) in Figure 18.9b can be derived using integral calculus. The result is

$$\left[\begin{array}{l}\textbf{Adiabatic} \\ \textbf{expansion or} \\ \textbf{compression of} \\ \textbf{an ideal gas}\end{array}\right] \qquad P_iV_i^{\gamma} = P_fV_f^{\gamma} \qquad (18.5)$$

where the exponent γ is the ratio of the specific heat capacities at constant pressure and constant volume, $\gamma = c_P/c_V$. Equation 18.5 applies in conjunction with the ideal gas law, for *each point* on the adiabatic curve satisfies the relation $PV = nRT$.

18.6 SPECIFIC HEAT CAPACITIES AND THE FIRST LAW OF THERMODYNAMICS

In this section the first law of thermodynamics is used to gain some understanding of the factors that determine the specific heat capacity of a material. Remember, when the temperature of a substance changes as a result of heat flow, the change in temperature ΔT and the amount of heat Q are related according to $Q = cm\,\Delta T$. In this expression c denotes the specific heat capacity in units of J/(kg·C°), and m is the mass in kilograms. Here, it is more convenient to express the amount of material as the number of moles n, rather than the number of kilograms. Therefore, we replace the expression $Q = cm\,\Delta T$ with the following analogous expression:

$$Q = Cn \, \Delta T \qquad (18.6)$$

where the capital letter C (as opposed to the lowercase c) refers to the **molar specific heat capacity** in units of J/(mol·K). In addition, the unit for measuring the temperature change ΔT is the Kelvin (K) rather than the Celsius degree (C°). For gases it is necessary to distinguish between the molar specific heat capacity C_P for conditions of constant pressure and C_V for conditions of constant volume. With the help of the first law and an ideal gas as an example, it is possible to see why C_P and C_V differ.

According to the first law, $Q = \Delta U + W$. Thus, the heat needed to raise the temperature of an ideal gas from an initial temperature T_i to a final temperature T_f can be calculated from ΔU and W. The change ΔU in internal energy is the same for conditions of constant pressure and for conditions of constant volume, because the temperature changes are identical and the internal energy of an ideal gas depends only on temperature. For a monatomic ideal gas $U = \frac{3}{2}nRT$ (Equation 17.7), and $\Delta U = U_f - U_i = \frac{3}{2}nR(T_f - T_i)$. When the heating process occurs at constant pressure, the work done is given by Equation 18.2: $W = P \, \Delta V = P(V_f - V_i)$. For an ideal gas $PV = nRT$, so the work becomes $W = nR(T_f - T_i)$. On the other hand, when volume is constant, $\Delta V = 0$, and the work done is zero, as it is in any isochoric process. The calculation of the heat is summarized below:

$$Q = \Delta U + W$$

$$Q_{\text{constant pressure}} = \tfrac{3}{2}nR(T_f - T_i) + nR(T_f - T_i)$$

$$Q_{\text{constant volume}} = \tfrac{3}{2}nR(T_f - T_i) + 0$$

The molar specific heat capacities can now be determined, since Equation 18.6 indicates that $C = Q/[n(T_f - T_i)]$:

Specific heat at constant pressure for a monatomic ideal gas		

$$C_P = \tfrac{3}{2}R + R = \tfrac{5}{2}R \qquad (18.7)$$

Specific heat at constant volume for a monatomic ideal gas		

$$C_V = \tfrac{3}{2}R \qquad (18.8)$$

The ratio γ of the specific heats is

Monatomic ideal gas	

$$\gamma = \frac{C_P}{C_V} = \frac{\tfrac{5}{2}R}{\tfrac{3}{2}R} = \frac{5}{3} \qquad (18.9)$$

For real monatomic gases near room temperature, experimental values of C_P and C_V give ratios very close to the theoretical value of $\frac{5}{3}$.

The difference between C_P and C_V arises because work is done when the gas expands in response to the addition of heat energy under conditions of constant pressure, whereas no work is done under conditions of constant volume. For a monatomic ideal gas, C_P exceeds C_V by an amount equal to R, the ideal gas constant:

$$C_P - C_V = R \qquad (18.10)$$

In fact, it can be shown that Equation 18.10 applies to any kind of ideal gas—monatomic, diatomic, etc.

18.7 LATENT HEATS AND THE FIRST LAW OF THERMODYNAMICS

Heat energy must be supplied or removed to enable a substance to change from one phase to another. As Section 15.3 explains, the energy per unit mass is called the latent heat of the phase change. The symbols L_f, L_v, and L_s denote, respectively, the latent heats of fusion, vaporization, and sublimation. The first law of thermodynamics can be applied to the phase change process and helps to explain the factors determining latent heat.

For example, Figure 18.10 shows the vaporization of a liquid of mass m. The heat Q needed to accomplish the change of phase is equal to the latent heat of vaporization L_v multiplied by the mass m of liquid that has been vaporized: $Q = mL_v$. According to the first law, however, $Q = \Delta U + W$. Equating these two expressions for Q reveals that $mL_v = \Delta U + W$. While the phase change is occurring, the pressure is constant, so the work is given by the isobaric work expression, $W = P\,\Delta V$. With this substitution we find that

$$mL_v = \underbrace{\Delta U}_{U_g - U_\ell} + \underbrace{P\,\Delta V}_{P(V_g - V_\ell)} \tag{18.11}$$

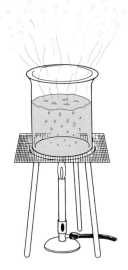

FIGURE 18.10 The heat provided by the burner causes the liquid to vaporize.

This equation indicates that two factors determine the latent heat of vaporization: (1) the difference ΔU between the internal energy U_g in the gas phase and the internal energy U_ℓ in the liquid phase; (2) the work $P\,\Delta V$ done as the substance expands from the smaller volume V_ℓ of the liquid phase to the larger volume V_g of the gas phase. Example 5 illustrates the relative importance of these two factors.

EXAMPLE 5

At a temperature of 373 K and a pressure of 1.01×10^5 Pa, heat is applied to one kilogram of liquid water until the water is converted entirely into vapor. Find (a) the heat energy added to the liquid, (b) the work done by the water as it changes phase, and (c) the difference in internal energy between water in the vapor phase and water in the liquid phase.

SOLUTION

(a) Since the latent heat of vaporization for water is $L_v = 2.26 \times 10^6$ J/kg, the heat required to change 1.00 kg of water from liquid to vapor is $\boxed{Q = 2.26 \times 10^6 \text{ J}}$.

(b) The isobaric work done by the water in changing phase is $W = P\,\Delta V = P(V_g - V_\ell)$. The volume V_ℓ of the liquid is the mass divided by the density of liquid water at 373 K ($\rho = 958$ kg/m³):

$$V_\ell = \frac{m}{\rho} = \frac{1.00 \text{ kg}}{958 \text{ kg/m}^3} = 1.04 \times 10^{-3} \text{ m}^3$$

The volume V_g of the gas can be estimated from the ideal gas

law. One kilogram of water vapor corresponds to $n = (1.00 \times 10^3 \text{ g})/(18.0 \text{ g/mol}) = 55.6$ mol. Therefore,

$$V_g = \frac{nRT}{P} = \frac{(55.6 \text{ mol})[8.31 \text{ J/(mol·K)}](373 \text{ K})}{1.01 \times 10^5 \text{ Pa}}$$
$$= 1.71 \text{ m}^3$$

Clearly, V_ℓ can be neglected compared to V_g. Therefore, the isobaric work is

$$W = P(V_g - V_\ell) = PV_g = (1.01 \times 10^5 \text{ Pa})(1.71 \text{ m}^3)$$
$$= \boxed{0.17 \times 10^6 \text{ J}}$$

(c) The results from parts (a) and (b) can be used in Equation 18.11 to show that the difference between the internal energies of the gas and liquid phases is

$$\Delta U = mL_v - P(V_g - V_\ell) = 2.26 \times 10^6 \text{ J} - 0.17 \times 10^6 \text{ J}$$
$$= \boxed{2.09 \times 10^6 \text{ J}}$$

Equations analogous to Equation 18.11 also exist for the solid-liquid and solid-gas phase changes. As in Example 5, the nature of the phases determines the relative importance of ΔU and $P\,\Delta V$, as far as the latent heat is concerned.

SUMMARY

Thermodynamics is the branch of physics that is built upon the laws that are obeyed by energy in the form of work and heat. A thermodynamic **system** is the collection of objects on which attention is being focused, and the **surroundings** are everything else. The **state of the system** is the physical condition of the system, as described by values for physical parameters, usually pressure, volume, and temperature.

Two systems are in **thermal equilibrium** if there is no net flow of heat energy between them when they are brought into thermal contact. The **zeroth law of thermodynamics** states that two systems individually in thermal equilibrium with a third system are in thermal equilibrium with each other. **Temperature** is the indicator of thermal equilibrium in the sense that there is no net flow of heat between two systems in thermal contact that have the same temperature.

The **first law of thermodynamics** states that when a system absorbs an amount of heat Q and performs an amount of work W, the internal energy of the system changes from its initial value of U_i to a final value of U_f, according to $\Delta U = U_f - U_i = Q - W$. The first law is an expression of the conservation of energy. The internal energy is called a **function of state,** because it depends only on the state of the system and not on the method by which the system came to be in a given state. Heat and work are not functions of state, because they depend on how the system is changed from one state to another.

Thermal processes are **quasi-static** when they occur slowly enough so that a uniform pressure and temperature exist throughout the system. An **isobaric process** is one that occurs at constant pressure. The work W done when a system changes from an initial volume V_i to a final volume V_f at a constant pressure P is $W = P\,\Delta V = P(V_f - V_i)$. An **isochoric process** is one that takes place at constant volume, and no work is done in such a process. An **isothermal process** is one that occurs at constant temperature. An **adiabatic process** is one that takes place without the transfer of heat. The work done in a quasi-static thermal process is given by the area under the pressure-versus-volume graph for the process.

When n moles of an ideal gas change quasi-statically from an initial volume V_i to a final volume V_f at a constant Kelvin temperature T, the work done is $W = nRT \ln(V_f/V_i)$. Since the internal energy of an ideal gas is proportional to the Kelvin temperature, the internal energy does not change during an isothermal process. When n moles of a monatomic ideal gas change quasi-statically and adiabatically from an initial temperature T_i to a final temperature T_f, the work done is $W = \frac{3}{2}nR(T_i - T_f)$. Along with the ideal gas law, an ideal gas also obeys the relation $P_iV_i^{\gamma} = P_fV_f^{\gamma}$ in an adiabatic process, where $\gamma = C_P/C_V$ is the ratio of the specific heat capacities at constant pressure and constant volume.

The **molar specific heat capacity** C of a substance determines how much heat energy Q is added or removed when the temperature of n moles of the substance changes by an amount ΔT: $Q = Cn\,\Delta T$. For a monatomic ideal gas, the molar specific heat capacities at constant pressure and constant volume are, respectively, $C_P = \frac{5}{2}R$ and $C_V = \frac{3}{2}R$, where R is the ideal gas constant. For any type of ideal gas $C_P - C_V = R$.

Two factors determine the latent heat L **of a phase change:** (1) the difference ΔU between the internal energies of the phases and (2) the work $P\,\Delta V$ done by the substance in changing phase at a constant pressure.

SOLVED PROBLEMS

SOLVED PROBLEM 1
Related Problems: *19 *20 **21 **22

The temperature of the atmosphere usually decreases with increasing altitude. Sometimes, however, meteorologists speak about a *temperature inversion*. In a temperature inversion, the temperature increases, rather than decreases, with increasing altitude. The drawing illustrates how a so-called subsidence inversion comes about. Cool, low-pressure air at a higher altitude subsides or drops to a lower altitude where the pressure is greater. In falling into the region of greater pressure, the air is compressed adiabatically. The compression is adiabatic, because the drop is rapid and there is little time for heat energy to flow into or out of the falling air. As a result of the adiabatic compression, the air becomes warmer, and its temperature can become greater than that of the air at a lower altitude. In the drawing, the air at the lower altitude of 0.5 km, for example, has a temperature of 285 K. Assume the air is a diatomic ideal gas for which $\gamma = 7/5$. Suppose that air from an altitude of two kilometers ($T_i = 280$ K, $P_i = 79\ 500$ Pa) drops to an altitude of one kilometer ($P_f = 89\ 900$ Pa). Determine the temperature T_f that results because of the drop.

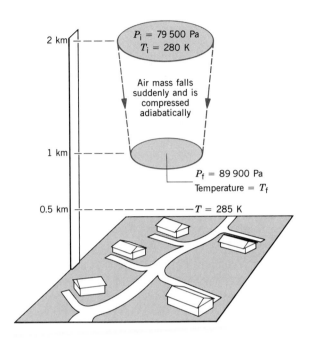

Solution This problem is done in two steps. First, the ratio of the final and initial volumes is determined using Equation 18.5

($P_i V_i^\gamma = P_f V_f^\gamma$):

$$\left(\frac{V_f}{V_i}\right)^{7/5} = \frac{P_i}{P_f} = \frac{79\ 500\ \text{Pa}}{89\ 900\ \text{Pa}} = 0.884$$

$$\frac{V_f}{V_i} = (0.884)^{5/7} = 0.916$$

Next, the ideal gas law is used to determine the ratio of final and initial temperatures. Then, the final temperature can be determined:

$$\frac{T_f}{T_i} = \frac{\dfrac{P_f V_f}{nR}}{\dfrac{P_i V_i}{nR}} = \left(\frac{P_f}{P_i}\right)\left(\frac{V_f}{V_i}\right) = \left(\frac{89\ 900\ \text{Pa}}{79\ 500\ \text{Pa}}\right)(0.916) = 1.04$$

$$T_f = (1.04)T_i = (1.04)(280\ \text{K}) = \boxed{290\ \text{K}}$$

The temperature of 290 K at an altitude of 1 km is greater than that of 285 K at an altitude of 0.5 km (see the drawing). Thus, as altitude increases from 0.5 km to 1 km, the temperature increases. This increase in temperature is the temperature inversion.

Summary of Important Points The main intent here is to illustrate how the initial pressure, volume, and temperature are related to the final pressure, volume, and temperature in an adiabatic process involving an ideal gas. Since the relation $P_i V_i^\gamma = P_f V_f^\gamma$ does not include temperature explicitly, problems that deal with the initial and final temperatures can be solved by bringing the ideal gas law into play.

QUESTIONS

1. Ignore friction and assume that air behaves as an ideal gas. The plunger of a bicycle tire pump is pushed down rapidly with the end of the pump sealed so no air escapes. Explain why the cylinder of the pump becomes warm to the touch.

2. One hundred joules of heat energy are added to a gas, and the gas expands at constant pressure. Is it possible that the internal energy increases by 200 J? Account for your answer with the aid of the first law of thermodynamics.

3. In an isobaric expansion of an ideal gas, is it possible for heat energy to flow out of the gas? Explain, using the first law of thermodynamics.

4. Listed below are five values of heat and work that result when a system interacts with its environment. In each case, state whether the internal energy of the system increases, decreases, or remains the same, and justify your choices:

(a) $W = -500$ J and $Q = 0$
(b) $W = 0$ and $Q = -200$ J
(c) $W = +100$ J and $Q = +100$ J
(d) $W = -100$ J and $Q = -100$ J
(e) $W = +300$ J and $Q = +500$ J

5. (a) Is it possible for the temperature of a substance to rise without heat energy flowing into the substance? (b) Does the temperature of a material necessarily have to change because heat energy flows into or out of it? In each case, give your reasoning and use the example of an ideal gas.

6. The drawing shows an arrangement for an adiabatic free expansion or "throttling" process. The process is adiabatic because the entire arrangement is contained within perfectly insu-

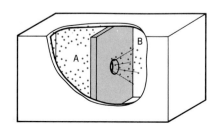

lating walls. The gas in chamber A rushes suddenly into chamber B through a hole in the partition. Chamber B is evacuated, so the gas expands there under zero external pressure and the work $W = P \Delta V$ is zero. Assume the gas is an ideal gas and explain how the final temperature of the gas after expansion compares to its initial temperature.

7. Suppose a material contracts when it is heated. Follow the same line of reasoning used in the text to reach Equations 18.7 and 18.8 and deduce which specific heat capacity for the material is larger, C_P or C_V.

8. The latent heat of fusion is an extremely good estimate of the change in internal energy per kilogram that occurs when a solid melts. An analogous statement cannot be made in connection with the latent heat of vaporization or sublimation. Explain why.

PROBLEMS

Section 18.3 The First Law of Thermodynamics

1. When one gallon of gasoline is burned in a car engine, 1.19×10^8 J of internal energy are released. Suppose that 1.00×10^8 J of this energy flow directly into the surroundings (engine block and exhaust system) in the form of heat. If 6.0×10^5 J of work are required to make the car go one mile, how many miles can the car travel on one gallon of gas?

2. One-half mole of a monatomic ideal gas absorbs 1200 J of heat energy while performing 2500 J of work. By how much does the temperature of the gas change? Is the change an increase or a decrease?

3. The drawing shows two objects, A and B. The arrows in the drawing symbolize the flow of heat energy into or out of an object, as well as the work done on or by an object. Find the change in the internal energy of the system when the system is (a) object A only, (b) object B only, and (c) objects A and B together. In each case, specify whether the change is a decrease or an increase.

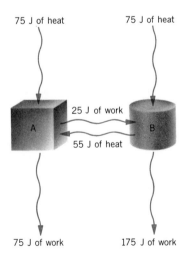

75 J of heat   75 J of heat

A   25 J of work   B

55 J of heat

75 J of work   175 J of work

* 4. In exercising, a weight lifter loses 0.100 kg of water through evaporation, the heat required to evaporate the water coming from the weight lifter's body. The work done in lifting weights is 1.00×10^5 J. (a) Assuming the latent heat of vaporization of per-spiration is 2.42×10^6 J/kg, find the amount by which the internal energy of the weight lifter decreases. (b) Determine the minimum number of nutritional calories of food that must be consumed to replace the loss of internal energy.

Section 18.4 Thermal Processes Involving Pressure, Volume, and Temperature

5. A gas is contained in a chamber such as that in Figure 18.4. Suppose the region outside the chamber is evacuated and the total mass of the block and the movable piston is 120 kg. When 1750 J of heat energy flow into the gas, the internal energy of the gas increases by 1550 J. What is the distance s through which the piston rises?

6. The specific heat capacity of a solid material is 1100 J/(kg·C°). The temperature of 2.0 kg of this material is raised by 6.0 C°. Ignoring the work that corresponds to the small change in the volume of the material, determine the change in the internal energy of the material.

7. The internal energy of a system increases by 1350 J when the system absorbs 1150 J of heat energy at a constant pressure of 1.01×10^5 Pa. By how much does the volume of the system change? Does the volume increase or decrease?

8. When a .22-caliber rifle is fired, the expanding gas from the burning gunpowder creates a pressure behind the bullet. This pressure causes the force that pushes the bullet through the barrel. The barrel has a length of 0.61 m and an opening whose radius is 2.8 mm. A 2.6-g bullet has a speed of 370 m/s after passing through this barrel. Ignore kinetic friction and determine the average pressure of the expanding gas.

* 9. (a) Using the data presented in the accompanying pressure-versus-volume graph, estimate the work done when the system

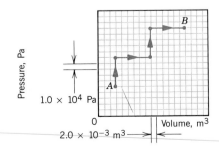

Pressure, Pa

1.0×10^4 Pa

A

B

0   Volume, m^3

2.0×10^{-3} m^3

changes from A to B along the path shown. (b) Is work done by the system or on the system? (c) Is the work positive or negative? (d) What would be the work, both magnitude and sign, if the system started at B and ended at A by moving along the same path?

* **10.** When a monatomic ideal gas expands at a constant pressure of 2.0×10^5 Pa, the volume of the gas increases by 5.0×10^{-3} m³. Determine how much heat energy flows into or out of the gas. Specify the direction of the flow.

* **11.** A cylindrical aluminum rod has a length of 0.50 m and a radius of 3.0 cm. The temperature of the rod is raised from 20 °C to 320 °C. How much work does the expanding rod do on the surrounding air, if the air pressure is 1.01×10^5 Pa?

* **12.** The pressure and volume of a gas are changed along the straight line from A to B in the accompanying graph. From this graph determine an expression for the work done by the gas in terms of the parameters P_A, P_B, V_A, and V_B.

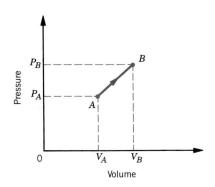

** **13.** The pressure and volume of an ideal gas change from A to B to C, as the drawing shows. Determine the total heat for the process and state whether the flow of heat energy is into or out of the gas.

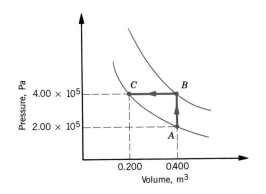

** **14.** The drawing refers to one mole of a monatomic ideal gas and shows a process that has four steps, two isobaric (A to B, C to D) and two isochoric (B to C, D to A). Complete the following table by calculating ΔU, W, and Q (including the algebraic signs) for each of the four steps. Note that the gas has returned to its

initial state at the end of the process, so the value for the total ΔU can be predicted in advance without any calculations.

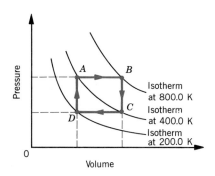

	ΔU	W	Q
A to B			
B to C			
C to D			
D to A			
Total			

Section 18.5 Thermal Processes That Utilize an Ideal Gas

15. A bubble from the tank of a scuba diver in a lake contains 3.5×10^{-4} moles of gas. The bubble expands as it rises to the surface from a freshwater depth of 10.3 m. Assuming the gas is an ideal gas and the temperature remains constant at 18 °C, find the amount of heat energy that flows into the bubble.

16. Two grams of helium (molecular mass = 4.0 u) expand isothermally at 77 °C and do 1600 J of work. Assuming helium is an ideal gas, determine the ratio of the final volume of the gas to the initial volume.

17. When 2.00×10^3 J of work are performed adiabatically to compress one-half mole of a monatomic ideal gas, the Kelvin temperature of the gas doubles. Determine the initial temperature of the gas.

18. A monatomic ideal gas ($\gamma = 5/3$) is compressed adiabatically and its volume is reduced by a factor of two. Determine the factor by which its pressure increases.

* **19.** Unlike a gasoline engine, a diesel engine does not use spark plugs to ignite the mixture of fuel and air in the cylinders. Instead, the temperature required to ignite the fuel occurs because the pistons compress the air in the cylinders. Suppose air at an initial temperature of 27 °C is compressed adiabatically to a temperature of 681 °C. Assume the air to be an ideal gas for which $\gamma = 7/5$.

Find the compression ratio, which is the ratio of the initial volume to the final volume. *(See Solved Problem 1 for a related problem.)*

* 20. A monatomic ideal gas ($\gamma = 5/3$) is contained within a perfectly insulated cylinder that is fitted with a movable piston. The initial pressure of the gas is 1.50×10^5 Pa. The piston is pushed so as to compress the gas, with the result that the Kelvin temperature doubles. What is the final pressure of the gas? *(See Solved Problem 1 for a related problem.)*

**21. The work done by one mole of a monatomic ideal gas ($\gamma = 5/3$) in expanding adiabatically is 825 J. The initial temperature and volume of the gas are 393 K and 0.100 m³. Obtain the final temperature and volume of the gas. *(See Solved Problem 1 for a related problem.)*

**22. The drawing shows an adiabatically isolated cylinder that is divided initially into two identical parts by an adiabatic partition. Both sides contain one mole of a monatomic ideal gas ($\gamma = 5/3$), with the initial temperature being 525 K on the left and 275 K on the right. The partition is then allowed to move slowly (i.e., quasi-statically) to the right, until the pressures on each side of the partition are the same. Find the final temperatures on the left and right. *(See Solved Problem 1 for a related problem.)*

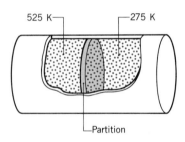

525 K   275 K

Partition

Section 18.6 Specific Heat Capacities and the First Law of Thermodynamics, Section 18.7 Latent Heats and the First Law of Thermodynamics

23. Argon is a monatomic gas whose molecular mass is 39.9 u. The temperature of eight grams of argon is raised by 75 K under conditions of constant pressure. Assuming that argon is an ideal gas, how many kilocalories of heat are required?

24. Heat energy Q is added to a monatomic ideal gas at constant pressure. As a result, the gas does work W. Find the ratio Q/W.

25. The latent heat of fusion for water is 3.35×10^5 J/kg. When one kilogram of ice melts, its volume decreases by 9.03×10^{-5} m³. Determine the increase in the internal energy of 2.50 kg of water when the solid-to-liquid phase change takes place at atmospheric pressure.

26. The latent heat of sublimation for zinc (atomic mass = 65.4 u) at 600 K is 1.99×10^6 J/kg. Assume that zinc vapor can be treated as a monatomic ideal gas and that the volume of one kilogram of solid is negligible compared to that of the vapor. What

percentage of the latent heat serves to change the internal energy during sublimation?

* 27. The temperature of two moles of an ideal gas increases, because 750 J of heat energy are absorbed under conditions of constant volume. Under conditions of constant pressure, the temperature increases by the same amount when 1.00×10^3 J of heat energy are absorbed. By how many kelvins does the temperature change?

* 28. Even at rest, the human body generates heat. The heat arises because of the body's metabolism, that is, the chemical reactions that are always occurring in the body to generate energy. In rooms designed for use by large groups, adequate ventilation or air conditioning must be provided to remove this heat. Consider a classroom containing 200 students. Assume that the metabolic rate of generating heat is 130 W for each student and that the heat accumulates during a fifty-minute lecture. In addition, assume that the air has a molar specific heat of $C_V = \frac{5}{2}R$ and that the room (volume = 1200 m³, initial pressure = 1.01×10^5 Pa, and initial temperature = 21 °C) is sealed shut. If all the heat energy generated by the students were absorbed by the air, by how much would the air temperature rise during a lecture?

* 29. A monatomic ideal gas expands at constant pressure. (a) What percentage of the heat energy being supplied to the gas is used to increase the internal energy of the gas? (b) What percentage is used for work of expansion? (c) Answer the same two questions assuming the gas expands at constant temperature.

* 30. One mole of neon, a monatomic gas, starts out at conditions of standard temperature and pressure. The gas is heated at constant volume until its pressure is tripled, then further heated at constant pressure until its volume is doubled. Assume that neon behaves as an ideal gas. For the entire process, find (a) the heat energy added to the gas, (b) the work done by the gas, and (c) the change in the internal energy of the gas.

ADDITIONAL PROBLEMS

31. Five moles of oxygen expand isothermally from 0.100 m³ to 0.400 m³. To maintain the constant temperature, 2.50×10^4 J of heat energy are added to the system. Assuming oxygen to be an ideal gas, determine the temperature.

32. The work done to compress one mole of a monatomic ideal gas is 6200 J. The temperature of the gas changes from 350 to 550 K. How much heat energy flows between the gas and its surroundings? Determine whether the energy flows into or out of the gas.

33. A system gains 1500 J of heat energy, while the internal energy of the system increases by 4500 J and the volume decreases by 0.010 m³. Assume the pressure is constant and find its value.

34. Suppose 550 J of heat energy are removed from two moles of a monatomic ideal gas. What drop in temperature occurs when the energy is removed under conditions of (a) constant volume and (b) constant pressure?

* **35.** Using the relationship for an adiabatic expansion or compression of an ideal gas ($P_iV_i^\gamma = P_fV_f^\gamma$) together with the ideal gas law, derive (a) an expression similar to the one above, but involving only volume, temperature, and γ and (b) an expression involving only pressure, temperature, and γ.

* **36.** A ten-watt heater is used to heat a monatomic ideal gas at a constant pressure of 2.50×10^5 Pa. During the process, the 1.00×10^{-3}-m³ volume of the gas increases by 20.0%. How long was the heater on?

* **37.** The volume of an ideal gas is changed along the *isothermal* curve in the drawing. (a) Estimate the work for the process that takes the gas from A to B. (b) Is the work done on the gas or by the gas? (c) Is the work positive or negative? (d) What is the heat for the process? (e) Does heat energy flow into or out of the gas?

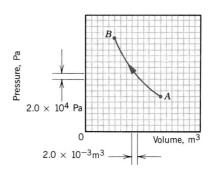

38. Suppose a monatomic ideal gas is contained within a vertical cylinder that is fitted with a movable piston. The piston is frictionless and has a negligible mass. The radius of the piston is 0.100 m, and the pressure outside the cylinder is 1.01×10^5 Pa. One-half of a kilocalorie of heat is removed from the gas. Through what distance does the piston drop?

CHAPTER 19

Thermodynamics: The Second Law and the Third Law

A ride on an old train is a nostalgic experience. The steam locomotive is a heat engine in that the locomotive uses thermal energy to do the work of moving the train. When a heat engine does work, the second law of thermodynamics limits the maximum efficiency that an engine can have.

19.1 THE SECOND LAW OF THERMODYNAMICS

If a cold can of soda is placed in a warm room, the soda warms up because heat energy flows spontaneously from the warmer surroundings into the cooler soda. On the other hand, the soda never becomes cooler as it sits in the room, for heat energy does not flow spontaneously from a cooler object into a warmer object. Observations such as these form the basis of the *second law of thermodynamics.*

> **THE SECOND LAW OF THERMODYNAMICS: THE HEAT FLOW STATEMENT**
> Heat energy flows spontaneously from a substance at a high temperature to a substance at a low temperature and does not flow spontaneously in the reverse direction.

In the form stated above, the second law of thermodynamics refers to the natural tendency of heat to flow from hot to cold. It should be noted, however, that heat can be *forced* to flow in the reverse direction, *if work is done* to make it do so. There are several familiar examples of such forced reverse flow. An air conditioner makes heat flow from the cool interior of a house to the hot outdoors with the aid of the work done by electrical energy. Refrigerators and heat pumps are also devices that use work to make heat flow from cold to hot, against its natural tendency.

The second law of thermodynamics has important implications for the manner in which many devices operate, and we now turn our attention to one of them, the heat engine.

19.2 HEAT ENGINES

INTRODUCTION

A *heat engine* is any device that uses heat energy to perform work. To illustrate the features that are common to all heat engines, Figure 19.1 shows a simplified steam engine. The boiler receives heat from a high-temperature source, such as an oil or gas burner. In the boiler, liquid water is converted into steam at a high temperature and pressure. The steam is then sent through the intake valve and enters the cylinder, where the steam expands against the piston and forces it to move. The piston may be coupled to the wheel of a locomotive, for example. The work done on the moving piston comes from the internal energy of the steam, so the temperature of the steam in the cylinder drops. As the piston returns to its original position, the cooler steam is forced out the exhaust valve and into the condenser, where the steam is changed back into a liquid. The heat of condensation is given up to a heat sink, which may be the atmosphere. The water is then pumped back to the boiler, and the cycle is repeated. The combination of steam and water is known as the *working substance* of the engine, for it is the agent that carries the heat energy and does the work.

The essential features of a steam engine are listed below, because they are shared by all types of heat engines:

1. Heat energy is supplied to the engine at a relatively high temperature.
2. Part of the input heat is used to perform work.

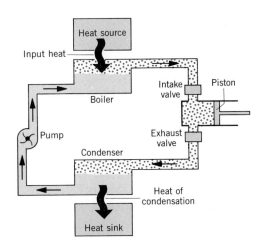

FIGURE 19.1 A steam engine.

3. The remainder of the input heat is rejected at a lower temperature than that at which the input heat is supplied.

Figure 19.2 emphasizes these essential features in a schematic fashion. In this drawing the symbol Q_H refers to the magnitude of the input heat. The subscript H stands for "hot," and the place from which the input heat comes is called the "hot reservoir." The symbol Q_C denotes the magnitude of the rejected heat. The subscript C means "cold," and the place in the environment where the rejected heat goes is known as the "cold reservoir." The symbol W denotes the magnitude of the work done by the engine. The symbols Q_H, Q_C, and W refer to magnitudes only, without reference to algebraic signs, so negative values are not used for these quantities in any of the equations that appear in this chapter.

Heat engines, such as the steam engine, operate in repeating cycles. In other words, the conclusion of one cycle is the start of another, so the working substance is in the same state at the end of one cycle and the beginning of the next.

EFFICIENCY

From a practical point of view an important characteristic of an engine is its **efficiency.** An engine that converts most of the input heat into work is efficient. Conversely, an engine that rejects most of its input energy and does relatively little work is inefficient. The efficiency of a heat engine is defined as the ratio of the work W done by the engine to the input heat Q_H:

$$\text{Efficiency} = \frac{\text{Work done}}{\text{Input heat}} = \frac{W}{Q_H} \tag{19.1}$$

If the input heat were converted entirely into work, the engine would have an efficiency of 1.00, since $W = Q_H$; such an engine would be 100% efficient. Even though efficiencies are often quoted in percentages, Equation 19.1 does not include the corresponding factor of 100. Thus, an efficiency of 68% would mean that a value of 0.68 is used for the value of the efficiency in the equation.

An engine, like any device, must obey the principle of conservation of energy. Some of an engine's input heat Q_H is converted into work W and the remainder Q_C is

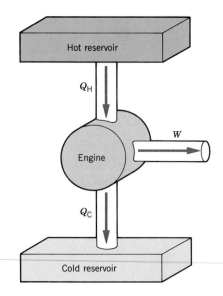

FIGURE 19.2 This schematic representation of a heat engine shows the input heat (magnitude = Q_H) that originates from the hot reservoir, the work (magnitude = W) done by the engine, and the heat (magnitude = Q_C) that the engine rejects to the cold reservoir.

rejected to the cold reservoir. If there are no other losses in the engine, the principle of energy conservation requires that

$$Q_H = W + Q_C \qquad (19.2)$$

By solving this equation for W and substituting the result into Equation 19.1, we arrive at the following alternative expression for the efficiency of a heat engine:

$$\text{Efficiency} = \frac{Q_H - Q_C}{Q_H} = 1 - \frac{Q_C}{Q_H} \qquad (19.3)$$

Example 1 illustrates how the concepts of efficiency and energy conservation are applied to a heat engine.

EXAMPLE 1

An automobile engine has an efficiency of 22.0% and produces 2510 J of work. How much heat is rejected by the engine?

SOLUTION

The amount Q_C of heat rejected to the cold reservoir can be calculated by using $W = 2510$ J in the expression $Q_C = Q_H - W$, provided the input heat Q_H is known. According to Equation 19.1,

$$Q_H = \frac{W}{\text{Efficiency}} = \frac{2510 \text{ J}}{0.220} = 11\,400 \text{ J}$$

The amount of rejected heat is

$$Q_C = Q_H - W = 11\,400 \text{ J} - 2510 \text{ J} = \boxed{8900 \text{ J}} \qquad (19.2)$$

In Example 1, less than one-quarter of the input heat is converted into work, because the efficiency of the automobile engine is only 22.0%. If the engine were 100% efficient, all the input heat would be converted into the work of moving the car. Unfortunately, nature does not permit 100% efficient heat engines to exist, as the next two sections discuss.

19.3 CARNOT'S PRINCIPLE

REVERSIBLE PROCESSES

Any heat engine has a maximum efficiency when the processes within the engine are reversible. Figure 19.3 reveals the essence of a *reversible process* by using a gas-filled

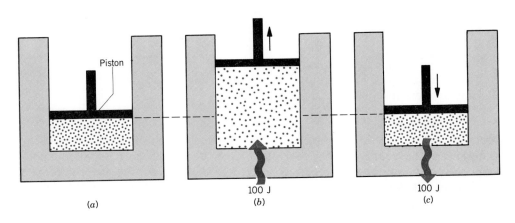

Piston

100 J
(a)   (b)   100 J   (c)

FIGURE 19.3 (a) The frictionless piston is supported by the gas. (b) When 100 J of heat are added to the gas, the gas expands and does work in lifting the piston. (c) If the process is reversible, the system (the gas) and the environment (the piston and the rest of the universe) can both be returned to the same initial states they were in before the process occurred.

cylinder and a frictionless piston. In part a of the drawing, the gas in the cylinder supports the piston and has a pressure, volume, and temperature of P_1, V_1, and T_1. When 100 J of heat energy are added to the gas, as in part b, the gas expands. As the gas expands, it pushes the piston upward, thereby doing work. When the gas stops expanding, its pressure, volume, and temperature are P_2, V_2, and T_2. In part c of the drawing, the piston is allowed to fall back to its original position, thereby doing work in compressing the gas. If the process is reversible, 100 J of heat energy flow out of the gas and back into the environment as the gas returns to its original state (P_1, V_1, and T_1). Thus, *a reversible process is one in which both the system (the gas) and its environment (the piston and the rest of the universe) can be returned to exactly the states they were in before the process occurred.*

In a reversible process, *both* the system and its environment can be returned to their initial states. Therefore, a process that involves an energy-dissipating mechanism, such as friction, cannot be reversible, because the energy wasted due to friction would alter either the system or the environment or both. For example, suppose there were friction between the piston and the wall of the cylinder in Figure 19.3. When the piston returns to its initial position (part c), the pressure and temperature would not be P_1 and T_1, so the gas would not be in the initial state it had in part a.

The presence of friction is one reason why a real process may not be reversible. There are also other reasons. For instance, the spontaneous flow of heat energy from a hot substance to a cold substance is irreversible, even though friction is not present. For heat to flow in the reverse direction, work must be done. The agent doing such work must be located in the environment of the hot and cold substances, and, therefore, the environment must change while the heat is moved back from cold to hot. Since the system and the environment cannot *both* be returned to their initial states, the process of spontaneous heat flow is irreversible. In fact, all spontaneous processes are irreversible, such as the explosion of an unstable chemical or the bursting of a bubble.

When the word "reversible" is used in connection with engines, it does not mean just a gear that allows the engine to operate a device in reverse. All cars have a reverse gear, for instance, but no automobile engine is thermodynamically reversible, since friction exists no matter which way the car moves.

THE STATEMENT OF CARNOT'S PRINCIPLE

The idea that the efficiency of a heat engine is a maximum when the engine operates reversibly originated with the French engineer Sadi Carnot (1796–1832) and is referred to as *Carnot's principle.*

CARNOT'S PRINCIPLE: AN ALTERNATIVE STATEMENT OF THE SECOND LAW OF THERMODYNAMICS
No irreversible engine operating between two reservoirs at constant temperatures can have a greater efficiency than a reversible engine operating between the same temperatures. Furthermore, all reversible engines operating between the same temperatures have the same efficiency.

Carnot's principle is quite remarkable, for no mention is made of the working substance of the engine. It does not matter if the working substance is a gas, a liquid, or a solid. As long as the process is reversible, the efficiency of the engine is a maximum. Carnot's principle does *not* state, or even imply, that a reversible engine has an efficiency of 100%.

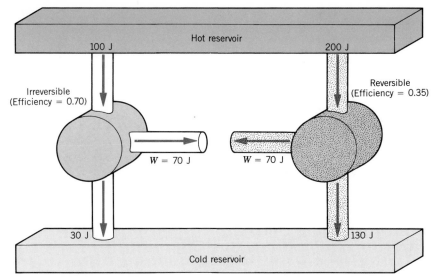

FIGURE 19.4 The drawing depicts a hypothetical situation that violates Carnot's principle, because the irreversible engine is more efficient than the reversible engine, each engine using the same hot and cold reservoirs. Being more efficient, the irreversible engine requires a smaller input heat to produce a given amount of work. The numbers shown are arbitrary.

If Carnot's principle did not apply, it would be possible to build an irreversible engine that uses the same hot and cold reservoirs as a reversible engine does, but is more efficient. For example, Figure 19.4 shows a reversible engine with efficiency of 0.35 and an irreversible engine with an efficiency of 0.70. The reversible engine uses 200 J of input heat to produce $(0.35)(200\ \mathrm{J}) = 70\ \mathrm{J}$ of work, while rejecting the remaining 130 J to the cold reservoir. Being twice as efficient, the irreversible engine needs only 100 J of input heat to produce the same 70 J of work $[(0.70)(100\ \mathrm{J}) = 70\ \mathrm{J}]$. This irreversible engine rejects only 30 J to the cold reservoir. According to Carnot's principle, a situation like this *cannot exist,* for, as we will see, it violates the second law of thermodynamics.

Suppose, however, that Carnot's principle were not valid and the situation depicted in Figure 19.4 could exist. Then, because it is reversible, the engine on the right could be run in reverse without changing the numbers. As Figure 19.5a shows, the 70 J of work put out by the irreversible engine could be used to drive the reversible engine in reverse. With the aid of the 70 J of work, the reversible engine would take 130 J of heat from the cold reservoir and deposit a total of 200 J into the hot reservoir (the 130 J of heat plus the 70 J of work).

Is there anything wrong with the two-engine tandem shown in Figure 19.5a? Yes, there is, and part b of the drawing helps to explain why. Here, a gray box encloses the engines to focus attention on the two-engine tandem as a single device in itself. The gray box constitutes a device that takes 100 J of heat from the cold reservoir $(130\ \mathrm{J} - 30\ \mathrm{J} = 100\ \mathrm{J})$ and deposits 100 J of heat into the hot reservoir $(200\ \mathrm{J} - 100\ \mathrm{J} = 100\ \mathrm{J})$. In performing the heat transfer, *the gray box receives no work from the environment.* Such a device violates the second law of thermodynamics, for the device transfers heat from cold to hot without the aid of work to counteract the natural tendency of heat to flow spontaneously the other way. Therefore, situations like those shown in Figures 19.4 and 19.5 cannot exist, and we are left with the conclusion that Carnot's principle must be valid. In effect, Carnot's principle is another way of expressing the second law of thermodynamics.

To show that all reversible engines operating between the same hot and cold reservoirs have the same efficiency, just substitute a reversible engine for the irreversible engine in Figure 19.5a and assume this reversible engine also needs only 100 J of input heat to produce the 70 J of work. Then, the reversible engine on the left in the

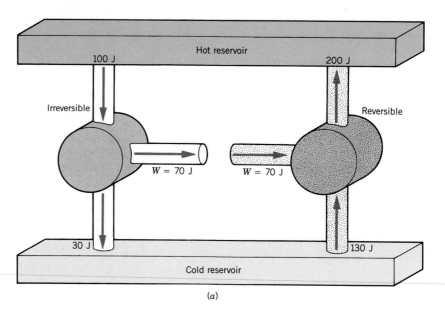

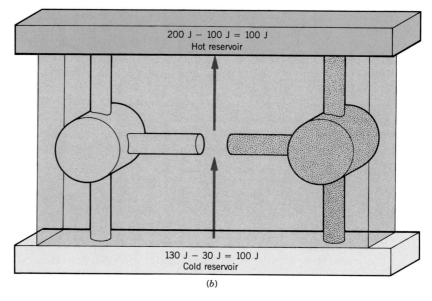

FIGURE 19.5 (*a*) If Carnot's principle were not valid, and the situation depicted in Figure 19.4 could exist, it would be possible to operate the reversible engine in that drawing in reverse and construct the two-engine tandem shown here. (*b*) The gray box stresses the fact that the two-engine tandem would function as a single device for moving heat from the cold reservoir to the hot reservoir without the aid of work from agents in the environment. This situation is a violation of the second law of thermodynamics.

drawing has a different efficiency than the reversible engine on the right. Once again, as part *b* of the drawing indicates, heat would flow spontaneously from cold to hot, in violation of the second law of thermodynamics. Thus, the assumption that the two reversible engines have different efficiencies must be false.

19.4 THE EFFICIENCY OF THE CARNOT HEAT ENGINE

According to Carnot's principle, a heat engine has maximum efficiency when the engine operates reversibly. A hypothetical engine that operates reversibly, therefore, provides a useful standard for judging how well real engines perform. Figure 19.6

shows a reversible engine, called a **Carnot engine,** that is particularly useful as an idealized model. An important feature of the Carnot engine is that all input heat Q_H originates from a hot reservoir *at a single temperature* T_H and all rejected heat Q_C goes into a cold reservoir *at a single temperature* T_C.

Carnot's principle implies that the efficiency of a reversible engine is independent of the working substance of the engine, and therefore can depend only on the temperatures of the hot and cold reservoirs. Since Efficiency $= 1 - Q_C/Q_H$ according to Equation 19.3, the ratio Q_C/Q_H can depend only on the reservoir temperatures. This observation led Kelvin to define a *thermodynamic temperature scale.* He proposed that the thermodynamic temperatures of the cold and hot reservoirs be defined such that the ratio of these temperatures is equal to Q_C/Q_H. Thus, the thermodynamic temperature scale is related to the heat energies absorbed and rejected by a Carnot engine, and is independent of the working substance. If a reference point on the thermodynamic temperature scale is properly chosen, it can be shown that the scale is identical to the Kelvin temperature scale introduced in Section 14.2 and used in the ideal gas law. Thus, the ratio of the rejected heat Q_C to the input heat Q_H can be written as

$$\frac{Q_C}{Q_H} = \frac{T_C}{T_H} \qquad (19.4)$$

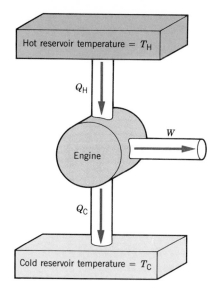

FIGURE 19.6 A Carnot engine is a reversible engine in which all input heat Q_H originates from a hot reservoir at a single temperature T_H, and all rejected heat Q_C goes into a cold reservoir at a single temperature T_C. The work done by the engine is W.

where the temperatures T_C and T_H *must be expressed in kelvins.*

The efficiency of a Carnot engine can be written in a particularly useful way by substituting Equation 19.4 into the relation Efficiency $= 1 - Q_C/Q_H$:

$$\text{Efficiency of a Carnot engine} = 1 - \frac{T_C}{T_H} \qquad (19.5)$$

This relation gives the maximum possible efficiency for a heat engine operating between two Kelvin temperatures T_C and T_H, and the next example illustrates its application.

EXAMPLE 2

Water near the surface of a tropical ocean has a temperature of 25.0 °C, while water 700 m beneath the surface has a temperature of 7.0 °C. It has been proposed that the warm water be used as the hot reservoir and the cool water as the cold reservoir of a heat engine. (a) Find the maximum possible efficiency for such an engine. (b) Determine the minimum input heat Q_H that would be needed if a number of these engines were to produce an amount of work equal to the 8.5×10^{19} J of energy that the United States consumed in 1980.

SOLUTION

(a) The maximum possible efficiency is given by Equation 19.5, in which the Kelvin temperatures 298.2 K (25.0 °C) and 280.2 K (7.0 °C) are used:

$$\begin{array}{c} \text{Efficiency} \\ \text{of the Carnot} \\ \text{engine} \end{array} = \begin{array}{c} \text{Maximum} \\ \text{possible} \\ \text{efficiency} \end{array}$$

$$= 1 - \frac{T_C}{T_H} = 1 - \frac{280.2 \text{ K}}{298.2 \text{ K}} = \boxed{0.060 \ (6\%)}$$

(b) With a knowledge of the maximum possible efficiency, it is a straightforward matter to use Equation 19.1 to calculate the input heat Q_H needed for the engines to produce 8.5×10^{19} J of work:

$$Q_H = \frac{W}{\text{Efficiency}} = \frac{8.5 \times 10^{19} \text{ J}}{0.060} = \boxed{1.4 \times 10^{21} \text{ J}}$$

Real engines, being less efficient than a Carnot engine, would require an input energy that is greater than 1.4×10^{21} J to produce the same amount of work.

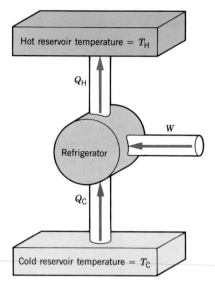

FIGURE 19.7 In a refrigeration process, work W is used to remove heat Q_C from the cold reservoir and deposit heat Q_H into the hot reservoir. A refrigeration process is the reverse of an engine process.

In Example 2 the maximum possible efficiency is only 6%. The small efficiency arises because the Kelvin temperatures of the hot and cold reservoirs are nearly the same. A greater efficiency is possible only when there is a greater difference between the reservoir temperatures. In any event, Equation 19.5 indicates that a *perfect heat engine has an efficiency that is always less than 1.0 or 100%.* We note in this regard that when T_C approaches absolute zero (0 K), the maximum possible efficiency approaches 1.0. However, experiments have shown that it is not possible to cool a substance to absolute zero, so nature does not permit the existence of a 100% efficient heat engine. Because such a heat engine does not exist, there is always heat rejected to a cold reservoir whenever a heat engine does work, even if friction and other irreversible processes are eliminated completely.

19.5 REFRIGERATORS, AIR CONDITIONERS, AND HEAT PUMPS

Refrigerators, air conditioners, and heat pumps are familiar devices that make heat flow from cold to hot, and each operates in a similar fashion. As Figure 19.7 illustrates, these devices use work W to "reach into" a cold reservoir, "grab onto" an amount of heat Q_C, and deposit an amount of heat Q_H into a hot reservoir. Generally speaking, such a process is called a *refrigeration process.* Often the work is provided by an electric motor, although the work can be done by other means. For example, the work needed to operate an automobile air conditioner comes from the engine of the car. A comparison of this drawing with Figure 19.6 shows that the directions of the arrows symbolizing heat and work in a refrigeration process are opposite to those in an engine process. Energy is conserved during a refrigeration process, just as it is in an engine process, so $Q_H = W + Q_C$. Moreover, if the process occurs reversibly, we have ideal devices that are called Carnot refrigerators, Carnot air conditioners, and Carnot heat pumps. For these ideal devices, the relation $Q_C/Q_H = T_C/T_H$ applies, just as it does for the Carnot engine.

Refrigerators, air conditioners, and heat pumps differ mainly in the nature of their hot and cold reservoirs. Figure 19.8 indicates that the interior of a refrigerator is the cold reservoir, while the warmer exterior is the hot reservoir. The refrigerator takes heat energy from the food inside and deposits it into the kitchen, along with the energy that is needed to do the work of making the heat flow from cold to hot. For this reason, the outside surfaces (usually the sides and back) of most refrigerators are warm to the touch while they are operating. Thus, a refrigerator warms the kitchen. An air conditioner is like a refrigerator, except the room itself is the cold reservoir and the outdoors is the hot reservoir.

In a sense, refrigerators and air conditioners operate like pumps. They pump heat "uphill" from a lower temperature to a higher temperature, just as a water pump forces water uphill from a lower elevation to a higher elevation. It would be appropriate to call them heat pumps. However, the name "heat pump" is reserved for the device illustrated in Figure 19.9, which is a kind of home heating appliance. The heat pump uses work W to make heat Q_C from the wintry outdoors (the cold reservoir) flow up the temperature "hill" into a warm house (the hot reservoir). According to the conservation of energy, the heat pump deposits inside the house an amount of heat equal to $Q_H = Q_C + W$. The air conditioner and the heat pump do closely related jobs. The air conditioner refrigerates the inside of the house and heats up the outdoors, while the heat pump refrigerates the outdoors and heats up the inside. In

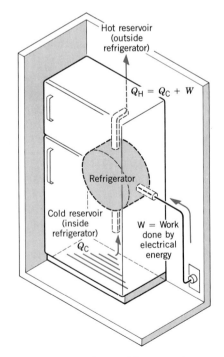

FIGURE 19.8 In a refrigerator the cold reservoir is the interior region where the food is kept, and the hot reservoir is the kitchen itself.

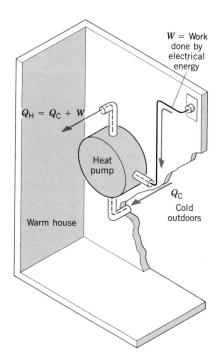

FIGURE 19.9 In a heat pump the cold reservoir is the wintry outdoors, and the hot reservoir is the inside of the house.

fact, these jobs are so closely related that most heat pump systems serve in a dual capacity, because they come equipped with a switch that converts them from heaters in the winter into air conditioners in the summer.

Heat pumps are becoming increasingly popular for home heating in today's energy-conscious world, and it is easy to understand why. Suppose 1000 J of energy are available to use for home heating. Figure 19.10a shows that a conventional electric heating system uses this 1000 J to heat a coil of wire, just as in a toaster. A fan blows

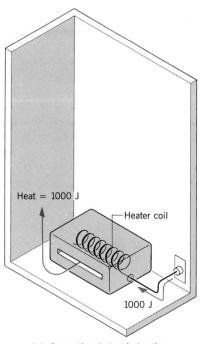

(a) Conventional electric heating

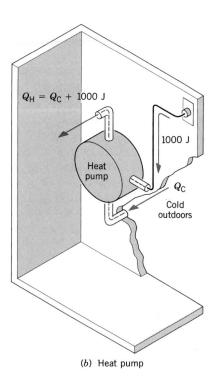

(b) Heat pump

FIGURE 19.10 (a) In a conventional electric heating system, 1000 J of energy are used to heat a coil of wire, and 1000 J of heat are delivered to the living room. (b) In a heat pump, 1000 J of energy are used as work to make heat Q_C flow from the outdoors into the room. More than 1000 J of heat are delivered to the room, the amount being $Q_H = Q_C + 1000$ J.

air across the hot coil, and forced convection carries the 1000 J of heat into the living room. Part *b* of the drawing shows that a heat pump does not use the 1000 J directly as heat. Instead, the heat pump uses the 1000 J to do the work of pumping heat Q_C from the cooler outdoors into the warmer room. In the process, the pump delivers to the inside an amount of energy $Q_H = Q_C + 1000$ J. Clearly, the heat pump puts more than 1000 J of heat into the room, whereas the conventional electric heating system provides only 1000 J. In principle, then, the heat pump is a superior system.

The next example illustrates how the basic relations $Q_H = W + Q_C$ and $Q_C/Q_H = T_C/T_H$ are used with heat pumps.

EXAMPLE 3

An ideal or Carnot heat pump is used to heat a house to a temperature of 21 °C (294 K). How much work must be done by the pump to deliver 3350 J of heat into the house when the outdoor temperature is (a) 0 °C (273 K) and (b) −21 °C (252 K)?

$$Q_C = Q_H \frac{T_C}{T_H} = (3350 \text{ J}) \left(\frac{273 \text{ K}}{294 \text{ K}} \right) = 3110 \text{ J}$$

The work W done by the heat pump is

$$W = 3350 \text{ J} - Q_C = 3350 \text{ J} - 3110 \text{ J} = \boxed{240 \text{ J}}$$

SOLUTION

(a) The conservation of energy (Equation 19.2) applies whether or not the heat pump is ideal and can be used to determine the work: $W = Q_H - Q_C = 3350 \text{ J} - Q_C$. To obtain a numerical answer for W, a value for Q_C is needed. A value for Q_C can be obtained from Equation 19.4, because a Carnot heat pump is a pump that operates reversibly:

(b) This solution is identical to that in part (a), except that it is now cooler outside, so $T_C = 252$ K. Consequently, more work must be done by the heat pump: $\boxed{W = 480 \text{ J}}$. It is not surprising that more work must be done, considering that the heat must be pumped up a greater temperature "hill" when the outside is colder than when it is warmer.

A measure of the performance of a heat pump can be obtained by specifying the ratio of the heat Q_H delivered into the house to the work W required to deliver it. This ratio is known as the **coefficient of performance:**

[Heat pump]
$$\begin{array}{c} \text{Coefficient} \\ \text{of} \\ \text{performance} \end{array} = \frac{Q_H}{W} \qquad (19.6)$$

The coefficient of performance depends on the indoor and outdoor temperatures. In Example 3, when the outdoor temperature is 273 K, the coefficient of performance is $Q_H/W = (3350 \text{ J})/(240 \text{ J}) = 14$. When the outdoor temperature cools to 252 K, the coefficient of performance becomes $Q_H/W = (3350 \text{ J})/(480 \text{ J}) = 7.0$. Such large coefficients of performance result because the heat pump is an ideal or Carnot heat pump. Commercially available heat pumps have coefficients of performance of about 2–4 under favorable temperature conditions.

It is also possible to specify a coefficient of performance for refrigerators and air conditioners. However, unlike a heat pump, the job of these two devices is to cool, not to heat. As a result, the coefficient of performance of a refrigerator or an air conditioner is the ratio of the heat Q_C removed from the cold reservoir to the work W needed to remove it:

$$\begin{bmatrix} \textbf{Refrigerator} \\ \textbf{or} \\ \textbf{air conditioner} \end{bmatrix} \qquad \begin{array}{c} \text{Coefficient} \\ \text{of} \\ \text{performance} \end{array} = \frac{Q_C}{W} \qquad (19.7)$$

Commercially available refrigerators and air conditioners have coefficients of performance in the range 2–6, depending on the temperatures involved, values that are less than those for ideal or Carnot devices.

19.6 ENTROPY AND THE SECOND LAW OF THERMODYNAMICS

INTRODUCTION

A Carnot engine has the maximum possible efficiency for its operating conditions. According to Carnot's principle, the reason for the maximum efficiency is that the processes occurring in a Carnot engine are reversible. However, irreversible processes, such as those involving friction, always cause real engines to operate at less than maximum efficiency.

When an irreversible process occurs, there is a loss in our ability to use heat for performing work. As an extreme example, imagine that a hot object is placed in thermal contact with a cold object, so heat flows spontaneously, and hence irreversibly, from hot to cold. Eventually both objects reach the same temperature, and $T_C = T_H$. A Carnot engine using these two objects as heat reservoirs is unable to do work, because the efficiency of the engine is zero [Efficiency $= 1 - (T_C/T_H) = 0$]. In general, irreversible processes cause us to lose some, but not necessarily all, of our ability to perform work. This partial loss can be expressed in terms of a concept called *entropy.*

ENTROPY

To introduce the idea of entropy, recall the relation $Q_C/Q_H = T_C/T_H$ that applies to a Carnot engine. This equation can be rearranged to read $Q_C/T_C = Q_H/T_H$, a result that focuses attention on the heat Q that is transferred divided by the Kelvin temperature T at which the transfer occurs. The quantity Q/T is called the *change in the entropy* ΔS:

$$\Delta S = \left(\frac{Q}{T}\right)_R \tag{19.8}$$

The subscript R emphasizes that this relation pertains to reversible processes. It can be shown that Equation 19.8 applies to any process in which heat Q enters or leaves a system reversibly, provided the temperature remains constant while the energy flows. Such is the case for the heat that flows into and out of the reservoirs of a Carnot engine. Equation 19.8 indicates that the SI unit for entropy is a joule per kelvin (J/K). If heat is expressed in kilocalories, the corresponding entropy unit is kilocalories per kelvin (kcal/K).

Entropy, like internal energy, is a function of the state or condition of the system. Only the state of a system determines the entropy S that a system has. Therefore, the change in entropy ΔS is equal to the entropy of the final state of the system minus the entropy of the initial state.

We can now describe what happens to the entropy of a Carnot engine. As the engine operates, the entropy of the hot reservoir decreases, since heat Q_H departs at a Kelvin temperature T_H. The change in the entropy of the hot reservoir is $\Delta S_H = -Q_H/T_H$, where the minus sign is needed to indicate a decrease in entropy, since the symbol Q_H denotes only the magnitude of the heat. In contrast, the entropy of the cold reservoir increases by an amount $\Delta S_C = +Q_C/T_C$, for the rejected heat enters the cold reservoir at a Kelvin temperature T_C. The total change in entropy is*

---

* Over one cycle, the change in entropy of the working substance is zero, because it is in the same state at the beginning and ending of each cycle.

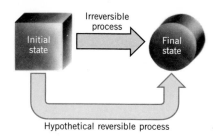

Hypothetical reversible process

ΔS for irreversible process $=$ ΔS for hypothetical reversible process

FIGURE 19.11 Although the relation $\Delta S = (Q/T)_R$ applies to reversible processes, it can be used as part of an indirect procedure to find the entropy change for an irreversible process. This drawing illustrates the procedure discussed in the text.

$$\Delta S_C + \Delta S_H = \frac{Q_C}{T_C} - \frac{Q_H}{T_H} = 0$$

because $Q_C/T_C = Q_H/T_H$, according to Equation 19.4.

The fact that the total change in entropy is zero for a Carnot engine is one specific illustration of a general result. It can be proved that when *any* reversible process occurs, the total change in the entropy of the universe is zero; $\Delta S_{universe} = 0$ for a reversible process. The word "universe" means that $\Delta S_{universe}$ takes into account the entropy changes of all parts of the system and all parts of the environment. **Reversible processes, then, do not alter the total entropy of the universe.** To be sure, the entropy of one part of the universe may change because of a reversible process, but if so, the entropy of another part must change in the opposite way by the same amount.

To understand what happens to the entropy of the universe when an *irreversible* process occurs is more difficult, for the expression $\Delta S = (Q/T)_R$ does not apply directly to such a process. However, if a system changes irreversibly from an initial state to a final state, this expression can be used to calculate ΔS indirectly, as Figure 19.11 indicates. We imagine a hypothetical reversible process that causes the system to change between *the same initial and final states* and then find ΔS for this reversible process. The value obtained for ΔS also applies to the irreversible process that actually occurs, since only the difference between the entropies of the initial and final states determines ΔS. Example 4 illustrates this indirect method and shows that spontaneous (irreversible) processes cause the entropy of the universe to increase.

EXAMPLE 4

Figure 19.12 shows that 1200 J of heat energy flow spontaneously from a hot reservoir at 650 K to a cold reservoir at 350 K. Determine the amount by which this irreversible process changes the entropy of the universe.

SOLUTION

The hot-to-cold heat flow is irreversible, so the relation $\Delta S = (Q/T)_R$ is applied to a hypothetical process whereby the 1200 J of heat are taken reversibly from the hot reservoir and added reversibly to the cold reservoir. The total entropy change of the

universe is the algebraic sum of the entropy changes for each reservoir:

$$\Delta S_{universe} = -\underbrace{\frac{1200 \text{ J}}{650 \text{ K}}}_{\substack{\text{Entropy lost} \\ \text{by the hot} \\ \text{reservoir}}} + \underbrace{\frac{1200 \text{ J}}{350 \text{ K}}}_{\substack{\text{Entropy gained} \\ \text{by the cold} \\ \text{reservoir}}} = \boxed{+1.6 \text{ J/K}}$$

The irreversible process causes the entropy of the universe to increase by 1.6 J/K.

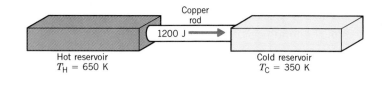

Copper rod

1200 J

Hot reservoir $T_H = 650$ K

Cold reservoir $T_C = 350$ K

FIGURE 19.12 Twelve hundred joules of heat flow spontaneously from a hot reservoir at a constant temperature of 650 K to a cold reservoir at a constant temperature of 350 K.

Example 4 is a specific illustration of a quite general result: **any irreversible process always increases the entropy of the universe.** In other words, $\Delta S_{universe} > 0$ for an irreversible process. Reversible processes do not alter the entropy of the universe, whereas irreversible processes cause the entropy to increase. Therefore, the entropy of the universe continually increases, like time itself. For this reason, entropy is sometimes called "time's arrow." It can be shown that the behavior of the entropy of the universe constitutes a completely general statement of the second law of thermodynamics, which applies not only to heat flow but also to all kinds of other processes.

THE SECOND LAW OF THERMODYNAMICS STATED IN TERMS OF ENTROPY
The total entropy of the universe does not change when a reversible process occurs ($\Delta S_{\text{universe}} = 0$) and always increases when an irreversible process occurs ($\Delta S_{\text{universe}} > 0$).

ENTROPY AND ENERGY THAT IS UNAVAILABLE FOR DOING WORK

Heat energy can be used to do work. However, when an irreversible process occurs and the entropy of the universe increases, some heat energy becomes unavailable for doing work, as the next example helps to illustrate.

EXAMPLE 5

Suppose that 1200 J of heat are used as input for an engine under two different conditions. In Figure 19.13a the heat is supplied by a hot reservoir whose temperature is 650 K. In part b of the drawing, the heat first flows irreversibly through a copper rod into a second reservoir whose temperature is 350 K, and then the heat enters the engine. In either case, a 150-K reservoir is at hand to use as the cold reservoir. For each condition, determine the maximum amount of work that can be obtained from the 1200 J of heat.

SOLUTION

The work obtained from the engine is the product of its efficiency and the input heat: $W = (\text{Efficiency})Q_{\text{H}} = (\text{Efficiency}) \times (1200 \text{ J})$. The maximum amount of work is obtained when the efficiency is a maximum, that is, when the engine is a Carnot engine. In each case, the efficiency can be determined

from Equation 19.5 and the Kelvin temperatures of the hot and cold reservoirs:

$$\begin{bmatrix} \text{Before} \\ \text{irreversible} \\ \text{heat flow} \end{bmatrix} \quad \text{Efficiency} = 1 - \frac{T_{\text{C}}}{T_{\text{H}}}$$

$$= 1 - \frac{150 \text{ K}}{650 \text{ K}} = 0.77$$

$$W = (\text{Efficiency})(1200 \text{ J}) = (0.77)(1200 \text{ J}) = \boxed{920 \text{ J}}$$

$$\begin{bmatrix} \text{After} \\ \text{irreversible} \\ \text{heat flow} \end{bmatrix} \quad \text{Efficiency} = 1 - \frac{T_{\text{C}}}{T_{\text{H}}}$$

$$= 1 - \frac{150 \text{ K}}{350 \text{ K}} = 0.57$$

$$W = (\text{Efficiency})(1200 \text{ J}) = (0.57)(1200 \text{ J}) = \boxed{680 \text{ J}}$$

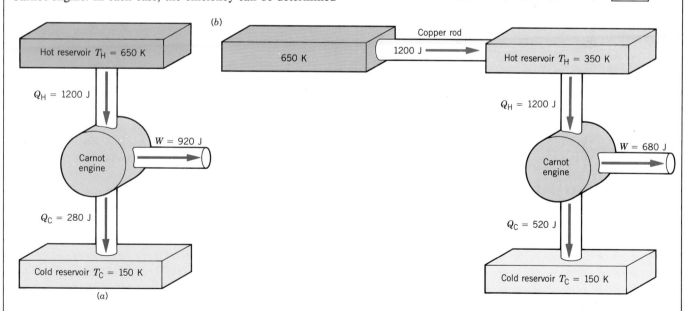

FIGURE 19.13 (a) A Carnot engine uses 1200 J of heat from a hot reservoir at 650 K to produce 920 J of work. The engine rejects 280 J of heat to a cold reservoir at 150 K. (b) After the 1200 J have been allowed to flow irreversibly through a copper rod from 650 to 350 K, only 680 J of work can be extracted using a Carnot engine that rejects heat into the same 150-K cold reservoir.

When the 1200 J of input heat energy are taken from the 350-K reservoir instead of the 650-K reservoir, the efficiency of the Carnot engine is smaller. As a result, less work (680 J versus 920 J) can be extracted from the 1200 J of input heat. In this sense, the input heat is of lower "quality" when it is taken from a 350-K reservoir compared to a 650-K reservoir.

Example 5 shows that 240 J less work (920 J − 680 J) can be performed when the 1200 J of input heat is obtained from the reservoir with the smaller temperature. In other words, the irreversible process of heat flow through the copper rod causes 240 J of heat energy to become unavailable for doing work. The irreversible process also causes the entropy of the universe to increase, and the change in entropy $\Delta S_{universe}$ is related to the energy that is unavailable for doing work $W_{unavailable}$. Example 4 deals with the same irreversible heat flow that is present in Example 5. If the change in entropy calculated in Example 4 ($\Delta S_{universe} = +1.6$ J/K) is multiplied by the lowest Kelvin temperature in Example 5 (150 K), an answer is obtained that is identical to the 240 J of "unavailable" energy just discussed. This is a specific illustration of the following general result:

$$W_{unavailable} = T_0 \Delta S_{universe} \qquad (19.9)$$

where T_0 is the Kelvin temperature of the coldest heat reservoir. Since irreversible processes always cause the entropy of the universe to increase, they cause energy to be degraded in the sense that part of the energy becomes unavailable for the performance of work. In contrast, there is no penalty when reversible processes occur, because for them $\Delta S_{universe} = 0$ and there is no loss of work, $W_{unavailable} = T_0 \Delta S_{universe} = 0$.

Since the entropy of the universe is always increasing, the amount of energy that is unavailable for work is also increasing. Therefore, at some time in the future, it is conceivable that all energy will be degraded to the point where none can be used to do work. Such an eventuality is referred to as the "heat death" of the universe. "Heat death" will arrive when irreversible heat flow from the hot to the cold parts of the universe has occurred to such an extent that all regions have a common temperature. Under such a condition, the hot and cold heat reservoirs that engines must have to produce work would not exist. And without the ability to convert energy into work, civilization could not survive.

ORDER AND DISORDER

Entropy can also be interpreted in terms of order and disorder. As an example, consider a block of ice, with each of its H_2O molecules fixed rigidly in place in a highly structured and ordered arrangement. In comparison, the puddle of water into which the ice melts is disordered and unorganized, for the molecules in a liquid are free to

Block of ice — an ordered system   Puddle of water—a disordered system

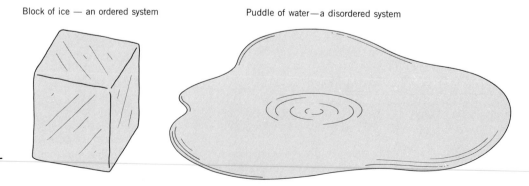

FIGURE 19.14 A block of ice is an example of an ordered system relative to a puddle of water, in which the H_2O molecules have considerable motional freedom.

move from place to place, as Figure 19.14 indicates. Heat is required to melt the ice and produce the disorder. Moreover, heat flow into a system increases the entropy of the system, according to $\Delta S = (Q/T)_R$. Thus, an increase in entropy is associated with an increase in disorder. Conversely, a decrease in entropy is associated with a decrease in disorder or a greater degree of order. Example 6 illustrates this kind of order-to-disorder change and the resulting increase of entropy that accompanies it.

EXAMPLE 6

Find the change in entropy that results when a 2.30-kg block of ice melts slowly (reversibly) at 0 °C (273 K).

SOLUTION

Since the phase change occurs reversibly at a constant temperature, the change in entropy can be found by using $\Delta S = (Q/T)_R$, where Q is the heat absorbed by the melting ice. This heat can be determined by using the latent heat of fusion of water L_f: $Q =$

$mL_f = +(2.30 \text{ kg})(80.0 \text{ kcal/kg}) = +184 \text{ kcal}$. The change in entropy is

$$\Delta S = \frac{+184 \text{ kcal}}{273 \text{ K}} = \boxed{+0.674 \text{ kcal/K}}$$

a result that is positive, since the ice absorbs heat as it melts.

According to the second law of thermodynamics, the entropy of the universe is continually increasing. In terms of order and disorder, this indicates that the universe is becoming an evermore disordered and chaotic place. Under such conditions, how is it that life has evolved? After all, the components of the human body are arranged in a highly structured, well-ordered fashion. To some, the low-entropy arrangement of the parts in the human body seems to fly in the face of the second law of thermodynamics. However, the second law states what happens to *the total entropy of the universe,* not to the entropy of any one single part of the universe. It is perfectly consistent with the second law for a low-entropy structure such as the human body to come into being, as long as there is a sufficient increase in entropy somewhere else to guarantee that the total entropy of the universe increases. Furthermore, work must be done to maintain order. The human body and our society as a whole accomplish this work by extracting the necessary energy from food and fuel, but always at the expense of an increase in entropy and disorder in the environment.

19.7 THE THIRD LAW OF THERMODYNAMICS

To the zeroth, first, and second laws of thermodynamics we add the third (and last) law. The **third law of thermodynamics** indicates that it is impossible to reach a temperature of absolute zero.

THE THIRD LAW OF THERMODYNAMICS

It is not possible to lower the temperature of any system to absolute zero in a finite number of steps.

This law, like the second law, can be expressed in a number of ways, but a discussion of them is beyond the scope of this text. There is nothing in the second law of thermodynamics that prohibits the temperature of a system from being lowered to absolute zero. The third law is needed to explain a number of experimental observations that cannot be explained by the other laws of thermodynamics.

SUMMARY

A **reversible process** is one in which both the system and its environment can be returned to exactly the initial states they were in before the process occurred. All spontaneous processes, such as the conduction of heat and any process involving friction, are irreversible.

The **second law of thermodynamics** can be stated in a number of equivalent forms. In terms of heat flow, the second law declares that heat energy flows spontaneously from a substance at a high temperature to a substance at a low temperature and does not flow spontaneously in the reverse direction. In the form known as **Carnot's principle,** the second law states that no irreversible engine operating between two reservoirs at constant temperatures can have a greater efficiency than a reversible engine operating between the same temperatures. Furthermore, all reversible engines operating between the same temperatures have the same efficiency. In terms of **entropy,** the second law states that the total entropy of the universe does not change when a reversible process occurs and always increases when an irreversible process occurs.

A **heat engine** operates in cycles and produces work W from input heat Q_H that is extracted from a heat reservoir at a relatively high temperature. The engine rejects heat Q_C into a reservoir at a relatively low temperature. The **efficiency** of a heat engine is defined as Efficiency = W/Q_H. In addition, the principle of the conservation of energy requires that $Q_H = W + Q_C$.

A **Carnot engine** is a reversible engine in which all input heat Q_H originates from a hot reservoir at a single Kelvin temperature T_H and all rejected heat Q_C goes into a cold reservoir at a single Kelvin temperature T_C. For a Carnot engine, $Q_C/Q_H = T_C/T_H$. The **efficiency of a Carnot engine** is the maximum efficiency that an engine operating be-

tween two fixed temperatures can have: Efficiency of a Carnot engine = $1 - T_C/T_H$. The maximum efficiency of a Carnot engine is less than one, so even an ideal engine has an efficiency less than 100%.

Refrigerators, air conditioners, and **heat pumps** are devices that utilize work W to make heat Q_C flow from a lower Kelvin temperature T_C to a higher Kelvin temperature T_H. In the process (the refrigeration process) they deposit an amount of heat Q_H at the higher temperature. The principle of the conservation of energy requires that $Q_H = W + Q_C$. If the refrigeration process is perfect, in the sense that it occurs reversibly, the devices are called Carnot devices and the relation $Q_C/Q_H = T_C/T_H$ holds. The **coefficient of performance of a heat pump** is given by Q_H/W, while the **coefficient of performance of a refrigerator or an air conditioner** is Q_C/W.

The **change in entropy** ΔS for a process in which heat Q enters or leaves a system reversibly at a constant Kelvin temperature T is $\Delta S = (Q/T)_R$, where R stands for "reversible." Irreversible processes cause energy to be degraded in the sense that part of the energy becomes unavailable for the performance of work. The **energy that is unavailable for doing work** because of an irreversible process is $W_{unavailable} = T_0 \Delta S_{universe}$, where $\Delta S_{universe}$ is the total entropy change of the universe and T_0 is the Kelvin temperature of the coldest reservoir into which heat is rejected. Increased entropy is associated with a greater degree of disorder and decreased entropy with a lesser degree of disorder (more order).

The **third law of thermodynamics** states that it is not possible to lower the temperature of any system to absolute zero in a finite number of steps.

SOLVED PROBLEMS

SOLVED PROBLEM 1
Related Problems: **6   **7   **41

An engine uses a monatomic ideal gas as a working substance. One cycle of the engine appears in the drawing. The cycle consists of an isochoric (constant volume) tripling of the pressure from A to B, an isothermal (constant temperature) tripling of the volume from B to C, and an isobaric (constant pressure) compression of the volume by a factor of three from C to A. Find the efficiency of the engine.

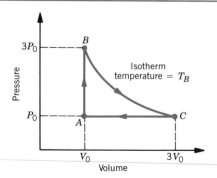

Solution The efficiency of the engine is given by Equation 19.1 as Efficiency $= W/Q_H$. In this expression W is the total work done by the engine, so we begin by calculating the work done during each of the three parts of the cycle, the intent being to compute the total work from the algebraic sum of the three individual values.

Step 1. From A to B on the pressure–volume plot the volume remains constant, so the work is zero, as in any isochoric process (see Section 18.4): $W_{AB} = 0$. From B to C the gas expands isothermally at temperature T_B, from an initial volume $V_i = V_0$ to a final volume $V_f = 3V_0$. Therefore, Equation 18.3 can be used to determine the positive work of expansion:

$$W_{BC} = nRT_B \ln\left(\frac{V_f}{V_i}\right)$$

$$= nRT_B \ln\left(\frac{3V_0}{V_0}\right) = (1.10)nRT_B$$

From C to A the gas is compressed isobarically at the pressure P_0, so Equation 18.2 can be used to determine the negative work of compression:

$$W_{CA} = P_0(V_f - V_i) = P_0(V_0 - 3V_0) = -2P_0V_0$$

The total work W is

$$W = W_{AB} + W_{BC} + W_{CA} = 0 + (1.10)nRT_B - 2P_0V_0$$

Step 2. A value for Q_H, the total input heat to the engine, is also needed if the efficiency is to be determined. This total heat is the sum of the individual input heats associated with the various parts of the cycle. From A to B the pressure increases at constant volume, which means, according to the ideal gas law, that the temperature increases from T_A to T_B. The heat energy that flows into the gas can be calculated by using the molar specific heat capacity at constant volume for a monatomic ideal gas, $C_V = \frac{3}{2}R$ (Equation 18.8):

$$Q_{AB} = C_V n \,\Delta T = \tfrac{3}{2}Rn(T_B - T_A) \qquad (18.6)$$

From B to C the temperature remains constant at T_B while the gas does work in expanding. Under this isothermal condition, the heat energy that flows into the ideal gas equals the work done by the gas (see Section 18.5):

$$Q_{BC} = W_{BC} = (1.10)nRT_B$$

From C to A the volume decreases at constant pressure, which means, according to the ideal gas law, that the temperature decreases from T_C to T_A. Thus, heat *flows out of the system* from C to A. Consequently, only the heat for steps $A \rightarrow B$ and $B \rightarrow C$ need be combined to obtain the total *input* heat Q_H:

$$Q_H = Q_{AB} + Q_{BC} = \tfrac{3}{2}nR(T_B - T_A) + (1.10)nRT_B$$

The efficiency, then, is

$$\text{Efficiency} = \frac{W}{Q_H} = \frac{(1.10)nRT_B - 2P_0V_0}{\tfrac{3}{2}nR(T_B - T_A) + (1.10)nRT_B}$$

Step 3. To complete the problem, note that the ideal gas law applies, so $PV/T = nR =$ constant. The volume is the same at A and B, and the pressure at B is three times that at A. Therefore, the temperature at B is three times that at A or $T_B = 3T_A$. In addition, using $P = P_0$ and $V = V_0$ at point A, it can be seen that $P_AV_A = P_0V_0 = nRT_A$. With these substitutions for T_B and P_0V_0, the expression for efficiency of the engine becomes

$$\text{Efficiency} = \frac{(1.10)nR(3T_A) - 2nRT_A}{\tfrac{3}{2}nR(3T_A - T_A) + (1.10)nR(3T_A)} = \boxed{0.206}$$

Summary of Important Points There are three steps in calculating the efficiency of an engine directly from its closed-loop, pressure–volume plot. First, determine the work associated with each segment of the loop and then combine all the contributions algebraically to get the total work. Second, calculate the heat associated with each segment of the loop for which heat *flows into the system* and combine all the contributions to get the total input heat. In one or more of the loop segments, heat may flow out of the system; the heat for each of these segments is not included with the total input heat. Lastly, when dividing the total work by the total input heat to determine the efficiency, take advantage of any relations between the pressure, volume, and temperature at various points in the cycle. For an ideal gas, $PV = nRT$ and, if expansion or compression occurs adiabatically, $P_iV_i^\gamma = P_fV_f^\gamma$.

QUESTIONS

1. Two *irreversible* engines operate between the same hot and cold reservoirs. Does Carnot's principle require that these two engines have the same efficiency? Justify your answer.

2. Consider a hypothetical engine that takes 10 000 J of heat from a hot reservoir and 5000 J of heat from a cold reservoir and produces 15 000 J of work. (a) Does this engine violate the first law of thermodynamics? Explain. (b) Does the engine violate the second law of thermodynamics? Explain.

3. The second law of thermodynamics, in the form of Carnot's principle, indicates that the most efficient heat engine operating between two temperatures is a reversible one. Does this mean that a reversible engine operating between the temperatures of 600 and 400 K must be more efficient than an *irreversible* engine operating between 700 and 300 K? Provide a reason for your answer.

4. Three reversible engines are shown in the drawing. Engine A operates between the temperatures of 1000 and 400 K. Engines B

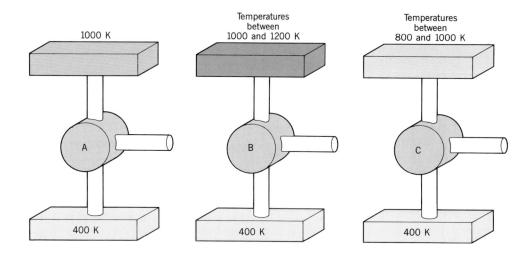

and C receive their input heat over a range of temperatures, as the drawing shows. Rank these engines in order of increasing efficiency (smallest efficiency first). Account for your answer.

5. Suppose you wish to improve the efficiency of a Carnot engine. Compare the improvement to be realized via each of the following alternatives: (1) lower the Kelvin temperature of the cold reservoir by a factor of four, (2) raise the Kelvin temperature of the hot reservoir by a factor of four, (3) cut the Kelvin temperature of the cold reservoir in half and double the Kelvin temperature of the hot reservoir. Give your reasoning.

6. Explain why you cannot (a) cool your kitchen by leaving the refrigerator door open and (b) cool your bedroom by putting a window air conditioner on the floor.

7. Is it possible for a Carnot heat pump to have a coefficient of performance that is less than one? Justify your answer.

8. Explain why the coefficient of performance of a heat pump cannot be determined from a statement such as "the pump delivers 15 000 J of heat per cycle."

9. Suppose the size of the Celsius degree remains unchanged, but absolute zero occurs at −73 °C instead of −273 °C. (a) Would this increase or decrease the efficiency of a Carnot engine that operates between two reservoirs whose temperatures are the steam point and the ice point of water? (b) Would this increase or decrease the coefficient of performance of a Carnot heat pump? Provide a reason for each answer.

10. An event happens somewhere in the universe and, as a result, the entropy of an object changes by −5 J/K. According to the second law of thermodynamics, which one (or more) of the following is a possible value for the entropy change for the rest of the universe: −5 J/K, 0 J/K, +5 J/K, +10 J/K? Account for your choice(s).

11. When water freezes from a less-ordered liquid to a more-ordered solid, its entropy decreases. Why doesn't this decrease in entropy violate the entropy version of the second law of thermodynamics?

12. In each of the following cases, which has the greater entropy: (a) a handful of popcorn kernels or the popcorn that results from them, (b) a salad before or after it has been tossed, (c) a messy apartment with clothes strewn all over or a neat apartment? Why?

13. A glass of water contains a teaspoon of dissolved sugar. After a while, the water evaporates, leaving behind sugar crystals. The entropy of the sugar crystals is less than the entropy of the dissolved sugar, because the sugar crystals are in a more ordered state. Explain why this process does not violate the entropy version of the second law of thermodynamics.

14. A builder uses lumber to construct a building, which is unfortunately destroyed in a fire. Thus, the lumber existed at one time or another in three different states: (1) as unused building material, (2) as a building, and (3) as a burned-out shell of a building. Rank these three states in order of decreasing entropy (largest first). Provide a reason for the ranking.

PROBLEMS

Section 19.2 Heat Engines

1. The input heat for an engine is 5.75 kcal, and the rejected heat is 1.40 kcal. Find (a) the work done by the engine and (b) the efficiency of the engine.

2. An engine has an efficiency of 71% and produces 4800 J of work. Determine (a) the input heat and (b) the rejected heat.

3. In doing 16 600 J of work, an engine rejects 9700 J of heat. What is the efficiency of the engine?

* **4.** Engine A rejects two times more heat than engine B, when each receives the same input heat. The efficiency of B is three times greater than that of A. Find the efficiency of each engine.

* **5.** Two engines are arranged so the rejected heat of one provides the input heat for the other, as the drawing shows. The efficiencies of the two engines are e_1 and e_2. The overall efficiency of this two-engine device is the total work done ($W_1 + W_2$) divided by the input heat Q_H. Prove that the overall efficiency e is given by $e = e_1 + e_2 - e_1 e_2$.

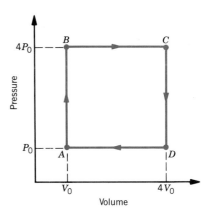

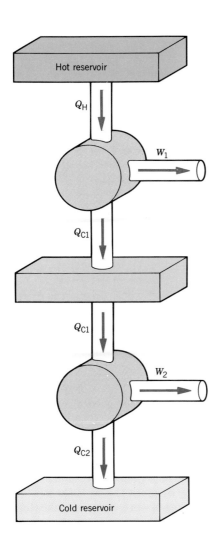

**6. One cycle of an engine appears on a pressure–volume plot as the rectangle $ABCD$ shown in the drawing. The working substance of the engine is a monatomic ideal gas. The cycle includes an isochoric pressure rise by a factor of four from A to B, an isobaric expansion of the volume by a factor of four from B to C, an isochoric pressure drop by a factor of four from C to D, and an isobaric compression of the volume by a factor of four from D to A. Determine the efficiency of the engine. *(See Solved Problem 1 for a related problem.)*

**7. The drawing shows the pressure–volume plot for one cycle of an engine that utilizes a monatomic ideal gas as the working substance. The cycle begins at point A and involves an isochoric rise in pressure from A to B, an adiabatic expansion from B to C, and an isobaric compression from C to A. The temperatures at B and C are given in the drawing. Find the efficiency of this engine. *(See Solved Problem 1 for a related problem.)*

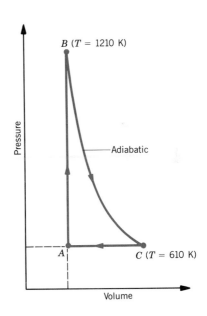

Section 19.4 The Efficiency of the Carnot Heat Engine

8. Five thousand joules of heat are put into a Carnot engine whose hot and cold reservoirs have temperatures of 500 and 200 K, respectively. How much heat is (a) converted into work and (b) rejected to the cold reservoir?

9. Electric motors can convert electrical energy into work with efficiencies approaching 95%. For a Carnot engine to have such an efficiency, what must be the ratio of the Kelvin temperatures of the hot and cold reservoirs?

10. The ratio of the input heat to the rejected energy of a Carnot engine is 1.5. (a) What is the efficiency of the engine? (b) What is the ratio of the Kelvin temperature of the hot reservoir to the Kelvin temperature of the cold reservoir?

11. A Carnot engine operates between reservoirs whose temperatures are 650 and 350 K. To improve the efficiency of the engine, it is decided either to raise the temperature of the hot reservoir by 40 K or to lower the temperature of the cold reservoir by 40 K. Which change gives the greatest improvement? Justify your answer by calculating the efficiency in each case.

12. A Carnot engine does 20 900 J of work and rejects 7330 J of heat into a cold reservoir at 25 °C. What is the Kelvin temperature of the hot reservoir?

* 13. A power plant taps steam superheated by geothermal energy to 232 °C (the temperature of the hot reservoir) and uses the steam to do work in turning the turbine of an electric generator. The steam is then converted back into water in a condenser at 50 °C (the temperature of the cold reservoir), after which the water is pumped back down into the earth where it is heated again. The output power of the plant is 84 000 kilowatts. Determine (a) the maximum efficiency at which this plant can operate and (b) the minimum amount of rejected heat that must be removed from the condenser every twenty-four hours.

* 14. Two Carnot engines A and B are connected as shown in the drawing. The 503-K cold reservoir of A also serves as the hot reservoir for B, so the rejected heat of A is the input heat for B. Each engine produces 1750 J of work when the input heat to engine A is 5550 J. Find the Kelvin temperatures, T_1 and T_2, of the hottest and coldest reservoirs. See Drawing 14.

* 15. Two Carnot engines are connected to heat reservoirs, as the drawing indicates. The input heat to each engine is the same. The overall efficiency of this two-engine device is the total work done ($W_1 + W_2$) divided by the total input heat $2Q_H$. What is the overall efficiency? See Drawing 15.

**16. The hot and cold reservoirs of a Carnot engine have temperatures of 905 and 405 K, respectively. The engine does the work of lifting a 10.0-kg block straight up from rest, so that at a height of 4.00 m the block has a speed of 8.00 m/s. How much heat must be put into the engine?

**17. A nuclear-fueled electric power plant utilizes a so-called "boiling water reactor." In this type of reactor, nuclear energy causes water under pressure to boil at 285 °C (the temperature of the hot reservoir). After the steam does the work of turning the turbine of an electric generator, the steam is converted back into water in a condenser at 40 °C (the temperature of the cold reservoir). To keep the condenser at 40 °C, the rejected heat must be carried away by some means, for example, by water from a river. The plant operates at three-fourths of its Carnot efficiency, and the electrical output power of the plant is 1.2×10^9 watts. A river with a water flow rate of 1.0×10^5 kg/s is available to remove the rejected heat from the plant. Find the number of Celsius degrees by which the temperature of the river rises.

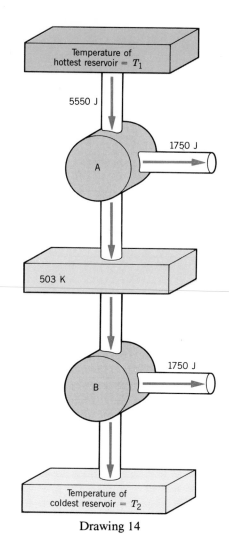

Drawing 14

Section 19.5 Refrigerators, Air Conditioners, and Heat Pumps

18. A Carnot air conditioner uses 25 500 J of electrical energy, and the temperatures indoors and outdoors are 27 °C and 39 °C, respectively. How much heat is deposited outdoors?

19. The water in a deep underground well is used as the cold reservoir of a Carnot heat pump that maintains the temperature of a house at 25 °C. To deposit 12 600 J of heat in the house, the heat pump requires 806 J of work. Determine the Kelvin temperature of the well water.

20. A Carnot refrigerator maintains the food inside it at 3 °C, while the temperature of the kitchen is 25 °C. The refrigerator removes 3.00×10^4 J of heat energy from a quantity of food. (a) How much heat is delivered to the kitchen? (b) Determine the work done to operate the refrigerator.

21. A Carnot engine has an efficiency of 0.70. If this engine were run backward as a heat pump, what would be the coefficient of performance?

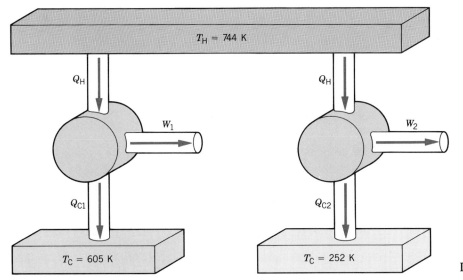

Drawing 15

22. A heat pump removes 0.50 kcal of heat from the outdoors and delivers 0.75 kcal of heat to the inside of a house. (a) How much work does the heat pump need? (b) What is the coefficient of performance of the heat pump?

* **23.** Two kilograms of water at 0 °C are put into the freezer compartment of a Carnot refrigerator. The temperature of the compartment is -15 °C and the temperature of the kitchen is 27 °C. If the cost of electrical energy is ten cents per kilowatt·hour, how much does it cost to make two kilograms of ice at 0 °C?

* **24.** An engine with an efficiency of 0.300 provides the work to run a Carnot air conditioner. The air conditioner removes 1930 kcal of heat from a room at 27 °C. Only 388 kcal of input heat are available for the engine. What is the maximum temperature of the hot reservoir of the air conditioner?

* **25.** How long would a 3.00-kW space heater have to run to put into a kitchen the same amount of heat as a refrigerator (coefficient of performance = 3.00) does when it freezes 1.50 kg of water at 20.0 °C into ice at 0.0 °C?

**26. The temperatures of the reservoirs of a Carnot engine are 1684 and 842 K. As the drawing shows, the work from this engine drives a Carnot heat pump. The heat pump removes from the 842-K reservoir an amount of heat that equals the input heat to the engine and pumps it into a hot reservoir of unknown temperature. What is the Kelvin temperature of this reservoir?

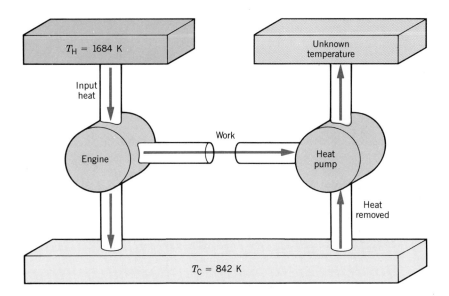

Section 19.6 Entropy and the Second Law of Thermodynamics

27. Suppose the entropy of a system decreases by 25 J/K because of some process. (a) Based on the second law of thermodynamics, what can you conclude about the entropy change of the environment that surrounds this system? (b) Interpret your answer to part (a) in terms of order and disorder of the environment.

28. Four kilograms of carbon dioxide sublime from solid "dry ice" to a gas at a pressure of 1.00 atm and a temperature of -78.5 °C. The latent heat of sublimation is 5.77×10^5 J/kg. Find the change in entropy of the carbon dioxide.

29. A hot reservoir consists of a large mass of water at its normal boiling point. A cold reservoir consists of a large mass of water at its normal freezing point. The temperature of each reservoir remains constant at all times. As the drawing shows, some heat Q from the hot reservoir flows spontaneously through a metal rod to a reservoir at an unknown temperature T, which is between the normal boiling and freezing points of water. The Carnot engine connected between this intermediate reservoir and the cold reservoir cannot produce work from twenty percent of the heat Q. What is the temperature T? See Drawing 29.

* 30. (a) Find the equilibrium temperature that results when one kilogram of water at 100.0 °C is added to two kilograms of water at 10.0 °C in a perfectly insulated container. (b) When heat is added to or removed from a solid or liquid of mass m and specific heat capacity c, the change in entropy can be shown to be $\Delta S = mc \ln(T_f/T_i)$, where T_i and T_f are the initial and final Kelvin temperatures. Use this equation to calculate the entropy change for each amount of water. Then combine the two entropy changes algebraically to obtain the total entropy change. Note that the

process is irreversible, so the total entropy change of the universe is greater than zero. (c) Assuming the coldest reservoir at hand has a temperature of 0 °C, determine the amount of energy that becomes unavailable for doing work because of the irreversible process.

* 31. Review Example 5 and Figure 19.13 in the text. Also review the discussion that follows the example concerning the 240 J of energy that become unavailable for doing work. This unavailable energy results, because 1200 J of heat energy flow irreversibly from a 650-K reservoir to a 350-K reservoir. An inventor claims that to recover this unavailable energy, we need only use a heat pump and return the 1200 J of heat to the 650-K reservoir, from which location we can then proceed as in Figure 19.13a. To see why this inventor is wrong, do the following. (a) Use a Carnot heat pump (the best possible kind) to remove the 1200 J from the 350-K reservoir and calculate the amount of energy the heat pump would deposit into the 650-K reservoir. (b) Then, with a Carnot engine operating between 650 and 350 K, determine how much work can be extracted from the energy calculated in part (a). (c) Finally, obtain the work needed to operate the heat pump in part (a) and compare this work to that determined in part (b). Notice that there is no point in doing what this inventor claims.

**32. (a) Five kilograms of water at 80.0 °C are mixed in a perfect thermos with 2.00 kg of ice at 0.0 °C, and the mixture is allowed to reach equilibrium. Using the expression $\Delta S = mc \ln(T_f/T_i)$ [see problem 30] and the change in entropy for melting, find the change in entropy that occurs. (b) Should the entropy of the universe increase or decrease as a result of the mixing process? Give your reasoning and state whether your answer in part (a) is consistent with your answer here.

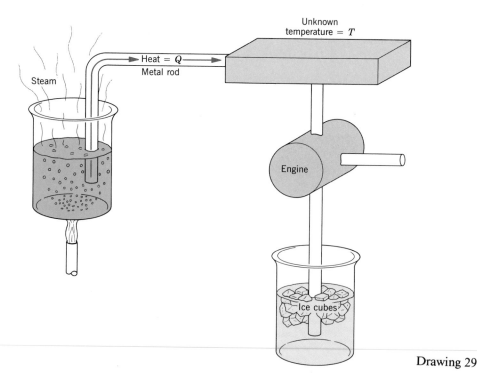

Drawing 29

ADDITIONAL PROBLEMS

33. Find the change in entropy of the H_2O molecules when (a) three kilograms of ice melt into water at 0.0 °C and (b) three kilograms of water change into steam at 100.0 °C. (c) On the basis of the answers to parts (a) and (b), discuss which change creates more disorder in the collection of H_2O molecules.

34. An engine rejects three times more heat than it converts into work. What is the efficiency of the engine?

35. A Carnot heat pump operates between an outdoor temperature of −8 °C and an indoor temperature of 25 °C. Find its coefficient of performance.

36. A process occurs in which the entropy of a system increases by 125 J/K. During the process, the energy that becomes unavailable for doing work is zero. (a) Is this process reversible or irreversible? Give your reasoning. (b) Determine the change in the entropy of the environment of the system.

37. A Carnot engine has an efficiency of 0.700, and the temperature of its cold reservoir is 105 °C. (a) Determine the Kelvin temperature of its hot reservoir. (b) If 1.25 kcal of heat are rejected to the cold reservoir, what amount of heat is put into the engine?

* 38. An engine is run in reverse as a heat pump. An identical engine (with the same values of Q_H, Q_C, and W) is run in reverse as a refrigerator. The coefficient of performance of the heat pump is three times greater than the coefficient of performance of the refrigerator. Obtain (a) the coefficient of performance of the refrigerator, (b) the coefficient of performance of the heat pump, and (c) the efficiency of the engine.

* 39. Engine A receives three times more input heat, produces five times more work, and rejects two times more heat than engine B. Find the efficiency of each engine.

**40. Suppose the gasoline in a car engine burns at 631 °C, while the exhaust temperature is 139 °C and the outdoor temperature is 27 °C. Assume the engine can be treated as a Carnot engine (a gross oversimplification). In an attempt to increase mileage performance, an inventor builds a second engine that functions between the exhaust and outdoor temperatures and uses the exhaust heat to produce additional work. Assume the inventor's engine can also be treated as a Carnot engine. By what percentage does the inventor's device increase the work obtained from the burning gasoline?

**41. An engine uses a monatomic ideal gas as a working substance, and the drawing illustrates one cycle of its operation, which consists of three steps. From A to B on the pressure–volume plot, an isothermal expansion at temperature T doubles the volume of the gas. From B to C an isobaric compression reduces the volume. From C to A an adiabatic compression returns the gas to its initial state. Find the efficiency of the engine. *(See Solved Problem 1 for a related problem.)*

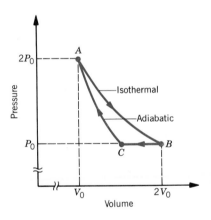

CHAPTER 20

An Introduction to Waves

A surfer catches a wave for a ride. Sound and light are also waves, and "catching" these waves gives us our senses of hearing and sight. The important features of waves are presented in the text.

20.1 THE NATURE OF WAVES

Water waves reveal two essential features that all waves have. First, *a wave is a traveling disturbance.* In Figure 20.1 the wave created by the motorboat travels across the lake and disturbs the fisherman's boat. It is important to realize that *there is no bulk flow of water* outward from the motorboat. The wave is not a bulk movement of water such as a river, but, rather, a disturbance traveling on the surface of the water. Second, *a wave carries energy from place to place.* Part of the energy of the wave in the drawing is used to do the work of moving the fisherman's boat.

There are many kinds of waves, depending on the kind of disturbance that is created by the source of the wave and the kind of material or medium the wave travels through. It is useful, therefore, to introduce two basic types of waves, transverse and longitudinal. Figure 20.2 illustrates how a transverse wave can be generated using a Slinky, a remarkable toy that is a long, loosely coiled spring. If one end of the Slinky is jerked up and down, an upward pulse is sent traveling toward the right. If the end is then jerked down and up, a downward pulse is generated and also moves to the right. If the end is continually moved up and down in simple harmonic motion, an entire

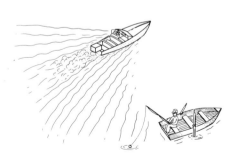

FIGURE 20.1 Waves created by the motorboat travel across the lake. When they reach the fisherman's boat, they disturb it.

wave is produced. The wave consists of a series of alternating upward and downward sections that propagate to the right, disturbing the vertical position of the Slinky in the process. To focus attention on the disturbance, a colored dot has been attached to the Slinky in the last drawing in the figure. As the wave advances to the right, the dot is displaced up and down in simple harmonic motion. The motion of the dot occurs perpendicular, or transverse, to the horizontal direction in which the wave advances. This example shows that *a transverse wave is one in which the disturbance occurs perpendicular to the direction of travel of the wave.* Radio waves, light waves, and microwaves are transverse waves.

A longitudinal wave can also be generated with a Slinky, and Figure 20.3 demonstrates how. When one end of the Slinky is pushed forward along its length (i.e., longitudinally) and then returned to its starting point, a compressed region where the coils are squeezed together is sent traveling to the right. If the end is pulled backward and then returned to its starting point, a stretched region where the coils are pulled apart is formed and also moves to the right. If the end is continually moved back and forth in simple harmonic motion, an entire wave is created. The wave consists of a series of alternating compressed and stretched regions that travel to the right and disturb the separation between adjacent coils. As before, a colored dot has been attached to the Slinky to emphasize the vibratory nature of the disturbance. In response to the wave, the dot is displaced back and forth in simple harmonic motion along the line of travel of the wave. This example reveals that *a longitudinal wave is one in which the disturbance is parallel to the line of travel of the wave.* A sound wave is a longitudinal wave.

There are waves that are neither completely transverse nor completely longitudinal. For instance, in a water wave the motion of the water particles is not strictly perpendicular or strictly parallel to the line along which the wave travels. Instead, the motion includes both transverse and longitudinal components, since the water parti-

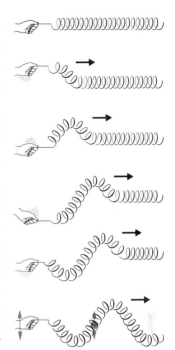

FIGURE 20.2 When the end of the Slinky is moved up and down in simple harmonic motion, a transverse wave is generated. The wave travels to the right.

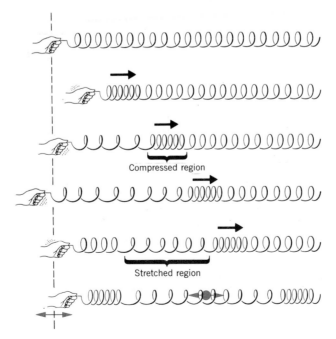

FIGURE 20.3 When the end of the Slinky is moved back and forth in simple harmonic motion parallel to the length of the Slinky, a longitudinal wave is created. The wave travels to the right.

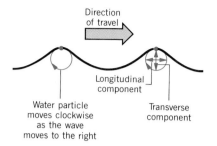

FIGURE 20.4 A water wave is neither transverse nor longitudinal, since water particles at the surface move clockwise on circular paths as the wave moves from left to right.

cles at the surface move on circular paths, as Figure 20.4 indicates. This text focuses primarily on transverse and longitudinal waves.

20.2 PERIODIC WAVES

Both the transverse and longitudinal waves that we have been discussing are ***periodic waves.*** A periodic wave consists of patterns that are produced over and over again by the source. In Figures 20.2 and 20.3 the repetitive patterns occur as a result of the simple harmonic motion of the left end of the Slinky. In fact, some of the terminology (amplitude, period, and frequency) used to describe periodic waves is the same as that used in connection with simple harmonic motion.

To introduce this terminology, Figure 20.5 presents a graphical representation of a transverse wave on a Slinky. In part *a* of the drawing the vertical position of the Slinky is plotted on the ordinate, while the corresponding location along the length of the Slinky is plotted on the abscissa. Such a graph is equivalent to a photograph of the wave taken at a particular instant in time and shows the disturbance that exists at each point along the Slinky's length. As marked on this graph, the ***amplitude*** A is the maximum excursion of a particle of the medium from the particle's undisturbed position. The amplitude is the distance between a crest, or highest point on the wave pattern, and the undisturbed position; the amplitude is also the distance between a trough, or lowest point on the wave pattern, and the undisturbed position. The ***wavelength*** λ is the horizontal length of one cycle of the wave, as shown in color in Figure 20.5*a*. The wavelength is also the horizontal distance between two successive crests, two successive troughs, or any two successive equivalent points on the wave. A wave is a series of many cycles or wavelengths.

Part *b* of Figure 20.5 shows a graph in which time, rather than distance, is plotted on the abscissa. This graph is obtained by observing a single point on the Slinky. As the wave passes, the point moves up and down as a function of time. As indicated on the graph, the ***period*** T is the time required for one cycle of the wave to pass an

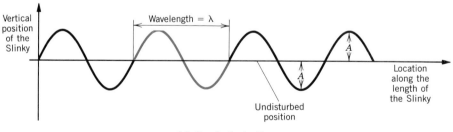

(*a*) At a Particular Time

FIGURE 20.5 In parts *a* and *b*, one cycle of the wave is shown in color, and the amplitude of the wave is denoted as A. The wavelength λ is the horizontal length (in meters) of one cycle, as part *a* illustrates. The period T is the time (in seconds) required for one cycle to pass an observer, as part *b* indicates.

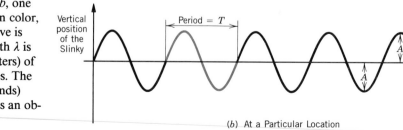

(*b*) At a Particular Location

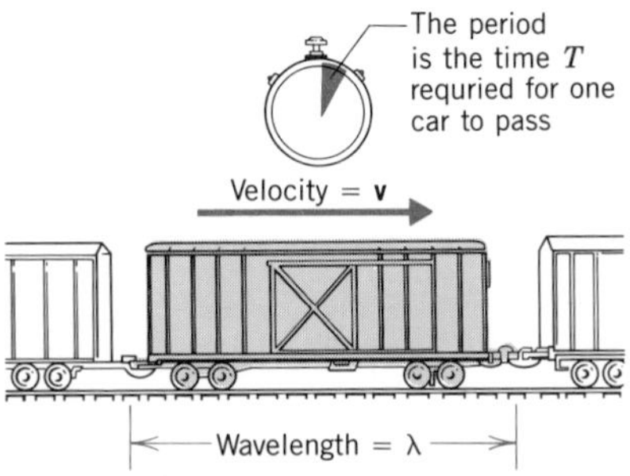

FIGURE 20.6 A train moving at a constant speed serves as an analogy for a traveling wave.

observer. Equivalently, the period is the time required for the wave to travel a distance of one wavelength. Moreover, since the point under observation moves up and down in simple harmonic motion, the period is also the time required for one complete up/down cycle, just as it is for an object on a spring.

As for any example of simple harmonic motion, the period T is related to the *frequency f* according to

$$f = \frac{1}{T} \tag{11.8}$$

Frequency is measured in cycles per second or hertz (Hz). If, for instance, one cycle of a wave takes one-tenth of a second to pass an observer, then ten cycles pass during each second, as Equation 11.8 indicates [$f = 1/(0.1 \text{ s}) = 10$ cycles/s].

A simple relation exists between the period, the wavelength, and the speed of a wave, a relation that Figure 20.6 helps to introduce. Imagine waiting at a railroad crossing, while a freight train passes by at a constant speed. The train consists of a long line of identical boxcars, each of which has a length λ and requires a time T to pass. You could easily calculate the speed v of the train by dividing the length by the time; that is, $v = \lambda/T$. This same equation applies for a wave and relates the speed of the wave to the wavelength λ and the period T. Since the frequency of a wave is $f = 1/T$, the expression for the speed is often written as

$$v = \lambda f \tag{20.1}$$

Example 1 illustrates how the wavelength of a wave is determined by the wave speed and the frequency established by the source.

EXAMPLE 1

AM and FM radio waves are transverse waves that consist of electric and magnetic disturbances. These waves travel at a speed of 3.00×10^8 m/s. A station broadcasts an AM radio wave whose frequency is 1230×10^3 Hz (1230 kHz on the dial) and an FM radio wave whose frequency is 91.9×10^6 Hz (91.9 MHz on the dial). Find the distance between adjacent crests in each wave.

SOLUTION

The distance between adjacent crests is the wavelength λ. Since the speed of each wave is known to be $v = 3.00 \times 10^8$ m/s and

the frequency is established at the broadcasting station in each case, $v = \lambda f$ can be used to determine the wavelength:

[AM] $\quad \lambda = \dfrac{v}{f} = \dfrac{3.00 \times 10^8 \text{ m/s}}{1230 \times 10^3 \text{ Hz}} = \boxed{244 \text{ m}}$

[FM] $\quad \lambda = \dfrac{v}{f} = \dfrac{3.00 \times 10^8 \text{ m/s}}{91.9 \times 10^6 \text{ Hz}} = \boxed{3.26 \text{ m}}$

Notice that the wavelength of an AM radio wave exceeds the length of two and one-half football fields!

The terminology discussed above and the fundamental relations $f = 1/T$ and $v = f\lambda$ apply to longitudinal as well as transverse waves. The next chapter deals with an important example of a longitudinal wave, namely, a sound wave.

20.3 THE SPEED OF A WAVE ON A STRING

THE DEPENDENCE OF WAVE SPEED ON PROPERTIES OF THE STRING

The properties of the material* or medium through which a wave travels determine the speed of the wave. To illustrate this important fact, let us consider a transverse wave on a string or a wire. Figure 20.7 shows such a wave and draws attention to seven string particles that have been drawn as colored dots. As the wave speeds along, each particle is displaced, one after the other, from its undisturbed position. For instance, particles 1–3 have already been displaced upward, while particles 4–7 are not yet affected by the wave, although particle 4 is just about to be displaced upward. Particle 4 will move upward because the section of string immediately to its left (i.e., particle 3) will pull it upward.

Figure 20.7 leads us to conclude that the speed with which the wave moves to the right depends on how quickly one particle of the string is accelerated upward in response to the net pulling force exerted by its adjacent neighbors. In accord with Newton's second law, a stronger net force results in a greater acceleration, and, thus, a faster-moving wave. The ability of one particle of the string to pull on its neighbors depends on how tightly the string is stretched, that is, on the tension in the string (see Section 4.9 for a review of tension). The greater the tension, the greater the pulling force the particles exert on each other. Thus, other things being equal, a wave travels faster on a string with a greater tension than on one with a smaller tension.

In addition to the tension, there is another factor that influences the wave speed. According to Newton's second law, the inertia or mass of particle 4 in Figure 20.7 also affects how quickly it responds to the upward pull of particle 3. For a given net pulling force, a smaller mass has a greater acceleration than a larger mass. Therefore, other things being equal, a wave travels faster on a string whose particles have a small mass, or, as it turns out, on a string that has a small mass per unit length. The mass per unit length is called the *linear density* of the string.

The effects of the tension F and the mass per unit length m/L are evident in the following expression for the speed v of a wave on a string:

$$v = \sqrt{\frac{F}{m/L}} \qquad (20.2)$$

Clearly, a larger tension F and a smaller linear density m/L lead to a larger value for the speed. This equation applies only to small-amplitude waves for a reason that is explained later.

The motion of transverse waves along a string is important for the operation of

FIGURE 20.7 As a transverse wave moves to the right with speed v, each string particle is displaced, one after the other, from its undisturbed position. At the time shown in the picture, particle 4 is just about to be pulled vertically upward.

* Electromagnetic waves can move through a vacuum, as well as through materials such as glass and water.

many musical instruments. In all instruments that utilize strings (guitar, violin, piano, etc.), the strings are either plucked, bowed, or struck to produce transverse waves. Example 2 discusses the speed of the waves on the strings of a guitar.

EXAMPLE 2

Transverse waves travel on the strings of an electric guitar after the strings are plucked, as Figure 20.8 illustrates. The length of each string between its two fixed ends is 0.629 m, and the mass is 0.208 g for the highest pitched E string and 3.32 g for the lowest pitched E string. Each string is under a tension of 226 N. Find the speeds of the waves on the two strings.

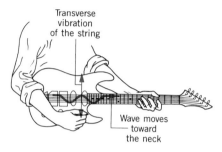

Transverse vibration of the string

Wave moves toward the neck

FIGURE 20.8 Transverse waves are generated on a guitar string by plucking it.

SOLUTION

The tension F, the mass m, and the length L are known for each string, and the speeds of the waves are given by Equation 20.2:

$$\begin{bmatrix} \textbf{High-} \\ \textbf{pitched} \\ \textbf{E} \end{bmatrix} \quad v = \sqrt{\frac{F}{m/L}}$$

$$= \sqrt{\frac{226 \text{ N}}{(0.208 \times 10^{-3} \text{ kg})/(0.629 \text{ m})}} = \boxed{827 \text{ m/s}}$$

$$\begin{bmatrix} \textbf{Low-} \\ \textbf{pitched} \\ \textbf{E} \end{bmatrix} \quad v = \sqrt{\frac{F}{m/L}}$$

$$= \sqrt{\frac{226 \text{ N}}{(3.32 \times 10^{-3} \text{ kg})/(0.629 \text{ m})}} = \boxed{207 \text{ m/s}}$$

Notice how fast the waves move; the speeds correspond to 1850 mi/h and 463 mi/h, respectively.

DERIVATION OF $v = \sqrt{F/(m/L)}$

Figure 20.9 shows one cycle of a wave traveling from left to right on a string. An application of Newton's second law to a small string segment of mass m and length L leads to an expression for the speed of the wave. One way of applying the second law is to envision the cycle at rest, as it would appear to you if you were moving toward the right, just matching its speed. From such a point of view, the string itself would appear to be moving toward the left at constant speed v, as if being pulled through a frictionless tube of exactly the same shape as the cycle. At the crest, the string would be moving along a path that is approximately the arc of a circle (see the drawing), and the speed v would be the tangential speed along the arc. The motion of the string at the crest can then be regarded as *uniform circular motion,* a kind of motion that Chapter 6 discusses. For uniform circular motion, Newton's second law is

$$F_c = ma_c = \frac{mv^2}{r} \tag{6.3}$$

where F_c and a_c are the centripetal force and centripetal acceleration, v is the tangential speed along the arc, and r is the radius of the circle.

To determine the centripetal force, it is necessary to consider the tension F in the

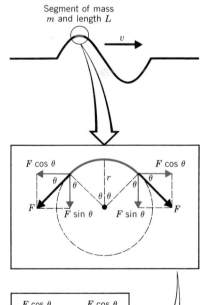

Segment of mass m and length L

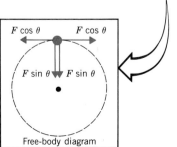

Free-body diagram

FIGURE 20.9 The speed v of a transverse wave on a string can be determined by applying Newton's second law to a string segment of mass m and length L. The segment is at the crest of the cycle. The tension in the string is denoted by F, and the derivation in the text shows that $v = \sqrt{F/(m/L)}$.

string. Because of the tension, a force of magnitude F acts on either end of the string segment and is tangent to the circle, as the drawing shows. The free-body diagram indicates that the two horizontal components of the tension cancel, while the two vertical components reinforce, both pointing toward the center of the circle. Thus,

$$F_c = 2F \sin \theta$$

where θ is one-half the angle subtended by the segment. If θ is small and expressed in radians, $\sin \theta = \theta$ to a good degree of approximation. Consequently,

$$F_c = 2F \theta = \frac{mv^2}{r}$$

The angle 2θ is given in radians as

$$2\theta = \frac{\text{Arc length}}{\text{Radius}} = \frac{L}{r} \qquad (9.1)$$

With this substitution for 2θ, the expression for the centripetal force becomes

$$F_c = F\left(\frac{L}{r}\right) = \frac{mv^2}{r}$$

Solving this expression for the speed v shows that $v = \sqrt{F/(m/L)}$. This result is applicable only for a wave whose amplitude is small compared to its wavelength. If the amplitude were of the same magnitude as or larger than the wavelength, the tension would be different at different points on the string, and it would be incorrect to speak of the string as having a single tension F.

*20.4 THE MATHEMATICAL DESCRIPTION OF A WAVE

It is useful to have a mathematical expression for a wave. For periodic waves that result from simple harmonic motion at the source, the expression involves a sine or cosine, a fact that is not surprising. After all, in Chapter 11 simple harmonic motion is described using sinusoidal equations, and the graphs for a wave in Figure 20.5 look like a plot of position versus time for an object oscillating on a spring (see Figure 11.12).

Our tack will be to present the expression for a wave and then show graphically that it gives a correct description. Equation 20.3 represents a wave that travels in the $+x$ direction (to the right) and has an amplitude A, frequency f, and wavelength λ. Equation 20.4 applies to a wave moving in the $-x$ direction (to the left).

$$\begin{bmatrix} \textbf{Wave motion} \\ \textbf{toward } +x \end{bmatrix} \qquad y = A \sin\left(2\pi ft - \frac{2\pi x}{\lambda}\right) \qquad (20.3)$$

$$\begin{bmatrix} \textbf{Wave motion} \\ \textbf{toward } -x \end{bmatrix} \qquad y = A \sin\left(2\pi ft + \frac{2\pi x}{\lambda}\right) \qquad (20.4)$$

The symbol y denotes the displacement of a particle of the medium from the particle's undisturbed position. These equations apply to transverse or longitudinal waves and assume that $y = 0$ when $x = 0$ and $t = 0$.

Consider a transverse wave moving in the $+x$ direction along a string. The term

$(2\pi ft - 2\pi x/\lambda)$ in Equation 20.3 is called the *phase angle* of the wave. A string particle located at the origin ($x = 0$) exhibits simple harmonic motion with a phase angle of $2\pi ft$, that is, $y = A \sin(2\pi ft)$. A particle located at a distance x also exhibits simple harmonic motion, but its phase angle is $2\pi f(t - x/f\lambda) = 2\pi f(t - x/v)$, where the quantity x/v is the time needed for the wave to travel the distance x. In other words, the simple harmonic motion that occurs at x is delayed by the time interval x/v compared to that occurring at the origin.

Figure 20.10 shows y plotted as a function of position x along the string at a series of time intervals separated by one-fourth of the period T ($t = 0, \frac{1}{4}T, \frac{2}{4}T, \frac{3}{4}T, T$). These graphs are constructed by substituting the corresponding value for t in Equation 20.3, remembering that $f = 1/T$, and then calculating y at a series of values for x. The graphs are like photographs taken at various times as the wave moves to the right. For reference, the colored square on each graph marks the place on the wave that is located at $x = 0$ when $t = 0$. As time passes, the colored square clearly moves to the right, along with the wave. It should be noted that the phase angle $(2\pi ft - 2\pi x/\lambda)$ is measured in *radians,* not degrees. Therefore, when using a calculator to evaluate the function $\sin(2\pi ft - 2\pi x/\lambda)$, the calculator must be set to its radian mode. In a similar manner, it can be seen that Equation 20.4 represents a wave moving in the $-x$ direction.

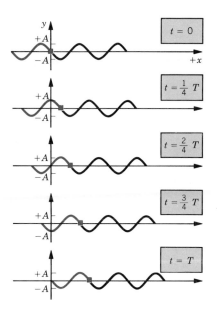

FIGURE 20.10 Equation 20.3 is plotted here at a series of times separated by one-fourth of the period T. The colored square in each of the graphs marks the place on the wave that is located at $x = 0$ when $t = 0$. As time passes, the colored square moves to the right, in the $+x$ direction.

SUMMARY

A **wave** is a traveling disturbance and carries energy from place to place. In a **transverse wave,** the disturbance occurs perpendicular to the direction of travel of the wave. In a **longitudinal wave,** the disturbance occurs parallel to the line along which the wave travels.

In a **periodic wave,** the pattern of the disturbance is produced over and over again by the source of the wave. The **amplitude** of the wave is the maximum excursion of a particle of the medium from the particle's undisturbed position. The **wavelength** λ is the distance along the length of the wave between two successive equivalent points, such as two crests or two troughs. The **period** T is the time required for the wave to travel a distance of one wavelength. The **frequency** f (in hertz) is the number of wave cycles per second that passes an observer and is the reciprocal of the period (in seconds): $f = 1/T$. The **speed** v of a wave is related to the wavelength and the frequency according to $v = \lambda f$.

The **speed of a wave** depends on the properties of the medium that the wave travels through. For a transverse wave on a string that has a tension F and a mass per unit length m/L, the wave speed is $v = \sqrt{F/(m/L)}$.

The **mathematical expression for a wave** traveling in the $+x$ direction is $y = A \sin(2\pi ft - 2\pi x/\lambda)$, where y denotes the displacement of a particle of the medium from the particle's undisturbed position and A is the amplitude of the wave. For wave motion in the $-x$ direction, the expression is $y = A \sin(2\pi ft + 2\pi x/\lambda)$.

QUESTIONS

1. Considering the nature of a water wave (see Figure 20.4), describe the motion of a fishing float on the surface of a lake when a wave passes beneath the float. Is it really correct to say that the float bobs "up and down"? Explain.

2. "Domino Toppling" is one entry in the *Guiness Book of World Records*. The event consists of lining up an incredible number of dominoes and then letting them topple, one after another. (See Photo 2.) Is the disturbance that propagates along the line of dominoes transverse, longitudinal, or partly both. Explain.

3. Suppose that a longitudinal wave moves along a Slinky at a speed of 5 m/s. Does one coil of the Slinky move through a distance of 5 m in one second? Justify your answer.

4. A transverse wave on a horizontal wire travels with a speed of 2000 mi/h. The particles of the wire move up and down in simple harmonic motion. (a) Explain whether the particles themselves are moving at a speed of 2000 mi/h, using Equation 11.11 to justify your answer. (b) If you say that the particles are *not* traveling at 2000 mi/h, then explain what factors determine the speed of the particles.

5. A wire is strung tightly between two immovable posts. Discuss how an increase and a decrease in temperature affect the speed of a transverse wave on this wire. Give your reasoning.

6. One end of each of two identical strings is attached to a wall. Each string is being pulled tight by someone at the other end. A transverse pulse is sent traveling along one of the strings. A bit later an identical pulse is sent traveling along the other string. What, if anything, can be done to make the second pulse catch up with and pass the first pulse? Account for your answer.

Photo 2

7. In Section 4.9 the hypothetical concept of a "massless" rope is discussed. How long would it take for a transverse wave to travel the length of a massless rope? Justify your answer.

PROBLEMS

Section 20.2 Periodic Waves

1. A person standing in the ocean notices that after a wave crest passes by, ten more crests pass in a time of 120 s. What is the frequency of the wave?

2. Sound travels at a speed of 343 m/s in air at 20 °C. The wavelength of a sound wave is 1.31 m. Find the period of the wave.

3. A longitudinal wave with a frequency of 3.0 Hz takes 1.7 s to travel the length of a 2.5-m Slinky (see Figure 20.3). Determine the wavelength of the wave.

4. The left-most and right-most keys on a piano produce sound waves with frequencies of 27.5 and 4185.6 Hz, respectively. Assuming the speed of sound in air is 343 m/s, find the corresponding wavelengths.

5. A wave has a frequency of 45 Hz and a speed of 22 m/s. Determine, if possible, (a) its period, (b) its wavelength, and (c) its amplitude. If it is not possible to determine any of these quantities from the data given, then so state.

6. A person fishing from a pier observes that four wave crests pass by in 7.0 s and estimates the distance between two successive crests as 4.0 m. The timing starts with the first crest and ends with the fourth. What is the speed of the wave?

7. Suppose the amplitude and frequency of the transverse wave in Figure 20.2 are, respectively, 1.3 cm and 5.0 Hz. Find the *total vertical distance* through which the colored dot moves in 3.0 s.

＊ 8. In Figure 20.3 the colored "dot" exhibits simple harmonic motion as the longitudinal wave passes. The wave has an amplitude of 5.4×10^{-3} m and a frequency of 4.0 Hz. Find the maximum acceleration of the dot.

＊＊9. A water-skier is moving at a speed of 12.0 m/s. When she skis in the same direction as a traveling wave, she springs upward every 0.500 s because of the wave crests. When she skis in the direction opposite to that in which the wave moves, she springs upward every 0.400 s in response to the crests. Determine (a) the speed and (b) the wavelength of the wave.

Section 20.3 The Speed of a Wave on a String

10. The linear density of the A string on a violin is 7.8×10^{-4} kg/m. A wave that travels on the string has a frequency of 440 Hz and a wavelength of 65 cm. What is the tension in the string?

11. A 0.200-kg wire is stretched between two posts 20.0 m apart and has a tension of 90.0 N. The wire is struck at one end, and a transverse pulse is sent traveling toward the other end. How long does the pulse take to travel the length of the wire?

12. The middle C string on a piano is under a tension of 944 N. The period and wavelength of a wave on this string are 3.82×10^{-3} s and 1.26 m, respectively. Find the linear density of the string.

＊ 13. To measure the acceleration due to gravity on a distant planet, an astronaut hangs a 0.085-kg lead ball from the end of a wire. The wire has a length of 1.5 m and a linear density of 3.1×10^{-4} kg/m. Using electronic equipment, the astronaut measures the time required for a transverse pulse to travel the length of the wire and obtains a value of 0.083 s. The mass of the wire is negligible compared to the mass of the ball. Determine the acceleration due to gravity.

＊ 14. A horizontal wire is under a tension of 315 N and has a mass per unit length of 6.50×10^{-3} kg/m. A transverse wave with an amplitude of 2.50 mm and a frequency of 585 Hz is traveling on this wire. As the wave passes, a particle of the wire moves up and down in simple harmonic motion. Obtain (a) the speed of the wave and (b) the maximum speed with which the particle moves up and down.

＊ 15. A wire has a diameter of 0.23 mm and is made from a material whose density is 7900 kg/m³. Determine the wire's *linear density* in kg/m.

＊＊16. A copper wire, 1.2 mm in diameter, has a linear density of 7.0×10^{-3} kg/m and is strung between two walls. At the ambient temperature, a transverse wave travels with a speed of 46 m/s on this wire. The coefficient of linear expansion for copper is 17×10^{-6} (C°)⁻¹, and Young's modulus for copper is 1.1×10^{11} Pa. What will be the speed of the wave when the temperature is lowered by 14 C°?

Section 20.4 The Mathematical Description of a Wave

17. The displacement (in meters) of a wave is $y = (0.26) \sin (\pi t - 3.7\pi x)$, where t is in seconds and x is in meters. (a) Is the wave traveling in the $+x$ or $-x$ direction? (b) What is the displacement y when $t = 38$ s and $x = 13$ m?

18. A wave has the following properties: amplitude $= 0.37$ m, period $= 0.80$ s, wave speed $= 12$ m/s. The wave is traveling in the $-x$ direction. What is the mathematical expression for the wave?

19. A wave has a displacement (in meters) of $y = (0.45) \sin (8.0\pi t + \pi x)$, where t and x are expressed in seconds and meters, respectively. (a) Find the amplitude, the frequency, the wavelength, and the speed of the wave. (b) Is this wave traveling in the $+x$ or $-x$ direction?

20. The drawing shows two graphs that represent a transverse wave on a string. The wave is moving in the $+x$ direction. Using the information contained in these graphs, write the mathematical expression for the wave.

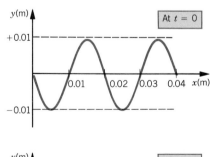

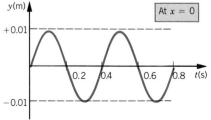

21. A transverse wave on a string is traveling in the $+x$ direction and has an amplitude of 2.00 mm, a wavelength of 0.300 m, and a speed of 15.0 m/s. Write the equation that gives the disturbance as a function of time for a point located on the string at $x = +0.400$ m.

ADDITIONAL PROBLEMS

22. A vibrator moves one end of a rope up and down to generate a wave. The tension in the rope is 58 N. The frequency is then doubled. To what value must the tension be adjusted, so the new wave has the same wavelength as the old one?

23. A person lying on an air mattress in the ocean is observed to rise and fall through one complete cycle once every five seconds. The crests of the wave causing the motion are 20.0 m apart. Determine (a) the period, (b) the frequency, and (c) the speed of the wave.

24. A wave is moving in the $+x$ direction. Assuming the wave has the following properties, write the equation of the wave:
(a) wavelength = 3.5 m,   frequency = 420 Hz,   amplitude = 0.71 m
(b) amplitude = 0.46 m, period = 0.010 s, speed = 18 m/s
(c) speed = 7.1 m/s,   amplitude = 0.15 m,   wavelength = 0.28 m.

25. In Figure 20.2 the hand moves the end of the Slinky up and down through two complete cycles in one second. The wave moves along the Slinky at a speed of 0.50 m/s. Find the distance between two adjacent crests on the wave.

* **26.** The drawing shows a frictionless incline and pulley. The two blocks are connected by a wire (mass per unit length = 0.0250 kg/m) and remain stationary. A transverse wave on the wire has a speed of 75.0 m/s. Neglecting the weight of the wire relative to the tension in the wire, find the masses m_1 and m_2 of the blocks.

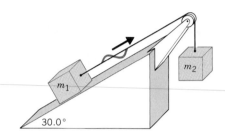

* **27.** The amplitude and frequency of a transverse wave on a string are, respectively, 7.8×10^{-3} m and 6.0 Hz. Remembering that a particle of the string exhibits simple harmonic motion as the wave passes, determine the maximum speed of a particle. *(Hint: This speed is not the speed of the wave.)*

CHAPTER 21

Sound

Dolphins are highly intelligent mammals that communicate with each other by transmitting audible "squeeks" and "clicks" through water. Sound can also be propagated through gases and solids, as is explored in this chapter.

21.1 THE NATURE OF SOUND

LONGITUDINAL SOUND WAVES

Sound is a longitudinal wave that is created by a vibrating object, such as a guitar string, the human vocal cords, or the diaphragm of a hi-fi speaker. Moreover, sound can be created only in a medium, which can be a gas, a liquid, or a solid. As we will see, the particles of the medium must be present for the disturbance of the wave to move from place to place. Consequently, sound cannot be created in a vacuum.

To see how sound waves are produced and why they are longitudinal, consider the vibrating diaphragm of a hi-fi speaker (sometimes called the speaker "cone"). When the diaphragm moves outward, it compresses the air directly in front of it, as in Figure 21.1*a*. This compression causes the air pressure to rise slightly. The region of increased pressure is called a **condensation,** and, once created, it travels away from the speaker at the speed of sound. After producing a condensation, the diaphragm reverses its motion and moves inward, as in part *b* of the drawing. The inward motion produces a region known as a **rarefaction,** where the air pressure is slightly less than normal. Following immediately behind the condensation, the rarefaction also travels away from the speaker at the speed of sound.

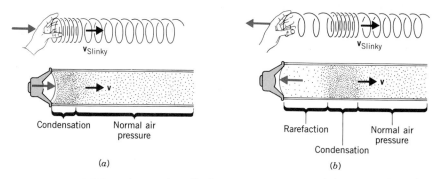

Condensation   Normal air pressure   Rarefaction   Normal air pressure

Condensation

(a)   (b)

FIGURE 21.1 (*a*) When the speaker diaphragm moves outward, it creates a condensation. (*b*) When the diaphragm moves inward, it creates a rarefaction. The condensation and rarefaction on the Slinky are included for comparison. In reality, the velocity of the wave on the Slinky v_{Slinky} is much smaller than the velocity of sound in air **v**. However, for the sake of comparison, the two waves are shown to have the same velocity.

For comparison, Figure 21.1 includes the condensation and rarefaction of a longitudinal wave on a Slinky. Figure 21.2 emphasizes that the sound wave, like the Slinky wave, is longitudinal. As the wave passes, the colored dots attached both to the Slinky and to an air molecule execute simple harmonic motion about their undisturbed positions. The colored arrows on either side of the dots indicate that the simple harmonic motion occurs parallel to the line of travel. The drawing also illustrates that the wavelength λ is the distance between the centers of two successive condensations; λ is also the distance between the centers of two successive rarefactions.

Usually, sound waves are not restricted to propagate inside a tube, as in the previous drawings. Figure 21.3 illustrates the more normal situation of a wave spreading out in space as it leaves a hi-fi speaker. When the condensations and rarefactions arrive at the ear, they force the eardrum to vibrate at the same frequency as that of the speaker diaphragm. The vibratory motion of the eardrum is interpreted by the brain as a tone of *constant loudness.*

It should be emphasized that sound is not a mass movement of air, such as occurs on a windy day. In Figure 21.3, for example, the air molecules on which the speaker diaphragm pushes are not the same ones that push on the eardrum. The condensations and rarefactions of the sound wave travel outward from the vibrating diaphragm. But the individual air molecules are not carried along with the wave. Rather, each molecule executes simple harmonic motion about a fixed location. In doing so, one molecule collides with a neighbor further down the line and passes the condensations and rarefactions on. The neighbor, in turn, repeats the process.

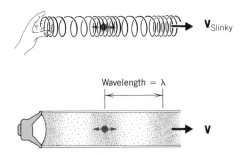

v_{Slinky}

Wavelength = λ

V

FIGURE 21.2 Both the sound wave and the wave on the Slinky are longitudinal waves. The colored dot attached to an air molecule, like the colored dot attached to the Slinky, vibrates back and forth parallel to the line of travel of the wave.

THE FREQUENCY OF A SOUND WAVE

Each complete cycle of a sound wave includes one condensation and one rarefaction, and the *frequency* is the number of cycles per second that passes by a given location. For example, if the diaphragm of a speaker vibrates back and forth at a frequency of 1000 Hz, then 1000 condensations, each followed by a rarefaction, are generated every second, thus forming a sound wave whose frequency is also 1000 Hz. A sound with a single frequency is called a *pure tone.* Experiments have shown that a healthy young person hears all sound frequencies in the range from approximately 20 Hz to approximately 20 000 Hz (20 kHz). The ability to hear the high frequencies decreases with age, however, and a normal middle-aged adult hears frequencies only up to 12–14 kHz.

Of course, sound can be generated whose frequency lies below 20 Hz and above 20 kHz, although humans normally do not hear it. In contrast, some dogs hear frequencies as large as 30 kHz, and, therefore, can respond to dog whistles that humans cannot hear. Bats can hear even higher frequencies and depend on high-frequency sound (up to a frequency of 100 kHz) for locating their prey and navigating. Sound waves whose frequencies lie above 20 kHz are called *ultrasonic waves,* while those that lie below 20 Hz are referred to as *infrasonic waves.*

Frequency is an objective property of a sound wave, because frequency can be measured with an electronic frequency counter. A listener's perception of frequency, however, is subjective. The brain interprets the frequency detected by the ear primarily in terms of the subjective quality called *pitch.* A pure tone with a large (high) frequency is interpreted as a high-pitched sound, while a pure tone with a small (low) frequency is interpreted as a low-pitched sound. A piccolo produces high-pitched sounds, and a tuba produces low-pitched sounds.

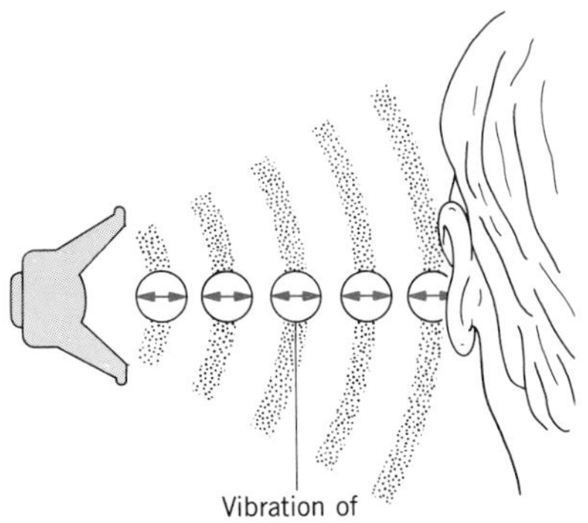

Vibration of of an individual air molecule

FIGURE 21.3 Although the condensations and rarefactions travel from the speaker to the listener, the individual air molecules do not move with the wave. A given molecule vibrates back and forth about a fixed location.

THE PRESSURE AMPLITUDE OF A SOUND WAVE

Figure 21.4 illustrates a pure-tone sound wave traveling in a tube. Attached to the tube is a series of pressure gauges that indicate the pressure variations along the wave. The graph shows that the air pressure varies sinusoidally along the length of the tube. Although this graph has the appearance of a transverse wave, remember that the sound wave itself is a longitudinal wave. The graph also shows the *pressure amplitude* of the wave, which is the maximum amount of change in pressure, measured relative to the undisturbed or atmospheric pressure. The pressure fluctuations in a sound wave are normally very small. For instance, in a conversation between two people the

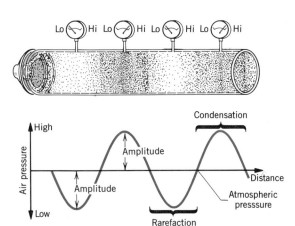

FIGURE 21.4 A sound wave is a series of alternating condensations and rarefactions. The graph shows that the condensations are regions of higher-than-normal air pressure, while the rarefactions are regions of lower-than-normal pressure.

pressure amplitude is about 3×10^{-2} Pa, certainly a small amount compared with the atmospheric pressure of $1.01 \times 10^{+5}$ Pa. The ear is remarkable in being able to detect such small changes.

Loudness is an important attribute of sound that depends primarily on the amplitude of the wave: the larger the amplitude the louder the sound. Since it can be measured with electronic equipment, the pressure amplitude is an objective property of a sound wave. Loudness, on the other hand, is subjective. Each individual determines what is loud, depending on the acuteness of his hearing.

21.2 THE SPEED OF SOUND

GASES

Sound travels through gases, liquids, and solids at considerably different speeds, as Table 21.1 reveals. Near room temperature, the speed of sound in air is 343 m/s (767 mi/h) and is markedly greater in liquids and solids. For example, sound travels more than four times faster in water and more than seventeen times faster in steel than it does in air. In general, sound travels slowest in gases, faster in liquids, and fastest in solids.

Like the speed of a wave on a string, the speed of sound depends on the properties of the medium. In a gas, it is only when molecules collide that the condensations and rarefactions of a sound wave can move from place to place. It is reasonable, then, to expect the speed of sound in a gas to have the same order of magnitude as the average molecular speed between collisions. For an ideal gas this average speed is the translational rms-speed given by Equation 17.6: $v_{rms} = \sqrt{3kT/m}$, where T is the Kelvin temperature, m is the mass of a molecule, and k is Boltzmann's constant. Although the expression for v_{rms} overestimates the speed of sound, it does give the correct dependence on Kelvin temperature and particle mass. Careful analysis shows that the speed of sound in an ideal gas is given by

TABLE 21.1 Speed of Sound in Gases, Liquids, and Solids

Substance	Temperature (°C)	Speed (m/s)
Gases		
Air	0	331
Air	20	343
Carbon dioxide	0	259
Oxygen	0	316
Helium	0	965
Liquids		
Chloroform	20	1004
Ethanol	20	1162
Mercury	20	1450
Fresh water	20	1482
Solids (bulk)		
Copper	—	5010
Glass (Pyrex)	—	5640
Lead	—	1960
Steel	—	5960

[Ideal gas]
$$v = \sqrt{\frac{\gamma k T}{m}}$$
(21.1)

where $\gamma = c_P/c_V$ is the ratio of the specific heat capacity at constant pressure c_P to the specific heat capacity at constant volume c_V.

The factor γ is introduced in Section 18.5, where the adiabatic compression and expansion of an ideal gas is discussed. The factor γ enters into Equation 21.1 because the condensations and rarefactions of a sound wave are formed by adiabatic compressions and expansions of the gas. When a sound wave travels through a gas, the regions that are compressed (the condensations) become slightly warmed, and the regions that are expanded (the rarefactions) become slightly cooled. However, no appreciable amount of heat energy flows from a condensation to an adjacent rarefaction, because the distance between the two (half a wavelength) is relatively large for most audible sound waves and a gas is a poor thermal conductor. Thus, the compression and expansion process is adiabatic. Example 1 uses Equation 21.1 to illustrate how camera technology takes advantage of the speed of sound.

EXAMPLE 1

Some cameras have a mechanism that automatically focuses the camera with the aid of sound waves. Figure 21.5 illustrates the central idea of this feature. To initiate the focusing process, the camera generates a pulse of ultrasonic sound that travels to the subject being photographed. Like an echo reflecting from a distant cliff, the pulse reflects off the subject and returns to the camera. By measuring the time it takes the sound to make the round-trip and using a preset value for the speed of sound, the camera can determine the distance to the subject and set the lens to its proper focus. Suppose the time required for the sound pulse to make the round-trip is 20.0×10^{-3} s on a day when the air temperature is 23 °C (296 K). Assuming air behaves as an ideal gas for which $\gamma = 1.40$ and assuming the average molecular mass of an air molecule is 28.9 u, find the distance s between the camera and the subject.

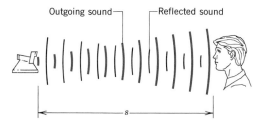

Outgoing sound   Reflected sound

FIGURE 21.5 To set its focus automatically, the camera determines the distance s with the aid of ultrasonic sound waves. The camera measures the time for the sound to travel to the subject and back and computes s from a knowledge of the speed of sound.

SOLUTION

The distance between the camera and the subject is $s = vt$, where v is the speed of sound and t is the time for the sound pulse to reach the subject. The time t is just one-half the round-trip time, so $t = 10.0 \times 10^{-3}$ s. The speed of sound in air can be obtained directly from Equation 21.1, once the mass of an air molecule is known. The mass of an air molecule is the average molecular mass of air (expressed in kilograms) divided by Avogadro's number N_A (see Section 17.2):

$$m = \frac{28.9 \times 10^{-3} \text{ kg/mol}}{N_A}$$

$$= \frac{28.9 \times 10^{-3} \text{ kg/mol}}{6.022 \times 10^{23} \text{ mol}^{-1}} = 4.80 \times 10^{-26} \text{ kg}$$

The speed of sound is

$$v = \sqrt{\frac{\gamma k T}{m}}$$
(21.1)

$$= \sqrt{\frac{(1.40)(1.38 \times 10^{-23} \text{ J/K})(296 \text{ K})}{4.80 \times 10^{-26} \text{ kg}}}$$

$$= 345 \text{ m/s}$$

The distance from the camera to the subject is

$$s = vt = (345 \text{ m/s})(10.0 \times 10^{-3} \text{ s}) = \boxed{3.45 \text{ m}}$$

LIQUIDS

In a liquid, the speed of sound is a function of the density ρ and the *adiabatic* bulk modulus B_{ad} of the liquid:

[Liquid]
$$v = \sqrt{\frac{B_{ad}}{\rho}} \qquad (21.2)$$

The concept of the bulk modulus is introduced in Section 11.2 in a discussion of the volume deformation of liquids and solids. In that discussion it is tacitly assumed that the temperature remains constant while the volume of the material changes; i.e., the compression or expansion is isothermal. The values for the bulk moduli in Table 11.3 reflect this experimental condition and, consequently, are known as isothermal bulk moduli. However, the condensations and rarefactions in a sound wave occur under *adiabatic* rather than isothermal conditions. Thus, the adiabatic bulk modulus must be used when calculating the speed of sound in liquids. Values of B_{ad} will be provided as needed in this text. The next example emphasizes that sound travels much faster in a liquid than in a gas.

EXAMPLE 2

For seawater the adiabatic bulk modulus is $B_{ad} = 2.31 \times 10^9$ Pa and the density is $\rho = 1025$ kg/m^3. Will the autofocusing mechanism of the camera in Example 1 function correctly when used under the sea to photograph marine life?

SOLUTION

The answer is no, for the ultrasonic sound pulse travels much faster in seawater than it does in air. Therefore, the pulse returns to the camera in a much shorter time. This quicker return "fools" the camera into believing the subject is much closer than it actually is. The speed of sound in seawater can be determined from Equation 21.2:

$$v = \sqrt{\frac{B_{ad}}{\rho}} = \sqrt{\frac{2.31 \times 10^9 \text{ Pa}}{1025 \text{ kg/m}^3}} = 1500 \text{ m/s}$$

In comparison, the speed of sound in air is 345 m/s in Example 1. Thus, the camera calculates, erroneously, that the subject is $1500/345 = 4.3$ times closer under water than it actually is.

SOLID BARS

When sound travels through a long slender solid bar, the speed of the sound depends on the properties of the medium according to

$$\begin{bmatrix} \text{Long slender} \\ \text{solid bar} \end{bmatrix} \qquad v = \sqrt{\frac{Y}{\rho}} \qquad (21.3)$$

where Y is Young's modulus (defined in Section 11.2) and ρ is the density. For example, Table 11.1 lists Young's modulus for steel as $Y = 2.0 \times 10^{11}$ Pa, and the density of steel is $\rho = 7860$ kg/m^3. Using these values in Equation 21.3 reveals that the speed of sound in a long, slender, solid bar of steel is $v = 5000$ m/s, considerably less than the value of 5960 m/s given for bulk steel in Table 21.1.

21.3 SOUND INTENSITY

Sound waves carry energy that can be used to do work, like forcing the eardrum into the oscillatory motion that initiates the hearing process. Or, in an extreme case such as a sonic boom, the energy can be sufficient to cause damage to windows and buildings. The amount of energy transported per second by a sound wave is called the *power* of the wave, and, like all power, is measured in SI units of joules per second (J/s) or watts (W).

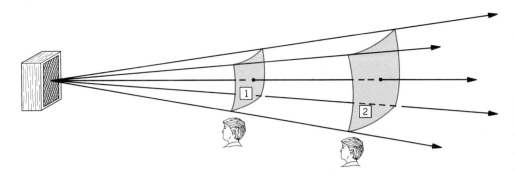

FIGURE 21.6 The intensity of a sound wave is the sound power that passes perpendicularly through a surface, such as surface 1 or surface 2, divided by the area of the surface.

When a sound wave leaves a source, such as the loudspeaker in Figure 21.6, the power spreads out and passes through increasingly larger areas. The *sound intensity I* is defined as the sound power *P* that passes perpendicularly through a surface divided by the area *A* of the surface:

$$I = \frac{P}{A} \qquad (21.4)$$

The unit of sound intensity is power per unit area, or W/m². For instance, suppose that 12×10^{-5} watts of power (a typical value under normal circumstances) pass perpendicularly through the two surfaces labeled 1 and 2 in Figure 21.6. These surfaces have areas of $A_1 = 4.0$ m² and $A_2 = 12$ m². The sound intensity is less at the more distant surface, where the same power passes through a threefold greater area:

[Surface 1] $I_1 = \dfrac{P}{A_1} = \dfrac{12 \times 10^{-5}\ \text{W}}{4.0\ \text{m}^2} = 3.0 \times 10^{-5}\ \text{W/m}^2$

[Surface 2] $I_2 = \dfrac{P}{A_2} = \dfrac{12 \times 10^{-5}\ \text{W}}{12\ \text{m}^2} = 1.0 \times 10^{-5}\ \text{W/m}^2$

The ear of a listener, with its fixed area, intercepts less power where the intensity, or power per unit area, is smaller. Thus, listener 2 intercepts less of the sound power than listener 1. With less power striking the ear, the sound is quieter.

For a 1000-Hz tone, the smallest sound intensity that the human ear can detect is about 1×10^{-12} W/m²; this intensity is often called the *threshold of hearing.* On the other extreme, continuous exposure to intensities greater than 1 W/m² can be painful and result in permanent hearing damage. The human ear is remarkable for the wide range of intensities to which it is sensitive.

If a source emits sound *uniformly in all directions,* the sound intensity depends on distance in a simple way. Figure 21.7 shows such a source at the center of an imaginary sphere of radius *r*. Since all the radiated sound power *P* passes through the spherical surface of area $A = 4\pi r^2$, the intensity at a distance *r* is

$$\begin{bmatrix} \textbf{Intensity of} \\ \textbf{spherically} \\ \textbf{uniform} \\ \textbf{radiation} \end{bmatrix} \qquad I = \frac{P}{4\pi r^2} \qquad (21.5)$$

The intensity of a source that radiates sound uniformly in all directions varies as $1/r^2$. For example, if the distance increases by a factor of two, the sound intensity decreases by a factor of $2^2 = 4$. Equation 21.5 is valid only when no walls, ceilings, floors, etc.,

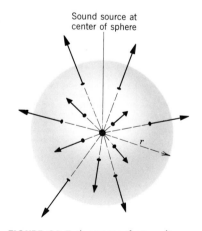

Sound source at center of sphere

FIGURE 21.7 A source that emits sound uniformly in all directions. The imaginary spherical surface drawn around the source has a radius *r* and an area $A = 4\pi r^2$.

are present to reflect the sound and cause it to pass through the same area more than once. Example 3 illustrates the effect of the $1/r^2$ dependence of intensity on distance.

EXAMPLE 3

During a fireworks display, a rocket explodes high in the air, as Figure 21.8 illustrates. Assume the sound spreads out uniformly in all directions and any reflections from the ground can be ignored. When the sound reaches listener 2, who is $r_2 = 640$ m away from the explosion, the sound has an intensity of $I_2 = 0.10$ W/m². What is the sound intensity detected by listener 1, who is only $r_1 = 160$ m away from the explosion?

SOLUTION

Listener 1 is four times closer to the explosion than listener 2. Therefore, the sound intensity detected by listener 1 is $4^2 = 16$ times greater than that detected by listener 2, as the calculation below indicates:

$$\frac{I_1}{I_2} = \frac{\dfrac{P}{4\pi r_1^2}}{\dfrac{P}{4\pi r_2^2}} = \frac{r_2^2}{r_1^2} = \frac{(640 \text{ m})^2}{(160 \text{ m})^2} = 16$$

As a result, $I_1 = (16)I_2 = (16)(0.10 \text{ W/m}^2) = \boxed{1.6 \text{ W/m}^2}$.

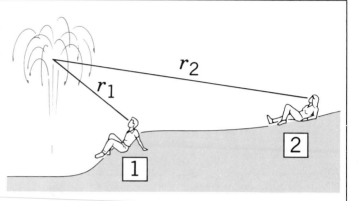

FIGURE 21.8 If an explosion in a fireworks display radiates sound uniformly in all directions, the intensity at any distance r is $I = P/(4\pi r^2)$, where P is the sound power of the explosion.

21.4 DECIBELS

COMPARING SOUND INTENSITIES USING DECIBELS

The **decibel** (*dB*) is a measurement unit encountered frequently in connection with audio equipment. For example, the decibel is used extensively on specification sheets to describe the performance characteristics of audio amplifiers, turntables, speakers, cassette decks, and FM tuners. The main application of the decibel concept is for comparing two sound intensities. Of course, the simplest method of comparison would be to compute the ratio of the intensities. For instance, we could compare $I = 8 \times 10^{-12}$ W/m² to $I_0 = 1 \times 10^{-12}$ W/m² by computing $I/I_0 = 8$ and stating that I is eight times greater than I_0. However, because of the way in which the human hearing mechanism responds to intensity, it is more convenient to use a logarithmic scale for the comparison. For this purpose, the **intensity level** β (expressed in decibels) is defined as follows:

$$\beta \text{ (in decibels)} = 10 \log\left(\frac{I}{I_0}\right) \tag{21.6}$$

where "log" denotes the logarithm to the base ten. I_0 is the intensity of the reference level to which I is being compared and is often the threshold of hearing, $I_0 = 1.00 \times 10^{-12}$ W/m². With the aid of a calculator, the intensity level can be evaluated for the values of I and I_0 given above:

$$\beta = 10 \log \left(\frac{8 \times 10^{-12} \text{ W/m}^2}{1 \times 10^{-12} \text{ W/m}^2} \right) = 10 \log 8 = 10(0.9) = 9 \text{ dB}$$

This result indicates that I is 9 decibels greater than I_0.

Although β is called the "intensity level," it is *not* an intensity and does *not* have intensity units of W/m^2. In fact, the decibel unit, like the radian, is dimensionless, since it is the product of the number 10 and a logarithm, both of which are pure numbers without any units.

Notice that when I is at the threshold of hearing, i.e., when $I = I_0$, the intensity level is 0 dB according to Equation 21.6:

$$\beta = 10 \log \left(\frac{I_0}{I_0} \right) = 10 \log 1 = 0$$

since $\log 1 = 0$. Thus, ***an intensity level of zero decibels does not mean that the sound intensity is zero; it means that $I = I_0$.***

Table 21.2 lists the intensities I and the associated intensity levels β for some common sounds, using the threshold of hearing as the reference level. The intensities can be measured with a sound level meter, such as the one in Figure 21.9a. While the sound level meter actually responds to the intensity I, it is the intensity level β that is displayed on its scale, as part b of the drawing indicates.

INTENSITY LEVEL CHANGES AND LOUDNESS CHANGES

When a sound wave reaches a listener's ear, the sound is interpreted by the brain as loud or soft, depending on the intensity of the wave. Greater intensities give rise to louder sounds. However, the relation between intensity and loudness is not a simple proportionality; doubling the intensity does *not* double the loudness. The correlation between intensity and loudness is a fascinating one, as we will now see.

Suppose you are sitting in front of a stereo system that is producing an intensity level of 90 dB. If the volume control on the amplifier is turned up slightly to produce a 91 dB level, you would just barely notice the accompanying change in loudness. ***Hearing tests have revealed that a one decibel (1 dB) change in the intensity level produces, approximately, the smallest change in loudness that an average listener can detect.*** Since 1 dB gives rise to the smallest perceivable increment in loudness, a change in the intensity level of only 3 dB, say, from 90 dB to 93 dB, still produces a rather small change in loudness. However, a 3-dB increase in loudness, while small, corresponds to a doubling of the sound intensity, as Example 4 illustrates.

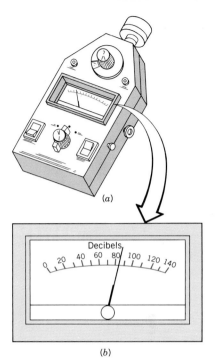

FIGURE 21.9 (*a*) A sound level meter and (*b*) a close-up view of its measurement scale. The scale is calibrated in terms of the sound intensity level, expressed in decibels.

TABLE 21.2 Typical Sound Intensities and Intensity Levels Relative to the Threshold of Hearing

	Intensity I (W/m^2)	Intensity level β (dB)
Threshold of hearing	1.0×10^{-12}	0
Rustling leaves	1.0×10^{-11}	10
Whisper	1.0×10^{-10}	20
Normal conversation (1 meter)	3.2×10^{-6}	65
Inside car in city traffic	1.0×10^{-4}	80
Car without muffler	1.0×10^{-2}	100
Live rock concert	1.0	120
Threshold of pain	10	130

EXAMPLE 4

Figure 21.10 shows two loudspeaker systems. System 1 produces an intensity level of $\beta_1 = 90.0$ dB, while system 2 produces an intensity level of $\beta_2 = 93.0$ dB. The corresponding intensities (in W/m²) are I_1 and I_2, respectively. Determine the ratio I_2/I_1.

SOLUTION

Subtracting the two intensity levels reveals that

$$\beta_2 - \beta_1 = 93.0 \text{ dB} - 90.0 \text{ dB}$$

$$= 10 \log\left(\frac{I_2}{I_0}\right) - 10 \log\left(\frac{I_1}{I_0}\right)$$

$$3.0 \text{ dB} = 10 \left[\log\left(\frac{I_2}{I_0}\right) - \log\left(\frac{I_1}{I_0}\right)\right]$$

$$= 10 \log\left(\frac{I_2}{I_1}\right)$$

The last step uses the fact that $(\log A - \log C) = \log (A/C)$. Thus, $0.30 = \log (I_2/I_1)$ and

$$\frac{I_2}{I_1} = 10^{0.30} = \boxed{2.0}$$

We see, then, that increasing the loudness by only a small amount (3 dB) corresponds to doubling the intensity. In other words, doubling the intensity does *not* double the loudness.

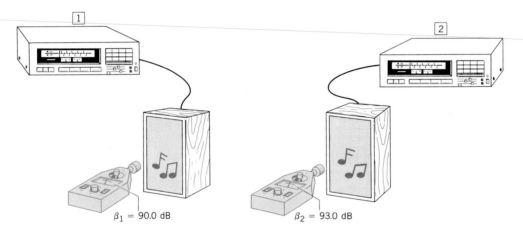

$\beta_1 = 90.0$ dB $\beta_2 = 93.0$ dB

FIGURE 21.10 System 2 sounds slightly louder than system 1, since β_2 is only 3.0 dB greater than β_1.

To double the loudness of a sound, the intensity must be increased by more than a factor of two. *Experiment shows that if the intensity level increases by 10 dB, the new sound appears approximately twice as loud as the original sound.* For instance, a 70-dB intensity level sounds about twice as loud as a 60-dB level, and an 80-dB intensity level sounds about twice as loud as a 70-dB level. The factor by which the sound intensity must be increased to double the loudness can be calculated by the method used in Example 4:

$$\beta_2 - \beta_1 = 10.0 \text{ dB} = 10 \left[\log\left(\frac{I_2}{I_0}\right) - 10 \log\left(\frac{I_1}{I_0}\right)\right]$$

Solving this equation reveals that $I_2/I_1 = 10.0$. Thus, increasing the sound intensity (in W/m²) by a factor of ten will only double the perceived loudness. Consequently, the 200-watt loudspeaker system in Figure 21.11 will sound only about twice as loud as the much cheaper 20-watt system.

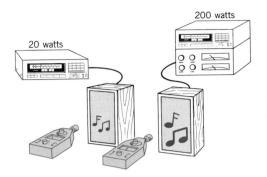

FIGURE 21.11 In spite of its tenfold greater power, the 200-watt loudspeaker system has only about double the loudness of the 20-watt system.

20 watts

200 watts

21.5 APPLICATIONS OF SOUND

SONAR

Sonar (**so**und **na**vigation **r**anging) is a technique for determining water depth and locating underwater objects, such as reefs, submarines, and schools of fish. The core of a sonar unit consists of an ultrasonic transmitter and receiver mounted on the bottom of a ship, as Figure 21.12 illustrates. The transmitter emits a pulse of ultrasonic sound, and a short time later the reflected pulse, or echo, returns and is detected by the receiver. The water depth can be determined from the electronically measured round-trip time of the pulse and a knowledge of the speed of sound in water; the depth registers automatically on an appropriate meter. Such a depth measurement is similar to the distance measurement discussed for the autofocusing camera in Examples 1 and 2.

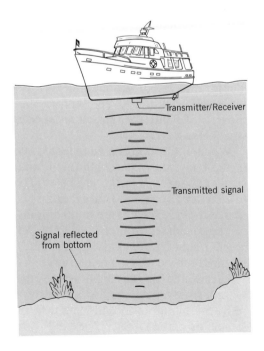

Transmitter/Receiver

Transmitted signal

Signal reflected from bottom

FIGURE 21.12 Ultrasonic sound can be used to measure water depth, as well as to detect the presence of underwater objects such as a school of fish.

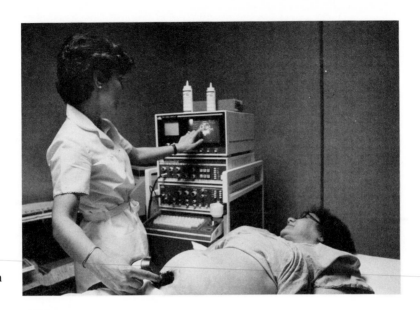

An ultrasound scanner produces an image of a 17-week-old fetus.

ULTRASOUND IN MEDICINE

Ultrasonic waves are used in medicine for diagnostic purposes. As in sonar, high-frequency sound pulses are produced by a transmitter and directed into the body. Reflections occur each time a pulse encounters a boundary between two tissues that have different densities or a boundary between a tissue and the adjacent fluid. By scanning ultrasonic waves across the body and detecting the echoes generated from various locations within the body, it is possible to obtain a "picture" of the inner anatomy. Ultrasonic waves have been employed extensively in the field of obstetrics to examine the developing fetus. The fetus, surrounded by the amniotic sac, can be distinguished from other anatomical features so that fetal structural abnormalities can be detected.

Ultrasound has also been applied in medically related areas other than obstetrics. For instance, malignancies in the liver, kidney, brain, and pancreas have been detected with ultrasound. In cases where internal hemorrhaging occurs, it is possible to identify the bleeding area and even obtain a gross estimate of blood loss using ultrasonic techniques. Yet another application involves monitoring the real-time movement of pulsating structures, such as the heart valves ("echocardiography") and large blood vessels.

When ultrasound is used to locate internal anatomical features or foreign objects in the body, the wavelength of the sound wave must be about the same size, or smaller, than the object to be located. Therefore, high frequencies in the range from 1 to 15 MHz (1 MHz = 1 megahertz = 1×10^6 Hz) are used in ultrasonic diagnosis, so that the wavelengths are small. For instance, the wavelength of 5 MHz ultrasound can be calculated from $\lambda = v/f$ to be 0.3 mm if a value of 1540 m/s is used for the speed of sound through tissue. A sound wave with a frequency higher than 5 MHz is required for locating objects smaller than 0.3 mm.

Ultrasound also has applications other than locating objects in the body. Neurosurgeons use a device called a **cavitron ultrasonic surgical aspirator** (CUSA) to remove brain tumors once thought to be inoperable. Ultrasonic sound waves cause the slender tip of the CUSA probe (see Figure 21.13) to vibrate at approximately 23 kHz. The probe shatters any section of the tumor that it touches, and the fragments are then flushed out of the brain with a saline solution. Because the tip of the probe is small, the surgeon can selectively remove small bits of malignant tissue without damaging the surrounding healthy tissue.

ULTRASONIC CLEANERS

Ultrasonic cleaners are especially popular with jewelers and scientists for cleaning delicate instruments and objects with hard-to-get-at places. Figure 21.14 illustrates the main features of an ultrasonic cleaner, which includes a metal tank filled with a cleaning fluid. Attached to the bottom of the tank is an ultrasonic transmitter that typically produces 40-kHz sound waves. The high-frequency sound causes a phenomenon called cavitation, in which small, partially evacuated spaces form in the cleaning fluid. The partial vacuum in these spaces causes dirt and grease particles to be pulled from even the tiniest cracks and crevices.

21.6 THE DOPPLER EFFECT

INTRODUCTION

Have you ever heard the siren of an approaching fire truck and noticed the distinct change in pitch of the siren as the truck passes? While the truck approaches, the pitch of the siren is relatively high, but as the truck passes and moves away, the pitch drops suddenly. A less familiar, but similar, phenomenon occurs when an observer moves toward or away from a stationary source of sound. Such phenomena were first identified in 1842 by the Austrian physicist Christian Doppler (1803–1853). The *Doppler effect,* then, is the change in pitch or frequency of the sound detected by an observer because the sound source and the observer have different velocities with respect to the medium of sound propagation.

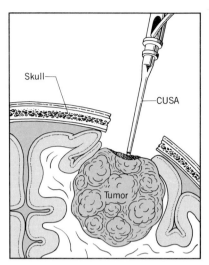

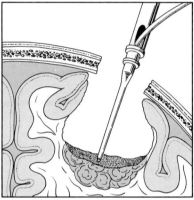

FIGURE 21.13 Neurosurgeons are now using an ultrasonic tool to "cut out" brain tumors without adversely affecting the surrounding healthy tissue. The instrument is called a cavitron ultrasonic surgical aspirator (CUSA).

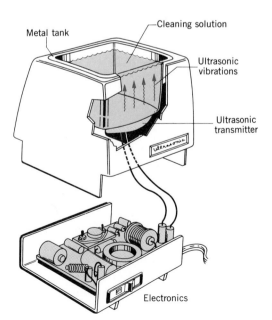

FIGURE 21.14 An ultrasonic cleaner.

MOVING SOURCE

In Figure 21.15a, sound is assumed to spread out in a spherical pattern after leaving the siren of a stationary fire truck. Like the truck, the air is also assumed to be stationary with respect to the earth. Each circular line in the drawing represents a condensation of the sound wave. Since the sound pattern is symmetrical, listeners standing in front of or behind the truck detect the same number of condensations per second and, consequently, hear the same frequency. Once the truck begins to move, however, the situation changes, as part b of the picture illustrates. Ahead of the truck the condensations are closer together, resulting in a decrease in the effective wavelength of the sound. This "bunching-up" effect occurs because the truck is moving with respect to the air. Thus, the truck "gains ground" on a previously emitted condensation before emitting the next one. Since the condensations are closer together, the observer standing in front of the truck senses more of them arriving per second than he does when the truck is stationary. The increased rate of arrival corresponds to a greater sound frequency, which the listener hears as a higher pitch.

Behind the moving truck, the condensations are farther apart than they are when the truck is stationary. This increase in the effective wavelength occurs because the truck pulls away from the condensations emitted toward the rear. Consequently, fewer condensations per second arrive at the ear of an observer behind the truck, corresponding to a smaller sound frequency or lower pitch.

Let us now determine how much the frequency changes because of the Doppler effect. If the stationary siren emits one condensation at the time $t = 0$, it will emit the next one at time T, where T is the period of the wave. The distance between these two condensations is the wavelength λ of the sound produced by the stationary source, as

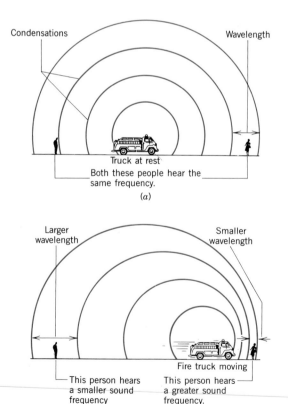

FIGURE 21.15 (a) The person in front of and the person behind a stationary fire truck hear the same frequency of 1000 Hz when the siren emits a 1000-Hz sound wave. (b) When the truck is moving forward, the observer in front hears a frequency greater than 1000 Hz, while the observer behind hears a frequency less than 1000 Hz.

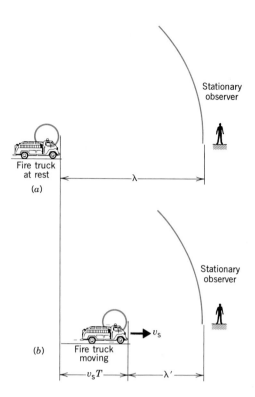

FIGURE 21.16 (*a*) When the fire truck is stationary, the distance between successive condensations is the wavelength λ. (*b*) When the truck moves forward with a speed v_s, the wavelength of the sound in front of the truck is shortened to λ'.

Figure 21.16*a* indicates. When the truck is moving with a speed v_s (the subscript s stands for the "source" of sound) toward a stationary observer, the siren also emits two successive condensations, one at $t = 0$ and one at time T. However, prior to emitting the second condensation, the truck moves closer to the observer by a distance v_sT, as Figure 21.16*b* shows. As a result, the distance between successive condensations is no longer the wavelength λ created by the stationary siren, but, rather, a shortened wavelength λ' given by

$$\lambda' = \lambda - v_sT$$

The frequency f' of the sound wave, as perceived by the stationary observer, is just the speed of sound v divided by the shortened wavelength λ', according to Equation 20.1:

$$f' = \frac{v}{\lambda'} = \frac{v}{\lambda - v_sT}$$

But it is also true for the stationary siren that $v = f\lambda$ and $T = 1/f$, where f is the frequency at which the source emits the sound (not the frequency f' perceived by the listener). With the aid of these two equations, the expression for f' can be arranged to give the following result:

$$\begin{bmatrix} \textbf{Sound source} \\ \textbf{moving toward} \\ \textbf{stationary} \\ \textbf{listener} \end{bmatrix} \qquad f' = f\left(\frac{1}{1 - \dfrac{v_s}{v}} \right) \qquad\qquad (21.7)$$

The denominator in Equation 21.7 is less than one, so the listener hears a frequency f' that is *greater* than the frequency f emitted by the source. The difference between these two frequencies, $f' - f$, is called the *Doppler shift,* and its magnitude depends on the speed of the source v_s relative to the speed of sound v.

When the siren moves away from the listener, rather than toward the listener, the wavelength λ' becomes *greater* than λ according to

$$\lambda' = \lambda + v_s T$$

Notice the presence of the "+" sign in this equation, in contrast to the "−" sign that appeared earlier. The same reasoning that led to Equation 21.7 can be used again to obtain an expression for the observed frequency f':

$$\begin{bmatrix} \textbf{Sound source} \\ \textbf{moving away from} \\ \textbf{stationary} \\ \textbf{listener} \end{bmatrix} \qquad f' = f\left(\dfrac{1}{1 + \dfrac{v_s}{v}}\right) \qquad (21.8)$$

The denominator in Equation 21.8 is greater than one, so the listener hears a frequency f' that is *less* than the frequency f emitted by the source. The next example illustrates the Doppler effect.

EXAMPLE 5

A high-speed train is traveling at a speed of 44.7 m/s (100 mi/h) when the engineer sounds the 415-Hz warning horn. The speed of sound is 343 m/s. What are the frequency and the wavelength of the sound, as perceived by a person standing at a crossing, when the train is (a) approaching and (b) leaving the crossing?

SOLUTION

(a) When the train approaches, the observed frequency is

$$f' = f\left(\dfrac{1}{1 - \dfrac{v_s}{v}}\right) = (415 \text{ Hz})\left(\dfrac{1}{1 - \dfrac{44.7 \text{ m/s}}{343 \text{ m/s}}}\right) \qquad (21.7)$$

$$= \boxed{477 \text{ Hz}}$$

According to Equation 20.1, the wavelength is

$$\lambda' = \dfrac{v}{f'} = \dfrac{343 \text{ m/s}}{477 \text{ Hz}} = \boxed{0.719 \text{ m}}$$

(b) When the train leaves the crossing, the observed frequency is

$$f' = f\left(\dfrac{1}{1 + \dfrac{v_s}{v}}\right) = (415 \text{ Hz})\left(\dfrac{1}{1 + \dfrac{44.7 \text{ m/s}}{343 \text{ m/s}}}\right) \qquad (21.8)$$

$$= \boxed{367 \text{ Hz}}$$

In this case, the wavelength is

$$\lambda' = \dfrac{v}{f'} = \dfrac{343 \text{ m/s}}{367 \text{ Hz}} = \boxed{0.935 \text{ m}}$$

MOVING OBSERVER

Figure 21.17 shows how the Doppler effect arises when the sound source is stationary and the observer moves, again assuming the air is stationary. In part *a* of the drawing, both the source of sound and the observer are stationary, and the observer hears the frequency f emitted by the source. Moving with a speed v_o ("o" stands for observer) toward the stationary source, the observer covers a distance $v_o t$ in a time t, as part *b* of the picture indicates. During this time, the moving observer encounters all the condensations detected by the stationary observer, *plus an additional number.* The additional number of condensations encountered is the distance $v_o t$ divided by the distance λ between successive condensations, or $v_o t/\lambda$. Thus, the additional number

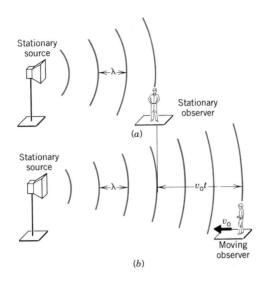

Stationary
source

Stationary
observer

(a)

Stationary
source

Moving
observer

(b)

FIGURE 21.17 An observer moving with a speed v_o toward a stationary source of sound, as in part b, intercepts more wave condensations per unit of time than does the stationary observer in part a. Consequently, the moving observer hears a higher frequency.

of condensations encountered per second is v_o/λ, and the moving observer hears a higher frequency f' given by

$$f' = f + \frac{v_o}{\lambda} = f\left(1 + \frac{v_o}{f\lambda}\right)$$

Using the fact that $v = f\lambda$, we find that

$$\begin{bmatrix} \text{Observer moving} \\ \text{toward stationary} \\ \text{sound source} \end{bmatrix} \qquad f' = f\left(1 + \frac{v_o}{v}\right) \qquad (21.9)$$

An observer moving *away from* a stationary source moves in the same direction as the sound wave, and, as a result, intercepts *fewer* condensations per second than a stationary observer does. In this case, the moving observer hears a smaller frequency f' that is given by

$$\begin{bmatrix} \text{Observer moving} \\ \text{away from stationary} \\ \text{sound source} \end{bmatrix} \qquad f' = f\left(1 - \frac{v_o}{v}\right) \qquad (21.10)$$

The physical mechanism producing the Doppler effect in the case of the moving observer is different from that in the case of the moving source. When the source moves and the observer is stationary, the wavelength of the sound changes, giving rise to the frequency f' heard by the observer. On the other hand, when the observer moves and the source is stationary, *the wavelength of the sound does not change.* Instead, a moving observer intercepts a different number of wave condensations per second than does a stationary observer and, therefore, detects a different frequency f'.

GENERAL CASE

It is possible for both the sound source and the observer to move with respect to the medium of sound propagation. If the medium is stationary, the expression for the detected frequency f' in such a situation is

$$\begin{bmatrix} \text{Sound source moving} \\ \text{and} \\ \text{observer moving} \end{bmatrix} \qquad f' = f\left(\dfrac{1 \pm \dfrac{v_o}{v}}{1 \mp \dfrac{v_s}{v}} \right) \qquad (21.11)$$

In the numerator, the plus sign applies when the observer moves toward the source, and the minus sign applies when the observer moves away from the source. In the denominator, the minus sign is used when the source moves toward the observer, and the plus sign is used when the source moves away from the observer. The symbols v_o, v_s, and v denote numbers without an algebraic sign, because the direction of travel has been taken into account by the plus and minus signs that appear directly in Equation 21.11 (or, alternatively, in Equations 21.7–21.10).

DOPPLER FLOW METER

The Doppler flow meter measures the speed of blood flow and is an interesting medical application of the Doppler effect. The device consists of transmitting and receiving elements that are placed directly on the skin, as in Figure 21.18. The transmitter emits a continuous sound wave whose frequency is typically about 5 MHz. When the sound is reflected from the red blood cells, its frequency is changed in a kind of Doppler effect, because the cells are moving. The receiving element detects the reflected sound, and an electronic counter measures its frequency, which is Doppler-shifted relative to the transmitter frequency. From the change in frequency the speed of the blood flow can be determined. Typically, the change in frequency is around 6000 Hz for flow speeds of about 0.1 m/s. The Doppler flow meter can be used to locate regions where blood vessels have narrowed, since greater flow speeds occur in the narrowed regions, according to the equation of continuity (see Section 13.2). In addition, the Doppler flow meter can be used to detect the motion of a fetal heart as early as 8–10 weeks after conception.

The Doppler effect is also employed in radar devices to measure the speed of moving vehicles. However, electromagnetic waves, rather than sound waves, are used for such purposes.

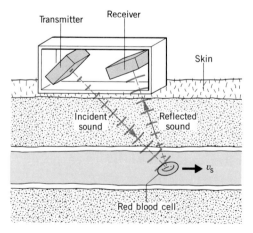

FIGURE 21.18 A Doppler flow meter measures the speed of red blood cells by utilizing the Doppler effect.

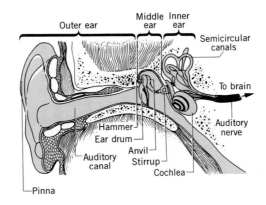

FIGURE 21.19 The human ear has three principal sections: the outer ear, the middle ear, and the inner ear.

*21.7 THE HUMAN EAR

PHYSIOLOGY OF THE EAR

Figure 21.19 shows the three principal sections of the ear: the outer ear, the middle ear, and the inner ear. The **outer ear** consists of three parts: (1) the pinna, which is the external, visible part of the ear; (2) the auditory canal, which is about 2.5 – 3 cm long; and (3) the eardrum, which stretches across the inner end of the auditory canal. The pinna funnels the sound waves into the auditory canal. When the waves reach the eardrum, the condensations and rarefactions cause the eardrum to vibrate.

The **middle ear** begins just inside the eardrum. It is a small, air-filled chamber that contains three tiny bones linked together: the hammer, the anvil, and the stirrup. The hammer, being firmly attached to the eardrum, vibrates along with it. The vibration of the hammer is passed to the anvil, which, in turn, transmits it to the stirrup. The inner end of the stirrup fits against the oval window (not shown in the figure), a membrane that stretches across an opening in the next chamber, the inner ear. As the stirrup vibrates, it pushes the oval window in and out. Thus, this delicate system of three bones transmits the vibrations from the outer ear to the inner ear.

The **inner ear** contains two general structures, the semicircular canals and the cochlea. The semicircular canals are responsible for our sense of balance and orientation and are not involved in the hearing process. It is within the cochlea that the hearing receptors are located. The cochlea is a liquid-filled cavity, resembling the coiled shell of a snail, and is partitioned into three canals by membranes that run along its length. One of these membranes, the basilar membrane, supports a cluster of sensitive cells that is known as the organ of Corti. The vibrations transmitted to the cochlear fluid through the oval window cause the basilar membrane, and hence the organ of Corti, to oscillate. In the organ of Corti about 30 000 sensory cells, called the hair cells, are connected to nerve fibers. When the basilar membrane vibrates, the hair cells stimulate the nerve fibers, producing nerve impulses that the auditory nerve transmits to the brain.

SENSITIVITY OF THE EAR

Although the ear is capable of detecting sound intensities as small as 1×10^{-12} W/m^2, it is *not* equally sensitive to all frequencies, as Figure 21.20 shows. This figure displays a series of graphs that are known as the **Fletcher-Munson curves,** after H.

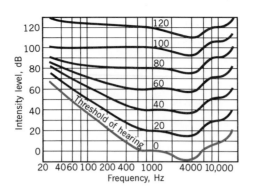

FIGURE 21.20 Each curve represents the intensity levels at which sounds of various frequencies have the same loudness. The curves are labeled by their intensity levels at 1000 Hz. These curves are known as the Fletcher-Munson curves.

Fletcher and M. Munson, who first determined them in 1933. In these graphs the audible sound frequencies are plotted on the horizontal axis, while the sound intensity levels (in decibels) are plotted on the vertical axis. Each curve is a *constant loudness* curve, in the sense that it shows the sound intensity level needed at each frequency to make the sound appear to have the same loudness. For example, the lowest curve represents the threshold of hearing. It shows the intensity levels at which sounds of different frequencies just become audible. The graph indicates that the intensity level of a 100-Hz sound must be about 37 dB greater than the intensity level of a 1000-Hz sound to be at the threshold of hearing. Therefore, the ear is *less sensitive* to a 100-Hz sound than it is to a 1000-Hz sound. In general, Figure 21.20 reveals that the ear is most sensitive in the range from about 1 to 5 kHz, and becomes progressively less sensitive at higher and lower frequencies.

Each curve in Figure 21.20 represents a different loudness, and each is labeled according to its intensity level at 1000 Hz. For instance, the curve labeled "60" represents all sounds that have the same loudness as that of a 1000-Hz sound whose intensity level is 60 dB. These constant-loudness curves become "flatter" as the loudness increases, the relative flatness indicating that the ear is nearly equally sensitive to all frequencies when the sound is loud. Thus, when you play your hi-fi at very loud levels, you hear the low frequencies, the middle frequencies, and the high frequencies equally well. However, when you turn down the volume control on the amplifier so the sound becomes quiet, the high and low frequencies seem to "disappear," for the ear is relatively insensitive to these frequencies under such conditions.

Many manufacturers of audio equipment include a *loudness switch* on their amplifiers (see Figure 21.21) to compensate for the loss of hearing sensitivity at quiet listening levels. When this switch is turned on, the intensities of quiet high- and low-frequency sounds are increased automatically. For loud sounds, the loudness function is designed to have little effect, consistent with the fact that our ears hear all frequencies more or less uniformly under such conditions.

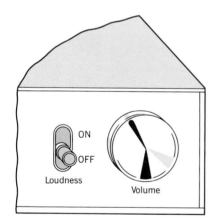

FIGURE 21.21 The loudness switch on a hi-fi amplifier.

SUMMARY

A **sound wave** is a longitudinal wave that consists of alternating regions of greater-than-normal pressure (condensations) and less-than-normal pressure (rarefactions). A sound wave can travel through a solid, a liquid, or a gas. Each cycle of a sound wave includes one condensation and one rarefaction. The **frequency** is the number of cycles per second that passes by a given location. A sound wave with a large frequency is interpreted by the brain as a high-pitched sound, while one with a small frequency is interpreted as a low-pitched sound. The **wavelength** is the length of one cycle, measured along the line of travel of the wave. The **pressure amplitude** of a sound wave is the maximum amount of change in pressure, measured relative to the undisturbed pressure. The larger the pressure amplitude, the louder the sound.

The **speed of sound** v depends on the properties of the medium in which the sound travels. In an ideal gas, the speed of sound is given by $v = \sqrt{\gamma k T/m}$, where $\gamma = c_P/c_V$ is the ratio of the specific heat capacities at constant pressure and constant volume, k is Boltzmann's constant, T is the Kelvin temperature, and m is the mass of a molecule of the gas. In a liquid, the speed of sound is given by $v = \sqrt{B_{ad}/\rho}$, where B_{ad} is the adiabatic bulk modulus and ρ is the density. For a solid in the shape of a long slender bar, the expression for the speed of sound is $v = \sqrt{Y/\rho}$, where Y is Young's modulus and ρ is the density.

The **intensity** I of a sound wave is the power P that passes perpendicularly through a surface divided by the area A of the surface: $I = P/A$. The SI unit of intensity is watts per square meter (W/m²). The smallest sound intensity that humans can detect is known as the **threshold of hearing** and is about 1×10^{-12} W/m² for a 1-kHz sound. When a source emits sound uniformly in all directions, the intensity of the sound is inversely proportional to the square of the distance from the source.

The **intensity level** β is used to compare a sound intensity I to the sound intensity I_0 of a reference level: $\beta = 10 \log (I/I_0)$, where β is expressed in **decibels** (dB). I_0 is often taken to be the threshold of hearing.

The **Doppler effect** is the change in frequency detected by an observer because the sound source and the observer have different velocities with respect to the medium of sound propagation. If the observer and source move with speeds v_o and v_s, respectively, and if the medium is stationary, the frequency f' detected by the observer is

$$f' = f\left(\frac{1 \pm \dfrac{v_o}{v}}{1 \mp \dfrac{v_s}{v}}\right)$$

where f is the frequency of the sound emitted by the source, and v is the speed of sound. In the numerator, the plus sign applies when the observer moves toward the source, and the minus sign applies when the observer moves away from the source. In the denominator, the minus sign is used when the source moves toward the observer, and the plus sign is used when the source moves away from the observer.

QUESTIONS

1. There is a widely known rule of thumb for estimating how far away a storm is. Following a lightning flash, count off the seconds until the thunder is heard. Divide the number of seconds by five. The result gives the approximate distance *in miles*. Explain why this rule works, noting that light waves from the lightning flash travel at a speed that is much greater than the speed of sound.

2. Do you expect an echo to return to you more quickly or less quickly on a hot day as compared to a cold day, other things being equal? Account for your answer.

3. A vibrating tuning fork produces a sound wave. Does the wavelength of the sound increase, decrease, or remain the same, when the wave travels from air into water? Justify your answer.

4. Some animals rely on an acute sense of hearing for survival, and the visible part of the ear on such animals is often relatively large. Explain how this anatomical feature helps to increase the sensitivity of the animal's hearing for low-intensity sounds.

5. The sound intensity I produced by a hi-fi speaker in your living room is *not* given by $I = P/4\pi r^2$, where P is the sound power emitted by the speaker and r is the distance from the speaker. Give reasons why this relation does not apply.

6. If the sound intensity level is measured relative to the threshold of hearing, what does it mean for an intensity level to be a negative number of decibels, e.g., -20 dB?

7. If two people talk simultaneously and each creates an intensity level of 65 dB at a certain point, does the total intensity level at this point equal 130 dB? Account for your answer.

8. Suppose you are swinging on a swing. Somewhere in the distance in front of you a stationary whistle is blowing. Specify where in the motional cycle of the swing you would hear the highest pitch, where you would hear the lowest pitch, and where you would hear the same pitch as that heard by a stationary observer. Note that there may be more than one place in the cycle where each of the above is heard. Give your reasoning.

9. Two cars, one behind the other, are traveling *in the same direction at the same speed*. Does either driver hear the other's horn at a frequency that is different than that heard when both cars are at rest? Justify your answer.

PROBLEMS

Section 21.2 The Speed of Sound

1. Suppose someone said: "Sound whose wavelength is larger than the size of your ear cannot be heard." (a) Assume the speed of sound is 343 m/s and compute the wavelength of sound at the limits of human hearing, 20 Hz and 20 kHz. (b) Compare these values with the (estimated) width of your ear. Based on this comparison, is the statement above correct?

2. The magnetic tape of a cassette deck moves with a speed of $1\frac{7}{8}$ inches per second, as the drawing indicates. The recording head records a 15 000-Hz tone on the tape. What is the "wavelength" λ of the magnetized regions?

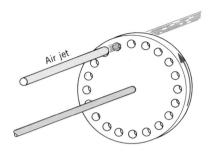

Drawing 8

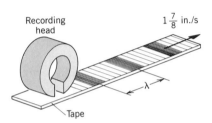

3. Argon (molecular mass = 39.9 u) is a monatomic gas. Assuming that it behaves like an ideal gas, find (a) the rms-speed of argon atoms at 25 °C, and (b) the speed of sound in argon at 25 °C.

4. The speed v of longitudinal waves in a liquid is given by $v = \sqrt{B_{ad}/\rho}$, where B_{ad} is the adiabatic bulk modulus and ρ is the density. Show that $\sqrt{B_{ad}/\rho}$ has the units of speed.

5. Aluminum has a Young's modulus of 6.9×10^{10} Pa and a density of 2700 kg/m³. How much time does it take for sound to travel the 1.4-m length of an aluminum rod?

6. The wavelength of a sound wave is 2.74 m in air at 20 °C. What is the wavelength of this same sound wave in fresh water at 20 °C?

7. At 20 °C the densities of fresh water and ethanol are, respectively, 998 and 789 kg/m³. Find the ratio of the adiabatic bulk modulus of fresh water to the adiabatic bulk modulus of ethanol at 20 °C.

8. As the drawing illustrates, a siren can be made by blowing a jet of air through 100 equally spaced holes in a rotating disk. If the siren is to produce a 2200-Hz tone, what must be the angular speed of the disk in (a) rad/s and (b) rev/s? See Drawing 8.

* 9. An explosion occurs at the end of a pier. The sound reaches the other end of the pier by traveling through three media: air, fresh water, and a slender handrail of solid steel. Assume that the air and water temperature is 20 °C and that the sound travels a distance of 125 m in each medium. (a) Through which medium does the sound arrive first, second, and third? (b) After the first sound arrives, how much later do the second and third sounds arrive? *(Hint: The speed of sound given for bulk steel in Table 21.1 does not apply here.)*

* 10. At a height of ten meters above the surface of a lake, a sound pulse is generated. The echo from the bottom of the lake returns to the point of origin 0.140 s later. The air and water temperatures are both 20 °C. How deep is the lake?

* 11. A long slender bar is made from an unknown material. The length of the bar is 0.83 m, its cross-sectional area is 1.3×10^{-4} m², and its mass is 2.1 kg. A sound wave travels from one end of the bar to the other end in 1.9×10^{-4} s. From which one of the materials listed in Table 11.1 is the bar most likely to be made?

* 12. A sonar unit on a boat is capable of detecting the return of an ultrasonic pulse during times up to 1.500 s after the pulse is transmitted. (a) Assuming the freshwater temperature is 20 °C, what is the maximum water depth that can be measured? (b) If the sonar unit measures time with an accuracy of ±0.004 s, what is the error (in meters) in the depth measurement?

* 13. Both krypton (Kr) and neon (Ne) can be approximated as ideal gases and are monatomic. The atomic mass of krypton is 83.8 u, while that of neon is 20.2 u. A tuning fork produces a sound whose wavelength in krypton is 1.25 m. If the tuning fork were used in neon at the same temperature, what would be the wavelength?

**14. As a prank, someone drops a water-filled balloon out of a hotel window. The balloon is released from rest at a height of 10.0 m above the ears of a man who is the target. Because of a guilty conscience, however, the prankster decides to shout a warning after the balloon is released. The warning will do no good, however, if shouted after the balloon reaches a certain point, even if the man could react infinitely quickly. Assuming the air temperature is 20 °C and ignoring the effect of air resistance on the balloon, determine how far above the man's ears this point is.

**15. In a mixture of argon (atomic mass = 39.9 u) and neon (atomic mass = 20.2 u), the speed of sound is 363 m/s at 27 °C. Assume that both monatomic gases behave as ideal gases. Find the percentage of the atoms that are argon and the percentage that are neon.

Section 21.3 Sound Intensity

16. A typical adult ear has a surface area of 21 cm². The sound intensity during a normal conversation is about 3.2×10^{-6} W/m² at the listener's ear. Assume the sound strikes the surface of the ear perpendicularly. How much power is intercepted by the ear?

17. At a distance of 3.8 m from a siren, the sound intensity is 3.6×10^{-2} W/m². Assuming the siren radiates sound uniformly in all directions, find the total power radiated.

* 18. When a helicopter is hovering 1100 m directly overhead, an observer on the ground measures a sound intensity I. Assume that sound is radiated uniformly from the helicopter and that ground reflections are negligible. How far must the helicopter fly in a straight line parallel to the ground before the observer measures a sound intensity of $\frac{1}{3}I$?

* 19. A source radiates 1.2×10^{-5} W of sound power uniformly in all directions. A cube and a sphere (radius = 0.56 m) are centered on this source. The sphere just fits within the cube. What is the sound intensity at a corner of the cube?

**20. A rocket, starting from rest, travels straight up with an acceleration of 58.0 m/s². When the rocket is at a height of 562 m, it produces sound that eventually reaches a ground-based monitoring station directly below. This station measures a sound intensity I. Later on, the station measures an intensity $\frac{1}{4}I$. Ignoring air resistance and assuming the speed of sound is 343 m/s, find the time that has elapsed between the two measurements.

Section 21.4 Decibels

21. The sound intensity level of a jet engine is 138 dB above the threshold of hearing. What is the sound intensity?

22. One of the important specifications of a cassette tape deck is its signal-to-noise rating. This specification indicates how much of the sound intensity created when playing a tape is due to the musical tones (the signal) and how much is due to the hissing sound produced by the moving tape (the noise). Suppose the sound intensity due to the musical tones is 1.2×10^{-4} W/m², while that due to the tape hiss is 3.2×10^{-11} W/m². What is the signal-to-noise rating, which is the number of *decibels* by which the signal exceeds the noise?

23. The intensity level of the sound produced at a rock concert often reaches 120 dB. The intensity level of a quiet flute is about 67 dB. What is the ratio of the sound intensity of a rock concert to the sound intensity of a quiet flute?

24. The equation $\beta = 10 \log (I/I_0)$, which defines the decibel, is sometimes written in terms of power P (in watts) rather than intensity I (in watts/meter²). The form $\beta = 10 \log (P/P_0)$ can be used to compare two power levels in terms of decibels. Suppose that stereo amplifier A is rated at 250 watts per channel, while

amplifier B has a rating of 45 watts per channel. (a) Expressed in decibels, how much more powerful is A compared to B? (b) Will A sound more than twice as loud as B? Justify your answer.

25. For information, read problem 24 before attempting this problem. Hi-fi manufacturers often express the power output of an audio amplifier using the decibel, abbreviated as dBW, where the "W" indicates that a reference level of one watt of power has been used. If an amplifier has a power rated at 17.5 dBW, how many watts of power can this component deliver?

* 26. The intensity level of sound A is 5.0 dB greater than that of sound B and 3.0 dB less than that of sound C. Determine the ratio of the intensity of sound C to the intensity of sound B; that is, determine I_C/I_B.

* 27. A source emits sound uniformly in all directions. A single radial line is drawn from this source. On this line, determine the positions of two points, 1.00 m apart, such that the intensity level at one point is 2.00 dB greater than that at the other.

* 28. The sound intensity level of a person speaking normally is about 65 dB above the threshold of hearing. What is the minimum number of people speaking simultaneously, each with this intensity level, that is necessary to produce sound at least 78 dB above the threshold of hearing?

Section 21.6 The Doppler Effect

29. At a football game, a stationary spectator is watching the halftime show. A trumpet player in the band is playing a 784-Hz tone while marching directly toward the spectator at a speed of 0.83 m/s. On a day when the speed of sound is 343 m/s, what frequency does the spectator hear?

30. A train is blowing its whistle while traveling at a speed of 33.0 m/s. The speed of sound is 343 m/s. Observer A is directly in front of the train, while observer B is directly behind it. Find the ratio of the whistle frequency heard by A to that heard by B.

* 31. Two trucks travel at the same speed. They are far apart on adjacent lanes and, thus, approach each other essentially head-on. One driver hears the horn of the other truck at a frequency that is 1.20 times the frequency he hears when the trucks are stationary. The speed of sound is 343 m/s. At what speed is each truck moving?

* 32. A motorcycle starts from rest and accelerates along a straight line at 2.81 m/s², on a day when the speed of sound is 343 m/s. A siren is located at the starting point and remains stationary. How far has the motorcycle gone when the driver hears the frequency of the siren at 90.0% of the value it has when the motorcycle is stationary?

**33. A microphone is attached to a spring that is suspended from the ceiling, as the drawing illustrates. Directly below on the floor is a stationary 440-Hz source of sound. The microphone vibrates up and down in simple harmonic motion with a period of 2.0 s. The difference between the maximum and minimum sound frequencies detected by the microphone is 2.1 Hz. Ignoring any reflections of sound in the room and using 343 m/s for the speed of sound, determine the amplitude of the simple harmonic motion.

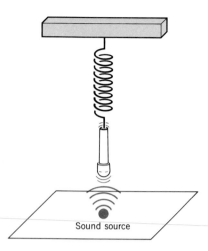

Sound source

ADDITIONAL PROBLEMS

34. The volume control on a stereo amplifier is adjusted so the sound intensity level increases from 23 to 61 dB. What is the ratio of the final sound intensity to the original sound intensity?

35. A bat emits sound whose frequency is 91 kHz. The air temperature is 35 °C. Find (a) the wavelength of the sound and (b) the distance between the centers of a condensation and an adjacent rarefaction in the sound wave.

36. An observer is moving away from a stationary source of sound. (a) In air at 20 °C, the observer hears the frequency at a value that is only 98.5% of the vibration frequency at the source. How fast is the observer moving? (b) Repeat part (a), assuming the observer and the source are in fresh water at 20 °C.

37. A source radiates 0.38 W of sound power uniformly in all directions and is at the center of a sphere. How much power passes through a 0.22-m² patch of surface on the sphere if the radius of the sphere is (a) 1.3 m and (b) 2.8 m?

38. The distance between a hi-fi speaker and the left ear of a listener is 2.70 m. (a) Calculate the time required for sound to travel this distance if the air temperature is 20 °C. (b) Assuming the sound frequency is 523 Hz, how many wavelengths of sound are contained in this distance? (c) The sound has to travel an additional 12.0 cm before reaching the right ear. Determine how much more time is required for the sound to reach this ear.

39. A listener doubles his distance from a source that emits sound uniformly in all directions. By how many decibels does the sound intensity level change, and does it increase or decrease?

* 40. At an outdoor party, Arnold is talking at a distance of 1.5 m from the beer keg. A number of other people at a distance of 3.5 m from the keg are also talking. Each individual, including Arnold, is producing the same sound power, which is assumed to spread out uniformly in all directions. The sound intensity at the beer keg due to all the others is at least as large as that produced by Arnold. What is the minimum number of other people talking?

* 41. A 33⅓-rpm record has a 5.00-kHz tone cut in the groove. (a) If the groove is located 10.0 cm from the center of the record (see drawing), what is the "wavelength" in the groove? (b) How many wavelengths are there per meter of groove length?

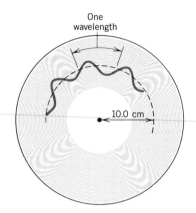

One wavelength

10.0 cm

* 42. A hi-fi speaker is generating sound in a room. At a certain point, the sound waves coming directly from the speaker (without reflecting from the walls) create an intensity level of 75.0 dB. The waves reflected from the walls create an intensity level of 72.0 dB at the same point. These levels are relative to the threshold of hearing. Relative to the threshold of hearing, what is the total intensity level? *(Hint: The answer is not 147.0 dB.)*

**43. The horn of a car produces a frequency *f*. When the car is driven perpendicularly toward the side of a tall building at a speed v_c, the sound of the horn reflects from the building. The driver hears the reflected sound at a frequency *f′* that is not equal to *f* because of the Doppler effect. Derive the equation that relates *f′* to *f*, v_c, and the speed of sound *v*. Carry out this derivation in the following sequence: (1) treat the building as a stationary observer and determine the frequency the building "observes," (2) treat the building as a stationary source that emits the reflected sound at the frequency determined in (1), the driver of the car now hearing the reflected sound at frequency *f′*.

**44. Civil engineers use a transit theodolite when surveying. A modern version of this device determines distance by measuring the time required for an ultrasonic pulse to reach a target, reflect from it, and return. Effectively, such a theodolite is calibrated properly when it is programmed with the speed of sound appropriate for the ambient air temperature. (a) Suppose the round-trip time for the pulse is 0.580 s on a day when the air temperature is 20.0 °C, the temperature for which the instrument is calibrated. How far is the target from the theodolite? (b) Assume that air behaves as an ideal gas. If the air temperature were 25.0 °C, rather than the calibration temperature of 20.0 °C, what percentage error would there be in the distance measured by the theodolite?

CHAPTER 22

The Principle of Linear Superposition and Interference Phenomena

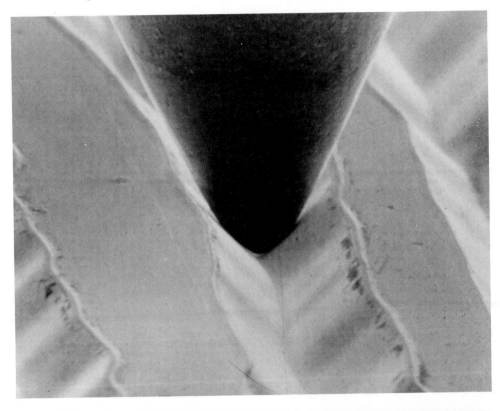

A diamond stylus rides in the groove of a record. Each wall of the groove contains the "music" for one channel of a stereo program. The sound we hear from the speakers is a linear superposition of the sound coming from each wall of the groove.

22.1 THE PRINCIPLE OF LINEAR SUPERPOSITION

Often, two or more waves pass through the same place simultaneously. For example, when two people speak at the same time, the sound waves they produce pass through each other easily. To see how to deal with such situations, examine Figures 22.1 and 22.2, which show two transverse pulses moving toward each other along a Slinky. In Figure 22.1 both pulses are "up," while in Figure 22.2 one pulse is "up" and the other is "down." Part *b* of each drawing shows an instant when the two pulses just begin to overlap, as the dashed lines indicate. The pulses merge, and the Slinky assumes a shape that is *the sum of the shapes of the individual pulses.* Thus, when the two "up" pulses overlap completely, as in Figure 22.1*c*, the Slinky has a pulse height that is twice the height of an individual pulse. Likewise, when the "up" pulse and the

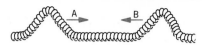

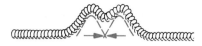

(a) The approaching pulses

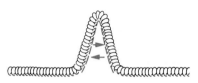

(b) Overlap begins

(c) Total overlap; notice that the slinky has twice the height of either pulse

FIGURE 22.1 Two transverse "up" pulses passing through each other.

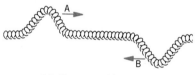

(d) The receding pulses

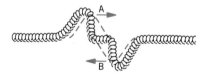

(a) The approaching pulses

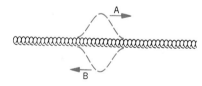

(b) Overlap begins

(c) Total overlap

FIGURE 22.2 Two transverse pulses, one "up" and one "down," passing through each other.

(d) The receding pulses

"down" pulse overlap exactly, as in Figure 22.2c, they momentarily cancel, and the Slinky becomes straight. In either case, the two pulses move apart after overlapping, and the Slinky once again conforms to the shape of the individual pulses.

The adding together of individual pulses to form a resultant pulse is an example of a more general concept called the ***principle of linear superposition.***

THE PRINCIPLE OF LINEAR SUPERPOSITION
When two or more waves are present simultaneously at the same place, the resulting disturbance is the sum of the disturbances of the individual waves at that point.

This principle can be applied to all types of waves, including sound waves, water waves, and electromagnetic waves such as light. It embodies one of the most important concepts in all of physics, and the remainder of this chapter deals with examples that are related to it.

22.2 CONSTRUCTIVE AND DESTRUCTIVE INTERFERENCE OF SOUND WAVES

CONSTRUCTIVE INTERFERENCE

Suppose that the sounds from two speakers overlap in the middle of a listening area, as in Figure 22.3, and that each speaker produces a sound wave of the same amplitude and the same frequency. For convenience, the wavelength of the sound is chosen to be $\lambda = 1.0$ m. In addition, assume the diaphragms of the speakers vibrate in and out synchronously; that is, they move outward together and inward together. If the distance of each speaker from the overlap point is the same (4.0 m in the drawing), the condensations of one wave always meet the condensations of the other when the waves come together; similarly, rarefactions always meet rarefactions. Figure 22.4 shows the pressure patterns of the individual sounds, as well as the combined pressure pattern at the overlap point. According to the principle of linear superposition, the combined pattern is the sum of the individual patterns. As a result, the pressure fluctuations at the overlap point have twice the amplitude that the individual waves

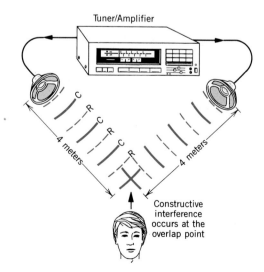

Constructive interference occurs at the overlap point

FIGURE 22.3 As a result of constructive interference between the two sound waves, a loud sound is heard at an overlap point located equally distant from two hi-fi speakers.

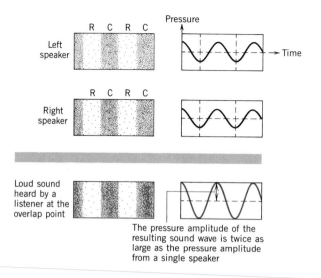

FIGURE 22.4 When sound waves from the left and right speakers in Figure 22.3 arrive at the overlap point, condensations always meet condensations, and rarefactions always meet rarefactions. When the two waves add together, constructive interference results, and a loud sound is heard, as the bottom drawings indicate.

The pressure amplitude of the resulting sound wave is twice as large as the pressure amplitude from a single speaker

have. Consequently, a listener at this spot hears a sound that is louder than the sound coming from either speaker alone. When two waves always meet condensation-to-condensation and rarefaction-to-rarefaction (or crest-to-crest and trough-to-trough), they are said to be *exactly in phase* and exhibit *constructive interference.*

DESTRUCTIVE INTERFERENCE

Now consider what happens if one of the speakers is moved. The result is surprising. Figure 22.5 shows the situation when the left speaker is moved away* from the overlap point by a distance equal to one-half of the wavelength, or 0.5 m. Therefore, to reach the overlap point, the wave from the left speaker travels one-half wavelength farther than the wave from the right speaker, and a condensation arriving from the left meets a rarefaction arriving from the right. Likewise, a rarefaction arriving from

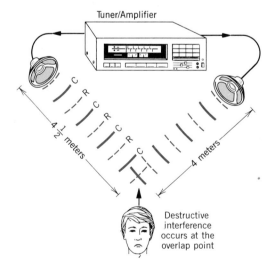

FIGURE 22.5 The left speaker is one-half of a wavelength (0.5 m) farther from the overlap point than the right speaker. Because of destructive interference, no sound is heard at the overlap point.

Destructive interference occurs at the overlap point

* When the left speaker is moved back, its sound intensity and, hence, its pressure amplitude decrease at the overlap point. Assume the power delivered to the left speaker by the amplifier is increased slightly to offset this effect and keep the amplitudes of the right and left sound waves equal at the overlap point.

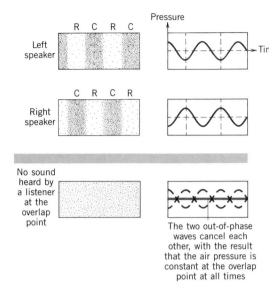

Left speaker

R C R C

Right speaker

C R C R

No sound heard by a listener at the overlap point

The two out-of-phase waves cancel each other, with the result that the air pressure is constant at the overlap point at all times

FIGURE 22.6 The sound waves from the left and right speakers in Figure 22.5 arrive at the overlap point in such a way that the condensations from one speaker always meet the rarefactions from the other. When the waves combine, the individual pressure variations exactly cancel. As a result of this destructive interference, no sound is heard, as the bottom drawings indicate.

the left meets a condensation arriving from the right. According to the principle of linear superposition, the net effect is a mutual cancellation of the two waves, as Figure 22.6 emphasizes. The condensations from one wave offset the rarefactions from the other wave, leaving only a *constant air pressure*. A constant air pressure, devoid of condensations and rarefactions, means that a listener detects no sound at this spot. When two waves always meet condensation-to-rarefaction (or crest-to-trough), they are said to be *exactly out of phase* and exhibit *destructive interference.*

INTERFERENCE — THE GENERAL PICTURE

It should be apparent that if the left speaker in Figure 22.5 were moved away from the overlap point by *another* one-half wavelength ($4\frac{1}{2}$ m + $\frac{1}{2}$ m = 5.0 m), the two waves would again be in phase, and the listener would again hear a loud sound. In such a case, constructive interference results because the left sound wave travels one whole wavelength (1.0 m) farther than the right wave and, at the overlap point, condensation meets condensation and rarefaction meets rarefaction. In general, whenever the left speaker in Figure 22.5 is moved a distance of one-half wavelength toward or away from the overlap point, the listener there experiences a dramatic change in loudness. The change occurs because the interference changes from being constructive to destructive or vice versa.

Interference effects can also be detected if the two speakers are fixed in position and the listener moves about the room. Consider Figure 22.7, where the sound waves spread outward from each speaker, as indicated by the concentric circular arcs. Each solid arc represents the center of a condensation, while each dashed arc represents the center of a rarefaction. Where the two waves overlap, there are places of constructive interference and places of destructive interference.* Constructive interference occurs at any spot where two condensations or two rarefactions intersect, and the drawing shows four such places as solid dots. A listener stationed at any one of these locations hears a loud sound. On the other hand, destructive interference occurs at any place where a condensation and a rarefaction intersect, such as the two open dots in the

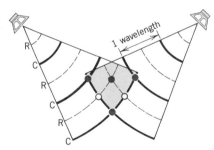

FIGURE 22.7 The overlapping of two sound waves produces interference effects in the shaded region. The solid lines denote the centers of condensations, while the dashed lines denote the centers of rarefactions. As the text explains, constructive interference occurs at each location marked by a solid dot, and destructive interference occurs at each location marked by an open dot.

---

* The sound intensity produced at a given point by each speaker depends on the distance from the speaker. As a first approximation, we ignore this dependence.

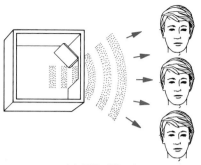

(a) With diffraction

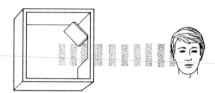

(b) Without diffraction

FIGURE 22.8 (a) The bending of sound waves around an obstacle, such as the edge of the doorway, is an example of diffraction. The source of sound waves within the room is not shown in the drawing. (b) If diffraction did not occur, the sound waves would not bend as they pass through the doorway.

picture. A listener situated at a point of destructive interference hears no sound. At locations where neither constructive nor destructive interference occurs, the two waves partially reinforce or partially cancel, depending on the position relative to the speakers. Listeners at such places hear a sound whose loudness is between that heard at the points of constructive and destructive interference.

Since there are many places of constructive and destructive interference within the overlap region in Figure 22.7, it is possible for a listener to walk about the area and hear marked variations in loudness. Interference redistributes the sound intensity by taking the intensity from places of destructive interference and giving it to places of constructive interference. In a sense, interference "robs Peter to pay Paul." The phenomenon is exhibited by all types of waves, not just sound waves.

22.3 DIFFRACTION

Diffraction is the bending of a wave around an obstacle or the edges of an opening. All kinds of waves exhibit diffraction. For instance, a sound wave produced by a stereo system bends around the edges of an open doorway, as Figure 22.8a illustrates. If such bending did not occur, sound could be heard outside the room only at locations directly in front of the doorway, as part b of the drawing suggests. (It is assumed that no sound is transmitted directly through the walls.)

To demonstrate how the bending of waves arises, Figure 22.9 shows an expanded view of Figure 22.8. When the sound wave reaches the doorway, the air in the doorway is set into longitudinal vibration. As far as the outside is concerned, each air molecule in the doorway is a source of a sound wave in its own right, and, for purposes of illustration, the drawing shows two of the molecules. Each molecule produces a sound wave that expands spherically outward, much like the water wave generated when a stone is dropped into a pond. The sound waves generated by all of the molecules in the doorway must be added together to obtain the total sound intensity at any location outside the room, as prescribed by the principle of linear superposition. However, even considering only the waves from the two molecules in the picture, it is clear that the expanding wave patterns reach locations off to either side of

FIGURE 22.9 Each vibrating air molecule in the doorway generates a sound wave that expands outward and bends, or diffracts, around the edges of the doorway. Because of interference effects from the sound waves produced by all the molecules in the doorway, the sound intensity is mostly confined to the region on either side of the doorway defined by the angle θ.

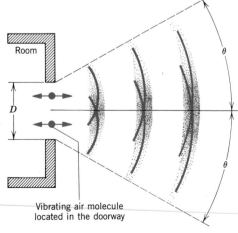

the doorway. The net effect is a "bending" or diffraction of the sound around the edges of the opening.

When the sound waves generated by every molecule in the doorway are added together, it is found that there are places where the sound intensity is a maximum and places where it is zero, in a fashion similar to that discussed in the previous section. Analysis shows that at a great distance from the doorway the sound intensity is a maximum directly opposite the center of the opening. As the distance to either side of center increases, the intensity decreases and reaches zero, then rises again to a maximum, falls again to zero, rises back to a maximum, and so on. The intensity pattern consists of a series of alternating maxima and minima. Only the maximum at the center is a strong one. The other maxima are weak and become progressively weaker at greater distances from the center. In Figure 22.9 the angle θ defines the location of the first minimum intensity point on either side of center. Equation 22.1 gives θ in terms of the wavelength λ and the width D of the doorway and assumes that the doorway can be treated like a slit whose height is very large compared to its width:

$$\begin{bmatrix} \text{Single slit—} \\ \text{first minimum} \end{bmatrix} \qquad \sin \theta = \frac{\lambda}{D} \qquad (22.1)$$

An important point to remember about Equation 22.1 is that the extent of the diffraction depends on the ratio of the wavelength to the size of the opening. For larger values of the ratio λ/D, the angle θ is larger, and the waves spread out over a larger region. On the other hand, if λ is small compared to D, θ is small. Little diffraction occurs, and the waves are beamed in the forward direction, much like the light from a flashlight.

Waves not only bend around the edges of a single slit, but they also bend around the edges of other types of openings. Particularly important is the diffraction of sound by a circular opening, such as that in most hi-fi speakers. In this case the angle θ is related to the wavelength and the diameter D of the opening by

$$\begin{bmatrix} \text{Circular opening} \\ \text{—first minimum} \end{bmatrix} \qquad \sin \theta = 1.22 \frac{\lambda}{D} \qquad (22.2)$$

As with a single slit, greater values of the ratio λ/D lead to greater values of θ and, hence, more diffraction. Therefore, low-frequency sound, with its relatively large wavelength, spreads out, or disperses, over a wide listening area as it leaves the speaker. Such sound is said to have a "wide dispersion." In contrast, high-frequency sound, with its relatively small wavelength, tends to be concentrated in a narrow beam and has a relatively "narrow dispersion." Most manufacturers strive to produce speakers that give the sound as wide a dispersion as possible, so the music "fills the room" after leaving the speaker. Example 1 illustrates one way in which wide dispersion is achieved in loudspeakers.

EXAMPLE 1

(a) A 1500-Hz sound and a 8500-Hz sound each come from a loudspeaker whose diameter is 0.30 m. Assuming the speed of sound in air is 343 m/s, find the diffraction angle θ for each sound. (b) If the 8500-Hz sound is to be produced by a second speaker, how small should this speaker be, so the sound has the same wide dispersion as does the low-pitched sound coming from the 0.30-m speaker?

SOLUTION

(a) The diffraction angle θ for each sound wave is given by $\sin \theta = 1.22(\lambda/D)$. However, to use this equation, we must first calculate the wavelengths from $\lambda = v/f$:

$$\lambda_{1500} = \frac{343 \text{ m/s}}{1500 \text{ Hz}} = 0.23 \text{ m}$$

and

$$\lambda_{8500} = \frac{343 \text{ m/s}}{8500 \text{ Hz}} = 0.040 \text{ m}$$

The diffraction angles can now be determined:

$$\begin{bmatrix} \textbf{1500-Hz} \\ \textbf{sound} \end{bmatrix} \quad \sin \theta = 1.22 \frac{\lambda_{1500}}{D} \quad (22.2)$$

$$= 1.22 \left(\frac{0.23 \text{ m}}{0.30 \text{ m}} \right) = 0.94$$

$$\theta = \sin^{-1} 0.94 = \boxed{70°}$$

$$\begin{bmatrix} \textbf{8500-Hz} \\ \textbf{sound} \end{bmatrix} \quad \sin \theta = 1.22 \frac{\lambda_{8500}}{D} \quad (22.2)$$

$$= 1.22 \left(\frac{0.040 \text{ m}}{0.30 \text{ m}} \right) = 0.16$$

$$\theta = \sin^{-1} 0.16 = \boxed{9.2°}$$

Figure 22.10 illustrates these results.

(b) To find the speaker diameter that will give a $\theta = 70°$ dispersion to the 8500-Hz sound, we again use the relation $\sin \theta = 1.22(\lambda/D)$:

$$D = \frac{1.22 \, \lambda_{8500}}{\sin \theta} = \frac{1.22 \, (0.040 \text{ m})}{\sin (70°)} = \boxed{0.052 \text{ m}}$$

This result shows that high-frequency sound can have a wide dispersion, *provided the diameter of the speaker is small enough.* Accordingly, loudspeaker designers use small diameter speakers (called "tweeters") to generate high-frequency sound, as Figure 22.11 indicates.

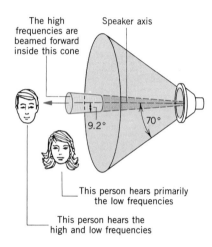

FIGURE 22.10 Because the high frequencies are not dispersed to the same extent as the low frequencies, you should be directly in front of the speaker to hear both the high and low frequencies equally well.

FIGURE 22.11 Small-diameter speakers, called tweeters, are used to produce high-frequency sound. The small diameter helps to promote a wider dispersion of the sound.

22.4 BEATS

Constructive or destructive interference can occur when two sound waves with *the same frequency* overlap. This section considers what happens when two waves with *slightly different frequencies* overlap. Once again, the principle of linear superposition will be our guide. In Figure 22.12 the waves come from two tuning forks placed side by side. A tuning fork has the property of producing a single-frequency sound wave when struck with a sharp blow. The two tuning forks in the drawing are identical, and each is designed to produce a 440-Hz tone. However, a small piece of putty has been attached to one fork, whose frequency is lowered to 438 Hz because of

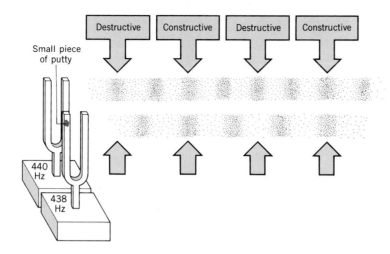

FIGURE 22.12 Two tuning forks have slightly different frequencies of 440 and 438 Hz. The phenomenon of beats occurs when the tuning forks are sounded simultaneously. For clarity, the frequencies of the sound waves displayed have a ratio of about 4/3 rather than 440/438.

the added mass. When the forks are sounded simultaneously, the loudness of the resulting sound rises and falls periodically—faint, then loud, then faint, then loud again, and so on. The periodic variations in loudness are called *beats* and result from the interference between two sound waves with slightly different frequencies.

For clarity, Figure 22.12 shows the condensations and rarefactions of the sound waves separately. In reality, however, the waves spread out and overlap. In accord with the principle of linear superposition, the ear detects the combined total of the two. Notice that there are places where the waves interfere constructively and places where they interfere destructively. When a region of constructive interference reaches the ear, a loud sound is heard. When a region of destructive interference arrives, the sound intensity drops to zero (assuming each of the waves has the same amplitude). The number of times per second that the loudness rises and falls is the *beat frequency* and is the *difference* between the two sound frequencies. Thus, in the situation illustrated in Figure 22.12, an observer hears the sound loudness rise and fall at the rate of 2 times per second (440 Hz − 438 Hz).

Figure 22.13 helps to explain why the beat frequency is the difference between the two frequencies. The drawing displays graphical representations of the pressure pat-

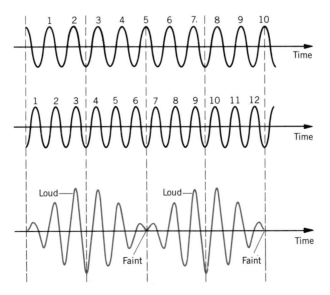

FIGURE 22.13 A 10-Hz sound wave and a 12-Hz sound wave, when added together, produce a wave with a beat frequency of 2 Hz. The drawings show the pressure patterns of the individual waves and the pressure pattern (in color) that results when the two overlap and are combined according to the principle of linear superposition. The time interval shown for each pattern is one second.

terns of a 10-Hz wave and a 12-Hz wave, along with the pressure pattern that results when the two overlap. These frequencies have been chosen for convenience, even though they lie below the audio range and are inaudible. Audible sound waves behave in exactly the same way. The top two drawings show the pressure variations in a 1-second interval of each wave. The third drawing shows the result of adding together the first two patterns according to the principle of linear superposition. Notice that the amplitude in the third drawing is not constant, as it is in the individual waves. Instead, the amplitude changes from a minimum to a maximum, back to a minimum, etc. When such pressure variations reach the ear and occur in the audible frequency range, they produce a loud sound when the amplitude is a maximum, and a faint sound when the amplitude is a minimum. Two loud-faint cycles, or beats, occur in the 1-second interval shown in the drawing, corresponding to a beat frequency of 2 Hz. This is consistent with our earlier statement that the beat frequency is the difference between the frequencies of the individual waves, that is, 12 Hz − 10 Hz = 2 Hz.

Musicians often tune their instruments by listening to a beat frequency. For instance, a guitar player sounds an out-of-tune string along with a tone from a source known to have the correct frequency. The guitarist adjusts the tension in the string until the beats vanish, ensuring that the string is vibrating at the correct frequency.

22.5 TRANSVERSE STANDING WAVES

GENERATING STANDING WAVES

A standing wave is another effect that can occur when two waves travel through the same place at the same time. Figure 22.14 shows some of the essential features of transverse standing waves. In this figure one end of a string is attached to a wall and the other end is vibrated back and forth. Regions of the string move so fast that they appear only as a "blur" in the photographs. Each of the patterns shown is called a *standing wave pattern.* Notice that there are special places along the patterns called nodes and antinodes. The *nodes* are points that do not vibrate at all, and *antinodes* are points where maximum vibration occurs. Figure 22.15 shows a series of superimposed drawings for each of the patterns in Figure 22.14. The drawings freeze the shape of the string at various times. The colored "dot" attached to the string emphasizes the maximum vibration that occurs at an antinode.

Each standing wave pattern in Figure 22.14 is produced at a unique frequency of vibration. These unique frequencies form a series, the smallest frequency f_1 corresponding to the one-loop pattern and the larger frequencies being integer multiples of f_1, as Figure 22.15 indicates. Thus, if f_1 is 10 Hz, the frequency of vibration needed to establish the 2-loop pattern is $2f_1$ or 20 Hz, while that needed to create the 3-loop pattern is $3f_1$ or 30 Hz, and so on. The frequencies in this series—f_1, $2f_1$, $3f_1$, etc.—are called *harmonics.* The lowest frequency f_1 is called the first harmonic, and the higher frequencies are labeled as the second harmonic ($2f_1$), the third harmonic ($3f_1$), and so forth. The harmonic number (1st, 2nd, 3rd, etc.) corresponds to the number of loops in the standing wave pattern. The frequencies in this series are also referred to as the fundamental frequency, the first overtone, the second overtone, and so on. Thus, the frequencies above the fundamental are called *overtones* (see Figure 22.15).

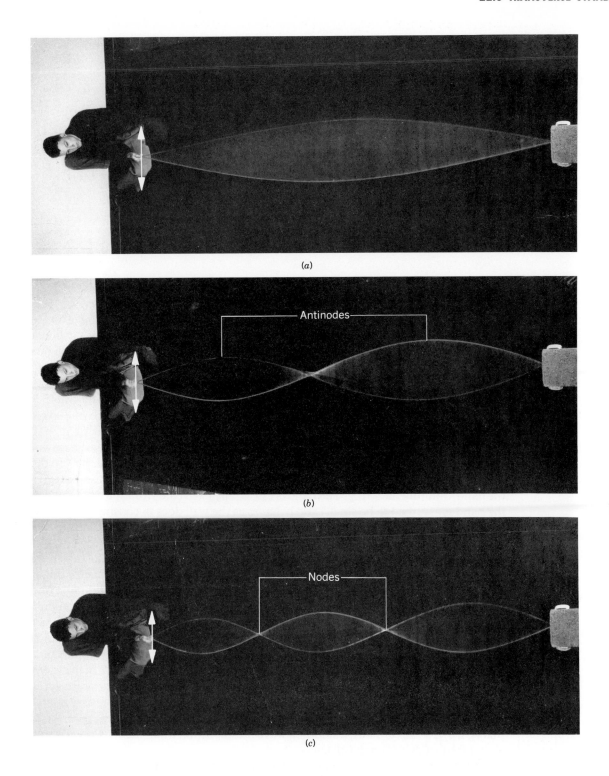

FIGURE 22.14 Vibrating a string at certain unique frequencies sets up standing wave patterns, such as the three shown here. Patterns (*b*) and (*c*) result, respectively, by vibrating the hand at two and three times the frequency used in (*a*).

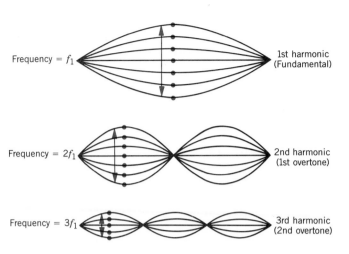

FIGURE 22.15 This illustration shows a series of superimposed drawings for each of the standing wave patterns in Figure 22.14. The colored dots attached to the string at an antinode focuses attention on the maximum vibration that occurs there. In addition, the illustration includes the frequencies and the associated nomenclature for the standing waves.

RESONANCE AND STANDING WAVES

The standing wave patterns in Figure 22.14 arise because *identical* waves travel on the string in *opposite directions* and combine in accord with the principle of linear superposition. A standing wave is said to be "standing," since it does not travel in one direction or the other, as do the individual waves that produce it.

Figure 22.16 shows why there are waves traveling in both directions on the string. At the top of the picture, one-half of a wave cycle (the remainder of the wave is omitted for clarity) is moving toward the wall on the right. When the half-cycle reaches the wall, it causes the string to pull upward on the wall. Consistent with Newton's action-reaction law, the wall pulls downward on the string, and a downward-pointing half-cycle is sent back toward the left. Thus, the wave reflects from the wall. Upon arriving back at the point of origin, the wave reflects again, this time from the hand vibrating the string. The hand is essentially fixed (the vibration amplitude of the hand is assumed to be small) and behaves as the wall does in causing reflections. Repeated reflections at the right and left ends of the string create a multitude of wave cycles traveling in both directions along the string.

As each new cycle is formed by the vibrating hand, previous cycles that have reflected from the wall arrive and reflect again from the hand. Unless the timing is right, however, the new cycles and the reflected cycles tend to offset one another, and the formation of a standing wave is inhibited. Think about pushing someone on a swing and timing your pushes so that the effect of one push reinforces that of another. Such reinforcement in the case of wave cycles on the string leads to a large amplitude standing wave. Suppose the string has a length L and its left end is being vibrated at a frequency f_1. The time required to create a new wave cycle is the period T of the wave, where $T = 1/f_1$. On the other hand, the time needed for a cycle to travel from the hand to the wall and back, a distance of $2L$, is $2L/v$, where v is the wave speed. Reinforcement between new and reflected cycles occurs if these two times are equal; that is, if $1/f_1 = 2L/v$. Thus, a standing wave is established when the string is vibrated with a frequency of $f_1 = v/(2L)$.

Repeated reinforcement between newly created and reflected cycles causes a large amplitude standing wave to develop on the string, *even though the hand itself vibrates with only a small amplitude.* Thus, the motion of the string is a resonance effect, analogous to that discussed in Section 11.9 for an object attached to a spring. The frequency f_1 at which resonance occurs is sometimes called a ***natural frequency*** of the string, similar to the frequency at which an object oscillates on a spring.

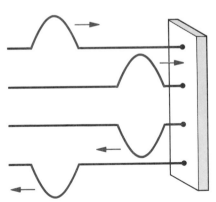

FIGURE 22.16 In reflecting from the wall, a forward-traveling half-cycle becomes a backward-traveling half-cycle that is inverted.

The resonance of the string is not completely analogous to the resonance of a spring system, however. An object on a spring has only a single natural frequency, whereas the string has a *series* of natural frequencies. The series arises because a reflected wave cycle need not return to its point of origin in time to reinforce *every* newly created cycle. Reinforcement can occur, for instance, on *every other* new cycle, as it does if the string is vibrated at twice the frequency f_1, or $f_2 = 2f_1$. Likewise, if the vibration frequency is $f_3 = 3f_1$, reinforcement occurs on *every third* new cycle. Similar arguments apply for any frequency $f_n = nf_1$, where n is an integer. As a result, the series of natural frequencies that lead to standing waves on a string fixed at both ends is given by

$$\begin{bmatrix} \text{String} \\ \text{fixed at} \\ \text{both ends} \end{bmatrix} \qquad f_n = n\left(\frac{v}{2L}\right) \qquad n = 1, 2, 3, 4, \ldots \qquad (22.3)$$

This series predicts the harmonics listed in Figure 22.15.

It is also possible to obtain Equation 22.3 in another way. Figure 22.17 illustrates the first four harmonics of the string in order to show that each "loop" in a standing wave pattern corresponds to one-half a wavelength. Since the two fixed ends of the string are nodes, the length L of the string must contain an integer number n of half-wavelengths: $L = n(\frac{1}{2}\lambda_n)$ or $\lambda_n = 2L/n$. Using this result for the wavelength in the relation $f_n\lambda_n = v$ shows that $f_n(2L/n) = v$, which can be rearranged to give Equation 22.3.

Standing waves on a string play an important role in the way many musical instruments produce sound. For instance, a guitar string is stretched between two supports and, when plucked, vibrates according to the series of natural frequencies given by Equation 22.3. The next example illustrates how this series of frequencies governs the design and playing of a guitar.

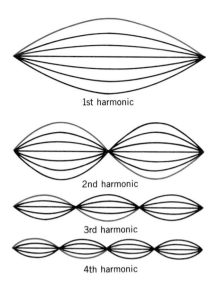

1st harmonic

2nd harmonic

3rd harmonic

4th harmonic

FIGURE 22.17 In any standing wave pattern, one of the "loops," or the distance between two successive nodes, corresponds to a half wavelength.

EXAMPLE 2

The heaviest string on an electric guitar has a linear density of $m/L = 5.28 \times 10^{-3}$ kg/m and is stretched with a tension $F = 226$ N. This string produces the musical note E when vibrating along its entire length in a standing wave at the fundamental frequency of 164.8 Hz. (a) Find the length L of the string between its two fixed ends (see Figure 22.18a). (b) A guitar player wants the string to vibrate at a fundamental frequency of 2×164.8 Hz = 329.6 Hz, as it must if the musical note E is to be sounded one octave higher in pitch. To accomplish this, he presses the string against the proper fret and then plucks the string (see part b of the drawing). Find the distance L between the fret and the bridge of the guitar.

SOLUTION

(a) The fundamental frequency f_1 is given by Equation 22.3 with $n = 1$: $f_1 = v/2L$. Since f_1 is known, the length L can be calculated directly from this expression once the speed v is known. The speed, in turn, is related to the tension F and the linear density m/L:

$$v = \sqrt{\frac{F}{m/L}} = \sqrt{\frac{226 \text{ N}}{5.28 \times 10^{-3} \text{ kg/m}}} = 207 \text{ m/s} \quad (20.2)$$

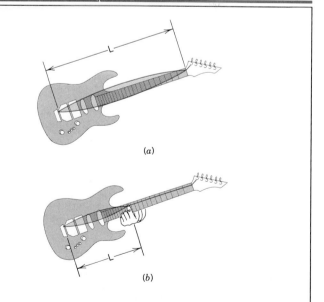

(a)

(b)

FIGURE 22.18 By pushing a string against a fret on the neck of the guitar, a player can adjust the frequency of the sound that is produced when the string is plucked.

According to $f_1 = v/2L$, the length of the string is

$$L = \frac{v}{2f_1} = \frac{207 \text{ m/s}}{2(164.8 \text{ Hz})} = \boxed{0.628 \text{ m}}$$

(b) The distance L that locates the fret can be determined exactly as in part (a) by using the wave speed $v = 207$ m/s

and noting that the frequency is now $f_1 = 329.6$ Hz: $\boxed{L = 0.314 \text{ m}}$. This length is exactly half that determined in part (a), because the frequencies have a ratio of $2:1$. Similarly, all the frets on the neck of a guitar have precise locations, according to the notes associated with them (see problem 40).

22.6 LONGITUDINAL STANDING WAVES

THE NATURE OF A LONGITUDINAL STANDING WAVE

Standing wave patterns can also be formed from longitudinal waves traveling in opposite directions. For example, when sound reflects from a wall, the forward- and backward-going waves can produce a standing wave. Figure 22.19 illustrates the vibrational motion in a longitudinal standing wave on a Slinky. As in a transverse standing wave, there are nodes and antinodes. At the nodes the coils of the Slinky do not vibrate at all, while at the antinodes the coils vibrate with maximum amplitude, as indicated by the colored dots attached to the Slinky in the picture. The vibration occurs along the line of travel of the individual waves, as is to be expected for longitudinal waves. In a standing wave of sound, the molecules behave as the colored dots do.

LONGITUDINAL STANDING WAVES IN AIR COLUMNS

Musical instruments in the wind family depend on longitudinal standing waves in producing sound. Since wind instruments (trumpet, flute, clarinet, pipe organ, etc.) are modified forms of air columns, it is useful to examine the standing waves that can be set up in such columns. Figure 22.20 shows three cylindrical columns or tubes of air that are open at both ends. Sound waves, originating from a tuning fork, travel up and down within each tube, since they reflect from the ends of the tubes, even though the ends are open. If the frequency f of the tuning fork matches one of the natural frequencies of the air column, the downward- and upward-traveling waves combine to form a standing wave, and the sound of the tuning fork becomes markedly louder. To emphasize the longitudinal nature of the standing wave patterns, the left side of the drawing replaces the air in the tubes with Slinkies, on which the nodes and

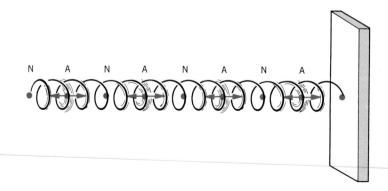

FIGURE 22.19 A longitudinal standing wave on a Slinky showing the displacement nodes (N) and antinodes (A).

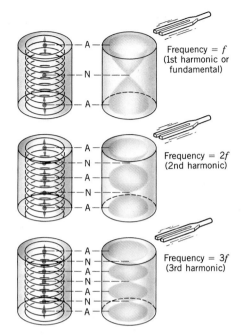

Frequency = f
(1st harmonic or fundamental)

Frequency = $2f$
(2nd harmonic)

Frequency = $3f$
(3rd harmonic)

FIGURE 22.20 A pictorial representation of longitudinal standing waves on a Slinky (left side) and in a tube (right side) open at both ends (A, antinode; N, node). The picture shows the standing waves that can be established at the three lowest frequencies and includes the associated nomenclature.

antinodes are indicated with colored dots. As an additional aid in visualizing the standing waves, the right side of the drawing shows "blurred" colored patterns within each tube. These patterns symbolize the amplitude of the vibrating air molecules at various locations. Wherever the pattern is widest, the amplitude of vibration is greatest (a displacement antinode), and wherever the pattern is narrowest the vibration is zero (a displacement node).

To determine the natural frequencies of the air columns in Figure 22.20, notice that there is a displacement antinode at each end of the tube, because the air molecules there are free to move.*

As in a transverse standing wave, the distance between two successive antinodes is one-half of a wavelength, so the length L of the tube must be an integer number n of half-wavelengths: $L = n(\frac{1}{2}\lambda_n)$ or $\lambda_n = 2L/n$. Using this wavelength in the relation $f_n = v/\lambda_n$ shows that the natural frequencies f_n of the tube are

$$\begin{bmatrix} \text{Tube open} \\ \text{at both} \\ \text{ends} \end{bmatrix} \qquad f_n = n\left(\frac{v}{2L}\right) \qquad n = 1, 2, 3, 4, \ldots \qquad (22.4)$$

At these frequencies, large amplitude standing waves develop within the tube, due to resonance. Example 3 illustrates how Equation 22.4 plays a role in the construction of a flute.

---

* In reality, the antinode does not occur exactly at the open end. However, if the diameter of the tube is small compared to the length of the tube, little error is made in assuming that the antinode is located right at the end.

EXAMPLE 3

When all the holes are closed on a standard flute, the lowest note it can sound is a middle C, whose fundamental frequency is 261.6 Hz. (a) The air temperature is 20 °C, and the speed of sound is 343 m/s. Assuming the flute is a cylindrical tube open at both ends, determine the distance L in Figure 22.21, that is, the distance from the mouthpiece to the end of the tube. (This distance is only approximate, since the antinode does not occur exactly at the mouthpiece.) (b) A flautist can alter the length of the flute by adjusting the extent to which the head joint is inserted into the main stem of the instrument. If the air temperature rises to 32 °C, to what length must a flute be adjusted to play a middle C?

SOLUTION

(a) The fundamental frequency f_1 is given by Equation 22.4 with

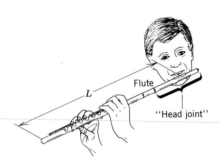

L

Flute

"Head joint"

FIGURE 22.21 When all the holes are closed on a flute, the frequency of the lowest note is determined by the length L of the tube between the mouthpiece and the end, as well as the speed of sound in air. Geometrical factors within the head joint also have an effect, but they can be ignored as a first approximation.

$n = 1: f_1 = v/2L$. This expression can be used to calculate the length L of the flute:

$$L = \frac{v}{2f_1} = \frac{343 \text{ m/s}}{2(261.6 \text{ Hz})} = \boxed{0.656 \text{ m}}$$

(b) When the temperature changes, the speed of sound changes. Therefore, at a given frequency, the wavelength changes, and the length of the flute must be adjusted to accommodate the new wavelength. Assuming air behaves as an ideal gas, we know from Equation 21.1 ($v = \sqrt{\gamma kT/m}$) that the speed of sound is directly proportional to the square root of the Kelvin temperature. Taking ratios gives

$$\frac{v(T = 305 \text{ K})}{v(T = 293 \text{ K})} = \frac{\sqrt{305 \text{ K}}}{\sqrt{293 \text{ K}}} = 1.02$$

Thus,

$$v(T = 305 \text{ K}) = 1.02(343 \text{ m/s}) = 3.50 \times 10^2 \text{ m/s}$$

The adjusted flute length is

$$L = \frac{v}{2f_1} = \frac{3.50 \times 10^2 \text{ m/s}}{2(261.6 \text{ Hz})} = \boxed{0.669 \text{ m}}$$

Thus, to play in tune at the higher temperature, a flautist must lengthen the flute by 1.3 cm.

Example 4 uses Equation 22.4 to compare two closely related wind instruments.

EXAMPLE 4

Assume that both a piccolo and a flute are cylindrical tubes with both ends open. The lowest fundamental frequency produced by a piccolo is 587.3 Hz, while that produced by a flute is 261.6 Hz. How long is a piccolo compared to a flute?

SOLUTION

The length of the piccolo can be compared to the length of the flute by noting that L is inversely proportional to the fundamen-

tal frequency, according to Equation 22.4: $L \propto 1/f_1$. Therefore,

$$\frac{L_{\text{piccolo}}}{L_{\text{flute}}} = \frac{(f_1)_{\text{flute}}}{(f_1)_{\text{piccolo}}} = \frac{261.6 \text{ Hz}}{587.3 \text{ Hz}} = \boxed{0.445}$$

The piccolo is slightly less than one-half the length of the flute.

Standing waves can also exist in a tube with only one end open, as the patterns in Figure 22.22 indicate. Note the difference between these patterns and those in Figure 22.20. Here the standing waves have a displacement antinode at the open end and a displacement node at the closed end, where the air molecules are not free to move. Since the distance between a node and an adjacent antinode is one-fourth of a wavelength, the length L of the tube must be an odd number of quarter-wavelengths: $L = 1(\frac{1}{4}\lambda)$, $L = 3(\frac{1}{4}\lambda)$, and $L = 5(\frac{1}{4}\lambda)$, for the three standing wave patterns in Figure 22.22. In general, then, $L = n(\frac{1}{4}\lambda_n)$, where n is any odd integer ($n = 1, 3, 5, \ldots$). From this result it follows that $\lambda_n = 4L/n$, and the natural frequencies f_n can be obtained from the relation $f_n = v/\lambda_n$:

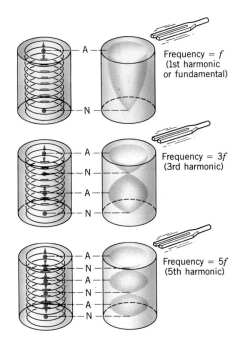

Frequency = f
(1st harmonic
or fundamental)

Frequency = $3f$
(3rd harmonic)

Frequency = $5f$
(5th harmonic)

FIGURE 22.22 A pictorial representation of the longitudinal standing waves that can be set up at the three lowest frequencies in a tube open only at one end. Because of the requirement that a displacement antinode (A) be at the open end and a displacement node (N) be at the closed end, only the odd harmonics are present.

$$\begin{bmatrix} \textbf{Tube open} \\ \textbf{at only} \\ \textbf{one end} \end{bmatrix} \qquad f_n = n\left(\frac{v}{4L}\right) \qquad n = 1, 3, 5, \ldots \qquad (22.5)$$

A tube open at only one end can develop standing waves only at the odd harmonic frequencies f_1, f_3, f_5, etc. In comparison, a tube open at both ends can develop standing waves at all harmonic frequencies f_1, f_2, f_3, etc. Moreover, the fundamental frequency f_1 of a tube is half as high when the tube has only one end open compared to when it has both ends open. In other words, a tube open at both ends must be *twice* as long as a tube open at only one end in order to produce the *same* fundamental frequency.

*22.7 COMPLEX SOUND WAVES

Musical instruments produce sound in a way that depends on standing waves. Example 2 illustrates the role of transverse standing waves on the string of an electric guitar, while Examples 3 and 4 stress the role of longitudinal standing waves in the air column within a flute and a piccolo. In each example sound is produced at the fundamental frequency of the instrument.

In general, however, a musical instrument does not produce just the fundamental frequency when it plays a note, but generates a number of harmonics or overtones as well. Since different instruments are constructed differently, each instrument generates overtones to a different extent, and they give an instrument its characteristic sound quality or timbre. Suppose, for instance, that a flute player and a trumpet player both sound concert A, a note whose fundamental frequency is 440 Hz. Even though both instruments are playing the same note, most people can distinguish the sound of the flute from that of the trumpet. The instruments sound different because

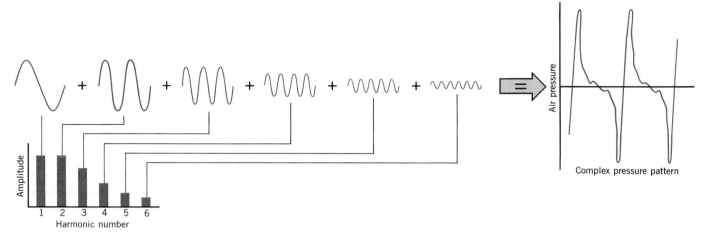

FIGURE 22.23 The graph on the right shows the pattern of pressure fluctuations such as a singer might produce in singing a note. The pattern is the sum of the first six harmonics. The relative amplitudes of the harmonics correspond to the heights of the vertical bars in the bar graph.

the relative amplitudes of the higher harmonics (880 Hz, 1320 Hz, etc.) that the instruments create are different.

The sound wave corresponding to a note on a musical instrument is called a **complex sound wave,** because it consists of a mixture of the fundamental and overtone frequencies. The pattern of pressure fluctuations in a complex wave can be obtained by using the principle of linear superposition, as Figure 22.23 indicates. This drawing shows a bar graph in which the heights of the bars give the relative amplitudes of the harmonics contained in a note such as a singer might produce. When the individual pressure patterns for each of the six harmonics are added together, they yield the complex pressure pattern shown on the right side of the picture.*

In practice, a bar graph such as that in Figure 22.23 is determined with the aid of an electronic instrument known as a spectrum analyzer. When the note is produced, the complex sound wave is detected by a microphone that converts the wave into an electrical signal. The electrical signal, in turn, is fed into the spectrum analyzer, as Figure 22.24 illustrates. The spectrum analyzer then determines the amplitude and frequency of each overtone present in the complex wave and displays the results on its screen.

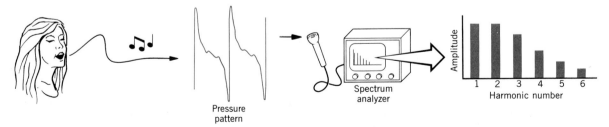

FIGURE 22.24 A microphone detects a complex sound wave, and a spectrum analyzer determines the amplitude and frequency of each harmonic present in the wave.

---

* In carrying out the addition, we assume that each individual pattern begins at zero at the origin when the time equals zero.

SUMMARY

The **principle of linear superposition** states that when two or more waves are present simultaneously at the same place, the resulting disturbance is the sum of the disturbances of the individual waves at that point. **Constructive interference** occurs at a point when two waves meet there crest-to-crest and trough-to-trough, thus reinforcing each other. **Destructive interference** occurs when the waves meet crest-to-trough and cancel each other. When the waves meet crest-to-crest and trough-to-trough, they are **exactly in phase.** When they meet crest-to-trough, they are **exactly out of phase.**

Diffraction is the bending of a wave around an obstacle or the edges of an opening. The angle through which the wave bends depends on the ratio of the wavelength λ of the wave to the width D of the opening; the greater the ratio λ/D, the greater the angle.

Beats are the periodic variations in amplitude that arise from the linear superposition of two waves that have slightly different frequencies. When the waves are sound waves, the variations in amplitude cause the loudness to vary at the **beat frequency,** which is the difference between the frequencies of the waves.

A transverse or longitudinal **standing wave** is the pattern of disturbance that results when oppositely traveling waves of the same frequency and amplitude pass through each other. A standing wave has points of minimum and maximum vibration called, respectively, **nodes** and **antinodes.** Under resonance conditions, standing waves can be established only at certain frequencies f_n, known as the **natural frequencies** of the vibrating medium. For a string that is fixed at both ends and has a length L, the natural frequencies are $f_n = n(v/2L)$, where v is the speed of the wave on the string and n is any positive integer, $n = 1, 2, 3, \ldots$. For a gas in a cylindrical tube open at both ends, the natural frequencies are given by the same expression, where v is the speed of sound in the gas and $n = 1, 2, 3, \ldots$. However, if the cylindrical tube is open only at one end, the natural frequencies are $f_n = n(v/4L)$, where $n = 1, 3, 5, \ldots$.

A **complex sound wave** is a sound wave that consists of a mixture of a fundamental frequency and overtone frequencies.

QUESTIONS

1. Does the principle of linear superposition imply that two sound waves, passing through the same place at the same time, always create a louder sound than either wave alone? Explain.

2. Suppose you are sitting at the overlap point between the two speakers in Figure 22.5. Because of destructive interference, you hear no sound, even though both speakers are emitting identical sound waves. One of the two speakers is suddenly shut off. Describe what you would hear.

3. Refer to Example 1 in Section 20.2. Which type of radio wave, AM or FM, diffracts more readily around a given obstacle? Give your reasoning.

4. The drawing shows water waves involved in one of the phenomena discussed in the chapter. Which phenomenon is it, diffraction, beats, or standing waves? Account for your choice.

5. Sometimes loudspeaker manufacturers use a "diffraction horn" for reproducing the middle frequencies and high frequen-

cies. The width w of this horn is chosen to be much smaller than the wavelengths of the sounds that emerge from the horn. If the horn is to diffract the sound into the widest possible listening area,

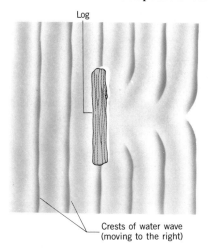

Log

Crests of water wave
(moving to the right)

Drawing for Question 4

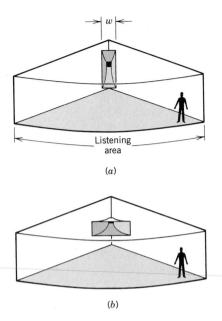

(a)

(b)

should the horn be mounted vertically, as in part *a* of the drawing, or should it be mounted horizontally, as in part *b*? Justify your answer.

6. There are three tuning forks. Two of the three are sounded together and produce beats. Is it possible for the beat frequency to be the same, no matter which tuning forks are chosen? Explain.

7. Take a rubber band and fasten one end to an immovable object. Slowly stretch the rubber band, all the while plucking it to produce a sound. Describe and explain the way in which the sound changes.

8. A string is attached to a wall and vibrated back and forth, as in Figure 22.14. The vibration frequency and length of the string are fixed. The tension in the string is changed, and it is observed that at certain values of the tension a standing wave pattern develops. Account for the fact that no standing waves are observed once the tension is increased beyond a certain value.

9. By blowing across the top of an empty soda bottle, a relatively loud sound can be created, because a standing wave develops in the bottle. Explain why the pitch of the sound produced is higher when the bottle is partially filled with water.

10. In terms of length, why is it an advantage to use a pipe closed at one end, rather than a pipe open at both ends, to generate the lowest audible frequency on a pipe organ?

11. Standing waves can ruin the acoustics of a concert hall if there is excessive reflection of the sound waves that the performers generate. For example, suppose a performer generates a 2093-Hz tone. If a large amplitude standing wave is present, it is possible for a listener to move a distance of only 8.2 cm and hear the loudness of the tone change from loud to faint. Account for this observation in terms of standing waves, pointing out why the distance is 8.2 cm.

12. The tones produced by a typical orchestra are complex sound waves, and most have fundamental frequencies less than 5000 Hz. However, a high-quality stereo system must be able to reproduce frequencies up to 20 000 Hz accurately. Explain why.

13. The sound of a normal voice is a complex wave, consisting of a fundamental frequency plus various overtones that create a characteristic voice quality. When the lungs are filled with helium instead of air, the voice has a high-pitched quality, somewhat like Donald Duck's voice. Assume the sound of the voice is generated by the vocal cords vibrating above a gas-filled tube closed at one end. Using data from Table 21.1, explain why helium has the effect of raising the pitch.

PROBLEMS

Section 22.2 Constructive and Destructive Interference of Sound Waves

1. The drawing shows a string on which two rectangular pulses are traveling at a constant speed of 1 cm/s at time *t* = 0. (a) Using the principle of linear superposition, draw the shape of the string at *t* = 1 s, 2 s, 3 s, 4 s, and 5 s. (b) What is the value of *t* when the centers of the pulses coincide?

2. Repeat problem 1, assuming the pulse on the right is pointing downward, rather than upward.

3. Repeat problem 1, assuming the pulses have the shape (half up and half down) shown in the drawing.

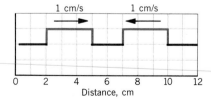

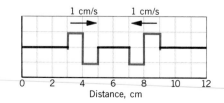

* 4. Suppose the two speakers in Figure 22.7 are separated by a distance of 3.000 m. Consider a point P that is in front of the speakers and whose perpendicular distance from the line joining the speakers is 2.200 m. The point P is between the speakers and is 2.500 m from the speaker on the right. The speed of sound is 343 m/s. Does constructive or destructive interference occur at P when the speakers produce identical sound waves whose frequency is (a) 1466 Hz and (b) 977 Hz?

Section 22.3 Diffraction

5. A speaker has a diameter of 0.30 m. (a) Assuming the speed of sound is 343 m/s, find the diffraction angle θ for a 2.0-kHz tone and for a 6.0-kHz tone. (b) What speaker diameter D should be used to generate a 6.0-kHz tone whose diffraction angle is as wide as that for the 2.0-kHz tone in part (a)?

6. A circular speaker produces a 1250-Hz tone, and the diffraction angle θ is 65°. The speed of sound is 343 m/s. Determine the diameter of the speaker.

* 7. A 4.00-kHz tone is being produced by a speaker with a diameter of 15.0 cm. The air temperature changes from 0 °C to 35 °C. Assuming air to be an ideal gas, find the *change* in the dispersion angle θ.

Section 22.4 Beats

8. Two pure tones are sounded together. The drawing shows the pressure variations of the two sound waves, measured with respect to atmospheric pressure. What is the beat frequency?

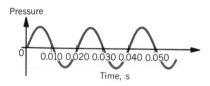

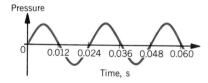

9. When a guitar string is sounded along with a 440-Hz tuning fork, a beat frequency of 5 Hz is heard. When the same string is sounded along with a 436-Hz tuning fork, a beat frequency of 9 Hz is heard. What is the frequency of the string?

10. A tuning fork vibrates at a frequency of 523 Hz. An out-of-tune piano string vibrates at 519 Hz. How much time separates successive beats?

* 11. Each of three tuning forks A, B, and C has a slightly different frequency. When A and B are sounded together, they produce a beat frequency of 2 Hz. When A and C are sounded together they produce a beat frequency of 5 Hz. What is the beat frequency that occurs when B and C are sounded together? There are two possible answers.

**12. Two hi-fi speakers are mounted on a merry-go-round whose radius is 9.00 m. The speakers both play a tone whose true frequency is 100.0 Hz. As the drawing illustrates, they are situated at opposite ends of a diameter. The speed of sound is 343 m/s, and the merry-go-round revolves once every 20.0 s. What is the beat frequency that is detected when the merry-go-round is near the position shown in the drawing?

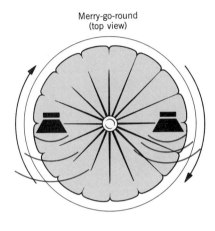

Merry-go-round
(top view)

Listener

Section 22.5 Transverse Standing Waves

13. If the end of the string in Figure 22.14 is vibrated at a frequency of 3.0 Hz and the distance between two successive nodes is 0.23 m, what is the speed of the waves on the string?

14. On a cello, the string with the largest linear density $(1.56 \times 10^{-2}$ kg/m$)$ is the C string. This string produces a fundamental frequency of 65.4 Hz and has a length of 0.800 m between two fixed ends. Find the tension in the string.

15. For the cello C string in problem 14, find the time required for a wave to travel the length of the string.

16. The lowest note on a piano has a fundamental frequency of 27.5 Hz and is produced by a wire that has a length of 1.18 m. The speed of sound in air is 343 m/s. Determine the ratio of the wavelength of the sound wave to the wavelength of the waves that travel on the wire.

17. The G string on a guitar has a fundamental frequency of 196 Hz and a length of 0.62 m. This string is pressed against the proper fret to produce the note C, whose fundamental frequency is 262 Hz. What is the distance between the fret and the end of the string at the top of the neck of the guitar?

18. Ideally, the four strings on a violin are stretched with the same tension. Each has the same length between its two fixed ends. The musical notes and corresponding fundamental frequencies of these strings are G (196.0 Hz), D (293.7 Hz), A (440.0 Hz), and E (659.3 Hz). The linear density of the E string is 3.47×10^{-4} kg/m. What is the linear density of each of the other strings?

* 19. A copper block is suspended in air from a wire. As the drawing shows, a container of mercury is then raised up around the block, until the fundamental frequency of the wire is reduced by a factor of two. Determine the ratio h/h_0 that gives the fraction of the block immersed in the mercury. *(Hint: See Table 12.1 for density values.)*

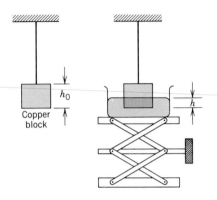

Copper block

* 20. A length of steel wire for a bass guitar has the following specifications: cross-sectional area = 5.0×10^{-7} m², density = 7860 kg/m³. (This is *not* the linear density; it is the mass per unit volume.) The wire is stretched between two fixed supports that are 0.90 m apart, and the second harmonic is 196 Hz. What is the tension in the wire?

**21. The drawing shows a uniform log supported at a 30.0° angle with respect to the ground by a wire that has a linear density of 6.00×10^{-3} kg/m. The wire has a length of 1.20 m, is perpendicular to the log, and has a fundamental frequency of 89.0 Hz. (a) Find the mass m of the log. (b) How much additional mass must be hung from the right end of the log to increase the fundamental frequency of the wire by fifty percent?

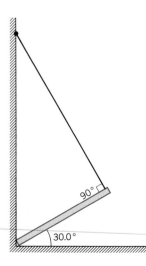

90°

30.0°

**22. The note that is three octaves above middle C is supposed to have a fundamental frequency of 2093 Hz. On a certain piano the steel wire that produces this note has a diameter of 1.00 mm. The wire is stretched between two pegs. When the piano is tuned properly to produce the correct frequency at 25.0 °C, the wire is under a tension of 818.0 N. Suppose the temperature drops to 20.0 °C. In addition, as an approximation, assume the wire is kept from contracting as the temperature drops. Consequently, the tension in the wire changes. What beat frequency is produced when this piano and another instrument (properly tuned) sound the note simultaneously?

Section 22.6 Longitudinal Standing Waves

23. The fundamental frequency of a vibrating system is 400 Hz. For each of the following systems, give the three lowest frequencies (excluding the fundamental) at which standing waves can occur: (a) a string fixed at both ends, (b) a cylindrical pipe with both ends open, and (c) a cylindrical pipe with only one end open.

24. A tube is open only at one end, has a length of 2.50 m, and is at a temperature of 37 °C. The speed of sound at 20 °C is 343 m/s. Assuming air is an ideal gas, find the frequencies of the fundamental and first two overtones produced by the tube.

25. An organ pipe, open at both ends, produces the middle C note (262 Hz) when sustaining a standing wave at its third harmonic. The speed of sound is 343 m/s. What is the length of the pipe?

26. One method for measuring the speed of sound uses standing waves. A cylindrical tube is open at both ends, and one end is placed against a loudspeaker. A movable plunger is inserted into the other end. The distance between the loudspeaker and the plunger is L. When the loudspeaker generates a 485-Hz tone, the smallest value of L for which a standing wave is formed is 0.264 m. What is the speed of sound in the gas that fills the tube?

27. A tube, open at both ends, contains an unknown ideal gas for which $\gamma = 1.40$. At 20 °C, the shortest tube in which a standing wave can be set up with a 294-Hz tuning fork has a length of 0.248 m. Find the molecular mass of the gas.

* 28. A cylindrical pipe is *closed at both ends.* Derive an expression for the frequencies of the allowed standing waves, similar in form to Equations 22.4 and 22.5, in terms of the speed of sound v, the length of the pipe L, and the standing wave number n. Be sure to state which integer values of n are allowed.

* 29. A tunnel leading straight through a hill makes tones at 135 Hz and 165 Hz especially loud because of standing waves. Assuming the speed of sound is 343 m/s, find the smallest length the tunnel could have.

**30. A tube, open at only one end, is cut into two shorter (nonequal) lengths. The piece open at both ends has a fundamental frequency of 425 Hz, while the piece open only at one end has a fundamental frequency of 675 Hz. What is the fundamental frequency of the original tube?

**31. Two loudspeakers are facing each other and are producing identical 440-Hz tones. A listener walks from one speaker toward the other at a constant speed and hears the loudness change at a frequency of 3.0 Hz. The air temperature is 20 °C. What is the walking speed?

ADDITIONAL PROBLEMS

32. The A string on a string bass is tuned to vibrate at a fundamental frequency of 55.0 Hz. If the tension in the string were changed so the fundamental frequency is doubled, what would be the ratio of the new tension to the original tension?

33. The range of human hearing is roughly from twenty hertz to twenty kilohertz. (a) Based on these limits and a value of 343 m/s for the speed of sound, what are the lengths of the longest and shortest pipes (open at both ends and producing sound at their fundamental frequencies) that you expect to find in a pipe organ? (b) Repeat part (a) for pipes open at only one end.

34. When a tuning fork is sounded together with a 492-Hz tone, a beat frequency of 2 Hz is heard. Then a small piece of putty is stuck to the tuning fork, and the tuning fork is again sounded along with the 492-Hz tone. The beat frequency decreases. What is the frequency of the tuning fork?

35. The sound produced by the loudspeaker in the drawing has a frequency of 5700 Hz and arrives at the microphone via two different paths. The sound travels through the left tube *SXM*, which has a fixed length. Simultaneously, the sound travels through the right tube *SYM*, the length of which can be changed by moving the sliding section. As the length of the path *SYM* is changed, the sound loudness detected by the microphone changes. When the sliding section is pulled out by 2.5 cm, the loudness changes from a maximum to a minimum. Find the speed at which sound travels through the gas in the tube.

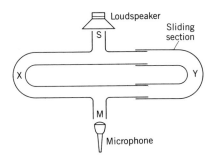

36. Both neon (Ne) and helium (He) are monatomic gases and can be assumed to be ideal gases. The fundamental frequency of a tube of neon is 268 Hz. What is the fundamental frequency of the tube if the tube is filled with helium, all other factors remaining the same?

37. Sometimes, when a wind blows across a long wire, a low-frequency "moaning" sound is produced. This sound arises be-

cause a standing wave is set up on the wire, like a standing wave on a guitar string. Assume that a wire (linear density = 0.029 kg/m) sustains a tension of 22 N, because the wire is stretched between two poles that are 45 m apart. The lowest frequency that a human ear can detect is about 20.0 Hz. What is the lowest harmonic number that could be responsible for the "moaning" sound?

* **38.** Two tuning forks X and Y have different frequencies and produce an 8-Hz beat frequency when sounded together. When X is sounded along with a 392-Hz tone, a 3-Hz beat frequency is detected. When Y is sounded along with the 392-Hz tone, a 5-Hz beat frequency is heard. What are the frequencies f_X and f_Y when (a) f_X is greater than f_Y and (b) f_X is less than f_Y?

* **39.** A vertical tube is closed at one end and open to air at the other end. The air pressure is 1.01×10^5 Pa. The tube has a length of 0.75 m. Mercury is poured into it to shorten the effective length for standing waves. What is the absolute pressure at the bottom of the mercury column, when the fundamental frequency of the shortened, air-filled tube is equal to the third harmonic of the original tube?

** **40.** As the drawing shows, the length of a guitar string is 62.8 cm. Note that the frets are numbered for convenience. A performer can play a musical scale on a single string, because the spacing *between the frets* is designed according to the following rule: When the string is pushed against any fret *j*, the fundamental frequency of the shortened string is larger by a factor of the twelfth root of two ($\sqrt[12]{2}$) than it is when the string is pushed against the fret *j* − 1. Assuming the tension in the string is the same for any note, find the spacing (a) between fret 1 and fret 0 and (b) between fret 7 and fret 6.

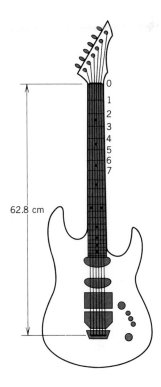

CHAPTER 23

Electric Forces and Electric Fields

This spectacular display of lightning occurred around the Kitt Peak National Observatory in Arizona. Lightning arises when electric charge is transferred between the clouds and the ground because of the electric forces between positive and negative charges.

23.1 THE ORIGIN OF ELECTRICITY

The discovery of electricity can be traced to the Greek philosopher Thales (640?–546 B.C.). He found that after amber (a petrified tree resin) is rubbed with wool or fur, the amber attracts small bits of leaves or straw. The origin of this attractive force is now known to be the electrical nature of matter itself. The word "electric" is, in fact, derived from the Greek word for amber (ēlektron). Despite the ancient roots of electricity, a correct understanding of its laws and the development of practical electrical devices is only a scant 200 years old.

The electrical nature of matter is inherent in the atoms of all substances. An atom consists of a small, relatively massive nucleus that contains particles called protons. Surrounding the nucleus is a cloud of orbiting particles known as electrons, as Figure 23.1 indicates. In a normal atom there are as many protons as there are electrons. A proton has a mass of 1.67×10^{-27} kg, while an electron has a mass of 9.11×10^{-31} kg. Like mass, *electric charge* is an intrinsic property of protons and electrons.

Only two types of electric charge have been discovered, positive and negative; a proton has a positive charge, and an electron has a negative charge.

The SI unit for measuring the magnitude of an electric charge is the **coulomb*** (**C**). Experiment reveals that the magnitude of the charge on the proton is *exactly equal* to the magnitude of the charge on the electron. Thus, the proton carries a charge $+e$, and the electron carries a charge $-e$, where e has been determined experimentally to be

$$e = 1.60 \times 10^{-19} \text{ C}$$

In a normal atom the numbers of protons and electrons are equal, so the algebraic sum of the positive charge of the nucleus and the negative charge of the electrons is zero. When an atom, or any object, carries no net charge, the object is said to be *electrically neutral.*

The charge on an electron or a proton is the *smallest* amount of free charge that has been discovered. Charges of larger magnitude are built up on an object by adding or removing electrons. Thus, any charge of magnitude q is an integer multiple of e, that is, $q = Ne$, where N is an integer. For example, the number N of electrons that must be added to a neutral object to give it a negative charge of one microcoulomb (1 μC = 10^{-6} C) is

$$N = \frac{q}{e} = \frac{1.00 \times 10^{-6} \text{ C}}{1.60 \times 10^{-19} \text{ C}} = 6.25 \times 10^{12} \text{ electrons}$$

Because any electric charge q occurs in integer multiples of elementary, indivisible charges e, electric charge is said to be *quantized.*

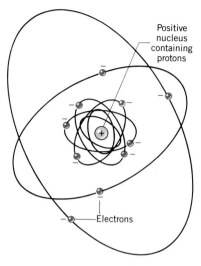

FIGURE 23.1 An atom contains a small, positively charged nucleus, about which the negatively charged electrons move. The closed-loop paths shown here are symbolic only. In reality, the electrons do not follow discreet paths, as Section 37.5 discusses.

23.2 CHARGED OBJECTS AND THE ELECTRIC FORCES BETWEEN THEM

THE SEPARATION OF CHARGES

It is possible to transfer electric charge from one object to another. Usually electrons are transferred, and the body that gains electrons acquires an excess of negative charge. The body that loses electrons has an excess of positive charge. Such separation of charge occurs often when two unlike materials are rubbed together. For example, when an ebonite rod (hard, black rubber) is rubbed with animal fur, some of the electrons from the fur are transferred to the rod. The ebonite becomes negatively charged, and the fur becomes positively charged, as Figure 23.2 indicates. Similarly, if a glass rod is rubbed with a silk cloth, some of the electrons are removed from the glass and deposited on the silk, leaving the glass positively charged and the silk negatively charged. There are many familiar examples of charge separation, as when someone walks across a nylon rug, vigorously runs a comb through dry hair, or removes a pullover sweater. In each case objects become "electrified" as surfaces rub against one another and a transfer of electrons occurs.

Charge separation plays a fundamental role in the operation of electrical equipment. For example, batteries, microphones, playback heads in cassette decks, alter-

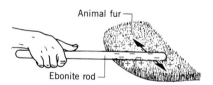

FIGURE 23.2 When an ebonite rod is rubbed against animal fur, electrons from the fur are transferred to the rod. This transfer gives the rod a negative charge (−) and leaves a positive charge (+) on the fur.

---

* At this time we omit a precise definition of the coulomb, since such a definition depends on electric currents and magnetic fields, concepts discussed in later chapters.

nators in automobile electrical systems, and electric power generators all depend on the separation of electric charges for their operation.

THE CONSERVATION OF CHARGE

When an ebonite rod is rubbed with animal fur, the rubbing process serves only to separate electrons and protons already present in the materials. No electrons or protons are created or destroyed during the process. Whenever an electron is transferred to the ebonite, a proton is left behind on the fur. Since the charges on the electron and proton have identical magnitudes but opposite signs, the algebraic sum of the two charges is zero, and the transfer does not change the net charge of the fur/rod system. If each material contains an equal number of protons and electrons to begin with, the net charge of the system is zero initially and remains zero at all times during the rubbing process.

Electric charges play a role in many processes other than rubbing two surfaces together. Chemical reactions, for instance, cause individual atoms to combine and form different molecules. And in radioactive decay, fission, or other nuclear reactions, nuclei disintegrate or combine to form new particles. A great number of experiments have verified that in any situation, the *law of conservation of electric charge* is obeyed: *During any process, the net electric charge of an entire isolated system remains constant (is conserved).*

THE ELECTRIC FORCE BETWEEN CHARGES

It is easy to demonstrate that two electrically charged objects exert a force on one another. Consider Figure 23.3*a*, which shows two small balls that have been *oppositely charged* and are light and free to move. The balls are attracted toward each other. On the other hand, balls with the *same* type of charge, either both positive or both negative, repel each other, as part *b* of the drawing illustrates. It is a fundamental characteristic of electric charges that *like charges repel and unlike charges attract each other.*

23.3 CONDUCTORS AND INSULATORS

Not only can electric charge exist *on an object,* but it can also move *through an object.* However, materials differ vastly in their ability to allow electric charge to move or be conducted through them. In other words, they differ in conductivity.

To help illustrate such differences in conductivity, Figure 23.4*a* recalls the con-

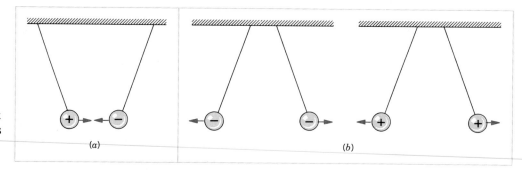

FIGURE 23.3 (*a*) A positive charge (+) and a negative charge (−) attract each other. (*b*) Two negative charges or two positive charges repel each other.

duction of heat through a bar of material whose ends are maintained at different temperatures. As Section 16.3 discusses, metals conduct heat readily and, therefore, are known as thermal conductors. On the other hand, substances that conduct heat poorly are referred to as thermal insulators.

A situation analogous to the conduction of heat arises when a metal bar is placed between two charged objects, as in Figure 23.4b. Electrons are conducted through the bar from the negatively charged object toward the positively charged object. Substances with high electrical conductivity are called **electrical conductors.** In general, materials that are good thermal conductors are also good electrical conductors. Metals such as copper, aluminum, silver, and gold are excellent electrical conductors and, therefore, are used in electrical wiring. Materials that conduct electric charge *poorly* are known as **electrical insulators.** In many cases, thermal insulators are also electrical insulators. Common electrical insulators are rubber, many plastics, and wood. Insulators, such as the rubber or plastic that coats electrical wiring, prevent electric charge from going where it is not wanted.

It is because of the atomic properties of the materials themselves that conductors and insulators have such different conductivities. As electrons orbit the nucleus, those in the outer orbits experience a weaker force of attraction to the nucleus than do the electrons in the inner orbits. Consequently, the outermost or valence electrons can be dislodged more easily than the inner electrons. In a good conductor, one or more of the valence electrons actually become detached from a parent atom and wander more or less freely throughout the material, belonging to no one atom in particular. The exact number of electrons detached from each atom depends on the nature of the material, but is usually between one and three. When one end of a conducting bar is placed in contact with a negatively charged object and the other end in contact with a positively charged object, as in Figure 23.4b, the "free" electrons are able to move readily away from the negative end and toward the positive end. The ready movement of electrons is the hallmark of a good conductor. In an insulator the situation is different, for there are very few electrons free to move throughout the material. Virtually every electron remains bound to its parent atom. Without the "free" electrons, there is very little flow of charge when the material is placed between two oppositely charged bodies, with the result that the material is an electrical insulator.

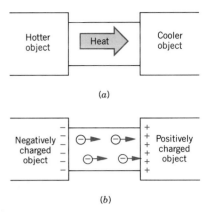

FIGURE 23.4 (*a*) Heat is conducted from the hotter end of the bar to the cooler end. (*b*) Electrons are conducted from the negatively charged end of the bar to the positively charged end.

23.4 CHARGING BY CONTACT AND BY INDUCTION

CHARGING BY CONTACT

When a negatively charged ebonite rod is rubbed along a metal object, such as the sphere in Figure 23.5a, some of the excess electrons from the rod are transferred to the object. Once the electrons are on the metal sphere, where they can move readily, their mutual repulsive forces cause them to spread out over the surface of the sphere. The insulated stand on which the sphere rests prevents the electrons from flowing to the earth, where they could spread out even more. When the rod is removed, as in part *b* of the picture, the sphere is left with a uniformly distributed negative charge. In a similar manner, the sphere would be left with a positive charge after being in contact with a positively charged rod. In this case, electrons from the sphere would be transferred to the rod. The process of giving one object a net electric charge by placing it in contact with a charged object is known as **charging by contact.**

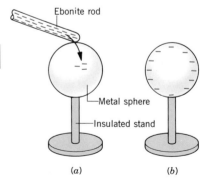

FIGURE 23.5 (*a*) Electrons are transferred by rubbing the negatively charged ebonite rod along the metal sphere. (*b*) When the rod is removed, the electrons distribute themselves uniformly over the surface of the sphere.

FIGURE 23.6 (*a*) When a charged rod is brought near the metal sphere without touching it, some of the negative charges in the sphere are separated from the positive charges. (*b*) If a ground wire is attached to the sphere, some of the electrons leave the sphere, with the result (*c*) that the sphere acquires a net positive charge.

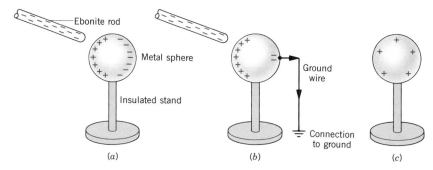

CHARGING BY INDUCTION

Figure 23.6 illustrates another method of charging a conductor. Here a negatively charged rod is brought close to, *but does not touch,* a metal sphere. In the sphere, the free electrons closest to the rod move to the other side, as part *a* of the drawing indicates. The resulting positive charge on the left of the sphere and the negative charge on the right have been "induced" or "persuaded" to form because of the repulsive force between the free electrons in the sphere and the negative rod. If the rod were removed, the free electrons would return to their original places, and the charges on either side of the sphere would disappear.

Under most conditions the earth is a good electrical conductor. So when a metal wire is attached between the sphere and the ground, as in Figure 23.6*b*, some of the free electrons leave the sphere and distribute themselves over the much larger earth. If the ground wire is removed first, followed by the ebonite rod, the sphere is left with a positive net charge, as part *c* of the picture shows. The process of giving an object a net electric charge *without* touching the object to a second charged object is called **charging by induction.** The process could also be used to give the sphere a negative net charge, if the rod were positively charged. Then, electrons would be drawn up from the ground, through the ground wire, and onto the sphere.

THE ELECTROSCOPE

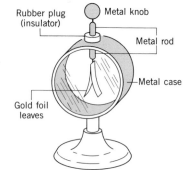

FIGURE 23.7 An electroscope.

The electroscope is a device for detecting small amounts of charge. Figure 23.7 shows that an electroscope consists of two thin gold foil strips, or leaves, mounted at the bottom end of a metal rod. A metal knob caps the top of the rod, and glass windows enclose the leaves to prevent the effects of air currents. An insulating rubber plug separates the metal rod from the metal case, so any charge on the rod does not leak away.

The electroscope can be used to determine if an insulator is charged and, if so, whether the charge is positive or negative. First, the electroscope is given a charge of known polarity, by touching the metal knob with a negatively charged ebonite rod, for example. As can be seen in Figure 23.8*a*, the negative charge spreads out over the leaves, and the repulsive force between the like charges causes the leaves to spread apart. Then, the unknown charge is brought near the electroscope *without touching it.* If the unknown charge is positive, as in part *b* of the drawing, some of the electrons are drawn off the leaves and onto the metal knob. The loss of negative charge causes a reduction in the repulsive force between the leaves and, as a result, they partially collapse. Conversely, if the unknown charge is negative, as in part *c*, it forces free

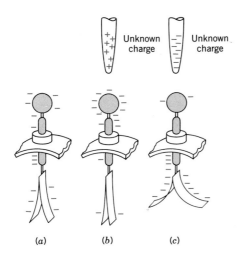

FIGURE 23.8 (*a*) A negative charge has been placed on the knob and leaves of the electroscope. (*b*) When a positively charged insulator is brought near the knob, electrons are drawn off the leaves, causing the leaves to collapse. (*c*) Conversely, a negatively charged insulator causes the leaves to diverge further.

electrons to leave the knob and increase the negative charge on the leaves. Consequently, the leaves spread apart even further. In either case, the unknown charge is not brought close enough to the electroscope to cause the charge on the leaves to change polarity (see question 6).

23.5 COULOMB'S LAW

THE FORCE BETWEEN TWO POINT CHARGES — COULOMB'S LAW

The electric force that charged objects exert on each other depends on the amount of charge on and the distance between the objects. Figure 23.9 shows two charged bodies whose physical sizes are small compared to the separation distance r between them, so small that the bodies can be regarded as mathematical points. The "point charges" have magnitudes q_1 and q_2. If the charges have *unlike* signs, as in part *a* of the picture, each charge is *attracted* to the other by a force that is directed along the line between them; $+\mathbf{F}$ is the electric force exerted on charge 1 by charge 2 and $-\mathbf{F}$ is the electric force exerted on charge 2 by charge 1. If the charges have the *same* sign (either both positive or both negative), as in part *b*, each charge is repelled from the other. The repulsive forces, like the attractive forces, act along the line between the charges. The two forces, whether attractive or repulsive, are equal in magnitude but opposite in direction. These forces always exist as a pair, each one acting on a different object, in accord with Newton's action–reaction law.

The French physicist Charles-Augustin Coulomb (1736–1806) carried out a number of experiments to determine how the electric force between two point charges depends on the amount of each charge and the separation between them. His result, now known as *Coulomb's law,* is stated below.

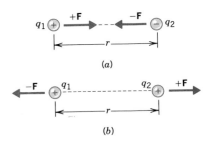

FIGURE 23.9 Each point charge exerts a force of magnitude F on the other. Regardless of whether the forces are (*a*) attractive or (*b*) repulsive, they are directed along the line between the charges and have equal magnitudes.

COULOMB'S LAW

The magnitude F of the electrical force exerted by one point charge on another point charge is directly proportional to the magnitudes q_1 and q_2 of the charges and inversely proportional to the square of the distance r between the charges:

$$F = k \frac{q_1 q_2}{r^2} \qquad (23.1)$$

where k is a proportionality constant whose value in SI units is $k = 8.99 \times 10^9 \ \text{N} \cdot \text{m}^2/\text{C}^2$.

It is common practice to express k in terms of another constant ϵ_0, by writing $k = 1/(4\pi\epsilon_0)$; ϵ_0 is called the **_permittivity of free space_** and has a value of $\epsilon_0 = 1/(4\pi k) = 8.85 \times 10^{-12} \ \text{C}^2/(\text{N} \cdot \text{m}^2)$. The force **F** is often called the electrostatic force. Equation 23.1 gives the magnitude of the electrostatic force between two point charges. When using this equation, then, it is important to remember to substitute only the charge magnitudes (without algebraic signs) for q_1 and q_2, as Example 1 illustrates.

EXAMPLE 1

Two very small objects, whose charges are $+1.0$ C and -1.0 C, are separated by 1.5 m. Find the magnitude of the attractive force that either charge exerts on the other.

SOLUTION

Coulomb's law may be used to find the magnitude of the force, provided that only the *magnitudes of the charges* are used in the calculation:

$$F = k \frac{q_1 q_2}{r^2} = \frac{(8.99 \times 10^9 \ \text{N} \cdot \text{m}^2/\text{C}^2)(1.0 \ \text{C})(1.0 \ \text{C})}{(1.5 \ \text{m})^2}$$

$$= \boxed{4.0 \times 10^9 \ \text{N}}$$

The force calculated in Example 1 corresponds to nearly 900 million pounds and is enormous. However, charges as large as one coulomb are usually encountered only in the most severe conditions, as in a lightning bolt, where as much as 25 coulombs can be transferred between the cloud and the ground. The typical charges produced in the laboratory are much smaller and are measured conveniently in microcoulombs.

Coulomb's law has a form remarkably similar to Newton's law of gravitation ($F = Gm_1m_2/r^2$). The force in both laws depends on the inverse square ($1/r^2$) of the distance between the two objects and is directed along the line between them. In addition, the force is proportional to the product of an intrinsic property of each of the objects, the charges q_1 and q_2 in Coulomb's law and the masses m_1 and m_2 in the gravitation law. On the other hand, there is a major difference between the two laws. The electrostatic force can be either repulsive or attractive, depending on whether the charges have the same sign or not; in contrast, the gravitational force is always an attractive force.

Section 6.5 discusses how the gravitational attraction between the earth and a satellite provides the centripetal force that keeps the satellite in orbit. Example 2 illustrates that an electrostatic force of attraction plays a similar role in a famous model of the atom created by the Danish physicist Niels Bohr (1885–1962).

EXAMPLE 2

In the Bohr model of the hydrogen atom, the electron is in orbit about the nuclear proton at a radius of 5.29×10^{-11} m, as Figure 23.10 shows. (a) Determine the force on the electron and (b) find the speed of the electron, assuming the orbit to be circular.

SOLUTION

(a) The electron experiences an electrostatic force of attraction because of the proton, and the magnitude of this force is

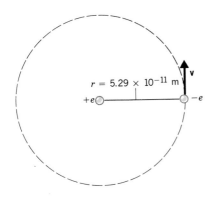

FIGURE 23.10 In the Bohr model of the hydrogen atom, the electron ($-e$) orbits the proton ($+e$) at a distance of 5.29×10^{-11} m. The velocity of the electron is **v**.

$$F = k \frac{q_1 q_2}{r^2} \tag{23.1}$$

$$= \frac{(8.99 \times 10^9 \ \text{N} \cdot \text{m}^2/\text{C}^2) \times (1.60 \times 10^{-19} \ \text{C})(1.60 \times 10^{-19} \ \text{C})}{(5.29 \times 10^{-11} \ \text{m})^2}$$

$$= \boxed{8.22 \times 10^{-8} \ \text{N}}$$

The electron is also pulled toward the proton by a gravitational force. However, the gravitational force due to the proton is negligible in comparison to the electrostatic force.

(b) Recall from Section 6.3 that any object moving with speed v on a circular path must experience a net force $\mathbf{F}_c$, called the centripetal force. This force is directed toward the center of the circle and has a magnitude of $F_c = mv^2/r$, where m is the mass of the object and r is the radius of the circle. For the electron ($m = 9.11 \times 10^{-31}$ kg) in the hydrogen atom, the centripetal force is provided by the electrostatic force. Thus, the speed of the electron is

$$v = \sqrt{\frac{F_c r}{m}}$$

$$= \sqrt{\frac{(8.22 \times 10^{-8} \ \text{N})(5.29 \times 10^{-11} \ \text{m})}{9.11 \times 10^{-31} \ \text{kg}}}$$

$$= \boxed{2.18 \times 10^6 \ \text{m/s}}$$

This orbital speed is almost five million miles per hour.

THE FORCE ON A POINT CHARGE DUE TO TWO OR MORE OTHER POINT CHARGES

Up to now, we have dealt with the problem of finding the electrostatic force on a point charge q_1 due to a point charge q_2. Suppose that another point charge q_3 were also present. What would be the net force on q_1 due to both q_2 and q_3? It is convenient to deal with such a problem in parts. First, find the magnitude and direction of the force exerted on q_1 by q_2 (ignoring q_3). Then, determine the force exerted on q_1 by q_3 (ignoring q_2). The *net force* on q_1 is the *vector sum* of these two forces. Examples 3 and 4 illustrate this procedure when the charges lie along a straight line and when they lie in a plane, respectively.

EXAMPLE 3

Figure 23.11a shows three point charges that lie along the x axis. Determine the magnitude and direction of the net electrostatic force on q_1.

SOLUTION

Let $\mathbf{F}_{12}$ be the force exerted on q_1 by q_2, and let $\mathbf{F}_{13}$ be the force exerted on q_1 by q_3. Part b of the drawing shows the directions of these forces, and their magnitudes can be determined from Coulomb's law:

$$F_{12} = k \frac{q_1 q_2}{r_{12}^2}$$

$$= \frac{(8.99 \times 10^9 \ \text{N} \cdot \text{m}^2/\text{C}^2)(3.0 \times 10^{-6} \ \text{C})(4.0 \times 10^{-6} \ \text{C})}{(0.20 \ \text{m})^2}$$

$$= 2.7 \ \text{N}$$

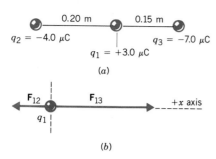

(a)

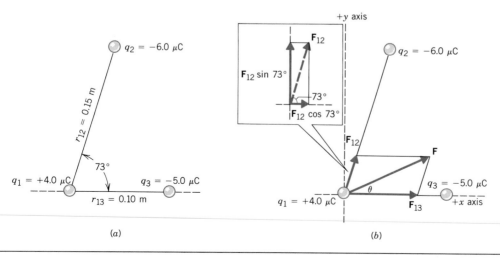

(b)

FIGURE 23.11 (a) Three charges lying along the x axis. (b) The force exerted on q_1 by q_2 is $\mathbf{F}_{12}$, while the force exerted on q_1 by q_3 is $\mathbf{F}_{13}$. The positive x axis points to the right.

$$F_{13} = k \frac{q_1 q_3}{r_{13}^2}$$

$$= \frac{(8.99 \times 10^9 \ \text{N} \cdot \text{m}^2/\text{C}^2)(3.0 \times 10^{-6} \ \text{C})(7.0 \times 10^{-6} \ \text{C})}{(0.15 \ \text{m})^2}$$

$$= 8.4 \ \text{N}$$

The net force on q_1 is the vector sum of $\mathbf{F}_{12}$ and $\mathbf{F}_{13}$. Since $\mathbf{F}_{12}$ points in the negative x direction, and $\mathbf{F}_{13}$ points in the positive x direction, the net force $\mathbf{F}$ is

$$\mathbf{F} = \mathbf{F}_{12} + \mathbf{F}_{13} = (-2.7 \ \text{N}) + (+8.4 \ \text{N}) = \boxed{+5.7 \ \text{N}}$$

The net force points along the positive x axis.

EXAMPLE 4

Find the magnitude and direction of the net electrostatic force on q_1 in Figure 23.12a.

SOLUTION

Let $\mathbf{F}_{12}$ be the force exerted on q_1 by q_2, and let $\mathbf{F}_{13}$ be the force exerted on q_1 by q_3. Part b of the picture indicates the directions of these forces, and their magnitudes can be obtained from Coulomb's law:

$$F_{12} = k \frac{q_1 q_2}{r_{12}^2}$$

$$= \frac{(8.99 \times 10^9 \ \text{N} \cdot \text{m}^2/\text{C}^2)(4.0 \times 10^{-6} \ \text{C})(6.0 \times 10^{-6} \ \text{C})}{(0.15 \ \text{m})^2}$$

$$= 9.6 \ \text{N}$$

$$F_{13} = k \frac{q_1 q_3}{r_{13}^2}$$

$$= \frac{(8.99 \times 10^9 \ \text{N} \cdot \text{m}^2/\text{C}^2)(4.0 \times 10^{-6} \ \text{C})(5.0 \times 10^{-6} \ \text{C})}{(0.10 \ \text{m})^2}$$

$$= 18 \ \text{N}$$

The net force $\mathbf{F}$ is the vector sum of $\mathbf{F}_{12}$ and $\mathbf{F}_{13}$, as part b of the drawing shows. The components of $\mathbf{F}$ that lie in the x and y directions are $\mathbf{F}_x$ and $\mathbf{F}_y$, respectively. Our approach to finding $\mathbf{F}$ is same as that used in Chapters 1 and 5. The forces $\mathbf{F}_{12}$ and $\mathbf{F}_{13}$ are resolved into x and y components. Then, the x components are combined to give $\mathbf{F}_x$, and the y components are combined to give $\mathbf{F}_y$. Once $\mathbf{F}_x$ and $\mathbf{F}_y$ are known, the magnitude and direction of $\mathbf{F}$ can be determined:

Force	*x component*	*y component*
$\mathbf{F}_{12}$	$+(9.6 \ \text{N}) \cos 73°$ $= +2.8 \ \text{N}$	$+(9.6 \ \text{N}) \sin 73°$ $= +9.2 \ \text{N}$
$\mathbf{F}_{13}$	$+18 \ \text{N}$	0
$\mathbf{F}$	$F_x = +21 \ \text{N}$	$F_y = +9.2 \ \text{N}$

The magnitude F and the angle θ of the net force are

$$F = \sqrt{F_x^2 + F_y^2} = \sqrt{(21 \ \text{N})^2 + (9.2 \ \text{N})^2} = \boxed{23 \ \text{N}}$$

$$\theta = \tan^{-1}\left(\frac{F_y}{F_x}\right) = \tan^{-1}\left(\frac{9.2 \ \text{N}}{21 \ \text{N}}\right) = \boxed{24°}$$

FIGURE 23.12 (a) Three charges lying in a plane. (b) The net force acting on q_1 is $\mathbf{F} = \mathbf{F}_{12} + \mathbf{F}_{13}$. The angle that $\mathbf{F}$ makes with the x axis is θ. The positive x axis points to the right, and the positive y axis points upward.

(a)

(b)

Examples 3 and 4 illustrate how to determine the net electrostatic force exerted on a point charge by two other point charges. The same techniques can be used for any number of point charges.

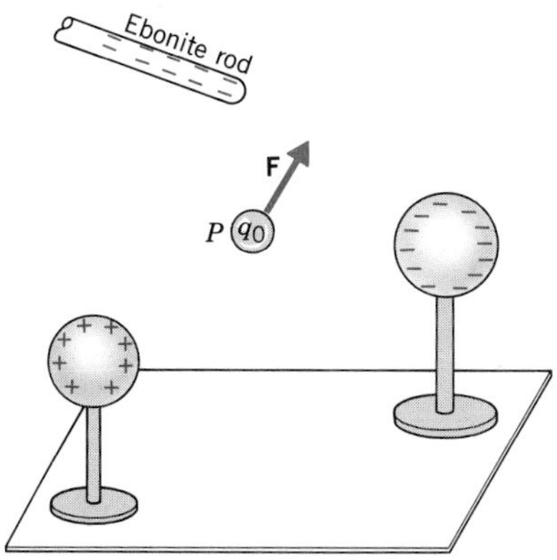

FIGURE 23.13 A positive charge q_0 placed at point P experiences an electrostatic force **F** due to the presence of the charges in its environment (i.e., the charges on the ebonite rod and the two spheres).

23.6 THE ELECTRIC FIELD

DEFINITION OF THE ELECTRIC FIELD

A charge can experience an electrostatic force due to the presence of other charges in the environment. For instance, when a positive charge q_0 is placed at the point P in Figure 23.13, the charge feels a force **F**, which is the vector sum of the forces exerted by the charges on the rod and the two spheres. It is useful to think of q_0 as a **test charge** for determining the extent to which the environmental charges generate a force at point P. However, in using a test charge, we must be careful to select a charge whose magnitude is very small, so that it does not alter the locations of the other charges in the environment. The next example illustrates how the concept of a test charge is applied.

EXAMPLE 5

A positive test charge ($q_0 = +3.0 \times 10^{-8}$ C) is placed at the point P in Figure 23.13 and experiences a force **F** $= 6.0 \times 10^{-8}$ N in the direction shown in the drawing. (a) Find the *force per coulomb* that the test charge experiences. (b) Using the result of part (a), predict the force that a charge of $+12 \times 10^{-8}$ C would experience if it were placed at P.

SOLUTION

(a) The force per coulomb of charge is

$$\frac{\mathbf{F}}{q_0} = \frac{6.0 \times 10^{-8}\text{ N}}{3.0 \times 10^{-8}\text{ C}} = \boxed{2.0\text{ N/C}}$$

(b) The result from part (a) indicates that the environmental charges can exert 2.0 newtons of force per coulomb of charge placed at the point P. Thus, a charge of $+12 \times 10^{-8}$ C would experience a force whose magnitude is

$$F = (2.0\text{ N/C})(12 \times 10^{-8}\text{ C}) = \boxed{24 \times 10^{-8}\text{ N}}$$

The direction of this force would be the same as that experienced by the test charge, since both have the same positive sign.

The force per coulomb, $\mathbf{F}/q_0$, calculated in Example 5(a) is one illustration of a quantity that is very important in the study of electricity. This quantity is called the *electric field.*

DEFINITION OF ELECTRIC FIELD

The electric field **E** that exists at a point is the electrostatic force **F** experienced by a small test charge q_0 placed at that point divided by the charge itself:

$$\mathbf{E} = \frac{\mathbf{F}}{q_0} \tag{23.2}$$

The electric field is a vector, and its direction is the same as the direction of the force **F** on a positive test charge.

SI unit of electric field: newton per coulomb (N/C)

Equation 23.2 indicates that the unit for the electric field is that of force divided by charge, which is a newton/coulomb (N/C) in SI units. The definition also emphasizes that the electric field is a vector with the same direction as the force on a *positive* test charge.*

It is the charges in the environment that create an electric field at a given point. The field exists in the sense that whenever a positive or negative charge is placed at the point, the field exerts a force on the charge, as the next example indicates.

EXAMPLE 6

In Figure 23.14a the environmental charges (the charges on the two metal spheres and the ebonite rod) create an electric field at point P. This field has a magnitude of 2.0 N/C and is directed as in the drawing. Determine the force on a charge placed at P, if the charge has a value of (a) $q_0 = +18 \times 10^{-8}$ C and (b) $q_0 = -24 \times 10^{-8}$ C.

SOLUTION

(a) The magnitude of the force is the product of the magnitudes of q_0 and E:

$$F = q_0 E = (18 \times 10^{-8} \text{ C})(2.0 \text{ N/C})$$
$$= \boxed{36 \times 10^{-8} \text{ N}} \qquad (23.2)$$

Since q_0 is positive, the force points in the same direction as the electric field, as part b of the drawing indicates.

(b) In this case, the magnitude of the force is

$$F = q_0 E = (24 \times 10^{-8} \text{ C})(2.0 \text{ N/C})$$
$$= \boxed{48 \times 10^{-8} \text{ N}} \qquad (23.2)$$

The force on the negative charge points in the direction *opposite* to the force on the positive charge, that is, opposite to the electric field (see part c of the drawing). Clearly, the electric field at point P can exert a variety of forces, depending on the magnitude and sign of the charge placed there. The charges placed at point P are assumed to be small enough that they do not alter the locations of the environmental charges creating the electric field.

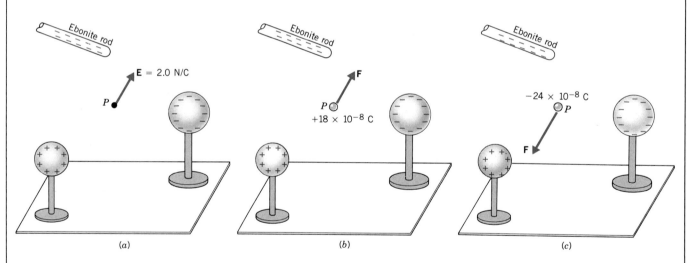

(a) (b) (c)

FIGURE 23.14 (a) The electric field **E** that exists at point P can exert a variety of forces at that spot, depending on the magnitude and sign of the charge placed there. (b) The force on a positive charge points in the same direction as **E**, while (c) the force on a negative charge points opposite to **E**.

Any charge in the environment contributes to the electric field that exists at a point. To determine the net electric field, it is necessary to determine the various

* It is not necessary to use a positive test charge. If a negative test charge were used, the force **F** in Equation 23.2 would point opposite to that found with a positive charge. However, since q_0 is negative, the electric field **F**/q_0 would have the same direction as that found with the positive charge.

contributions separately and then find the vector sum of the contributions to get the net field. This kind of approach is an illustration of the principle of linear superposition, as applied to electric fields. (This principle is introduced in Section 22.1, in connection with waves.) Example 7 emphasizes the vector nature of the electric field.

EXAMPLE 7

Figure 23.15 shows two charged objects, A and B. Each contributes as follows to the net electric field at point P: $\mathbf{E}_A = 3.00$ N/C directed to the right, and $\mathbf{E}_B = 2.00$ N/C directed downward. Thus, $\mathbf{E}_A$ and $\mathbf{E}_B$ are perpendicular. What is the net field at P?

SOLUTION

The net field $\mathbf{E}$ is the vector sum of $\mathbf{E}_A$ and $\mathbf{E}_B$: $\mathbf{E} = \mathbf{E}_A + \mathbf{E}_B$. Since $\mathbf{E}_A$ and $\mathbf{E}_B$ are perpendicular, the magnitude of $\mathbf{E}$ is given by the Pythagorean theorem:

$$E = \sqrt{E_A{}^2 + E_B{}^2} = \sqrt{(3.00 \text{ N/C})^2 + (2.00 \text{ N/C})^2}$$

$$= \boxed{3.61 \text{ N/C}}$$

The direction of $\mathbf{E}$ is given by the angle θ in the drawing:

$$\theta = \tan^{-1}\left(\frac{E_B}{E_A}\right) = \tan^{-1}\left(\frac{2.00 \text{ N/C}}{3.00 \text{ N/C}}\right)$$

$$= \boxed{33.7°}$$

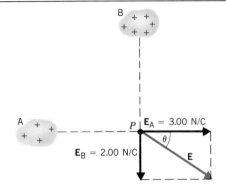

FIGURE 23.15 The electric field is a vector. Therefore, the electric field contributions $\mathbf{E}_A$ and $\mathbf{E}_B$, which come from the two charge distributions, must be added vectorially to obtain the net field $\mathbf{E}$.

ELECTRIC FIELDS PRODUCED BY POINT CHARGES

A more complete understanding of the electric field can be gained by considering the field created by a point charge, as in the following example.

EXAMPLE 8

There is an isolated point charge of $q = +15$ μC at the left in Figure 23.16a. Using a test charge of $q_0 = +0.80$ μC, determine the electric field at point P, which is 0.20 m away.

SOLUTION

Following the definition of the electric field, we place the test charge q_0 at point P, determine the force acting on the test charge, and then divide the force by the test charge. The magnitude of the force can be found from Coulomb's law:

$$F = \frac{kq_0 q}{r^2} \tag{23.1}$$

$$= \frac{(8.99 \times 10^9 \text{ N} \cdot \text{m}^2/\text{C}^2)(0.80 \times 10^{-6} \text{ C})(15 \times 10^{-6} \text{ C})}{(0.20 \text{ m})^2}$$

$$= 2.7 \text{ N}$$

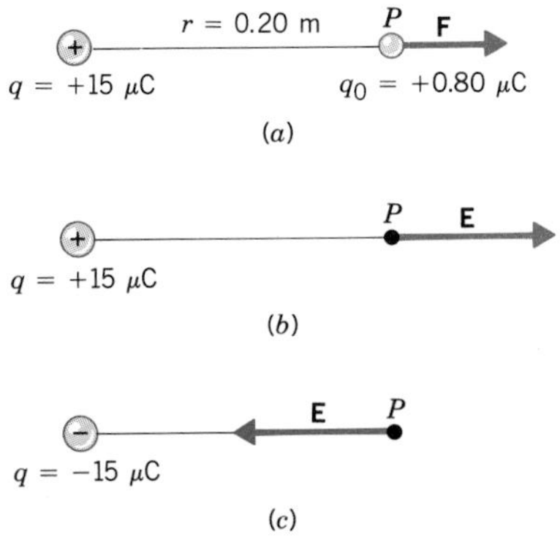

FIGURE 23.16 (a) When a positive test charge q_0 is placed at location P, the test charge experiences a repulsive force $\mathbf{F}$ due to the positive point charge q. (b) The electric field at P is represented by the arrow labeled $\mathbf{E}$ and is directed to the right. (c) If the charge q were negative rather than positive, the electric field would have the same magnitude as in (b) but would be directed to the left.

The magnitude of the electric field is

$$E = \frac{F}{q_0} = \frac{2.7 \text{ N}}{0.80 \times 10^{-6} \text{ C}} = \boxed{3.4 \times 10^6 \text{ N/C}} \quad (23.2)$$

The electric field **E** points in the *same direction* as the force **F** on the positive test charge. Since the test charge experiences a force of repulsion directed to the right, the electric field vector also points to the right, as Figure 23.16*b* shows.

The electric field produced by a point charge q can be obtained in general terms from Coulomb's law. First, note that the magnitude of the force on a test charge q_0 is $F = kqq_0/r^2$ and then divide this value by q_0 to obtain the magnitude of the field. Since q_0 is eliminated algebraically from the result, *the electric field does not depend on the test charge:*

$$\begin{bmatrix} \textbf{Electric field of} \\ \textbf{a point charge } q \end{bmatrix} \qquad E = \frac{kq}{r^2} \qquad (23.3)$$

As in Coulomb's law, only the magnitude of q is used in Equation 23.3, without regard to whether q is positive or negative. If q is positive, **E** is directed away from q, as in Figure 23.16*b*. On the other hand, if q is negative, **E** is directed toward q, since a negative charge attracts a positive test charge. For instance, Figure 23.16*c* shows the electric field that would exist at point P if there were a charge of $-15 \ \mu$C instead of $+15 \ \mu$C at the left of the drawing.

The last example in this section reemphasizes the fact that all the charges in the environment make a contribution to the electric field that exists at a given location.

EXAMPLE 9

Two positive point charges, $q_1 = +16 \ \mu$C and $q_2 = +4.0 \ \mu$C, are separated by a distance of 3.0 m, as Figure 23.17 illustrates. Find the spot on the line between the charges where the net electric field is zero.

SOLUTION

Between the charges the two field contributions have opposite directions, and the electric field is zero at the place where the magnitude of $\mathbf{E}_1$ equals that of $\mathbf{E}_2$. However, since q_2 is smaller than q_1, this location must be *closer* to q_2, in order that the field

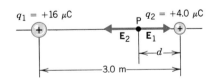

FIGURE 23.17 The two point charges q_1 and q_2 create electric fields $\mathbf{E}_1$ and $\mathbf{E}_2$ that cancel at a location P on the line between the charges. The spot where cancellation occurs is closer to the smaller charge.

of the smaller charge can balance the field of the larger charge. In the drawing, the cancellation spot is labeled P, and its distance from q_2 is d. At P, $E_1 = E_2$, and using the expression $E = kq/r^2$, we have

$$\frac{k(16 \times 10^{-6} \text{ C})}{(3.0 \text{ m} - d)^2} = \frac{k(4.0 \times 10^{-6} \text{ C})}{d^2}$$

Rearranging this expression shows that $4.0d^2 = (3.0 \text{ m} - d)^2$, and taking the square root reveals that

$$\pm 2.0d = 3.0 \text{ m} - d$$

The plus and minus signs on the left occur because either the positive or negative root can be taken. Therefore, there are two possible values for d: $+1.0$ and -3.0 m. The negative value corresponds to a location off to the right of both charges, where the magnitudes of $\mathbf{E}_1$ and $\mathbf{E}_2$ are equal, but where the directions are the same. Thus, $\mathbf{E}_1$ and $\mathbf{E}_2$ do not cancel at this spot. The positive value corresponds to the location shown in the drawing and is the zero-field location: $\boxed{d = +1.0 \text{ m}}$.

THE ELECTRIC FIELD PRODUCED BY A PARALLEL PLATE CAPACITOR

Figure 23.18 shows that a *parallel plate capacitor* consists of two parallel metal plates, each with area A. A charge $+q$ is spread uniformly over one plate, while a charge $-q$ is spread uniformly over the other plate. In the region between the plates and away from the edges, the electric field points from the positive plate toward the negative plate, is perpendicular to both plates, and has a magnitude of

$$\begin{bmatrix} \textbf{Electric field} \\ \textbf{between the} \\ \textbf{plates of a} \\ \textbf{parallel plate} \\ \textbf{capacitor} \end{bmatrix} \qquad E = \frac{q}{\epsilon_0 A} = \frac{\sigma}{\epsilon_0} \qquad (23.4)$$

In this expression the Greek symbol sigma (σ) denotes the charge per unit area ($\sigma = q/A$) and is sometimes called the charge density. Except in the region near the edges, the field has the same value at all places between the plates. The field does *not* depend on the distance from the charges, in distinct contrast to the field created by an isolated point charge. It should be noted, however, that Equation 23.4 can be derived by applying Coulomb's law to each tiny part of the charge on each plate and using integral calculus and the definition of the electric field.

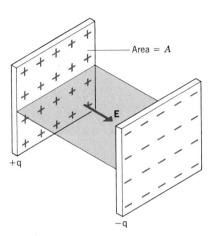

FIGURE 23.18 A parallel plate capacitor.

23.7 ELECTRIC FIELD LINES

A visual representation of the electric field can be obtained in terms of *electric field lines,* an idea proposed by the great English physicist Michael Faraday (1791–1867). Electric field lines can be thought of as a "map" that provides qualitative (and even quantitative) information about the direction and strength of the electric field in various places. Because electric field lines provide information about the electric field, which, in turn, leads to information about the electric force exerted on a charge, the lines are sometimes called *lines of force.*

To introduce the electric-field-line concept, Figure 23.19a shows a positive point charge $+q$. At the locations numbered 1–8, a positive test charge $+q_0$ would experience a repulsive force, as the arrows in the drawing indicate. Therefore, the electric field created by the charge $+q$ is directed radially outward. The electric field lines are lines drawn to show the field direction, as part b of the drawing illustrates. The lines

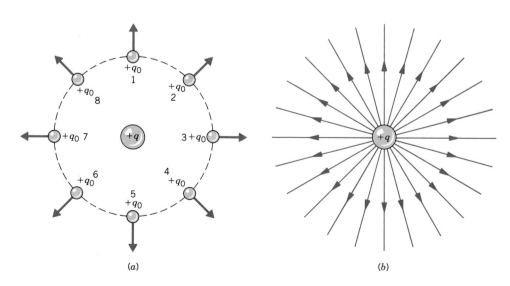

(a)

(b)

FIGURE 23.19 (*a*) A positive test charge $+q_0$, placed anywhere in the vicinity of a positive point charge $+q$, experiences a repulsive force directed radially outward. (*b*) The electric field lines are directed radially outward from the positive point charge $+q$.

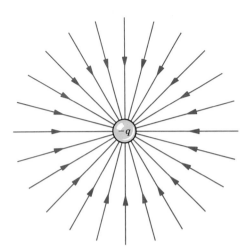

FIGURE 23.20 The electric field lines are directed radially inward toward a negative point charge $-q$.

begin on the charge $+q$ and point radially *outward*. Figure 23.20 shows the electric field lines in the vicinity of a negative charge $-q$. In this case the lines are directed radially *inward,* because the force on a positive test charge is one of attraction, indicating that the electric field points inward. In general, ***electric field lines are always directed away from positive charges and toward negative charges.***

The electric field lines in Figures 23.19 and 23.20 are drawn in only two dimensions, as a matter of convenience. Electric field lines radiate from the charges in three dimensions, and an infinite number of lines could be drawn. However, for clarity only a small number is ever included in pictures. The number is chosen to be proportional to the magnitude of the charge; thus, five times as many lines would emerge from a $+5q$ charge as from a $+q$ charge.

The pattern of electric field lines also provides information about the strength of the field. Notice in Figures 23.19 and 23.20 that near the charges, where the electric field is strongest, the field lines are close together. Conversely, at distances far from the charges, where the electric field is weaker, the electric field lines are more spread out. In other words, the electric field is strongest in regions where the field lines are closest together. In fact, no matter how many charges are present, ***the number of lines per unit area passing perpendicularly through an area is proportional to the magnitude of the electric field.*** Thus, the lines are drawn according to an arbitrary scale factor such as 1000 lines/m^2 equals a field magnitude of 1 N/C.

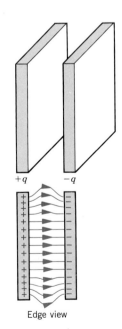

FIGURE 23.21 In the central region of a parallel plate capacitor the electric field lines are parallel and evenly spaced, indicating that the electric field there has the same magnitude at all points.

In regions where the electric field lines are equally spaced, there is the same number of lines per unit area everywhere, and the electric field has the same strength at all points. For example, Figure 23.21 shows the field lines between the plates of a parallel plate capacitor. The lines are parallel and equally spaced, except near the edges where they bulge outward. The equally spaced, parallel lines indicate that the electric field has the same magnitude at all points in the central region of the capacitor.

Electric field lines are not always straight. More often they are curved, as in the case of an ***electric dipole.*** An electric dipole consists of two separated point charges that have the same magnitude but opposite signs. The electric field of a dipole is proportional to the product of the magnitude of one of the charges and the separation between the charges. This product is called the ***dipole moment.*** Many molecules, such as H_2O and HCl, have permanent dipole moments.

Figure 23.22 depicts the curved electric field lines in the vicinity of a dipole. For a curved field line, the electric field vector at a point is *tangent* to the line at that point

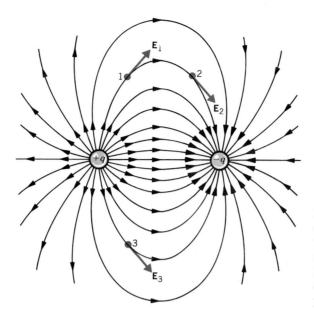

FIGURE 23.22 The electric field lines in the vicinity of an electric dipole are curved and extend from the positive charge to the negative charge. At any point, such as 1, 2, or 3, the electric field created by the dipole is tangent to the line that passes through the point.

(see points 1, 2, and 3 in the drawing). The pattern of the lines for the dipole indicates that the electric field is greatest in the region between and immediately surrounding the two charges, since the lines are closest together there. Especially notice that a given field line starts on the positive charge and ends on the negative charge. In general, *electric field lines always begin on a positive charge and end on a negative charge and do not start or stop in mid-space; furthermore, the lines begin and end on equal amounts of charge.*

The electric field lines are also curved in the case of two identical separated charges. Figure 23.23 shows the pattern of lines associated with two positive point charges and reveals that the lines in the region between the two like charges seem to repel each other, just as the charges themselves do. The pattern indicates that the electric field is relatively weak in the region between the two like charges, in contrast to the dipole.

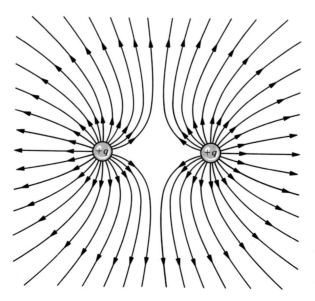

FIGURE 23.23 The electric field lines for two identical positive point charges. If the charges were both negative, the directions of the lines would be reversed.

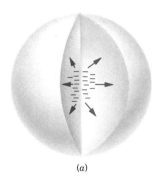

(a)

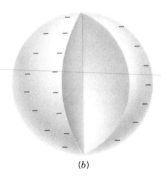

(b)

FIGURE 23.24 (a) Under electrostatic conditions, any excess charge within a conducting material quickly moves to the surface where (b) it resides at equilibrium.

23.8 THE ELECTRIC FIELD INSIDE A CONDUCTOR: SHIELDING

In conducting materials such as copper, electric charges move readily in response to the forces that electric fields exert. This characteristic property of conducting materials has a major effect on the electric field that can exist within and around them.

Suppose that a piece of copper carries a number of excess electrons somewhere within it, as in Figure 23.24a. Each electron would experience a force of repulsion because of the electric field of its neighbors. And, since copper is a conductor, the excess electrons move readily in response to that force. In fact, as a consequence of the $1/r^2$ dependence on distance in Coulomb's law, they rush to the surface of the material. Once static equilibrium is established with all of the excess charge on the surface, no further movement of charge occurs, as part b of the drawing indicates. Similarly, excess positive charge also moves to the surface of a conductor. In general, then, *at equilibrium under electrostatic conditions, excess charge of any type resides on the surface of a conductor.*

Now consider the interior of the copper in Figure 23.24b. The interior is electrically neutral, although there are still free electrons that can move under the influence of an electric field. The absence of a net movement of these free electrons indicates that there is no net electric field present within the conductor. If a net field were present, the free electrons would be moving in response to it. In fact, the excess charges arrange themselves on the conductor surface precisely in the manner needed to *make* the total field zero within the material. Thus, *at equilibrium under electrostatic conditions, the electric field at any point within a conducting material is zero.* This fact has some fascinating and important implications.

Figure 23.25a shows an uncharged, solid, cylindrical conductor at equilibrium in the central region of a parallel plate capacitor. Induced charges on the surface of the cylinder alter the electric field lines of the capacitor. Since an electric field cannot exist within the conductor under these conditions, the electric field lines do not penetrate the cylinder and end or begin on the induced charges. Consequently, a test charge placed *inside* the conductor would feel no force due to the presence of the charges on the capacitor. In other words, *the conductor shields any charge within it from electric fields created outside the conductor.* The shielding results from the induced charges on the conductor surface.

FIGURE 23.25 (a) A cylindrical conductor (shown in cross section) is placed between the two oppositely charged plates of a parallel plate capacitor. The electric field lines do not penetrate the conductor. (b) The electric field is zero in a cavity that is inside the conductor. (c) Just outside the conductor, the electric field lines are perpendicular to its surface.

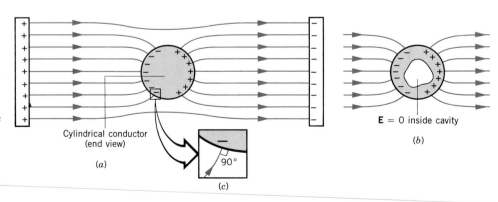

Since the electric field is zero inside the conductor, nothing is disturbed if a cavity is cut from the interior of the material, as in part *b* of the drawing. Thus, the interior of the cavity is also shielded from external electric fields, a fact that has important applications, particularly in electronics. "Stray" electric fields are produced by various electrical appliances, and these fields can interfere with the operation of sensitive electronic circuits, such as those in stereo amplifiers, televisions, and computers. To eliminate such interference, circuits are often enclosed within metal boxes that provide shielding.

Figure 23.25*c* shows another aspect of the way conducting materials alter the electric field lines created by external charges. The lines are altered because *the electric field just outside the surface of a conductor is perpendicular to the surface at equilibrium under electrostatic conditions.* If the field were not perpendicular, there would be a component of the field parallel to the surface. Since the free electrons in the conductor can move in this direction, they would move under the force exerted by the parallel component. But, in reality, no electron flow occurs at equilibrium. Therefore, there can be no parallel component, and the electric field is perpendicular to the surface.

The preceding discussion deals with aspects of the electric field within and around a conductor at equilibrium under electrostatic conditions. These features are related to the fact that conductors contain electrons that can move freely. Since insulators do not contain free electrons, these features *do not apply to insulators.*

This section concludes with an example illustrating the behavior of a conducting material in the presence of an electric field.

EXAMPLE 10

A charge $+q$ is suspended at the center of a hollow, electrically neutral, spherical conductor, as Figure 23.26 illustrates. Show that this charge induces (a) a charge of $-q$ on the interior surface and (b) a charge of $+q$ on the exterior surface of the conductor.

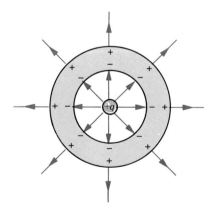

FIGURE 23.26 A positive charge $+q$ is suspended at the center of a hollow spherical conductor that is electrically neutral. Induced charges appear on the inner and outer surfaces of the conductor. The electric field within the conductor itself is zero.

SOLUTION

(a) Electric field lines emanate from the positive charge $+q$. Since the electric field inside the metal conductor must be zero at equilibrium under electrostatic conditions, each field line ends when it reaches the conductor, as the picture shows. Consequently, there is an induced *negative* charge on the interior surface of the conductor, since field lines terminate only on negative charges. Furthermore, the lines begin and end on equal amounts of charge, so the magnitude of the total induced charge is the same as the magnitude of the charge at the center. Thus, the induced charge on the interior surface is $-q$.

(b) Since an induced charge of $-q$ appears on the interior surface and since the conductor carries no excess charge to begin with, a charge $+q$ must be induced on the outer surface. As we have seen, there can be no excess charge within the metal. The positive charge on the outer surface generates electric field lines that radiate outward (see drawing) as if they originated from the central charge and the conductor were absent. The conductor does not shield the outside from the electric field produced by the central charge.

23.9 APPLICATIONS OF ELECTROSTATICS

XEROGRAPHY

Figure 23.27a illustrates an office copier. The copying process is called *xerography,* from the Greek *xeros* and *graphos,* meaning "dry writing." The heart of the machine is the xerographic drum, an aluminum cylinder coated with a layer of selenium. Aluminum is an excellent electrical conductor. Selenium, on the other hand, is a photoconductor; it is an insulator in the dark but becomes a conductor when exposed to light. Consequently, if a positive charge is deposited on the selenium surface, the

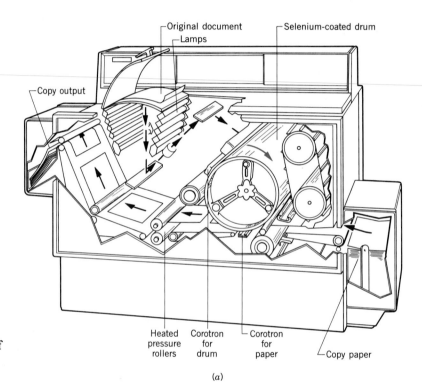

(a)

FIGURE 23.27 (a) This cutaway view shows the essential elements of a copying machine. (b) The five steps in the xerographic process.

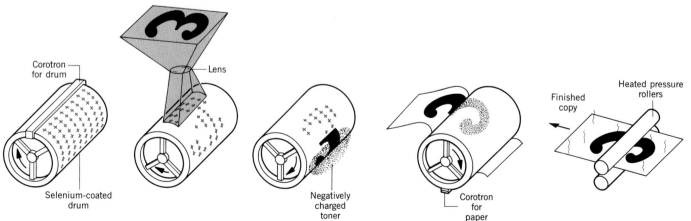

1. Charging the drum
2. Imaging the document on the drum
3. Fixing the toner to the drum
4. Transferring the toner to the paper
5. Melting the toner into the paper

(b)

charge will remain there—provided the selenium is kept in the dark. If the drum is exposed to light, however, electrons from the aluminum pass through the conducting selenium, and neutralize the positive charge. The photoconductive property of selenium lies at the core of the xerographic process.

Figure 23.27b illustrates the five steps of the process. First, an electrode called a *corotron* gives the entire selenium surface a positive charge in the dark. Second, a series of lenses and mirrors focuses an image of a document onto the revolving drum. The dark and light areas of the document produce corresponding areas on the drum. The dark areas retain their positive charge, but the light areas become conducting and lose their positive charge, ending up neutralized. Thus, a positive-charge image of the document remains on the selenium surface. In the third step, a special dry, black powder, called the *toner,* is given a negative charge and then spread onto the drum, where it adheres selectively to the positively charged areas. The fourth step involves transferring the toner onto a blank piece of paper. However, the attraction of the positive-charge image holds the toner to the drum. To transfer the toner, the paper is given a *greater positive charge* than that of the image, with the aid of another corotron. Lastly, the paper and adhering toner pass through heated pressure rollers. As a result of the heat, the toner melts into the fibers of the paper and produces the finished copy.

LASER PRINTER

A laser printer is used with computers to provide high-quality copies of text and graphics. The laser printer is very similar in operation to the xerographic machine, except the information to be reproduced is not on paper. Instead, the information

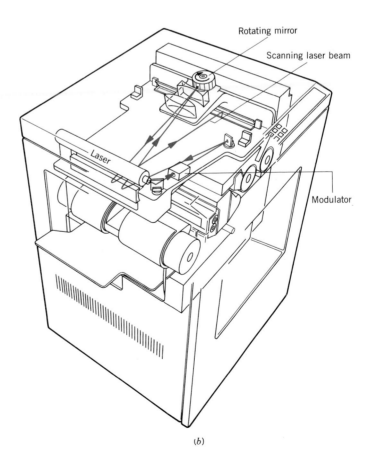

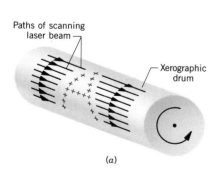

FIGURE 23.28 (*a*) As the laser beam scans across the surface of the xerographic drum, a positive-charge image of the letter "A" is created. (*b*) A laser printer showing the laser, modulator, and rotating mirror.

originates from the computer's memory and is transferred to the printer through an electrical cable. Laser light is used to copy the information onto the selenium–aluminum drum. A laser beam, focused to a fine point, is scanned rapidly from side to side across the rotating drum, as Figure 23.28*a* indicates. While the light remains on, the positive charge on the drum is neutralized. As the laser beam moves, the computer turns the beam off at the right moments during each scan to produce the desired positive-charge image, which is the letter "A" in the picture.

Figure 23.28*b* shows the general mechanism that turns the laser beam off and on and scans it across the xerographic drum. The light from the laser is sent through a device called a "modulator." The modulator also receives the information to be printed from the computer and, accordingly, either allows the light to pass or blocks it. Thus, the laser output beam from the modulator is turned off and on, at rates often exceeding one million times per second. The output beam is then directed by a series of mirrors and lenses to a rotating polygonal mirror that causes the reflected beam to sweep from side to side. When the beam reflected from the rotating mirror is directed onto the xerographic drum, the beam scans across the drum and produces the positive-charge image.

INKJET PRINTERS

An inkjet printer is another type of printer that uses electric charges in its operation. While shuttling back and forth across the paper, the inkjet printhead ejects a thin stream of ink. Figure 23.29 illustrates the elements of one type of printhead. The ink is forced out of a small nozzle and breaks up into extremely small droplets, whose diameters are less than 0.004 inch. Typically, about 150 000 droplets leave the nozzle each second and travel with a speed of approximately 40 mph toward the paper. During their flight, the droplets pass through two electrical components, a *charging electrode* and the *deflection plates* (a parallel plate capacitor). When the printhead moves over regions of the paper that are not to be inked, the charging electrode is left on and gives the ink droplets a net charge. The deflection plates divert the charged droplets into a gutter and thus prevent them from reaching the paper. Whenever ink is to be placed on the paper, the charging control, responding to the computer, turns off the charging electrode. The uncharged droplets fly straight through the deflection plates and strike the paper.

Inkjet printers are popular, because they can produce color copies. Full-color printheads use at least three nozzles for different colors (usually cyan or blue, magenta, and yellow). Often they also use a fourth nozzle for black ink. By adjusting the amount of ink that each nozzle produces, the printhead can deposit any color on the paper.

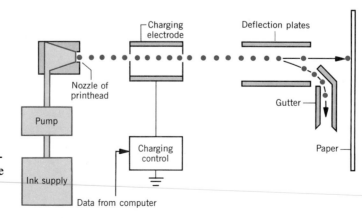

FIGURE 23.29 An inkjet printhead ejects a steady flow of ink droplets. The charging electrode is used to charge the droplets that are not needed on the paper. Charged droplets are deflected into a gutter by the deflection plates, while uncharged droplets fly straight onto the paper.

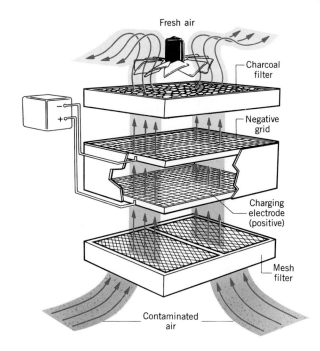

Fresh air

Charcoal filter

Negative grid

Charging electrode (positive)

Mesh filter

Contaminated air

FIGURE 23.30 An electrostatic air cleaner. Unwanted airborne particles are given a positive charge as they pass the charging electrode. The positively charged particles are removed from the air stream, because they become stuck to the negative grid.

ELECTROSTATIC AIR CLEANER

The electrostatic air cleaner is a great aid to people who suffer from respiratory problems, for it can remove up to 95% of all airborne particles, such as dust and pollen. Figure 23.30 shows the essential components of an electrostatic air cleaner. A fan draws the contaminated air into the cleaner. The air first passes through an ordinary mesh filter that removes the larger particles, and then through a positively charged wire grid, known as the *charging electrode.* This electrode gives contaminant particles a positive charge. The positively charged particles continue upward to the negatively charged grid, where they stick because of the electrostatic force of attraction. Usually the clean air leaving the negative grid is passed through an activated charcoal filter before being discharged into the room. Activated charcoal is a first-rate odor absorber and leaves the air smelling fresh.

SUMMARY

There are two kinds of **electric charge,** positive and negative; the SI unit of electric charge is the **coulomb (C).** The magnitude of the charge on an electron or a proton is $e = 1.60 \times 10^{-19}$ C. The electron carries a charge of $-e$, while the proton carries a charge of $+e$. The charge on any object, whether positive or negative, is **quantized,** in the sense that the charge consists of an integral number of protons or electrons. **The law of conservation of electric charge** states that the net electric charge of an entire isolated system remains constant during any process.

An **electrical conductor** is a material, such as copper, that conducts electric charge readily. An **electrical insulator** is a material, such as rubber, that conducts electric charge poorly. **Charging by contact** is the process of giving one object a net electric charge by placing it in contact with another, already charged object. **Charging by induction** is the process of giving an object a net electric charge without touching it to another already charged object.

One electric charge exerts a force on another electric charge. **For like charges the force is a repulsion, while for** unlike charges the force is an attraction. **Coulomb's law** gives the magnitude F of the electric force between two point charges as $F = kq_1q_2/r^2$, where q_1 and q_2 are the magnitudes of the charges, r is the distance between the charges, and $k = 8.99 \times 10^9$ N·m²/C². The force acts along the line between the two point charges. The permittivity of free space ϵ_0 is defined by the relation $k = 1/(4\pi\epsilon_0)$.

The **electric field E** at a given point is a vector and is the electrostatic force **F** experienced by a small test charge q_0 placed at that point divided by the charge itself: $\mathbf{E} = \mathbf{F}/q_0$. The direction of the electric field is the same as the direction of the force on a positive test charge. The SI unit for the electric field is the newton per coulomb (N/C). The source of the electric field at any spot is the charged objects in the environment. The magnitude of the electric field created by a point charge q is $E = kq/r^2$, where r is the distance between the charge and the point of interest. The magnitude of the electric field between the plates of a parallel plate capacitor is $E = q/(\epsilon_0 A)$, where q is the magnitude of the charge on either plate, A is the area of either plate, and ϵ_0 is the permittivity of free space.

Electric field lines are lines that can be thought of as a "map" providing information about the direction and strength of the electric field in various regions. Electric field lines are always directed away from positive charges and toward negative charges. The direction of the lines gives the direction of the electric field, since the electric field vector at a point is tangent to the line at that point. The electric field is strongest in regions where the number of lines per unit area passing perpendicularly through the area is the greatest, that is, where the lines are closest together.

Excess negative or positive charge resides on the surface of a conductor at equilibrium under electrostatic conditions. In such a situation, the electric field at any point within the conducting material is zero, and the electric field just outside the surface of the conductor is perpendicular to the surface.

QUESTIONS

1. There are three identical metal spheres, A, B, and C. Sphere A carries a charge of $+5q$. Sphere B carries a charge of $-q$. Sphere C carries no net charge. Spheres A and B are touched together and then separated. Sphere C is then touched to sphere A and separated from it. Lastly, sphere C is touched to sphere B and separated from it. How much charge ends up on sphere C? Give your reasoning.

2. In Figure 23.6 the ground wire is removed first, followed by the rod, with the result that the sphere is left with a positive charge. If the rod were removed first, followed by the ground wire, would the sphere be left with a charge? Account for your answer.

3. A rod made from insulating material carries a net charge, while a copper sphere is neutral. Is it possible for the rod and the sphere to (a) attract one another and (b) repel one another? Justify your answers.

4. An electroscope is charged so that its leaves are spread apart. The spread between the leaves decreases slightly when an electrically neutral copper object is brought near the metal knob of the electroscope without touching it. Explain.

5. The leaves of an electroscope are given a positive charge. As an insulator is brought from far away and moved toward the knob of the electroscope, the leaves are observed to spread apart even further. Is the insulator charged positively or negatively? Give your reasoning.

6. A negatively charged insulator is brought near, but does not touch, the metal knob of a charged electroscope. As the insulator approaches, the leaves collapse. When the insulator is brought even closer, the leaves are observed to diverge. Deduce the sign of the original charge on the electroscope and account for the observed behavior.

7. Suppose an electroscope is electrically neutral, rather than being negatively charged as in Figure 23.8a. Can this neutral electroscope be used to determine (a) whether an object carries a net charge and (b) whether the net charge on the object is negative or positive? Justify your answers.

8. A particle is attached to a spring and is pushed so that the spring is compressed more and more. As a result, the spring exerts a greater and greater force on the particle. Similarly, a charged particle experiences a greater and greater force when pushed closer and closer to another particle that is fixed in position and has a charge of the same polarity. In spite of the similarity, the charged particle will *not* exhibit simple harmonic motion upon being released, as will the particle on the spring. Explain why not.

9. On a thin, nonconducting rod, positive charges are spread evenly, so that there is the same amount of charge per unit length at every point. On another identical rod, positive charges are spread evenly over only the left half, and the same amount of negative charges are spread evenly over the right half. For each rod, deduce the *direction* of the electric field at a point that is located directly above the midpoint of the rod. Give your reasoning.

10. Suppose there is an electric field at point P. A very small charge is placed at this point and experiences a force. Another very small charge is then placed at this point and experiences a force that differs in both magnitude and direction from that experi-

enced by the first charge. How can these two different forces result from the single electric field that exists at point P?

11. The three drawings show four charges fixed to the corners of a rectangle in various ways. Consider the net electric field at the center C of the rectangle in each case. Rank the field magnitudes in decreasing order (largest first). Justify your ranking.

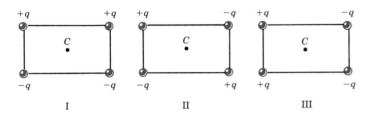

I   II   III

12. Drawings I and II show two examples of electric field lines. Decide which of the following statements are true and which are false, defending your choice in each case. (a) In both I and II the electric field is the same everywhere. (b) As you move from left to right in each case, the electric field becomes stronger. (c) The electric field in I is the same everywhere, but in II the electric field becomes stronger as you move from left to right. (d) The electric fields in both I and II are created by negative charges located somewhere on the left and positive charges somewhere on the right. (e) Both I and II arise from a single positive point charge located somewhere to the left.

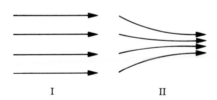

I   II

13. A positively charged particle is moving horizontally when it enters the region between the plates of a capacitor, as the drawing

illustrates. (a) Draw the trajectory that the particle follows in moving through the capacitor. (b) When the particle is within the capacitor, which of the following four vectors, if any, are *parallel* to the electric field lines: the particle's displacement, its velocity, its linear momentum, its acceleration? For each vector explain why the vector is, or is not, parallel to the electric field.

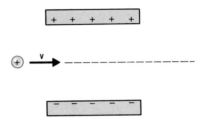

14. The drawing is supposed to show the electric field lines in the region between two charged metal conductors. Give two reasons why this drawing is incorrect.

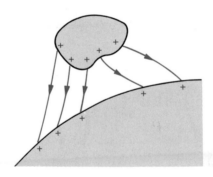

15. In all the pictures of electric field lines in the text there are no examples of two field lines that intersect. In fact, two field lines never intersect. Suppose that it were possible for two field lines to intersect. Discuss what would happen to a test charge placed at the crossover point.

PROBLEMS

Note: All charges are point charges, unless specified otherwise.

Section 23.5 Coulomb's Law

1. How many electrons must be removed from an electrically neutral silver dollar to give it a charge of $+3.8\ \mu C$?

2. A metal sphere has a charge of $+8.0\ \mu C$. What is the net charge after 6.0×10^{13} electrons have been placed on the sphere?

3. The nucleus of the helium atom contains two protons that

are separated by about 3.0×10^{-15} m. Find the magnitude of the electrostatic force between the protons. (The protons remain together in the nucleus because the repulsive electrostatic force is balanced by an attractive force called the strong nuclear force.)

4. Two very small spheres are initially neutral and separated by a distance of 0.50 m. Suppose that 3.0×10^{13} electrons are removed from one sphere and placed on the other. (a) What is the magnitude of the electrostatic force that acts on each sphere? (b) Is the force attractive or repulsive? Why?

5. An object has a mass of 215 kg. Suppose this object and the earth each carried an identical positive charge q. Assuming the earth's charge is located at the center of the earth (radius $= 6.38 \times 10^6$ m), determine q such that the electrostatic force exactly cancels the gravitational force.

6. Three charges are located on the $+x$ axis as follows: $q_1 = +25\ \mu$C at $x = 0$, $q_2 = +11\ \mu$C at $x = +2.0$ m, and $q_3 = +45\ \mu$C at $x = +3.5$ m. (a) Find the electrostatic force (magnitude and direction) acting on q_2. (b) Suppose q_2 were $-11\ \mu$C, rather than $+11\ \mu$C. Without performing any detailed calculations, specify the magnitude and direction of the force exerted on q_2. Give your reasoning.

7. A charge of $-3.00\ \mu$C is fixed at the center of a compass. Two additional charges are fixed on the circle of the compass (radius $= 0.100$ m). The charges on the circle are $-4.00\ \mu$C at the position due north and $+5.00\ \mu$C at the position due east. What is the magnitude and direction of the net electrostatic force acting on the charge at the center?

8. An equilateral triangle has sides of 0.15 m. Charges of $-9.0\ \mu$C, $+8.0\ \mu$C, and $+2.0\ \mu$C are located at the corners of the triangle. Find the magnitude and direction of the net electrostatic force exerted on the 2.0-μC charge.

* 9. Two particles, with identical positive charges and a separation of 2.60×10^{-2} m, are released from rest. Immediately after the release, particle 1 has an acceleration $\mathbf{a}_1$ whose magnitude is 4.60×10^3 m/s^2, while particle 2 has an acceleration $\mathbf{a}_2$ whose magnitude is 8.50×10^3 m/s^2. Particle 1 has a mass of 6.00×10^{-6} kg. Find (a) the charge on each particle and (b) the mass of particle 2.

* 10. A charge $+q$ is fixed to each of three corners of a square. On the empty corner a charge is placed, such that there is no net electrostatic force acting on the diagonally opposite charge. What charge (magnitude and sign) is placed on the empty corner? Express your answer in terms of q.

* 11. Two small objects A and B are fixed in place and separated by 2.00 cm. Object A has a charge of $+1.00\ \mu$C, and object B has a charge of $-1.00\ \mu$C. How many electrons must be removed from A and put onto B to make the electrostatic force between the two objects an attractive force whose magnitude is 45.0 N?

* 12. Two spheres are mounted on identical horizontal springs and rest on a frictionless table, as in the drawing. When the spheres are uncharged, the spacing between them is 5.0 cm, and the springs are unstressed. When each sphere has a charge of $+1.6\ \mu$C, the spacing doubles. Assuming the spheres have a negligible diameter, determine the spring constant of the springs.

**13. There are four charges, each with a magnitude of 2.0 μC. Two are positive, and two are negative. The charges are fixed to the corners of a 0.30-m square, one to a corner, in such a way that the force on any charge is directed toward the center of the square. Find the magnitude of the net electrostatic force experienced by any charge.

**14. Two identical, very small insulating balls are suspended by separate 0.25-m threads that are attached to a common point on the ceiling. Each ball has a mass of 8.0×10^{-4} kg. Initially the balls are uncharged and hang straight down. They are then given identical positive charges and, as a result, spread apart with an angle of 36° between the threads. Determine (a) the magnitude of the charge on each ball and (b) the tension in the threads.

**15. Two charges, when combined, give a total charge of $+9.00\ \mu$C. When the charges are separated by 3.00 m, the force exerted by one charge on the other has a magnitude of 8.00×10^{-3} N. Find the amount of each charge, assuming (a) that each is positive and (b) that they have opposite polarities.

**16. Two objects are identical and small enough that their sizes can be ignored relative to the distance between them, which is 0.200 m. Each object carries a different charge, and they attract each other with a force of 1.20 N. The objects are brought into contact, so the net charge is shared equally, and then they are returned to their initial positions. Now it is found that the objects repel one another with a force whose magnitude is equal to that of the initial attractive force. What is the initial charge on each object? Note that there are two answers.

Section 23.6 The Electric Field, Section 23.7 Electric Field Lines

17. An electric field with a magnitude of 160 N/C exists at a spot that is 0.15 m away from a charge. At a place that is 0.45 m away from this charge, what is the electric field strength?

18. The electric field at a distance of 0.50 m from a charge is 9.0×10^5 N/C, directed toward the charge. Find the magnitude and polarity of the charge.

19. A charge of 3.0×10^{-5} C is located at a spot where the environment produces an electric field that points due east and has a magnitude of 15 000 N/C. What are the magnitude and direction of the force acting on the charge, assuming the charge is (a) negative and (b) positive?

20. The helium atom contains two protons in its nucleus. When helium is singly ionized, one of its two electrons is completely removed. In the Bohr model of the singly ionized helium atom, the remaining electron orbits the nucleus at a radius of 2.65×10^{-11} m. (a) What is the electric field (magnitude and direction) created at the location of the electron by the positively charged nucleus? (b) What is the electrostatic force (magnitude and direction) acting on the electron? (c) Find the speed of the electron.

21. The magnitude of the electric field between the plates of a parallel plate capacitor is 2.4×10^5 N/C. Each plate carries a charge whose magnitude is 0.15 μC. What is the area of each plate?

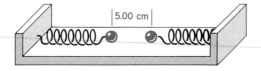

|5.00 cm|

22. A charge of $+3.5 \mu C$ is fixed on the x axis at $x = +0.55$ m, while a charge of $-15 \mu C$ is fixed at the origin. (a) Determine the net electric field (magnitude and direction) on the x axis at $x = +0.80$ m. (b) What force (magnitude and direction) would act on a charge of $-8.0 \mu C$ placed on the x axis at $x = +0.80$ m?

23. Two charges each have a magnitude of $0.36 \mu C$, one being positive and the other negative. These charges are fixed to the corners of an equilateral triangle, 0.75 m on a side. Determine the net electric field (magnitude and direction) that exists at the empty corner of the triangle.

* 24. A small object has a mass of 2.0×10^{-3} kg and a charge of $-25 \mu C$. It is placed at a certain spot where there is an electric field. When released, the object experiences an acceleration of 3.5×10^3 m/s^2 in the direction of the $+x$ axis. Determine the magnitude and direction of the electric field at this spot.

* 25. A negative charge $-q$ is fixed to one corner of a rectangle, as in the drawing. What positive charge must be fixed to corner A and what positive charge must be fixed to corner B, so the total electric field at the remaining corner is zero? Express your answers in terms of q.

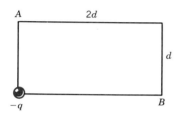

* 26. The magnitude of the electric field between the plates of a parallel plate capacitor is 480 N/C. A silver dollar is placed between the plates and oriented parallel to the plates. (a) Ignoring the edges of the coin, find the induced charge density on each face of the coin. (b) Assuming the coin has a radius of 1.9 cm, find the total charge on each face of the coin.

* 27. The drawing shows two positive charges q_1 and q_2 fixed to a circle. At the center of the circle they produce a net electric field that is directed upward along the vertical axis. Determine the ratio q_2/q_1.

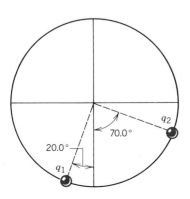

**28. A rectangle has sides L and $3L$. Two charges, with magnitudes q_1 and q_2, are located on opposite corners of this rectangle, as in the drawing. The electric field at corner A is directed along the diagonal. (a) Decide whether the charges have the same or different polarities. (b) Find the ratio q_2/q_1 of the charge magnitudes.

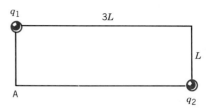

**29. A small plastic ball of mass 6.50×10^{-3} kg and charge $+0.150 \mu C$ is suspended from an insulating thread and hangs between the plates of a capacitor (see the drawing). The ball is in equilibrium, with the thread making an angle of 30.0° with respect to the vertical. The area of each plate is 0.0150 m^2. What is the magnitude of the charge on each plate?

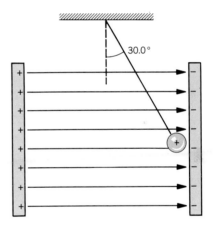

ADDITIONAL PROBLEMS

30. Two charges attract each other with a force of 1.5 N. What will be the force if the distance between them is (a) reduced to one-ninth of its original value and (b) increased to five times its original value?

31. Three positive charges are fixed along a line. From left to right they are q_1, q_2, and q_3. The charge q_2 is situated one-fourth of the way between q_1 and q_3 and experiences no net electrostatic force. Find the ratio q_3/q_1.

32. A tiny 12-g ball carries a charge of $-18 \mu C$. What electric field (magnitude and direction) is needed to cause the ball to float above the ground?

33. Three charges are fixed to an xy coordinate system. Charge $q_1 = +18 \ \mu C$ is on the y axis at $y = +3.0$ m. Charge $q_2 = -12 \ \mu C$ is at the origin. Charge $q_3 = +45 \ \mu C$ is on the x axis at $x = +3.0$ m. (a) Determine the magnitude and direction of the net electrostatic force on q_3. (b) Without doing any detailed calculations, find the net force (magnitude and direction) that q_3 alone exerts on the combination of q_1 and q_2.

34. Two charges, -16 and $+4.0 \ \mu C$, are fixed in place and separated by 3.0 m. (a) At what spot along the line between the charges is the net electric field zero? Locate this spot relative to the positive charge. *(Hint: The spot does not necessarily lie between the two charges.)* (b) What would be the force on a charge of $+14 \ \mu C$ placed at this spot?

35. Two tiny spheres have the same mass and carry charges of the same magnitude. The mass of each sphere is 2.0 mg. The gravitational force acting between the spheres is exactly balanced by the electric force. (a) What polarities do the charges have? (b) Determine the charge magnitude.

* 36. A particle of mass 3.8×10^{-5} kg and charge $+12 \ \mu C$ is released from rest in a region where there is a constant electric field of 470 N/C. (a) What is the acceleration of the particle due to the electric field? (b) How fast is the particle moving after traveling 2.0×10^{-2} m?

* 37. In the rectangle in the drawing, a charge is to be placed at the empty corner to make the net force on the charge at corner A point along the vertical direction. What charge (magnitude and sign) must be placed at the empty corner?

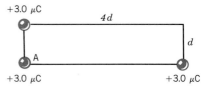

+3.0 μC

4 d

d

A

+3.0 μC

+3.0 μC

**38. A small spherical insulator of mass 8.00×10^{-2} kg and charge $+0.600 \ \mu C$ is hung by a thin wire of negligible mass. A charge of $-0.900 \ \mu C$ is held 0.150 m away from the sphere and directly to the right of it, so the wire makes an angle θ with the vertical (see drawing). Find (a) the angle θ and (b) the tension in the wire.

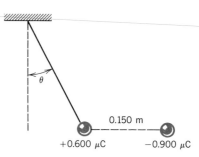

θ

0.150 m

+0.600 μC

$-0.900 \ \mu$C

Electric Potential Energy and the Electric Potential

Electricity is transmitted over long distances by high-voltage lines. Voltage and its relation to electric energy are discussed in this chapter.

24.1 POTENTIAL ENERGY

GRAVITATIONAL POTENTIAL ENERGY

In Section 7.5 it is pointed out that the gravitational force is a conservative force and that a potential energy can always be associated with a conservative force. The electric force between two point charges, $F = kq_1q_2/r^2$, is identical in form to the gravitational force between two point masses, $F = Gm_1m_2/r^2$, and the electric force is also a conservative force. Thus, an electric potential energy can be defined whose properties are similar to those of the gravitational potential energy. To set the stage for a discussion of electric potential energy, let us review some of the important aspects of gravitational potential energy.

Figure 24.1 shows a ball of mass m located at point A, which is at a height h_A above the surface of the earth. Relative to the surface, the gravitational potential energy

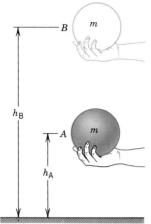

FIGURE 24.1 The gravitational potential energy of the ball at point A is $\text{GPE}_A = mgh_A$. Work W_{AB} (done by the hand) is required to raise the ball from A to B, where the gravitational potential energy is $\text{GPE}_B = mgh_B$. The work is $W_{AB} = \text{GPE}_B - \text{GPE}_A$.

GPE* of the ball is given by Equation 7.6 as $\text{GPE}_A = mgh_A$, where g is the acceleration due to gravity. This expression assumes the height is small compared to the radius of the earth, so g is essentially constant at all heights. If the ball is to be lifted to a higher point B, an external force must be applied to the ball to compensate for the gravitational force. The hand in the drawing provides the external force. When the ball is lifted at a constant speed, there is no change in its kinetic energy KE, and the work W_{AB} done on the ball by the external force goes entirely into changing the gravitational potential energy:

$$W_{AB} = \underbrace{(\text{KE}_B - \text{KE}_A)}_{= 0} + (\text{GPE}_B - \text{GPE}_A) = mgh_B - mgh_A \qquad (7.7a)$$

If the ball is released from rest at B, it is accelerated toward the earth by the gravitational force. As the ball falls, its potential energy decreases and its kinetic energy increases. In the absence of dissipative forces such as friction, the total mechanical energy is conserved; that is, the sum of the gravitational potential energy and the kinetic energy remains constant at all times during the fall.

ELECTRIC POTENTIAL ENERGY

Electric potential energy is analogous to gravitational potential energy. Consider Figure 24.2, which shows a positive test charge $+q_0$ situated between two oppositely charged plates. Because of the charges on the plates, the test charge experiences an electric force that is directed toward the lower plate (the gravitational force is being neglected here). The hand in the drawing provides the external force needed to compensate for the electric force and move the test charge from A to B. When q_0 is moved at constant speed, its kinetic energy does not change, and the work W_{AB} done on q_0 by the external force goes into changing the electric potential energy. The work equals the difference between the electric potential energy EPE at B and that at A:

$$W_{AB} = \text{EPE}_B - \text{EPE}_A \qquad (24.1)$$

Because the electric force is a conservative force, the path along which the charge is moved from A to B is of no consequence, for the work W_{AB} is the same for all paths.

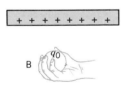

FIGURE 24.2 At locations A and B, the test charge $+q_0$ has electric potential energies of EPE_A and EPE_B. The work done in moving the test charge from A to B at a constant speed is $W_{AB} = \text{EPE}_B - \text{EPE}_A$.

24.2 THE ELECTRIC POTENTIAL DIFFERENCE

The work done to move the charge from A to B in Figure 24.2 depends on the magnitude of the charge, because the strength of the electric force opposing the motion depends on the magnitude of the charge. It is useful to express the work on a per-unit-charge basis by dividing both sides of Equation 24.1 by the charge q_0:

$$\frac{W_{AB}}{q_0} = \frac{\text{EPE}_B - \text{EPE}_A}{q_0} \qquad (24.2)$$

The electric potential energy per unit charge is an important concept in electricity and is known as the *electric potential,* or, simply, the *potential.*

---

* The gravitational potential energy is denoted by GPE to distinguish it from the electric potential energy EPE.

DEFINITION OF ELECTRIC POTENTIAL
The electric potential V at a given point is the electric potential energy EPE of a small test charge q_0 situated at that point divided by the charge itself:

$$V = \frac{EPE}{q_0} \qquad (24.3)$$

SI unit of electric potential: joule/coulomb = volt (V)

The SI unit of electric potential is a joule per coulomb, a quantity known as a *volt,* so named for Alessandro Volta (1745–1827) who invented the voltaic pile, the forerunner of the battery. Note that, in spite of the similarity in names, the electric potential energy EPE and the electric potential V are *not* the same. The electric potential energy, as its name implies, is an *energy* and, therefore, is measured in joules. In contrast, the electric potential is an *energy per unit charge,* and is measured in joules per coulomb, or volts.

According to Equations 24.2 and 24.3, the electric potential difference between two points A and B is related to the work per unit charge in the following manner:

$$V_B - V_A = \frac{EPE_B - EPE_A}{q_0} = \frac{W_{AB}}{q_0} \qquad (24.4)$$

Often, the "delta" notation is used to express the difference in potentials and the difference in potential energies: $\Delta V = V_B - V_A$ and $\Delta(EPE) = EPE_B - EPE_A$. In terms of this notation, Equation 24.4 takes the form

$$\Delta V = \frac{\Delta(EPE)}{q_0} = \frac{W_{AB}}{q_0} \qquad (24.4)$$

Equation 24.4 indicates that the work W_{AB} is related to the *difference* in potential, $V_B - V_A$, or to the *difference* in potential energy, $EPE_B - EPE_A$. Neither the potential V nor the potential energy EPE can be determined in an absolute sense, for only the differences are measurable. The gravitational potential energy has this same characteristic, since only the value at one height relative to that at some reference height has any significance. Example 1 emphasizes the relative nature of the electric potential.

EXAMPLE 1
In Figure 24.2, $+5.0 \times 10^{-5}$ J of work are performed in moving the test charge ($q_0 = +2.0 \times 10^{-6}$ C) at a steady speed from A to B. (a) Find the difference in the electric potential energies of the charge between the two points. (b) Determine the potential difference between the two points.

SOLUTION
(a) The difference in potential energy between points A and B is equal to the work done in moving the charge from A to B. Therefore, $EPE_B - EPE_A = W_{AB} = \boxed{+5.0 \times 10^{-5} \text{ J}}$.

(b) The potential difference between A and B is the difference in potential energy divided by the charge

$$V_B - V_A = \frac{EPE_B - EPE_A}{q_0} = \frac{+5.0 \times 10^{-5} \text{ J}}{+2.0 \times 10^{-6} \text{ C}} \qquad (24.4)$$

$$= \boxed{+25 \text{ V}}$$

The electric potential at B exceeds that at A by 25 V. It is not possible to determine separate values for V_B and V_A.

In Example 1, energy in the form of work is expended in moving the positive charge from the lower potential at A to the higher potential at B. According to the principle of the conservation of energy, this energy does not disappear. If the positive charge is released at B, it accelerates toward A because of the repulsion from the upper plate and the attraction to the lower plate. The speed of the charge increases as electric potential energy is converted into kinetic energy. Thus, *a positive charge accelerates from a region of higher potential toward a region of lower potential.* A negative charge behaves in the opposite fashion, since the electric force acting on it is directed opposite to that on a positive charge. *A negative charge accelerates from a region of lower potential toward a region of higher potential.* The next example illustrates the way positive and negative charges behave.

EXAMPLE 2

In Figure 24.2, point B has an electric potential that is 25 V greater than that at point A, so $V_B - V_A = 25$ V. A particle has a mass of 1.8×10^{-5} kg and a charge whose magnitude is 3.0×10^{-5} C. The effects of gravity and friction are negligible. (a) If the particle has a positive charge and is released from rest at B, what speed v_A does the particle have when it arrives at A? (b) If the particle has a negative charge and is released from rest at A, what speed v_B does the particle have at B?

SOLUTION

(a) The only force acting on the moving charge is the conservative electric force. Therefore, the total energy of the charge (the sum of the kinetic energy KE and the electric potential energy EPE) is the same at points A and B:

$$KE_A + EPE_A = KE_B + EPE_B$$

Since $KE_A = \frac{1}{2}mv_A^2$ and the particle is at rest at point B, it follows that $\frac{1}{2}mv_A^2 = EPE_B - EPE_A$. The difference in po-

tential energies is related to the difference in potentials by Equation 24.4, $EPE_B - EPE_A = q_0(V_B - V_A)$. As a result,

$$\frac{1}{2}mv_A^2 = q_0(V_B - V_A)$$

$$v_A = \sqrt{\frac{2q_0(V_B - V_A)}{m}} = \sqrt{\frac{2(3.0 \times 10^{-5} \text{ C})(25 \text{ V})}{1.8 \times 10^{-5} \text{ kg}}}$$

$$= \boxed{9.1 \text{ m/s}}$$

(b) A negative charge accelerates from a region of lower potential toward a region of higher potential. Therefore, when the negatively charged particle is released from rest at point A, it accelerates toward point B. The magnitude of the electric force causing this acceleration is the same as it is in part (a), but here the direction of the force is reversed. Thus, the speed v_B of the particle at point B is the same as that calculated in part (a), $\boxed{v_B = 9.1 \text{ m/s}}$.

As a familiar application of electric potential energy and electric potential, Figure 24.3 shows a 12-V automobile battery with a headlight connected between the battery terminals. The positive terminal has a potential that is 12 V higher than the potential at the negative terminal. Positive charges are repelled from the positive terminal and travel through the wires and headlight toward the negative terminal.* As the charges

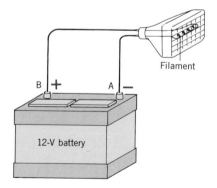

FIGURE 24.3 A headlight connected to a 12-V battery.

---

* Historically, it was believed that positive charges flow in the wires of an electric circuit. Today, it is known that negative charges flow in wires from the negative toward the positive terminal. However, it is customary to describe the flow of negative charges by specifying the opposite but equivalent flow of positive charges.

pass through the headlight, virtually all their potential energy is converted into thermal energy that heats up the filament, causing it to glow "white hot" and emit light. When the charges reach the negative terminal, they no longer have any potential energy. The battery then gives the charges an additional "shot" of potential energy by moving them to the higher-potential positive terminal, and the cycle is repeated. In raising the potential energy of the charges, the battery does work W_{AB} on them, and draws from its reserve of chemical energy to do so. Example 3 illustrates the concept of the electric potential difference as applied to a battery.

EXAMPLE 3

Determine the number of particles, each carrying a charge of 1.60×10^{-19} C (the magnitude of the charge on an electron), that passes between the terminals of a 12-V car battery when a 60.0-W headlight burns for one hour.

SOLUTION

To obtain the number of charged particles, we determine the total charge needed to provide the energy consumed by the headlight in one hour. Dividing the total charge by the charge on each particle gives the number of particles. At a rate of 60.0 joules per second (60.0 watts) for one hour, the headlight consumes a total energy of

$$\text{Energy} = \text{Power} \times \text{Time} = (60.0 \text{ W})(3600 \text{ s}) \quad (7.10)$$
$$= 2.2 \times 10^5 \text{ J}$$

Equation 24.4 gives the total amount of charge that delivers this energy upon passing through a 12-V potential difference:

$$q_0 = \frac{\Delta(\text{EPE})}{\Delta V} = \frac{2.2 \times 10^5 \text{ J}}{12 \text{ V}} = 1.8 \times 10^4 \text{ C}$$

The number of particles whose individual charges combine to provide this total charge is $(1.8 \times 10^4 \text{ C})/(1.60 \times 10^{-19} \text{ C}) = \boxed{1.1 \times 10^{23}}$.

As used in connection with batteries, the volt is a familiar unit for measuring electric potential difference. The word "volt" also appears in another context, as part of a unit that is used to measure energy, particularly the energy of an atomic particle, such as an electron or a proton. This energy unit is called the *electron volt* (eV). **One electron volt is the change in potential energy of an electron ($q_0 = 1.60 \times 10^{-19}$ C) when the electron moves through a potential difference of one volt.** Since the change in potential energy equals $q_0 \Delta V$, one electron volt is equal to $(1.60 \times 10^{-19}$ C) $\times (1.00 \text{ V}) = 1.60 \times 10^{-19}$ J; thus

$$1 \text{ eV} = 1.60 \times 10^{-19} \text{ J}$$

One million or 10^{+6} electron volts of energy is referred to as one MeV, and one billion or 10^{+9} electron volts of energy is one GeV, where the "G" stands for the prefix "giga."

24.3 THE ELECTRIC POTENTIAL DIFFERENCE CREATED BY POINT CHARGES

A SINGLE POINT CHARGE

A point charge creates an electric potential in a fashion that Figure 24.4 helps to explain. This picture shows a positive point charge q and two locations A and B, situated at distances r_A and r_B from the charge. An electrostatic force of repulsion acts on a positive test charge $+q_0$ located at any position between these spots, the magnitude of the force being given by Coulomb's law as $F = kq_0q/r^2$. Thus, to move the test charge from A to B at a constant speed, some external agent must apply a force $F = kq_0q/r^2$ to balance the electric force. Since r varies between r_A and r_B, F also

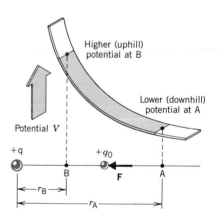

FIGURE 24.4 The positive test charge $+q_0$ experiences a repulsive force due to the positive point charge q. As a result, a force **F** must be applied to move the test charge from A to B. Since **F** points in the same direction as the displacement, positive work is done by the force in moving the test charge from A to B. Consequently, the electric potential is higher (uphill) at B and lower (downhill) at A.

varies, and therefore the work done on q_0 is not simply the product of F and the distance between the points. However, the work can be found by using integral calculus. According to Equation 24.4, dividing the work W_{AB} by q_0 gives the potential difference between B and A. The result is

$$V_B - V_A = \frac{W_{AB}}{q_0} = \frac{kq}{r_B} - \frac{kq}{r_A} \qquad (24.5)$$

As point A is located farther and farther away from the charge q, r_A becomes larger and larger, and the term kq/r_A becomes zero when r_A becomes infinitely large; thus, $V_B = kq/r_B$. In this limit, it is customary to omit the subscripts and write the potential in the following form:

$$\begin{bmatrix} \text{Potential of} \\ \text{a point} \\ \text{charge} \end{bmatrix} \qquad V = \frac{kq}{r} \qquad (24.6)$$

The symbol V in Equation 24.6, while often called the potential, does not refer to the potential in any absolute sense. Rather, $V = kq/r$ stands for the amount by which the potential at a distance r from a point charge differs from the potential at an infinite distance away. In other words, V refers to a potential with the arbitrary assumption that the potential at infinity is zero. Unless otherwise specified, we will always make this assumption.

With the aid of Equation 24.6, it is possible to describe the effect that a point charge q has on the surrounding space. When q is positive, the value of $V = kq/r$ is also positive, indicating the positive charge has everywhere raised the potential above the zero reference value. Conversely, when q is negative, the potential V is also negative, indicating the negative charge has everywhere decreased the potential below the zero reference value. The next example deals with these effects quantitatively.

EXAMPLE 4

Using a zero reference potential at infinity, determine the amount by which a point charge of 4.0×10^{-8} C alters the electric potential at a spot 1.2 m away when the charge is (a) positive and (b) negative.

SOLUTION

(a) Figure 24.5a shows the potential when the charge is positive:

$$V = \frac{kq}{r} = \frac{(8.99 \times 10^9 \text{ N} \cdot \text{m}^2/\text{C}^2)(+4.0 \times 10^{-8} \text{ C})}{1.2 \text{ m}} \qquad (24.6)$$

$$= \boxed{+300 \text{ V}}$$

(b) Part b of the drawing illustrates the results when the charge is negative. A calculation similar to that in part (a) shows the potential is now negative: $\boxed{-300 \text{ V}}$.

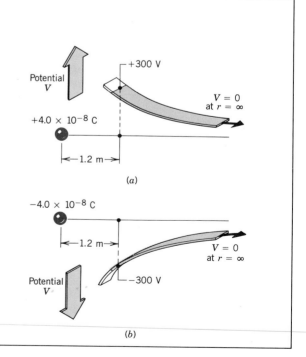

FIGURE 24.5 A point charge of 4.0×10^{-8} C alters the potential at a spot 1.2 m away. The potential is (a) increased by 300 V when the charge is positive and (b) decreased by 300 V when the charge is negative, assuming a zero reference potential at infinity.

The electric potential created by a point charge plays an important role in the structure of atoms. Example 5 illustrates this role in the case of the hydrogen atom.

EXAMPLE 5

In the Bohr model of the hydrogen atom, the electron is in an orbit around the nuclear proton at a distance of 5.29×10^{-11} m, as Figure 24.6 shows. Find (a) the electric potential that the proton creates at this distance, (b) the total energy of the atom, and (c) the ionization energy for the atom. The ionization energy is the energy that must be put into the atom to remove the electron and place it at rest infinitely far from the proton. Express the answers to parts (b) and (c) in electron volts.

SOLUTION

(a) The electric potential created by the proton charge of $+1.60 \times 10^{-19}$ C is

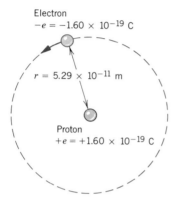

Electron
$-e = -1.60 \times 10^{-19}$ C

$r = 5.29 \times 10^{-11}$ m

Proton
$+e = +1.60 \times 10^{-19}$ C

FIGURE 24.6 In the Bohr model of the hydrogen atom, the electron orbits the proton at a distance of 5.29×10^{-11} m. The total energy of the atom is the sum of the electric potential energy and the kinetic energy of the electron.

$$V = \frac{kq}{r} = \frac{(8.99 \times 10^9 \ \text{N} \cdot \text{m}^2/\text{C}^2)(+1.60 \times 10^{-19} \ \text{C})}{5.29 \times 10^{-11} \ \text{m}}$$ (24.6)

$$= \boxed{+27.2 \ \text{V}}$$

(b) The total energy of the hydrogen atom is the sum of the electric potential energy and the kinetic energy of the electron. The potential energy, relative to a zero reference value at infinity, can be found from Equation 24.4 by using the charge on the electron and the potential of $+27.2$ V determined in part (a):

$$\text{EPE} = qV = (-1.60 \times 10^{-19} \ \text{C})(+27.2 \ \text{V})$$

$$= -4.35 \times 10^{-18} \ \text{J}$$

From Example 2 in Section 23.5, we know that the speed of the electron is 2.18×10^6 m/s. Since the mass of the electron is 9.11×10^{-31} kg, its kinetic energy is $\text{KE} = \frac{1}{2}mv^2 = 2.17 \times 10^{-18}$ J. The total energy E of the hydrogen atom is

$$E = \text{EPE} + \text{KE} = -4.35 \times 10^{-18} \ \text{J} + 2.17 \times 10^{-18} \ \text{J}$$

$$= -2.18 \times 10^{-18} \ \text{J}$$

The total energy in electron volts (1 eV $= 1.60 \times 10^{-19}$ J) is $\boxed{-13.6 \ \text{eV}}$.

(c) Removing the electron from its orbit and placing the electron at rest at infinity requires $+13.6$ eV of energy. Then, the total energy will be zero, the assumed reference value for the potential energy. Thus, the ionization energy of the hydrogen atom is $\boxed{+13.6 \ \text{eV}}$, which agrees with the experimental value.

POTENTIAL OF MULTIPLE POINT CHARGES

A single point charge raises or lowers the potential at a given location, depending on whether the charge is positive or negative. When two or more charges are present, the potential due to all the charges is obtained by adding together the individual potentials, as Example 6 shows.

EXAMPLE 6

At the locations A and B in Figure 24.7, find the total electric potential due to the two point charges.

SOLUTION

At every location, each of the charges contributes to the total electric potential. The individual contributions are determined

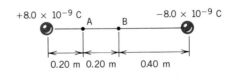

$+8.0 \times 10^{-9}$ C -8.0×10^{-9} C

0.20 m 0.20 m   0.40 m

FIGURE 24.7 Both the positive and negative charges affect the electric potential at locations A and B.

by using $V = kq/r$. To obtain the total potential, the individual contributions are added algebraically.

Location	Contribution from + charge	Contribution from − charge	Total potential
A	$\dfrac{(8.99 \times 10^9 \text{ N} \cdot \text{m}^2/\text{C}^2)(+8.0 \times 10^{-9} \text{ C})}{0.20 \text{ m}}$ +	$\dfrac{(8.99 \times 10^9 \text{ N} \cdot \text{m}^2/\text{C}^2)(-8.0 \times 10^{-9} \text{ C})}{0.60 \text{ m}}$ =	$\boxed{+240 \text{ V}}$
B	$\dfrac{(8.99 \times 10^9 \text{ N} \cdot \text{m}^2/\text{C}^2)(+8.0 \times 10^{-9} \text{ C})}{0.40 \text{ m}}$ +	$\dfrac{(8.99 \times 10^9 \text{ N} \cdot \text{m}^2/\text{C}^2)(-8.0 \times 10^{-9} \text{ C})}{0.40 \text{ m}}$ =	$\boxed{0}$

At A the total potential is positive, because this spot is closer to the positive charge, whose effect dominates over that of the more distant negative charge. At B, midway between the charges, the total potential is zero, since the potential of one charge exactly offsets that of the other.

24.4 EQUIPOTENTIAL SURFACES AND THEIR RELATION TO THE ELECTRIC FIELD

EQUIPOTENTIAL SURFACES

An *equipotential surface* is a surface on which the electric potential is the same everywhere. The easiest equipotential surfaces to visualize are those that surround an isolated point charge. According to Equation 24.6, the potential at a distance r from a point charge q is $V = kq/r$. Thus, wherever r is the same, the potential is the same, and the equipotential surfaces are spherical surfaces centered on the charge. There are an infinite number of such surfaces, one for every value of r, and Figure 24.8 illustrates two of them. The larger the distance r, the smaller the potential of the equipotential surface.

No work is required to move a charge at constant speed on an equipotential surface. This important characteristic arises because when work is done on an object, either the potential energy or the kinetic energy of the object changes. Since both types of energy remain unchanged when a charge is moved at constant speed on an equipotential surface, no work is needed to accomplish the motion. In Figure 24.8, for instance, no work is done in moving a test charge at constant speed along the circular arc ABC. The only force that must be applied to the test charge is that needed to counteract the electric force produced by q, and this applied force, being perpendicular to the path ABC, does no work. (See Section 7.1 for a review of work.) In contrast,

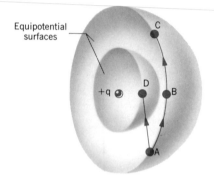

Equipotential surfaces

FIGURE 24.8 The equipotential surfaces that surround the point charge $+q$ are spherical, as are the two shown here. No work is required to move a charge at a constant speed on a path that lies on an equipotential surface, such as the path ABC. However, work is required to move a charge between two equipotential surfaces, as along the path AD.

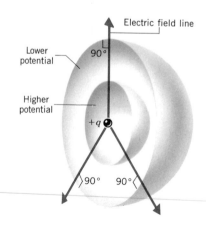

FIGURE 24.9 The radially directed electric field of a point charge is perpendicular to the spherical equipotential surfaces that surround the charge. The electric field points in the direction of *decreasing* potential, from a region of higher potential to one of lower potential.

work must be done to move a charge at a constant speed *between* equipotential surfaces, as from A to D in the picture. The work is the product of the charge and the difference between the potentials of the surfaces, in accord with Equation 24.4.

The spherical equipotential surfaces that surround an isolated point charge illustrate another characteristic of such surfaces. Figure 24.9 shows two equipotential surfaces around a positive point charge, along with some electric field lines. The electric field lines give the direction of the electric field, and for a positive point charge, the electric field is directed radially outward. Therefore, at each location on an equipotential sphere the electric field is perpendicular to the surface and points outward in the direction of decreasing potential, as the drawing emphasizes. This perpendicular relation is valid whether or not the shape of the equipotential surface is spherical; *the electric field created by any group of charges is always perpendicular to the associated equipotential surfaces and points in the direction of decreasing potential.* For example, Figure 24.10 shows the electric field lines around an electric dipole,

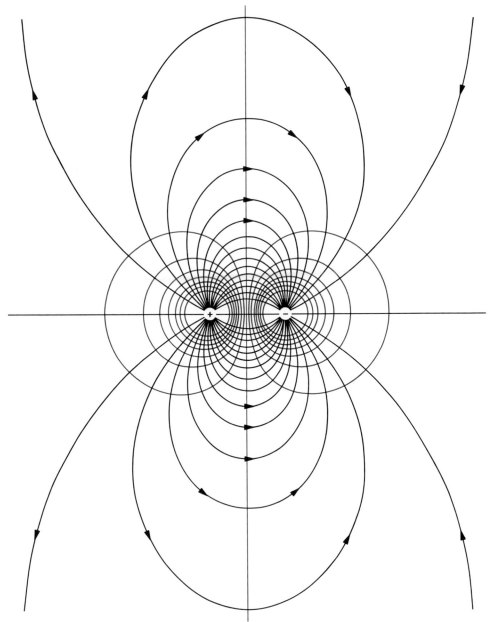

FIGURE 24.10 A cross-sectional view of the equipotential surfaces (in color) of an electric dipole. The surfaces are drawn so that at every point they are perpendicular to the electric field lines (in black) of the dipole.

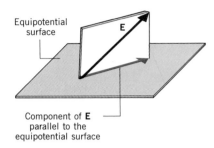

Equipotential surface

E

Component of **E** parallel to the equipotential surface

FIGURE 24.11 In the hypothetical situation shown here, the electric field **E** is not perpendicular to the equipotential surface. As a result, there would be a component of **E** parallel to the surface.

along with some equipotential surfaces. Since the field lines are not simply radial, the equipotential surfaces are no longer spherical, but, instead, have the necessary shape so as to be everywhere perpendicular to the field lines.

To see why an equipotential surface must be perpendicular to the electric field, consider Figure 24.11, which shows a hypothetical situation in which the perpendicular relation does *not* hold. If **E** were not perpendicular to the equipotential surface, there would be a component of **E** parallel to the surface. This field component would exert an electric force on a test charge placed on the surface, and work would have to be done to counteract the effect of this force to move the test charge at a constant speed along the surface. Thus, the surface could not be an equipotential surface as assumed. The only way out of the dilemma is for the electric field to be perpendicular to the surface, so there is no component of the field parallel to the surface.

The surface of any *conductor* is an equipotential surface when the conductor is at equilibrium under electrostatic conditions. Then, as discussed in Section 23.8, the direction of the electric field just outside the conductor is perpendicular to the surface of the conductor. Consequently, there is no component of the electric field parallel to the surface, so the potential on the surface is the same everywhere. In fact, since the electric field is zero everywhere inside a conductor whose charges are in equilibrium, the entire conductor can be regarded as an equipotential volume.

THE RELATION BETWEEN THE ELECTRIC FIELD AND THE ELECTRIC POTENTIAL

There is a quantitative relation between the electric field and the equipotential surfaces. To understand this relation, consider the parallel plate capacitor in Figure 24.12. As Section 23.6 discusses, the electric field **E** between the metal plates is perpendicular to the plates and is the same everywhere, ignoring fringe fields at the edges. To be perpendicular to the electric field, the equipotential surfaces must be planes that are parallel to the plates, which themselves are equipotential surfaces. The potential difference between the plates is given by Equation 24.4 as $V_B - V_A = \Delta V = W_{AB}/q_0$, where A is a point on the negative plate and B is a point on the positive plate. The work required to move a positive test charge q_0 from A to B is $W_{AB} = F \Delta s$, where F is the magnitude of the applied force, directed to the left in opposition to the electric force, and Δs is the magnitude of the displacement along a line perpendicular to the plates. If the test charge is moved at a constant speed, the magnitude of the applied force equals the magnitude of the electric force, so $F = q_0 E$, and the work becomes $W_{AB} = F \Delta s = q_0 E \Delta s$. The potential difference between the capacitor plates is the work per unit charge, so

Equipotential surfaces

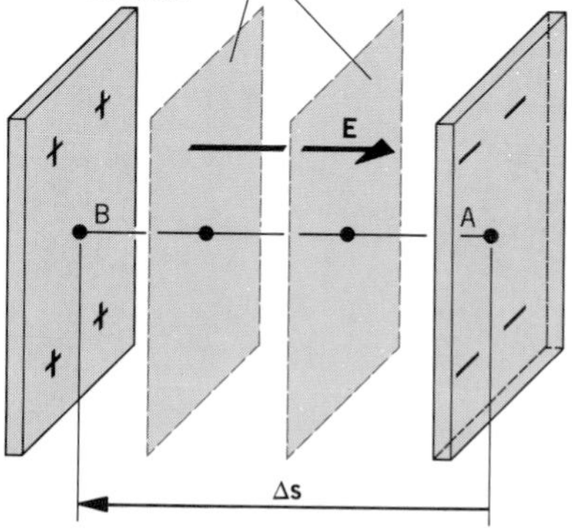

FIGURE 24.12 The metal plates of a parallel plate capacitor are equipotential surfaces. Two additional equipotential surfaces (planes) are shown in color between the plates. These two equipotential surfaces are parallel to the plates and are perpendicular to the electric field between the plates.

$$\Delta V = \frac{W_{AB}}{q_0} = \frac{q_0 E \Delta s}{q_0} \quad \text{or} \quad E = \frac{\Delta V}{\Delta s}$$

As the test charge is moved from the negative plate to the positive plate, the potential difference increases with distance ($\Delta V/\Delta s$ is positive). However, the electric field is in the opposite direction, for it points from the positive plate toward the negative plate. To account for the fact that the electric field is directed opposite to the direction in which the potential increases, Equation 24.7 includes a minus sign:

$$E = -\frac{\Delta V}{\Delta s} \tag{24.7}$$

The quantity $\Delta V/\Delta s$ is referred to as the *potential gradient* and has units of volts per meter. The next example deals further with the equipotential surfaces between the plates of a capacitor.

EXAMPLE 7

The plates of the capacitor in Figure 24.12 are separated by a distance of 3.2 cm, and the potential difference between them is 64 V. Between the two equipotential surfaces shown in color there is a potential difference of 3.0 V. Find the spacing between the two colored surfaces.

value for E can be obtained, since the distance and potential difference between the plates are known:

$$E = -\frac{\Delta V}{\Delta s} = -\frac{64 \text{ V}}{0.032 \text{ m}} = -2.0 \times 10^3 \text{ V/m}$$

The spacing between the colored equipotential surfaces can now be determined:

SOLUTION

To determine the spacing between the two colored equipotential surfaces, we solve $E = -\Delta V/\Delta s$ for Δs, with $\Delta V = 3.0$ V and E equal to the electric field between the plates of the capacitor. A

$$\Delta s = -\frac{\Delta V}{E} = -\frac{3.0 \text{ V}}{-2.0 \times 10^3 \text{ V/m}} = \boxed{1.5 \times 10^{-3} \text{ m}}$$

24.5 CAPACITORS AND DIELECTRICS

THE CAPACITANCE OF A CAPACITOR

A *capacitor* is a device that can store charge and consists of two conductors placed near one another, but not touching. If the conductors have the shape of parallel plates, as in Figure 24.13, the capacitor is a parallel plate capacitor. For a reason that will become clear later on, it is common practice to fill the region between the plates with an electrically insulating material called a *dielectric.* Each plate carries a charge of the *same magnitude,* one plate being positively charged and the other negatively charged. Because of the charges, the electric potential of the positive plate exceeds that of the negative plate by an amount V. Experiment shows that when the magnitude q of the charge on each plate is doubled, the electric potential difference V is also doubled, so q is proportional to V: $q \propto V$. Equation 24.8 expresses this proportionality with the aid of a proportionality constant C, which is the *capacitance* of the capacitor.

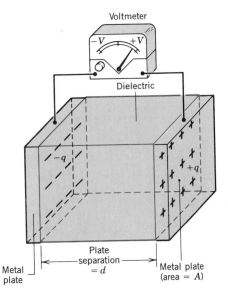

FIGURE 24.13 A parallel plate capacitor consists of two metal plates, one carrying a charge $+q$ and the other carrying a charge $-q$. The potential of the positively charged plate exceeds that of the negatively charged plate by an amount V. Often the region between the plates is filled with a dielectric.

THE RELATION BETWEEN CHARGE AND POTENTIAL DIFFERENCE FOR A CAPACITOR

The magnitude q of the charge on each plate of a capacitor is directly proportional to the magnitude V of the potential difference between the plates:

$$q = CV \tag{24.8}$$

where C is the capacitance.

SI unit of capacitance: coulomb/volt = farad (F)

Equation 24.8 shows that the SI unit of capacitance is the coulomb per volt (C/V). This unit is called the *farad* (F), named after the English scientist Michael Faraday (1791–1867). One farad is an enormous amount of capacitance. Usually smaller amounts, such as a microfarad (1 μF = 10^{-6} F) or a picofarad (1 pF = 10^{-12} F), are used in electric circuits. The capacitance reflects the ability of the capacitor to store charge, in the sense that a larger capacitance C allows more charge q to be put onto the plates for a given value of the potential difference V. Figure 24.14 shows some commercially available capacitors that have various geometries and dielectrics between their plates.

THE DIELECTRIC CONSTANT

Figure 24.15a shows the electric field lines inside an empty capacitor. The capacitor is assumed to be isolated in the sense that it is not connected to anything else, such as a battery, so the charge on each plate remains constant. If a dielectric is inserted between the plates, the capacitance can increase markedly, because of the way in which the dielectric alters the electric field between the plates. Figure 24.15b indicates how this alteration arises by showing what happens to the molecules of the dielectric. In many substances the molecules, although electrically neutral, possess a permanent dipole moment, for one part of the molecule has a slight excess of negative charge and

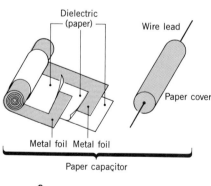

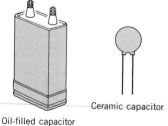

FIGURE 24.14 Various types of commercial capacitors.

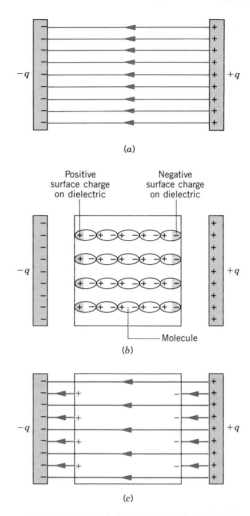

FIGURE 24.15 (*a*) The electric field lines inside an empty capacitor. (*b*) The electric field produced by the charges on the plates of the capacitor produces an end-to-end alignment of the molecular dipoles within the dielectric. (*c*) The resulting positive and negative surface charges that appear on the dielectric cause a reduction in the electric field within the dielectric. The empty space between the dielectric and the plates is intended only for clarity in showing the electric field lines. In reality, the dielectric completely fills the region between the plates.

another part has a slight excess of positive charge. When the molecules are placed between the charged plates of the capacitor, the negative ends are attracted to the positive plate and the positive ends are attracted to the negative plate, with the result that the dipolar molecules tend to orient themselves as in part *b* of the drawing. Whether or not a molecule has a permanent dipole moment, the electric field can cause the electrons to shift position within the molecule, making one end slightly negative, while leaving the opposite end slightly positive. Once again, the result is similar to that in Figure 24.15*b*; the dipoles tend to be oriented end to end. As a result of the end-to-end orientation, the extreme left surface of the dielectric becomes positively charged, and the extreme right surface becomes negatively charged. The surface charges are shown in color in the picture.

Because of the surface charges on the dielectric, not all the electric field lines generated by the charges on the plates pass through the dielectric. As Figure 24.15*c* shows, some of the field lines end on the negative surface charges and begin again on the positive surface charges. Thus, the electric field inside the dielectric is less than the electric field inside an empty capacitor, assuming the charge on the plates remains constant. This reduction in the electric field is described by the ***dielectric constant*** κ, which is the ratio of the field magnitude E_0 without the dielectric to the field magnitude E inside the dielectric:

$$\kappa = \frac{E_0}{E} \qquad (24.9)$$

Being a ratio of two field strengths, the dielectric constant is a number without units. Moreover, since the field $\mathbf{E}_0$ without the dielectric is greater than the field $\mathbf{E}$ inside the dielectric, the dielectric constant is greater than unity. The value of κ depends on the nature of the dielectric material, as Table 24.1 indicates.

THE CAPACITANCE OF A PARALLEL PLATE CAPACITOR

The capacitance of a capacitor is affected by the geometry of the plates and the dielectric constant of the material between them. For example, Figure 24.13 shows a parallel plate capacitor in which the area of each plate is A and the separation between the plates is d. The magnitude of the electric field inside the dielectric is given by Equation 24.7 (without the minus sign) as $E = V/d$, where V is the potential difference between the plates. If the charge on each plate is kept fixed, the electric field inside the dielectric is related to the electric field in the absence of the dielectric via $E = E_0/\kappa$. Therefore, $E_0/\kappa = V/d$. Since the electric field within an empty capacitor is $E_0 = q/(\epsilon_0 A)$ (see Equation 23.4), it follows that $q/(\epsilon_0 A\kappa) = V/d$, which can be solved for q to give

$$q = \left(\frac{\kappa \epsilon_0 A}{d}\right) V$$

A comparison of this expression with $q = CV$ (Equation 24.8) reveals that the capacitance C is

$$\left[\begin{array}{l}\textbf{Parallel plate}\\\textbf{capacitor filled}\\\textbf{with a dielectric}\end{array}\right] \qquad C = \frac{\kappa \epsilon_0 A}{d} \qquad (24.10)$$

Notice that only the geometry of the plates, that is, A and d, and the dielectric constant κ affect the capacitance. With C_0 representing the capacitance of the empty capacitor ($\kappa = 1$), Equation 24.10 shows that $C = \kappa C_0$. In other words, the capacitance with the dielectric present is increased by a factor of κ over the capacitance without the dielectric. It can be shown that the relation $C = \kappa C_0$ applies to any capacitor, not just to a parallel plate capacitor. One reason, then, that capacitors are filled with dielectric materials is to increase the capacitance. Example 8 deals with an important device that uses capacitors.

TABLE 24.1 Dielectric Constants of Some Common Substances[a]

Substance	Dielectric Constant
Vacuum	1
Air	1.00054
Teflon	2.1
Benzene	2.28
Paper (royal gray)	3.3
Ruby mica	5.4
Neoprene rubber	6.7
Methyl alcohol	33.6
Water	80.4

[a] Near room temperature.

EXAMPLE 8

Each key on the keyboard of a personal computer acts like a tiny electrical switch. By depressing the appropriate keys, the user closes the switches needed to enter information into the computer. One common kind of keyboard uses capacitance switching, in which each key is mounted on one end of a plunger, the other end being attached to a movable metal plate (see Figure 24.16). When a key is pressed, the movable plate is pushed closer to the fixed metal plate, the two plates forming a capacitor whose capacitance increases as a result of the reduced plate separation. Electronic circuitry enables the computer to detect the *change* in capacitance. The separation of the plates is normally 5.00×10^{-3} m, but decreases to 0.150×10^{-3} m when a key is pressed. The plate area is 9.50×10^{-5} m² and the capacitor is filled with a material whose dielectric constant is 3.50. Determine the change in capacitance that must be detected by the computer.

SOLUTION

When the key is pressed, the capacitance is

$$C = \frac{\kappa \epsilon_0 A}{d} = \frac{(3.50)[8.85 \times 10^{-12} \text{ C}^2/(\text{N}\cdot\text{m}^2)](9.50 \times 10^{-5} \text{ m}^2)}{0.150 \times 10^{-3} \text{ m}}$$

$$= 19.6 \times 10^{-12} \text{ F} \quad (19.6 \text{ pF})$$

A similar calculation shows that when the key is *not* pressed, the capacitance is 0.589×10^{-12} F (0.589 pF). The *change* in capacitance is $\boxed{19.0 \times 10^{-12} \text{ F} (19.0 \text{ pF})}$. The presence of the dielectric increases the *change* in the capacitance. The greater the change in capacitance, the easier it is for the circuitry within the computer to recognize it.

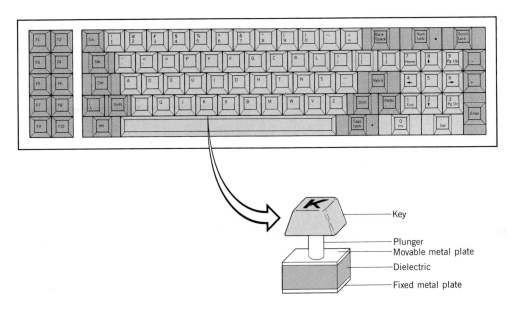

FIGURE 24.16 One common kind of computer keyboard uses capacitance switching, in which each key, when pressed, changes the separation between the plates of a capacitor.

ENERGY STORAGE IN A CAPACITOR

A capacitor is a device for storing charge. Alternatively, it is possible to view the capacitor as a device for storing electric energy. After all, the charge on the plates possesses electric potential energy, which arises because work must be done to deposit the charge on the plates. In fact, as each small increment of charge is deposited during the charging process, the potential difference between the plates increases, and a larger amount of work is needed to bring up the next increment of charge. The total work W done in charging the capacitor, and hence the total electric potential energy EPE, can be obtained from Equation 24.4 by using an average potential difference $\overline{V}$; thus, $W = \text{EPE} = q\overline{V}$. As the capacitor becomes fully charged, the potential of the positive plate relative to the negative plate increases from 0 to V. The average potential difference is $\overline{V} = \frac{1}{2}V$, so the electric potential energy stored is EPE $= \frac{1}{2}qV$. Since $q = CV$, the energy stored becomes

$$\text{Energy} = \frac{1}{2}(CV)V = \frac{1}{2}CV^2 \qquad (24.11)$$

It is also possible to regard the energy as being stored in the electric field between the plates, rather than in the potential energy of the charge on the plates. Such a viewpoint can be useful when the strength of the electric field is known, instead of the magnitude of the charge that creates the field. The relation between energy and field strength can be obtained for a parallel plate capacitor by substituting $V = Ed$ and $C = \kappa\epsilon_0 A/d$ into Equation 24.11:

$$\text{Energy} = \frac{1}{2}\left(\frac{\kappa\epsilon_0 A}{d}\right)(Ed)^2$$

Since the area A times the separation d is the volume between the plates, the energy per unit volume or *energy density* is

$$\text{Energy density} = \frac{\text{Energy}}{\text{Volume}} = \frac{1}{2}\kappa\epsilon_0 E^2 \qquad (24.12)$$

It can be shown that this expression is valid for any electric field strength, not just that between the plates of a capacitor.

The energy-storing capability of a capacitor is often put to good use in electronic circuits. Recently, for example, state-of-the-art capacitors with large capacitances (1.0–3.3 F) have been employed to provide backup power in computers in the event of a power line failure. The energy stored in such capacitors can maintain certain kinds of random access computer memory (RAM) for up to 48 hours after the main power source is disrupted. The use of a capacitor with a large capacitance and a relatively small physical size is attractive, because it eliminates the need for a battery pack.

*24.6 MEDICAL APPLICATIONS OF ELECTRIC POTENTIAL DIFFERENCES

Several important medical diagnostic techniques depend on the fact that the surface of the human body is *not* an equipotential surface. Between various points on the human body there are small potential differences (approximately 30–500 μV), and these differences provide the basis for electrocardiography, electroencephalography, and electroretinography.

The reason that potential differences exist between various points on the surface of the body can be traced to the electrical characteristics of muscle cells and nerve cells. In carrying out their biological functions, these cells utilize charged atoms and molecules. For example, they use positively charged sodium and potassium ions and negatively charged chlorine ions that exist within the cells and in the intercellular fluid outside the cells. As a result of such charged particles, electric fields are generated that extend to the surface of the body and lead to small potential differences.

Figure 24.17 shows some locations on the body where electrodes are placed to measure potential differences in electrocardiography. The potential difference be-

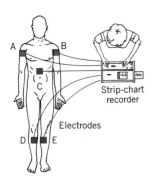

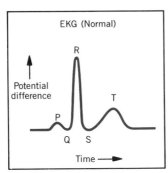

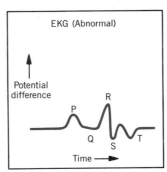

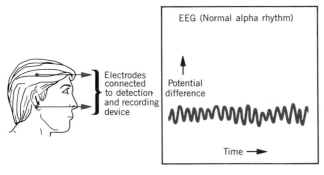

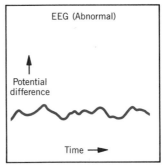

FIGURE 24.17 The potential differences generated by heart muscle activity provide the basis for electrocardiography. Each EKG pattern corresponds to one heartbeat.

tween two points changes as the heart beats and, therefore, the potential difference forms a repetitive pattern in time. The recorded pattern of potential difference versus time is called an electrocardiogram (ECG or EKG), and its shape depends on which pair of points in the picture (A and B, B and C, etc.) is used to locate the electrodes. The figure also shows some EKGs and indicates the regions (P, Q, R, S, and T) that can be associated with specific parts of the heart's beating cycle. The distinct differences between the EKGs of healthy and damaged hearts provide physicians with a valuable diagnostic tool.

In electroencephalography the electrodes are placed at specific locations on the head, as Figure 24.18 indicates, and they record the potential differences that characterize brain behavior. The graph of potential difference versus time is known as an electroencephalogram (EEG). The various parts of the patterns in an EEG are often referred to as "waves" or "rhythms." The drawing shows an example of the main resting rhythm of the brain, the so-called alpha rhythm, and also illustrates the distinct differences that are found between the EEGs generated by healthy and diseased (abnormal) tissue.

The electrical characteristics of the retina of the eye lead to the potential differences measured in electroretinography. Figure 24.19 shows a typical electrode placement used to record the pattern of potential difference versus time that occurs when the eye is stimulated by a flash of light. One electrode is mounted on a contact lens, while the other is often placed on the forehead. The recorded pattern is called an electroretinogram (ERG), parts of the pattern being referred to as the "A wave" and the "B wave." As the graphs show, the ERGs of normal and diseased eyes can differ markedly.

FIGURE 24.18 In electroencephalography the potential differences created by the electrical activity of the brain are used for diagnosing abnormal behavior.

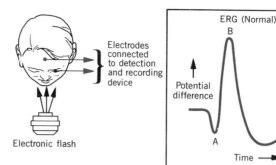

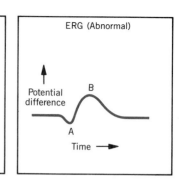

FIGURE 24.19 The electrical activity of the retina of the eye generates the potential differences used in electroretinography.

SUMMARY

When work W_{AB} is done to move a positive test charge $+q_0$ at constant speed from point A to point B, the work equals the difference between the **electric potential energy** EPE at B and that at A: $W_{AB} = \text{EPE}_B - \text{EPE}_A$. The **electric potential** V is the electric potential energy per unit charge, so the electric potential difference between two points is $V_B - V_A = (\text{EPE}_B - \text{EPE}_A)/q_0 = W_{AB}/q_0$. A positive charge accelerates from a region of higher potential toward a region of lower potential. Conversely, negative charge accelerates from a region of lower potential toward a region of higher potential. The electric potential at a distance r from a **point charge** q is $V = kq/r$. This expression for V assumes the potential is zero at an infinite distance away from the charge.

The **electron volt** (eV) is a unit of energy. One electron volt corresponds to 1.60×10^{-19} J and is the change in potential energy of an electron when the electron moves through a potential difference of one volt.

An **equipotential surface** is a surface on which the electric potential is the same everywhere. No work is needed to move a charge at a constant speed on an equipotential surface. However, work is done when a charge moves between equipotential surfaces. The electric field created by any group of charges is always perpendicular to the associated equipotential surfaces and points in the direction of decreasing potential. The electric field is given by $E = -\Delta V/\Delta s$, where ΔV is the potential difference and Δs is the magnitude of the displacement perpendicular to the potential surfaces in the direction of increasing potential.

A **capacitor** is a device that can store charge and consists of two conductors that are near one another, but not touching. The magnitude q of the charge on each plate is given by $q = CV$, where V is the magnitude of the potential difference between the plates and C is the **capacitance**. The insulating material included between the plates is called a **dielectric**. The dielectric constant κ of the material is $\kappa = E_0/E$, where E_0 and E are, respectively, the magnitudes of the electric fields between the plates without and with a dielectric, assuming the charge on the plates is kept fixed. The capacitance of a parallel plate capacitor is $C = \kappa \epsilon_0 A/d$, where A is the area of each plate, and d is the distance between the plates. The **electric potential energy** stored in a capacitor is $\frac{1}{2}CV^2$. The **energy density** or energy stored per unit volume is $\frac{1}{2}\kappa \epsilon_0 E^2$.

SOLVED PROBLEMS

SOLVED PROBLEM 1
Related Problems: *16 *17 **18

The drawing shows three point charges, initially very far apart, that are brought together and placed at the corners of an equilateral triangle. Determine the electric potential energy of the group. In other words, determine the amount by which the electric potential energy of the triangular group differs from that of the three charges in their initial, widely separated locations.

Solution This problem is done by adding the charges to the triangle one at a time and calculating the energy needed to bring up each charge. The total potential energy of the triangular group

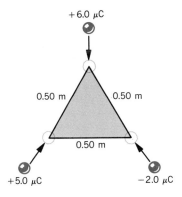

+6.0 μC

0.50 m · · · 0.50 m

0.50 m

+5.0 μC · · · −2.0 μC

is the sum of the energies needed to bring together the three charges.

The order in which the charges are put on the triangle does not matter; we begin with the charge of $+5.0\ \mu C$. No energy is required to place this charge at a corner of the triangle, since the two other charges remain very far away. Once the charge is in place, the potential it creates at either empty corner ($r = 0.50$ m) is

$$V = \frac{kq}{r} = \frac{(8.99 \times 10^9 \text{ N·m}^2/\text{C}^2)(+5.0 \times 10^{-6} \text{ C})}{0.50 \text{ m}} \quad (24.6)$$

$$= +9.0 \times 10^4 \text{ V}$$

The energy required to bring up the charge of $+6.0\ \mu C$ and place it at one of the empty corners is

$$\text{EPE} = qV = (+6.0 \times 10^{-6} \text{ C})(+9.0 \times 10^4 \text{ V}) \quad (24.4)$$

$$= +0.54 \text{ J}$$

The electric potential at the remaining empty corner is the sum of the potentials due to the charges at the other two corners:

$$V = \frac{(8.99 \times 10^9 \text{ N·m}^2/\text{C}^2)(+5.0 \times 10^{-6} \text{ C})}{0.50 \text{ m}}$$

$$+ \frac{(8.99 \times 10^9 \text{ N·m}^2/\text{C}^2)(+6.0 \times 10^{-6} \text{ C})}{0.50 \text{ m}} = +2.0 \times 10^5 \text{ V}$$

The energy needed to bring up the charge of $-2.0\ \mu C$ and place it at the third corner is

$$\text{EPE} = qV = (-2.0 \times 10^{-6} \text{ C})(+2.0 \times 10^5 \text{ V}) \quad (24.4)$$

$$= -0.40 \text{ J}$$

The total potential energy of the triangular group differs from that of the widely separated charges by an amount that is the sum of the potential energies calculated above:

$$\text{Total potential energy} = 0.54 \text{ J} - 0.40 \text{ J} = \boxed{+0.14 \text{ J}}$$

This potential energy originates in the work that is done to bring the charges together.

Summary of Important Points This problem shows how to find the electric potential energy of a group of point charges. Initially, the charges are widely separated. Then, each charge is brought up, one at a time, to form the group. The energy needed to add each charge to the group is determined. The total electric potential energy of the group is the sum of the individual energies. The value of the total energy does not depend on the order in which the individual charges are added to the group.

QUESTIONS

1. Three points, X, Y, and Z, are located from left to right on a line. A positive test charge is released from rest at X and moves toward and reaches Y. Subsequently, a positive test charge is released from rest at Y and moves toward and reaches Z. Assuming that only motion along the line is possible, what will a negative test charge do when it is released from rest at Y? Explain.

2. What charges, all having the same magnitude, would you place at the corners of a rectangle, so that both the electric field and the electric potential (assuming a zero reference value at infinity) are zero at the center of the rectangle? Account for the fact that the charge distribution gives rise to *both* a zero field and a zero potential.

3. The electric field at a single point is zero. Does this fact necessarily mean that the electric potential at the same point is zero? Use a spot on the line between two identical point charges as an example to support your reasoning.

4. To measure the potential at the midpoint between the positive and negative charges of a dipole, a positive test charge is brought in from infinity at a constant speed. (a) When the path followed is along the perpendicular bisector of the dipole, no work is required along any portion of the path. Why? (b) When the path followed is not along the perpendicular bisector, the total work required to reach the midpoint is still zero, but for any one portion of the path the work is not necessarily zero. Explain.

5. The drawing shows two points, A and B, on an equipotential surface. A positive test charge is moved from A to B along two different paths. In the first case, the path lies entirely on the surface

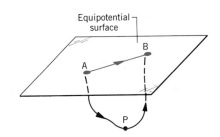

Equipotential surface

B

A

P

(see colored line). In the second case, the path includes the point P, which does *not* lie on the surface. Compare the work done in moving the test charge from A to B in both cases. Justify your answer.

6. The drawing shows two equipotential surfaces. On a separate sheet of paper, draw the electric field line (including its direction) that passes through the following points: (a) A and A', (b) B and B', and (c) C and C'.

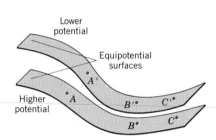

7. The potential is constant throughout a given region of space. Is the electric field zero or nonzero in this region? Justify your answer.

8. A positive test charge is placed in an electric field. In what direction should the charge be moved relative to the field, such that the charge experiences a constant electric potential? Account for your answer.

9. The drawing shows three electric field lines. On a separate sheet of paper, draw the equipotential surface that passes through the points (a) A, A', and A" and (b) B, B', and B". Which surface is at the higher potential? Why?

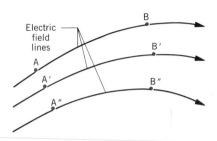

10. The electric field can be expressed in either of two units: newtons/coulomb or volts/meter. Show that these units for the electric field are equivalent.

11. A proton and an electron are released from rest at the midpoint between the plates of a charged parallel plate capacitor. Except for these particles, nothing else is between the plates. Which particle strikes a capacitor plate first? Give your reasoning.

PROBLEMS

Note: All charges are assumed to be point charges unless specified otherwise.

Section 24.1 Potential Energy, Section 24.2 The Electric Potential Difference

1. The anode (positive terminal) of an x-ray tube is at a potential of $+125\,000$ V with respect to the cathode (negative terminal). (a) How much work (in joules) is done on an electron that is accelerated from the cathode to the anode? (b) If the electron is initially at rest, what kinetic energy does the electron have when it arrives at the anode?

2. An agent moves a charge of $+18.0 \times 10^{-5}$ C at a constant speed from point A to point B and performs 5.80×10^{-3} J of work on the charge. (a) What is the difference between the electric potential energies of the charge at the two points? (b) Determine the potential difference between the two points. (c) State which point is at the higher potential.

3. The potential of point A relative to point B is $V_A - V_B = +95$ V, while the potential of point C relative to point B is $V_C - V_B = +23$ V. How much work is needed to move a charge of $+45\ \mu$C from C to A?

4. In a television picture tube, electrons strike the screen after being accelerated from rest through a potential difference of 25 000 V. The speeds of the electrons are quite large, and in accurate calculations of the speeds, the effects of special relativity must be taken into account. Ignoring such effects and assuming an electron starts from rest, find the electron speed just before the electron strikes the screen.

5. A particle with a charge of $-1.5\ \mu$C and a mass of 2.5×10^{-6} kg is released from rest at point X and accelerates toward point Y, arriving there with a speed of 42 m/s. (a) What is the potential difference between X and Y? (b) Which point is at the higher potential? Give your reasoning.

6. A proton, released from rest, accelerates from one point to another and gains kinetic energy, because there is a potential difference of $1.5\ \mu$V between the two points. To acquire the same amount of kinetic energy, through what height would the proton have to fall freely, after being released from rest in the presence of the earth's gravity?

* 7. The energy in a lightning bolt is enormous. Consider, for example, a lightning bolt in which 25 C of charge move through a potential difference of 1.2×10^8 V. With the amount of energy in this bolt, how many kilograms of water at 100 °C could be boiled into steam at 100 °C?

Section 24.3 The Electric Potential Difference Created by Point Charges

8. Three point charges are located at the corners of a square whose sides are 2.0 m in length. The charges are $+2.0, +14,$ and $+5.0\ \mu C$. The empty corner of the square is opposite the 14-μC charge. How much work is required to bring up a fourth charge of $+8.0\ \mu C$ and place it at the empty corner?

9. Two point charges are fixed in place with a separation d. One charge is positive and has twice the magnitude of the other charge, which is negative. The positive charge lies to the left of the negative charge. Relative to the negative charge, locate the two spots on the line through the charges where the total potential is zero. One spot is (a) between the charges, and the other spot is (b) to the right of the negative charge. (c) Why is there no such spot to the left of the positive charge?

10. Two point charges, each with a magnitude of $q = 2.00\ \mu C$, are fixed to adjacent corners of an isosceles right triangle, as the drawing shows. One charge is positive and one is negative. What charge (both magnitude and sign) must be fixed to the midpoint of the hypotenuse, so the electric potential at the empty corner is zero?

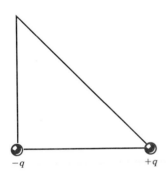

11. Location A is 2.00 m from a charge of -3.00×10^{-8} C, while location B is 3.00 m from the charge. Find the potential difference $V_B - V_A$ between the two points, and state which point is at the higher potential.

12. Charges of $+2q$ and $-q$ are fixed in place and separated by a distance of 2.0 m. A dashed line is drawn through the negative charge, perpendicular to the line between the charges. Relative to the negative charge, where on the dashed line is the total potential equal to zero? There are two places.

* 13. A positive charge of $+q_1$ is located 3.00 m to the left of a negative charge $-q_2$. The charges have different magnitudes. On the line through the charges, the net *electric field* is zero at a spot 1.00 m to the right of the negative charge. On this line there are also two spots where the potential is zero. Locate these two spots relative to the negative charge.

* 14. According to the Bohr model, a singly ionized helium atom contains one electron in orbit at a distance of 2.65×10^{-11} m from a nucleus that contains two protons and two neutrons. (The neutrons are electrically neutral and play no role in this problem.) (a) Calculate the ionization energy (in joules) for the atom. (b) Express the answer in electron volts.

* 15. A charge of $-3.00\ \mu C$ is fixed in place. From a distance of 4.50 cm, a particle of mass 7.20 g and charge $-8.00\ \mu C$ is fired with an initial speed of 65.0 m/s directly toward the fixed charge. How far does the particle travel before its speed is zero?

* 16. A square is 0.50 m on a side. How much work is done to bring in four identical charges ($5.0\ \mu C$) from infinity and place each at a corner of the square? *(See Solved Problem 1 for a related problem.)*

* 17. Determine the electric potential energy for the array of three charges shown in the drawing, relative to its value when the charges are infinitely far away. *(See Solved Problem 1 for a related problem.)*

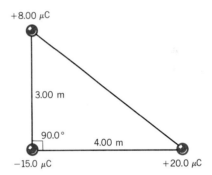

**18. Charges q_1 and q_2 are fixed in place, q_2 being located at a distance d to the right of q_1. A third charge q_3 is then fixed to the line joining q_1 and q_2 at a distance d to the right of q_2. The third charge is chosen so the potential energy of the group is zero; that is, the potential energy has the same value as that of the three charges when they are widely separated. Determine q_3, assuming that (a) $q_1 = q_2 = q$ and (b) $q_1 = q$ and $q_2 = -q$. *(See Solved Problem 1 for a related problem.)*

**19. A positive charge $+q_1$ is located to the left of a negative charge $-q_2$. On a line passing through the two charges, there are two places where the total potential is zero. The first place is between the charges and is 4.00 cm to the left of the negative charge. The second place is 7.00 cm to the right of the negative charge. (a) What is the distance between the charges? (b) Find q_1/q_2, the ratio of the magnitudes of the charges.

Section 24.4 Equipotential Surfaces and Their Relation to the Electric Field

20. The inner and outer surfaces of a cell membrane carry a negative and positive charge, respectively. Because of these charges, a potential difference of about 95 mV exists across the membrane. The thickness of the membrane is 7.5×10^{-9} m. What is the magnitude of the electric field in the membrane?

21. Two points, A and B, are separated by 0.016 m. The potential at A is $+28$ V, and that at B is $+95$ V. (a) What is the magnitude of the potential gradient? (b) Determine the magnitude and direction of the electric field between the two points.

22. Consider the equipotential surfaces that surround a point charge of $+1.50 \times 10^{-8}$ C. How far from the 150-V surface is the 75-V surface?

23. It is possible to receive a shock when you reach out to touch a metal surface, after walking across a nylon rug on a dry day. Because your feet rub on the rug, your body becomes electrically charged, and the potential difference between your body and the metal surface increases. When the potential difference becomes large enough, a spark jumps between your hand and the metal. A spark occurs when the electric field strength created by the charges reaches the dielectric strength of the air. The dielectric strength of the air is 3.0×10^6 N/C and is the field strength at which the air suffers electrical breakdown. Suppose a spark 3.0 mm long jumps between your hand and a metal doorknob. Assuming the electric field is uniform, find the potential difference between your hand and the doorknob.

* 24. The drawing shows the potential at five points in space. From the data shown, find the magnitude and direction of the electric field in the vicinity of the center point.

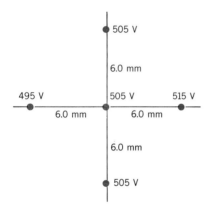

* 25. Two equipotential surfaces surround a point charge of $+1.00 \times 10^{-9}$ C. The surfaces are separated by a distance of 1.00 m and have a potential difference of 1.00 V between them. (a) What are the radii of the surfaces? (b) What is the potential of each surface?

Section 24.5 Capacitors and Dielectrics

26. One farad is a large amount of capacitance. To see just how large, determine the area of each plate of an empty, one-farad parallel plate capacitor whose plate separation is one meter. Express your answer in square miles.

27. An axon is the relatively long tail-like part of a neuron, or nerve cell. The outer surface of the axon membrane (dielectric constant = 5, thickness = 1×10^{-8} m) is charged positively, and the inner portion is charged negatively. Thus, the membrane is a kind of capacitor. Assuming an axon can be treated like a parallel plate capacitor with a plate area of 5×10^{-6} m², what is its capacitance?

28. A defibrillator is a device used in emergency situations to stimulate the heart muscle and start the heart beating again. Two "paddles" are placed on the body near the heart, and the energy stored in a capacitor is discharged through them. The energy discharged is 500 J and the capacitance is 100 μF. What is the potential difference across the capacitor plates?

29. A capacitor has a capacitance of 2.5×10^{-8} F. In the charging process, electrons are removed from one plate and placed on the other plate. When the potential difference between the plates is 450 V, how many electrons have been transferred?

30. What is the potential difference between the plates of a 3.3-F capacitor that stores sufficient energy to operate a 75-W light bulb for one minute?

31. A parallel plate capacitor is filled with ruby mica, and the effective area of each plate is 3.8 m². The capacitor stores 2.7 μC of charge when a 1.5-V flashlight battery provides the potential difference between the plates. What is the plate separation?

32. Each plate of a parallel plate capacitor has an area of 2.2×10^{-4} m² and stores a charge whose magnitude is 4.8×10^{-9} C. Determine the magnitude of the electric field between the plates when the capacitor is (a) empty and (b) filled with Teflon.

33. The electronic flash attachment for a camera contains a capacitor for storing the energy used to produce the flash. In one such unit, the potential difference between the plates of a 750-μF capacitor is 330 V. (a) Determine the energy that is used to produce the flash in this unit. (b) Assuming the flash lasts for 5.0×10^{-3} s, find the effective "wattage" of the flash.

* 34. The membrane that surrounds a certain type of living cell has a surface area of about 5×10^{-9} m² and a thickness of about 1×10^{-8} m. Assume that the membrane behaves like a parallel plate capacitor and has a dielectric constant of 5. (a) If the potential on the outer surface of the membrane is $+60$ mV greater than that on the inside surface, how much charge resides on the outer surface? (b) If the charge in part (a) is due to K$^+$ ions, how many such ions are present on the outer surface? (c) What is the electrical energy, in electron volts, stored in this biological capacitor?

* 35. The equipotential surfaces within the plates of an empty parallel plate capacitor are such that two surfaces, 2.00 mm apart, have a potential difference of 1.20 mV. The area of each plate is 7.50×10^{-4} m². How much charge is on each plate?

* 36. The dielectric strength of an insulating material is the maximum electric field strength to which the material can be subjected without electrical breakdown occurring. Suppose a parallel plate capacitor is filled with a material whose dielectric constant is 3.5 and whose dielectric strength is 1.4×10^7 N/C. If this capacitor is to store 1.7×10^{-7} C of charge on its plate without suffering breakdown, what must be the radius of its circular plates?

* 37. The capacitance of an empty capacitor is 1.2 μF. The capacitor is connected to a 12-V battery and charged up. (a) With the capacitor connected to the battery, so the 12-V potential difference between the plates is maintained, a slab of dielectric material is inserted between the plates. As a result, 2.6×10^{-5} C of *additional* charge flow from one plate, through the battery, and on to the other plate. What is the dielectric constant of the material? (b) Suppose the capacitor in part (a) is initially charged up and then disconnected from the battery. Therefore, no charge can flow between the plates as the slab is inserted. Find the amount by which the potential difference across the plates changes. Specify whether the change is an increase or a decrease.

**38.   The plate separation of a charged capacitor is 8.00 cm. A proton and an electron are released from rest at the midpoint between the plates. How far has the proton traveled by the time the electron strikes the positive plate of the capacitor?

ADDITIONAL PROBLEMS

39.   A charge of $+125\ \mu$C is fixed in place at the center of a square that is 0.64 m on a side. How much work is required to move a charge of $+7.0\ \mu$C from one corner of the square to any other empty corner?

40.   A parallel plate capacitor has a capacitance of $7.0\ \mu$F when filled with a dielectric. The effective area of each plate is 1.5 m² and the separation between the plates is 1.0×10^{-5} m. What is the dielectric constant of the dielectric?

41.   Point A is at a potential of $+250$ V, and point B is at a potential of -150 V. An α-particle is a helium nucleus that contains two protons and two neutrons; the neutrons are electrically neutral. An α-particle starts from rest at A and accelerates toward B. When the α-particle arrives at B, what kinetic energy (in electron volts) does it have?

42.   A charge of $+9q$ is fixed to one corner of a square, while a charge of $-8q$ is fixed to the opposite corner. Expressed in terms

of q, what charge should be fixed to the center of the square, so the potential is zero at each of the two empty corners?

* 43.   The electric field has a constant value of 3.0×10^3 V/m and is directed downward. The field is the same everywhere. The potential at a point P within this region is 135 V. Find the potential at the following points: (a) 8.0 mm directly above P, (b) 3.3 mm directly below P, (c) 5.0 mm directly to the right of P.

* 44.   The group of charges shown in the drawing is called a *quadrupole*. Assuming the distance R to be much larger than the distance d and assuming the usual zero reference value for the potential at infinity, derive an expression for the electric potential at point P. *(Hint: The potential is proportional to $1/R^3$.)*

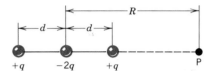

**45.   The potential difference between the plates of a capacitor is 175 V. Midway between the plates, a proton and an electron are released. The electron is released from rest. The proton is projected perpendicularly toward the negative plate with an initial speed. The proton strikes the negative plate at the same instant the electron strikes the positive plate. Find the initial speed of the proton.

**46.   One particle has a mass of 3.00 g and a charge of $+8.00\ \mu$C. A second particle has a mass of 6.00 g and the same charge. The two particles are initially held in place and then released. The particles fly apart, and when the separation between them is 10.0 cm, the speed of the 3.00-g particle is 125 m/s. Find the initial separation between the particles. *(Hint: Both energy and momentum are conserved in this problem.)*

CHAPTER 25

Electric Circuits: Basic Concepts

Electric circuits are used to provide the power for the lights and rides that make the midway of a fair such a memorable place on a summer's evening.

FIGURE 25.1 In an electric circuit, energy is transferred from a source to a device that needs it.

Labels in figure: Conducting wire · Device · Moving charges · Source of electrical energy

25.1 THE PURPOSE OF ELECTRIC CIRCUITS

Applications of electricity take advantage of the ease with which electrical energy can be transferred from a source to the devices that need it. Such devices may include the components of a stereo system, appliances, motors, lights, and the like. The transfer takes place via electric circuits, in which the source of energy and the energy-consuming device are connected by conducting wires, as Figure 25.1 illustrates. Electric charges move in the wires.

Many types of sources of electrical energy are used in circuits, the battery being the most familiar. In a battery, a chemical reaction occurs and provides the energy needed for separating electric charges onto the positive and negative terminals. Figure 25.2 shows the terminals of a car battery, as well as those of a flashlight battery. The drawing also shows the symbol ($\dashv\vdash$) used to represent a battery in circuit drawings. The energy of the charges on the terminals is reflected in the electric potential difference that the chemical reaction maintains between the terminals. The maximum potential difference is called the ***electromotive force* (emf)*** of the battery, for which the symbol $\mathscr{E}$ is often used. In a typical car battery, the chemical reaction maintains the potential of the positive terminal at a maximum of 12 volts (12 joules/coulomb) higher than the potential of the negative terminal, so the emf is 12 V. Thus, one coulomb of charge emerging from the battery and entering a circuit has at most 12 joules of energy. In a typical flashlight battery the emf is 1.5 V. In reality, the potential difference between the terminals of a battery is somewhat less than the maximum value indicated by the emf, for reasons that Section 26.4 discusses.

In a circuit such as that in Figure 25.2, the battery creates an electric field within† and parallel to the wire, directed from the positive toward the negative terminal. The field exerts a force on the free electrons in the wire, and they respond by moving. The resulting flow of charge is known as an ***electric current.*** The current is the amount of charge flowing per unit time, much in the same sense that a river current is the amount of water flowing per unit time. If the rate is constant, the current I is given by

$$I = \frac{q}{t} \tag{25.1}$$

where q is the magnitude of the charge that flows past a given point in the wire in a time t. If charge does not flow at a constant rate, then Equation 25.1 gives the average current. Since the units for charge and time are the coulomb (C) and the second (s), the SI unit for current is a coulomb per second (C/s). One coulomb per second is referred to as an ***ampere*** (A), after the French mathematician André-Marie Ampère (1775–1836).

If the charges move around a circuit in the same direction at all times, the current is said to be ***direct current (dc).*** Batteries, for example, create direct current. In contrast, the current is said to be ***alternating current (ac)*** when the charges move first one way and then the opposite way, changing direction from moment to moment. Many energy sources create alternating current, e.g., generators at power companies, microphones, and tape playback heads. Example 1 deals with the direct current produced by the battery in a pocket calculator.

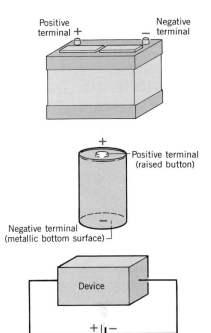

FIGURE 25.2 Typical batteries and the symbol ($\underset{\vdash}{+}\blacksquare\underset{\dashv}{-}$) used to represent them in electric circuits.

EXAMPLE 1

The current from the 3.0-V battery of a pocket calculator is 0.17 mA (milliamperes). In one hour of operation, (a) how much charge flows in the circuit and (b) how much energy does the battery deliver to the circuit?

SOLUTION

(a) Equation 25.1 gives the charge. Note that one milliampere is 10^{-3} ampere.

$$q = It = (0.17 \times 10^{-3} \text{ A})(3600 \text{ s}) = \boxed{0.61 \text{ C}}$$

(b) In a 3.0-V battery, the charges separated onto the terminals have 3.0 joules of energy per coulomb of charge. To find the energy delivered to the circuit, we multiply the amount of charge delivered to the circuit by the energy per unit charge:

$$\text{Energy} = \text{Charge} \times \underbrace{\frac{\text{Energy}}{\text{Charge}}}_{\text{Battery emf}}$$

$$= (0.61 \text{ C})(3.0 \text{ V}) = \boxed{1.8 \text{ J}}$$

Today, it is known that electrons flow in any kind of conducting wire. Thus, Figure 25.3 shows the negative electrons emerging from the negative terminal of the battery

---

* The word "force" appears in this context for historical reasons, even though it is incorrect; as we know, electric potential is energy per unit charge, which is not force.

† Here, an electric field *can* exist inside a conductor, in contrast to the situation in electrostatics. The field can exist here because the battery keeps the charges moving and prevents them from coming to equilibrium on the outer surface of the conductor, where they would cause the net electric field on the interior to be zero.

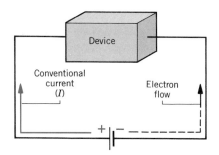

FIGURE 25.3 In a circuit, electrons actually flow through the conducting wires. However, it is customary to use a so-called conventional current to describe the flow of charges (see text).

and moving around the circuit toward the positive terminal. It is customary, however, *not* to refer to the flow of electrons when discussing circuits. Instead, a so-called **conventional current** is used, for reasons that date back to the time when it was believed that positive charges moved through metal wires. Conventional current is the hypothetical flow of positive charges that would have the same effect in the circuit as the movement of negative charges that actually does occur. In Figure 25.3, negative electrons *arrive* at the positive terminal. The same effect could have been achieved if an equivalent amount of positive charge had *left* the positive terminal. Therefore, the drawing shows the conventional current (in color) originating from the positive terminal. A conventional current of hypothetical positive charges is consistent with our earlier use of a positive test charge for defining electric fields and potentials. The direction of conventional current is always from a point of higher potential toward a point of lower potential, that is, from the positive terminal toward the negative terminal. In this text, the symbol I stands for conventional current.

25.2 OHM'S LAW

The current that a battery can push through a wire is analogous to the water flow that a pump can push through a pipe. Greater pump pressures lead to larger water flow rates, and, similarly, greater battery voltages* lead to larger electric currents. In the simplest case, the current I is directly proportional to the voltage V; that is, $I \propto V$. Thus, a 12-V battery leads to twice as much current as a 6-V battery, when each is connected to the same circuit.

In a water pipe, the flow rate is not only determined by the pump pressure but is also affected by the length and diameter of the pipe. Longer and narrower pipes offer higher resistance to the moving water and lead to smaller flow rates for a given pump pressure. A similar situation exists in electrical circuits. In the electrical case, the **resistance** (R) is defined as the ratio of the voltage V applied across a piece of material to the current I through the material, or $R = V/I$. When only a small current results from a large voltage, there is a high resistance to the moving charge. For many materials (e.g., metals), the ratio V/I is the same for a given piece of material over a wide range of voltages and currents. In such a case, the resistance is a constant, and the relation $R = V/I$ is referred to as **Ohm's law,** after the German physicist Georg Simon Ohm (1789–1854), who discovered it.

OHM'S LAW
The ratio V/I is a constant, where V is the voltage applied across a piece of material (such as a wire) and I is the resulting current through the material:

$$\frac{V}{I} = R = \text{constant} \quad \text{or} \quad V = IR \qquad (25.2)$$

R is the resistance of the piece of material.

    SI unit of resistance: volt/ampere (V/A) = ohm (Ω)

---

* The potential difference between two points, such as the terminals of a battery, is commonly called the voltage between the points.

The SI unit for resistance is a volt per ampere, which is called an *ohm* and is represented by the Greek letter omega (Ω). Note that Ohm's law is not a fundamental law of nature like Newton's laws. Rather, it is only a statement of the way certain materials behave in electric circuits.

To the extent that a wire or an electrical device offers resistance to the flow of charges, it is called a **resistor**. The resistance can take on a wide range of values. The copper wires in a television set, for instance, have a negligibly small resistance. On the other hand, commercial resistors can have resistances up to many kiloohms (kΩ) or megaohms (MΩ). Such resistors play an important role in electric circuits, where they are used to limit the amount of current (see Figure 25.4).

In drawing electric circuits we follow the usual conventions: (1) a zigzag line (–⋀⋀–) represents a resistor and (2) a straight line (——) represents an ideal conducting wire, or one with a negligible resistance. Example 2 illustrates an application of Ohm's law to the circuit in a flashlight.

Resistors

FIGURE 25.4 A printed circuit board.

EXAMPLE 2

The filament in a light bulb is a resistor in the form of a thin piece of wire. The wire becomes hot enough to glow brightly because of the current in it. Figure 25.5 shows a flashlight that uses two 1.5-V batteries (effectively a single 3.0-V battery) to provide a current of 0.40 A in the filament. Determine the resistance of the glowing filament.

SOLUTION

The resistance of the filament is

$$R = \frac{V}{I} = \frac{3.0 \text{ V}}{0.40 \text{ A}} = \boxed{7.5 \ \Omega} \qquad (25.2)$$

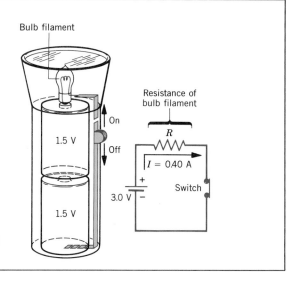

FIGURE 25.5 The circuit in many flashlights consists of a resistor (the filament of the light bulb) connected to a 3.0-V battery.

25.3 RESISTANCE AND RESISTIVITY

In a water pipe, the length and cross-sectional area of the pipe determine the resistance the pipe offers to the flow of water. Longer pipes with smaller cross-sectional areas offer greater resistance. Analogous effects are found in the electrical case. For a wide range of materials, the resistance of a piece of material of length L and cross-sectional area A is

$$R = \rho \frac{L}{A} \tag{25.3}$$

where ρ is a proportionality constant known as the ***resistivity*** of the material. It can be seen from Equation 25.3 that the unit for resistivity is the ohm · meter ($\Omega \cdot m$), and Table 25.1 lists values for various materials. All the conductors in Table 25.1 are metals and have small resistivities. Insulators such as rubber have large resistivities. Materials like germanium and silicon have intermediate resistivity values and are, accordingly, called ***semiconductors.***

Resistivity is an inherent property of a material, inherent in the same sense that the density of a material is an inherent property. Resistance, on the other hand, depends on both the resistivity and the geometry of the material. Thus, two wires can be made from copper, which has a resistivity of $1.72 \times 10^{-8}\ \Omega \cdot m$, but Equation 25.3 indicates that a short wire with a large cross-sectional area offers a smaller resistance to current than a long, thin wire. Wires that carry large currents, such as main power cables, are thick rather than thin so that the resistance of the wires is kept as small as possible. For the same reason, the instructions for electric lawn mowers recommend using a thicker extension cord when the length exceeds about 100 ft. The larger

TABLE 25.1 Resistivities[a] of Various Materials

Material	Resistivity ρ ($\Omega \cdot m$)
Conductors	
Aluminum	2.82×10^{-8}
Copper	1.72×10^{-8}
Gold	2.44×10^{-8}
Iron	9.7×10^{-8}
Mercury	95.8×10^{-8}
Nichrome (alloy)	100×10^{-8}
Silver BEST CONDUCTOR	1.59×10^{-8}
Tungsten	5.6×10^{-8}
Semiconductors	
Carbon	3.5×10^{-5}
Germanium	0.5^b
Silicon	$20 - 2300^b$
Insulators	
Mica	$10^{11} - 10^{15}$
Rubber (hard)	$10^{13} - 10^{16}$
Teflon	10^{16}
Wood (maple)	3×10^{10}

[a] The values pertain to temperatures near 20 °C.

[b] Depending on purity.

TABLE 25.2 Temperature Coefficients[a] of Resistivity for Various Materials

Material	Temperature Coefficient of Resistivity α [$(C°)^{-1}$]
Aluminum	0.0039
Carbon	−0.0005
Copper	0.00393
Germanium	−0.05
Gold	0.0034
Iron	0.0050
Mercury	0.00089
Nichrome (alloy)	0.0004
Silicon	−0.07
Silver	0.0038
Tungsten	0.0045

[a] The values pertain to a temperature of 20 °C.

diameter offsets the effect of the greater length and keeps the resistance to a small value.

The resistivity of a material depends on temperature. In metals, the resistivity increases with increasing temperature, whereas in semiconductors the reverse is true. Certain materials have the property that their resistivity drops to zero at very low temperatures. Such materials are called *superconductors*, because, with zero resistivity, they offer no resistance to electric current.

For many materials and limited temperature ranges it is possible to express the temperature dependence of the resistivity as follows:

$$\rho = \rho_0[1 + \alpha(T - T_0)] \qquad (25.4)$$

In this expression ρ and ρ_0 are the resistivities at temperatures T and T_0, respectively. The term α has the unit of reciprocal temperature and is the *temperature coefficient of resistivity*. Table 25.2 gives values of α for various materials. When the resistivity increases with increasing temperature, α is positive, as it is for metals. When the resistivity decreases with increasing temperature, α is negative, as it is for carbon, germanium, and silicon. Since resistance is given by $R = \rho L/A$, both sides of Equation 25.4 can be multiplied by L/A to show that resistance depends on temperature according to

$$R = R_0[1 + \alpha(T - T_0)] \qquad (25.5)$$

This section concludes with an example to illustrate the role of the resistivity and its temperature coefficient in determining the electrical resistance of a piece of material.

EXAMPLE 3

Figure 25.6 shows a heating element from an electric stove. The element contains a wire (length = 1.1 m, cross-sectional area = 3.1×10^{-6} m^2) through which electric charge flows. This wire is imbedded within an electrically insulating material that is contained within a metal casing. The casing becomes hot in response to the flowing charge. The material of the wire has a resistivity of $\rho_0 = 6.8 \times 10^{-5}$ $\Omega \cdot$m at $T_0 = 320$ °C and a temperature coefficient of resistivity of $\alpha = 2.0 \times 10^{-3}$ $(C°)^{-1}$. Determine the resistance of the heater wire at an operating temperature of 420 °C.

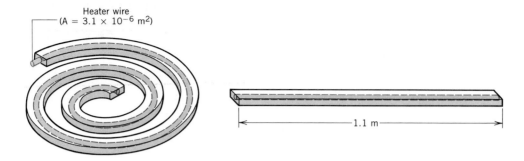

Heater wire
$(A = 3.1 \times 10^{-6}$ m$^2)$

1.1 m

FIGURE 25.6 A heating element from an electric stove.

SOLUTION

The relation $R = \rho L/A$ provides a straightforward solution, provided we first use Equation 25.4 to find the resistivity of the metal at the 420 °C operating temperature:

$$\rho = \rho_0[1 + \alpha(T - T_0)]$$
$$= (6.8 \times 10^{-5} \ \Omega \cdot m)[1$$
$$+ (2.0 \times 10^{-3} \ (C°)^{-1})(420 \ °C - 320 \ °C)]$$
$$= 8.2 \times 10^{-5} \ \Omega \cdot m$$

This value of the resistivity can be used along with the length and cross-sectional area shown in Figure 25.6 to find the resistance of the heating element:

$$R = \frac{\rho L}{A} = \frac{(8.2 \times 10^{-5} \ \Omega \cdot m)(1.1 \ m)}{3.1 \times 10^{-6} \ m^2}$$
$$= \boxed{29 \ \Omega}$$

(25.3)

25.4 ELECTRIC POWER

If the potential difference between the terminals of a battery is V, then a charge q emerging from the battery has an energy qV, according to the definition of potential given in Equation 24.3. Clearly, qV has units of energy (joules) since q is measured in coulombs and V is measured in volts or joules/coulomb. Since the amount of energy per second is the power, dividing qV by the time t gives the electric power qV/t provided to the circuit. The charge flowing per second q/t is the current I, so the electric power is IV, the product of current and voltage.

ELECTRIC POWER

When there is a current I in a circuit as a result of a voltage V, the electric power P delivered to the circuit is

$$P = IV \qquad (25.6)$$

SI unit of power: watt (W)

Power is measured in watts, and Equation 25.6 indicates, therefore, that the product of an ampere and a volt is equal to a watt.

Many electrical devices are essentially resistors that become hot when provided with sufficient electric power: toasters, irons, space heaters, heating elements on electric stoves, and incandescent light bulbs, to name just a few. In such cases, it is possible to obtain two additional, but equivalent, expressions for the power. These two expressions follow directly upon substituting $V = IR$ into the relation $P = IV$:

$$P = IV \qquad\qquad (25.6a)$$

$$P = I(IR) = I^2R \qquad\qquad (25.6b)$$

$$P = \left(\frac{V}{R}\right)V = \frac{V^2}{R} \qquad\qquad (25.6c)$$

Example 4 deals with the electric power utilized by the bulb of a flashlight.

EXAMPLE 4

In the flashlight in Figure 25.5, the current is 0.40 A, and the voltage is 3.0 V. Find (a) the power delivered to the bulb and (b) the energy dissipated in the bulb in 5.5 minutes of operation.

SOLUTION

(a) The power is the product of current and voltage:

$$P = IV = (0.40 \text{ A})(3.0 \text{ V}) \qquad (25.6a)$$

$$= \boxed{1.2 \text{ W}}$$

The wattage rating of this bulb would therefore be 1.2 W.

(b) The energy consumed in 5.5 minutes follows from the definition of power as energy per unit time:

$$\text{Energy} = Pt = (1.2 \text{ W})(330 \text{ s}) = \boxed{4.0 \times 10^2 \text{ J}}$$

Monthly electric bills specify the cost for the amount of energy consumed during the month. Energy is the product of power and time, and electric companies compute your energy consumption using power in kilowatts and time in hours. Therefore, a commonly used unit for energy is the *kilowatt-hour* (kWh). For instance, if you used an average power of 2200 watts (2.2 kW) for thirty days (720 h), your energy consumption would be 1600 kWh. At a cost of $0.10 per kWh, your monthly bill would be $160. One kilowatt-hour equals 3.60×10^6 J of energy.

Figure 25.7 shows an interesting application of the heat generated when electric charge flows through a resistance. The application involves the use of a bimetallic strip formed by fastening together two strips of *dissimilar* metals. As Section 14.4 discusses, the metals expand by different amounts when heated, causing the strip itself to bend. The drawing shows a bimetallic strip with a resistance heater wire wrapped around it. While the strip is cool, its end touches the contact point. Charges from the battery pass directly through the strip and cause the light bulb to glow. However, as charges continue to flow, the resistance heater becomes hot, and the bimetallic strip bends away from the contact point, shutting off the current in the circuit. The light goes out. As the bimetallic strip cools, it bends back and touches the contact point again, turning the light back on. The on–off cycle repeats itself every second or so, and the result is a flashing light that can be used as a warning device.

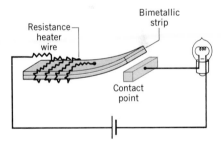

FIGURE 25.7 A bimetallic strip flasher.

25.5 ALTERNATING CURRENT

Many electric circuits use batteries and involve direct current (dc). However, there are considerably more circuits that operate with alternating current (ac), in which the direction of charge flow reverses periodically. For example, the electrical outlets in a house provide alternating current, so all of us use ac circuits routinely. This section deals with ac circuits that contain only resistance.

The common generators that create ac electricity depend on magnetic forces for their operation and are discussed in Chapter 28, after the basic principles of magne-

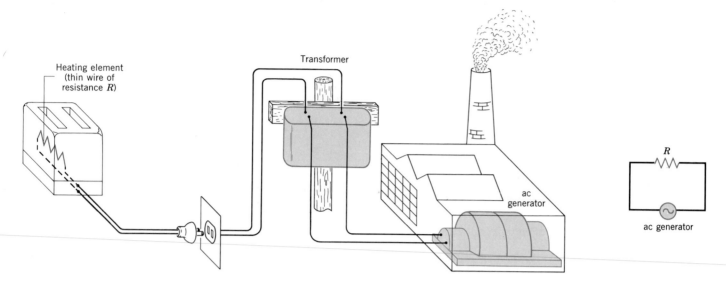

FIGURE 25.8 A circuit consisting of a toaster and the ac generator (including transformer) provided by the electric power company.

tism have been presented. In an ac circuit, these generators serve the same purpose as a battery serves in a dc circuit, that is, they give energy to the moving electric charges.

Figure 25.8 shows the ac circuit that is formed when a toaster is plugged into a wall socket. The heating element of a toaster is essentially a thin wire of resistance R and becomes red hot when energy is dissipated in it. The circuit schematic in the picture introduces the symbol $\odot$ that is used to represent the generator. In this case, the generator (including the transformer) is provided by the electric power company.

Figure 25.9 shows a graph that records the voltage produced between the terminals

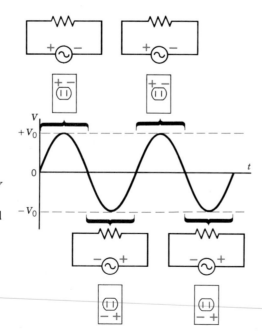

FIGURE 25.9 The voltage produced between the terminals of an ac generator fluctuates from moment to moment. A graph of the voltage V versus time t is sinusoidal in the most common case. The circuits and wall sockets indicate the relative polarity of the generator terminals during the positive and negative parts of the sinusoidal graph. For clarity, the wall sockets do not include the hole for the third prong of a three-prong plug.

of the most common kind of ac generator at each moment of time. The voltage fluctuates sinusoidally between positive and negative values as time passes. The voltage V that exists at time t is

$$V = V_0 \sin 2\pi ft \qquad (25.7)$$

where V_0 is the maximum or peak value of the voltage, and f is the frequency at which the voltage oscillates. In the United States, the voltage present at most home wall outlets has a peak value of approximately $V_0 = 170$ volts and oscillates with a frequency of 60 Hz. Thus, the period of each cycle is 1/60 s, and the polarity of the generator terminals reverses twice during each cycle, as Figure 25.9 indicates.

The current in an ac circuit also oscillates. In circuits that contain only resistance, the current reverses direction each time the polarity of the generator terminals reverses. Thus, the current in the circuit shown in Figure 25.9 has a frequency of 60 Hz and changes direction twice during each cycle. Substituting $V = V_0 \sin 2\pi ft$ into $V = IR$ shows that the current can be represented as

$$I = \left(\frac{V_0}{R}\right) \sin 2\pi ft = I_0 \sin 2\pi ft \qquad (25.8)$$

The peak current is given by $I_0 = V_0/R$, so it can be calculated if the peak voltage and the resistance are known.

The power delivered to an ac circuit by the generator is given by $P = IV$, just as it is in a dc circuit. However, since both I and V depend on time, the power fluctuates as time passes. Substituting Equations 25.7 and 25.8 for V and I into $P = IV$ gives

$$P = I_0 V_0 \sin^2 2\pi ft \qquad (25.9)$$

This expression is plotted in Figure 25.10.

Since the power fluctuates in an ac circuit, it is customary to consider the average power $\overline{P}$, which is just one-half the peak power, as Figure 25.10 indicates:

$$\overline{P} = \tfrac{1}{2} I_0 V_0 \qquad (25.10)$$

On the basis of this expression, a kind of average current and average voltage can be introduced that are very useful when discussing ac circuits. A slight rearrangement of Equation 25.10 reveals that

$$\overline{P} = \left(\frac{I_0}{\sqrt{2}}\right)\left(\frac{V_0}{\sqrt{2}}\right) = I_{rms} V_{rms} \qquad (25.11)$$

I_{rms} and V_{rms} are called the **root mean square (rms)** current and voltage, respectively, and may be calculated from the peak values by dividing them by $\sqrt{2}$*:

$$I_{rms} = \frac{I_0}{\sqrt{2}} \qquad (25.12)$$

$$V_{rms} = \frac{V_0}{\sqrt{2}} \qquad (25.13)$$

For instance, in the United States the maximum ac voltage at a home wall socket is typically $V_0 = 170$ volts, and the corresponding rms voltage is $V_{rms} = 170$ volts/$\sqrt{2} = 120$ volts. When the instructions for an electric appliance specify 120 V, it is an

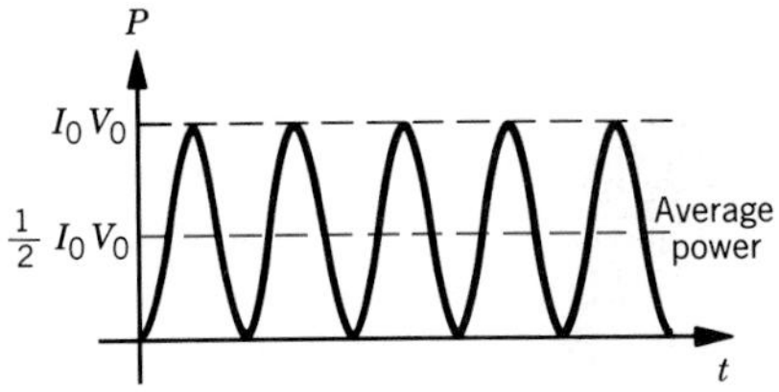

---

* Equations 25.12 and 25.13 apply only for sinusoidal voltage and current.

FIGURE 25.10 The power P in an ac circuit oscillates between zero and a peak value of $I_0 V_0$, where I_0 and V_0 are the peak current and the peak voltage, respectively. The average power is $\tfrac{1}{2} I_0 V_0$.

rms voltage that is indicated. Similarly, when we specify an ac voltage or current in this text, it is an rms value, unless indicated otherwise. Likewise, when we specify ac power, it is an average power, unless stated otherwise.

Except for dealing with average quantities, the relation $\overline{P} = I_{rms}V_{rms}$ has the same form as Equation 25.6a ($P = IV$). Moreover, Ohm's law can be written conveniently in terms of rms quantities:

$$V_{rms} = I_{rms}R \qquad (25.14)$$

Substituting Equation 25.14 into $\overline{P} = I_{rms}V_{rms}$ shows that the average power can be expressed in one of the following ways:

$$\overline{P} = I_{rms}V_{rms} \qquad (25.11a)$$

$$\overline{P} = I_{rms}^2 R \qquad (25.11b)$$

$$\overline{P} = \frac{V_{rms}^2}{R} \qquad (25.11c)$$

These expressions are completely analogous to $P = IV = I^2R = V^2/R$ for dc circuits. Example 5 deals with the average power in one familiar ac circuit.

EXAMPLE 5

A stereo receiver applies a peak ac voltage of 45 V to a speaker in a stereo system. The speaker is an 8.0-Ω speaker, in the sense that it behaves approximately* as an 8.0-Ω resistance. Figure 25.11 shows the circuit. Determine (a) the rms voltage, (b) the rms current, and (c) the average power for this circuit.

SOLUTION

(a) The peak value of the voltage is 45 V, so the corresponding rms value is

$$V_{rms} = \frac{V_0}{\sqrt{2}} = \frac{45 \text{ V}}{\sqrt{2}} = \boxed{32 \text{ V}} \qquad (25.13)$$

(b) The rms current can be obtained from Ohm's law:

$$I_{rms} = \frac{V_{rms}}{R} = \frac{32 \text{ V}}{8.0 \ \Omega} = \boxed{4.0 \text{ A}} \qquad (25.14)$$

(c) The average power is

$$\overline{P} = I_{rms}V_{rms} = (4.0 \text{ A})(32 \text{ V}) = \boxed{130 \text{ W}} \qquad (25.11)$$

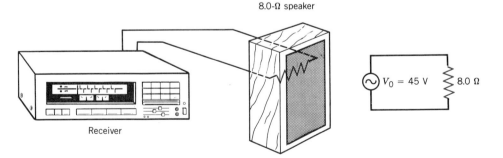

8.0-Ω speaker

$V_0 = 45$ V   8.0 Ω

Receiver

FIGURE 25.11 A stereo receiver applies an ac voltage (peak value = 45 V) to an 8.0-Ω speaker.

---

* Other factors besides resistance can affect the current and voltage in ac circuits; they are discussed in Chapter 29.

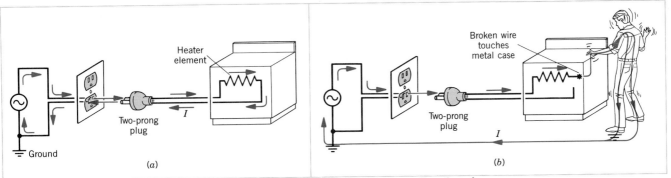

FIGURE 25.12 (*a*) A normally operating clothes drier that is connected to a wall socket via a two-prong plug. (*b*) The drier malfunctions, because an internal wire touches the metal casing. A person who touches the drier can receive an electrical shock.

*25.6 SAFETY AND THE PHYSIOLOGICAL EFFECTS OF CURRENT

Electric circuits, while very useful, can also be hazardous. To reduce the danger inherent in using circuits, proper *electrical grounding* is necessary. Figures 25.12 and 25.13 help to illustrate what electrical grounding means and how it is achieved.

Figure 25.12*a* shows a clothes drier connected to a wall socket via an ordinary two-prong plug. The drier is operating normally, that is, the wires inside are insulated from the metal casing of the drier, so no charge flows through the casing itself. Notice that one terminal of the ac generator is customarily connected to ground (⏚) by the electric power company. Part *b* of the drawing shows the hazardous result that occurs if a wire comes loose and contacts the metal casing of the drier. A person touching the casing receives a shock, since electric charge flows from the generator, through the casing, the person's body, and the ground, arriving back at the generator, as the picture illustrates.

Figure 25.13*a* shows the same clothes drier connected to a wall socket via a three-prong plug that provides safe electrical grounding. The third prong on the plug connects the metal casing of the drier directly to a copper rod driven into the ground

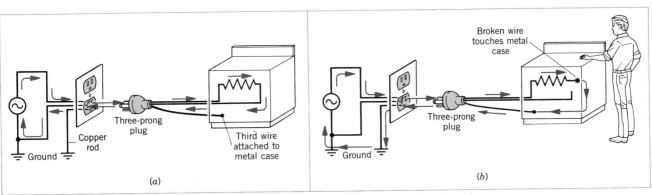

FIGURE 25.13 (*a*) A normally operating clothes drier that is connected to a wall socket via a three-prong plug. (*b*) When the drier malfunctions, a person touching it receives no shock, since electric charge flows through the third prong and into the ground, rather than into the person's body.

or to a copper water pipe that is in the ground. This arrangement protects against electrical shock in the event that a broken wire contacts the metal casing, as in part *b* of the drawing. The charge flows from the generator, through the casing, through the third prong of the plug, and into the ground, returning eventually to the generator. No charge flows into the person's body, since the copper rod connected to the third prong of the plug provides much less electrical resistance than the body does.

Serious and sometimes fatal injuries can result from electric shock. The severity of the injury depends on the magnitude of the current and the parts of the body that the moving charges pass through. The amount of current that the body senses as a mild tingling sensation is about 0.001 A. Currents on the order of 0.01 – 0.02 A can cause muscle spasms, in which a person "can't let go" of the object causing the shock. Currents of approximately 0.2 A are potentially fatal, because they can make the heart fibrillate, or beat in an uncontrolled manner. Substantially larger currents stop the heart completely. However, since the heart often begins beating normally again after the current ceases, the larger currents can be less dangerous than the smaller currents that cause fibrillation.

Since it is the current that determines the severity of an electric shock, one must worry about more than just high voltages when dealing with electrical equipment. Current is determined by electrical resistance, as well as voltage. The resistance of the human body ranges between about 100 and 10^6 Ω, depending on a variety of factors. The dampness of the skin in contact with an electric circuit plays an important role. For example, a person in bare feet on a wet floor presents much less resistance to the current in Figure 25.12*b*, and runs a greater risk of being shocked than does a person wearing a pair of shoes with dry, thick rubber soles. The dry rubber presents a very large resistance, and thus a smaller amount of current is created for a given voltage. If the resistance presented by the body is as small as 100 Ω, even a voltage as small as 20 V can lead to fatal currents.

SUMMARY

There must be at least one source or generator of electrical energy in an electric circuit. The **electromotive force (emf)** of a generator, such as a battery, is the maximum electric potential difference (in volts) that exists between the terminals of the generator.

The rate of flow of charge is called the **electric current** and is given by $I = q/t$, where q is the magnitude of the charge flowing past a given point in a time t. The SI unit for current is the coulomb per second (C/s), and one coulomb per second is referred to as an ampere (A). When the charges flow only in one direction around a circuit, the current is called **direct current (dc).** When the direction of charge flow changes from moment to moment, the current is known as **alternating current (ac). Conventional current** is the hypothetical flow of positive charges that would accomplish the same effect in the circuit as the movement of negative electrons that actually does occur in the wires. The direction of conventional current is always from a point at a higher potential toward a point at a lower potential.

The definition of **electrical resistance** is $R = V/I$, where V is the voltage applied across a piece of material and I is the current through the material. If the ratio V/I is constant for all values of V and I, the relation $R = V/I$ or $V = IR$ is referred to as **Ohm's law.** Resistance is measured in volts per ampere, a unit that is called an ohm (Ω).

The resistance of a piece of material of length L and cross-sectional area A is $R = \rho L/A$, where ρ is the **resistivity** of the material. The resistivity of a conductor is small; in comparison, the resistivity of an insulator is large. The resistivity of a material depends on the temperature. For many materials and limited temperature ranges, the temperature dependence is given by $\rho = \rho_0[1 + \alpha(T - T_0)]$, where ρ and ρ_0 are the resistivities at temperatures T and T_0, respectively, and α is the **temperature coefficient of resistivity.**

In a circuit in which a current I results from a voltage V, the **electric power** delivered to the circuit is $P = IV$. Since $V = IR$, the power dissipated in a resistance R is also given by $P = I^2R$ or $P = V^2/R$.

The **alternating voltage between the terminals of an ac generator** can be represented by $V = V_0 \sin 2\pi ft$, where V_0 is the peak value of the voltage, t is the time, and f is the frequency at which the voltage oscillates. Correspondingly, in a circuit containing only resistance, the **ac current** is $I = I_0 \sin 2\pi ft$, where I_0 is the peak value of the current and is related to the peak voltage via $I_0 = V_0/R$. The **root mean square (rms)** voltage and current are related to the peak values according to $V_{rms} = V_0/\sqrt{2}$ and $I_{rms} = I_0/\sqrt{2}$. The power in an ac circuit is the product of the current and the voltage and oscillates in time. The **average power** is $\overline{P} = I_{rms}V_{rms}$. Since $V_{rms} = I_{rms}R$, the average power can also be written as $\overline{P} = I_{rms}^2 R$ or $\overline{P} = V_{rms}^2/R$.

QUESTIONS

1. The drawing shows a circuit in which a light bulb is connected to the household ac voltage via two switches S_1 and S_2. This is the kind of wiring, for example, that allows you to turn a carport light on and off from either inside the house or out in the carport. Explain which position (A or B) of S_2 turns the light on when S_1 is set to (a) position A and (b) position B.

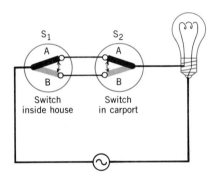

Switch inside house   Switch in carport

2. The resistance of a light bulb is not the same when the bulb is off as it is when the bulb is on. Why?

3. Two materials have different resistivities. Two wires of the same length are made, one from each of the materials. Is it possible for each wire to have the same resistance? Explain.

4. Does the resistance of a copper wire increase or decrease when both the length and the diameter of the wire are doubled? By what factor does the resistance change? Justify your answers.

5. One electrical appliance operates with a voltage of 120 V, while another operates with 240 V. Based on this information alone, is it correct to say that the second device uses more power than the first? Give your reasoning.

6. A long extension cord is used to connect a light to an electrical outlet. The rms current in the bulb is slightly less than that calculated using Ohm's law with the resistance of the light bulb and the rms voltage at the socket. Why?

PROBLEMS

Note: For problems in this set that involve ac conditions, the current and voltage are rms values and the power is an average value, unless indicated otherwise.

Section 25.1 The Purpose of Electric Circuits, Section 25.2 Ohm's Law

1. The wiring in a typical house can safely handle about 15 A of current. At this current level, how much charge flows through a wire in one hour?

2. A battery charger is connected to a dead battery and delivers a current of 8.0 A for 3.0 hours, keeping the voltage across the battery terminals at 12 V in the process. How much energy is delivered to the battery?

3. The heating element of a clothes drier has a resistance of 11 Ω and is connected across a 240-V electrical outlet. What is the current in the heating element?

4. In the arctic, electric socks are useful. A battery-operated pair of socks uses a 9.0-V battery pack for each sock. A current of 0.11 A is drawn from each battery pack by wire woven into the socks. Find the resistance of the wire in one sock.

5. A car battery has a rating of 220 ampere·hours (A·h). This rating is one indication of the *total charge* that the battery can provide to a circuit before failing. (a) Show that an ampere·hour is a unit for measuring charge. Determine the maximum current that the battery can provide for (b) 15 minutes and (c) 38 minutes. (d) What is the total charge that this battery can provide?

* 6. As problem 5 discusses, the "ampere·hour" rating of a battery indicates the total charge that a battery can deliver before failing. Equivalently, this rating also indicates the number of hours that a battery would last if the current that it produced were one ampere. The 3.0-V battery pack in a pocket calculator is rated at 0.080 A·h. When it is operating, the calculator draws 0.15 mA of current from the battery pack. (a) For how many hours can the calculator be operated on a single battery pack? (b) Find the total energy stored in the battery pack.

Section 25.3 Resistance and Resistivity

7. High-voltage power lines are a familiar sight throughout the country. The aluminum wire used for some of these lines has a diameter of 2.5 cm. What is the resistance of ten kilometers of this wire?

8. Two wires have the same length and the same resistance. One is made from copper and the other from aluminum. Obtain the ratio of the diameter of the aluminum wire to that of the copper wire.

9. The drawing shows a resistance-vs.-temperature graph for a piece of wire. Using this graph, and the relation $R = R_0[1 + \alpha(T - T_0)]$, determine α at a temperature of 400 °C.

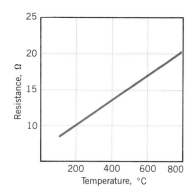

10. A copper wire has a diameter of 1.00 mm. Find the resistance *per unit length* for this wire.

11. A wire of unknown composition has a resistance of $R_0 = 35.0$ Ω when immersed in water at 20.0 °C. When the wire is placed in boiling water, its resistance rises to 47.6 Ω. (a) Based on Table 25.2, what is the material from which the wire is made? (b) What is the temperature of a hot summer day when the wire has a resistance of 37.8 Ω?

* 12. A wire has a resistance of 21 Ω. It is melted down, and from the metal a new wire is made that is three times as long as the original wire. What is the resistance of the new wire?

* 13. Liquid mercury in one rectangular container is poured into another rectangular container that has one-half the cross-sectional area. The resistance between the top and bottom surfaces of the mercury is measured in each case. Find the ratio of the resistance of the mercury in the new container to that of the mercury in the original container.

* 14. An iron wire has a resistance of 5.90 Ω at 20.0 °C and a gold wire has a resistance of 6.70 Ω at the same temperature. At what temperature do the wires have the same resistance?

**15. A digital thermometer uses a thermistor as the tempera-

ture sensing element. A thermistor is a kind of semiconductor and has a large negative temperature coefficient of resistivity α. Suppose $\alpha = -0.0600$ (C°)$^{-1}$ for the thermistor in a digital thermometer used to measure the temperature of a sick patient. The resistance of the thermistor decreases by 15% relative to its value at the normal body temperature of 37.0 °C. What is the patient's temperature?

Section 25.4 Electric Power

16. A stove is connected to a 240-V outlet and receives power P_{240}. When a restaurant owner uses the same stove with a 208-V outlet, the stove receives power P_{208}. Ignoring any change of resistance with temperature, find the ratio P_{240}/P_{208}.

17. An electric alarm clock uses a 5.0-W motor and runs all day, every day. If electricity costs $0.10 per kWh, determine the yearly cost of running the clock.

18. The heating element in a toaster has a resistance of 14 Ω. (a) The toaster is plugged into a 120-V outlet. What is the power dissipated by the toaster? (b) What is the current in the toaster?

19. A cigarette lighter in a car is just a resistor that, when activated, is connected across the 12-V battery. Suppose a lighter dissipates 33 W of power. Find (a) the resistance of the lighter and (b) the current that the lighter draws from the battery.

20. A commercial resistor can safely dissipate power only up to a certain rated level. Beyond this level, the resistor becomes excessively hot and often cracks apart. What is the largest voltage to which a 680-Ω resistor can be connected, when the resistor is rated at (a) 0.25 W and (b) 2.0 W?

* 21. An electric motor rotates the magnetic disks in a personal computer. The motor produces a maximum output power of 0.010 hp when operated with an input current and voltage of 0.65 A and 12 V. One horsepower equals 746 W. Determine the efficiency of the motor, that is, the percentage of the input electrical power converted into output power.

**22. An iron wire has a resistance of 12 Ω at 20.0 °C and a mass of 1.3×10^{-3} kg. A constant current of 0.10 A is sent through the wire for one minute and causes the wire to become hot. Assuming all the electrical energy is dissipated in the wire and remains there, find the final temperature of the wire. *(Hint: Consider the average resistance of the wire during the heating process.)*

Section 25.5 Alternating Current

23. The current in a circuit is ac and has a peak value of 2.50 A. Determine the rms current.

24. A blow-drier and a vacuum cleaner each operate with an ac voltage of 120 V. The current rating of the blow-drier is 11 A, while that of the vacuum cleaner is 4.0 A. Determine the power consumed by (a) the blow-drier and (b) the vacuum cleaner. (c) Determine the ratio of the energy used by the blow-drier in 15 minutes to the energy used by the vacuum cleaner in one-half an hour. (d) The electric power company charges you for energy. Which device is more expensive to operate in part (c)?

25. The heating element in an iron has a resistance of 16 Ω and is connected to a 120-V wall socket. (a) What is the average power consumed by the iron, and (b) the peak power?

26. Birds sit on uninsulated ac power lines without being harmed. Suppose these lines are aluminum (diameter = 1.3 cm) and carry a current of 4200 A. The feet of a bird are 6.0 cm apart as it sits on a wire. (a) Find the voltage between the feet of this bird. (b) Considering that the resistance of the bird is much greater than the resistance of the wire between the bird's feet, explain why the bird is not harmed.

＊ 27. The *recovery time* of a hot water heater is the time required to heat all the water in the unit to the desired temperature. Suppose that a 42-gal (1.00 gal = 3.79×10^{-3} m³) unit starts with cold water at 11 °C and delivers hot water at 55 °C. The unit is electric and utilizes a resistance heater (120 V ac, 3.2 Ω) to heat the water. Assuming no heat is lost to the environment, determine the recovery time of the unit.

＊ 28. The heating element in an oven is operated with an ac voltage of 240 V and carries a current of 19 A. A stereo receiver is operated with an ac voltage of 120 V and a current of 0.80 A. Electricity costs $0.10 per kWh. (a) What is the cost of running the oven for two hours? (b) For the amount of money calculated in part (a), how long can the stereo receiver be operated?

＊ 29. On its highest setting, a heating element on an electric stove (see Figure 25.6) is connected to an ac voltage of 240 V. This element has a resistance of 29 Ω. Find (a) the power dissipated in the element and (b) the current in the element. (c) Assuming only three-fourths of the heat produced by the element is used to heat a pot of water (the rest being wasted), find the time required to bring half a gallon of water (1.9 kg, 15 °C) to a boil.

ADDITIONAL PROBLEMS

30. The two headlights of a car consume a total power of 120 W. A driver parks the car but leaves the lights on. The 12-V battery is rated at 95 A·h. (See problems 5 and 6 for an explanation of this rating.) How long does it take for the battery to lose its charge?

31. The drawing shows the cross sections of three wires, two square and one circular. Each wire has the same length and is made from the same material. (a) Find the ratio R_A/R_B for the resistances of wires A and B. (b) Find the ratio R_A/R_C for the resistances of wires A and C.

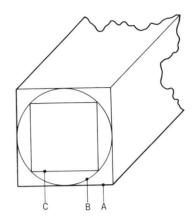

32. An electric blanket is connected to a 120-V outlet and consumes 140 W of power. What is the resistance of the wire in the blanket?

33. The filament of a light bulb has a resistance of 580 Ω. A voltage of 120 V is connected across the filament. How much current is in the filament?

＊ 34. An electric furnace runs nine hours a day to heat a house during January (31 days). The heating element has a resistance of 5.3 Ω and carries a current of 44 A. The cost of electricity is $0.10 per kWh. Find the monthly cost of running the furnace.

＊ 35. A piece of nichrome wire has a diameter of 1.3 mm. It is used in a laboratory to make a heater that dissipates 400 W of power when connected to a voltage source of 120 V. Ignoring the effect of temperature on resistance, estimate the necessary length of wire.

＊＊36. Two wires have the same cross-sectional area and are joined end to end to form a single wire. One is tungsten and the other carbon. The total resistance of the composite wire is the sum of the resistances of the pieces. The total resistance of the composite does *not change with temperature*. What is the ratio of the lengths of the tungsten and carbon sections?

CHAPTER 26

Electric Circuits: Additional Concepts

Commander John Young (left) and pilot Robert Crippen go over a checklist on the spaceship *Columbia* during a mission simulation. The numerous lights and switches on the instrument panel hint that many complex electric circuits are used to operate the vehicle.

26.1 SERIES WIRING

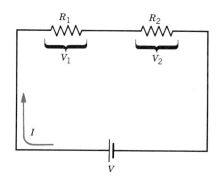

FIGURE 26.1 When two resistors are connected in series, the same current is in both of them.

In the previous chapter all the applications deal with circuits that include only a single device, such as a light bulb or a loudspeaker. There are, however, many important circuits in which more than one device is connected to a voltage source. This section introduces one method by which such connections may be made, namely, series wiring. *Series wiring means that the devices are connected in such a way that there is the same electric current through each device.* Figure 26.1 shows a circuit in which two different devices, represented by the resistors R_1 and R_2, are connected in series with a battery. Note that if the current in one resistor is interrupted, the current in the other is also interrupted. Such an interruption could occur, for example, if two light bulbs were connected in series, and the filament of one bulb broke. Because of the series wiring, the voltage V supplied by the battery is divided between the two resistors. Figure 26.1 indicates that the portion of the voltage across R_1 is V_1, while the portion across R_2 is V_2, so $V = V_1 + V_2$. Applying the definition of resistance to each resistor individually shows that

$$V = IR_1 + IR_2 = I(R_1 + R_2) = IR_S$$

where R_S is called the **equivalent resistance** of the series circuit. Thus, two resistors connected in series are equivalent to a single resistor whose resistance is $R_S = R_1 + R_2$, in the sense that there is the same current through R_S as there is through the series combination of R_1 and R_2. This line of reasoning can be extended to any number of resistors in series, with the result that

$$\begin{bmatrix} \text{Series} \\ \text{resistors} \end{bmatrix} \qquad\qquad R_S = R_1 + R_2 + R_3 + \cdots \qquad\qquad (26.1)$$

Example 1 illustrates the concept of the equivalent resistance in a series circuit.

EXAMPLE 1

A 6.00-Ω resistor and a 3.00-Ω resistor are connected in series with a 12.0-V battery, as Figure 26.2 indicates. Assuming the battery contributes no resistance to the circuit, find (a) the current, (b) the power dissipated in each resistor, and (c) the total power delivered to the resistors by the battery.

SOLUTION

(a) Since the resistors are wired in series, the current I can be determined by applying Ohm's law to an equivalent circuit containing an equivalent resistance R_S, as in the drawing:

$$R_S = 6.00\ \Omega + 3.00\ \Omega = 9.00\ \Omega \qquad (26.1)$$

$$I = \frac{V}{R_S} = \frac{12.0\ \text{V}}{9.00\ \Omega} = \boxed{1.33\ \text{A}} \qquad (25.2)$$

(b) Now that the current is known, the power dissipated in each resistor can be obtained from $P = I^2R$ (Equation 25.6b):

$$\begin{bmatrix} \text{6.00-}\Omega \\ \text{resistor} \end{bmatrix} \quad P = I^2R = (1.33\ \text{A})^2(6.00\ \Omega) = \boxed{10.6\ \text{W}}$$

$$\begin{bmatrix} \text{3.00-}\Omega \\ \text{resistor} \end{bmatrix} \quad P = I^2R = (1.33\ \text{A})^2(3.00\ \Omega) = \boxed{5.31\ \text{W}}$$

(c) The total power delivered by the battery is the sum of the two contributions in part (b): $P = 10.6\ \text{W} + 5.31\ \text{W} = 15.9\ \text{W}$. Alternatively, the total power can be obtained directly by using the equivalent resistance $R_S = 9.00\ \Omega$ and the current from part (a):

$$P = I^2R_S = (1.33\ \text{A})^2(9.00\ \Omega) = \boxed{15.9\ \text{W}}$$

In general, the total power delivered to any number of resistors in series is equal to the power delivered to the equivalent resistor.

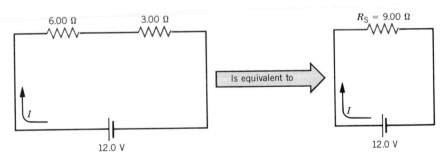

FIGURE 26.2 A 6.00-Ω and a 3.00-Ω resistor connected in series are equivalent to a single 9.00-Ω resistor.

26.2 PARALLEL WIRING

Parallel wiring is another method of connecting multiple devices together. *Parallel wiring means that the devices are connected in such a way that the same voltage is*

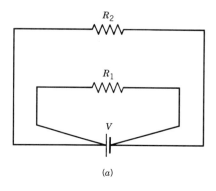

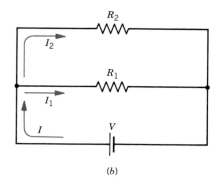

FIGURE 26.3 (a) When two resistors are connected in parallel, the same voltage V is applied across each resistor. (b) This circuit drawing is equivalent to that in part a. I_1 and I_2 are, respectively, the currents in R_1 and R_2.

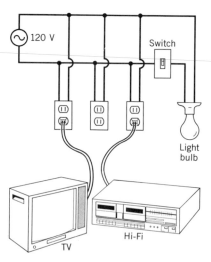

FIGURE 26.4 This drawing shows some of the parallel connections found in a typical home. Each wall socket provides 120 V to the appliance connected to it. In addition, 120 V is applied to the light bulb when the switch is turned on.

applied across each device. Figure 26.3 shows two resistors connected in parallel between the terminals of a battery. Part *a* of the picture is drawn so as to emphasize that the entire voltage of the battery is applied across each resistor. Actually, parallel connections are rarely drawn in this manner; instead they are drawn as in part *b*, where the dots indicate the points where the wires for the two branches are joined together. Parts *a* and *b* are equivalent representations of the same circuit.

Parallel wiring is quite common. In fact, when an electrical appliance is plugged into a wall socket, the appliance is connected in parallel with other appliances already operating, as in Figure 26.4, where the entire voltage of 120 V is applied across the television, the hi-fi, and the light bulb. The presence of the unused socket or other devices that are turned off does not affect the operation of those devices that are turned on. Moreover, if the current in one device is interrupted (perhaps by an opened switch or a broken wire), the current in the other devices is not interrupted. In contrast, if household appliances were connected in series, there would be no current through any appliance if the current at any point in the circuit were halted.

When two resistors R_1 and R_2 are connected as in Figure 26.3, each receives current from the battery as if the other were not present. Therefore, R_1 and R_2 together draw more current from the battery than does either resistor alone. According to the definition of resistance, a larger current arises from a smaller resistance. Thus, the two parallel resistors behave as a single equivalent resistance that is *smaller* than either R_1 or R_2. Figure 26.5 returns to the water flow analogy to provide additional insight into this important feature of parallel wiring. In part *a*, two sections of pipe that have the same length are connected in parallel with a pump. In part *b*

FIGURE 26.5 (a) Two equally long pipe sections, with cross-sectional areas A_1 and A_2, are connected in parallel to a water pump. (b) The two pipe sections in parallel are equivalent to a single pipe of the same length whose cross-sectional area is $A_1 + A_2$. However, the radius of the larger pipe does not equal the sum of the radii of the smaller pipes.

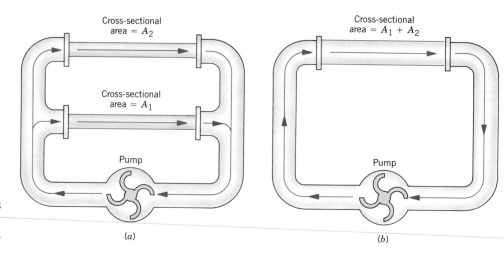

these two sections have been replaced with a single pipe of the same length, whose cross-sectional area equals the combined cross-sectional areas of section 1 and section 2. The pump can push more water per second through the wider pipe in part *b* than it can through *either* of the narrower pipes in part *a*. In effect, the wider pipe (sections 1 and 2 acting together) offers less resistance to the flow than either of the narrower pipes offers individually.

As in a series circuit, it is possible to replace a parallel combination of resistors with an equivalent resistor that results in the same total current and power for a given voltage as the original combination. The equivalent resistance of two resistors connected in parallel can be determined in the following fashion. First, notice in Figure 26.3 that the total current I from the battery is the sum of I_1 and I_2, where I_1 is the current in resistor R_1 and I_2 is the current in resistor R_2: $I = I_1 + I_2$. Since the same voltage V is applied across each resistor, the definition of resistance indicates that $I_1 = V/R_1$ and $I_2 = V/R_2$. Therefore,

$$I = I_1 + I_2 = \frac{V}{R_1} + \frac{V}{R_2} = V\left(\frac{1}{R_1} + \frac{1}{R_2}\right) = V\left(\frac{1}{R_P}\right)$$

where R_P is the equivalent resistance. Hence, when two resistors are connected in parallel, they are equivalent to a single resistor whose resistance R_P is given by $1/R_P = 1/R_1 + 1/R_2$. For any number of resistors in parallel, a similar line of reasoning shows that

$$\begin{bmatrix} \text{Parallel} \\ \text{resistors} \end{bmatrix} \qquad \frac{1}{R_P} = \frac{1}{R_1} + \frac{1}{R_2} + \frac{1}{R_3} + \cdots \qquad (26.2)$$

The next example deals with a parallel combination of resistors that occurs in a hi-fi system.

EXAMPLE 2

Most hi-fi amplifiers allow the user to connect a pair of "remote" speakers (to play music in another room, for instance) in addition to the main speakers. Figure 26.6 shows that the remote speaker and the main speaker for the right stereo channel are connected to the amplifier in parallel (for clarity, the speakers for the left channel are not shown). At the instant shown in the picture, the ac voltage across the speakers is 6.00 V. The main speaker has a resistance of 8.00 Ω, and the remote speaker has a resistance of 4.00 Ω.* Determine (a) the equivalent resistance of the two speakers, (b) the total current supplied by the amplifier, (c) the current in each speaker, (d) the power dissipated in each speaker, and (e) the total power delivered by the amplifier.

SOLUTION

(a) Since the speakers are in parallel, the equivalent resistance is

$$\frac{1}{R_P} = \frac{1}{8.00 \ \Omega} + \frac{1}{4.00 \ \Omega} = \frac{3}{8.00 \ \Omega} \qquad (26.2)$$

$$R_P = \frac{8.00 \ \Omega}{3} = \boxed{2.67 \ \Omega}$$

This result is illustrated in part *b* of the drawing.

(b) The total current supplied by the amplifier can be obtained from Ohm's law and the equivalent resistance:

$$I_{\text{rms}} = \frac{V_{\text{rms}}}{R_P} = \frac{6.00 \ \text{V}}{2.67 \ \Omega} = \boxed{2.25 \ \text{A}}$$

(c) The current in each speaker can be determined from Ohm's law and the resistance of each speaker:

$$\begin{bmatrix} 8.00\text{-}\Omega \\ \text{speaker} \end{bmatrix} \quad I_{\text{rms}} = \frac{V_{\text{rms}}}{R} = \frac{6.00 \ \text{V}}{8.00 \ \Omega} = \boxed{0.750 \ \text{A}}$$

$$\begin{bmatrix} 4.00\text{-}\Omega \\ \text{speaker} \end{bmatrix} \quad I_{\text{rms}} = \frac{V_{\text{rms}}}{R} = \frac{6.00 \ \text{V}}{4.00 \ \Omega} = \boxed{1.50 \ \text{A}}$$

---

* In reality, frequency-dependent characteristics of the speaker (see Chapter 29) play a role in the operation of a loudspeaker. We assume here, however, that the frequency of the sound is low enough that the speakers behave as pure resistances.

The sum of these currents is equal to the current from the amplifier, as determined in part (b).

(d) The average power dissipated in each speaker can be calculated using $\bar{P} = I_{rms}V_{rms}$ (Equation 25.11):

$\begin{bmatrix} \textbf{8.00-}\Omega \\ \textbf{speaker} \end{bmatrix}$ $\bar{P} = I_{rms}V_{rms} = (0.750 \text{ A})(6.00 \text{ V}) = \boxed{4.50 \text{ W}}$

$\begin{bmatrix} \textbf{4.00-}\Omega \\ \textbf{speaker} \end{bmatrix}$ $\bar{P} = I_{rms}V_{rms} = (1.50 \text{ A})(6.00 \text{ V}) = \boxed{9.00 \text{ W}}$

(e) The total power delivered by the amplifier is the sum of the individual values found in part (d), $\bar{P} = 4.50 \text{ W} + 9.00 \text{ W} = 13.5 \text{ W}$. Alternatively, the total power can be obtained using the total current supplied by the amplifier, as determined in part (b):

$$\bar{P} = I_{rms}V_{rms} = (2.25 \text{ A})(6.00 \text{ V}) = \boxed{13.5 \text{ W}}$$

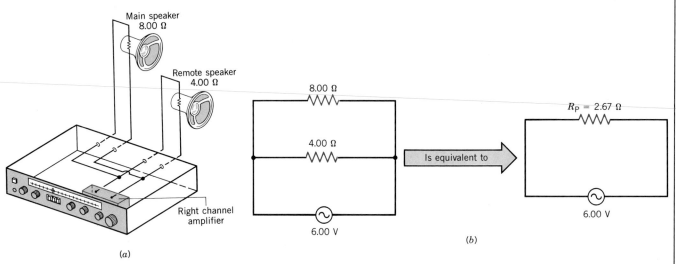

(a)

FIGURE 26.6 (*a*) The main and remote speakers in a hi-fi system are connected in parallel to the amplifier. Only the main and remote speakers for the right stereo channel are shown. (*b*) The circuit schematic shows the situation when the ac voltage across the speakers is 6.00 V.

When a number of resistors are connected in parallel, the equivalent resistance is *less than* any of the individual resistances [see part (a) of Example 2]. In fact, it is the *smallest* resistance that has the largest impact in determining the equivalent resistance. If one resistance approaches zero, then according to Equation 26.2, the equivalent resistance also approaches zero. In such a case, the zero resistance is said to *short out* the other resistances, by providing a zero resistance path for the current to follow as a shortcut around the other resistances.

26.3 CIRCUITS THAT ARE WIRED PARTIALLY IN SERIES AND PARTIALLY IN PARALLEL

Often an electric circuit is wired partially in series and partially in parallel. The key to determining current, voltage, and power in such a case is to deal with the circuit in parts, with the resistances in each part being either in series or in parallel with each other. Example 3 shows how this analysis is carried out.

EXAMPLE 3
Figure 26.7 shows a circuit composed of a 24-V battery and four resistors, whose resistances are 110, 180, 220, and 250 Ω. Find (a) the total current supplied by the battery and (b) the voltage between points A and B in the circuit.

SOLUTION
(a) The total current supplied by the battery can be obtained from $I = V/R$, where R is the equivalent resistance of the four resistors. The equivalent resistance can be calculated by dealing with the circuit in parts. The 220-Ω resistor and the 250-Ω resistor are in series, so they are equivalent to a single resistor whose resistance is $220\ \Omega + 250\ \Omega = 470\ \Omega$ (see Figure 26.7). The 470-Ω resistor is in parallel with the 180-Ω resistor. Their equivalent resistance is given by Equation 26.2:

$$\frac{1}{R_P} = \frac{1}{470\ \Omega} + \frac{1}{180\ \Omega} = 0.0077\ \Omega^{-1}$$

$$R_P = \frac{1}{0.0077\ \Omega^{-1}} = 130\ \Omega$$

The circuit is now equivalent to a circuit containing a 110-Ω resistor in series with a 130-Ω resistor (see the drawing). This combination acts like a single resistor whose resistance is $R = 110\ \Omega + 130\ \Omega = 240\ \Omega$. The total current from the battery is, then,

$$I = \frac{V}{R} = \frac{24\ \text{V}}{240\ \Omega} = \boxed{0.10\ \text{A}}$$

(b) Ohm's law indicates that the voltage across the 130-Ω resistor between points A and B is

$$V = IR = (0.10\ \text{A})(130\ \Omega) = \boxed{13\ \text{V}}$$

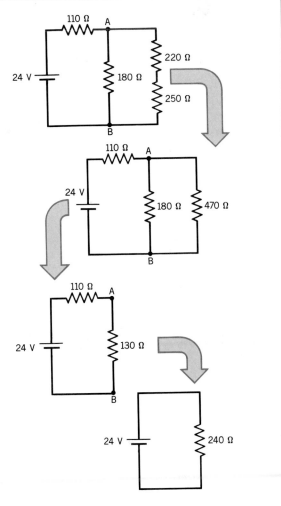

FIGURE 26.7 The four circuits shown in this picture are equivalent.

26.4 INTERNAL RESISTANCE

So far, the circuits considered in this chapter and in Chapter 25 have used batteries or generators that add only their emfs to a circuit. In reality, however, such devices also add some resistance. This resistance is called the **internal resistance** of the battery or generator, because it is located inside the device. In a battery, the internal resistance is the resistance encountered by the current due to the chemicals within the battery. In a generator, the internal resistance is the resistance of wires and other components in the generator.

Figure 26.8 shows a schematic representation of the internal resistance r of a battery. The drawing emphasizes that when a resistance R is connected to the battery, the resistance is connected *in series* with the internal resistance. The internal resis-

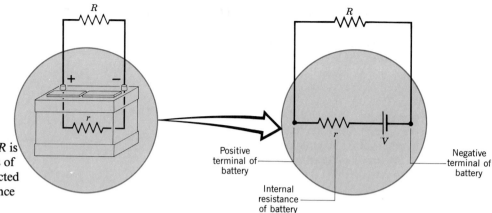

FIGURE 26.8 When a resistance R is connected between the terminals of a battery, the resistance is connected in series with the internal resistance r of the battery.

tance of a functioning battery is typically small (several thousandths of an ohm for a new car battery). Nevertheless, the effect of the internal resistance may not be negligible. Example 4 illustrates that when current is drawn from a battery, the internal resistance causes the voltage between the terminals to drop below the maximum value specified by the battery's emf. The actual voltage between the terminals of a battery is known as the ***terminal voltage.***

EXAMPLE 4

Figure 26.9 shows a car battery whose emf is 12.0 V and whose internal resistance is 0.010 Ω. This resistance is relatively large because the battery is old and the terminals are corroded. What is the terminal voltage when the current I drawn from the battery is (a) 10.0 A and (b) 100.0 A?

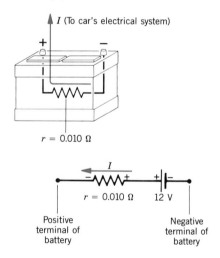

FIGURE 26.9 A 12.0-V car battery, being old and having corroded terminals, has a rather large internal resistance of $r = 0.010$ Ω.

SOLUTION

(a) The voltage between the terminals is not the entire 12.0-V emf, because part of the emf is needed to make the current go through the internal resistance. The amount of voltage needed can be determined from $V = Ir$, since the current through the battery is known: $V = (10.0 \text{ A})(0.010 \text{ Ω}) = 0.10$ V. To find the terminal voltage, remember that the direction of conventional current is always from high potential toward low potential. To emphasize this fact in the drawing, plus and minus signs have been included at the right and left ends, respectively, of the resistance r. The terminal voltage can be calculated by starting at the negative terminal of the battery and following how the voltage increases and decreases as we move toward the positive terminal. The voltage rises by 12.0 V due to the battery's emf. However, the voltage drops by 0.10 V because of the potential difference across the internal resistance. Therefore, the terminal voltage is 12.0 V − 0.10 V = $\boxed{11.9 \text{ V}}$.

(b) When the current through the battery is 100.0 A, the amount of voltage needed to make the current go through the internal resistance is $V = (100.0 \text{ A})(0.010 \text{ Ω}) = 1.0$ V. The terminal voltage of the battery now decreases to 12.0 V − 1.0 V = $\boxed{11.0 \text{ V}}$.

Example 4 indicates that the terminal voltage of a battery is smaller when the current drawn from the battery is larger, an effect that any car owner can demonstrate. Turn the headlights on before starting your car, so the current through the battery is about 10 A, as in part (a) of Example 4. Then start the car. The starter motor draws a large amount of additional current from the battery, momentarily increasing the total current by an appreciable amount. Consequently, the terminal voltage of the battery decreases, causing the headlights to dim.

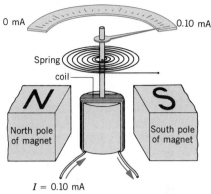

26.5 THE MEASUREMENT OF CURRENT, VOLTAGE, AND RESISTANCE

THE GALVANOMETER

Current and voltage can be measured with devices known, respectively, as ammeters and voltmeters. There are two types of such devices, those that use digital electronics and those that do not. The essential feature of nondigital devices is the dc *galvanometer*. As Figure 26.10 illustrates, a galvanometer consists of a coil of wire, a pointer, a spring, a magnet, and a calibrated scale. The coil is mounted in such a way that it can rotate, the rotation causing the pointer to move in relation to the scale. As Section 27.6 will discuss, the coil rotates in response to the torque applied by the magnet when there is a current in the coil. The coil stops rotating when this torque is counteracted by the torque of the spring.

A galvanometer has two important characteristics that must be considered when it is used as part of a measurement device. First, the amount of dc current that causes a full-scale deflection of the pointer indicates the sensitivity of the galvanometer. For instance, Figure 26.10 shows an instrument that deflects full scale when the current in the coil is 0.10 mA. The second important characteristic is the resistance R_C of the wire in the coil. Figure 26.11 shows how a galvanometer with a coil resistance of $R_C = 50 \ \Omega$ is represented in a circuit diagram. Both the full-scale current and the coil resistance depend on the details of the galvanometer construction.

FIGURE 26.10 A dc galvanometer consists of a coil of wire, a pointer, a spring, a magnet, and a measurement scale. The coil rotates when there is a current in the wire. In the example shown, the pointer deflects full scale when the current is 0.10 mA.

THE AMMETER

An *ammeter* (*am*pere *meter*) is a device for measuring current and must be inserted into the circuit so the current passes directly through the ammeter, as Figure 26.12

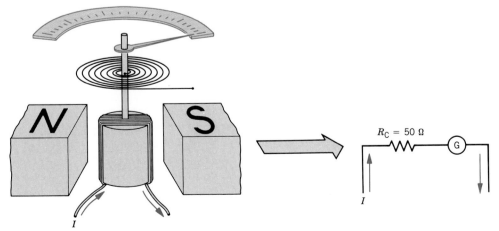

FIGURE 26.11 This picture shows how a galvanometer with a coil resistance of $R_C = 50 \ \Omega$ is represented in a circuit diagram.

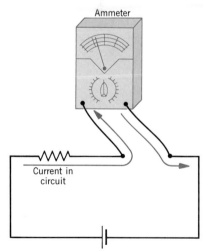

FIGURE 26.12 An ammeter measures the current in a circuit. The ammeter is inserted into the circuit in such a way that the current passes directly through the ammeter.

shows. An ammeter includes a galvanometer and one or more *shunt resistors*. The purpose of a shunt resistor is to allow excess current to bypass the galvanometer coil, so the ammeter can be used to measure a current that exceeds the full-scale amount of the galvanometer. A shunt resistor is connected in parallel with the galvanometer coil, and the next example illustrates how the value of the shunt resistance is selected.

EXAMPLE 5

A galvanometer has a full-scale current of 0.100 mA and a coil resistance of $R_C = 50.0\ \Omega$. This galvanometer is used with a shunt resistor to form an ammeter that will register a full-scale deflection for a current of 60.0 mA. Determine the shunt resistance R.

SOLUTION

Since only 0.100 mA out of the available 60.0 mA is needed to cause a full-scale deflection of the galvanometer, the shunt resistor must allow the excess current of 59.9 mA to detour around the meter coil, as Figure 26.13 indicates. The value for the shunt resistance can be obtained by recognizing that the 50.0-Ω coil resistance and the shunt resistance are in parallel, both being connected between points A and B in the drawing. Thus, the voltage across each resistance is the same. Expressing voltage as the product of current and resistance, we find that

$$\underbrace{(59.9 \times 10^{-3}\ A)(R)}_{\substack{\text{Voltage across} \\ \text{shunt resistance}}} = \underbrace{(0.100 \times 10^{-3}\ A)(50.0\ \Omega)}_{\substack{\text{Voltage across} \\ \text{coil resistance}}}$$

$$R = \frac{(0.100 \times 10^{-3}\ A)(50.0\ \Omega)}{(59.9 \times 10^{-3}\ A)} = \boxed{0.0835\ \Omega}$$

Typically, an ammeter includes a number of shunt resistors that provide several selectable current ranges.

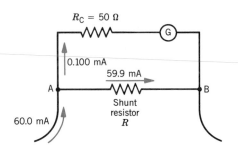

FIGURE 26.13 If a galvanometer with a full-scale deflection of 0.100 mA is to be used to measure a current of 60.0 mA, a shunt resistance R must be used, so the excess current of 59.9 mA can skirt around the galvanometer coil.

When an ammeter is inserted into a circuit, the equivalent resistance of the coil and the shunt resistor adds to the circuit resistance. Any increase in circuit resistance causes a reduction in current, and this is a problem, for an ammeter should only measure the current, not change it. Therefore, an *ideal* ammeter would have zero resistance. In practice, a good ammeter is designed with a sufficiently small equivalent resistance, so there is only a negligible reduction of the current in the circuit when the ammeter is inserted.

THE VOLTMETER

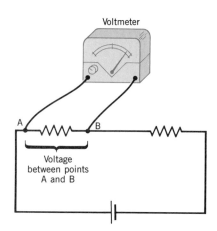

FIGURE 26.14 To measure the voltage between two points, A and B, in a circuit, a voltmeter is connected between the points. The voltmeter is not inserted into the circuit as an ammeter is.

A *voltmeter* is an instrument that measures the voltage between two points, A and B, in a circuit. Figure 26.14 shows that the voltmeter must be connected between the points and is *not* inserted into the circuit as an ammeter is. A voltmeter includes a galvanometer whose scale is calibrated in volts. Suppose, for instance, that a galvanometer has a full-scale current of 0.1 mA and a coil resistance of 50 Ω. Under full-scale conditions, the voltage across the coil would be $V = IR_C = (0.1 \times 10^{-3}\ A)(50\ \Omega) = 0.005\ V$, as Figure 26.15 indicates. Thus, this galvanometer could be used to register voltages in the range 0–0.005 V. A voltmeter, then, is a galvanometer used in this fashion, along with some provision for adjusting the range of voltages to be measured. Example 6 illustrates how the range of voltages can be extended by the simple expedient of connecting a resistor in series with the coil.

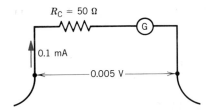

FIGURE 26.15 The galvanometer shown has a full-scale deflection of 0.1 mA and a coil resistance of 50 Ω. Under full-scale conditions, the voltage across the coil is $V = (0.1 \times 10^{-3}\ \text{A})(50\ \Omega) = 0.005\ \text{V}$.

EXAMPLE 6

A galvanometer has a full-scale current of 0.100 mA and a coil resistance of $R_C = 50\ \Omega$. Determine the resistance R that must be connected in series with the coil to produce a voltmeter that will register a full-scale voltage of 0.500 V.

SOLUTION

Figure 26.16 shows the galvanometer connected in series with the resistance R. The resistance R is chosen to ensure the following result: When a voltage of 0.500 V is applied between points A and B in the drawing, the full-scale current of 0.100 mA will be in the galvanometer coil. The equivalent resistance of the series combination is $R + 50\ \Omega$, and according to Ohm's law, $V = I(R + 50\ \Omega)$. Therefore, the resistance R is

$$R = \frac{V}{I} - 50\ \Omega = \frac{0.500\ \text{V}}{0.100 \times 10^{-3}\ \text{A}} - 50\ \Omega = \boxed{4950\ \Omega}$$

Usually a voltmeter includes a number of additional series resistors that provide a variety of selectable voltage ranges.

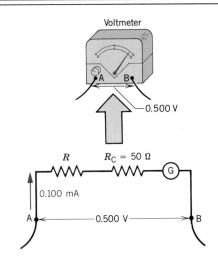

Voltmeter

0.500 V

R $R_C = 50\ \Omega$

0.100 mA

A   0.500 V   B

FIGURE 26.16 A galvanometer can be used to measure voltages over a range of values. The range can be adjusted by connecting a resistance R in series with the galvanometer coil. Example 6 shows how to determine the value of R.

Ideally, the voltage registered by a voltmeter should be the same as the voltage that exists when the voltmeter is not connected. However, a voltmeter takes current from a circuit and, thus, alters the circuit voltage to some extent. An ideal voltmeter would have infinite resistance and draw away only an infinitesimal amount of current. In reality, a good voltmeter is designed with a resistance that is large enough so the unit does not appreciably alter the voltage in the circuit to which it is connected.

THE WHEATSTONE BRIDGE

One way to measure resistance is to use a circuit called a *Wheatstone bridge,* after Charles Wheatstone (1802–1875), the English physicist who established its usefulness. The circuit illustrates a method of measurement known as the *null method.* In addition to the unknown resistance R, a Wheatstone bridge includes three other resistances R_1, R_2, and R_v, as Figure 26.17 shows. A galvanometer records any current between points A and B. To measure the unknown resistance, the variable resistance R_v is adjusted until the galvanometer registers zero or null current, in which case the Wheatstone bridge is said to be "balanced." Example 7 shows how the unknown resistance can be obtained from the value of R_v and the ratio R_1/R_2 in a balanced Wheatstone bridge.

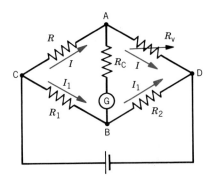

A

R R_v

C I R_C I D

I_1 I_1

R_1 G R_2

B

FIGURE 26.17 A Wheatstone bridge circuit. The resistor marked with the black arrow is a variable resistor.

EXAMPLE 7

When the Wheatstone bridge in Figure 26.17 is balanced, the variable resistance is $R_v = 173 \, \Omega$, and the ratio R_1/R_2 is 0.100. Determine the value of the unknown resistance R.

SOLUTION

The key to solving this problem lies in the information that the bridge is balanced. In other words, there is no current through the galvanometer. The direction of conventional current is from high potential toward low potential. But since there is no current through the galvanometer, points A and B in the circuit *must be at the same potential*. Furthermore, R_1 and R are connected at point C, so the voltages across R_1 and R must be the same. Since voltage is current times resistance and I_1 and I represent the currents in R_1 and R, it follows that

$$I_1 R_1 = IR$$

In a similar fashion, the voltage across R_2 must be the same as that across R_v:

$$I_1 R_2 = I R_v$$

Dividing these two equations shows that $R_1/R_2 = R/R_v$ or

$$R = \frac{R_1}{R_2} R_v = (0.100)(173 \, \Omega) = \boxed{17.3 \, \Omega}$$

26.6 KIRCHHOFF'S RULES

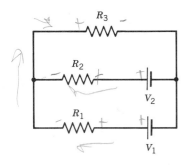

FIGURE 26.18 No two of the three resistors in this circuit are in series, and no two are in parallel. If the two batteries were not present, the three resistors would be in parallel.

Electric circuits that contain a number of resistors can often be analyzed by combining individual groups of resistors in series and parallel, as Section 26.3 discusses. However, there are circuits, such as that in Figure 26.18, in which no two resistors are in series or in parallel. To deal with such circuits it is necessary to employ methods other than the series–parallel method. One alternative is to take advantage of Kirchhoff's rules, named after their developer Gustav Kirchhoff (1824–1887). There are two of these rules, one dealing with the currents in a circuit and the other dealing with the voltages.

Figure 26.19 illustrates the basic idea behind Kirchhoff's first rule, or *junction rule,* as it is called. The picture shows a junction where several wires are connected together. As Section 23.2 discusses, electric charge is conserved. Therefore, since there is no continual accumulation of charges at the junction itself, the total charge per second flowing into the junction must equal the total charge per second flowing out of the junction. In other words, the junction rule states that *the total current directed into a junction must equal the total current directed out of the junction,* or 7 A = 5 A + 2 A for the specific case shown in the picture.

Figure 26.20 illustrates the basic idea behind Kirchhoff's second rule, or *loop rule,* as it is known. The drawing shows a circuit in which a 12-V battery is connected to a series combination of a 5-Ω and a 1-Ω resistor. The plus and minus signs associated with each resistor remind us that conventional current is directed from a higher potential toward a lower potential. Thus, from left to right, there is a potential drop of 10 V across the first resistor and another drop of 2 V across the second resistor. Keeping in mind that potential is the electric potential energy per unit charge, let us follow a positive test charge clockwise* around the circuit. Starting at the negative terminal of the battery, we see that the test charge gains potential energy because of the 12-V rise in potential due to the battery. The test charge then loses potential energy because of the 10-V drop in potential across the first resistor and the 2-V drop across the second resistor, ultimately arriving back at the negative terminal. In traversing the closed circuit loop, the test charge is like a skier gaining gravitational

* The choice of the clockwise direction is arbitrary.

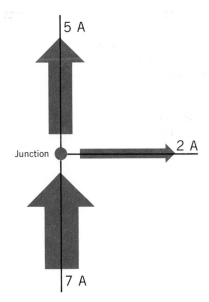

FIGURE 26.19 A junction is a point in a circuit where a number of wires are connected together. If 7 A of current are directed into the junction, then a total of 7 A of current must be directed out of the junction, assuming there is no continual accumulation of charges at the junction.

potential energy in going up a hill on a chair lift and then losing it to friction in coming down. When the skier returns to the starting point, the gain equals the loss, so there is no net change in potential energy. Similarly, when the test charge arrives back at its starting point, there is no net change in electric potential energy, the gains matching the losses. This behavior of the test charge is an example of energy conservation, which the loop rule expresses in terms of the electric potential energy per unit charge. The loop rule states that *for a closed circuit loop, the total of all the potential rises* (**12 V**) *is the same as the total of all the potential drops* (**10 V + 2 V**).

Kirchhoff's rules can be applied to any circuit, even when the resistors are not in series or in parallel. The two rules are summarized below, and Example 8 illustrates how to use them.

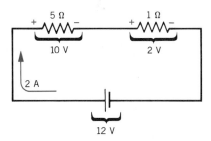

FIGURE 26.20 The plus and minus signs marking the ends of the resistors emphasize that the conventional current of 2 A is directed from a higher potential (+) toward a lower potential (−). Following a positive test charge clockwise around the circuit, we see that the total voltage drop of 10 V + 2 V across the two resistors equals the voltage rise of 12 V provided by the battery.

KIRCHHOFF'S RULES

Junction rule. The sum of the magnitudes of the currents directed into a junction equals the sum of the magnitudes of the currents directed out of the junction.

Loop rule. Around any closed circuit loop, the sum of the potential drops equals the sum of the potential rises.

EXAMPLE 8

In a car, the headlights are connected to the battery and would deplete the battery if it were not for the alternator. The alternator is run by the engine. Figure 26.21 indicates how the headlights and the alternator are connected to the battery. The picture also gives a circuit schematic in which the alternator is approximated as an additional 14.00-V battery for the sake of simplicity. The circuit includes an internal resistance of 0.0100 Ω for the battery and its leads, an internal resistance of 0.100 Ω for the alternator, and a resistance of 1.20 Ω for the headlights. Determine the current through the headlights, the battery, and the alternator.

SOLUTION

The first step in applying Kirchhoff's rules is to label the currents in the headlights (I_H), the battery (I_B), and the alternator (I_A). The drawing shows the directions chosen for these currents. The directions are *arbitrary,* and if any of them is incorrect, then the analysis will show that the corresponding value for the current is negative.

The second step in the solution is to mark the resistors with the plus and minus signs that serve as an aid in identifying the potential drops and rises for the loop rule. In marking the resis-

[Handwritten annotations:]
BEFA
I(6-Ω) -12V + 3 V - I(3-Ω) + I_B(3-Ω)
I(6-Ω) -12V + 3V = 12V + I_B(3-Ω)
I_B(6-Ω) + 3V = 12V
I_B(6-Ω) + I_B(6-Ω) = 12V
I_H(6-Ω) + I_B(6-Ω) =

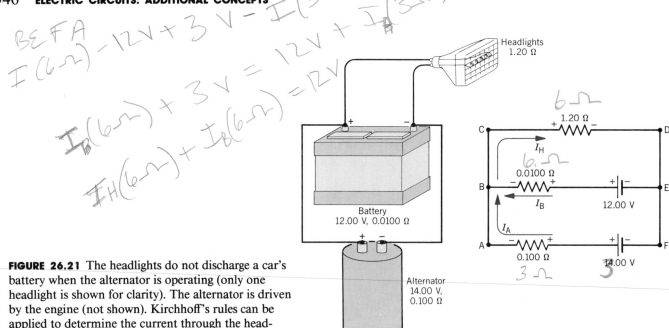

FIGURE 26.21 The headlights do not discharge a car's battery when the alternator is operating (only one headlight is shown for clarity). The alternator is driven by the engine (not shown). Kirchhoff's rules can be applied to determine the current through the headlights, the battery, and the alternator, as Example 8 shows.

tors we remember that conventional current is always directed from a higher potential (+) toward a lower potential (−). Thus, given the directions selected for I_H, I_B, and I_A, the plus and minus signs *must* be those indicated in Figure 26.21.

Kirchhoff's rules can now be used. The junction rule can be applied to junction B or junction E. In either case, the same equation results:

$$\left[\begin{array}{c}\text{Junction rule}\\\text{applied to}\\\text{junction B}\end{array}\right]\quad \underbrace{I_A + I_B}_{\substack{\text{Current}\\\text{into}\\\text{junction}}} = \underbrace{I_H}_{\substack{\text{Current}\\\text{out of}\\\text{junction}}}$$

In applying the loop rule to the lower loop BEFA, we start at point B, move clockwise around the loop, and identify the potential rises and drops. The clockwise direction is arbitrary, and the same result is obtained with a counterclockwise path. There is a potential rise of $I_B(0.0100 \ \Omega)$ across the 0.0100-Ω resistor. This rise is followed by a drop of 12.00 V as we proceed from the positive terminal to the negative terminal of the battery. Continuing on around the loop, we find a 14.00-V rise across the alternator, followed by a drop of $I_A(0.100 \ \Omega)$ across the 0.100-Ω resistor. Setting the sum of the potential drops equal to the sum of the potential rises gives the following result:

$$\left[\begin{array}{c}\text{Loop rule}\\\text{applied}\\\text{clockwise}\\\text{around BEFA}\end{array}\right]\quad \underbrace{I_A(0.100 \ \Omega) + 12.00 \text{ V}}_{\text{Potential drops}}$$
$$= \underbrace{I_B(0.0100 \ \Omega) + 14.00 \text{ V}}_{\text{Potential rises}}$$

Since there are three unknown variables in this problem, I_A, I_B, and I_H, a third equation is needed for a solution. To obtain the third equation, we apply the loop rule to the upper loop CDEB, choosing a clockwise path for convenience. The result is

$$\left[\begin{array}{c}\text{Loop rule}\\\text{applied}\\\text{clockwise}\\\text{around CDEB}\end{array}\right]\quad \underbrace{I_B(0.0100 \ \Omega) + I_H(1.20 \ \Omega)}_{\substack{\text{Potential}\\\text{drops}}} = \underbrace{12.00 \text{ V}}_{\substack{\text{Potential}\\\text{rises}}}$$

These three equations can be solved simultaneously to show that

$$\boxed{I_H = 10.1 \text{ A}, \ I_B = -9.0 \text{ A}, \ I_A = 19.1 \text{ A}}$$

The negative answer for I_B indicates that the current through the battery is not directed from right to left, as drawn in Figure 26.21. Instead, the 9.0-A current is directed from left to right, opposite to the way current would be directed if the alternator were not connected. It is the left-to-right current created by the alternator that keeps the battery charged.

26.7 CAPACITORS IN SERIES AND PARALLEL

Capacitors, like resistors, can be connected in series and in parallel. Consider a combination of parallel capacitors first, since it is easier to understand than a series

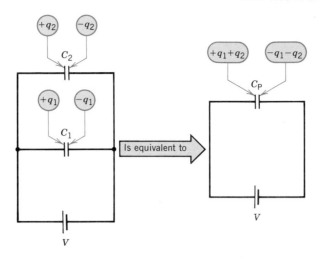

FIGURE 26.22 In a parallel combination of capacitances C_1 and C_2, the voltage V across each capacitor is the same, but the charges q_1 and q_2 on each capacitor are different. The combination of capacitors can be replaced by an equivalent capacitor that stores the charge $q = q_1 + q_2$. The equivalent capacitance is $C_P = C_1 + C_2$.

combination. Figure 26.22 shows two capacitors connected in parallel to a battery. Since the capacitors are in parallel, they have the same voltage V across their plates. However, the capacitors *contain different amounts of charge.* The charge stored by a capacitor is $q = CV$ (Equation 24.8), so $q_1 = C_1V$ and $q_2 = C_2V$.

As with resistors, it is always possible to replace a parallel combination of capacitors with an *equivalent capacitor* that stores the same charge and energy for a given voltage as the combination does. To determine the equivalent capacitance C_P, note that the total charge q stored by the two capacitors is $q = q_1 + q_2$. Consequently,

$$q = q_1 + q_2 = C_1V + C_2V = (C_1 + C_2)V = C_PV$$

This result indicates that two capacitors in parallel can be replaced by an equivalent capacitor whose capacitance is $C_P = C_1 + C_2$. For any number of capacitors in parallel, the equivalent capacitance is

$$\begin{bmatrix} \textbf{Parallel} \\ \textbf{capacitors} \end{bmatrix} \qquad C_P = C_1 + C_2 + C_3 + \cdots \qquad (26.3)$$

Capacitances in parallel simply add together to give an equivalent capacitance. This behavior contrasts with that of resistors in parallel, which combine as reciprocals, according to Equation 26.2.

The equivalent capacitor not only stores the same amount of charge as the parallel combination of capacitors, but it also stores the same amount of energy. For instance, the energy stored in a single capacitor is $\frac{1}{2}CV^2$ (Equation 24.11), so the total energy U stored by two capacitors in parallel is

$$U = \tfrac{1}{2}C_1V^2 + \tfrac{1}{2}C_2V^2 = \tfrac{1}{2}(C_1 + C_2)V^2 = \tfrac{1}{2}C_PV^2$$

which is equal to the energy stored in the equivalent capacitor C_P.

When capacitors are connected in series, the equivalent capacitance is calculated differently than when they are in parallel. As an example, Figure 26.23 shows two capacitors in series and reveals the following important fact: All capacitors in series, regardless of their capacitances, *contain charges of the same magnitude, $+q$ and $-q$, on their plates.* The battery places a charge of $+q$ on plate a of capacitor C_1, and this charge induces a charge of $+q$ to depart from the opposite plate a', leaving behind a charge $-q$. The $+q$ charge that leaves plate a' is deposited on plate b of capacitor C_2 (since these two plates are connected by a wire), where it induces a $+q$ charge to move

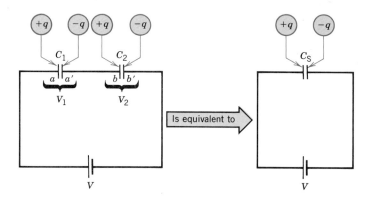

FIGURE 26.23 In a series combination of capacitances C_1 and C_2, the same amount of charge q is on the plates of each capacitor, but the voltages V_1 and V_2 across each capacitor are different. The combination can be replaced by an equivalent capacitor that stores a charge q. The equivalent capacitance C_S is given by $1/C_S = 1/C_1 + 1/C_2$.

away from the opposite plate b′, leaving behind a charge of $-q$. Thus, all capacitors in series contain charges of the same magnitude on their plates.

Note the difference between charging capacitors in parallel and in series. In charging parallel capacitors, the battery moves a charge q that is the *sum* of the charges moved for each of the capacitors: $q = q_1 + q_2 + q_3 + \cdots$. In contrast, in a series combination of n capacitors, the battery only moves a charge q, not nq, because the charge q passes by induction from one capacitor directly to the next one in line.

The equivalent capacitance C_S for the series connection in Figure 26.23 can be determined by observing that the battery voltage V is shared by the two capacitors, as it is in a series combination of resistors. The drawing indicates that the voltages across C_1 and C_2 are V_1 and V_2 and that $V = V_1 + V_2$. The voltage across each capacitor is related to the magnitude of the charge on each capacitor according to $V_1 = q/C_1$ and $V_2 = q/C_2$, the charge q being the same for each. Therefore,

$$V = \frac{q}{C_1} + \frac{q}{C_2} = q\left(\frac{1}{C_1} + \frac{1}{C_2}\right) = q\left(\frac{1}{C_S}\right)$$

where C_S is the equivalent capacitance. Thus, two capacitors in series can be replaced by a single capacitor whose capacitance C_S is given by $1/C_S = 1/C_1 + 1/C_2$. For any number of capacitors connected in series the result is

$$\begin{bmatrix} \textbf{Series} \\ \textbf{capacitors} \end{bmatrix} \qquad \frac{1}{C_S} = \frac{1}{C_1} + \frac{1}{C_2} + \frac{1}{C_3} + \cdots \qquad (26.4)$$

Equation 26.4 indicates that capacitances in series combine as reciprocals and do not simply add together as resistors in series do. It is left as an exercise (problem 41) to show that the equivalent capacitance stores the same electrostatic energy as the sum of the energies of the individual capacitors in the series combination.

It is possible to simplify circuits containing a number of capacitors in the same general fashion as that outlined for resistors in Example 3 and Figure 26.7. The capacitors in a parallel grouping can be combined according to Equation 26.3, and those in a series grouping can be combined according to Equation 26.4.

26.8 RC CIRCUITS

Many electric circuits contain both resistors and capacitors. Figure 26.24 illustrates an example of a resistor–capacitor or *RC* circuit. In part *a* of the drawing a switch

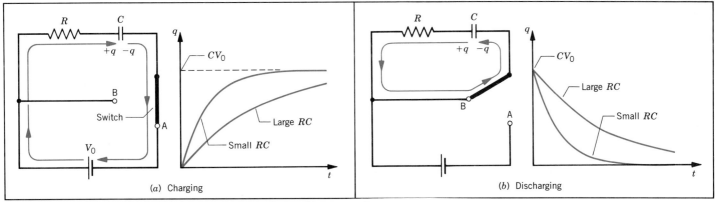

FIGURE 26.24 (*a*) Charging a capacitor and (*b*) discharging a capacitor.

completes the circuit at point A, so that the battery can charge up the capacitor plates. When the switch is closed, the capacitor does not charge up immediately. Rather, the charge builds up gradually to its equilibrium value of $q = CV_0$, where V_0 is the voltage of the battery. Assuming the capacitor is uncharged at time $t = 0$, it can be shown that the magnitude q of the charge on the plates at time t is given by

$$\begin{bmatrix} \textbf{Capacitor} \\ \textbf{charging} \end{bmatrix} \qquad q = CV_0[1 - e^{-t/(RC)}] \qquad (26.5)$$

where the exponential e has the value of 2.718. . . . Figure 26.24*a* also shows a graph of this expression, which indicates that the charge is $q = 0$ when $t = 0$ and increases gradually toward the equilibrium value of $q = CV_0$. The voltage V across the capacitor at any time can be obtained from Equation 26.5 by dividing the charge by C, since $V = q/C$.

The term RC in the exponent in Equation 26.5 is called the ***time constant τ*** of the circuit:

$$\tau = RC \qquad (26.6)$$

The time constant is measured in seconds; verification of the fact that an ohm times a farad is equivalent to a second is left as an exercise (see question 8). The time constant is the amount of time required for the capacitor to accumulate 63.2% of its equilibrium charge, as can be seen by substituting $t = \tau = RC$ in Equation 26.5; $q = CV_0(1 - e^{-1}) = CV_0(1 - 0.368) = CV_0(0.632)$. The charge approaches its equilibrium value rapidly when the time constant is small and slowly when the time constant is large.

In Figure 26.24*b* the switch completes the circuit at point B, so the charge $+q$ on the left plate of the capacitor can flow counterclockwise through the resistor and neutralize the charge $-q$ on the right plate. Assuming the fully charged capacitor begins discharging at time $t = 0$, it can be shown that

$$\begin{bmatrix} \textbf{Capacitor} \\ \textbf{discharging} \end{bmatrix} \qquad q = CV_0e^{-t/(RC)} \qquad (26.7)$$

where q is the amount of charge remaining on either plate at time t. The graph of this expression in part *b* of the drawing shows that the charge begins at $q = CV_0$ when $t = 0$ and decreases gradually toward zero. Smaller values of the time constant RC lead to a more rapid discharge. Equation 26.7 indicates that when $t = \tau = RC$, the magnitude of the charge remaining on each plate is $q = CV_0e^{-1} = CV_0(0.368)$.

Therefore, the time constant is also the amount of time required for a fully charged capacitor to *lose* 63.2% of its charge.

The charging/discharging of a capacitor has many applications. For example, some automobiles come equipped with a feature that allows the windshield wipers to be used intermittently during a light drizzle. In this mode of operation the wipers remain off for a while and then turn on briefly. The timing of the on–off cycle is determined by the time constant of a resistor–capacitor combination.

SUMMARY

When devices are connected **in series,** there is the same electric current through each device. The **equivalent resistance of a series combination of resistances** (R_1, R_2, R_3, etc.) is $R_S = R_1 + R_2 + R_3 + \cdots$. The equivalent resistance dissipates the same total power as the series combination.

Connecting devices **in parallel** means that they are connected in such a way that the same voltage is applied across each device. In general, devices wired in parallel carry different currents. The **equivalent resistance of a parallel combination of resistances** is $1/R_P = 1/R_1 + 1/R_2 + 1/R_3 + \cdots$. The equivalent resistance dissipates the same total power as the parallel combination.

The **internal resistance** of a battery or generator is the electrical resistance encountered in the battery or generator by the current. The **terminal voltage** is the voltage between the terminals of a battery or generator and is equal to the emf only when there is no current through the device. Because of the internal resistance, the terminal voltage is less than the emf when there is current.

A **galvanometer** is a device that responds to electric current and is used in nondigital ammeters and voltmeters. An **ammeter** is an instrument that measures current and must be inserted into a circuit so the current passes directly through the ammeter. An ideal ammeter has zero resistance, so the current is not altered when the ammeter is inserted into a circuit. A **voltmeter** is an instrument for measuring the voltage between two points in a circuit. For this measurement a voltmeter must be connected between the two points and is not inserted into a circuit as an ammeter is. An ideal voltmeter has an infinite resistance, so the

voltmeter will not draw any current from the circuit to which it is attached. A **Wheatstone bridge** is a device that uses the **null method** for measuring resistance.

Kirchhoff's rules may be used to analyze the currents and potential differences in electric circuits. The **junction rule** states that the sum of the magnitudes of the currents directed into a junction equals the sum of the magnitudes of the currents directed out of the junction. The **loop rule** states that, around any closed circuit loop, the sum of the potential drops equals the sum of the potential rises.

The **equivalent capacitance for a parallel combination of capacitances** (C_1, C_2, C_3, etc.) is $C_P = C_1 + C_2 + C_3 + \cdots$. In general, each capacitor in a parallel combination carries a different amount of charge. The equivalent capacitor carries the same total charge and stores the same total energy as the parallel combination.

The **equivalent capacitance for a series combination of capacitances** is given by $1/C_S = 1/C_1 + 1/C_2 + 1/C_3 + \cdots$. Each capacitor in the combination carries the same amount of charge. The equivalent capacitor carries the same amount of charge as *any one* of the capacitors in the combination and stores the same total energy as the entire combination.

The **charging or discharging of a capacitor** in a dc series circuit (resistance R, capacitance C, and voltage V_0) does not occur instantaneously. Charge builds up gradually according to the relation $q = CV_0[1 - e^{-t/(RC)}]$, where q is the charge on the capacitor at time t. The discharging of the capacitor through the same resistor is described by $q = CV_0e^{-t/(RC)}$. The term RC is the **time constant** of the circuit.

QUESTIONS

1. The power rating of a 1000-W heater specifies the power consumed when the heater is connected to an ac voltage of 120 V. Explain why the power consumed by two of these heaters connected in series to a voltage of 120 V is not 2×1000 W = 2000 W.

2. A car has two headlights. The filament of one burns out, so charges can no longer flow out of the battery and through the headlight. However, the other headlight stays on. Draw a circuit diagram that shows how the headlights are connected to the battery.

3. A normal light bulb has a single filament and a power rating that corresponds to the power dissipated in the filament resistance when the bulb is connected to an ac voltage of 120 V. Some bulbs, however, contain two separate filaments. These are the familiar three-way bulbs that, in the proper socket, can be switched to provide three different wattages. Observe that in the following three-way bulbs the highest wattage is the sum of the other two choices: 30 W/70 W/100 W and 50 W/200 W/250 W. With this observation in mind, explain how the two filaments can be connected to the voltage source to give three wattage ratings.

4. For the circuit shown in the drawing, account for the fact that there is no current in the resistor R_1. No calculation is required.

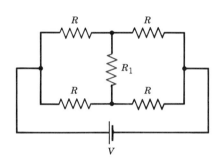

5. One of the circuits shown in the drawing contains resistors that are neither in series nor in parallel. Which is it?

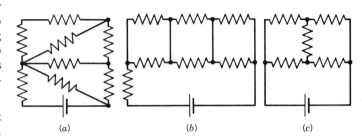

(a)   (b)   (c)

6. Compare the resistance of an ideal ammeter with the resistance of an ideal voltmeter and explain why the resistances are different.

7. Describe what would happen to the current in a circuit if a voltmeter, inadvertently mistaken for an ammeter, were inserted into the circuit.

8. The time constant of a series RC circuit is $\tau = RC$. Verify that an ohm times a farad is equivalent to a second.

PROBLEMS

Note: For problems that involve ac conditions, the current and voltage are rms values and the power is an average power, unless indicated otherwise.

Section 26.1 Series Wiring

1. A 16.0-Ω resistor and an 8.0-Ω resistor are connected in series across a 12.0-V battery. What is the voltage across each resistor?

2. The current in a series circuit is 15.0 A. When an additional 8.00-Ω resistor is inserted in series, the current drops to 12.0 A. What is the resistance in the original circuit?

3. Three resistors (9.0, 5.0, and 1.0 Ω) are connected in series across a 24-V battery. Find (a) the current in, (b) the voltage across, and (c) the power dissipated in each resistor.

4. A battery dissipates 2.50 W of power in each of two 47.0-Ω resistors connected in series. What is the voltage of the battery?

* 5. A 47-Ω resistor can dissipate up to 0.25 W of power without burning up. What is the smallest number of such resistors that can be connected in series across a 9.0-V battery?

* 6. Three resistors are connected in series across a battery. The value of each resistance and its maximum power rating are as follows: 5.0 Ω and 20.0 W, 30.0 Ω and 10.0 W, and 15.0 Ω and 10.0 W. (a) What is the greatest voltage that the battery can have without one of the resistors burning up? (b) How much power does the battery deliver to the circuit in (a)?

Section 26.2 Parallel Wiring

7. A 16-Ω loudspeaker and an 8.0-Ω loudspeaker are connected in parallel across the terminals of an amplifier. Assuming the speakers behave as resistors, determine the equivalent resistance of the two speakers.

8. Four resistors, 5.0, 8.0, 12.0, and 15.0 Ω, are connected in parallel. The current through the 15.0-Ω resistor is 4.0 A. (a) Determine the currents in the other three resistors. (b) What is the total power consumed by the four resistors?

9. How many 4.0-Ω resistors must be connected in parallel to create an equivalent resistance of one-sixteenth of an ohm?

10. A wire whose resistance is R is cut into three equally long pieces, which are then connected in parallel. (a) In terms of R, what is the resistance of the parallel combination? (b) Repeat part (a), assuming there are n pieces. Express your answer in terms of R and n.

* 11. The total current delivered to a number of devices connected in parallel is the sum of the individual currents in each device. Circuit breakers are resettable automatic switches that protect against dangerously large totals by opening to stop the current at a specified safe value. A 1650-W toaster, a 1090-W iron, and a 1250-W microwave oven are turned on in a kitchen. As the drawing shows, they are all connected through a 20-A circuit breaker to an ac voltage of 120 V. (a) Find the equivalent resistance of the three devices. (b) Obtain the total current delivered by the source and determine whether the breaker will open to prevent an accident.

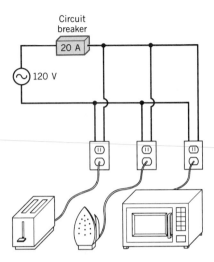

**12. The rear window defogger of a car consists of thirteen thin wires (resistivity = $88.0 \times 10^{-8} \ \Omega \cdot$m) imbedded in the glass. The wires are connected in parallel to the 12.0-V battery, and each has a length of 1.30 m. The defogger can melt 21.0 g of ice at 0 °C in two minutes. Assume that all the power dissipated in the wires is used immediately to melt the ice. Find the radius of the wires.

Section 26.3 Circuits That Are Wired Partially in Series and Partially in Parallel

13. Find the equivalent resistance between points A and B in the drawing.

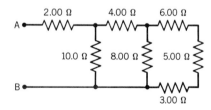

14. Circuit A has three resistors connected in series ($R_1 = 13 \ \Omega$, $R_2 = 25 \ \Omega$, and $R_3 = 320 \ \Omega$). Circuit B has three resistors (different from any of those in circuit A) connected in parallel. In circuit B each resistor has the same resistance. What is the resistance of each resistor in circuit B, such that the total resistance in B equals the total resistance in A?

15. Two 25.0-Ω resistors are connected in series. This combination is connected between the terminals of a 75.0-V battery. A 50.0-Ω resistor is also connected across the battery, in parallel with the series combination. (a) How much current is supplied by the battery? (b) What is the current in the 50.0-Ω resistor? (c) How much power is dissipated in one of the 25.0-Ω resistors?

* 16. Determine the power dissipated in the 2.0-Ω resistor in the circuit shown in the drawing.

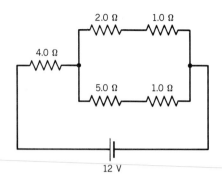

* 17. Three identical resistors are connected in parallel. The equivalent resistance increases by 700 Ω when one resistor is removed and connected in series with the remaining two, which are still in parallel. What is the resistance of each resistor?

* 18. A resistor (resistance = R) is connected first in parallel and then in series with a 2.00-Ω resistor. A battery delivers five times more current to the parallel combination than it does to the series combination. Determine the two possible values for R.

**19. Determine the equivalent resistance between the open terminals for each of the three groups of resistors in the drawing.

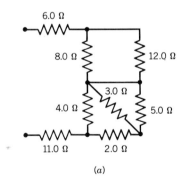

(a)

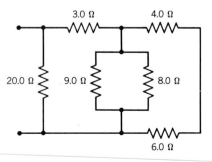

(b)

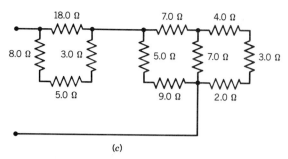

(c)

20. The current in the 8.00-Ω resistor in the drawing is 0.500 A. Find the current in the 20.0-Ω resistor and in the 9.00-Ω resistor.

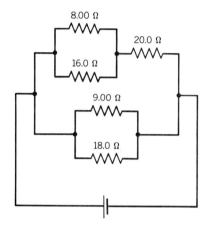

Section 26.4 Internal Resistance

21. A battery has an emf of 12.0 V and an internal resistance of 0.15 Ω. (a) What is the terminal voltage when the battery is connected to a 1.50-Ω resistor? (b) What is the power dissipated in the 1.50-Ω resistor?

22. A battery has an internal resistance of 0.50 Ω. A number of identical light bulbs, each with a resistance of 15 Ω, are connected in parallel across the battery terminals. The terminal voltage of the battery is observed to be one-half the emf of the battery. How many bulbs are connected?

23. A 2.00-Ω resistor is connected across a 6.00-V battery. The voltage between the terminals of the battery is observed to be only 4.90 V. Find the internal resistance of the battery.

* 24. A 75.0-Ω and a 45.0-Ω resistor are connected in parallel. When this combination is connected across a battery, the current delivered by the battery is 0.294 A. When the 45.0-Ω resistor is disconnected, the current from the battery drops to 0.116 A. Determine (a) the emf and (b) the internal resistance of the battery.

Section 26.5 The Measurement of Current, Voltage, and Resistance

25. A galvanometer has a coil resistance of 250 Ω and requires a current of 1.5 mA for full-scale deflection. This device is

to be used in an ammeter that has a full-scale current of 25.0 mA. What is the value of the necessary shunt resistance?

26. A galvanometer with a coil resistance of 32.0 Ω and a full-scale current of 25.0 μA is used with a shunt resistor to make an ammeter. The ammeter registers a maximum current of 15.0 mA. Find the equivalent resistance of the ammeter.

27. The resistance of a voltmeter is 140 000 Ω. The voltmeter uses a galvanometer that has a full-scale deflection of 180 μA. What is the maximum voltage that can be measured by the voltmeter?

* 28. Two scales on a voltmeter measure voltages up to 20.0 and 30.0 V, respectively. The resistance connected in series with the galvanometer is 1680 Ω for the 20.0-V scale and 2930 Ω for the 30.0-V scale. Determine the coil resistance and the full-scale current of the galvanometer that is used in the voltmeter.

* 29. One way to measure an unknown resistance is to use an ammeter (resistance = 36.0 Ω) to determine the current in the resistor and a voltmeter (resistance = 5870 Ω) to determine the voltage across the resistor. According to the definition, resistance is the voltage divided by the current. Parts a and b of the drawing show the two methods by which the meters can be connected. Find the current and the voltage registered by the meters and the *apparent* value for R calculated from these measurements for both connection methods.

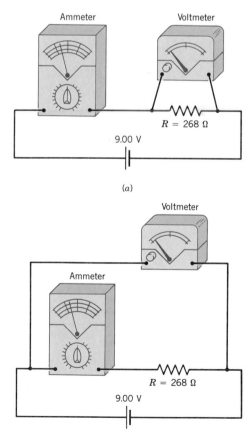

* 30. In measuring a voltage, a voltmeter uses some current from the circuit. Consequently, the voltage measured is only an approximation to the voltage present when the voltmeter is not connected. Consider a circuit consisting of two 1550-Ω resistors connected in series across a 60.0-V battery. (a) Find the voltage across one of the resistors. (b) A voltmeter has a full-scale voltage of 60.0 V and uses a galvanometer with a full-scale deflection of 5.00 mA. Determine the voltage that this voltmeter registers when it is connected across the resistor used in part (a).

Section 26.6 Kirchhoff's Rules

31. The drawing shows resistors that are partly in series and partly in parallel. (a) Find the current in the 4.0-Ω resistor without using Kirchhoff's rules. (b) Redetermine the current in the 4.0-Ω resistor, this time using Kirchhoff's rules. Verify that the answer obtained is the same as that in part (a).

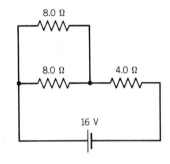

32. Two batteries, each with an internal resistance of 0.015 Ω, are connected as in the drawing. In effect, the 9.0-V battery is being used to charge the 8.0-V battery. What is (a) the current in the circuit and (b) the terminal voltage of each battery?

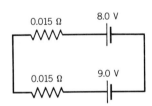

33. (a) Find the magnitude and direction of the current in the 2.0-Ω resistor in the drawing. (b) How much power is dissipated in the 3.0-Ω resistor?

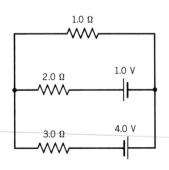

* 34. Determine the voltage across the 5.0-Ω resistor in the drawing. Which end of the resistor is at the higher potential?

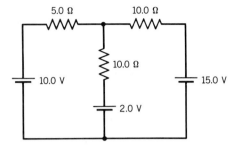

* 35. For the circuit in the drawing, find the current in the 10.0-Ω resistor. Specify the direction of the current.

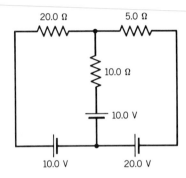

**36. Suppose the resistors in Figure 26.17 have the following resistances: $R = 10.0\ \Omega$, $R_1 = 20.0\ \Omega$, $R_2 = 30.0\ \Omega$, $R_v = 40.0\ \Omega$, and $R_C = 50.0\ \Omega$. If the battery is a 10.0-V battery, what is the voltage between points A and B in the circuit? State which point is at the higher potential.

Section 26.7 Capacitors in Series and Parallel

37. Determine the equivalent capacitance between A and B for the group of capacitors in the drawing.

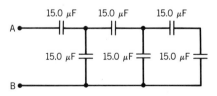

38. A 2.0-μF and a 4.0-μF capacitor are connected to a 60.0-V battery. How much charge is supplied by the battery in charging the capacitors when the wiring is (a) parallel and (b) series?

39. Three capacitors (4.0, 6.0, and 12.0 μF) are connected in series across a 50.0-V battery. Find the voltage across the 4.0-μF capacitor.

40. Three capacitors have identical geometries. One is filled with a material whose dielectric constant is 3.00. Another is filled with a material whose dielectric constant is 5.00. The third capacitor is filled with a material whose dielectric constant κ is such that this single capacitor has the same capacitance as the series combination of the two other capacitors. Determine κ.

41. Suppose three capacitors (C_1, C_2, and C_3) are connected in series. Show that the sum of the energies stored in these capacitors is equal to the energy stored in the equivalent capacitor.

* 42. A 16.0-μF and a 4.0-μF capacitor are connected in parallel and charged by a 22-V battery. What voltage is required to charge a series combination of the two capacitors with the same total *energy?*

* 43. A 3.00-μF capacitor and a 5.00-μF capacitor are connected in series across a 30.0-V battery. A 7.00-μF capacitor is then connected in parallel across the 3.00-μF capacitor. Determine (a) the voltage across and (b) the energy stored in the 7.00-μF capacitor.

**44. The drawing shows two fully charged capacitors (C_1 = 2.00 μF, q_1 = 6.00 μC; C_2 = 8.00 μF, q_2 = 12.0 μC). The switch is closed, and charge flows until equilibrium is reestablished. Find the resulting voltage across each capacitor.

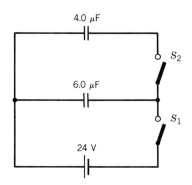

**45. Consider the circuit in the drawing. The 6.0-μF capacitor is first charged by closing switch S_1. Switch S_1 is then opened, and the charged capacitor is connected to the uncharged 4.0-μF capacitor by closing switch S_2. Charge flows until equilibrium is reestablished. What charge results on the 4.0-μF capacitor?

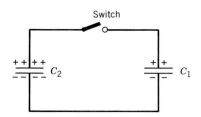

Section 26.8 *RC* Circuits

46. An electronic flash attachment for a camera produces a flash by using the energy stored in a 750-μF capacitor. Between flashes, the capacitor recharges through a resistor whose resistance

is chosen so that the capacitor will recharge with a time constant of 3.0 s. Determine the value of the resistance.

47. The capacitor in the drawing is fully charged. When the switch is opened, the capacitor begins to discharge. (a) What is the time constant for the discharge? (b) What is the voltage across the capacitor after a discharge period of one time constant has elapsed?

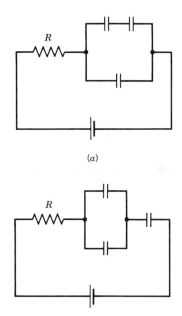

* 48. Three identical capacitors are connected with a resistor in two different ways. When they are connected as in part *a* of the drawing, the time constant is 0.020 s. What is the time constant when they are connected with the same resistor as in part *b*?

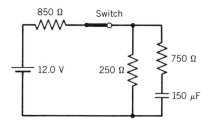

(a)

(b)

**49. How many time constants must elapse before a capacitor in a series *RC* circuit is charged to within 0.10% of its equilibrium charge?

ADDITIONAL PROBLEMS

50. A 3.0-μF capacitor and a 4.0-μF capacitor are connected in series across a 40.0-V battery. A 10.0-μF capacitor is also connected directly across the battery terminals. Find the total charge that the battery delivers to the capacitors.

51. The drawing shows a variable resistor used as a so-called voltage divider. This arrangement provides a way to obtain a voltage that is only a fraction of the battery voltage V. Derive an expression [involving V and the resistances R_1 and R_2 on either side of the movable contact ($\downarrow$)] for the voltage between points A and B.

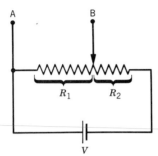

52. Suppose the resistances and the battery voltages in Figure 26.18 have the following values: $R_1 = 3.00 \ \Omega$, $R_2 = 6.00 \ \Omega$, $R_3 = 6.00 \ \Omega$, $V_1 = 3.00 \ \text{V}$, and $V_2 = 12.0 \ \text{V}$. (a) Find the current in the 3.0-Ω resistor. Be sure to specify the direction of the current. (b) What is the potential difference between the ends of R_2? Specify which end of the resistor is at the higher potential.

53. A voltmeter utilizes a galvanometer that has a 180-Ω coil resistance and a full-scale current of 8.30 mA. The voltmeter measures voltages up to 30.0 V. Determine the resistance that is connected in series with the galvanometer.

54. Eight different values of resistance can be obtained by connecting together three resistors (1.00, 2.00, and 3.00 Ω) in all possible ways. What are they?

* 55. When an ammeter is inserted into a circuit, its resistance is added to the circuit resistance and causes the current to drop. Therefore, the current measured by an ammeter is only an approximation for the current that exists when the ammeter is absent. Consider a circuit consisting of a 212-Ω resistor and a 134-Ω resistor connected in series with a 9.00-V battery. (a) Find the current in the circuit. (b) An ammeter uses a galvanometer with a full-scale current of 2.00 mA and a coil resistance of 285 Ω. The ammeter has a full-scale current of 40.0 mA. What percentage error is made when this ammeter is used to measure the current in the circuit?

* 56. A 46-m extension cord is used with an electric weed trimmer that has a resistance of 15.0 Ω. The extension cord is made of two copper wires, the wire diameter being 1.3 mm. (a) Determine the resistance of the extension cord. (b) The extension is plugged into a 120-V socket. What voltage is applied to the trimmer itself?

* 57. A resistor has a resistance R, and a battery has an internal resistance r. When the resistor is connected across the battery, ten percent less power is dissipated in the resistor than there would be if the battery had no internal resistance. Find the ratio r/R.

* 58. Three resistors are connected in series to a battery. From left to right, the resistances are R_1, $R_2 = 5.0 \ \Omega$, and R_3. The voltage across R_1 and R_2 together is 8.0 V, while the voltage across R_2 and R_3 together is 4.0 V. The equivalent resistance of the three resistors is 22.0 Ω. Determine R_1, R_3, and the battery voltage.

* 59. In the circuit in the drawing, there is no current in the 4.0-Ω resistor. Find the ratio R_1/R_2 of the unknown resistances.

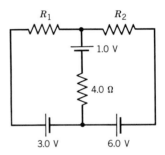

**60. A sheet of gold foil (negligible thickness) is placed between the plates of a capacitor and has the same area as each of the plates. The foil is parallel to the plates, at a position one-third of the way from one to the other. Before the foil is inserted, the capacitance is C_0. What is the capacitance after the foil is in place?

**61. Each resistor in the drawing is a 5.0-W resistor. In other words, each can dissipate at most 5.0 W of power without burning up. What is the maximum possible value for the current I?

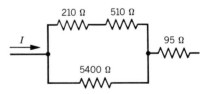

CHAPTER 27

Magnetic Forces and Magnetic Fields

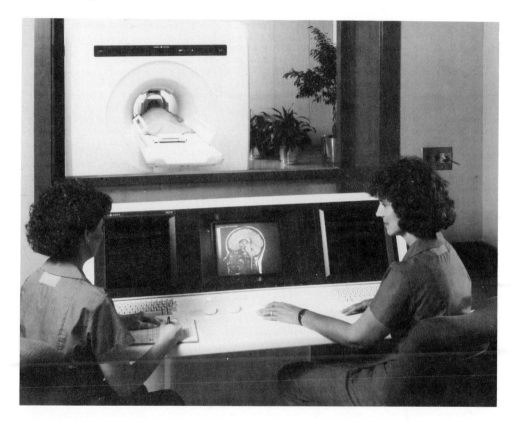

A patient is inside a magnetic resonance imaging machine that shows an image of internal organs on a comupter screen. Magnetism is important in many areas of our lives.

27.1 MAGNETS AND MAGNETIC FIELDS

PERMANENT MAGNETS

Historically, permanent magnets have been widely used in navigational compasses. As Figure 27.1 illustrates, the compass needle is a permanent magnet supported so it can rotate freely in a plane. When the compass is placed on a horizontal surface, the needle rotates until one end points approximately to the north. The end of the needle that points north is labeled the *north pole;* the opposite end is the *south pole.*

Magnets can exert forces on each other. Figure 27.2 shows that the magnetic forces between north and south poles have the property that *like poles repel each other, and unlike poles attract each other.* This behavior is similar to that of like and unlike electric charges. However, there is a significant difference between magnetic poles and electric charges. It is possible to separate positive from negative electric charges and produce isolated charges of either kind. In contrast, no one to date has found a magnetic monopole (an isolated north or south pole). For example, any attempt to

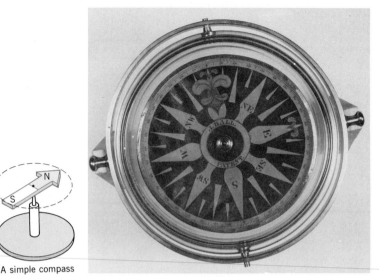

A mariner's compass.

A simple compass

FIGURE 27.1 The needle of a compass is a permanent magnet that has a north magnetic pole (N) at one end and a south magnetic pole (S) at the other end. The needle is supported so it can rotate freely in a plane.

separate north and south poles by cutting a bar magnet in half fails, because each piece becomes a smaller magnet with its own north and south poles. Repeated cutting only produces more bar magnets, without yielding isolated magnetic poles.

THE MAGNETIC FIELD

An electric field exists in the space around electric charges. In a similar manner, a *magnetic field* exists in the region around a magnet. The magnetic field, like the electric field, is a vector and has both a magnitude and a direction. We postpone a discussion of the magnitude until Section 27.2, concentrating our attention here only on the direction of the field.

The direction of the magnetic field at any point in space is the direction indicated by the north pole of a small compass needle placed at that point. Figure 27.3 shows how compasses can be used to map out the magnetic field in the space surrounding a bar magnet. Since like poles repel and unlike poles attract, the needle of each compass becomes aligned relative to the bar magnet in the manner shown in the picture. The compass needles provide a visual picture of the magnetic field that the bar magnet creates in the surrounding space.

As an aid in visualizing the electric field, Section 23.7 introduces the notion of electric field lines. In a similar fashion, it is possible to draw magnetic field lines in the vicinity of a magnet. Figure 27.4a illustrates some magnetic field lines around a bar magnet. The lines appear to originate from the north pole and to end on the south pole; the lines do not start or stop in midspace. A visual image of the magnetic field lines in a plane can be created by sprinkling finely ground iron filings on a piece of paper that covers a bar magnet. Iron filings in a magnetic field behave like tiny magnets and align themselves along the magnetic field lines, as part b of the drawing shows.

As is the case with electric field lines, the magnetic field at any point is tangent to the magnetic field line at that point. Furthermore, the strength of the magnetic field is

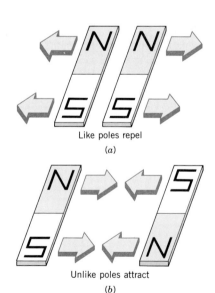

Like poles repel
(a)

Unlike poles attract
(b)

FIGURE 27.2 Bar magnets have a north magnetic pole at one end and a south magnetic pole at the other end. (a) Like poles repel each other, and (b) unlike poles attract each other.

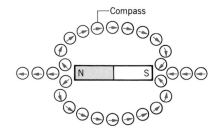

FIGURE 27.3 At any location in the vicinity of a magnet, the north pole of a small compass needle points in the direction of the magnetic field at that location.

proportional to the number of lines per unit area that passes through a surface oriented perpendicular to the lines. Thus, the magnetic field is stronger in regions where the field lines are relatively close together and weaker where they are relatively far apart. For instance, in Figure 27.4a the lines are closest together near the north and south poles, so the strength of the magnetic field is greatest in these regions. Away

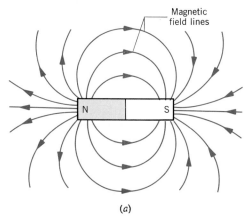

Magnetic
field lines

(a)

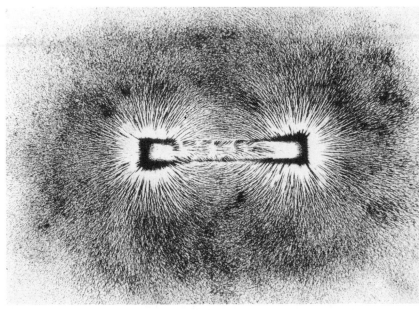

(b)

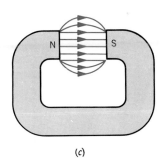

(c)

FIGURE 27.4 (a) The magnetic field lines and (b) the pattern of iron filings are shown in the vicinity of a bar magnet. (c) The magnetic field lines of a horseshoe magnet. The magnetic field lines are directed from the north pole to the south pole.

from the poles, the magnetic field becomes weaker. Notice in part *c* that the magnetic field lines in the central region between the poles of the horseshoe magnet are nearly parallel and equally spaced, indicating the magnetic field is approximately constant there.

GEOMAGNETISM

Although the north pole of a compass needle points northward, it does not in general point directly at the north geographic pole of the earth. The north geographic pole is that point where the earth's axis of rotation crosses the surface in the northern hemisphere (see Figure 27.5). Measurements of the magnetic field surrounding the earth show that the earth behaves magnetically almost as if it were a bar magnet. As the drawing illustrates, the orientation of this fictitious magnet defines a magnetic axis for the earth. The location where the magnetic axis crosses the surface in the northern hemisphere is known as the north magnetic pole. The north magnetic pole does not coincide with the north geographic pole but, instead, lies in Hudson Bay, Canada, some 1300 km to the south. It is interesting to note that the position of the north magnetic pole is not fixed, but moves over the years. For example, the current location of the north magnetic pole is about 770 km northwest of its position in 1904.

The north magnetic pole is so named because it is the location toward which the north end of a compass needle points. Since unlike poles attract, the south pole of the earth's fictitious magnet must lie beneath the north magnetic pole, as Figure 27.5 indicates.

For any location on the surface of the earth, the angle that a compass needle deviates from the north geographic pole is called the *angle of declination* for that location. For New York City, the angle of declination is about 12° west, meaning that a compass needle points 12° west of geographic north. The angle of declination is about 0° at Chicago, Illinois, and 15° east at San Diego, California. Since the north magnetic pole moves, the angle of declination varies from year to year; on average, the angle changes by several degrees per century.

Figure 27.5 shows that the earth's magnetic field lines are not parallel to the surface at all points. For instance, near the north magnetic pole the field lines are almost perpendicular to the surface of the earth. The angle that the magnetic field makes with respect to the surface at any point is known as the *angle of dip*.

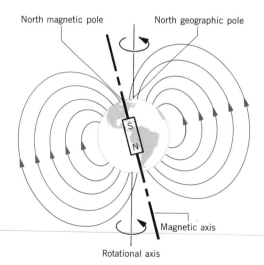

FIGURE 27.5 The earth behaves magnetically as if a fictitious bar magnet were located near its center. The earth's magnetic axis does not coincide with the earth's rotational axis; the two axes are currently about 11.5° apart.

27.2 THE FORCE THAT A MAGNETIC FIELD EXERTS ON A MOVING CHARGE

THE NATURE OF THE MAGNETIC FORCE

When a charge is placed in an electric field, the charge experiences an electric force. It is natural to ask, therefore, whether a charge placed in a magnetic field experiences a **magnetic force.** The answer is yes, provided two conditions are met:

1. The charge must be moving, for no magnetic force acts on a stationary charge.
2. The velocity of the moving charge must have a component that is perpendicular to the direction of the magnetic field.

To examine the second condition more closely, consider Figure 27.6, which shows a positive test charge $+q_0$ moving with a velocity **v** through a magnetic field labeled by the symbol **B**. The magnetic field is produced by an arrangement of magnets not shown in the drawing and is assumed to be constant in both magnitude and direction. If the charge moves *parallel or antiparallel* to the magnetic field, as in part *a* of the drawing, the charge experiences *no magnetic force.* If, on the other hand, the charge moves *perpendicular* to the magnetic field, as in part *b*, the charge feels the *maximum possible force* **F**. In general, if a charge moves at an angle $\theta*$ with respect to the magnetic field (see part *c* of the drawing), only the component of the velocity that is perpendicular to the magnetic field, $v \sin \theta$, gives rise to a magnetic force. This force is smaller than the maximum possible force. The component of the velocity that is parallel to the magnetic field yields no force.

Figure 27.6 shows that the direction of the magnetic force **F** is perpendicular to

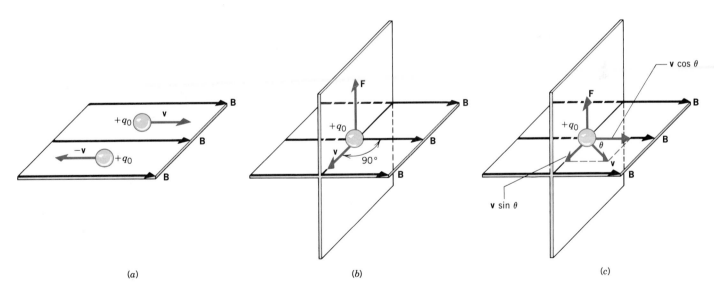

(a)   (b)   (c)

FIGURE 27.6 (*a*) No magnetic force acts on a charge moving with a velocity **v** that is parallel or antiparallel to a magnetic field **B**. (*b*) The charge experiences a maximum force **F** when the charge moves perpendicular to a magnetic field. (*c*) If the charge travels at an angle θ with respect to **B**, only the component of the velocity that is perpendicular to the magnetic field $v \sin \theta$, gives rise to a magnetic force.

---

* The angle θ between the velocity of the charge and the magnetic field is chosen so that it lies in the range $0 \leq \theta \leq 180°$.

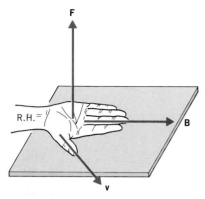

FIGURE 27.7 Right-Hand Rule No. 1 is illustrated. When the right hand (R.H.) is oriented so the fingers point along the magnetic field **B** and the thumb points along the velocity **v** of a positively charged particle, the palm faces in the direction of the magnetic force **F** applied to the particle.

both **v** and **B**; in other words, **F** is perpendicular to the plane defined by **v** and **B**. As an aid in remembering the direction of the force, it is convenient to use *Right-Hand Rule No. 1 (RHR-1),* as Figure 27.7 illustrates:

> **Right-Hand Rule No. 1.** Extend the right hand so the fingers point along the direction of the magnetic field **B** and the thumb points along the velocity **v** of the charge. The palm of the hand then faces in the direction of the magnetic force **F** that acts on a positive charge.

In this rule, it is as if the open palm of the right hand is pushing on the positive charge in the direction of the magnetic force. If the moving charge is *negative* instead of positive, the direction of the magnetic force is *opposite* to that predicted by RHR-1. Thus, there is an easy method for finding the force on a moving negative charge. First, assume the charge is positive and use RHR-1 to find a force direction. Then, reverse this direction to find the direction of the force that acts on the negative charge.

DEFINITION OF THE MAGNETIC FIELD

It is observed experimentally that when a charge moves through a magnetic field, the charge experiences a magnetic force whose magnitude is directly proportional to (1) the magnitude of the charge and (2) the component of its velocity that is perpendicular to the magnetic field. Because of these facts, it is possible to define the magnitude of the magnetic field in a manner that is similar in spirit to that used for the electric field, although the details differ.

Recall that the electric field at any point in space is the force per unit charge that acts on a test charge q_0 placed at that point. In other words, to determine the electric field **E**, we divide the electric force **F** by the charge q_0: $\mathbf{E} = \mathbf{F}/q_0$. However, the magnetic force depends not only on the charge magnitude q_0, but also on the velocity component $v \sin \theta$ that is perpendicular to the magnetic field. Therefore, to determine the magnitude of the magnetic field, we divide the magnitude of the magnetic force not only by q_0, but also by $v \sin \theta$, according to the following definition:

DEFINITION OF THE MAGNETIC FIELD

The magnitude B of the magnetic field at any point in space is defined as

$$B = \frac{F}{q_0(v \sin \theta)} \tag{27.1}$$

where F is the magnitude of the magnetic force on a positive test charge q_0 whose velocity **v** makes an angle θ ($0 \le \theta \le 180°$) with the field. The magnetic field **B** is a vector, and its direction can be determined by using a small compass needle.

SI unit of magnetic field: $\dfrac{\text{newton} \cdot \text{second}}{\text{coulomb} \cdot \text{meter}} = 1$ tesla (T)

The unit of magnetic field strength that follows from Equation 27.1 is the $\text{N} \cdot \text{s}/(\text{C} \cdot \text{m})$. This unit is called the *tesla* (T), in tribute to the Croatian-born, American engineer Nikola Tesla (1856–1943). Thus, one tesla is the strength of the magnetic field in which a unit test charge, traveling perpendicular to the magnetic field with a speed of one meter per second, experiences a force of one newton. Because a coulomb per second is called an ampere, 1 C/s = 1 A, the tesla is often written as $1 \text{ T} = 1 \text{ N}/(\text{A} \cdot \text{m})$.

In many situations the magnetic field has a value that is considerably less than one tesla. For example, the strength of the magnetic field near the earth's surface is approximately 10^{-4} T. In such circumstances, a magnetic field unit called the *gauss* (G) is sometimes used. Although not an SI unit, the gauss is a convenient size for many applications involving magnetic fields. The relation between the gauss and the tesla is

$$1 \text{ gauss} = 10^{-4} \text{ tesla}$$

Example 1 deals with the magnetic force exerted on a moving proton and on a moving electron.

EXAMPLE 1

A proton in a particle accelerator has a speed of 5.0×10^6 m/s. The proton encounters a magnetic field whose magnitude is 0.40 T and whose direction makes an angle of $\theta = 30.0°$ with respect to the proton's velocity (see Figure 27.6c). Find (a) the magnitude and direction of the magnetic force on the proton and (b) the acceleration of the proton. (c) What would be the force and acceleration if the particle were an electron instead of a proton?

SOLUTION

(a) Since the positive charge on a proton is 1.60×10^{-19} C, the magnitude of the magnetic force is

$$F = q_0 Bv \sin \theta \qquad (27.1)$$
$$= (1.60 \times 10^{-19} \text{ C})(0.40 \text{ T})(5.0 \times 10^6 \text{ m/s})(\sin 30.0°)$$
$$= \boxed{1.6 \times 10^{-13} \text{ N}}$$

The direction of the magnetic force is given by RHR-1 and is directed upward in Figure 27.6c, with the magnetic field pointing to the right.

(b) The acceleration of the proton follows directly from Newton's second law as the magnetic force divided by the mass m_p of the proton:

$$a = \frac{F}{m_p} = \frac{1.6 \times 10^{-13} \text{ N}}{1.67 \times 10^{-27} \text{ kg}} = \boxed{9.6 \times 10^{13} \text{ m/s}^2} \quad (4.1)$$

(c) The magnitude of the force does not change when the proton is replaced by an electron, since both have the same charge magnitude. However, the direction of the force on the electron is opposite to that on the proton, since the charge on the electron is negative. Furthermore, the electron has a smaller mass m_e and, therefore, experiences a significantly greater acceleration:

$$a = \frac{F}{m_e} = \frac{1.6 \times 10^{-13} \text{ N}}{9.11 \times 10^{-31} \text{ kg}} = \boxed{1.8 \times 10^{17} \text{ m/s}^2}$$

27.3 THE MOTION OF A CHARGED PARTICLE IN A MAGNETIC FIELD

COMPARING THE MOTION IN ELECTRIC AND MAGNETIC FIELDS

The motion of a charged particle in an electric field is noticeably different than the motion in a magnetic field. For example, Figure 27.8a shows a positive charge moving between the plates of a parallel plate capacitor. Initially, the charge is moving perpendicular to the direction of the electric field. Since the direction of the electric force on a positive charge is in the same direction as the electric field, the particle is deflected sideways in the drawing. Part b of the drawing shows the same particle traveling initially at right angles to a magnetic field. An application of RHR-1 shows that when the charge enters the field, the charge is deflected upward (not sideways) by the magnetic force. As the charge moves upward, the direction of the magnetic force

FIGURE 27.8 (*a*) The electric force **F** that acts on a positive charge is in the same direction as the electric field **E**. Here the electric force causes the particle's trajectory to bend in a horizontal plane. (*b*) The magnetic force **F** is perpendicular to both the magnetic field **B** and the particle's velocity **v**. In this instance, the magnetic force causes the particle's trajectory to bend in a vertical plane.

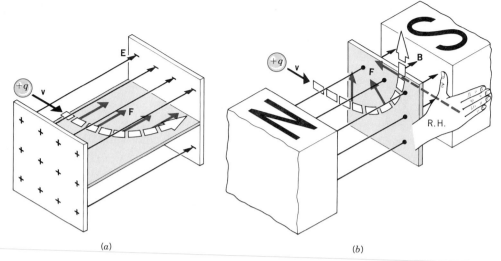

(*a*)   (*b*)

changes, always remaining perpendicular to both the magnetic field and the velocity. When a charged particle travels in a magnetic field, then, the charge never experiences a force that is parallel to the field, as it does in the electric case. Because of the difference in the way that electric and magnetic fields exert forces on charges, the work done on the particle by each field is different, as we will now see.

THE WORK DONE ON A CHARGED PARTICLE MOVING THROUGH ELECTRIC AND MAGNETIC FIELDS

In Figure 27.8*a* an electric field applies a force to a positively charged particle, and, consequently, the path of the particle bends in the direction of the force. Because there is a component of the particle's displacement in the direction of the electric force, the force does work on the particle. This work increases the kinetic energy, and hence speed, of the particle, as specified by the work–energy theorem presented in Section 7.2.

In contrast to an electric field, *a constant magnetic field does no work on a moving charged particle.** As Figure 27.8*b* indicates, this fact arises because the magnetic force always acts in a direction that is perpendicular to the motion of the charge.

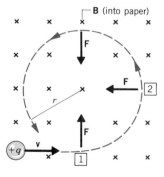

FIGURE 27.9 A positively charged particle is moving perpendicular to a constant magnetic field. The magnetic force **F** causes the particle to move on a circular path.

---

* This statement is valid as long as the magnetic field is constant and no agent other than the magnetic field exerts a force on the charge.

Consequently, the displacement of the moving charge never has a component in the direction of the magnetic force. As a result, the magnetic force cannot change the kinetic energy of the charge, although the force can alter the direction of motion.

THE CIRCULAR TRAJECTORY

To describe the motion of a charged particle in a magnetic field more completely, and to emphasize that the field does no work, we now discuss the special case in which the velocity of the particle is perpendicular to a uniform magnetic field.* As Figure 27.9 illustrates, the magnetic force serves to move the particle in a circular path. To understand why the path is circular, consider two points on the circumference labeled 1 and 2. When the positively charged particle is at point 1, the magnetic force **F** is perpendicular to the velocity **v** and points directly upward in the drawing. This force causes the trajectory to bend upward. When the particle reaches point 2, the magnetic force still remains perpendicular to the velocity, but is now directed to the left in the drawing. *The magnetic force always remains perpendicular to the velocity and is directed toward the center of the circular path.* This situation is very similar to that of an object being whirled around at the end of a string at constant speed, in which case the force acting on the object is provided by the tension in the string. The tension always remains perpendicular to the velocity and is directed along the string toward the center of the circle. With its "center-seeking" direction, the tension keeps the object moving in a circular path, but does not change the kinetic energy of the object. In a similar manner, the magnetic force is always perpendicular to the particle's velocity and serves only to change the direction of the motion. The magnetic force does no work on the particle, and so the kinetic energy and hence the speed of the particle remain constant.

To find the radius of the circular path in Figure 27.9, we recall the concept of centripetal force from Section 6.3. The centripetal force is the net force, directed toward the center of the circle, that is needed to keep a particle moving in a circular path. The magnitude F_c of the centripetal force depends on the speed v and mass m of the particle, as well as the radius r of the circle:

$$F_c = \frac{mv^2}{r} \tag{6.3}$$

In the present situation, the magnetic force furnishes the centripetal force needed to keep the charge $+q$ on the circular path. According to Equation 27.1, the magnetic force is $qvB \sin 90°$, so $qvB = mv^2/r$ or

$$r = \frac{mv}{qB} \tag{27.2}$$

---

* In many instances it is convenient to orient the magnetic field **B** so its direction is perpendicular to the page. In these cases it is customary to use a dot to symbolize the magnetic field pointing out of the page (toward the reader); this dot symbolizes the tip of the arrow representing the **B** vector. Therefore, a region where a constant magnetic field is directed *out of the page* is drawn as

$$\cdot \quad \cdot \quad \cdot \quad \cdot$$
$$\cdot \quad \cdot \quad \cdot \quad \cdot$$

A region where a constant magnetic field is directed *into the page* is drawn as a series of crosses that indicate the tail feathers of the arrows representing the **B** vectors. Such a region looks like

$$\times \quad \times \quad \times \quad \times$$
$$\times \quad \times \quad \times \quad \times$$
$$\times \quad \times \quad \times \quad \times$$

Equation 27.2 shows that the radius of the circle is inversely proportional to the magnitude of the magnetic field, with stronger fields producing "tighter" circular paths.

Example 2 illustrates the fact that an electric field can change the kinetic energy of a particle, but a magnetic field cannot.

EXAMPLE 2

A proton starts from rest at the positive plate of a parallel plate capacitor and is accelerated toward the negative plate by the electric force. The potential difference between the plates is $V = 2100$ volts. The high-speed proton leaves the capacitor through a small hole in the negative plate. Once outside the capacitor, the proton travels at a constant velocity until it enters a region of constant magnetic field of magnitude 0.10 T. The velocity and magnetic field are perpendicular, as in Figure 27.9. Find (a) the speed of the proton when it leaves the capacitor, (b) the change in the proton's kinetic energy due to the magnetic field, and (c) the radius of the circular path on which the proton moves in the magnetic field.

SOLUTION

(a) Initially, when the proton (charge $= +e$) is at the positive plate, the electric potential energy relative to the negative plate is eV. As the proton approaches the negative plate, the potential energy is converted into kinetic energy. When the proton reaches the negative plate, $eV = \frac{1}{2}mv^2$, and the speed of the proton is

$$v = \sqrt{\frac{2eV}{m}} = \sqrt{\frac{2(1.60 \times 10^{-19} \text{ C})(2100 \text{ V})}{1.67 \times 10^{-27} \text{ kg}}}$$

$$= \boxed{6.3 \times 10^5 \text{ m/s}}$$

(b) The magnetic field does no work on the moving proton. Therefore, according to the work–energy theorem, the kinetic energy of the proton does not change. The speed of the proton remains constant at $v = 6.3 \times 10^5$ m/s while the proton travels along a circular path within the magnetic field.

(c) The radius of the circle can be found from Equation 27.2:

$$r = \frac{mv}{qB} = \frac{(1.67 \times 10^{-27} \text{ kg})(6.3 \times 10^5 \text{ m/s})}{(1.60 \times 10^{-19} \text{ C})(0.10 \text{ T})}$$

$$= \boxed{6.6 \times 10^{-2} \text{ m}}$$

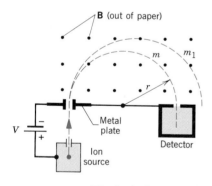

FIGURE 27.10 The basic features of a mass spectrometer. After being accelerated through a potential difference V, positive ions enter a constant magnetic field **B** that is directed out of the page, toward the reader. The dashed lines are the paths traveled by two ions of different masses. Ions with mass m follow the path of radius r and enter the detector. Ions with the larger mass m_1 follow the path that has a larger radius and, consequently, miss the detector.

27.4 THE MASS SPECTROMETER AND THE HALL EFFECT

THE MASS SPECTROMETER

Physicists use mass spectrometers for determining the relative masses and abundances of isotopes (atoms that have the same atomic number, but different atomic masses due to the presence of different numbers of neutrons in the nuclei). Chemists use these instruments to help identify unknown molecules produced in chemical reactions. Mass spectrometers are also used during surgery, where they give the anesthesiologist information on the gases, including the anesthetic, in the patient's lungs.

In the type of mass spectrometer illustrated in Figure 27.10, the atoms or molecules are first vaporized and then ionized by the ion source. The ionization process removes one electron from the particle, leaving it with a net positive charge of $+e$. The positive ions are then accelerated through the potential difference V that is applied between the ion source and the metal plate. The ions pass through a hole in the plate with a speed v and enter a region of constant magnetic field **B**, where they are deflected in semicircular paths. Only those ions following a path with the proper radius r strike the detector, which records the number of ions arriving per second.

The mass m of the detected ions can be expressed in terms of r, B, and v by recalling that the radius of the path followed by a particle of charge $+e$ is $r = mv/eB$ (Equation 27.2). In addition, the results of Example 2 show that the ion speed v can be expressed in terms of the accelerating potential V as $v = \sqrt{2eV/m}$. Algebraically eliminating v from these two equations and solving for the mass gives

$$m = \left(\frac{er^2}{2V}\right) B^2$$

This result shows that the mass of each ion reaching the detector is proportional to B^2. By experimentally changing the value of B, and keeping the term in the parentheses constant, ions of different masses are allowed to enter the detector. A plot of the detector output as a function of B^2 then gives an indication of what masses are present and the abundance of each mass.

Figure 27.11 shows a record obtained by a mass spectrometer for naturally occurring neon gas. The results show that the element neon has three isotopes whose atomic mass numbers are 20, 21, and 22. These isotopes occur because neon atoms exist with different numbers of neutrons in the nucleus. Notice that the isotopes have different abundances, with neon-20 being the most abundant.

THE HALL EFFECT

The current in a metal conductor is due to the motion of electrons, and electrons carry a negative charge. However, there are important materials in which the electric current is not necessarily caused by the motion of negative charge carriers. For example, semiconductors—most notably silicon and germanium—are important in the technology of integrated circuits. In contrast to the situation in metals, the charge carriers in semiconductors can be either negative or positive, depending on how the semiconductors are fabricated. When silicon is mixed with a small amount of phosphorus, the composite is known as an n-type semiconductor, because the charge carriers are negative. On the other hand, when a small amount of boron is added to silicon, a p-type semiconductor results in which the charge carriers behave as if they were positive. Both types of semiconductors will be discussed in Section 29.6.

When new types of semiconductors are developed, it is important to identify whether the charge carriers are negative or positive. One of the few experimental methods for unambiguously determining the type of carrier was devised by Edwin H. Hall in 1879. Figure 27.12 illustrates Hall's method, which is widely used today. A thin, flat, conducting slab is placed in a constant magnetic field, such that the field is oriented perpendicular to the wide face of the slab. Suppose that a current I of positive

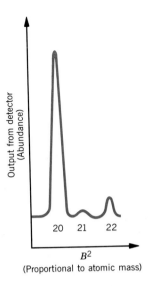

FIGURE 27.11 The mass spectrum of naturally occurring neon, showing three isotopes whose atomic mass numbers are 20, 21, and 22. The larger the peak, the more abundant the isotope, so neon-20 is the most abundant isotope of neon.

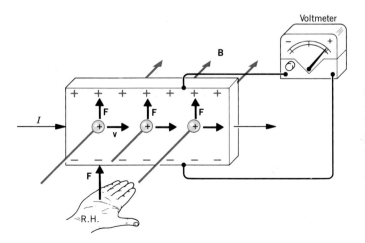

FIGURE 27.12 Positive charges moving to the right are deflected upward by the magnetic force **F**, giving the top surface of the slab a positive charge and the bottom surface a negative charge. The potential difference between the top and bottom surfaces is called the Hall emf, and is registered by the voltmeter.

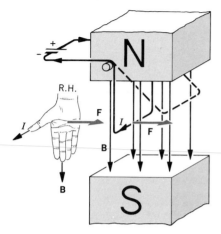

FIGURE 27.13 The wire carries a current I and is oriented perpendicular to a magnetic field **B**. A magnetic force **F** deflects the wire to the right. If the current were reversed, the directions of the force and the deflection would be reversed.

charges is driven to the right by a battery (not shown in the drawing). According to RHR-1, the charges are deflected upward by the magnetic force **F**. Thus, positive charges accumulate at the top edge of the slab, while corresponding negative charges accumulate at the bottom edge. Because of the buildup of positive and negative charges, an emf, called the *Hall emf* (or *Hall voltage*), appears across the slab, with the top of the slab being at a higher potential relative to the bottom. The Hall emf can be measured with a voltmeter, such as the one shown in the drawing. The emf builds up until the electric field produced by the separated positive and negative charges exerts an electric force on the current I that is equal and opposite to the magnetic force. Therefore, a current of positively charged carriers produces a situation in which the top of the slab becomes positively charged and the bottom becomes negatively charged.

On the other hand, the *same* current I could also have been caused by negative charge carriers moving to the *left* in Figure 27.12. The same current arises if the negative carriers possess both an opposite charge and an opposite velocity compared to the moving positive charges in the drawing. An application of RHR-1 (with a reversal of the direction of the predicted force, since the moving charges are negative) shows that the top of the slab now becomes negatively charged and the bottom becomes positively charged. Therefore, negative charges moving to the left generate a Hall emf of opposite polarity to that produced by positive charges moving to the right. Thus, the polarity of the Hall emf reveals whether the charge carriers are positive or negative.

Another use of the Hall effect is found in an instrument known as a Hall probe, which measures the strength of a magnetic field. It has been determined experimentally that the Hall emf is directly proportional to the strength of the magnetic field into which the conducting slab is placed. A Hall probe is a convenient, hand-held instrument that has been calibrated to register the strength of the magnetic field, rather than the Hall emf.

27.5 THE FORCE ON A CURRENT IN A MAGNETIC FIELD

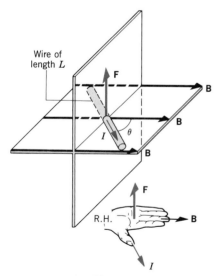

FIGURE 27.14 The current I in the wire, oriented at an angle θ with respect to a magnetic field **B**, is acted upon by a magnetic force **F**. Note the similarity between this drawing and Figure 27.6c.

As we have seen, a charge moving through a magnetic field can experience a magnetic force. Since an electric current is a collection of moving charges, a current in the presence of a magnetic field can also experience a magnetic force. In Figure 27.13, for instance, a current-carrying wire is placed between the poles of a magnet. When the direction of the current I is as shown, the moving charges, and hence the wire itself, feel a magnetic force that pushes the wire to the right in the drawing. The direction of the force is determined in the usual manner by using RHR-1, with the minor modification that the velocity of a positive charge is replaced by the direction of the conventional current I. If the direction of the current in the drawing were reversed by switching the leads to the battery, the direction of the force would be reversed, and the wire would be pushed to the left.

When a charge moves through a magnetic field, the magnitude of the force that acts on the charge is $F = qvB \sin \theta$. With the aid of Figure 27.14, this expression can be put into a form that is more suitable for use with an electric current. The drawing shows a wire of length L that carries a current I. The wire is oriented at an angle θ with respect to a magnetic field **B**. This picture is nearly identical to Figure 27.6c, except that now the charges move in a wire. The magnitude F of the magnetic force exerted on this length of wire is the net force acting on the total charge q moving in the wire.

Multiplying and dividing the right side of $F = qvB \sin \theta$ by t, the time needed for the total charge to travel the length of the wire, gives

$$F = \left(\frac{q}{t}\right)(vt)B \sin \theta$$

The term q/t is the current I in the wire, and the term vt is the length L of the wire. With these two substitutions, we arrive at the following expression for the magnetic force exerted on a current-carrying wire:

$$\begin{bmatrix} \textbf{Magnetic force on} \\ \textbf{a current-carrying} \\ \textbf{wire of length } L \end{bmatrix} \qquad F = ILB \sin \theta \qquad (27.3)$$

As in the case of a single charge traveling in a magnetic field, the maximum magnetic force on a current-carrying wire results when the wire is oriented perpendicular to the field ($\theta = 90°$), and the force vanishes when the current is parallel or antiparallel to the field ($\theta = 0°$ or $180°$). The direction of the magnetic force is given by RHR-1.

Most hi-fi speakers operate on the principle that a magnetic field exerts a force on a current-carrying wire. Figure 27.15a shows a speaker design that consists of three principal parts: a cone, a voice coil, and a permanent magnet. The cone is usually made from specially treated, stiff paper and is mounted so it can vibrate back and forth freely. In vibrating, the cone pushes and pulls on the air in front of it, thereby creating sound waves. Attached to the apex of the cone is a hollow cardboard cylinder, around which many turns of wire are wound. This cylinder and its coils of wire are collectively called the "voice coil"; the voice coil is slipped over one pole of the permanent magnet, which is the north pole in the drawing. The permanent magnet itself does not move, but the voice coil is designed to move freely over the north pole. The two ends of the voice-coil wire are connected to the speaker terminals located at the rear panel of an audio amplifier.

The amplifier acts as an ac generator, sending an alternating current to the voice coil. The ac current interacts with the magnetic field to generate an alternating force that pushes and pulls on the voice coil and the attached cone. To see how the magnetic force arises, consider Figure 27.15b, which illustrates a cross-sectional view of the voice coil and the magnet. In the cross section shown in the drawing, the

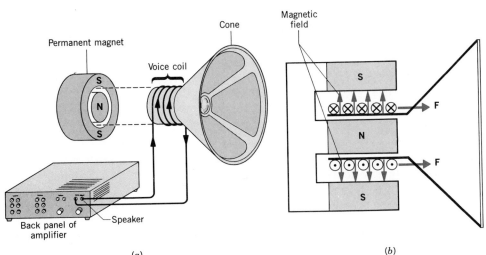

(a)

(b)

FIGURE 27.15 (a) An "exploded" view of one type of speaker design, which shows a cone, a voice coil, and a permanent magnet. (b) Because of the current in the voice coil (shown as $\otimes$ and $\odot$), the magnetic field causes a force **F** to be exerted on the voice coil and cone. This force accelerates the voice coil and cone.

current is directed into the page in the upper half of the voice coil ($\otimes\otimes\otimes$) and out of the page in the lower half of the voice coil ($\odot\odot\odot$). In both cases the magnetic field is perpendicular to the wire, so the maximum possible force is exerted on the wire. An application of RHR-1 to both the upper and lower halves of the voice coil shows that the magnetic force **F** in the drawing is directed to the right, causing the cone to accelerate in that direction. One-half of a cycle later when the current is reversed, the direction of the magnetic force is also reversed, and the cone accelerates to the left. If, for example, the ac current from the amplifier has a frequency of 1000 Hz, the alternating magnetic force causes the cone to vibrate back and forth at the same frequency, and a 1000-Hz sound wave is produced. Thus, it is the magnetic force on a current-carrying wire that is responsible for converting an electrical signal into a sound wave. In Example 3 a typical force and acceleration for a loudspeaker are determined.

EXAMPLE 3

The voice coil of a speaker has a diameter of $d = 2.5$ cm, contains 55 turns of wire, and is placed in a 0.10-T magnetic field. The current in the voice coil is 2.0 A. (a) Determine the magnetic force that acts on the coil and cone. (b) If the voice coil and cone have a combined mass of 0.020 kg, find their acceleration.

SOLUTION

(a) To find the magnetic force, we need a value for the effective length of the voice coil wire. Since the magnetic field acts perpendicular to all parts of the wire, the effective length is the number of turns N times the circumference of one turn: $L = N\pi d = 55\pi(2.5 \times 10^{-2} \text{ m}) = 4.3$ m. The force on the voice coil is

$$F = ILB \sin\theta \qquad (27.3)$$

$$= (2.0 \text{ A})(4.3 \text{ m})(0.10 \text{ T}) \sin 90°$$

$$= \boxed{0.86 \text{ N}}$$

(b) The acceleration of the voice coil and cone can be found from Newton's second law:

$$a = \frac{F}{m} = \frac{0.86 \text{ N}}{0.020 \text{ kg}} = \boxed{43 \text{ m/s}^2} \qquad (4.1)$$

This acceleration is more than four times the acceleration due to gravity.

The voice coil of a hi-fi speaker moves when current is sent to it by an audio amplifier. The same basic idea plays a role in some personal computer systems that incorporate a hard disk drive. In these drives the element that reads information from or writes information on the spinning disk is the read/write head (see Figure 27.16). Hard disk drives often use "voice-coil positioners" to move the read/write head to the proper location on the disk. In response to instructions from the user, current is sent to the voice-coil positioner, and a magnetic force causes the head to move across the surface of the disk to the appropriate location.

27.6 THE TORQUE ON A CURRENT-CARRYING COIL

THE GENERAL PICTURE

The last section shows that a current-carrying wire can experience a force when placed in a magnetic field. If a loop of wire is suspended properly in a magnetic field, the magnetic force produces a torque that tends to rotate the loop. This torque is responsible for the operation of a number of useful devices, including galvanometers and electric motors.

Figure 27.17*a* shows a rectangular loop of wire attached to a vertical shaft. The

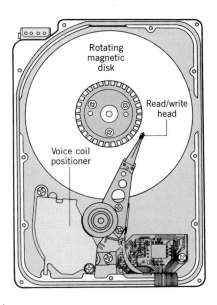

FIGURE 27.16 Many hard disk drives use a voice-coil positioner to move the read/write head to the appropriate location over the rotating disk.

shaft is mounted in bearings (not shown in the drawing) and is free to rotate within a uniform magnetic field. When there is a current in the loop, the loop rotates because a magnetic force is exerted on each of the two vertical sides, labeled 1 and 2 in the drawing. Part *b* of the drawing shows a top view of the loop and the magnetic forces on the two sides. These two forces have the same magnitude, but an application of RHR-1 shows that they point in opposite directions, so the loop experiences no net force. The loop does, however, experience a net torque that tends to rotate the loop in

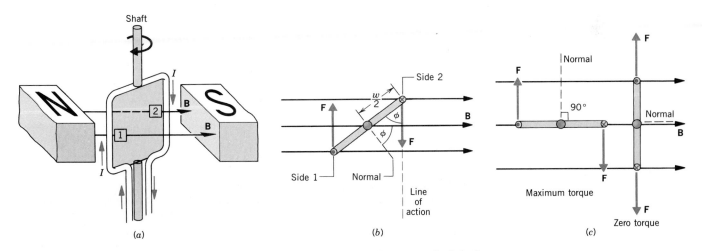

FIGURE 27.17 (*a*) A current-carrying loop of wire, which can rotate about a vertical shaft, is situated in a magnetic field. (*b*) A top view of the loop. The current in side 1 is directed out of the page (⊙), while the current in side 2 is directed into the page (⊗). The current in side 1 experiences a force **F** that is opposite in direction to the force exerted on side 2. The two forces produce a clockwise torque about the shaft. (*c*) Maximum torque occurs when the normal to the plane of the loop is perpendicular to the magnetic field, while no torque exists when the normal is parallel to the field.

a clockwise fashion about the vertical shaft. Part c of the drawing shows that the torque is maximum when the normal to the plane of the loop is perpendicular to the field. In contrast, the torque is zero when the normal is parallel to the field. *When a current-carrying loop is placed in a magnetic field, the loop tends to rotate such that its normal becomes aligned with the magnetic field.* In this respect, a current loop behaves like a magnet (e.g., a compass needle) suspended in a magnetic field, since a magnet also rotates to line itself up with the magnetic field.

It is possible to determine the magnitude of the torque on the loop. From Equation 27.3, the magnetic force on each vertical side is $F = ILB \sin 90°$, where L is the length of side 1 or side 2, and $\theta = 90°$ because the current I always remains perpendicular to the magnetic field as the loop rotates. As Section 10.1 discusses, the torque produced by a force is the product of the force and the lever arm. In Figure 27.17b the lever arm is the perpendicular distance from the line of action of the force to the shaft. This distance is given by $(w/2) \sin \phi$, where w is the width of the loop, and ϕ is the angle between the normal to the plane of the loop and the direction of the magnetic field. The net torque is the sum of the torques on the two sides, so

$$\text{Net torque} = \tau = ILB(\tfrac{1}{2}w \sin \phi) + ILB(\tfrac{1}{2}w \sin \phi) = IAB \sin \phi$$

where the product Lw has been replaced by the area A of the loop. If the wire is wrapped so as to form a coil containing N loops, each of area A, the force on each side is N times larger, and the torque becomes proportionally greater:

$$\tau = NIAB \sin \phi \qquad (27.4)$$

This relation has been derived for a rectangular coil, but, in fact, it is valid for any shape of flat coil, such as a circular coil. It is apparent that the torque depends on (1) the geometric properties of the coil itself and the current in it (NIA), (2) the magnitude B of the magnetic field, and (3) the orientation of the normal to the coil with respect to the direction of the field ($\sin \phi$). The quantity NIA is known as the *magnetic moment* of the coil, and its units are ampere·meter². The greater the magnetic moment of a current-carrying coil, the greater the torque that the coil experiences when placed in a magnetic field. Example 4 discusses the torque that a magnetic field applies to such a coil.

EXAMPLE 4

A coil of wire has an area of 2.0×10^{-4} m², consists of 100 loops or turns, and contains a current of 45 mA. The coil is placed in a uniform magnetic field of magnitude 0.15 T. (a) Determine the magnetic moment of the coil. (b) Find the maximum torque that the magnetic field can exert on the coil.

SOLUTION

(a) The magnetic moment of the coil is

$$\text{Magnetic moment} = NIA$$
$$= (100 \text{ turns})(45 \times 10^{-3} \text{ A})$$
$$\times (2.0 \times 10^{-4} \text{ m}^2)$$
$$= \boxed{9.0 \times 10^{-4} \text{ A·m}^2}$$

The magnetic moment depends only on the properties of the coil (N and A) and the current I in it; the magnetic moment does not depend on the orientation of the coil in the magnetic field or on the strength of the field.

(b) According to Equation 27.4, the torque is the product of the magnetic moment NIA and $B \sin \phi$. However, the maximum torque occurs when $\phi = 90°$ so

$$\tau = (\text{Magnetic moment})(B \sin 90°)$$
$$= (9.0 \times 10^{-4} \text{ A·m}^2)(0.15 \text{ T})$$
$$= \boxed{1.4 \times 10^{-4} \text{ N·m}}$$

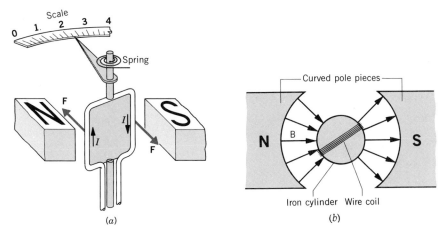

FIGURE 27.18 (*a*) The greater the current *I*, the greater the torque that tends to rotate the coil and pointer. The pointer rotates until the torque exerted on the coil by the magnetic forces is counterbalanced by the torque due to the spring. (*b*) Top view of a galvanometer mechanism showing the curved pole pieces, the iron cylinder, and the coil. For clarity, the scale, pointer, and spring have been omitted.

THE GALVANOMETER

The galvanometer is the basic component of nondigital ammeters and voltmeters. In measuring the current, a galvanometer relies on the fact that a current-carrying coil can rotate when placed in a magnetic field. Figure 27.18*a* shows the coil (only one turn is shown) of a galvanometer suspended in a magnetic field, the coil being able to rotate about a vertical shaft. Attached to the shaft is a pointer and a spring. When a current exists in the coil, the magnetic torque causes the coil to rotate. As the coil rotates, the spring winds up and produces a countertorque. The coil comes to rest when the magnetic torque is counterbalanced by the spring torque. The greater the current, the greater the torque, and the further the coil rotates. In a properly designed instrument, the deflection of the coil and pointer is directly proportional to the current, so the measurement scale can be calibrated to indicate the magnitude of the current.

It is common practice to use a galvanometer coil that consists of many turns of wire, so as to generate a larger torque. The enhanced torque produces a greater deflection of the pointer for a given current, thus improving the ability of the instrument to detect a small current. Moreover, the permanent magnet is typically made with curved pole pieces, and the galvanometer coil surrounds a stationary iron cylinder, as part *b* of the drawing illustrates. The curved pole pieces and iron cylinder tend to orient the magnetic field lines so they are radial, rather than parallel. Because the magnetic field is radial, the direction of the field at any orientation of the coil is perpendicular to the normal of the coil. This configuration ensures that a maximum magnetic torque acts on the coil as the coil rotates.

THE DIRECT-CURRENT ELECTRIC MOTOR

The electric motor is found in many devices, such as tape decks, turntables, automobiles, washing machines, and air conditioners. Figure 27.19*a* shows the essential parts of a direct-current (dc) motor. The elements of a motor are similar to those of a galvanometer, except the spring is removed so the coil can rotate continuously in one direction. The coil of wire contains many turns and is wrapped around a movable iron cylinder, although these features have been omitted to simplify the drawing. The coil and iron cylinder assembly is known as the armature. Each end of the wire coil is attached to a metallic half-ring. Rubbing against each of the half-rings is a graphite

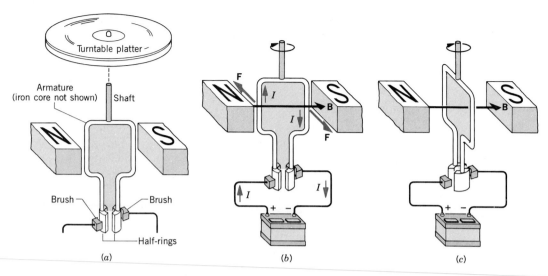

FIGURE 27.19 (*a*) The basic components of a dc motor. Note the similarity to Figure 27.18*a*. The platter of a turntable is shown as it might be attached to the motor. (*b*) When a current exists in the coil, the coil experiences a torque. (*c*) Because of its rotational inertia, the coil continues to rotate when there is no current.

contact called a brush. The two half-rings and the associated brushes are referred to as a split-ring commutator, the purpose of which will be explained shortly.

The operation of a motor can be understood by considering Figure 27.19*b*. The current from the battery enters the coil through one of the brushes and the corresponding half-ring, goes around the coil, and then leaves through the other half-ring and brush. According to RHR-1, the directions of the forces on the two sides of the coil are as shown in the drawing, and these forces produce the torque that turns the coil. The coil rotates until it reaches the position shown in part *c* of the drawing. In this position the half-rings momentarily lose electrical contact with the brushes, and, as a result, there is no current in the coil and no applied torque. However, like any moving object, the rotating coil does not stop immediately, for its rotational inertia carries it onward. When the rings reestablish contact with the brushes, there again is a current in the coil, and a magnetic torque again rotates the coil in the same direction. The split-ring commutator ensures that the current is always in the proper direction to yield a torque that produces a continuous rotation of the coil.

27.7 MAGNETIC FIELDS PRODUCED BY CURRENTS

The previous two sections deal with the magnetic force that a current-carrying wire experiences when placed in a magnetic field. The magnetic field is assumed to be produced by some external source, such as a permanent magnet. In this section we consider the phenomena in which *a current-carrying wire produces a magnetic field.* Hans Christian Oersted (1777–1851) first discovered this effect in 1820 when he observed that a current-carrying wire influenced the orientation of a nearby compass needle. The compass needle aligns itself with the net magnetic field produced by the current and the magnetic field of the earth. Oersted's discovery, which linked the motion of electric charges with the creation of a magnetic field, marked the beginning of a technologically important discipline called **electromagnetism.**

THE MAGNETIC FIELD PRODUCED BY A LONG, STRAIGHT, CURRENT-CARRYING WIRE

Figure 27.20a illustrates the essence of Oersted's discovery with a very long, straight wire. When a current is present, the compass needles are observed to point in a circular pattern about the wire. The pattern indicates that the magnetic field lines produced by the current are circles centered on the wire. If the direction of the current is reversed, the needles also reverse their directions, indicating that the direction of the magnetic field has reversed. The direction of the magnetic field can be obtained by using Right-Hand Rule No. 2 (RHR-2), as part b of the drawing indicates:

Right-Hand Rule No. 2. Curl the fingers of the right hand into the shape of a half-circle. Point the thumb in the direction of the current I, and the tips of the fingers will point in the direction of the magnetic field **B**.

With a Hall-effect probe, or some other device that measures the magnetic field, the magnitude of **B** can be measured as a function of the current I in the wire and the radial distance r from the wire. It is found that the magnitude of the field is directly proportional to the current and inversely proportional to the radial distance: $B \propto I/r$. The proportionality constant is determined by experiment and is written as $\mu_0/2\pi$. Thus, the magnitude of the magnetic field created by the current in a very long, straight wire is

$$\begin{bmatrix} \textbf{Long straight} \\ \textbf{wire} \end{bmatrix} \qquad B = \frac{\mu_0 I}{2\pi r} \qquad (27.5)$$

The constant μ_0 is known as the **permeability of free space,** and its value is $\mu_0 = 4\pi \times 10^{-7}$ T·m/A. Since the magnetic field becomes stronger nearer the wire where r is smaller, the magnetic field lines near the wire are closer together than those located farther away, where the field is weaker. Figure 27.20c shows the pattern of field lines.

Many factories use industrial robots to carry materials and other parts from one place to another. One type of robot follows a current-carrying cable buried in the

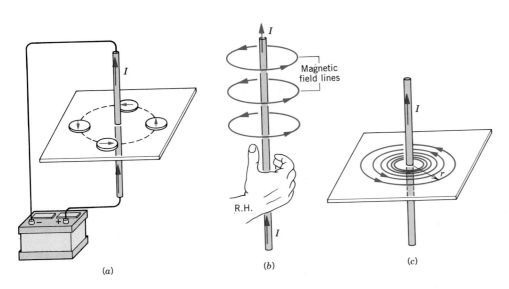

(a) (b) (c)

FIGURE 27.20 (a) A long, straight, current-carrying wire produces magnetic field lines that are circular about the wire. One such circular line is indicated by the compass needles. (b) If the thumb of the right hand (R.H.) is directed along the current I, the curled fingers point in the direction of the magnetic field, according to RHR-2. (c) The magnetic field becomes stronger as the radial distance r decreases, so the field lines are closer together near the wire.

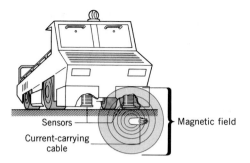

FIGURE 27.21 The robot follows the buried current-carrying cable by sensing the magnetic field that surrounds the cable.

floor. As Figure 27.21 suggests, the robot follows the cable by using special sensors to detect the magnetic field around the cable.

An electric current can create a magnetic field of its own, as well as experience a force created by another magnetic field. Therefore, the magnetic field that a current creates can cause a force to act on another nearby current. Example 5 illustrates this magnetic interaction between two currents.

EXAMPLE 5

Figure 27.22 shows two parallel straight wires. The wires are separated by a distance of 0.065 m and carry currents of $I_1 = 15$ A and $I_2 = 7.0$ A. Find the magnitude and direction of the force that the magnetic field of wire 1 applies to a 1.5-m length of wire 2 when the currents are (a) in the same direction and (b) in opposite directions.

SOLUTION

(a) To find the force on wire 2, we first determine the magnetic field that wire 1 produces at the location of wire 2. At wire 2, the magnitude of the magnetic field created by wire 1 is given by Equation 27.5 with $r = 0.065$ m:

$$B = \frac{\mu_0 I_1}{2\pi r} = \frac{(4\pi \times 10^{-7}\ \text{T}\cdot\text{m/A})(15\ \text{A})}{2\pi(0.065\ \text{m})}$$

$$= 4.6 \times 10^{-5}\ \text{T}$$

The direction of this field is upward at the location of wire 2, as part a of the figure shows. The direction can be obtained using RHR-2 (thumb of right hand along I_1, curled fingers

point upward at wire 2 and indicate the direction of **B**). The field is perpendicular to wire 2 ($\theta = 90°$), so the magnitude of the force on a 1.5-m length of wire 2 is

$$F = I_2 L B \sin \theta \qquad (27.3)$$

$$= (7.0\ \text{A})(1.5\ \text{m})(4.6 \times 10^{-5}\ \text{T}) \sin 90°$$

$$= \boxed{4.8 \times 10^{-4}\ \text{N}}$$

The direction of the magnetic force on wire 2 is toward wire 1, as part a of the drawing indicates; the force direction is found by using RHR-1 (fingers of the right hand extended upward along **B**, thumb points along I_2, palm pushes to the left).

In a like manner, the current in wire 2 also creates a magnetic field that produces a force on wire 1. Reasoning similar to that above shows that wire 1 is also attracted to wire 2 with a force of 4.8×10^{-4} N. Thus, each wire generates a force on the other and, if the currents are in the *same* direction, the wires *attract* each other. The fact that the two wires exert equal, but oppositely directed forces on each

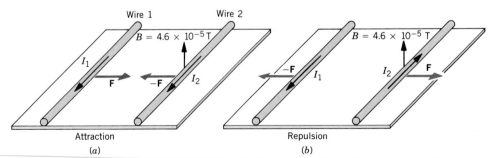

FIGURE 27.22 (a) Two long, parallel wires carrying currents I_1 and I_2 in the same direction attract each other. (b) The wires repel each other when the currents are in opposite directions.

other is consistent with Newton's third law, the action-reaction law.

(b) If the current in wire 2 is reversed, as part *b* of the drawing indicates, wire 2 is repelled from wire 1, because the direction of the magnetic force is reversed. However, the magnitude of the force is the same as that calculated in part *a* of this example. Likewise, wire 1 is also repelled from wire 2. Two parallel wires carrying currents in *opposite* directions *repel* each other.

THE MAGNETIC FIELD PRODUCED BY A LOOP OF WIRE

If a current-carrying wire is bent into a circular loop, the magnetic field lines around the loop have the pattern shown in Figure 27.23a. At the *center* of a loop of radius R, the magnetic field is perpendicular to the plane of the loop and has the value

$$\begin{bmatrix} \textbf{Center of a} \\ \textbf{circular loop} \end{bmatrix} \qquad\qquad B = \frac{\mu_0 I}{2R} \qquad\qquad (27.6)$$

where I is the current in the loop. Often, the loop consists of N turns of wire that are wound sufficiently close together that they form a flat coil with a single radius. In this case, the magnetic fields of the individual turns add together to give a net field that is N times greater than that of a single loop. For such a coil the magnetic field at the center is $B = N(\mu_0 I/2R)$.

The direction of the magnetic field at the center of the loop can be determined with the help of RHR-2. If the thumb of the right hand is pointed along the current and the curled fingers are placed at the center of the loop, the fingers indicate that the magnetic field is directed from right to left in Figure 27.23b.

A comparison of the magnetic field lines around a current loop with those in the vicinity of the short bar magnet in Figure 27.24a shows that the two patterns are quite similar. Not only are the patterns similar, but the loop itself behaves as a bar magnet with a "north pole" on one side and a "south pole" on the other side. Part *b* of the drawing emphasizes this point by including a "phantom" magnet at the center of the loop to symbolize that the loop may be imagined to be a bar magnet. The side of the loop that acts like a north pole can be determined with the aid of RHR-2, for the

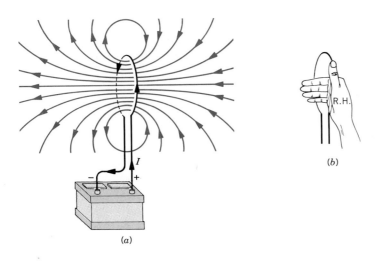

(a)

(b)

FIGURE 27.23 (a) The magnetic field lines in the vicinity of a current-carrying circular loop. (b) The direction of the magnetic field at the center of the loop is given by RHR-2.

FIGURE 27.24 (*a*) Note how the field lines around the bar magnet resemble those around the loop in Figure 27.23*a*. (*b*) The current loop can be imagined to be a phantom bar magnet with a north pole and a south pole.

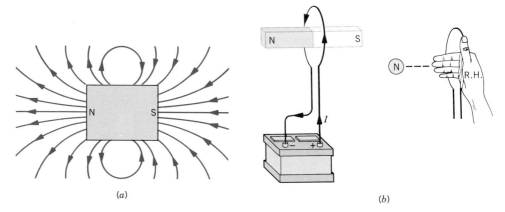

(*a*)

(*b*)

fingers of the right hand not only point in the direction of **B**, but they also point toward the north pole.

Example 6 shows how the magnetic fields produced by the current in a loop of wire and the current in a long, straight wire combine to form a net magnetic field.

EXAMPLE 6

A long, straight wire carries a current of 8.0 A. A portion of the wire is then bent into a circular loop of radius 0.020 m, as Figure 27.25 illustrates. Find the magnitude and direction of the net magnetic field at the center *C* of the loop.

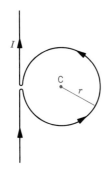

FIGURE 27.25 Part of a long, straight wire is bent into a circular loop that carries a current *I*.

SOLUTION

The net magnetic field at the point *C* is the sum of two contributions: (1) the field that the circular loop produces at its center, and (2) the field that the long, straight wire generates. An application of RHR-2 shows that the magnetic field generated by the circular loop at *C* is directed out of the plane of the paper, toward the reader. Similarly, RHR-2 shows that the magnetic field created at *C* by the long, straight wire is directed into the plane of the paper. Therefore, the directions of the two magnetic fields are opposite, and the net field is

$$B = \underbrace{\frac{\mu_0 I}{2r}}_{\substack{\text{Center of} \\ \text{a circular} \\ \text{loop}}} - \underbrace{\frac{\mu_0 I}{2\pi r}}_{\substack{\text{Long,} \\ \text{straight} \\ \text{wire}}} = \frac{\mu_0 I}{2r}\left(1 - \frac{1}{\pi}\right)$$

$$= \frac{(4\pi \times 10^{-7}\ \text{T}\cdot\text{m/A})(8.0\ \text{A})}{2\,(0.020\ \text{m})}\left(1 - \frac{1}{\pi}\right) = \boxed{1.7 \times 10^{-4}\ \text{T}}$$

The net field is directed perpendicularly out of the plane of the paper.

Because a current-carrying loop acts like a bar magnet, two adjacent loops can be either attracted to or repelled from each other, depending on the relative directions of the currents. Figure 27.26 includes a "phantom" magnet for each loop and shows that the loops are attracted to each other when the currents are in the same direction and repelled from each other when the currents are in opposite directions. This behavior is analogous to that of the two long, straight wires discussed earlier in Example 5.

As Section 27.6 mentions, a current loop tends to align itself such that the normal to the loop becomes parallel to an external magnetic field. This behavior can now be

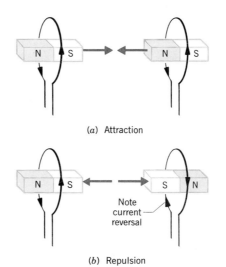

(a) Attraction

(b) Repulsion

Note current reversal

FIGURE 27.26 (a) The two current loops attract each other if the direction of the current is the same in both loops. (b) The loops repel each other if the currents are in opposite directions. The "phantom" magnet included for each loop helps explain the attraction and repulsion between the loops.

interpreted from another point of view, by thinking of the loop as a "phantom" bar magnet suspended in the field. The imaginary magnet tends to align itself with the field in the same manner that a real magnet or a compass needle does.

THE SOLENOID

A solenoid is a long coil of wire wound in the shape of a helix (see Figure 27.27). If the wire is wound so the turns are packed close to each other and the solenoid is long compared to its diameter, the magnetic field lines have the appearance shown in the drawing. Notice that the field inside the solenoid and away from its ends is nearly constant in magnitude and directed parallel to the axis. The direction of the magnetic field inside the solenoid is given by RHR-2, just as it is for a circular current loop. The magnitude of the magnetic field in the interior of a long solenoid is

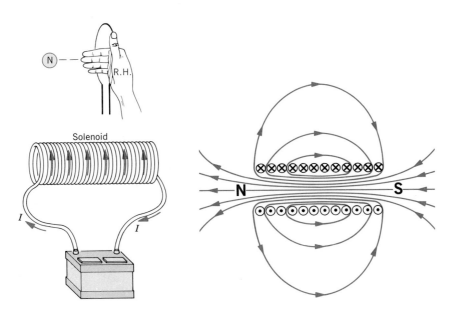

FIGURE 27.27 A solenoid and a cross-sectional view of it, showing the magnetic field lines and the north and south poles.

$$\begin{bmatrix} \text{Interior of a} \\ \text{long solenoid} \end{bmatrix} \qquad\qquad B = \mu_0 nI \qquad\qquad (27.7)$$

where n is the number of turns per unit length of the solenoid and I is the current. If, for example, the solenoid contains 100 turns and has a length of 0.05 m, the number of turns per unit length is $n = 2000$ turns/m.

As with a single loop of wire, a solenoid can also be imagined to be a bar magnet, for the solenoid is just an array of connected current loops. And, just as in the case of a circular current loop, the location of the north pole can be determined with RHR-2. Figure 27.27 shows that the left end of the solenoid acts as a north pole, and the right end behaves as a south pole. Solenoids are often referred to as *electromagnets,* and they have several advantages over permanent magnets. For one thing, the strength of the magnetic field can be altered by changing the current and/or the number of turns per unit length. Furthermore, the north and south poles of an electromagnet can be readily switched by reversing the current. An important application of electromagnetism in tape recording will be considered in the next section.

27.8 MAGNETIC MATERIALS

FERROMAGNETISM

The similarity between the magnetic field lines in the neighborhood of a bar magnet and those around a current loop suggests that the magnetism in each case arises from a common cause. The field that surrounds the loop is created by the charges moving in the wire. In fact, the magnetic field around a bar magnet is also due to the motion of charges, but the motion is not that of a bulk current through the magnetic material. Instead, the motion responsible for the magnetism is that which the electrons exhibit within the atoms of the material.

The magnetism produced by electrons within an atom can arise from two motional effects. First, each electron orbiting the nucleus behaves like an atomic-sized loop of current that generates a small magnetic field; this situation is similar to the field created by the current loop in Figure 27.23. Second, each electron possesses a spin that also gives rise to a magnetic field. The net magnetic field created by the electrons within an atom is due to the combined fields created by their orbital and spin motions.

In most substances the magnetism produced at the atomic level tends to cancel out, with the result that the substance is nonmagnetic overall. However, there are some materials, known as *ferromagnetic materials,* in which the cancellation does not occur for groups of approximately $10^{16} – 10^{19}$ neighboring atoms, because they have electron spins that are naturally aligned parallel to each other. This alignment results from a special type of quantum mechanical* interaction between the spins. The result of the interaction is a small but highly magnetized region of about 0.01 to 0.1 mm in size, depending on the nature of the material; this region is called a **magnetic domain.** Each domain behaves as a small magnet with its own north and south poles.

Ferromagnetic materials are important technologically because they can be permanently magnetized and used, for example, as the magnetic medium in tape decks

---

* The branch of physics called quantum mechanics is mentioned in Section 36.6, although a detailed discussion of quantum mechanics is beyond the scope of this book.

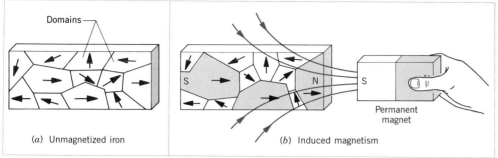

FIGURE 27.28 (*a*) Each magnetic domain is a highly magnetized region that behaves like a small magnet (represented by an arrow whose head indicates a noth pole). An unmagnetized piece of iron consists of many domains that are randomly aligned. The size of each domain is exaggerated for clarity. (*b*) The external magnetic field of the permanent magnet causes those domains that are parallel or nearly parallel to the field to grow in size (shown in color). Magnetism is thus induced into the iron.

and computer disks. Common ferromagnetic materials are iron (and certain types of steel), nickel, cobalt, chromium dioxide, and alnico (an *al*uminum-*ni*ckel-*co*balt alloy).

INDUCED MAGNETISM

Often the magnetic domains in a ferromagnetic material are arranged randomly, as Figure 27.28*a* illustrates for a piece of iron. In such a situation, the magnetic fields of the domains cancel each other, so the iron displays little, if any, overall magnetism. The unmagnetized iron represents the normal state of affairs in a piece of "erased" magnetic recording tape, for example. However, an unmagnetized piece of iron can be magnetized by placing it in an external magnetic field provided by a permanent magnet or an electromagnet. The external magnetic field penetrates the unmagnetized iron and *induces* (or "brings about") a state of magnetism in the iron by causing two effects on the domains. Those domains whose magnetism is parallel or nearly parallel to the external magnetic field grow in size at the expense of other domains that are not so oriented. Part *b* of the drawing shows the growing domains in color. In addition, the magnetic alignment of some domains may rotate and become more oriented in the direction of the external field. The resulting preferred alignment of the domains gives the iron an overall magnetism, so the iron behaves like a magnet with associated north and south poles. When the external field is removed, the domains remain aligned for the most part, and the iron becomes a permanent magnet.

The magnetism induced into a ferromagnetic material can be surprisingly large, even in the presence of a weak external field. For instance, it is not unusual for the induced magnetic field to be a hundred to a thousand times greater than the strength of the external field that causes the alignment. For this reason, high-field electromagnets are constructed by wrapping the current-carrying wire around a solid core made from iron or other ferromagnetic material.

Induced magnetism explains why a permanent magnet sticks to a refrigerator door and why an electromagnet can pick up scrap steel at a junkyard. Notice in Figure 27.28*b* that there is a north pole at the end of the iron that is closest to the south pole of the permanent magnet. This north pole arises because the north poles of the magnetic domains within the iron tend to line up so they face the south pole of the permanent magnet. The net result is that the two opposite poles give rise to an attraction between the iron and the permanent magnet. Conversely, the north pole of the permanent magnet would also attract the piece of iron by inducing a south pole in the nearest side of the iron. In nonferromagnetic materials, such as aluminum and copper, the formation of magnetic domains does not occur, so magnetism cannot be induced into these substances. Consequently, magnets do not stick to aluminum cans or copper pennies.

MAGNETIC TAPE RECORDING

Magnetic tape recording uses a fascinating application of induced magnetism, and Figure 27.29 illustrates the essential features of the recording process. The weak electrical signal from a microphone is routed to an amplifier where it is amplified. The current from the output of the amplifier is then sent to the recording head, which is essentially a wire wrapped around an iron core. The iron core has the approximate shape of a horseshoe with a small gap between the two ends. The ferromagnetic iron substantially enhances the magnetic field produced by the current in the wire.

When there is a current in the coil, the coil becomes an electromagnet with a north pole at one end and a south pole at the other end. The magnetic field lines pass through the iron core and cross the gap. Within the gap, the lines are directed from the north pole to the south pole. Some of the field lines in the gap "bow outward," as Figure 27.29 indicates, the bowed region of magnetic field being called the *fringe field*. The fringe field penetrates the magnetic coating on the tape and induces magnetism in the coating. This induced magnetism is retained when the tape leaves the vicinity of the recording head and, thus, provides a means for storing audio information. Audio information can be stored, because at any instant in time the extent to

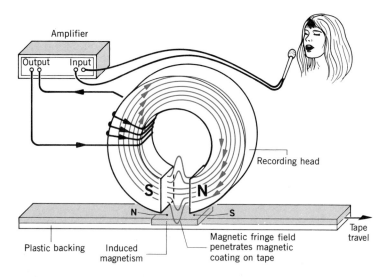

FIGURE 27.29 The magnetic fringe field of the recording head penetrates the magnetic coating on the tape and causes the coating to become magnetized.

which the tape is magnetized depends on the amount of current in the recording head. The current, in turn, depends on the sound intensity picked up by the microphone, so that changes in the sound intensity that occur from moment to moment are preserved as changes in the tape's magnetization.

27.9 OPERATIONAL DEFINITIONS OF THE AMPERE AND THE COULOMB

In Section 25.1 electric current is defined as the rate at which charge flows, or $I = q/t$ when the current is constant. Therefore, one way of measuring current is to determine the amount of charge q that flows in a time t. According to this procedure, one ampere of current exists when one coulomb of charge flows for one second. In practice, however, it is difficult to measure an ampere precisely by measuring the amount of charge flowing in a known time interval. It would be far superior if the ampere could be measured in terms of force and distance, quantities that can be measured with a high degree of precision. Such a measurement is possible in terms of the magnetic force that two current-carrying wires exert on each other.

Suppose the same current I is sent through two long, straight, parallel wires that are separated by a distance r. According to Equation 27.5, the magnetic field $\mathbf{B}$ produced at the location of one wire by the other wire has a magnitude of $B = \mu_0 I/2\pi r$. Since this magnetic field is perpendicular to the wire (see Figure 27.22), the field exerts a force $\mathbf{F}$ on a length L of the wire:

$$F = ILB \sin 90° = \frac{\mu_0 I^2 L}{2\pi r} \qquad (27.3)$$

With special instruments, this force can be measured accurately. The wire length L and the separation r can also be determined accurately, and μ_0 has been assigned the value of $4\pi \times 10^{-7}$ T·m/A. With these values, the equation above can be solved for the current I. For instance, suppose $F = 2.000 \times 10^{-7}$ N, $r = 1.000$ m, and $L = 1.000$ m. The current is

$$I = \sqrt{\frac{2\pi r F}{\mu_0 L}} = \sqrt{\frac{2\pi(1.000 \text{ m})(2.000 \times 10^{-7} \text{ N})}{(4\pi \times 10^{-7} \text{ T·m/A})(1.000 \text{ m})}} = 1.000 \text{ A}$$

Therefore, one ampere of current is defined as the amount of electric current in each of two long, parallel wires that gives rise to a magnetic force per unit length of 2×10^{-7} N/m on each wire when the wires are separated by one meter. This definition provides a means for measuring current in terms of force and distance and obviates the need to define the ampere in terms of the amount of moving charge per unit time.

With the ampere defined in terms of force and distance, the coulomb can now be defined as the quantity of electrical charge that passes a given point in one second when the current is one ampere, or $1 \text{ C} = 1 \text{ A·s}$. This definition is preferred, since scientists can measure electric current and time more accurately than they can measure the amount of moving charge.

SUMMARY

A magnet has a north pole and a south pole. The north pole is the end that points toward the north magnetic pole of the earth when the magnet is freely suspended. **Like poles repel each other and unlike poles attract each other.**

A **magnetic field** exists in the space around a magnet. The magnetic field is a vector whose direction at any point is the direction indicated by the north pole of a small compass needle placed at that point. The magnitude B of the magnetic field at any point in space is defined as $B = F/(q_0 v \sin \theta)$, where F is the magnitude of the magnetic force that acts on a charge q_0 whose velocity $\mathbf{v}$ makes an angle θ with respect to the magnetic field. The SI unit for the magnitude of the magnetic field is the tesla (T). The direction of the magnetic force is perpendicular to both $\mathbf{v}$ and $\mathbf{B}$, and for a positive charge the direction can be determined with the aid of Right-Hand Rule No. 1 (RHR-1, see Section 27.2). The magnetic force on a moving negative charge is opposite to the force on a moving positive charge.

A **magnetic force does no work on a charged particle,** provided the magnetic field is constant and no agent other than the magnetic field exerts a force on the charge. The magnetic force does no work because the direction of the force is always perpendicular to the motion of the particle. Consequently, the magnetic force cannot change the kinetic energy of the particle; however, the magnetic force can change the direction in which the particle moves. If a particle of charge q and mass m moves with speed v perpendicular to a uniform magnetic field $\mathbf{B}$, the magnetic force causes the charge to move on a circular path of radius $r = mv/(qB)$.

A **mass spectrometer** is an instrument that can determine the masses of atoms and molecules. The Hall effect occurs when an emf, called the **Hall emf,** develops across a current-carrying metal or semiconductor that has been placed in a magnetic field. The Hall emf arises because of the deflection of the moving charges by the magnetic force. The polarity of the Hall emf indicates whether the charge carriers are positive or negative.

An **electric current, being composed of moving charges, can experience a magnetic force when placed in a magnetic field.** For a straight wire that has a length L and carries a current I, the magnetic force has a magnitude of $F = ILB \sin \theta$, where θ is the angle between the directions of I and $\mathbf{B}$. The direction of the force is perpendicular to both I and $\mathbf{B}$ and is given by RHR-1.

Magnetic forces can exert a **torque** on a current-carrying loop of wire and thus cause the loop to rotate. If a current I exists in a coil of wire with N turns, each of area A, in the presence of a magnetic field $\mathbf{B}$, the coil experiences a torque whose magnitude is $\tau = NIAB \sin \phi$, where ϕ is the angle between the direction of the magnetic field and the normal to the plane of the coil. The operation of a galvanometer and an electric motor is based on the torque exerted on a current-carrying coil.

An **electric current produces a magnetic field,** with different current geometries giving rise to different magnetic field patterns. For a **long, straight wire,** the magnetic field lines are circles centered on the wire, and their direction is given by RHR-2 (see Section 27.7). The magnitude of the field at a radial distance r from the wire is $B = \mu_0 I/(2\pi r)$, where I is the current and μ_0 is the **permeability of free space** ($\mu_0 = 4\pi \times 10^{-7}$ T·m/A). The magnetic field produced by a **flat coil** consisting of N closely spaced turns of wire resembles that of a permanent magnet. The coil has associated with it a north pole on one side and a south pole on the other side. The side of the coil that behaves like a north pole can be predicted by using RHR-2. The magnetic field at the center of a circular coil consisting of N turns, each of radius R, is $B = \mu_0 NI/(2R)$. A **solenoid** is a coil of wire wound in the shape of a helix. Inside a long solenoid the magnetic field is nearly constant and has the value $B = \mu_0 nI$, where n is the number of turns per unit length of the solenoid. Solenoids are often known as electromagnets, especially if the wire is wrapped around a ferromagnetic core.

Ferromagnetic materials, such as iron, are made up of tiny regions called domains, each of which behaves as a small magnet. In an unmagnetized piece of iron, the domains are randomly aligned. In a permanent magnet, many of the domains are aligned, and a high degree of magnetism results. An unmagnetized piece of iron can be induced into becoming magnetized by placing the iron in an external magnetic field.

QUESTIONS

1. Consider the magnetic field lines shown in the drawing. At which point, P_1, P_2, or P_3, is the magnetic field (a) the strongest, (b) the weakest, and (c) nearly constant? Justify your answers. See Drawing 1.

2. In all the drawings of magnetic field lines in this chapter there are no instances of two field lines crossing each other. Magnetic field lines, like electric field lines, never intersect. Suppose it were possible for two field lines to intersect at a point in space. Discuss

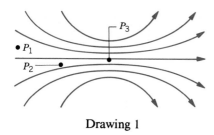

Drawing 1

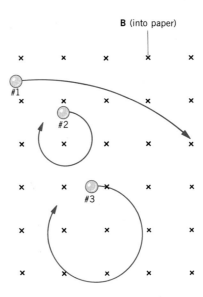

what this would imply about the force(s) that act on a charge moving through such a point, thereby ruling out the possibility of field lines crossing each other.

3. Suppose you accidentally use your left hand, instead of your right hand, to determine the direction of the magnetic force on a positive charge moving in a magnetic field. Do you get the correct answer? If not, what direction do you get?

4. A charged particle, passing through a certain region of space, has a velocity whose magnitude and direction remain constant. (a) If it is known that the magnetic field is zero everywhere, can you conclude that the electric field is also zero? (b) If it is known that the electric field is zero everywhere, can you conclude that the magnetic field is also zero? Justify your answers.

5. A stationary charge is located between the poles of a horseshoe magnet. Is a magnetic force exerted on the charge? Why?

6. Three particles, each with the same mass, move through a constant magnetic field and follow the paths shown in the drawing. Determine whether each particle is positively charged, negatively charged, or neutral. Give a reason for each answer.

8. The drawing shows a top view of four interconnected chambers. A negative charge is fired into chamber 1. By turning on separate magnetic fields in each chamber, the charge can be made to exit from chamber 4. (a) Describe how the magnetic field in each chamber should be directed. (b) If the speed of the charge is v when it enters chamber 1, what is the speed of the charge when it exits chamber 4? Why?

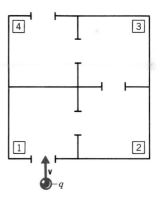

9. A positive charge moves along a circular path under the influence of a magnetic field. The magnetic field is perpendicular to the plane of the circle, as in Figure 27.9. If the velocity of the particle is reversed at some point along the path, will the particle retrace its path? If not, draw the new path. Justify your answer.

10. A positively charged particle travels on a circular path in the presence of a magnetic field, as in Figure 27.9. A uniform electric field is then turned on. Draw the path of the particle when the electric field is directed (a) perpendicular to the magnetic field and pointing to the right and (b) parallel to the magnetic field.

11. The drawing shows a positive charge $+q$ located at the coordinate origin and a target located in the third quadrant. A magnetic field is directed perpendicularly into the plane of the

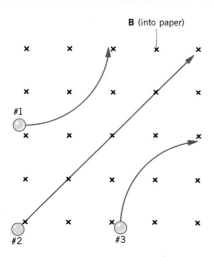

7. Three particles have identical charges and masses. They enter a constant magnetic field and follow the paths shown in the picture. Which particle is moving the fastest, and which is moving the slowest? Justify your answers.

paper. The charge can be projected in the plane of the paper only, along the positive or negative x or y axis. Thus, there are four possible directions along which the charge can be projected. The charge can be made to hit the target for only two of the four directions. Which two are they? Give your reasoning, along with the two paths that the charge can follow on its way to the target.

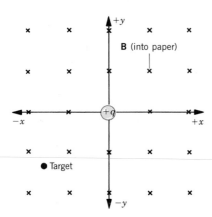

12. Refer to Figure 27.13. (a) What happens to the direction of the magnetic force if the current is reversed? (b) What happens to the direction of the force if *both* the current and the magnetic poles are reversed? Explain your answers for parts (a) and (b).

13. Suppose that the permanent magnet in a galvanometer has lost some of its magnetism. Would the reading on the scale of the galvanometer be greater than or less than the actual value of the current? Give a reason for your answer.

14. The drawing shows an end-on view of three parallel wires that are perpendicular to the plane of the paper. In two of the wires the current is directed into the paper, while in the remaining wire the current is directed out of the paper. The two outermost wires are held rigidly in place. Which way will the middle wire move? Explain.

15. For each electromagnet at the left of the drawing, explain whether it will be attracted to or repelled from the adjacent magnet at the right.

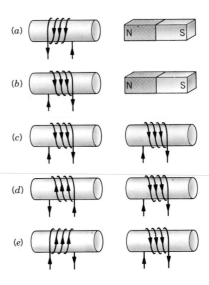

16. Refer to Figure 27.5. If the earth's magnetism is assumed to originate from a large circular loop of current within the earth, how is the plane of this current loop oriented relative to the magnetic axis, and what is the direction of the current around the loop?

17. Suppose you have two bars, one of which is a permanent magnet and the other is not a magnet, but is made from a ferromagnetic material like iron. The two bars look exactly alike. (a) Using a third bar, which is known to be a magnet, how can you determine which of the look-alike bars is the magnet and which is not? (b) Can you determine the identities of the look-alike bars with the aid of a third bar that is not a magnet, but made from a ferromagnetic material? Give a reason for your answer.

PROBLEMS

Section 27.1 Magnets and Magnetic Fields, Section 27.2 The Force That a Magnetic Field Exerts on a Moving Charge

1. A charge of 12 μC, traveling with a speed of 9.0×10^6 m/s in a direction perpendicular to a magnetic field, experiences a magnetic force of 8.7×10^{-3} N. What is the magnitude of the field?

2. A particle with a charge of 6.0 μC and a speed of 25 m/s

enters a uniform magnetic field whose magnitude is 0.15 T. For each of the eight cases shown in the drawing, find the magnitude and direction of the magnetic force on the charge. See Drawing 2.

3. An electron, traveling with a velocity of 4.5×10^6 m/s due east, experiences a maximum magnetic force of 8.0×10^{-14} N due south. (a) What is the magnitude and direction of the magnetic field? (b) Answer part (a), assuming the electron is replaced by a proton.

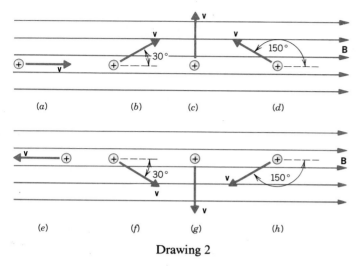

(a)   (b)   (c)   (d)

(e)   (f)   (g)   (h)

Drawing 2

4. A charged body, moving with a velocity of 8.0×10^4 m/s at an angle of $30.0°$ with respect to a magnetic field of 5.6×10^{-5} T, feels a force of 2.0×10^{-4} N. What is the magnitude of the charge?

* 5. In a television set, electrons are accelerated from rest through a potential difference of 15 kV. The electrons then pass through a magnetic field that deflects them to the appropriate spot on the screen. (a) If the magnetic field has a magnitude of 3500 gauss, find the maximum force that an electron can experience. (b) Find the ratio of the maximum magnetic force to the gravitational force that acts on an electron.

Section 27.3 The Motion of a Charged Particle in a Magnetic Field, Section 27.4 The Mass Spectrometer and the Hall Effect

6. An electron moves at a speed of 6.0×10^6 m/s perpendicular to a constant magnetic field. The path is a circle of radius 1.3×10^{-3} m. (a) What is the magnitude of the magnetic field? (b) Determine the magnitude of the electron's acceleration. (c) Draw a sketch, showing the magnetic field and the electron's path.

7. The solar wind is a thin hot gas given off by the sun. Charged particles in this gas enter the magnetic field of the earth and can experience a magnetic force. Suppose a charged particle traveling with a speed of 9.0×10^6 m/s encounters the earth's magnetic field at an altitude where the field has a magnitude of 1.2×10^{-7} T. Assuming the particle's velocity is perpendicular to the magnetic field, find the radius of the circular path on which the particle would move if it were (a) an electron and (b) a proton.

8. A beam of protons moves in a circle of radius 0.25 m. The beam moves perpendicular to a 0.30-T magnetic field. (a) What is the speed of the protons? (b) Determine the magnitude of the centripetal force that acts on each proton.

9. A mass spectrometer uses a potential difference of 2.00 kV to accelerate a singly charged ion to the proper speed. A 0.400-T magnetic field then bends the ion into a circular path of radius 0.226 m. What is the mass of the ion?

10. An ion source in a mass spectrometer produces a mixture of protons and deuterons (a deuteron is a particle that has twice

the mass of a proton, but the same charge). These particles are accelerated from rest through a potential difference of 2.00×10^3 V, after which they enter a 0.600-T magnetic field. Find the radii of the circular paths of these particles.

11. Two isotopes of carbon, carbon-12 and carbon-13, have masses of 19.92×10^{-27} kg and 21.59×10^{-27} kg, respectively. These two isotopes are singly ionized and each is given a speed of 6.667×10^5 m/s. The ions then enter the bending region of a mass spectrometer where the magnetic field is 0.8500 T. Determine the spatial separation between the two isotopes after they have traveled through a half-circle.

* 12. A positively charged particle of mass 7.2×10^{-8} kg is traveling due east with a speed of 85 m/s. The particle enters a 0.31-T uniform magnetic field, and 2.2×10^{-3} s later the particle leaves the field heading due south with a speed of 85 m/s. All during the motion the particle moves perpendicular to the magnetic field. (a) What is the magnitude of the magnetic force acting on the particle? (b) Determine the charge of the particle.

* 13. An electron moves in a circular orbit of radius 1.7 m in a magnetic field of 2.2×10^{-5} T. The electron moves perpendicular to the magnetic field. Determine the kinetic energy of the electron in electron volts.

* 14. A *velocity selector* is a device for measuring the speed of a charged particle. The drawing shows that a velocity selector consists of a cylindrical tube located within a constant magnetic field **B**. Inside the tube there is a parallel plate capacitor that produces an electric field **E**. The magnetic and electric fields are perpendicular to each other. A positive charge enters the left end of the tube and has a velocity that is perpendicular to both **B** and **E**. The charge experiences both a magnetic force and an electric force. However, if the forces are adjusted so as to cancel each other, the net force acting on the charge is zero. The charge then moves down the tube at a constant speed v in a straight line. For such a situation, derive an expression for the speed of the particle in terms of B and E.

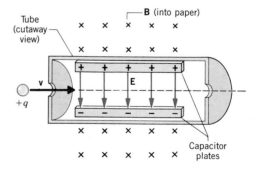

* 15. Work problem 14 before attempting to solve this problem. A charged particle moves through a velocity selector at a constant speed in a straight line. The electric field of a velocity selector is 5.65×10^3 N/C, while the magnetic field is 0.114 T. When the electric field is turned off, the charged particle travels on a circular path whose radius is 2.90 cm. Find the charge-to-mass ratio of the particle.

* **16.** Suppose that an ion source in a mass spectrometer produces *doubly* ionized gold ions (Au^{2+}) whose masses are 3.270×10^{-25} kg. The ions are accelerated from rest through a potential difference of 1.000 kV. A 0.5000-T magnetic field causes the ions to follow a circular path. Determine the radius of curvature of the path.

** **17.** The drawing shows one arrangement that could be used to measure the speed of blood in an artery. As the blood flows with a speed v, positive and negative ions are carried perpendicularly into a magnetic field. The positive ions are deflected by the magnetic force to one side of the artery, while negative ions are deflected to the other side. Charge builds up on either side until an electric field develops, such that the electric force acting on the moving charges balances the magnetic force. At the balance point, charge buildup ceases, and the positive and negative ions pass through the fields undeflected. The charge that has accumulated can then be detected with electrodes placed on either side of the artery. For the purpose of this problem, these electrodes can be treated as a parallel plate capacitor, the plates being located on opposite sides of the artery. For a magnetic field of 0.080 T, an artery diameter of 2.5 mm, and a measured voltage of 6.3×10^{-5} V between the electrodes, find the speed of the blood.

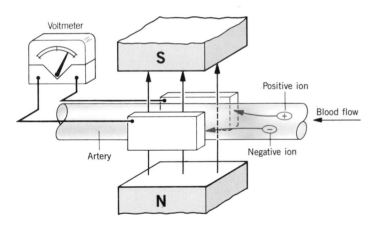

** **18.** An α-particle is the nucleus of a helium atom; the orbiting electrons are missing. The α-particle contains two protons and two neutrons, and has a mass of 6.64×10^{-27} kg. Suppose an α-particle is accelerated from rest through a potential difference and then enters a region where its velocity is perpendicular to a 2.10×10^{-2}-T magnetic field. With what angular speed ω does the α-particle move on its circular path?

** **19.** A singly ionized atom of neon-20 follows a circular path in the bending region of a mass spectrometer. (a) For the same magnetic field in the bending region, how much larger or smaller (in percent) is the radius of the circular path followed by a singly charged ion of neon-22? (b) Repeat part (a) for a doubly ionized atom of neon-22.

Section 27.5 The Force on a Current in a Magnetic Field

20. The drawing shows three examples of a wire with length L

and current I, lying in a plane that is perpendicular to a magnetic field **B**. In all cases $B = 25 \times 10^{-2}$ T, $L = 0.60$ m, and $I = 15$ A. Find the magnitude and direction of the magnetic force on the current in each situation.

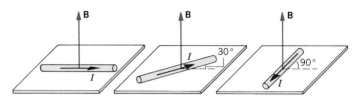

21. An electric power line carries a current of 1400 A in a location where the earth's magnetic field is 0.50×10^{-4} T. The line makes an angle of 75° with respect to the field. Determine the magnitude of the magnetic force on a 120-m length of line.

22. Near the equator in South America the earth's magnetic field has a strength of about 0.30 gauss; the field is parallel to the surface of the earth and points due north. A straight wire, 25 m in length, has an east–west orientation and experiences a magnetic force of 4.1×10^{-2} N, directed vertically down (toward the earth). What is the magnitude and direction of the current in the wire?

23. A wire of length 0.655 m carries a current of 21.0 A. In the presence of a 0.470-T magnetic field, the wire experiences a force of 5.46 N. What is the angle between the wire and the magnetic field?

24. At New York City, the earth's magnetic field has a vertical (downward) component of 0.52×10^{-4} T and a horizontal component of 0.18×10^{-4} T that is directed toward geographic north. What is the magnitude of the magnetic force on a long straight wire, 8.0 m in length, that carries a 35-A current due east?

* **25.** A 125-turn rectangular coil of wire is hung from one arm of a balance, as the drawing shows. With the magnetic field turned

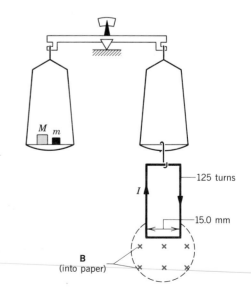

off, a mass M is added to the pan on the other arm in order to balance the mass of the coil. When a constant magnetic field of magnitude 0.200 T is turned on and there is a current of 8.50 A in the coil, how much *additional* mass m must be added to regain the balance?

* 26. A copper rod of length 0.85 m is lying on a frictionless table (see the drawing). Each end of the rod is attached to a fixed wire by an unstretched spring whose spring constant is $k = 75$ N/m. A magnetic field with a strength of 0.16 T is oriented perpendicular to the surface of the table. (a) What must be the direction of the current in the copper rod that causes the springs to stretch? (b) If the current is 12 A, by how much does each spring stretch?

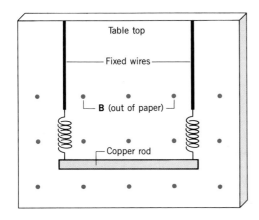

**27. A 0.20-kg aluminum rod is lying on top of two conducting rails that are separated by 1.6 m. A 0.050-T magnetic field has the direction shown in the drawing. The coefficient of static friction between the rod and a rail is $\mu_s = 0.45$. (a) How much current must be sent through the rod before the rod begins to move? (b) In what direction will the rod move, toward the battery or away from it? Explain.

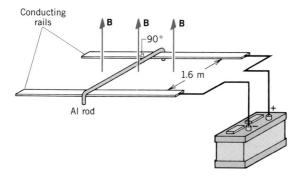

**28. A horizontal wire of length 0.20 m and mass 0.080 kg is hung from the ceiling of a room by two massless strings. A 0.070-T magnetic field is directed from the ceiling to the floor. When a current of 42 A passes through the wire, the wire swings upward through an angle ϕ, as the drawing shows. Find (a) the angle ϕ and (b) the tension in each of the two strings.

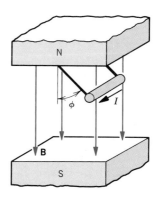

Section 27.6 The Torque on a Current-Carrying Coil

29. A circular coil of wire has a radius of 0.10 m. The coil has 50 turns and a current of 15 A, and is placed in a magnetic field whose magnitude is 0.20 T. (a) Determine the magnetic moment of the coil. (b) What is the maximum torque the coil can experience in this field? (c) What is the orientation of the coil relative to the magnetic field under the condition of maximum torque?

30. A coil carries a current and experiences a torque due to a magnetic field. The value of the torque is 80.0% of the maximum possible torque. (a) What is the angle between the magnetic field and the normal to the plane of the coil? (b) Make a drawing, showing how this coil would be oriented relative to the magnetic field. Be sure to include the angle in the drawing.

31. A 0.50-m length of wire is formed into a single-turn, planar loop in which there is a current of 12 A. The loop is placed in a magnetic field of 0.12 T, as in Figure 27.17a. What is the maximum torque that the loop can experience when it is (a) square and (b) circular?

* 32. In the model of the hydrogen atom due to Niels Bohr, the electron moves around the proton at a speed of 2.2×10^6 m/s in a circle of radius 0.53×10^{-10} m. Considering the orbiting electron to be a small current loop, determine the magnetic moment associated with this motion.

* 33. A charge of 4.0×10^{-6} C is placed on a small conducting sphere that is located at the end of a thin insulating rod whose length is 0.20 m. The rod rotates with an angular speed of 150 rad/s about an axis that passes perpendicularly through its other end. Find the magnetic moment of the rotating charge.

Section 27.7 Magnetic Fields Produced by Currents

34. In a lightning bolt, about 15 C of charge flows in a time of 1.5×10^{-3} s. Assuming that the lightning bolt can be represented as a long, straight line of current, what is the magnetic field at a distance of 25 m from the bolt?

35. A 6.00-μC charge is moving with a velocity of 7.50×10^6 m/s parallel to a long, straight wire. The wire carries a current of 67.0 A in the same direction as the moving charge, and is 5.00×10^{-2} m away from the charge. Find the magnitude and direction of the force on the charge.

36. What must be the radius of a circular loop of wire so the magnetic field at its center is 1.4×10^{-4} T when the loop carries a current of 6.0 A?

37. A long solenoid consists of 1400 turns of wire and has a length of 0.65 m. There is a current of 4.7 A in the wire. What is the magnitude of the magnetic field within the solenoid?

38. Two straight, parallel wires are separated by 0.15 m. The first wire carries a current of 125 A, and the magnetic field produced by this current exerts a force of 3.0×10^{-3} N on a 2.1-m length of the second wire. What is the current in the second wire?

39. A 123-m length of copper wire has a resistance per unit length of 5.90×10^{-3} Ω/m. The wire is wound into a thin, flat coil of many turns that has a radius of 0.140 m. The ends of the wire are connected to a 12.0-V battery. Find the magnetic field at the center of the coil.

* 40. Two rigid rods are oriented parallel to each other and to the ground. The rods carry the same current in the same direction. The length of each rod is 0.85 m, while the mass of each is 0.073 kg. One rod is held in place above the ground, and the other floats beneath it at a distance of 8.2 mm. Determine the current in the rods.

* 41. Two long, straight, parallel wires A and B are separated by a distance of one meter. They carry currents in opposite directions, and the current in wire A is one-third of that in wire B. On a line drawn perpendicular to the wires and passing through them, find the point where the net magnetic field is zero. Determine this point relative to wire A.

** 42. A rectangular current loop is located near a long, straight wire that carries a current of 12 A (see the drawing). The current in the loop is 25 A. (a) Determine the net magnetic force that acts on the loop. (b) Is the loop attracted to or repelled from the straight wire? Why?

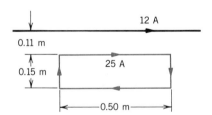

* 43. Two circular coils are concentric and lie in the same plane. The inner coil contains 120 turns of wire, has a radius of 0.012 m, and carries a current of 6.0 A. The outer coil contains 150 turns and has a radius of 0.017 m. What must be the magnitude and direction (relative to the current in the inner coil) of the current in the outer coil, such that the net magnetic field at the common center of the two coils is zero?

**44. The drawing shows two long, straight wires that are suspended from a ceiling. Each of the four strings suspending the wires has a length of 1.2 m. The mass per unit length of each wire

is 0.050 kg/m. When the wires carry identical currents in opposite directions, the angle between the strings holding the two wires is 15°. What is the current in each wire?

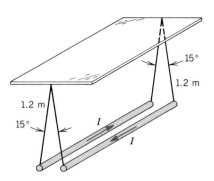

ADDITIONAL PROBLEMS

45. Two long, straight wires are separated by 0.120 m. The wires carry currents of 8.0 A in opposite directions, as the drawing indicates. Find the magnitude of the net magnetic field at the points labeled A, B, and C.

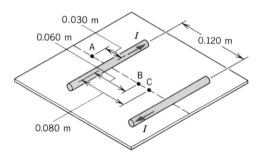

46. A charge $q_1 = 25.0$ μC moves with a speed of 4.50×10^3 m/s perpendicular to a uniform magnetic field. The charge experiences a magnetic force of 7.31×10^{-3} N. A second charge $q_2 = 5.00$ μC travels at an angle of 40.0° with respect to the same magnetic field and experiences a 1.90×10^{-3}-N force. Determine (a) the magnitude of the magnetic field and (b) the speed of q_2.

47. A semicircular closed loop is made from a wire of length 5.00×10^{-2} m. There is a current of 2.00 A in the wire. (a) In the presence of a 1.50-T magnetic field, what is the largest torque that this loop can experience? (b) What should be the orientation of the loop with respect to the magnetic field to yield the largest torque?

48. A long, straight wire carrying a current of 305 A is placed in a uniform magnetic field whose magnitude is 7.00×10^{-3} T. The wire is perpendicular to the field. Find a point in space where the net magnetic field is zero. Locate this point by specifying its perpendicular distance from the wire.

49. A charged particle with a charge-to-mass ratio of 5.7×10^8 C/kg travels on a circular path that is perpendicular to a magnetic field whose magnitude is 0.72 T. How much time does it take for the particle to complete one revolution?

50. A 45-m length of wire is stretched horizontally between two vertical posts. The wire carries a current of 75 A and experiences a magnetic force of 0.15 N. Find the magnitude of the earth's magnetic field at the location of the wire, assuming the field makes an angle of 60.0° with respect to the wire.

* **51.** A circular loop of wire and a long straight wire each carry the same current, as the drawing shows. The loop and wire lie in the same plane. The net magnetic field at the center of the loop is zero. Find the distance H, expressing your answer in terms of R, the radius of the loop.

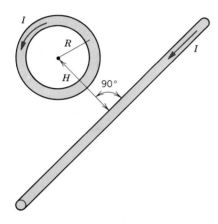

* **52.** The electrons in the beam of a television tube have a kinetic energy of 2.40×10^{-15} J. Initially, the electrons move horizontally from west to east. The vertical component of the earth's magnetic field points down, toward the surface of the earth, and has a magnitude of 2.00×10^{-5} T. (a) In what direction are the electrons deflected by this component of the magnetic field? (b) What is the acceleration of an electron in part (a)?

* **53.** A charge of $+3.5 \times 10^{-5}$ C is distributed uniformly around a thin ring of insulating material. The ring has a radius of 0.25 m and rotates with an angular speed of $\omega = 6500$ rad/s about an axis perpendicular to the plane of the ring and passing through its center. Determine the magnitude of the magnetic field produced at the center of the ring.

* **54.** Two charged particles have the same *linear momentum,* but particle 1 has three times the charge of particle 2. Both particles travel perpendicular to a uniform magnetic field. What is the ratio r_1/r_2 of the radii of the circles on which they move?

* **55.** Suppose you have two identical lengths of very thin insulated wire with which to make two planar coils. Coil A consists of a single circular turn, while coil B consists of two circular turns. The coils are located in the same magnetic field and each carries the same current. What is the ratio τ_B/τ_A of the maximum torques that these coils can experience?

** **56.** The drawing shows two wires that carry the same current of $I = 85.0$ A and are oriented perpendicular to the plane of the paper. The current in one wire is directed out of the paper, while the current in the other is directed into the paper. Find the magnitude and direction of the net magnetic field at the point P.

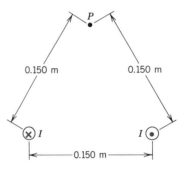

** **57.** A particle of charge $q = +25 \times 10^{-10}$ C is fired horizontally with a speed of 4300 m/s at the equator of the earth. The particle moves west to east (see the drawing) and has a mass of 1.2×10^{-6} kg. The radius of the earth is 6.4×10^6 m. What should be the direction and magnitude of the smallest magnetic field that would cause the particle to orbit the earth?

CHAPTER 28

Electromagnetic Induction

These large electric generators are in the Nevada Wing of the Hoover Dam powerplant. They produce electrical energy in a manner that is described by Faraday's law of electromagnetic induction, the central theme of this chapter.

28.1 INDUCED EMF AND INDUCED CURRENT

Since an electric current produces a magnetic field, it is natural to ask whether a magnetic field can produce an electric current. In this chapter we will see that the use of magnetic fields to produce electricity is, in fact, widespread. For example, large generators at power plants utilize magnetic fields in the production of electricity for home use. On a completely different scale, a phono cartridge and a cassette deck also use magnetic fields to generate electrical signals. These signals are sent to the remainder of the hi-fi system for amplification and, ultimately, conversion into sound by the speakers.

Figure 28.1 illustrates one way that a magnetic field can be used to generate a current. This drawing shows a bar magnet and a helical coil of wire to which an ammeter is connected. When there is no relative motion between the bar magnet and the coil, as in part *a* of the drawing, the ammeter reads zero, indicating there is no current in the coil. However, when the magnet moves toward the coil, as in part *b*, a current appears in the coil. As the magnet approaches, the magnetic field that it creates at the location of the coil becomes stronger and stronger, and it is this *changing* magnetic field that produces the current. When the magnet moves away from the coil, as in part *c*, a current also exists, but the direction of the current is

A turntable.

A double cassette deck.

reversed. In this instance, the magnetic field at the coil becomes weaker as the magnet moves away. Once again, though, it is the *changing* magnetic field at the coil that generates the current.

A current would also be created in Figure 28.1 if the magnet were held stationary and the coil were moved, because the magnetic field at the coil would be changing as the coil approached or receded from the magnet. Only relative motion between the magnet and the coil is needed to generate a current; it does not matter which one moves.

The current in the coil is called an **induced current,** because the current is brought about (or "induced") by a changing magnetic field. Since a source of emf is always needed to produce a current, the coil itself behaves as if it were a source of emf. The emf is known as an **induced emf,** and its value can be measured in the usual manner by connecting a voltmeter in parallel with the coil. Thus, a changing magnetic field induces an emf in the coil, and the emf leads to an induced current.

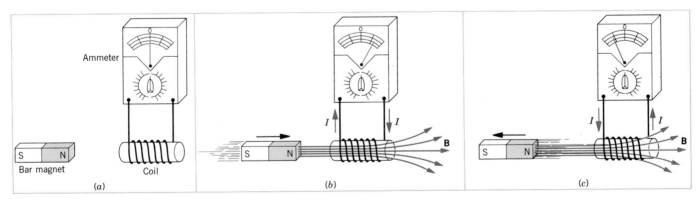

FIGURE 28.1 (*a*) When there is no relative motion between the coil of wire and the bar magnet, there is no current in the coil. (The ammeter indicates zero current when the pointer is in the center of the scale; the pointer deflects right or left when there is a current, depending on the direction of the current.) (*b*) A current is created in the coil when the magnet moves toward the coil. (*c*) A current also exists when the magnet moves away from the coil, but the direction of the current is opposite to that in (*b*).

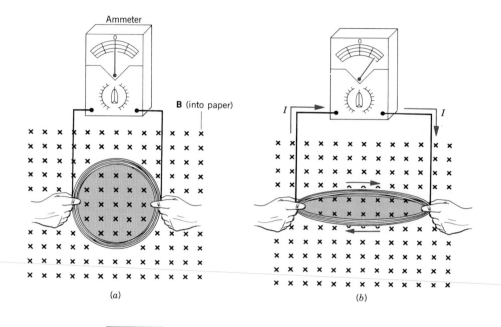

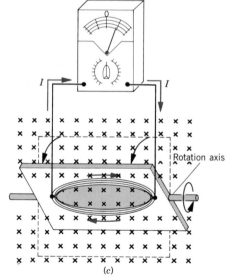

FIGURE 28.2 (*a*) No current exists in a coil of constant area (the shaded region) that is located in a constant magnetic field. (*b*) While the area of the coil is changing, an induced emf and current are generated. (*c*) An induced emf and current are also produced while the coil is rotating about an axis perpendicular to the magnetic field.

Figure 28.2 shows other ways to induce an emf and a current in a coil. Parts *a* and *b* of the drawing illustrate that an emf can be induced by *changing the area* of a coil in a constant magnetic field. Here the shape of the coil is being distorted so as to reduce the area. As long as the area is changing, an induced emf and current exist; they vanish when the area is no longer changing. If the distorted coil is returned to its original circular shape, thereby increasing the area, an oppositely directed current is generated as long as the area is changing.

Part *c* of Figure 28.2 indicates that an induced emf is also generated when a coil of constant area is rotated in a constant magnetic field and the *orientation* of the coil *changes* with respect to the field. When the rotation stops, the emf, and hence the current, vanishes.

In each of the examples above, both an emf and current are induced in the coil because the coil is part of a complete, or closed, circuit. If the circuit were open—

perhaps because of an open switch—there would be no induced current. However, an emf would still be induced in the coil, whether the current exists or not.

Changing the magnetic field, changing the area of a coil, and changing the orientation of a coil are all methods that can be used to create an induced emf. The phenomenon of producing an induced emf with the aid of a magnetic field is called *electromagnetic induction.* The next section discusses how an induced emf arises when a conducting rod moves through a magnetic field.

28.2 MOTIONAL EMF

THE EMF INDUCED IN A MOVING CONDUCTOR

When a conducting rod moves through a constant magnetic field, an emf is induced in the rod. This special case of electromagnetic induction arises as a result of the magnetic force that acts on a moving charge, the same force discussed in Section 27.2. Consider the metal rod of length L moving to the right in Figure 28.3*a*. The velocity **v** of the rod is constant and is perpendicular to a uniform magnetic field **B**. Each charge within the rod also moves with a velocity **v** and experiences a magnetic force of magnitude $F = qvB$, according to Equation 27.1. By using RHR-1 as it pertains to negative charges, it can be seen that the mobile, free electrons are driven to the bottom of the rod, leaving behind an equal amount of positive charge at the top. The positive and negative charges continue to accumulate, until the attractive electric force between the two charge distributions becomes equal in magnitude to the magnetic force. When the electric force balances the magnetic force, equilibrium is reached and no further charge separation occurs.

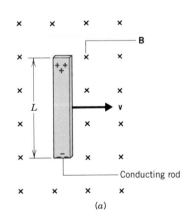

(a)

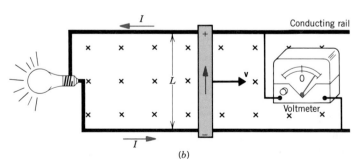

(b)

FIGURE 28.3 (*a*) When a conducting rod moves at right angles to a constant magnetic field, the magnetic force causes opposite charges to appear at the ends of the rod, giving rise to an induced emf. (*b*) The induced emf causes an induced current I to appear in the circuit. The voltmeter measures the induced emf.

The separated charges on the ends of the moving conductor give rise to an induced emf, called a *motional emf,* because it originates from the motion of charges through a magnetic field. The motional emf exists as long as the rod moves. If the rod is brought to a halt, the magnetic force vanishes, with the result that the attractive electric force reunites the positive and negative charges and the emf disappears. The emf of the moving rod is similar to that between the terminals of a battery. The difference between a battery and the rod is that the emf of a battery is produced by chemical reactions, whereas the emf of the rod is created by the agent that moves the rod through the magnetic field.

Figure 28.3*b* shows the rod sliding on conducting rails that form part of a closed circuit. Electrons flow in a clockwise direction around the circuit as long as the rod continues to move. Positive charge would flow in the direction opposite to the electron flow, so the conventional current *I* is drawn counterclockwise in the picture.

An expression for the magnitude of the motional emf $\mathcal{E}$ can be obtained by recalling that the emf is the work per unit charge needed to bring a positive charge from the negative end of the rod to the positive end: $\mathcal{E} = \text{Work}/q$ (see Sections 24.2 and 25.1). According to Equation 7.1, the work done is the product of the magnetic force *F* and the distance *L* (the length of the rod) over which the charge is moved:

$$\mathcal{E} = \frac{\text{Work}}{q} = \frac{FL}{q} = \frac{(qvB)L}{q} = vBL \qquad (28.1)$$

This result shows that the magnitude of the motional emf is the product of the magnetic field strength *B*, the length *L* of the rod between the rails, and the speed *v* with which the rod moves at right angles to the magnetic field. As expected, $\mathcal{E} = 0$ when $v = 0$, for no motional emf is developed in a stationary rod. As with batteries, $\mathcal{E}$ is expressed in volts and can be measured by a voltmeter connected between the ends of the rod, as Figure 28.3*b* indicates. Example 1 illustrates how to determine the electrical energy that the motional emf delivers to a device.

EXAMPLE 1

Suppose the rod in Figure 28.3*b* is moving at a speed of 5.0 m/s in a direction perpendicular to a 0.80-T magnetic field. The rod has a length of 1.6 m and has negligible electrical resistance. (a) What is the emf produced by the rod? Assuming the rails also have negligible resistance and the light bulb has a resistance of 96 Ω, find (b) the induced current in the circuit, (c) the electrical power delivered to the bulb, and (d) the energy consumed by the bulb in 60.0 s.

SOLUTION

(a) The motional emf is given by Equation 28.1 as

$$\mathcal{E} = vBL = (5.0 \text{ m/s})(0.80 \text{ T})(1.6 \text{ m}) = \boxed{6.4 \text{ V}}$$

(b) Ohm's law is applicable to metals, so the induced current is equal to the motional emf divided by the resistance of the circuit:

$$I = \frac{\mathcal{E}}{R} = \frac{6.4 \text{ V}}{96 \ \Omega} = \boxed{0.067 \text{ A}} \qquad (25.2)$$

(c) The electrical power *P* delivered to the light bulb is the product of the current *I* and the potential difference across the bulb:

$$P = I\mathcal{E} = (0.067 \text{ A})(6.4 \text{ V}) = \boxed{0.43 \text{ W}} \qquad (25.6)$$

(d) Since power is energy per unit time, the energy *E* consumed in 60.0 s is the product of the power and the time:

$$E = Pt = (0.43 \text{ W})(60.0 \text{ s}) = \boxed{26 \text{ J}}$$

MOTIONAL EMF AND ELECTRICAL ENERGY

Motional emf arises because a magnetic force acts on the charges in a conductor that is moving through a magnetic field. Whenever this emf causes a current, as it does in

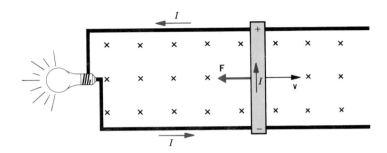

FIGURE 28.4 A magnetic force **F** is exerted on the current *I* in the moving rod. The magnetic force is opposite to the velocity **v** of the rod. The rod will slow down unless a counterbalancing force (not shown) is applied to the rod.

Figure 28.3*b*, a second magnetic force acts on the current. This second force plays an important role in the manner by which the motional emf can be used to supply energy to any electrical device connected to the circuit.

In Figure 28.3*b* the second force arises because the current *I* in the rod is perpendicular to the magnetic field. The current, and hence the rod, experiences a magnetic force **F** whose magnitude is given by Equation 27.3 as $F = ILB \sin 90°$. Using the values of *I*, *L*, and *B* from Example 1, we see that $F = (0.067 \text{ A})(1.6 \text{ m})(0.80 \text{ T}) = 0.086 \text{ N}$. The direction of **F** is specified by RHR-1 and is *opposite* to the velocity **v** of the rod, as Figure 28.4 shows. Hence, **F** tends to *slow down* the rod. To keep the rod moving to the right with a constant velocity, a counterbalancing force must be applied to the rod by an external agent. This agent (not shown in the drawing) could be somebody pushing on the rod. Whatever the agent, the counterbalancing force must have a magnitude of 0.086 N and must be directed to the right in the drawing. If the counterbalancing force of 0.086 N were removed, the rod would decelerate under the influence of the magnetic force **F** and eventually come to rest. During the deceleration, the motional emf would decrease and the light bulb would eventually go out.

In Example 1 the light bulb consumes 26 J of electrical energy in sixty seconds. We can now answer an important question—Where does this energy come from? The energy originates with the external agent that supplies the 0.086-N counterbalancing force needed to keep the rod moving at a constant velocity. This agent does work, and Example 2 shows that the work done by the agent is equal to the electrical energy consumed by the light bulb.

EXAMPLE 2

Determine the work done in 60.0 s by the external agent supplying the 0.086-N force that keeps the rod in Example 1 moving at a constant speed of 5.0 m/s.

SOLUTION

The work *W* done by the external agent is equal to the magnitude of the external force times the distance the rod moves. The distance *x* traveled by the rod in a time *t* is $x = vt$, so the work is

$$W = Fx = F(vt) = (0.086 \text{ N})(5.0 \text{ m/s})(60.0 \text{ s}) = \boxed{26 \text{ J}}$$

The 26 J of work done on the rod by the external agent is the same as the 26 J of energy consumed by the light bulb. Hence, the moving rod converts mechanical work into electrical energy, much as a battery converts chemical energy into electrical energy.

It is interesting to speculate what would happen if the direction of the current in Figure 28.4 were *reversed*. If the current were reversed, the magnetic force **F** would also be reversed and point in the same direction as **v**. Then the force **F** would cause the rod to accelerate, rather than decelerate. There would be no need for an external force to keep the rod moving. This hypothetical electric generator would be able to run itself and supply energy to the light bulb. In effect, such a device would create energy out of nothing. Such a device cannot exist, because it would violate the principle of conservation of energy. This principle states that "energy cannot be created or destroyed, but can only be converted from one form to another."

28.3 MAGNETIC FLUX

MOTIONAL EMF AND MAGNETIC FLUX

Any method used to create an induced emf can be described in terms of a concept called magnetic flux. This concept can be introduced by using the example of motional emf. According to Equation 28.1, the emf induced in a rod moving perpendicular to a magnetic field is $\mathcal{E} = vBL$. We now show how this expression can be written in terms of magnetic flux.

Figure 28.5a shows the position of the rod at a time $t = 0$ and at a later time t_0. During this time interval, the rod moves a distance x_0 to the right. At an even later time t, the rod has moved an even greater distance x to the right, as part b of the drawing indicates. The speed v of the rod is the distance traveled divided by the elapsed time: $v = (x - x_0)/(t - t_0)$. Substituting this expression for v into $\mathcal{E} = vBL$ gives

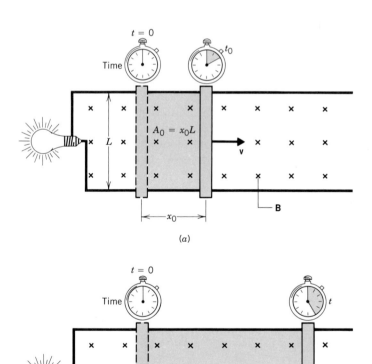

FIGURE 28.5 (a) In a time t_0, the moving rod sweeps out an area $A_0 = x_0 L$. (b) The area swept out in a time t is $A = xL$. In both parts of the figure the areas are shaded in color.

$$\mathcal{E} = \left(\frac{x - x_0}{t - t_0}\right)BL = \left(\frac{xL - x_0L}{t - t_0}\right)B$$

As the drawing indicates, the term x_0L is the area A_0 swept out by the rod in moving a distance x_0, while xL is the area A swept out in moving a distance x. In terms of these areas, the emf is

$$\mathcal{E} = \left(\frac{A - A_0}{t - t_0}\right)B = \frac{(BA) - (BA)_0}{t - t_0}$$

Notice how the product BA of the magnetic field strength and the area appears in the numerator of this expression. The quantity BA is given the name *magnetic flux* and is represented by the symbol Φ (Greek letter *phi*); thus, $\Phi = BA$. The magnitude of the induced emf is the *change* in flux $\Delta\Phi = \Phi - \Phi_0$ divided by the time interval $\Delta t = t - t_0$ during which the change occurs:

$$\mathcal{E} = \frac{\Phi - \Phi_0}{t - t_0} = \frac{\Delta\Phi}{\Delta t}$$

In other words, the induced emf equals the time rate of change of the magnetic flux.

You will almost always see the equation above written with a minus sign, namely, $\mathcal{E} = -\Delta\Phi/\Delta t$. The minus sign is introduced for the following reason: The direction of the current induced in the circuit is such that the magnetic force $\mathbf{F}$ acts on the rod to *oppose* its motion, thereby tending to slow down the rod (see Figure 28.4). The presence of the minus sign reminds us that the polarity of the induced emf sends the induced current in the proper direction so as to give rise to this opposing magnetic force.

The advantage of writing the induced emf as $\mathcal{E} = -\Delta\Phi/\Delta t$ is that this relation is far more general than our present discussion suggests. In Section 28.4 we will see that $\mathcal{E} = -\Delta\Phi/\Delta t$ can be applied to *all possible ways of generating induced emfs.*

A GENERAL EXPRESSION FOR MAGNETIC FLUX

In Figure 28.5 the direction of the magnetic field $\mathbf{B}$ is perpendicular to the surface swept out by the moving rod, or, alternatively, $\mathbf{B}$ is parallel to the normal to the surface. In general, however, $\mathbf{B}$ may not always be parallel to the normal, as Figure 28.6 shows. In such a situation, the flux is computed by using only the component of the magnetic field that is parallel to the normal. The drawing shows that this component is $B \cos\theta$, where θ is the angle between $\mathbf{B}$ and the normal to the surface. The general expression for magnetic flux is

$$\Phi = (B \cos\theta)A = BA \cos\theta \qquad (28.2)$$

If the magnetic field is not constant over the surface, an average magnetic field must be used in computing the flux. Equation 28.2 shows that the unit of magnetic flux is the tesla $\cdot$ meter^2 (T$\cdot$m^2). This unit is called a *weber* (Wb), after the German physicist Wilhelm Weber (1804–1891): 1 Wb = 1 T$\cdot$m^2. Example 3 illustrates how to determine the magnetic flux for three different orientations of the surface of a coil relative to the magnetic field.

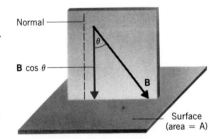

FIGURE 28.6 When computing the magnetic flux, the component of the magnetic field that is parallel to the normal to the surface must be used; this component is $B \cos\theta$. The flux is $\Phi = (B \cos\theta)A$.

EXAMPLE 3

A rectangular coil of wire is situated in a constant magnetic field whose magnitude is 0.50 T. The coil has an area of 2.0 m^2. Determine the magnetic flux for the three orientations, $\theta = 0°$, 60.0°, and 90.0°, shown in Figure 28.7.

SOLUTION

The flux in the three cases can be computed using $\Phi = BA \cos \theta$:

$\theta = 0°$: $\Phi = (0.50 \text{ T})(2.0 \text{ m}^2) \cos 0° = \boxed{1.0 \text{ Wb}}$

$\theta = 60.0°$: $\Phi = (0.50 \text{ T})(2.0 \text{ m}^2) \cos 60.0° = \boxed{0.50 \text{ Wb}}$

$\theta = 90.0°$: $\Phi = (0.50 \text{ T})(2.0 \text{ m}^2) \cos 90.0° = \boxed{0}$

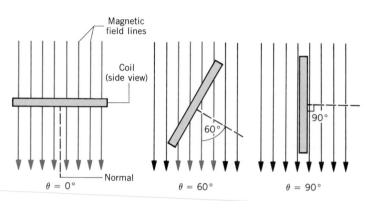

FIGURE 28.7 This picture shows three orientations of a rectangular coil (drawn as a side view), relative to the magnetic field lines. The magnetic field lines that pass through the coil are drawn in color; those that do not pass through the coil are drawn in black.

GRAPHICAL INTERPRETATION OF MAGNETIC FLUX

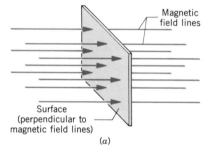

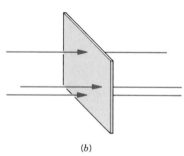

FIGURE 28.8 The magnitude of the magnetic field in (a) is three times greater than that in (b), because the number of magnetic field lines crossing the surfaces is in the ratio of 3:1.

It is possible to interpret the magnetic flux graphically by noting that *the flux is proportional to the number of magnetic field lines that passes through a surface.* This useful interpretation stems from the fact that the magnitude of the magnetic field **B** in any region of space is proportional to the number of magnetic field lines per unit area that passes through a surface perpendicular to the field lines (see Section 27.1). For instance, the magnitude of **B** in Figure 28.8a is three times larger than it is in part b of the drawing, since the number of field lines drawn through the identical surfaces is in the ratio of 3:1. Because Φ is directly proportional to B for a given area, the flux in part a is also three times larger than that in part b. Therefore, we can say that the magnetic flux is proportional to the number of magnetic field lines that passes through a surface.

The graphical interpretation of flux as being proportional to the number of magnetic field lines passing through a surface also applies when the surface is oriented at an angle with respect to **B**. For example, as the coil in Figure 28.7 is rotated from $\theta = 0°$ to $60°$ to $90°$, the number of magnetic field lines passing through the surface (see the colored field lines) changes in the ratio of 8:4:0 or 2:1:0. The results of Example 3 show that the flux in the three orientations changes by the same ratio. Because the magnetic flux is proportional to the number of field lines passing through a surface, one often encounters phrases like "the flux that passes through a surface bounded by a loop of wire."

28.4 FARADAY'S LAW OF ELECTROMAGNETIC INDUCTION

Two scientists are given credit for the discovery of electromagnetic induction: the Englishman Michael Faraday (1791–1867) and the American Joseph Henry (1797–

1878). Although Henry was actually the first to observe electromagnetic induction, Faraday investigated it in more detail and published his findings first. Consequently, the law that describes electromagnetic induction bears his name.

Faraday discovered that whenever there is a *change in flux* through a loop of wire, a emf is induced in the loop. A constant flux creates no emf. Only when the flux changes does an emf appear. In fact, Faraday found that the magnitude of the induced emf is equal to the time rate of change of the magnetic flux. This is just the relation we obtained in Section 28.3 for the specific case of motional emf: $\mathcal{E} = -\Delta\Phi/\Delta t$.

Often the magnetic flux passes through a coil of wire containing more than one loop (or turn). If the coil consists of N loops, and if the same flux passes through each loop, it is found experimentally that the total induced emf is N times that induced in a single loop. An analogous situation occurs within a flashlight when two 1.5-V batteries are stacked in series on top of one another to give a total emf of 3.0 volts. For the general case of N loops, the total induced emf is described by *Faraday's law of electromagnetic induction* in the following manner.

FARADAY'S LAW OF ELECTROMAGNETIC INDUCTION

The average emf $\mathcal{E}$ induced in a coil of N loops during a time interval Δt is N times the change in magnetic flux $\Delta\Phi$ through each loop divided by the time interval:

$$\mathcal{E} = -N\left(\frac{\Phi - \Phi_0}{t - t_0}\right) = -N\frac{\Delta\Phi}{\Delta t} \qquad (28.3)$$

The term $\Delta\Phi/\Delta t$ is the average time rate of change of the flux that passes through one loop.

SI unit of induced emf: volt (V)

Faraday's law states that an emf is generated if the flux changes for any reason. Since the flux is given by Equation 28.2 as $\Phi = BA \cos\theta$, the flux depends on three factors, B, A, and θ, any of which may change. The examples in the remainder of this section illustrate how such changes can lead to an induced emf. Example 4 considers a changing magnetic field.

EXAMPLE 4

A coil of wire consists of 20 turns, each of which has an area of 1.5×10^{-3} m^2. A constant magnetic field is perpendicular to the surface of each loop, so $\theta = 0°$. At time $t_0 = 0$, the magnitude of the magnetic field at the location of the coil is $B_0 = 0.050$ T. At a later time of $t = 0.10$ s, the magnitude of the field at the coil has increased to $B = 0.060$ T. (a) Find the average emf induced in the coil during this time. (b) What would be the value of the induced emf if the magnitude of the magnetic field decreased from 0.060 T to 0.050 T in 0.10 s?

SOLUTION

(a) To find the induced emf, we use Faraday's law of electromagnetic induction, combining it with the definition of magnetic flux from Equation 28.2:

$$\mathcal{E} = -N\left(\frac{\Phi - \Phi_0}{t - t_0}\right) = -N\left(\frac{BA \cos\theta - B_0 A \cos\theta}{t - t_0}\right)$$

$$= -NA \cos\theta\left(\frac{B - B_0}{t - t_0}\right)$$

Writing $\mathcal{E}$ in the manner above makes it clear in this case that it is the change in the magnitude of the field with time, $(B - B_0)/(t - t_0)$, that gives rise to the induced emf. Substituting numbers into this equation yields

$$\mathcal{E} = -(20)(1.5 \times 10^{-3} \text{ m}^2)(\cos 0°)\left(\frac{0.060 \text{ T} - 0.050 \text{ T}}{0.10 \text{ s} - 0}\right)$$

$$= \boxed{-3.0 \times 10^{-3} \text{ V}}$$

(b) The reasoning here proceeds in the same manner as in part (a), except the initial and final values of B are interchanged. This interchange reverses the sign of the emf, so

$\boxed{\mathscr{E} = +3.0 \times 10^{-3} \text{ V}}$. Because the polarity of the emf is reversed, the direction of the induced current is opposite to that in part (a).

The next example demonstrates that an emf can be created when a coil is rotated in a magnetic field.

EXAMPLE 5

A flat coil of wire has an area of 2.0×10^{-2} m² and consists of 50 turns. At $t_0 = 0$ the coil is oriented so its surface is perpendicular ($\theta_0 = 0°$) to a constant magnetic field of magnitude 0.18 T. The coil is then rotated through an angle of $\theta = 30.0°$ in a time of 0.10 s (see Figure 28.2c). (a) Determine the average induced emf. (b) What would be the induced emf if the coil were returned to its initial orientation in the same time of 0.10 s?

SOLUTION

(a) Faraday's law yields

$$\mathscr{E} = -N\left(\frac{\Phi - \Phi_0}{t - t_0}\right) = -N\left(\frac{BA \cos\theta - BA \cos\theta_0}{t - t_0}\right)$$

$$= -NBA\left(\frac{\cos\theta - \cos\theta_0}{t - t_0}\right)$$

$$\mathscr{E} = -(50)(0.18 \text{ T})(2.0 \times 10^{-2} \text{ m}^2)\left(\frac{\cos 30.0° - \cos 0°}{0.10 \text{ s} - 0}\right)$$

$$= \boxed{+0.24 \text{ V}}$$

(b) When the coil is rotated back to its initial orientation in a time of 0.10 s, the induced emf has the same magnitude, but opposite polarity, so $\boxed{\mathscr{E} = -0.24 \text{ V}}$.

28.5 LENZ'S LAW

THE POLARITY OF THE INDUCED EMF

An induced emf drives current around a circuit just as the emf of a battery does. With a battery, conventional current is directed out of the positive terminal, through the attached device, and into the negative terminal. The same is true for an induced emf, although the location of the positive and negative terminals is generally not as obvious. Therefore, a method is needed for determining the polarity of the induced emf, so the terminals can be identified. The method that we will use is based on a discovery made by the Russian physicist Heinrich Lenz (1804–1865), which is stated below as *Lenz's law.*

> **LENZ'S LAW**
> The polarity of an induced emf is such that the emf would produce a current whose own magnetic field opposes the change in flux that causes the induced emf.

An induced current—like any current—creates a magnetic field; this field is called the *induced magnetic field.* During the following discussion, it will be helpful to keep in mind that the net magnetic field penetrating a coil of wire is the sum of the fields from two sources: (1) the magnetic field that produces the flux whose change leads to the induced emf and (2) the induced magnetic field that is created by the induced current. According to Lenz's law, the induced magnetic field points in such a direction so as to oppose the **change** in the flux (not necessarily the flux itself) that causes the induced emf in the first place.

Lenz's law is best illustrated with examples. Each of the examples is worked out according to the following procedure:

1. First, we determine whether the magnetic flux that penetrates a coil is increasing or decreasing.
2. Second, we find what the direction of the induced magnetic field must be so that it opposes the **change** in the flux.
3. Third, having found the direction of the induced magnetic field, we use RHR-2 (see Section 27.7) to determine the direction of the induced current. Then the polarity of the induced emf can be assigned, because the conventional current is directed out of the positive terminal, through the external circuit, and into the negative terminal.

EXAMPLE 6

Figure 28.9a shows a permanent magnet approaching a loop of wire. The external circuit attached to the loop consists of the resistance R, which could be the resistance of the filament in a light bulb, for instance. Find the direction of the induced current and the polarity of the induced emf.

SOLUTION

The magnetic flux through the loop is increasing, since the magnitude of the magnetic field at the loop is increasing as the magnet approaches. To oppose the increase in the flux, the direction

of the induced magnetic field must be opposite to that of the bar magnet. Since the field of the bar magnet passes through the loop from left to right in part a of the drawing, the induced magnetic field must pass through the loop from right to left, as in part b. To create such an induced magnetic field, the induced current must be directed *counterclockwise* around the loop, when viewed from the side nearest the magnet. (See the application of RHR-2 in the drawing.) The loop behaves as a source of emf, just like a battery, with the positive and negative terminals as shown in Figure 28.9b.

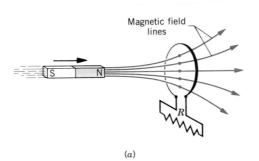

(a)

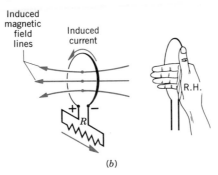

(b)

FIGURE 28.9 (a) As the magnet moves to the right, the magnetic flux through the loop increases. The external circuit attached to the loop has a resistance R. (b) The polarity of the induced emf is indicated by the + and − symbols.

In Example 6 the direction of the induced magnetic field is opposite to the direction of the external field of the bar magnet. However, the induced field does not always have to oppose the external field, for Lenz's law requires only that the induced magnetic field opposes the **change** in the flux that generates the emf. Example 7 illustrates this point.

EXAMPLE 7

In Figure 28.10 there is a constant magnetic field in a rectangular region of space. This field is directed perpendicularly into the plane of the paper. Outside this region there is no magnetic field. A copper ring slides through the region, from position 1 to position 5. For each of the five positions, determine if an induced current exists in the ring and, if so, find the direction of the current.

SOLUTION

Position 1: No flux passes through the moving ring, because the magnetic field is everywhere zero outside the rectangular region. Consequently, there is no change in flux and no induced emf or current in the ring.

Position 2: As the ring moves into the region of the magnetic field, the flux increases, and there is an induced emf and an induced current. To determine the direction of the current, we require that the induced magnetic field point opposite to the external field, so as to oppose the increase in the flux, in accord with Lenz's law. With the induced field pointing perpendicularly out of the plane of the paper, RHR-2 indicates that the direction of the induced current is counterclockwise, as the drawing indicates.

Position 3: Even though a flux passes through the moving ring, there is no induced emf or current, because the flux remains constant within the rectangular region. To induce an emf, it is not sufficient just to have a flux. The flux must ***change*** to generate an emf.

Position 4: As the ring leaves the magnetic field, the flux decreases. Once again, the induced magnetic field must be in such a direction so as to oppose this change. Since the change is a decrease in flux, the induced field must point in the *same* direction as the external field, namely, into the paper. With this orientation, the induced field increases the net magnetic field through the ring and thereby increases the flux. With the induced field pointing into the paper, RHR-2 indicates that the induced current is clockwise around the ring, opposite to

what it was in position 2. By comparing the results for positions 2 and 4, it should be clear that the induced magnetic field does not always oppose the external field.

Position 5: As in position 1, there is no induced current, since the magnetic field is everywhere zero.

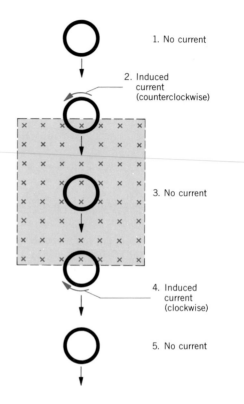

FIGURE 28.10 A copper ring passes through a rectangular region where a constant magnetic field is directed into the page. The picture shows the current induced in the ring at five different locations.

Lenz's law should not be thought of as an independent law, for it is a consequence of the law of conservation of energy. The connection between energy conservation and induced emf has already been discussed in Section 28.2 for the specific case of motional emf. However, the connection is valid for any type of induced emf. In fact, the polarity of the induced emf, as specified by Lenz's law, ensures that energy is conserved.

28.6 APPLICATIONS OF ELECTROMAGNETIC INDUCTION RELATED TO THE REPRODUCTION OF SOUND

Electromagnetic induction plays an important role in the technology used for the reproduction of sound. Figure 28.11 shows an audio system, to which an electric

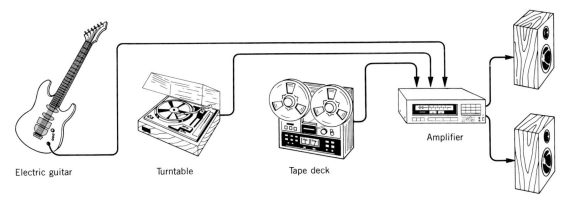

FIGURE 28.11 The operation of an electric guitar, a turntable, and a tape deck is based on electromagnetic induction.

guitar, a turntable, and a tape deck are connected. As we will now see, each of these generates an induced emf. In general, however, the emf is rather small, so it is strengthened by an amplifier before being sent to the speakers.

THE ELECTRIC GUITAR PICKUP

Virtually all electric guitars use electromagnetic pickups in which an induced emf is generated in a coil of wire by a vibrating string. Most guitars have at least two pickups for each string and some, as Figure 28.12 illustrates, have three pickups. These pickups are positioned at different locations under the string, so that each pickup is sensitive to different harmonics produced by the vibrating string. The drawing also shows an enlarged side view of one pickup.

The guitar string is made from a magnetizable material, such as steel. The pickup itself consists of a coil of wire with a permanent magnet located inside the coil. The permanent magnet produces a magnetic field that penetrates the guitar string, causing the string to become magnetized with an associated north and south pole. When the magnetized string is plucked, it oscillates above the coil, thereby changing the magnetic flux that passes through the coil. The change in flux induces an emf in the coil. The polarity of the emf reverses with the vibratory motion of the string, so a string vibrating at 440 Hz, for example, induces a 440-Hz ac emf in the coil. This 440-Hz signal, after being amplified, is sent to the speakers, which produce a 440-Hz sound wave (concert A).

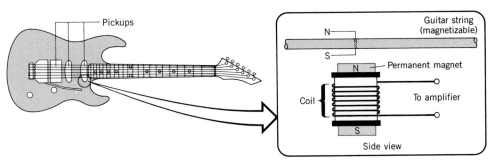

FIGURE 28.12 When the string of an electric guitar vibrates, an emf is induced in the coil of the pickup. The two ends of the coil are connected to the input of an amplifier.

FIGURE 28.13 (*a*) A magnetic phono cartridge and the stylus that rides in the record groove. (*b*) A simplified diagram of a monaural moving-coil phono cartridge. As the stylus vibrates, the magnetic flux through the coil changes, and an emf is induced in the coil.

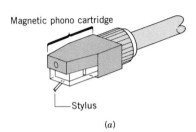

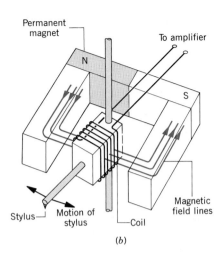

(*a*)

(*b*)

THE MAGNETIC PHONO CARTRIDGE

Most high-quality turntables use a magnetic phono cartridge to generate an output electrical signal (see Figure 28.13*a*). As the stylus rides in the record groove, the stylus is forced to vibrate by the undulations cut into the groove. Part *b* of the drawing illustrates the moving-coil type of magnetic phono cartridge. (The version shown is monaural; the additional complexity of a stereo version is omitted for clarity.) The stylus is attached to a small coil of wire that moves back and forth in a uniform magnetic field as the stylus rides in the groove. When the coil moves, its orientation relative to the magnetic field lines changes. Consequently, the flux through the coil changes, and an emf is induced in the coil. As with an electric guitar pickup, the ends of the coil are connected to the input of an amplifier. Another popular type of magnetic phono cartridge is the moving-magnet cartridge. In this design, a tiny magnet is attached to the stylus and the coil remains stationary.

THE PLAYBACK HEAD OF A TAPE DECK

The playback head of a cassette or open-reel tape deck utilizes a moving tape to generate an emf in a coil of wire. Figure 28.14 shows a section of magnetized tape in which a series of "tape magnets" have been created in the magnetic layer of the tape

FIGURE 28.14 The magnetic playback head of a tape deck. As each tape magnet passes by the gap in the iron core, some of the magnetic field lines are routed through the core and the coil. The change in flux through the coil causes an induced emf to appear in the coil. Only one field line is shown for clarity, and the width of the gap has been exaggerated.

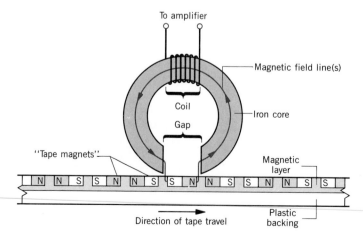

during the recording process. The tape moves beneath the playback head, which consists of a coil of wire wrapped around an iron core. The iron core has the approximate shape of a horseshoe with a small gap between the two ends. The drawing shows an instant when a tape magnet is adjacent to the gap. Some of the magnetic field lines of the tape magnet are routed through the highly magnetizable iron core, and hence through the coil, as they proceed from the north pole to the south pole. Consequently, the flux through the coil changes as the tape magnet moves past the playback head. The change in flux leads to an emf, with each pair of oppositely directed tape magnets inducing one cycle of the ac signal in the playback head. The signal, as usual, is then amplified and sent to the speakers.

28.7 THE ELECTRIC GENERATOR

HOW A GENERATOR PRODUCES AN EMF

The importance of electric generators stems from the fact that they produce virtually all the electrical energy consumed in the world. A generator produces electrical energy from mechanical work, just the opposite of what a motor does. In a motor, an *input* electric current causes the coil to rotate, thereby doing mechanical work on any object attached to the shaft of the motor. In a generator, the shaft is rotated by some mechanical means, such as by an engine or a turbine, and an emf is induced in the coil. If the generator is connected to an external circuit, an electric current is the *output* of the generator.

In its simplest form, an ac generator consists of a coil of wire that is rotated in a uniform magnetic field, as Figure 28.15a indicates. Although not shown in the

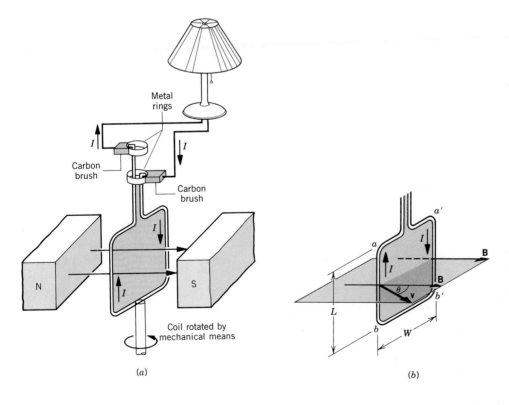

(a)

(b)

FIGURE 28.15 (*a*) This electric generator consists of a coil (only one loop is shown) of wire that is rotated in a uniform magnetic field by some mechanical means. (*b*) The current *I* arises because of the magnetic force exerted on the charges in the wire, which is moving with a velocity **v** in the magnetic field **B**.

illustration, the wire is usually wound around an iron core, and, as in an electric motor, the coil/core combination is called the armature. The drawing indicates that each end of the wire forming the coil is connected to the external circuit by means of a metal ring that rotates with the coil. Each ring slides against a stationary carbon brush, to which the external circuit (the lamp in the drawing) is connected.

To see how the current is produced by the generator, consider the two vertical sides of the coil, ab and $a'b'$, shown in Figure 28.15b. Since each side is moving with a velocity **v** in a magnetic field **B**, the magnetic force exerted on the charges in the wire causes them to flow, thus creating a current. With the aid of RHR-1 (fingers of extended right hand point along **B**, thumb along **v**, palm pushes in the direction of the force on a positive charge), it can be seen that the direction of the current in side ab is from b toward a, while in side $a'b'$ it is from a' toward b'. Thus, charge flows around the loop. The top and bottom segments of the loop, aa' and bb', are also moving. However, the magnetic force on the charges in these segments can be ignored, because the force is toward the sides of the wire and not along the length. The emf generated in the coil results only from the magnetic force on the charges in the sides ab and $a'b'$.

The magnitude of the motional emf developed in a conductor moving through a magnetic field is given by Equation 28.1. To apply this expression to side ab, whose length is L, we need to use the velocity component $v_\perp$ that is perpendicular to **B**. Letting θ be the angle between **v** and **B**, it follows that $v_\perp = v \sin \theta$, and the emf can be written as

$$\mathcal{E} = BLv_\perp = BLv \sin \theta$$

The emf induced in side $a'b'$ has the same magnitude and polarity as that of side ab, with the result that the emf developed in a complete loop is $\mathcal{E} = 2BLv \sin \theta$. If the coil consists of N loops, the net induced emf is N times greater than that of one loop, so

$$\mathcal{E} = N(2BLv \sin \theta)$$

It is convenient to express the variables v and θ in terms of the angular speed ω (in radians per second) of the coil. Equation 9.2 shows that the angle θ is the product of the angular speed and the time, $\theta = \omega t$, if it is assumed that $\theta = 0$ when $t = 0$. Furthermore, any point on each vertical side moves on a circular path of radius $r = W/2$, where W is the width of the coil (see the drawing), so the tangential speed v of each side is related to the angular speed ω via Equation 9.9 as $v = r\omega = (W/2)\omega$. Substituting these expressions for θ and v in the equation above for $\mathcal{E}$, and recognizing that the product LW is the area A of the coil, we can write the induced emf as

$$\begin{bmatrix} \textbf{Emf induced} \\ \textbf{in a rotating} \\ \textbf{planar coil} \end{bmatrix} \qquad \mathcal{E} = NAB\omega \sin \omega t = \mathcal{E}_0 \sin \omega t \qquad (28.4)$$

This expression shows that the emf varies sinusoidally with time. The peak, or maximum, emf $\mathcal{E}_0$ occurs when $\sin \omega t = 1$ and has the value $\mathcal{E}_0 = NAB\omega$. Although Equation 28.4 was derived for a rectangular coil, the result is valid for any planar shape of area A, such as a circle.

The emf of Equation 28.4 is plotted in Figure 28.16, which shows that the emf changes polarity as the coil rotates. This changing polarity is exactly the same as that discussed for an ac voltage in Section 25.5 and illustrated in Figure 25.9. If the external circuit connected to the generator is a closed circuit, an alternating current results that changes direction at the same rate as the emf changes polarity. Therefore, this electric generator is also called an *alternating current (ac) generator.*

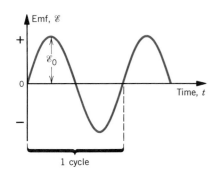

FIGURE 28.16 An ac generator produces an alternating emf $\mathcal{E}$ that varies as $\mathcal{E} = \mathcal{E}_0 \sin \omega t$ (see Equation 28.4).

EXAMPLE 8

In Figure 28.15 the armature of the ac generator rotates at a frequency of 60.0 Hz and develops an emf of 120 V(rms). The coil has an area of $A = 3.0 \times 10^{-3}$ m^2 and consists of $N = 500$ turns. Find the magnitude of the magnetic field in which the coil rotates.

SOLUTION

The magnetic field can be found from the relation $\mathscr{E}_0 = NAB\omega$, provided $\mathscr{E}_0$ and ω can be determined. $\mathscr{E}_0$ is the *peak* emf, and is related to the rms emf by $\mathscr{E}_{rms} = \mathscr{E}_0/\sqrt{2}$ (Equation 25.13). Therefore, $\mathscr{E}_0 = \sqrt{2}\,\mathscr{E}_{rms} = \sqrt{2}\,(120 \text{ V}) = 170 \text{ V}$. Since one revolution corresponds to 2π radians, the angular speed of the coil is $\omega = 2\pi(60.0 \text{ Hz}) = 377$ rad/s. The magnitude of the magnetic field is

$$B = \frac{\mathscr{E}_0}{NA\omega} = \frac{170 \text{ V}}{(500)(3.0 \times 10^{-3} \text{ m}^2)(377 \text{ rad/s})} = \boxed{0.30 \text{ T}}$$

THE ELECTRICAL ENERGY DELIVERED BY A GENERATOR AND THE COUNTERTORQUE

Some power-generating stations burn fossil fuel (coal, gas, or oil) to heat water and produce pressurized steam for turning the blades of a turbine. Others use nuclear fuel or falling water as a source of energy. The shaft of the turbine is linked to that of the generator, so as the generator coil rotates, mechanical work is transformed into electrical energy.

The devices to which the generator supplies electricity are known collectively as the "load," because they place a burden or load on the generator by taking electrical energy from it. If all the devices are switched off, the generator runs under a no-load condition. In this instance there is no current in the external circuit, and the generator does not supply electrical energy. The only work that the turbine must do is to overcome friction and other mechanical losses within the generator itself. Thus, the consumption of fuel is at a minimum under a no-load condition.

Figure 28.17a illustrates a situation in which a load is connected to a generator. Because there is now a current I in the coil, and the coil is situated in a magnetic field, the current experiences a magnetic force **F**. Part b of the drawing shows the magnetic force acting on the left side of the coil, the direction of **F** being given by RHR-1. A force of equal magnitude but opposite direction acts on the right side of the coil, although this force is not shown in the drawing. The magnetic force **F** retards the motion of the coil. We encountered such a retarding force in Section 28.2 when discussing the motional emf of a rod sliding along two conducting rails (see Figure 28.4). In the ac generator, the magnetic force **F** gives rise to a *countertorque* that opposes the rotational motion. The greater the current drawn from the generator, the greater the countertorque, and the harder it is for the turbine to turn the coil. To compensate for this countertorque and to keep the coil rotating at a constant angular speed, work must be done on the coil by the turbine, which means more fuel must be burned. This is another example of the law of conservation of energy, since the electrical energy consumed by the load must ultimately come from the energy source used to drive the turbine.

THE BACK EMF GENERATED BY AN ELECTRIC MOTOR

A generator converts mechanical work into electrical energy; in contrast, an electric motor converts electrical energy into mechanical work. Both devices are similar, for each consists of a coil of wire that rotates in a magnetic field. In fact, as the armature of a motor rotates, the magnetic flux passing through the coil changes and an emf is induced in the coil. Thus, when a motor is operating, two sources of emf are present: (1) the applied emf V, for instance, from a 120-V outlet that supplies current to drive

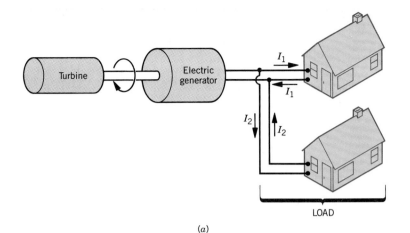

(a)

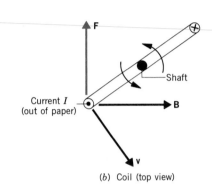

(b) Coil (top view)

FIGURE 28.17 (a) Current is drawn when a load is connected to the generator. (b) The current *I* in the coil experiences a magnetic force **F** because the charges are moving in a magnetic field **B**. This magnetic force retards the motion of the coil.

the motor, and (2) the emf $\mathcal{E}$ induced by the generator-like action of the rotating coil. The circuit diagram in Figure 28.18 shows these two emfs.

Consistent with Lenz's law, the induced emf $\mathcal{E}$ acts to oppose the applied emf V and is called the ***back emf*** or the ***counter emf*** developed by the motor. The greater the speed of the motor, the greater the flux change through the coil, and the greater is the back emf. Because V and $\mathcal{E}$ have opposite polarities, the net emf in the circuit is $V - \mathcal{E}$. If R in Figure 28.18 is the resistance of the wire in the coil, the current I drawn by the motor is determined from Ohm's law as the net emf divided by the resistance:

$$I = \frac{V - \mathcal{E}}{R} \qquad (28.5)$$

The next example uses Equation 28.5 to illustrate that the current in a motor depends on both the applied emf V and the back emf $\mathcal{E}$.

FIGURE 28.18 The applied emf V supplies the current I to drive the motor. The circuit shows V, along with the electrical equivalent of the motor (in color) that includes the resistance R of its coil and the back emf $\mathcal{E}$. The net emf in the circuit is $V - \mathcal{E}$.

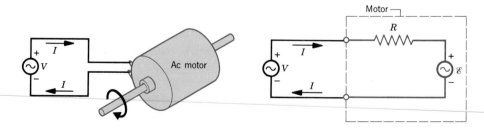

EXAMPLE 9

The coil of an ac motor has a resistance of $R = 4.1 \, \Omega$. The motor is plugged into an outlet where $V = 120.0$ volts, and the coil develops a back emf of $\mathcal{E} = 118.0$ volts when rotating at normal speed. Find (a) the current when the motor first starts up and (b) the current when the motor is operating at normal speed.

SOLUTION

(a) When the motor just starts up, the coil is not rotating, so there is no back emf induced in the coil and $\mathcal{E} = 0$. The start-up current drawn by the motor is

$$I = \frac{V - \mathcal{E}}{R} = \frac{120.0 \text{ V}}{4.1 \, \Omega} = \boxed{29 \text{ A}} \qquad (28.5)$$

(b) At normal speed, the motor develops a back emf of $\mathcal{E} = 118.0$ volts, so the current is

$$I = \frac{V - \mathcal{E}}{R} = \frac{120.0 \text{ V} - 118.0 \text{ V}}{4.1 \, \Omega} = \boxed{0.49 \text{ A}}$$

Example 9 illustrates that when a motor is just starting, there is little back emf and, consequently, a relatively large current exists in the coil. As the motor speeds up, the back emf increases until it reaches a maximum value when the motor is rotating at normal speed. The back emf becomes almost equal to the applied emf, and the current is reduced to a relatively small value. This limiting value of the current is sufficient to provide the torque on the coil to drive the load (such as a fan) and to overcome frictional losses.

28.8 MUTUAL INDUCTANCE AND SELF-INDUCTANCE

MUTUAL INDUCTANCE

Figure 28.19 illustrates an important method of inducing an emf in a coil. Here, two coils of wire are placed close to each other. Coil 1 is connected to an ac generator that sends an alternating current I_1 through the coil. Coil 2 is not attached to an ac generator, although a voltmeter is connected between the ends of coil 2 to register any induced emf. It is customary to call coil 1, the coil connected to the ac generator, the *primary coil* and coil 2 the *secondary coil.*

The current-carrying primary coil is an electromagnet and creates a magnetic field in the surrounding region. If the two coils are close to each other, a significant fraction of this magnetic field penetrates the secondary coil and produces a magnetic flux. The flux is changing in time, since the current in the primary coil and its associated

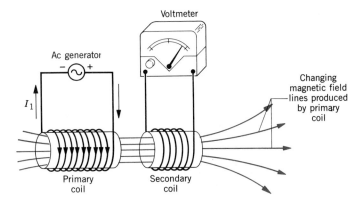

FIGURE 28.19 An alternating current I_1 in the primary coil creates an alternating magnetic field. This changing magnetic field induces an emf in the secondary coil. The effect in which an ac current in one coil induces an emf in another coil is known as mutual induction.

magnetic field are changing in time. Because of the change in the flux, an emf is induced in the secondary coil.

The effect in which a changing current in one circuit induces an emf in another circuit is called **mutual induction.** According to Faraday's law of electromagnetic induction, the emf $\mathcal{E}_2$ induced in the secondary coil is proportional to the change in flux $\Delta\Phi_2$ passing through it. However, as mentioned above, $\Delta\Phi_2$ is produced by the change in current ΔI_1 in the primary coil. For this reason, Faraday's law is recast into a more convenient form that relates $\mathcal{E}_2$ to ΔI_1. To see how this recasting is accomplished, note that the net magnetic flux passing through the secondary coil is $N_2\Phi_2$, where N_2 is the number of loops in the secondary coil and Φ_2 is the flux through each loop (assumed to be the same for all loops). The net flux is proportional to the magnetic field, which, in turn, is proportional to the current I_1 in the primary. Thus, we can write $N_2\Phi_2 \propto I_1$. This proportionality can be converted into an equation in the usual manner by introducing a proportionality constant M, known as the **mutual inductance:**

$$N_2\Phi_2 = MI_1 \quad \text{or} \quad M = \frac{N_2\Phi_2}{I_1} \tag{28.6}$$

Substituting this equation into Faraday's law, we find that

$$\mathcal{E}_2 = -N_2\frac{\Delta\Phi_2}{\Delta t} = -\frac{\Delta(N_2\Phi_2)}{\Delta t} = -\frac{\Delta(MI_1)}{\Delta t} = -M\frac{\Delta I_1}{\Delta t}$$

$$\begin{bmatrix} \textbf{Emf due to} \\ \textbf{mutual} \\ \textbf{induction} \end{bmatrix} \qquad \mathcal{E}_2 = -M\frac{\Delta I_1}{\Delta t} \tag{28.7}$$

Writing Faraday's law in this manner makes it clear that the emf $\mathcal{E}_2$ induced in the secondary coil is due to the change in the current ΔI_1 in the primary coil. Equation 28.7 shows that the measurement unit for the mutual inductance M is $V\cdot s/A$, which is called a *henry* (H) after Joseph Henry: $1\ V\cdot s/A = 1\ H$.

The mutual inductance links the rate of change in current in one circuit to the emf induced in another circuit and depends on factors related to the geometry of the coils and the nature of any ferromagnetic core material that might be present. Although M can be calculated for some highly symmetrical arrangements, it is usually measured experimentally. In most situations, values of M are less than 1 H, and they are often on the order of millihenries (mH) or microhenries (μH).

The importance of mutual induction is that it permits the transfer of electrical energy from one circuit to another without any physical contact between them. Consequently, there are many applications of mutual induction. As an example, Figure 28.20a shows an induction ammeter, which is a device for measuring alternating current in situations where it would be too time-consuming or too dangerous to disconnect a wire and insert a standard ammeter in the circuit. Consider a person who needs to know quickly whether a wire in a broken appliance carries a 60-Hz current, and, if so, how much. Part b of the figure indicates that the "jaw" of the induction ammeter is slipped around the wire in question. The alternating current produces a changing magnetic field in the space around the wire. Part of this field passes through a coil wrapped around the jaw of the ammeter, as part c of the drawing shows. The changing magnetic field induces an emf that is registered by the meter connected to the coil. The jaw is often made of iron to enhance the magnetic field and the emf. Since the induced emf is proportional to the current in the appliance wire, the meter can be calibrated to read this current.

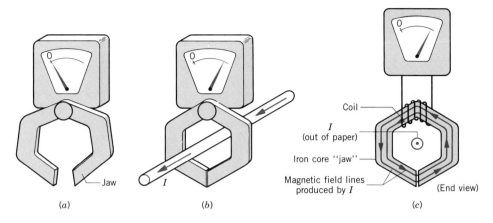

Coil

I
(out of paper)

Iron core "jaw"

Magnetic field lines
produced by I

(End view)

FIGURE 28.20 An induction ammeter with its jaws (a) open and (b) closed around a wire carrying an alternating current I. (c) Some of the magnetic field lines around a current-carrying wire are routed through the coil by the iron core. Since the magnetic field produced by the alternating current is changing, an emf is induced in the coil. The meter detects the induced emf and is calibrated to display the amount of current in the wire.

SELF-INDUCTANCE

In all the examples of induced emfs presented so far, the magnetic field has been produced by an external source, such as a permanent magnet or an electromagnet. However, it is not necessary that the magnetic field arise from an external source. An emf can be induced in a current-carrying coil by a change in the magnetic field that the current itself produces. For instance, Figure 28.21 shows a coil connected to an ac generator. The alternating current creates an alternating magnetic field that, in turn, creates a changing flux through the coil. The change in flux induces an emf in the coil, in accord with Faraday's law. The effect in which a changing current in a circuit induces an emf in the same circuit is referred to as **self-induction.**

When dealing with self-induction, as with mutual induction, it is customary to recast Faraday's law into a form in which the induced emf is proportional to the changing current in the coil, rather than to the changing flux. If Φ is the magnetic flux that passes through each turn of the coil, then $N\Phi$ is the net flux through a coil of N turns. Since Φ is proportional to the magnetic field, and the magnetic field is proportional to the current I, it follows that $N\Phi \propto I$. By inserting a constant L, called the **self-inductance** or simply the **inductance** of the coil, we can convert this proportionality into Equation 28.8:

$$N\Phi = LI \quad \text{or} \quad L = \frac{N\Phi}{I} \qquad (28.8)$$

Faraday's law of induction now gives the induced emf as

$$\mathscr{E} = -N\frac{\Delta\Phi}{\Delta t} = -\frac{\Delta(N\Phi)}{\Delta t} = -\frac{\Delta(LI)}{\Delta t} = -L\frac{\Delta I}{\Delta t}$$

$$\begin{bmatrix} \textbf{Emf due to} \\ \textbf{self-} \\ \textbf{induction} \end{bmatrix} \qquad \mathscr{E} = -L\frac{\Delta I}{\Delta t} \qquad (28.9)$$

Like mutual inductance, L is measured in henries. The magnitude of L depends on the geometry of the coil and on the core material; L does not depend on the current, however. By wrapping the coil around a ferromagnetic (iron) core, the magnetic flux — and therefore the inductance — can be increased substantially relative to that for an air core. Because of their self-inductance, coils are known as **inductors** and are widely used in electronics. Inductors come in all sizes, typically in the range between

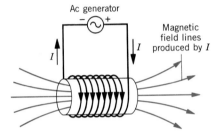

Ac generator

Magnetic
field lines
produced by I

FIGURE 28.21 The alternating current in the coil generates an alternating magnetic field that induces an emf in the coil. The effect in which a changing current in a coil induces an emf in the same coil is known as self-induction.

millihenries and microhenries. The next example shows how to determine the inductance of a solenoid.

EXAMPLE 10

A solenoid of length $\ell = 8.0 \times 10^{-2}$ m and cross-sectional area $A = 5.0 \times 10^{-5}$ m² contains $n = 6500$ turns per meter of length. (a) Find the self-inductance of the solenoid, assuming the core is air. (b) Determine the emf induced in the solenoid when the current increases from 0 to 1.5 A in a time interval of 0.20 s.

SOLUTION

(a) The self-inductance can be found by using Equation 28.8, $L = N\Phi/I$, provided the flux Φ can be determined. The flux is given by Equation 28.2 as $\Phi = BA \cos \theta$. In the case of a solenoid, the interior magnetic field is directed perpendicular to the plane of the loops, so $\theta = 0°$. According to Equation 27.7, the magnetic field inside the solenoid has the value $B = \mu_0 nI$, where n is the number of turns per unit length. Therefore, the self-inductance of the solenoid is

$$L = \frac{N\Phi}{I} = \frac{N(BA)}{I} = \frac{N(\mu_0 nI)A}{I} = \mu_0 nNA = \mu_0 n^2 A\ell$$

where we have replaced N by $n\ell$. Substituting the given values into this result yields

$$L = \mu_0 n^2 A\ell = (4\pi \times 10^{-7} \text{ T·m/A})(6500 \text{ turns/m})^2$$
$$\times (5.0 \times 10^{-5} \text{ m}^2)(8.0 \times 10^{-2} \text{ m})$$
$$= \boxed{0.21 \text{ mH}}$$

(b) The induced emf that results from the increasing current is

$$\mathcal{E} = -L\frac{\Delta I}{\Delta t} = -(0.21 \times 10^{-3} \text{ H})\left(\frac{1.5 \text{ A}}{0.20 \text{ s}}\right) \quad (28.9)$$
$$= \boxed{-1.6 \text{ mV}}$$

The negative sign reminds us that the induced emf opposes the increasing current that induces the emf.

THE ENERGY STORED IN AN INDUCTOR

An inductor, like a capacitor, can store energy. To see how this stored energy arises, let us compute the work a generator does to establish a current in an inductor. Suppose an inductor is connected to a generator whose terminal voltage can be varied continuously from zero to some final value. As the voltage is increased, the current I in the circuit rises continuously from zero to its final value. While the current is rising, an induced emf $\mathcal{E} = -L(\Delta I/\Delta t)$ appears across the inductor. Conforming with Lenz's law, the polarity of the induced emf $\mathcal{E}$ is opposite to that of the generator voltage, so as to oppose the increase in the current. Thus, the generator must do work to push the charges through the inductor against this induced emf. The increment of work ΔW done by the generator in moving a small amount of charge ΔQ through the inductor is $\Delta W = (\Delta Q)\mathcal{E} = (\Delta Q) L (\Delta I/\Delta t)$, according to Equation 24.4. To ensure that the work done by the generator is positive, as it must be since the generator is driving charge against an opposing emf, the minus sign in front of the $L(\Delta I/\Delta t)$ term has been removed. Since $\Delta Q/\Delta t$ is the current I, the work done by the generator is

$$\Delta W = LI (\Delta I)$$

In this expression ΔW represents the work done by the generator to increase the current in the inductor by an amount ΔI. To determine the total work W done during the time interval when the current is changed from zero to its final value, all the small increments of work ΔW must be added together. This summation is left as an exercise at the end of this chapter (see problem 39). The result is $W = \frac{1}{2}LI^2$, where I represents the final current in the inductor. This work is stored as energy in the inductor, so that

$$\begin{bmatrix} \textbf{Energy stored} \\ \textbf{in an} \\ \textbf{inductor} \end{bmatrix} \qquad \text{Energy} = \tfrac{1}{2}LI^2 \qquad (28.10)$$

For instance, an 85-mH inductor carrying a current of 5.0 A stores an amount of energy given by Energy $= \frac{1}{2}(85 \times 10^{-3}$ H$)(5.0$ A$)^2 = 1.1$ J. This equation for the energy stored in an inductor is analogous to Equation 24.11 for the energy stored in a capacitor: Energy $= \frac{1}{2}CV^2$, where C is the capacitance and V is the potential difference between the charged plates of the capacitor.

As Section 24.5 discusses, it is possible to regard the energy in a capacitor as being stored in the electric field between the plates. Similarly, it is possible to regard the energy in an inductor as being stored in its magnetic field. For the special case of a long solenoid, Example 10 shows that the self-inductance is $L = \mu_0 n^2 A \ell$, where n is the number of turns per meter, A is the cross-sectional area, and ℓ is the length of the solenoid. As a result, the energy stored in a solenoid is

$$\text{Energy} = \tfrac{1}{2}LI^2 = \tfrac{1}{2}\mu_0 n^2 A \ell I^2$$

Since $B = \mu_0 nI$ at the interior of a long solenoid (Equation 27.7), this energy can be expressed as

$$\text{Energy} = \frac{1}{2\mu_0} B^2 A \ell$$

The term $A\ell$ is the volume inside the solenoid in which the magnetic field exists, so the energy per unit volume or **energy density** is

$$\text{Energy density} = \frac{\text{Energy}}{\text{Volume}} = \frac{1}{2\mu_0} B^2 \tag{28.11}$$

This result applies only to magnetic fields in air (or vacuum) or in nonmagnetic materials. Although Equation 28.11 was obtained for the special case of a long solenoid, the relation is valid in general for any region of space where a magnetic field exists. Thus, energy is stored in a magnetic field, just as it is in an electric field.

28.9 TRANSFORMERS

One of the most important applications of mutual induction and self-induction takes place in a transformer. A **transformer** is a device for increasing or decreasing an ac voltage. For example, whenever a portable radio or calculator is plugged into a wall receptacle to recharge the batteries, a transformer plays a role in reducing the 120-V ac voltage to a much smaller value; typically, between 3 and 9 V are needed to energize batteries. In another instance, a picture tube in a television set needs about 15 000 V to accelerate the electron beam, and a transformer is used to obtain this high voltage from the relatively low voltage provided by a wall receptacle.

Figure 28.22 shows a drawing of a transformer. The transformer consists of an iron core on which two coils are wound: a primary coil with N_p turns, and a secondary coil with N_s turns. As the drawing indicates, the primary coil is connected to an ac generator. For the moment, suppose the switch in the secondary circuit is open, so there is no current in this circuit.

The alternating current in the primary coil establishes a changing magnetic field in the iron core. Because iron is easily magnetized, it greatly enhances the magnetic field relative to that in an air core and guides the field lines to the secondary coil. In a well-designed core, nearly all the magnetic flux Φ that passes through each turn of the primary coil also goes through each turn of the secondary coil. Since the magnetic

FIGURE 28.22 A transformer consists of a primary coil with N_p turns and a secondary coil with N_s turns; both coils are wound on the same iron core. The changing magnetic flux produced by the current in the primary coil induces an emf in the secondary coil. The drawing at the far right shows the symbol used for a transformer in circuit diagrams.

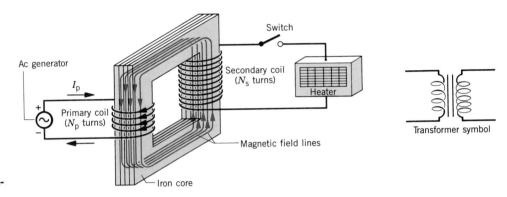

field is changing, the flux through the primary and secondary coils is also changing, and consequently an emf is induced in both coils. In the secondary coil the induced emf $\mathcal{E}_s$ arises from mutual induction and is given by Faraday's law of electromagnetic induction as

$$\mathcal{E}_s = -N_s \frac{\Delta \Phi}{\Delta t}$$

In the primary coil the induced emf $\mathcal{E}_p$ is due to self-induction and is specified by Faraday's law as

$$\mathcal{E}_p = -N_p \frac{\Delta \Phi}{\Delta t}$$

The term $\Delta\Phi/\Delta t$ is the same in both of these equations, since the same flux penetrates each turn of both coils. Dividing the two equations shows that

$$\frac{\mathcal{E}_s}{\mathcal{E}_p} = \frac{N_s}{N_p}$$

In a high-quality transformer the resistances of the coils are negligible, so the magnitudes of the emfs, $\mathcal{E}_s$ and $\mathcal{E}_p$, are nearly equal to the terminal voltages, V_s and V_p, across the coils (see Section 26.4 for a discussion of terminal voltage). The relation $\mathcal{E}_s/\mathcal{E}_p = N_s/N_p$ is called the ***transformer equation*** and is usually written in terms of the terminal voltages:

$$\left[\begin{array}{c} \textbf{Transformer} \\ \textbf{equation} \end{array}\right] \qquad \frac{\text{Secondary voltage}}{\text{Primary voltage}} = \frac{V_s}{V_p} = \frac{N_s}{N_p} \qquad (28.12)$$

According to the transformer equation, if N_s is greater than N_p, the secondary (output) voltage is greater than the primary (input) voltage. In this case we have a *step-up* transformer. On the other hand, if N_s is less than N_p, the secondary voltage is less than the primary voltage, and we have a *step-down* transformer. The ratio N_s/N_p is referred to as the *turns ratio* of the transformer. A turns ratio of 8/1 (often written as 8 : 1) means, for example, that the secondary coil has eight times as many turns as the primary coil. Conversely, a turns ratio of 1 : 8 implies that the secondary coil has one-eighth as many turns as the primary coil.

A transformer operates with ac electricity and not with steady direct current. A steady direct current in the primary coil produces a flux that does not change, and thus no emf is induced in the secondary coil. The ease with which transformers can be used to change voltages from one value to another is the principal reason why ac is preferred over dc.

If the switch in the secondary circuit of Figure 28.22 is closed, a current I_s exists in the circuit and electrical energy is fed to the heater. This energy comes from the ac generator connected to the primary coil. Although the secondary voltage V_s may be larger or smaller than the primary voltage V_p, energy is not being created or destroyed by the transformer. Energy conservation requires that the energy delivered to the secondary coil must be the same as the energy delivered to the primary coil, provided no energy is dissipated in heating these coils or is otherwise lost. In a well-designed transformer, less than 1% of the input energy is lost in the form of heat. Noting that power is energy per unit time, and assuming 100% energy transfer, the average power $\overline{P}_p$ delivered to the primary coil is equal to the average power $\overline{P}_s$ delivered to the secondary coil: $\overline{P}_p = \overline{P}_s$. But $P = IV$ (Equation 25.6), so $I_pV_p = I_sV_s$, or

$$\frac{I_p}{I_s} = \frac{V_s}{V_p} = \frac{N_s}{N_p} \qquad (28.13)$$

Observe that V_s/V_p is equal to the turns ratio N_s/N_p, while I_s/I_p is equal to the inverse turns ratio N_p/N_s. Consequently, a transformer that steps up the voltage simultaneously steps down the current. Conversely, a transformer that steps down the voltage steps up the current. However, in an ideal transformer, the power is neither stepped up nor stepped down, since $\overline{P}_p = \overline{P}_s$. Example 11 emphasizes this fact.

EXAMPLE 11

A step-down transformer inside a stereo receiver has 330 turns in the primary coil and 25 turns in the secondary coil. The plug connects the primary coil to a 120-V wall receptacle, and there is a current of 0.83 A in the primary coil while the receiver is turned on. Connected to the secondary coil are the transistor circuits of the receiver. Find (a) the voltage across the secondary coil, (b) the current in the secondary coil, and (c) the average electrical power delivered to the transistor circuits.

SOLUTION

(a) The voltage across the secondary coil can be found from the transformer equation:

$$V_s = V_p \frac{N_s}{N_p} = (120 \text{ V})\left(\frac{25}{330}\right) = \boxed{9.1 \text{ V}}$$

(b) The current in the secondary coil follows from Equation 28.13 as

$$I_s = I_p \frac{N_p}{N_s} = (0.83 \text{ A})\left(\frac{330}{25}\right) = \boxed{11 \text{ A}}$$

(c) The average power $\overline{P}_s$ delivered to the secondary is the product of I_s and V_s:

$$\overline{P}_s = I_sV_s = (11 \text{ A})(9.1 \text{ V}) = \boxed{1.0 \times 10^2 \text{ W}} \quad (25.6)$$

As a check on our calculation, we verify that the power delivered to the secondary coil is the same as that sent to the primary coil from the wall receptacle: $\overline{P}_p = I_pV_p = (0.83 \text{ A})(120 \text{ V}) = 1.0 \times 10^2 \text{ W}$.

Transformers play an important role in the transmission of power between electrical generating plants and the communities they serve. Whenever electricity is transmitted, there is always some loss of power in the transmission lines themselves due to resistive heating. Since the resistance of the wires is proportional to their length, the longer the wires the greater the power loss. Power companies reduce this loss by using transformers that step up the voltage to high levels, while reducing the current. A smaller current means less power loss, since $P = I^2R$, where R is the resistance of the transmission wires. Figure 28.23 shows one possible scenario for transmitting power. The power plant produces a voltage of 12 000 V. This voltage is

FIGURE 28.23 Transformers are important in the transmission of electric power. The voltage on the transmission line has been stepped up to a very high value (240 000 V in this instance) so as to reduce the current, thereby minimizing power losses due to resistive heating of the wires.

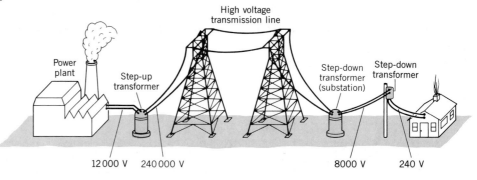

then raised to 240 000 V by a 20:1 step-up transformer. The high-voltage power is sent over the long-distance transmission line. Upon arrival at the city, the voltage is reduced to about 8000 V at a substation using a 1:30 step-down transformer. The power is then distributed to users. However, before any domestic use, the voltage is further reduced to 240 V (or possibly 120 V) by another step-down transformer that is often mounted near the top of a utility pole.

SUMMARY

The **magnetic flux** Φ that passes through a surface is $\Phi = BA \cos \theta$, where A is the area of the surface, B is the magnitude of the magnetic field at the surface, and θ is the angle between **B** and the normal to the surface.

Electromagnetic induction is the phenomenon in which an emf is induced in a coil of wire by a change in the magnetic flux that passes through the coil.

Faraday's law of electromagnetic induction states that the average emf $\mathcal{E}$ induced in a coil of N loops during a time interval Δt is N times the change in magnetic flux $\Delta \Phi$ through each loop divided by the time interval:

$$\mathcal{E} = -N \left(\frac{\Phi - \Phi_0}{t - t_0} \right) = -N \frac{\Delta \Phi}{\Delta t}$$

For the special case of a conductor of length L moving with speed v perpendicular to a magnetic field **B**, the induced emf is called motional emf and its value is given by $\mathcal{E} = vBL$.

Lenz's law provides a way to determine the polarity of an induced emf. Lenz's law states that the polarity of an induced emf is such that the emf would produce a current whose own magnetic field opposes the change in flux that causes the induced emf. Lenz's law is a consequence of the law of conservation of energy.

In its simplist form, an **electric generator** consists of a coil of N loops that rotates in a uniform magnetic field **B**. The emf produced by this generator is $\mathcal{E} = NAB\omega \sin \omega t = \mathcal{E}_0 \sin \omega t$, where A is the area of the coil, ω is the angular

speed (in rad/s) of the coil, and $\mathcal{E}_0$ is the peak emf produced by the generator.

When an electric motor is running, it exhibits a generator-like behavior by producing an induced emf, called a **back emf**. The current I needed to keep the motor running at a constant speed is $I = (V - \mathcal{E})/R$, where V is the emf applied to the motor by an external source, $\mathcal{E}$ is the back emf generated by the motor, and R is the resistance of the coil.

Mutual induction is the effect in which a changing current in the primary coil induces an emf in the secondary coil. The emf $\mathcal{E}_2$ induced in the secondary coil by a change in current ΔI_1 in the primary coil is $\mathcal{E}_2 = -M(\Delta I_1/\Delta t)$, where Δt is the time interval during which the change occurs. The constant M is the **mutual inductance** between the two coils and is measured in henries (H).

Self-induction is the effect in which a change in current ΔI in a coil induces an emf $\mathcal{E} = -L(\Delta I/\Delta t)$ in the same coil. The constant L is the **self-inductance** or **inductance** of the coil and is measured in henries.

To establish a current I in an inductor, work must be done by an external agent. This work is stored as energy in the inductor, the amount being Energy $= \frac{1}{2}LI^2$, where L is the self-inductance. The energy stored in an inductor can be regarded as being stored in its magnetic field. At any point in space where a magnetic field **B** exists, the **energy density,** or the energy stored per unit volume, is Energy density $= B^2/(2\mu_0)$.

A **transformer** consists of a primary coil of N_p turns and a secondary coil of N_s turns. When an emf $\mathcal{E}_p$ is applied to the primary coil, an emf $\mathcal{E}_s$ is induced in the secondary coil according to the relation $\mathcal{E}_s/\mathcal{E}_p = N_s/N_p$. A transformer functions with ac electricity, not with steady dc electricity.

If the transformer is 100% efficient in transferring power from the primary coil to the secondary coil, the ratio of the primary current I_p to the secondary current I_s is $I_p/I_s = N_s/N_p$.

QUESTIONS

1. A uniform magnetic field points due east. A horizontal copper rod is perpendicular to this field and is oriented in the north–south direction. The rod falls freely to the earth. (a) Which end of the rod, north or south, becomes positively charged? (b) Which end of the rod becomes positively charged if the rod is initially oriented parallel to the magnetic field? Account for your answers.

2. In the discussion concerning Figure 28.4, we saw that a force of 0.086 N from an external agent was required to keep the rod moving at a constant speed. Suppose the light bulb in the figure is unscrewed from its socket. How much force would now be needed to keep the rod moving at a constant speed? Justify your answer.

3. Eddy currents are electric currents that can arise in a piece of metal when it moves through a region where the magnetic field is not the same everywhere. The picture shows, for example, a metal sheet moving to the right at a velocity **v** and a magnetic field **B** that is directed perpendicular to the sheet. At the instant represented, the magnetic field only extends over the left half of the sheet. An emf is induced in the sheet and leads to the eddy current shown. Explain why this current causes the metal sheet to slow down. This action of eddy currents is used in various devices as a brake to damp out unwanted motion.

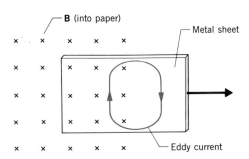

4. Suppose the magnetic flux through a 1-m² flat surface is known to be 2 Wb. From this data alone, is it possible to determine the average magnetic field at the surface? If it is not possible to determine the magnitude of the field, what can be ascertained about the field?

5. A square loop of wire is moving (but not rotating) through a uniform magnetic field. The normal to the loop is oriented parallel to the magnetic field. Is an emf induced in the loop? Give a reason for your answer.

6. Explain how a bolt of lightning can produce a current in the circuit of an electrical appliance, even when the lightning does not directly strike the appliance.

7. A robot is designed to move in a line parallel to a cable hidden under the floor. The cable carries a steady direct current I. A sensor mounted on the robot consists of a coil of wire. The coil is near the floor and parallel to it. As long as the robot moves parallel to the cable, as in the drawing, no emf is induced in the coil, since the magnetic flux through the coil does not change. But when the robot deviates from the parallel path, an induced emf appears in the coil. The emf is sent to electronic circuits that provide instructions to bring the robot back to the path. Explain why an emf would be induced in the sensor coil.

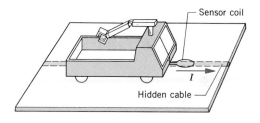

8. In Figure 28.2*b* a coil of wire is being stretched. (a) Using Lenz's law, verify that the induced current in the coil has the direction shown in the drawing. (b) Deduce the direction of the induced current if the direction of the external magnetic field in the figure were reversed.

9. (a) When the switch in the circuit in the drawing is closed, a current is established in the coil and the metal ring "jumps" upward. Explain this behavior. (b) Describe what would happen to the ring if the battery polarity were reversed.

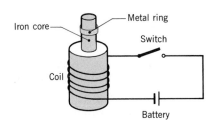

10. The string of an electric guitar vibrates in a standing wave pattern that consists of nodes and antinodes (see Section 22.5 for a discussion of standing waves). Where should an electromagnetic pickup be located in the standing wave pattern to produce a maximum emf, at a node or an antinode? Why?

11. An electric motor in a hair dryer is running at normal speed and, thus, is drawing a relatively small amount of current, as in part (b) of Example 9. What happens to the current drawn by the motor if the shaft is prevented from turning, so the back emf is suddenly reduced to zero? Remembering that the wire in the coil of the motor has some resistance, what happens to the temperature of the coil? Justify your answers.

12. Would a steady direct current in a wire register on the induction ammeter shown in Figure 28.20? Explain.

13. The drawing shows a plot of the current as a function of time in the primary coil of a transformer. During which intervals of time—Δt_1, Δt_2, Δt_3, Δt_4, Δt_5—would you expect an emf to be induced in the secondary coil? Why?

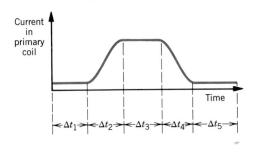

PROBLEMS

Section 28.2 Motional Emf

1. An emf of 0.35 V is generated between the ends of a bar moving through a magnetic field of 0.11 T, as in Figure 28.3a. What field strength would be needed to produce an emf of 1.5 V between the ends of the bar, assuming all other factors remain the same?

2. The wingspan (tip-to-tip) of a Boeing 747 jetliner is 59 m. The plane is flying horizontally at a speed of 220 m/s. The vertical component of the earth's magnetic field is 5.0×10^{-6} T. Find the emf induced between the wing tips.

3. Near San Francisco, where the vertically downward component of the earth's magnetic field is 0.48 G, a car is traveling at 25 m/s. An emf of 2.4×10^{-3} V is induced between the sides of the car. (a) Which side of the car is positive, the driver's side or the passenger's side? (b) What is the width of the car? (c) How would the answer in part (a) change if the car were turned around and traveled in the opposite direction?

4. A spark can jump between two nontouching conductors if the potential difference between them is sufficiently large. Approximately, a potential difference of 940 V is required to produce a spark in a 0.010-cm air gap. Suppose the light bulb in Figure 28.3b is replaced by such a gap. How fast would a 1.6-m rod have to be moving in a magnetic field of 0.80 T to cause a spark to jump across the gap?

* 5. Suppose the light bulb in Figure 28.3b is replaced by a 6.0-Ω electric heater that consumes 15 W of power. The conducting bar moves to the right at a constant speed, the field strength is 2.4 T, and the length of the rod between the rails is 1.2 m. (a) How fast is the bar moving? (b) What force must be applied to the bar to keep it moving to the right at the constant speed found in part (a)?

**6. A conducting rod slides down between two frictionless vertical copper tracks at a constant speed of 5.4 m/s perpendicular to

a 0.30-T magnetic field (see the drawing). The rod maintains electrical contact with the tracks at all times and has a length of 1.2 m. A 0.50-Ω resistor is attached between the tops of the tracks. (a) What is the mass of the rod? (b) Find the change in gravitational potential energy that occurs in a time of 0.20 s. (c) Determine the electrical energy dissipated in the resistor in 0.20 s.

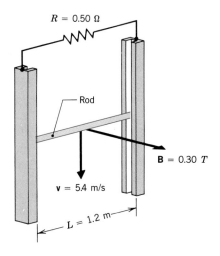

Section 28.3 Magnetic Flux

7. A hand is held flat and placed in a uniform magnetic field of magnitude 0.35 T. The hand has an area of 160 cm² and negligible thickness. Determine the magnetic flux that passes through the hand when the normal to the hand is (a) parallel and (b) perpendicular to the magnetic field.

8. A rectangle (0.60 m $\times$ 0.30 m) lies in the xy plane. An identical rectangle lies in the xz plane. A uniform 0.17-T magnetic field points along the positive z direction. Find the flux through each rectangle.

9. A house has a floor area of 112 m² and an outside wall that has an area of 28 m². The earth's magnetic field here has a horizontal component of 2.6×10^{-5} T that points due north and a vertical component of 4.2×10^{-5} T that points straight down, toward the earth. Determine the magnetic flux through the wall if the wall faces (a) north and (b) east. (c) Calculate the magnetic flux that passes through the floor.

* 10. A five-sided object, whose dimensions are shown in the drawing, is placed in a uniform magnetic field. The magnetic field has a magnitude of 0.25 T and points along the positive y direction. Determine the magnetic flux through each of the five sides.

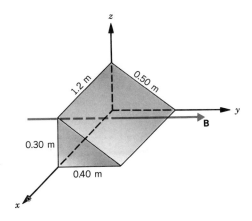

Section 28.4 Faraday's Law of Electromagnetic Induction

11. A circular loop of wire is placed in a uniform magnetic field that is parallel to the normal to the loop. The strength of the magnetic field is 3.0 T. The area of the loop begins shrinking at a constant rate of 0.40 m²/s. What is the magnitude of the emf induced in the loop while the area is shrinking?

12. A circular coil (950 turns, radius = 6.0 cm) is rotating in a uniform magnetic field. At $t = 0$ the normal to the coil is perpendicular to the magnetic field and at $t = 0.010$ s the normal makes an angle of 45° with the field. An average emf of magnitude 65 mV is induced in the coil. Find the magnitude of the magnetic field at the location of the coil.

13. A 75-turn conducting coil has an area of 8.5×10^{-3} m² and the normal to the coil is parallel to a magnetic field **B**. The coil has a resistance of 14 Ω. At what rate (in T/s) must the magnitude of **B** change for an induced current of 7.0 mA to exist in the coil?

14. Magnetic resonance imaging (MRI) is a new medical technique for producing "pictures" of the interior of the body. The patient is placed inside the coil of a large electromagnet that produces a strong magnetic field aligned along the length of the body. One safety concern is what would happen to the positively and negatively charged particles in the body fluids if an equipment failure caused the magnetic field to be suddenly shut off. The fear

is that an induced emf would cause these particles to flow, producing an electric current within the body. Consider a loop of diameter 20.0 cm as defining the largest area of the body through which flux passes, and suppose the magnetic field has a strength of 1.5 T. If the normal to the loop is parallel to the magnetic field, determine the smallest time period during which the field can be allowed to vanish if the induced emf within the body is to be kept at a value less than 0.010 V.

15. A 1.8-m-long aluminum rod is rotating about an axis that is perpendicular to one end. A 0.27-T magnetic field is directed parallel to the axis. The rod rotates through one-fourth of a circle in 2.0 s. What is the magnitude of the average emf generated between the ends of the rod during this time?

* 16. A copper rod is sliding on two conducting rails that make an angle of 15° with respect to each other, as in the drawing. The rod is moving to the right with a constant speed of 0.40 m/s. A 0.42-T uniform magnetic field is perpendicular to the plane of the paper. Determine the magnitude of the average emf induced in the triangle ABC during the 5.0-s period after the rod has passed point A.

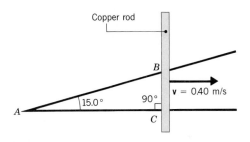

* 17. A conducting coil of 1850 turns is connected to a galvanometer, and the total resistance of the circuit is 45.0 Ω. The area of each turn is 4.70×10^{-4} m². This coil is moved from a region where the magnetic field is zero into a region where it is nonzero, the normal to the coil being kept parallel to the magnetic field. The amount of charge that is induced to flow around the circuit is measured to be 8.87×10^{-3} C. Find the magnitude of the magnetic field. (Such a device can be used to measure the magnetic field strength and is called a *flux meter.*)

** 18. Two 0.50-m-long conducting rods are rotating at the same speed in opposite directions, and both are perpendicular to a

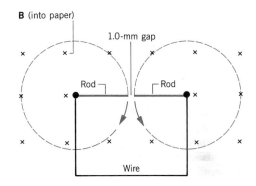

0.40-T magnetic field. As the drawing shows, the ends of these rods come to within 1.0 mm of each other as they rotate. Moreover, the fixed ends about which the rods are rotating are connected by a wire, and, therefore, these ends are at the same electric potential. If a potential difference of 4.5×10^3 V is required to cause a 1.0-mm spark in air, what is the angular speed (in rad/s) of the rods when a spark jumps across the gap?

Section 28.5 Lenz's Law

19. (a) What is the direction of the induced current through the resistor R in part a of the drawing just after the switch is closed? (b) What is the direction of the induced current through R in part b of the drawing as the current I decreases to zero? Provide a reason for each of your answers.

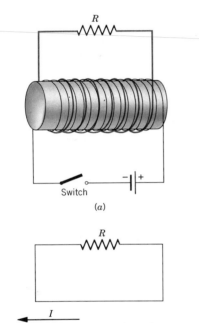

(a)

(b)

20. As the picture shows, a loop of copper wire is lying flat on a table and is attached to a battery via a switch. The current I in the loop establishes the magnetic field lines shown in color. There are also two smaller conducting loops A and B lying flat on the table, but not connected to batteries. Determine the direction of the induced current in loops A and B when the switch is (a) opened

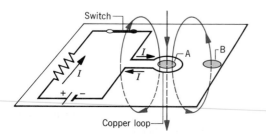

and (b) closed again. Specify the direction of the currents to be clockwise or counterclockwise when viewed from above the table.

21. The drawing shows that a uniform magnetic field is directed perpendicularly out of the plane of the paper and fills the entire region to the left of the y axis. There is no magnetic field to the right of the y axis. A rigid right triangle ABC is made of copper wire. The triangle rotates counterclockwise about the origin at point C. What is the direction (clockwise or counterclockwise) of the induced current when the triangle is crossing (a) the $+y$ axis, (b) the $-x$ axis, (c) the $-y$ axis, and (d) the $+x$ axis? For each case, provide a reason for your answer.

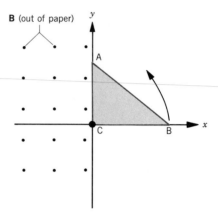

＊ 22. Indicate the direction of the electric field between the plates of the parallel plate capacitor shown in the drawing if the magnetic field is decreasing in time. Explain your reasoning.

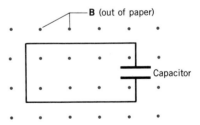

＊ 23. A long, straight wire lies on a table and carries a current I. As the drawing shows, a small circular loop of wire is pushed across the top of the table from position 1 to position 4. Determine the direction of the induced current, clockwise or counterclockwise, as the loop moves past each of the four positions. Justify your answers. See Drawing 23.

＊＊24. A wire loop is suspended from a string that is attached to point P in the drawing. When released, the loop swings downward, from left to right, through a uniform magnetic field, with the plane of the loop remaining perpendicular to the plane of the paper at all times. (a) Determine the direction of the current induced in the loop as it swings past the locations labeled I and II. Specify the direction of the current in terms of the points x, y, and z on the loop (e.g., $x \rightarrow y \rightarrow z$ or $z \rightarrow y \rightarrow x$). The points x, y, and z

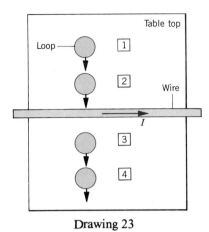

Drawing 23

lie behind the plane of the paper. (b) What is the direction of the induced current at the locations II and I when the loop swings back, from right to left? Provide reasons for your answers.

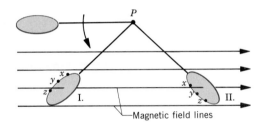

Section 28.7 The Electric Generator

25. A 200-turn rectangular coil has a cross-sectional area of 0.040 m². The coil is rotating at an angular speed of 15 rad/s about an axis that is perpendicular to a magnetic field of 1.5 T. (a) Plot one cycle of the induced emf as a function of time, displaying on the graph numerical values for the maximum emf and the period. (b) On the same graph, plot the emf induced in the coil for one rotation if the angular speed is doubled.

26. Suppose you are requested to design a 60.0-Hz ac generator whose maximum emf is to be 5500 V. The generator is to contain a 150-turn coil whose area is 0.85 m². What should be the magnitude of the magnetic field in which the coil rotates?

27. A generator produces a peak emf of 12.0 V when the armature rotates at 750 rev/min. What is the peak emf when the armature rotates at 2250 rev/min, assuming everything else remains the same?

28. The maximum strength of the earth's magnetic field is about 7.0×10^{-5} T near the south magnetic pole. In principle, this field could be used with a rotating coil to generate 60.0-Hz ac electricity. What is the minimum number of turns (area per turn = 0.016 m²) that the coil must have so as to produce an rms voltage of 120 V?

29. A generator has a square coil consisting of 248 turns. The coil rotates at 755 rev/min in a 0.170-T magnetic field. The rms output of the generator is 53.0 V. What is the length of one side of the coil?

30. The current in the electric motor of a vacuum cleaner is 2.0 A when the cleaner is plugged into a 120.0-V receptacle and is running at normal speed. The coil resistance of the motor is 2.5 Ω. Determine the back emf generated by the motor.

* 31. A motor is designed to operate on 117 V and draws a current of 37.5 A when it first starts up. At its normal operating speed, the motor draws a current of 4.10 A. Obtain (a) the resistance of the armature coil, (b) the back emf developed at normal speed, and (c) the current drawn by the motor at one-third normal speed.

**32. A generator is mounted on a bicycle to power a headlight. A wheel of the generator is pressed against the bike tire and turns the armature 44 times for each revolution of the tire. The tire has a radius of 0.33 m. The armature has 75 turns, each with an area of 2.6×10^{-3} m², and rotates in a 0.10-T magnetic field. When the peak emf being generated is 6.0 V, what is the translational speed (in m/s) of the bike?

Section 28.8 Mutual Inductance and Self-Inductance

33. The mutual inductance between two coils is $M = 8.0$ mH. The current in the primary coil changes at a constant rate from 2.0 to 5.5 A in 0.020 s. Determine (a) the magnitude of the emf induced in the secondary coil, (b) the magnitude of the current in the secondary coil if the resistance of the secondary circuit is 1.5 Ω, and (c) the electrical energy transferred to the secondary circuit during this time.

34. Mutual induction can be used as the basis for a metal detector. A typical setup uses two large coils that are parallel to each other and have a common axis. Because of mutual induction, the ac generator connected to the primary coil causes an emf of 0.46 V to be induced in the secondary coil. When someone without metal objects walks through the coils, the mutual inductance and, consequently, the induced emf do not change much. But when a person carrying a handgun walks through, the mutual inductance increases. If the mutual inductance increases by a factor of three, find the new value of the induced emf. This change in emf can be used to trigger an alarm.

35. A coil consists of 275 turns and has a self-inductance of 15.0 mH. The coil carries a current of 17.0 mA. Obtain the magnetic flux through one turn of the coil.

36. How much energy is stored in an 85-mH inductor that carries a current of (a) 2.5 A and (b) 5.0 A?

37. The earth's magnetic field, like any magnetic field, stores energy. Its maximum strength is about 0.70 G. Find the maximum magnetic energy stored in the space above a city if the space occupies an area of 5.0×10^{8} m² and has a height of 1500 m.

* 38. A long current-carrying solenoid with an air core has 1750 turns per meter of length and a radius of 1.80 cm. A coil of 125 turns is wrapped tightly around the outside of the solenoid. What is the mutual inductance of this system?

* 39. The purpose of this problem is to show that the work W needed to establish a final current I_f in an inductor is $W = \frac{1}{2}LI_f^2$ (Equation 28.10). In Section 28.8 we saw that the amount of work ΔW needed to change the current through an inductor by an amount ΔI is $\Delta W = LI(\Delta I)$, where L is the inductance. The drawing shows a graph of LI versus I. Notice that $LI(\Delta I)$ is the area of the shaded vertical rectangle whose height is LI and whose width is ΔI. Use this fact to show that the total work W needed to establish a current I_f is $W = \frac{1}{2}LI_f^2$.

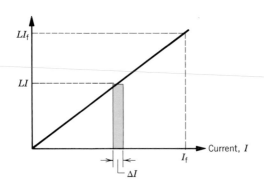

Section 28.9 Transformers

40. In some parts of the country, insect "zappers," with their blue lights, are a familiar sight on a summer's night. These devices use a high voltage to electrocute insects. One such device uses an ac voltage of 4150 V, which is obtained from a standard 120.0-V outlet by means of a transformer. If the primary coil has 300 turns, how many turns are in the secondary coil?

41. Electric doorbells found in many homes require 10.0 V to operate. To obtain this voltage from the standard 120-V supply, a transformer is used. Is a step-up or a step-down transformer needed, and what is its turns ratio?

42. A transformer plays a role in establishing the proper voltage for the operation of a fluorescent light. The input to the primary coil is 120 V, while the current in the secondary coil is 0.40 A. (a) When 60.0 W of power are being delivered to the fluorescent light, determine the voltage across the secondary coil. (b) Is the transformer a step-up or a step-down unit, and what is its turns ratio?

43. A step-down transformer (turns ratio = 1:8) is used with an electric train to reduce the voltage from the wall receptacle to a value needed to operate the train. When the train is running, the current in the secondary coil is 3.4 A. What is the current in the primary coil?

44. The secondary coil of a transformer provides the voltage that operates an electrostatic air filter. The turns ratio of the transformer is 100:1. The primary coil is plugged into a standard 120-V outlet. The current in the secondary coil is 2.0 mA. Find the power consumed by the air filter.

* 45. A generating station is producing 1.0×10^6 W of power that is to be sent to a small town located 7.0 km away. Each of the two wires that comprise the transmission line has a resistance per unit of length of 5.0×10^{-2} Ω/km. (a) Find the power lost in heating the wires if the power is transmitted at 1200 V. (b) A 100:1 step-up transformer is used to raise the voltage before the power is transmitted. How much power is now lost in heating the wires?

**46. A generator is connected across the primary coil (N_p turns) of a transformer, while a resistance R_2 is connected across the secondary coil (N_s turns). This circuit is equivalent to a circuit in which a single resistance R_1 is connected directly across the generator, without the transformer. Show that $R_1 = (N_p/N_s)^2 R_2$, by starting with Ohm's law as applied to the secondary coil.

ADDITIONAL PROBLEMS

47. A 300-turn rectangular loop of wire has an area of 50.0×10^{-4} m^2. At $t_0 = 0$ a magnetic field is turned on, and its magnitude increases to 0.40 T when $t = 0.80$ s. The field is directed at an angle of 30.0° with respect to the normal of the loop. (a) Find the magnitude of the average emf induced in the loop. (b) If the loop is a closed circuit whose resistance is 6.0 Ω, determine the average induced current.

48. Suppose in Figure 28.1 that the bar magnet is held stationary, but the coil of wire is free to move. Which way will current be directed through the ammeter, left-to-right or right-to-left, when the coil is moved (a) to the left and (b) to the right? Explain.

49. The coil of an electromagnet carries a steady direct current of 8.0 A and has a self-inductance of 150 mH. Suddenly a switch is opened and the current decreases to zero in 7.0×10^{-3} s. Obtain the magnitude of the average emf induced in the coil during this time.

50. The resistance of the primary coil of a transformer is measured to be 56 Ω, while the resistance of the secondary coil is found to be 14 Ω. Using these data, determine the turns ratio N_s/N_p of the transformer. Explain the rationale for your calculation.

51. The back emf in a motor is 115 V when the motor is turning at 1800 rev/min. What is the back emf when the motor turns at 3600 rev/min, assuming all other factors remain the same?

* 52. A magnetic field has a magnitude of 12 T. What is the magnitude of an electric field that stores the same energy per unit volume as this magnetic field?

* 53. The armature of an electric-drill motor has a resistance of 3.0 Ω. When connected to a 120.0-V outlet, the motor rotates at its normal speed and develops a back emf of 108 V. (a) What is the current through the motor? (b) If the armature "freezes up" due to a lack of lubrication in the bearings and can no longer rotate, what is the current in the stationary armature? (c) What is the current when the motor runs at only half speed?

* 54. A large circular loop carries a current I. A much smaller circular loop is held above the center of the large loop, with its plane parallel to the plane of the large loop. The small loop is released, so it falls downward through the large loop, all the while maintaining its parallel orientation. The center of the small loop remains in line with the center of the large loop at all times. Is the direction of the current induced in the small loop the same as I or opposite to I when (a) the small loop is above the large loop and (b) the small loop has fallen below the large loop? *(Hint: With the aid of Figure 27.23, first identify the direction of the magnetic field along the axis of the large loop.)* Justify your answers.

* 55. A magnetic field is passing through a loop of wire whose area is 1.8×10^{-2} m^2. The direction of the magnetic field is parallel to the normal to the loop, and the magnitude of the field is increasing at the rate of 0.20 T/s. (a) Determine the magnitude of the emf induced in the loop. (b) Suppose the area of the loop can be enlarged or shrunk. If the magnetic field is increasing as in part (a), at what rate (in m^2/s) should the area be changed at the instant when $B = 1.8$ T if the induced emf is to be zero? Explain whether the area is to be enlarged or shrunk.

**56. Coil 1 is a flat circular coil that has N_1 turns and a radius R_1. At its center is a much smaller flat, circular coil that has N_2 turns and radius R_2. The planes of the coils are parallel. Assume coil 2 is so small that the magnetic field at its location due to coil 1

is nearly constant. Determine an expression for the mutual inductance between these two coils in terms of N_1, R_1, N_2, and R_2.

**57. The drawing shows a copper wire bent into a circular shape with a radius of 0.50 m. The radial section BC is fixed in place, while the copper bar AC sweeps around at an angular speed of 15 rad/s. The bar makes electrical contact with the wire at all times. A uniform magnetic field exists everywhere, is perpendicular to the plane of the circle, and has a magnitude of 3.8×10^{-3} T. Find the magnitude of the current induced in the loop ABC.

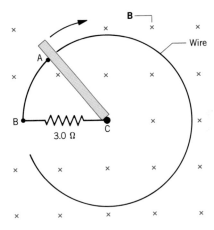

CHAPTER 29

Alternating Current Circuits

The Cray computer is one of today's supercomputers, because it can perform a large number of highly complex mathematical operations at a very high speed. Computers, as well as home appliances and hi-fi systems, typically use alternating current to provide the energy they need to operate.

The basic elements of ac electricity (frequency, rms-voltage and current, and power) have been discussed in Section 25.5, with an emphasis on circuits that contain only resistors. However, circuits often include other components. This chapter deals with a number of these additional components, including capacitors, inductors, diodes, and transistors. Capacitors and inductors are familiar from Sections 24.5 and 28.8, respectively, while the diode and the transistor will be introduced in Section 29.6.

29.1 CAPACITORS AND CAPACITIVE REACTANCE

In an ac circuit, a capacitor plays a different role than it does in a dc circuit. In a dc circuit, charge flows only while the capacitor is charging up for the brief period following the application of the voltage, as we have seen in Section 26.8. After the capacitor becomes fully charged, however, no further flow of charge occurs. In an ac circuit the polarity of the voltage applied to the capacitor continually switches back and forth, and, in response, charges flow first one way around the circuit and then the other way. This flow of charge, surging back and forth in the wires leading to the capacitor, constitutes an alternating current. Thus, charge flows continuously in an ac circuit containing a capacitor.

To help set the stage for the present discussion, recall that $V_{rms} = I_{rms}R$ for a purely resistive ac circuit. The resistance R has the same value for any frequency. Figure 29.1 emphasizes this fact by showing that a graph of resistance versus frequency is a horizontal straight line.

For the rms-voltage across a capacitor the following expression applies, which is analogous to $V_{rms} = I_{rms}R$:

$$V_{rms} = I_{rms}X_C \qquad (29.1)$$

The term X_C appears in place of the resistance R and is called the ***capacitive reactance.*** The capacitive reactance, like resistance, is measured in *ohms* and determines how much rms-current exists in a capacitor in response to a given rms-voltage across the capacitor.

The capacitive reactance depends on the frequency of the ac voltage applied to the capacitor. After all, a larger frequency means that the voltage changes polarity more times per second. And charge flows to and from the capacitor plates as the voltage changes. Since current is the charge flowing per second, there is more current when the voltage is changing rapidly, as it is when the frequency is larger. A greater current for a given voltage means that the capacitive reactance $X_C = V_{rms}/I_{rms}$ is smaller, so the reactance decreases as frequency increases. In fact, it can be shown that the capacitive reactance is inversely proportional to both the frequency f and the capacitance C:

$$X_C = \frac{1}{2\pi fC} \qquad (29.2)$$

For a fixed value of the capacitance C, Figure 29.2 gives a plot of X_C versus frequency, according to Equation 29.2. A comparison of this drawing with Figure 29.1 reveals that a capacitor behaves differently than a resistor. As the frequency becomes very large, Figure 29.2 shows that X_C approaches zero, signifying that a capacitor offers only a negligibly small opposition to the alternating current. In contrast, in the limit of zero frequency (i.e., dc current), X_C becomes infinitely large, and a capacitor provides so much opposition to the motion of charges that there is no alternating current. Example 1 illustrates how frequency and capacitance determine the amount of current in an ac circuit.

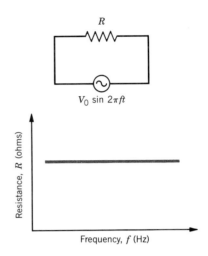

FIGURE 29.1 The resistance in a purely resistive circuit has the same value at all frequencies. Here we use V_0 for the maximum emf of the generator, rather than $\mathscr{E}_0$, which was used in Section 28.7; this choice simplifies our notation.

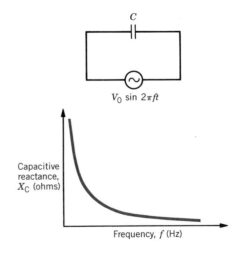

FIGURE 29.2 The capacitive reactance X_C is measured in ohms and is inversely proportional to the frequency f according to $X_C = 1/(2\pi fC)$.

EXAMPLE 1

For the circuit in Figure 29.2, the capacitance of the capacitor is 1.50 μF and the rms-voltage of the generator is 25.0 V. What is the rms-current in the circuit when the frequency of the generator is (a) 1.00×10^2 Hz and (b) 5.00×10^3 Hz?

SOLUTION

(a) The current can be found from $I_{rms} = V_{rms}/X_C$, once the capacitive reactance is determined:

$$X_C = \frac{1}{2\pi fC} = \frac{1}{2\pi(1.00 \times 10^2 \text{ Hz})(1.50 \times 10^{-6} \text{ F})}$$

$$= 1060 \ \Omega \qquad (29.2)$$

$$I_{rms} = \frac{V_{rms}}{X_C} = \frac{25.0 \text{ V}}{1060 \ \Omega} = \boxed{0.0236 \text{ A}} \qquad (29.1)$$

(b) When the frequency is 5.00×10^3 Hz, the calculations are similar:

$$X_C = \frac{1}{2\pi fC} = \frac{1}{2\pi(5.00 \times 10^3 \text{ Hz})(1.50 \times 10^{-6} \text{ F})} = 21.2 \ \Omega$$

$$I_{rms} = \frac{V_{rms}}{X_C} = \frac{25.0 \text{ V}}{21.2 \ \Omega} = \boxed{1.18 \text{ A}}$$

As noted earlier, the capacitive reactance becomes smaller at the higher frequency, giving rise to a larger current.

To gain a more complete understanding of the effects that capacitors have in circuits, we now consider the behavior of the instantaneous (not rms) voltage and current. To provide background material, Figure 29.3 shows graphs of voltage and current versus time in a resistive circuit. These graphs indicate that, when only resistance is present, the voltage and current are proportional to each other at every moment. For example, when the voltage increases from A to B on the graph, the current follows along exactly in step, increasing from A' to B' during the same time. Likewise, when the voltage decreases from B to C, the current decreases from B' to C'. For this reason, the current in a resistance R is said to be **in phase** with the voltage across the resistance. For a capacitor, this in-phase relation between instantaneous voltage and current does *not* exist.

Figure 29.4 shows graphs of the ac voltage and current versus time for a circuit that contains only a capacitor. As the voltage increases from A to B, the charge on the capacitor increases and reaches its full value at B. The current, or rate of flow of charge, has a maximum positive value at the start of the charging process at A', when there is no charge on the capacitor and hence no capacitor voltage to oppose the generator voltage. When the capacitor is fully charged at B, the capacitor voltage has a magnitude equal to that of the generator and completely opposes the generator

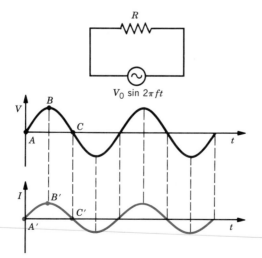

FIGURE 29.3 The instantaneous voltage V and current I in a resistive circuit are *in phase,* which means that they increase and decrease exactly in step with one another.

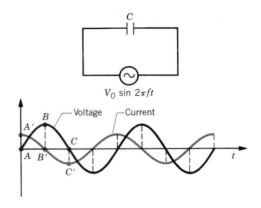

FIGURE 29.4 In a circuit containing only a capacitor, the instantaneous voltage and current are not in phase, as they are in a purely resistive circuit. Instead, the current *leads* the voltage by a phase angle of 90° (one quarter of a cycle).

voltage. The result is that the current decreases to zero at B'. While the capacitor voltage decreases from B to C, the charges flow out of the capacitor in a direction opposite to that of the charging current, as indicated by the negative current from B' to C'. Thus, voltage and current are not in phase but are, in fact, one-quarter wave cycle out of step, or out of phase. More specifically, assuming the voltage fluctuates as $V_0 \sin (2\pi ft)$, the current varies as $I_0 \sin (2\pi ft + \pi/2) = I_0 \cos (2\pi ft)$. Since $\pi/2$ radians correspond to 90° and since the current reaches its maximum value *before* the voltage does, it is said that the current through a capacitor *leads* the voltage across the capacitor by a phase angle of 90°.

The fact that the current and voltage for a capacitor are 90° out of phase has an important consequence from the point of view of electric power, since power is the product of current and voltage. For the time interval between points A and B (or A' and B') in Figure 29.4, both current and voltage are positive. Therefore, the instantaneous power is also positive, meaning that the generator is delivering energy to the capacitor. However, during the period between B and C (or B' and C'), the current is negative while the voltage remains positive, and the power, being the product of the two, is negative. During this period, the capacitor is returning energy to the generator. Thus, the power alternates between positive and negative values for equal periods of time. In other words, the capacitor alternately absorbs and releases energy. Consequently, *on the average, the power is zero and a capacitor uses no energy in an ac circuit.*

It will prove useful later on to use a model for the voltage and current in ac circuits. In this model, voltage and current are represented by rotating arrows, often called *phasors*, whose lengths correspond to the maximum voltage V_0 and maximum current I_0, as Figure 29.5 indicates. These phasors rotate counterclockwise at a fre-

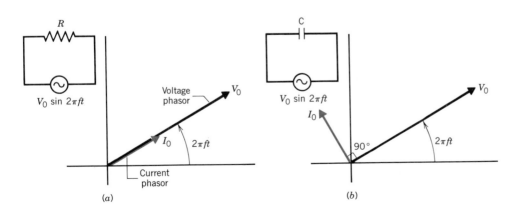

FIGURE 29.5 These rotating arrow (phasor) models represent the voltage and the current in ac circuits that contain (*a*) only a resistor and (*b*) only a capacitor.

quency f. For a resistor, the phasors are colinear as they rotate (see part a of the drawing), because voltage and current are in phase. For a capacitor (see part b), the phasors remain perpendicular while rotating, because the phase angle between the current and the voltage is $90°$. Since current leads voltage for a capacitor, the current phasor is ahead of the voltage phasor in the direction of rotation. In both cases in Figure 29.5, the instantaneous voltage and current are given by the vertical components of the phasors.

29.2 INDUCTORS AND INDUCTIVE REACTANCE

As Section 28.8 discusses, an inductor is usually a coil of wire, and the basis of its operation is Faraday's law of electromagnetic induction. According to Faraday's law, an inductor develops a voltage that opposes a change in the current. This voltage V* is given by $V = -L(\Delta I/\Delta t)$ (see Equation 28.9), where $\Delta I/\Delta t$ is the rate at which the current changes and L is the inductance of the inductor. In an ac circuit the current is always changing and, consequently, the current is affected by the presence of an inductor.

The expression $V = -L(\Delta I/\Delta t)$ can be used to show that the rms-voltage across an inductor is

$$V_{rms} = I_{rms} X_L \qquad (29.3)$$

In this result, which is analogous to $V_{rms} = I_{rms} R$, the term X_L appears in place of the resistance R and is called the ***inductive reactance.*** The inductive reactance is measured in ohms and determines how much rms-current exists in an inductor for a given rms-voltage across the inductor.

The opposition of an inductor to ac current depends on frequency. At higher frequencies the current changes more rapidly than at lower frequencies. As described by Faraday's law, the rapid changes in the current lead to a greater induced emf that opposes the change. Thus, it is to be expected that the inductive reactance increases with increasing frequency. The inductive reactance is related to the frequency f and the inductance L in the following way:

$$X_L = 2\pi f L \qquad (29.4)$$

This relation indicates that the larger the inductance, the larger the inductive reactance. Moreover, the inductive reactance is directly proportional to the frequency, $X_L \propto f$. In contrast, the capacitive reactance is inversely proportional to the frequency, $X_C \propto 1/f$.

Figure 29.6 shows a graph of the inductive reactance versus frequency for a fixed value of the inductance according to Equation 29.4. As frequency becomes very large, X_L also becomes very large. In such a situation, an inductor provides a large opposition to the ac current. In the limit of zero frequency (i.e., dc current), X_L becomes zero, a result indicating that an inductor does not oppose dc current at all. The next example demonstrates the effect of inductive reactance on the current in an ac circuit.

---

* When an inductor is used in a circuit, the notation is simplified if we designate the potential difference across the inductor as the voltage V, rather than the emf $\mathscr{E}$.

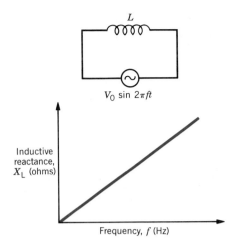

FIGURE 29.6 In an ac circuit the inductive reactance X_L is directly proportional to the frequency f, according to $X_L = 2\pi fL$.

EXAMPLE 2

The circuit in Figure 29.6 contains a 3.60-mH inductor. The rms-voltage of the generator is 25.0 V. Find the rms-current in the circuit when the generator frequency is (a) 1.00×10^2 Hz and (b) 5.00×10^3 Hz.

SOLUTION

(a) The current can be calculated from $I_{rms} = V_{rms}/X_L$, provided the inductive reactance is obtained first:

$$X_L = 2\pi fL = 2\pi(1.00 \times 10^2 \text{ Hz})(3.60 \times 10^{-3} \text{ H}) \quad (29.4)$$

$$= 2.26 \ \Omega$$

$$I_{rms} = \frac{V_{rms}}{X_L} = \frac{25.0 \text{ V}}{2.26 \ \Omega} = \boxed{11.1 \text{ A}} \quad (29.3)$$

(b) The calculation is similar when the frequency is 5.00×10^3 Hz:

$$X_L = 2\pi fL = 2\pi(5.00 \times 10^3 \text{ Hz})(3.60 \times 10^{-3} \text{ H}) = 113 \ \Omega$$

$$I_{rms} = \frac{V_{rms}}{X_L} = \frac{25.0 \text{ V}}{113 \ \Omega} = \boxed{0.221 \text{ A}}$$

By virtue of its inductive reactance, an inductor affects the amount of current that exists in an ac circuit. The inductor also influences the current in another way, as Figure 29.7 shows. This figure displays graphs of voltage and current versus time for a circuit containing only an inductor. At a maximum or minimum on the current graph, the current does not change much with time, so the voltage generated by the inductor to oppose the ac current is zero. At the points on the current graph where the current is zero, the graph is at its steepest, and the current has the largest rate of

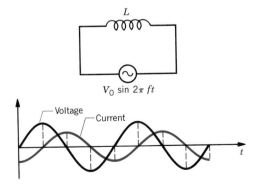

FIGURE 29.7 The instantaneous voltage and current in an inductive circuit are not in phase. The current *lags behind* the voltage by a phase angle of 90° when only an inductor is present.

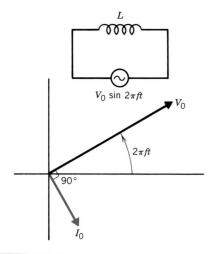

FIGURE 29.8 This phasor model represents the voltage and current in a circuit that contains only an inductor.

increase or decrease. Correspondingly, the voltage generated by the inductor to oppose the ac current has the largest positive or negative value. Thus, current and voltage are not in phase but are one-quarter of a wave cycle out of phase. If the voltage varies as $V_0 \sin(2\pi ft)$, the current fluctuates as $I_0 \sin(2\pi ft - \pi/2) = -I_0 \cos(2\pi ft)$. The current reaches its maximum *after* the voltage does, and it is said that the current *lags behind* the voltage by a phase angle of 90° ($\pi/2$ radians). In a purely capacitive circuit, in contrast, the current leads the voltage by 90°.

In an inductor the 90° phase difference between current and voltage leads to the same result for average power that it does in a capacitor. An inductor alternately absorbs and releases energy for equal periods of time, so **on the average, the power is zero and an inductor uses no energy in an ac circuit.**

As an alternative to the graphs in Figure 29.7, Figure 29.8 uses phasors to describe the instantaneous voltage and current in a circuit containing only an inductor. The voltage and current phasors remain perpendicular as they rotate, for there is a 90° phase angle between them. The current phasor lags behind the voltage phasor, relative to the direction of rotation, in contrast to the equivalent picture for a capacitor. Once again, the instantaneous values for voltage and current are given by the vertical components of the phasors.

29.3 THE SERIES RCL-CIRCUIT

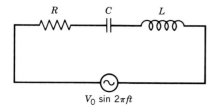

FIGURE 29.9 A series RCL-circuit contains a resistor, a capacitor, and an inductor.

Capacitors and inductors can be combined along with resistors in a single circuit. The simplest combination is the series RCL-circuit, which contains a resistor, a capacitor, and an inductor, as Figure 29.9 shows. In a series RCL-circuit the total opposition to the flow of charge is called the *impedance* of the circuit, and the impedance comes partially from the resistance R, the capacitive reactance X_C, and the inductive reactance X_L. Figure 29.10 shows a graph of impedance versus frequency and emphasizes the frequency regions where each circuit component dominates. At low frequencies X_C becomes very large, and so does the impedance, with X_C making a much greater contribution than either X_L or R. At high frequencies X_L becomes very large, leading once again to a large impedance. However, in this case X_L dominates over X_C and R. At intermediate frequencies the impedance is smaller than it is at either extreme. In

FIGURE 29.10 In a series RCL-circuit the impedance varies with frequency, as this graph shows.

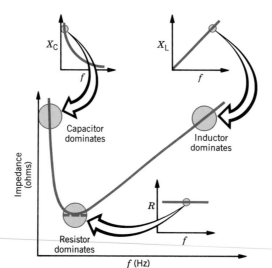

fact, we will see that there is a single frequency where the capacitive and inductive reactances cancel, leaving the frequency-independent resistance to dominate the circuit in this region.

Because the resistor, the capacitor, and the inductor are wired in series, it is tempting to follow the analogy of a series combination of resistors and calculate the impedance by simply adding together R, X_C, and X_L. However, such a procedure is not correct. Instead, the phasors shown in Figure 29.11 must be used. The lengths of the voltage phasors in this drawing represent the maximum voltages V_R, V_C, and V_L across the resistor, the capacitor, and the inductor, respectively. The current is the same for each device, since the circuit is wired in series. The length of the current phasor represents the maximum current I_0. Notice that the drawing shows the current phasor to be (1) in phase with the voltage phasor for the resistor, (2) ahead of the voltage phasor for the capacitor by 90°, and (3) behind the voltage phasor for the inductor by 90°.

The basis for dealing with the voltage phasors in Figure 29.11 is Kirchhoff's loop rule. In an ac circuit this rule applies to the *instantaneous* voltages across each circuit component and the generator. Therefore, it is necessary to take into account the fact that these voltages do not have the same phase, that is, the phasors V_R, V_C, and V_L point in different directions in the drawing. Kirchhoff's loop rule indicates that the phasors add together to give the total voltage V_0 that is supplied to the circuit by the generator. The addition, however, must be a vector addition to take into account the different directions. Since V_L and V_C point in opposite directions, they combine to give a resultant phasor of $V_L - V_C$, as Figure 29.12 shows. In this drawing the resultant $V_L - V_C$ is perpendicular to V_R and may be combined with it according to the Pythagorean theorem to give the total voltage V_0:

$$V_0^2 = V_R^2 + (V_L - V_C)^2$$

In this equation each of the symbols stands for a maximum voltage and when divided by $\sqrt{2}$ gives the corresponding rms-voltage. Therefore, it is possible to divide both sides of the equation by $(\sqrt{2})^2$ and obtain a result for $V_{rms} = V_0/\sqrt{2}$. This result has exactly the same form as that above, but involves the rms-voltages $V_{R\text{-rms}}$, $V_{C\text{-rms}}$, and $V_{L\text{-rms}}$. However, to avoid using such awkward symbols, we simply interpret V_R, V_C, and V_L as rms-quantities in the following expression:

$$V_{rms}^2 = V_R^2 + (V_L - V_C)^2 \qquad (29.5)$$

The last step in determining the impedance of the circuit is to remember that $V_R = I_{rms}R$, $V_C = I_{rms}X_C$, and $V_L = I_{rms}X_L$. With these substitutions the voltage expression above becomes

$$V_{rms} = I_{rms}\sqrt{R^2 + (X_L - X_C)^2}$$

Therefore, for the entire RCL-circuit, it follows that

$$V_{rms} = I_{rms}Z \qquad (29.6)$$

$$\begin{bmatrix} \text{Series} \\ \text{RCL-combination} \end{bmatrix} \qquad Z = \sqrt{R^2 + (X_L - X_C)^2} \qquad (29.7)$$

The quantity Z is the impedance of the circuit and, like R, X_C, and X_L, it is measured in ohms. In Equation 29.7, $X_L = 2\pi fL$ and $X_C = 1/(2\pi fC)$, and a plot of Z versus frequency f gives the graph shown earlier in Figure 29.10. The minimum in the graph can now be seen to occur when $X_L = X_C$, so that at this point $Z = R$, the resistance in the circuit.

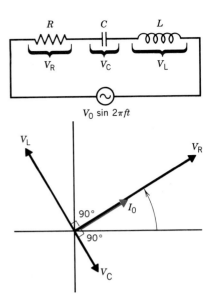

FIGURE 29.11 This drawing illustrates the relation between the three voltage phasors (V_R, V_C, and V_L) and the current phasor (I_0) for a series RCL-circuit.

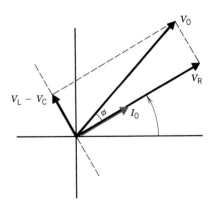

FIGURE 29.12 This simplified version of Figure 29.11 results when the phasors V_L and V_C, which point in opposite directions, are combined to give a resultant of $V_L - V_C$.

The phase angle between the current and the voltage in a series RCL-circuit is the angle ϕ between the current phasor I_0 and the voltage phasor V_0 in Figure 29.12. According to the drawing, this angle is

$$\tan \phi = \frac{V_L - V_C}{V_R} = \frac{I_{rms}X_L - I_{rms}X_C}{I_{rms}R}$$

$$\begin{bmatrix} \text{Series} \\ \text{RCL-combination} \end{bmatrix} \qquad \tan \phi = \frac{X_L - X_C}{R} \qquad (29.8)$$

The phase angle ϕ is important, because it has a major effect on the power dissipated by the circuit. Remember, on the average, only the resistance consumes power, that is, $\overline{P} = I_{rms}^2 R$ (Equation 25.11b). According to Figure 29.12, $\cos \phi = V_R/V_0 = (I_{rms}R)/(I_{rms}Z) = R/Z$, so that $R = Z \cos \phi$. Therefore,

$$\overline{P} = I_{rms}^2 Z \cos \phi = I_{rms}(I_{rms}Z) \cos \phi$$

$$\overline{P} = I_{rms}V_{rms} \cos \phi \qquad (29.9)$$

The term $\cos \phi$ is called the **power factor** of the circuit. As a check on the validity of Equation 29.9, note that if no resistance is present, $R = 0$, and $\cos \phi = R/Z = 0$. Consequently, $\overline{P} = I_{rms}V_{rms} \cos \phi = 0$, a result that is expected since neither a capacitor nor an inductor uses energy on the average. Conversely, if only resistance is present, $Z = \sqrt{R^2 + (X_L - X_C)^2} = R$, and $\cos \phi = R/Z = 1$. In this case, $\overline{P} = I_{rms}V_{rms} \cos \phi = I_{rms}V_{rms}$, which is the correct expression for the average power dissipated in a resistor. Example 3 deals with the current, voltages, and power for a series RCL-circuit.

EXAMPLE 3

A series RCL-circuit contains a 148-Ω resistor, a 1.50-μF capacitor, and a 35.7-mH inductor. The generator has a frequency of 512 Hz and an rms-voltage of 35.0 V. Obtain (a) the rms-voltage across each circuit element and (b) the electrical power consumed by the circuit.

SOLUTION

(a) The rms-voltages across each circuit element can be determined from $V_R = I_{rms}R$, $V_C = I_{rms}X_C$, and $V_L = I_{rms}X_L$, as soon as the rms-current and the reactances X_C and X_L are known. Since the rms-current can be found from $I_{rms} = V_{rms}/Z$, the first step in the solution is to find the impedance Z from the individual reactances:

$$X_C = \frac{1}{2\pi f C} = \frac{1}{2\pi(512 \text{ Hz})(1.50 \times 10^{-6} \text{ F})} \qquad (29.2)$$

$$= 207 \ \Omega$$

$$X_L = 2\pi f L = 2\pi(512 \text{ Hz})(35.7 \times 10^{-3} \text{ H}) \qquad (29.4)$$

$$= 115 \ \Omega$$

$$Z = \sqrt{R^2 + (X_L - X_C)^2} = \sqrt{(148 \ \Omega)^2 + (115 \ \Omega - 207 \ \Omega)^2}$$

$$= 174 \ \Omega \qquad (29.7)$$

The current through each circuit element is

$$I_{rms} = \frac{V_{rms}}{Z} = \frac{35.0 \text{ V}}{174 \ \Omega} = 0.201 \text{ A} \qquad (29.6)$$

The rms-voltages across each device now follow immediately:

$$V_R = I_{rms}R = (0.201 \text{ A})(148 \ \Omega) = \boxed{29.7 \text{ V}} \quad (25.14)$$

$$V_C = I_{rms}X_C = (0.201 \text{ A})(207 \ \Omega) = \boxed{41.6 \text{ V}} \quad (29.1)$$

$$V_L = I_{rms}X_L = (0.201 \text{ A})(115 \ \Omega) = \boxed{23.1 \text{ V}} \quad (29.3)$$

Observe that these three rms-voltages do not add up to give the generator's rms-voltage, which is 35.0 V. Instead, the rms-voltages satisfy Equation 29.5. It is the sum of the *instantaneous* voltages across R, C, and L, rather than the sum of the rms-voltages, that add up to give the generator's *instantaneous* voltage, according to Kirchhoff's loop rule.

(b) The dissipated power is given by $\overline{P} = I_{rms}V_{rms} \cos \phi$. Therefore, a value for the phase angle ϕ is needed and can be obtained as follows:

$$\tan \phi = \frac{X_L - X_C}{R} = \frac{115 \ \Omega - 207 \ \Omega}{148 \ \Omega} \qquad (29.8)$$

$$= -0.62$$

$$\phi = \tan^{-1}(-0.62) = -32°$$

The phase angle is negative since the circuit is more capacitive than inductive (X_C is greater than X_L), and the current leads the voltage. The average power consumed is

$$\overline{P} = I_{rms}V_{rms}\cos\phi \qquad (29.9)$$
$$= (0.201 \text{ A})(35.0 \text{ V})\cos(-32°)$$
$$= \boxed{6.0 \text{ W}}$$

29.4 THE ROLE OF CAPACITORS AND INDUCTORS IN THE DESIGN OF HI-FI LOUDSPEAKERS

The application of capacitors and inductors in the design of hi-fi loudspeakers is a fascinating one. The use of these circuit components is related to the fact that a loudspeaker usually contains a number of speakers within its enclosure. Technically, each of these speakers is known as a driver. For instance, Figure 29.13 shows a so-called two-way loudspeaker with its front cover removed; the presence of two separate drivers is apparent. In such a design the driver that produces primarily low-frequency tones is called the woofer, and the driver that produces mainly high-frequency tones is called the tweeter. In a three-way loudspeaker a woofer, a tweeter, and a separate midrange driver for intermediate frequencies are present.

Since musical sound contains a mixture of frequencies that lie in the range from 20 Hz to 20 kHz, the receiver delivers to the loudspeaker a mixture of ac voltages whose frequencies also lie in this range. Each individual voltage in the mixture has a frequency that is one of the frequencies present in the musical sound. When an individual voltage with a given frequency causes a large amount of current to exist in one of the drivers, then that driver produces a loud sound. Notice in Figure 29.13 that the receiver sends all frequencies to the loudspeaker simultaneously. The question is—How does the loudspeaker route the low-frequency voltages to the woofer and the high-frequency voltages to the tweeter? The answer lies in the clever use of inductors and capacitors.

An inductor can be used so that low-frequency voltages create more current in the woofer than high-frequency voltages do. Figure 29.14 shows how. In this application

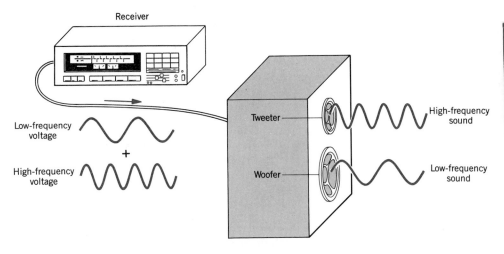

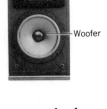

FIGURE 29.13 A two-way loudspeaker contains two drivers, a woofer that generates the lower frequencies and a tweeter that generates the higher frequencies.

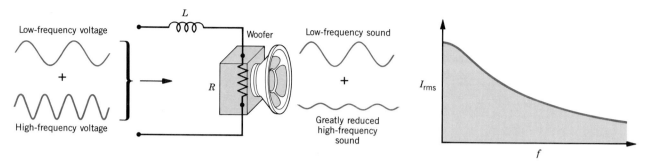

FIGURE 29.14 An inductor wired in series with a woofer significantly reduces high-frequency current, with the result that the woofer generates mainly low-frequency sound.

the inductor L is wired in series with the woofer. For an ac voltage with frequency f, the inductive reactance is $X_L = 2\pi f L$ and becomes larger as the frequency becomes larger. The inductor, therefore, offers greater opposition to high-frequency current than to low-frequency current. In other words, the inductor selectively discriminates against high-frequency current, while allowing low-frequency current to reach the woofer relatively unaffected. The plot of rms-current versus frequency included in the figure shows that the current in the woofer is larger at lower frequencies. Thus, it is primarily low-frequency current that causes the woofer to create mostly low-frequency sound.

A capacitor can be used in a similar fashion to ensure that primarily high-frequency current reaches the tweeter of a two-way loudspeaker (see Figure 29.15). The capacitive reactance is $X_C = 1/(2\pi f C)$ and becomes larger at lower frequencies and smaller at higher frequencies. Thus, the capacitor opposes low-frequency current, keeping it to a small level, while permitting the high-frequency current to reach the tweeter relatively unattenuated. The graph in the figure shows that the current in the tweeter is larger at higher frequencies. The result is that primarily high-frequency current causes the tweeter to produce mainly high-frequency sound.

Figure 29.16 presents one possible arrangement of inductors and capacitors within a loudspeaker. Here, the woofer circuit is connected in parallel with the tweeter circuit. The points A and B in the drawing denote the places on the back panel of the cabinet where the wires from the amplifier are connected. In addition, the picture includes the rms-current-versus-frequency graphs for the two drivers superimposed on a single plot. The frequency at which the two graphs intersect is called the crossover frequency f_c, and the inductor/capacitor combination is called a crossover network. Frequencies above f_c produce relatively large currents in the tweeter and small currents in the woofer, while frequencies below f_c create the opposite effect.

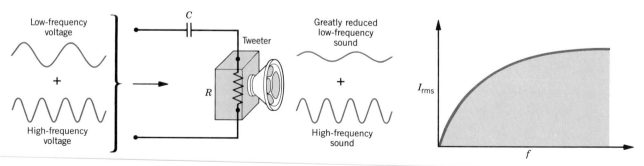

FIGURE 29.15 A capacitor wired in series with a tweeter significantly reduces low-frequency current, so the tweeter creates mostly high-frequency sound.

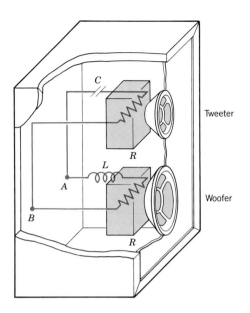

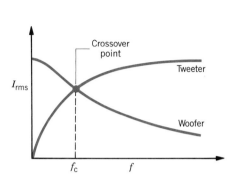

FIGURE 29.16 This two-way loudspeaker contains a capacitor and an inductor arranged to form a crossover network whose crossover frequency is f_c.

29.5 RESONANCE IN ELECTRIC CIRCUITS

The behavior of current and voltage in a series RCL-circuit can give rise to a condition of **resonance.** Resonance occurs when the frequency of a vibrating force exactly matches a natural frequency of the object to which the force is applied, as discussed in Section 11.9. In the electric case the vibrating force is provided by the oscillating electric field that is related to the voltage supplied by the generator. For a generator whose voltage is V_{rms}, the resulting current is $I_{rms} = V_{rms}/Z$, Z being the impedance of the circuit. Therefore, as Figure 29.17 illustrates, the current is a maximum when the impedance $Z = \sqrt{R^2 + (X_L - X_C)^2}$ is a minimum, assuming a given generator voltage. The minimum impedance occurs when the frequency is such that $X_L = X_C$, and this frequency f_0 is called the **resonance frequency** of the circuit; thus, $2\pi f_0 L = 1/(2\pi f_0 C)$. Therefore,

$$f_0 = \frac{1}{2\pi \sqrt{LC}} \tag{29.10}$$

The resonance frequency is determined by the inductance and the capacitance, but not the resistance.

Figure 29.18 helps us to understand why there is a resonance frequency for an ac circuit. This drawing presents an analogy between the electrical case (ignoring resistance) and the mechanical case of a mass on a horizontal spring (ignoring friction). Part a shows a fully stretched spring that has just been released, with the speed v of the mass being zero. All the energy is stored in the form of elastic potential energy. When the mass begins to move, it gradually loses potential energy and picks up kinetic energy, until in part b the mass moves with maximum kinetic energy through the position where the spring is unstretched (zero potential energy). Because of its inertia, the moving mass coasts through this position and eventually comes to a halt in part c when the spring is fully compressed and all kinetic energy has been converted back into elastic potential energy. Part d of the picture is just like part b, except the direction of motion is reversed.

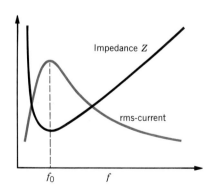

FIGURE 29.17 In a series RCL-circuit the impedance is a minimum, and the current is a maximum, when the frequency f equals the resonance frequency f_0 of the circuit.

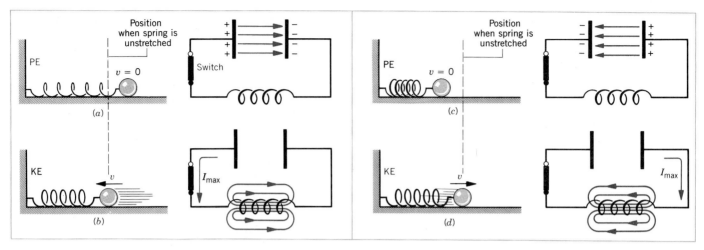

FIGURE 29.18 The oscillation of a mass on a spring is analogous to the oscillation of the electric and magnetic fields that occur, respectively, in a capacitor and in an inductor.

In the electrical case, Figure 29.18a begins with a fully charged capacitor that has just been connected to an inductor by the closing of a switch. At this instant the energy is stored in the electric field between the capacitor plates. As the capacitor discharges, the electric field between the plates decreases, while a magnetic field builds up around the inductor because of the increasing current in the circuit. The maximum current and the maximum magnetic field exist at the instant the capacitor is completely discharged, as in part b of the figure. Energy is now stored entirely in the magnetic field of the inductor. The voltage induced in the inductor keeps the charges flowing until the capacitor again becomes fully charged, but now with reverse polarity, as in part c. Once again, the energy is stored in the electric field between the plates and no energy resides in the magnetic field of the inductor. Part d of the cycle repeats part b, but with reversed directions of current and magnetic field. Thus, we see that an ac circuit can have a resonance frequency, because there is a natural tendency for energy to shuttle back and forth [at frequency $f_0 = 1/(2\pi \sqrt{LC})$] between the electric field of the capacitor and the magnetic field of the inductor.

The effect of resistance on electrical resonance is to make the "sharpness" of the circuit response less pronounced, as Figure 29.19 indicates. When the resistance is small, the current-versus-frequency graph falls off suddenly on either side of the maximum current. When the resistance is large, the falloff is more gradual, and there is less current at the maximum.

The following example deals with one application of resonance in electrical circuits. In this example the focus is on the oscillation of energy between a capacitor and an inductor. Once a capacitor/inductor combination is energized, the energy will oscillate indefinitely as in Figure 29.18, provided there is some provision to replace any dissipative losses that occur because of resistance. Circuits that include this type of provision are called oscillator circuits.

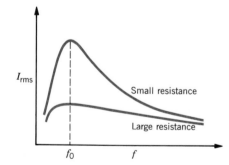

FIGURE 29.19 This graph illustrates the effect that resistance has on the current in a series RCL-circuit.

EXAMPLE 4

Figure 29.20 illustrates a heterodyne metal detector, which utilizes two capacitor/inductor oscillator circuits, A and B. Each circuit produces its own resonance frequency, $f_{0A} =$ $1/(2\pi \sqrt{L_A C})$ and $f_{0B} = 1/(2\pi \sqrt{L_B C})$. Any difference between these two frequencies is detected through earphones as a beat frequency $f_{0B} - f_{0A}$, similar to the beat frequency that two musical tones produce. In the absence of any nearby metal object, the inductances L_A and L_B are the same, and f_{0A} and f_{0B} are identical. There is no beat frequency. When inductor B (the search coil) comes near a piece of metal, the inductance L_B decreases, the corresponding oscillator frequency f_{0B} increases, and a beat frequency is heard. Suppose that initially each inductor is adjusted so $L_B = L_A$, and each oscillator has a resonance frequency of 855.5 kHz. Assuming the inductance of search coil B decreases by 1.00% due to a nearby piece of metal, determine the beat frequency heard through the earphones.

SOLUTION

The solution to this problem depends on finding the amount by which the resonance frequency f_{0B} changes because of a 1.00% change in the inductance L_B. The ratio of f_{0B} to f_{0A} is

$$\frac{f_{0B}}{f_{0A}} = \frac{\dfrac{1}{2\pi\sqrt{L_B C}}}{\dfrac{1}{2\pi\sqrt{L_A C}}} = \sqrt{\frac{L_A}{L_B}}$$

But due to the nearby piece of metal $L_B = 0.9900 L_A$, so that

$$\frac{f_{0B}}{f_{0A}} = \sqrt{\frac{L_A}{0.9900 L_A}} = 1.005$$

Therefore, the new value for f_{0B} is $f_{0B} = 1.005 f_{0A} = 1.005 \times (855.5 \text{ kHz}) = 859.8 \text{ kHz}$. As a result, the detected beat frequency is

$$f_{0B} - f_{0A} = 859.8 \text{ kHz} - 855.5 \text{ kHz} = \boxed{4.3 \text{ kHz}}$$

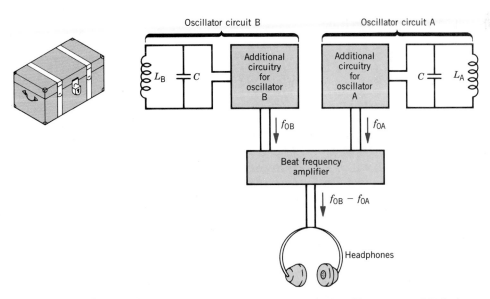

FIGURE 29.20 A heterodyne metal detector uses two electrical oscillators, A and B, in its operation. When the resonance frequency of oscillator B is changed due to the proximity of a piece of metal, a beat frequency, whose value is $f_{0B} - f_{0A}$, is heard in the headphones.

29.6 SEMICONDUCTOR DEVICES

Semiconductor devices are widely used in modern electronics. This section discusses two of these important devices, the diode and the transistor, and Figure 29.21 illustrates one way in which they are used. The drawing shows an audio system in which small ac voltages (originating in a compact disc player, turntable, etc.) are amplified so they can drive the speaker(s). The electric circuits that accomplish the amplification depend on the power provided by the power supply, which is simply a battery in portable units. In nonportable units, however, the power supply is a separate electric circuit containing diodes, along with other elements. As we will see, the diodes convert the 60-Hz ac voltage present at a wall outlet into the dc voltage needed by the amplifier, which, in turn, performs its job of amplification with the aid of transistors.

n-TYPE AND p-TYPE SEMICONDUCTORS

The materials used in diodes and transistors are semiconductors, such as silicon and germanium. However, they are not pure materials, because small amounts of "impurity" atoms (about one part in a million) have been added to them. For instance, Figure 29.22a shows an array of atoms that symbolizes the crystal structure in pure silicon. Each silicon atom has four outer-shell* electrons, and each electron participates with electrons from neighboring atoms in forming the bonds that hold the crystal together. Since they participate in forming bonds, these electrons generally do not move throughout the crystal. Consequently, pure silicon and germanium are poor conductors of electricity. It is possible, however, to make them conductors by adding tiny amounts of impurity atoms, such as phosphorus or arsenic, whose atoms have five outer-shell electrons. For example, when a phosphorus atom replaces a silicon atom in the crystal, only four of the five outer-shell electrons of phosphorus fit into the crystal structure. The extra fifth electron does not fit in and is relatively free to diffuse throughout the crystal, as part *b* of the drawing suggests. A semiconductor

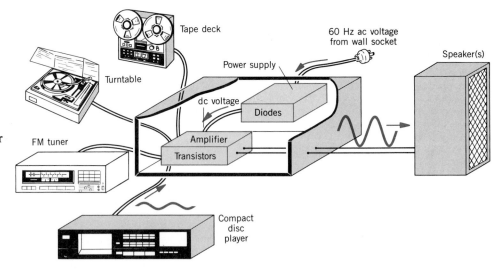

FIGURE 29.21 In a typical audio system, diodes are used in the power supply to create a dc voltage from the ac voltage present at the wall socket. This dc voltage is necessary so the transistors in the amplifier can perform their task of enlarging the small ac voltages originating in the compact disc player, the turntable, etc.

* Section 37.6 discusses the electronic structure of the atom in terms of "shells."

containing small amounts of phosphorus can, therefore, be envisioned as containing immobile, positively charged phosphorus atoms and a pool of electrons that are free to wander throughout the material. These mobile electrons allow the semiconductor to conduct electricity and are analogous to the free electrons that give a metal like copper its ability to conduct electricity.

The process of adding impurity atoms is called *doping.* A semiconductor doped with an impurity that contributes mobile electrons is called an ***n-type semiconductor,*** since the mobile charge carriers have a **n**egative charge. Note that an *n*-type semiconductor is overall electrically neutral, since it contains equal numbers of positive and negative charges.

It is also possible to dope a silicon crystal with an impurity whose atoms have only three outer-shell electrons (e.g., boron or gallium). Because of the missing fourth electron, there is a "hole" in the lattice structure at the boron atom, as part *c* of Figure 29.22 illustrates. An electron from a neighboring silicon atom can jump into this hole, in which event the region around the boron atom, having acquired the electron, becomes negatively charged. Of course, when a nearby electron does jump, it leaves behind a hole. This hole is positively charged, since it results from the removal of an electron from the vicinity of a neutral silicon atom. The vast majority of atoms in the lattice are silicon, so the hole is almost always next to another silicon atom. Consequently, an electron from one of these adjacent atoms can jump into the hole, with the result that the hole moves to yet another location. In this fashion, a positively charged hole can wander through the crystal. This type of semiconductor can, therefore, be viewed as containing immobile, negatively charged boron atoms and an equal number of positively charged, mobile holes. Because of the mobile holes, the semiconductor can conduct electricity. In this case the charge carriers are positive, as can be verified by measuring the Hall emf (see Section 27.4). A semiconductor doped with an impurity that introduces mobile positive holes is called a ***p-type semiconductor.***

THE SEMICONDUCTOR DIODE

A ***p-n junction diode*** is a device formed from a *p*-type semiconductor and an *n*-type semiconductor. The *p-n* junction between the two materials is of fundamental importance to the operation of diodes and transistors. Figure 29.23*a* shows separate *p*-type and *n*-type semiconductors, each electrically neutral, while part *b* of the drawing shows them joined together to form a diode. Electrons from the *n*-type semiconductor and holes from the *p*-type semiconductor flow across the junction and combine. This process leaves the *n*-type material with a positive charge layer and the *p*-type material with a negative charge layer, as part *c* of the drawing indicates. The positive and negative charge layers on the two sides of the junction set up an electric field **E**, much like that in a parallel plate capacitor. This electric field tends to prevent any further movement of charge across the junction, and all charge flow quickly stops.

Suppose now that a battery is connected across the *p-n* junction, as in Figure 29.24*a*, where the negative terminal of the battery is attached to the *n*-material, and the positive terminal is attached to the *p*-material. In this situation the junction is said to be in a condition of ***forward bias,*** and as will now be seen, a current exists in the circuit. The mobile electrons in the *n*-material are repelled by the negative terminal of the battery and move toward the junction. Likewise, the positive holes in the *p*-material are repelled by the positive terminal of the battery and also move toward the junction. At the junction the electrons fill the holes. In the meantime, the negative

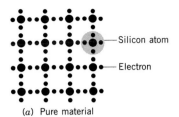
(a) Pure material

— Silicon atom

— Electron

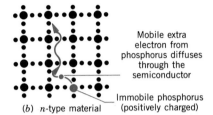
(b) *n*-type material

Mobile extra electron from phosphorus diffuses through the semiconductor

Immobile phosphorus (positively charged)

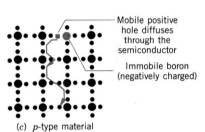
(c) *p*-type material

Mobile positive hole diffuses through the semiconductor

Immobile boron (negatively charged)

FIGURE 29.22 The drawing illustrates a silicon crystal that is (*a*) undoped or pure, (*b*) doped with phosphorus to produce an *n*-type material, and (*c*) doped with boron to produce a *p*-type material.

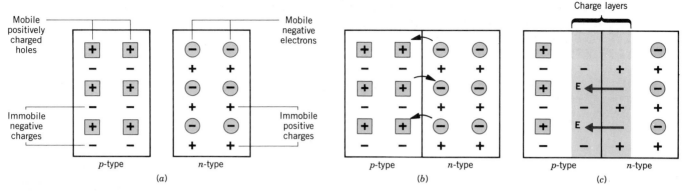

FIGURE 29.23 (*a*) An *n*-type semiconductor contains mobile electrons and immobile, positive charges at the impurity atoms (phosphorus). A *p*-type semiconductor contains mobile, positive holes and immobile, negative charges at the impurity atoms (boron). (*b*) At the junction between *n*- and *p*-materials, electrons and holes combine and (*c*) create positive and negative charge layers. The electric field produced by the charge layers is **E**.

terminal of the battery provides a fresh supply of electrons to the *n*-material, and the positive terminal pulls off electrons from the *p*-material, forming new holes in the process. Consequently, a continual flow of charge, and hence a current, is maintained in the circuit.

In Figure 29.24*b* the battery polarity has been reversed, and the *p-n* junction is in a condition known as **reverse bias.** The battery forces electrons in the *n*-material and holes in the *p*-material away from the junction. As a result, the potential across the junction builds up until it opposes the battery potential, and very little current can be sustained through the diode. The diode, then, is a unidirectional device in the sense that it allows current to pass only in one direction.

The graph in Figure 29.25 shows the dependence of the current on the magnitude and polarity of the voltage applied across a *p-n* junction diode. The exact values of the current depend on the nature of the semiconductor and the extent of the doping. Also shown in the drawing is the symbol used for a diode (—▶︎|—). The direction of the arrowhead in the symbol indicates the direction of the conventional current in the diode under a forward bias condition.

Because diodes are unidirectional devices, they are commonly used in **rectifier circuits,** which convert an ac voltage into a dc voltage. For instance, Figure 29.26 shows a circuit in which charges flow through the resistance *R* only while the ac generator biases the diode in the forward direction. Since current occurs only during one-half of every generator voltage cycle, the circuit is called a half-wave rectifier. A plot of the output voltage applied to the resistor reveals that only the positive halves of each cycle are present. If a capacitor is added across the resistor, as indicated in the drawing, the capacitor charges up and keeps the voltage from dropping to zero between each positive half-cycle. It is also possible to construct full-wave rectifier

FIGURE 29.24 (*a*) There is an appreciable current through the diode when the diode is forward biased. (*b*) Under a reverse bias condition, there is almost no current through the diode.

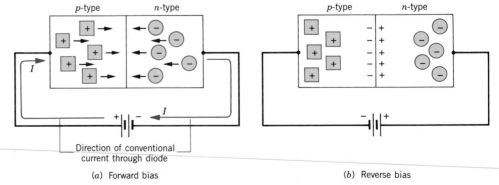

circuits, in which both halves of every cycle of the generator voltage drive current through the load resistor in the same direction (see question 11).

When a circuit such as that in Figure 29.26 includes a capacitor and also a transformer to establish the desired voltage level, the circuit is called a power supply. In the hi-fi system in Figure 29.21, the power supply receives the 60-Hz ac voltage from a wall socket and produces a dc output voltage that is used for the transistors within the amplifier. Power supplies using diodes are also found in virtually all electronic appliances, such as televisions and microwave ovens.

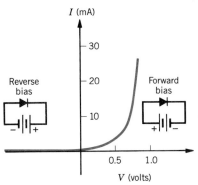

FIGURE 29.25 The current-versus-voltage characteristics of a typical *p-n* junction diode. The polarity of the battery in the forward bias condition is opposite to the polarity in the reverse bias condition.

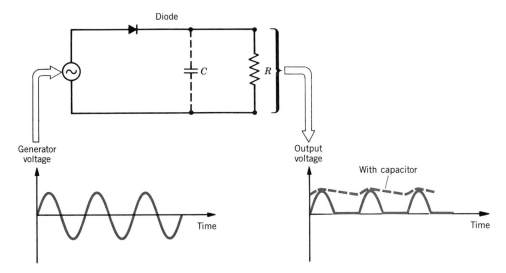

FIGURE 29.26 A half-wave rectifier circuit, together with a capacitor, constitutes a dc power supply, because the rectifier converts an ac voltage into a dc voltage.

TRANSISTORS

A number of different kinds of transistors are in use today. One common type is the *bipolar junction transistor,* which consists of two *p-n* junctions formed by three layers of doped semiconductors. As Figure 29.27 indicates, there are *pnp* and *npn* transistors. In either case, the middle region is made very thin compared to the outer regions.

A transistor is useful because it is capable of amplifying a smaller voltage into one that is much greater. A transistor plays the same kind of role in an amplifier circuit that a valve does when it controls the flow of water through a pipe. A small change in the valve setting produces a large change in the amount of water per second that flows through the pipe. In other words, a small change in the voltage applied as input to a transistor produces a large change in the output from the transistor.

Figure 29.28 shows a *pnp* transistor connected to two batteries, labeled V_E and V_C. The voltage V_E is applied in such a way that the *p-n* junction on the left has a forward bias, while the *p-n* junction on the right has a reverse bias. Moreover, the voltage V_C is usually much larger than V_E for a reason to be discussed shortly. The drawing also shows the standard symbol and nomenclature for the three sections of the transistor, namely, the *emitter,* the *base,* and the *collector.* The arrowhead in the symbol points in the direction of the conventional current through the emitter.

The positive terminal of V_E pushes the mobile positive holes in the *p*-type material of the emitter toward the emitter/base junction. And since this junction has a forward bias, the holes enter the base region readily. Once in the base region, the holes come under the strong influence of V_C and are attracted to its negative terminal. Since the base is so thin (about 10^{-6} m or so), approximately 98% of the holes are drawn

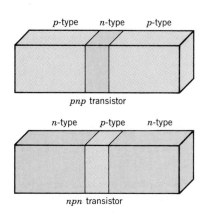

FIGURE 29.27 There are two kinds of bipolar junction transistors, *pnp* and *npn.*

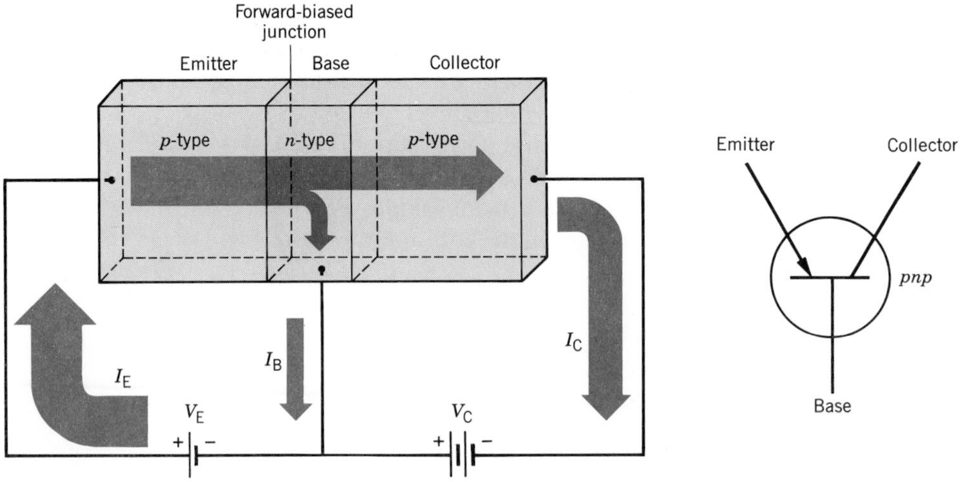

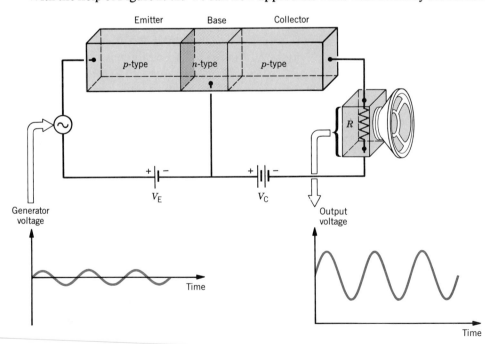

FIGURE 29.28 This drawing shows a *pnp* transistor, along with its bias voltages V_E and V_C. On the symbol for the *pnp* transistor, the emitter is marked with an arrowhead that denotes the direction of conventional current through the emitter.

directly through the base and on into the collector. The remaining 2% of the holes combine with free electrons in the base region, thereby giving rise to a small base current I_B. As the drawing shows, the moving holes in the emitter and collector constitute currents that are labeled I_E and I_C, respectively. From Kirchhoff's junction rule it follows that $I_C = I_E - I_B$.

Because the base current I_B is small, the collector current is determined primarily by current from the emitter ($I_C = I_E - I_B \approx I_E$). This means that a change in I_E will cause a change in I_C of nearly the same amount. Furthermore, a substantial change in I_E can be caused by only a small change in the forward bias voltage V_E. To see that this is the case, look back at Figure 29.25 and notice how steep the current-versus-voltage curve is for a *p-n* junction; small changes in the forward bias voltage give rise to large changes in the current.

With the help of Figure 29.29 we can now appreciate what was meant by the earlier

FIGURE 29.29 The basic *pnp* transistor amplifier in this drawing amplifies a small generator voltage to produce an enlarged voltage across the resistance R.

statement that a small change in the voltage applied as input to a transistor leads to a large change in the output. This picture shows an ac generator connected in series with the battery V_E, and a resistance R connected in series with the collector. The generator voltage could originate from an electric guitar pickup or the phono cartridge of a turntable, while the resistance R could represent a loudspeaker. The generator introduces small voltage changes in the forward bias across the emitter/base junction and, thus, causes large corresponding changes in the current I_C leaving the collector and passing through the resistance R. As a result, the output voltage across R is an enlarged or amplified version of the input voltage of the generator.

The operation of an *npn* transistor is similar to that of a *pnp* transistor. The main difference is that the bias voltages (and current directions) are reversed, as Figure 29.30 indicates.

It is important to realize that the increased power available at the output of a transistor amplifier does *not* come from the transistor itself. Rather, it comes from the power provided by the voltage source V_C. The transistor, acting like an automatic valve, merely allows the small, weak signals from the input generator to control the power taken from the voltage source V_C and delivered to the resistance R.

Today it is possible to combine arrays of thousands of transistors, diodes, resistors, and capacitors on a tiny "chip" of silicon that usually measures less than a centimeter on a side. These arrays are called integrated circuits (ICs), and they can be designed to perform almost any desired electronic function. Integrated circuits have revolutionized the electronics industry and now lie at the heart of computers, hand-held calculators, digital watches, and programmable appliances.

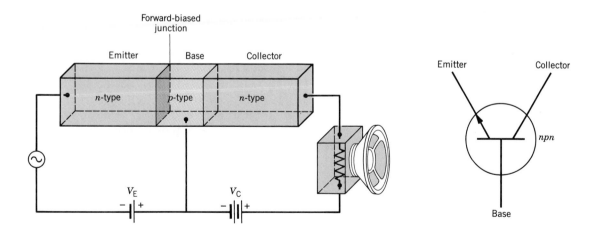

FIGURE 29.30 An *npn* transistor amplifier can also be used to convert a small input voltage from a generator into a large output voltage across the resistance R. On the symbol for the *npn* transistor, the emitter is marked with an arrowhead to denote the direction of conventional current through the emitter.

This dime-sized integrated circuit is a computer microprocessor that contains about 150 000 transistors.

SUMMARY

In an ac circuit the rms-voltage across a capacitor is related to the rms-current according to $V_{rms} = I_{rms}X_C$, where X_C is the **capacitive reactance.** The capacitive reactance is measured in ohms and is given by $X_C = 1/(2\pi f C)$ for a capacitance C and a frequency f. The current in a capacitor leads the voltage across the capacitor by a phase angle of 90°, and as a result, a capacitor consumes no power, on the average.

For an inductor the rms-voltage and the rms-current are related by $V_{rms} = I_{rms}X_L$, where X_L is the **inductive reactance.** For an inductance L and frequency f the inductive reactance is given in ohms as $X_L = 2\pi f L$. The ac current in an inductor lags behind the voltage by a phase angle of 90°. Consequently, an inductor, like a capacitor, consumes no power, on the average.

When a resistor, a capacitor, and an inductor are connected in series, the rms-voltage across the combination is related to the rms-current according to $V_{rms} = I_{rms}Z$, where Z is the **impedance** of the combination. The impedance (in ohms) for the series combination is $Z = \sqrt{R^2 + (X_L - X_C)^2}$. The phase angle ϕ between current and voltage for a series RCL-combination is given by $\tan \phi =$ $(X_L - X_C)/R$. Only the resistor in the combination dissipates power on the average, according to the relation $P = I_{rms}V_{rms} \cos \phi$. The term $\cos \phi$ is the **power factor** of the circuit.

A series RCL-circuit exhibits the phenomenon of resonance. The **resonance frequency** f_0 of the circuit is $f_0 = 1/(2\pi \sqrt{LC})$. At resonance the impedance of the circuit has a minimum value equal to the resistance R, and the rms-current has a maximum value.

The two basic types of semiconductor materials are called n-type and p-type. In an **n-type material,** mobile negative electrons carry the current. An n-type material is produced by doping a semiconductor such as silicon with a small amount of impurity such as phosphorus. In a **p-type material,** mobile positive holes in the crystal structure carry the current. A p-type material is made by doping a semiconductor with an impurity such as boron. These two types of semiconductor materials are used in the **p-n junction diode** and in **pnp and npn bipolar junction transistors.**

QUESTIONS

1. A light bulb is connected directly to the 60-Hz ac voltage present at a wall outlet. Suppose a parallel plate capacitor (without a dielectric between the plates) is inserted in series between the light bulb and the wall outlet. (a) Describe what would happen to the brightness of the bulb. (b) Suppose the capacitor is replaced with another capacitor, one that is identical, except the space between the plates is filled with a dielectric material. Describe what now happens to the brightness. In both (a) and (b) explain your reasoning.

2. The ends of a long, straight wire are connected to the terminals of an ac generator, and the current is measured. The wire is then disconnected, wound into the shape of a multiple-turn coil, and reconnected to the generator. In which case does the generator deliver a larger current? Explain.

3. An air-core inductor is connected in series with a light bulb and this circuit is plugged into an electrical outlet. What happens to the brightness of the bulb when a piece of iron is inserted inside the inductor? Give a reason for your answer.

4. The drawing shows three circuits. The generators in each are identical. (a) In the limit of very small generator frequencies, rank the circuits in ascending order, according to the amount of current delivered by the generator (i.e., smallest current first). (b) Repeat part (a) for the limit of very large frequencies. Explain your reasoning in each case.

I
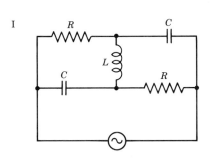

II
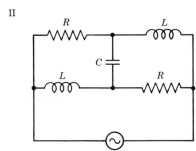

III
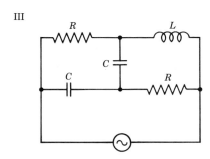

5. A light bulb is connected to an ac generator. When an inductor is added in series with the bulb, the brightness decreases, no matter what the value of the inductance is. However, if a capacitor is now added in series with the bulb and the inductor, the brightness may either increase or decrease, depending on the value of the capacitance. Explain these observations.

6. In a series circuit a resistor and an inductor are connected to a generator whose rms-voltage is 160 V. It is determined that the rms-voltage across the resistor is 110 V, while the rms-voltage across the inductor is also 110 V. Notice that 110 V + 110 V is greater than the 160 V provided by the generator. Does this situation involving rms-voltages violate Kirchhoff's loop rule? Explain.

7. An inductor and a capacitor are connected in parallel across the terminals of a generator. What happens to the current from the generator as the frequency becomes (a) very large and (b) very small?

8. Is it possible for two series RCL-circuits to have the same resonance frequencies and yet have (a) different R values and (b) different C and L values? Justify your answers.

9. Both *p-n* junction diodes and bipolar junction transistors utilize the junctions between *p*-type and *n*-type semiconductors. A transistor can be used as an amplifying device. Can a diode also be used in this fashion? Explain.

10. It is possible to describe the action of a diode in terms of resistance. Qualitatively, how does the resistance of a forward-biased diode at point A in the drawing compare to the resistance of a reverse-biased diode at point B?

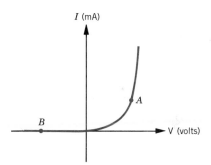

11. The drawing shows a full-wave rectifier circuit (a so-called bridge-rectifier) that uses four diodes. The direction of the current in the load resistance R is the *same* for both positive and negative halves of the voltage cycle of the generator. (a) When the generator causes the potential at A to be positive relative to that at B, through which two diodes do charges pass, and what is the direction of the conventional current in the load resistance R? (b) Repeat part (a) when the potential at B becomes positive relative to that at A.

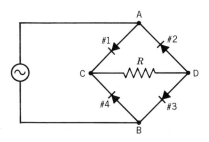

PROBLEMS

Note: For problems in this set, the ac current and voltage are rms-values and the power is an average value, unless indicated otherwise.

Section 29.1 Capacitors and Capacitive Reactance

1. At what frequency does a 7.50-μF capacitor have a reactance of 168 Ω?

2. What voltage is needed to create a current of 29.0 mA in a circuit containing only a 0.565-μF capacitor, when the frequency is 2.60 kHz?

3. Three capacitors are connected in parallel across the terminals of a 440-Hz generator. The capacitances are 2.0, 4.0, and 7.0 μF. (a) Find the equivalent capacitance of these capacitors. (b) If the generator supplies a total current of 0.62 A, what is the voltage of the generator? (c) What is the current supplied to each of the three capacitors?

4. A capacitor is attached to a 5.00-Hz generator. The current is observed to reach a maximum value at a certain time. What is the least amount of time that passes before the voltage across the capacitor reaches its maximum value?

* 5. A capacitor consists of two square metal plates that are parallel and are one centimeter on a side. This capacitor is connected to a generator that has a frequency of 11 kHz and a voltage of 150 V. The current in the circuit is measured to be 9.4 μA. Assuming air exists between the plates, determine the distance between them.

Section 29.2 Inductors and Inductive Reactance

6. What is the inductance of an inductor that has a reactance of 1.8 kΩ at a frequency of 4.2 kHz?

7. The transformer for an electric toy train has a primary winding whose inductance is 2.4 H. A voltage of 120 V (frequency = 60.0 Hz) is applied to the primary coil when the transformer is plugged into an electrical outlet in the United States. In Europe the corresponding values are 240 V and 50.0 Hz. Assuming the train is not connected to the secondary of the transformer, find the current in the primary in (a) the United States and (b) Europe.

* 8. A 0.313-H and a 0.127-H inductor are connected in series across the terminals of a generator. A 0.508-H inductor is also connected across the terminals of the generator. The generator has a voltage of 9.00 V and a frequency of 266 Hz. What is the total current that the generator delivers?

* 9. A 47-mH inductor and a 75-mH inductor are wired in parallel across the terminals of a generator that has a voltage of 2.1 V and supplies a current of 23 mA. Find the frequency of the generator.

* 10. The current in a solenoid is 36 mA when the solenoid is connected to an 18-kHz generator. The solenoid has a cross-sectional area of 3.1×10^{-5} m² and a length of 2.5 cm. The solenoid has 135 turns. Determine the *peak voltage* of the generator.

Section 29.3 The Series RCL-Circuit, Section 29.4 The Role of Capacitors and Inductors in the Design of Hi-Fi Loudspeakers

11. A 2700-Ω resistor and a 1.1-μF capacitor are connected in series across a generator (60.0 Hz, 120 V). Determine the power dissipated in the circuit.

12. The purpose of this problem is to verify the shapes of the graphs of capacitive and inductive reactance versus frequency, which are shown in Figures 29.2 and 29.6. Plot these graphs for a 20.0-μF capacitor and a 5.00-mH inductor. Use a frequency of 10 Hz and ten equally spaced frequencies between 100 and 1000 Hz.

13. An ac generator has a frequency of 5.60 kHz and produces a current of 53.0 mA in a series circuit that contains a 218-Ω resistor and a 0.100-μF capacitor. Obtain (a) the voltage of the generator and (b) the phase angle between the current and the voltage.

14. A series circuit has an impedance of 192 Ω, and the current leads the voltage by 75.0°. The circuit contains two different elements. (a) From the phase angle between current and voltage, decide which elements are present, R and C, R and L, or C and L. (b) Find values for the appropriate quantities, R and X_C, R and X_L, or X_C and X_L.

15. A circuit consists of a 215-Ω resistor and a 0.200-H inductor. These two elements are connected in series across a generator that has a frequency of 106 Hz and a voltage of 234 V. (a) What is the current in the circuit? (b) Determine the phase angle between the current and the voltage.

16. Figure 29.14 shows the part of a crossover network that involves the woofer in a hi-fi loudspeaker. Suppose the inductor has an inductance of 2.50 mH and the resistor has a resistance of 8.00 Ω. A generator that has a voltage of 8.00 V is connected to this inductor/resistor combination. Obtain the current in the woofer when the frequency has the following values: 100.0, 300.0, 500.0, 700.0, 1000.0, and 2000.0 Hz. Plot a graph of current versus frequency.

17. In a crossover network a tweeter is wired in series with a capacitor, as Figure 29.15 shows. In one case, the capacitance is 40.0 μF, and the tweeter resistance is 8.00 Ω. (a) Assume the generator has a voltage of 8.00 V and find the current in the tweeter for the following frequencies: 100.0, 300.0, 500.0, 700.0, 1000.0, and 2000.0 Hz. Plot a graph of current versus frequency. (b) Compare the results here with those in problem 16 and estimate the crossover frequency from the graphs.

* 18. For the circuit shown in the drawing, find the current provided by the generator when the frequency is (a) very large and (b) very small.

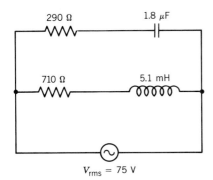

290 Ω 1.8 μF

710 Ω 5.1 mH

V_{rms} = 75 V

* 19. In reality, there is some resistance R in the wire from which an inductor is made. Therefore, an actual inductor should be represented as a resistor in series with an ideal (resistanceless) inductor. With this in mind, suppose the current in a 2.8-mH inductor is I_0 when the inductor is connected to a 12-V battery. However, when the battery is replaced with a 1500-Hz generator whose voltage is 12 V, the current is $I_0/3$. What is the resistance R of the wire?

* 20. The graph in Figure 29.14 shows a plot of current versus frequency for a series circuit containing a resistor and an inductor. The plot applies for a fixed voltage from a generator. If $R = 16$ Ω and $L = 4.0$ mH, find the frequency at which the current drops to one-half of its value at zero frequency.

**21. When a resistor is connected by itself to an ac generator, the average power dissipated in the resistor is 1.000 W. When a capacitor is added in series with the resistor, the power dissipated is 0.500 W. When an inductor is added in series with the resistor (without the capacitor), the power dissipated is 0.250 W. Determine the power dissipated when both the capacitor and the inductor are added in series with the resistor.

Section 29.5 Resonance in Electric Circuits

22. A series RCL-circuit has a capacitance of 1.20 μF and an inductance of 2.00 mH. What is the resonance frequency of the circuit?

23. A series RCL-circuit has a resonance frequency of 690 kHz. If the value of the inductance is 26 μH, what is the value of the capacitance?

24. The resistor in a series RCL-circuit has a resistance of 92 Ω, while the voltage of the generator is 3.0 V. At resonance, what is the average power dissipated in the circuit?

25. A series RCL-circuit is at resonance and contains a variable resistor that is set to 175 Ω. The power dissipated in the circuit is 2.6 W. How much power is dissipated when the variable resistor is set to 562 Ω?

* 26. The ratio of the inductive reactance to the capacitive reactance is observed to be 5.36 in a series RCL-circuit. The resonance frequency of the circuit is 225 Hz. What is the frequency of the generator that is connected to the circuit?

* 27. The power dissipated in a series RCL-circuit is 65.0 W, and the current is 0.530 A. The circuit is at resonance. Determine the *peak voltage* of the generator.

* 28. Suppose you have a number of capacitors. Each of these capacitors is identical to the capacitor that is already in a series RCL-circuit. How many of these additional capacitors must be inserted in series in the circuit, so the resonance frequency triples?

* 29. The resonance frequency of a series RCL-circuit that contains an 18 μH inductor is 13 MHz. The capacitor is an empty parallel plate capacitor with a plate separation of 1.5 mm. (a) Find the plate area. (b) What is the resonance frequency when the capacitor is filled with a material whose dielectric constant is 5.2?

**30. In a series RCL-circuit the dissipated power drops by a factor of two when the frequency of the generator is changed from the resonance frequency to a nonresonance frequency. The peak voltage is held constant while this change is made. Determine the power factor of the circuit at the nonresonance frequency.

**31. When the frequency is twice the resonance frequency, the impedance of a series RCL-circuit is twice the value of the impedance at resonance. Obtain the ratios of the inductive and capacitive reactances to the resistance, that is, obtain X_L/R and X_C/R.

ADDITIONAL PROBLEMS

32. In a series circuit, a generator (1350 Hz, 15.0 V) is connected to a 16.0-Ω resistor, a 4.10-μF capacitor, and a 5.30-mH inductor. Determine (a) the voltage across each element in the circuit and (b) the power dissipated in the circuit.

33. A circuit consists of a 3.00-μF and a 6.00-μF capacitor connected in series across the terminals of a 510-Hz generator. The voltage of the generator is 120 V. (a) Determine the equivalent capacitance of the two capacitors. (b) Find the current in the circuit. (c) What is the voltage across each capacitor?

34. A simple metal detector consists of a series circuit formed by a 1.70-mH inductor, a 3.00-μF capacitor, and a generator with a voltage of 9.00 V. The inductor has the shape of a large coil (see the drawing). The resistance of the wire in the coil is 3.50 Ω. The circuit is set up to be at resonance when there is no metal passing through the coil. When a person with a metal object walks through the coil, the inductance increases and, consequently, the current in the circuit changes. The change in current can be used to sound a warning. Determine the current in the circuit (a) when no metal is present and (b) when a metal object causes a 4.0% increase in the inductance. (c) Find the change in current and specify whether it is an increase or a decrease.

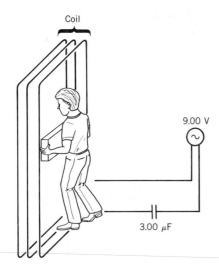

Coil

9.00 V

3.00 μF

case interpret the algebraic sign of the phase angle, remembering that a positive sign means that the voltage "leads" the current and the circuit is more inductive than capacitive.

**38. The drawing shows the current-versus-frequency graph for a series RCL-circuit. The drawing also shows f_1 and f_2, the two frequencies at which the current has a value that is one-half of the maximum value. The width of the graph is indicated by the quantity $f_2 - f_1$. Derive an expression for this quantity (sometimes called the full width at half the maximum height) in terms of R and L.

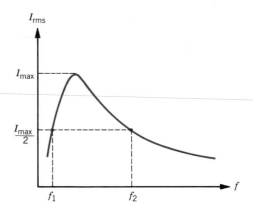

35. An 8.2-mH inductor is connected to an ac generator (10.0 V, 620 Hz). Determine the *peak value* of the current supplied by the generator.

* 36. A series RCL-circuit contains a 5.10-μF capacitor and a generator whose voltage is 11.0 V. At a resonance frequency of 1.30 kHz the power dissipated in the circuit is 25.0 W. Find the values of (a) the inductance and (b) the resistance. (c) Calculate the power factor when the generator frequency is 2.31 kHz.

* 37. The elements in a series RCL-circuit are a 106-Ω resistor, a 3.30-μF capacitor, and a 31.0-mH inductor. What is the impedance of the circuit and the phase angle between the current and the voltage when the frequency is (a) 215 Hz and (b) 609 Hz? In each

**39. A 108-Ω resistor, a 0.200-μF capacitor, and a 5.42-mH inductor are connected in series to a generator whose voltage is 26.0 V. The current in the circuit is 0.141 A. Because of the shape of the current-versus-frequency graph (see Figure 29.17), there are two possible values for the frequency that correspond to this current. Obtain these two values.

CHAPTER 30

Electromagnetic Waves

Some of the 27 dishes that comprise the Very Large Array (VLA) radio telescope near Socorro, NM are shown here. The VLA receives radio waves, a type of electromagnetic waves, from distant stars or galaxies and is the largest effective radio telescope in the world. Other important types of electromagnetic waves discussed in the text are infrared waves, light waves, X rays, and gamma rays.

30.1 THE NATURE OF ELECTROMAGNETIC WAVES

It was the great Scottish physicist James Clerk Maxwell (1831–1879) who predicted that electric and magnetic fields fluctuating together can form a propagating wave, appropriately called an ***electromagnetic wave.*** Visible light is just such a wave. Other familiar electromagnetic waves are radio waves, microwaves, and X rays. This chapter is devoted to explaining the general properties of electromagnetic waves.

Figure 30.1 illustrates one way to create an electromagnetic wave. The setup consists of two straight copper wires that are connected to the terminals of an ac generator and serve as an antenna. The potential difference between the terminals changes sinusoidally with time and has a period T and a frequency $f = 1/T$. Part a shows the moment when there is no charge at the ends of either wire. Since there is no charge, there is no electric field at the point P just to the right of the antenna. Part b shows the situation a little while later at time $t = \frac{1}{8}T$, when the generator has gone through one-eighth of its cycle. Now the bottom wire is negatively charged and the top wire is positively charged, with the result that the electric field at P (drawn as a

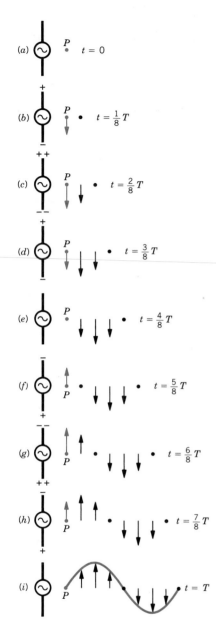

FIGURE 30.1 The ac generator causes the positive and negative charges on the antenna wires to change from moment to moment. In each part of the drawing, the colored arrow represents the electric field produced at point P by the charges on the antenna at the indicated time. The black arrows represent the electric fields created at earlier times. All the fields propagate to the right. Fields also are produced and propagate in other directions, but these fields are omitted for simplicity.

colored arrow) points downward.* As time passes, the charges on the wires increase until they attain their maximum values at a time $t = \frac{2}{8}T$, as in part c. Correspondingly, the electric field at point P has increased to its maximum strength in the downward direction. Part c also indicates that the electric field created in part b has not disappeared, but has moved to the right. Here lies the crux of the matter. At points far removed from the wire, the effect of the charges in creating the electric field is not felt immediately. Instead, the field is created first near the wires and then, like the effect of a pebble dropped into a pond, the field moves outward as a wave in all directions. Only the field moving to the right is shown in the picture for the sake of clarity. Eventually, after a time determined by the speed at which the wave travels, the field reaches distant points.

Parts d–i of Figure 30.1 show the creation of the electric field at point P (colored arrow) at later times during the generator cycle. In each part, the fields produced earlier in the sequence (black arrows) continue propagating toward the right. Parts f–h show the charges on the wires when the polarity of the generator has reversed, so the top wire is negative and the bottom wire is positive. As a result, the electric field at P has reversed its direction and points upward. In part i of the sequence, a complete sine wave has been drawn through the tips of the electric field vectors to emphasize that the field changes sinusoidally.

So far, our focus has been on the electric field that is created by the charges on the antenna wires. However, a magnetic field is also created. After all, the charges flowing in the antenna constitute an electric current, and an electric current creates a magnetic field. Figure 30.2 illustrates the magnetic field direction at point P at the instant when the current in the antenna wire is upward. With the aid of Right-Hand Rule No. 2 (thumb of right hand points along I, fingers curl in the direction of $\mathbf{B}$), the magnetic field at P can be seen to point perpendicularly into the page. As the oscillating current changes in magnitude and direction, the magnetic field at P changes accordingly. As with electric fields, however, the magnetic fields created at earlier times propagate outward as a wave. Moreover, (1) the magnetic field is zero when the electric field has its maximum positive or negative value and (2) the magnetic field has its maximum positive or negative value when the electric field is zero. In other words, the two fields are 90° out of phase with one another.

It is important to notice that the magnetic field in Figure 30.2 is perpendicular to the page, whereas the electric field in Figure 30.1 lies in the plane of the page. Thus, the electric and magnetic fields created by the antenna wires are mutually perpendicular, and they remain so as they move away from the antenna. Moreover, both fields are perpendicular to the direction of travel. These perpendicular electric and magnetic fields, moving together, constitute an electromagnetic wave.

The electric and magnetic fields illustrated in Figures 30.1 and 30.2 decrease to zero rapidly with increasing distance from the antenna. Therefore, they exist mainly near the antenna and together are called the **near field.** Electric and magnetic fields do form an electromagnetic wave at large distances from the antenna, however. These fields arise from an effect that is different than that which produces the near field and are referred to as the **radiation field.** Faraday's law of electromagnetic induction provides part of the basis for the radiation field. As Section 28.4 discusses, this law describes the emf or potential difference produced by a changing magnetic field. And, as Section 24.4 explains, a potential difference can be related to an electric field. Thus, a changing magnetic field produces an electric field. Maxwell predicted that the reverse effect also occurs; namely, that a changing electric field produces a magnetic

* The direction of the electric field can be obtained by imagining a positive test charge at P and determining the direction in which it would be pushed because of the charges on the wires.

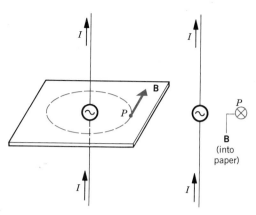

FIGURE 30.2 The ac generator produces an oscillating current I in the antenna wires. At point P the current creates a magnetic field **B** whose direction is given by the tangent to a circle centered on the wire. When the current is upward, the magnetic field at point P is directed into the page, as shown on the right. At a later time, when the current is downward, the field is directed out of the page at point P.

field. The radiation field arises, then, from both effects, the changing magnetic field creating an electric field that fluctuates in time, and the changing electric field creating the magnetic field.

Figure 30.3 shows the electromagnetic wave of the radiation field far from the antenna. The picture shows only the part of the wave traveling along the $+x$ axis. The part traveling in the other directions has been omitted for clarity. Note that for the radiation field, the electric and magnetic parts of the wave reach a maximum together. In other words, the fluctuating electric and magnetic fields are in phase, in contrast to the 90° phase relation for the near field.

It should be clear from Figure 30.3 that *an electromagnetic wave is a transverse wave,* because the electric and magnetic fields are both perpendicular to the direction in which the wave travels. Moreover, this kind of transverse wave, unlike a wave on a string, does not require a medium in which to propagate. *Electromagnetic waves can travel through a vacuum or a material substance,* since electric and magnetic fields can exist in either one.

Electromagnetic waves can be produced in situations that do not involve a wire antenna such as that in Figures 30.1–30.3. In general, any electric charge that is accelerating emits an electromagnetic wave, whether the charge is inside a wire or not. An electron oscillating in simple harmonic motion along the length of a wire is one example of an accelerating charge.

All electromagnetic waves move through a vacuum at the same speed, and a special symbol c is used to denote its value. This speed is called the *speed of light in a*

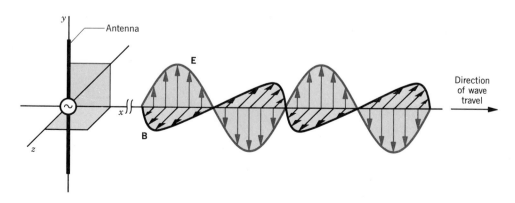

FIGURE 30.3 The ac generator at the origin and the antenna along the y axis generate an electromagnetic wave that travels outward in all directions. For clarity, only the part of the wave traveling along the positive x axis is shown. The electric field **E** points along the y axis, and the magnetic field **B** points along the z axis. The picture shows the wave of the radiation field far away from the antenna. Observe that **E** and **B** are perpendicular to each other, and both are perpendicular to the direction of travel.

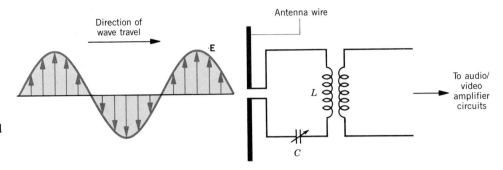

FIGURE 30.4 A radio wave can be detected with a receiving antenna wire that is parallel to the electric field of the wave. The magnetic field of the radio wave has been omitted for simplicity.

vacuum and is $c = 3.00 \times 10^8$ m/s. In air, electromagnetic waves travel at very nearly the same speed as they do in a vacuum, but, in general, they move through a substance such as glass at a speed that is less than c.

The frequency of an electromagnetic wave is determined by the vibration frequency of the electric charges at the source of the wave. In Figures 30.1–30.3 the wave frequency would equal the frequency of the ac generator. Suppose, for example, that the antenna is broadcasting the electromagnetic waves known as radio waves. The frequencies of AM radio waves lie between 545 and 1605 kHz, these numbers corresponding to the limits of the AM broadcast band on the dial. In contrast, the frequencies of FM radio waves lie between 88 and 108 MHz on the dial. Television channels 2–6, on the other hand, utilize electromagnetic waves with frequencies between 54 and 88 MHz, while channels 7–13 use frequencies between 174 and 216 MHz.

Radio and television reception involves a process that is the reverse of that outlined earlier for the creation of electromagnetic waves. When broadcasted waves reach a receiving antenna, they interact with the electric charges in the antenna wires. Either the electric field or the magnetic field of the waves can be used. To take full advantage of the electric field, the wires of the receiving antenna must be parallel to the electric field, as Figure 30.4 indicates. The electric field acts on the electrons in the wire, forcing them to oscillate back and forth along the length of the wire. Consequently, an ac current exists in the antenna and the circuit connected to it. The variable-capacitor C and the inductor L in the circuit provide one way to select the frequency of the desired electromagnetic wave. By adjusting the value of the capacitance, it is possible to adjust the corresponding resonance frequency f_0 of the circuit [$f_0 = 1/(2\pi\sqrt{LC})$, Equation 29.10] to match the frequency of the wave. Under the condition of resonance there will be a maximum oscillating current in the inductor. Because of mutual inductance, this current creates a maximum voltage in the second coil in the drawing, and this voltage can then be amplified and processed by the remaining radio or television circuitry.

To detect the magnetic field of a broadcasted radio wave, a receiving antenna in the form of a loop can be used, as Figure 30.5 shows. For best reception, the normal to

FIGURE 30.5 With a receiving antenna in the form of a loop, the magnetic field of a broadcasted radio wave can be detected. The normal to the plane of the loop should be parallel to the magnetic field for best reception. For clarity, the electric field of the radio wave has been omitted.

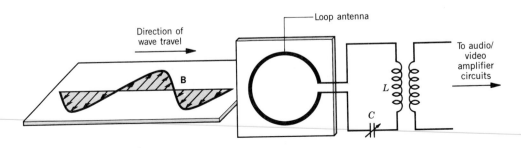

the plane of the wire loop is oriented parallel to the magnetic field. Then, as the wave sweeps by, the magnetic field penetrates the loop, and the changing magnetic flux induces a voltage and a current in the loop, in accord with Faraday's law. Once again, the resonance frequency of a capacitor/inductor combination can be adjusted to match the frequency of the desired electromagnetic wave.

Radio waves are only one part of the broad spectrum of electromagnetic waves that has been discovered. The next section discusses the entire spectrum.

30.2 THE ELECTROMAGNETIC SPECTRUM

An electromagnetic wave, like any wave, has a frequency f and a wavelength λ that are related to the speed v of the wave according to Equation 20.1 as $v = \lambda f$. For electromagnetic waves traveling through a vacuum or, to a good approximation, through air, the speed is $v = c$, so $c = \lambda f$.

As Figure 30.6 shows, electromagnetic waves exist with an enormous range of frequencies, from values less than 10^4 Hz to greater than 10^{22} Hz. The series of electromagnetic waves depicted in the drawing, arranged in order of their frequencies, is called the *electromagnetic spectrum.* Since all these waves travel through a vacuum at the same speed of $c = 3.00 \times 10^8$ m/s, Equation 20.1 can be used to find the correspondingly wide range of wavelengths that the picture also displays. Historically, regions of the electromagnetic spectrum have been given names such as radio waves and infrared waves. Although the boundary between two regions is drawn as a sharp line in Figure 30.6, the boundary is not so well defined in practice, and the regions often overlap.

Beginning on the left in Figure 30.6, we find radio waves. The lower-frequency radio waves are generally produced by electric oscillator circuits, while the higher-frequency radio waves (called microwaves) are usually generated using electron tubes called klystrons. Infrared radiation, sometimes loosely called heat waves, originates with the vibration and rotation of molecules within a material. Visible light is emitted by hot objects, such as the sun, a burning log or the filament of an incandescent light

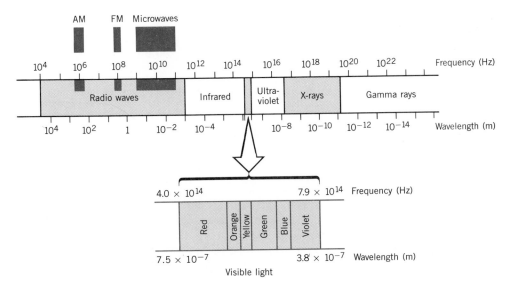

FIGURE 30.6 The electromagnetic spectrum and the common names associated with regions of the spectrum.

bulb, when the temperature is high enough to excite the electrons within an atom. Ultraviolet frequencies can be produced from the discharge of an electric arc. X rays are produced by the sudden deceleration of high-speed electrons. And, finally, gamma rays are radiation from nuclear emissions.

Of all the frequency ranges in the electromagnetic spectrum, the most familiar is that of visible light, although it is the smallest range indicated in Figure 30.6. Only waves with frequencies between about 4.0×10^{14} Hz and 7.9×10^{14} Hz are perceived by the human eye as visible light. Usually visible light is discussed in terms of wavelengths (in vacuum) rather than frequencies. As Example 1 indicates, the wavelengths of visible light are extremely small and, therefore, they are normally expressed in *nanometers* (nm); 1 nm = 10^{-9} m. An obsolete (non-SI) unit occasionally used for wavelengths is the *angstrom* (Å); 1 Å = 10^{-10} m.

EXAMPLE 1

Find the range in wavelengths (in vacuum) for visible light in the frequency range between 4.0×10^{14} Hz and 7.9×10^{14} Hz. Express the answers in nanometers.

SOLUTION

The wavelength corresponding to a frequency of 4.0×10^{14} Hz is given by Equation 20.1 as

$$\lambda = \frac{c}{f} = \frac{3.00 \times 10^8 \text{ m/s}}{4.0 \times 10^{14} \text{ Hz}} = 7.5 \times 10^{-7} \text{ m}$$

Since 1 nm = 10^{-9} m, it follows that $\boxed{\lambda = 750 \text{ nm}}$.

The calculation for a frequency of 7.9×10^{14} Hz is similar:

$$\lambda = \frac{c}{f} = \frac{3.00 \times 10^8 \text{ m/s}}{7.9 \times 10^{14} \text{ Hz}} = 3.8 \times 10^{-7} \text{ m} \quad \text{or} \quad \boxed{\lambda = 380 \text{ nm}}$$

The eye recognizes light of different frequencies as different colors. A wavelength of 750 nm (in vacuum) is approximately the longest wavelength of red light, whereas 380 nm (in vacuum) is approximately the shortest wavelength of violet light. Between these limits are found the other familiar colors, as Figure 30.6 indicates. In this drawing the boundaries between colors are only approximate. In reality, one color changes gradually into the next.

The picture of light as a wave is supported by experiments that will be discussed in Chapter 34. However, it must be noted that there are also experiments indicating that light can behave as if it were composed of discrete particles, rather than waves. These experiments will be discussed in Chapter 36. Wave theories and particle theories of light have been around for hundreds of years, and it is now widely accepted that light, as well as other electromagnetic radiation, exhibits a dual nature. Either wave-like or particle-like behavior can be observed, depending on the kind of experiment being performed.

30.3 THE SPEED OF LIGHT

EXPERIMENTAL DETERMINATION OF THE SPEED OF LIGHT

At a speed of 3.00×10^8 m/s, light travels from the earth to the moon in a little over a second, so the time required for light to move normal distances on the earth is very short. In fact, early attempts at measuring the speed of light ran into just this problem. Galileo (1564–1642), for example, attempted to measure the speed of light at night by stationing a helper at a distant point. The helper was to uncover a lamp as soon as

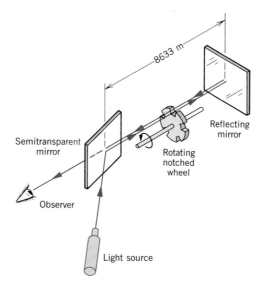

FIGURE 30.7 This drawing shows a version of the experiment that Fizeau carried out to measure the speed of light. His results were reported in 1849.

he saw the light from a lamp that Galileo had uncovered. Galileo hoped to measure the time it took for light to travel from his lamp to his assistant and for the reverse trip. Of course, the time was so short that the light appeared to move instantaneously from one point to the other, leaving only the conclusion that light travels very rapidly indeed.

In the first measurement of the speed of light that did not depend on astronomical observations, the French scientist Armand Fizeau (1819–1896) used a rotating notched wheel, as Figure 30.7 illustrates. The light passes through one notch on its way to a mirror located some distance away (8633 m in Fizeau's experiment). After reflection, the light travels back and passes through another notch in the wheel only if the rotational speed of the wheel is just right. The rotational speed must be such that the time it takes for the light to travel the round-trip must match the time it takes for the second (or third or fourth, etc.) notch to rotate into the position occupied by the first notch. From a knowledge of the rotational speed at which light passes back through the second notch, Fizeau was able to determine the speed of light as 3.13×10^8 m/s.

More accurate measurements of the speed of light were performed later using a rotating mirror instead of a notched wheel. Figure 30.8 shows a simplified version of this setup. It was used first by the French scientist Jean Foucault (1819–1868) and later in a more refined version by the American physicist Albert Michelson (1852–1931), who obtained the value of $c = (2.997\ 96 \pm 0.000\ 04) \times 10^8$ m/s in 1926. The next example illustrates the method.

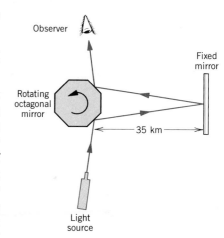

FIGURE 30.8 Between 1878 and 1931 Michelson used a rotating eight-sided mirror to measure the speed of light. The picture shown here presents a simplified version of the setup.

EXAMPLE 2

If the speed of the rotating eight-sided mirror in Figure 30.8 is adjusted correctly, light reflected from one side travels to the fixed mirror, reflects, and can be detected after reflecting from another side that has rotated into place at just the right time. For one of his experiments, Michelson placed mirrors on Mt. San Antonio and Mt. Wilson in California, a distance of 35 km apart. Knowing that $c = 3.00 \times 10^8$ m/s, obtain the minimum rotational speed for the eight-sided mirror in Michelson's experiment.

SOLUTION

The minimum rotational speed is that at which one side of the mirror rotates one-eighth of a revolution during the time it takes for the light to make the round-trip between Mt. San Antonio and Mt. Wilson. The round trip travel time is $t = 2(35 \times 10^3 \text{ m})/(3.00 \times 10^8 \text{ m/s}) = 2.3 \times 10^{-4}$ s. This time corresponds to a rotational speed of

$$\text{Rotational speed} = \frac{\frac{1}{8} \text{ revolution}}{2.3 \times 10^{-4} \text{ s}} = \boxed{540 \text{ rev/s}}$$

The best experimental value to date for the speed of light in a vacuum is

$$c = 299\ 792\ 458\ \text{m/s} \pm 1\ \text{m/s}$$

The experiment that yielded this value was carried out using a laser method developed at the National Bureau of Standards, and the small uncertainty of ± 1 m/s indicates the remarkable precision of the method.

THEORETICAL PREDICTION OF THE SPEED OF LIGHT

In 1865 Maxwell calculated that electromagnetic waves propagate through a vacuum at a speed given by

$$c = \frac{1}{\sqrt{\epsilon_0 \mu_0}} \tag{30.1}$$

where $\epsilon_0 = 8.85 \times 10^{-12}$ C^2/(N·m^2) is the (electric) permittivity of free space and $\mu_0 = 4\pi \times 10^{-7}$ T·m/A is the (magnetic) permeability of free space. Originally ϵ_0 was introduced in Section 23.5 as an alternative way of writing the proportionality constant k in Coulomb's law [$k = 1/(4\pi\epsilon_0)$] and, hence, plays a basic role in determining the strengths of the electric fields created by point charges. The role of μ_0 is similar for magnetic fields; it was introduced in Section 27.7 as part of a proportionality constant in the expression for the magnetic field created by the current in a long straight wire. Substituting the values for ϵ_0 and μ_0 into Equation 30.1 shows that

$$c = \frac{1}{\sqrt{[8.85 \times 10^{-12}\ \text{C}^2/\text{N}\cdot\text{m}^2][4\pi \times 10^{-7}\ \text{T}\cdot\text{m/A}]}} = 3.00 \times 10^8\ \text{m/s}$$

The experimental and predicted values for c agree. Maxwell's success in predicting c provided a basis for inferring that light behaves as a wave consisting of oscillating electric and magnetic fields.

30.4 THE ENERGY CARRIED BY ELECTROMAGNETIC WAVES

Electromagnetic waves, like water waves or sound waves, carry energy. The energy is carried by the electric and magnetic fields that comprise the wave. It is because of this energy, for example, that sunlight can warm you, microwaves can cook a dinner, and radio waves can make a radio work.

Suppose an electric field **E** exists in a certain region of space. As we saw in Section 24.5, the electric energy stored in the field per unit volume of space is the electric energy density:

$$\text{Electric energy density} = \frac{\text{Electric energy}}{\text{Volume}} = \tfrac{1}{2}\epsilon_0 E^2 \tag{24.12}$$

where the dielectric constant κ has been set equal to unity, since we are dealing with an electric field in a vacuum (or air). From Section 28.8, the analogous expression for the magnetic energy density is

$$\begin{matrix} \text{Magnetic} \\ \text{energy} \\ \text{density} \end{matrix} = \frac{\text{Magnetic energy}}{\text{Volume}} = \frac{1}{2\mu_0} B^2 \qquad (28.11)$$

Therefore, the **total energy density** u of an electromagnetic wave in a vacuum is the sum of these two energy densities:

$$u = \frac{\text{Total energy}}{\text{Volume}} = \tfrac{1}{2}\epsilon_0 E^2 + \frac{1}{2\mu_0} B^2 \qquad (30.2)$$

In an electromagnetic wave propagating through a vacuum or air, the electric field and the magnetic field carry equal amounts of energy per unit volume of space. To see that this is indeed the case, it is necessary to realize that the fields of the wave are related by the speed of light according to

$$E = cB \qquad (30.3)$$

a result that we state without proof. Now, if the fields carry energy equally, the ratio of the two terms on the right in Equation 30.2 should be unity. Let's see if it is, using the fact that $c = \dfrac{1}{\sqrt{\epsilon_0 \mu_0}}$:

$$\frac{\tfrac{1}{2}\epsilon_0 E^2}{\frac{1}{2\mu_0} B^2} = \epsilon_0 \mu_0 \frac{E^2}{B^2} = \epsilon_0 \mu_0 c^2 = \epsilon_0 \mu_0 \left(\frac{1}{\sqrt{\epsilon_0 \mu_0}} \right)^2 = 1$$

Since $\tfrac{1}{2}\epsilon_0 E^2 = (1/2\mu_0)B^2$, it is possible to rewrite Equation 30.2 in two additional, but equivalent forms:

$$u = \tfrac{1}{2}\epsilon_0 E^2 + \frac{1}{2\mu_0} B^2 \qquad (30.2a)$$

$$u = \epsilon_0 E^2 \qquad (30.2b)$$

$$u = \frac{1}{\mu_0} B^2 \qquad (30.2c)$$

The total energy density is related to the **intensity** of the wave. Recall from Section 21.3 that the intensity of a sound wave is the sound power that passes perpendicularly through a surface divided by the area of the surface (see Equation 21.4). In other words, intensity is power per unit area or energy per unit time per unit area. To help us apply this definition of intensity in the present situation, Figure 30.9 shows an

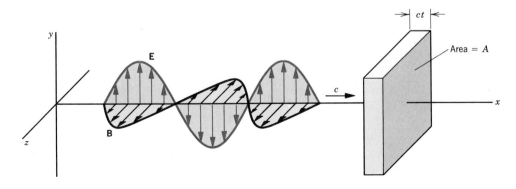

FIGURE 30.9 In a time t an electromagnetic wave moves a distance ct along the x axis and passes through a surface of area A.

electromagnetic wave traveling in a vacuum along the x axis. In a time t the wave travels the distance ct, passing through the area A. Consequently, the volume of space through which the wave passes is ctA. The total (electric and magnetic) energy in this volume is

$$\text{Total energy} = (\text{Total energy density}) \times \text{Volume} = u(ctA)$$

Using this expression, it can be seen that the intensity S and the total energy density u are related as follows:

$$S = \frac{\text{Total energy}}{\text{Time} \cdot \text{Area}} = \frac{uctA}{tA} = cu \qquad (30.4)$$

Substituting Equations 30.2 into Equation 30.4 shows that the intensity of an electromagnetic wave depends on the electric and magnetic fields according to the following equivalent relations:

$$S = cu = \tfrac{1}{2} c\epsilon_0 E^2 + \frac{c}{2\mu_0} B^2 \qquad (30.5a)$$

$$S = c\epsilon_0 E^2 \qquad (30.5b)$$

$$S = \frac{c}{\mu_0} B^2 \qquad (30.5c)$$

In the waves that have been discussed in this chapter, the electric and magnetic fields fluctuate sinusoidally in time. Equations 30.2 and 30.5, then, give the energy density and the intensity of the wave at any instant in time. If average values for u and S are desired, average values are needed for E^2 and B^2. In Section 25.5 we faced a similar situation for ac currents and voltages and introduced rms quantities. Using an analogous procedure here, it follows that the rms values for the electric and magnetic fields, E_{rms} and B_{rms}, are related to the maximum values of these fields, E_0 and B_0, by

$$E_{rms} = \frac{1}{\sqrt{2}} E_0 \quad \text{and} \quad B_{rms} = \frac{1}{\sqrt{2}} B_0$$

Equations 30.2 and 30.5 can now be interpreted as giving the average energy density $\overline{u}$ and the average intensity $\overline{S}$, provided the symbols E and B are interpreted to mean the rms values given above.

30.5 POLARIZATION

POLARIZED ELECTROMAGNETIC WAVES

One of the essential features of electromagnetic waves is that they are transverse waves, and because of this feature they can be polarized. Figure 30.10 illustrates the idea of polarization by showing a transverse wave as it travels along a rope toward a slit. The wave is said to be *linearly polarized,* which means that its vibrations always occur along one direction. This direction is called the direction of polarization. In part a of the picture, the slit is oriented parallel to the direction of polarization, and the wave passes through easily. However, when the slit is turned perpendicular to the direction of polarization, as in part b, the wave cannot pass, because the slit prevents the rope from oscillating. Note that for longitudinal waves, such as sound waves, the notion of polarization has no meaning. In a longitudinal wave the direction of vibration is along the direction of travel, and, thus, the orientation of the slit would have no effect on the wave.

In an electromagnetic wave such as that in Figure 30.3, the electric field oscillates along the y axis everywhere. Similarly the magnetic field oscillates along the z axis. Therefore, the wave is linearly polarized. The direction of polarization is taken arbitrarily to be that along which the electric field oscillates. If the wave is a radio wave generated by a straight-wire transmitting antenna, the direction of polarization is determined by the orientation of the antenna. In comparison, however, the electromagnetic waves given off as light by an incandescent light bulb are completely unpolarized. In this case the light waves are emitted by an exceedingly large number of atoms in the hot filament of the bulb. When an electron in an atom oscillates, the atom behaves as a miniature antenna that broadcasts light for brief periods of time, about 10^{-8} seconds, or so. However, the directions of these atomic antennas change randomly as time passes, for the direction in which an electron oscillates changes randomly as a result of collisions. Unpolarized light, then, consists of many individual waves, emitted in short bursts by many atomic antennas, each with its own direction of polarization. Figure 30.11 compares polarized and unpolarized light. In the unpolarized case, the arrows shown around the direction of wave travel symbolize the random directions of polarization of the individual waves that comprise the light.

Linearly polarized light can be produced from unpolarized light with the aid of certain materials. One commercially available polarizing material goes under the name of Polaroid. Such materials allow only the component of the electric field along one direction to pass through, while absorbing the field component perpendicular to this direction. As Figure 30.12 indicates, the direction of polarization that a polarizing material allows through is called the ***transmission axis*** of the material. No matter how the transmission axis is oriented, the intensity of the polarized light transmitted to the photocell is one-half that of the unpolarized incident light. The reason for this fact is that the unpolarized light contains all polarization directions to an equal extent. Moreover, the electric field for each direction can be resolved into components perpendicular and parallel to the transmission axis, with the result that the average component perpendicular to the transmission axis equals the average component parallel to the transmission axis. As a result, the polarizing material absorbs as much of the electric field strength as it transmits, and the intensity of the transmitted polarized light is one-half the intensity of the incident unpolarized light.

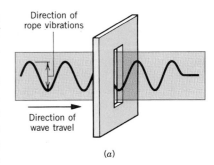

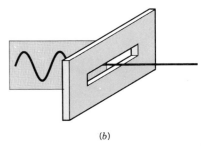

FIGURE 30.10 A transverse wave is linearly polarized when its vibrations always occur along one direction. (*a*) A linearly polarized wave on a rope can pass through a slit that is parallel to the direction of the rope vibrations, but (*b*) cannot pass through a slit that is perpendicular to the vibrations. The rope is vibrating in the vertical plane (see colored surface).

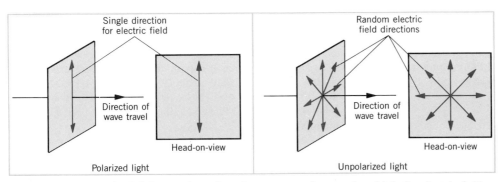

FIGURE 30.11 Polarized light consists of an electromagnetic wave in which the electric field fluctuates along a single direction. Unpolarized light consists of short bursts of electromagnetic waves emitted by many different atoms, each burst having its own direction of polarization. Thus, in unpolarized light, the electric field directions of these bursts are perpendicular to the direction of wave travel but are distributed randomly about it.

FIGURE 30.12 With the aid of a piece of polarizing material, polarized light may be produced from unpolarized light. The transmission axis of the material is the direction of polarization of the light that passes through the material. Each gray surface is perpendicular to the direction of travel of the light and emphasizes the plane in which the electric field oscillates.

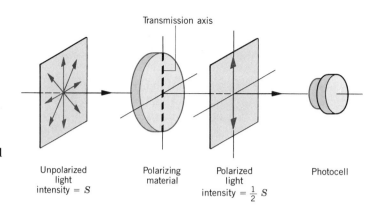

Transmission axis

Unpolarized light intensity = S

Polarizing material

Polarized light intensity = $\frac{1}{2} S$

Photocell

MALUS' LAW

qual

Once polarized light has been produced with one piece of polarizing material, it is possible to use a second piece to change the polarization direction and to adjust the intensity of the light. Figure 30.13 shows how. In this picture the first piece of polarizing material is called the ***polarizer,*** while the second piece is referred to as the ***analyzer.*** The transmission axis of the analyzer is oriented at an angle θ relative to the transmission axis of the polarizer. If the electric field strength of the polarized light incident on the analyzer is E, the field strength passing through is the component parallel to the transmission axis of the analyzer, or $E \cos \theta$. According to Equation 30.5b, the intensity is proportional to the square of the electric field strength. Consequently, the average intensity of polarized light passing through the analyzer is proportional to $\cos^2 \theta$. Thus, both the polarization direction and the intensity of the light can be adjusted by rotating the transmission axis of the analyzer relative to that of the polarizer. The average intensity $\overline{S}$ of the light leaving the analyzer, then, is

[Malus' law] $$\overline{S} = \overline{S}_0 \cos^2 \theta \qquad (30.6)$$

where $\overline{S}_0$ is the average intensity of the light entering the analyzer. Equation 30.6 is sometimes called ***Malus' law,*** for it was discovered by the French engineer Etienne-Louis Malus (1775–1812). Example 3 illustrates the use of Malus' law.

EXAMPLE 3
What value of θ should be used in Figure 30.13, so the average intensity of the polarized light reaching the photocell is one-tenth the average intensity of the unpolarized light?

SOLUTION
It is important to realize that both the polarizer and the analyzer reduce the intensity of the light. The polarizer reduces the intensity by a factor of one-half, as discussed earlier. Therefore, if the

average intensity of the unpolarized light is I, the average intensity of the polarized light leaving the polarizer and striking the analyzer is $\overline{S}_0 = I/2$. The angle θ must now be selected so the average intensity of the light leaving the analyzer is $\overline{S} = I/10$. It follows, then, from Malus' law that $I/10 = (I/2) \cos^2 \theta$. Consequently,

$$\cos \theta = \sqrt{\tfrac{1}{5}} = 0.447 \quad \text{and} \quad \theta = \cos^{-1}(0.447) = \boxed{63.4°}$$

When $\theta = 90°$ in Figure 30.13, the polarizer and analyzer are said to be ***crossed,*** and no light is transmitted by the polarizer/analyzer combination. As an illustration of this effect, Figure 30.14 shows two pairs of Polaroid sunglasses in uncrossed and crossed configurations.

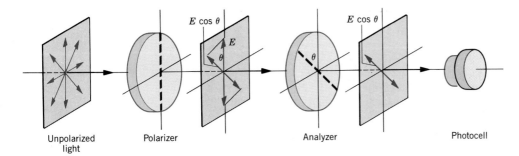

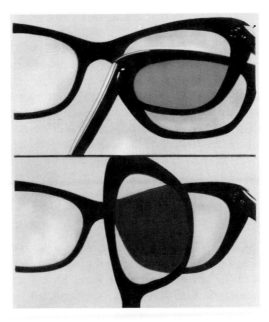

FIGURE 30.13 Two sheets of polarizing material, called the polarizer and the analyzer, may be used together to adjust the polarization direction and the intensity of the light that reaches the photocell. This can be done by changing the angle θ between the transmission axes of the polarizer and analyzer.

FIGURE 30.14 When Polaroid sunglasses are uncrossed (top photograph), the transmitted light is dimmed due to the extra thickness of tinted plastic. However, when they are crossed (lower photograph), the transmitted light is reduced to zero because of the effects of polarization.

A clever application of a crossed polarizer/analyzer combination occurs in one kind of liquid crystal display (LCD). LCDs are widely used in pocket calculators and digital watches. The display usually consists of blackened numbers and letters set against a light gray background. As Figure 30.15 indicates, each number or letter is

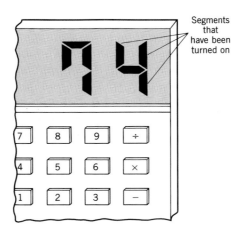

FIGURE 30.15 Liquid crystal displays use liquid crystal "segments" to form the numbers. When a segment is turned on, it appears black.

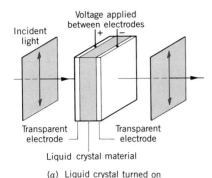

(a) Liquid crystal turned on

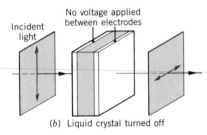

(b) Liquid crystal turned off

FIGURE 30.16 A liquid crystal is shown here as it functions in its (a) "on" state and (b) "off" state.

formed from a combination of liquid crystal segments that have been "turned on." Let us now see what it means for a liquid crystal to be turned on and how polarized light is used.

The liquid crystal part of an LCD segment consists of the liquid crystal material sandwiched between two transparent electrodes, as in Figure 30.16. When a voltage is applied between the electrodes, the liquid crystal is said to be "on." Part a of the picture shows that linearly polarized incident light passes through the "on" material without having its direction of polarization affected. When the voltage is removed, as in part b, the liquid crystal is said to be "off" and now rotates the direction of polarization by 90°.

A working LCD segment includes a crossed polarizer/analyzer combination, as Figure 30.17 illustrates. The polarizer, analyzer, electrodes, and liquid crystal material are packaged as a complete unit. The polarizer produces polarized light from incident unpolarized light. With the display segment turned on, the polarized light emerges from the liquid crystal only to be absorbed by the analyzer, since the light is polarized perpendicular to the transmission axis of the analyzer (part a). Since no light emerges from the analyzer, an observer sees a black segment against a light gray background, as in Figure 30.15. On the other hand, the segment is turned off when the voltage is removed, in which case the liquid crystal rotates the direction of polarization by 90° to coincide with the axis of the analyzer (part b). The light now passes through the analyzer and enters the eye of the observer. However, the light coming from the segment has been designed to have the same color and shade (light gray) as the background of the display, so the segment becomes indistinguishable from the background.

FIGURE 30.17 A liquid crystal display (LCD) incorporates a crossed polarizer/analyzer combination. (a) The LCD segment is turned on when a voltage is applied to the electrodes. In this case, no light is transmitted through the analyzer, and an observer sees a black segment. (b) Removing the voltage turns the LCD segment off. Now, light from the segment, which is the same color as the background light, reaches the observer. The observer cannot see the segment, however, due to a lack of contrast between it and the background.

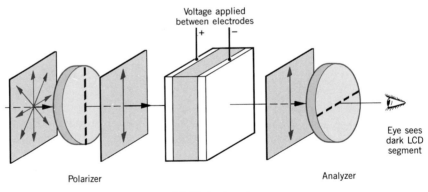

(a) Display turned on

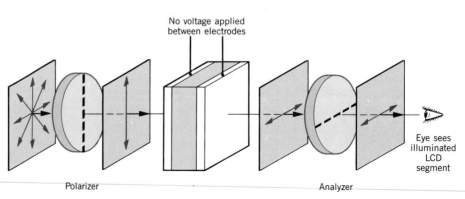

(b) Display turned off

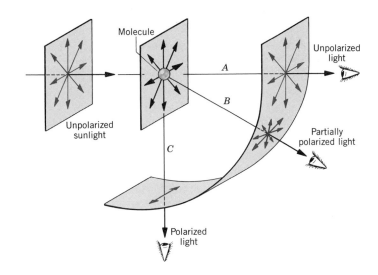

FIGURE 30.18 In the process of being scattered from atmospheric molecules, unpolarized light from the sun becomes polarized.

THE OCCURRENCE OF POLARIZED LIGHT IN NATURE

Polaroid is a familiar material because of its widespread use in sunglasses. Such sunglasses are designed so that the axis of the Polaroid is oriented vertically when the glasses are worn in the usual fashion. Thus, the glasses prevent any light that is polarized horizontally from reaching the eye. Light from the sun is unpolarized, but a considerable amount of horizontally polarized sunlight originates by reflection from horizontal surfaces such as that of a lake. Section 32.4 discusses this effect. Polaroid sunglasses reduce glare by preventing the horizontally polarized reflected light from reaching the eyes.

Polarized sunlight also originates from the scattering* of light by molecules in the atmosphere. Figure 30.18 shows light being scattered by a single atmospheric molecule. What happens is that the electric fields in the unpolarized sunlight set the electrons in the molecule into vibrations that are perpendicular to the direction in which the light is traveling. The electrons, in turn, reradiate the electromagnetic waves in different directions, as the drawing illustrates. The light radiated straight ahead in direction A is unpolarized, just like the incident light. But light radiated perpendicular to the incident light in direction C is polarized. Light radiated in the intermediate direction B is partially polarized.

SUMMARY

An *electromagnetic wave* in a vacuum consists of mutually perpendicular and oscillating electric and magnetic fields. The wave is a transverse wave, since both the electric and magnetic fields are perpendicular to the direction in which the wave travels. All electromagnetic waves, regardless of their frequency, travel through a vacuum at the same speed, namely, the speed of light c ($c = 3.00 \times 10^8$ m/s).

The frequency f and wavelength λ of an electromagnetic

---

* The scattering of light from molecules in the atmosphere (primarily oxygen and nitrogen) is also responsible for the fact that the sky appears blue. Normal white light is a mixture of colors or wavelengths, and the shorter blue wavelengths scatter most efficiently. During the daytime, you see this scattered blue light. At sunset the sun's rays come in from the horizon and follow a long path through the atmosphere before reaching your eyes. Blue light has been scattered all along this path, so the sunlight, minus the blue wavelengths, has a red hue.

wave in a vacuum are related to its speed through the relation $c = \lambda f$. The series of electromagnetic waves, arranged in order of their frequencies, is called the **electromagnetic spectrum.** The electromagnetic spectrum is composed of groups of waves that are known as radio waves, infrared radiation, visible light, ultraviolet radiation, X rays, and gamma rays.

Visible light has frequencies between about 4.0×10^{14} Hz and 7.9×10^{14} Hz. The human eye and brain perceive different frequencies as different colors. Colors are often discussed in terms of the wavelengths (in vacuum) that correspond to the various frequencies, the wavelengths being expressed in nanometers (1 nm = 10^{-9} m).

Maxwell calculated that the speed of light in a vacuum is $c = 1/\sqrt{\epsilon_0 \mu_0}$, where ϵ_0 is the (electric) permittivity of free space and μ_0 is the (magnetic) permeability of free space.

The **total energy density** u of an electromagnetic wave is the total energy per unit volume of the wave and, in a vacuum, is given by $u = \frac{1}{2}\epsilon_0 E^2 + (1/2\mu_0)B^2$, where E and B are the magnitudes of the electric and magnetic fields. Since E and B are related by $E = cB$ in a vacuum, the electric and magnetic parts of the total energy density are equal.

The **intensity** of an electromagnetic wave is the power that the wave carries perpendicularly through a surface divided by the area of the surface. In a vacuum, the intensity S is related to the total energy density u according to $S = cu$.

A **linearly polarized** electromagnetic wave is one in which all oscillations of the electric field occur along one direction, which is taken to be the direction of polarization. In **unpolarized light** the direction of polarization does not remain fixed, but fluctuates randomly in time.

Polarizing materials allow only the component of the wave's electric field along one direction to pass through them. The preferred transmission direction for the electric field is called the **transmission axis** of the material. When unpolarized light is incident on a piece of polarizing material, the transmitted polarized light has an intensity that is one-half that of the incident light. When two pieces of polarizing material are used one after the other, the first one is called the polarizer, while the second one is referred to as the analyzer. If the average intensity of polarized light falling on an analyzer is $\bar{S}_0$, the average intensity $\bar{S}$ of the light leaving the analyzer is given by **Malus' law** as $\bar{S} = \bar{S}_0 \cos^2 \theta$, where θ is the angle between the transmission axes of the polarizer and analyzer. When $\theta = 90°$, the polarizer and the analyzer are said to be "crossed," and no light passes through the analyzer.

QUESTIONS

1. Compare the properties of electromagnetic waves and sound waves by filling in the table below.

Property	Electromagnetic Wave	Sound Wave
Transverse or longitudinal?		
Can be polarized?		
Can travel through a vacuum?		
Can travel through a material such as glass?		
Involves electric and magnetic fields?		
Involves pressure oscillations?		

2. A transmitting antenna is located at the origin of an x, y, z axis system and broadcasts an electromagnetic wave whose electric field oscillates along the y axis. The wave travels along the $+x$ axis. There are three possible wire loops that can be used with an LC-tuned circuit to detect this wave: one loop lies in the xy plane, another in the xz plane, and the third in the yz plane. Which of the loops will detect the wave? Justify your answers.

3. In Section 22.3 we discussed the diffraction of sound waves, that is, the ability of the waves to bend around obstacles. Electromagnetic waves also have the same ability. On the basis of this earlier discussion and the fact that AM radio waves have larger wavelengths than FM radio waves, would you expect AM or FM radio waves to exhibit a greater ability to bend around an obstacle such as a building? Explain.

4. The speed of sound is about one-fifth of a mile per second, and there is a rule of thumb based on this fact. The rule specifies that if you divide the number of seconds between a lightning flash and the associated thunder clap by five, you get the approximate distance (in miles) to the lightning. Explain why this rule works, using what you know about the speed of sound compared to the speed of light. Would this rule be useful if the speed of light were nearly equal to the speed of sound? Why?

5. Refer to Figure 30.8 and Example 2 in the text. Would light be detected by the observer if the eight-sided mirror were to rotate at an angular speed of 1080 rev/s, instead of 540 rev/s? Explain.

6. Why is it said that astronomers looking at distant stars are "looking back in time"?

7. Malus' law applies to the setup in Figure 30.13, which shows the analyzer rotated through an angle θ and the polarizer held fixed. Does Malus' law apply when the analyzer is held fixed and the polarizer is rotated? Give your reasoning.

8. Light is incident from the left on two pieces of polarizing material, 1 and 2. As part *a* of the drawing illustrates, the transmission axis of material 1 is along the vertical direction, while that of material 2 makes an angle of θ with respect to the vertical. In part *b* of the drawing the two polarizing materials are interchanged. (a) Assume the incident light is unpolarized and determine whether the intensity of the transmitted light in part *a* is greater than, equal to, or less than, that in part *b*. (b) Repeat part (a), assuming the incident light is linearly polarized along the vertical direction. Justify your answers to both parts (a) and (b).

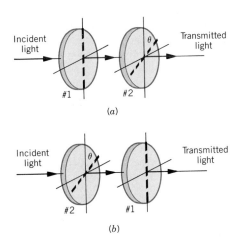

(a)

(b)

PROBLEMS

Section 30.1 The Nature of Electromagnetic Waves

1. The mean distance from the sun to the earth is about 1.50×10^{11} m. How long (in minutes) does it take for sunlight to reach the earth?

2. In astronomy, distances are often expressed in light-years. One light-year is the distance traveled by light in one year. The distance to Alpha Centauri, the closest star to us other than our own sun, is 4.3 light-years. Express this distance in meters.

3. In Figure 30.4 the value of the inductance is 2.6×10^{-4} H. For an AM radio station broadcasting at 1200 kHz, find the value to which the capacitor must be adjusted.

4. For an FM radio station broadcasting at 88.0 MHz, the capacitance in Figure 30.4 must be adjusted to a value of 23.0×10^{-12} F. Assuming the inductance does not change, determine the value of the capacitance for an FM station broadcasting at 108.0 MHz.

* 5. Equation 20.3, $y = A \sin (2\pi ft - 2\pi x/\lambda)$, gives the mathematical representation of a wave oscillating in the y direction and traveling in the positive x direction. Let y in this equation equal the electric field of an electromagnetic wave traveling in a vacuum. Assuming that the maximum electric field is $A = 156$ N/C and that the frequency is $f = 1.50 \times 10^8$ Hz, plot a graph of the electric field strength versus position, using for x the following values: 0, 0.25, 0.50, 0.75, 1.00, 1.25, 1.50, 1.75, and 2.00 m. Plot this graph for (a) a time $t = 0$ and (b) a time t that is one-fourth of the wave's period.

Section 30.2 The Electromagnetic Spectrum

6. Some of the X rays produced in a X-ray machine have a wavelength of 2.1 nm. What is the frequency of these electromagnetic waves?

7. A truck driver is broadcasting at a frequency of 26.965 MHz with a CB (citizen's band) radio. Determine the wavelength of the electromagnetic wave being used.

8. A radio station broadcasts a radio wave whose wavelength is 274 m. (a) What is the frequency of the wave? (b) What kind of radio wave is this? (See Figure 30.6.)

9. Obtain the wavelengths in vacuum for blue light with a frequency of 6.34×10^{14} Hz and for orange light with a frequency of 4.95×10^{14} Hz. Express your answers in nanometers (nm).

* 10. Sections 22.5 and 22.6 deal with standing waves. Electromagnetic waves also can form standing waves. In a standing wave pattern formed from microwaves, the distance between a node and an adjacent antinode is 0.50 cm. What is the microwave frequency?

Section 30.3 The Speed of Light

11. (a) Neil A. Armstrong was the first person to walk on the moon. The distance between the earth and the moon is 3.85×10^8 m. Find the time it took for his voice to reach earth via radio waves. (b) Similarly, determine the communication time for the first person who will some day walk on Mars, which is 5.6×10^{10} m from earth at the point of closest approach.

12. The distance between earth and the moon can be determined from the length of time it takes for a laser beam to travel from earth to a reflector on the moon and back. If the round-trip time can be measured to an accuracy of one-tenth of a nanosecond (1 ns = 10^{-9} s), what is the corresponding error in the earth–moon distance?

13. In Fizeau's experiment (see Figure 30.7) the rotating wheel contained 720 notches. Knowing that the speed of light in air is 3.00×10^8 m/s, obtain the rotational speed (in rev/s) of the wheel if the incident light is to pass through one notch and the reflected light is to pass through an adjacent notch.

* 14. A mirror faces a cliff, located some distance away. Mounted on the cliff is a second mirror, directly opposite the first mirror and facing toward it. A gun is fired very close to the first mirror. The speed of sound is 343 m/s. How many times does the flash of the gunshot travel the round-trip distance between the mirrors before the echo of the gunshot is heard?

* 15. The President of the United States holds a press conference. The conference is televised live, and a television viewer hears the sound picked up by a microphone placed directly in front of the president. This viewer is seated 2.0 m from the television set. A reporter at the press conference is located 5.0 m from the stage and hears the president's words directly *at the very same instant* that the television viewer hears them. Using a value of 343 m/s for the speed of sound, determine the maximum distance (in miles) between the television viewer and the president.

Section 30.4 The Energy Carried by Electromagnetic Waves

16. Suppose the electric field in an electromagnetic wave has a maximum strength of 2140 N/C. What is the maximum strength of the magnetic field of the wave?

17. The maximum value of the electric field in an electromagnetic wave is 174 N/C. The wave passes perpendicularly through a surface of area 0.35 m². How much energy does this wave carry across the surface in one minute?

18. Show that, in addition to Equations 30.2a–30.2c, the total energy density for an electromagnetic wave can be expressed as $u = (\sqrt{\epsilon_0/\mu_0})EB$.

* 19. An argon-ion laser produces a cylindrical beam of light whose diameter is 2.00 mm. The average power produced by the laser is 0.75 W. (a) Determine the maximum value of the electric field in the laser light. (b) How much energy is contained in a 2.5-m length of the beam?

* 20. In an electromagnetic wave the magnetic field strength has an rms value of 9.11×10^{-8} T. Find (a) the maximum value of the magnetic field strength, (b) the rms value of the electric field strength, (c) the maximum value of the electric field strength, (d) the average total energy density of the wave, and (e) the average intensity of the wave.

**21. The mean distance between earth and the sun is 1.50×10^{11} m. The average intensity of solar radiation incident on the upper atmosphere of the earth is 1390 W/m². Assuming the sun emits radiation uniformly in all directions, determine the total power radiated by the sun.

Section 30.5 Polarization

22. Linearly polarized light is incident on a piece of polarizing material. What is the ratio of the transmitted light intensity to the incident light intensity when the angle between the transmission axis and the incident electric field is (a) 25°, (b) 45°, (c) 65°, and (d) 85°?

23. What should be the angle between the transmission axes of the polarizer and the analyzer in Figure 30.13, so the polarized light reaching the photocell has an intensity that is (a) one-half and (b) one-fourth of the intensity of the incident unpolarized light?

24. In the polarizer/analyzer combination in Figure 30.13, 90.0% of the light falling on the analyzer is absorbed. Determine the angle between the transmission axes of the polarizer and the analyzer.

* 25. The orientation of the transmission axis for each of the three sheets of polarizing material in the drawing is labeled relative to the vertical. A beam of light, polarized in the vertical direction, is incident on the first sheet. The intensity of the incident beam is 1550 W/m². Obtain the intensity of the beam transmitted through the three sheets when: (a) $\theta_1 = 0°$, $\theta_2 = 40.0°$, $\theta_3 = 75.0°$; (b) $\theta_1 = 30.0°$, $\theta_2 = 30.0°$, $\theta_3 = 70.0°$.

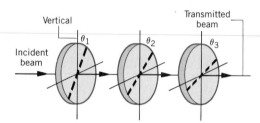

* 26. The polarizer and the analyzer in Figure 30.13 are crossed ($\theta = 90.0°$), and no light falls on the photocell. Then, a third piece of polarizing material is put between the polarizer and the analyzer, with its transmission axis oriented at 45.0° relative to the transmission axes of the polarizer and the analyzer. If the unpolarized light intensity incident on the polarizer is I, what fraction of I now falls on the photocell?

**27. More than one analyzer can be used in a setup like that in Figure 30.13, each analyzer following the previous one. Suppose that the transmission axis of the first analyzer is rotated 27° relative to the transmission axis of the polarizer, and that the transmission axis of each additional analyzer is rotated 27° relative to the transmission axis of the previous one. What is the minimum number of analyzers needed, so the light reaching the photocell has an intensity that is reduced by at least a factor of one hundred relative to that striking the first analyzer?

ADDITIONAL PROBLEMS

28. Determine the range of wavelengths for (a) AM radio waves with frequencies between 535 and 1605 kHz and (b) FM radio waves with frequencies between 88.0 and 108.0 MHz.

29. An industrial laser is used to burn a hole through a piece of metal. The average intensity of the light is 1.23×10^9 J/(m²·s). What is the maximum value of (a) the electric field and (b) the magnetic field in the electromagnetic wave emitted by the laser?

30. In 1980 and 1981, two Voyager spacecraft sent back beautiful photographs of Saturn via radio transmission. If the distance between earth and Saturn was 1277×10^9 m, how much travel time (in minutes) was required for the radio waves?

31. TV channel 3 (VHF) broadcasts at a frequency of 63.0 MHz. TV channel 23 (UHF) broadcasts at a frequency of 527 MHz. Find the ratio (VHF/UHF) of the wavelengths for these channels.

32. Polarized light strikes a piece of polarizing material. The incident light is polarized at an angle of 30.0° relative to the transmission axis of the material. What percentage of the light intensity is transmitted?

* 33. As Section 30.3 discusses, Maxwell calculated that the speed of light in a vacuum is given by $c = 1/\sqrt{\epsilon_0 \mu_0}$. The unit for ϵ_0 is $C^2/(N \cdot m^2)$ and the unit for μ_0 is $T \cdot m/A$. Show that the unit for $1/\sqrt{\epsilon_0 \mu_0}$ is meters per second.

* 34. In experiment 1, unpolarized light falls on the polarizer in Figure 30.13, with $\theta = 60.0°$. In experiment 2, the unpolarized light is replaced with light of the same intensity, but which is polarized along the direction of the polarizer transmission axis. By how many *additional* degrees and in what direction must the analyzer be rotated, so the light falling on the photocell has the same intensity as it did in experiment 1?

* 35. A flat coil of wire is used with an LC-tuned circuit as a receiving antenna. The coil has a radius of 25 cm and consists of 450 turns. The transmitted radio wave has a frequency of 1.2 MHz. The magnetic field of the wave is parallel to the normal to the coil and has a maximum value of 2.0×10^{-13} T. Using Faraday's law of electromagnetic induction and the fact that the magnetic field changes from zero to its maximum value in one-quarter of a wave period, *estimate* the magnitude of the average emf induced in the antenna during this time.

* 36. The average intensity of sunlight reaching the earth is 1390 $J/(s \cdot m^2)$. A charge of 2.6×10^{-8} C is placed in the path of this electromagnetic wave. (a) What is the maximum electric force that the charge experiences? (b) If the charge is moving at a speed of 3.7×10^4 m/s, what is the maximum magnetic force that the charge could experience?

**37. Suppose that the light falling on the polarizer in Figure 30.13 is partially polarized (average intensity $= \overline{S}_P$) and partially unpolarized (average intensity $= \overline{S}_U$). The total incident intensity is $\overline{S}_P + \overline{S}_U$, and the percentage polarization is $100\overline{S}_P/(\overline{S}_P + \overline{S}_U)$. When the analyzer is rotated in such a situation, the intensity reaching the photocell varies between a minimum value of $\overline{S}_{min}$ and a maximum value of $\overline{S}_{max}$. Show that the percentage polarization can be expressed as $100(\overline{S}_{max} - \overline{S}_{min})/(\overline{S}_{max} + \overline{S}_{min})$.

**38. A tiny source of light emits light uniformly in all directions. The total energy emitted per second is 60.0 W. For a point located 8.00 m away from this source, determine the rms electric and magnetic field strengths in the light waves.

CHAPTER 31

The Reflection of Light and Mirrors

The Taj Mahal, located near New Delhi, India, sits beside a beautiful reflecting pool. The reflection of light is also responsible for the images produced by plane and curved mirrors.

31.1 WAVE FRONTS AND RAYS

When we "see" an object, light from it enters our eyes and evokes the sensation of vision. Some objects emit light, like the sun, a flame, or a light bulb. More commonly, however, objects reflect light into our eyes. This chapter deals with reflected light and pays particular attention to reflection by mirrors. In our discussion of reflection the concepts of a wave front and a ray play important roles.

Consider a small spherical object whose surface is pulsating in simple harmonic motion. A spherical sound wave is emitted that moves outward from the object at a constant speed. To represent this wave, we draw surfaces through all points of the wave that are in the same phase of motion. These surfaces of constant phase are called *wave fronts.* Figure 31.1 shows a two-dimensional view of the wave fronts. In this view the wave fronts appear as concentric circles about the vibrating object. If the wave fronts are drawn through the condensations, or crests, of the sound wave, as they are in Figure 31.1, the distance between adjacent wave fronts equals the wavelength λ of the sound. The radial lines pointing outward from the source and perpen-

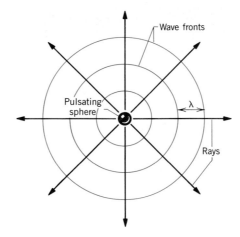

FIGURE 31.1 A cross-sectional view of a spherical sound wave emitted by a pulsating spherical object. Here, the wave fronts are drawn through the condensations of the wave, so the distance between two successive wave fronts is the wavelength λ of the sound. The rays are perpendicular to the wave fronts and point in the direction of the velocity of the wave.

dicular to the wave fronts are called **rays.** The rays point in the direction of the velocity of the wave.

Figure 31.2a shows a small section of two adjacent spherical wave fronts. At large distances from the source, the wave fronts become less and less curved and approach the shape of flat surfaces, as part b of the drawing shows. Waves whose wave fronts are flat surfaces (i.e., planes) are known as **plane waves** and are important in understanding the properties of mirrors and lenses. Since rays are perpendicular to the wave fronts, the rays for a plane wave are parallel to each other.

The concepts of wave fronts and rays can also be applied to light waves. For light waves, the ray concept is particularly convenient for showing the path taken by the light. We will make frequent use of light rays, and they can be regarded essentially as a narrow beam of light.

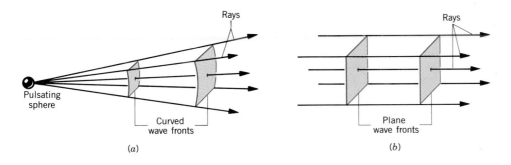

FIGURE 31.2 (a) Portions of the spherical wave fronts are shown. The rays are perpendicular to the wave fronts and point in the direction of the wave's velocity. (b) For a plane wave, the wave fronts are flat surfaces, and the rays are parallel to each other.

31.2 THE REFLECTION OF LIGHT

Most objects reflect a certain portion of the light falling on them, and it is usually this reflected light that enables us to see them. Suppose a ray of light is incident on a flat, shiny surface, such as the mirror in Figure 31.3. As part b of the drawing shows, the *angle of incidence* θ_i is the angle that the incident ray makes with respect to the normal to the surface, while the *angle of reflection* θ_r is the angle that the reflected ray makes with the normal. The **law of reflection** describes the behavior of the incident and reflected rays.

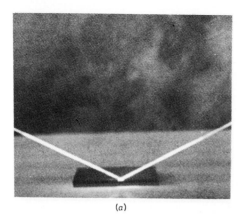

(a)

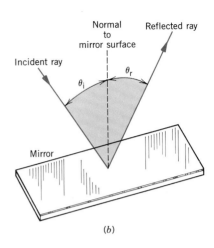

(b)

FIGURE 31.3 (a) A ray of light is reflected by a flat mirror. (b) The angle of reflection θ_r equals the angle of incidence θ_i.

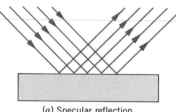

(a) Specular reflection

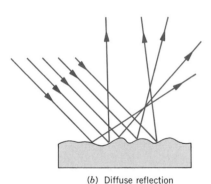

(b) Diffuse reflection

FIGURE 31.4 (a) The drawing shows specular reflection from a polished plane surface, such as a mirror. The reflected rays are parallel to each other. (b) The rough surface reflects the light in all directions; this type of reflection is known as diffuse reflection.

LAW OF REFLECTION
The incident ray, the reflected ray, and the normal to the surface all lie in the same plane, and the angle of reflection θ_r equals the angle of incidence θ_i:

$$\theta_r = \theta_i$$

When parallel light rays strike a smooth, plane surface, such as that in Figure 31.4a, the reflected rays are parallel to each other. This type of reflection is known as *specular reflection* and is important in determining the properties of mirrors. Most surfaces, however, are not perfectly smooth, for they contain irregularities the sizes of which are equal to or greater than the wavelength of light. The irregular surface reflects the light rays in various directions, as part b of the drawing suggests. This type of reflection is known as *diffuse reflection.* Common surfaces that give rise to diffuse reflection are most papers, wood, nonpolished metals, and walls covered with a flat paint.

31.3 THE FORMATION OF IMAGES BY A PLANE MIRROR

When you look into a plane (flat) mirror, you see an image of yourself that has four properties:

1. The image is upright.
2. The image has left–right reversal. That is, if you wave your *right* hand, it is the *left* hand of the image that waves back, as Figure 31.5a illustrates. Therefore, letters and words held up to a mirror are reversed. In fact, to account for this reversal, ambulances and other emergency vehicles are often lettered in reverse, as in part b of the drawing, so the letters will appear normal when seen in the rearview mirror of a car.
3. The image is located as far behind the mirror as you are in front of it.
4. The image is the same size as you are.

To illustrate how an image appears to originate from behind a plane mirror, Figure 31.6a shows a light ray leaving the top of an object. This ray reflects from the mirror (angle of reflection equals angle of incidence) and enters the eye. To the eye, it seems

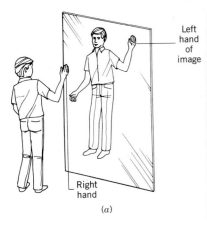

Left
hand
of
image

Right
hand

(a)

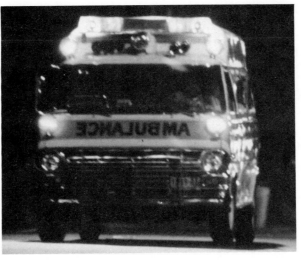

(b)

FIGURE 31.5 (a) The person's right hand becomes the image's left hand when viewed in a plane mirror. (b) Many emergency vehicles are reverse-lettered so the lettering appears normal when viewed through the rearview mirror of a car.

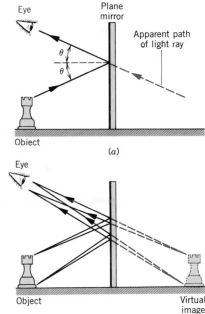

FIGURE 31.6 (a) A ray of light from the top of the chess piece reflects from the mirror. To the eye, the ray seems to come from behind the mirror. (b) Two bundles of rays, one originating from the top of the object and the other from the bottom, appear to originate from the image behind the mirror.

that the ray originates from behind the mirror, somewhere back along the dashed colored line in the picture. Actually, rays going in all directions leave each point on the object. But only a small bundle of such rays is intercepted by the eye. Part b of the figure shows a bundle of two rays leaving the top of the object and a similar bundle leaving the bottom. All the rays that enter the eye appear to originate from the image behind the mirror, as the dashed colored lines indicate. For each point on the object, there is a corresponding point on the image.

Although rays of light *seem* to come from the image, it is apparent from Figure 31.6b that no light exists behind the mirror where the image appears to be. Because the rays of light do not actually emanate from the image, it is called a ***virtual image.*** In this text the parts of the light rays that appear to come from a virtual image are represented by dashed lines. *Curved* mirrors, on the other hand, can produce images from which light rays actually do emanate. Such images are known as ***real images,*** and are discussed in later sections.

With the aid of the law of reflection, it is possible to show that the image is located as far behind a plane mirror as the object is in front of it. In Figure 31.7 the object distance is d_o and the image distance is d_i. A ray of light leaves the base of the object,

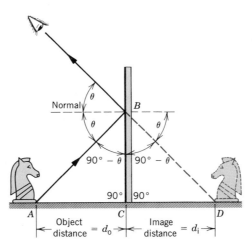

FIGURE 31.7 This drawing illustrates the geometry that is used to show that for a plane mirror the image distance d_i equals the object distance d_o.

strikes the mirror at an angle of incidence θ, and is reflected at the same angle. To the eye, this ray appears to come from the base of the image. In the drawing there are two right triangles, ABC and DBC, that share a common side BC. The angles at the top of each triangle are both equal to $90° - \theta$, as the drawing shows. Since the triangles share a common side BC and have two identical angles, $90°$ and $90° - \theta$, the triangles are identical (congruent). It then follows that the object distance d_o equals the image distance d_i.

By starting with a light ray from the top of the object, rather than from the bottom, we can extend the line of reasoning given above to show that the height of the image also equals the height of the object.

Example 1 discusses an interesting feature of plane mirrors.

EXAMPLE 1

A woman stands in front of a plane mirror to see her full height. She is 1.68 m tall, and her eyes are 0.08 m below the top of her head. Show that the mirror need only be *half as tall* as the woman.

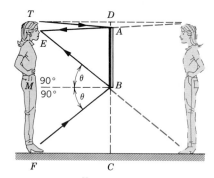

FIGURE 31.8 For the woman to see her full-sized image, only a half-sized mirror AB is needed.

SOLUTION

In Figure 31.8 the mirror is labeled AB. A ray from the woman's foot F strikes the mirror at B with an angle of incidence equal to θ. The ray is reflected with an angle of reflection equal to θ and proceeds to the woman's eye E. The two right triangles EBM and MBF are identical, since they share the common side MB and have two angles, θ and $90°$, that are identical. Therefore,

$$EM = MF = \tfrac{1}{2}EF = \tfrac{1}{2}(1.68 \text{ m} - 0.08 \text{ m}) = 0.80 \text{ m}$$

which is also the distance BC. Similarly, a ray from the top of the woman's head T strikes the mirror at A and proceeds to her eye. The same line of reasoning as above leads us to the conclusion that

$$DA = \tfrac{1}{2}TE = \tfrac{1}{2}(0.08 \text{ m}) = 0.04 \text{ m}$$

Thus, the length AB of the mirror is given by $AB = DC - BC - DA = 1.68 \text{ m} - 0.80 \text{ m} - 0.04 \text{ m} = 0.84 \text{ m}$. The mirror, then, need only be *half as tall* as the woman, if its bottom edge is located 0.80 m off the ground.

Although many "full-length" mirrors are sold today, the bottom part BC and the top part DA of these mirrors are useless, because light rays that strike these parts do not have the proper angle of incidence to reach the person's eyes. Note that the conclusions here are valid regardless of how far the person stands from the mirror.

31.4 SPHERICAL MIRRORS

The most common type of curved mirror is a spherical mirror. As Figure 31.9 shows, a spherical mirror has the shape of a section from the surface of a sphere. If the inside or concave surface of the mirror is polished, we have a **concave mirror.** If the outside or convex surface is polished, we have a **convex mirror.** The drawing shows both types of mirrors, with a light ray reflecting from the polished surface. Each type has a radius of curvature R, and the center of curvature is located at point C. The **principal axis** of the mirror is a straight line drawn through C and the midpoint of the mirror.

Figure 31.10 shows a point on a tree from which light rays are emanating. This point lies on the principal axis of the mirror and is beyond the center of curvature C. Those rays that are near the principal axis are reflected from the mirror and cross the axis at a point called the image point. The rays continue to diverge from the image

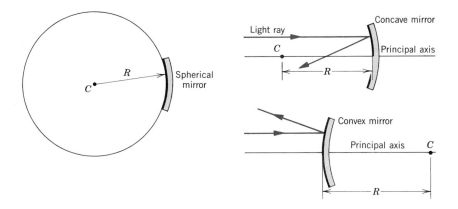

FIGURE 31.9 A spherical mirror has the shape of a segment of a spherical surface. The radius of curvature is R, and the center of curvature is point C. For a concave mirror, the inner surface of the spherical segment is the reflecting surface, while for a convex mirror the outer surface is the reflecting surface.

point as if there were an object there. Since light rays actually come from the image point, the image is a real image.

If the tree in Figure 31.10 is infinitely far from the mirror, the rays are parallel to each other and to the principal axis as they approach the mirror. Figure 31.11 shows how rays parallel to the principal axis are reflected from the mirror and pass through an image point. In this special case the image point is referred to as the **focal point** F of the mirror. Therefore, an object infinitely far away on the principal axis gives rise to an image at the focal point of the mirror. The distance between the focal point and the middle of the mirror is the **focal length** f of the mirror.

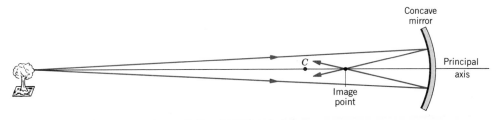

FIGURE 31.10 A point on the tree lies on the principal axis of the concave mirror. Rays from this point that are near the principal axis are reflected from the mirror and cross the axis at the image point.

We can use the geometry shown in Figure 31.12 to show that the focal point F lies halfway between the center of curvature C and the middle of the mirror. A light ray parallel to the principal axis strikes the concave mirror at point A. The line CA is the radius of the mirror and, therefore, is normal to the spherical surface. The ray reflects from the mirror such that the angle of reflection θ equals the angle of incidence. Furthermore, the angle ACF is also θ, because the radial line CA is a transversal of two parallel lines. The triangle CAF is an isosceles triangle, because two of its angles are equal; thus, sides CF and FA are equal. When the incoming ray lies close to the principal axis, the angle of incidence θ is small. When θ is small, the distance FA does

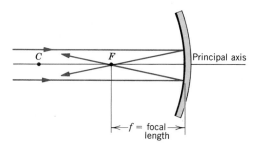

FIGURE 31.11 Parallel light rays that are near the principal axis are reflected from a concave mirror and converge at the focal point F. The focal length f of the mirror is the distance between F and the middle of the mirror.

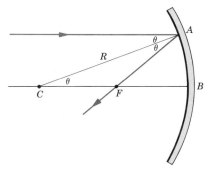

FIGURE 31.12 This drawing is used to show that the focal point F of a concave mirror is halfway between the center of curvature C and the mirror at point B. In the proof, it is necessary to assume that the angle of incidence θ is small for a ray that is parallel to the principal axis and strikes the mirror at A (the angle θ has been exaggerated for clarity).

not differ appreciably from the distance FB, so the focal point F lies halfway between the mirror and the center of curvature. Thus, the focal length f is one-half of the radius of curvature R:

$$\left[\begin{array}{l}\textbf{Focal length of}\\ \textbf{a concave mirror}\end{array}\right] \qquad f=\tfrac{1}{2}R \qquad (31.1)$$

Rays that lie close to the principal axis are known as **paraxial rays,** and Equation 31.1 is valid only for such rays. Rays that are far from the principal axis do not converge to a single point after reflection from the mirror, as Figure 31.13 shows. The result is a blurred image. The fact that a spherical mirror does not bring all rays parallel to the axis to a single image point is known as **spherical aberration.** Spherical aberration can be minimized by using a mirror whose height is small compared to the radius of curvature.

A sharp image point can be obtained with a large mirror, if the mirror is parabolic in shape instead of spherical. The shape of a parabolic mirror is such that all light rays parallel to the principal axis, regardless of their distance from the axis, are reflected through a single image point. However, parabolic mirrors are costly to manufacture and are only used where the sharpest images are required, as in telescopes.

Figure 31.14 shows parallel rays incident on a convex mirror. Clearly, the rays diverge after being reflected. If the incident rays are paraxial, the reflected rays seem to come from a single point F behind the mirror. This point is the focal point of the convex mirror, and its distance from the midpoint of the mirror is the focal length f. The magnitude of the focal length of a convex mirror is also one-half of the radius of curvature, just as it is for a concave mirror. However, by convention, the focal length of a convex mirror is considered to be a negative number, a convention that will prove useful later on:

$$\left[\begin{array}{l}\textbf{Focal length of}\\ \textbf{a convex mirror}\end{array}\right] \qquad f=-\tfrac{1}{2}R \qquad (31.2)$$

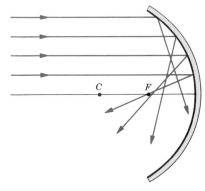

FIGURE 31.13 Rays that are farthest from the principal axis have the greatest angle of incidence and miss the focal point F after reflection from the mirror.

31.5 THE FORMATION OF IMAGES BY SPHERICAL MIRRORS

IMAGE FORMATION BY A CONCAVE MIRROR

As we have seen, some of the light rays emitted from an object in front of a mirror strike the mirror, reflect from it, and form an image. For a concave mirror, three paraxial rays are particularly helpful in determining the location and size of the

FIGURE 31.14 When paraxial light rays that are parallel to the principal axis strike a convex mirror, the reflected rays appear to originate from the focal point F. The radius of curvature is R and the focal length is $f=-\tfrac{1}{2}R$.

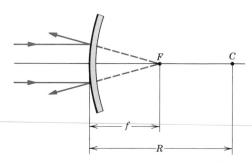

image. Figure 31.15 shows these rays—labeled 1, 2, and 3—leaving a point on the top of an object:

Ray 1. This ray is initially parallel to the principal axis and, therefore, passes through the focal point F after reflection from the mirror.

Ray 2. This ray passes through the focal point F and is reflected parallel to the principal axis. Ray 2 is analogous to ray 1, except the order of the incident and reflected rays is interchanged.

Ray 3. This ray travels along a line that passes through the center of curvature C and follows a radius of the spherical mirror; as a result, the ray strikes the mirror perpendicularly and reflects back on itself.

If the three rays above are superimposed on a scale drawing, they converge at a point on the top of the image, as can be seen in Figure 31.16a.* Although three rays have been used here to locate the image, only two are really needed; the third ray is usually drawn to serve as a check. In a similar fashion, rays from all other points on the object locate corresponding points on the image, and the mirror forms a complete image of the object. If you place your eye as shown in the drawing, you will see an image that is *larger* and *inverted* relative to the object. The image is real, because the light rays actually pass through the image.

If the object and image in Figure 31.16a are interchanged, the situation in part b of the drawing results. The three rays in part b are the same as those in part a, except the directions are reversed. These drawings illustrate the ***principle of reversibility,*** which states that ***if the direction of a light ray is reversed, the light retraces its original path.***

When the object is placed between the focal point F and the mirror, as in Figure 31.17a, three rays can again be drawn to find the image. But now ray 2 does not go through the focal point on its way to the mirror, since the object is inside the focal point. However, when projected backward, ray 2 appears to come from the focal point. Therefore, after ray 2 is reflected it is directed parallel to the principal axis. In

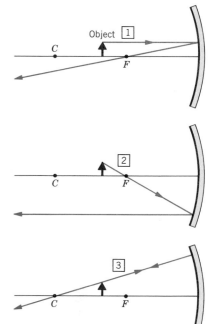

FIGURE 31.15 The rays labeled 1, 2, and 3 are useful in locating the image of an object that is located in front of a concave spherical mirror. The object is represented as a vertical arrow.

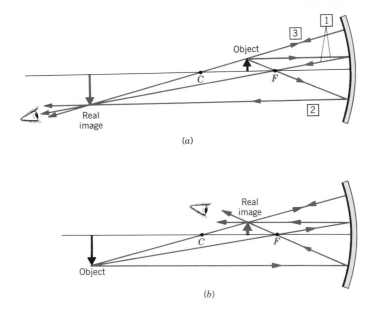

(a)

(b)

FIGURE 31.16 (*a*) When an object is placed between the focal point F and the center of curvature C of a concave mirror, a real image is formed. The image is enlarged and inverted with respect to the object. (*b*) When the object is located beyond the center of curvature C, a real image is created that is reduced in size and inverted with respect to the object.

---

* In the drawings that follow, we assume the rays are paraxial, although the distance between the rays and the principal axis is often exaggerated for clarity.

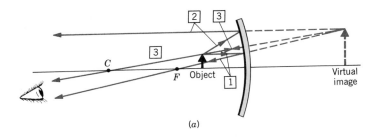

(a)

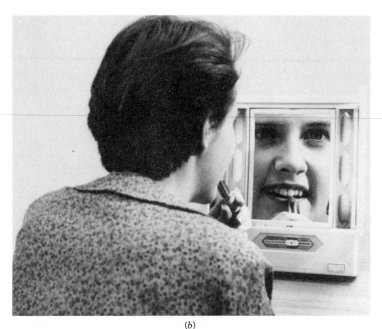

FIGURE 31.17 (a) When an object is located between the focal point F and a concave mirror, an enlarged, upright, virtual image is produced. (b) A makeup mirror is concave, and normally the image is formed when the object is within the focal point of the mirror.

(b)

this case the three reflected rays diverge from each other and do not converge to a common point. However, when projected behind the mirror, the three rays appear to come from a point behind the mirror; thus, a virtual image is formed. This virtual image is larger than the object and upright. Makeup and shaving mirrors are concave mirrors. When you place your face between the mirror and its focal point, you see an enlarged virtual image of yourself, as part b of the drawing shows.

IMAGE FORMATION BY A CONVEX MIRROR

The procedure for determining the location and size of an image in a convex mirror is similar to that for a concave mirror. The same three rays are used. However, the focal point and center of curvature of a convex mirror lie behind the mirror, not in front of it. Figure 31.18a shows the three rays that are summarized below:

Ray 1. This ray is initially parallel to the principal axis and, therefore, appears to originate from the focal point F after reflection from the mirror.

Ray 2. This ray heads toward F, emerging parallel to the principal axis after reflection. Ray 2 is analogous to ray 1, except the order of the incident and reflected rays is interchanged.

Ray 3. This ray travels toward the center of curvature C; as a result, the ray strikes the mirror perpendicularly and reflects back on itself.

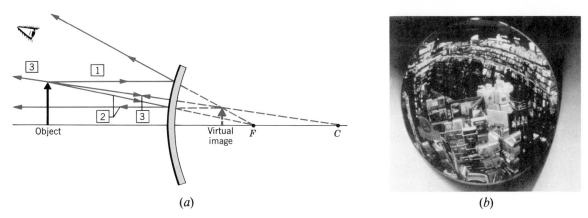

(a)

(b)

FIGURE 31.18 (*a*) An object placed in front of a convex mirror produces a virtual image behind the mirror. The virtual image is reduced in size and upright. (*b*) Convex mirrors are often used for security purposes, for the shape of the mirror gives a wider field of view than is obtainable with other types of mirrors.

These three rays appear to come from a point on a virtual image that is behind the mirror. The virtual image is diminished in size and upright, relative to the object. A convex mirror *always* forms a virtual image of the object in Figure 31.18*a*, no matter where the object is placed. Part *b* of the figure shows a convex mirror used for security purposes.

31.6 THE MIRROR EQUATION AND THE MAGNIFICATION EQUATION

CONCAVE MIRRORS

Ray diagrams drawn to scale are useful for determining the location and size of the image formed by a mirror. However, for a precise description of the image, a more analytical technique is needed. It is possible to derive an equation, called the mirror equation, that gives the image distance if the object distance and the focal length of the mirror are known. The image distance is the distance between the image and the mirror, while the object distance is the distance between the object and the mirror. In Figure 31.19 the image and object distances are labeled d_i and d_o, respectively. The height of the image is h_i, and the height of the object is h_o. Part *a* of the drawing shows

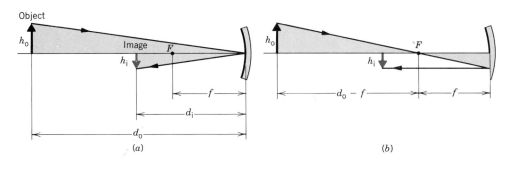

(a)   *(b)*

FIGURE 31.19 These diagrams are used to derive the mirror equation and the magnification equation. (*a*) The triangle shaded in gray is similar to the triangle shaded in color. (*b*) If the ray is close to the principal axis, the two shaded triangles are almost similar triangles.

a ray leaving the top of the object and striking the mirror at the point where the principal axis intersects the mirror. Since the principal axis is perpendicular to the mirror, this ray reflects at an equal angle and passes through the image. The gray and the colored triangles are similar because they have equal angles, so

$$\frac{h_o}{h_i} = \frac{d_o}{d_i}$$

In part *b* another ray leaves the top of the object, this time passing through the focal point *F*, reflecting parallel to the principal axis, and then passing through the image. Provided the ray remains close to the axis, the gray triangle and the colored area can be considered to be similar triangles, with the result that

$$\frac{h_o}{h_i} = \frac{d_o - f}{f}$$

Setting the two equations above equal to each other yields $d_o/d_i = (d_o - f)/f$. Rearranging this result gives the **mirror equation:**

$$\left[\begin{array}{c}\textbf{Mirror}\\\textbf{equation}\end{array}\right] \qquad \frac{1}{d_o} + \frac{1}{d_i} = \frac{1}{f} \qquad\qquad (31.3)$$

When using the mirror equation, it is useful to construct a ray diagram to guide your thinking and check on your calculation.

The **magnification** *m* of a mirror is defined as the ratio of the image height to the object height: $m = h_i/h_o$. If the image height is less than the object height, *m* is less than unity. Conversely, if the image is larger than the object, *m* is greater than unity. Since $h_i/h_o = d_i/d_o$, it follows that

$$\left[\begin{array}{c}\textbf{Magnification}\\\textbf{equation}\end{array}\right] \qquad m = \frac{\text{Image height}}{\text{Object height}} = -\frac{d_i}{d_o} \qquad\qquad (31.4)$$

By convention, the minus sign is inserted into Equation 31.4 to help in determining whether the image is upright or inverted. As Examples 2 and 3 show, the value of *m* is positive if the image is upright and negative if the image is inverted.

EXAMPLE 2

A 2.0-cm-high object is placed 7.10 cm from a concave mirror whose radius of curvature is 10.20 cm. Find (a) the position of the image and (b) its size.

SOLUTION

(a) Since $f = \frac{1}{2}R = (10.20 \text{ cm})/2 = 5.10 \text{ cm}$, the object is located between the focal point *F* and the center of curvature *C* of the mirror. This is the situation illustrated in Figure 31.16a. With $d_o = 7.10$ cm and $f = 5.10$ cm, the mirror equation can be used to find the image distance:

$$\frac{1}{d_i} = \frac{1}{f} - \frac{1}{d_o} = \frac{1}{5.10 \text{ cm}} - \frac{1}{7.10 \text{ cm}}$$

$$= 0.055 \text{ cm}^{-1} \boxed{d_i = 18 \text{ cm}}$$

In this calculation, *f* and d_o are positive numbers, indicating that the focal point and the object are in front of the mirror.

Thus, the positive answer for d_i means that the image is also in front of the mirror, and the reflected rays actually pass through the image, as Figure 31.16a shows. In other words, the positive value for d_i indicates that the image is a real image.

(b) The height of the image can be determined once the magnification *m* of the mirror is known. The magnification equation can be used to find *m*:

$$m = -\frac{d_i}{d_o} = -\frac{18 \text{ cm}}{7.10 \text{ cm}} = -2.5$$

The image height is $h_i = mh_o = (-2.5)(2.0 \text{ cm}) = \boxed{-5.0 \text{ cm}}$. The image is 2.5 times larger than the object, the negative values for *m* and h_i indicating that the image is inverted with respect to the object, as in Figure 31.16a.

EXAMPLE 3

An object is placed 6.00 cm in front of a concave mirror that has a 10.0-cm focal length, as in Figure 31.17a. (a) Determine the location of the image. (b) If the object is 1.2 cm high, find the image height. Note that the setup in this problem is analogous to a person using a shaving or makeup mirror.

SOLUTION

(a) As Figure 31.17a illustrates, the image is a virtual image that lies behind the mirror, and our answer here will make this evident. Using the mirror equation with $d_o = 6.00$ cm and $f = 10.0$ cm, we have

$$\frac{1}{d_i} = \frac{1}{f} - \frac{1}{d_o} = \frac{1}{10.0 \text{ cm}} - \frac{1}{6.00 \text{ cm}}$$

$$= -0.067 \text{ cm}^{-1} \boxed{d_i = -15 \text{ cm}}$$

The answer for d_i is negative, indicating that the image is *behind* the mirror. Thus, as expected, the image is a virtual image.

(b) The image height h_i can be found from the magnification and the object height h_o:

$$m = -\frac{d_i}{d_o} = -\frac{(-15 \text{ cm})}{6.00 \text{ cm}} = 2.5$$

The image height is $h_i = mh_o = (2.5)(1.2 \text{ cm}) = \boxed{3.0 \text{ cm}}$. The image is larger than the object, and the positive values for m and h_i indicate that the image is upright (see Figure 31.17a).

CONVEX MIRRORS

The mirror equation and the magnification equation can also be used with convex mirrors, provided the focal length f is taken to be a *negative number*. Using a negative focal length is plausible, because the focal point for a convex mirror lies behind the mirror. Example 4 deals with a convex mirror.

EXAMPLE 4

A convex mirror is used to reflect light from an object placed 65 cm in front of the mirror, as in Figure 31.18a. The focal length of the mirror is $f = -45$ cm (note the minus sign). Find (a) the location of the image and (b) its size relative to the object.

SOLUTION

(a) It is evident in Figure 31.18a that the image lies behind the mirror (a virtual image) and is smaller than the object. These features should also result from our analysis here and in part (b). With $d_o = 65$ cm and $f = -45$ cm, the mirror equation gives

$$\frac{1}{d_i} = \frac{1}{f} - \frac{1}{d_o} = \frac{1}{-45 \text{ cm}} - \frac{1}{65 \text{ cm}}$$

$$= -0.038 \text{ cm}^{-1} \boxed{d_i = -26 \text{ cm}}$$

The negative sign for d_i indicates that the image is behind the mirror and, therefore, is a virtual image.

(b) The size of the image relative to the object is given by the magnification equation:

$$m = -\frac{d_i}{d_o} = -\frac{(-26 \text{ cm})}{65 \text{ cm}} = \boxed{0.40}$$

The image is smaller (m is less than unity) and upright (m is positive) with respect to the object.

SUMMARY OF SIGN CONVENTIONS

We conclude this section by summarizing the sign conventions that are used with the mirror equation and the magnification equation. These conventions apply to both concave and convex mirrors:

Object distance

d_o is + if the object is in front of the mirror (real object).

d_o is — if the object is behind the mirror (virtual object).*

Image distance

d_i is + if the image is in front of the mirror (real image).

d_i is — if the image is behind the mirror (virtual image).

Focal length

f is + for a concave mirror.

f is — for a convex mirror.

Magnification

m is + for an image that is upright with respect to the object.

m is — for an image that is inverted with respect to the object.

SUMMARY

Wave fronts are surfaces on which all points of a wave are in the same phase of motion. If the wave fronts are flat surfaces, the wave is called a **plane wave. Rays** are lines that are perpendicular to the wave fronts and point in the direction of the velocity of the wave.

When light reflects from a smooth surface, the reflected light obeys the **law of reflection,** which states that (a) the incident ray, the reflected ray, and the normal to the surface all lie in the same plane, and (b) the angle of reflection equals the angle of incidence. The law of reflection explains how mirrors form images. A **virtual image** is one from which rays of light do not actually come, but only appear to do so. A **real image** is one from which rays of light actually emanate. A **plane mirror** forms an upright, virtual image that is located as far behind the mirror as the object is in front of the mirror. In addition, the heights of the image and the object are equal.

A **spherical mirror** has the shape of a section from the surface of a sphere. The **principal axis** of a mirror is a straight line drawn through the center of curvature and the middle of the mirror's surface. Rays that lie close to the principal axis are known as **paraxial rays.** The **radius of curvature** R of the mirror is the distance from the center of curvature to the mirror. The **focal point** of a concave spherical mirror is a point on the principal axis, in front of the mirror. Incident paraxial rays that are parallel to the principal axis converge to the focal point after being reflected from the concave mirror. The focal point of a convex spherical mirror is a point on the principal axis behind the mirror. For a convex mirror, paraxial rays that are parallel to the principal axis seem to diverge from the focal point after reflecting from the mirror. The **focal length** f of a mirror is the distance from the focal point to the middle of the mirror. The focal length and the radius of curvature are related by $f = \frac{1}{2}R$ for a concave mirror and $f = -\frac{1}{2}R$ for a convex mirror.

The image produced by a concave mirror can be located by the **ray technique,** using the three rays shown in Figure 31.15. Similarly, the image produced by a convex mirror can be found using the three rays shown in Figure 31.18.

The **mirror equation** can be used with either concave or convex mirrors and specifies the relation between the image distance d_i, the object distance d_o, and the focal length f of the mirror: $1/d_o + 1/d_i = 1/f$. The **magnification** m of a mirror is the ratio of the image height h_i to the object height h_o. The magnification is also related to d_i and d_o by the **magnification equation:** $m = h_i/h_o = -d_i/d_o$. The algebraic sign conventions for the variables appearing in these equations are summarized at the end of Section 31.6.

QUESTIONS

1. A sign painted on a store window is reversed when viewed from inside the store. If a person inside the store views the reversed sign in a plane mirror, does the sign appear as it would when viewed from outside the store? (Try it by writing some letters on a

---

* Sometimes optical systems use two (or more) mirrors, and the image formed by the first mirror serves as the object for the second mirror. Occasionally, such an object falls *behind* the second mirror. In this case the object distance is negative, and the object is said to be a virtual object.

transparent sheet of paper and then holding the back side of the paper up to a mirror.)

2. Which kind of spherical mirror, concave or convex, can be used to start a fire with sunlight? For the best results, how far from the mirror should the paper to be ignited be placed? Explain.

3. Why is your image fuzzy when you look at yourself in a small shiny silver sphere, such as a Christmas tree ornament?

4. (a) Can the image formed by a concave mirror ever be projected directly onto a screen, without the help of other mirrors or lenses? If so, specify where the object should be placed relative to the mirror. (b) Repeat part (a) assuming the mirror is convex.

5. When you look at the back side of a shiny teaspoon, held at arm's length, you see yourself upright. When you look at the other side of the spoon, you see yourself upside down. Explain.

6. The drawing shows a person standing in front of a trick mirror that has the shape of an "S." The top part of the mirror, opposite the person's head, is approximately a convex mirror with a small focal length, whereas the bottom part opposite the feet is approximately a concave mirror with a large focal length. The picture shows the focal points. Sketch the distorted image that the person sees and explain your drawing. See Drawing 6.

7. Suppose you wish to design a searchlight that produces a parallel beam of light. The searchlight consists of a light bulb in

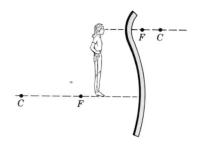

Drawing 6

front of a concave spherical mirror. Where should the bulb be positioned along the principal axis of the mirror? Give your reasoning.

8. When you see the image of yourself formed by a mirror, it is because (1) light rays actually coming from a real image enter your eyes or (2) light rays appearing to come from a virtual image enter your eyes. If light rays from the image do not enter your eyes, you do not see yourself. Are there any places on the principal axis where you cannot see yourself when you are in front of a mirror that is (a) convex and (b) concave? If so, where are these places?

9. Suppose you stand in front of a spherical mirror (concave or convex). Is it possible for your image to be (a) real and upright or (b) virtual and inverted? Justify your answer in each case.

PROBLEMS

Section 31.2 The Reflection of Light, Section 31.3 The Formation of Images by a Plane Mirror

1. Two plane mirrors are separated by 120°, as the drawing illustrates. If a ray strikes mirror M_1, at a 65° angle of incidence, at what angle θ does it leave mirror M_2?

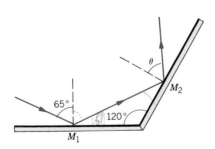

2. A person whose eyes are 1.50 m above the floor stands in front of a plane mirror. The top of his head is 0.11 m above his eyes. (a) What is the height of the shortest mirror in which he can see his entire image? (b) How far above the floor should the bottom edge of the mirror be placed?

3. Two diverging light rays, originating from the same point, have an angle of 10° between them. After the rays reflect from a plane mirror, what is the angle between them? Construct one possible ray diagram that supports your answer.

4. Suppose you walk with a speed of 0.90 m/s toward a plane mirror. What is the speed of your image *relative to you*, when your velocity is (a) perpendicular to the mirror and (b) at an angle of 50.0° with respect to the normal to the mirror?

* 5. A person, trying on a new suit in a clothing store, stands in front of two mirrors to examine the "fit." The mirrors intersect at a 90° angle. When the person looks into the mirrors, he can see three images of himself. Draw the rays and show where the images are located.

* 6. A ray of light strikes a plane mirror at a 45° angle of incidence. The mirror is then rotated by 15° into the position shown in color in the drawing, while the incident ray is kept fixed. (a) Through what angle ϕ does the reflected ray rotate? (b) What is the answer to part (a) if the angle of incidence is 60°? See Drawing 6.

**7. The drawing shows a square room. One wall is missing, and the other three are each mirrors. From point P in the center of the open side, a laser is fired, with the intent of hitting a small target

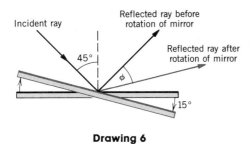

Drawing 6

located at the center of one wall. Identify five directions in which the laser can be fired and score a hit, assuming the light does not strike any mirror more than once. Draw the rays to confirm your choices.

Section 31.4 Spherical Mirrors, Section 31.5 The Formation of Images by Spherical Mirrors

8. A 2.0-cm-high object is situated 15.0 cm in front of a concave mirror that has a radius of curvature of 10.0 cm. Using a ray diagram drawn to scale on a piece of paper, measure the location and the height of the image. The radius of curvature must be drawn to scale, and the scale chosen for the horizontal and vertical directions must be the same.

9. Repeat problem 8 for a concave mirror with a focal length of 20.0 cm, an object distance of 12.0 cm, and an 8.0-cm-high object.

10. Repeat problem 8 for a convex mirror with a radius of curvature of 1.00 m, an object distance of 25 cm, and a 10.0-cm-high object.

11. Repeat problem 8 for a concave mirror with a focal length of 7.50 cm, an object distance of 11.0 cm, and a 2.0-cm-high object.

Section 31.6 The Mirror Equation and the Magnification Equation

12. The image behind a convex mirror (radius of curvature = 68 cm) is located 22 cm from the mirror. (a) Where is the object located and (b) what is the magnification of the mirror? State whether the image is (c) upright or inverted and (d) larger or smaller than the object.

13. The image of a very distant car is located 12 cm behind a convex mirror. (a) What is the radius of curvature of the mirror? (b) Draw a ray diagram to scale showing this situation.

14. A small postage stamp is placed in front of a concave mirror (radius = R), such that the image distance equals the object distance. (a) In terms of R, what is the object distance? (b) What is the magnification of the mirror? (c) State whether the image is upright or inverted relative to the object. Draw a ray diagram to guide your thinking.

15. Convex mirrors are being used to monitor the aisles in a store. The mirrors have a radius of curvature of 4.0 m. (a) What is the image distance if a customer is 15 m in front of the mirror? (b) Is the image real or virtual? (c) If a customer is 1.6 m tall, how tall is the image? (d) Determine d_i and h_i from an accurately drawn ray diagram and compare your answers with those obtained above.

16. A concave mirror ($R = 64.0$ cm) is used to project a transparent slide onto a wall. The slide is located at a distance of 38.0 cm from the mirror, and a small flashlight shines light through the slide and onto the mirror. (a) How far from the wall should the mirror be located? (b) The height of the object on the slide is 1.20 cm. What is the height of the image? (c) State whether the image is upright or inverted relative to the object.

17. A dentist's mirror is placed 1.5 cm from a tooth. The enlarged image is located 4.3 cm behind the mirror. (a) What kind of mirror (plane, concave, or convex) is being used? (b) Determine the focal length of the mirror. (c) What is the magnification? (d) How is the image oriented relative to the object?

18. The radius of curvature of a plane mirror is infinite ($R = \infty$). With this in mind, show that the mirror equation and the magnification equation correctly predict the location, magnification, orientation (upright or inverted), and nature (real or virtual) of an image formed by a plane mirror.

* 19. A concave makeup mirror is designed so the virtual image is twice the size of the object, when the distance between the object and the mirror is 15 cm. (a) Determine the radius of curvature of the mirror. (b) Draw a ray diagram to scale, showing this situation.

* 20. Show that to produce an image with magnification m, using a mirror whose focal length is f, the object must be placed at a distance d_o from the mirror, where $d_o = (m - 1)f/m$.

* 21. A gemstone is placed 20.0 cm in front of a concave mirror and is within the focal point. (a) When the concave mirror is replaced with a plane mirror, the image moves 15.0 cm toward the mirror. Find the focal length of the concave mirror. (b) Where would the plane mirror have to be placed relative to the gem so the image did not move?

**22. A spherical mirror is polished on both sides. When used as a convex mirror, the magnification is $+1/4$. What is the magnification when used as a concave mirror, the object remaining the same distance from the mirror?

**23. Using the mirror equation and the magnification equation, show that for a convex mirror the image is always (a) virtual (i.e., d_i is always negative) and (b) upright and smaller, relative to the object (i.e., m is positive and less than unity).

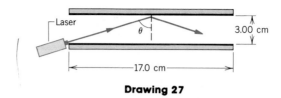

Drawing 27

ADDITIONAL PROBLEMS

24. A coin is placed 8.0 cm in front of a concave mirror. The mirror produces a real image that has a diameter 4.0 times larger than that of the coin. Find the radius of curvature of the mirror.

25. The intent of this problem is to verify the phenomenon of spherical aberration. Draw a semicircle with a radius of 10 cm to represent a concave spherical mirror. Recall that a radial line drawn between the center of curvature and any point on the arc is perpendicular to the arc and, hence, is a normal to the mirror. (a) Draw a ray parallel to the principal axis at a distance of 5 cm from the axis. Where the ray strikes the mirror, draw the normal and the reflected ray, such that the angle of reflection equals the angle of incidence. Extend the reflected ray until it intersects the principal axis (this ray should cross the axis just inside the focal point). (b) Repeat (a) with a ray drawn at a distance of 7.5 cm from the principal axis and note how much farther from the focal point this ray crosses the axis.

26. (a) Where should a diamond ring be placed in front of a concave mirror, such that the image is twice the size of the ring? (*Note:* The image may be inverted.) Express your answer in terms of the radius of curvature R. (b) Draw a ray diagram to confirm your answer.

* 27. Two plane mirrors are facing each other. They are parallel, 3.00 cm apart, and 17.0 cm in length, as the picture indicates. A laser beam is directed at the top mirror, from the left edge of the bottom mirror. What is (a) the smallest and (b) largest angle of incidence θ that allows the beam to hit each mirror once or less? See Drawing 27.

* 28. Complete each row in the following table. As an example, the first row of the table shows a completed set of data.

* 29. The drawing shows the top view of a *corner reflector* that is formed by placing two plane mirrors at right angles to each other. The incident ray lies in a plane that is perpendicular to the mirrors. The corner reflector has the property that any such ray that strikes the reflector is reflected back parallel to the incident ray. For the ray shown in the drawing, prove that this property is true for any angle of incidence θ.

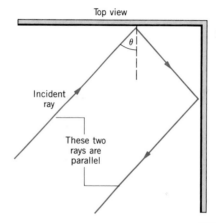

**30. A concave mirror has a focal length of 30.0 cm. The distance between an object and its image is 45.0 cm. (a) Assuming d_o is greater than d_i, find the object and image distances. There are two solutions to this problem. (b) Construct a ray diagram corresponding to each solution.

Type of Mirror	Focal Length	Radius of Curvature	Object Distance	Image			Magnification
				Distance	Real or Virtual?	Upright or Inverted?	
Concave	+18.0 cm	36.0 cm	+62.0 cm	+25.4 cm	Real	Inverted	−0.410
Plane	—	—	+30.0 cm				
Concave		64.0 cm					+8.00
				+85.0 cm			−2.00
	+27.0 cm			−71.0 cm			
			+16.0 cm	−12.0 cm			
	−49.0 cm		+7.50 cm				

CHAPTER 32

The Refraction of Light and Lenses

A diamond sparkles because it is cut to enhance light that is refracted into the diamond and then shines back out because of a phenomenon known as total internal reflection. Four common cuts are (clockwise, from upper left) oval, emerald, pear, and brilliant.

32.1 THE INDEX OF REFRACTION

Many materials, such as air, water, and glass are transparent to light. When a ray of light enters a transparent material at an angle other than normal incidence, the ray deviates from its incident direction. This change in the direction of travel as light passes from one medium into another is called *refraction.* Refraction plays a central role in determining the properties of the lenses used in a wide variety of optical instruments, including eyeglasses, cameras, microscopes, telescopes, and even the human eye itself.

As we will see, refraction depends on the speed of light in a material, and when light travels through a solid, a liquid, or a gas, its speed is different than that in a vacuum. The *index of refraction* (or *refractive index*) n of a material is the ratio of the speed of light c in a vacuum to the speed of light v in the material:

$$\left[\begin{array}{c} \textbf{Index of} \\ \textbf{refraction} \end{array} \right] \qquad n = \frac{\text{Speed of light in a vacuum}}{\text{Speed of light in the material}} = \frac{c}{v} \qquad (32.1)$$

TABLE 32.1 Index of Refraction[a]
for Various Substances

Substance	Index of Refraction, n
Solids at 20 °C	
Diamond	2.419
Glass, crown	1.52
Ice (0 °C)	1.309
Sodium chloride	1.544
Quartz	
Crystalline	1.544
Fused	1.458
Liquids at 20 °C	
Benzene	1.501
Carbon	
tetrachloride	1.461
Ethyl alcohol	1.362
Water	1.333
Gases at 0 °C, 1 atm	
Air	1.000 293
Carbon dioxide	1.000 45
Oxygen, O_2	1.000 271
Hydrogen, H_2	1.000 139

[a] Measured with light whose wavelength in a vacuum is 589 nm.

Table 32.1 lists the refractive indices for some common substances. As the table indicates, the values of n are greater than unity, so the speed of light in a material is less than it is in a vacuum. For example, the index of refraction for diamond is $n = 2.42$, so the speed of light in diamond is $v = c/n = (3.00 \times 10^8 \text{ m/s})/2.42 = 1.24 \times 10^8$ m/s. In contrast, the index of refraction for air (and other gases as well) is so close to unity that $n_{air} = 1$ for most purposes. The index of refraction depends slightly on the wavelength of the light, and the values in Table 32.1 correspond to a wavelength of $\lambda = 589$ nm in a vacuum.

32.2 SNELL'S LAW AND THE REFRACTION OF LIGHT

SNELL'S LAW

When light strikes the interface between two transparent materials, such as air and water, the light divides into two parts, as Figure 32.1a illustrates. Part of the light is reflected, with the angle of reflection equaling the angle of incidence. The remainder of the light is transmitted across the interface. If the incident ray does not strike the interface at normal incidence, the transmitted ray has a different direction than the incident ray. The ray that enters the second material is said to be refracted.

In Figure 32.1a the light travels from a medium where the refractive index is smaller (air) into a medium where it is larger (water), and the refracted ray is bent *toward* the normal. Both the incident and refracted rays obey the principle of revers-

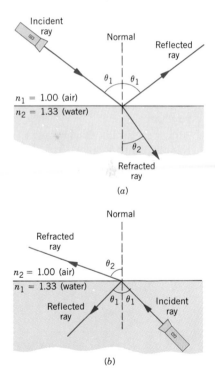

FIGURE 32.1 (a) When light is directed from air into water, part of the light is reflected at the surface and the remainder is refracted into the water. The refracted ray is bent *toward* the normal ($\theta_2 < \theta_1$). (b) When light is directed from water into air, the refracted ray in air is bent *away* from the normal ($\theta_2 > \theta_1$).

ibility, so their directions can be reversed to give a situation like that in part *b* of the drawing. Here light travels from a material with a greater refractive index into one with a smaller refractive index, and the refracted ray is bent *away* from the normal. In this case the reflected ray lies in the water, rather than in the air. In both parts of the drawing the angles of incidence, refraction, and reflection are measured relative to the normal. In addition, the index of refraction of air is labeled n_1 in part *a* of the figure, while it is labeled n_2 in part *b*, because *we label all variables associated with the incident (and reflected) ray with a subscript 1 and all variables associated with the refracted ray with a subscript 2.*

The angle of refraction θ_2 depends on the angle of incidence θ_1 and on the indices of refraction, n_2 and n_1, of the two media. The relation between these quantities is known as ***Snell's law of refraction,*** after the Dutch mathematician Willebrord Snell (1591–1626) who discovered it experimentally. A proof of Snell's law is presented at the end of this section.

SNELL'S LAW OF REFRACTION

When light travels from a material with refractive index n_1 into a material with refractive index n_2, the refracted ray, the incident ray, and the normal to the surface all lie in the same plane; furthermore, the angle of refraction θ_2 is related to the angle of incidence θ_1 by

$$n_1 \sin \theta_1 = n_2 \sin \theta_2 \qquad (32.2)$$

Example 1 illustrates Snell's law.

EXAMPLE 1

A light ray strikes an air/water surface at an angle of 46° with respect to the normal. Find the angle of refraction when the direction of the ray is (a) from air to water and (b) from water to air.

SOLUTION

(a) The incident ray is in air, so $\theta_1 = 46°$ and $n_1 = 1.00$. The refracted ray is in water, so $n_2 = 1.33$. Snell's law can be used to find the angle of refraction:

$$\sin \theta_2 = \frac{n_1 \sin \theta_1}{n_2} = \frac{(1.00) \sin 46°}{1.33} = 0.54$$

$$\theta_2 = \sin^{-1}(0.54) = \boxed{33°}$$

The refracted ray is bent *toward* the normal, since θ_2 is less than θ_1, as Figure 32.1*a* shows.

(b) Now the incident ray propagates in water ($\theta_1 = 46°$, $n_1 = 1.33$), and the refracted ray propagates in air ($n_2 = 1.00$). Snell's law yields

$$\sin \theta_2 = \frac{n_1 \sin \theta_1}{n_2} = \frac{(1.33) \sin 46°}{1.00} = 0.96$$

$$\theta_2 = \sin^{-1}(0.96) = \boxed{74°}$$

Since θ_2 is greater than θ_1, the refracted ray is bent *away* from the normal, as Figure 32.1*b* indicates.

The simultaneous reflection and refraction of light at an interface finds applications in a number of devices. For instance, many cars come equipped with an interior rearview mirror that has an adjustment lever. One position of the lever sets the mirror for day driving, while another position sets it for night driving. The night setting is useful for reducing glare from the headlights of the car behind. As Figure 32.2*a* indicates, this kind of mirror is a glass wedge, the back side of which is silvered and highly reflecting. Part *b* of the picture shows the day setting. Light from the car behind follows the path *ABCE* in reaching the driver's eye. At points *A* and *C*, where the light strikes the front air–glass surface, there are both reflected and refracted rays. The reflected rays are drawn as thin colored lines, the thinness denoting that only a small

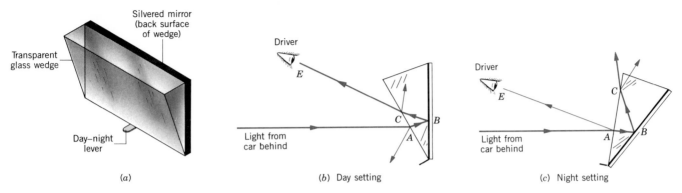

FIGURE 32.2 An interior rearview mirror with a day–night adjustment lever operates by taking advantage of the simultaneous reflection and refraction that occurs at an air–glass surface.

percentage (about 10%) of the light is reflected. The weak reflected rays at *A* and *C* do not reach the driver's eye. In contrast, almost all the light reaching the silvered back surface at *B* is reflected toward the driver. Since most of the light follows the path *ABCE*, the driver sees a bright image of the car behind during the day.

During the night, the adjustment lever can be used to rotate the mirror clockwise (see part *c* of the drawing), away from the driver. Now, most of the light from the headlights behind follows the path *ABC* and does not reach the driver. Only the light that is weakly reflected from the front surface along path *AE* is seen. As a result, there is significantly less glare.

APPARENT DEPTH

One of the interesting consequences of refraction is that an object lying beneath the surface of water appears to be closer to the surface than it actually is. Example 2 sets the stage for explaining why, by showing what must be done to shine a light on such an object.

EXAMPLE 2

A searchlight on a yacht is being used at night to illuminate a sunken chest, as in Figure 32.3. At what angle of incidence θ_1 should the light be aimed?

SOLUTION

The angle of incidence θ_1 can be determined from Snell's law, provided the angle of refraction θ_2 can be found. From the data in the drawing it follows that $\tan \theta_2 = (2.0 \text{ m})/(3.3 \text{ m})$, so $\theta_2 = 31°$. With $n_1 = 1.00$ for air and $n_2 = 1.33$ for water, Snell's law gives

$$\sin \theta_1 = \frac{n_2 \sin \theta_2}{n_1} = \frac{(1.33) \sin 31°}{1.00} = 0.69$$

$$\theta_1 = \sin^{-1}(0.69) = \boxed{44°}$$

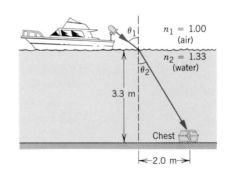

FIGURE 32.3 The beam from the searchlight is refracted when it enters the water and, consequently, does not follow a single straight-line path to the chest.

When the sunken chest in Example 2 is viewed from the boat, light rays from the chest pass upward through the water, refract away from the normal when they enter

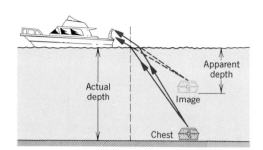

FIGURE 32.4 Because light from the chest is refracted away from the normal when the light enters the air, the apparent depth of the image is less than the actual depth.

the air, and then travel to the observer, as Figure 32.4 suggests. This picture is similar to Figure 32.3, except the direction of the rays is reversed and the searchlight is replaced by an observer. The rays entering the air are extended back into the water (see dashed colored lines) and indicate that the observer sees an image of the chest at an *apparent depth* that is less than the actual depth. When the observer is *directly above* the submerged object, the apparent depth d' is related to the actual depth d by

$$\begin{bmatrix} \textbf{Apparent depth with} \\ \textbf{observer directly} \\ \textbf{above object} \end{bmatrix} \qquad d' = d\left(\frac{n_2}{n_1}\right) \qquad\qquad (32.3)$$

In this result, n_1 is the refractive index of the medium associated with the incident ray (the medium in which the object is located), while n_2 refers to the medium associated with the refracted ray (the medium in which the observer is situated). Thus, if you look straight down into a 3.0-m-deep pool, an object on the bottom appears to be only $d' = (3.0 \text{ m})(1.00/1.33) = 2.3$ m away. Equation 32.3 also applies when the observer is beneath the water and is looking at an object in the air directly above. In this case n_2 is greater than n_1, so the object appears to be farther away (not closer) than it actually is. The proof of Equation 32.3 is left as problem 17 at the end of the chapter.

THE DISPLACEMENT OF LIGHT BY A TRANSPARENT SLAB OF MATERIAL

A common use of a transparent material, such as glass, is for windows. A window pane consists of a plate of glass with parallel surfaces. When a ray of light passes through the glass, the emergent ray is parallel to the incident ray, but displaced from it, as Figure 32.5 shows. This result can be verified by applying Snell's law to each of the two glass surfaces, with the result that $n_1 \sin \theta_1 = n_2 \sin \theta_2 = n_3 \sin \theta_3$. Since air surrounds the glass, $n_1 = n_3$, and the relation above shows that $\sin \theta_1 = \sin \theta_3$. Therefore, $\theta_1 = \theta_3$, and the emergent ray is parallel to the incident ray. However, the drawing shows that the emergent ray is displaced laterally relative to the incident ray. The extent of the lateral displacement depends on the angle of incidence and on the thickness and refractive index of the glass.

DERIVATION OF SNELL'S LAW

Snell's law can be derived by considering what happens to the wave fronts when the light passes from one medium into another. Figure 32.6a shows light propagating from medium 1, where the speed is relatively large, into medium 2, where the speed is smaller; therefore, n_1 is less than n_2. The plane wave fronts in this picture are drawn perpendicular to the incident and refracted rays. Since the part of each wave front that penetrates medium 2 slows down, the wave fronts in medium 2 are rotated

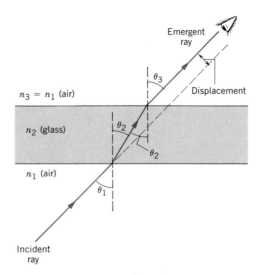

FIGURE 32.5 When a ray of light passes through a pane of glass that has parallel surfaces and is surrounded by air, the emergent ray is parallel to the incident ray ($\theta_3 = \theta_1$), but is displaced from it.

clockwise relative to those in medium 1. Correspondingly, the ray in medium 2 is bent toward the normal, as the drawing shows.

Although the incident and refracted waves have different speeds, *they have the same frequency f.* Each wave front crosses the boundary between the two media. Therefore, the number of wave fronts per second arriving at the boundary equals the number of wave fronts per second leaving the boundary, and the frequencies of the incident and refracted waves are the same.

The distance between successive wave fronts in Figure 32.6a has been chosen to be the wavelength λ. Since the frequencies are the same in both media, but the speeds are different, it follows from Equation 20.1 that $\lambda_1 = v_1/f$ and $\lambda_2 = v_2/f$. Since v_1 is assumed to be larger than v_2, λ_1 is larger than λ_2, and the wave fronts are farther apart in medium 1.

Figure 32.6b shows an enlarged view of the incident and refracted wave fronts at the surface. The angles θ_1 and θ_2 within the shaded right triangles are, respectively, the angles of incidence and refraction. In addition, the triangles share the same hypotenuse h. Therefore,

$$\sin \theta_1 = \frac{\lambda_1}{h} = \frac{(v_1/f)}{h} \quad \text{and} \quad \sin \theta_2 = \frac{\lambda_2}{h} = \frac{(v_2/f)}{h}$$

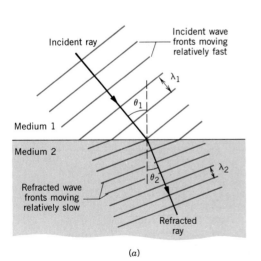

(a)

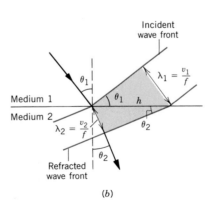

(b)

FIGURE 32.6 (a) The wave fronts are refracted as the light ray passes from medium 1 into medium 2. The distance between successive wave fronts has been chosen to be the wavelength λ of the light. The angles θ_1 and θ_2 are, respectively, the angles of incidence and refraction. (b) An enlarged view of the incident and refracted wave fronts at the surface.

Combining these two equations into a single equation by eliminating the common term hf gives

$$\frac{\sin \theta_1}{v_1} = \frac{\sin \theta_2}{v_2}$$

By multiplying each side of this result by c, the speed of light in a vacuum, and recognizing that the ratio c/v is the index of refraction n, we arrive at Snell's law of refraction: $n_1 \sin \theta_1 = n_2 \sin \theta_2$.

32.3 TOTAL INTERNAL REFLECTION

THE CRITICAL ANGLE AND TOTAL INTERNAL REFLECTION

When light passes from a medium of larger refractive index into one of smaller refractive index—for example, from water to air—the refracted ray bends *away* from the normal, as in Figure 32.7a. In fact, at a particular angle of incidence, called the *critical angle* θ_c, the angle of refraction is 90°, and the refracted ray points along the surface. Figure 32.7b illustrates the critical angle. When the angle of incidence exceeds the critical angle, as in part c of the drawing, there is no refracted light. All the incident light is reflected back into the medium from which it came, a phenomenon known as *total internal reflection.*

Total internal reflection occurs only when light travels from a higher-index medium toward a lower-index medium. Total internal reflection does not occur when light propagates in the reverse direction—for example, from air to water. In this situation, the refracted ray bends toward the normal, rather than away from it, so there is always a refracted ray, regardless of the angle of incidence.

An expression for the critical angle θ_c can be obtained directly from Snell's law by setting $\theta_1 = \theta_c$ and $\theta_2 = 90°$:

$$\begin{bmatrix} \text{Critical} \\ \text{angle} \end{bmatrix} \qquad \sin \theta_c = \frac{n_2 \sin 90°}{n_1} = \frac{n_2}{n_1} \qquad (n_1 > n_2) \qquad (32.4)$$

For example, the critical angle for light traveling from water ($n_1 = 1.33$) to air ($n_2 = 1.00$) is $\theta_c = \sin^{-1}(1.00/1.33) = 48.8°$. For incident angles greater than 48.8°, Snell's

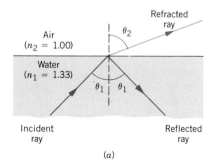

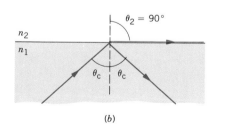

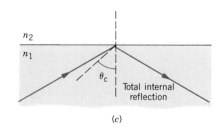

FIGURE 32.7 (a) When light travels from a higher-index medium (water) into a lower-index medium (air), the refracted ray is bent away from the normal. (b) When the angle of incidence is equal to the critical angle θ_c, the angle of refraction is 90°. (c) If θ_1 is greater than θ_c, there is no refracted ray, and total internal reflection occurs.

law predicts that $\sin \theta_2$ is greater than unity, a value that is not possible. Thus, for all light rays with incident angles exceeding 48.8° there is no refracted light, and the light is totally reflected back into the water. The next example illustrates how the critical angle changes when the indices of refraction change.

EXAMPLE 3

A beam of light is propagating through diamond ($n_1 = 2.42$) and strikes a diamond–air interface at an angle of incidence of 28°. (a) Will part of the beam enter the air or will the beam be totally reflected at the interface? (b) Repeat part (a), assuming the diamond is surrounded by water ($n_2 = 1.33$).

SOLUTION

(a) The critical angle θ_c for total internal reflection at the diamond–air interface is given by Equation 32.4 as

$$\theta_c = \sin^{-1}\left(\frac{n_2}{n_1}\right) = \sin^{-1}\left(\frac{1.00}{2.42}\right) = 24.4°$$

Because the angle of incidence of 28° is greater than the critical angle, the light is totally reflected.

(b) If water, rather than air, surrounds the diamond, the critical angle for total internal reflection becomes larger:

$$\theta_c = \sin^{-1}\left(\frac{n_2}{n_1}\right) = \sin^{-1}\left(\frac{1.33}{2.42}\right) = 33.3°$$

Now a ray of light that has an angle of incidence of 28° (less than the critical angle) at the diamond–water interface is refracted into the water.

The critical angle determines how well a diamond sparkles. For instance, Figure 32.8 shows the critical angle for light incident on a diamond–air interface. Light with a large range of incident angles strikes a bottom facet. Those rays whose angle of incidence exceeds the critical angle are totally reflected back into the diamond, eventually exiting out the top to give the diamond its sparkle. If the diamond is placed in water, the critical angle increases. Therefore, more rays now strike the bottom facet at an angle less than the critical angle, and some of this light is lost out the bottom. As a result, the gem sparkles less.

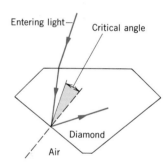

FIGURE 32.8 When a light ray strikes the bottom facet of a diamond and the angle of incidence is greater than the critical angle, the light is totally reflected.

PRISMS AND TOTAL INTERNAL REFLECTION

Many optical instruments, such as binoculars, periscopes, and telescopes, use glass prisms to turn a beam of light through 90° or 180°. Figure 32.9a shows a light ray striking a 45°–45°–90° glass prism ($n_1 = 1.5$). Most of the light enters the prism and is directed toward the hypotenuse of the prism with a 45° angle of incidence. The critical angle for a glass–air interface is $\theta_c = \sin^{-1}(n_2/n_1) = \sin^{-1}(1.0/1.5) = 42°$. Since the angle of incidence is greater than the critical angle, the light is totally reflected at the hypotenuse and is directed vertically upward in the drawing, having been turned through an angle of 90°. Part b of the picture shows how the same prism can turn the beam through 180° if total internal reflection occurs twice. Prisms can

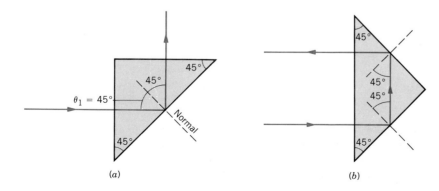

FIGURE 32.9 Total internal reflection at a glass–air interface can be used to turn a ray of light through an angle of (a) 90° or (b) 180°.

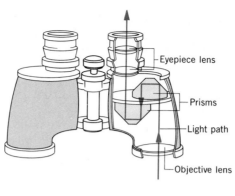

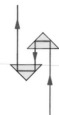

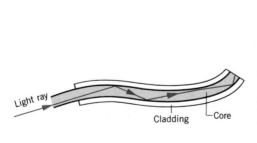

FIGURE 32.10 Two prisms, each reflecting the light by total internal reflection, are sometimes used in binoculars.

also be used in tandem to produce a lateral displacement of a light beam, while leaving its initial direction unaltered. Figure 32.10 illustrates such an application in binoculars.

FIBER OPTICS

Another application of total internal reflection is in the exciting field of fiber optics. In fiber optics, hair-thin threads of glass, called optical fibers, "pipe" light from one place to another. Figure 32.11 shows that an optical fiber consists of a cylindrical inner *core* that carries the light and an outer concentric shell, the *cladding*. The core is made from transparent glass or plastic that has a relatively high index of refraction.

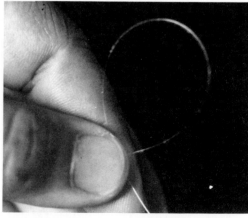

FIGURE 32.11 Light can travel with little loss in a curved optical fiber, because the light is totally reflected whenever it strikes the core–cladding interface and because the absorption of light by the glass core itself is small. The thickness of the optical fiber (core plus cladding) is about the thickness of a human hair.

The cladding, also made of glass, has a relatively low index of refraction. Light enters one end of the core, strikes the core/cladding surface at an angle of incidence greater than the critical angle, and, therefore, is reflected back into the core. Light, then, travels inside the optical fiber along a zigzag path. In a well-designed fiber, little light is lost as a result of absorption by the core, so light can travel many kilometers before its intensity diminishes appreciably. Fibers are often bundled together to produce cables that usually contain 72 fibers. Because the fibers themselves are so thin, the cables are relatively small and flexible, and they can fit into places inaccessible to larger wire cables.

Optical fibers are revolutionizing the manner in which video, telephone, and computer-data communications are transmitted, because a light beam can carry information through an optical fiber just as electricity carries information through copper wires and radio waves carry information through space. The information-carrying capacity of light, however, is thousands of times greater than that of electricity or radio waves. A laser beam traveling through a single optical fiber can carry tens of thousands of telephone conversations and several TV programs simultaneously. Optical fiber cables are the medium of choice for high-quality telecommunications, because the cables are relatively immune to external electrical interference.

Flexible, fiber optic cables are also used in medicine. For instance, such a cable can be passed through the esophagus into the stomach to search for ulcers and other abnormalities. Light is carried into the stomach by the outer fibers of the cable, reflected back by the stomach wall, and transmitted out of the stomach by the inner fibers of the same cable. The image can be displayed on a TV monitor or recorded on film. In arthroscopic surgery, a small surgical instrument, several millimeters in diameter, is mounted at the end of an optical fiber cable. The surgeon can insert the instrument and cable into a joint, such as the knee, with only a tiny incision and minimal damage to the surrounding tissue.

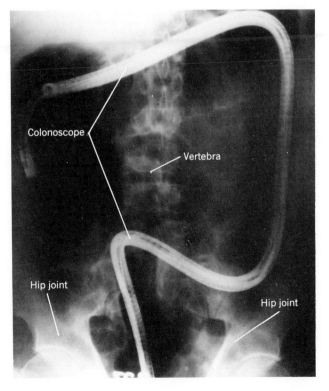

Fiber optics has led to the development of the colonoscope, which is used to examine the interior of the colon.

32.4 POLARIZATION AND THE REFLECTION AND REFRACTION OF LIGHT

For incident angles other than 0°, unpolarized light becomes partially polarized in reflecting from a nonmetallic surface, such as from water. To demonstrate this fact, simply rotate a pair of Polaroid sunglasses in the sunlight reflected from a lake. You will see that the light intensity transmitted through the glasses is a minimum when the glasses are oriented as they are normally worn. Since sunglasses are built with the transmission axis aligned in the vertical direction, it follows that the light reflected from the lake is preferentially polarized in the horizontal direction.

There is one special angle of incidence at which the reflected light is completely polarized parallel to the surface, the refracted ray being only partially polarized. This angle is called the **Brewster angle** θ_B. Figure 32.12 summarizes what happens when unpolarized light strikes a nonmetallic surface at the Brewster angle. The value of θ_B is given by **Brewster's law,** in which n_1 and n_2 are, respectively, the refractive indices of the materials in which the incident and refracted rays propagate:

$$\left[\begin{array}{c}\textbf{Brewster's}\\\textbf{law}\end{array}\right] \qquad\qquad \tan \theta_B = \frac{n_2}{n_1} \qquad\qquad (32.5)$$

This relation is named after the Scotsman David Brewster (1781–1868) who discovered it empirically.

Figure 32.12 also indicates that the reflected and refracted rays are perpendicular to each other when light strikes the surface at the Brewster angle. This result follows directly from Snell's law and Brewster's law:

$$\sin \theta_B = \left(\frac{n_2}{n_1}\right) \sin \theta_2 = \tan \theta_B \sin \theta_2 = \left(\frac{\sin \theta_B}{\cos \theta_B}\right) \sin \theta_2$$

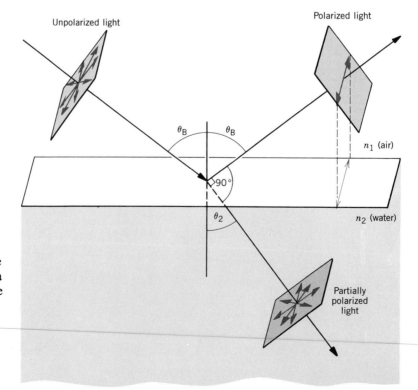

FIGURE 32.12 When unpolarized light is incident on a nonmetallic surface at the Brewster angle θ_B, the reflected light is 100% polarized in a direction parallel to the surface. The refracted light, however, is only partially polarized. The angle between the reflected and refracted rays is 90°.

In other words, $\cos \theta_B = \sin \theta_2$. The cosine of an angle can be equal to the sine of another angle only if the two angles are complementary, so $\theta_B + \theta_2 = 90°$, and the reflected and refracted rays are perpendicular.

32.5 THE DISPERSION OF LIGHT: PRISMS AND RAINBOWS

Figure 32.13a shows a ray of light passing through a glass prism. When the light enters the prism at the left face, the refracted ray is bent toward the normal, for the refractive index of glass is greater than that of air. Conversely, when the light leaves the prism at the right face and enters the air, the light is refracted away from the normal. Thus, the net effect of the prism is to change the direction of the ray. Because the refractive index of the glass depends on wavelength (see Table 32.2), the rays corresponding to different colors are bent by different amounts by the prism and depart traveling in different directions. The greater the index of refraction for a given color, the greater the bending, and part b of the drawing shows the refractions for the colors red and violet, which are at opposite ends of the visible spectrum. If a beam of sunlight, which contains all colors, is sent through a prism, the sunlight is separated into the spectrum of colors, as part c shows. The spreading of light into its color components is called *dispersion.*

Another example of dispersion occurs in rainbows, in which refraction by water droplets gives rise to the colors. Rainbows are often seen just as a storm is leaving, and we look at the departing rain with the sun at our backs. When light from the sun enters

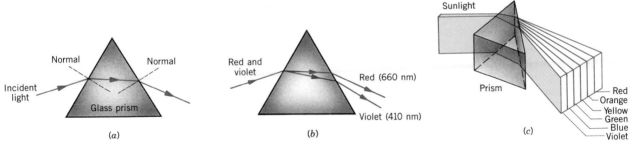

FIGURE 32.13 (a) A ray of light is refracted as it passes through the prism. (b) Two different colors are refracted by different amounts. For clarity, the amount of refraction has been exaggerated. (c) Because the refractive index of glass varies with wavelength, the prism disperses the sunlight into its color components.

TABLE 32.2 Indices of Refraction n of Selected Materials at Various Wavelengths

Color	Wavelength in Vacuum (nm)	Crown Glass	Flint Glass	Diamond
Red	660	1.520	1.662	2.410
Orange	610	1.522	1.665	2.415
Yellow	580	1.523	1.667	2.417
Green	550	1.526	1.674	2.426
Blue	470	1.531	1.684	2.444
Violet	410	1.538	1.698	2.458

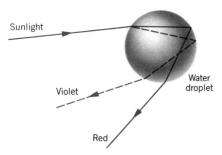

FIGURE 32.14 When sunlight enters a water droplet, the light is refracted at the front surface and then reflected from the rear surface. A second refraction occurs when the light reenters the air. As a result, the light that emerges from the droplet is dispersed into its constituent colors, of which only two are shown.

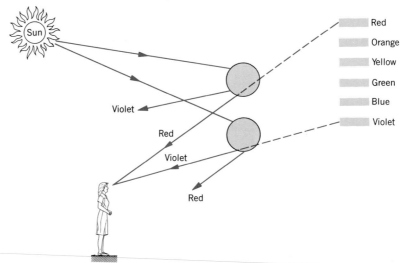

FIGURE 32.15 The different colors seen in a rainbow originate from water droplets at different angles of elevation.

a spherical raindrop, as in Figure 32.14, light of each color is refracted or bent by an amount that depends on the refractive index of water for that wavelength. After reflection from the back surface of the droplet, the different colors are again refracted as they reenter the air. Although each droplet disperses the light into its full spectrum of colors, the observer in Figure 32.15 sees only one color of light coming from any given droplet, since only one color travels in the right direction to reach the observer's eyes. However, all colors are visible in a rainbow, because each color originates from different droplets at different angles of elevation.

32.6 CONVERGING AND DIVERGING LENSES

CONVERGING LENSES

A most useful application of the refraction of light is in lenses for optical instruments. A lens is a piece of transparent material that refracts light in such a way that an image of the source of the light is formed. Figure 32.16a shows a crude lens formed from two glass prisms. Suppose an object, centered on the principal axis, is infinitely far from the lens so the rays from the object are parallel to the principal axis. In passing through the prisms, these rays are bent toward the axis because of refraction. Unfortunately, the rays do not all cross the axis at the same place, and, therefore, such a crude lens gives rise to a "blurred" image of the object.

A better lens can be constructed from a single piece of transparent material with properly curved surfaces, often spherical, as in part b of the drawing. With this improved lens, rays that are near the principal axis (paraxial rays) and parallel to it converge to a single point on the axis after emerging from the lens. This point is called

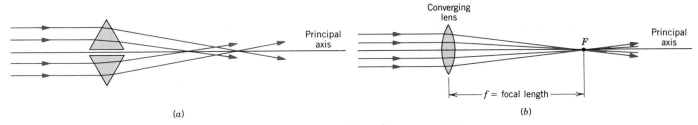

FIGURE 32.16 (*a*) This two-prism, converging lens causes rays of light that are parallel to the principal axis to cross the axis at different points after passing through the lens. (*b*) With a well-designed converging lens, paraxial rays that are parallel to the axis converge to a single point *F* after passing through the lens. This point is the focal point of the lens, and its distance *f* from the lens is the focal length.

the *focal point* *F* of the lens. Thus, an object located infinitely far away on the principal axis leads to an image at the focal point of the lens. The distance between the focal point and the lens is the *focal length f*. In what follows, we assume the lens is sufficiently thin compared to *f* that it makes no difference whether *f* is measured between the focal point and either surface of the lens or the center of the lens. The type of lens shown in Figure 32.16*b* is known as a *converging lens,* because the lens causes incident parallel rays to converge at the focal point.

DIVERGING LENSES

Another type of lens commonly used in optical instruments is a *diverging lens,* which causes incident parallel rays to diverge after exiting the lens. Two prisms can also be used to form a crude diverging lens, as in Figure 32.17*a*. In a properly designed diverging lens, such as that in part *b* of the picture, paraxial rays that are parallel to the axis appear to originate from a single point on the axis after passing through the lens. This point is the focal point *F* of the diverging lens, and its distance *f* from the lens is the focal length. Again, we assume that the lens is thin compared to the focal length.

Converging and diverging lenses come in a variety of shapes, as Figure 32.18 illustrates. Observe, however, that converging lenses are always thickest at the center, whereas diverging lenses are always thinnest at the center.

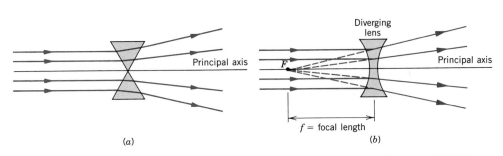

FIGURE 32.17 (*a*) Two prisms can be arranged to form a crude diverging lens. (*b*) With a properly designed diverging lens, paraxial rays that are parallel to the axis appear to originate from a single point *F* after passing through the lens. This point is the focal point of the lens, and its distance from the lens is the focal length *f*.

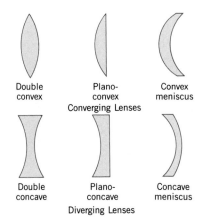

FIGURE 32.18 Converging and diverging lenses come in a variety of shapes.

32.7 THE FORMATION OF IMAGES BY CONVERGING AND DIVERGING LENSES

RAY DIAGRAMS

Each point on an object emits light rays in all directions, and when some of these rays pass through a lens, they form an image. As with mirrors, ray diagrams can be drawn to determine the location and size of the image. Lenses differ from mirrors, however, in that light can pass through a lens from left to right or from right to left. Therefore, when constructing ray diagrams, begin by locating a focal point F on *each side of the lens;* each point lies on the principal axis at the same distance f from the lens. The lens is assumed to be a thin lens, in that its thickness is small compared with the focal length and the distances of the object and the image from the lens. For convenience, it is also assumed that the object is located to the left of the lens and is oriented perpendicular to the principal axis. There are three paraxial rays that leave a point on the top of the object and are especially helpful in drawing ray diagrams. They are labeled 1, 2, and 3 in Figure 32.19 and are as follows:

Converging Lens	Diverging Lens
Ray 1	
This ray initially travels parallel to the principal axis. In passing through a converging lens, the ray is refracted toward the axis and travels through the focal point on the right side of the lens, as Figure 32.19a shows.	This ray initially travels parallel to the principal axis. In passing through a diverging lens, the ray is refracted away from the axis, and *appears* to have originated from the focal point on the left of the lens. See part *b*, where the dashed line represents the apparent path of the ray.
Ray 2	
This ray first passes through the focal point on the left and then is refracted by the lens in such a way that it leaves traveling parallel to the axis, as in part *c*.	This ray leaves the object and moves toward the focal point on the right of the lens. Before reaching the focal point, however, the ray is refracted by the lens so as to exit parallel to the axis. See part *d*, where the dashed line indicates the ray's path in the absence of the lens.
Ray 3	
This ray travels directly through the center of the thin lens without any appreciable bending, as in part *e*.	This ray travels directly through the center of the thin lens without any appreciable bending, as in part *f*.

Ray 3 does not bend as it proceeds through the lens, because the front and back surfaces of the lens are nearly parallel at the center. Thus, the lens behaves as a transparent slab. As Figure 32.5 shows, the rays incident on and exiting from a slab travel in the same direction with only a lateral displacement. If the lens is sufficiently thin, the displacement is negligibly small.

IMAGE FORMATION BY A CONVERGING LENS

Figure 32.20a illustrates the formation of a real image by a converging lens. Here the object is located at a distance from the lens that is greater than twice the focal length

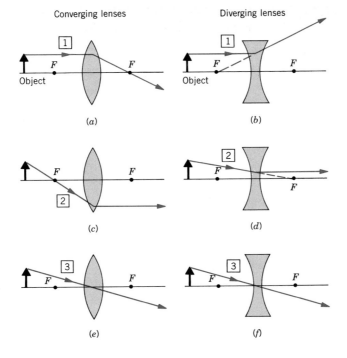

FIGURE 32.19 The rays shown here are useful in determining the nature of the images formed by converging and diverging lenses.

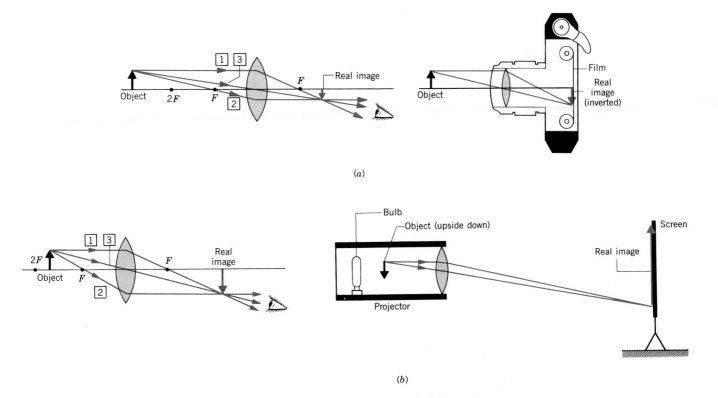

FIGURE 32.20 (a) This optical system gives rise to a real, inverted, and smaller image. The system is like that used in a camera. (b) Here the image is real, inverted, and larger than the object. This arrangement is found in projectors.

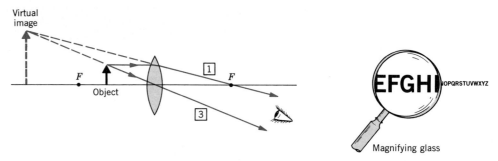

FIGURE 32.21 When an object is placed inside the focal point F of a converging lens, an upright, enlarged, virtual image is created. Such an image is seen when looking through a magnifying glass.

(beyond the point labeled $2F$). Any two of the three special rays can be drawn from the tip of the object. The point on the right side of the lens where these rays intersect locates the tip of the image. The ray diagram indicates that the image is real, inverted, and smaller than the object. This optical arrangement is similar to that used in a camera, where a piece of film records the image.

When the object is placed between $2F$ and F, as in Figure 32.20b, the image is still real and inverted; however, the image is now larger than the object. This optical system is used in a projector in which a small piece of film (a slide) is the object and the enlarged image falls on a screen. However, to obtain an image that is right-side up, the slide must be placed in the projector upside down.

When the object is located between the focal point and the lens, as in Figure 32.21, the rays diverge after leaving the lens. To a person viewing the diverging rays, they appear to come from an image behind and to the left of the lens. Because the rays do not actually come from the image, it is a virtual image. The ray diagram shows that the virtual image is upright and enlarged. A magnifying glass uses this arrangement.

IMAGE FORMATION BY A DIVERGING LENS

Light rays diverge upon leaving a diverging lens, as Figure 32.22 shows, and the ray diagram indicates that a virtual image is formed on the left side of the lens. In fact, regardless of the position of a real object, a diverging lens always forms a virtual image that is upright and smaller relative to the object.

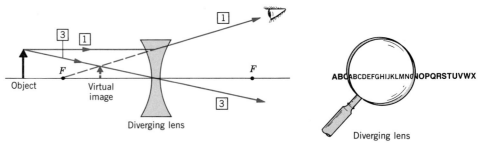

FIGURE 32.22 A diverging lens always forms a virtual image of a real object. The image is upright and smaller relative to the object. The picture at the right shows the image seen through a diverging lens.

32.8 THE THIN-LENS EQUATION AND THE MAGNIFICATION EQUATION

For an object in front of a mirror, it is possible to determine the location, size, and nature of the image with the mirror equation and the magnification equation. A similar analysis can be carried out for lenses, and, fortunately, the equations are identical to those used with mirrors. With lenses, however, the mirror equation becomes the *thin-lens equation.* These equations are

$$\left[\begin{array}{c}\textbf{Thin-lens}\\\textbf{equation}\end{array}\right] \qquad \frac{1}{d_\text{o}} + \frac{1}{d_\text{i}} = \frac{1}{f} \qquad (32.6)$$

$$\left[\begin{array}{c}\textbf{Magnification}\\\textbf{equation}\end{array}\right] \qquad m = \frac{\text{Image height}}{\text{Object height}} = \frac{h_\text{i}}{h_\text{o}} = -\frac{d_\text{i}}{d_\text{o}} \qquad (32.7)$$

Figure 32.23 defines the symbols in these expressions with the aid of a converging lens, but the expressions also apply to a diverging lens. The only restriction is that the lenses are thin lenses. Equations 32.6 and 32.7 are used extensively in the design of optical devices such as cameras, eyeglasses, and laser-based compact disc players. The derivations of these equations are presented at the end of this section.

Certain sign conventions accompany the use of the thin-lens and magnification equations, and the conventions are similar to those used with mirrors in Section 31.6. These conventions allow the equations to convey information about whether the image is real or virtual, upright or inverted, and enlarged or reduced with respect to the object. The issue of real-versus-virtual images, however, is slightly different with lenses than with mirrors. With a mirror, a real image is formed on the *same side* of the mirror as the object, in which case the image distance d_i is a positive number (see Figure 31.16). With a lens, a positive value for d_i also means the image is real. But, starting with an actual object, a real image is formed on the *opposite side* of the lens as the object (see Figure 32.23). In what follows, it is assumed that a real object is placed to the left of the lens, and the *sign conventions* listed below apply to light rays traveling from left to right.

Object distance

d_o is + if the object is to the left of the lens (real object), as is usually the case.
d_o is − if the object is to the right of the lens (virtual object).*

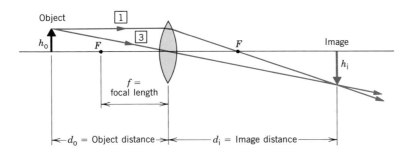

FIGURE 32.23 The drawing shows the focal length f, the object distance d_o, and the image distance d_i, for a converging lens. The object and image heights are, respectively, h_o and h_i.

* This situation arises in systems containing more than one lens, where the image formed by the first lens becomes the object for the second lens. In such a case, the object of the second lens may lie to the right of that lens, in which event d_o is assigned a negative value and the object is called a virtual object.

Image distance

d_i is + for an image (real) formed to the right of the lens by a real object.

d_i is − for an image (virtual) formed to the left of the lens by a real object.

Focal length

f is + for a converging lens.

f is − for a diverging lens.

Magnification

m is + for an image that is upright with respect to the object.

m is − for an image that is inverted with respect to the object.

Examples 4 and 5 illustrate the use of the thin-lens and magnification equations.

EXAMPLE 4

A 2.0-cm-tall object is placed 7.50 cm to the left of a converging lens whose focal length is 11.8 cm. (a) Find the image distance and determine whether the image is real or virtual. (b) Determine the magnification and the image height.

SOLUTION

(a) This optical arrangement is similar to that in Figure 32.21, where the object lies between the focal point and the lens. Using the thin-lens equation with $d_o = 7.50$ cm and $f = 11.8$ cm, we find that

$$\frac{1}{d_i} = \frac{1}{f} - \frac{1}{d_o} = \frac{1}{11.8 \text{ cm}} - \frac{1}{7.50 \text{ cm}} = -0.048 \text{ cm}^{-1}$$

$$\boxed{d_i = -21 \text{ cm}}$$

The negative sign for d_i indicates that the image is virtual and lies to the left of the lens.

(b) The magnification follows from Equation 32.7:

$$m = -\frac{d_i}{d_o} = -\frac{(-21 \text{ cm})}{7.50 \text{ cm}} = \boxed{2.8}$$

The image is 2.8 times larger than the object and is upright (m is +). Since the object height is $h_o = 2.0$ cm, the image height is

$$h_i = mh_o = (2.8)(2.0 \text{ cm}) = \boxed{5.6 \text{ cm}}$$

EXAMPLE 5

An object is placed 7.10 cm to the left of a diverging lens whose focal length is $f = -5.08$ cm (a diverging lens has a negative focal length). (a) Find the image distance and determine whether the image is real or virtual. (b) Obtain the magnification.

SOLUTION

(a) This situation is similar to that in Figure 32.22, and, once again, the thin-lens equation can be used to find the image distance d_i:

$$\frac{1}{d_i} = \frac{1}{f} - \frac{1}{d_o} = \frac{1}{-5.08 \text{ cm}} - \frac{1}{7.10 \text{ cm}} = -0.338 \text{ cm}^{-1}$$

$$\boxed{d_i = -2.96 \text{ cm}}$$

The image distance is negative, indicating that the image is virtual and located to the left of the lens.

(b) Since d_i and d_o are known, the magnification can be determined:

$$m = -\frac{d_i}{d_o} = -\frac{(-2.96 \text{ cm})}{7.10 \text{ cm}} = \boxed{0.417}$$

The image is upright (m is +) and smaller ($m < 1$) than the object.

The thin-lens and magnification equations can be derived by considering rays 1 and 3 in Figure 32.24a. Ray 1 is shown separately in part b of the drawing, where the angle θ is the same in each of the two colored triangles. Thus, tan θ is the same for each triangle, so

$$\tan \theta = \frac{h_o}{f} = \frac{h_i}{d_i - f}$$

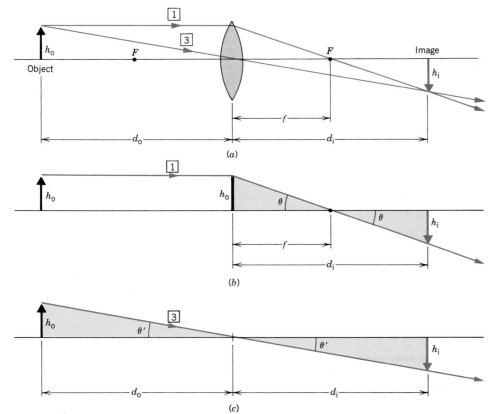

(a)

(b)

(c)

FIGURE 32.24 These ray diagrams are used for deriving the thin-lens and magnification equations.

Ray 3 is shown separately in part *c* of the drawing, where the angles θ' are the same. Therefore,

$$\tan \theta' = \frac{h_o}{d_o} = \frac{h_i}{d_i}$$

The first equation gives $h_i/h_o = (d_i - f)/f$, while the second equation yields $h_i/h_o = d_i/d_o$. Equating these two expressions for h_i/h_o and rearranging the result produces the thin-lens equation, $1/d_o + 1/d_i = 1/f$.

The magnification equation follows directly from the equation $h_i/h_o = d_i/d_o$, if we recognize that h_i/h_o is the magnification m of the lens. As with mirrors, a minus sign is inserted in front of the ratio d_i/d_o, so a positive value for m indicates an image that is upright relative to the object, while a negative value indicates the opposite.

32.9 LENSES IN COMBINATION

Many optical instruments, such as microscopes and telescopes, use a number of lenses together to produce an image, as will be seen in Chapter 33. Among other things, a multiple-lens system can produce an image that is magnified more than is possible with a single lens. For instance, Figure 32.25a shows a two-lens system used in a microscope. The first lens, the lens closest to the object, is referred to as the *objective*. The second lens is known as the *eyepiece* (or *ocular*). The object is placed

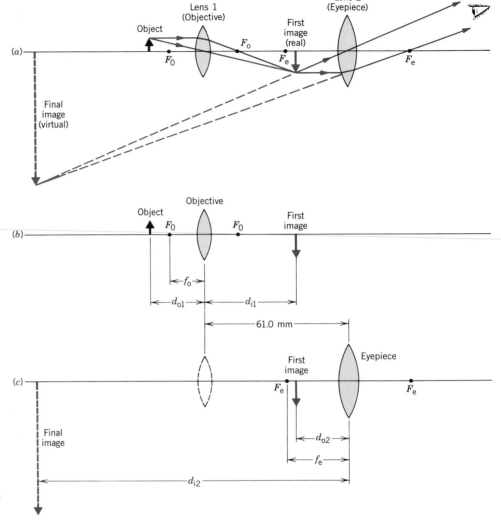

FIGURE 32.25 (a) This two-lens system can be used as a compound microscope to produce a virtual, enlarged, and inverted final image. (b) The objective forms the first image and (c) the eyepiece forms the final image.

just outside the focal point F_o of the objective. The image formed by the objective—called the "first image" in the drawing—is real, inverted, and enlarged compared to the object. This first image then serves as the object for the eyepiece. Since the first image falls between the eyepiece and its focal point F_e, the eyepiece forms an enlarged, virtual, final image, which is what the observer sees.

The nature and overall magnification of the final image in a multiple-lens system can be determined by applying the thin-lens and magnification equations to each lens separately. The key point to remember in such situations is that ***the image produced by one lens serves as the object for the next lens,*** as the next example illustrates.

EXAMPLE 6

The objective and eyepiece lenses of the compound microscope in Figure 32.25 are both converging lenses and have focal lengths of $f_o = 15.0$ mm and $f_e = 25.5$ mm. A distance of 61.0 mm separates the lenses. The microscope is being used to examine an object placed $d_{o1} = 24.1$ mm in front of the objective. Find (a)

the final image distance and (b) the overall magnification of the instrument.

SOLUTION

(a) We begin by calculating the "first image" distance d_{i1}, using the thin-lens equation with $d_{o1} = 24.1$ mm and $f_o = 15.0$ mm (see Figure 32.25b):

$$\frac{1}{d_{i1}} = \frac{1}{f_o} - \frac{1}{d_{o1}} = \frac{1}{15.0 \text{ mm}} - \frac{1}{24.1 \text{ mm}} = 0.0252 \text{ mm}^{-1}$$

$$d_{i1} = 39.7 \text{ mm}$$

The first image now becomes the object for the eyepiece (see part c of the drawing). Since the distance between the lenses is 61.0 mm, the object distance for the eyepiece is $d_{o2} = 61.0$ mm $- d_{i1} = 61.0$ mm $- 39.7$ mm $= 21.3$ mm. Noting that the focal length of the eyepiece is $f_e = 25.5$ mm, we can determine the final image distance with the aid of the thin-lens equation:

$$\frac{1}{d_{i2}} = \frac{1}{f_e} - \frac{1}{d_{o2}} = \frac{1}{25.5 \text{ mm}} - \frac{1}{21.3 \text{ mm}} = -0.0077 \text{ mm}^{-1}$$

$$\boxed{d_{i2} = -130 \text{ mm}}$$

The fact that d_{i2} is negative indicates that the final image is virtual and lies to the left of the eyepiece, as the drawing shows.

(b) Since the objective produces a first image that is m_o times larger than the object and since the eyepiece magnifies this already enlarged image by another factor of m_e, the overall magnification of the two-lens system is $m = m_o m_e$. From the magnification equation we find that the magnification produced by the objective is

$$m_o = -\frac{d_{i1}}{d_{o1}} = -\frac{39.7 \text{ mm}}{24.1 \text{ mm}} = -1.65$$

Similarly, the magnification for the eyepiece is

$$m_e = -\frac{d_{i2}}{d_{o2}} = -\frac{(-130 \text{ mm})}{21.3 \text{ mm}} = 6.1$$

The overall magnification of the two-lens system is $m = m_o m_e = (-1.65)(6.1) = \boxed{-10}$. With respect to the original object, the final image is 10 times larger and inverted (m is negative).

SUMMARY

The **index of refraction** n of a material is the ratio of the speed of light c in a vacuum to the speed of light v in the material: $n = c/v$.

When light strikes the interface between two media, part of the light is reflected and the remainder is transmitted across the interface. The change in the direction of travel as light passes from one medium into another is called **refraction. Snell's law of refraction** states that (1) the refracted ray, the incident ray, and the normal to the interface all lie in the same plane, and (2) the angle of refraction θ_2 is related to the angle of incidence θ_1 by $n_1 \sin \theta_1 = n_2 \sin \theta_2$, where n_1 and n_2 are the indices of refraction of the incident and refracting media. The angles are measured relative to the normal.

Because of refraction, a submerged object has an **apparent depth** that is different than its actual depth. If the observer is directly above the object, the apparent depth d' is related to the actual depth d by $d' = d(n_2/n_1)$, where n_1 and n_2 are the refractive indices of the media in which the object and the observer, respectively, are located.

When light passes from a medium of larger refractive index n_1 into one of smaller refractive index n_2, the refracted ray is bent away from the normal. If the incident ray is at the **critical angle** θ_c, the angle of refraction is 90°. The critical angle is determined from Snell's law and is given by

$\sin \theta_c = n_2/n_1$. When the angle of incidence exceeds the critical angle, all the incident light is reflected back into the medium from which it came, a phenomenon known as **total internal reflection.**

When light is incident on a nonmetallic surface at the **Brewster angle** θ_B, the reflected light is completely polarized parallel to the surface. The Brewster angle is given by $\tan \theta_B = n_2/n_1$, where n_1 and n_2 are the refractive indices of the incident and refracting media, respectively. At the Brewster angle, the reflected and refracted rays are perpendicular to each other.

A glass prism can spread a beam of sunlight into a spectrum of colors, because the index of refraction of the glass depends on the wavelength of the light. The spreading of light into its color components is known as **dispersion.**

There are two kinds of lenses, **converging lenses** and **diverging lenses,** and each depends on the phenomenon of refraction in forming an image. With a converging lens, paraxial rays that are parallel to the principal axis are focused to a point on the axis by the lens. This point is called the **focal point** of the lens. The distance between the focal point and the lens is the **focal length** f. Light rays that are parallel to the principal axis of a diverging lens appear to originate from its focal point after passing through the lens.

The image produced by a converging or a diverging lens can be located with the help of a **ray diagram,** which can be constructed using the rays shown in Figure 32.19.

The **thin-lens equation** can be used with either converging or diverging lenses that are thin, and it relates the image distance d_i, the object distance d_o, and the focal length f of the lens: $1/d_o + 1/d_i = 1/f$. The magnification m of a lens is the ratio of the image height h_i to the object height h_o. The magnification is also related to d_i and d_o by the **magnification equation:** $m = -(d_i/d_o)$. The algebraic sign conventions for the variables appearing in the thin-lens and magnification equations are summarized in Section 32.8. When two or more lenses are used in combination, the image produced by one lens serves as the object for the next lens.

QUESTIONS

1. Two slabs with parallel faces are made from different types of glass. A ray of light enters each slab at the same angle of incidence, as the drawing shows. Which slab has the greater index of refraction? Why?

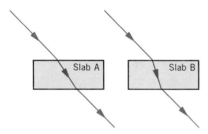

2. Two blocks, made from the same transparent material, are immersed in different liquids. A ray of light strikes each at the same angle of incidence. From the drawing, determine which liquid has the greater index of refraction. Justify your answer.

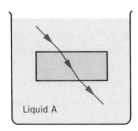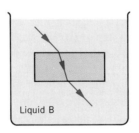

3. When an observer peers over the edge of a deep bowl, he does not see the entire bottom surface, so a small object lying on the bottom remains hidden from view. However, when the bowl is filled with water, the object can be seen. Explain this effect.

4. To a pearl diver swimming under water, does an object suspended in the air above the water appear to be at the actual height above the surface (assuming the diver is not wearing goggles)? If not, does the object appear to be higher or lower than the actual height? Why?

5. When you look through an aquarium window at a fish, is the fish as close as it seems? Explain.

6. At night, when it's dark outside and you are standing in a brightly lit room, it is easy to see your reflection in a window. During the day it is not so easy. Account for these facts.

7. A man is fishing from a dock. (a) If he is using a bow and arrow, should he aim above the fish, at the fish, or below the fish, to strike it? (b) How would he aim if he were using a laser gun?

8. Two rays of light converge to a point on a screen. A plane-parallel plate of glass is placed in the path of this converging light. Will the point of convergence remain on the screen? If not, will the point move toward the glass or away from it? Justify your answer by drawing a diagram and showing how the rays are affected by the glass.

9. A person sitting at the beach is wearing a pair of Polaroid sunglasses and notices little discomfort due to the glare from the water on a bright sunny day. When she lies on her side, however, she notices that the glare increases. Why?

10. Suppose a narrow beam of sunlight passes through a plate of glass with parallel sides, as in Figure 32.5. When the light leaves the glass, is the light dispersed into colors, as when light leaves a glass prism? Justify your answer, commenting on whether the separation of the colors (if they are separated) is larger or smaller with thicker plates.

11. Figure 32.13c illustrates the dispersion of sunlight by a prism. Would the dispersion be enhanced or reduced if the prism were immersed in water? Explain.

12. Suppose you wanted to make a rainbow by spraying water from a garden hose into the air. (a) Where must you stand relative to the water and the sun to see the rainbow? (b) Why can't you ever walk under the rainbow?

13. A beam of blue light is propagating in glass. When the light reaches the boundary between the glass and the surrounding air, the beam is totally reflected back into the glass. However, red light with the same angle of incidence is not totally reflected, and some of the light is refracted into the air. Account for the difference in the behavior of these two colors.

14. A beacon in a lighthouse is to produce a parallel beam of light. The beacon consists of a bulb and a converging lens. Should the bulb be placed outside the focal point, at the focal point, or inside the focal point of the lens? State your reason.

15. Sooner or later, most people discover that a lens can be used in bright sunlight to start a fire. (a) What kind of lens must be used? (b) Relative to the lens, where should a piece of paper be placed so the fire can be started as quickly as possible?

16. The drawing shows cross-sectional views of three styles of glass earrings. Conceivably, each might act as a lens and cause a small area on the wearer's neck to become hotter than normal on a sunny day. Rank the styles according to how much heating they might be expected to produce ("hottest" earring first). State your reasoning.

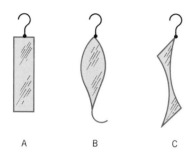

A   B   C

17. A spherical mirror and a lens are immersed in water. Compared to the way they work in air, which one do you expect will be more affected? Why?

18. The focal length of a thin lens is determined in part by the refractive index of the material from which it was made. Is the focal length the same for all colors of light? Explain.

19. In a TV mystery program, a photographic negative is introduced as evidence in a court trial. The negative shows an image of a house (now burned down) that was the scene of the crime. At the trial the defendant's acquittal depends on knowing exactly how far above the ground a window was. An expert called by the defense claims that this height can be calculated from only two pieces of information: (1) the measured height on the film, and (2) the focal length of the camera lens. Explain whether the expert is making sense, using the thin-lens and magnification equations to guide your thinking.

PROBLEMS

Section 32.1 The Index of Refraction

1. What is the speed of light in (a) benzene and (b) ice?

2. Find the ratio of the speed of light in diamond to the speed of light in ice.

3. A light wave has a frequency of 5.09×10^{14} Hz in air, water, and diamond. What is the wavelength of this light in each medium?

4. The speed of light is fifty percent larger in material A than it is in material B. Determine the ratio n_A/n_B of the refractive indices of these materials.

✱ **5.** In a certain time, light travels 3.50 km in air. During the same time, light travels only 2.50 km in a liquid. What is the refractive index of the liquid?

Section 32.2 Snell's Law and the Refraction of Light

6. A light ray in air is incident on a water surface at a 43° angle of incidence. Find (a) the angle of reflection and (b) the angle of refraction.

7. A ray of light is propagating in water and strikes a plate of fused quartz. The angle of refraction in the quartz is measured to be 36.7°. What is the angle of incidence?

8. A ray of sunlight hits a frozen lake at a 45° angle of incidence. At what angle of refraction does the ray penetrate (a) the ice and (b) the water beneath the ice?

9. A beam of light is traveling in air and strikes a material. The angles of incidence and refraction are measured to be 50.0° and 30.3°, respectively. Obtain the speed of light in the material.

10. A spotlight on a boat is 2.5 m above the water, and the light strikes the water at a point that is 8.0 m horizontally from the spotlight (see the drawing). The depth of the water is 4.0 m. Determine the distance d, which locates the point where the light strikes the bottom.

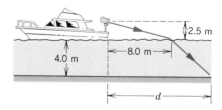

11. A block of crown glass is placed on top of a printed page. The block is 6.00 cm thick. When viewed directly from above, how far *above* the page does the printing appear to be?

✱ **12.** A prism is made from ice and is surrounded by air. The cross section of the prism is an isosceles right triangle. As the

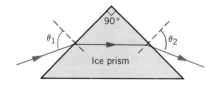

drawing indicates, a ray of light hits the prism, and once inside the prism the ray travels parallel to the hypotenuse of the right triangle. Find the angle of incidence θ_1 of the entering ray and the angle of refraction θ_2 of the ray that leaves the prism.

* 13. A silver medallion is sealed within a transparent block of plastic. An observer in air, viewing the medallion from directly above, sees the medallion at an apparent depth of 1.6 cm beneath the top surface of the block. How far below the top surface would the medallion appear if the observer (not wearing goggles) and the block were under water?

* 14. Three slabs of different material, each with parallel faces, are stacked on top of one another. A light ray originates in the bottom material. Prove that the angle of refraction at which a ray emerges into the top material is the same, whether or not the middle material is present.

** 15. Suppose a small logo is embedded in a thick block of crown glass, 3.20 cm beneath the top surface of the glass. The block is put into a container of water, so there is 1.50 cm of water above the top surface of the block. The logo is viewed from directly above by an observer in air. How far beneath the top surface of the water does the logo appear to be?

** 16. Light travels from point A to point B in medium 1 (refractive index $= n_1$). As the picture indicates, the light then refracts into medium 2 (refractive index $= n_2$) where the light travels from B to C. The travel time of the light in each medium is the same, and $n_2 = 2n_1$. Determine the ratio d_1/d_2 of the thicknesses.

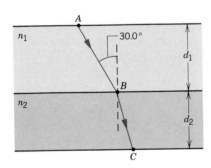

** 17. Refer to Figure 32.4 and assume the observer is directly above the submerged object. For this situation, derive the expression for the apparent depth: $d' = d(n_2/n_1)$, Equation 32.3. *(Hint: Use Snell's law of refraction and the fact that the angles of incidence and refraction are small, so $\tan \theta \approx \sin \theta$.)*

Section 32.3 Total Internal Reflection

18. What is the critical angle for light emerging from (a) ice into air and (b) water into ice?

19. One method of determining the refractive index of a transparent solid is to measure the critical angle when the solid is in air. If θ_c is found to be 40.5°, what is the index of refraction of the solid?

20. A point source of light is submerged 2.2 m below the surface of a lake and emits rays in all directions. On the surface of the lake, directly above the source, the area illuminated is a circle. What is the maximum radius that this circle could have?

21. The drawing shows a crown glass slab with a rectangular cross section. As illustrated, a laser beam strikes the upper surface at an angle of 60°. After reflecting from the upper surface, the beam reflects from the side and bottom surfaces. (a) If the glass is surrounded by air, determine where part of the beam first exits the glass, at point A, B, or C. (b) Repeat part (a), assuming the glass is surrounded by water.

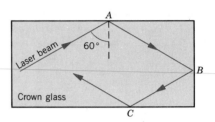

* 22. Three materials, A, B, and C, have refractive indices n_A, n_B, and n_C. The materials are in the form of parallel plates and are stacked on top of one another with A on the bottom and B in the middle. A ray of light originates in material A and strikes the A–B surface with an angle of incidence θ_A. It is observed that the light penetrates into material B only when θ_A is less than 50.0° and penetrates into material C only when θ_A is less than 30.0°. Find n_B/n_A and n_B/n_C.

* 23. A glass block ($n = 1.60$) is immersed in a liquid. A ray of light within the glass hits a glass–liquid surface at a 65.0° angle of incidence. Some of the light enters the liquid. What is the smallest possible value for the refractive index of the liquid?

* 24. The picture shows the core ($n_1 = 1.470$, diameter $= 1.0 \times 10^{-2}$ cm) of an optical fiber. The cladding ($n_2 = 1.455$) that surrounds the core is not included in the picture for the sake of clarity. A ray of light originates inside the fiber at point A on the core–cladding interface and strikes point B. (a) Calculate the critical angle for the core–cladding interface. (b) Determine the angle of incidence at B and decide whether total internal reflection occurs. (c) The dashed line in the drawing indicates a sharp bend (angle $= \alpha$) in the fiber. How large can α become before total internal reflection ceases to occur and light escapes the fiber at B?

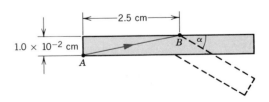

** 25. The back wall of a home aquarium is a mirror that is 30.0 cm away from the front wall. The walls of the tank are negligibly thin. A fish is swimming midway between the front and back walls. (a) How far from the front wall does the fish seem to be located? (b) How far behind the mirror does the image of the fish seem to be located? (c) Would the refractive index of the liquid have to be larger or smaller in order for the image of the fish to appear in *front* of the mirror, rather than behind it? Why?

Section 32.4 Polarization and the Reflection and Refraction of Light

26. At what angle of incidence is the light from the sun completely polarized upon being reflected from the surface of a lake (a) in the summer and (b) in the winter when the water is frozen?

27. Light is incident from air onto a beaker of benzene. If the reflected light is 100% polarized, what is the angle of refraction of the light that penetrates into the benzene?

28. When light strikes the surface between two materials from above, the Brewster angle is 65.0°. What is the Brewster angle when the light encounters the same surface from below?

**29. For a surface between two nonconducting materials, prove that the Brewster angle is never larger than the critical angle, so that light can never be 100% reflected from such a surface and simultaneously be 100% polarized.

Section 32.5 The Dispersion of Light: Prisms and Rainbows

30. A beam of sunlight encounters a plate of crown glass at a 45.00° angle of incidence. Using the data in Table 32.2, find the angle between the violet ray and the red ray in the glass.

* 31. The intent of this problem relates to Figure 32.13*b*, which illustrates the dispersion of light by a prism. The prism is made from flint glass (see Table 32.2), and its cross section is an equilateral triangle. The angle of incidence for both the red and violet light is 60.0°. Find the angles of refraction at which the red and violet rays emerge into the air from the prism.

Section 32.6 Converging and Diverging Lenses, Section 32.7 The Formation of Images by Converging and Diverging Lenses, Section 32.8 The Thin-Lens Equation and the Magnification Equation

32. A converging lens is used to focus light from a small bulb onto a book. The lens has a focal length of 5.0 cm, and the lens is located 25 cm from the book. How far is the lens from the light bulb?

33. A figurine is placed 15.0 cm in front of a converging lens ($f = 40.0$ cm). Using a ray diagram drawn to scale, find (a) the image distance and (b) the magnification.

34. Standard 35-mm slides have a usable area on the film of about 24 mm × 35 mm. The lens in a camera has a focal length of 50.0 mm and can be adjusted so objects as close as 0.50 m will be in focus on the film. At a distance of 0.50 m, what is the largest height the object can have, if the image is to fit into the long dimension of the film area?

35. A magnifying glass uses a converging lens whose focal length is 15 cm. The magnifying glass produces a virtual and upright image that is 3.0 times larger than the object. (a) How far is the object from the lens? (b) What is the image distance? (c) Confirm your answers to parts (a) and (b) by drawing a ray diagram to scale.

36. (a) For a diverging lens ($f = -20.0$ cm) [handwritten: *10.0*], construct a ray diagram to scale and find the image distance for an object that is 20.0 cm from the lens. (b) Determine the magnification of the system from the diagram.

37. A diverging lens has a focal length of -38 cm. An object is placed 28 cm in front of this lens. Calculate (a) the image distance and (b) the magnification. (c) Is the image real or virtual, (d) upright or inverted, and (e) enlarged or reduced in size?

* 38. Outdoors on a sunny day, a piece of paper rapidly catches fire when placed 15 cm from a lens. (a) What kind of lens is it, and what is its focal length? (b) What is the image distance of a flower placed 45 cm from the lens? (c) The height of the flower is 2.0 cm. What is the height of the image, and is it upright or inverted relative to the flower?

* 39. On a roll of movie film, each picture has a width of 70.0 mm. The projector lens has a focal length of 305 mm. If the screen in a theater is 60.0 m from the projector lens, what is the width of the image projected on the screen?

* 40. The moon's diameter is 3.48×10^6 m and its mean distance from the earth is 3.85×10^8 m. The moon is being photographed by a camera whose lens has a focal length of 50.0 mm. (a) Find the diameter of the moon's image on the slide film. (b) When the slide is projected onto a screen that is 15.0 m from the lens of the projector ($f = 110.0$ mm), what is the diameter of the moon's image on the screen? (c) Repeat parts (a) and (b), assuming the camera uses a telephoto lens with a 135-mm focal length and the projector lens remains the same.

**41. An object is 20.0 cm from a converging lens, and the image falls on a screen. When the object is moved 4.00 cm closer to the lens, the screen must be moved 2.70 cm farther away from the lens to register a sharp image. Determine the focal length of the lens.

**42. A converging lens ($f = 25.0$ cm) is used to project an image of a 3.00-cm-tall object onto a screen. The object and the screen are 125 cm apart, and between them the lens can be placed at either of two locations. Find (a) the two object distances and (b) the corresponding sizes of the images.

Section 32.9 Lenses in Combination

43. A converging lens ($f = 12.0$ cm) is located 30.0 cm to the left of a diverging lens ($f = -6.00$ cm). A postage stamp is placed 36.0 cm to the left of the converging lens. (a) Locate the final image of the stamp relative to the diverging lens. (b) Find the overall magnification. (c) Is the final image real or virtual? With respect to the original object, is the final image (d) upright or inverted, and is it (e) larger or smaller?

44. A converging lens ($f = 12.0$ cm) is 28.0 cm to the left of a diverging lens ($f = -14.0$ cm). An object is located 6.00 cm to the left of the converging lens. Draw an accurate ray diagram and from it find (a) the final image distance, measured from the diverging lens, and the overall magnification. (b) Confirm your answers to part (a) by using the thin-lens and magnification equations.

* 45. A coin is located 15.00 cm to the left of a converging lens ($f = 10.00$ cm). A second, identical lens is placed to the right of the first lens, such that the image formed by the combination has the same size and orientation as the original coin. Find the separation between the lenses.

**46. Two converging lenses ($f_1 = 9.00$ cm and $f_2 = 6.00$ cm) are separated by 18.0 cm. The lens on the left has the longer focal length. An object stands 12.0 cm to the left of the combination. (a) Locate the final image relative to the lens on the right. (b) Obtain the overall magnification. (c) Is the final image real or virtual? With respect to the original object, is the final image (d) upright or inverted and is it (e) larger or smaller?

ADDITIONAL PROBLEMS

47. When a diverging lens is held 13 cm above a line of print, as in Figure 32.22, the image is 5.0 cm beneath the lens. (a) What is the focal length of the lens? Is the image (b) real or virtual, (c) upright or inverted, and (d) enlarged or reduced?

48. A beam of light impinges from air onto a block of ice at a 60.0° angle of incidence. Assuming this angle remains the same, find the percentage by which the angle of refraction changes when the ice turns to water, and state whether the change is an increase or a decrease.

49. An object, 0.75 cm tall, is placed 12.0 cm to the left of a diverging lens ($f = -8.00$ cm). A converging lens is placed 8.00 cm to the right of the diverging lens. The final image is virtual and is 29.0 cm to the left of the diverging lens. Determine (a) the focal length of the converging lens and (b) the height of the final image.

50. The drawing shows a beam of light passing through three

media: oil, glass, and water. The light starts at point A, passes through the media, reflects from a plane mirror, and returns to A. Find the time for the round-trip.

51. To focus a camera on objects at different distances, the converging lens is moved toward or away from the film, so a sharp image always falls on the film. A camera with a telephoto lens ($f = 200.0$ mm) is to be focused on an object located first at a distance of 3.5 m and then at 50.0 m. Over what distance must the lens be movable?

52. Amber ($n = 1.546$) is a transparent brown-yellow fossil resin. An insect, trapped and preserved within the amber, appears to be 2.5 cm beneath the surface, when viewed directly from above. How far below the surface is the insect really located?

53. An object is located 30.0 cm to the left of a converging lens whose focal length is 50.0 cm. (a) Draw a ray diagram to scale and from it determine the image distance and the magnification. (b) Use the thin-lens and magnification equations to verify your answers to part (a).

* 54. An office copier uses a lens to place an image of a document onto a rotating drum. The copy is made from this image. (a) What kind of lens is used? (b) If the document and its copy are to have the same size, how far from the document is the lens located and how far from the lens is the image located? Express your answers in terms of the focal length f of the lens.

* 55. A 15.0° wedge is made from fused quartz and is surrounded by air. A beam of light strikes the wedge at a 30.0° angle of incidence, as the drawing indicates. At what angle θ does the beam emerge from the wedge?

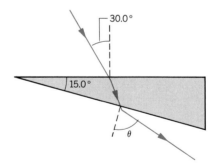

* 56. When a converging lens is used in a camera (as in Figure 32.20a), the film must be placed at a distance of 0.35 m from the lens to record an image of an object that is 2.0 m from the lens. The same lens is then used in a projector (illustrated in Figure 32.20b), with the screen located 1.2 m from the lens. How far from the projector lens should the film be placed?

**57. A glass prism, surrounded by air, has a cross section that is a 30.0°–60.0° right triangle. Two light rays, labeled 1 and 2 in the drawing, strike adjacent faces of the prism perpendicularly. (a) For ray 1, no light exits through the face represented by the hypotenuse, while for ray 2 light does exit through that face. What range of values can the refractive index of the glass have? (b) What

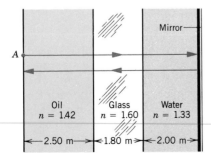

additional information, if any, can you deduce about the refractive index of the glass, if the information in part (a) is also true when the prism is surrounded by water?

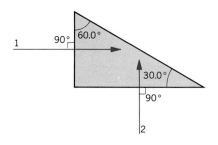

∗∗58. A number of slabs, each with parallel faces, are stacked one on top of the other. The air above the stack has a refractive index of 1.00. Suppose each slab has a refractive index that is ten percent larger than the previous one; that is, the first slab has $n = (1.00)(1.10) = 1.10$, the second slab has $n = (1.10)(1.10) = 1.21$, etc. Light strikes the top slab at a 75° angle of incidence. What is the minimum number of slabs needed so a ray of light will emerge into the bottom slab with an angle of refraction of 25° or smaller?

CHAPTER 33

Optical Instruments

This photograph of the spiral nebula in the constellation Canes Venatici was taken with the 200-in. reflecting telescope at Mount Wilson Observatory. Telescopes, cameras, the human eye, and microscopes are considered in the following pages.

A large number of optical instruments utilize mirrors and lenses, and the present chapter describes the camera, the human eye, the microscope, and the telescope. Man-made optical instruments often incorporate compound lenses. A *compound lens* consists of a number of individual lenses designed to act together as a single lens without introducing the aberrations of a single lens. Section 33.6 discusses some of the aberrations that a single lens has. In the interest of simplicity, we will treat such compound lenses as single thin lenses in our drawings and discussions of optical instruments.

33.1 THE CAMERA

As Figure 33.1 indicates, a camera consists of a converging lens, a diaphragm with a variable aperture (opening) for regulating the amount of light reaching the film, a shutter for controlling the exposure time, and the film itself. When the shutter is opened, light from the object being photographed is focused by the lens to form a real image on the film. The image is inverted and smaller than the object. The popular

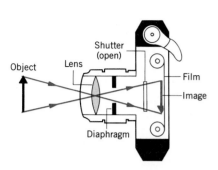

FIGURE 33.1 A cross-sectional view of a camera. The film is exposed when the shutter is opened for a fraction of a second. The diaphragm and its adjustable aperture (opening) control the amount of light that strikes the film. The aperture must be adjusted to suit the shutter speed and the type of film.

35-mm camera, for instance, gets its name from the small film area within which the image must fit, namely, a 35-mm × 24-mm rectangle.

The separation between the lens and the film (the image distance) is adjustable, so objects located at different distances from the camera can produce sharply focused images on the film. Example 1 illustrates that the lens-to-film distance depends on the focal length of the lens, as well as the object distance.

EXAMPLE 1

A 50.0-mm lens ($f = 50.0$ mm) is mounted on a 35-mm camera. (a) Determine the separation between the lens and the film (the image distance) when the object is as close as 0.45 m and as far as 10.0 m. (b) From the results in part (a), obtain the range over which the lens position must be adjustable. (c) What is the image height on the film when the camera is used to photograph a 1.7-m-tall person who is standing 10.0 m away?

SOLUTION

(a) With an object distance of $d_o = 450$ mm and a focal length of $f = 50.0$ mm, the thin-lens equation can be used to find the image distance d_i:

$$\frac{1}{d_i} = \frac{1}{f} - \frac{1}{d_o} = \frac{1}{50.0 \text{ mm}} - \frac{1}{450 \text{ mm}} \qquad (32.6)$$

$$= 0.0178 \text{ mm}^{-1}$$

$$\boxed{d_i = 56.2 \text{ mm}}$$

Repeating this calculation with $d_o = 10.0 \times 10^3$ mm shows that the corresponding image distance is $\boxed{d_i = 50.3 \text{ mm}}$.

(b) The lens must be capable of moving back and forth over the range between the two image distances calculated above, or

Range $= 56.2$ mm $- 50.3$ mm $= \boxed{5.9 \text{ mm}}$.

(c) The image height h_i can be obtained from the object height $h_o = 1.7$ m and the magnification m, since $h_i = mh_o$. The magnification is given by the magnification equation in terms of the image distance $d_i = 50.3$ mm and the object distance $d_o = 10.0 \times 10^3$ mm:

$$m = -\frac{d_i}{d_o} = -\frac{50.3 \text{ mm}}{10.0 \times 10^3 \text{ mm}} \qquad (32.7)$$

$$= -5.03 \times 10^{-3}$$

The image height is then

$$h_i = mh_o = (-5.03 \times 10^{-3})(1.7 \times 10^3 \text{ mm}) = \boxed{-8.6 \text{ mm}}$$

The fact that h_i is negative indicates that the image is inverted.

The amount of light that strikes the film depends on a number of factors. One of them is the **shutter speed,** which controls the length of time that the shutter is open.

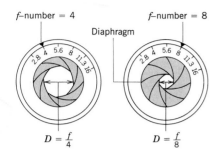

f-number = 4   f-number = 8

Diaphragm

$D = \frac{f}{4}$ $D = \frac{f}{8}$

FIGURE 33.2 The smaller the f-number, the larger the diameter D of the aperture in the diaphragm.

This time is often called the exposure time. The shutter is a mechanical device usually located just in front of the film. Shorter opening times require faster shutter speeds. However, with fast shutter speeds the subject must be well lit, so sufficient light to expose the film passes through the shutter during the brief time the shutter is open. The opening times that may be selected are usually marked on the camera. A setting of 125, for instance, means that the shutter remains open for $\frac{1}{125}$ s.

Another factor that affects the amount of light striking the film is the *size of the aperture* in the diaphragm. This opening can be enlarged to let more light reach the film or reduced to restrict the amount of light. The size of the aperture is selected to be compatible with lighting conditions and the chosen shutter speed. The size of the aperture is indicated by its diameter, and often there are numbers on the barrel of a camera lens that are related to the various diameters that may be selected. These numbers are called *f-numbers,* and Figure 33.2 shows the front view of a lens with two different settings for the f-numbers. From the picture it is clear that larger f-numbers correspond to smaller aperture diameters. This inverse relation occurs because the f-number is defined as the focal length f of the lens divided by the diameter D of the aperture:

$$f\text{-number} = \frac{\text{Focal length of lens}}{\text{Diameter of aperture}} = \frac{f}{D} \qquad (33.1)$$

Thus, when a lens is set to an f-number of 4, as in the figure, the lens is said to have an aperture of $f/4$. An aperture of $f/4$ means that the aperture diameter is $D = f/4$, or one-fourth the focal length.

Each f-number setting labeled on the barrel of a camera lens is often called an *f-stop* by photographers. These f-stops are calibrated such that *if you change from one f-stop to the next, the amount of light reaching the film changes by a factor of 2.* For instance, in Figure 33.2 the 2.8 setting admits twice as much light as the 4 setting, and in turn, the 4 setting admits twice as much light as the 5.6 setting. The reason that each change in f-stop changes the amount of light by a factor of 2 is as follows. According to Equation 33.1, changing the f-number from 4 to 2.8 changes the aperture diameter by a factor of $4/2.8 = 1.4 = \sqrt{2}$. But the amount of light admitted by the camera is proportional to the area of the aperture or to the square of the diameter. Thus, changing the f-number from 4 to 2.8 increases the aperture area and the amount of light by a factor of $(\sqrt{2})^2 = 2$. Similar reasoning applies to all changes between adjacent f-stops. Example 2 deals with the diameter of the aperture in a standard lens used with a 35-mm camera.

EXAMPLE 2

(a) A 35-mm camera is fitted with a 50.0-mm lens ($f = 50.0$ mm). The aperture is adjusted to have an f-number of 2. What is the diameter of the aperture? (b) How much more light reaches the film when the aperture has an f-number of 2 compared to an f-number of 16?

SOLUTION

(a) Equation 33.1 gives the aperture diameter D directly:

$$D = \frac{f}{f\text{-number}} = \frac{50.0 \text{ mm}}{2} = \boxed{25.0 \text{ mm}}$$

(b) The ratio of the amounts of light reaching the film when the lens is set for aperture diameters of $f/2$ and $f/16$ is equal to

the ratio of the aperture areas or the squared diameters, so

$$\frac{\text{Ratio of amounts of light}}{} = \frac{\text{Area with } f\text{-number set at 2}}{\text{Area with } f\text{-number set at 16}} = \frac{D_2^2}{D_{16}^2}$$

In part (a) we found for the $f/2$ aperture that $D_2 = (50.0 \text{ mm})/2$. Similarly, the diameter of the $f/16$ aperture is $D_{16} = (50.0 \text{ mm})/16$. Thus,

$$\frac{\text{Ratio of amounts of light}}{} = \frac{[(50.0 \text{ mm})/2]^2}{[(50.0 \text{ mm})/16]^2} = \left(\frac{16}{2}\right)^2 = \boxed{64}$$

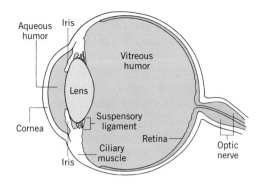

FIGURE 33.3 A cross-sectional view of the human eye.

33.2 THE HUMAN EYE

THE ANATOMY OF THE EYE

Without doubt, the human eye is the most remarkable of all optical instruments, and Figure 33.3 shows some of its main anatomical features. The eyeball is approximately spherical with a diameter of about 25 mm. Light enters the eye through a transparent membrane (the *cornea*). This membrane covers a clear liquid region (the *aqueous humor*), behind which is a variable-aperture diaphragm (the *iris*), the *lens,* a region filled with a jelly-like substance (the *vitreous humor*), and, finally, the *retina.* The retina is the light-sensitive part of the eye, consisting of millions of structures called *rods* and *cones.* When stimulated by light, these structures send electrical impulses via the *optic nerve* to the brain, which interprets the image on the retina.

The iris is the colored portion of the eye and controls the amount of light reaching the retina. The iris acts as a controller, because it is a muscular diaphragm with a variable opening at its center, through which the light passes. The opening is called the *pupil.* The diameter of the pupil varies from about 2 to 7 mm, decreasing in bright light and increasing (dilating) in dim light.

Of prime importance to the operation of the eye is the fact that the lens is flexible, and its shape can be altered by the action of the *ciliary muscle.* The lens is connected to the ciliary muscle by the *suspensory ligaments* (see the drawing). We will see shortly how the shape-changing ability of the lens affects the focusing property of the eye.

THE OPTICS OF THE EYE

Optically, the eye and the camera are similar; both have a lens system and a variable-aperture diaphragm. Moreover, the retina of the eye and the film in the camera serve similar functions, for both record the image formed by the lens system. In the eye, the image formed on the retina is real, inverted, and smaller than the object, just as it is in a camera. Although the image on the retina is inverted, it is interpreted by the brain as being right-side up.

For clear vision, the eye must refract the incoming light rays, so as to form a sharp image on the retina. In reaching the retina, the light travels through five different media, each with a different index of refraction n: air ($n = 1.00$), the cornea ($n = 1.38$), the aqueous humor ($n = 1.33$), the lens ($n = 1.40$, on the average), and the vitreous humor ($n = 1.34$). Each time light passes from one medium into another, it is refracted at the boundary. Collectively, all the boundaries participate in refracting the light to form the image on the retina. However, the greatest amount of refraction, about 70% or so, occurs at the air/cornea boundary. According to Snell's law, the large refraction at this interface occurs primarily because the refractive index of air ($n = 1.00$) is so different from that of the cornea ($n = 1.38$). The refraction at all the other boundaries is relatively small, because the indices of refraction on either side of these boundaries are nearly equal. The lens itself contributes only about 20–25% of the total refraction, since the surrounding aqueous and vitreous humors have indices of refraction that are nearly the same as that of the lens.

Even though the lens contributes only a quarter of the total refraction or less, the function of the lens is an important one. Unlike a camera, the eye has a fixed image distance, that is, the distance between the retina and the lens is constant. Therefore, the only way that objects at different distances can be made to produce images on the retina is for the focal length of the lens to be adjustable. And it is the ciliary muscle that adjusts the focal length. When the eye looks at a very distant object, the ciliary muscle is not tensed. The lens has its least curvature and, consequently, its longest focal length. Under this condition the eye is said to be "fully relaxed," and the rays form a sharp image on the retina, as in Figure 33.4a. When the object moves closer to the eye, the ciliary muscle automatically tenses, thereby increasing the curvature of the lens, shortening the focal length, and permitting a sharp image to form again on the retina (Figure 33.4b). When a sharp image of an object is formed on the retina, we say the eye is "focused" on the object. The process in which the lens changes its focal length to focus on objects at different distances is called **accommodation** and occurs so swiftly that we are usually unaware of it.

When you hold a book too close, the print is blurred because the lens cannot be adjusted enough to bring the book into focus. The point nearest the eye at which an object can be placed and still produce a sharp image on the retina is called the **near point** of the eye. The ciliary muscle is fully tensed when an object is placed at the near point. For people in their early twenties with normal vision, the near point is located about 25 cm from the eye. It increases to about 50 cm at age 40 and to roughly 500 cm at age 60. Since most reading material is held at a distance of 45 cm or so from the eye, older adults typically need eyeglasses to overcome the loss of accommodation. The **far point** of the eye is the location of the farthest object on which the fully relaxed eye can focus. A person with normal eyesight can see objects very far away, such as the planets and stars, and thus has a far point located nearly at infinity.

NEARSIGHTEDNESS

A person who is **nearsighted (myopic)** can focus on nearby objects but cannot clearly see objects far away. For such a person, the far point of the fully relaxed eye is not at

FIGURE 33.4 (a) When the lens of the eye is fully relaxed, the lens has its longest focal length, and an image of a distant object (infinitely far away) is formed on the retina. (b) When the ciliary muscle is tensed, the lens is thicker and has a shorter focal length. Because of the shorter focal length, an image of a closer object is also formed on the retina.

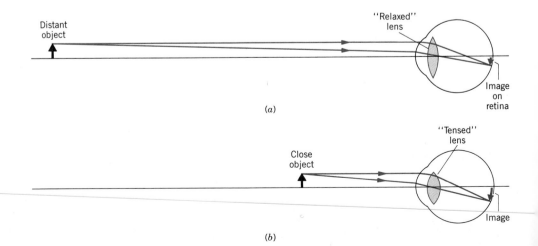

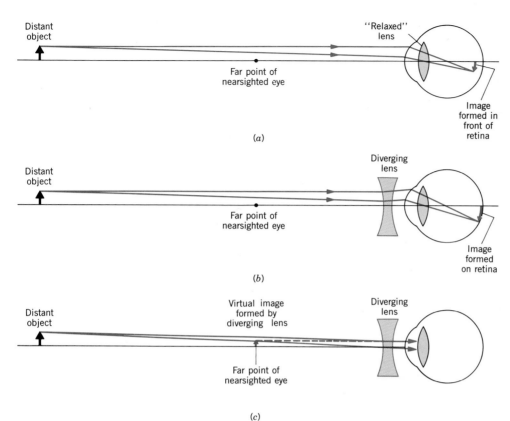

FIGURE 33.5 (a) When a nearsighted person views a distant object, the image is formed in front of the retina. The result is blurred vision. (b) With a suitable diverging lens placed in front of the eye, the image is moved onto the retina and clear vision results. (c) The diverging lens is designed to form a virtual image at the far point of the nearsighted eye.

infinity and may even be as close to the eye as three or four meters. When a nearsighted eye tries to focus on a distant object, the eye is fully relaxed, like a normal eye. However, the nearsighted eye has a focal length that is shorter than it should be, so rays from the distant object form a sharp image in front of the retina, as Figure 33.5a shows, and blurred vision results.

The nearsighted eye can be corrected with glasses or contacts that use *diverging* lenses, as Figure 33.5b suggests. The rays from the object diverge after leaving the eyeglass lens. Therefore, when they are subsequently refracted toward the principal axis by the eye, a sharp image is formed farther back and falls on the retina. Since the relaxed (but nearsighted) eye can focus on an object at the eye's far point—but not on objects farther away—the diverging lens is designed to transform a very distant object into an image located at the far point. Part c of the drawing shows this transformation, and the next example illustrates how to determine the focal length of the necessary diverging lens.

EXAMPLE 3

A nearsighted person has a far point located only 521 cm from the eye. Assuming that eyeglasses are to be worn 2 cm in front of the eye, find the focal length needed for the diverging lenses of the glasses so the person can see distant objects.

SOLUTION

In Figure 33.5c the far point is 521 cm away from the eye. Since the glasses are worn 2 cm from the eye, the far point is 519 cm to the left of the diverging lens. The image distance, then, is $d_i = -519$ cm, the negative sign indicating that the image is a virtual

image formed to the left of the lens. The object is far away from the diverging lens, so $d_o \approx \infty$. The thin-lens equation gives

$$\frac{1}{f} = \frac{1}{d_o} + \frac{1}{d_i} = \frac{1}{\infty} + \frac{1}{-519 \text{ cm}} \qquad (32.6)$$

$$\boxed{f = -519 \text{ cm}}$$

The value for f is negative, indicating the lens is a diverging lens.

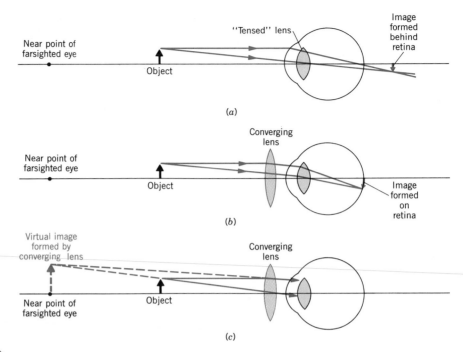

FIGURE 33.6 (*a*) When a farsighted person views an object located inside the near point of the eye, an image is formed behind the retina, causing blurred vision. (*b*) With a suitable converging lens in front of the eye, the image is moved onto the retina and clear vision results. (*c*) The converging lens is designed to form a virtual image at the near point of the farsighted eye.

FARSIGHTEDNESS

A *farsighted (hyperopic)* person can usually see distant objects clearly, but cannot focus on those nearby. Whereas the near point of a "normal" eye is located about 25 cm from the eye, the near point of a farsighted eye may be considerably farther away than that, perhaps as far as several hundred centimeters. When a farsighted eye tries to focus on a book held closer than the near point, the eye accommodates and shortens its focal length as much as it can. However, even at its shortest, the focal length of a farsighted eye is longer than it should be. Therefore, the light rays from the book form a sharp image behind the retina, as Figure 33.6*a* indicates, leading to blurred vision.

Figure 33.6*b* shows that farsightedness can be corrected by placing a *converging* lens in front of the eye. The lens refracts the light rays more toward the principal axis before they enter the eye. Consequently, when the rays are refracted even further by the eye, they converge to form an image on the retina. Part *c* of the figure illustrates what the eye sees when it looks through the converging lens. The lens is designed so that the eye perceives the light to be coming from a virtual image located at the near point. Example 4 shows how the focal length of the converging lens is determined to correct for farsightedness.

EXAMPLE 4

A farsighted person has a near point located 210 cm from the eyes. Obtain the focal length of the converging lenses in a pair of contacts that can be used to read a book held 25.0 cm from the eyes.

SOLUTION

A contact lens is placed directly against the eye. Thus, the object distance, which is the distance from the book to the lens, is $d_o = 25.0$ cm. The lens forms an image of the book at the near point of the eye, so the image distance is $d_i = -210$ cm. The minus sign indicates that the image is a virtual image formed to the left of the lens, as in Figure 33.6c. The focal length f can now be obtained from the thin-lens equation:

$$\frac{1}{f} = \frac{1}{d_o} + \frac{1}{d_i} = \frac{1}{25.0 \text{ cm}} + \frac{1}{-210 \text{ cm}} \qquad (32.6)$$

$$= 0.0352 \text{ cm}^{-1}$$

$$\boxed{f = 28.4 \text{ cm}}$$

THE REFRACTIVE POWER OF A LENS — THE DIOPTER

The degree to which rays of light are refracted by a lens depends on the focal length of the lens. However, the optometrists who prescribe correctional lenses and the opticians who make the lenses do not specify the focal length directly in the prescriptions. Instead, they use the concept of **refractive power** to describe the degree to which a lens refracts the light:

$$\begin{array}{c} \text{Refractive power} \\ \text{of a lens} \\ \text{(in diopters)} \end{array} = \frac{1}{f \text{(in meters)}} \qquad (33.2)$$

The refractive power of a lens is measured in units of *diopters*. One diopter is 1 m^{-1}.

Equation 33.2 shows that a converging lens has a refractive power of 1 diopter if it focuses parallel light rays to a focal point 1 m beyond the lens. If a lens refracts parallel rays even more and converges them to a focal point only 0.25 m beyond the lens, the lens has four times more refractive power, or 4 diopters. Since a converging lens has a positive focal length and a diverging lens has a negative focal length, the refractive power of a converging lens is positive while that of a diverging lens is negative. For instance, the eyeglass in Example 3 would be described in a prescription from an optometrist in the following way: Refractive power $= 1/(-5.19 \text{ m}) = -0.193$ diopters. The contact lenses in Example 4 would be described in a similar fashion: Refractive power $= 1/(0.284 \text{ m}) = 3.52$ diopters.

33.3 ANGULAR MAGNIFICATION AND THE MAGNIFYING GLASS

ANGULAR SIZE

If you hold a penny at arm's length, the penny looks larger than the moon. The reason is that the penny, being so close, forms a larger image on the retina of the eye than does the more distant moon. The brain interprets the larger image of the penny as arising from a larger object. As far as the brain is concerned, then, the size of the image on the retina determines how large an object appears to be. Usually, however, the image size itself is not discussed directly, because it is difficult to measure. Alternatively, the angle θ subtended by the image can be used as a measure of the image size. Figure 33.7 shows this alternative, which has the advantage that θ is also the angle subtended by the object and, hence, can be measured easily. The angle θ is called the **angular size** of both the image and the object. The larger the angular size, the larger the image on the retina, and the larger the object appears to be.

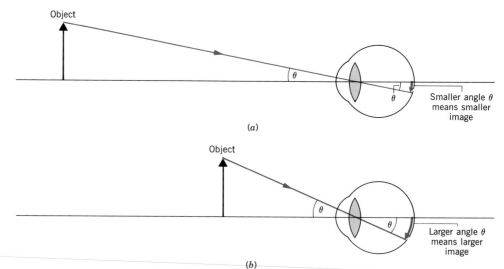

(a)

Object

Smaller angle θ
means smaller
image

Object

Larger angle θ
means larger
image

(b)

FIGURE 33.7 In both (a) and (b) the object is the same size, but in (b) the image on the retina is larger because the object is closer to the eye. The brain thinks that the object in (b) is larger than that in (a). The angle θ is the angular size of both the image and the object.

According to Equation 9.1, the angle θ (measured in radians) is the length of the circular arc subtended by the angle divided by the radius of the arc, as Figure 33.8a indicates. Part b of the drawing shows the situation when we view an object of height h_o at a distance d_o from the eye. Comparing part a with part b, it is evident that when θ is small, h_o is approximately equal to the arc length and d_o is nearly equal to the radius. Thus, for small angles, θ can be expressed in terms of the measurable quantities h_o and d_o:

$$\theta \text{ (in radians)} = \text{Angular size} \approx \frac{h_o}{d_o}$$

The approximation is good to within one percent for angles of $9°$ or smaller. It is now possible to compare the angular size of a penny ($h_o = 1.9$ cm) held at arm's length ($d_o = 71$ cm) with that of the moon ($h_o = 3.5 \times 10^6$ m, $d_o = 3.9 \times 10^8$ m):

[Penny] $\theta = \dfrac{h_o}{d_o} = \dfrac{1.9 \text{ cm}}{71 \text{ cm}} = 0.027$ rad $(1.5°)$

[Moon] $\theta = \dfrac{h_o}{d_o} = \dfrac{3.5 \times 10^6 \text{ m}}{3.9 \times 10^8 \text{ m}} = 0.0090$ rad $(0.52°)$

The penny thus appears to be about three times larger than the moon.

FIGURE 33.8 (a) The angle θ, measured in radians, is the arc length divided by the radius. (b) If θ is less than $9°$, its value in radians is approximately equal to h_o/d_o to within one percent, where h_o and d_o are the object height and distance.

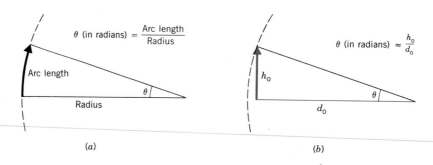

θ (in radians) $= \dfrac{\text{Arc length}}{\text{Radius}}$

Arc length

Radius

θ

(a)

θ (in radians) $\approx \dfrac{h_o}{d_o}$

h_o

d_o

θ

(b)

ANGULAR MAGNIFICATION

An optical instrument, such as a magnifying glass, allows us to view small or distant objects, because it produces a larger image on the retina than would be possible otherwise. In other words, an optical instrument magnifies the angular size of the object. The **angular magnification** (or **magnifying power**) M is the angular size θ' of the final image produced by the instrument divided by a reference angular size θ. The reference angular size is taken to be the angular size of the object when seen without the instrument:

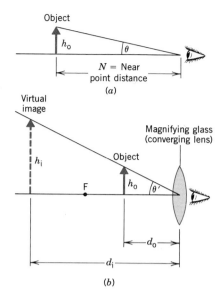

$$\begin{bmatrix} \textbf{Angular} \\ \textbf{magnification} \end{bmatrix} \quad M = \frac{\begin{array}{c} \text{Angular size of} \\ \text{final image produced} \\ \text{by optical instrument} \end{array}}{\begin{array}{c} \text{Reference angular size} \\ \text{of object seen without} \\ \text{optical instrument} \end{array}} = \frac{\theta'}{\theta} \qquad (33.3)$$

A magnifying glass is the simplest device that provides angular magnification. To compute its angular magnification, we first determine the reference angular size θ. In this case, θ is chosen to be the angular size that the object has when placed at the near point of the eye and is seen without the magnifying glass. Since an object cannot be brought closer than the near point of the eye and still produce a sharp image on the retina, θ represents the largest angular size obtainable without the magnifying glass. Figure 33.9a indicates that θ, measured in radians, is $\theta \approx h_o/N$, where N is the distance from the eye to the near point. To compute θ', recall from Section 32.7 and Figure 32.21 that a magnifying glass is usually a single converging lens used with the object located inside the focal point. In this situation, Figure 33.9b indicates that the lens produces a virtual image that is enlarged and upright with respect to the object. Assuming the eye is next to the magnifying glass, the angular size θ' seen by the eye is $\theta' \approx h_o/d_o$, where d_o is the object distance. The angular magnification is

$$M = \frac{\theta'}{\theta} \approx \frac{h_o/d_o}{h_o/N} = \frac{N}{d_o}$$

FIGURE 33.9 (a) Without a magnifying glass, the largest angular size θ occurs when the object is placed at the near point of the eye. (b) A magnifying glass is a converging lens. It produces an enlarged, virtual image of an object placed inside the focal point F of the lens. The angular size of both the image and the object is θ'.

According to the thin-lens equation, d_o is related to the image distance d_i and the focal length f of the lens by

$$\frac{1}{d_o} = \frac{1}{f} - \frac{1}{d_i} \qquad (32.6)$$

Substituting this expression for $1/d_o$ into the expression above for M leads to the following result:

$$\begin{bmatrix} \textbf{Angular magni-} \\ \textbf{fication of a} \\ \textbf{magnifying glass} \end{bmatrix} \quad M = \frac{\theta'}{\theta} \approx \left(\frac{1}{f} - \frac{1}{d_i} \right) N \qquad (33.4)$$

Two special cases of Equation 33.4 are of interest, depending on whether the image is located as close to the eye as possible or as far away as possible. To be seen clearly, the closest the image can be relative to the eye is at the near point; for this situation, $d_i = -N$. The minus sign indicates that the image lies to the left of the lens and is virtual. Under this condition, Equation 33.4 becomes $M \approx N/f + 1$. The farthest the image can be from the eye is at infinity ($d_i = -\infty$); this occurs when the object is placed at the focal point of the lens. When the image is at infinity, Equation 33.4 simplifies to $M \approx N/f$. Clearly, the angular magnification is greater when the image is

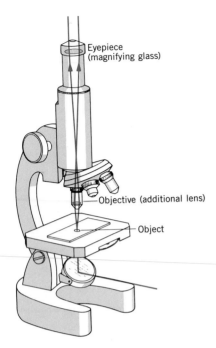

Eyepiece (magnifying glass)

Objective (additional lens)

Object

FIGURE 33.10 A compound microscope.

at the near point of the eye rather than at infinity. In either case, however, the greatest magnification is achieved by using a magnifying glass with the shortest possible focal length.

33.4 THE COMPOUND MICROSCOPE

To increase the angular magnification beyond that possible with a magnifying glass, an additional converging lens can be included to "premagnify" the object before the magnifying glass comes into play. The result is an optical instrument known as the **compound microscope** (Figure 33.10). The magnifying glass is called the **eyepiece** of the microscope, and the additional lens is referred to as the **objective.**

In obtaining the angular magnification M of the compound microscope, we follow the same approach used in the last section and begin with $M = \theta'/\theta$ (Equation 33.3), where θ' is the angular size of the final image and θ is the reference angular size. As with the magnifying glass in Figure 33.9, the reference angular size is determined by the height h_o of the object when the object is located at the near point of the unaided eye: $\theta \approx h_o/N$, where N is the distance between the eye and the near point.

To find the angular size θ' of the final image produced by the microscope, note from Figure 33.11a that the objective lens produces a "first image," which then serves as the object for the eyepiece. The eyepiece, in turn, produces the final image. Part b of the drawing shows that the angular size θ' of the final image is equal to that of the

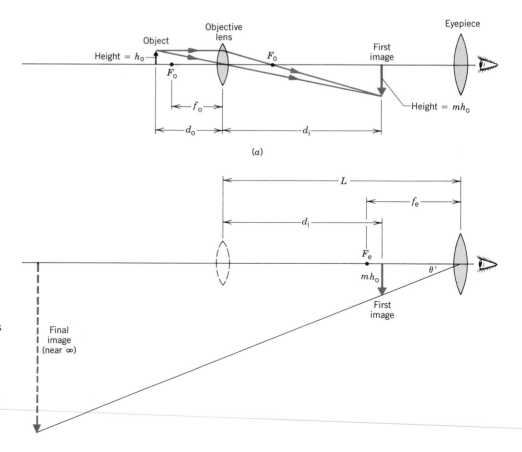

(a)

(b)

FIGURE 33.11 (a) In a two-lens microscope, the objective lens forms a "first image" of the object. This first image is enlarged, real, and inverted. (b) The first image becomes the object for the eyepiece, which behaves as a magnifying glass and produces an enlarged, virtual, final image near infinity.

first image, so we need only to determine θ' from the first image. The first image is inverted and its height is mh_o, where m is the magnification of the objective lens and h_o is the height of the object. The magnification equation (Equation 32.7) specifies m in terms of the object distance d_o and the image distance d_i, according to $m = -d_i/d_o$. If, as in part a of the drawing, the object is placed just outside the focal point F_o of the objective lens to achieve a large magnification, then $d_o \approx f_o$, where f_o is the focal length of the objective. Furthermore, if the microscope is designed so the eye is fully relaxed when viewing the final image, the final image must be very far from the eyepiece, or near "infinity." This location of the final image implies that the first image must fall just inside the focal point F_e of the eyepiece, as part b of the drawing indicates. Therefore, $d_i \approx L - f_e$, where L is the separation between the two lenses and f_e is the focal length of the eyepiece. With these approximations for d_o and d_i, the magnification of the objective lens becomes

$$m = -\frac{d_i}{d_o} \approx -\frac{(L - f_e)}{f_o}$$

Since the distance between the first image and the eyepiece is nearly f_e, the angular size θ' of the first image (and also the final image) is

$$\theta' \approx \frac{\text{Height of first image}}{\text{Distance of first image from eyepiece}} \approx \frac{mh_o}{f_e} \approx -\left(\frac{L - f_e}{f_o}\right)\frac{h_o}{f_e}$$

Using the expressions above for θ and θ', we find that the angular magnification of the microscope is

$$M = \frac{\theta'}{\theta} \approx \frac{-\left(\dfrac{L - f_e}{f_o}\right)\dfrac{h_o}{f_e}}{\dfrac{h_o}{N}}$$

$$\boxed{\begin{array}{c}\text{Angular magni-}\\\text{fication of a}\\\text{compound}\\\text{microscope}\end{array}} \qquad M \approx -\frac{(L - f_e)N}{f_o f_e} \qquad\qquad (33.5)$$

Equation 33.5 shows that the angular magnification is greatest when f_o and f_e are as small as possible (since they are in the denominator) and when L is as large as possible. Furthermore, L must be greater than the sum of f_o and f_e for this equation to be valid. Example 5 deals with the angular magnification of a compound microscope.

EXAMPLE 5

The objective of a compound microscope has a focal length of $f_o = 0.40$ cm, while that of the eyepiece is $f_e = 3.0$ cm. The two lenses are separated by a distance of $L = 20.0$ cm. A person with normal eyes ($N = 25$ cm) is using the microscope. (a) Determine the angular magnification of the microscope. (b) Compare the answer in part (a) with the largest angular magnification obtainable by using the eyepiece alone as a magnifying glass.

SOLUTION

(a) Equation 33.5 gives the angular magnification of the microscope:

$$M \approx -\frac{(L - f_e)N}{f_o f_e} = -\frac{(20.0 \text{ cm} - 3.0 \text{ cm})(25 \text{ cm})}{(0.40 \text{ cm})(3.0 \text{ cm})}$$

$$= \boxed{-350}$$

The minus sign indicates that the final image is inverted relative to the initial object.

(b) When the eyepiece is used alone as a magnifying glass, as in Figure 33.9b, the largest angular magnification occurs when the image seen through the eyepiece is as close as possible to the eye. As Section 33.3 discusses, the image in this case is at the near point and the angular magnification is

$$M \approx \frac{N}{f_e} + 1 = \frac{25 \text{ cm}}{3.0 \text{ cm}} + 1 = \boxed{9.3}$$

Clearly, the effect of the objective lens is to increase the angular magnification of the microscope by a factor of $350/9.3 = 38$.

33.5 THE TELESCOPE

THE ASTRONOMICAL TELESCOPE

A telescope is an instrument for magnifying distant objects, such as stars and planets. Like a microscope, a telescope consists of an objective lens and an eyepiece lens (also called the *ocular*). Since the object is usually far away, the light rays striking the telescope are nearly parallel, and the "first image" is formed just beyond the focal point F_o of the objective, as Figure 33.12a illustrates. The first image is real and inverted. Unlike that in the microscope, however, this image is *smaller* than the object. If, as in part b of the drawing, the telescope is constructed so the first image lies just inside the focal point F_e of the eyepiece, the eyepiece acts like a magnifying glass. It forms a final image that is greatly enlarged, virtual, and located near infinity. This final image can then be viewed with a fully relaxed eye.

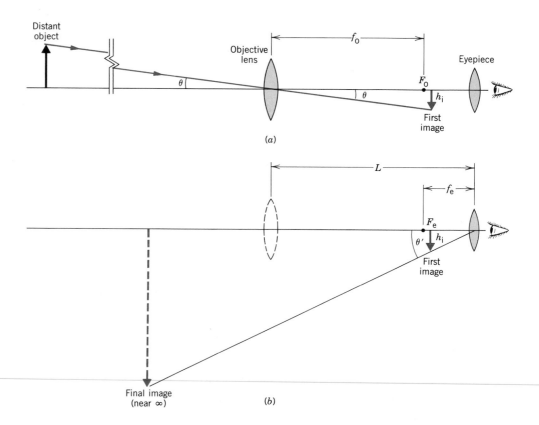

FIGURE 33.12 (a) An astronomical telescope is used to view distant objects (note the "break" in the principal axis, between the object and the objective). The objective lens produces a real, inverted, first image. (b) The eyepiece magnifies the first image to produce the final image near infinity.

The angular magnification M of a telescope, like that of a magnifying glass or a microscope, is the angular size θ' subtended by the final image of the telescope divided by the reference angular size θ of the object as seen without the telescope. For an astronomical object, such as a planet, it is convenient to use as a reference the angular size of the object seen in the sky with the unaided eye. Since the object is far away, the angular size seen by the unaided eye is nearly the same as the angle θ subtended at the objective of the telescope in Figure 33.12a. Moreover, θ is also the angle subtended by the first image, so $\theta \approx h_i/f_o$, where h_i is the height of the first image and f_o is the focal length of the objective. To obtain an expression for θ', note from part b of the figure that the first image is located very near the focal point F_e of the eyepiece, so $\theta' \approx -h_i/f_e$, where f_e is the focal length of the eyepiece lens. The minus sign is present because the first image is inverted. The angular magnification of the telescope is, then,

$$
\begin{bmatrix}
\textbf{Angular} \\
\textbf{magnification of} \\
\textbf{an astronomical} \\
\textbf{telescope}
\end{bmatrix}
\qquad
M = \frac{\theta'}{\theta} \approx \frac{-h_i/f_e}{h_i/f_o} = -\frac{f_o}{f_e}
\qquad (33.6)
$$

The angular magnification is determined by the ratio of the focal length of the objective lens to the focal length of the eyepiece lens. For large angular magnifications, the objective lens should have a long focal length and the eyepiece a short one. Some of the design features of a telescope are the topic of the next example.

EXAMPLE 6

The telescope shown in Figure 33.13 has the following specifications: $f_o = 1.0 \times 10^3$ mm and $f_e = 5.0$ mm. From these data, find (a) the angular magnification of the telescope and (b) the length of the telescope.

SOLUTION

(a) The angular magnification follows directly from Equation 33.6:

$$
M = -\frac{f_o}{f_e} = -\frac{1.0 \times 10^3 \text{ mm}}{5.0 \text{ mm}} = \boxed{-200}
$$

(b) Since the first image is located just beyond the focal point F_o of the objective (Figure 33.12a) and just inside the focal point F_e of the eyepiece (Figure 33.12b), these two points are close together and the distance L between the two lenses is approximately

$$
L \approx f_o + f_e = 1000 \text{ mm} + 5 \text{ mm} \approx \boxed{1000 \text{ mm}}
$$

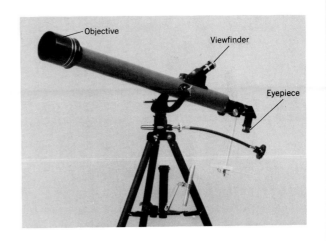

FIGURE 33.13 An astronomical telescope. The viewfinder is a separate small telescope with low magnification and serves as an aid in locating the object. Once the object has been found, the viewer looks through the eyepiece to obtain the full magnification of the astronomical telescope.

Two types of astronomical telescopes are used today, the ***refracting telescope*** and the ***reflecting telescope.*** Both include a lens for the eyepiece. The refracting telescope

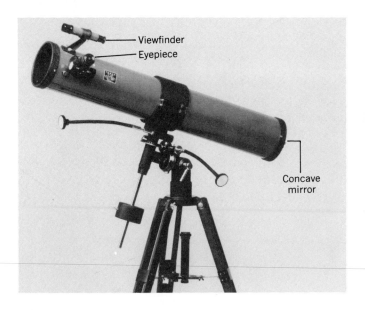

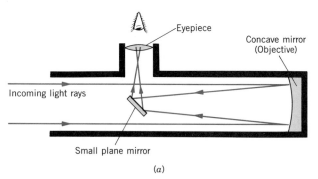

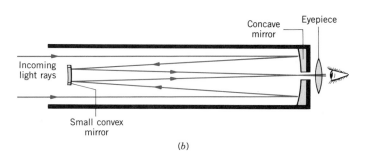

FIGURE 33.14 (*a*) A reflecting telescope with a Newtonian focus. (*b*) A reflecting telescope with a Cassegrainian focus.

also uses a lens for the objective and is the type illustrated in Figure 33.12. The reflecting telescope, however, uses a concave mirror for the objective, as Figure 33.14 indicates. In part *a* of the drawing the incoming light rays reflect from a concave mirror at the right end. A small plane mirror intercepts the reflected, converging rays and directs them toward the eyepiece. This particular design was developed by Isaac Newton and is said to have a Newtonian focus. Figure 33.14*b* shows another design, which is said to have a Cassegrainian focus.

Because the light from distant stars is so faint, telescopes need a great light-gathering ability, so their objectives have large diameters. Since a lens can be supported only at its edge, a large lens begins to sag under its own weight and produce distorted images. Consequently, large concave mirrors are preferable, because they can be supported over the entire back surface. Also, a mirror has only one surface to be ground and polished rather than two, and a mirror is free from the chromatic aberration that exists in all lenses.* All the larger astronomical telescopes in the world are reflectors.

* Section 33.6 discusses chromatic aberration.

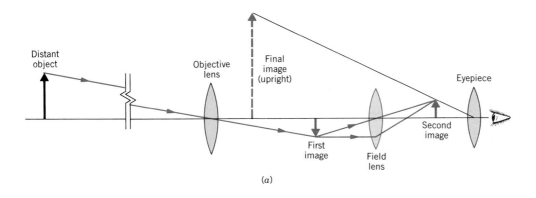

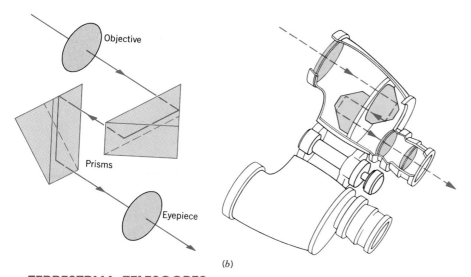

FIGURE 33.15 (*a*) In a terrestrial telescope a field lens produces an upright second image. This second image serves as the object for the eyepiece, which produces an upright final image. (*b*) A pair of prisms is used in binoculars to produce an upright image of the object. By folding the light path, the prisms also allow the instrument to be shorter than a normal telescope.

TERRESTRIAL TELESCOPES

As you can see in Figure 33.12, an astronomical telescope gives an inverted image of a distant object. It makes little difference whether a star or planet is seen upside down, but for viewing objects on earth, this inversion is undesirable. ***Terrestrial telescopes,*** therefore, contain an extra lens or prisms to produce a final image that is upright with respect to the object. Figure 33.15*a* suggests one possible design in which a third lens (the *field lens*) produces an upright second image that leads to an upright final image. This type of terrestrial telescope is sometimes referred to as a spyglass. Often telescopes used by amateur astronomers can be converted into terrestrial telescopes by removing the eyepiece and replacing it with one that contains a field lens or prisms. Prisms are frequently used in binoculars, as Figure 33.15*b* illustrates. The prisms serve both to invert the image so it is upright relative to the object and to fold the light path. With a folded path, the instrument can be relatively short, and yet the objective can have the long focal length required for large magnification.

33.6 LENS ABERRATIONS

SPHERICAL ABERRATION

Rather than forming a sharp image, a single lens typically forms an image that is slightly out of focus. This lack of sharpness arises because the rays originating from a

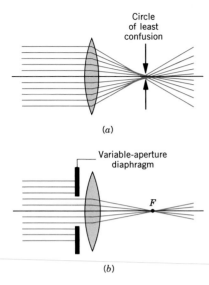

FIGURE 33.16 (*a*) Spherical aberration results when light rays parallel to the principal axis do not all cross at the same point after passing through the lens. Instead, they converge to a minimum cross-sectional area at the circle of least confusion. (*b*) Spherical aberration can be reduced by allowing only rays near the principal axis to pass through the lens. The refracted rays now converge more nearly to a single focal point *F*.

single point on the object are not focused to a single point on the image; as a result, each point on the image becomes a small "blur." The lack of point-to-point correspondence between object and image is called an aberration.

One common type of aberration is ***spherical aberration,*** and it ocurs with converging and diverging lenses made with spherical surfaces. Figure 33.16*a* shows how it arises with a converging lens. Ideally, all rays traveling parallel to the principal axis are refracted so they cross the axis at the same point after passing through the lens. However, rays far from the principal axis are refracted more by the lens than those closer in. Consequently, the outer rays cross the axis closer to the lens than do the inner rays, and spherical aberration prevents a lens from having a unique focal point. Instead, as the drawing suggests, there is a location along the principal axis where the light converges to the smallest cross-sectional area. This area is circular and is known as the ***circle of least confusion.*** The circle of least confusion is where the most satisfactory image can be formed by the lens.

Spherical aberration can be reduced substantially by using a variable-aperture diaphragm to allow only those rays close to the principal axis to pass through the lens. Figure 33.16*b* indicates that a reasonably sharp focal point can be achieved by this method, although less light now passes through the lens. Lenses with parabolic surfaces are also used to reduce this type of aberration, but they are difficult to make and thus expensive.

CHROMATIC ABERRATION

Chromatic aberration also causes blurred images. It arises because the index of refraction of the material from which the lens is made varies with wavelength. Section 32.5 discusses how this variation leads to the phenomenon of dispersion, whereby different colors refract by different amounts.

Figure 33.17*a* shows sunlight incident on a converging lens. Within the lens, the light spreads into its color spectrum because of dispersion, but for clarity the picture shows only the colors at the opposite ends of the spectrum—red and violet. Violet is refracted more than red, so the violet ray crosses the principal axis closer to the lens than does the red ray. Thus, the focal length of the lens is shorter for violet than for red, with intermediate values of the focal length corresponding to the colors in between. As a result of chromatic aberration, an undesirable color fringe surrounds the image.

Chromatic aberration can be greatly reduced by using a compound lens, such as the combination of a converging lens and a diverging lens shown in Figure 33.17*b*. Each lens is made from a different glass. With this lens combination the red and violet rays almost come to a common focus and, thus, chromatic aberration is reduced. A lens combination designed to reduce chromatic aberration is called an *achromatic*

FIGURE 33.17 (*a*) Chromatic aberration arises when different colors are focused at different points along the principal axis: F_V = focal point for violet light, F_R = focal point for red light. (*b*) Chromatic aberration can be minimized by using an achromatic lens (a converging and a diverging lens in tandem). The achromatic lens brings different colors more nearly to the same focal point *F*.

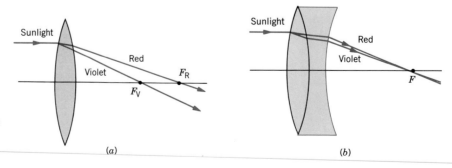

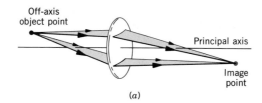

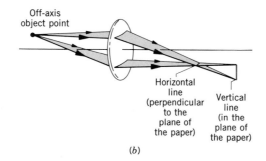

FIGURE 33.18 (*a*) An ideal converging lens focuses light rays from an off-axis point on the object to a single point on the image. (*b*) When a lens exhibits off-axis astigmatism, it focuses rays from an off-axis point on the object to an image that consists of a short horizontal line and a short vertical line, instead of a single point.

lens (from the Greek "achromatos," meaning "without color"). All high-quality cameras use achromatic lenses.

OFF-AXIS ASTIGMATISM

Ideally, when a point on an object lies off the principal axis, a lens should focus this point to a corresponding single point on the image. Figure 33.18*a* illustrates this ideal situation for a converging lens. *Off-axis astigmatism* is a troublesome kind of aberration in which an object point lying off the principal axis is focused to two short *lines,* instead of a single image point. Figure 33.18*b* illustrates the effect. The line closest to the lens is horizontal and perpendicular to the plane of the paper. The other line is vertical and lies in the plane of the paper.

To reduce off-axis astigmatism, a photographer can center the subject in the field of view to make sure that the important part of the picture is as close as possible to the principal axis. Specially designed lens systems can also be used to reduce this form of aberration.

SUMMARY

A **camera** uses a converging lens system to produce real, inverted images on film. The amount of light reaching the film is controlled by the **shutter speed** and the size of the **aperture.** Aperture sizes are often indicated by *f*-numbers on the barrel of the lens. The **f-number** is the ratio of the focal length *f* of the lens to the diameter *D* of the aperture: *f*-number = *f*/*D*. More light strikes the film when the *f*-number is smaller than when it is larger.

The **human eye,** like a camera, forms a real, inverted image on a light-sensitive surface, called the retina. **Accommodation** is the process by which the focal length of the eye is automatically adjusted, so that objects at different distances can be made to produce focused images on the retina. The **near point** of the eye is the point nearest the eye at which an object can be placed and still produce a sharp image on the retina. The **far point** of the eye is the location of the farthest object on which the fully relaxed eye can focus. For a normal eye, the near point is located 25 cm from the eye and the far point is located at infinity.

A **nearsighted (myopic)** eye is one that can focus on nearby objects, but not on distant ones. Nearsightedness can be corrected by wearing eyeglasses or contacts made from diverging lenses. A **farsighted (hyperopic)** eye can see distant objects clearly, but not those close up. Farsightedness can be corrected by using converging lenses.

The **refractive power** of a lens is measured in diopters and is given by $1/f$, where f is the focal length of the lens in meters. A converging lens has a positive refractive power, while a diverging lens has a negative refractive power.

The **angular size** of an object is the angle that it subtends at the eye of the viewer. For small angles, the angular size in radians is $\theta \approx h_o/d_o$, where h_o is the height of the object and d_o is the object distance. The **angular magnification** M of an optical instrument is the angular size θ' of the final image produced by the instrument divided by the reference angular size θ of the object, which is that seen without the instrument: $M = \theta'/\theta$.

A **magnifying glass** is usually a single converging lens that forms an enlarged, upright, and virtual image of an object placed at or inside the focal point of the lens. For a magnifying glass held close to the eye, the angular magnification is approximately $M \approx (1/f - 1/d_i)N$, where f is the focal length of the lens, d_i is the image distance, and N is the distance of the viewer's near point from the eye.

A **compound microscope** consists of an objective lens and an eyepiece lens. The final image is enlarged, inverted, and virtual. The angular magnification of a microscope is $M \approx -(L - f_e)N/(f_o f_e)$, where f_o and f_e are, respectively, the focal lengths of the objective and eyepiece, L is the distance between the two lenses, and N is the distance of the viewer's near point from the eye.

An **astronomical telescope** magnifies distant objects with the aid of objective and eyepiece lenses and produces a final image that is inverted and virtual. The angular magnification of a telescope is $M \approx -f_o/f_e$, where f_o and f_e are the focal lengths of the objective and eyepiece lenses. A **terrestrial telescope** uses one or more extra lenses or prisms to produce a final image that is upright relative to the object.

Lens aberrations limit the formation of perfectly focused or sharp images by optical instruments. **Spherical aberration** occurs because rays that pass through the outer edge of a lens with spherical surfaces are not focused at the same point as those that pass through the center of the lens. **Chromatic aberration** arises because a lens focuses different colors at different points. When a lens exhibits an aberration known as **off-axis astigmatism,** an off-axis point on the object is not focused to a corresponding point on the image. Instead, the image of the off-axis point consists of a horizontal line followed by a vertical line, as Figure 33.18b illustrates.

QUESTIONS

1. Suppose two people who wear glasses are camping. One of them is nearsighted and the other is farsighted. Whose glasses may be useful in starting a fire with the sun's rays? Give your reasoning.

2. Suppose that a 21-year-old with normal vision is standing in front of a plane mirror. How close can he stand and still see himself in focus?

3. If we read for a long time, our eyes become "tired." When this happens, it helps to stop reading and look at a distant object. From the point of view of the ciliary muscle, why does this refresh the eyes?

4. To a swimmer under water, distant objects look blurred and out of focus. However, when the swimmer wears goggles that keep the water away from the eyes, the objects appear sharp and in focus. Why do goggles improve a swimmer's underwater vision?

5. The refractive power of the lens of the eye is 15 diopters when surrounded by the aqueous and vitreous humors. If this lens is removed from the eye and then surrounded by air, its refractive power increases to about 150 diopters. Why is the refractive power of the lens so much greater outside the eye?

6. Three lenses have refractive powers of 1, 2, and 4 diopters, respectively. Draw each of the three lenses and locate their focal points to scale. Parallel rays of light are incident on each lens. (a) Draw the refracted rays as they pass through the focal point of each lens. (b) From your drawings, decide which lens bends the rays to the greatest extent.

7. By means of a ray diagram, show that the eyes of a person wearing glasses appear to be (a) smaller when the glasses use diverging lenses to correct for nearsightedness and (b) larger when the glasses use converging lenses to correct for farsightedness.

8. Can a diverging lens be used as a magnifying glass? Justify your answer with a ray diagram.

9. Who benefits more from using a magnifying glass, a person whose near point is located 25 cm away from the eyes or a person whose near point is located 75 cm away from the eyes? Provide a reason for your answer.

10. Two lenses, whose focal lengths are 3.0 cm and 45 cm, are used to build a telescope. Which lens should be the objective? Why?

11. Two refracting telescopes have identical eyepieces, although one telescope is twice as long as the other. Which telescope has the greater angular magnification? Provide a reason for your answer.

12. Suppose a well-designed optical instrument is composed of two converging lenses separated by 14 cm. The focal lengths of the lenses are 6.0 mm and 45 mm. Is the instrument a microscope or a telescope? Why?

13. It is often thought that virtual images are somehow less important than real images. To show that this is not true, identify which of the following instruments normally produce final images that are virtual: (a) the projector, (b) the camera, (c) the magnifying glass, (d) eyeglasses (or contact lenses), (e) the compound microscope, (f) the astronomical telescope, (g) the terrestrial telescope.

14. Would you expect the image produced by a camera to have less spherical aberration at higher *f*-number settings or at lower *f*-number settings? Why?

15. Why does chromatic aberration occur in lenses, but not in mirrors?

16. When taking pictures outdoors using the same lens, would you expect chromatic aberration to cause more problems at sunset or at noon? Explain your reasoning.

PROBLEMS

Section 33.1 The Camera

1. A macroscopic (or macro) lens is usually a lens of normal focal length built into a lens barrel that can be adjusted to provide the additional lens-to-film distance needed when focusing at very close range. (a) Suppose that a 50.0-mm macro lens has a maximum lens-to-film distance of 275 mm. How close can the object be located in front of the lens? (b) For comparison, determine the object distance for a normal 50.0-mm lens that has a maximum lens-to-film distance of 65.0 mm.

2. A camera with a 50.0-mm lens is focused on a distant mountain. By how much and in what direction must the lens be moved to focus on a kitten 0.65 m away?

3. By what factor does the amount of light admitted to a camera increase when the aperture is changed from *f*/8 to *f*/2?

4. A 4.0×10^2-mm telephoto lens is used with an aperture diameter of 71 mm. What is the corresponding *f*-number?

* 5. From a distance of sixty meters, a photographer uses a 5.00×10^2-mm telephoto lens to take a picture of a charging rhinoceros. How far from the rhinoceros would the photographer have to be to record an image of the same size using a 50.0-mm lens?

* 6. A camera takes a picture with an exposure time of 1/125 s and its lens set to an *f*-number of 16. Assuming that the amount of light reaching the film is to remain the same, determine the correct exposure time if the *f*-number is set to 8.

**7. To produce a "trick" photograph, a small-scale model (1 : 375) of an intergalactic spaceship is photographed with a 90.0-mm lens. The image of the model on the film has the same size as the image of the actual spaceship when photographed at a distance of 10.0 m using a 50.0-mm lens. At what distance from the model was the trick photograph taken?

Section 33.2 The Human Eye

8. A nearsighted person is diagnosed to have a far point lo-

cated only 220 cm from his eyes. Determine the focal length of contact lenses that will enable him to see distant objects clearly.

9. Suppose your friend wears contact lenses that have a focal length of 35.1 cm. The lenses are designed so she can read a magazine held as close as 25.0 cm. Where is the near point of her unaided eyes?

10. A person has far points of 5.0 m from the right eye and 6.5 m from the left eye. Write a prescription for the refractive power of the corrective contact lenses.

11. A farsighted person cannot focus clearly on objects that are less than 145 cm from his eyes. To correct this problem, the person wears eyeglasses that are located 2.0 cm in front of his eyes. Determine the focal length that will permit this person to read a newspaper at a distance of 32.0 cm from his eyes.

* 12. As a person grows older, the lens of the eye loses most of its elasticity, and the focal length of the lens can be changed by only a relatively small amount. Consequently, the eye loses much of its power of accommodation, a condition known as presbyopia. Bifocals (see drawing) are necessary if a person with presbyopia is to see both near and distant objects clearly. The upper segment of the

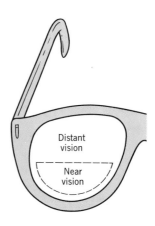

lens has a focal length chosen for distant vision, while the lower segment has a focal length suitable for near vision. Suppose a presbyopic eye has a far point located at a distance of 4.30 m and a near point located at a distance of 1.25 m from the eye. The glasses are to be worn 2.0 cm in front of the eyes, and reading material is to be held 25.0 cm from the eyes. Obtain the focal lengths of the upper and lower segments of the bifocal lens.

* 13. A nearsighted person wears contacts to correct for a far point that is only 3.62 m from his eyes. The near point of his unaided eyes is 25.0 cm from his eyes. If he does not remove the lenses when reading, how close can he hold a book and see it clearly?

**14. The far point of a nearsighted person is 4.37 m from her eyes, and she wears contacts that enable her to see distant objects clearly. A tree is 12.0 m away and 3.00 m high. (a) When she looks through the contacts at the tree, what is its image distance? (b) How high is the image formed by the contacts?

**15. The contacts worn by a farsighted person allow her to see objects clearly that are as close as 25.0 cm, even though her uncorrected near point is 79.0 cm from her eyes. When she is looking at a poster, the contacts form an image of the poster at a distance of 217 cm from her eyes. (a) How far away is the poster actually located? (b) If the poster is 0.350 m tall, how tall is the image formed by the contacts?

Section 33.3 Angular Magnification and the Magnifying Glass

16. A spectator, seated in the left field stands, is watching a 1.9-m-tall baseball player who is 75 m away. On a TV screen, the same player has a 12-cm image. To a viewer located 3.0 m from the screen, does the ball player appear to be larger or smaller than what the spectator sees? Give your reasoning.

17. A converging lens whose focal length is 11.0 cm is being used to examine a small caterpillar by a person whose near point is located 45 cm from his eyes. What is (a) the maximum angular magnification of the lens and (b) the angular magnification when the lens is positioned so the eye is fully relaxed?

18. A magnifying glass is held above a magazine such that the image is located at the near point of the eye. The near point is 0.30 m away from the eye, and the angular magnification is 3.4. Find the focal length of the magnifying glass.

19. A butterfly collector is examining a rare specimen and uses a magnifying glass with a refractive power of 10.0 diopters. The magnifying glass is held close to the eye, and the butterfly-to-lens distance is adjusted so a virtual image is formed at infinity. The angular magnification is 4.0. What is the distance between the collector's eyes and the near point?

* 20. A person holds a book 25 cm in front of her eyes; the print in the book is 2.0 mm high. If the effective lens of the eye is located 1.7 cm from the retina, what is the size of the print image on the retina?

* 21. A stamp collector is viewing a special stamp whose letter-ing is 1.0 mm high. The collector's near point is 25 cm from his eyes. (a) What is the refracting power of a magnifying glass that makes the letters appear to be 6.0 mm high when the image of the stamp is located at the near point of the eye? (b) What is the apparent size of the lettering when the image of the stamp is 45 cm from the eye?

**22. A farsighted person can read printing as close as 25.0 cm when she wears contacts that have a focal length of 45.4 cm. One day, however, she forgets her contacts and uses a magnifying glass, which has a maximum angular magnification of 7.50 for a young person with normal vision. What is the maximum angular magnification that the magnifying glass can provide for her?

Section 33.4 The Compound Microscope

23. An insect subtends an angle of only 4.0×10^{-3} rad at the unaided eye when placed at the near point. What is the angular size when the insect is viewed through a microscope whose angular magnification has a magnitude of 160?

24. A compound microscope has three interchangeable objective lenses whose focal lengths are 2.0 mm, 5.0 mm, and 8.0 mm. The eyepiece has a focal length of 2.8 cm, and the distance between the eyepiece and the objective is 18.0 cm. What are the angular magnifications of the microscope for a viewer with a near point 25 cm from her eyes?

25. A compound microscope has a barrel whose length is 16.0 cm and an eyepiece whose focal length is 1.4 cm. The viewer has a near point located 25 cm from his eyes. What focal length must the objective have so the angular magnification of the microscope is -320?

* 26. The maximum angular magnification of a magnifying glass is 12.0 when a person uses it who has a near point that is 25.0 cm from his eyes. The same person finds that a microscope, using this magnifying glass as the eyepiece, has an angular magnification of -525. The separation between the eyepiece and the objective lenses of the microscope is 23.0 cm. Obtain the focal length of the objective lens.

* 27. It is possible to interchange the eyepiece and the objective of a microscope without changing the angular magnification of the instrument, provided the separation L between the two lenses is suitably adjusted. Derive an equation that gives the new separation L' in terms of the original separation L, the focal length f_o of the original objective, and the focal length f_e of the original eyepiece.

Section 33.5 The Telescope

28. An astronomical telescope has an objective with a focal length of 92 cm and interchangeable eyepieces whose focal lengths are 3.0 and 0.80 cm. What angular magnifications are possible?

29. A refracting telescope has an objective and an eyepiece that have refractive powers of 1.25 diopters and 250 diopters, respectively. Find the angular magnification of the telescope.

30. A refracting astronomical telescope for hobbyists has an angular magnification of −155. The eyepiece has a focal length of 5.0 mm. (a) Determine the focal length of the objective lens. (b) About how long is the telescope?

31. An astronomical telescope has an angular magnification of −184 and uses an objective with a focal length of 48.0 cm. What is the focal length of the eyepiece?

* 32. A refracting telescope has an angular magnification of −109. The length of the barrel is 1.10 m. What are the focal lengths of the objective and eyepiece lenses?

* 33. The refracting telescope at Yerkes Observatory in Wisconsin has an objective whose diameter is 102 cm and whose focal length is 19.4 m. Its eyepiece has a focal length of 10.0 cm. (a) What is the *f*-number of the objective? (b) What is the angular magnification of the telescope? (c) If the telescope is used to look at a lunar crater (diameter = 1500 m), what is the size of the first image, assuming the moon is 3.85×10^8 m from the earth? (d) How close does the crater appear to be when seen through the telescope?

**34. An astronomical telescope is being used to examine a relatively close object that is only 114 m away from the objective of the telescope. The objective and eyepiece have focal lengths of 1.500 m and 0.070 m, respectively. Noting that the expression $M \approx -f_o/f_e$ is no longer applicable because the object is so close, use the thin-lens and magnification equations to find the angular magnification of the telescope. *(Hint: See Figure 33.12 and note that the focal points F_o and F_e are so close together that the distance between them may be ignored.)*

ADDITIONAL PROBLEMS

35. Suppose the focal length of the objective in a microscope is 1.5 cm. The distance between the lenses is 18 cm. If the microscope is to provide an angular magnification of −83 when used by a person with a normal near point (25 cm from the eye), what must be the focal length of the eyepiece?

36. A camera is supplied with two interchangeable lenses, whose focal lengths are 35.0 and 150.0 mm. A woman whose height is 1.80 m stands 8.00 m in front of the camera. What is the height (including sign) of her image that each lens produces on the film?

37. A jeweler whose near point is 65 cm from his eye uses a small magnifying glass (called a loupe) to examine a watch held 5.0 cm from the lens. Find the angular magnification of the magnifying glass.

38. An amateur astronomer decides to build a telescope from a discarded pair of eyeglasses. One of the lenses has a refractive power of 11 diopters, while the other has a refractive power of 1.3 diopters. (a) Which lens should be the objective? (b) How far apart should the lenses be separated? (c) What is the angular magnification of the telescope?

39. An optometrist prescribes contact lenses that have a focal length of 55.0 cm. (a) Are the lenses converging or diverging, and (b) is the person who wears them nearsighted or farsighted? (c) Where is the unaided near point of the person located, if the lenses are designed so that objects no closer than 35.0 cm can be seen clearly?

* 40. A person using a magnifying glass observes that for clear vision its maximum angular magnification is 25% larger than its minimum angular magnification. Assuming that the person has a near point located 25 cm from her eye, what is the focal length of the magnifying glass?

* 41. A reflecting telescope uses a concave mirror whose radius of curvature is 2.4 m. If the angular magnification of the telescope is −360, what is the focal length of the eyepiece?

* 42. A racing fan is photographing a car being refueled in the pit area. The correct exposure is obtained with a shutter speed setting of 125 and a lens aperture of *f*/4. When the car begins to move, the photographer must use a shutter speed setting of 1000 to freeze the motion. Determine the correct *f*-number that must be used, assuming the lighting conditions are the same.

* 43. Often, the angular magnification M of a microscope is written as the product of the angular magnification M_o of the objective and the angular magnification M_e of the eyepiece: $M = M_o M_e$. Which part of the angular magnification $M \approx -(L - f_e)N/(f_o f_e)$ corresponds to M_o and which part to M_e? Justify your answer.

* 44. At age forty, a man requires contact lenses ($f = 65.0$ cm) to read a book held 25.0 cm from his eyes. At age forty-five, he finds that while wearing these contacts he must now hold a book 29.0 cm from his eyes. (a) By what distance has his near point changed? (b) What focal length lenses does he require at age forty-five to read a book at 25.0 cm?

**45. The angular magnification of a refracting telescope is 32 800 times larger when you look through the correct end of the telescope than when you look through the wrong end. What is the angular magnification of the telescope?

**46. Bill is farsighted and has a near point located 125 cm from his eyes. Anne is also farsighted, but her near point is 75.0 cm from her eyes. Both have glasses that correct their vision to a normal near point (25.0 cm from the eyes), and both wear the glasses 2.0 cm from the eyes. Relative to the eyes, what is the closest object that can be seen clearly (a) by Anne when she wears Bill's glasses and (b) by Bill when he wears Anne's glasses?

CHAPTER 34

Interference and the Wave Nature of Light

Saw-whet owls have marvelous night vision. In part because of their relatively large pupils, owl eyes have good resolving power and are able to distinguish between two objects that are close together.

Up to now, we have assumed that light travels through a uniform medium in a straight line and only changes direction when it reflects or refracts at a boundary between two different media. The study of the straight-line motion of light and reflection and refraction is referred to as *geometrical optics.* Because of its wave nature, however, light can also exhibit effects due to interference and diffraction, which geometrical optics cannot explain. The study of the interference and diffraction of light is referred to as *wave optics* or *physical optics* and is the topic of the present chapter. Light exhibits the same interference and diffraction phenomena that sound waves do (see Chapter 22). Therefore, Section 34.1 begins with a review of the principle of linear superposition, which lies at the heart of wave interference and applies to electromagnetic waves, as well as to sound waves.

34.1 THE PRINCIPLE OF LINEAR SUPERPOSITION

The principle of linear superposition states that when two or more waves are present simultaneously at the same place, the resulting disturbance is the sum of the distur-

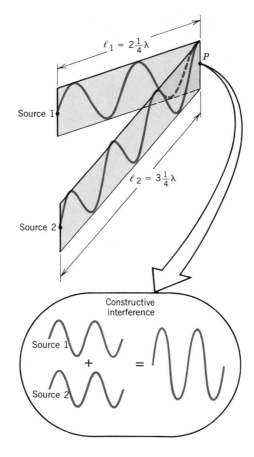

FIGURE 34.1 The waves emitted by source 1 and source 2 arrive at point P in phase, leading to constructive interference at that point.

bances of the individual waves. For a light wave the disturbance is electromagnetic; thus, the electric fields of two light waves passing through a given point combine to give the total electric field at that point. And the square of the electric field strength is proportional to the intensity of the light, or the light energy per second per square meter. The intensity, in turn, is related to the brightness of the light.

Figure 34.1 illustrates what happens when two identical waves arrive at the point P in phase with one another, that is, crest to crest and trough to trough. The waves have the same wavelength λ. According to the principle of linear superposition, the waves reinforce each other and *constructive interference* occurs. The resulting total wave at P has an amplitude that is twice the amplitude of either individual wave, and in the case of light waves, the brightness at P is greater than that of either wave alone. The reason the waves arrive at P in phase is that the distance ℓ_1 between source 1 and this spot is $2\frac{1}{4}$ wavelengths, whereas the corresponding distance ℓ_2 for source 2 is $3\frac{1}{4}$ wavelengths, these distances differing by exactly one wavelength. Constructive interference will result at P whenever the distances are the same or differ by any integer number of wavelengths; in other words, assuming ℓ_2 is the larger distance, whenever $\ell_2 - \ell_1 = m\lambda$, where $m = 0, 1, 2, 3, \ldots$.

Figure 34.2 shows what occurs when two identical waves arrive at the point P out of phase with one another, or crest to trough. Now the waves mutually cancel, according to the principle of linear superposition, and *destructive interference* results. With light waves this would mean that there is no brightness. The waves are out of phase at P, because the distances through which they travel in reaching this spot differ by one-half of a wavelength ($\ell_1 = 2\frac{3}{4}\lambda$ and $\ell_2 = 3\frac{1}{4}\lambda$ in the drawing). Destructive interference will take place at P whenever the distances differ by any half-integer

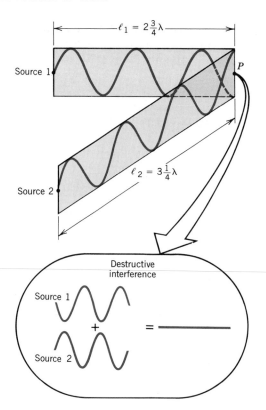

FIGURE 34.2 The waves emitted by the two sources arrive at point P out of phase, and destructive interference occurs at P.

number of wavelengths, that is, whenever $\ell_2 - \ell_1 = (m + \frac{1}{2})\lambda$, where $m = 0, 1, 2, 3, \ldots$, and ℓ_2 is the larger distance.

If constructive or destructive interference is to continue occurring at a point, the sources of the waves must be ***coherent sources.*** Two sources are coherent if the waves they emit maintain a constant phase relation. Effectively, this means that the waves do not shift relative to one another as time passes. For instance, suppose that the wave pattern of source 1 in Figure 34.2 shifted forward or backward by random amounts at random moments. Then, on average, neither constructive nor destructive interference would be observed at point P, because there would be no stable relation between the two wave patterns.

FIGURE 34.3 Source 1 is a TV transmitting antenna that emits electromagnetic waves. One wave reaches the house directly from the transmitting antenna, while another arrives via an indirect path after being reflected from a passing airplane. The airplane acts as wave source 2. As the plane flies by, conditions of constructive and destructive interference come and go at the receiving antenna, causing the TV picture to flutter.

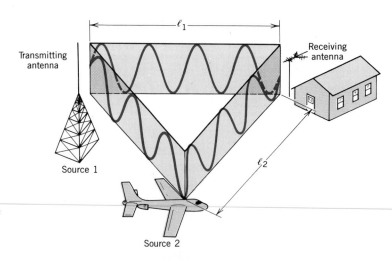

Many people have observed the effects of interference between two electromagnetic waves while watching TV. Television programming is carried by electromagnetic waves that are emitted by a transmitting antenna and detected by a receiving antenna connected to the TV set, as Figure 34.3 suggests. Sometimes these waves reflect from an airplane, and then the receiving antenna may detect waves coming from two sources. Source 1 is the transmitting antenna of the station, which is located at a fixed distance ℓ_1 from your house. The airplane acts as source 2, and its distance ℓ_2 from the house is changing as the plane moves. As a result, the difference between ℓ_1 and ℓ_2 is also changing, and the conditions of constructive and destructive interference come and go at the receiving antenna. Correspondingly, a signal that increases and decreases is delivered to the TV set, and one sees a picture that flutters as the plane passes overhead.

34.2 YOUNG'S DOUBLE-SLIT EXPERIMENT

The English scientist Thomas Young (1773–1829) demonstrated interference between light waves, and Figure 34.4 shows one arrangement of his experiment. Light passes through a single narrow slit S_0 and falls on two closely spaced slits S_1 and S_2. These two slits act as coherent sources of light waves that interfere constructively and destructively at different points on the screen to produce a pattern of alternating bright and dark fringes. The purpose of the single slit S_0 is to ensure that only light from one direction falls on the double slit. Without it, light coming from different points on the light source would strike the double slit from different directions and cause the pattern on the screen to be washed out. The slits S_1 and S_2 act as coherent sources of light waves, because the light from each originates from the same primary source, namely, the single slit S_0.

To help explain the origin of the bright and dark fringes, Figure 34.5 presents three top views of the double slit and the screen. Part *a* illustrates how a bright fringe arises directly opposite the midpoint between the two slits. At this location on the screen, the distances ℓ_1 and ℓ_2 to the slits are equal, each containing the same number of wavelengths. Therefore, constructive interference results, leading to the bright fringe. Part *b* indicates that constructive interference produces another bright fringe on one

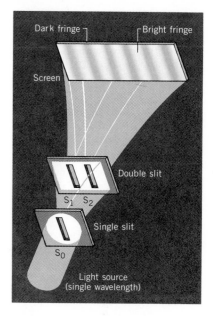

FIGURE 34.4 In the Young's double-slit experiment illustrated here, the two slits S_1 and S_2 act as coherent sources of light. The light waves from these slits interfere constructively and destructively at different places on the screen to produce, respectively, the bright and dark fringes. The slit widths and the distance between the slits have been exaggerated for clarity.

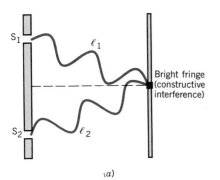
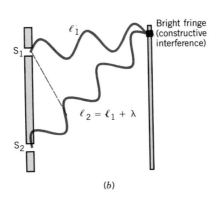
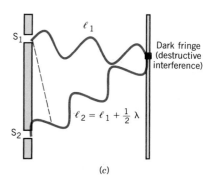

FIGURE 34.5 The waves emitted by slits S_1 and S_2 interfere constructively (parts *a* and *b*) or destructively (part *c*) on the screen, depending on the difference in distances between the slits and the screen. The slit widths and the distance between the slits have been exaggerated for clarity.

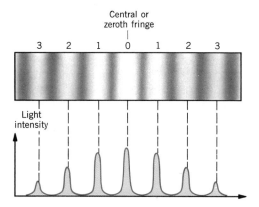

Central or
zeroth fringe

3   2   1   0   1   2   3

Light
intensity

FIGURE 34.6 In a Young's double-slit experiment, the central fringe is the brightest, that is, it has the greatest light intensity. The intensities of the other bright fringes decrease to either side of the center, as the graph indicates.

side of the midpoint when the distance ℓ_2 is larger than ℓ_1 by exactly one wavelength. A bright fringe also occurs symmetrically on the other side of the midpoint when the distance ℓ_1 exceeds ℓ_2 by one wavelength; for clarity, however, this bright fringe is not shown. Constructive interference produces additional bright fringes on both sides of the middle wherever the difference between ℓ_1 and ℓ_2 is an integer number of wavelengths. Part c shows how the first dark fringe arises. Here the distance ℓ_2 is larger than ℓ_1 by exactly one-half a wavelength, so the waves interfere destructively, giving rise to the dark fringe. Destructive interference creates additional dark fringes on both sides of the center wherever the difference between ℓ_1 and ℓ_2 equals a half-integer number of wavelengths.

The brightness of the fringes in Young's experiment varies. As an indication of the brightness, Figure 34.6 gives a graph of the light intensity for the fringe pattern. The central fringe is labeled with a zero, while the other bright fringes are numbered in ascending order on either side of center. It can be seen that the central fringe has the greatest intensity. To either side of center, the intensities of the other fringes decrease in a way that depends on how small the slit widths are relative to the wavelength of the light. Figure 34.7 shows a photograph of the fringe pattern observed in a typical Young's experiment.

The position of the fringes observed on the screen in Young's experiment can be calculated with the aid of Figure 34.8. If the screen is located very far away compared with the separation d of the slits, then the lines labeled ℓ_1 and ℓ_2 in part a are nearly parallel. Being nearly parallel, these lines make approximately equal angles θ with the horizontal. When constructive interference occurs, the distances ℓ_1 and ℓ_2 differ by an integer number of wavelengths $m\lambda$, and it can be seen in part b that $m\lambda$ is one of the sides of the colored right triangle whose hypotenuse is the separation d between the slits. Therefore, the angle θ of the interference maxima can be determined from the following expression:

$$\left[\begin{array}{c}\textbf{Bright fringes}\\\textbf{of a double}\\\textbf{slit}\end{array}\right] \qquad \sin\theta = m\frac{\lambda}{d} \qquad m = 0, 1, 2, 3, \dots \qquad (34.1)$$

FIGURE 34.7 This photograph shows the fringe pattern obtained in a typical Young's double-slit experiment. The arrows indicate the central bright fringe.

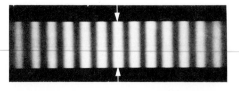

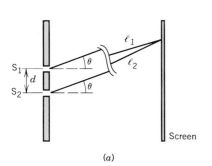
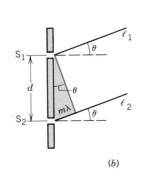
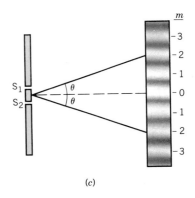

(a) (b) (c)

FIGURE 34.8 With the help of these pictures, the expression $\sin\theta = m\lambda/d$ ($m = 0, 1, 2, \ldots$) is derived in the text. This expression gives the angles θ at which the bright fringes occur in terms of the order m of the fringe, the wavelength λ of the light, and the separation d between the slits S_1 and S_2. The distances between slits S_1 and S_2 and a point on the screen are ℓ_1 and ℓ_2, respectively.

The value of m specifies the *order* of the fringe, as in the phrase "second-order bright fringe" for $m = 2$. Part c of the drawing stresses that the angle θ given by Equation 34.1 locates bright fringes on either side of the midpoint between the slits. A similar line of reasoning leads to the conclusion that the dark fringes, which lie between the bright fringes, are located according to

$$\begin{bmatrix} \textbf{Dark fringes} \\ \textbf{of a double} \\ \textbf{slit} \end{bmatrix} \qquad \sin\theta = (m + \tfrac{1}{2})\frac{\lambda}{d} \qquad m = 0, 1, 2, 3, \ldots \qquad (34.2)$$

Example 1 illustrates the application of these expressions and shows how to determine the distance of a higher-order bright fringe from the central bright fringe.

EXAMPLE 1

Red light ($\lambda = 713$ nm in vacuum) is used in Young's experiment with the slits separated by a distance $d = 0.120$ mm. The screen is located at a distance from the slits given by $L = 2.75$ m. Find the distance y on the screen between the central bright fringe and the third-order bright fringe (see Figure 34.9).

SOLUTION

This problem can be solved by first using Equation 34.1 to determine the value of θ that locates the third-order bright fringe. Then trigonometry can be used to obtain the distance y:

$$\sin\theta = m\frac{\lambda}{d} = 3\left(\frac{713 \times 10^{-9}\text{ m}}{0.120 \times 10^{-3}\text{ m}}\right) = 1.78 \times 10^{-2}$$

$$\theta = \sin^{-1}(1.78 \times 10^{-2}) = 1.02°$$

According to Figure 34.9, the distance y can be calculated from $\tan\theta = y/L$:

$$y = L\tan\theta = (2.75\text{ m})\tan 1.02° = \boxed{0.0490\text{ m}}$$

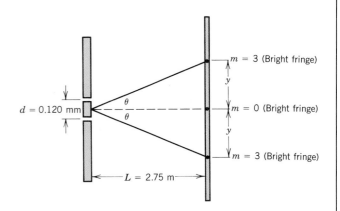

FIGURE 34.9 The third-order bright fringe ($m = 3$) is observed on the screen at a distance y from the central bright fringe ($m = 0$).

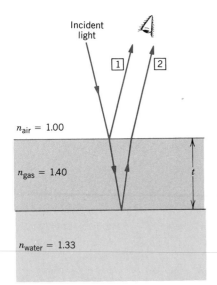

$n_{air} = 1.00$

$n_{gas} = 1.40$ t

$n_{water} = 1.33$

FIGURE 34.10 Because of reflection and refraction, two light waves, represented by rays 1 and 2, enter the eye when light shines on a thin film of gasoline floating on water. Constructive or destructive interference can result between these waves.

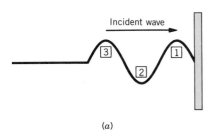

(a)

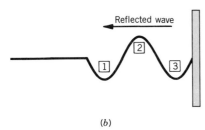

(b)

FIGURE 34.11 When a wave on a string reflects from a wall, the wave undergoes a phase change that is equivalent to one-half of a wave cycle or one-half of a wavelength. Thus, after reflection an upward-pointing half-cycle of the wave becomes a downward-pointing half-cycle, and vice versa, as the numbered labels in the drawing indicate.

Historically, Young's experiment provided strong evidence that light has a wave-like character. If light behaved only as a stream of "tiny particles," as others believed at the time,* then the two slits would deliver the light energy into only two bright fringes located directly opposite the slits on the screen. Instead, Young's experiment shows that wave interference redistributes the energy from the two slits into many bright fringes.

34.3 THIN FILM INTERFERENCE

Young's double-slit experiment is one example of interference between light waves. Interference also occurs in more common circumstances. For instance, Figure 34.10 shows a thin film, such as gasoline floating on water. The film is assumed to have a constant thickness. Consider what happens when light of a single wavelength (monochromatic light) strikes the film nearly perpendicularly. At the top surface of the film reflection occurs and produces the light wave represented by ray 1. However, refraction also occurs, and some light enters the film. Part of this light reflects from the bottom surface and passes back up through the film, eventually reentering the air. Thus, a second light wave, which is represented by ray 2, also exists. Moreover, this wave, having traversed the film twice, has traveled farther than wave 1. Because of the extra travel distance, there can be interference between the two waves. If constructive interference occurs, an observer, whose eyes detect the superposition of waves 1 and 2, would see a uniformly bright film. If destructive interference occurs, an observer would see a uniformly dark film. The controlling factor is whether the extra distance for wave 2 is an integer or half-integer number of wavelengths.

The wavelength that is important for thin film interference is the wavelength within the film, as opposed to the wavelength in vacuum. The wavelength within the film can be calculated from the wavelength in vacuum by using the index of refraction n, since $n = c/v = (c/f)/(v/f) = \lambda_{vacuum}/\lambda_{film}$. In other words,

$$\lambda_{film} = \frac{\lambda_{vacuum}}{n} \qquad (34.3)$$

In explaining the interference that can occur in Figure 34.10, we need to add one more important part to the story. Whenever waves reflect at a boundary, it is possible for them to change phase. As Section 22.5 discusses, for example, a wave on a string is inverted when it reflects from the end of a string that is tied to a wall (see Figure 34.11). This inversion is equivalent to a half-cycle of the wave, as if the wave had traveled an additional distance of one-half wavelength. In contrast, a phase change does not occur when a wave on a string reflects from the end of a string that is hanging free. When light waves undergo reflection, similar phase changes occur as follows:

1. When light travels through a material with a smaller refractive index toward a material with a larger refractive index (e.g., air to gasoline), reflection at the boundary occurs along with a phase change that is equivalent to one-half wavelength.

---

* It is now known that the particle or corpuscular theory of light, which Isaac Newton promoted, does indeed explain some experiments that the wave theory cannot explain. Today, light is regarded as having both particle and wave characteristics. Chapter 36 discusses this dual nature of light.

2. When light travels from larger toward smaller refractive index, there is no phase change upon reflection at the boundary.

The next example indicates how the phase change that can accompany reflection is taken into account when dealing with thin film interference.

EXAMPLE 2

(a) A thin film of gasoline floats on a puddle of water. Sunlight falls almost perpendicularly on the film and reflects into your eyes. Although sunlight is white, since it contains all colors, the film has a yellow hue, because destructive interference eliminates the color of blue ($\lambda_{\text{vacuum}} = 469$ nm) from the reflected light. If the refractive indices of the blue light in gasoline and water are 1.40 and 1.33, respectively, determine the minimum nonzero thickness t of the film. (b) Repeat part (a) assuming the gasoline is on glass ($n_{\text{glass}} = 1.52$) instead of water.

SOLUTION

(a) To solve this problem, we must express the condition for destructive interference in terms of the film thickness t. In the process, it is necessary to consider any phase changes that occur upon reflection. In Figure 34.10 the phase change for wave 1 is equivalent to one-half wavelength, since this light travels from a smaller refractive index ($n_{\text{air}} = 1.00$) toward a larger refractive index ($n_{\text{gas}} = 1.40$). In contrast, there is no phase change when wave 2 reflects from the bottom surface of the film, since this light travels from a larger refractive index ($n_{\text{gas}} = 1.40$) toward a smaller refractive index ($n_{\text{water}} = 1.33$). The net phase change between waves 1 and 2 due to reflection is, thus, equivalent to one-half wavelength, $\frac{1}{2}\lambda_{\text{film}}$. This half wavelength must be combined with the extra distance traveled by wave 2 relative to wave 1, to determine the condition for destructive interference. For destructive interference, the combined total must be a half-integer number of wavelengths. Since wave 2 travels back and forth through the film and since light strikes the film nearly perpendicularly, the extra distance traveled by wave 2 is twice the film thickness, or $2t$. Thus, the condition for destructive

interference is $2t + \frac{1}{2}\lambda_{\text{film}} = \frac{1}{2}\lambda_{\text{film}}, 1\frac{1}{2}\lambda_{\text{film}}$, and so forth. This condition is satisfied when

$$2t = m\lambda_{\text{film}} \qquad m = 0, 1, 2, 3, \ldots$$

With $m = 1$, the expression above gives the minimum nonzero film thickness for which the blue color is missing in the reflected light: $t = \frac{1}{2}\lambda_{\text{film}}$. Equation 34.3 gives the wavelength of blue light in the film as $\lambda_{\text{film}} = (469 \text{ nm})/1.40 = 335$ nm. Therefore, the minimum film thickness is

$$t = \frac{1}{2}\lambda_{\text{film}} = \frac{1}{2}(335 \text{ nm}) = \boxed{168 \text{ nm}}$$

(b) When the water in Figure 34.10 is replaced by glass, the phase change that accompanies the reflection of wave 2 from the bottom surface of the film is no longer zero. Instead, the phase change is equivalent to one-half wavelength, because the wave now travels from a smaller refractive index ($n_{\text{gas}} = 1.40$) toward a larger refractive index ($n_{\text{glass}} = 1.52$). In other words, wave 2 now behaves exactly as wave 1. Consequently, there is no net phase change between the waves. As a result, destructive interference occurs when the extra distance traveled by wave 2 is a half-integer number of wavelengths, which is a different condition than that in part (a):

$$2t = (m + \tfrac{1}{2})\lambda_{\text{film}} \qquad m = 0, 1, 2, 3, \ldots$$

The minimum nonzero thickness for destructive interference corresponds to $m = 0$ in this equation, so

$$t = \frac{(m + \frac{1}{2})\lambda_{\text{film}}}{2} = \frac{\lambda_{\text{film}}}{4} = \frac{335 \text{ nm}}{4} = \boxed{83.8 \text{ nm}}$$

Under natural conditions a thin film does not have a uniform thickness. Consequently, destructive interference eliminates different colors from the light reflected at different points on the film, depending on the thickness. Thus, a gasoline film floating on water looks multicolored, as does a soap bubble. The colors also depend on the viewing angle. At an oblique angle, the light corresponding to ray 2 in Figure 34.10 would travel a greater distance within the film than it does at nearly perpendicular incidence. The greater distance would lead to destructive interference for a different wavelength.

Thin film interference can be beneficial in optical instruments. A well-designed camera lens, for instance, is constructed so a maximum amount of light passes through the lens and reaches the film. Light reflected from the lens is wasted and reduces the quality of the photograph. To achieve minimum reflection, high-quality lenses are often covered with a thin nonreflective coating of magnesium fluoride ($n = 1.38$). This situation is like that in part (b) of Example 2, except the thickness is

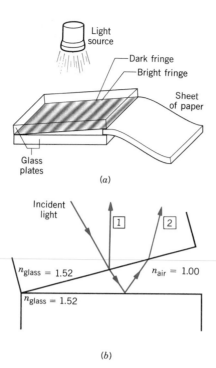

FIGURE 34.12 (*a*) The wedge of air formed between two flat glass plates causes an interference pattern of alternating dark and bright fringes to appear in reflected light. (*b*) A side view of the glass plates.

usually chosen to ensure that destructive interference eliminates the reflection of green light, which is in the middle of the visible spectrum.

Another interesting illustration of thin film interference is the air wedge. As Figure 34.12*a* shows, an air wedge is formed when two flat plates of glass are separated along one side, perhaps by a thin sheet of paper. The thickness of this film of air varies between zero, where the plates touch, and the thickness of the paper. When monochromatic light reflects from this arrangement, alternate bright and dark fringes are formed by constructive and destructive interference, as the picture indicates. Example 3 deals with the interference caused by an air wedge.

EXAMPLE 3

(a) Assuming that green light ($\lambda_{\text{vacuum}} = 552$ nm) strikes the glass plates nearly perpendicularly in Figure 34.12, determine the number of bright fringes that occurs between the place where the plates touch and the edge of the sheet of paper (thickness = 4.10×10^{-5} m). (b) Explain why there is a dark fringe where the plates touch.

SOLUTION

(a) First we consider the phase changes due to reflection for the waves corresponding to rays 1 and 2 in Figure 34.12*b*. There is no phase change for wave 1, since this light travels from larger (glass) toward smaller (air) refractive index in the process of reflecting. In contrast, there is a half-wavelength phase change for wave 2, since the ordering of the refractive indices is reversed at the air/glass boundary where reflection occurs. The net phase change due to reflection for waves 1 and 2, then, is equivalent to a half wavelength. The second stage in the solution is to combine any extra distance trav-

eled by ray 2 with this half wavelength and determine the condition for the constructive interference that creates the bright fringes. Constructive interference occurs whenever the *combination* yields an integer number of wavelengths. Therefore, the extra travel distance must be a half-integer number of wavelengths if it is to cause constructive interference. At nearly perpendicular incidence, the extra travel distance is nearly twice the thickness t of the wedge at any point, so the condition for constructive interference can be written as

$$2t = (m + \tfrac{1}{2})\lambda_{\text{film}} \qquad m = 0, 1, 2, 3, \ldots$$

In this expression, note that the "film" is a film of air. Since the refractive index of air is nearly unity, λ_{film} is virtually the same as that in vacuum, so $\lambda_{\text{film}} = 552$ nm. When t equals the thickness of the paper holding the plates apart, the corresponding value of m can be obtained from the equation above:

$$m = \frac{2t}{\lambda_{film}} - \frac{1}{2} = \frac{2(4.10 \times 10^{-5} \text{ m})}{552 \times 10^{-9} \text{ m}} - \frac{1}{2} = 148$$

Since the first bright fringe occurs when $m = 0$, the number of bright fringes is $m + 1 = \boxed{149}$.

(b) Where the plates touch, there is a dark fringe because of destructive interference between the light waves represented by rays 1 and 2. Destructive interference occurs, since the thickness of the wedge is zero here and the only difference between the rays is the half-wavelength phase change due to reflection from the lower plate.

Figure 34.13a shows a photograph of the fringes observed for an air wedge between two ultraflat, or optically flat, plates. Part b shows the fringe pattern observed when the plates are not optically flat. In fact, one way to test a plate for flatness is to use it to form an air wedge with a second reference plate that is known to be flat. When the fringes are straight as in part a, rather than wavy as in part b, no further polishing is needed to flatten the plate being tested.

Another type of air wedge can also be used to determine the degree to which the surface of a lens or mirror is spherical. When an accurate spherical surface is put in contact with an optically flat plate, as in Figure 34.14a, the circular interference fringes shown in part b of the figure can be observed. The circular fringes are called *Newton's rings*. They arise in the same way that the straight fringes arise in Figure 34.12. When the curved surface is irregular, the interference fringes are no longer circular, as in Figure 34.14c.

34.4 THE MICHELSON INTERFEROMETER

An interferometer is an apparatus that can be used to measure the wavelength of light by utilizing interference between two light waves. One particularly famous interferometer is that developed by Albert A. Michelson (1852–1931). Among other things, its fame rests upon a series of experiments that provided support for Einstein's theory of special relativity.

(a)   (b)

FIGURE 34.13 In reflected light a pattern of interference fringes can be observed due to the air wedge between two glass plates. (a) When the plates are ultraflat, or optically flat, the pattern consists of straight fringes. (b) When the plates are not flat, the pattern is wavy.

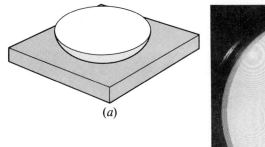

(a)

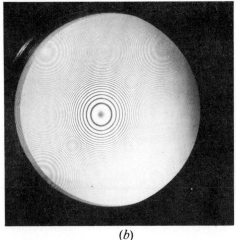

(b)

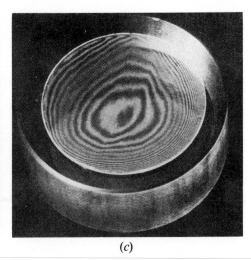

(c)

FIGURE 34.14 (a) The air wedge between an accurate spherical glass surface and an optically flat plate leads to (b) a pattern of circular interference fringes that is known as Newton's rings. (c) When the curved surface is irregular, the interference fringes are not circular.

The Michelson interferometer uses reflection to set up conditions where two light waves interfere. Figure 34.15 presents a schematic drawing of the instrument. Waves emitted by the light source strike a *beam splitter,* so called because it splits the beam of light into two parts. The beam splitter is a glass plate, the far side of which is coated with a thin layer of silver that reflects part of the beam upward as wave A in the drawing. The coating is so thin, however, that it also allows the remainder of the beam to pass directly through as wave F. Wave A strikes an adjustable mirror M_A and reflects back on itself. It again crosses the beam splitter and then enters the viewing telescope. Wave F strikes a fixed mirror M_F and returns, to be partly reflected into the viewing telescope by the beam splitter. Note that wave A passes through the glass

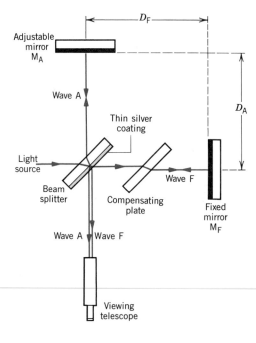

FIGURE 34.15 A schematic drawing of a Michelson interferometer.

plate of the beam splitter three times in reaching the viewing scope, while wave F passes through it only once. The compensating plate in the path of wave F has the same thickness as the beam splitter plate and ensures that wave F also passes three times through the same thickness of glass on the way to the viewing scope. Thus, an observer who views the combination of waves A and F through the telescope sees constructive or destructive interference, depending only on the difference in path lengths D_A and D_F traveled by the two waves.

Now suppose the mirrors M_A and M_F are perpendicular to each other, the beam splitter makes a 45° angle with each, and the distances D_A and D_F are equal. The waves A and F travel the same distance, and the field of view in the telescope is uniformly bright due to constructive interference. However, if the adjustable mirror is moved a distance of $\frac{1}{4}\lambda$, one wave travels back and forth over the quarter wavelength and moves through an extra distance of $\frac{1}{2}\lambda$. Consequently, the two waves are out of phase when they reach the viewing scope, destructive interference occurs, and the viewer sees a dark field. As mirror M_A is moved further, brightness again returns as soon as the waves are in phase and interfere constructively. The in-phase condition occurs when one of the waves travels a total extra distance of λ relative to the other. Thus, as the mirror is continuously moved, the viewer sees the field of view change from bright to dark, then back to bright, and so on.

It is also possible to describe the interference between waves A and F by regarding the silvered surface of the beam splitter as a mirror that forms a virtual image of the fixed mirror M_F. When the distances D_A and D_F are equal, the image of M_F coincides exactly with the adjustable mirror M_A. When D_A is made greater than D_F, the image of M_F lies in front of M_A, as Figure 34.16a illustrates. The light that appears to come from the image of M_F combines with the light coming from M_A to create the interference effects seen through the telescope.

When the adjustable mirror is rotated, it lies at an angle with respect to the image of M_F, as Figure 34.16b shows. Under this condition, M_A and the image of M_F are analogous to the surfaces of an air wedge, and the field of view in the telescope contains a pattern of bright and dark fringes like that in Figure 34.13a. As with the air wedge, these fringes arise because the distance D_A is different for light striking the adjustable mirror at different points along its surface. Each fringe corresponds to a different location on the mirror. When M_A is moved toward or away from the beam splitter, the fringe pattern appears to move across the field of view in the telescope, as a dark fringe changes into a bright fringe in response to the changing distance D_A. The distance by which D_A has been changed can be measured and related to the wavelength of the light, since a bright fringe changes into a dark fringe and back again each time D_A is changed by a half wavelength. (The back-and-forth change in distance is λ.) If a sufficiently large number of wavelengths are counted in this manner, the Michelson interferometer can be used to obtain a very accurate value for the wavelength from the measured changes in D_A.

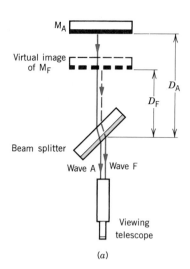

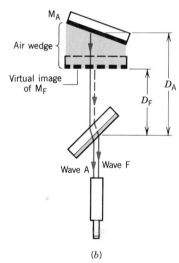

FIGURE 34.16 (a) In a Michelson interferometer, a virtual image of mirror M_F is formed in front of mirror M_A when the distance D_A is made greater than D_F. (b) When mirror M_A is rotated with respect to the virtual image of M_F, a kind of air wedge is created, and the field of view in the telescope contains a pattern of bright and dark fringes.

34.5 DIFFRACTION

As Section 22.3 discusses, **diffraction** is a bending of waves around obstacles or the edges of an opening. In Figure 34.17, for example, sound waves are leaving a room through an open doorway. Because the exiting sound waves bend, or diffract, around the edges of the opening, a listener outside the room can hear the sound even when standing "around the corner" from the doorway.

Diffraction is an interference effect, and the Dutch scientist Christian Huygens

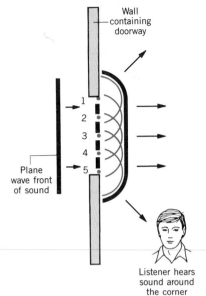

FIGURE 34.17 Sound bends or diffracts around the edges of a doorway, so a person can hear the sound even though he is not standing directly in front of the opening. The five colored points within the doorway act as sources and emit the five Huygens wavelets shown in color.

(1629–1695) developed a principle that is useful in explaining why diffraction arises. *Huygens' principle* describes how a wave front that exists at one instant gives rise to the wave front that exists at a later instant. The principle states that *every point on a wave front acts as a source of tiny wavelets that move forward with the same speed as the wave; the wave front at a later instant is the surface that is tangent to the wavelets.*

We begin by using Huygens' principle to explain the diffraction of sound waves in Figure 34.17. The drawing shows a plane wave front of sound approaching a doorway and identifies five representative points on the wave front when it is just leaving the opening. According to Huygens' principle, each of these points acts as a source of wavelets. The picture shows the wavelets as colored circular arcs at some moment after they are emitted. The tangent to the wavelets from points 2, 3, and 4 indicates that the wave front at this later instant is flat in the region directly in front of the doorway. But near the edges of the opening, points 1 and 5 are the last points that produce wavelets. Thus, Huygens' principle suggests that in conforming to the curved shape of the wavelets emitted near the edges, the new wave front propagates into regions that it would not reach otherwise. The sound wave, then, bends or diffracts around the edges of the doorway.

Huygens' principle applies not only to sound waves, but to all kinds of waves. For instance, light has a wavelike nature and, consequently, exhibits diffraction. Therefore, you may ask, "Since I can hear around the edges of a doorway, why can't I also see around them?" As a matter of fact, light waves do bend around the edges of a doorway. However, the degree of bending is extremely small, so the diffraction of light is not enough to allow you to "see around the corner."

As we will see, the extent to which a wave bends around the edges of an opening is determined by the ratio λ/W, where λ is the wavelength of the wave and W is the width of the opening. The photographs in Figure 34.18 illustrate the effect of this ratio on the diffraction of water waves. In part *a*, the ratio λ/W is small, because the wavelength (as indicated by the distance between the wave fronts) is small relative to the width of the opening. The wave fronts move through the opening with little bending or diffraction into the regions behind the edges. In part *b*, the wavelength is larger and the width of the opening is smaller. As a result, the ratio λ/W is larger, and the degree of bending becomes more pronounced, with the wave fronts penetrating more into the regions behind the edges.

FIGURE 34.18 These photographs show the plane wave fronts (horizontal lines) of water waves approaching an opening whose width W is smaller in (*b*) than it is in (*a*). In addition, the wavelength λ of the waves is larger in (*b*) than in (*a*). Therefore, the ratio λ/W increases from (*a*) to (*b*) and so does the extent of the diffraction. The waves bend around the edges of the opening more in (*b*) than in (*a*).

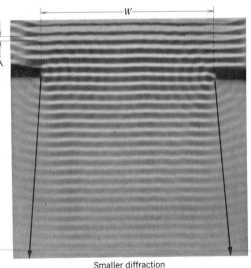

Smaller diffraction
(*a*) Smaller value for λ/W

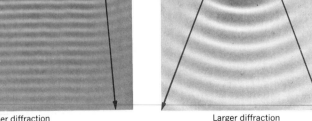

Larger diffraction
(*b*) Larger value for λ/W

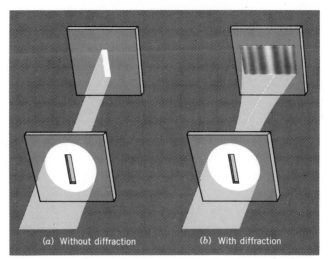

(a) Without diffraction   (b) With diffraction

FIGURE 34.19 (*a*) If light were to pass through a very narrow slit *without* diffraction, only the region on the screen directly opposite the slit would be illuminated. (*b*) Diffraction causes the light to bend around the edges of the slit into regions it would not otherwise reach and to form a pattern of alternating bright and dark fringes on the screen. The bright fringes to either side of the central bright fringe have greatly reduced intensities relative to that of the central fringe. Here, the intensities of the bright fringes to either side of center have been exaggerated for the sake of clarity. The slit width has also been exaggerated for clarity.

Based on the pictures in Figure 34.18, it might be expected that light waves of wavelength λ will bend or diffract appreciably when they pass through an opening whose width W is small enough to make the ratio λ/W sufficiently large. This is indeed the case, as Figure 34.19 illustrates. In this picture it is assumed that parallel rays (or plane wave fronts) of light fall on a very narrow slit and illuminate a viewing screen that is located far from the slit. Part *a* of the drawing shows what would happen if there were no diffraction. Light would pass through the slit without bending around the edges and produce an image of the slit on the screen. Part *b* shows what actually happens. The light diffracts around the edges of the slit and brightens regions on the screen that are not directly opposite the slit. The diffraction pattern on the screen consists of a bright central band, accompanied by a series of narrower faint fringes that are parallel to the slit itself.

To help explain how the pattern of diffraction fringes arises in Figure 34.19*b*, Figure 34.20 shows an enlarged top view of a plane wave front approaching the slit and singles out five sources of Huygens wavelets. Consider how the light reaches the midpoint on the screen. To simplify things, the screen is assumed to be far from the slit. Then, the rays from each Huygens source are nearly parallel,* and the wavelets all travel the same distance to the midpoint, arriving there in phase. As a result, constructive interference creates a bright central fringe on the screen, directly opposite the slit.

The wavelets emitted by the Huygens sources in the slit can also interfere destructively on the screen, as Figure 34.21 illustrates. Part *a* shows the light traveling from each source toward the first dark fringe. The angle θ gives the position of this dark fringe relative to the line between the midpoint of the slit and the central bright fringe. Since the screen is far from the slit, the rays of light from each Huygens source are nearly parallel and are all oriented at nearly the same angle θ, as in part *b* of the drawing. The wavelet from source 1 travels the shortest distance to the screen, while that from source 5 travels the farthest. Destructive interference creates the first dark fringe when the extra distance traveled by the wavelet from source 5 is exactly one wavelength, as the colored right triangle in the drawing indicates. Under this condi-

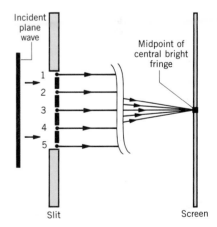

FIGURE 34.20 A plane wave front is incident on a single slit. This enlarged top view of the slit shows five sources of Huygens wavelets. The wavelets travel toward the midpoint of the central bright fringe on the screen, as the rays indicate. The screen is assumed to be far from the slit.

---

* When the rays are parallel, the diffraction is called Fraunhofer diffraction in tribute to the German optician Josef Fraunhofer (1787–1826). When the rays are not parallel, the diffraction is referred to as Fresnel diffraction, named for the French physicist Augustin Jean Fresnel (1788–1827).

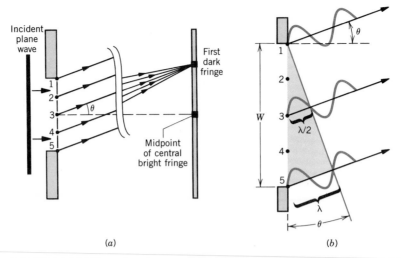

(a)   (b)

FIGURE 34.21 These drawings pertain to single-slit diffraction and show the condition for the destructive interference that creates the first dark fringe on either side of the central bright fringe. Only one of the dark fringes is included here for the sake of clarity. The screen is assumed to be far from the slit.

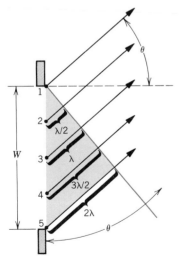

FIGURE 34.22 In a single-slit diffraction pattern there are multiple dark fringes on either side of the central bright fringe. This drawing shows the condition for the destructive interference that creates the second dark fringe on a very distant screen.

tion, the extra distance traveled by the wavelet from source 3 at the center of the slit is exactly one-half a wavelength. Therefore, wavelets from sources 1 and 3 are exactly out of phase and interfere destructively when they reach the screen. Similarly, a wavelet that originates slightly below source 1 cancels a wavelet that originates the same distance below source 3. Thus, each wavelet from the upper half of the slit cancels a corresponding wavelet from the lower half, and no light reaches the screen. As can be seen from the colored right triangle, the angle θ locating the first dark fringe is given by $\sin\theta = \lambda/W$, where W is the width of the slit.

Figure 34.22 shows the condition that leads to destructive interference at the second dark fringe on either side of the midpoint on the screen. In reaching the screen, the light from source 5 now travels a distance of two wavelengths farther than the light from source 1. To see that this extra distance leads to another dark fringe, notice that in the top half of the slit the wavelet from source 3 travels one wavelength farther than the wavelet from source 1. Furthermore, in the bottom half, the wavelet from source 5 travels one wavelength farther than the wavelet from source 3. Therefore, each half of the slit can be treated as the entire slit was in the previous paragraph: all the wavelets from the top half interfere destructively with each other, and all the wavelets in the bottom half do likewise. As a result, no light from either half reaches the screen. The colored triangle in the drawing shows that the second dark fringe occurs when $\sin\theta = 2\lambda/W$. Similar arguments hold for the third- and higher-order dark fringes, with the general result being

$$\begin{bmatrix}\textbf{Dark fringes}\\\textbf{for single-slit}\\\textbf{diffraction}\end{bmatrix}\qquad \sin\theta = m\frac{\lambda}{W}\qquad m = 1, 2, 3, \ldots \qquad (34.4)$$

Between each pair of dark fringes there is a bright fringe due to constructive interference. The brightness of the fringes is related to the light intensity, just as loudness is related to sound intensity. The intensity of the light at any location on the screen is the amount of light energy per second per unit area that strikes the screen there. Figure 34.23 gives a graph of the light intensity along with a photograph of a single-slit diffraction pattern. The central fringe, which is approximately twice as

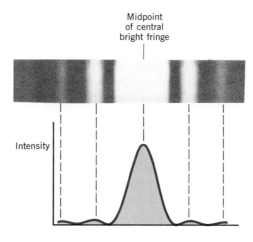

FIGURE 34.23 The photograph shows a single-slit diffraction pattern, with a bright and wide central fringe. The higher-order bright fringes are much less intense than the central fringe, as the graph indicates.

wide as the other bright fringes, has by far the greatest intensity. The width of the central fringe provides one indication of the extent of diffraction, as Example 4 illustrates.

EXAMPLE 4

Suppose that red light ($\lambda = 690$ nm in vacuum) passes through a slit and falls on a screen located at a distance of $L = 0.40$ m from the slit (see Figure 34.24). The distance between the middle of the central bright fringe and the first dark fringe is y. Determine the width $2y$ of the central bright fringe when the slit width is (a) $W = 4.0 \times 10^{-6}$ m and (b) $W = 80.0 \times 10^{-6}$ m.

SOLUTION

(a) The angle θ in Equation 34.4 locates the first dark fringe when $m = 1$; $\sin \theta = (1)\lambda/W$. Therefore,

$$\theta = \sin^{-1} \frac{\lambda}{W} = \sin^{-1} \left(\frac{690 \times 10^{-9} \text{ m}}{4.0 \times 10^{-6} \text{ m}} \right) = 9.9°$$

Now that θ is known, trigonometry can be used to determine the distance y. According to the picture, $\tan \theta = y/L$, so that

$$y = L \tan \theta = (0.40 \text{ m}) \tan 9.9° = 0.070 \text{ m}$$

The width of the central fringe, then, is $\boxed{2y = 0.14 \text{ m}}$.

(b) Repeating the calculation above with $W = 80.0 \times 10^{-6}$ m shows that $\theta = 0.49°$ and $\boxed{2y = 0.0068 \text{ m}}$. Notice how the spreading of the diffracted light decreases from 9.9° to 0.49° as the width of the slit increases and the ratio λ/W decreases. In the limit that the slit width becomes very large relative to the wavelength, the angle θ approaches zero, and the spreading effect of diffraction becomes unobservable.

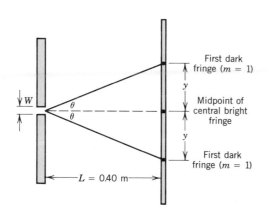

FIGURE 34.24 The distance $2y$ is the width of the central bright fringe.

Another example of diffraction can be seen when light from a point source falls on a solid disk, such as a coin (Figure 34.25). The effects of diffraction modify the dark shadow cast by the disk in several ways. First, the light waves diffracted around the circular edge of the disk interfere constructively at the center of the shadow to produce a small bright spot. There are also circular bright fringes in the shadow area. In addition, the boundary between the circular shadow and the lighted screen is not

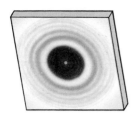

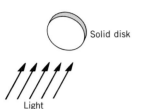

FIGURE 34.25 The diffraction pattern formed by a solid disk consists of a small bright spot in the middle of the shadow, circular bright fringes (not visible in the drawing) within the shadow, and concentric bright and dark fringes surrounding the shadow.

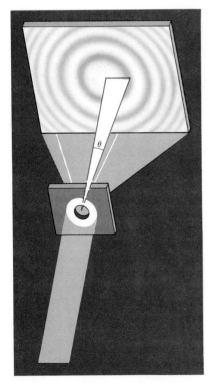

sharply defined, but consists of concentric bright and dark fringes. The various fringes are analogous to those produced by a single slit, and they are due to the interference between Huygens wavelets that originate from different points near the edge of the disk.

34.6 RESOLVING POWER

In a fashion similar to that for a single slit, diffraction also occurs when light passes through circular or nearly circular openings, such as those that admit light into a camera, a microscope, a telescope, or the human eye itself. The resulting diffraction pattern places a natural limit on the resolving power of such optical instruments. The *resolving power* is the ability of the instrument to distinguish between two closely spaced objects.

Figure 34.26 shows the diffraction pattern created by a small circular opening when the viewing screen is far from the opening. The pattern consists of a central bright circular region, surrounded by alternating bright and dark circular fringes. These fringes are analogous to the rectangular fringes that a single slit produces. The angle θ in the picture locates the first circular dark fringe relative to the central bright region and is given by

$$\sin \theta = 1.22 \frac{\lambda}{D} \tag{34.5}$$

where λ is the wavelength of the light and D is the diameter of the circular opening. This expression is similar to Equation 34.4 for a slit ($\sin \theta = \lambda/D$, when $m = 1$) and is valid when the distance to the screen is much larger than the diameter of the aperture.

An optical instrument with the ability to resolve two closely spaced objects can produce images of them that can be identified separately. For instance, think about the images on the film when light from two widely separated point objects passes through the circular aperture of a camera. As Figure 34.27 illustrates, each image is a circular diffraction pattern, but the two patterns do not overlap and are completely resolved. On the other hand, if the objects are sufficiently close together, the intensity patterns created by the diffraction overlap, as Figure 34.28a suggests. In fact, if the overlap is extensive it may no longer be possible to distinguish the patterns separately. In such a case, the picture from a camera would show a single "smeared-out" object instead of two separate objects. In Figure 34.28b the diffraction patterns overlap, but not enough to prevent us from seeing that two objects are present. Ultimately, then, diffraction limits the ability of an optical instrument to produce distinguishable images of objects that are close together.

It is useful to have a criterion for judging whether two closely spaced objects are resolved by an optical instrument. Figure 34.28a presents the *Rayleigh criterion* for resolution, which is named for Lord Rayleigh (1842–1919), who first proposed it:
Two point objects are just resolved when the first dark fringe in the diffraction pattern

FIGURE 34.26 When light passes through a small circular opening, a circular diffraction pattern is observed on a viewing screen. The angle θ locates the first dark fringe relative to the central bright region. In reality, the intensities of the bright fringes are greatly reduced relative to the intensity of the central bright region. Here the intensities of the bright fringes, as well as the diameter of the opening, have been exaggerated in the interest of clarity.

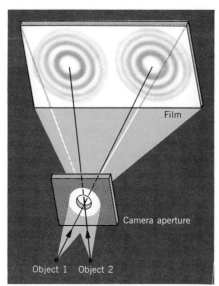

FIGURE 34.27 When light from two point objects passes through the circular aperture of a camera, two circular diffraction patterns are formed as images on the film. The images here are completely separated or resolved, because the objects themselves are widely separated.

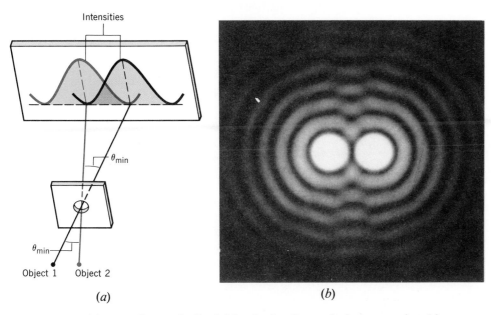

(a)   (b)

FIGURE 34.28 (a) According to the Rayleigh criterion for resolution, two point objects are just resolved when the first dark fringe (zero intensity) of one of the images falls on the central bright fringe (maximum intensity) of the other. (b) The photograph shows the resulting diffraction pattern when the images of two objects are just resolved.

of one falls directly on the central bright fringe in the diffraction pattern of the other. The minimum angle $\theta_{\min}$ between the two objects in the drawing is that given by Equation 34.5. If $\theta_{\min}$ is small and is expressed in radians, $\sin \theta_{\min} \approx \theta_{\min}$. Then, Equation 34.5 can be rewritten as

$$\theta_{min} \approx 1.22 \frac{\lambda}{D} \tag{34.6}$$

For a given wavelength λ and aperture diameter D, this result specifies the smallest angle that two point objects can subtend at the aperture and still be resolved. According to Equation 34.6, optical instruments designed to resolve closely spaced objects (small values of θ_{min}) must utilize the smallest possible wavelength and the largest possible aperture diameter.

To conclude this section, Example 5 compares the resolving power of the human eye with that of an eagle's eye.

EXAMPLE 5

(a) A hang glider is flying at an altitude of $H = 120$ m. Green light (wavelength = 555 nm in vacuum) enters the pilot's eye through a pupil that has a diameter $D = 2.5$ mm. The average index of refraction of the material in the eye is approximately $n = 1.36$. Determine how far apart two point objects must be on the ground if the person on the hang glider is to have any hope of distinguishing between them (see Figure 34.29). (b) The pupil of an eagle's eye has a diameter $D = 6.2$ mm. Assume that the material within its eyes has about the same refractive index as that of human eyes. Repeat part (a) for an eagle flying at the same altitude as the glider.

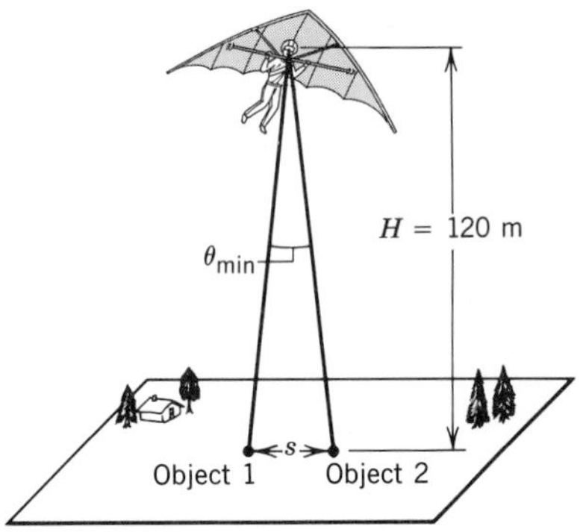

SOLUTION

(a) The minimum angle that the two objects can subtend at the pupil of the eye is given in radians by $\theta_{min} \approx 1.22\lambda/D$; from a knowledge of this angle and the altitude H, it is possible to obtain the separation s. Before using this expression, however, it is important to remember that the wavelength within the eye is given by Equation 34.3 as $\lambda = \lambda_{vacuum}/n = (555$ nm$)/1.36 = 408$ nm. Therefore,

$$\theta_{min} \approx 1.22 \frac{\lambda}{D} = 1.22 \left(\frac{408 \times 10^{-9}\ m}{2.5 \times 10^{-3}\ m} \right) \tag{34.6}$$

$$= 2.0 \times 10^{-4}\ rad$$

According to Equation 9.1, θ_{min} in radians is $\theta_{min} \approx s/H$, so

$$s \approx \theta_{min} H = (2.0 \times 10^{-4}\ rad)(120\ m) = \boxed{0.024\ m}$$

The two objects must be separated by at least 2.4 cm for them to be distinguished separately by the person on the hang glider.

(b) Since the diameter of the pupil of an eagle's eye is larger than that in a human eye, diffraction creates less of a limitation for the eagle. A calculation like that above, using $D = 6.2$ mm, reveals that the diffraction limit for the eagle's eye is $\boxed{s = 0.0096\ m}$.

FIGURE 34.29 The Rayleigh criterion for resolution can be used to estimate the smallest distance s that can separate two objects on the ground, if a person on a hang glider is to have any hope of distinguishing one object from the other.

34.7 THE DIFFRACTION GRATING

THE INTERFERENCE PATTERN OF A DIFFRACTION GRATING

Diffraction patterns of bright and dark fringes occur when monochromatic light passes through a single or double slit. Fringe patterns also result when light passes through more than two slits, and an arrangement consisting of a large number of parallel, closely spaced slits is called a ***diffraction grating.*** Gratings with as many as

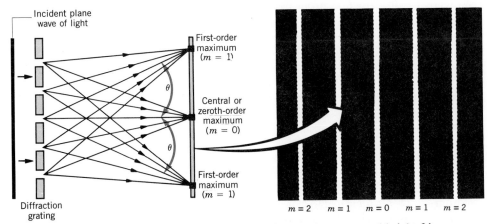

FIGURE 34.30 When light passes through a diffraction grating, a central bright fringe ($m = 0$) and higher-order bright fringes ($m = 1, 2, \ldots$) form when the light falls on a viewing screen.

40 000 slits per centimeter can be made, depending on the production method. In one method a diamond cutting tool is used to inscribe closely spaced parallel lines on a glass plate, the spaces between the lines serving as the slits. In fact, the number of slits per centimeter is often quoted as the number of lines per centimeter.

Figure 34.30 illustrates how light travels to a distant viewing screen from each of five slits in a grating and forms the central bright fringe and the first-order bright fringes on either side. Higher-order bright fringes also form, as the photograph in the figure shows. Each bright fringe is located by an angle θ relative to the central fringe. These bright fringes are sometimes called the *principal fringes* or *principal maxima*, since they are places where the light intensity is a maximum. Moreover, the term "principal" distinguishes them from other, much less bright fringes that are referred to as secondary fringes or secondary maxima. These secondary fringes are not visible in Figure 34.30.

Constructive interference creates the principal fringes, and to show how, we assume the screen is far from the grating. Thus, the rays remain nearly parallel while the light travels toward the screen, as in Figure 34.31. In reaching the place on the screen where the first-order maximum is located, light from slit 2 travels a distance of one wavelength farther than light from slit 1. Similarly, light from slit 3 travels one wavelength farther than light from slit 2, and so forth, as emphasized by the colored

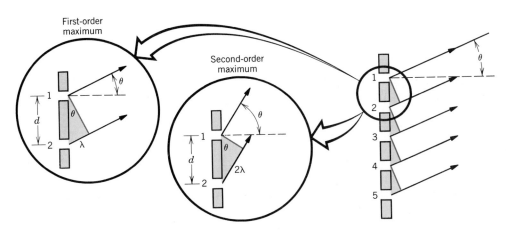

FIGURE 34.31 The conditions shown here lead to the first- and second-order intensity maxima in the diffraction pattern produced by a diffraction grating.

right triangles. For the first-order maximum, the enlarged view of one of these right triangles shows that constructive interference occurs when $\sin \theta = \lambda/d$, where d is the separation between the slits. The second-order maximum forms when the extra distance traveled by light from adjacent slits is two wavelengths, so that $\sin \theta = 2\lambda/d$. The general result is

$$
\begin{bmatrix} \textbf{Principal maxima} \\ \textbf{of a diffraction} \\ \textbf{grating} \end{bmatrix}
\qquad \sin \theta = m\frac{\lambda}{d} \qquad m = 0, 1, 2, 3, \ldots \qquad (34.7)
$$

The separation d between the slits can be calculated from the number of slits per centimeter of grating; for instance, a grating with 2500 slits per centimeter has a slit separation of $d = 1/2500$ cm.

Equation 34.7 is identical to Equation 34.1 for the double slit. A grating, however, produces bright fringes that are much *narrower* or *sharper* than those from a double slit, as the intensity patterns in Figure 34.32 reveal. Consider a grating with 100 slits/cm, for instance. The extra distance traveled by light from adjacent slits in forming the first-order maximum is exactly one wavelength λ. For an extra distance slightly larger than one wavelength, the light reaches the screen at a point slightly displaced from the maximum. If the extra distance is $\lambda + \lambda/100$, this slightly displaced point is already a place where complete destructive interference occurs, according to the following reasoning. The extra distance traveled by light from slit 51 compared to that from slit 1 is $50(\lambda + \lambda/100) = 50.5\lambda$. The additional half wavelength means that crests and troughs from these two slits combine to create destructive interference. The same result applies to light from slits 52 and 2, 53 and 3, and so on; in other words, to all the light from the grating. If a grating contains 10 000 slits/cm instead of 100 slits/cm, then destructive interference occurs when the extra distance traveled by light from adjacent slits is $\lambda + \lambda/10\,000$, instead of $\lambda + \lambda/100$. Thus, for a greater number of slits per centimeter, a smaller displacement from the maximum is required to produce the adjacent point of destructive interference on the screen, with the result that the principal fringes are even narrower. Note in Figure 34.32 that between the principal fringes there are secondary maxima with much smaller intensities.

The next example illustrates the ability of a grating to separate the components in a mixture of colors.

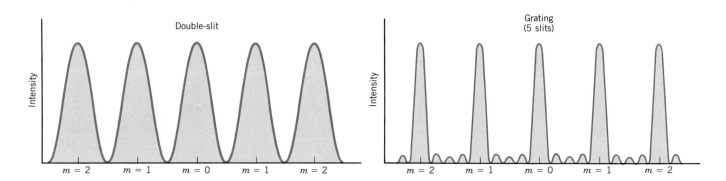

FIGURE 34.32 The bright fringes produced by a diffraction grating are much narrower than those produced by a double slit, as the intensity patterns in this picture indicate. Note the three small secondary bright fringes between the principal bright fringes. For a large number of slits, these secondary fringes become very small.

EXAMPLE 6

A mixture of violet light ($\lambda = 410$ nm in vacuum) and red light ($\lambda = 660$ nm in vacuum) falls on a grating that contains 9800 lines/cm. For each wavelength, find the angle θ that locates the first-order maxima.

SOLUTION

Before Equation 34.7 can be used here, a value for the separation d between the slits is needed: $d = 1/(9800$ lines/cm$) = 1.0 \times 10^{-4}$ cm, or 1.0×10^{-6} m. For violet light, the angle θ_{violet} for the first-order maxima ($m = 1$) is given by $\sin \theta_{violet} = m\lambda/d = \lambda/d$. Consequently,

$$\theta_{violet} = \sin^{-1}\frac{\lambda}{d} = \sin^{-1}\left(\frac{410 \times 10^{-9} \text{ m}}{1.0 \times 10^{-6} \text{ m}}\right) = \boxed{24°}$$

For red light, a similar calculation with $\lambda = 660 \times 10^{-9}$ m shows that $\boxed{\theta_{red} = 41°}$. Because θ_{violet} and θ_{red} are different, separate first-order bright fringes are seen for violet and red light on a viewing screen.

If the light in Example 6 had been sunlight, the angles for the first-order maxima would cover all values in the range between 24° and 41°, since sunlight contains all colors or wavelengths between violet and red. Consequently, a rainbowlike dispersion of the colors would be observed to either side of the central fringe on a screen. The central bright fringe would be white, however, since all the colors overlap there.

THE GRATING SPECTROSCOPE

A grating spectroscope is an instrument designed to measure the angles at which the principal maxima of a grating occur. With a measured value of the angle, calculations such as those in Example 6 can be turned around to provide values of the corresponding wavelengths. As Chapter 37 will point out, the atoms in a hot gas emit discrete wavelengths, and determining the values of these wavelengths is one important technique that can be used to identify the atoms. Figure 34.33 contains a sketch of a grating spectroscope. The slit that admits light from the source (e.g., a hot gas) is located at the focal point of the collimating lens, so the light rays striking the grating are parallel. The telescope is used to detect the bright fringes and, hence, to measure the angle θ.

COMPACT DISC PLAYERS

The compact disc (CD) has revolutionized stereo sound reproduction, and a diffraction grating is an essential feature of CD players. A CD contains a spiral track that holds the audio information. This spiral track is analogous to the spiral groove on a conventional LP record. However, the audio information on the CD track is detected using a laser beam, as Figure 34.34 illustrates. The information is encoded in the form of raised areas (which appear as "pits" when viewed from the top side of the CD) separated by flat areas (called "land"). As the disc rotates, the laser beam must be able to follow or track the pits and land along the spiral.

One type of CD player utilizes a three-beam tracking method to ensure that the laser beam follows the spiral properly. Figure 34.35 shows that a diffraction grating is the key element in this method. Before the laser beam strikes the CD, the beam passes through a grating that produces a central maximum and two first-order maxima, one on either side of the central maximum. As the picture indicates, the central maximum beam falls directly on the spiral track. This beam reflects into a detector, and the reflected light intensity fluctuates as the pits and land areas pass by, the fluctuations conveying the audio information as a series of binary numbers (zeros and ones).

The two first-order maxima beams are called tracking beams. They hit the CD between the arms of the spiral and also reflect into detectors of their own. Under

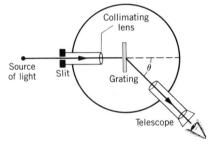

FIGURE 34.33 A grating spectroscope.

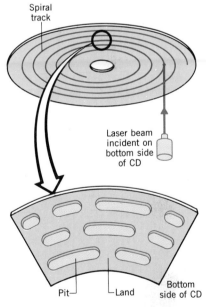

FIGURE 34.34 The bottom surface of a compact disc (CD) carries the audio information in the form of pits and land areas along a spiral track. A CD is played by using a laser beam that strikes the bottom surface of the disc and reflects from it.

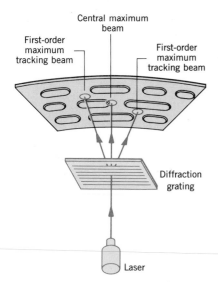

FIGURE 34.35 A three-beam tracking method is often used in CD players to ensure that the laser follows the spiral track correctly. The three beams are derived from a single laser beam by using a diffraction grating that produces a central maximum and two first-order maxima. The bottom surface of the disc is shown here.

perfect conditions, the intensities of the two reflected tracking beams do not fluctuate, since they originate from the smooth surface between the arms of the spiral where there are no pits. As a result, each tracking beam detector puts out the same constant electrical signal. However, when the tracking drifts to either side due to mechanical inaccuracies, the reflected intensity of each tracking beam changes because of the pits. In response, the tracking beam detectors produce different electrical signals. The difference between the signals is used in a "feedback" circuit to correct for the drift and put the three beams back into their proper positions.

34.8 X-RAY DIFFRACTION

Not all diffraction gratings are commercially made. Nature also creates diffraction gratings, although these gratings do not look like an array of closely spaced slits. Instead, nature's gratings are the arrays of regularly spaced atoms that exist in crystalline solids. For example, Figure 34.36 shows a crystal of ordinary salt (NaCl). Typically, the atoms in a crystalline solid are separated by distances of about 1.0×10^{-10} m, so we might expect a crystalline array of atoms to act like a grating with roughly this "slit" spacing for electromagnetic waves of the appropriate wavelength. Assuming that $\sin \theta = 0.5$ and that $m = 1$ in Equation 34.7, it follows that $0.5 = \lambda/d$. A value of $d = 1.0 \times 10^{-10}$ m, then, corresponds to a wavelength of approximately $\lambda = 0.5 \times 10^{-10}$ m. This wavelength is much shorter than that of visible light and, in fact, falls in the X-ray region of the electromagnetic spectrum. (See Figure 30.6.)

It is indeed true that a diffraction pattern results when X rays are directed onto a crystalline material, as Figure 34.37 illustrates for a crystal of DNA. The pattern consists of a complicated arrangement of spots, because a crystal has a complex three-dimensional structure. It is from patterns such as these that the spacing between atoms and the nature of the crystal structure can be determined.

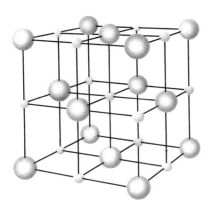

FIGURE 34.36 In this drawing of the crystalline structure of sodium chloride, the small colored spheres represent positive sodium ions, while the large spheres represent negative chloride ions.

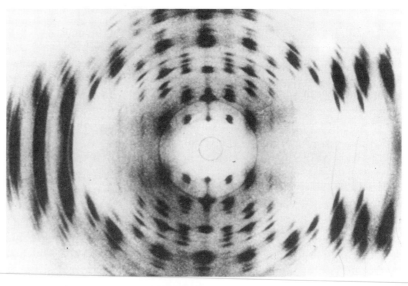

FIGURE 34.37 The X-ray diffraction pattern from crystalline DNA.

SUMMARY

The **principle of linear superposition** states that when two or more waves are present simultaneously in the same region of space, the resulting disturbance at any point is the sum of the individual wave disturbances at that point. According to this principle, two or more light waves can interfere constructively or destructively when they exist at the same place at the same time, provided they originate from mutually **coherent sources**. Two sources are coherent if they emit waves that have a constant phase relationship.

In **Young's double-slit experiment,** light passes through a pair of closely spaced slits and produces a pattern of alternating bright and dark fringes on a viewing screen. The fringes arise because of constructive and destructive interference. The angle θ for the mth higher-order bright fringes on either side of the central bright fringe is given by $\sin \theta = m\lambda/d$, where d is the spacing between the narrow slits and $m = 0, 1, 2, 3, \ldots$. Similarly, the angle for the dark fringes is given by $\sin \theta = (m + \frac{1}{2})\lambda/d$.

Constructive and destructive interference of light waves can occur with **thin films** of transparent materials. The interference occurs between light waves that reflect from the top and bottom surfaces of the film. One important factor in thin film interference is the thickness of the film relative to the wavelength of the light within the film. The wavelength within the film is $\lambda_{\text{film}} = \lambda_{\text{vacuum}}/n$, where n is the refractive index of the film. A second important factor is the phase change that can occur when light undergoes reflection at each surface of the film; the exact nature of the phase change is discussed in Section 34.3.

Diffraction is a bending of waves around obstacles or the edges of an opening. Diffraction is an interference effect that can be explained with the aid of **Huygens' principle.**

This principle states that every point on a wave front acts as a source of tiny wavelets that move forward with the same speed as the wave; the wave front at a later instant is the surface that is tangent to the wavelets. When light passes through a single slit and falls on a viewing screen, a pattern of bright and dark fringes is formed because of the superposition of such wavelets. The angle θ for the mth dark fringe on either side of the central bright fringe is given by $\sin \theta = m\lambda/W$, where W is the slit width and $m = 1, 2, 3, \ldots$.

The **resolving power** of an optical instrument is the ability of the instrument to distinguish between two closely spaced objects. Resolving power is limited by the diffraction that occurs when light waves enter an instrument, often through a circular opening. Consideration of the diffraction fringes leads to the **Rayleigh criterion** for resolution. This criterion specifies that two point objects are just resolved when the first dark fringe in the diffraction pattern of one falls directly on the central bright fringe in the diffraction pattern of the other. According to this specification, the minimum angle that two point objects can subtend at an aperture of diameter D and still be resolved as separate objects is $\theta_{\text{min}} \approx 1.22\lambda/D$, where λ is the wavelength of the light.

A **diffraction grating** is a device consisting of a large number of parallel, closely spaced slits. When light passes through a diffraction grating and falls on a viewing screen, the light forms a pattern of bright and dark fringes. The bright fringes are referred to as principal maxima and are found at an angle θ, such that $\sin \theta = m\lambda/d$, where d is the separation between two successive slits and $m = 0, 1, 2, 3, \ldots$.

QUESTIONS

1. A Young's double-slit experiment is performed using sunlight. The zeroth-order or central bright fringe on the screen is observed to be white, while rainbowlike patterns of color are observed on either side of the central fringe. Explain.

2. (a) How would the pattern of bright and dark fringes produced in a Young's double-slit experiment change if the light waves coming from *both* of the slits had their phases shifted by a half wavelength? (b) How would the pattern change, if the light coming from *only one* of the slits had its phase shifted by a half wavelength?

3. Replace slits S_1 and S_2 in Figure 34.4 with identical loudspeakers and use the same ac electrical signal to drive them. The two sound waves produced will then be identical and you will have the audio equivalent of Young's double-slit experiment. In terms of loudness and softness, describe what you would hear as you walked along the screen from one end to the other.

4. A camera lens is covered with a nonreflective coating that eliminates the reflection of perpendicularly incident green light. Recalling Snell's law, would you expect the reflected green light to be eliminated if it were incident on the lens at an angle of 45° rather than perpendicularly? Justify your answer.

5. When white light reflects from a soap bubble of nonuniform thickness, the bubble appears multicolored, because destructive interference removes different wavelengths from the light reflected at different places, depending on the thickness of the soap film. As a soap bubble becomes thinner and thinner, it looks darker and darker, appearing black just before it bursts. This black appearance means that destructive interference removes *all* wavelengths from the reflected light when the soap film is very thin. Explain why.

6. In Figure 34.14 there is a dark spot at the center of the pattern of Newton's rings. By considering the phase changes that occur when light reflects from the upper curved surface and the lower flat surface, account for the dark spot.

7. A transparent coating is deposited on a glass plate and has a refractive index that is *larger than that of the glass,* not smaller, as it is for a typical nonreflective coating. For a certain wavelength within the coating, the thickness of the coating is a quarter wavelength. The coating *enhances* the reflection of the light corresponding to this wavelength. Explain why, referring to part (a) of Example 2 in the text to guide your thinking.

8. The drawing shows an apparatus (Pohl's apparatus) in which interference effects arise because of reflection. A point source S_0 shines light on a thin sheet of mica. Some light reflects from the front surface and forms the virtual image S_F. In addition, some light penetrates the mica and reflects from the back surface to form a second virtual image S_B. Light waves that appear to come from these images travel different distances in reaching the screen, where they combine to create interference fringes. What is the shape of these fringes? Give your reasoning.

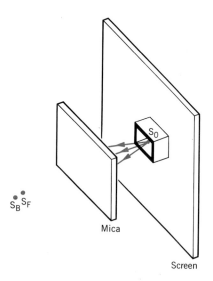

9. Ultrasound, or sound waves with a frequency above the audible range, is used today in the treatment of kidney stones. In the medical procedures, the ultrasound has a frequency of approximately 5.0×10^6 Hz and must be able to impinge on the kidney stones; it must not diffract or bend around them. Diffraction will be a minor effect to the extent that the wavelength of the waves is less than the diameter of a stone. Suppose that a typical stone has a diameter of 1.0 cm. Assuming the speed of sound within the kidney is 1560 m/s, explain why the frequency of the sound waves is chosen to be so far above the audible range.

10. Do any dark fringes appear in a single-slit diffraction pattern (see Figure 34.19b) when the wavelength of the light is greater than the width of the slit? Why?

11. Account for the fact that a sound wave diffracts much more than a light wave does when the two pass through the same doorway.

12. The French postimpressionist artist Georges Seurat developed a technique of painting in which dots of color are placed close together on the canvas. From sufficiently far away the individual dots are no longer distinguishable, and the images in the picture take on a more normal appearance. Explain why this is so, utilizing what you have learned about the resolving power of the eye.

13. Suppose the pupil of your eye were elliptical instead of circular in shape, with the long axis of the ellipse oriented in the vertical direction. (a) Would the resolving power of your eye be the same in the horizontal and vertical directions? (b) In which direction would the resolving power be greatest? Justify your answers by discussing how the diffraction of light waves would differ in the two directions.

14. The cameras used for taking photographs of the earth from satellites have large-diameter lenses. From the point of view of photographing two objects that are close together, account for this feature.

15. Suppose you were designing an eye and could select the size of the pupil and the wavelength of the electromagnetic waves to which the eye is sensitive. As far as the limitation created by diffraction is concerned, rank the following design choices in order of decreasing resolving power (greatest first): (a) large pupil and ultraviolet wavelengths, (b) small pupil and infrared wavelengths, and (c) small pupil and ultraviolet wavelengths.

16. What would happen to the distance between the bright fringes produced by a diffraction grating if the entire interference apparatus (light source, grating, and screen) were immersed in water? Why?

Section 34.1 The Principle of Linear Superposition, Section 34.2 Young's Double-Slit Experiment

1. Two sources are in phase and emit waves that have a wavelength of 0.44 m. Determine whether constructive or destructive interference occurs at a point whose distances from the two sources are as follows: (a) 1.32 and 3.08 m; (b) 2.67 and 3.33 m; (c) 2.20 and 3.74 m; (d) 1.10 and 4.18 m; (e) 1.10 and 1.10 m.

2. A Young's double-slit experiment is performed using light that has a wavelength of 630 nm. The separation between the slits is 0.053 mm. Find the angles that locate the first-, second-, and third-order bright fringes on the screen.

3. A flat observation screen is placed at a distance of 4.5 m from a pair of slits. The separation on the screen between the central bright fringe and the first-order bright fringe is 3.7 cm. The light illuminating the slits has a wavelength of 490 nm. Determine the slit separation.

4. In a Young's double-slit experiment, the seventh dark fringe is located 2.5 cm to the side of the central bright fringe on a flat screen, which is 1.1 m away from the slits. The slits are 0.14 mm apart. What is the wavelength of the light being used?

* 5. At most, how many bright fringes can be formed on either side of the central bright fringe when light of wavelength 625 nm falls on a double slit whose slit separation is 3.76×10^{-6} m?

* 6. In a Young's double-slit experiment the separation y between the first-order bright fringe and the central bright fringe on a flat screen is 2.40 cm, when light is used that has a wavelength of 475 nm. Assume the angles that locate the fringes on the screen are small enough so that $\sin \theta \approx \tan \theta$. Find the separation y when the light has a wavelength of 611 nm.

* 7. Two parallel slits are illuminated by light composed of two wavelengths, one of which is 486 nm. On a viewing screen, the light whose wavelength is known produces its third dark fringe at the same place where the light whose wavelength is unknown produces its fourth-order bright fringe. The fringes are counted relative to the central or zeroth-order bright fringe. What is the unknown wavelength?

Section 34.3 Thin Film Interference

8. A nonreflective coating of magnesium fluoride ($n = 1.38$) covers the glass ($n = 1.52$) of a camera lens. Assuming the coating prevents reflection of yellow-green light (wavelength in vacuum = 565 nm), determine the minimum nonzero thickness that the coating can have.

9. A transparent film ($n = 1.43$) is deposited on a glass plate ($n = 1.52$) to form a nonreflective coating. The film has a thickness of 1.07×10^{-7} m. What is the longest possible wavelength of light (in vacuum) for which this film has been designed?

10. A layer of transparent plastic ($n = 1.61$) on glass ($n = 1.52$) looks dark when illuminated by light whose wavelength is 589 nm in vacuum. Find the three smallest possible nonzero values for the thickness of the layer.

11. Example 3(a) in the text deals with the air wedge formed between two plates of glass ($n = 1.52$). Repeat this example, assuming the wedge of air is replaced by water ($n = 1.33$).

* 12. A film of oil lies on wet pavement. The refractive index of the oil exceeds that of the water. The film has the minimum nonzero thickness such that it appears dark due to destructive interference when viewed in red light (wavelength = 660 nm in vacuum). Assuming the visible spectrum extends from 380 nm to 750 nm, what are the visible wavelength(s) (in vacuum) for which the film will appear bright due to constructive interference?

* 13. A layer of glycerol ($n = 1.47$) on glass ($n = 1.52$) has a thickness of 1.02×10^{-7} m. This is the minimum nonzero thickness for which the layer will look dark in a certain monochromatic light. What is the next largest thickness that the layer could have and still look dark in the same light?

** 14. A uniform layer of water ($n = 1.33$) lies on a glass plate ($n = 1.52$). Light shines perpendicularly on the layer, which looks maximally bright when the wavelength of the light is 432 nm in vacuum and when it is 648 nm in vacuum. (a) Obtain the minimum thickness of the film. (b) Assuming that the film has the minimum thickness and that the visible spectrum extends from 380 nm to 750 nm, determine the visible wavelength(s) (in vacuum) for which the film appears completely dark.

Section 34.5 Diffraction

15. Light shines through a single slit whose width is 5.6×10^{-4} m. A diffraction pattern is formed on a flat screen located 4.0 m away. The distance between the middle of the central bright fringe and the first dark fringe is 3.4 mm. What is the wavelength of the light?

16. A diffraction pattern forms when light passes through a single slit. The wavelength of the light is 575 nm. Determine the angle that locates the first dark fringe when the width of the slit is (a) 0.18 mm, (b) 0.018 mm, and (c) 0.0018 mm.

17. A slit whose width is 0.0450 mm is located 1.23 m from a flat observation screen. Light shines through the slit and falls on the screen. Find the width of the central fringe of the diffraction pattern when the wavelength of the light is (a) 415 nm and (b) 655 nm.

18. A doorway is 0.91 m wide. (a) Obtain the angle that locates the first dark fringe in the Fraunhofer diffraction pattern formed when red light (wavelength = 660 nm) passes through the doorway. (b) Repeat part (a) for a 440-Hz tone (concert A), assuming the speed of sound is 343 m/s. (c) Which diffraction is so small that it is unnoticeable?

* 19. A single slit (width $= 4.85 \times 10^{-7}$ m) produces *no* dark fringes on an observation screen, even though light shines on the slit. Determine a minimum value for the wavelength of the light being used.

* 20. A loudspeaker produces an 1100-Hz tone and a 3100-Hz tone. The sound waves are emitted through a vertically oriented slit. The angle locating the first diffraction minimum, or audio "dark fringe," is 15° for the 3100-Hz tone. What is the corresponding angle for the 1100-Hz tone?

**21. In a single-slit diffraction pattern, the central fringe is 450 times as wide as the slit. The screen is 18 000 times as far from the slit as the slit is wide. What is the ratio λ/W, where λ is the wavelength of the light shining through the slit and W is the width of the slit? Assume the angle that locates a dark fringe on the screen is small, so that $\sin \theta \approx \tan \theta$.

Section 34.6 Resolving Power

22. In a dot matrix printer, an array of dots is used to form the printed characters. If the dots are close enough together, they cannot be resolved individually by the eye and, therefore, appear to form solid lines. Suppose that the pupil of the eye has a diameter of 2.0 mm in bright yellow-green light (wavelength = 565 nm in vacuum), that the material in the eye has an average refractive index of $n = 1.36$, and that the printed page is to be read at a distance of 31 cm. Considering the limit created by diffraction, find the smallest separation between the dots that the eye can see.

23. Late one night on a highway, a car speeds by you and fades into the distance. Under these conditions the pupils of your eyes (average refractive index = 1.36) have diameters of about 7.0 mm. The taillights of this car are separated by a distance of 1.2 m and emit red light (wavelength = 660 nm in vacuum). How far away from you is this car when its taillights appear to merge into a single spot of light because of the effects of diffraction?

24. In an experiment, red light from a ruby laser (wavelength = 694.3 nm) is passed through a telescope in reverse and is sent on its way to the moon. At the surface of the moon, which is 3.77×10^8 m away, the light strikes a reflector left there by astronauts. The reflected light returns to the earth, where it is detected. When it leaves the telescope, the circular beam of light has a diameter of about 0.20 m, and diffraction causes the beam to spread as the light travels to the moon. In effect, the first circular dark fringe in the diffraction pattern defines the size of the central bright spot on the moon. Determine the diameter (not the radius) of the central bright spot on the moon.

25. You are looking down at the earth from inside a commercial jetliner flying at an altitude of 8690 m, and the pupil of your eye has a diameter of 2.00 mm. The average refractive index of the material in the eye is 1.36. Determine how far apart two cars must be on the ground if you are to have any hope of distinguishing between them in (a) red light (wavelength = 665 nm in vacuum) and (b) violet light (wavelength = 405 nm in vacuum).

* 26. The largest refracting telescope in the world is at the Yerkes Observatory in Williams Bay, Wisconsin. The objective lens of the telescope has a diameter of 40.0 inches. With light whose wavelength is 565 nm, it is said that this telescope can resolve two objects separated 1.00 inch apart at a distance of 23.3 miles. Verify this statement with a calculation of your own.

**27. You are using a microscope to examine a blood sample. Recall from Section 33.4 that the sample should be placed just outside the focal point of the objective lens of the microscope. (a) If the specimen is being illuminated with light of wavelength λ and the f-number of the objective is 1, determine the closest distance between two blood cells that can just be resolved. Express your answer in terms of λ. (b) Based on your answer to (a), should you use light with a longer wavelength or a shorter wavelength if you wish to resolve two blood cells that are even closer together?

Section 34.7 The Diffraction Grating

28. The diffraction gratings discussed in the text are transmission gratings, because light *passes through* them. There are also gratings in which the light *reflects from* the grating to form a pattern of fringes. Equation 34.7 also applies to a reflection grating with straight parallel lines when the incident light shines perpendicularly on the grating. The surface of a compact disc (CD) has a multicolored appearance because it acts like a reflection grating and spreads sunlight into its colors. The arms of the spiral track on the CD are separated by 1.1×10^{-6} m. Using Equation 34.7, estimate the angle that corresponds to the first-order maximum for a wavelength of (a) 660 nm (red) and (b) 410 nm (violet).

29. The wavelength of the laser beam used in a compact disc player is 790 nm. Suppose that a diffraction grating produces first-order tracking beams that are 1.2 mm apart at a distance of 3.0 mm from the grating. Estimate the spacing between the slits of the grating.

30. A diffraction grating produces a first-order bright fringe that is 8.94 cm away from the central bright fringe on a flat screen. The separation between the slits of the grating is 4.17×10^{-6} m, and the distance between the grating and the screen is 0.625 m. What is the wavelength of the light shining on the grating?

31. The number of higher-order bright fringes that can be seen with a grating is limited by the fact that the value for $\sin \theta$ can never exceed unity in Equation 34.7. A diffraction grating contains 4820 lines/cm and is used with blue light (wavelength = 470 nm). What is the highest-order bright fringe that can be seen with this grating?

32. When a grating is used with light that has a wavelength of 575 nm, a second-order maximum is formed at an angle of 11.2°. How many lines per centimeter does this grating have?

* 33. There are 5620 lines per centimeter in a grating that is used with light whose wavelength is 471 nm. A flat observation screen is located at a distance of 0.750 m from the grating. What is the minimum width that the screen must have so the *centers* of all the principal maxima formed on either side of the central maximum fall on the screen?

**34. The separation between the slits of a grating is 2.2×10^{-6} m. This grating is used with light that contains all wavelengths between 410 and 660 nm. Rainbowlike spectra form on a screen 3.2 m away. How wide (in meters) is (a) the first-order spectrum and (b) the second-order spectrum?

ADDITIONAL PROBLEMS

35. A single slit has a width of 2.1×10^{-6} m and is used to form a diffraction pattern. Find the angle that locates the second dark fringe when the wavelength of the light is (a) 430 nm, (b) 570 nm, and (c) 660 nm.

36. A rock concert is being held in an open field. Two loudspeakers are separated by 7.00 m. As an aid in arranging the seating, a test is conducted in which both speakers vibrate in phase and produce an 80.0-Hz bass tone simultaneously. The speed of sound is 343 m/s. A reference line is marked out in front of the speakers, perpendicular to the midpoint of the line between the speakers. Relative to either side of this reference line, what is the smallest angle that locates the places where destructive interference occurs? People seated in these places would have trouble hearing the 80.0-Hz bass tone.

37. The first-order maximum produced by a grating is located by an angle of $\theta = 28°$. What is the angle for the second-order maximum with the same light?

38. A mixture of yellow light (wavelength = 580 nm in vacuum) and violet light (wavelength = 410 nm in vacuum) falls perpendicularly on a film of gasoline that is floating on a puddle of water. For both wavelengths, the refractive index in gasoline is $n = 1.40$ and in water is $n = 1.33$. What is the minimum nonzero thickness of the film in a spot that looks (a) yellow and (b) violet?

39. It is claimed that some professional baseball players can see which way the ball is spinning as it travels toward home plate. One way to judge this claim is to estimate the distance at which a batter can first hope to distinguish two points on opposite sides of a baseball, which has a diameter of 7.38 cm. (a) Estimate this distance, assuming the pupil of the eye has a diameter of 2.0 mm, the material within the eye has a refractive index of 1.36, and the wavelength of the light is 550 nm in vacuum. (b) Considering that the distance between the pitcher's mound and home plate is 18.4 m, can you rule out the claim based on your answer to part (a)?

* 40. The same diffraction grating is used with two different wavelengths of light, λ_A and λ_B. The fourth-order principal maximum of light A falls exactly on top of the third-order principal maximum of light B. Find the ratio λ_A/λ_B.

* 41. A film of gasoline ($n = 1.40$) floats on water ($n = 1.33$). Yellow light (wavelength = 580 nm in vacuum) shines perpendicularly on this film. (a) Determine the minimum nonzero thickness of the film, such that the film appears bright yellow due to constructive interference. (b) Repeat part (a), assuming the gasoline film is on glass ($n = 1.52$) instead of water.

* 42. A telescope is being used to view two objects that are separated by 480 m on the moon's surface. The surface of the moon is 3.77×10^8 m away from the surface of the earth. Assume that diffraction effects, rather than atmospheric turbulence, limits the resolving power of the telescope and that the wavelength of the light being used is 550 nm. Determine the diameter that the objective lens of this telescope must have if the two objects on the moon are to be resolved.

* 43. Violet light (wavelength = 410 nm) and red light (wavelength = 660 nm) lie at opposite ends of the visible spectrum. (a) For each wavelength, find the angle θ that locates the first-order maximum produced by a grating with 3300 lines/cm. This grating converts a mixture of all colors between violet and red into a rainbowlike dispersion between the two angles. Repeat the calculation above for (b) the second-order maximum and (c) the third-order maximum. (d) From your results, decide whether there is any overlap between any of the "rainbows" and specify which orders overlap.

* 44. Two slits are 0.158 mm apart. A mixture of red light (wavelength = 665 nm) and yellow-green light (wavelength = 565 nm) falls on the slits. A flat observation screen is located 2.24 m away. What is the distance on the screen between the third-order red fringe and the third-order yellow-green fringe?

**45. Two gratings A and B have slit separations d_A and d_B, respectively. They are used with the same light and the same observation screen. When grating A is replaced with grating B, it is observed that the first-order maximum of A is exactly replaced by the second-order maximum of B. (a) Determine the ratio d_B/d_A of the spacings between the slits of the gratings. (b) Find the next two principal maxima of grating A and the principal maxima of B that exactly replace them when the gratings are switched. Identify these maxima by their order numbers.

**46. A piece of curved glass has a radius of curvature of 10.0 m and is used to form Newton's rings, as in Figure 34.14. Not counting the dark spot at the center of the pattern, there are one hundred dark fringes, the last one being at the outer edge of the curved piece of glass. The light being used has a wavelength of 654 nm in vacuum. What is the radius of the outermost dark ring in the pattern?

CHAPTER 35

Special Relativity

Albert Einstein made many important contributions to science. The equation $E = mc^2$, which expresses the equivalence between energy and mass, is one of the most famous in all of physics and is a consequence of his theory of special relativity.

35.1 EVENTS AND INERTIAL REFERENCE FRAMES

Although Albert Einstein (1879–1955) made a number of important contributions to physics, he is best known for the theory of special relativity. This theory, published in 1905 when Einstein was 26 years old, significantly altered notions about time and space.

The theory of special relativity deals with the way that an event is measured by observers who are moving relative to the event. An *event*, such as the launching of the space shuttle in Figure 35.1, is a physical "happening" that occurs at a certain place and time. In this drawing two observers are watching the lift-off, one standing on the earth and one seated in an airplane that is flying at a constant velocity relative to the earth. To record the event, each observer uses a *reference frame* that consists of a set of x, y, z axes (called a *coordinate system*) and a clock. The coordinate system is used to establish where the event occurs, and the clock is used to specify when it happens. Each observer is at rest relative to his own reference frame. Since the earth-based observer and the airborne observer are moving relative to each other, their respective reference frames are also in relative motion.

The theory of special relativity deals with a "special" kind of reference frame,

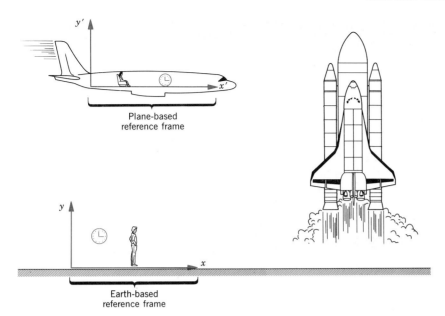

FIGURE 35.1 Using an earth-based reference frame, an observer standing on the earth records the location and time of an event (the lift-off). Likewise, an observer in the airplane uses a plane-based reference frame to describe the event.

called an *inertial reference frame.* As Section 4.2 discusses, an inertial reference frame is one in which Newton's law of inertia is valid. That is, if the net force acting on a body is zero, the body either remains at rest or moves at a constant velocity. In other words, the acceleration of such a body is zero when measured in an inertial reference frame. Rotating and otherwise accelerating reference frames are not inertial reference frames. The earth-based reference frame in Figure 35.1 is not quite an inertial frame, because it is subjected to centripetal accelerations as the earth spins on its axis and revolves around the sun. In most situations, however, the effects of these accelerations are small, so we can consider an earth-based reference frame to be an inertial one. To the extent that the earth-based reference frame is an inertial frame, so is the reference frame attached to the plane, for the plane moves at a constant velocity relative to the earth. The next section discusses why inertial reference frames are important in relativity.

35.2 THE POSTULATES OF SPECIAL RELATIVITY

Einstein based his theory of special relativity on two fundamental assumptions or postulates about the way nature behaves.

THE POSTULATES OF SPECIAL RELATIVITY

1. **The Relativity Postulate.** The laws of physics are the same in every inertial reference frame.

2. **The Speed of Light Postulate.** The speed of light in a vacuum, measured in any inertial reference frame, always has the same value of c, no matter how fast the source of light and the observer are moving relative to each other.

It is not difficult to accept the relativity postulate. For instance, in Figure 35.1 each observer, using his own inertial reference frame, can make measurements on the motion of the shuttle. The relativity postulate asserts that both observers find their data to be consistent with Newton's laws of motion. Similarly, both observers find that the behavior of the electronics on board the shuttle is described by the laws of electromagnetism, such as Faraday's law of electromagnetic induction. The relativity postulate states that *any inertial reference frame is as good as any other for expressing the laws of physics, because the laws are the same in all such frames.* In other words, with regard to inertial reference frames, nature does not play favorites.

Since the laws of physics are the same in all inertial reference frames, there is no experiment that can distinguish between an inertial frame that is at rest and one that is moving at a constant velocity. When you are seated on the aircraft in Figure 35.1, for instance, it is just as valid to say that you are at rest and the earth is moving as it is to say the reverse. It is not possible to single out one particular inertial reference frame as being at "absolute rest." Consequently, it is meaningless to talk about the "absolute velocity" of an object—that is, its velocity measured relative to a reference frame at "absolute rest." Thus, the earth moves relative to the sun, which itself moves relative to the center of our galaxy. And the galaxy moves relative to other galaxies, and so on. According to Einstein, only the relative velocity between objects, not their absolute velocities, can be measured and is physically meaningful.

While the relativity postulate is not too difficult to accept, the speed of light postulate defies common sense. For instance, Figure 35.2 illustrates a person standing on the bed of a truck that is moving at a constant speed of 15 m/s relative to the ground. Now, suppose you are standing on the ground and the person on the truck shines a flashlight at you. The person on the truck observes the speed of light to be c. What do you measure for the speed of light? You might guess that the speed of light would be $c + 15$ m/s. However, this guess is inconsistent with the speed of light postulate, which states that all observers in inertial reference frames measure the speed of light to be c—nothing more, nothing less. Therefore, you must also measure the speed of light to be c, the same as that measured by the person on the truck. According to the speed of light postulate, the fact that the flashlight is moving toward you has no influence whatsoever on the speed of the light approaching you. This property of light, although surprising, has been verified many times by experiment.

Since waves, such as water waves and sound waves, require a medium through which to propagate, it was natural for scientists before Einstein to assume that light did too. This hypothetical medium was called the *luminiferous ether* and was assumed to fill all of space. Furthermore, it was believed that light traveled only at the speed c when measured with respect to the ether. According to this view, an observer moving relative to the ether would measure a speed for light that was slower or faster than c, depending on whether the observer moved with or against the light. During the years 1883–1887, however, the American scientists A. A. Michelson and E. W. Morley carried out a series of famous experiments whose results were not consistent with the ether theory. Their results indicated that the speed of light is indeed the *same*

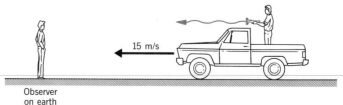

FIGURE 35.2 Both the person on the truck and the observer on the earth measure the speed of light to be c, regardless of the speed of the truck.

15 m/s

Observer
on earth

in all inertial reference frames and does not depend on the motion of the observer relative to the source of the light. These experiments, and others, led eventually to the demise of the ether theory and the acceptance of the theory of special relativity.

35.3 THE RELATIVITY OF TIME: TIME DILATION

TIME DILATION

Common experience indicates that time passes just as fast for a person standing on the ground as it does for an astronaut in a spacecraft. In contrast, the theory of special relativity reveals that the bystander sees time passing more slowly for the astronaut than for himself. We can see how this curious effect arises with the help of the clock illustrated in Figure 35.3. This clock uses a pulse of light to mark time. A short pulse of light is emitted by a light source, reflects from a mirror, and then strikes a detector that is situated next to the source. Each time a pulse reaches the detector, a "tick" registers on the chart recorder, another short pulse of light is emitted, and the cycle repeats. Thus, the time interval between successive "ticks" is marked by a beginning event (the firing of the light source) and an ending event (the pulse striking the detector). The source and detector are so close to each other that the two events can be considered to occur at the same place.

Suppose two identical clocks are built. One clock is kept on earth, and the other is placed aboard a spacecraft that travels at a constant velocity relative to the earth. The astronaut is at rest with respect to the clock on the spacecraft and, therefore, sees the light pulse move on the up/down path shown in Figure 35.4a. According to the astronaut, the time interval Δt_0 required for the light to follow this path is the distance $2D$ divided by the speed of light c; $\Delta t_0 = 2D/c$. To the astronaut, Δt_0 is the time interval between the "ticks" of the spacecraft clock, that is, the time interval between the beginning and ending events of the clock. An earth-based observer, however, does *not* measure Δt_0 as the time interval between these two events. Since the spacecraft is moving, the earth-based observer sees the light pulse follow the diagonal path shown

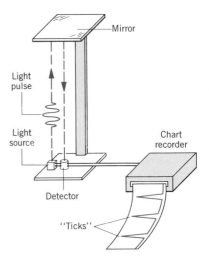

FIGURE 35.3 A light clock.

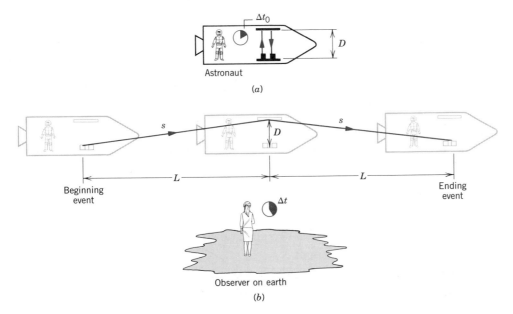

FIGURE 35.4 (a) The astronaut measures the time interval Δt_0 between successive "ticks" of the light clock on board the spaceship. (b) To an observer on earth who is watching the astronaut's clock, the light pulse travels a greater distance between "ticks" than it does in part a. Consequently, an earth-based observer measures a time interval Δt between "ticks" that is greater than Δt_0.

in part *b* of the drawing. This diagonal path is longer than the up/down path seen by the astronaut. But light travels at the *same speed c* for both observers, in accord with the speed of light postulate. Therefore, the earth-based observer measures a time interval Δt between the two events that is *greater* than the time interval Δt_0 measured by the astronaut. This result of the theory of special relativity is known as **time dilation.** (To *dilate* means to expand, and the time interval Δt is "expanded" relative to Δt_0.) From the point of view of the earth-based observer, the moving clock is measured to run slowly, because the time interval between successive ticks is larger when the spacecraft is moving than when it is at rest relative to the earth.

The time interval Δt that the earth-based observer measures in Figure 35.4*b* can be determined as follows. While the light pulse travels from the source to the detector, the spacecraft moves a distance $2L = v\,\Delta t$ to the right, where v is the speed of the spacecraft relative to the earth. From the drawing it can be seen that the light pulse travels a total diagonal distance of $2s$ during the time interval Δt. Applying the Pythagorean theorem, we find that

$$2s = 2\sqrt{D^2 + L^2} = 2\sqrt{D^2 + \left(\frac{v\,\Delta t}{2}\right)^2}$$

But the distance $2s$ is also equal to the speed of light times the time interval Δt, so $2s = c\,\Delta t$. Therefore,

$$c\,\Delta t = 2\sqrt{D^2 + \left(\frac{v\,\Delta t}{2}\right)^2}$$

Squaring this result and solving for Δt gives

$$\Delta t = \frac{2D}{c}\,\frac{1}{\sqrt{1 - \dfrac{v^2}{c^2}}}$$

But $2D/c = \Delta t_0$, the time interval between successive "ticks" of the spacecraft's clock as measured by the astronaut. With this substitution, the equation above can be expressed as

$$\left[\begin{array}{c}\textbf{Time} \\ \textbf{dilation}\end{array}\right] \qquad\qquad \Delta t = \frac{\Delta t_0}{\sqrt{1 - \dfrac{v^2}{c^2}}} \qquad\qquad (35.1)$$

The symbols in this formula are summarized below:

Δt_0 = time interval between two events, as measured by an observer who is at rest with respect to the events and who views the events as occurring at the same place.

Δt = time interval measured by an observer who is in motion with respect to the events

v = relative speed between the two observers

c = speed of light in a vacuum

Since v is less than c, the term $\sqrt{1 - v^2/c^2}$ in Equation 35.1 is less than 1, and Δt is greater than Δt_0. Example 1 illustrates the time dilation effect.

EXAMPLE 1

The spacecraft in Figure 35.4 is moving past the earth at a constant speed of $v = 0.92c$. The astronaut measures the time interval between successive "ticks" of the spacecraft clock to be $\Delta t_0 = 1.0$ s. What is the time interval Δt that an earth observer measures on the astronaut's clock?

SOLUTION

The time-dilation relation provides an answer to this question:

$$\Delta t = \frac{\Delta t_0}{\sqrt{1 - \dfrac{v^2}{c^2}}} = \frac{1.0 \text{ s}}{\sqrt{1 - \left(\dfrac{0.92c}{c}\right)^2}} = \boxed{2.6 \text{ s}}$$

From the point of view of the earth-based observer, the astronaut is using a clock that is running slowly, for the earth-based observer measures a time between "ticks" that is longer than what the astronaut measures. The earth observer measures the clock on the spacecraft to lose 1.6 s every second.

Example 1 shows that time dilation is appreciable when the speed v is comparable to the speed of light. The speeds we experience in everyday life are far too small for time dilation to be noticeable; for a clock aboard a jetliner traveling at $v = 0.000\ 000\ 75c$ (about 500 miles per hour), the time intervals Δt and Δt_0 in Example 1 would differ by only 2.8×10^{-13} s. This small difference in time means it would take about 110 000 years for the two clocks to differ by only 1 second.

PROPER TIME

In Figure 35.4 both the astronaut and the person standing on the earth are measuring the time interval between a beginning event (the firing of the light source) and an ending event (the light pulse striking the detector). For the astronaut, who is at rest with respect to the light clock, the two events occur at the same place. Being at rest with respect to a clock is the usual or "proper" situation, so the time interval Δt_0 measured by the astronaut is called the ***proper time interval.*** In general, the proper time interval Δt_0 between two events is the time interval measured by an observer who is at rest relative to the events and sees the events at the *same point* in space. On the other hand, the earth-based observer does not see the two events occurring at the same point in space, since the spacecraft is in motion. The time interval Δt that this observer measures is, therefore, not a proper time interval in the sense that we have defined it.

To understand situations involving time dilation, it is essential to distinguish between Δt_0 and Δt. In such situations it is helpful if one first identifies the two events that define the time interval. These events may be something other than the firing of a light source and the light pulse striking a detector. Then determine the reference frame in which the two events occur at the same place. For an observer at rest in this reference frame, the time interval is the proper time interval Δt_0.

SPACE TRAVEL

One of the intriguing aspects of time dilation occurs in conjunction with space travel. Since enormous distances are involved, travel to even the closest star outside our solar system would take a long time. However, as the following example shows, the time for such a trip can be considerably less for the passengers than one might guess.

EXAMPLE 2

The star closest to our solar system is Alpha Centauri, which is 4.3 light-years away. This means that, as measured by a person on earth, it would take light 4.3 years to reach this star. If a rocket leaves for Alpha Centauri at a speed of $v = 0.95c$ relative to the earth, by how much will the passengers have aged, according to their own clock, when they reach their destination?

SOLUTION

The two events in this problem are the departure from earth and the arrival at Alpha Centauri. At departure, earth is just outside the spaceship. Upon arrival at the destination, Alpha Centauri is just outside. Therefore, relative to the passengers, the two events occur at the same place, namely, just outside the spaceship. Thus, the passengers measure the proper time interval Δt_0 on their clock. For a person left behind on earth, the events occur at *different places,* so such a person measures the dilated time interval Δt rather than the proper time interval. Since it takes 4.3 years for light to traverse the distance and the rocket is moving at $v = 0.95c$ relative to the earth, a person on earth measures the time interval to be $\Delta t = (4.3 \text{ years})/0.95 = 4.5$ years. A person's

physiological aging is meaningful only when the aging process is measured by a clock at rest with respect to the person, so the passengers age according to the proper time interval Δt_0 indicated by their own clock on board the rocket. Using the time-dilation equation, we find that

$$\Delta t_0 = \Delta t \sqrt{1 - \frac{v^2}{c^2}} = (4.5 \text{ years}) \sqrt{1 - \left(\frac{0.95c}{c}\right)^2} = \boxed{1.4 \text{ years}}$$

Thus, the people aboard the rocket have aged by only 1.4 years when they reach Alpha Centauri, and not the 4.5 years an earthbound observer has calculated.

VERIFICATION OF TIME DILATION

A striking confirmation of time dilation was achieved in 1971 by an experiment carried out by J. C. Hafele and R. E. Keating.* They transported very precise cesium-beam atomic clocks around the world on commercial jets. Since the speed of a jet plane is considerably less than c, the time-dilation effect is extremely small. However, the atomic clocks were accurate to about $\pm 10^{-9}$ s, so the effect could be measured. The clocks were in the air for 45 hours, and their times were compared to reference atomic clocks kept on earth. The experimental results revealed that, within experimental error, the clocks on board the planes ran slower than those on earth by an amount that agreed with the prediction of relativity.

Time dilation has also been confirmed with experiments using subatomic particles called *muons.* These particles are created high in the atmosphere, at altitudes of about 10 000 m. When at rest, muons are short-lived, existing for a time of about 2.2×10^{-6} s before disintegrating into other particles. With such a short lifetime, these particles could never make it down to the earth's surface, even if they traveled close to the speed of light. However, *a large number of muons do reach the earth.* The only way they can do so is to live longer because of time dilation, as Example 3 illustrates.

EXAMPLE 3

A muon created in the upper atmosphere travels toward the earth at a speed of $v = 0.998c$. Find, on the average, (a) how long a muon lives according to an observer on earth, and (b) how far the muon travels before disintegrating.

SOLUTION

(a) The two events of interest are the generation and subsequent disintegration of the muon. When the muon is at rest, these events occur at the same place, so the muon's average (at rest) lifetime of 2.2×10^{-6} s is a proper time interval Δt_0. When the muon moves at a speed $v = 0.998c$ relative to the earth, an observer on the earth measures a dilated lifetime of

$$\Delta t = \frac{\Delta t_0}{\sqrt{1 - \frac{v^2}{c^2}}} = \frac{2.2 \times 10^{-6} \text{ s}}{\sqrt{1 - \left(\frac{0.998c}{c}\right)^2}} \quad (35.1)$$

$$= \boxed{35 \times 10^{-6} \text{ s}}$$

(b) The average distance x traveled by a muon, as measured by an earth observer, is equal to the muon's speed times the dilated time interval:

$$x = v \, \Delta t = (0.998)(3.00 \times 10^8 \text{ m/s})(35 \times 10^{-6} \text{ s})$$

$$= \boxed{1.0 \times 10^4 \text{ m}}$$

Thus, the dilated, or extended, lifetime of the muon allows it sufficient time to reach the surface of the earth. If its lifetime were only 2.2×10^{-6} s, a muon would travel only 660 m before disintegrating and could never reach the earth.

---

*J. C. Hafele and R. E. Keating, Around the world atomic clocks: Relativistic time gains observed, *Science,* 168 (July 14, 1972).

35.4 THE RELATIVITY OF LENGTH: LENGTH CONTRACTION

Because of time dilation, observers moving at a constant velocity relative to each other measure different time intervals between two events. For instance, Example 2 in the previous section illustrates that a trip from earth to Alpha Centauri takes 4.5 years according to a clock on earth, but only 1.4 years according to a clock in the rocket. These two times differ by the factor $\sqrt{1 - v^2/c^2}$. Since the times for the trip are different, one might ask if the observers measure different distances between earth and Alpha Centauri. The answer, according to special relativity, is yes. For the rocket passenger, the distance is not as far as that measured by the earth-based observer. This shortening of the distance between two points is one example of a phenomenon known as **length contraction.**

Length contraction happens for the rocket trip to Alpha Centauri, because both the earth-based observer and the rocket passenger agree that the relative speed between the rocket and earth is $v = 0.95c$. Since speed is distance divided by time, and the time is different for the two observers, it follows that the distances must also be different. Thus, the earth observer determines the distance to Alpha Centauri to be $L_0 = v\,\Delta t = (0.95c)(4.5 \text{ years}) = 4.3$ light-years. On the other hand, a passenger aboard the rocket finds the distance is only $L = v\,\Delta t_0 = (0.95c)(1.4 \text{ years}) = 1.3$ light-years. The passenger, measuring the shorter time, also measures the shorter distance.

The relation between the distances measured by two observers in relative motion at a constant velocity can be obtained with the aid of Figure 35.5. Part a of the drawing shows the situation from the point of view of the earth-based observer. This person measures the time of the trip to be Δt, the distance to be L_0, and the relative

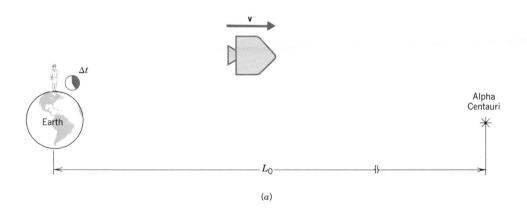

(a)

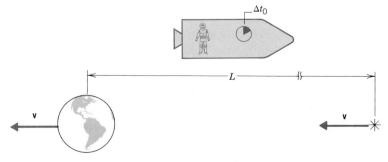

(b)

FIGURE 35.5 (a) As measured by an observer on the earth, the distance to Alpha Centauri is L_0 and the time required to make the trip is Δt. (b) According to the passenger on the spacecraft, the earth and Alpha Centauri move with speed v relative to the craft. The passenger measures the distance and time of the trip to be L and Δt_0, respectively, both quantities less than those in part a.

speed of the rocket to be $v = L_0/\Delta t$. Part b of the drawing presents the point of view of the passenger, for whom the rocket is at rest, and the earth and Alpha Centauri appear to move by at a speed v. The passenger determines the distance of the trip to be L, the time to be Δt_0, and the relative speed to be $v = L/\Delta t_0$. Since the relative speed computed by the passenger equals that computed by the earth-based observer, it follows that $v = L/\Delta t_0 = L_0/\Delta t$. Using this result and the time-dilation equation, Equation 35.1, we obtain the following relation between L and L_0:

$$\boxed{\begin{array}{c} \textbf{Length} \\ \textbf{contraction} \end{array}} \qquad\qquad L = L_0\sqrt{1 - \frac{v^2}{c^2}} \qquad\qquad (35.2)$$

The length L_0 is called the ***proper length;*** it is the length (or distance) between two points *as measured by an observer at rest with respect to them*. Since v is less than c, the term $\sqrt{1 - v^2/c^2}$ is less than 1, and L is less than L_0. It is important to note that this length contraction occurs only along the direction of the motion. Those dimensions that are perpendicular to the motion are not shortened, as the next example discusses.

EXAMPLE 4

An astronaut, using a meter stick that is at rest relative to a cylindrical spacecraft, measures the length and diameter of the spacecraft to be 82 and 21 m, respectively. The spacecraft moves with a constant speed of $v = 0.95c$ relative to the earth, as in Figure 35.5. What are the dimensions of the spacecraft, as measured by an observer on earth?

SOLUTION

The length of 82 m is a proper length L_0, since it is measured using a meter stick that is at rest relative to the spacecraft. The length L measured by the observer on earth can be determined from the length-contraction formula:

$$L = L_0\sqrt{1 - \frac{v^2}{c^2}} = (82\ \text{m})\sqrt{1 - \left(\frac{0.95c}{c}\right)^2} = \boxed{26\ \text{m}}$$

The diametric dimension is perpendicular to the motion, so the earth-observer does not measure any change in the diameter: $\boxed{\text{Diameter} = 21\ \text{m}}$. Figure 35.5$a$ shows the size of the spacecraft as measured by the earth observer, while part b shows the size measured by the astronaut.

When dealing with relativistic effects, be sure to distinguish between the criteria for defining the proper time interval and the proper length. The proper time interval Δt_0 between two events is the time interval measured by an observer who is at rest relative to the events and who sees them occurring at the *same place*. All other moving inertial observers will measure a larger value for this interval. The proper length L_0 of an object is the length measured by an observer who is *at rest* with respect to the object. All other moving inertial observers will measure a shorter value for this length. The observer who measures the proper time interval may not be the same one who measures the proper length. For instance, Figure 35.5 shows that the astronaut measures the proper time interval for the trip between earth and Alpha Centauri, while the earth-based observer measures the proper length (or distance) for the trip.

It should be emphasized that the word "proper" in the phrases proper time or proper length does *not* mean that these quantities are the "correct" or "preferred" quantities. If this were so, the observer measuring these quantities would be using a preferred reference frame for making the measurement, a situation that is prohibited by the relativity postulate. According to this postulate, all inertial reference frames are equivalent. Therefore, when two observers are moving relative to each other at a constant velocity, each measures the other person's clock to run more slowly than his own, and each measures the other person's length to be contracted.

35.5 THE RELATIVITY OF MASS: MASS INCREASE

MASS INCREASE

According to special relativity, time intervals and lengths depend on the relative motion between observers and events. Another surprising result of relativity is that the mass of a body also depends on its motion relative to an observer. When an object is at rest with respect to an observer, the mass of the object is called the **rest mass** m_0. When the same object moves with speed v relative to an observer, the observer does not measure a mass equal to the rest mass. Instead, the observer measures a mass m, which is called the **relativistic mass** to distinguish it from the rest mass. Einstein showed that the relativistic mass is related to the rest mass by the following equation:

$$\begin{bmatrix} \text{Mass} \\ \text{increase} \end{bmatrix} \qquad m = \frac{m_0}{\sqrt{1 - \dfrac{v^2}{c^2}}} \qquad (35.3)$$

Note that the same factor of $\sqrt{1 - v^2/c^2}$ appears in Equation 35.3 and in the time-dilation and length-contraction equations. Since this factor is always less than 1 and occurs in the denominator of Equation 35.3, the relativistic mass m is larger than the rest mass m_0.

Relativistic mass increase is significant only at speeds approaching that of light, as Figure 35.6 shows. For speeds attained by ordinary vehicles, such as planes and cars, the increase in mass is negligibly small. In Example 5, the relativistic mass of an electron is determined when the electron is moving close to the speed of light.

EXAMPLE 5

The rest mass of an electron is $m_0 = 9.11 \times 10^{-31}$ kg. The electron is accelerated to a speed of $v = 0.999\,99c$. Find the relativistic mass of the electron.

SOLUTION

The relativistic mass m of the moving electron can be determined directly from Equation 35.3:

$$m = \frac{m_0}{\sqrt{1 - \dfrac{v^2}{c^2}}} = \frac{m_0}{\sqrt{1 - \left(\dfrac{0.999\,99c}{c}\right)^2}} = 224m_0$$

The relativistic mass is 224 times greater than the rest mass of the electron: $m = 224m_0 = \boxed{2.04 \times 10^{-28} \text{ kg}}$.

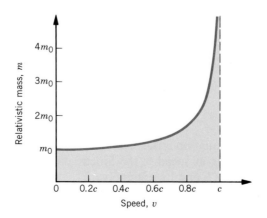

FIGURE 35.6 This graph shows how the mass m of an object increases as its speed approaches the speed of light. The rest mass of the object is m_0.

THE SPEED OF LIGHT IS THE ULTIMATE SPEED

The mass-increase formula $m = m_0/\sqrt{1 - v^2/c^2}$ implies that an object with a finite rest mass cannot have a speed that is equal to or greater than the speed of light. Observe from the graph in Figure 35.6 that as the speed of a particle approaches the speed of light, the relativistic mass of the particle becomes greater and greater. When $v = c$, the relativistic mass becomes infinite, and to accelerate the mass up to the speed of light would require infinite energy. An object with a finite rest mass can therefore never have a speed that equals or exceeds the speed of light. The speed of light represents the "ultimate" speed for such an object.

35.6 THE EQUIVALENCE OF ENERGY AND MASS: $E = mc^2$

THE TOTAL ENERGY OF AN OBJECT

One of the most astonishing results of special relativity is that mass is another form of energy. In this sense, mass and energy are equivalent. Consider, for example, an object traveling at a speed v relative to an observer. The equivalence between energy and mass is expressed by Einstein's famous equation $E = mc^2$:

$$\left[\begin{array}{l} \textbf{Total energy} \\ \textbf{of an object} \end{array}\right] \qquad E = mc^2 = \frac{m_0 c^2}{\sqrt{1 - \dfrac{v^2}{c^2}}} \qquad (35.4)$$

The symbols in this expression have the following meanings:

E = total energy of an object that is moving at a speed v relative to the observer

m = relativistic mass of the moving object, as measured by the observer

m_0 = rest mass of the object

To gain some understanding of $E = mc^2$, consider the special case when the object is at rest. When $v = 0$, the total energy is called the ***rest energy*** E_0, and Equation 35.4 reduces to $E_0 = m_0 c^2$. The rest energy represents the energy equivalent of the rest mass. For example, 1 kg of rest mass is equivalent to $E_0 = m_0 c^2 = (1 \text{ kg})(3 \times 10^8 \text{ m/s})^2 = 9 \times 10^{16}$ J of energy. This is an enormous amount of energy, for the entire amount of energy consumed by the United States in 1980 was only about 9×10^{19} J.

When an object is accelerated from rest to a speed v, the object acquires kinetic energy in addition to its rest energy. The total energy E is the sum of the rest energy E_0 and the kinetic energy KE, or $E = E_0 + \text{KE}$. Using Equation 35.4, we can write the kinetic energy as

$$\text{KE} = E - E_0 = m_0 c^2 \left(\frac{1}{\sqrt{1 - \dfrac{v^2}{c^2}}} - 1 \right) \qquad (35.5)$$

This equation is the relativistically correct expression for the kinetic energy of a mass moving at speed v; the kinetic energy is the difference between the object's total energy E and its rest energy E_0.

Equation 35.5 looks nothing like the kinetic energy expression used in Chapter 7, namely, $\text{KE} = \frac{1}{2}m_0 v^2$. However, for speeds much less than the speed of light ($v \ll c$), the relativistic equation for the kinetic energy reduces to $\text{KE} = \frac{1}{2}m_0 v^2$, as can be seen by using the binomial expansion* to represent the square root term in Equation 35.5:

$$\frac{1}{\sqrt{1 - \dfrac{v^2}{c^2}}} = 1 + \frac{1}{2}\left(\frac{v^2}{c^2}\right) + \frac{3}{8}\left(\frac{v^2}{c^2}\right)^2 + \cdots$$

Suppose v is much smaller than c, say $v = 0.01c$. The second term in the binomial expansion has the value $\frac{1}{2}(v^2/c^2) = 5.0 \times 10^{-5}$, while the third term has the much smaller value $\frac{3}{8}(v^2/c^2)^2 = 3.8 \times 10^{-9}$. The additional terms are even smaller than the third term, so if $v \ll c$, we can neglect the third and additional terms in comparison with the first and second terms. Substituting the first two terms of the binomial expansion into Equation 35.5 gives

$$\text{KE} \approx m_0 c^2 \left(1 + \frac{1}{2}\frac{v^2}{c^2} - 1 \right) = \tfrac{1}{2} m_0 v^2$$

which is the familiar form for the kinetic energy. However, Equation 35.5 gives the correct kinetic energy for all speeds and must be used for speeds near the speed of light, as in Example 6.

EXAMPLE 6

An electron is accelerated from rest to a speed of $v = 0.9995c$ in a particle accelerator. (a) Determine the rest energy E_0 of an electron (rest mass $= 9.109 \times 10^{-31}$ kg). (b) Find the total energy of the high-speed electron. (c) What is the kinetic energy of the electron?

SOLUTION

(a) The electron's rest energy is

$$E_0 = m_0 c^2 = (9.109 \times 10^{-31} \text{ kg})(2.998 \times 10^8 \text{ m/s})^2$$

$$= \boxed{8.187 \times 10^{-14} \text{ J}}$$

Energy is often expressed in units of electron volts (eV). Since 1 eV $= 1.602 \times 10^{-19}$ J, the electron's rest energy is

$$E_0 = \frac{8.187 \times 10^{-14} \text{ J}}{1.602 \times 10^{-19} \text{ J/eV}} = 5.11 \times 10^5 \text{ eV} \quad (0.511 \text{ MeV})$$

(b) The total energy of an electron traveling at $v = 0.9995c$ is

$$E = \frac{m_0 c^2}{\sqrt{1 - \dfrac{v^2}{c^2}}} = \frac{(9.109 \times 10^{-31} \text{ kg})(2.998 \times 10^8 \text{ m/s})^2}{\sqrt{1 - \left(\dfrac{0.9995c}{c}\right)^2}}$$

$$= \boxed{2.59 \times 10^{-12} \text{ J} \quad (16.2 \text{ MeV})} \qquad (35.4)$$

(c) The kinetic energy is the difference between the total energy and the rest energy:

$$\text{KE} = E - E_0 = 2.59 \times 10^{-12} \text{ J} - 8.2 \times 10^{-14} \text{ J}$$

$$= \boxed{2.51 \times 10^{-12} \text{ J} \quad (15.7 \text{ MeV})}$$

THE TRANSFORMATION BETWEEN MASS AND OTHER FORMS OF ENERGY

Einstein suggested that, since mass is a form of energy, it might be possible to transform mass into other forms of energy, just as potential energy can be transformed into kinetic energy. In other words, a change in mass Δm should be accompanied by a corresponding change in energy ΔE, where $\Delta E = (\Delta m)c^2$. For instance, life on earth is dependent on the sun for energy. The sun produces energy by transforming mass into electromagnetic energy in a thermonuclear reaction called *fusion* (see Section 39.5). The following example deals with the amount of mass per second transformed by the sun.

---

* The binomial expansion states that $(1 - x)^n = 1 - nx + n(n - 1)x^2/2 + \cdots$. In our case, $x = v^2/c^2$ and $n = -1/2$.

EXAMPLE 7

The sun radiates electromagnetic energy at the rate of 3.92×10^{26} W. (a) How much mass is transformed into energy each second? (b) The mass of the sun is 1.99×10^{30} kg. What fraction of the sun's mass is used during a human lifetime of 75 years?

SOLUTION

(a) Since power is energy per unit time, the amount of energy radiated by the sun in one second is $\Delta E = 3.92 \times 10^{26}$ J. The amount of mass that is equivalent to this much energy is

$$\Delta m = \frac{\Delta E}{c^2} = \frac{3.92 \times 10^{26} \text{ J}}{(3.00 \times 10^8 \text{ m/s})^2} = \boxed{4.36 \times 10^9 \text{ kg}}$$

Over 4 billion kilograms of mass are lost by the sun during each second.

(b) The amount of mass lost by the sun in 75 years is

$$\Delta m = (4.36 \times 10^9 \text{ kg/s})(3.16 \times 10^7 \text{ s/year})(75 \text{ years})$$
$$= 1.0 \times 10^{19} \text{ kg}$$

While this is an enormous amount of mass, it represents only a tiny fraction of the sun's mass:

$$\frac{\Delta m}{m_{\text{sun}}} = \frac{1.0 \times 10^{19} \text{ kg}}{1.99 \times 10^{30} \text{ kg}} = \boxed{5.0 \times 10^{-12}}$$

Nuclear power plants also generate energy by transforming some of the mass of the uranium fuel into heat energy in a process called *fission* (see Section 39.3). The heat energy is then used to generate electrical energy.

Another example of a mass-to-energy transformation arises when matter collides with antimatter. The positron (see Section 38.4), created in high-energy accelerators, has the same mass as an electron but an opposite electrical charge. The positron is the antiparticle of the electron. Because of their opposite charges, the two particles are pulled toward each other. If they collide, the electron and the positron are completely annihilated, and a burst of high-energy electromagnetic waves is produced. Their combined mass is transformed into electromagnetic energy according to $\Delta E = (\Delta m)c^2$.

The transformation of electromagnetic energy into mass also happens. In one experiment, an extremely high-energy electromagnetic wave, called a gamma ray (see Section 38.4), passes close to the nucleus of an atom. If the gamma ray has sufficient energy, it can create an electron and a positron. The gamma ray disappears and the electron and positron appear in its place; except for picking up some momentum, the nearby nucleus remains unchanged. The process in which the energy of the gamma ray is transformed into the masses of the two antiparticles is known as *pair production*.

Any change in the energy of a system causes a change in the mass of the system according to $\Delta E = (\Delta m)c^2$. It does not matter whether the change in energy is due to a change in kinetic energy, potential energy, thermal energy, electromagnetic energy, or so on. While any change in energy gives rise to a change in mass, in most instances the change in mass is too small to be noticed. For instance, if 4186 J of heat is used to raise the temperature of 1 kg of water by 1 C°, the mass can be expected to change by only $\Delta m = \Delta E/c^2 = (4186 \text{ J})/(3.00 \times 10^8 \text{ m/s})^2 = 4.7 \times 10^{-14}$ kg.

35.7 THE RELATIVISTIC ADDITION OF VELOCITIES

The velocity of an object relative to an observer plays a central role in special relativity, for the effects on time, length, and mass depend on how fast the relative motion is compared to the speed of light. To determine the velocity of an object relative to one or more observers, it is sometimes necessary to add two or more

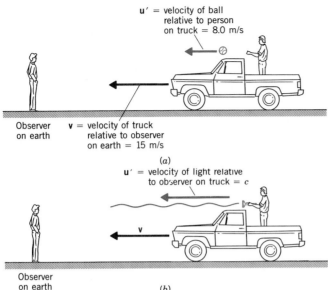

u' = velocity of ball relative to person on truck = 8.0 m/s

Observer on earth

v = velocity of truck relative to observer on earth = 15 m/s

(a)

u' = velocity of light relative to observer on truck = c

v

Observer on earth

(b)

FIGURE 35.7 (a) The truck is approaching the earth-based observer at a relative velocity of $v = 15$ m/s. The velocity of the baseball relative to the truck is $u' = 8.0$ m/s. (b) The speed of the light emitted by the flashlight is c relative to both the truck and the observer on earth.

velocities together. For instance, Figure 35.7 illustrates a truck moving at a constant velocity of $v = 15$ m/s toward an observer standing on the earth. Suppose someone on the truck throws a baseball toward the observer at a velocity of $u' = 8.0$ m/s relative to the truck. We might conclude that the observer on earth sees the ball approaching at a velocity of $u = u' + v = 23$ m/s. Although this conclusion seems reasonable, careful measurements would show that this is not quite right. The equation $u = u' + v$ is not valid, because if the velocity of the truck were close to the speed of light, the equation would predict that the observer on earth would see the baseball moving at a velocity greater than the speed of light. This is an impossibility, since no object with a finite rest mass can move faster than the speed of light.

For the case where the truck and ball are moving along the same direction, the theory of special relativity states that the velocities are related according to the *velocity-addition formula:*

$$\begin{bmatrix} \textbf{Velocity} \\ \textbf{addition} \end{bmatrix} \qquad u = \frac{u' + v}{1 + \dfrac{u'v}{c^2}} \qquad\qquad (35.6)$$

In this equation the symbols have the following meanings:

u = the velocity of the object as measured by the observer on the earth

u' = the velocity of the object measured by the person on the truck, which itself is moving at a velocity v relative to the earth.

When the motion occurs along a straight line, the velocities in Equation 35.6 can have either positive or negative values, depending on whether they are directed along the positive or negative direction. In Figure 35.7a, $u' = 8.0$ m/s and $v = 15$ m/s, assuming that the direction to the left is positive. Equation 35.6 differs from the nonrelativistic formula ($u = u' + v$) by the presence of the $u'v/c^2$ term in the denominator. When u' and v are small compared to c, the $u'v/c^2$ term is small compared to 1, so the velocity-addition formula reduces to $u \approx u' + v$. However, when either u' or v is comparable to c, the results can be quite different, as Example 8 illustrates.

EXAMPLE 8

Imagine a hypothetical situation in which the truck in Figure 35.7a is approaching the observer on the earth at a relative velocity of $v = 0.8c$. A person riding on the truck throws a baseball toward the observer at a velocity of $u' = 0.5c$ relative to the truck. At what velocity does the observer on earth see the ball approaching?

SOLUTION

The observer on earth does *not* see the baseball approaching at

$u = 0.5c + 0.8c = 1.3c$. This cannot be, because the velocity of the ball exceeds the speed of light. The velocity-addition formula gives the correct velocity, which is less than the speed of light:

$$u = \frac{u' + v}{1 + \dfrac{u'v}{c^2}} = \frac{0.5c + 0.8c}{1 + \dfrac{(0.5c)(0.8c)}{c^2}} = \frac{1.3c}{1 + 0.4} = \boxed{0.93c}$$

The velocity-addition formula is consistent with the speed of light postulate, which states that all inertial observers must measure the speed of light to be c. Consider Figure 35.7b, which shows the person riding on the truck and holding a flashlight. The speed of the light, as measured by this person, is $u' = c$. According to the observer standing on the earth, the speed of this light is given by the velocity-addition formula as

$$u = \frac{u' + v}{1 + \dfrac{u'v}{c^2}} = \frac{c + v}{1 + \dfrac{cv}{c^2}} = \frac{(c + v)c}{(c + v)} = c$$

Thus, consistent with the speed of light postulate, the velocity-addition formula indicates that the observer on earth and the person on the truck both measure the speed of light to be c, independent of the relative velocity v between them.

SUMMARY

The special theory of relativity is based on two postulates. The **relativity postulate** states that the laws of physics are the same in every inertial reference frame. The **speed of light postulate** says that the speed of light in a vacuum, measured in any inertial reference frame, always has the same value of c, no matter how fast the source of light and the observer are moving relative to each other.

The **proper time interval** Δt_0 between two events is the time interval measured by an observer who is at rest relative to the events and views them occurring at the same place. A moving observer who does *not* see the two events occurring at the same place measures a dilated time interval Δt. The dilated time interval is greater than the proper time interval, according to the **time-dilation equation:** $\Delta t = \Delta t_0 / \sqrt{1 - v^2/c^2}$. In this expression, v is the relative speed between the observer who measures Δt_0 and the observer who measures Δt.

The **proper length** L_0 between two points is the length measured by an observer who is at rest relative to the points. An observer moving with a relative speed v parallel to the line between the two points does not measure the proper length. Instead, such an observer measures a contracted length L given by the **length-contraction formula:** $L = L_0 \sqrt{1 - v^2/c^2}$.

The **rest mass** m_0 of an object is its mass as measured by an observer who is at rest relative to the object. The **relativistic mass** m of the same object moving with speed v relative to the observer is greater than the rest mass according to the **mass-increase equation:** $m = m_0 / \sqrt{1 - v^2/c^2}$.

Energy and mass are equivalent. The total energy E of an object moving at speed v is related to its relativistic mass m (which is greater than the rest mass m_0) according to $E = mc^2$. The total energy of an object is the sum of its rest energy, $E_0 = m_0c^2$, and its kinetic energy KE: $E = E_0 +$ KE. The kinetic energy is, therefore, $\text{KE} = E - E_0 = mc^2 - m_0c^2$.

When an object is moving with respect to a reference frame that itself is moving relative to an observer, the **velocity-addition formula** (Equation 35.6) gives the velocity of the object as measured by the observer.

QUESTIONS

1. A baseball player at home plate hits a pop fly straight up (the beginning event) that is caught by the catcher at home plate (the ending event). Which of the following observers record the proper time interval between the two events: (a) a spectator sitting in the stands, (b) a spectator watching the game on TV, and (c) an astronaut passing overhead? Explain your answers.

2. The drawing shows one face of a cubical die. Draw pictures to show how the face would appear to you if you moved past it at a constant velocity along each of the three indicated directions. Assume your speed is an appreciable fraction of the speed of light.

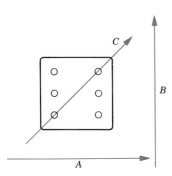

3. Suppose you are standing at a railroad crossing, watching a train go by. (a) Both you and a passenger in the train are looking at a clock on the train. Which of you measures the proper time interval? (b) Who measures the proper length of the train car? (c) Who measures the proper distance between the railroad ties under the track? Justify your answers.

4. The speed limit on interstate highways is 65 miles per hour. If the speed of light were 65 miles per hour, would you be able to drive at the speed limit? Give a reason for your answer.

5. There are tables that list data for the various particles of matter that physicists have discovered. Often, such tables list the masses of the particles in units of energy, such as in MeV (million electron volts), rather than in kilograms. Why is this possible?

6. (a) Does a compressed spring with elastic potential energy have more mass than a noncompressed spring (assume the spring is not vibrating)? (b) Do two positive, electric charges separated by a finite distance have more mass than when they are infinitely far apart (assume the charges remain stationary)? Provide a reason for each answer.

7. A person is approaching you in a truck that is traveling very close to the speed of light. This person throws a baseball toward you. Relative to the truck, the ball is also thrown with a speed nearly equal to the speed of light, so the person on the truck sees the baseball move away from the truck at a very high speed. Yet you see the baseball move away from the truck very slowly. Why? Use the velocity-addition formula to guide your thinking.

8. Which of the following quantities will an observer always measure to be the *same,* regardless of the relative velocity of the observer: (a) the time interval between two events; (b) the length of an object; (c) the mass of an object; (d) the speed of light; (e) the speed between the observer and another observer. In each case, give a reason for your answer.

9. If the speed of light were infinitely large instead of 3.0×10^8 m/s, would there be such effects as time dilation, length contraction, and mass increase? Explain, using the equations presented in the text to support your reasoning.

PROBLEMS

Before doing any calculations involving time dilation, length contraction, or mass increase, it is useful to identify which observer measures the proper time interval Δt_0, the proper length L_0, or the rest mass m_0.

Section 35.3 The Relativity of Time: Time Dilation

1. A law enforcement officer in an intergalactic "police car" turns on a red flashing light and sees it generate 10 flashes every 15 seconds. A person on earth measures that the 10 flashes occur in 25 seconds. How fast is the "police car" moving relative to the earth?

2. In 1980, Barbara Krause set the world's record for the 100-m freestyle. Suppose that this race had been monitored from a spaceship traveling at a speed of $0.900c$ relative to the earth and that the space travelers measured the time interval of the race to be 125.7 s. What was the time recorded on earth?

3. A spacecraft is passing through the solar system at a speed of $0.850c$ relative to the earth. What does the captain measure for the number of hours in an earth day if the spacecraft is (a) moving toward the earth or (b) away from the earth?

4. An astronaut travels at a speed of 7800 m/s relative to the earth. According to a clock on the earth, the trip lasts 15 days. Determine the *difference* (in seconds) between the time recorded by the earth clock and the astronaut's clock. [*Hint: When $v \ll c$, the following approximation is valid: $\sqrt{1 - v^2/c^2} \approx 1 - \frac{1}{2}(v^2/c^2)$.]*

* **5.** A 5.00-kg mass oscillates back and forth at the end of a spring whose spring constant is 49.3 N/m. An observer is traveling at a speed of 2.80×10^8 m/s relative to the fixed end of the spring. What does this observer measure for the period of oscillation?

** **6.** A certain type of bacteria is known to double in number every 24.0 hours. Two cultures of these bacteria are prepared, each consisting initially of one bacterium. One culture is left on earth and the other placed on a rocket that travels at a speed of $0.866c$ relative to the earth. At a time when the earthbound culture has grown to 256 bacteria, how many bacteria are in the culture on the rocket?

Section 35.4 The Relativity of Length: Length Contraction

7. The land speed record for a jet-propelled car was set by Craig Breedlove when his car attained an average speed of 274 m/s (613 mi/h) over a distance of 604 m. If the speed of light were 355 m/s, what distance would Breedlove have measured while driving the car?

8. Suppose the straight-line distance between New York and San Francisco is 4.2×10^6 m (neglecting the curvature of the earth). A UFO is flying between these two cities at a speed of $0.70c$ relative to the earth. What do the voyagers aboard the UFO measure for this distance?

9. Suppose you are traveling in space and pass a rectangular landing pad on a planet. The spacecraft has a speed of $0.85c$ relative to the planet and moves in a direction parallel to the length of the pad. While moving, you measure the length to be 18 m and the width to be 15 m. What are the dimensions of the landing pad according to the engineer who built it?

* **10.** As the drawing shows, a carpenter on a space station has constructed a 30.0° ramp. A rocket moves past the space station with a relative speed of $0.85c$ in a direction parallel to side x. What does a person aboard the rocket measure for the angle of the ramp?

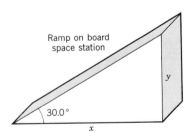

** **11.** A rectangle has the dimensions of 3.0 m × 2.0 m when viewed by someone at rest with respect to it. When you move past the rectangle along one of its sides, the rectangle looks like a square. What dimensions do you observe when you move at the same speed along the adjacent side of the rectangle?

Section 35.5 The Relativity of Mass: Mass Increase

12. A small rock, moving through the solar system at a speed of $0.70c$ relative to the earth, has a relativistic mass of 10.2 kg. What is the rest mass of the rock?

13. A car, whose rest mass is 1550 kg, is traveling at 15.0 m/s. If the speed of light were 25.0 m/s, what would be the mass of the car as measured by a person standing on the ground?

14. Intergalactic travelers observe earth astronauts constructing a space station in orbit. The astronauts are moving a structural member whose rest mass is 350 kg. The travelers, however, observe the mass to be 470 kg. How fast are the travelers moving?

Section 35.6 The Equivalence of Energy and Mass: $E = mc^2$

15. The total amount of energy consumed in the United States during 1980 is estimated to be about 9.0×10^{19} J. One penny has a rest mass of 2.9 g. How many pennies have the equivalent of this amount of energy?

16. The amount of heat required to melt 1 kg of ice at 0 °C is 3.35×10^5 J. By how many kilograms is the mass of the water greater than the mass of the ice?

17. A nuclear power reactor generates about 3.0×10^9 W of power. The energy comes from a fission process in which rest mass is transformed into other forms of energy. How much rest mass is transformed each year?

18. An elementary particle called a pion has been observed to decay completely into electromagnetic radiation. The pion has a rest mass of 2.4×10^{-28} kg. (a) What is its kinetic energy when traveling at a speed of $0.850c$? (b) How much energy in the form of electromagnetic radiation is released when the high-speed pion decays?

* **19.** In a TV picture tube, an electron is accelerated from rest through a potential difference of 24 000 V before striking the screen. (a) What is the kinetic energy (in joules) of the electron just before the electron hits the screen? (b) What is the ratio m/m_0 of this fast-moving electron?

Section 35.7 The Relativistic Addition of Velocities

20. A rocket ship is moving directly toward the earth with a velocity of $0.80c$ relative to the earth. The ship sends out a pulse of light that is aimed at the earth. What is the velocity that a person on earth sees for the approaching pulse?

21. An observer on the earth sees a spaceship approaching at a velocity of $0.50c$. The spacecraft then launches an exploration vehicle that, according to the earth observer, approaches at $0.70c$. What is the velocity of the exploration vehicle relative to the spaceship?

22. It has been proposed that spaceships of the future will be powered by ion propulsion engines. In one such engine the ions are to be ejected with a speed of $0.80c$ relative to the engine. If the ship were traveling away from the earth with a velocity of $0.70c$, what would be the velocity of the ions relative to the earth?

* **23.** An intergalactic cruiser has two types of guns: a photon cannon that fires a beam of laser light, and an ion gun that shoots atomic ions at a velocity of $0.950c$ relative to the cruiser. The cruiser closes in on a hostile spacecraft at a velocity of $0.800c$ relative to the enemy. The captain fires both types of guns. With

what velocity do the aliens aboard the hostile spacecraft see (a) the photons and (b) the ions approach them? With what velocity do the aliens see (c) the photons and (d) the ions move away from the cruiser?

* **24.** A person on earth notices a rocket approaching from the right at a speed of $0.75c$ and another rocket approaching from the left at $0.65c$. What is the relative velocity between the two rockets, as measured by a passenger on one of them?

25. Two atomic particles approach each other in a head-on collision. Each particle has a relativistic mass of 3.60×10^{-25} kg and a speed of 2.40×10^8 m/s when measured by an observer standing in the laboratory. (a) What is the speed of one particle as seen by the other particle? (b) Determine the relativistic mass of one particle, as would be observed by the other.

ADDITIONAL PROBLEMS

26. An electron and a positron collide and annihilate each other, their rest masses being transformed into electromagnetic energy. If each particle is moving at a speed of $0.20c$ relative to the laboratory before the collision, determine the energy released when they are annihilated. Express your answer in joules and in electron volts. *(Hint: Use the relativistic mass for each particle.)*

27. A particle known as a pion lives for a proper time of $2.6 \times$ 10^{-8} s before breaking apart into other particles. How long does this particle live according to a laboratory observer if the particle moves past the observer at a speed of $0.67c$?

28. A woman is 1.7 m tall and has a rest mass of 49 kg. She moves past an observer with the direction of the motion parallel to her height. The observer measures her mass to be 110 kg. What does the observer measure for her height?

29. How fast must a meter stick be moving if its length is observed to shrink to one-half a meter?

* **30.** A rocket is moving away from the earth with a speed of $0.75c$. An escape pod of length 45 m (as measured by the rocket crew) is launched from the rocket toward the earth with a speed of $0.55c$ relative to the rocket. What is the length of the escape pod as determined by an observer on earth?

* **31.** Four kilograms of water are heated from 20.0 °C to 60.0 °C. (a) How much heat is required to produce this change in temperature? [The specific heat capacity of water is 4186 J/ (kg·C°).] (b) By how much does the mass of the water increase?

32. Twins who are 19.0 years of age leave the earth and travel to a distant planet 12.0 light-years away. Assume the planet and earth are at rest with respect to each other. The twins depart at the same time on different spaceships. One twin travels at a speed of $0.900c$, while the other twin travels at $0.500c$. (a) According to the theory of special relativity, what is the difference between their ages when they meet again at the earliest possible time on the new planet? (b) Which twin is older? Why?

CHAPTER 36

Particles and Waves

This highly magnified view of a wasp's head was made with a scanning electron microscope. Such high resolution is possible because high-energy electrons possess small de Broglie wavelengths.

36.1 THE WAVE–PARTICLE DUALITY

The ability to exhibit interference effects is an essential characteristic of waves. For instance, Section 34.2 discusses Young's famous experiment in which light passes through two closely spaced slits and produces a pattern of bright and dark fringes on a screen (see Figure 34.4). The fringe pattern is a direct indication that interference is occurring between the light waves coming from each slit.

One of the most incredible discoveries of twentieth-century physics is that particles can also behave like waves and exhibit interference effects. For instance, Figure 36.1 shows a version of Young's experiment performed by directing *a beam of electrons* onto a double slit. In this experiment, the screen is like a television screen and glows wherever a electron strikes it. Part *a* of the picture indicates the pattern that would be seen on the screen if each electron, behaving strictly as a particle, were to pass through one slit or the other and strike the screen. The pattern would consist of an image of each slit. Part *b* shows the pattern actually observed, which consists of bright and dark fringes, reminiscent of that obtained when light waves pass through the double slit. The fringe pattern indicates that the electrons are somehow exhibiting the interference effects associated with waves.

But how can electrons behave like waves in the experiment shown in Figure 36.1*b*?

And what kind of waves are they? The answers to these profound questions will be discussed later in this chapter. For the moment, we intend only to emphasize that the picture of an electron as a tiny discrete particle of matter does not account for the fact that the electron can behave as a wave in some circumstances. In other words, the electron exhibits a dual nature, with both particle-like characteristics and wave-like characteristics.

There is another interesting question: If a particle can exhibit wave-like properties, can waves exhibit particle-like behavior? As the next three sections reveal, the answer is yes. In fact, experiments that demonstrated the particle-like behavior of waves were performed near the beginning of the twentieth century, before the experiments that demonstrated the wave-like properties of the electron. In any event, scientists now accept the *wave–particle duality* as an essential part of nature: ***Waves can exhibit particle-like characteristics and particles can exhibit wave-like characteristics.***

Section 36.2 begins the remarkable story of the wave-particle duality by discussing the electromagnetic waves that are radiated by a perfect blackbody. It is appropriate to begin with blackbody radiation, because it provided the first link in the chain of experimental evidence leading to our present understanding of the wave–particle duality.

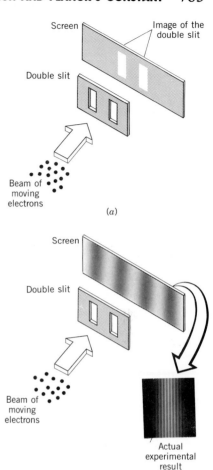

(a)

(b)

36.2 BLACKBODY RADIATION AND PLANCK'S CONSTANT

All bodies, no matter how hot or cold, continuously radiate electromagnetic waves. For instance, we see very hot objects glow, because they emit electromagnetic waves in the visible region of the spectrum. A temperature of about 1700 K produces the white-hot appearance of the filament in an incandescent light bulb, while a temperature near 1000 K creates the characteristic cherry red color of burning charcoal. However, at relatively low temperatures, cooler objects emit visible light waves only weakly and, as a result, do not appear to be glowing. Certainly the human body, at only 310 K, does not emit enough visible light to be seen in the dark with the unaided eye. But the body does emit electromagnetic waves in the infrared region of the spectrum, and these can be detected with infrared sensitive detectors.

At a given temperature, the intensities of the electromagnetic waves emitted by an object vary from wavelength to wavelength throughout the visible, the infrared, and other regions of the spectrum. Figure 36.2 illustrates how the intensity per unit wavelength depends on wavelength for a perfect blackbody emitter. As Section 16.4 discusses, a perfect blackbody at a constant temperature absorbs and reemits all the electromagnetic radiation that falls on it. The two curves in the drawing show that at the higher temperature the maximum emitted intensity increases and shifts toward shorter wavelengths. In accounting for the shape of these curves, the German physicist Max Planck (1858–1947) took the first step toward our present understanding of the wave–particle duality.

In 1900 Planck calculated the blackbody radiation curves, using a model that represents a blackbody as a large number of atomic vibrators, each of which emits electromagnetic waves. To obtain agreement between the theoretical and experimental curves, Planck assumed that the energy E of an atomic vibrator could have only the discrete values of $E = 0$, hf, $2hf$, $3hf$, and so on. In other words, he assumed that

$$E = nhf \qquad n = 0, 1, 2, 3, \ldots \qquad (36.1)$$

where n is a positive integer, f is the frequency of vibration (in hertz), and h is a

FIGURE 36.1 (a) If electrons behaved as discrete particles with no wave properties, they would pass through one or the other of the two slits and strike the screen, causing it to glow and produce exact images of the slits. (b) In reality, the screen reveals a pattern of bright and dark fringes, similar to the pattern produced when a beam of light is used and interference occurs between the light waves coming from each slit. The experimental result was obtained by Dr. Clauss Jönsson. The slit widths and the distance between the slits have been exaggerated for clarity.

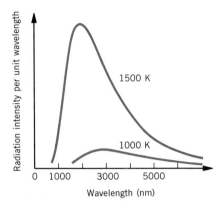

FIGURE 36.2 The intensity per unit wavelength of the electromagnetic radiation emitted by a perfect blackbody varies from wavelength to wavelength, as each curve indicates. At the higher temperature, the intensity of the radiation is not only greater, but the maximum intensity occurs at a shorter wavelength.

constant now called **Planck's constant.** * Experiment has shown that Planck's constant has a value of

$$h = 6.626\ 0755 \times 10^{-34}\ \text{J} \cdot \text{s}$$

The radical feature of Planck's assumption was that the energy of an atomic vibrator could have only discrete values (hf, $2hf$, $3hf$, etc.), with energies in between these values being forbidden. Whenever the energy of a system can have only certain definite values, and nothing in between, the energy is said to be *quantized.* This quantization of the energy was unexpected on the basis of the traditional physics of the time. However, it was soon realized that energy quantization had wide-ranging implications.

Conservation of energy requires that the energy carried off by the electromagnetic waves must equal the energy lost by the atomic vibrators. Suppose, for example, that a vibrator with an energy of $3hf$ emits an electromagnetic wave. According to Equation 36.1, the next smallest allowed value for the energy of the vibrator is $2hf$. In such a case, the energy carried off by the electromagnetic wave would have the value of hf, equaling the amount of energy lost by the vibrator. Thus, Planck's model for blackbody radiation sets the stage for the idea that electromagnetic energy occurs as a collection of discrete amounts or packets of energy, the energy of a packet being equal to hf. As the next section discusses, it remained for Einstein to make the specific proposal that light consists of such energy packets.

36.3 PHOTONS AND THE PHOTOELECTRIC EFFECT

If light with a sufficiently high frequency shines on a metal plate, electrons are emitted from the plate. As Figure 36.3 shows, the electrons move toward a positive electrode called the collector and cause a current to register on the ammeter. Because the electrons are ejected with the aid of light, the electrons are called **photoelectrons** and the phenomenon is known as the **photoelectric effect.** In 1905 Einstein presented an explanation of the photoelectric effect that took advantage of Planck's work concerning blackbody radiation. It was primarily for his theory of the photoelectric effect that Einstein was awarded the Nobel prize in physics in 1921.

In his photoelectric theory, Einstein proposed that light of frequency f could be regarded as a collection of discrete packets of energy, each packet containing an amount of energy E given by

$$\begin{bmatrix} \textbf{Energy of} \\ \textbf{a photon} \end{bmatrix} \qquad\qquad E = hf \qquad\qquad (36.2)$$

where h is Planck's constant. Today these energy packets are called **photons.** The light energy given off by a light bulb, for instance, is carried by photons. The brighter the light shining on a given area, the greater is the number of photons per second that strike the area. Example 1 estimates the number of photons emitted each second by a typical light bulb.

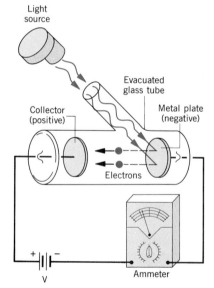

FIGURE 36.3 In the photoelectric effect, light shines on a metal surface, and if the frequency of the light is sufficiently high, electrons are ejected from the surface. These photoelectrons, as they are called, are drawn to the positive collector, thus producing a current.

* It is now known that the energy of a harmonic oscillator is $E = (n + \frac{1}{2})hf$, the extra term of $\frac{1}{2}$ being unimportant to the present discussion.

EXAMPLE 1

In converting electrical energy into light energy, a sixty-watt incandescent light bulb operates at about 2.1% efficiency. Assuming that all the light is green light (vacuum wavelength = 555 nm), determine the number of photons given off per second by the bulb.

SOLUTION

The number of photons emitted per second can be found by dividing the amount of light energy emitted per second by the energy of one photon. At an efficiency of 2.1%, the number of joules of light energy emitted per second by a sixty-watt bulb is $(0.021)(60.0 \text{ W}) = 1.3 \text{ J/s}$. The energy of a single photon is given

by $E = hf$, where the frequency f can be found from the wavelength according to $f = c/\lambda = (3.00 \times 10^8 \text{ m/s})/(555 \times 10^{-9} \text{ m}) = 5.41 \times 10^{14} \text{ Hz}$:

$$E = hf = (6.63 \times 10^{-34} \text{ J·s})(5.41 \times 10^{14} \text{ Hz}) \quad (36.2)$$

$$= 3.59 \times 10^{-19} \text{ J}$$

Therefore,

$$\text{Number of photons emitted per second} = \frac{1.3 \text{ J/s}}{3.59 \times 10^{-19} \text{ J/photon}}$$

$$= \boxed{3.6 \times 10^{18} \text{ photons/s}}$$

According to Einstein, when light shines on a metal, a photon can give up its energy to an electron in the metal. If the photon has enough energy to do the work of removing the electron from the metal, the electron can be ejected. The work required depends on how strongly the electron is held. For the *least strongly* held electrons, the necessary work has a minimum value W_0 and is called the ***work function*** of the metal. If a photon has energy in excess of the work needed to remove an electron, the excess energy appears as kinetic energy of the ejected electron. Thus, the least strongly held electrons are ejected with the maximum kinetic energy KE_{max}. Einstein applied the conservation of energy principle and proposed the following relation to describe the photoelectric effect:

$$\underbrace{hf}_{\substack{\text{Photon} \\ \text{energy}}} = \underbrace{\text{KE}_{max}}_{\substack{\text{Maximum} \\ \text{kinetic energy} \\ \text{of ejected} \\ \text{electron}}} + \underbrace{W_0}_{\substack{\text{Minimum} \\ \text{work needed to} \\ \text{eject electron} \\ \text{from metal}}} \quad (36.3)$$

According to this equation, $\text{KE}_{max} = hf - W_0$, which is plotted in Figure 36.4. The intercept at $f = f_0$ identifies the frequency f_0 at which the energy of an incident photon equals the work function W_0.

The photon picture provides an explanation for a number of features of the photoelectric experiment that are difficult to explain without using the concept of photons. It is known, for instance, that only light with a frequency above a certain minimum value will eject electrons. If the frequency of the light is below this value, no electrons are ejected, regardless of how intense the light is. The next example illustrates how Einstein's theory accounts for this minimum value of the frequency.

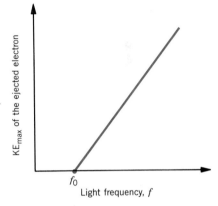

FIGURE 36.4 According to Einstein's theory of the photoelectric effect, photons of light can eject electrons from a metal when the light frequency is above a minimum value f_0. For frequencies above this minimum value, the ejected electrons have a maximum kinetic energy KE_{max} that is linearly related to the frequency of the light, as the graph shows.

EXAMPLE 2

The work function for a silver surface is $W_0 = 4.73 \text{ eV}$. Find the minimum frequency that light must have in order to eject electrons from this surface.

SOLUTION

The minimum frequency f_0 is that frequency at which the photon energy equals the work function W_0 of the metal, so the electron is ejected with zero kinetic energy. Since $1 \text{ eV} = 1.60 \times 10^{-19} \text{ J}$, the work function expressed in joules is $W_0 =$

$(4.73 \text{ eV})(1.60 \times 10^{-19} \text{ J/1 eV}) = 7.57 \times 10^{-19} \text{ J}$. The minimum frequency of the light can now be found:

$$hf_0 = \underbrace{\text{KE}_{max}}_{= 0} + W_0 \quad (36.3)$$

$$f_0 = \frac{W_0}{h} = \frac{7.57 \times 10^{-19} \text{ J}}{6.63 \times 10^{-34} \text{ J·s}} = \boxed{1.14 \times 10^{15} \text{ Hz}}$$

Since $\lambda_0 = c/f_0$, the wavelength of this light is $\lambda_0 = 263$ nm, which is in the ultraviolet region of the electromagnetic spectrum. Photons with frequencies less than f_0 do not have enough energy to eject electrons from a silver surface.

Another significant feature of the photoelectric effect is that the maximum kinetic energy of the ejected electrons remains the same when the intensity of the light increases, provided the light frequency remains the same. As the light intensity increases, more photons per second strike the metal, and consequently more electrons per second are ejected. However, since the frequency is the same for each photon, the energy of each photon is also the same. Thus, the ejected electrons always have the same maximum kinetic energy.

Whereas the photon model of light explains the photoelectric effect satisfactorily, the electromagnetic wave picture of light does not. Certainly, it is possible to imagine that the electric field of an electromagnetic wave would cause electrons in the metal to oscillate and tear free from the surface when the amplitude of oscillation becomes large enough. However, were this the case, higher intensity light would eject electrons with a greater maximum kinetic energy, a fact that experiment does not confirm. Moreover, in the electromagnetic wave picture, a relatively long time would be required with low-intensity light before the electrons would build up a sufficiently large oscillation amplitude to tear free. Instead, experiment shows that even the weakest light intensity causes electrons to be ejected almost instantaneously, provided the frequency of the light is above the minimum value f_0. The failure of the electromagnetic wave picture to explain the photoelectric effect does not mean that the wave model should be abandoned. But we must recognize that the wave picture does not account for all the characteristics of light. The photon model also makes an important contribution to our understanding of the way light behaves when it interacts with matter.

Because a photon has energy, the photon can eject an electron from a metal surface when it interacts with the electron. However, a photon is different than a normal particle. A normal particle has a rest mass, as defined in Einstein's theory of special relativity, but a photon does not. The fact that a photon does not have a rest mass follows from Equation 35.3, which gives the relativistic mass m of a particle moving at a speed v as $m = m_0/\sqrt{1 - (v/c)^2}$, where m_0 is the rest mass of the particle. In a vacuum, a photon moves at the speed of light, $v = c$, so if the photon had a finite rest mass, its relativistic mass would be infinite. According to Equation 35.4, $E = mc^2$, the energy of the photon would also be infinite. Since the energy of a photon is not infinitely large, a photon must have a zero rest mass.

36.4 THE MOMENTUM OF A PHOTON AND THE COMPTON EFFECT

Although Einstein presented his photon model for the photoelectric effect in 1905, it was not until 1923 that the photon picture began to achieve widespread acceptance. It was then that the American physicist Arthur H. Compton (1892–1962) used the photon model to explain his research on the scattering of X rays by the electrons in graphite. X rays are high-frequency electromagnetic waves and, like light, they are composed of photons.

Figure 36.5 illustrates what happens when an X-ray photon strikes an electron in a piece of graphite. Like two billiard balls colliding on a pool table, the X-ray photon

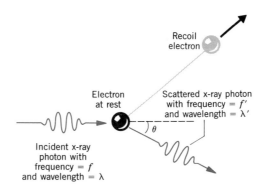

FIGURE 36.5 In an experiment performed by Arthur H. Compton, an X-ray photon (frequency $= f$) collides with an electron. The scattered photon and the recoil electron depart the collision in different directions. The frequency f' of the scattered photon is found to be less than the frequency f of the incident photon, indicating that the photon has given energy to the electron.

scatters in one direction and the electron recoils in another direction after the collision. Compton observed that the scattered photon has a frequency f' that is smaller than the frequency f of the incident photon. In addition, he found that the difference between the two frequencies depends on the angle θ at which the scattered photon leaves the collision. The phenomenon in which an X-ray photon is scattered from an electron, the scattered photon having a smaller frequency than the incident photon, is called the **Compton effect.**

In Section 8.4 the collision between two objects is analyzed using the fact that the total kinetic energy and the total linear momentum of the objects are the same before and after the collision. Similar analysis can be applied to the collision between a photon and an electron. The electron is assumed to be initially at rest and essentially free, that is, not bound to the atoms of the material. For the total energy, it follows that

$$\underbrace{hf}_{\substack{\text{Energy of} \\ \text{incident} \\ \text{photon}}} = \underbrace{hf'}_{\substack{\text{Energy of} \\ \text{scattered} \\ \text{photon}}} + \underbrace{\text{KE}}_{\substack{\text{Kinetic energy} \\ \text{of recoil} \\ \text{electron}}} \qquad (36.4)$$

where the relation $E = hf$ has been used for the photon energies. Equation 36.4 shows that the energy and corresponding frequency f' of the scattered photon are less than the energy and frequency of the incident photon, just as Compton observed. Since $\lambda = c/f$, the wavelength of the scattered X rays is larger than that of the incident X rays.

For an initially stationary electron, conservation of total linear momentum requires that

$$\begin{array}{cccc} \text{Momentum of} \\ \text{incident photon} \end{array} = \begin{array}{c} \text{Momentum of} \\ \text{scattered photon} \end{array} + \begin{array}{c} \text{Momentum of} \\ \text{recoil electron} \end{array} \qquad (36.5)$$

An expression for the magnitude p of the momentum of the photon can be obtained from the definition of momentum as mass times velocity, $p = mc$. The equivalent mass m of a photon is related to its total energy E by $E = mc^2$ (Equation 35.4) with the result that $p = mc = (E/c^2)c = E/c$. But the energy of a photon is $E = hf$, while the wavelength is $\lambda = c/f$. Therefore, the magnitude of the momentum is

$$p = \frac{hf}{c} = \frac{h}{\lambda} \qquad (36.6)$$

Compton was able to use Equations 36.4–36.6 to show that the difference between the wavelength λ' of the scattered photon and the wavelength λ of the incident photon is related to the scattering angle θ by

$$\lambda' - \lambda = \frac{h}{m_0 c}(1 - \cos\theta) \qquad (36.7)$$

In this equation m_0 is the rest mass of the electron. The quantity $h/(m_0 c)$ is referred to as the **Compton wavelength of the electron**, and has the value $h/(m_0 c) = 2.43 \times 10^{-12}$ m. Since $\cos\theta$ varies between $+1$ and -1, the shift $\lambda' - \lambda$ in the wavelength can vary between zero and $2h/(m_0 c)$, depending on the value of θ, a fact observed by Compton.

The photoelectric effect and the Compton effect provided compelling evidence that electromagnetic waves can exhibit particle-like characteristics attributable to energy packets called photons. Light and other electromagnetic radiation act as waves when they move from place to place, but behave as particles or photons when interacting with material substances.

36.5 THE DE BROGLIE WAVELENGTH AND THE WAVE NATURE OF MATTER

As a graduate student in 1923, Louis de Broglie (1892–1987) made the astounding suggestion that since light waves could exhibit particle-like behavior, particles of matter should exhibit wave-like behavior. De Broglie proposed that the wavelength λ of a particle is given by the same relation (Equation 36.6) that applies to a photon:

$$\left[\begin{array}{c}\textbf{De Broglie}\\\textbf{wavelength}\end{array}\right] \qquad \lambda = \frac{h}{p} = \frac{h}{mv} \qquad (36.8)$$

where h is Planck's constant and p is the magnitude of the momentum of the particle. The momentum is the product of the particle's relativistic mass and velocity, so that $p = mv$. Today, λ is known as the **de Broglie wavelength** of the particle.

Confirmation of de Broglie's suggestion came in 1927 from the experiments of the American physicists Clinton J. Davisson (1881–1958) and Lester H. Germer (1896–1971) and, independently, the English physicist George P. Thomson (1882–1975). Davisson and Germer directed a beam of electrons onto a crystal of nickel and observed that the electrons exhibited a diffraction behavior, analogous to that seen when X rays are diffracted by a crystal (see Section 34.8 for a discussion of X-ray diffraction). The wavelength of the electrons revealed by the diffraction pattern matched that predicted by de Broglie's hypothesis, $\lambda = h/mv$. More recently, Young's double-slit experiment has been performed with electrons, and they exhibit the effects of wave interference shown in Figure 36.1.

Particles other than electrons can also exhibit wave-like properties. For instance, neutrons are sometimes used in diffraction studies of crystal structure. Figure 36.6 compares the neutron diffraction pattern and the X-ray diffraction pattern caused by a crystal of rock salt (NaCl).

Although all moving particles have a de Broglie wavelength, the effects of this wavelength are observable only for particles whose masses are very small, on the order of the mass of an electron or a neutron, for instance. Example 3 illustrates the reason for this.

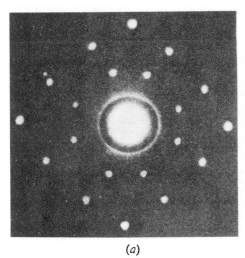

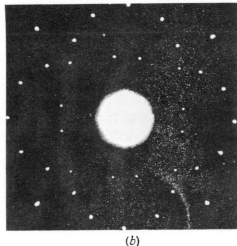

(a)   (b)

FIGURE 36.6 Neutrons can exhibit wave-like properties, as indicated by the neutron diffraction pattern shown in part a for a crystal of sodium chloride (NaCl). Compare this pattern to the X-ray diffraction pattern in part b for the same crystal.

EXAMPLE 3

Determine the de Broglie wavelength for (a) an electron (mass $= 9.1 \times 10^{-31}$ kg) moving at a speed of 6.0×10^6 m/s and (b) a baseball (mass $= 0.15$ kg) moving at a speed of 13 m/s.

SOLUTION

(a) The momentum of the electron is

$$p = mv = (9.1 \times 10^{-31} \text{ kg})(6.0 \times 10^6 \text{ m/s})$$

$$= 5.5 \times 10^{-24} \text{ kg} \cdot \text{m/s}$$

Consequently, the de Broglie wavelength of the electron is

$$\lambda = \frac{h}{p} = \frac{6.63 \times 10^{-34} \text{ J} \cdot \text{s}}{5.5 \times 10^{-24} \text{ kg} \cdot \text{m/s}} \qquad (36.8)$$

$$= \boxed{1.2 \times 10^{-10} \text{ m}}$$

A de Broglie wavelength of 1.2×10^{-10} m is about the size of the interatomic spacing in a solid, such as the nickel crystal

used by Davisson and Germer, and, therefore, leads to the observed diffraction effects.

(b) Calculations similar to those in part (a) show that the momentum and de Broglie wavelength of the baseball are $p = 2.0$ kg $\cdot$ m/s and $\boxed{\lambda = 3.3 \times 10^{-34} \text{ m}}$. The de Broglie wavelength of the baseball is incredibly small, even by comparison with the dimensions of an atom (10^{-10} m) or a nucleus (10^{-14} m). A wavelength of 3.3×10^{-34} m is so small that the wave characteristics of a baseball cannot be observed.

It is the size of Planck's constant h that determines the de Broglie wavelength for a given value of the momentum, and Planck's constant is very small. Moreover, the wavelength is inversely proportional to the momentum, so the wavelength for an object with a large momentum, like the baseball, is small.

The de Broglie equation for particle wavelength provides no hint as to what kind of wave is associated with a particle of matter. We turn our attention to the nature of these particle waves in the next section.

36.6 PARTICLE WAVES AND PROBABILITY

The de Broglie particle wavelength suggests that there is a wave associated with a moving particle, and Figure 36.7 helps to explain the nature of this wave. This picture

Screen

Double slit

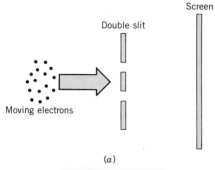

Moving electrons

(a)

(b) After 48 electrons.

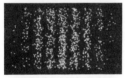

(c) After 1000 electrons.

(d) After 10,000 electrons

FIGURE 36.7 This electron version of Young's double-slit experiment was simulated on a computer by Dr. Elisha Huggins. The characteristic fringe pattern becomes recognizable only after a sufficient number of electrons have struck the screen.

shows a computer simulation of how the fringe pattern emerges on the screen when electrons are used in a version of Young's double-slit experiment. The bright fringes occur in places on the screen where particle waves coming from each slit interfere constructively, while the dark fringes occur in places where the waves interfere destructively.

When an electron passes through the double-slit arrangement and strikes a spot on the screen, the screen glows at that spot, and Figure 36.7 illustrates how the spots accumulate in time. As more and more electrons strike the screen, the spots eventually form the fringe pattern that is evident in part *d* of the drawing. Bright fringes occur where there is a high probability of electrons striking the screen, and dark fringes occur where there is a low probability. Here lies the key to understanding particle waves. *Particle waves are waves of probability,* waves whose magnitude at a point in space gives an indication of the probability that the particle will be found at that point. At the place where the screen is located, the pattern of probabilities conveyed by the particle waves causes the fringe pattern to emerge. The fact that no fringe pattern is apparent in part *b* of the picture does not mean that there are no probability waves present; it just means that too few electrons have struck the screen for the fringe pattern to be recognizable.

The pattern of probabilities that leads to the fringes in Figure 36.7 is analogous to the pattern of light intensities that creates the fringes in Young's original experiment with light waves (see Figure 34.4). Section 30.4 discusses the fact that the intensity of the light is proportional to either the square of the electric field strength or the square of the magnetic field strength of the wave. In the case of particle waves, the probability is proportional to the square of the magnitude Ψ of the wave; thus, probability $\propto \Psi^2$. Ψ is referred to as the *wave function* of the particle.

In 1925 the Austrian physicist Erwin Schrödinger (1887–1961) and the German physicist Werner Heisenberg (1901–1976) independently developed theoretical frameworks for determining the wave function. In so doing, they established a new branch of physics called *quantum mechanics.* The word "quantum" refers to the fact that in the world of the atom, where particle waves must be considered, the particle energy is quantized, so only certain energies are allowed. To understand the structure of the atom and the phenomena related to it, quantum mechanics is essential, and the Schrödinger equation for calculating the wave function is now widely used. A discussion of the Schrödinger equation is beyond the scope of this text. But in the next chapter, we will explore the structure of the atom based on the ideas of quantum mechanics.

36.7 THE UNCERTAINTY PRINCIPLE

As the previous section discusses, the bright fringes in Figure 36.7 indicate the places where there is a high probability of an electron striking the screen. And since there are a number of bright fringes, there is more than one place where each electron has some probability of hitting. Yet, any given electron can strike the screen in only one place after passing through the double slit. As a result, it is not possible to specify in advance exactly where on the screen an individual electron will fall. All we can do is speak of the probability that the electron may end up in a number of different places. No longer is it possible to say, as Newton's laws would suggest, that a single electron, fired through the double slit, will travel directly forward in a straight line and strike the screen. This simple picture just does not apply when a particle as small as an electron passes through a pair of narrow slits. Because the wave nature of particles is important in such circumstances, we lose the ability to predict with 100% certainty the path that a single particle will follow. Instead, only the average behavior of large numbers of particles is predictable, and the behavior of an individual particle is uncertain.

To see more clearly into the nature of the uncertainty, consider electrons passing through the single slit in Figure 36.8. After a sufficient number of electrons strike the screen, a diffraction pattern emerges. The electron diffraction pattern consists of alternating bright and dark fringes and is analogous to that for light waves shown in Figure 34.23. Figure 36.8 shows the slit and locates the first dark fringe on either side of the central bright fringe. The central fringe is bright because electrons strike the screen over the entire region between the dark fringes. If the electrons striking the screen outside the central bright fringe can be neglected, the extent to which the electrons are diffracted is given by the angle θ in the drawing. To reach locations within the central bright fringe, some electrons must have acquired momentum in the y direction, despite the fact that they enter the slit traveling along the x direction, and thus have no momentum in the y direction to start with. The figure illustrates that the y component of the momentum may be as large as Δp_y. The notation Δp_y indicates the difference between the maximum value of the y component of the momentum after the electron passes through the slit and its value of zero before the electron passes through the slit. Δp_y represents the *uncertainty* in the y component of the momentum, in that a diffracted electron may have any value from zero to Δp_y.

It is possible to relate Δp_y to the width W of the slit. To do this, we assume that Equation 34.4, which applies to light waves, also applies to particle waves whose de Broglie wavelength is λ. This equation, $\sin \theta = \lambda/W$, specifies the angle θ that locates the first dark fringe. If θ is small, $\sin \theta \approx \tan \theta$. Moreover, Figure 36.8 indicates that $\tan \theta = (\Delta p_y)/p_x$, where p_x is the x component of the momentum of the electron. Therefore, $(\Delta p_y)/p_x \approx \lambda/W$. But $p_x = h/\lambda$ according to de Broglie's equation, so it follows that

$$\frac{\Delta p_y}{h/\lambda} \approx \frac{\lambda}{W}$$

As a result,

$$\Delta p_y \approx \frac{h}{W} \qquad (36.9)$$

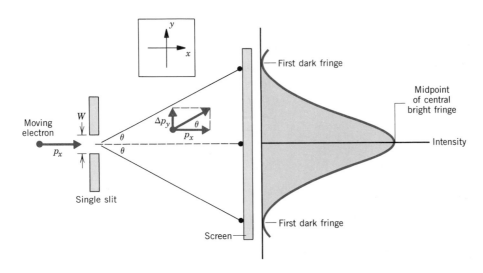

FIGURE 36.8 When a sufficient number of electrons pass through a single slit and strike the screen, a diffraction pattern of bright and dark fringes emerges. This pattern is due to the wave nature of the electrons and is analogous to that produced by light waves.

which indicates that a smaller slit width leads to a larger uncertainty in the y component of the electron's momentum.

It was Heisenberg who first suggested that the uncertainty Δp_y in the y component of the momentum is related to the uncertainty in the y position of the electron as the electron passes through the slit. Since the electron can pass through anywhere over the width W, the uncertainty in the y position of the electron is $\Delta y = W$. Substituting Δy for W in Equation 36.9 shows that $\Delta p_y \approx h/\Delta y$ or $(\Delta p_y)(\Delta y) \approx h$. The result of Heisenberg's more complete analysis is given below in Equation 36.10 and is known as the **Heisenberg uncertainty principle.**

THE HEISENBERG UNCERTAINTY PRINCIPLE

$$(\Delta p_y)(\Delta y) \geq \frac{h}{2\pi} \tag{36.10}$$

Δy = uncertainty in our knowledge of a particle's position along the y direction
Δp_y = uncertainty in our knowledge of the y component of the linear momentum of the particle

$$(\Delta E)(\Delta t) \geq \frac{h}{2\pi} \tag{36.11}$$

ΔE = uncertainty in our knowledge of the energy of a particle when the particle is in a certain state
Δt = time interval during which the particle is in the state

The uncertainty principle can also be expressed in terms of energy and time, as in Equation 36.11. According to the uncertainty principle, the shorter the lifetime of a particle in a given energy state, the greater is the uncertainty in the energy of that state.

Example 4 shows that the uncertainty principle has significant consequences for the motion of tiny particles such as electrons but has little effect on the motion of macroscopic objects, even those with as little mass as a Ping-Pong ball.

EXAMPLE 4

Assume that the position of an object is known so precisely that the uncertainty in the position is only $\Delta y = 1.5 \times 10^{-11}$ m. (a) Determine the minimum uncertainty in the momentum of the object. Find the corresponding minimum uncertainty in the speed of the object, if the object is (b) an electron (mass $= 9.1 \times 10^{-31}$ kg) and (c) a Ping-Pong ball (mass $= 2.2 \times 10^{-3}$ kg).

SOLUTION

(a) The minimum uncertainty in the y component of the momentum follows directly from the uncertainty principle:

$$\Delta p_y \geq \frac{h}{2\pi\,\Delta y} = \frac{6.63 \times 10^{-34}\text{ J}\cdot\text{s}}{2\pi(1.5 \times 10^{-11}\text{ m})} \tag{36.10}$$

$$= \boxed{7.0 \times 10^{-24}\text{ kg}\cdot\text{m/s}}$$

This small uncertainty in the momentum applies for both the electron and the Ping-Pong ball. However, the uncertainties in the speeds of these objects are vastly different, because, for a given momentum, the mass determines the speed.

(b) Since $\Delta p_y = m\,\Delta v_y$, the minimum uncertainty in the speed of the electron is

$$\Delta v_y = \frac{\Delta p_y}{m} = \frac{7.0 \times 10^{-24}\text{ kg}\cdot\text{m/s}}{9.1 \times 10^{-31}\text{ kg}}$$

$$= \boxed{7.7 \times 10^{6}\text{ m/s}}$$

Thus, the small uncertainty in the y position of the electron gives rise to a large uncertainty in the speed of the electron.

(c) The uncertainty in the speed of the Ping-Pong ball is

$$\Delta v_y = \frac{\Delta p_y}{m} = \frac{7.0 \times 10^{-24}\text{ kg}\cdot\text{m/s}}{2.2 \times 10^{-3}\text{ kg}}$$

$$= \boxed{3.2 \times 10^{-21}\text{ m/s}}$$

Because the mass of the Ping-Pong ball is relatively large compared to that of the electron, the uncertainty in the speed of the ball is unobservable.

The uncertainty principle places limits on the accuracy with which the behavior of particles can be predicted, and these limits are not just limits due to faulty measuring techniques. They are fundamental limits imposed by nature, in the same sense that the second law of thermodynamics places a natural limit on the efficiency of a heat engine. There are no ways to circumvent such limits. Equation 36.10, for instance, indicates that it is not possible for both Δy and Δp_y to be zero. In other words, it is impossible to specify simultaneously *both* the position and the momentum of a particle with 100% accuracy. If the position is known exactly, so that Δy is zero, then Δp_y is an infinitely large number, and the momentum of the particle is completely uncertain. In such a circumstance it is meaningless to specify a value for the momentum. Conversely, if we assume that Δp_y is zero, then Δy is an infinitely large number, and the position of the particle is completely uncertain. Similar comments apply to energy and time, as Equation 36.11 shows.

SUMMARY

The **wave–particle duality** refers to the fact that a wave can exhibit particle-like characteristics and a particle can exhibit wave-like characteristics.

At a constant temperature, a perfect blackbody absorbs and reemits all the electromagnetic radiation that falls on it. Max Planck calculated the emitted radiation intensity per unit wavelength as a function of wavelength. In his theory, Planck assumed that a blackbody consists of atomic vibrators that could have only quantized energies. Planck's quantized energies are given by $E = nhf$, where $n = 0, 1, 2, 3, \ldots$, h is **Planck's constant** $(6.63 \times 10^{-34} \text{ J} \cdot \text{s})$, and f is the vibration frequency.

All electromagnetic radiation consists of **photons,** which are packets of energy. The energy of a photon is $E = hf$, where h is Planck's constant and f is the frequency of the light. A photon in a vacuum always travels at the speed of light c and has no rest mass. The **photoelectric effect** is the phenomenon in which light shining on a metal surface causes electrons to be ejected from the surface. The **work function W_0** of a metal is the minimum work that must be done to eject an electron from the metal. In accordance with the conservation of energy, the electrons ejected from a metal have a maximum kinetic energy KE_{max} that is related to the energy hf of the incident photon by $hf = KE_{max} + W_0$.

The **Compton effect** is the scattering of a photon by an electron in a material, the scattered photon having a smaller frequency than the incident photon. The difference between the wavelength λ' of the scattered photon and the wavelength λ of the incident photon is related to the scattering angle θ by $\lambda' - \lambda = [h/(m_0 c)](1 - \cos\theta)$, where m_0 is the rest mass of the electron and the quantity $h/(m_0 c)$ is known as the Compton wavelength of the electron.

The **de Broglie wavelength** of a particle is $\lambda = h/p$, where $p = mv$ is the magnitude of the momentum of the particle (m = relativistic mass, v = speed). Because of its de Broglie wavelength, a particle can exhibit wave-like characteristics. The wave associated with a particle is a wave of probability.

The **Heisenberg uncertainty principle** places limits on our knowledge about the behavior of a particle. The uncertainty principle indicates that $(\Delta p_y)(\Delta y) \geq h/2\pi$, where Δy and Δp_y are, respectively, the uncertainties in the position and momentum of the particle. The uncertainty principle also states that $(\Delta E)(\Delta t) \geq h/2\pi$, where ΔE is the uncertainty in the energy of a particle when the particle is in a certain state and Δt is the time interval during which the particle is in the state.

QUESTIONS

1. Radiation with a given wavelength causes electrons to be emitted from the surface of one metal but not from the surface of another metal. Explain why this could be.

2. Which of the colored lights on a Christmas tree transmits, on the average, photons with (a) the least energy and (b) the greatest energy? Account for your answers.

3. When a sufficient number of visible light photons strike a piece of photographic film, the film becomes exposed. An X-ray photon is more energetic than a visible light photon. Yet, most photographic films are not exposed by the X-ray machines used at airport security checkpoints. Explain what these observations imply about the number of photons emitted by the X-ray machines.

4. In a Compton scattering experiment, an electron is accelerated straight ahead in the direction of the incident X-ray photon. Which way does the scattered photon move? Explain your reasoning, using the principle of conservation of momentum.

5. Photons can undergo Compton scattering from a molecule such as nitrogen, just as they do from an electron. However, the change in photon wavelength is much less than when an electron is scattered. Explain why, using Equation 36.7 for a nitrogen molecule instead of an electron.

6. In Section 17.4 the impulse-momentum theorem is used to analyze how gas molecules exert a force on a wall. (a) Use similar reasoning to discuss why a beam of light exerts a force on a surface it strikes. (b) Do you think a beam of light exerts more force on a mirror that reflects the light or on a black surface that absorbs the light? In your discussion assume the beam is perpendicular to both the mirror and the black surface.

7. A stone is dropped from rest from the top of a building. Explain what happens to the de Broglie wavelength of the stone as the stone falls.

8. A bullet leaving the barrel of a gun is analogous to an electron passing through the single slit in Figure 36.8. With this analogy in mind, explain whether the uncertainty principle is likely to have any effect on your success as a hunter.

PROBLEMS

In working these problems, ignore relativistic effects.

Section 36.3 Photons and the Photoelectric Effect

1. Ultraviolet light is responsible for sun tanning. Find the wavelength of an ultraviolet photon whose energy is 4.0 eV.

2. The wavelengths (in vacuum) of visible light occur between 380 and 750 nm. Determine the range of photon energies (in joules) to which this range of wavelengths corresponds.

3. An FM radio station broadcasts at a frequency of 98.1 MHz. The power radiated from the antenna is 5.0×10^4 W. How many photons per second does the antenna emit?

4. The work function for a sodium surface is 2.28 eV. What is the maximum wavelength that an electromagnetic wave can have and still eject electrons from this surface?

5. Light is shining perpendicularly on the surface of the earth with an intensity of 680 W/m². Assuming all the photons in the light have a wavelength of 730 nm, determine the number of photons per second per square meter that reach the earth.

6. A magnesium surface has a work function of 3.68 eV. Electromagnetic waves with a wavelength of 215 nm strike the surface and eject electrons. Find the maximum kinetic energy of the ejected electrons. Express your answer (a) in joules and (b) in electron volts.

7. An AM radio station broadcasts an electromagnetic wave at a frequency of 665 kHz, while an FM station broadcasts at a frequency of 91.9 MHz. How many AM photons are needed to have a total energy equal to that of one FM photon?

* 8. An owl has good night vision because its eyes can detect a light intensity as small as 5.0×10^{-13} W/m². What is the minimum number of photons per second that an owl eye can detect if

its pupil has a diameter of 8.5 mm and the light has a wavelength of 510 nm?

* 9. Radiation with a wavelength of 281 nm shines on a metal surface and ejects electrons that have a maximum velocity of 3.48×10^5 m/s. Which one of the following metals is present, the values in parentheses being the work functions: potassium (2.24 eV), calcium (2.71 eV), uranium (3.63 eV), aluminum (4.08 eV), and gold (4.82 eV)?

* 10. The maximum wavelength for which an electromagnetic wave can eject electrons from a platinum surface is 196 nm. When radiation with a wavelength of 141 nm shines on the surface, what is the maximum speed of the ejected electrons?

* 11. At night, approximately 530 photons per second must enter an unaided human eye for an object to be seen, assuming the light is green. The light bulb in Example 1 in the text emits green light uniformly in all directions and the diameter of the pupil of the eye is 7.0 mm. What is the maximum distance from which the bulb could be seen?

**12. (a) How many photons (wavelength = 620 nm) must be absorbed to melt a 2.0-kg block of ice at 0 °C? (b) On the average, how many H_2O molecules does one photon convert from the ice phase to the water phase?

**13. A laser emits 1.30×10^{18} photons per second in a beam of light that has a diameter of 2.00 mm and a wavelength of 514.5 nm. Determine (a) the average electric field strength and (b) the average magnetic field strength for the electromagnetic wave that constitutes the beam.

Section 36.4 The Momentum of a Photon and the Compton Effect

14. The microwaves used in a microwave oven have a wavelength of about 0.33 m. What is the momentum of a microwave photon?

15. Determine the *change* in the photon's wavelength that occurs when an electron scatters an X-ray photon (a) straight back at an angle of $\theta = 180.0°$, (b) at an angle $\theta = 90.0°$, and (c) at an angle of $\theta = 30.0°$. All angles are measured as in Figure 36.5.

16. In a Compton scattering experiment, the incident X rays have a wavelength of 0.2685 nm, while the scattered X rays have a wavelength of 0.2702 nm. At what angle θ in Figure 36.5 are the scattered X rays detected?

* 17. The X rays detected at a scattering angle of $\theta = 163°$ in Figure 36.5 have a wavelength of 0.1867 nm. Find (a) the wavelength of an incident photon, (b) the energy of an incident photon, (c) the energy of a scattered photon, and (d) the kinetic energy of the recoil electron. (For accuracy, use $h = 6.626 \times 10^{-34}$ J·s and $c = 2.998 \times 10^8$ m/s.)

Section 36.5 The de Broglie Wavelength and the Wave Nature of Matter

18. A honeybee (mass $= 1.3 \times 10^{-4}$ kg) is crawling at a speed of 0.020 m/s. What is the de Broglie wavelength of the bee?

19. The de Broglie wavelength of a proton in a particle accelerator is 1.30×10^{-14} m. Determine the kinetic energy of the proton.

20. How fast does a proton have to be moving to have the same de Broglie wavelength as an electron does when the electron moves at 4.5×10^6 m/s?

21. Recall from Section 17.4 that the average kinetic energy of an atom in an ideal gas is given by $\overline{\text{KE}} = \frac{3}{2}kT$, where $k = 1.38 \times 10^{-23}$ J/K and T is the Kelvin temperature of the gas. Determine the de Broglie wavelength of a helium atom (mass $= 6.65 \times 10^{-27}$ kg) that has the average kinetic energy at room temperature (20.0 °C).

* 22. In a Young's double-slit experiment performed with electrons, the two slits are separated by a distance of 2.0×10^{-6} m. The first-order bright fringes are located on the observation screen at a position given by $\theta = 1.6 \times 10^{-4}$ degrees in Equation 34.1. Find (a) the wavelength, (b) the momentum, and (c) the kinetic energy of the electrons.

* 23. In a television picture tube, electrons are accelerated from rest through a potential difference of 21 000 V. What is the de Broglie wavelength of the electrons?

**24. The kinetic energy of a particle is equal to the energy of a photon. The particle moves at 5.0% of the speed of light. Find the ratio of the photon wavelength to the particle wavelength.

Section 36.7 The Uncertainty Principle

25. The speed of a golf ball (mass $= 0.045$ kg) and of an electron is 71 m/s. If the uncertainty in the speed is 1.0%, estimate the minimum uncertainty in the position of each object.

26. A prisoner (75 kg) paces back and forth in his cell, and the uncertainty in his speed is 0.10 m/s. (a) Use the uncertainty principle to estimate the minimum uncertainty in his position. (b) Repeat part (a), assuming Planck's constant has a value of 663 J·s instead of 6.63×10^{-34} J·s.

27. When electrons pass through a single slit, as in Figure 36.8, they form a diffraction pattern. As Section 36.7 discusses, the central bright fringe extends to either side of the midpoint, according to an angle θ given by $\sin \theta = \lambda/W$, where λ is the de Broglie wavelength of the electron and W is the width of the slit. When λ is the same size as W, $\theta = 90°$, and the central fringe fills the entire observation screen. In this case, an electron passing through the slit has roughly the same probability of hitting the screen either straight ahead or anywhere off to one side or the other. Now, imagine yourself in a world where Planck's constant is large enough so you exhibit similar effects when you walk through a 0.90-m-wide doorway. If your mass is 82 kg and you walk at a speed of 0.50 m/s, how large would Planck's constant have to be in this hypothetical world?

ADDITIONAL PROBLEMS

28. What is the wavelength of (a) a 1.0-eV photon and (b) a 1.0-eV electron?

29. Photons A, B, and C have energies of 3.3×10^{-16} J, 2.0×10^{-23} J, and 1.3×10^{-20} J, respectively. Using Figure 30.6, identify the appropriate region in the electromagnetic spectrum for each of these photons.

30. As Section 34.5 discusses, sound waves diffract or bend around the edges of a doorway. Larger wavelengths diffract more than smaller wavelengths. (a) The speed of sound is 343 m/s. With what speed would a 55.0-kg person have to move through a doorway to diffract to the same extent as a 128-Hz bass tone? (b) At the speed calculated in part (a), how long (in years) would it take the person to move a distance of one meter?

31. Incident X rays have a wavelength of 0.3365 nm and are scattered by the "free" electrons in carbon. The scattering angle in Figure 36.5 is $\theta = 125°$. What is the magnitude of the momentum of (a) the incident photon and (b) the scattered X-ray photon?

32. Radiation of a certain wavelength causes electrons with a maximum kinetic energy of 0.68 eV to be ejected from a metal whose work function is 2.75 eV. What will be the maximum kinetic energy (in eV) with which this same radiation ejects electrons from another metal whose work function is 2.17 eV?

* 33. The width of the central bright fringe in a diffraction pattern on a screen is identical when either electrons or red light (vacuum wavelength $= 661$ nm) pass through a single slit. The distance between the screen and the slit is the same in each case and is large compared to the slit width. (a) How fast are the electrons moving? (b) To judge whether the speed in (a) is fast or slow for an electron, determine the speed acquired by an electron in accelerating from rest through a potential difference of one volt.

* **34.** Example 1 in the text calculates the number of photons per second given off by a sixty-watt incandescent light bulb. The photons are emitted uniformly in all directions. From a distance of 3.1 m you glance at this bulb for 0.10 s. The light from the bulb travels directly to your eye and does not reflect from anything. The pupil of the eye has a diameter of 2.0 mm. How many photons enter your eye?

* **35.** An electron and a proton have the same kinetic energy.

Determine the ratio of the de Broglie wavelength of the electron to that of the proton.

****36.** A beam of visible light has a wavelength of 395 nm and shines perpendicularly on a surface. As a result, there are 3.0×10^{18} photons per second striking the surface. By using the impulse-momentum theorem (Section 8.2), obtain the average force that this beam applies to the surface when (a) the surface is a mirror, so the momentum of each photon is reversed after reflection and (b) the surface is black, so each photon is absorbed and the momentum of the photon is reduced to zero in the process.

CHAPTER 37

The Nature of the Atom

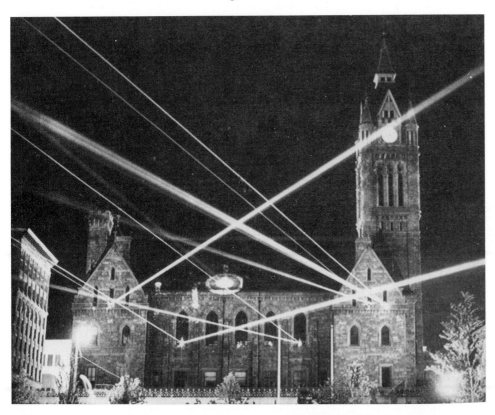

A laser can produce a narrow beam of light. Mirrors placed at various locations reflect the light to produce this laser light show. Lasers are used extensively in medicine, communications technology, and many other areas. The operation of a laser is rooted in the principles of modern physics.

37.1 RUTHERFORD SCATTERING AND THE NUCLEAR ATOM

An atom contains a small, positively charged nucleus (radius $\approx 10^{-14}$ m), which is surrounded at relatively large distances (radius $\approx 10^{-10}$ m) by a number of electrons, as Figure 37.1a illustrates. In the natural state, an atom is electrically neutral. The

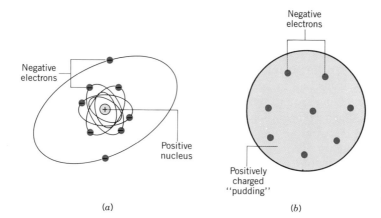

(a)

(b)

FIGURE 37.1 (*a*) The nuclear atom. (*b*) The "plum pudding" model of the atom (now discredited).

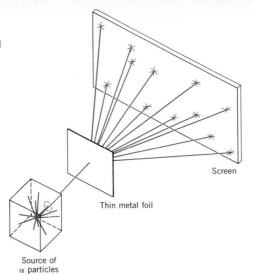

Screen

Thin metal foil

FIGURE 37.2 An illustration of a Rutherford scattering experiment where alpha particles (α particles) are scattered by a thin metal foil. The entire apparatus is placed in a vacuum chamber (not shown).

Source of α particles

neutrality arises because the nucleus contains a number of protons (each with a charge of $+e$) that exactly equals the number of electrons (each with a charge of $-e$). This model of the atom is universally accepted now and is referred to as the "nuclear atom."

The nuclear atom is a relatively recent idea. In the early part of the twentieth century a widely accepted model, due to the English physicist Joseph J. Thomson (1856–1940), pictured the atom very differently. In Thomson's view there was no nucleus at the center of an atom. Instead, the positive charge was assumed to be spread throughout the atom, forming a kind of "paste" or "pudding," in which the negative electrons were suspended like "plums." Figure 37.1 compares this "plum-pudding" model with the currently accepted view of the atom.

The "plum-pudding" model was discredited in 1911 when the New Zealand physicist Ernest Rutherford (1871–1937) published experimental results that the model could not explain. As Figure 37.2 indicates, Rutherford and his co-workers directed a beam of alpha particles (α particles) at a thin metal foil made of gold. Alpha particles are positively charged particles (the nuclei of helium atoms) emitted by some radioactive materials. If the "plum-pudding" model were correct, the α particles would be expected to pass nearly straight through the foil. After all, there is nothing in this model to deflect the relatively massive α particles, since the electrons have a comparatively small mass and the positive charge is spread out in a diluted "pudding." Using a zinc sulfide screen, which flashed briefly when struck by an α particle, Rutherford and his co-workers were able to determine that not all the α particles passed straight through the foil. Instead, some of the α particles were deflected at large angles, some even backward. Rutherford himself said, "It was almost as incredible as if you had fired a fifteen inch shell at a piece of tissue and it came back and hit you." Rutherford concluded that the positive charge, instead of being distributed throughout the atom, was concentrated in a small region called the nucleus. Thus, the idea of a nuclear atom was born.

But how could the electrons in a nuclear atom remain separated from the positively charged nucleus? If the electrons were stationary, they would be pulled inward by the attractive electric force of the nuclear charge. Therefore, it was realized that the electrons had to be moving around the nucleus in some fashion, like the planets revolving around the sun.

Such a planetary model of the atom has its own difficulties, however. For instance, an electron moving on a curved path has a centripetal acceleration, as Section 6.2 discusses. And when an electron is accelerating, it radiates electromagnetic waves. The difficulty is that the waves carry away energy, which decreases the energy of the electrons. With their energy constantly being depleted, the electrons would spiral inward and eventually collapse into the nucleus. Since such a collapse does not occur, the planetary model, while providing a more realistic picture of the atom than the "plum-pudding" model, must be telling only part of the story. The full story of atomic structure is fascinating, and the next section describes another aspect of it.

37.2 LINE SPECTRA

We have seen in Sections 16.4 and 36.2 that all objects emit electromagnetic waves. For a solid object, such as the hot filament of a light bulb, these waves have a continuous range of wavelengths, some of which are in the visible region of the spectrum. The continuous range of wavelengths is characteristic of the entire collection of atoms that make up the solid. In contrast, individual atoms, free from the strong interatomic interactions that are present in a solid, emit only certain specific wavelengths, rather than a continuous range. These wavelengths are characteristic of the atom and provide important clues about its structure. To study the behavior of individual atoms, low-pressure gases are used, in which the atoms are relatively far apart.

A low-pressure gas contained within a sealed tube can be made to emit electromagnetic waves by applying a sufficiently large potential difference between two electrodes located within the tube. With a grating spectroscope like that in Figure 34.33, the individual wavelengths emitted by the gas can be separated and identified as a series of bright fringes or lines. The series of lines is called a *line spectrum*. The simplest line spectrum is that of the hydrogen atom (H).* Figure 37.3 (between pages 812 and 813) shows the visible part of the line spectrum of atomic hydrogen, along with the visible parts of the line spectra of more complicated atoms such as neon and mercury. The specific visible wavelengths emitted by neon and mercury are familiar, because they give neon signs and mercury vapor street lamps their characteristic colors.

Much effort has been devoted to understanding the pattern of wavelengths observed in the line spectrum of atomic hydrogen. In addition to the series of lines found in the visible region, analogous series have been found in shorter and longer wavelength regions of the electromagnetic spectrum. In schematic form, Figure 37.4 illustrates some of the series of lines for atomic hydrogen. The group of lines in the visible region is known as the *Balmer series,* in recognition of Johann J. Balmer (1825–1898), a Swiss schoolteacher who found an empirical equation that gave the values for the observed wavelengths. This equation is given below, along with similar

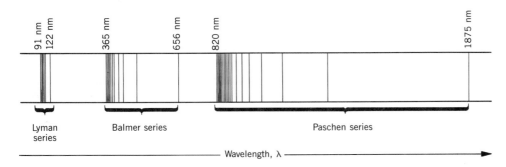

FIGURE 37.4 Line spectrum of atomic hydrogen. Only the Balmer series lies in the visible region of the electromagnetic spectrum.

* Molecular hydrogen (H$_2$) consists of two hydrogen atoms bonded together and has a more complicated line spectrum than that of atomic hydrogen.

equations that apply to the **Lyman series** and **Paschen series,** which are also shown in the drawing:

[Lyman series] $$\frac{1}{\lambda} = R\left(\frac{1}{1^2} - \frac{1}{n^2}\right) \qquad n = 2, 3, 4, \ldots \qquad (37.1)$$

[Balmer series] $$\frac{1}{\lambda} = R\left(\frac{1}{2^2} - \frac{1}{n^2}\right) \qquad n = 3, 4, 5, \ldots \qquad (37.2)$$

[Paschen series] $$\frac{1}{\lambda} = R\left(\frac{1}{3^2} - \frac{1}{n^2}\right) \qquad n = 4, 5, 6, \ldots \qquad (37.3)$$

In these equations, the constant term R has the value of $R = 1.097 \times 10^7$ m^{-1} and is called the *Rydberg constant.* An essential feature of each of these groups of lines is that they have both a long and a short wavelength limit, with the lines being increasingly crowded together toward the short wavelength limit. The drawing gives the wavelength limits for each of the three series, and Example 1 determines them for the Balmer series.

EXAMPLE 1

Find (a) the longest and (b) the shortest wavelengths of the Balmer series.

SOLUTION

(a) According to Balmer's empirical relation, Equation 37.2, the longest wavelength of the Balmer series occurs when $n = 3$:

$$\frac{1}{\lambda} = R\left(\frac{1}{2^2} - \frac{1}{n^2}\right) = (1.097 \times 10^7 \text{ m}^{-1})\left(\frac{1}{2^2} - \frac{1}{3^2}\right)$$

$$= 1.524 \times 10^6 \text{ m}^{-1}$$

$$\boxed{\lambda = 656 \text{ nm}}$$

(b) The shortest wavelength in the Balmer series arises when the integer n in Balmer's equation has a very large value, so that $1/n^2$ is essentially zero:

$$\frac{1}{\lambda} = (1.097 \times 10^7 \text{ m}^{-1})\left(\frac{1}{2^2} - 0\right) = 2.743 \times 10^6 \text{ m}^{-1}$$

$$\boxed{\lambda = 365 \text{ nm}}$$

Figure 37.4 shows the wavelength limits found in part (a) and part (b).

Equations 37.1–37.3 are useful, because they reproduce the wavelengths that hydrogen atoms radiate. However, these equations are empirical, and they provide no insight as to *why* certain wavelengths are radiated and others are not. It was the Danish physicist Niels Bohr (1885–1962) who provided the first theoretical model of the atom that predicted the discrete wavelengths emitted by atomic hydrogen. Bohr's model started us on the way toward understanding how the structure of the atom restricts the radiated wavelengths to certain values.

37.3 THE BOHR MODEL OF THE HYDROGEN ATOM

THE MODEL

In 1913 Bohr presented a model that led to equations such as Balmer's for predicting the specific wavelengths that the hydrogen atom radiates. Bohr's theory begins with Rutherford's picture of an atom as a nucleus surrounded by electrons moving in circular orbits. In analyzing this picture, Bohr made a number of assumptions in order to combine the new quantum ideas of Planck and Einstein with the traditional description of a particle in uniform circular motion.

Adopting Planck's idea of quantized energy levels, Bohr hypothesized that in a hydrogen atom there can be only certain values of the total energy (electron kinetic energy plus potential energy). These allowed energy levels correspond to different orbits for the electron as it moves around the nucleus, the larger orbits being associated with larger total energies. Figure 37.5 illustrates two of the allowed orbits. In addition, Bohr assumed that an electron in one of these orbits *does not* radiate electromagnetic waves. For this reason, the orbits are sometimes called **stationary orbits** or **stationary states.** Bohr recognized that radiationless orbits violated the laws of physics, as they were then known. But the assumption of such orbits was necessary, because the traditional laws indicated that an electron radiates electromagnetic waves as it accelerates around a circular path, and the loss of the energy carried by the waves would lead to the collapse of the orbit.

To incorporate Einstein's photon concept, Bohr theorized that a photon is emitted only when the electron *changes* orbits from a larger one with a higher energy to a smaller one with a lower energy, as Figure 37.5 indicates. But how do electrons get into the higher-energy orbits in the first place? They get there by picking up energy when two atoms collide, which happens more and more often when a gas is heated, or by acquiring energy when a high voltage is applied to a gas.

When an electron in an initial orbit with a larger energy E_i changes to a final orbit with a smaller energy E_f, the emitted photon has an energy that is $E_i - E_f$, consistent with the law of conservation of energy. But according to Einstein, the energy of a photon is hf, where f is the frequency and h is Planck's constant. Thus,

$$E_i - E_f = hf \qquad (37.4)$$

Since the frequency of an electromagnetic wave is related to the wavelength by $f = c/\lambda$, Bohr could use Equation 37.4 to determine the wavelengths radiated by a hydrogen atom. First, however, he had to derive expressions for the energies E_i and E_f.

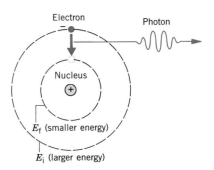

FIGURE 37.5 In the Bohr model of the hydrogen atom, a photon is emitted when the electron drops from a larger, higher energy orbit (energy $= E_i$) to a smaller, lower energy orbit (energy $= E_f$). The energy of the photon is equal to the difference between the energies, $E_i - E_f$.

THE ALLOWED ENERGIES AND RADII OF THE BOHR ORBITS

For an electron of mass m and speed v in an orbit of radius r, the total energy is the kinetic energy (KE $= \frac{1}{2}mv^2$) of the electron plus the electric potential energy. The potential energy is the product of the charge ($-e$) on the electron and the electric potential produced by the positive nuclear charge, in accord with Equation 24.3. We assume that the nucleus contains Z* protons, for a total nuclear charge of $+Ze$. The electric potential at a distance r from a point charge of $+Ze$ is given as $+kZe/r$ by Equation 24.6, where $k = 8.99 \times 10^9 \, \text{N} \cdot \text{m}^2/\text{C}^2$. The electric potential energy is, then, EPE $= (-e)(+kZe/r)$. Consequently, the total energy E of the atom is

$$E = \text{KE} + \text{EPE} = \frac{1}{2}mv^2 - \frac{kZe^2}{r} \qquad (37.5)$$

But a centripetal force of magnitude mv^2/r (Equation 6.3) must act on a particle in uniform circular motion. As Figure 37.6 indicates, the centripetal force is provided by the electrostatic force of attraction between the orbiting electron and the protons in the nucleus. According to Coulomb's law, the magnitude of the electrostatic force is kZe^2/r^2. Therefore, $mv^2/r = kZe^2/r^2$, or

* For hydrogen, $Z = 1$, but we also wish to consider situations in which Z is greater than 1.

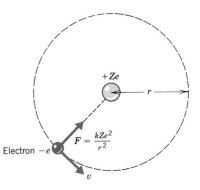

FIGURE 37.6 In the Bohr model, the electron is in uniform circular motion around the nucleus. The centripetal force is provided by the electrostatic force of attraction between the negative electron and the positive nuclear charge. According to Coulomb's law, the magnitude of this force is $F = kZe^2/r^2$.

$$mv^2 = \frac{kZe^2}{r} \tag{37.6}$$

We can use this relation to eliminate mv^2 from Equation 37.5, with the result that

$$E = \frac{1}{2}\left(\frac{kZe^2}{r}\right) - \frac{kZe^2}{r} = -\frac{kZe^2}{2r} \tag{37.7}$$

The total energy of the atom is negative, because the negative electric potential energy is larger in magnitude than the positive kinetic energy.

A value for the radius r is needed, if Equation 37.7 is to be useful. To determine r, Bohr made an assumption about the angular momentum of the electron. The angular momentum L is given by Equation 10.10 as $L = I\omega$, where $I = mr^2$ is the moment of inertia of the electron moving on its circular path and $\omega = v/r$ is the angular speed of the electron in radians per second. Thus, the angular momentum is $L = (mr^2)$ $\times (v/r) = mvr$. Bohr conjectured that the angular momentum of the electron has only certain discrete values; in other words, L is quantized. According to Bohr, the allowed values are integer multiples of Planck's constant divided by 2π:

$$L_n = mv_n r_n = n\frac{h}{2\pi} \qquad n = 1, 2, 3, \ldots \tag{37.8}$$

Solving this equation for v_n and substituting the result into Equation 37.6 leads to the following expression for the radius r_n of the nth Bohr orbit:

$$r_n = \left(\frac{h^2}{4\pi^2 mke^2}\right)\frac{n^2}{Z} \qquad n = 1, 2, 3, \ldots \tag{37.9}$$

With $h = 6.626 \times 10^{-34}$ J·s, $m = 9.109 \times 10^{-31}$ kg, $k = 8.988 \times 10^9$ N·m²/C², and $e = 1.602 \times 10^{-19}$ C, this expression reveals that

$$\begin{bmatrix}\text{Radii for} \\ \text{Bohr orbits}\end{bmatrix} \qquad r_n = (5.29 \times 10^{-11} \text{ m})\frac{n^2}{Z} \qquad n = 1, 2, 3, \ldots \tag{37.10}$$

Therefore, in the hydrogen atom ($Z = 1$) the smallest Bohr orbit ($n = 1$) has a radius of $r_1 = 5.29 \times 10^{-11}$ m. This particular value is sometimes called the **Bohr radius.** Figure 37.7 shows the first three Bohr orbits for the hydrogen atom.

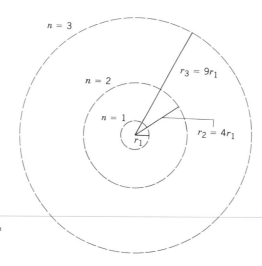

FIGURE 37.7 The first Bohr orbit in the hydrogen atom has a radius $r_1 = 5.29 \times 10^{-11}$ m. The second and third Bohr orbits have radii $r_2 = 4r_1$ and $r_3 = 9r_1$, respectively.

The expression for the radius of a Bohr orbit can be substituted into Equation 37.7 to show that the corresponding total energy for the *n*th orbit is

$$E_n = -\left(\frac{2\pi^2 mk^2 e^4}{h^2}\right)\frac{Z^2}{n^2} \qquad n = 1, 2, 3, \ldots \qquad (37.11)$$

Substituting values for h, m, k, and e into this expression yields

$$\begin{bmatrix} \text{Bohr energy} \\ \text{levels in} \\ \text{joules} \end{bmatrix} \qquad E_n = -(2.18 \times 10^{-18}\ \text{J})\frac{Z^2}{n^2} \qquad n = 1, 2, 3, \ldots \qquad (37.12)$$

Often, atomic energies are expressed in electron volts rather than joules. Since $1.60 \times 10^{-19}\ \text{J} = 1\ \text{eV}$, the result above can be rewritten as

$$\begin{bmatrix} \text{Bohr energy} \\ \text{levels in} \\ \text{electron volts} \end{bmatrix} \qquad E_n = -(13.6\ \text{eV})\frac{Z^2}{n^2} \qquad n = 1, 2, 3, \ldots \qquad (37.13)$$

ENERGY LEVEL DIAGRAMS

It is useful to represent the energy values given by Equation 37.13 on an *energy level diagram,* as in Figure 37.8. In this diagram, which applies to the hydrogen atom ($Z = 1$), the highest energy level corresponds to $n = \infty$ in Equation 37.13 and has an energy of 0 eV. This is the energy of the atom when the electron is completely removed ($r = \infty$) from the nucleus and is at rest. In contrast, the lowest energy level

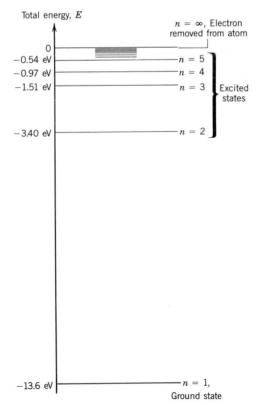

FIGURE 37.8 Energy level diagram for the hydrogen atom.

corresponds to $n = 1$ and has a value of -13.6 eV. The lowest energy level is called the **ground state,** to distinguish it from the higher levels, which are called **excited states.** Observe how the energies of the excited states come closer and closer together as n increases.

The electron in a hydrogen atom at room temperature spends most of its time in the ground state. To raise the electron from the ground state ($n = 1$) to the highest possible excited state ($n = \infty$), 13.6 eV of energy must be supplied. Supplying this amount of energy removes the electron from the atom, producing the positive hydrogen ion H^+. The energy needed to remove the electron is called the **ionization energy.** Thus, the Bohr model predicts that the ionization energy of atomic hydrogen is 13.6 eV, in excellent agreement with the experimental value. Example 2 applies the Bohr model to doubly ionized lithium.

EXAMPLE 2

The Bohr model does not apply when more than one electron orbits the nucleus, because the model does not account for the electrostatic forces that one electron exerts on another. For instance, an electrically neutral lithium atom (Li) contains three electrons in orbit around a nucleus that includes three protons ($Z = 3$), and Bohr's analysis does not apply. However, the Bohr model can be used for the doubly charged positive ion of lithium (Li^{2+}) that results when two electrons are removed from the neutral atom, leaving only one electron to orbit the nucleus. Obtain the ionization energy that is needed to remove the remaining electron from Li^{2+}.

SOLUTION

The Bohr energy levels for Li^{2+} are given by Equation 37.13 with

$Z = 3$; $E_n = -(13.6 \text{ eV})(3^2/n^2)$. Therefore, the ground state ($n = 1$) energy is

$$E_1 = -(13.6 \text{ eV}) \frac{3^2}{1^2} = -122 \text{ eV}$$

To remove the electron from Li^{2+}, 122 eV of energy must be supplied: $\boxed{\text{Ionization energy} = 122 \text{ eV}}$. This value for the ionization energy agrees well with the experimental value of 122.4 eV. It is larger than the 13.6 eV for atomic hydrogen because of the greater nuclear charge in Li^{2+}, which attracts the electron more strongly.

PREDICTION OF THE LINE SPECTRA OF THE HYDROGEN ATOM

To determine the wavelengths radiated by the atom, Bohr substituted Equation 37.11 for the energies into Equation 37.4 and used $f = c/\lambda$. He obtained the following result:

$$\frac{1}{\lambda} = \frac{2\pi^2 mk^2 e^4}{h^3 c} (Z^2) \left(\frac{1}{n_f^2} - \frac{1}{n_i^2} \right) \tag{37.14}$$

$$n_i, n_f = 1, 2, 3, \ldots \quad \text{and} \quad n_i > n_f$$

With the known values for h, m, k, e, and c, it can be seen that $2\pi^2 mk^2 e^4/(h^3 c) = 1.097 \times 10^7$ m^{-1}, in agreement with the Rydberg constant R that appears in the empirical Equations 37.1–37.3 for the Lyman, Balmer, and Paschen series of wavelengths. The agreement between the theoretical and experimental values of the Rydberg constant was a major accomplishment of Bohr's theory.

With $Z = 1$ and $n_f = 1$, Equation 37.14 reproduces Equation 37.1 for the Lyman series. Thus, Bohr's model shows that the Lyman series of lines occurs when electrons make transitions from higher energy levels with $n_i = 2, 3, 4, \ldots$ to the first energy level where $n_f = 1$. Figure 37.9 shows these transitions. Notice that when an electron makes a transition from $n_i = 2$ to $n_f = 1$, the longest wavelength photon in the series is emitted, since the energy change is the smallest possible. When an electron makes a transition from the highest level where $n_i = \infty$ to the lowest level where $n_f = 1$, the

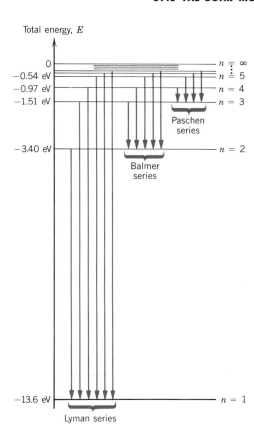

FIGURE 37.9 The Lyman, Balmer, and Paschen series of lines in the hydrogen atom spectrum correspond to transitions that the electron makes between higher and lower energy levels, as indicated here.

shortest wavelength is emitted, since the energy change is the largest possible. Since the higher energy levels are increasingly close together, the lines in the series become more and more crowded together toward the short wavelength limit, as observed. The drawing also shows the energy level transitions for the Balmer and Paschen series. In the Balmer series $n_i = 3, 4, 5, \ldots$, while $n_f = 2$. In the Paschen series $n_i = 4, 5, 6, \ldots$, while $n_f = 3$. The next example deals further with the line spectrum of the hydrogen atom.

EXAMPLE 3

In the line spectrum of atomic hydrogen there is also a group of lines known as the Brackett series. These lines are produced when electrons, excited to high energy levels, make transitions to the $n = 4$ level. Determine (a) the longest wavelength in this series and (b) the wavelength that corresponds to the transition from $n_i = 6$ to $n_f = 4$. (c) Refer to Figure 30.6 and identify the spectral region in which these lines are found.

SOLUTION

(a) The longest wavelength corresponds to the transition that has the smallest energy change. This would be between the $n_i = 5$ and $n_f = 4$ levels in Figure 37.9. Using Equation 37.14 with $Z = 1$, we find that

$$\frac{1}{\lambda} = (1.097 \times 10^7 \text{ m}^{-1})(1^2)\left(\frac{1}{4^2} - \frac{1}{5^2}\right) = 2.47 \times 10^5 \text{ m}^{-1}$$

$$\boxed{\lambda = 4050 \text{ nm}}$$

(b) The calculation here is similar to that above:

$$\frac{1}{\lambda} = (1.097 \times 10^7 \text{ m}^{-1})(1^2)\left(\frac{1}{4^2} - \frac{1}{6^2}\right) = 3.81 \times 10^5 \text{ m}^{-1}$$

$$\boxed{\lambda = 2620 \text{ nm}}$$

(c) According to Figure 30.6, these lines lie in the infrared region of the spectrum.

The various lines in the hydrogen atom spectrum are produced when electrons change from higher to lower energy levels. During these energy-level transitions photons are emitted, and, consequently, the spectral lines are called *emission lines.* Electrons can also make transitions from lower energy levels to higher energy levels, in a process known as absorption. In this case, an atom absorbs a photon that has precisely the energy needed to produce a transition from a lower energy level to a higher energy level. Thus, if photons with a continuous range of wavelengths pass through a gas and then are analyzed with a grating spectroscope, a series of dark *absorption lines* appear in the continuous spectrum. Such absorption lines can be seen in Figure 37.3 (between pages 812 and 813) in the spectrum of the sun, where they are called Fraunhofer lines, after their discoverer. They are due to atoms, located in the outer and cooler layers of the sun, that absorb radiation coming from within the sun. The inner and hotter portion of the sun emits a continuous spectrum of wavelengths, since it is too hot for individual atoms to retain their structures.

The Bohr model provides a great deal of insight into atomic structure. However, this model is now known to be oversimplified and has been superseded by a more detailed picture provided by quantum mechanics and the Schrödinger equation. More will be said about the present view of the atom in Section 37.5.

37.4 DE BROGLIE'S EXPLANATION OF BOHR'S ASSUMPTION ABOUT ANGULAR MOMENTUM

Of all the assumptions Bohr made in his model of the hydrogen atom, perhaps the most puzzling is the one about the angular momentum of the electron ($L_n = mv_n r_n = nh/2\pi$; $n = 1, 2, 3, \ldots$). Why should the angular momentum have only those values that are integer multiples of Planck's constant divided by 2π? In 1923, ten years after Bohr's work, de Broglie pointed out that his own theory for the wavelength of a moving particle could provide an answer to this question.

In de Broglie's way of thinking, the electron in its circular Bohr orbit must be pictured as a particle wave. And like waves traveling on a string, particle waves can lead to standing waves under resonance conditions. Section 22.5 discusses these conditions for a string: Standing waves form when the total distance traveled by a wave down the string and back is one wavelength, two wavelengths, or any integer number of wavelengths. The total distance around a Bohr orbit of radius r is the circumference of the orbit or $2\pi r$. By the same reasoning, then, the condition for standing particle waves for the electron in a Bohr orbit would be

$$2\pi r = n\lambda \qquad n = 1, 2, 3, \ldots$$

where n is the number of whole wavelengths that fit into the circumference of the circle. But according to Equation 36.8, the de Broglie wavelength is $\lambda = h/(mv)$, so that $2\pi r = nh/(mv)$. A rearrangement of this result gives

$$mvr = n\frac{h}{2\pi} \qquad n = 1, 2, 3, \ldots$$

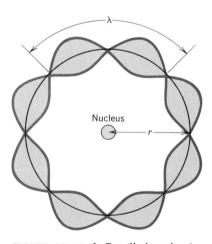

FIGURE 37.10 de Broglie imagined standing particle waves as an explanation for Bohr's angular momentum assumption. Here, a standing particle wave is illustrated on a Bohr orbit where four de Broglie wavelengths fit into the circumference of the orbit ($2\pi r = 4\lambda$).

which is just what Bohr assumed for the angular momentum of the electron. As an example, Figure 37.10 illustrates the standing particle wave on a Bohr orbit for which $2\pi r = 4\lambda$.

De Broglie's explanation of Bohr's assumption about angular momentum emphasizes an important fact, namely, that particle waves play a central role in the structure of the atom. Moreover, the theoretical framework of quantum mechanics provides the basis for determining the wave function Ψ that represents a particle wave. The next section deals with the picture that quantum mechanics gives for atomic structure, a picture that supersedes the Bohr model. In any case, the Bohr model can be applied when a single electron orbits the nucleus, while the theoretical framework of quantum mechanics can be applied, in principle, to atoms that contain an arbitrary number of electrons.

37.5 THE QUANTUM MECHANICAL PICTURE OF THE HYDROGEN ATOM

QUANTUM NUMBERS

The picture of the hydrogen atom that quantum mechanics and the Schrödinger equation provide differs in a number of ways from the Bohr model. The Bohr model uses a single integer number n to identify the various electron orbits and the associated energies. Because this number can have only discrete values, rather than a continuous range of values, n is called a *quantum number.* In contrast, quantum mechanics reveals that four different quantum numbers are needed to describe each state of the hydrogen atom. These four are described below.

1. **The principal quantum number n.** As in the Bohr model, this number determines the total energy of the atom and can have only integer values: $n = 1, 2, 3, \ldots$. In fact, the Schrödinger equation predicts that the energy of the hydrogen atom is identical to that obtained from the Bohr model*: $E_n = -(13.6 \text{ eV})Z^2/n^2$.

2. **The orbital quantum number ℓ.** This number determines the angular momentum of the electron due to its orbital motion. The values that ℓ can have depend on the value of n, and only the following integers are allowed:

$$\ell = 0, 1, 2, \ldots, (n - 1)$$

For instance, if $n = 1$, the orbital quantum number can have only the value $\ell = 0$, but if $n = 4$, the values $\ell = 0, 1, 2,$ and 3 are possible. According to the Schrödinger equation, the magnitude L of the angular momentum of the electron is

$$L = \sqrt{\ell(\ell + 1)} \, \frac{h}{2\pi} \tag{37.15}$$

3. **The magnetic quantum number m_ℓ.** The word "magnetic" is used here because this number has an effect on the energy of the atom when an external magnetic field is applied to the atom. The effect was discovered by the Dutch physicist Pieter Zeeman (1865–1943) and, hence, is known as the *Zeeman effect.* When there is no external magnetic field, m_ℓ does not affect the energy. In either event, the magnetic quantum number determines the component of the angular momentum along a specific direction, which is called the z direction by conven-

---

* This prediction requires that small relativistic effects and small interactions within the atom be ignored, and assumes that the hydrogen atom is not located in an external magnetic field.

TABLE 37.1 Quantum Numbers for the Hydrogen Atom

Name	Symbol	Allowed Values
Principal quantum number	n	1, 2, 3, . . .
Orbital quantum number	ℓ	0, 1, 2, . . . , $(n-1)$
Magnetic quantum number	m_ℓ	0, ± 1, ± 2, . . . , $\pm \ell$
Spin quantum number	m_s	$\pm \frac{1}{2}$

tion. The values that m_ℓ can have depend on the value of ℓ, with only the following positive and negative integers being permitted:

$$m_\ell = 0, \pm 1, \pm 2, \ldots, \pm \ell$$

For example, if the angular momentum quantum number is $\ell = 2$, then the magnetic quantum number can have the values $m_\ell = -2, -1, 0, +1,$ and $+2$. The Schrödinger equation shows that the component L_z of the angular momentum in the z direction is

$$L_z = m_\ell \frac{h}{2\pi} \qquad (37.16)$$

4. **The spin quantum number m_s.** This number is needed because the electron has an intrinsic property called spin angular momentum. Loosely speaking, we can view the electron as spinning while it orbits the nucleus, analogous to the way the earth spins as it moves around the sun. There are only two possible values for the spin quantum number of the electron:

$$m_s = +\tfrac{1}{2} \quad \text{or} \quad m_s = -\tfrac{1}{2}$$

Sometimes the phrases "spin up" and "spin down" are used to refer to the directions of the spin angular momentum that are associated with the two values for m_s.

Table 37.1 summarizes the four quantum numbers that are needed to describe each state of the hydrogen atom. One set of values for $n, \ell, m_\ell,$ and m_s corresponds to one state. As the principal quantum number n increases, the number of possible combinations of the four quantum numbers rises rapidly, as Example 4 illustrates.

EXAMPLE 4

Determine the number of possible states for the hydrogen atom when the principal quantum number is (a) $n = 1$ and (b) $n = 2$.

SOLUTION

(a) Each different combination of the four quantum numbers summarized in Table 37.1 corresponds to a different state. The diagram below shows the possibilities when $n = 1$:

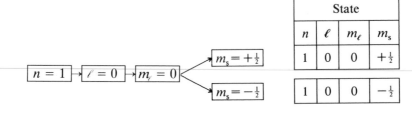

Thus, there are two different states for the hydrogen atom. These two states have the same energy, since they have the same value of n.

(b) When $n = 2$, there are eight possible combinations for the values of n, ℓ, m_ℓ, and m_s, as the diagram below indicates:

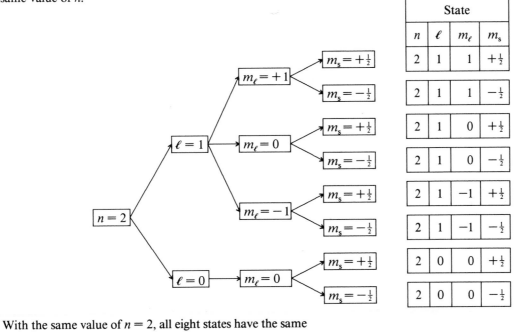

State			
n	ℓ	m_ℓ	m_s
2	1	1	$+\frac{1}{2}$
2	1	1	$-\frac{1}{2}$
2	1	0	$+\frac{1}{2}$
2	1	0	$-\frac{1}{2}$
2	1	-1	$+\frac{1}{2}$
2	1	-1	$-\frac{1}{2}$
2	0	0	$+\frac{1}{2}$
2	0	0	$-\frac{1}{2}$

With the same value of $n = 2$, all eight states have the same energy.

ELECTRON PROBABILITY CLOUDS

According to the Bohr model, the nth orbit is a circle of radius r_n, and every time the position of the electron in this orbit is measured, the electron is found exactly a distance r_n away from the nucleus. However, in the quantum mechanical picture of the atom, the spatial distribution of an electron is quite different than in the Bohr model. Suppose the electron is in a quantum mechanical state for which $n = 1$, and we imagine making a number of measurements of the electron's position with respect to the nucleus. We would find that the position of the electron is uncertain, in the sense that even in a state for which $n = 1$, there is a probability of finding the electron sometimes very near the nucleus, sometimes very far from the nucleus, and sometimes at intermediate locations. The probability is determined by the wave function Ψ, as Section 36.6 discusses. We can make a three-dimensional picture of our findings by marking a dot at each location where the electron is found. After a sufficient number of measurements are made, a picture of the quantum mechanical state emerges. A greater number of dots occur at places where the probability of finding the electron is higher. Figure 37.11 shows the spatial distribution for an electron in a state

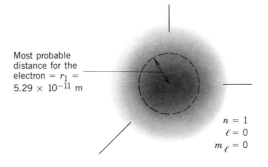

Most probable distance for the electron $= r_1 = 5.29 \times 10^{-11}$ m

$n = 1$
$\ell = 0$
$m_\ell = 0$

FIGURE 37.11 The electron probability cloud for the ground state ($n = 1$, $\ell = 0$, $m_\ell = 0$) of the hydrogen atom.

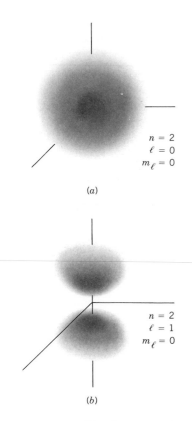

$n = 2$
$\ell = 0$
$m_\ell = 0$

(a)

$n = 2$
$\ell = 1$
$m_\ell = 0$

(b)

FIGURE 37.12 The electron probability clouds for the hydrogen atom when (a) $n = 2$, $\ell = 0$, $m_\ell = 0$ and (b) $n = 2$, $\ell = 1$, $m_\ell = 0$.

characterized by the quantum numbers $n = 1$, $\ell = 0$, and $m_\ell = 0$. This picture is constructed from so many measurements that the individual dots are no longer visible, but have merged to form a kind of probability "cloud" whose density changes gradually from place to place. The dense regions indicate places where the probability of finding the electron is higher, while the less dense regions indicate places where the probability is lower. Also indicated in Figure 37.11 is the radius where quantum mechanics predicts the greatest probability per unit radial distance of finding the $n = 1$ electron. This radius matches exactly the radius of 5.29×10^{-11} m found for the first Bohr orbit.

For a principal quantum number of $n = 2$, the probability clouds are different than for $n = 1$. In fact, more than one cloud shape is possible, because with $n = 2$ the orbital quantum number can be either $\ell = 0$ or $\ell = 1$. While the value of ℓ does not affect the energy of the hydrogen atom, the value does have a significant effect on the shape of the probability clouds. Figure 37.12a shows the cloud for $n = 2$, $\ell = 0$, and $m_\ell = 0$. Part b of the drawing shows that when $n = 2$, $\ell = 1$, and $m_\ell = 0$, the cloud has a two-lobe shape with the nucleus at the center between the lobes. For larger values of n, the probability clouds become increasingly more complex and are spread out over larger volumes of space.

37.6 THE PAULI EXCLUSION PRINCIPLE AND THE PERIODIC TABLE OF THE ELEMENTS

MULTIPLE-ELECTRON ATOMS

Except for hydrogen, all electrically neutral atoms contain more than one electron, the number being given by the atomic number Z of the element. In addition to being attracted by the nucleus, the electrons repel each other. This repulsion contributes to the total energy of a multiple-electron atom. As a result, the one-electron energy expression $[E_n = -(13.6 \text{ eV})Z^2/n^2]$ provided by both the Bohr model and by quantum mechanics does not apply to any other neutral atom except hydrogen. However, the simplest approach for dealing with a multiple-electron atom still uses the four quantum numbers n, ℓ, m_ℓ, and m_s to describe the quantum states of an atom.

Detailed quantum mechanical calculations reveal that the energy level of each state of a multielectron atom depends on both the principal quantum number n and the orbital quantum number ℓ. Figure 37.13 illustrates that the energy generally increases as n increases. Furthermore, for a given n, the energy also increases as ℓ increases, but there are some exceptions, as the drawing indicates.

In a multiple-electron atom, all electrons with the same value of n are said to be in the same **shell**. Electrons with $n = 1$ are in a single shell (sometimes called the K shell), electrons with $n = 2$ are in another shell (the L shell), those with $n = 3$ are in a third shell (the M shell), and so on. Those electrons with the same values for both n and ℓ are often referred to as being in the same **subshell**. The $n = 1$ shell consists of a single $\ell = 0$ subshell. The $n = 2$ shell has two subshells, one with $\ell = 0$ and one with $\ell = 1$. Similarly, the $n = 3$ shell has three subshells.

In the hydrogen atom near room temperature, the electron spends most of its time in the lowest energy level or ground state, namely, in the $n = 1$ shell. Similarly, when an atom contains more than one electron and is near room temperature, the electrons spend most of their time in the lowest energy levels possible. The lowest energy state for an atom is called the **ground state.** However, when a multiple-electron atom is in its ground state, not every electron is crowded into the $n = 1$ shell. The reason the

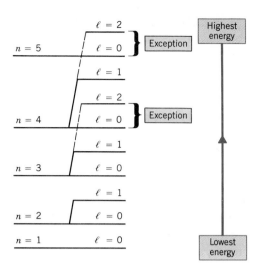

FIGURE 37.13 When more than one electron is present in an atom, the total energy of a given state depends on the principal quantum number n and the orbital quantum number ℓ. Generally, the energy increases with increasing n and, for a fixed n, with increasing ℓ. There are exceptions to the general rule, however, as indicated here. Levels for $n = 6$ and higher are not shown for clarity.

electrons are not all in the same shell is that they obey a principle discovered by the Austrian physicist Wolfgang Pauli (1900–1958).

THE PAULI EXCLUSION PRINCIPLE
No two electrons in an atom can have the same set of values for the four quantum numbers n, ℓ, m_ℓ, and m_s.

For instance, suppose two electrons in an atom have three quantum numbers that are identical: $n = 3$, $m_\ell = 1$, and $m_s = -\frac{1}{2}$. According to the exclusion principle, it is not possible for each to have $\ell = 2$, for example, since each would then have the same four quantum numbers. Each electron must have a different value for ℓ ($\ell = 1$ and $\ell = 2$, for instance) and, consequently, be in a different subshell. With the aid of the Pauli exclusion principle, we can determine which energy levels are occupied by the electrons in an atom in its ground state, as the next example demonstrates.

EXAMPLE 5

Determine which of the energy levels in Figure 37.13 are occupied by the electrons in the ground state of hydrogen (1 electron), helium (2 electrons), lithium (3 electrons), beryllium (4 electrons), and boron (5 electrons).

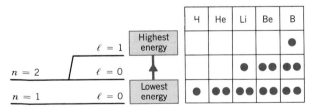

FIGURE 37.14 The electrons (●) in the ground state of an atom fill the available energy levels "from the bottom up," that is, from the lowest to the highest energy, consistent with the Pauli exclusion principle. The ranking of the energy levels in this figure is meant to apply for a given atom, not between one atom and another.

SOLUTION

As the colored dot in Figure 37.14 indicates, the electron in the hydrogen atom (H) is in the $n = 1$, $\ell = 0$ subshell, which has the lowest possible energy. A second electron is present in the helium atom (He) and both electrons can have the quantum numbers $n = 1$, $\ell = 0$, and $m_\ell = 0$. However, in accord with the Pauli exclusion principle, each electron must have a different spin quantum number, $m_s = +\frac{1}{2}$ for one electron and $m_s = -\frac{1}{2}$ for the other. Thus, the drawing shows both electrons in the lowest energy level.

The third electron that is present in the lithium atom (Li) would violate the exclusion principle if it were also in the $n = 1$, $\ell = 0$ subshell, no matter what the value for m_s. Thus, the $n = 1$, $\ell = 0$ subshell is filled when occupied by two electrons. With this level filled, the $n = 2$, $\ell = 0$ subshell becomes the next lowest energy level available and is where the third electron of lithium is found (see Figure 37.14). In the beryllium atom (Be), the fourth electron is in the $n = 2$, $\ell = 0$ subshell, along with the third

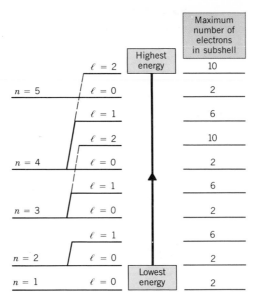

FIGURE 37.15 The maximum number of electrons that the ℓth subshell can hold is $2(2\ell + 1)$.

the middle of the table, the lanthanide series, and the actinide series. The similar chemical properties within a group can be explained on the basis of the configurations of the outer electrons of the elements in the group. Thus, quantum mechanics and the Pauli exclusion principle offer an explanation for the chemical behavior of the atoms.

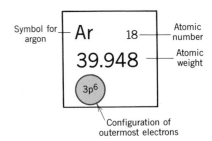

FIGURE 37.16 The entries in the periodic table of the elements often include the ground state configuration of the outermost electrons.

TABLE 37.3 Ground State Electronic Configurations of Atoms

Element	Number of Electrons	Configuration of the Electrons
Hydrogen (H)	1	$1s^1$
Helium (He)	2	$1s^2$
Lithium (Li)	3	$1s^2\,2s^1$
Beryllium (Be)	4	$1s^2\,2s^2$
Boron (B)	5	$1s^2\,2s^2\,2p^1$
Carbon (C)	6	$1s^2\,2s^2\,2p^2$
Nitrogen (N)	7	$1s^2\,2s^2\,2p^3$
Oxygen (O)	8	$1s^2\,2s^2\,2p^4$
Fluorine (F)	9	$1s^2\,2s^2\,2p^5$
Neon (Ne)	10	$1s^2\,2s^2\,2p^6$
Sodium (Na)	11	$1s^2\,2s^2\,2p^6\,3s^1$
Magnesium (Mg)	12	$1s^2\,2s^2\,2p^6\,3s^2$
Aluminum (Al)	13	$1s^2\,2s^2\,2p^6\,3s^2\,3p^1$
Silicon (Si)	14	$1s^2\,2s^2\,2p^6\,3s^2\,3p^2$
Phosphorus (P)	15	$1s^2\,2s^2\,2p^6\,3s^2\,3p^3$
Sulfur (S)	16	$1s^2\,2s^2\,2p^6\,3s^2\,3p^4$
Chlorine (Cl)	17	$1s^2\,2s^2\,2p^6\,3s^2\,3p^5$
Argon (Ar)	18	$1s^2\,2s^2\,2p^6\,3s^2\,3p^6$

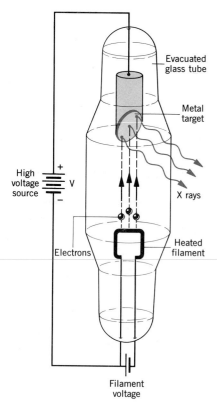

The full periodic table can be found in Appendix H. Group 0, the last column of elements on the right side of the table, consists of the noble gases, such as helium (He), neon (Ne), and argon (Ar). Chemically, these elements are relatively inert, because their outermost electrons form a shell or subshell that is completely full. Being full, the shell or subshell is very stable, not readily forming chemical bonds by accepting electrons from or donating electrons to other elements.

Group I is made up of the alkali metals. Sodium (Na) and potassium (K) are familiar members of this group. These elements are chemically reactive, because they have only a single electron in an outermost s subshell. This electron can be easily lost to other elements in a chemical reaction. Thus, singly charged positive ions, such as the sodium ion Na^+, are often formed by elements in group I.

Group VII consists of the halogens and includes fluorine (F), chlorine (Cl), and bromine (Br). These elements have outermost electrons in a p subshell that is only one electron shy of being full. The halogens are highly reactive. Their chemistry is characterized by reactions in which they accept a single electron from other elements to form a stable, filled subshell. For this reason, the halogens readily form ions such as the chloride ion Cl^-, which carry a single negative charge.

By looking at the other groups (II–VI) in the periodic table, you can see that elements within a group have outermost electrons in either s or p subshells. Within a group, the subshells are filled to the same extent. The transition elements are elements formed when electrons fill out primarily 3d, 4d, 5d, and 6d subshells. The lanthanide series mainly involves completing the 5d and 4f subshells. And finally, the actinide series corresponds primarily to filling out the 5f and 6d subshells.

FIGURE 37.17 In an X-ray tube, electrons are emitted by a heated filament, accelerate through a large potential difference, and strike a metal target. The X rays originate from the interaction between the electrons and the metal target.

37.7 X RAYS

X rays were discovered by Wilhelm K. Roentgen (1845–1923), a Dutch physicist who performed much of his work in Germany. X rays can be produced when electrons, accelerated through a large potential difference, collide with a metal target made from molybdenum or platinum, for example. The target is contained within an evacuated glass tube, as Figure 37.17 shows. A plot of X-ray intensity per unit wavelength versus the wavelength looks similar to Figure 37.18 and consists of sharp peaks or lines superimposed on a broad continuous spectrum. The sharp peaks are called characteristic lines or ***characteristic X rays,*** because they are characteristic of the target material. The broad continuous spectrum is referred to as ***Bremsstrahlung***

FIGURE 37.18 When a molybdenum target is bombarded with electrons that have been accelerated from rest through a potential difference of 45 000 V, the X-ray spectrum shown here is produced. The vertical axis is not drawn to scale.

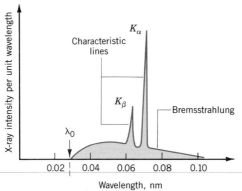

(German for "braking radiation"). Bremsstrahlung X rays are emitted when the electrons decelerate or "brake" upon hitting the target.

In Figure 37.18 the characteristic lines are marked K_α and K_β, because they involve the $n = 1$ or K shell of a metal atom. If an electron with enough energy strikes the target, one of the K-shell electrons can be knocked entirely out of a target atom. An electron in one of the outer shells can then fall into the K shell, and an X-ray photon is emitted in the process. Example 6 shows that large potential differences are needed to operate X-ray tubes, so the electrons impinging on the metal target have sufficient energy to generate the characteristic X rays.

EXAMPLE 6

Strictly speaking, the Bohr model does not apply to multiple-electron atoms, but it can be used to make estimates. Use the Bohr model to estimate the minimum energy that an incoming electron must have to knock a K-shell electron entirely out of an atom in a platinum ($Z = 78$) target of an X-ray tube.

SOLUTION

According to the Bohr model, the energy of a K-shell electron is given by Equation 37.13 with $n = 1$:

$$E_n = -(13.6 \text{ eV})\frac{Z^2}{n^2} = -(13.6 \text{ eV})\frac{77^2}{1^2} = -8.1 \times 10^4 \text{ eV}$$

In this calculation we have used 77 rather than 78 for the value of

Z, to account approximately for the fact that one of the two K-shell electrons shields the other from the effect of one nuclear proton by neutralizing the attractive force of the proton. When striking a platinum target, then, an incoming electron must have at least enough energy to raise the K-shell electron from an energy level of -8.1×10^4 eV up to the 0-eV level that corresponds to a very large distance from the nucleus. Therefore, the minimum energy for an incoming electron is $\boxed{8.1 \times 10^4 \text{ eV}}$.

One electron volt is the kinetic energy acquired when an electron accelerates from rest through a potential difference of one volt. Thus, the answer here means that a potential difference of 81 000 V must be applied to the X-ray tube.

The K_α line in Figure 37.18 arises when an electron in the $n = 2$ level falls into the vacancy that the impinging electron has created in the $n = 1$ level. Similarly, the K_β line arises when an electron in the $n = 3$ level falls to the $n = 1$ level. Example 7 determines an estimate for the K_α wavelength of platinum.

EXAMPLE 7

Use the Bohr model to estimate the wavelength of the K_α line in the X-ray spectrum of platinum.

SOLUTION

This example is very similar to Example 3, which deals with the emission line spectrum of the hydrogen atom. As in that example, we use Equation 37.14, this time with the initial value of n being $n_i = 2$ and the final value being $n_f = 1$. As in Example 6, a value of 77 rather than 78 should be used for Z, to account approximately for the shielding effect of the single K-shell electron in canceling out the attraction of one nuclear proton:

$$\frac{1}{\lambda} = \underbrace{\frac{2\pi^2 m k^2 e^4}{h^3 c}}_{\text{Rydberg constant}} (Z^2) \left(\frac{1}{n_f^2} - \frac{1}{n_i^2}\right)$$

$$\frac{1}{\lambda} = (1.097 \times 10^7 \text{ m}^{-1})(77^2)\left(\frac{1}{1^2} - \frac{1}{2^2}\right)$$

$$= 4.9 \times 10^{10} \text{ m}^{-1}$$

$$\boxed{\lambda = 2.0 \times 10^{-11} \text{ m}}$$

This answer is close to an experimental value of 1.9×10^{-11} m.

Another interesting feature of the X-ray spectrum in Figure 37.18 is the sharp cutoff that occurs at a wavelength of λ_0 on the short wavelength side of the Bremsstrahlung. This cutoff wavelength is independent of the target material, but depends on the energy of the impinging electrons. An impinging electron cannot give up any more than all of its kinetic energy (KE) when decelerated by the metal target in an X-ray tube. Thus, at most, an emitted X-ray photon can have an energy equal to KE and a frequency given by Equation 36.2 as $f = (\text{KE})/h$, where h is Planck's constant.

But the kinetic energy acquired by an electron in accelerating from rest through a potential difference V is eV, according to earlier discussions in Section 24.2; V is the potential difference applied across the X-ray tube. Thus, the maximum photon frequency is $f_0 = (eV)/h$. Since $f_0 = c/\lambda_0$, a maximum frequency corresponds to a minimum wavelength, which is the cutoff wavelength λ_0:

$$\lambda_0 = \frac{hc}{eV} \tag{37.17}$$

Figure 37.18, for instance, assumes a potential difference of 45 000 V, which corresponds to a cutoff wavelength of

$$\lambda_0 = \frac{(6.63 \times 10^{-34} \text{ J} \cdot \text{s})(3.00 \times 10^8 \text{ m/s})}{(1.60 \times 10^{-19} \text{ C})(45\ 000 \text{ V})} = 2.8 \times 10^{-11} \text{ m}$$

37.8 THE LASER

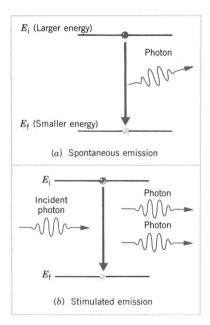

FIGURE 37.19 (a) Spontaneous emission of a photon occurs in a completely unprovoked manner at random moments, with the photon departing in a random direction when the electron (●) makes a transition from a higher to a lower energy level. (b) Stimulated emission of a photon occurs when an incoming photon with the correct energy induces an electron (●) to change energy levels, the emitted photon traveling in the same direction as the incoming photon.

When an electron makes a transition from a higher energy state to a lower energy state, a photon is emitted. The emission process can be one of two types, spontaneous emission or stimulated emission. In *spontaneous emission* (see Figure 37.19a), the photon is emitted spontaneously, in a random direction, without external provocation. In *stimulated emission* (see Figure 37.19b), an incoming photon induces or stimulates the electron to change energy levels. To produce stimulated emission, however, the incoming photon must have an energy that exactly matches the difference between the energies of the two levels, namely, $E_i - E_f$. Stimulated emission is similar to a resonance process, in which the incoming photon "jiggles" the electron at just the frequency to which it is particularly sensitive and, thus, causes the change between levels. This frequency is given by Equation 37.4 as $f = (E_i - E_f)/h$. The operation of lasers depends on stimulated emission.

Stimulated emission has three important features. First, one photon goes in and two photons come out. In this sense, the process amplifies the number of photons. In fact, this is the origin of the word "laser," which is an acronym for **L**ight **A**mplification by the **S**timulated **E**mission of **R**adiation. Second, the emitted photon travels in the same direction as the incoming photon that stimulates the electron to make the energy level transition. And third, the emitted photon is exactly in step with or has the same phase as the incoming photon. In other words, the two electromagnetic waves that these two photons represent are coherent and are locked in step with one another. In contrast, two photons emitted by the filament of an incandescent light bulb are not coherent, since one does not stimulate the emission of the other. They are emitted independently.

While stimulated emission plays a pivotal role in a laser, other factors are also important. For instance, an external source of energy must be provided to excite electrons into higher energy levels. The energy can be provided in a number of ways, including intense flashes of ordinary light and high-voltage discharges. If sufficient energy is delivered to the atoms, more electrons will be excited to a higher energy level than remain in a lower energy level, a condition known as a *population inversion.* Figure 37.20 compares the normal way energy levels are populated with a population inversion created artificially. The population inversions used in lasers involve a higher energy state that is *metastable,* in the sense that electrons remain in it for a much longer period of time than they do in an ordinary excited state (10^{-3} s versus

A beam of light from a laser.

10^{-8} s, for example). The requirement of a metastable higher energy state is essential in a laser, so that there is more time to enhance the population inversion.

Figure 37.21 shows the widely used helium/neon laser. To sustain the necessary population inversion a high voltage is discharged across a low-pressure mixture of 15% helium and 85% neon contained in a glass tube. The laser process begins when an atom, via spontaneous emission, emits a photon parallel to the axis of the tube. This photon, via stimulated emission, causes another atom to emit a photon parallel to the tube axis. These two photons, in turn, stimulate two more atoms, yielding four photons. Four yield eight, and so on, in a kind of avalanche effect. To ensure that more and more photons are created by stimulated emission, both ends of the tube are silvered to form mirrors that reflect the photons back and forth through the helium/neon mixture. One end is only partially silvered, however, so that some of the

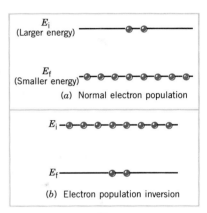

E_i (Larger energy)

E_f (Smaller energy)

(a) Normal electron population

E_i

E_f

(b) Electron population inversion

FIGURE 37.20 (a) In a normal situation at room temperature, most of the electrons in atoms are found in a lower or ground state energy level. (b) If an external energy source is provided to excite electrons into a higher energy level, a population inversion can be created, in which more electrons are in the higher level than in a lower level.

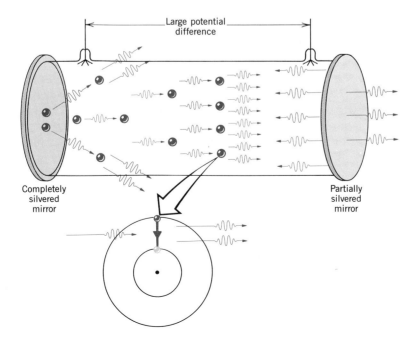

Large potential difference

Completely silvered mirror

Partially silvered mirror

FIGURE 37.21 A schematic drawing of a helium/neon laser.

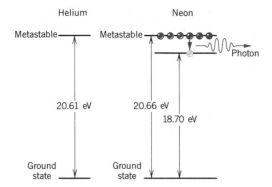

FIGURE 37.22 These energy levels are involved in the operation of a helium/neon laser.

photons can escape from the tube to form the laser beam. When the stimulated emission in a laser involves only a single pair of energy levels, the output beam has a single frequency or wavelength; that is, the radiation in the laser beam is monochromatic.

In addition to being monochromatic, a laser beam is also exceptionally narrow. The width is determined by the size of the opening through which the beam exits, and very little spreading-out occurs, except that due to diffraction around the edges of the opening. A laser beam does not spread much, because any photons emitted at an angle with respect to the tube axis are quickly reflected out the sides of the tube by the silvered ends (see Figure 37.21). These ends are carefully arranged to be perpendicular to the tube axis. Since all of the power in a laser beam can be confined to a narrow region, the intensity, or power per unit area, can be quite large.

Figure 37.22 shows the pertinent energy levels for a helium/neon laser. By coincidence, helium and neon have nearly identical metastable higher energy states, respectively located 20.61 and 20.66 eV above the ground state. The high-voltage discharge across the gaseous mixture excites electrons in helium atoms to the 20.61-eV state. Then, when an excited helium atom collides inelastically with a neon atom, the 20.61 eV of energy is given to an electron in the neon atom, along with 0.05 eV of kinetic energy from the moving atoms. As a result, the electron in the neon atom is raised to the 20.66-eV state. In this fashion, a population inversion is sustained in the neon, relative to an energy level that is 18.70 eV above the ground state. In producing the laser beam, stimulated emission causes electrons in neon to drop from the 20.66-eV level to the 18.70-eV level. The energy change of 1.96 eV corresponds to a wavelength of 633 nm, which is in the red region of the visible spectrum.

The helium/neon laser is not the only kind of laser. There are many different types, including the ruby laser, the argon-ion laser, the carbon dioxide laser, the gallium arsenide solid-state laser, and chemical dye lasers. Depending on the type and whether the laser operates continuously or in pulses, the available beam power ranges from milliwatts to megawatts. Since lasers provide coherent monochromatic electromagnetic radiation that can be confined to an intense narrow beam, they are useful in a wide variety of situations. Today they are used to reproduce music in compact disc players, to weld parts of automobile frames together, to perform delicate eye surgery, to measure distances accurately in surveying, to transmit telephone conversations and other forms of communication over long distances, and to study molecular structure. These are only a few of the applications that have become commonplace since the laser was invented in 1960.

SUMMARY

The idea of a **nuclear atom** originated in 1911, as a result of experiments performed by Ernest Rutherford in which α particles were scattered by a thin metal foil. The phrase "nuclear atom" refers to the fact that an atom consists of a small, positively charged nucleus surrounded at relatively large distances by a number of electrons, whose negative charge equals the positive nuclear charge when the atom is electrically neutral.

A **line spectrum** is a series of discrete electromagnetic wavelengths emitted by the atoms of a low-pressure gas that is subjected to a sufficiently high potential difference. Certain groups of discrete wavelengths are referred to as "series." The line spectrum of atomic hydrogen includes the **Lyman series,** the **Balmer series,** and the **Paschen series** of wavelengths.

The Bohr model applies to atoms that have only a single electron orbiting a nucleus containing Z protons. This model assumes that the electron exists in circular orbits that are called **stationary orbits,** because the electron does not radiate electromagnetic waves while in them. According to this model, a photon is emitted only when an electron changes from a higher energy orbit to a lower energy orbit. The model also assumes that the orbital angular momentum L_n of the electron can only have values that are integer multiples of Planck's constant divided by 2π: $L_n = n(h/2\pi)$; $n = 1, 2, 3, \ldots$. With the assumptions above, it can be shown that the nth Bohr orbit has a radius of $r_n = (5.29 \times 10^{-11}$ m$)(n^2/Z)$ and that the total energy associated with this orbit is $E_n = -(13.6$ eV$)(Z^2/n^2)$. The **ionization energy** is the energy needed to remove an electron completely from an atom. The Bohr model predicts that the wavelengths comprising the line spectrum emitted by an atom are given by Equation 37.14.

Quantum mechanics describes the hydrogen atom in terms of four quantum numbers: (1) **the principal quantum number** n, which can have the integer values $n = 1, 2, 3, \ldots$; (2) **the orbital quantum number** ℓ, which can have the integer values $\ell = 0, 1, 2, \ldots , (n - 1)$; (3) **the magnetic quantum number** m_ℓ, which can have the positive and negative integer values $m_\ell = 0, \pm 1, \pm 2, \ldots , \pm \ell$; and (4) **the spin quantum number** m_s, which, for an electron, can be either $m_s = +\frac{1}{2}$ or $m_s = -\frac{1}{2}$. According to quantum mechanics, an electron does not reside in a circular orbit but, rather, has some probability of being found at various distances from the nucleus.

The Pauli exclusion principle states that no two electrons in an atom can have the same set of values for the four quantum numbers n, ℓ, m_ℓ, and m_s. This principle determines the way in which the electrons in multiple-electron atoms are distributed into shells (defined by the value of n) and subshells (defined by the values of n and ℓ). Table 37.2 summarizes the conventional notation for atomic subshells. The arrangement of the periodic table of the elements is related to the exclusion principle.

X rays are electromagnetic waves emitted when high-energy electrons strike a metal target contained within an evacuated glass tube. The emitted X-ray spectrum of wavelengths consists of sharp "peaks" or "lines," called **characteristic X rays,** superimposed on a broad continuous range of wavelengths called **Bremsstrahlung.** The minimum wavelength, or cutoff wavelength, of the Bremsstrahlung is determined by the maximum kinetic energy of the electrons striking the target in the X-ray tube.

A **laser** is a device that generates electromagnetic waves via a process known as **stimulated emission.** In this process, one photon stimulates the production of another photon, by causing an electron in an atom to fall from a higher energy level to a lower energy level. Because of this mechanism of photon production, the electromagnetic waves generated by a laser are coherent and may be confined to a very narrow beam.

QUESTIONS

1. At room temperature, most of the atoms of atomic hydrogen contain electrons that are in the ground state or $n = 1$ energy level. A tube is filled with atomic hydrogen. Electromagnetic radiation with a continuous spectrum of wavelengths, including those in the Lyman, Balmer, and Paschen series, enters one end of this tube and leaves the other end. The exiting radiation is found to contain absorption lines. To which one (or more) of the series do the wavelengths of these absorption lines correspond? Explain.

2. When the outermost electron in an atom is in an excited state, the atom is more easily ionized than when the outermost electron is in the ground state. Why?

3. For a given value of the principal quantum number n, is it possible for the electron to have zero orbital angular momentum L in (a) the Bohr model and (b) quantum mechanics? Give your reasoning in each case.

4. In the Bohr model for the hydrogen atom, the closer the electron is to the nucleus, the smaller is the total energy of the atom. Is this also true in the quantum mechanical picture of the hydrogen atom? Justify your answer.

5. Consider two different hydrogen atoms. The electron in each atom is in an excited state. Is it possible for the two electrons to have different energies but the same orbital angular momentum (a) according to the Bohr model and (b) according to quantum mechanics? Is it possible for the two electrons to have different orbital angular momenta but the same energy (c) according to the Bohr model and (d) according to quantum mechanics? Account for your answer in each case.

6. Can a 5g subshell contain (a) 22 electrons and (b) 17 electrons? Why?

7. Explain why you would not expect hydrogen and helium atoms in their ground state to emit characteristic X rays.

8. The drawing shows the X-ray spectra produced by an X-ray tube when the tube is operated at two different potential differences. Explain why the characteristic lines occur at the same wave-

lengths in the two spectra, while the cutoff wavelength λ_0 shifts to the right when a smaller potential difference is used to operate the tube.

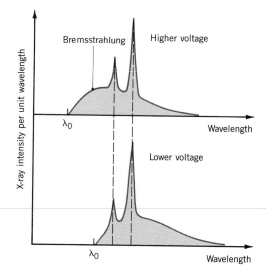

9. The short wavelength side of the spectrum of X rays ends abruptly at a cutoff wavelength λ_0 (see Figure 37.18). Does this cutoff wavelength depend on the target material used in the X-ray tube? Give your reasoning.

10. Explain why a laser beam focused to a small spot can cut through a piece of metal.

PROBLEMS

In working these problems, ignore relativistic effects.

Section 37.1 Rutherford Scattering and the Nuclear Atom

1. The nucleus of the hydrogen atom has a radius of about 1×10^{-15} m. The electron is normally at a distance of about 5.3×10^{-11} m from the nucleus. Assuming the hydrogen atom is a sphere with a radius of 5.3×10^{-11} m, find (a) the volume of the atom, (b) the volume of the nucleus, and (c) the percentage of the volume of the atom that is occupied by the nucleus.

2. There are Z protons in the nucleus of an atom, where Z is the atomic number of the element. An α particle carries a charge of $+2e$. In a scattering experiment, an α particle, heading directly toward a nucleus in a metal foil, will come to a halt when all the particle's kinetic energy is converted to electric potential energy. In such a situation, how close will an α particle with a kinetic energy of 5.0×10^{-13} J come to (a) a silver nucleus ($Z = 47$) and (b) a gold nucleus ($Z = 79$)?

3. The mass of an α particle is 6.64×10^{-27} kg. An α particle used in a scattering experiment has a kinetic energy of 7.00×10^{-13} J. What is the de Broglie wavelength of the particle?

* 4. The nucleus of an aluminum atom contains 13 protons and has a radius of 3.6×10^{-15} m. How much work (in electron volts) must be done to bring a proton from infinity, where it is at rest, to the "surface" of an aluminum nucleus?

Section 37.2 Line Spectra, Section 37.3 The Bohr Model of the Hydrogen Atom

5. If the line with the longest wavelength in the Balmer series for atomic hydrogen is counted as the first line, what is the wavelength of the third line?

6. On a piece of graph paper, make a copy of Figure 37.8, which shows the energy level diagram for the hydrogen atom ($Z = 1$) determined from Equation 37.13. On the same piece of graph paper, along side the hydrogen atom drawing, draw to scale the energy level diagram that Equation 37.13 predicts for (a) singly ionized helium He$^+$ ($Z = 2$), and for (b) doubly ionized lithium Li^{2+} ($Z = 3$). In these drawings include only the first two energy levels for hydrogen, the first four levels for He$^+$, and the first six levels for Li^{2+}. Which levels have the same energy in all three diagrams?

7. In the line spectrum of atomic hydrogen there is also a group of lines known as the Pfund series. These lines are produced when electrons, excited to high energy levels, make transitions to the $n = 5$ level. Determine (a) the longest wavelength and (b) the shortest wavelength in this series. (c) Refer to Figure 30.6 and identify the region of the electromagnetic spectrum in which these lines are found.

8. Determine the ionization energy that is needed to remove the remaining electron from a singly ionized helium atom He^+ ($Z = 2$). Express your answer in electron volts and in joules.

9. Find the energy of the photon that is emitted when the electron in a hydrogen atom undergoes a transition from the $n = 7$ energy level to produce a line in the Paschen series. Express your answer in electron volts and in joules.

10. What is the radius for the $n = 5$ Bohr orbit in a doubly ionized lithium atom Li^{2+} ($Z = 3$)?

11. Using the Bohr model, compare the nth orbit of a triply ionized beryllium atom Be^{3+} ($Z = 4$) to the nth orbit of a hydrogen atom (H) by calculating the ratio (Be^{3+}/H) of the following quantities: (a) the energies, (b) the radii, and (c) the angular momenta.

12. The electron in a hydrogen atom is in the first excited state, when the electron acquires an additional 2.86 eV of energy. What is the quantum number of the state into which the electron moves?

* 13. For atomic hydrogen, the Paschen series of lines occurs when $n_f = 3$, while the Brackett series occurs when $n_f = 4$ in Equation 37.14. Using this equation, show that the ranges of wavelengths in these two series overlap.

* 14. In the Bohr model, Equation 37.12 or 37.13 gives the total energy (kinetic plus potential). In a certain Bohr orbit, the total energy is -4.90 eV. For this orbit, determine the kinetic energy of the electron and the value of the electric potential energy.

* 15. In an unidentified ionized atom, only one electron moves about the nucleus. The radius of the $n = 3$ Bohr orbit is 2.38×10^{-10} m. What is the energy (in electron volts) of the $n = 7$ orbit for this atom?

* 16. The Bohr model can be applied to singly ionized helium He^+ ($Z = 2$). Using this model, consider the series of lines that is produced when the electron makes a transition from higher energy levels into the $n_f = 4$ level. Some of the lines in this series lie in the visible region of the spectrum (380 nm–750 nm). What are the values of n_i for the energy levels from which the electron makes the transitions corresponding to these lines?

**17. (a) Derive an expression for the velocity of the electron in the nth Bohr orbit, in terms of Z, n, and the constants k, e, and h. For the hydrogen atom, determine the velocity in (b) the $n = 1$ orbit and (c) the $n = 2$ orbit. (d) Generally, when speeds are less than one-tenth the speed of light, the effects of special relativity can be ignored. Do the speeds found in (b) and (c) justify ignoring relativistic effects in the Bohr model?

**18. A diffraction grating is used in the first order to separate the wavelengths in the Balmer series of atomic hydrogen (Section 34.7 discusses diffraction gratings). The grating and an observation screen are separated by a distance of 75.0 cm, as Figure 34.30 illustrates. You may assume that θ is small, so $\sin \theta \approx \theta$ when radian measure is used for θ. How many lines per centimeter should the grating have, so the longest and the next-to-the-longest wavelengths in the series are separated by 5.00 cm on the screen?

Section 37.5 The Quantum Mechanical Picture of the Hydrogen Atom

19. Write down the eighteen possible sets of the four quantum numbers that exist when the principal quantum number is $n = 3$.

20. The principal quantum number for an electron in an atom is $n = 6$, while the magnetic quantum number is $m_\ell = 2$. What possible values for the orbital quantum number ℓ could this electron have?

21. An electron in an atom has a value for the magnetic quantum number of $m_\ell = 4$. What are the *minimum* values that (a) the orbital quantum number and (b) the principal quantum number can have?

* 22. For an electron in a hydrogen atom, the z component of the angular momentum has a *maximum* value of $L_z = 2.11 \times 10^{-34}$ J·s. Find the three smallest possible values for the total energy (in electron volts) that this atom could have.

* 23. For the hydrogen atom, the Bohr model and quantum mechanics both give the same value for the energy of the nth state. However, they do not give the same value for the orbital angular momentum L. (a) For $n = 1$, determine the values of L (in units of $h/2\pi$) predicted by the Bohr model and quantum mechanics. (b) Repeat part (a) for $n = 2$, noting that quantum mechanics permits more than one value of ℓ when the electron is in the $n = 2$ state.

Section 37.6 The Pauli Exclusion Principle and the Periodic Table of the Elements

24. In the style indicated in Table 37.3, write down the ground state electronic configuration of (a) calcium Ca ($Z = 20$) and (b) arsenic As ($Z = 33$). For arsenic, note from Figure 37.15 that the 4s subshell fills before the 3d subshell.

25. Figure 37.15 was constructed using the Pauli exclusion principle and indicates that the $n = 1$ shell holds 2 electrons, the $n = 2$ shell holds 8 electrons, and the $n = 3$ shell holds 18 electrons. These numbers can be obtained by adding the numbers given in the figure for the subshells contained within a given shell. How many electrons can be put into (a) the $n = 4$ shell and (b) the $n = 5$ shell, neither of which is completely shown in the figure?

26. When an electron makes a transition between energy levels of an atom, there are no restrictions on the initial and final values of the principal quantum number n. According to quantum mechanics, however, there is a rule that restricts the initial and final values of the orbital quantum number ℓ. This rule is called a *selection rule* and states that $\Delta \ell = \pm 1$. In other words, when an electron makes a transition between energy levels, the value of ℓ can only increase or decrease by one. The value of ℓ may not remain the same or increase or decrease by more than one. According to this rule, which of the following energy level transitions are allowed: (a) 2s → 1s, (b) 2p → 1s, (c) 4p → 2p, (d) 4s → 2p, (e) 3d → 3s, and (f) 5d → 3d?

Section 37.7 X Rays

27. Suppose an X-ray machine in a doctor's office uses a potential difference of 61 kV to operate the X-ray tube. What is the shortest X-ray wavelength emitted by this machine?

28. By using the Bohr model to estimate the atomic number Z, decide which element is likely to emit a K_α X ray with a wavelength of 4.5×10^{-9} m.

29. Molybdenum has an atomic number of $Z = 42$. Using the Bohr model, estimate the wavelength of the K_α X ray.

* 30. An X-ray tube is being operated at a potential difference of 45.0 kV. What is the Bremsstrahlung wavelength that corresponds to a 25.0% loss of kinetic energy when an electron collides with the metal target in the tube?

Section 37.8 The Laser

31. A laser is used in eye surgery to weld a detached retina back into place. The wavelength of the laser beam is 514 nm, while the power is 2.0 W. During surgery, the laser beam is turned on for 0.10 s. During this time, how many photons are emitted by the laser?

32. In the helium/neon laser, there is an energy difference of 1.96 eV between the levels that participate in stimulated emission. Verify, by performing a calculation, that the laser produces a wavelength of 633 nm.

33. A carbon dioxide laser produces a wavelength of 1.06×10^{-5} m. A semiconductor laser produces a wavelength of 7.90×10^{-7} m. (a) Which laser produces the more energetic photons? Explain. (b) By what factor are they more energetic?

* 34. In a laser, the process of stimulated emission produces a kind of photon avalanche, one photon producing two photons, two producing four, four producing eight, and so on. The drawing illustrates three steps in this developing avalanche. What is the minimum number of such steps needed to produce at least a billion photons?

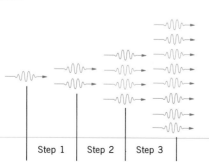

Step 1   Step 2   Step 3

35. What is the minimum potential difference that must be applied to an X-ray tube to knock a K-shell electron completely out of an atom in a copper ($Z = 29$) target? Use the Bohr model as needed.

36. It is possible to use electromagnetic radiation to ionize atoms. To do so, the atoms must absorb the radiation, the photons of which must have enough energy to remove an electron from an atom. What is the longest radiation wavelength that can be used to ionize the ground state hydrogen atom?

37. Write down the fourteen sets of the four quantum numbers that correspond to the electrons in a completely filled 4f subshell.

38. The K_β characteristic X-ray line for tungsten has a wavelength of 1.84×10^{-11} m. What is the difference in energy between the two energy levels that gives rise to this line? Express the answer in joules and in electron volts.

39. A 790-kg synchronous communications satellite has a period of one day. The radius of the orbit is 4.23×10^7 m. Suppose that Bohr's assumption about the angular momentum of the electron in orbit about the nucleus applies to this satellite. What is the quantum number n for the orbit of the satellite?

* 40. The total orbital angular momentum of the electron in a hydrogen atom has a magnitude of $L = 3.66 \times 10^{-34}$ J·s. What values can the angular momentum component L_z have?

* 41. In the line spectrum emitted by doubly ionized lithium atoms Li^{2+} ($Z = 3$), the shortest wavelength in one series of lines is 162.1 nm. This line is produced when an electron makes a transition from an initial energy level to a final energy level whose principal quantum number is n_f. What is the value of n_f?

* 42. Consider a particle of mass m that can exist only between $x = 0$ and $x = +L$ on the x axis. We could say that this particle is confined to a "box" of length L. In this situation, imagine the standing de Broglie waves that can fit into the box. For example, the drawing shows the first three possibilities. Note in this picture

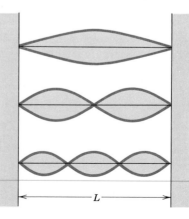

L

that there are either one, two, or three half-wavelengths that fit into the distance L. Use Equation 36.8 for the de Broglie wavelength of a particle and derive an expression for the allowed energies (only kinetic energy) that the particle can have. This expression involves m, L, Planck's constant, and a quantum number n that can have only the values 1, 2, 3, ...

* **43.** A wavelength of 410.2 nm is emitted by the hydrogen atoms in a high-voltage discharge tube. What are the initial and final values of the quantum number n for the energy level transition that produces this wavelength? (*Hint: The wavelength lies in the visible region of the electromagnetic spectrum.*)

**44. (a) Derive an expression for the time it takes the electron in the nth Bohr orbit to make one complete revolution around the nucleus. Express your answer in terms of Z, n, and the constants k, e, h, and m. For a hydrogen atom, determine this time for the (b) $n = 1$ orbit and (c) the $n = 2$ orbit.

**45. A certain species of ionized atoms produces an emission line spectrum according to the Bohr model, but the number of protons Z in the nucleus is unknown. A group of lines in the spectrum forms a series in which the shortest wavelength is 22.79 nm and the longest wavelength is 41.02 nm. Find the next-to-the-longest wavelength in the series of lines.

CHAPTER 38

Nuclear Physics and Radioactivity

Neutrinos are subatomic particles that travel at or near the speed of light. This neutrino detector is a tank of ultra-pure water that is located 700 meters beneath Lake Erie and is lined with 2048 light sensors. When a neutrino from outer space enters the tank and interacts with the protons and electrons that make up the water, a brief flash of light is emitted that is picked up by the light sensors. The diver here is inspecting the sensors.

38.1 NUCLEAR STRUCTURE

The nucleus of an atom consists of protons and neutrons, which are collectively referred to as *nucleons.* The *neutron* was discovered in 1932 by the English physicist James Chadwick (1891–1974). As Table 38.1 indicates, the neutron carries no electrical charge and has a mass that is slightly larger than that of a proton.

The number of protons in the nucleus is different in different elements and is given by the *atomic number* Z. In an electrically neutral atom, the number of nuclear protons equals the number of electrons in orbit around the nucleus. The number of neutrons in the nucleus is N. The total number of protons and neutrons is referred to as the *atomic mass number* A, because the total nuclear mass is approximately equal to A times the mass of a single nucleon:

$$A = Z + N \qquad (38.1)$$

Sometimes, A is also called the *nucleon number.* A shorthand notation is often used to specify Z and A along with the chemical symbol for the element. For instance, the

TABLE 38.1 Properties of Particles in the Atom

Particle	Electric Charge (C)	Rest Mass	
		Kilograms (kg)	Atomic Mass Units (u)
Electron	-1.60×10^{-19}	$9.109\ 390 \times 10^{-31}$	$5.485\ 799 \times 10^{-4}$
Proton	$+1.60 \times 10^{-19}$	$1.672\ 623 \times 10^{-27}$	$1.007\ 276$
Neutron	0	$1.674\ 929 \times 10^{-27}$	$1.008\ 665$

nuclei of all naturally occurring aluminum atoms have $A = 27$, and the atomic number for aluminum is $Z = 13$. In shorthand notation, then, the aluminum nucleus is specified as $^{27}_{13}\text{Al}$. The number of neutrons in an aluminum nucleus is $N = 14$. In general, for an element whose chemical symbol is X, the symbol for the nucleus is ^A_ZX. For a proton the symbol is ^1_1H, since the proton is the nucleus of a hydrogen atom. For a neutron the symbol is ^1_0n.

Nuclei that contain the same number of protons, but a different number of neutrons are known as *isotopes.* Carbon, for example, occurs in nature in two stable forms. In most carbon atoms (98.90%), the nucleus is the $^{12}_6\text{C}$ isotope, consisting of six protons and six neutrons. A small fraction (1.10%), however, contains nuclei that have six protons and seven neutrons, namely, the $^{13}_6\text{C}$ isotope. The percentages given above are the natural abundances of the isotopes. The atomic masses in the periodic table are average atomic masses, taking into account the abundances of the various isotopes.

Figure 38.1 illustrates that protons and neutrons are clustered together in the nucleus to form an approximately spherical region. Experiment shows that the radius r of the nucleus depends on the atomic mass number A and is given approximately in meters by

$$r \approx (1.2 \times 10^{-15}\ \text{m})A^{1/3} \tag{38.2}$$

The radius of the aluminum nucleus ($A = 27$), for example, is $r \approx (1.2 \times 10^{-15}\ \text{m})27^{1/3} = 3.6 \times 10^{-15}\ \text{m}$. Equation 38.2 indicates that r^3 is proportional to the

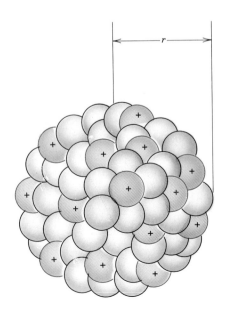

FIGURE 38.1 The nucleus is approximately spherical (radius $= r$) and contains protons ($\oplus$) clustered closely together with neutrons ($\bigcirc$).

nucleon number A. But $\frac{4}{3}\pi r^3$ is the volume of a sphere, so the volume of the nucleus is proportional to the number of nucleons it contains, the nucleons being clustered together as incompressible pieces of matter. The nucleon number, in turn, is nearly proportional to the total nuclear mass, since all nucleons have roughly the same mass. Thus, the volume and mass of a nucleus are nearly proportional, and the nuclear mass per unit volume, or the nuclear density, is approximately the same for all nuclei.

38.2 THE STRONG NUCLEAR FORCE AND THE STABILITY OF THE NUCLEUS

Two positive charges that are as close together as they are in a nucleus repel one another with a very strong electrostatic force. What, then, keeps the nucleus from flying apart? Clearly, some kind of attractive force must hold the nucleus together, since many kinds of naturally occurring atoms contain stable nuclei. The gravitational force of attraction between nucleons is too weak to counteract the repulsive electric force, so it must be recognized that a different type of force exists within the nucleus. This force is the *strong nuclear force* and is one of only four fundamental forces that have been discovered, fundamental in the sense that all forces in nature can be explained in terms of these four. We have already encountered two of the other three fundamental forces, the gravitational force and the electromagnetic force. The remaining force will be mentioned in Section 38.5.

The strong nuclear force is not yet completely characterized, but some of its features are known. The strong nuclear force is independent of electric charge. At a given separation distance, the same nuclear force of attraction exists between two protons, two neutrons, or between a proton and a neutron. The range of action of the strong nuclear force is extremely short, with the force of attraction being very strong when two nucleons are as close as 10^{-15} m and essentially zero at larger distances. In contrast, the electric force between two protons decreases to zero only gradually as the separation distance increases to large values and, therefore, has a relatively long range of action.

The limited range of action of the strong nuclear force plays an important role in the stability of the nucleus. Stability depends on a balance of forces. For a nucleus to be stable, the electrostatic repulsion between the protons must be balanced by the attraction between the nucleons due to the strong nuclear force. But one proton repels all other protons within the nucleus, since the electrostatic force has such a large range of action. In contrast, a proton or a neutron attracts only its nearest neighbors via the strong nuclear force, whose range is so limited. As the number Z of protons in the nucleus increases under these conditions, the number N of neutrons has to increase even more, if stability is to be maintained. Figure 38.2 shows a plot of N versus Z for naturally occurring elements that have stable nuclei. For reference, the plot also includes the straight line that represents the condition $N = Z$. With few exceptions, the points representing stable nuclei fall above this reference line, reflecting the fact that the number of neutrons becomes greater than the number of protons as the atomic number Z increases.

As more and more protons occur in a nucleus, there comes a point when a balance of repulsive and attractive forces cannot be achieved by an increased number of neutrons. Eventually, the limited range of action of the strong nuclear force prevents extra neutrons from balancing the long range electric repulsion of extra protons. The stable nucleus with the largest number of protons ($Z = 83$) is that of bismuth, $^{209}_{83}\text{Bi}$.

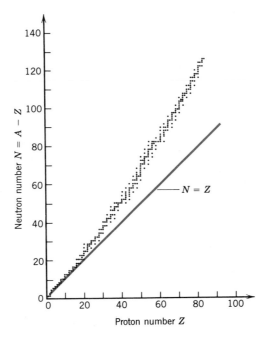

FIGURE 38.2 With few exceptions, the naturally occurring stable nuclei have a number N of neutrons that equals or exceeds the number Z of protons.

Nuclei with more than 83 protons [e.g., uranium ($Z = 92$)] are unstable and spontaneously break apart or rearrange their internal structures as time passes. This spontaneous disintegration or rearrangement of internal structure is called *radioactivity* and was first discovered in 1896 by the French physicist Henri Becquerel (1852–1908).

When an unstable nucleus disintegrates, certain kinds of particles and/or high-energy photons are released. These particles and photons are collectively called "rays." Section 38.4 will discuss three kinds of rays produced by naturally occurring radioactivity, namely, *α rays, β rays,* and *γ rays.* They are named according to the first three letters of the Greek alphabet, alpha (α), beta (β), and gamma (γ), to indicate the extent of their ability to penetrate matter. α rays are the least penetrating, being blocked by a thin (≈ 0.01-mm) sheet of lead, while β rays penetrate into lead much further (≈ 0.1 mm). γ rays are the most penetrating and can pass through an appreciable thickness (≈ 100 mm) of lead.

38.3 THE MASS DEFECT OF THE NUCLEUS AND NUCLEAR BINDING ENERGY

Because of the strong nuclear force, the nucleons in a stable nucleus are held tightly together. Therefore, energy is required to separate a stable nucleus into its constituent protons and neutrons, as Figure 38.3 illustrates. The more stable the nucleus is, the greater the amount of energy needed to tear it apart. The required energy is called the *binding energy* of the nucleus.

In Einstein's theory of special relativity, energy and mass are equivalent. A change Δm in the mass of a system is equivalent to a change in the total energy of the system by an amount $\Delta E = (\Delta m)c^2$, where c is the speed of light in a vacuum. Thus, in Figure 38.3, the binding energy ΔE used to disassemble the nucleus appears as extra mass of

FIGURE 38.3 Energy must be supplied to break the nucleus apart into its constituent protons and neutrons. Each of the separated nucleons is at rest and out of the range of the forces of the other nucleons.

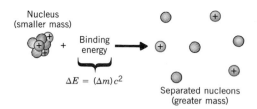

the separated nucleons; the sum of the individual masses of the separated protons and neutrons is greater by an amount Δm than the mass of the stable nucleus. The difference in mass Δm is known as the **mass defect** of the nucleus. As Example 1 shows, the binding energy of a nucleus can be determined from the mass defect by using $\Delta E = (\Delta m)c^2$:

$$\text{Binding energy} = (\text{Mass defect})c^2 = (\Delta m)c^2 \qquad (38.3)$$

EXAMPLE 1

The most abundant isotope of helium has a $_2^4\text{He}$ nucleus whose mass is 6.6447×10^{-27} kg. For this nucleus, find (a) the mass defect and (b) the binding energy.

SOLUTION

(a) To obtain the mass defect, it is first necessary to determine the sum of the individual masses of the separated protons (two) and neutrons (two) that comprise the $_2^4\text{He}$ nucleus. Using the masses from Table 38.1, we find that this sum is

$$\underbrace{2(1.6726 \times 10^{-27} \text{ kg})}_{\text{Two protons}} + \underbrace{2(1.6749 \times 10^{-27} \text{ kg})}_{\text{Two neutrons}}$$

$$= 6.6950 \times 10^{-27} \text{ kg}$$

This value is greater than the mass of the intact $_2^4\text{He}$ nucleus, and the mass defect is

$$\Delta m = 6.6950 \times 10^{-27} \text{ kg} - 6.6447 \times 10^{-27} \text{ kg}$$

$$= \boxed{0.0503 \times 10^{-27} \text{ kg}}$$

(b) The binding energy of the $_2^4\text{He}$ nucleus is

$$\frac{\text{Binding}}{\text{energy}} = (\Delta m)c^2 = (0.0503 \times 10^{-27} \text{ kg})(3.00 \times 10^8 \text{ m/s})^2$$

$$= 4.53 \times 10^{-12} \text{ J} \qquad (38.3)$$

Usually, binding energies are expressed in energy units of electron volts instead of joules ($1 \text{ eV} = 1.60 \times 10^{-19}$ J):

$$\frac{\text{Binding}}{\text{energy}} = (4.53 \times 10^{-12} \text{ J}) \left(\frac{1 \text{ eV}}{1.60 \times 10^{-19} \text{ J}} \right)$$

$$= 2.83 \times 10^7 \text{ eV} = \boxed{28.3 \text{ MeV}}$$

This value is more than two million times greater than the energy required to remove an orbital electron from an atom. For instance, in Section 37.3 the energy needed to separate the electron from the nucleus in a hydrogen atom is found to be 13.6 eV.

In calculations such as that in Example 1, it is customary to use the **atomic mass unit** (u) instead of the kilogram. As introduced in Section 17.2, the atomic mass unit is one-twelfth of the mass of a $_6^{12}\text{C}$ atom of carbon. In terms of this unit, the mass of a $_6^{12}\text{C}$ atom is exactly 12 u. Table 38.1 also gives the masses of the electron, the proton, and the neutron in atomic mass units. For future use, the energy equivalent of one atomic mass unit can be calculated by observing that the mass of a proton is 1.6726×10^{-27} kg or 1.0073 u, so that

$$1 \text{ u} = (1.6726 \times 10^{-27} \text{ kg}) \left(\frac{1 \text{ u}}{1.0073 \text{ u}} \right) = 1.6605 \times 10^{-27} \text{ kg}$$

and

$$\Delta E = (\Delta m)c^2 = (1.6605 \times 10^{-27}\ \text{kg})(2.9979 \times 10^8\ \text{m/s})^2 = 1.4924 \times 10^{-10}\ \text{J}$$

In electron volts, therefore, one atomic mass unit is equivalent to

$$(1.4924 \times 10^{-10}\ \text{J})\left(\frac{1\ \text{eV}}{1.6022 \times 10^{-19}\ \text{J}}\right) = 9.315 \times 10^8\ \text{eV} = 931.5\ \text{MeV}$$

A table of the isotopes, such as that in Appendix G, gives masses in atomic mass units. Typically, however, the given masses are not nuclear masses. They are *atomic masses,* that is, the masses of neutral atoms, including the mass of the orbital electrons. Example 2 deals again with the ^4_2He nucleus and shows how to take into account the effect of the orbital electrons when using data from a table of isotopes to determine energies.

EXAMPLE 2

Using atomic mass units instead of kilograms, obtain the binding energy of the ^4_2He nucleus.

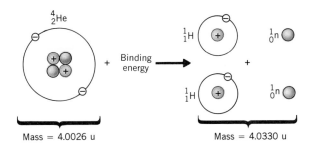

FIGURE 38.4 Tables of isotopes usually give the mass of the neutral atom (including the orbital electrons), rather than the mass of the nucleus. When using data from such tables to determine the mass defect of a nucleus, the mass of the orbital electrons must be taken into account, as this drawing illustrates for the ^4_2He isotope of helium.

SOLUTION

To determine the binding energy, we calculate the mass defect in atomic mass units and then use the fact that one atomic mass unit is equivalent to 931.5 MeV of energy. The table in Appendix G gives a mass of 4.0026 u for ^4_2He, *which includes the mass of the two electrons in the neutral helium atom.* To calculate the mass defect, we must subtract 4.0026 u from the sum of the individual masses of the nucleons, including the mass of the electrons. As Figure 38.4 illustrates, the electron mass will be included if the masses of two hydrogen atoms are used in the calculation instead of the masses of two protons. The mass of a ^1_1H hydrogen atom is 1.0078 u according to Appendix G, and the mass of a neutron is given in Table 38.1. The sum of the individual masses is

$$\underbrace{2(1.0078\ \text{u})}_{\substack{\text{Two hydrogen}\\\text{atoms}}} + \underbrace{2(1.0087\ \text{u})}_{\text{Two neutrons}} = 4.0330\ \text{u}$$

The mass defect is $\Delta m = 4.0330\ \text{u} - 4.0026\ \text{u} = 0.0304\ \text{u}$. Since 1 u is equivalent to 931.5 MeV, the binding energy is $\boxed{\text{Binding energy} = 28.3\ \text{MeV}}$. Considering round-off error, this result matches that obtained earlier in Example 1.

To see how the nuclear binding energy varies from nucleus to nucleus, it is necessary to compare the binding energy for each nucleus on a per-nucleon basis. Figure 38.5 shows a graph in which the binding energy divided by the nucleon number A is plotted against the nucleon number itself. In the graph, the peak for the ^4_2He isotope of helium indicates that the ^4_2He nucleus is particularly stable. The binding energy per nucleon increases rapidly for nuclei with small masses and reaches a maximum of approximately 8.7 MeV/nucleon for a nucleon number of about $A = 60$. For greater nucleon numbers, the binding energy per nucleon decreases gradually. Eventually, the binding energy per nucleon decreases enough so there is insufficient binding energy to hold the nucleus together. Nuclei more massive than the $^{209}_{83}\text{Bi}$ nucleus of bismuth are unstable and hence radioactive.

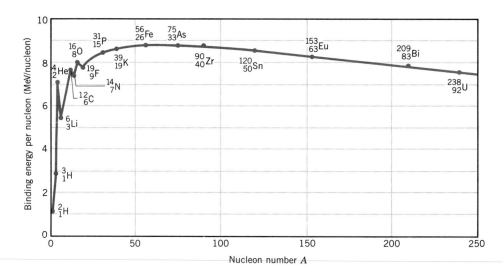

FIGURE 38.5 A plot of binding energy per nucleon versus the nucleon number A.

38.4 RADIOACTIVITY

When an unstable (radioactive) nucleus disintegrates spontaneously and produces α, β, or γ rays, the process must obey the laws of physics that are summarized below:

1. The conservation of **mass/energy** (see Sections 7.7 and 35.6)
2. The conservation of **electric charge** (see Section 23.2)
3. The conservation of **linear momentum** (see Section 8.3)
4. The conservation of **angular momentum** (see Section 10.6)
5. The conservation of **nucleon number**

Except for the conservation of nucleon number, these laws have been discussed earlier. As applied to the disintegration of a nucleus, the laws require that the energy, electric charge, linear momentum, angular momentum, and nucleon number that a nucleus possesses must remain unchanged when the nucleus disintegrates into nuclear fragments and accompanying α, β, or γ rays.

The three types of radioactivity that occur naturally can be observed in a relatively simple experiment. A piece of radioactive material is placed at the bottom of a narrow hole in a lead cylinder. The cylinder is located within an evacuated chamber, as Figure 38.6 illustrates. A magnetic field is directed perpendicular to the plane of the paper, and a photographic plate is positioned above the hole. Three spots appear on the developed plate, which are associated with the radioactivity of the nuclei in the material. Since moving particles are deflected by a magnetic field only when they are electrically charged, this experiment reveals that two types of radioactivity (α and β rays, as it turns out) consist of charged particles, while the third type (γ rays) does not.

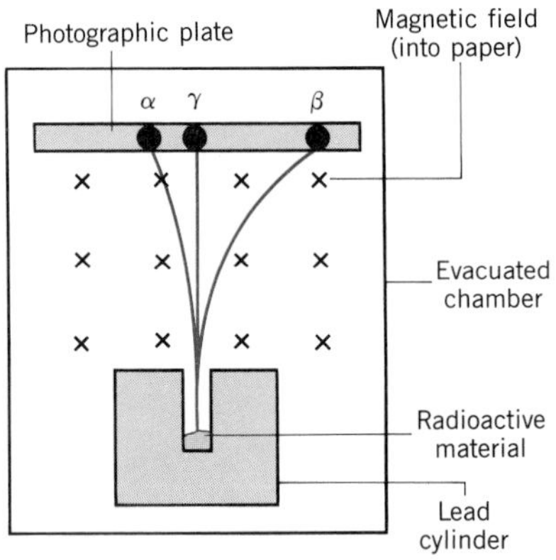

Photographic plate

Magnetic field (into paper)

Evacuated chamber

Radioactive material

Lead cylinder

FIGURE 38.6 α and β rays are deflected by a magnetic field and, therefore, consist of moving charged particles. γ rays are not deflected by a magnetic field and, consequently, must be uncharged.

α DECAY

When a nucleus disintegrates and produces α rays, it is said to undergo **α decay.** Experimental evidence shows that α rays consist of positively charged particles, each particle being the $^4_2$He nucleus of helium. Thus, an α particle has a charge of $+2e$ and a nucleon number of $A = 4$. Since the grouping of 2 protons and 2 neutrons in a $^4_2$He

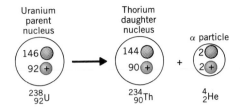

FIGURE 38.7 α decay occurs when an unstable parent nucleus emits an α particle and in the process is converted into a different or daughter nucleus.

nucleus is particularly stable, as we have seen in connection with Figure 38.5, it is not surprising that an α particle can be ejected as a unit from a more massive unstable nucleus.

Figure 38.7 shows the disintegration process for one example of α decay:

$$\underset{\substack{\text{Parent}\\\text{nucleus}\\\text{(uranium)}}}{^{238}_{92}\text{U}} \rightarrow \underset{\substack{\text{Daughter}\\\text{nucleus}\\\text{(thorium)}}}{^{234}_{90}\text{Th}} + \underset{\substack{\alpha\text{ particle}}}{^{4}_{2}\text{He}}$$

The original nucleus is referred to as the parent (P) nucleus, and the nucleus remaining after disintegration is called the daughter (D) nucleus. Upon emission of an α particle, the uranium $^{238}_{92}\text{U}$ parent is converted into the $^{234}_{90}\text{Th}$ daughter, which is an isotope of thorium. The parent and daughter nuclei are different, so α decay converts one element into another, a process known as **transmutation.**

Electric charge is conserved during α decay. In Figure 38.7, for instance, 90 of the 92 protons in the uranium nucleus end up in the thorium nucleus, and the remaining 2 protons are carried off by the α particle. The total number of 92, however, is the same before and after disintegration. α decay also conserves the number of nucleons, for the number is the same before (238) and after (234 + 4) disintegration. Consistent with the conservation of electric charge and nucleon number, the general form for α decay is

[α decay]
$$\underset{\substack{\text{Parent}\\\text{nucleus}}}{^{A}_{Z}\text{P}} \rightarrow \underset{\substack{\text{Daughter}\\\text{nucleus}}}{^{A-4}_{Z-2}\text{D}} + \underset{\substack{\alpha\text{ particle}}}{^{4}_{2}\text{He}}$$

When a nucleus releases an α particle, the nucleus also releases energy. In fact, the energy released by radioactive decay is responsible, in part, for keeping the interior of the earth hot and, in some places, even molten. The following example shows how the conservation of mass/energy can be used to determine the amount of energy released in α decay.

EXAMPLE 3

Determine the energy released when α decay converts $^{238}_{92}\text{U}$ into $^{234}_{90}\text{Th}$.

SOLUTION

Energy is released during α decay, because the combined mass of the $^{234}_{90}\text{Th}$ daughter nucleus and the α particle is less than the mass of the $^{238}_{92}\text{U}$ parent nucleus. The difference in mass is equivalent to the energy released. Appendix G gives the masses shown below:

$$\underset{\substack{238.0508 \text{ u}}}{^{238}_{92}\text{U}} \rightarrow \underset{\substack{234.0436 \text{ u}}}{^{234}_{90}\text{Th}} + \underset{\substack{4.0026 \text{ u}}}{^{4}_{2}\text{He}}$$
$$\overline{238.0462 \text{ u}}$$

The decrease in mass is 238.0508 u − 238.0462 u = 0.0046 u. As usual, the masses from Appendix G are atomic masses and include the mass of the orbital electrons. But this causes no error here, because the same total number of electrons is included for $^{238}_{92}\text{U}$, on the one hand, and for $^{234}_{90}\text{Th}$ plus $^{4}_{2}\text{He}$, on the other hand.

Since 1 u is equivalent to 931.5 MeV, the released energy is $\boxed{4.3 \text{ MeV}}$. Primarily,* this energy appears as kinetic energy of the α particle and the recoiling $^{234}_{90}$Th nucleus. However, $^{234}_{90}$Th is much more massive than the α particle, so that $^{234}_{90}$Th recoils with only a small velocity and correspondingly small kinetic energy. The law of conservation of momentum can be applied (see problem 20) to determine the individual velocities and, hence, kinetic energies.

* A small portion of the energy is also carried away as a γ ray.

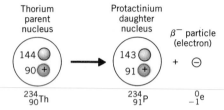

FIGURE 38.8 β decay occurs when a neutron in an unstable parent nucleus decays into a proton and an electron, the electron being emitted as the β^- particle. In the process, the parent nucleus is transformed into the daughter nucleus.

β DECAY

The β rays in Figure 38.6 are deflected by the magnetic field in a direction opposite to that of the positively charged α rays. Consequently, these β rays, which are the most common kind, consist of negatively charged particles or β^- particles. Experiment shows that β^- particles are electrons. As an illustration of β^- decay, consider the thorium $^{234}_{90}$Th nucleus, which decays by emitting a β^- particle, as in Figure 38.8:

$$\underset{\substack{\text{Parent} \\ \text{nucleus} \\ \text{(thorium)}}}{^{234}_{90}\text{Th}} \quad \rightarrow \quad \underset{\substack{\text{Daughter} \\ \text{nucleus} \\ \text{(protactinium)}}}{^{234}_{91}\text{Pa}} \quad + \quad \underset{\substack{\beta^- \text{ particle} \\ \text{(electron)}}}{^{0}_{-1}\text{e}}$$

β^- decay, like α decay, causes a transmutation of one element into another. In this case, thorium $^{234}_{90}$Th is converted into protactinium $^{234}_{91}$Pa. The law of conservation of charge is obeyed, since the net number of positive charges is the same before (90) and after (91 − 1) the β^- emission. The law of conservation of nucleon number is obeyed, since the nucleon number remains at $A = 234$. The general form for β^- decay is

[β^- decay]
$$\underset{\substack{\text{Parent} \\ \text{nucleus}}}{^A_Z\text{P}} \quad \rightarrow \quad \underset{\substack{\text{Daughter} \\ \text{nucleus}}}{^{A}_{Z+1}\text{D}} \quad + \quad \underset{\substack{\beta^- \text{ particle} \\ \text{(electron)}}}{^{0}_{-1}\text{e}}$$

The electron emitted in β^- decay does *not* actually exist within the parent nucleus and is *not* one of the orbital electrons. Instead, the electron is created when a neutron decays into a proton and an electron; when this occurs, the proton number of the parent nucleus increases from Z to $Z + 1$ and the nucleon number remains unchanged.

Example 4 illustrates that energy is released during β^- decay, just as it is during α decay, and that the conservation of mass/energy applies.

EXAMPLE 4

Find the energy released when β^- decay changes $^{234}_{90}$Th into $^{234}_{91}$Pa.

SOLUTION

To find the energy released, we follow the usual procedure of determining how much the mass has decreased because of the decay and then calculating the equivalent energy. The masses (see Appendix G) are shown below:

$$\underset{234.043\,59 \text{ u}}{^{234}_{90}\text{Th}} \quad \rightarrow \quad \underset{234.043\,30 \text{ u}}{^{234}_{91}\text{Pa}} \quad + \quad ^{0}_{-1}\text{e}$$

When the $^{234}_{90}$Th nucleus of a thorium atom is converted into a

$^{234}_{91}$Pa nucleus, the number of orbital electrons remains the same, so the resulting protactinium atom is missing one orbital electron. However, the mass taken from Appendix G includes all 91 electrons of a neutral protactinium atom. In effect, then, the value of 234.043 30 u for $^{234}_{91}$Pa already includes the mass of the β^- particle. The mass decrease that accompanies the β^- decay is 234.043 59 u − 234.043 30 u = 0.000 29 u. The equivalent energy (1 u = 931.5 MeV) is $\boxed{0.27 \text{ MeV}}$. This is the maximum kinetic energy that the emitted electron can have.

A second kind of β decay sometimes occurs.* In this process the particle emitted by the nucleus is a **positron**, rather than an electron. A positron, also called a β^+ particle, has the same mass as an electron, but carries a charge of $+e$ instead of $-e$. The disintegration process for β^+ decay is

[β^+ decay]

$$\underset{\substack{\text{Parent} \\ \text{nucleus}}}{^A_Z P} \rightarrow \underset{\substack{\text{Daughter} \\ \text{nucleus}}}{^A_{Z-1} D} + \underset{\substack{\beta^+ \text{ particle} \\ \text{(positron)}}}{^0_1 e}$$

The emitted positron does *not* exist within the nucleus but, rather, is created when a nuclear proton is transformed into a neutron. In the process, the proton number of the parent nucleus decreases from Z to $Z - 1$, and the nucleon number remains the same. As with β^- decay, the laws of conservation of charge and nucleon number are obeyed, and there is a transmutation of one element into another.

γ DECAY

The nucleus, like the orbital electrons, exists only in discrete energy states or levels. When a nucleus changes from an excited energy state (denoted by an asterisk *) to a lower energy state, a photon is emitted. The process is similar to the one discussed in Section 37.3 for the photon emission that leads to the hydrogen atom line spectrum. With nuclear energy levels, however, the photon has a much greater energy and is called a γ ray. The γ decay process is written as follows:

[γ decay]

$$\underset{\substack{\text{Parent nucleus} \\ \text{in excited} \\ \text{energy state}}}{^A_Z P^*} \rightarrow \underset{\substack{\text{Parent nucleus} \\ \text{in lower} \\ \text{energy state}}}{^A_Z P} + \underset{\gamma \text{ ray}}{\gamma}$$

γ decay does *not* cause a transmutation of one element into another. Often, γ-ray emission accompanies α or β decay, as the next example indicates.

EXAMPLE 5

When uranium $^{238}_{92}U$ is converted into thorium $^{234}_{90}Th$ via α decay (see Example 3), a γ ray is also emitted that has an energy of 0.0496 MeV. What is the wavelength of the γ-ray photon?

SOLUTION

The photon energy is the separation between two nuclear energy levels, and a value of 0.0496 MeV corresponds to

$$(0.0496 \times 10^6 \text{ eV}) \left(\frac{1.60 \times 10^{-19} \text{ J}}{1 \text{ eV}} \right) = 7.94 \times 10^{-15} \text{ J}$$

Equation 37.4 gives the relation between the energy level separation ΔE and frequency f of the photon as $\Delta E = hf$. Since $f\lambda = c$, the wavelength of the photon is

$$\lambda = \frac{hc}{\Delta E} = \frac{(6.63 \times 10^{-34} \text{ J} \cdot \text{s})(3.00 \times 10^8 \text{ m/s})}{7.94 \times 10^{-15} \text{ J}}$$

$$= \boxed{2.51 \times 10^{-11} \text{ m}}$$

38.5 THE NEUTRINO

When a β particle is emitted by a radioactive nucleus, energy is simultaneously released, as Example 4 illustrates. Experimentally, however, it is found that most β

---

* A third kind of β decay also occurs, in which a nucleus pulls in or captures one of the orbital electrons from outside the nucleus. The process is called **electron capture**, or **K capture**, since the electron normally comes from the innermost or K shell.

particles do not have enough kinetic energy to account for all the energy released. If a β particle carries away only part of the released energy, where does the remainder go? The question puzzled physicists until 1930, when Wolfgang Pauli proposed that part of the energy released during β decay is carried away by another particle that is emitted along with the β particle. This additional particle is called the ***neutrino,*** and its existence was verified experimentally in 1956. The emission of β particles and neutrinos involves a fundamental force that has not been mentioned before in this text. This force is much weaker than the strong nuclear force, and, hence, is referred to as the ***weak nuclear force.***

The neutrino has zero electrical charge. Moreover, at present there is no convincing experimental evidence to indicate that the neutrino has any rest mass. A particle with zero rest mass, like a photon, travels at the speed of light. The neutrino, therefore, travels near or at the speed of light. The Greek letter nu (v) is used to symbolize the neutrino. For instance, the β^- decay of thorium $^{234}_{90}\text{Th}$ (see Section 38.4) is more correctly written as

$$^{234}_{90}\text{Th} \rightarrow {}^{234}_{91}\text{Pa} + {}^{0}_{-1}\text{e} + \bar{v}$$

The bar above the v is included, because the neutrino emitted in this particular decay process is an antimatter neutrino or antineutrino. A normal neutrino (v without the bar) is emitted when β^+ decay occurs.

38.6 RADIOACTIVE DECAY AND ACTIVITY

The question of which radioactive nucleus in a group disintegrates at a given instant is decided like the drawing of numbers in a state lottery; individual disintegrations occur randomly. As time passes, the number N of parent nuclei decreases, as Figure 38.9 shows. This graph of N versus time indicates that the decrease occurs in a smooth fashion, with N approaching zero after enough time has passed. To help describe the

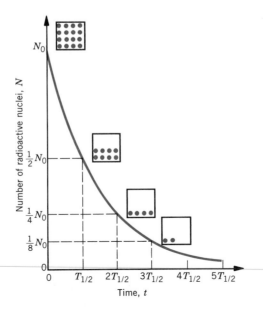

FIGURE 38.9 The half-life $T_{1/2}$ of a radioactive decay is the time in which one-half of the radioactive nuclei disintegrate.

TABLE 38.2 Some Half-lives for Radioactive Decay

Isotope	Half-life	Decay Mode
Polonium $^{214}_{84}$Po	1.64×10^{-4} s	α, γ
Krypton $^{89}_{36}$Kr	3.16 min	β^-, γ
Radon $^{222}_{86}$Rn	3.83 days	α, γ
Strontium $^{90}_{38}$Sr	28.5 yr	β^-
Radium $^{226}_{88}$Ra	1.6×10^3 yr	α, γ
Carbon $^{14}_{6}$C	5.73×10^3 yr	β^-
Uranium $^{238}_{92}$U	4.47×10^9 yr	α, γ
Indium $^{115}_{49}$In	4.41×10^{14} yr	β^-

graph, it is useful to define the *half-life* $T_{1/2}$ of a radioactive isotope as the time required for one-half of the nuclei present to disintegrate. For example, radium $^{226}_{88}$Ra has a half-life of 1600 years, for it takes this amount of time for one-half of a given quantity of this isotope to disintegrate into radon $^{222}_{86}$Rn. In another 1600 years, one-half of the remaining radium atoms will disintegrate, leaving only one-fourth of the original number intact. In Figure 38.9, the number of nuclei present at time $t = 0$ is $N = N_0$, while the number present at $t = T_{1/2}$ is $N = \frac{1}{2}N_0$, the number present at $t = 2\,T_{1/2}$ is $N = \frac{1}{4}N_0$, and so forth. The value of the half-life depends on the nature of the radioactive nucleus. Values ranging from a fraction of a second to billions of years have been found, as Table 38.2 indicates. Example 6 deals with the half-life of radon $^{222}_{86}$Rn.

EXAMPLE 6

Radon $^{222}_{86}$Rn is a radioactive gas produced when radium $^{226}_{88}$Ra undergoes α decay. There is growing concern about radon as a health hazard, because it can become trapped in houses, entering through walls and floors and in the drinking water. Suppose 3.0×10^7 radon atoms are trapped in a basement at the time the basement is sealed against further entry of the gas. The half-life of radon is 3.83 days. How many radon atoms remain after 31 days?

SOLUTION

During each half-life, the number of radon atoms is reduced by a factor of two. In a period of 31 days there are 31 days/3.83 days = 8.1 half-lives. In 8 half-lives the number of radon atoms is reduced by a factor of $2^8 = 256$. Ignoring the difference between 8 and 8.1 half-lives, we find that the number of atoms remaining is $3.0 \times 10^7/256 = \boxed{1.2 \times 10^5}$.

The *activity* of a radioactive sample is the number of disintegrations per second that occur. Each time a disintegration occurs, the number N of radioactive nuclei decreases. As a result, the activity can be obtained by dividing ΔN, the change in the number of nuclei, by Δt, the time interval during which the change takes place; the average activity over the time interval Δt is the magnitude of $\Delta N/\Delta t$. The number of disintegrations per second that occur in a sample is proportional to the number of radioactive nuclei present, so that

$$\frac{\Delta N}{\Delta t} = -\lambda N \qquad (38.4)$$

where λ is a proportionality constant referred to as the *decay constant.* The minus sign is present in this equation because each disintegration decreases N.

The SI unit for activity is the *becquerel* (Bq); one becquerel equals one disintegration per second. Activity is also measured in terms of a unit called the *curie* (Ci), in honor of Marie (1867–1934) and Pierre (1859–1906) Curie, the discoverers of radium and polonium. Historically, the curie was chosen as a unit because it is roughly the activity of one gram of pure radium. In terms of becquerels,

$$1 \text{ Ci} = 3.70 \times 10^{10} \text{ Bq}$$

The activity of the radium put into the dial of a watch to make it glow in the dark is about 4×10^4 Bq, and the activity used in radiation therapy for cancer treatment is approximately 4×10^{13} Bq.

The mathematical expression for the graph of N versus t shown in Figure 38.9 can be obtained from Equation 38.4 with the aid of calculus. The result is that the number N of radioactive nuclei present at time t is given by

$$N = N_0\, e^{-\lambda t} \tag{38.5}$$

assuming that the number at $t = 0$ is N_0. The exponential e has the value $e = 2.718 \ldots$, and many calculators provide the value of e^x. By substituting $N = \frac{1}{2}N_0$ and $t = T_{1/2}$ into Equation 38.5, we find that $\frac{1}{2} = e^{-\lambda T_{1/2}}$. Taking the natural logarithm of both sides of this equation reveals that

$$T_{1/2} = \frac{\ln 2}{\lambda} = \frac{0.693}{\lambda} \tag{38.6}$$

The following example illustrates the use of Equations 38.5 and 38.6.

EXAMPLE 7

As in Example 6, suppose there are 3.0×10^7 radon atoms ($T_{1/2} = 3.83$ days or 3.31×10^5 s) trapped in a basement. (a) How many radon atoms remain after 31 days? Find the activity (b) just after the basement is sealed against further entry of radon and (c) 31 days later.

SOLUTION

(a) The answer can be obtained directly from Equation 38.5, provided the decay constant is first determined from the half-life:

$$\lambda = \frac{0.693}{T_{1/2}} = \frac{0.693}{3.83 \text{ days}} = 0.181 \text{ days}^{-1}$$

$$N = N_0\, e^{-\lambda t} = (3.0 \times 10^7)e^{-(0.181 \text{ days}^{-1})(31 \text{ days})}$$

$$= \boxed{1.1 \times 10^5}$$

This value is slightly less than that found in Example 6, because there we ignored the difference between 8.0 and 8.1 half-lives.

(b) The activity can be obtained from Equation 38.4, provided the decay constant is expressed in reciprocal seconds: $\lambda = 0.693/(3.31 \times 10^5 \text{ s}) = 2.09 \times 10^{-6} \text{ s}^{-1}$. According to Equation 38.4,

$$\frac{\Delta N}{\Delta t} = -\lambda N = -(2.09 \times 10^{-6} \text{ s}^{-1})(3.0 \times 10^7)$$

$$= -63 \text{ disintegrations/s}$$

The activity is the magnitude of $\Delta N/\Delta t$, so initially $\boxed{\text{Activity} = 63 \text{ Bq}}$.

(c) From part (a), the number of radioactive nuclei remaining at the end of 31 days is $N = 1.1 \times 10^5$, and reasoning similar to that in part (b) reveals that $\boxed{\text{Activity} = 0.23 \text{ Bq}}$.

38.7 RADIOACTIVE DATING

One important application of radioactivity is the determination of the age of archeological or geological samples. If an object contains radioactive nuclei when it is formed, then the decay of these nuclei marks the passage of time like a clock, half of

Radioactive dating is often used to determine the age of artifacts obtained from an archeological dig.

the nuclei disintegrating during each half-life. If the half-life is known, a measurement of the number of nuclei present today relative to the number present initially can give the age of the sample. According to Equation 38.4, the activity of a sample is proportional to the number of radioactive nuclei present, so the age is frequently obtained by comparing current activity to initial activity.

The current activity of a sample can be measured using detectors such as a Geiger counter. But how is it possible to know what the original activity was, perhaps thousands of years ago? Radioactive dating methods entail certain assumptions that make it possible to estimate the original activity. For instance, the radiocarbon technique utilizes the $^{14}_{6}C$ isotope of carbon, which undergoes β^- decay with a half-life of 5730 yr. This isotope is present in the earth's atmosphere at an equilibrium concentration of about one atom for every 8.3×10^{11} atoms of normal carbon $^{12}_{6}C$. This value remains constant in time, because $^{14}_{6}C$ is created when cosmic rays interact with the earth's upper atmosphere, a production method that offsets the loss via β^- decay. Moreover, nearly all living organisms ingest the equilibrium concentration of $^{14}_{6}C$. However, once an organism dies, metabolism no longer sustains the input of $^{14}_{6}C$, and β^- decay causes half of the $^{14}_{6}C$ nuclei to disintegrate every 5730 yr. That the concentration of $^{14}_{6}C$ has always been at its present equilibrium value is, of course, an assumption, but the available evidence indicates that it is a reasonable one.

It is possible to calculate the $^{14}_{6}C$ activity of one gram of carbon in a living organism. One gram of carbon (atomic mass = 12 u) is 1.0/12 mol, and since there are 6.02×10^{23} atoms per mole (Avogadro's number), the number of $^{14}_{6}C$ atoms present is

$$\left(\frac{1.0}{12}\text{ mol}\right)\left(6.02 \times 10^{23}\ \frac{\text{atoms}}{\text{mol}}\right)\left(\frac{1}{8.3 \times 10^{11}}\right) = 6.0 \times 10^{10}\text{ atoms}$$

Since the half-life is 5730 yr (1.81×10^{11} s), the decay constant of $^{14}_{6}C$ is $\lambda = 0.693/T_{1/2} = 0.693/(1.81 \times 10^{11}\text{ s}) = 3.83 \times 10^{-12}\text{ s}^{-1}$. Therefore, Equation 38.4 indicates that the activity, or the magnitude of $\Delta N/\Delta t$, is

Activity of one
gram of carbon in = $\lambda N = (3.83 \times 10^{-12}\text{ s}^{-1})(6.0 \times 10^{10}) = 0.23$ Bq
a living organism

An organism that lived thousands of years ago, presumably had an activity of 0.23 Bq per gram of carbon. When the organism died, the activity began decreasing. From a sample of the remains, the current activity per gram of carbon can be measured and compared to the value of 0.23 Bq to determine the time that has transpired since death. This procedure is illustrated in Example 8.

EXAMPLE 8

The Dead Sea scrolls are famous ancient manuscripts, discovered in 1947. They were dated by applying the radiocarbon method to a sample of the linen in which they were wrapped. Linen is made from the flax plant. A $^{14}_{6}C$ activity of about 0.18 Bq per gram of carbon was measured. Determine the age of the scrolls.

SOLUTION

According to Equation 38.5, the number of nuclei remaining at time t is $N = N_0 e^{-\lambda t}$. Multiplying both sides of this expression by the decay constant λ and recognizing that the product of λ and N is the activity A, we find that

$$A = A_0 e^{-\lambda t}$$

where A_0 is the activity at time $t = 0$. For $^{14}_{6}C$, the decay constant is $\lambda = 0.693/T_{1/2} = 0.693/(5730 \text{ yr}) = 1.21 \times 10^{-4} \text{ yr}^{-1}$. Since $A = 0.18$ Bq and $A_0 = 0.23$ Bq, the age can be determined from

$$A = 0.18 \text{ Bq} = (0.23 \text{ Bq})e^{-(1.21 \times 10^{-4} \text{ yr}^{-1})t}$$

Taking the natural logarithm of both sides of this result gives

$$\ln\left(\frac{0.18 \text{ Bq}}{0.23 \text{ Bq}}\right) = -(1.21 \times 10^{-4} \text{ yr}^{-1})t$$

The age of the sample is $\boxed{t = 2.0 \times 10^3 \text{ yr}}$.

Radiocarbon dating is not the only radioactive dating method. For example, other methods utilize uranium $^{238}_{92}U$, potassium $^{40}_{19}K$, and lead $^{210}_{82}Pb$. And the related technique of thermoluminescence is being increasingly used.

38.8 RADIOACTIVE DECAY SERIES

When an unstable parent nucleus decays, the resulting daughter nucleus is sometimes also unstable. If so, the daughter then decays and produces its own daughter, and so on, until a completely stable nucleus is produced. This sequential decay of one nucleus after another is called a *radioactive decay series.* Examples 3 and 4 discuss the first two steps of a series that begins with uranium $^{238}_{92}U$:

Uranium   Thorium
$$^{238}_{92}U \longrightarrow \, ^{234}_{90}Th + \, ^{4}_{2}He$$
$$\longrightarrow \, ^{234}_{91}Pa + \, ^{0}_{-1}e$$
Protactinium

Furthermore, Example 6 deals with radon $^{222}_{86}Rn$, which is formed down the line in the $^{238}_{92}U$ radioactive decay series. Figure 38.10 shows the entire series. At several points in the series, branches occur, because more than one kind of decay is possible for an intermediate species. Ultimately, however, the series ends with lead $^{206}_{82}Pb$, which is stable.

The $^{238}_{92}U$ series and other such series are the only sources of some of the radioactive elements found in nature. Radium $^{226}_{88}Ra$, for instance has a half-life of 1600 yr, which is short enough that all the $^{226}_{88}Ra$ created when the earth was formed billions of years

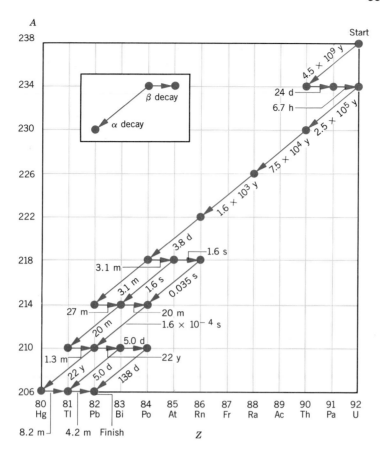

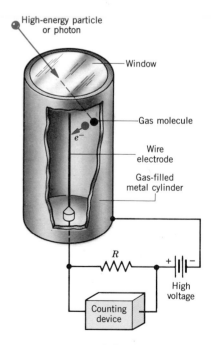

FIGURE 38.10 A radioactive decay series that begins with uranium $^{238}_{92}U$ and ends with lead $^{206}_{82}Pb$. The half-lives are given in seconds (s), minutes (m), hours (h), days (d), or years (y). The insert in the upper left corner of the graph identifies the type of decay that each nucleus undergoes.

ago has now disappeared. The $^{238}_{92}U$ series provides a continuing supply of $^{226}_{88}Ra$, however.

38.9 DETECTORS OF RADIATION

There are a number of devices that can be used to detect the particles and photons (γ rays) emitted when a radioactive nucleus decays. Such devices detect the ionization that these particles and photons cause as they pass through matter.

The most familiar detector is the **Geiger counter,** which Figure 38.11 illustrates. The Geiger counter consists of a gas-filled metal cylinder. The α, β, or γ rays enter the cylinder through a thin window at one end. γ rays can also penetrate directly through the metal. A wire electrode runs along the center of the tube and is kept at a high positive voltage (1000–3000 V) relative to the outer cylinder. When a high-energy particle or photon enters the cylinder, it collides with and ionizes a gas molecule. The electron produced from the gas molecule accelerates toward the positive wire, ionizing other molecules in its path. Additional electrons are formed, and an avalanche of electrons rushes toward the wire, leading to a pulse of current through the resistor R. This pulse can be counted or made to produce a "click" in a loudspeaker. The number of counts or clicks is related to the number of high-energy particles or photons present, or equivalently, to the number of disintegrations that produced the particles or photons.

FIGURE 38.11 A Geiger counter.

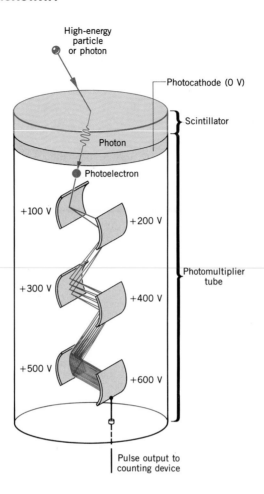

FIGURE 38.12 A scintillation counter.

The *scintillation counter* is another important radiation detector. As Figure 38.12 indicates, this device consists of a scintillator mounted on a photomultiplier tube. Often the scintillator is a crystal (e.g., cesium iodide) containing a small amount of impurity (thallium), but plastic, liquid, and gaseous scintillators are also used. In response to ionizing radiation, the scintillator emits a flash of visible light. The photons of the flash then strike the photocathode of the photomultiplier tube. The photocathode is made of a compound that emits electrons because of the photoelectric effect. These photoelectrons are then attracted to a special electrode kept at a voltage of about +100 V relative to the photocathode. The electrode is coated with a substance that emits several additional electrons for every electron striking it. The additional electrons are attracted to a second similar electrode (voltage = +200 V) where they generate even more electrons. Commercial photomultiplier tubes contain as many as 15 of these special electrodes, so photoelectrons resulting from the light flash of the scintillator lead to a cascade of electrons and a pulse of current. As in a Geiger tube, the current pulses can be counted.

Ionizing radiation can also be detected with several types of *semiconductor detectors.* Such devices utilize *n*- and *p*-type materials, and their operation depends on the electrons and holes formed in the materials as a result of the radiation. One of the main advantages of semiconductor detectors is their ability to discriminate between two particles with only slightly different energies.

A number of instruments provide a pictorial representation of the path that high-energy particles follow after they are emitted from unstable nuclei. In a *cloud*

chamber, a gas is cooled just to the point where it will condense into droplets, provided nucleating agents are available on which the droplets can form. When a high-energy particle, such as an α particle or a β particle, passes through the gas, the ions it leaves behind serve as nucleating agents, and droplets form along the path of the particle. A **bubble chamber** works in a similar fashion, except it contains a liquid that is just at the point of boiling. Tiny bubbles form along the trail of a high-energy charged particle passing through the liquid. The paths revealed in a cloud or bubble chamber can be photographed to provide a permanent record of the event. Figure 38.13 shows a photograph of the tracks in a bubble chamber. A **photographic emulsion** also can be used directly to produce a record of the path taken by a particle of ionizing radiation. The ions formed as the particle passes through the emulsion cause silver to be deposited along the track when the emulsion is developed.

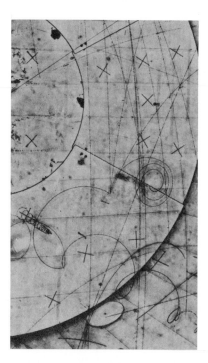

FIGURE 38.13 Particle tracks in a bubble chamber.

SUMMARY

The nucleus of an atom consists of protons and neutrons, which are collectively referred to as **nucleons.** A **neutron** is an electrically neutral particle whose mass is slightly larger than that of a proton. The **atomic number** Z is the number of protons in the nucleus. The **atomic mass number** or **nucleon number** A is the total number of protons and neutrons in the nucleus: $A = Z + N$, where N is the number of neutrons. Nuclei that contain the same number of protons, but a different number of neutrons, are called **isotopes.**

The **strong nuclear force** is the force of attraction between nucleons and is one of the four fundamental forces of nature. This force balances the electrostatic force of repulsion between protons and holds the nucleus together. The strong nuclear force has a very short range of action and does not depend on electric charge.

The **binding energy** of a nucleus is the energy required to separate the nucleus into its constituent protons and neutrons. The binding energy is equal to $(\Delta m)c^2$, where c is the speed of light in a vacuum and Δm is the mass defect of the nucleus. The **mass defect** is the amount by which the sum of the individual masses of the protons and neutrons exceeds the mass of the stable nucleus.

When specifying nuclear masses, it is customary to use the **atomic mass unit** (u), which is one-twelfth of the mass of a $^{12}_{6}C$ atom. One atomic mass unit is equivalent to an energy of 931.5 MeV.

Unstable nuclei spontaneously decay by breaking apart or rearranging their internal structures in a process called **radioactivity.** Naturally occurring radioactivity produces α rays, β rays, and γ rays. α **rays** consist of positively charged particles, each particle being the $^{4}_{2}He$ nucleus of helium. The most common kind of β **ray** consists of negatively charged particles, or β^- particles, which are electrons. Another kind of β-ray consists of positively charged particles, or β^+ particles. A β^+ particle, also called a **positron,** has the same mass as an electron but carries a charge of $+e$ instead of $-e$. γ **rays** are high-energy photons. If a radioactive parent nucleus disintegrates into a daughter nucleus that has a different atomic number, one element has been converted into another element, the conversion being referred to as a **transmutation.**

The **neutrino** is an electrically neutral particle that has near zero or zero rest mass. The neutrino travels near or at the speed of light and is emitted along with β particles.

The **half-life** $T_{1/2}$ of a radioactive isotope is the time required for one-half of the nuclei present to disintegrate or decay. The **activity** is the number of disintegrations per second that occur. In other words, the activity is the magnitude of $\Delta N/\Delta t$, where ΔN is the change in the number N of radioactive nuclei and Δt is the time interval during which the change occurs. The SI unit for activity is the becquerel (Bq), one becquerel being one disintegration per second.

Radioactive decay obeys the following relation: $\Delta N/\Delta t = -\lambda N$, where λ is the **decay constant**. This equation can be solved to show that $N = N_0 e^{-\lambda t}$, where N_0 is the number of nuclei present initially. The decay constant is related to the half-life according to $T_{1/2} = 0.693/\lambda$.

The sequential decay of one nucleus after another is called a **radioactive decay series,** and Figure 38.10 illustrates one such series.

A number of devices can be used to detect α, β, and γ rays. These devices include the **Geiger counter,** the **scintillation counter,** and others.

QUESTIONS

1. A material is known to be an isotope of lead, although the particular isotope is not known. From such limited information, which of the following quantities can you specify: (a) its atomic number, (b) its neutron number, and (c) its atomic mass number? Explain.

2. The density of an atom as a whole is much less than the density of the nucleus of the atom. Considering what Section 37.1 discusses about the structure of the atom, explain why.

3. Using Figure 38.5, rank the following nuclei in ascending order according to the binding energy per nucleon (smallest first): phosphorus $^{31}_{15}P$, cobalt $^{59}_{27}Co$, tungsten $^{184}_{74}W$, and thorium $^{232}_{90}Th$.

4. Describe how the radius of the daughter nucleus compares to that of the parent nucleus for (a) α decay and (b) β decay. Justify your answers in terms of Equation 38.2.

5. Uranium $^{238}_{92}U$ decays into thorium $^{234}_{90}Th$ by means of α decay, as Example 3 in the text discusses. A reasonable question to ask is "Why doesn't the $^{238}_{92}U$ nucleus just emit a single proton, instead of an α particle?" This hypothetical decay scheme is shown below, along with the pertinent atomic masses:

$$^{238}_{92}U \rightarrow {}^{237}_{91}Pa + {}^{1}_{1}H$$

Uranium	Protactinium	Proton
238.050 78 u	237.051 14 u	1.007 83 u

For a decay to be possible, it must bring the parent nucleus toward a more stable state by allowing the release of energy. Compare the total mass of the products of this hypothetical decay with the mass of $^{238}_{92}U$ and decide whether the emission of a single proton is possible for $^{238}_{92}U$. Explain.

6. Explain why unstable nuclei with short half-lives typically have only a small or zero natural abundance.

7. On the basis of the half-lives given in the isotope table in Appendix G, decide which isotope might be of use to date the pure H_2O in a sealed bottle. The water is thought to be about 5–15 yr old. Account for your choice.

8. To which of the following objects, each about 2000 yr old, can the radiocarbon dating technique *not* be applied: a glass vial, a wooden box, a gold statue, and a skeleton? Explain.

9. Suppose there were a greater number of carbon $^{14}_{6}C$ atoms in an animal living 5000 yr ago than is currently believed. When the bones of this animal are tested today using radiocarbon dating, is the age obtained too small or too large? Give your reasoning.

10. Tritium is an isotope of hydrogen and undergoes β^- decay with a half-life of 12.33 yr. Like carbon $^{14}_{6}C$, tritium is produced in the atmosphere because of cosmic rays and can be used in a radioactive dating technique. In any such technique, there must be a sufficient number of radioactive nuclei left in a sample to detect if the technique is to be useful. Can tritium dating be used to determine a reliable date for a sample that is about 700 yr old? Account for your answer.

PROBLEMS

The data given for atomic masses in these problems include the mass of the electrons orbiting the nucleus of the electrically neutral atom.

Section 38.1 Nuclear Structure, Section 38.2 The Strong Nuclear Force and the Stability of the Nucleus

1. How many protons and neutrons are there in the nucleus of (a) oxygen $^{18}_{8}O$, (b) chlorine $^{35}_{17}Cl$, (c) iron $^{56}_{26}Fe$, and (d) tin $^{120}_{50}Sn$?

2. What is the radius of a nucleus of uranium $^{238}_{92}U$?

3. Two isotopes of chlorine occur in nature. The $^{35}_{17}Cl$ isotope has an atomic mass of 34.968 85 u and a natural abundance of 75.77%. The $^{37}_{17}Cl$ isotope has an atomic mass of 36.965 90 u and a natural abundance of 24.23%. By a calculation of your own, verify that the value of 35.45 u listed in the periodic table is a weighted average of these individual atomic masses.

4. By what factor does the nucleon number of a nucleus have to increase in order for the nuclear radius to double?

5. In the nucleus of gold $^{197}_{79}Au$, what is the electrostatic force of repulsion between two protons, assuming the centers of the protons are located at opposite ends of a diameter of the gold nucleus?

* **6.** Two naturally occurring isotopes of carbon are $^{12}_{6}C$ (atomic mass = 12.000 000 u) and $^{13}_{6}C$ (atomic mass = 13.003 355 u). In one gram of each of these isotopes there are different numbers of atoms. Which contains more atoms, and how many more?

** **7.** (a) Determine an approximate value for the density (in kg/m^3) of the nucleus. (b) If a BB (radius = 2.3 mm) from an air rifle had a density equal to the nuclear density, what mass would the BB have? (c) Assuming the mass of a supertanker is about 1.5×10^8 kg, how many "supertankers" of mass would this hypothetical BB have?

Section 38.3 The Mass Defect of the Nucleus and Nuclear Binding Energy

8. Determine the mass defect of the nucleus for cobalt $^{59}_{27}Co$, which has an atomic mass of 58.933 198 u. Express your answer in (a) atomic mass units and (b) kilograms.

9. What is the binding energy (in MeV) for oxygen $^{16}_{8}O$ (atomic mass = 15.994 915 u)?

10. For radium $^{226}_{88}Ra$ (atomic mass = 226.025 402 u) obtain (a) the mass defect in atomic mass units, (b) the binding energy in MeV, and (c) the binding energy per nucleon.

* **11.** (a) Energy is required to separate a nucleus into its constituent nucleons, as Figure 38.3 indicates; this energy is the *total* binding energy of the nucleus. In a similar way one can speak of the energy that binds a single nucleon to the remainder of the nucleus. For example, separating nitrogen $^{14}_{7}N$ into nitrogen $^{13}_{7}N$ and a neutron takes energy equal to the binding energy of the neutron, as shown below:

$$^{14}_{7}N + Energy \rightarrow \, ^{13}_{7}N + \, ^{1}_{0}n$$

Find the energy that binds the neutron to the $^{14}_{7}N$ nucleus by considering the mass of $^{13}_{7}N$ and the mass of $^{1}_{0}n$, as compared to the mass of $^{14}_{7}N$ (see Appendix G for masses). (b) Similarly, one can speak of the energy that binds a single proton to the $^{14}_{7}N$ nucleus:

$$^{14}_{7}N + Energy \rightarrow \, ^{13}_{6}C + \, ^{1}_{1}H$$

Following the procedure outlined in part (a), determine the energy that binds the proton to the $^{14}_{7}N$ nucleus. (c) Which nucleon is more tightly bound, the neutron or the proton?

Section 38.4 Radioactivity

12. α decay occurs for each of the nuclei given below. Write the decay process for each, including the appropriate chemical symbols and values for Z and A for the daughter nuclei: (a) $^{228}_{90}Th$, (b) $^{231}_{91}Pa$, (c) $^{235}_{92}U$, and (d) $^{239}_{94}Pu$.

13. Carbon $^{14}_{6}C$ (atomic mass = 14.003 241 u) is converted into nitrogen $^{14}_{7}N$ (atomic mass = 14.003 074 u) via β^- decay. (a) Write this process in symbolic form, giving Z and A for the parent and daughter nuclei and the β^- particle. (b) Determine the energy (in MeV) released.

14. For the following nuclei, each undergoing β^- decay, write the decay process, identifying each daughter nucleus with its chemical symbol and values for Z and A: (a) $^{14}_{6}C$, (b) $^{35}_{16}S$, (c) $^{60}_{27}Co$, and (d) $^{212}_{82}Pb$.

15. What is the wavelength of the 0.186-MeV γ ray that is emitted by radium $^{226}_{88}Ra$?

16. In the form $^{A}_{Z}X$, identify the daughter nucleus that results when (a) plutonium $^{242}_{94}Pu$ undergoes α decay, (b) sodium $^{24}_{11}Na$ undergoes β^- decay, and (c) nitrogen $^{13}_{7}N$ undergoes β^+ decay.

* **17.** Determine the symbol $^{A}_{Z}X$ for the parent nucleus whose α decay produces the same daughter as the β^- decay of thallium $^{208}_{81}Tl$.

* **18.** Find the energy (in MeV) released when β^+ decay converts sodium $^{22}_{11}Na$ (atomic mass = 21.994 434 u) into neon $^{22}_{10}Ne$ (atomic mass = 21.991 383 u). Notice that the atomic mass for $^{22}_{11}Na$ includes the mass of 11 electrons, whereas the atomic mass for $^{22}_{10}Ne$ includes the mass of only 10 electrons.

* **19.** Thorium $^{232}_{90}Th$ undergoes α decay to produce a daughter nucleus that itself undergoes β^- decay. In the form $^{A}_{Z}X$, identify the nucleus that ultimately results.

** **20.** Example 3 in the text deals with the α decay of uranium $^{238}_{92}U$, which produces thorium $^{234}_{90}Th$ (atomic mass = 234.0436 u). The energy released in the decay is determined in this example to be 4.3 MeV. Use the conservation of linear momentum (see Example 3 in Chapter 8), and determine how much of this energy is carried away by the recoiling $^{234}_{90}Th$ daughter nucleus and how much by the α particle. Ignore the small amount of energy carried away by the γ ray that is also emitted.

** **21.** Sodium $^{24}_{11}Na$ emits a γ ray that has an energy of 0.423 MeV. Assuming the $^{24}_{11}Na$ nucleus is initially at rest, use the conservation of linear momentum to find the speed with which the nucleus recoils.

Section 38.6 Radioactive Decay and Activity

22. The number of radioactive nuclei present at the start of an experiment is 4.60×10^{15}. The number present twenty days later is 8.14×10^{14}. What is the half-life (in days) of the nuclei?

23. The $^{3}_{1}H$ isotope of hydrogen is called tritium and has a half-life of 12.33 yr. What is its decay constant in units of s^{-1}?

24. Strontium $^{90}_{38}Sr$ has a half-life of 28.5 yr. It is chemically similar to calcium, enters the body through the food chain, and collects in the bones. Consequently, $^{90}_{38}Sr$ is a particularly serious health hazard. How long (in years) will it take 99.99% of the $^{90}_{38}Sr$ released in a nuclear reactor accident to disappear?

25. To make the dial of a watch glow in the dark, 1.00×10^{-6} g of radium $^{226}_{88}Ra$ is used. The half-life of this isotope is 1.6×10^3 yr. How many grams of radium *disappear* while the watch is in use for fifty years?

* **26.** A device used in radiation therapy for cancer contains 0.50 g of cobalt $^{60}_{27}Co$. The half-life of $^{60}_{27}Co$ is 5.27 yr. Determine the activity of the radioactive material.

* 27. Two waste products from nuclear reactors are strontium $^{90}_{38}$Sr and cesium $^{134}_{55}$Cs. The half-life of $^{90}_{38}$Sr is 28.5 yr, while that of $^{134}_{55}$Cs is 2.06 yr. If these two species are initially present in a ratio of Sr/Cs = 7.80 × 10^{-3}, what is this ratio fifteen years later?

* 28. If the activity of a radioactive substance is initially 398 disintegrations/min and two days later it is 285 disintegrations/min, what is the activity four days later still? Give your answer in disintegrations/min.

* 29. A sample of ore containing a radioactive element has an activity of 4.0 × 10^4 Bq. How many grams of the element are in the sample, assuming the element is (a) radium $^{226}_{88}$Ra ($T_{1/2}$ = 1.6 × 10^3 yr) and (b) uranium $^{238}_{92}$U ($T_{1/2}$ = 4.47 × 10^9 yr)?

**30. Outside the nucleus, the neutron decays into a proton, an electron, and an antineutrino. The half-life for the neutron is 10.4 min. Over what distance will a beam of 5.00-eV neutrons travel before the number of neutrons per unit volume of the beam decreases by 25.0%?

Section 38.7 Radioactive Dating, Section 38.8 Radioactive Decay Series

31. Bones of the woolly mammoth have been found in North America. The youngest of these bones has a $^{14}_{6}$C activity per gram of carbon that is about 21% of what was present in the live animal. How long ago did this animal disappear from North America?

32. The practical limit to ages that can be determined by radiocarbon dating is about 41 000 yr. In a 41 000-yr-old sample, what percentage of the original $^{14}_{6}$C atoms remains?

33. The half-life for the α decay of uranium $^{238}_{92}$U is 4.47 × 10^9 yr. Determine the age of a rock that contains sixty percent of its original $^{238}_{92}$U atoms.

* 34. Using the isotope table in Appendix G, construct a plot like that in Figure 38.10, showing the radioactive series that begins with thorium $^{232}_{90}$Th and ends with lead $^{208}_{82}$Pb. You need not include half-lives.

**35. When any radioactive dating method is used, experimental error in the measurement of the sample's activity leads to error in the estimated age. In an application of the radiocarbon dating technique to certain fossils, an activity of 0.10 Bq per gram of carbon is measured to within an accuracy of ±ten percent. Find the age of the fossils and the maximum error (in years) in the value obtained. Assume that there is no error in the 5730-year half-life of $^{14}_{6}$C.

ADDITIONAL PROBLEMS

36. The $^{208}_{82}$Pb isotope of lead has an atomic mass of 207.976 627 u, while the $^{214}_{82}$Pb isotope has an atomic mass of 213.999 798 u. Obtain the binding energy per nucleon (in MeV) for each isotope.

37. According to the periodic table in Appendix H, what element does each symbol "X" represent: $^{195}_{78}$X, $^{32}_{16}$X, $^{63}_{29}$X, $^{11}_{5}$X, and $^{239}_{94}$X?

38. Find the energy (in MeV) released when α decay converts radium $^{226}_{88}$Ra (atomic mass = 226.025 40 u) into radon $^{222}_{86}$Rn (atomic mass = 222.017 57 u).

39. How many half-lives are required for the number of radioactive nuclei to decrease to one one-millionth of the initial number?

40. Write the β^+ decay process for each of the following nuclei, being careful to include Z and A and the proper chemical symbol for each daughter nucleus: (a) $^{18}_{9}$F, (b) $^{15}_{8}$O, and (c) $^{11}_{6}$C.

* 41. The photomultiplier tube in a commercial scintillator counter contains 15 of the special electrodes or dynodes. Each dynode produces 3 electrons for every electron that strikes it. One photoelectron strikes the first dynode. What is the maximum number of electrons that strikes the 15th dynode?

* 42. To see why one curie of activity was chosen to be 3.70 × 10^{10} Bq, determine the activity (in disintegrations per second) of one gram of radium $^{226}_{88}$Ra ($T_{1/2}$ = 1.6 × 10^3 yr).

* 43. Plutonium $^{239}_{94}$Pu (atomic mass = 239.052 16 u) undergoes α decay. Assuming all the released energy is in the form of kinetic energy of the α particle and ignoring the recoil of the daughter nucleus, find the speed of the α particle.

**44. Both gold $^{198}_{79}$Au ($T_{1/2}$ = 2.69 days) and iodine $^{131}_{53}$I ($T_{1/2}$ = 8.04 days) are used in diagnostic medicine related to the liver. At the time laboratory supplies are monitored, the activity of the gold is observed to be five times greater than the activity of the iodine. How many days later will the two activities be equal?

CHAPTER 39

Ionizing Radiation, Nuclear Energy, and Elementary Particles

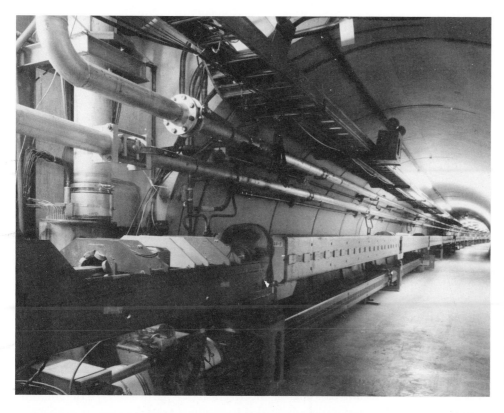

A portion of the particle accelerator at the Fermi National Accelerator Laboratory at Batavia, Illinois. The accelerator is 6.3 kilometers in circumference and is capable of accelerating protons to extremely high energies. The high-energy protons interact with matter to produce a variety of elementary particles.

39.1 BIOLOGICAL EFFECTS OF IONIZING RADIATION

TERMS AND UNITS

Ionizing radiation consists of photons and/or moving particles that have sufficient energy to knock an electron out of an atom or molecule, thus forming an ion. The photons usually lie in the ultraviolet, X-ray, or γ-ray regions of the electromagnetic spectrum, while the moving particles can be the α and β particles emitted during radioactive decay. An energy of roughly 1 to 35 eV is needed to ionize a molecule, and the particles and γ rays emitted during nuclear disintegration often have energies of several million eV. Therefore, a single α particle, β particle, or γ ray can ionize thousands of molecules.

Nuclear radiation is potentially harmful to humans, because the ionization it produces can significantly alter the structure of molecules within a living cell. The alterations cause the cell to malfunction and, if severe enough, can lead to the death of the cell and even the organism itself. Despite the potential hazards, ionizing radiation can be used in medicine for diagnostic and therapeutic purposes, such as locating bone fractures and treating cancer. Since radiation may produce deleterious effects, however, it is important to learn the fundamentals of radiation exposure, including dose units, and the biological effects of radiation.

Exposure is a measure of the ionization produced in air by X rays or γ rays, and it is defined in the following manner. A beam of X rays or γ rays is sent through a mass m of dry air at standard temperature and pressure (STP: 0 °C, 1 atm pressure). In passing through the air, the beam produces positive ions whose total charge is q. Exposure is defined as the total charge per unit mass of air: exposure $= q/m$. The SI unit of exposure is C/kg. However, the first radiation unit to be defined was the *roentgen* (R), and it is still used today. The exposure in roentgens is given by

$$\text{Exposure (in roentgens)} = \left(\frac{1}{2.58 \times 10^{-4}}\right)\frac{q}{m} \qquad (39.1)$$

Thus, an exposure of one roentgen produces $q = 2.58 \times 10^{-4}$ coulombs of positive charge in $m = 1$ kg of dry air, so 1 R $= 2.58 \times 10^{-4}$ C/kg [dry air, at STP].

Being defined in terms of the ionizing abilities of X rays and γ rays in air, exposure does not specify the effect of radiation on living tissue. For biological purposes, the **absorbed dose** is a more suitable quantity, because it is the energy absorbed from the radiation per unit mass of absorbing material:

$$\text{Absorbed dose} = \frac{\text{Energy absorbed}}{\text{Mass of absorbing material}} \qquad (39.2)$$

The SI unit of absorbed dose is the *gray* (Gy), which is a unit of energy divided by a unit of mass: 1 Gy $= 1$ J/kg. The absorbed dose is a concept that is applicable to all types of radiation and absorbing media.

Another unit is often used for absorbed dose, namely, the *rad* (rd). The word "rad" is an acronym for radiation absorbed dose. The rad and the gray are related by 1 rad $= 0.01$ gray.

The amount of biological damage produced by ionizing radiation is different for different kinds of radiation. For instance, a 1-rad dose of neutrons is far more effective in producing eye cataracts than a 1-rad dose of X rays. To compare the damage caused by different types of radiation, the **relative biological effectiveness** (RBE) is used.* The relative biological effectiveness of a particular type of radiation compares the dose of that radiation needed to produce a certain biological effect to the dose of 200-keV X rays needed to produce the same biological effect:

$$\begin{array}{l}\text{Relative biological} \\ \text{effectiveness (RBE)}\end{array} = \frac{\begin{array}{c}\text{The dose of 200-keV X rays that} \\ \text{produces a certain biological effect}\end{array}}{\begin{array}{c}\text{The dose of radiation that} \\ \text{produces the same biological effect}\end{array}} \qquad (39.3)$$

The RBE depends on the nature of the ionizing radiation and its energy, as well as the type of tissue being irradiated. Table 39.1 lists some typical RBE values for different kinds of radiation, assuming an "average" biological tissue is being irradiated. The values of RBE = 1 indicate that γ rays and β^- particles produce the same biological damage as do 200-keV X rays. The larger RBE values indicate that pro-

TABLE 39.1 Relative Biological Effectiveness (RBE) for Various Types of Radiation on Average Tissue

Type of Radiation	RBE
200-keV X rays	1
γ rays	1
β^- particles	1
Protons	10
α particles	10–20
Neutrons	
Slow	2
Fast	10

* The RBE is sometimes called the *quality factor* (QF).

tons, α particles, and fast neutrons cause substantially more damage. The RBE is often used in conjunction with the absorbed dose to reflect the damage-producing character of the radiation on tissue. The product of the absorbed dose in rads (not in grays) and the RBE is the *biologically equivalent dose:*

$$\begin{matrix} \text{Biologically equivalent dose} \\ \text{(in rem)} \end{matrix} = \begin{matrix} \text{Absorbed dose} \\ \text{(in rad)} \end{matrix} \times \text{RBE} \qquad (39.4)$$

The unit for the biologically equivalent dose is the *rem,* short for roentgen equivalent, man. Example 1 illustrates the use of the biologically equivalent dose.

EXAMPLE 1

A biological tissue is irradiated with γ rays that have an RBE of 0.70 for this type of specimen. The absorbed dose of γ rays is 850 rd. The tissue is then exposed to neutrons whose RBE is 3.5. The biologically equivalent dose of neutrons is the same as that of γ rays. What is the absorbed dose of neutrons?

SOLUTION

The biologically equivalent dose of γ rays is the product of the absorbed dose in rads and the RBE:

$$\begin{matrix} \text{Biologically} \\ \text{equivalent dose} \\ \text{of } \gamma \text{ rays} \end{matrix} = (850 \text{ rad})(0.70) = 6.0 \times 10^2 \text{ rem} \qquad (39.4)$$

For neutrons (RBE = 3.5), the biologically equivalent dose is the same. Therefore, 6.0×10^2 rem = (Absorbed dose of neutrons)(3.5) and

$$\begin{matrix} \text{Absorbed dose} \\ \text{of neutrons} \end{matrix} = \frac{6.0 \times 10^2 \text{ rem}}{3.5} = \boxed{170 \text{ rd}}$$

This result can also be obtained in another way. Because the RBE of neutrons is five times greater than that of γ rays, neutrons are five times more effective in damaging the tissue. Therefore, to produce the same damage as γ rays, the absorbed dose of neutrons needs to be only one-fifth as great as that of γ rays: $\frac{1}{5}(850 \text{ rad}) = 170 \text{ rad}$.

THE EFFECTS OF RADIATION ON HUMANS

Everyone is continually exposed to background radiation from natural sources, such as cosmic rays (high-energy particles that come from outside the solar system), radioactive materials in the environment, and radioactive nuclei—primarily carbon $^{14}_{6}$C and potassium $^{40}_{19}$K—within our own bodies. Table 39.2 lists the average biologically equivalent doses received from these sources by a person in the United States

TABLE 39.2 Average Biologically Equivalent Dose of Radiation Received in 1970 by a U.S. Resident

Source of Radiation	Biologically Equivalent Dose (mrem/yr)[a]
Natural background	
Cosmic rays	44
Radioactive earth and air	40
Internal radioactive nuclei	18
Man-made radiation	
Medical/dental diagnostics	74
Nuclear fallout	4
Total:	180

[a] 1 mrem = 10^{-3} rem.

during 1970. To the natural background of radiation, a significant amount of artificially produced radiation has been added, mostly from medical/dental diagnostic X rays. Table 39.2 indicates an average total dose of 180 mrem/yr from all sources.

The effects of radiation on humans can be grouped into two categories, according to the time span between initial exposure and the appearance of physiological effects: (1) short-term or acute effects that appear within a matter of minutes, days, or weeks, and (2) long-term or latent effects that appear years, decades, or even generations later.

Radiation sickness is the general term applied to the acute effects of radiation. Depending on the severity of the dose, a person with radiation sickness can exhibit nausea, vomiting, fever, diarrhea, and loss of hair. Ultimately, death can occur. The severity of radiation sickness is related to the dose received, and in the following discussion the biologically equivalent doses quoted are whole-body, single doses. A dose less than 50 rem causes no short-term ill effects. A dose between 50 and 300 rem brings on radiation sickness, the severity increasing with increasing dosage. A whole-body dose in the range of 400 – 500 rem is classified as an LD_{50} dose, meaning that it is a lethal dose (LD) for about 50% of the people so exposed; death occurs within a few months. Whole-body doses greater than 600 rem result in death for almost all individuals.

The long-term or latent effects of radiation may appear as a result of high-level, brief exposure or low-level exposure over a long period of time. Some long-term effects are loss of hair, eye cataracts, and various kinds of cancer. In addition, genetic defects caused by mutated genes may be passed on from one generation to the next.

Because of the hazards of radiation, the federal government has established dose limits. The permissible dose for an individual is defined as the dose, accumulated over a long period of time or resulting from a single exposure, that carries negligible probability of a severe health hazard. Federal standards state that an individual in the general population should not receive more than 170 mrem of man-made radiation each year, *exclusive* of medical sources. A person exposed to radiation in the workplace (e.g., a radiation therapist) should not receive more than 5 rem per year from work-related sources.

39.2 INDUCED NUCLEAR REACTIONS

Section 38.4 discusses how a radioactive parent nucleus disintegrates spontaneously into a daughter nucleus. It is also possible to bring about or "induce" the disintegration of an otherwise stable nucleus by striking it with another nucleus, an atomic or subatomic particle, or a γ-ray photon. A *nuclear reaction* is said to occur whenever the incident nucleus, particle, or photon causes a change to occur in a target nucleus.

In 1919 Ernest Rutherford observed that when an α particle strikes a nitrogen nucleus, an oxygen nucleus and a proton are produced. This nuclear reaction is written as

$$\underbrace{{}_{2}^{4}\text{He}}_{\substack{\text{Incident} \\ \alpha \text{ particle}}} + \underbrace{{}_{7}^{14}\text{N}}_{\substack{\text{Nitrogen} \\ \text{(target)}}} \rightarrow \underbrace{{}_{8}^{17}\text{O}}_{\text{Oxygen}} + \underbrace{{}_{1}^{1}\text{H}}_{\text{Proton}}$$

Because the incident α particle induces the transmutation of nitrogen into oxygen, this reaction is an example of an *induced nuclear transmutation.*

Nuclear reactions are often written in a shorthand form. For example, the reaction above is designated by $^{14}_{7}N(\alpha, p)^{17}_{8}O$. The first and last symbols represent the initial and final nuclei, respectively. The symbols inside the parentheses denote the incident particle (on the left) and the small emitted particle (on the right). Some other induced nuclear transmutations are listed below, together with the corresponding shorthand notations:

Nuclear Reaction	Notation
$^{1}_{0}n + {}^{10}_{5}B \rightarrow {}^{7}_{3}Li + {}^{4}_{2}He$	$^{10}_{5}B(n, \alpha)^{7}_{3}Li$
$\gamma + {}^{25}_{12}Mg \rightarrow {}^{24}_{11}Na + {}^{1}_{1}H$	$^{25}_{12}Mg(\gamma, p)^{24}_{11}Na$
$^{1}_{1}H + {}^{13}_{6}C \rightarrow {}^{14}_{7}N + \gamma$	$^{13}_{6}C(p, \gamma)^{14}_{7}N$

In any nuclear reaction, both the total electric charge of the nucleons and the total number of nucleons are conserved during the process. The fact that these quantities are conserved makes it possible to identify the nucleus produced in a nuclear reaction, as the next example illustrates.

EXAMPLE 2

An α particle strikes an aluminum $^{27}_{13}Al$ nucleus, and a nucleus $^{A}_{Z}X$ and a neutron are produced:

$$^{4}_{2}He + {}^{27}_{13}Al \rightarrow {}^{A}_{Z}X + {}^{1}_{0}n$$

Identify the nucleus produced, including its atomic number Z (the number of protons) and its atomic mass number A (the number of nucleons).

SOLUTION

Since the total electric charge of the nucleons and the total number of nucleons are conserved, it is possible to write the equations listed below:

Conserved Quantity	Before Reaction		After Reaction
Total electric charge (number of protons)	$2 + 13$	$=$	$Z + 0$
Total number of nucleons	$4 + 27$	$=$	$A + 1$

Solving these equations for Z and A gives $Z = 15$ and $A = 30$. Since $Z = 15$ identifies the element as phosphorus, the nucleus is $\boxed{^{30}_{15}P}$.

Induced nuclear transmutations can be used to produce isotopes that are not found naturally. In 1934, Enrico Fermi suggested a method for producing elements with a higher atomic number than uranium ($Z = 92$). These elements—neptunium ($Z = 93$), plutonium ($Z = 94$), americium ($Z = 95$), and so on—are known as *transuranium elements.* None of the transuranium elements occurs naturally. They are created in a nuclear reaction between a suitably chosen lighter element and a small incident particle, usually a neutron or an α particle. For example, Figure 39.1

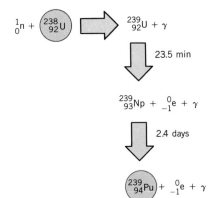

FIGURE 39.1 A nuclear reaction in which $^{238}_{92}U$ is transmuted into the transuranium element plutonium $^{239}_{94}Pu$.

shows a reaction that produces plutonium from uranium. A neutron is captured by a uranium $^{238}_{92}U$ nucleus, producing $^{239}_{92}U$ and a γ ray. The $^{239}_{92}U$ nucleus is radioactive and decays with a half-life of 23.5 min into neptunium $^{239}_{93}Np$. Neptunium is also radioactive and disintegrates with a half-life of 2.4 days into plutonium $^{239}_{94}Pu$. Plutonium is the final product and has a half-life of 24 100 yr.

39.3 NUCLEAR FISSION

THE FISSION PROCESS

In 1939 four German scientists, Otto Hahn, Lise Meitner, Fritz Strassmann, and Otto Frisch, made an important discovery that ushered in the atomic age. They found that a uranium nucleus, after absorbing a neutron, splits into two fragments, each with a smaller mass than the original nucleus. The splitting of a massive nucleus into two less-massive fragments is known as *nuclear fission.*

Figure 39.2 shows a fission reaction in which a uranium $^{235}_{92}U$ nucleus is split into barium $^{141}_{56}Ba$ and krypton $^{92}_{36}Kr$ nuclei. The reaction begins when $^{235}_{92}U$ absorbs a slowly moving neutron, creating a so-called "compound nucleus," $^{236}_{92}U$. The compound nucleus disintegrates quickly into $^{141}_{56}Ba$, $^{92}_{36}Kr$, and three neutrons according to the following reaction:

$$\underset{}{^{1}_{0}n} + ^{235}_{92}U \rightarrow \underbrace{^{236}_{92}U}_{\substack{\text{Compound}\\\text{nucleus}\\\text{(unstable)}}} \rightarrow \underbrace{^{141}_{56}Ba}_{\text{Barium}} + \underbrace{^{92}_{36}Kr}_{\text{Krypton}} + 3^{1}_{0}n$$

This reaction is only one of the many possible reactions that can occur when uranium fissions. For example, another reaction is

$$^{1}_{0}n + ^{235}_{92}U \rightarrow \underbrace{^{236}_{92}U}_{\substack{\text{Compound}\\\text{nucleus}\\\text{(unstable)}}} \rightarrow \underbrace{^{140}_{54}Xe}_{\text{Xenon}} + \underbrace{^{94}_{38}Sr}_{\text{Strontium}} + 2^{1}_{0}n$$

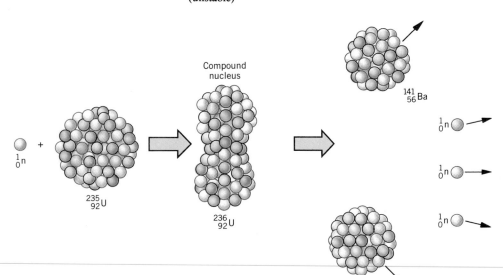

FIGURE 39.2 The slow neutron causes the uranium nucleus $^{235}_{92}U$ to fission into barium $^{141}_{56}Ba$, krypton $^{92}_{36}Kr$, and three neutrons.

Some reactions produce as many as 5 neutrons; however, the average number produced per fission is 2.5.

The fission of uranium is accompanied by the release of an enormous amount of energy, which appears primarily as kinetic energy of the fission fragments. An average of roughly 200 MeV of energy is released per fission. This energy is approximately 10^8 times greater than the energy released per molecule in an ordinary chemical reaction, such as the combustion of gasoline or coal. Example 3 demonstrates how to estimate the energy released during the fission of a nucleus.

EXAMPLE 3

Estimate the amount of energy released when a massive nucleus ($A = 240$) fissions.

SOLUTION

Figure 38.5 shows that the binding energy of a nucleus with $A = 240$ is about 7.6 MeV per nucleon. We assume that this nucleus fissions into two fragments, each with $A \approx 120$. According to Figure 38.5, the binding energy of the fragments increases to about 8.5 MeV per nucleon. Consequently, when a massive nucleus fissions, there is a release of about 8.5 MeV − 7.6 MeV = 0.9 MeV of energy per nucleon. Since there are 240 nucleons involved in the fission process, the total energy released per fission is approximately (0.9 MeV/nucleon)(240 nucleons) $\approx$ $\boxed{200 \text{ MeV}}$.

Virtually all naturally occurring uranium is composed of two isotopes. These isotopes and their natural abundances are $^{238}_{92}\text{U}$ (99.275%) and $^{235}_{92}\text{U}$ (0.720%). Although $^{238}_{92}\text{U}$ is by far the most abundant isotope, the probability that it will capture a neutron and fission is very small. For this reason, $^{238}_{92}\text{U}$ is not the isotope of choice for generating nuclear energy. In contrast, the isotope $^{235}_{92}\text{U}$ readily captures a neutron and fissions, *provided the neutron has a kinetic energy of about 0.04 eV or less.* A neutron with such an energy is called a ***thermal neutron,*** since the energy is comparable to the average thermal energy of a molecule at room temperature. The probability of a thermal neutron causing $^{235}_{92}\text{U}$ to fission is about five hundred times greater than a neutron whose energy is relatively high, say 1 MeV. Thermal neutrons can also be used to fission other nuclei, such as plutonium $^{239}_{94}\text{Pu}$.

CHAIN REACTION

The fact that the uranium fission reaction releases 2.5 neutrons, on the average, makes it possible for a self-sustaining series of fissions to occur. As Figure 39.3 illustrates, each neutron released by a fission can initiate another fission process, resulting in the emission of still more neutrons, followed by more fissions, and so on. A ***chain reaction*** is a series of nuclear fissions whereby the neutrons produced by each fission cause additional fissions. During an uncontrolled chain reaction, it would not be unusual for the number of fissions to increase a thousandfold within a few millionths of a second. With an average energy of about 200 MeV being released per fission, an uncontrolled chain reaction can generate an incredible amount of energy in a very short time, as happens in an atomic bomb (which is actually a *nuclear bomb*).

By limiting the number of neutrons in the environment of the fissioning nuclei, it is possible to establish a condition whereby each fission event contributes, on average, only *one neutron* that fissions another nucleus (see Figure 39.4). In this manner, the chain reaction and the rate of energy production are *controlled*. The controlled fission chain reaction is the principle behind the nuclear reactors used in the commercial generation of electric power.

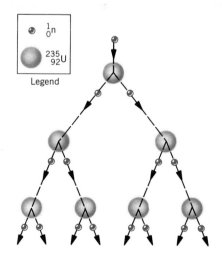

Legend: ^1_0n, $^{235}_{92}\text{U}$

FIGURE 39.3 A chain reaction. For clarity, it is assumed that each fission generates two neutrons (2.5 neutrons are actually liberated on the average, and the fission fragments are not shown).

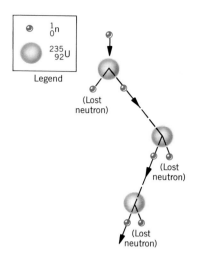

FIGURE 39.4 In a *controlled* chain reaction, only *one neutron,* on average, from each fission event causes another nucleus to fission. As a result, energy is released at a steady or controlled rate.

A nuclear explosion.

39.4 NUCLEAR REACTORS

BASIC COMPONENTS

A nuclear reactor is a kind of furnace in which energy is generated by a controlled fission chain reaction. The first nuclear reactor was built by Enrico Fermi in 1942, on the floor of a squash court under the west stands of Stagg Field at the University of Chicago. Today, there are many kinds and sizes of reactors, but they all have three basic components: fuel elements, a neutron moderator, and control rods. Figure 39.5 illustrates these components.

The *fuel elements* contain the fissionable fuel and, for example, may be in the

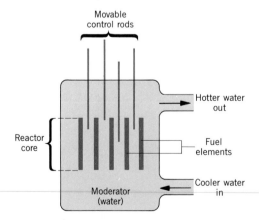

FIGURE 39.5 A nuclear reactor consists of fuel elements, control rods, and a moderator (in this case, water).

shape of thin rods about 1 cm in diameter. In a large power reactor there may be thousands of fuel elements placed close together, the entire region of fuel elements being known as the **reactor core.**

Uranium $^{235}_{92}U$ is a common reactor fuel. Since the natural abundance of this isotope is only about 0.7%, there are special uranium-enrichment plants to increase the percentage. Most commercial reactors use uranium in which the amount of $^{235}_{92}U$ is enriched to about 3%.

While neutrons with energies of about 0.04 eV or less readily fission $^{235}_{92}U$, the neutrons released during the fission process have significantly greater energies of several MeV or so. Consequently, a nuclear reactor must contain some type of material that will decrease or moderate the speed of such energetic neutrons so they can readily fission additional $^{235}_{92}U$ nuclei. The material that slows down the neutrons is called a **moderator.** One commonly used moderator is ordinary water. When an energetic neutron leaves a fuel element, the neutron enters the surrounding water and collides with water molecules. With each collision, the neutron loses an appreciable fraction of its energy and slows down. Once slowed down to thermal energy by the moderator, a process that takes less than 10^{-3} s, the neutron is capable of initiating a fission event upon reentering a fuel element.

If the output power from a reactor is to remain constant, only one neutron from each fission event must trigger a new fission, as Figure 39.4 suggests. When each fission leads to one additional fission — no more or no less — the reactor is said to be *critical.* A reactor normally operates in a critical condition, for then it produces a steady output of energy. The reactor is *subcritical* when, on average, the neutrons from each fission trigger *less than one* subsequent fission. In a subcritical reactor, the chain reaction is not self-sustaining and eventually dies out. When the neutrons from each fission trigger *more than one* additional fission, the reactor is *supercritical.* During a supercritical condition, the energy released by a reactor increases. If left unchecked, the increasing energy could lead to a partial or total meltdown of the reactor core, with the possible release of radioactive material into the environment.

Clearly, a control mechanism is needed to keep the reactor in its normal or critical state. This control is accomplished by a number of **control rods** that can be moved into and out of the reactor core (see Figure 39.5). The control rods contain an element, such as boron or cadmium, that readily absorbs neutrons without fissioning. If the reactor becomes supercritical, the control rods are automatically moved farther into the core to absorb the excess neutrons causing the condition. In response, the reactor returns to its critical state. Conversely, if the reactor becomes subcritical, the control rods are partially withdrawn from the core, so fewer neutrons are absorbed. Thus, more neutrons are available for fission, and the reactor returns to its critical state.

THE PRESSURIZED WATER REACTOR

Figure 39.6 illustrates a pressurized water reactor. In such a reactor, the heat generated within the fuel rods is carried away by water that surrounds the rods. To remove as much heat as possible, the water is heated above 300 °C. To prevent boiling, which occurs at 100 °C at 1 atmosphere of pressure, the water is pressurized in excess of 150 atmospheres. The hot water is pumped through a heat exchanger, where heat is transferred to water flowing in a second, closed system. The heat transferred to the second system produces steam that drives a turbine. The turbine is coupled to an electric generator, whose output electrical power is delivered to consumers via high-voltage transmission lines. After exiting the turbine, the steam is condensed back into water that is returned to the heat exchanger.

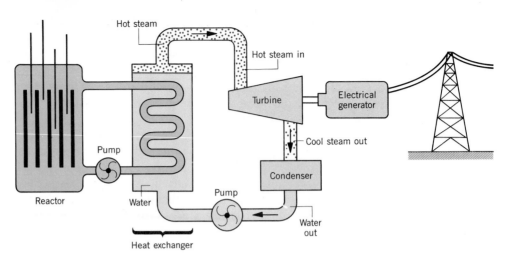

FIGURE 39.6 Diagram of a nuclear power plant that uses a pressurized water reactor.

THE BREEDER REACTOR

In a uranium reactor, only the less-abundant $^{235}_{92}\text{U}$ isotope produces nuclear energy, because the more abundant $^{238}_{92}\text{U}$ isotope does not fission with thermal neutrons. However, $^{238}_{92}\text{U}$ can be converted into plutonium $^{239}_{94}\text{Pu}$ by the nuclear reaction shown in Figure 39.1. This isotope of plutonium can be fissioned with thermal neutrons and is as good a nuclear fuel as $^{235}_{92}\text{U}$. Reactors have been built that produce energy and, at the same time, create or "breed" *more* plutonium than is consumed in the operation of the reactor itself; such a reactor is a ***breeder reactor.***

In a breeder reactor, there are two components in the core material. One component is fissionable (like $^{239}_{94}\text{Pu}$), and the other is nonfissionable (like $^{238}_{92}\text{U}$) but can be converted into the fissionable material. When a thermal neutron causes a plutonium nucleus to fission, three neutrons are released on the average. One of these neutrons is used to fission another plutonium nucleus, so as to sustain the chain reaction. The remaining neutrons can be captured by uranium $^{238}_{92}\text{U}$ nuclei, leading to the production of more plutonium nuclei. In this fashion, the reactor can breed more plutonium fuel than is being consumed. For example, at the Phénix demonstration breeder reactor in France, the breeding ratio is about 1.05, meaning that 21 fissionable nuclei are created for every 20 nuclei that are fissioned. Currently, a demonstration breeder reactor is being constructed at Clinch River, Tennessee. The breeding ratio of this reactor is expected to be about 1.3.

Despite the breeder's potential advantage, not a single *commercial* breeder is producing electricity in the United States, and it is unlikely that any will do so in the foreseeable future. There are many reasons for this situation. Perhaps the paramount reason is the substantial public concern about the safety of nuclear reactors in general, a concern that has been amplified by the accidents involving the nonbreeder reactors at Three Mile Island and Chernobyl. Moreover, plutonium is a very dangerous material, and the widespread use of breeder reactors would certainly increase the chances of weapons-grade material coming into the hands of terrorists.

39.5 NUCLEAR FUSION

In Example 3 of Section 39.3, the binding-energy-per-nucleon curve is used to estimate the amount of energy released in the fission process. As summarized in Figure

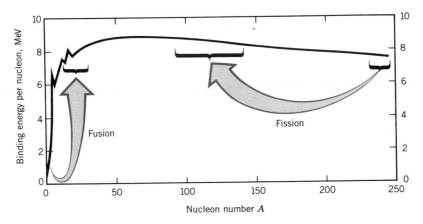

FIGURE 39.7 When fission occurs, a massive nucleus divides into two fragments whose binding energy per nucleon is greater than that of the original nucleus. When fusion occurs, two low-mass nuclei combine to form a more massive nucleus whose binding energy per nucleon is greater than that of the original nuclei.

39.7, the massive nuclei at the right end of the curve have a binding energy of about 7.6 MeV per nucleon, whereas the less-massive fragments are near the center of the curve and have a binding energy of approximately 8.5 MeV per nucleon. The energy released per nucleon by fission is the difference between these two values, or about 0.9 MeV per nucleon.

A glance at the far left end of the diagram suggests another means of generating energy. Two very low-mass nuclei with relatively small binding energies per nucleon, could be combined or "fused" into a single, more massive nucleus that has a greater binding energy per nucleon. This process is called **nuclear fusion.** A substantial amount of energy can be released during a fusion reaction, as Example 4 shows.

EXAMPLE 4

Two isotopes of hydrogen, $^2_1$H (deuterium) and $^3_1$H (tritium), fuse to form $^4_2$He according to the following reaction:

$$^2_1\text{H} + ^3_1\text{H} \rightarrow ^4_2\text{He} + ^1_0\text{n}$$

Determine the energy released by this fusion reaction.

SOLUTION

The masses of the initial and final nuclei in this reaction are given in the isotope table in Appendix G:

The mass defect is $\Delta m = (5.030 \text{ u} - 5.012 \text{ u}) = 0.018$ u. Since 1 u is equivalent to 931.5 MeV, the energy released is $\boxed{17 \text{ MeV}}$.

There are five nucleons that participate in the fusion, so the energy released per nucleon is about 3.4 MeV. This energy per nucleon is greater than that released in a fission process (≈ 0.9 MeV per nucleon). Thus, for a given mass of fuel, a fusion reaction yields more energy than a fission reaction.

Initial Masses		Final Masses	
$^2_1$H	2.014 u	$^4_2$He	4.003 u
$^3_1$H	3.016 u	$^1_0$n	1.009 u
Total:	5.030 u	Total:	5.012 u

Because fusion reactions release substantial amounts of energy, there is considerable interest in fusion reactors, although to date no commercial units have been constructed. The difficulties in building a fusion reactor arise mainly because the two low-mass nuclei must be brought sufficiently near each other so that the short-range strong nuclear force can pull them together, leading to fusion. But each nucleus has a positive charge and repels the other electrically. For the nuclei to get sufficiently close in the presence of the repulsive electric force, they must have large kinetic energies to start with. Experiments show that kinetic energies of about 0.01 MeV are needed to start a fusion reaction.

In Chapter 17, we saw that the average translational kinetic energy $\overline{KE}$ of an atom in an ideal gas is directly proportional to the Kelvin temperature T of the gas according to $\overline{KE} = \frac{3}{2}kT$ (Equation 17.6), where $k = 1.38 \times 10^{-23}$ J/K is the Boltzmann constant. The kinetic energy of 0.01 MeV (1.6×10^{-15} J) needed to start a fusion reaction corresponds to a gas temperature of

$$T = \frac{2(\overline{KE})}{3k} = \frac{2(1.6 \times 10^{-15} \text{ J})}{3(1.38 \times 10^{-23} \text{ J/K})} = 80 \times 10^6 \text{ K}$$

This is 80 million kelvins! Reactions that require such extremely high temperatures are called **thermonuclear reactions.** The most important thermonuclear reactions occur in stars, such as our own sun. The energy radiated by the sun comes from such reactions deep within its core, where the temperature is high enough to initiate the fusion process. One group of reactions thought to occur in the sun is the *proton–proton* cycle. This cycle is a series of reactions whereby six protons form a helium nucleus, two positrons, two γ rays, and two protons. The energy released by the proton–proton cycle is about 27 MeV.

Man-made fusion reactions have been carried out in a fusion-type nuclear bomb —commonly called a hydrogen bomb. In a hydrogen bomb, the fusion reaction is ignited by a fission bomb using uranium or plutonium. The temperature produced by the fission bomb is sufficiently high to initiate a thermonuclear reaction where, for example, hydrogen isotopes are fused into helium, releasing even more energy.

For fusion to be useful as a commercial energy source, the energy must be released in a steady, controlled manner—unlike the uncontrolled energy released by a hydrogen bomb. To date, scientists have not succeeded in constructing a fusion device that produces more energy on a continual basis than is expended in operating the device. A fusion device uses a high temperature to start a reaction, and under such a condition, all the atoms are completely ionized to form a *plasma* (a gas composed of charged particles, like $^2_1\text{H}^+$ and e^-). The problem is to confine the hot plasma for a long enough time so that collisions among the ions can lead to fusion.

One ingenious method of confining the plasma, called *magnetic confinement,* uses a magnetic field. Charges moving in a magnetic field are subject to magnetic forces,

The Tokamak Fusion Test Reactor at Princeton University uses the method of magnetic confinement to contain the hot plasma during nuclear fusion.

The NOVA Laser Facility at the Lawrence Livermore National Laboratory at Lawrence, California. Short pulses of laser light travel simultaneously through each of the cylindrical tubes shown here. The pulses converge inside the spherical chamber where they strike a small pellet containing the hydrogen isotopes deuterium and tritium. The laser pulses heat the pellet to about 100 million degrees Celsius, causing the isotopes to fuse and release energy.

and it is hoped that the plasma can be confined to a region of space by these forces. The problem of magnetic confinement is a difficult one, although steady progress is being made.

Another type of confinement scheme, known as *inertial confinement,* is also being developed. Tiny, solid pellets of fuel are dropped into a container. As each pellet reaches the center of the container, a number of high-intensity lasers or electron beams strike the pellet simultaneously. The heating causes almost instantaneous vaporization of the pellet. However, the inertia of the vaporized atoms keeps them from expanding outward as fast as the vapor is being formed. As a result, high pressures, high densities, and high temperatures are achieved at the center of the pellet, thus causing fusion.

When compared to fission, fusion has some attractive features as an energy source. One type of fuel, ^2_1H (deuterium), is found in the waters of the oceans and is plentiful, cheap, and easy to separate from the common ^1_1H isotope of hydrogen. Fissionable materials like naturally occurring uranium $^{235}_{92}U$ are much less available and supplies could be depleted within a century or two. Another attractive feature of fusion is that it produces a minimal amount of radioactive waste when compared to the waste produced by fission reactors. Unfortunately, the commercial use of fusion to provide cheap, pollution-free energy remains in the future.

39.6 ELEMENTARY PARTICLES

SETTING THE STAGE

By 1932 the electron, the proton, and the neutron had been discovered and were thought to be nature's three *elementary particles,* in the sense that they were the basic

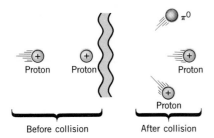

FIGURE 39.8 When an energetic proton collides with a stationary proton, a neutral pion (π^0) is produced. Part of the energy of the incident proton is converted into the pion's mass.

building blocks from which all matter is constructed. Experimental evidence obtained since then, however, shows that several hundred additional particles exist, and scientists now doubt that the proton and the neutron are elementary particles.

Most of these new particles have masses greater than the electron's mass, and many are more massive than protons or neutrons. Virtually all the new particles are unstable and decay with times between about 10^{-6} and 10^{-23} s.

Often, new particles are produced by accelerating protons or electrons to high energies and letting them collide with a target nucleus. For example, Figure 39.8 shows a collision between an energetic proton and a stationary proton. If the incoming proton has sufficient energy, the collision produces an entirely new particle, the *neutral pion* (π^0). The π^0 particle lives for only about 0.8×10^{-16} s before it decays into two γ-ray photons. Since the pion did not exist before the collision, the pion was created from part of the incident proton's energy. Because a new particle such as the neutral pion is often created from energy, it is customary to report the *rest-mass energy* of the particle in energy units of MeV. For instance, detailed analyses of experiments reveal that the rest mass of the π^0 particle is equivalent to an energy of 135.0 MeV. For comparison, the more massive proton has a rest-mass energy of 938.3 MeV. Analyses of experiments also provide the electric charge and other properties of particles created in high-energy collisions.

In the limited space available here, it is not possible to describe all the new particles that have been found. However, we will highlight some of the more significant discoveries.

NEUTRINOS

In 1930 Wolfgang Pauli suggested that a particle called the **neutrino** (now known as the electron neutrino) should accompany the β decay of a radioactive nucleus. As Section 38.5 discusses, the neutrino has no electric charge, has a very small (possibly zero) rest mass, and travels near or at the speed of light. Neutrinos were finally discovered in 1956. Today, neutrinos are created in abundance in nuclear reactors and particle accelerators and are thought to be plentiful in the universe.

POSITRONS AND ANTIPARTICLES

The year 1932 saw the discovery of the *positron* (a contraction for "positive electron"). The positron has the same mass as the electron, but carries an opposite charge of $+e$. A collision between a positron and an electron is likely to annihilate both particles, converting mass into electromagnetic energy in the form of γ rays. For this reason, positrons never coexist with ordinary matter for any appreciable length of time.

The positron is an example of an antiparticle, and after its discovery, scientists came to realize that for every particle there is an antiparticle. The antiparticle is a form of matter that has the same mass as the particle, but carries an opposite electric charge (e.g., the electron–positron pair) or a magnetic moment that is oriented in an opposite direction relative to the spin (e.g., the neutrino–antineutrino pair). A few electrically neutral particles, like the photon and the neutral pion (π^0), are their own antiparticles.

Antimatter would consist of positrons and antinucleons, such as antiprotons and antineutrons. While antimatter cannot coexist with matter, there is speculation that regions of the universe might consist entirely of antimatter. Presently, however, there is no experimental evidence to support such speculation.

MUONS AND PIONS

In 1937 the American physicists S. H. Neddermeyer and C. D. Anderson discovered a new charged particle whose mass was roughly 200 times greater than the mass of the electron. The particle is designated by the Greek letter μ (mu) and is known as a *muon.* There are two muons that have the same mass but opposite charge: the particle μ^- and its antiparticle μ^+. The μ^- muon has the same charge as the electron, while the μ^+ muon has the same charge as the positron. Both muons are unstable, with a lifetime of 2.2×10^{-6} s. The μ^- muon decays into an electron (β^-), a muon neutrino (ν_μ), and an electron antineutrino ($\bar{\nu}_e$), according to the following reaction:

$$\mu^- \rightarrow \beta^- + \nu_\mu + \bar{\nu}_e$$

The μ^+ muon decays into a positron (β^+), a muon antineutrino ($\bar{\nu}_\mu$), and an electron neutrino (ν_e):

$$\mu^+ \rightarrow \beta^+ + \bar{\nu}_\mu + \nu_e$$

Muons interact with protons and neutrons via the weak nuclear force.

The Japanese physicist Hidekei Yukawa (1907–1981) predicted in 1935 that *pions* exist, but they were not discovered until 1947. Pions come in three varieties: one that is positively charged, the negatively charged antiparticle with the same mass, and the neutral pion, mentioned earlier, which is its own antiparticle. The symbols for these pions are, respectively, π^+, π^-, and π^0. The charged pions are unstable and have a lifetime of 2.6×10^{-8} s. The decay of a charged pion produces a muon:

$$\pi^- \rightarrow \mu^- + \bar{\nu}_\mu$$
$$\pi^+ \rightarrow \mu^+ + \nu_\mu$$

As mentioned earlier, the neutral pion π^0 is also unstable and decays into two γ-ray photons, the lifetime being 0.8×10^{-16} s.

The pions are of great interest because, unlike the muons, the pions interact with protons and neutrons via the strong nuclear force. When pions are brought sufficiently close to a nucleus, their interaction with the nucleons causes the breakup of the nucleus, leading to the creation of new particles.

CLASSIFICATION OF PARTICLES

It is useful to group the known particles into three families, the photons, the leptons, and the hadrons, as Table 39.3 summarizes. This grouping is made according to the nature of the force by which a particle interacts with other particles. The *photon family,* for instance, has only one member, the photon. The photon interacts only with charged particles, and the interaction is only via the *electromagnetic force.* No other particle behaves in this manner.

The *lepton family* consists of particles that interact by means of the *weak nuclear force.* Leptons can also exert gravitational and electromagnetic (if the leptons are charged) forces on other particles. The four better-known leptons are the electron, the muon, the electron neutrino ν_e, and the muon neutrino ν_μ. Table 39.3 lists these particles together with their antiparticles. Recently, two other leptons have been discovered, the tau particle (τ) and its neutrino (ν_τ), bringing the number of particles in the lepton family to six.

The *hadron family* contains the particles that interact by means of the *strong*

TABLE 39.3 Some Elementary Particles

Family	Particle	Particle Symbol	Antiparticle Symbol	Rest-Mass Energy (MeV)	Lifetime (s)
Photon	Photon	γ	Self[a]	0	Stable
Lepton	Electron	e^- (or β^-)	e^+ (or β^+)	0.511	Stable
	Muon	μ^-	μ^+	105.7	2.2×10^{-6}
	Tau	τ^-	τ^+	1784	10^{-13}
	Electron neutrino	ν_e	$\bar{\nu}_e$	≈ 0	Stable
	Muon neutrino	ν_μ	$\bar{\nu}_\mu$	≈ 0	Stable
	Tau neutrino	ν_τ	$\bar{\nu}_\tau$	≈ 0	Stable
Hadron					
Mesons					
	Pion	π^+	π^-	139.6	2.6×10^{-8}
		π^0	Self[a]	135.0	0.8×10^{-16}
	Kaon	K^+	K^-	493.7	1.2×10^{-8}
		K^0_S	$\bar{K}^0_S$	497.7	0.9×10^{-10}
		K^0_L	$\bar{K}^0_L$	497.7	5.2×10^{-8}
	Eta	η^0	Self[a]	548.8	$< 10^{-18}$
	.				
	.				
	.				
	Plus other mesons				
Baryons					
	Proton	p	$\bar{p}$	938.3	Stable
	Neutron	n	$\bar{n}$	939.6	900
	Lambda	Λ^0	$\bar{\Lambda}^0$	1116	2.6×10^{-10}
	Sigma	Σ^+	$\bar{\Sigma}^-$	1189	0.8×10^{-10}
		Σ^0	$\bar{\Sigma}^0$	1192	6×10^{-20}
		Σ^-	$\bar{\Sigma}^+$	1197	1.5×10^{-10}
	Omega	Ω^-	Ω^+	1672	0.82×10^{-10}
	.				
	.				
	.				
	Plus other baryons				

[a] The particle is its own antiparticle.

nuclear force. Hadrons can also interact by gravitational and electromagnetic forces, but at short distances ($\leq 10^{-15}$ m) the strong nuclear force dominates. Among the hadrons are the proton, the neutron, and the pions. As Table 39.3 indicates, most hadrons are short-lived. The hadrons are subdivided into two groups, the **mesons** and the **baryons,** for a reason that will be discussed in connection with the idea of quarks.

QUARKS

As more and more hadrons were discovered, it became clear that they were not all elementary particles. The suggestion was made that the hadrons are made up of smaller, more elementary particles called **quarks.** In 1963 a quark theory was advanced independently by M. Gell-Mann (1929–   ) and G. Zweig (1937–   ).

TABLE 39.4 Quarks and Antiquarks

| Name | Quarks | | Antiquarks | |
	Symbol	Charge	Symbol	Charge
Up	u	$+\frac{2}{3}e$	$\bar{u}$	$-\frac{2}{3}e$
Down	d	$-\frac{1}{3}e$	$\bar{d}$	$+\frac{1}{3}e$
Strange	s	$-\frac{1}{3}e$	$\bar{s}$	$+\frac{1}{3}e$
Charm	c	$+\frac{2}{3}e$	$\bar{c}$	$-\frac{2}{3}e$
Top	t	$+\frac{2}{3}e$	$\bar{t}$	$-\frac{2}{3}e$
Bottom	b	$-\frac{1}{3}e$	$\bar{b}$	$+\frac{1}{3}e$

The theory proposed that there are three quarks and three corresponding antiquarks, and that hadrons are constructed from combinations of these. Thus, the quarks are elevated to the status of elementary particles for the hadron family. The particles in the photon and lepton families are considered to be elementary, and as such they are not composed of quarks.

The three quarks were named *up* (u), *down* (d), and *strange* (s), and were assumed to have, respectively, fractional charges of $+\frac{2}{3}e$, $-\frac{1}{3}e$, and $-\frac{1}{3}e$. In other words, a quark possesses a charge smaller than the charge of an electron. Table 39.4 lists the symbols and electric charges of these quarks. Experimentally, quarks should be recognizable by their fractional charges, but in spite of an extensive search for them, free quarks have never been found.

According to the original quark theory, the mesons are different from the baryons, for each meson consists of only two quarks—a quark and an antiquark—while a baryon contains three quarks. For instance, the π^- pion (a meson) is composed of a d quark and a $\bar{u}$ antiquark, $\pi^- = d + \bar{u}$, as Figure 39.9 shows. These two quarks combine to give the π^- pion a net charge of $-e$. Similarly, the π^+ pion is a combination of the $\bar{d}$ and u quarks, $\pi^+ = \bar{d} + u$. In contrast, protons and neutrons, being baryons, consist of three quarks. A proton contains the combination $d + u + u$, while a neutron contains the combination $d + d + u$ (see Figure 39.9). These groups of three quarks give the correct charge for the proton and neutron.

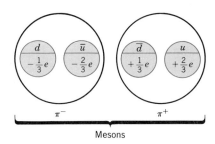

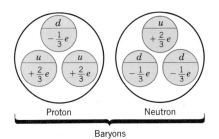

FIGURE 39.9 According to the original quark model of hadrons, all mesons consist of a quark and an antiquark, while baryons contain three quarks.

The original quark model was extremely successful in predicting not only the correct charges for the hadrons, but other properties as well. Except for the fact that no one had succeeded in isolating a quark, the quark theory was a phenomenal success. However, in 1974 a new particle, the J/ψ meson, was discovered. This meson has a rest-mass energy of 3100 MeV, far higher than other known mesons. The existence of the J/ψ meson could be explained only if a new quark–antiquark pair existed; this new quark was named *charm* (*c*). With the discovery of more and more particles, it has been necessary to postulate a fifth and a sixth quark; their names are *top* (*t*) and *bottom* (*b*), although some scientists prefer to call these quarks *truth* and *beauty*. Today, there is a grand total of six quarks (up, down, strange, charm, top, and bottom), and each has a corresponding antiquark. All the hundreds of the known hadrons can be accounted for in terms of these six quarks and their antiquarks. Whether the story is complete, however, remains to be seen.

SUMMARY

Ionizing radiation consists of photons and/or moving particles that have enough energy to ionize an atom or molecule. **Exposure** is a measure of the ionization produced in air by X rays or γ rays. An exposure of one roentgen (R) produces 2.58×10^{-4} coulombs of positive charge in 1 kg of dry air at STP conditions.

The **absorbed dose** is the amount of energy absorbed from the radiation per unit mass of absorbing material. The SI unit of absorbed dose is the gray (Gy); 1 Gy = 1 J/kg. However, the rad (rd) is another unit that is often used; 1 rd = 0.01 Gy.

The amount of biological damage produced by ionizing radiation is different for different types of radiation. The **relative biological effectiveness** (RBE) compares the dose of a given type of radiation needed to produce a certain biological effect to the dose of 200-keV X rays required to produce the same effect. The **biologically equivalent dose** is the product of the absorbed dose (in rads) and the RBE. The unit for the biologically equivalent dose is the rem.

An **induced nuclear transmutation** is the process whereby an incident particle or photon strikes a nucleus and causes the production of a new element.

Nuclear fission occurs when a massive nucleus splits into two less-massive fragments. Fission can be induced by the absorption of a thermal neutron. When a massive nucleus fissions, mass is transformed into energy when the binding energy per nucleon is greater for the fragments than for the original nucleus. Neutrons are also released during nuclear fission. These neutrons can, in turn, induce other nuclei to fission and lead to a process known as a **chain reaction**. A **fission reactor** is a device that generates energy by a controlled chain reaction.

In a **fusion** process, two nuclei with smaller masses combine to form a single nucleus with a larger mass. Energy is released by fusion when the binding energy per nucleon is greater for the larger nucleus than for the smaller nuclei. For fusion to occur, the temperature of the nuclei must be on the order of eighty million kelvins.

Subatomic particles are divided into three families: the **photon** family, the **lepton** family, and the **hadron** family. **Elementary particles** are the basic building blocks of matter. It is believed that all members of the photon and lepton families are elementary particles. The quark theory proposes that the hadrons are not elementary particles, but are composed of elementary particles called **quarks**. Currently, the hundreds of hadron particles can be accounted for in terms of six quarks and their antiquarks.

QUESTIONS

1. When a dentist x-rays your teeth, a lead apron is placed over your chest and lower body. What is the purpose of this apron?

2. State whether the two quantities in the following cases are related and, if so, give the relation between them: (a) rads and rems, (b) rads and grays, and (c) rads and roentgens.

3. Explain why the following reactions are *not* allowed: (a)

$^{60}_{28}$Ni(α, p)$^{62}_{29}$Cu, (b) $^{27}_{13}$Al(n, n)$^{28}_{13}$Al, (c) $^{39}_{19}$K(p, α)$^{36}_{17}$Cl.

4. Why is it possible for a thermal neutron (i.e., one with a small amount of energy) to penetrate a nucleus, whereas a proton or an α particle would need a large amount of energy to penetrate the same nucleus?

5. Would a release of energy accompany the fission of a nucleus of mass number 50 into two fragments of about equal mass? Using the curve in Figure 39.7, account for your answer.

6. In the fission of $^{235}_{92}$U there are, on the average, 2.5 neutrons

released per fission. Suppose a *different* element is being fissioned and, on the average, only 1.0 neutron is released per fission. If a small fraction of the thermal neutrons absorbed by the nuclei does *not* produce a fission, can a self-sustaining chain reaction be produced using this element? Justify your answer.

7. The mass of coal consumed in a coal-burning electric power plant is about two million times greater than the mass of $^{235}_{92}$U used to fuel a comparable nuclear power plant. Why?

8. Explain the difference between fission and fusion and why each process produces energy.

PROBLEMS

Section 39.1 Biological Effects of Ionizing Radiation

1. A beam of γ rays passes through 4.0×10^{-3} kg of dry air and generates 1.7×10^{12} singly charged ions. What is the exposure (in roentgens)?

2. A film badge worn by a radiologist indicates that she has received an absorbed dose of 2.5×10^{-3} Gy. The mass of the radiologist is 65 kg. How much energy has she absorbed?

3. What absorbed dose of α particles (RBE = 20) causes as much biological damage as a 60-rad dose of protons (RBE = 10)?

4. A person who receives a 500-rem dose of proton radiation (RBE = 10) has a 50% chance of dying within a few months or so. What is the absorbed dose (in rads) of this radiation?

* 5. Someone stands near a radioactive source and receives doses of the following types of radiation: γ rays (20 mrad, RBE = 1), electrons (30 mrad, RBE = 1), protons (4 mrad, RBE = 10), and slow neutrons (5 mrad, RBE = 2). What is the total biologically equivalent dose received by this person?

* 6. A water sample receives a 750-rad dose of radiation. Find the rise in the water temperature.

Section 39.2 Induced Nuclear Reactions

7. What is the nucleon number A in the reaction $^{27}_{13}$Al $(\alpha, n)^{A}_{15}$P?

8. Write the reactions below in the shorthand form discussed in the text.

(a) $^{27}_{13}$Al + $^{1}_{0}$n $\rightarrow$ $^{27}_{12}$Mg + $^{1}_{1}$H
(b) $^{16}_{8}$O + $^{2}_{1}$H $\rightarrow$ $^{14}_{7}$N + $^{4}_{2}$He
(c) $^{40}_{18}$Ar + $^{4}_{2}$He $\rightarrow$ $^{43}_{19}$K + $^{1}_{1}$H

9. Write the equation for the reaction $^{17}_{8}$O(γ, αn)$^{12}_{6}$C. The notation "αn" means that an α particle and a neutron are produced by the reaction.

10. Complete the following nuclear reactions, assuming the unknown quantity signified by the question mark is a single entity:

(a) $^{43}_{20}$Ca(α, ?)$^{46}_{21}$Sc
(b) $^{9}_{4}$Be(?, n)$^{12}_{6}$C
(c) $^{9}_{4}$Be(p, α) ?
(d) ? $(\alpha, p)^{17}_{8}$O
(e) $^{55}_{25}$Mn(n, γ) ?

* 11. During a nuclear reaction, an unknown particle is absorbed by a copper $^{63}_{29}$Cu nucleus, and the reaction products are $^{62}_{29}$Cu, a neutron, and a proton. What is the name, atomic number, and nucleon number of the compound nucleus?

Section 39.3 Nuclear Fission, Section 39.4 Nuclear Reactors

12. Determine the number of neutrons released during the following fission reaction: $^{1}_{0}$n + $^{235}_{92}$U $\rightarrow$ $^{133}_{51}$Sb + $^{99}_{41}$Nb + neutrons.

13. $^{235}_{92}$U absorbs a thermal neutron and fissions into rubidium $^{93}_{37}$Rb and cesium $^{141}_{55}$Cs. What other *nucleons* are produced by the fission, and how many are there?

14. During an underground nuclear test, an atomic bomb is detonated. The bomb produces an amount of energy equivalent to 36 kilotons of TNT (1 kiloton of TNT releases about 5×10^{12} J of energy). How much of the original mass of $^{235}_{92}$U is transformed into energy?

15. Uranium $^{235}_{92}$U fissions into two fragments plus three neutrons: $^{1}_{0}$n + $^{235}_{92}$U $\rightarrow$ (2 fragments) + 3^{1}_{0}n. The rest mass of a neutron is 1.008 665 u and that of $^{235}_{92}$U is 235.043 924 u. If 225.0 MeV of energy is released during the fission, what is the combined rest mass of the two fragments?

16. A particular fission reaction produces an energy of 210 MeV per fission. How many fissions occur per second if a reactor is generating 130 MW of power?

* 17. When 1 kg of coal is burned, about 3×10^7 J of energy is released. If the energy released per $^{235}_{92}$U fission is 200 MeV, how many kilograms of coal must be burned to produce the same energy as 1 kg of $^{235}_{92}$U?

* 18. (a) If each fission of a $^{235}_{92}$U nucleus releases about 200 MeV of energy, determine the energy (in joules) released by the complete fissioning of 1 gram of $^{235}_{92}$U. (b) How many grams of

$^{235}_{92}$U are consumed in one year, in order to supply the energy needs of a household that uses 30 kWh of energy per day, on the average?

* **19.** The water that cools a reactor core enters the reactor at 216 °C and leaves at 287 °C. (The water is pressurized, so it does not turn to steam.) The core is generating 5.6×10^9 W of power. Assume the specific heat of water is 4420 J/(kg·C°) over the temperature range stated above, and find the mass of water that passes through the core each second.

** **20.** A 20-kiloton atomic bomb releases as much energy as 20 kilotons of TNT (about 1×10^{14} J). Recall that about 200 MeV of energy is released when each $^{235}_{92}$U nucleus fissions. (a) How many $^{235}_{92}$U nuclei are fissioned to produce the bomb's energy? (b) How many grams of uranium are fissioned? (c) How many kilograms of rest mass are transformed into energy when the bomb explodes?

** **21.** A nuclear power plant is 25% efficient, meaning that 25% of the power it generates goes into producing usable electricity. The remaining 75% is wasted as heat. The plant generates 8.0×10^8 watts of usable electric power. If each fission releases 2.0×10^2 MeV of energy, how many kilograms of $^{235}_{92}$U are fissioned per year?

Section 39.5 Nuclear Fusion

22. Two deuterium ($^2_1$H) nuclei fuse and form $^3_2$He and a neutron. The atomic masses are $^2_1$H (2.0141 u), $^3_2$He (3.0161 u), and $^1_0$n (1.0087 u). Find the energy (in MeV) released by this process.

23. In one type of fusion reaction a proton fuses with a neutron to form a deuterium nucleus: $^1_1$H + $^1_0$n → $^2_1$H. The masses are $^1_1$H (1.0078 u), $^1_0$n (1.0087 u), and $^2_1$H (2.0141 u). How much energy (in MeV) is released by this reaction?

* **24.** Imagine your car is powered by a fusion engine in which the following reaction occurs: 3^2_1H → $^4_2$He + $^1_1$H + $^1_0$n. The masses are $^2_1$H (2.0141 u), $^4_2$He (4.0026 u), $^1_1$H (1.0078 u), and $^1_0$n (1.0087 u). The engine uses 6.1×10^{-6} kg of deuterium $^2_1$H fuel. If one gallon of gasoline produces 2.1×10^9 J of energy, how many gallons of gasoline would have to be burned to equal the energy released by all the deuterium fuel?

* **25.** Deuterium ($^2_1$H) is an attractive fuel for fusion reactions because it is abundant in the waters of the oceans. In the oceans, deuterium makes up about 0.015% of the hydrogen in the water (H_2O). (a) How many deuterium atoms are there in one kilogram of water? (b) If each deuterium nucleus produces about 7.5 MeV in a fusion reaction, how many kilograms of ocean water would be needed to supply the energy needs of the United States for one year (estimated to be 1×10^{20} J)?

Section 39.6 Elementary Particles

26. A high-energy proton collides with a stationary proton, and the reaction $p + p \rightarrow n + p + \pi^+$ occurs. The rest-mass energy of the π^+ pion is 139.6 MeV. Ignore momentum conservation and find the minimum energy the incident proton must have.

27. A collision between two protons produces three new parti-

cles: $p + p \rightarrow p + \pi^+ + \Lambda^0 + K^0$. The rest-mass energies of the new particles are π^+ (139.6 MeV), Λ^0 (1116 MeV), and K^0 (497.7 MeV). Note that one proton disappears during the reaction. How much energy (in MeV) is transformed into rest mass during this reaction?

28. An electron and its antiparticle annihilate each other, producing two γ-ray photons. The kinetic energies of the particles are negligible. For each photon, determine its (a) energy (in MeV), (b) wavelength, and (c) momentum.

* **29.** Suppose a neutrino is created and has an energy of 35 MeV. (a) If the neutrino, like the photon, has no rest mass and travels at the speed of light, find the momentum of the neutrino. (b) Determine the de Broglie wavelength of the neutrino.

* **30.** An energetic proton is fired at a stationary proton. For the reaction to produce new particles, the two protons must approach each other to within a distance of about 8.0×10^{-15} m. The moving proton must have a sufficient speed to overcome the repulsive Coulomb force. What must be the minimum initial kinetic energy (in MeV) of the proton?

ADDITIONAL PROBLEMS

31. During an X-ray examination, a person is exposed to radiation at a rate of 110 milligrays per hour. The exposure time is 0.10 s, and the mass of the exposed tissue is 1.2 kg. Determine the energy absorbed.

32. A Σ^+ particle (see Table 39.3) decays into a π^0 particle and a proton: $\Sigma^+ \rightarrow \pi^0 + p$. Ignore the kinetic energy of the Σ^+ particle, and determine how much energy is released in the process.

33. A nitrogen $^{14}_7$N nucleus absorbs a deuterium $^2_1$H nucleus during a nuclear reaction. What is the name, atomic number, and nucleon number of the compound nucleus?

34. What energy (in MeV) is liberated by the following fission reaction?

$$\underbrace{^1_0\text{n}}_{1.009 \text{ u}} + \underbrace{^{235}_{92}\text{U}}_{235.044 \text{ u}} \rightarrow \underbrace{^{141}_{56}\text{Ba}}_{140.914 \text{ u}} + \underbrace{^{92}_{36}\text{Kr}}_{91.926 \text{ u}} + 3^1_0\text{n}$$

35. Within the core of a nuclear reactor there are 3×10^{19} nuclei fissioning each second. The energy released by each fission is about 200 MeV. Determine the power (in watts) being generated.

* **36.** During an X-ray examination of the chest, a person receives an exposure of 15 mR. How many singly charged ions would be produced if the X rays passed through 2.0 m^3 of dry air at STP conditions?

* **37.** One proposed fusion reaction combines lithium $^6_3$Li (6.015 u) with deuterium $^2_1$H (2.014 u) to give helium $^4_2$He (4.003 u): $^2_1$H + $^6_3$Li → 2^4_2He. How many kilograms of lithium

would be needed to supply the energy needs of one household for a year (estimated to be 3.8×10^{10} J)?

* **38.** The energy consumed in one year in the United States is about 1×10^{20} J. When each $^{235}_{92}$U nucleus fissions, about 200 MeV of energy is released. How many kilograms of $^{235}_{92}$U would be needed to generate this energy if all the nuclei fissioned?

* **39.** Breeder reactors produce $^{239}_{94}$Pu. Suppose this nucleus fissions into two fragments whose mass ratio is $0.32 : 0.68$. With the aid of Figure 39.7, estimate the energy released during this fission.

** **40.** One kilogram of dry air at STP conditions is exposed to 1.0 R of X rays. One roentgen is defined by Equation 39.1. An equivalent definition can be based on the fact that an exposure of one roentgen deposits 0.83×10^{-2} J of energy per kilogram of dry air. Using the two definitions of the roentgen, determine the average energy (in eV) needed to produce a single ion in air.

Powers of Ten and Scientific Notation

In science, very large and very small decimal numbers are conveniently expressed in terms of powers of ten, some of which are listed below:

$$10^3 = 10 \times 10 \times 10 = 1000 \qquad 10^{-3} = \frac{1}{10 \times 10 \times 10} = 0.001$$

$$10^2 = 10 \times 10 = 100 \qquad 10^{-2} = \frac{1}{10 \times 10} = 0.01$$

$$10^1 = 10 \qquad 10^{-1} = \frac{1}{10} = 0.1$$

$$10^0 = 1$$

Using powers of ten, we can write the radius of the earth in the following way, for example:

$$\text{Earth radius} = 6\ 380\ 000\ \text{m} = 6.38 \times 10^6\ \text{m}$$

The factor of ten raised to the sixth power is ten multiplied by itself six times, or one million, so the earth's radius is 6.38 million meters. Alternatively, the factor of ten raised to the sixth power indicates that the decimal point in the term 6.38 is to be moved six places *to the right* to obtain the radius as a number without powers of ten.

For numbers less than one, negative powers of ten are used. For instance, the Bohr radius of the hydrogen atom is

$$\text{Bohr radius} = 0.000\ 000\ 000\ 0529\ \text{m} = 5.29 \times 10^{-11}\ \text{m}$$

The factor of ten raised to the minus eleventh power indicates that the decimal point in the term 5.29 is to be moved eleven places *to the left* to obtain the radius as a number without powers of ten. Numbers expressed with the aid of powers of ten are said to be in *scientific notation.*

Calculations that involve the multiplication and division of powers of ten are carried out as in the following examples:

$$(2.0 \times 10^6)(3.5 \times 10^3) = (2.0 \times 3.5) \times 10^{6+3} = 7.0 \times 10^9$$

$$\frac{9.0 \times 10^7}{2.0 \times 10^4} = \left(\frac{9.0}{2.0}\right) \times 10^7 \times 10^{-4} = \left(\frac{9.0}{2.0}\right) \times 10^{7-4} = 4.5 \times 10^3$$

The general rules for such calculations are:

$$\frac{1}{10^n} = 10^{-n} \qquad\qquad\qquad (\text{A-1})$$

$$10^n \times 10^m = 10^{n+m} \tag{A-2}$$

$$\frac{10^n}{10^m} = 10^{n-m} \tag{A-3}$$

where n and m are any positive or negative number.

Scientific notation is convenient because of the ease with which it can be used in calculations. Moreover, scientific notation provides a convenient way to express the significant figures in a number, as Appendix B discusses.

APPENDIX B

Significant Figures

The number of *significant figures* in a number is the number of digits whose values are known with certainty. For instance, a person's height is measured to be 1.78 m, with the measurement error being in the third decimal place. All three digits are known with certainty, and the number contains three significant figures. If a zero is given as the last digit to the right of the decimal point, the zero is presumed to be significant, so the number 1.780 m contains four significant figures. As another example, consider a distance of 1500 m. This number contains only two significant figures, the one and the five. The zeros immediately to the left of the unexpressed decimal point are not counted as significant figures. However, zeros located between significant figures are significant, so a distance of 1502 m contains four significant figures.

Scientific notation is particularly convenient from the point of view of significant figures. Suppose it is known that a certain distance is fifteen hundred meters, to four significant figures. Writing the number as 1500 m presents a problem, because it implies that only two significant figures are known. In contrast, the scientific notation of 1.500×10^3 m has the advantage of indicating that the distance is known to four significant figures.

When two or more numbers are used in a calculation, the number of significant figures in the answer is limited by the number of significant figures in the original data. For instance, a rectangular garden with sides of 9.8 m and 17.1 m has an area of (9.8 m)(17.1 m). A calculator gives 167.58 m² for this product. However, one of the original lengths is known only to two significant figures, so the final answer is limited to only two significant figures and should be rounded off to 170 m². In general, *when numbers are multiplied or divided, the final answer has a number of significant figures that equals the smallest number of significant figures in any of the original factors.*

The number of significant figures in the answer to an addition or subtraction is also limited by the original data. Consider the total distance covered by a biker's trail that consists of three segments with the distances shown below:

$$
\begin{array}{rl}
2.5 & \text{km} \\
11 & \text{km} \\
\underline{5.26} & \text{km} \\
\text{Total} \quad 18.76 & \text{km}
\end{array}
$$

The distance of 11 km contains no significant figures to the right of the decimal point. Therefore, neither does the sum of the three distances, so the total distance should not be reported as 18.76 km. Instead, the answer is rounded off to 19 km. In general, *when numbers are added or subtracted, the last significant figure in the answer occurs in the last column (counting from left to right) containing a number that results from a sum of digits that are all significant.* Thus, in the answer of 18.76 km, the eight is the sum of 2 + 1 + 5, each digit being significant. However, the seven is the sum of 5 + 0 + 2, and the zero is not significant, since it comes from the 11-km distance, which contains no significant figures to the right of the decimal point.

APPENDIX C

Algebra

C1 PROPORTIONS AND EQUATIONS

Physics deals with physical variables and the relations between them. Typically, variables are represented by the letters of the English and Greek alphabets. Sometimes, the relation between variables is expressed as a proportion or inverse proportion. Other times, however, it is more convenient or necessary to express the relation by means of an equation, which is governed by the rules of algebra.

If two variables are ***directly proportional*** and one of them doubles, then the other variable also doubles. Similarly, if one variable is reduced to one-half its original value, then the other is also reduced to one-half its original value. In general, if x is directly proportional to y, then increasing or decreasing one variable by a given factor causes the other variable to change in the same way by the same factor. This kind of relation is expressed as $x \propto y$, where the symbol $\propto$ means "is proportional to."

Since proportional variables x and y always increase and decrease by the same factor, the ratio of x to y must have a constant value, or $x/y = k$, where k is a constant, independent of the values for x and y. Consequently, a proportionality such as $x \propto y$ can also be expressed in the form of an equation: $x = ky$. The constant k is referred to as a ***proportionality constant.***

If two variables are ***inversely proportional*** and one of them increases by a given factor, then the other decreases by the same factor. An inverse proportion is written as $x \propto 1/y$. This kind of proportionality is equivalent to the following equation: $xy = k$, where k is a proportionality constant, independent of x and y.

C2 SOLVING EQUATIONS

Some of the variables in an equation typically have known values, and some do not. It is often necessary to solve the equation so that a variable whose value is unknown is expressed in terms of the known quantities. ***In the process of solving an equation, it is permissible to manipulate the equation in any way, as long as a change made on one side of the equals sign is also made on the other side.*** For example, consider the equation $v = v_0 + at$. Suppose values for v, v_0, and a are available, and the value of t is required. To solve the equation for t, we begin by subtracting v_0 from *both* sides:

$$\begin{array}{rcl} v & = & v_0 + at \\ -v_0 & & -v_0 \\ \hline v - v_0 & = & at \end{array}$$

Next, we divide *both* sides of $v - v_0 = at$ by the quantity a:

$$\frac{v - v_0}{a} = \frac{at}{a} = (1)t$$

On the right, the a in the numerator divided by the a in the denominator equals unity, so that

$$t = \frac{v - v_0}{a}$$

It is always possible to check the correctness of the algebraic manipulations performed in solving an equation by substituting the answer back into the original equation. In the previous example such a substitution proceeds as shown below:

$$v = v_0 + at$$

$$v = v_0 + a\left(\frac{v - v_0}{a}\right) = v_0 + (v - v_0) = v$$

The result $v = v$ implies that our algebraic manipulations were done correctly.

Algebraic manipulations other than addition, subtraction, multiplication, and division may play a role in solving an equation. The same basic rule applies, however: Whatever is done to the left side of an equation must also be done to the right side. As another example, suppose it is necessary to express v_0 in terms of v, a, and s, where $v^2 = v_0^2 + 2as$. By subtracting $2as$ from both sides, we isolate v_0^2 on the right:

$$
\begin{array}{rcl}
v^2 & = & v_0^2 + 2as \\
-2as & & -2as \\
\hline
v^2 - 2as & = & v_0^2
\end{array}
$$

To solve for v_0, we can take the positive or negative square root of *both* sides of $v^2 - 2as = v_0^2$. Taking the positive square root gives

$$v_0 = +\sqrt{v^2 - 2as}$$

C3 SIMULTANEOUS EQUATIONS

When more than one variable in a single equation is unknown, additional equations are needed if solutions are to be found for all of the unknown quantities. Thus, the equation $3x + 2y = 7$ cannot be solved by itself to give values for both x and y. However, if x and y simultaneously obey the equation $x - 3y = 6$, then both unknowns can be found.

There are a number of methods by which such simultaneous equations can be solved. A useful method is to solve one equation for x in terms of y and substitute the result into the other equation to obtain an expression containing only the single unknown variable y. The equation $x - 3y = 6$, for instance, can be solved for x by adding $3y$ to each side:

$$
\begin{array}{rcl}
x - 3y & = & 6 \\
+3y & & +3y \\
\hline
x & = & 6 + 3y
\end{array}
$$

The substitution into the equation $3x + 2y = 7$ is shown below:

$$3x + 2y = 7$$

$$3(6 + 3y) + 2y = 7$$

$$18 + 9y + 2y = 7$$

We find, then, that $18 + 11y = 7$, a result that can be solved for y:

$$\begin{array}{rr} 18 + 11y = & 7 \\ -18 & -18 \\ \hline 11y = & -11 \end{array}$$

Dividing both sides of this result by 11 shows that $y = -1$. The value of $y = -1$ can be substituted in either of the original equations to obtain a value for x:

$$\begin{array}{rcl} x - 3y & = & 6 \\ x - 3(-1) & = & 6 \end{array}$$

$$\begin{array}{rcl} x + 3 & = & 6 \\ -3 & & -3 \\ \hline x & = & 3 \end{array}$$

C4 THE QUADRATIC FORMULA

Equations occur in physics that include the square of a variable. Such equations are said to be *quadratic* in that variable and often can be put into the following form:

$$ax^2 + bx + c = 0 \tag{C-1}$$

where a, b, and c are constants independent of x. This equation can be solved to give the **quadratic formula,** which is

$$x = \frac{-b \pm \sqrt{b^2 - 4ac}}{2a} \tag{C-2}$$

The $\pm$ in the quadratic formula indicates that, in general, there are two solutions. For instance, if

$$2x^2 - 5x + 3 = 0$$

then $a = 2$, $b = -5$, and $c = 3$. According to the quadratic formula, the two solutions for x are as given below:

$\begin{bmatrix} \textbf{Solution 1:} \\ \textbf{Plus sign} \end{bmatrix}$
$$x = \frac{-b + \sqrt{b^2 - 4ac}}{2a}$$
$$= \frac{-(-5) + \sqrt{(-5)^2 - 4(2)(3)}}{2(2)}$$
$$= \frac{+5 + \sqrt{1}}{4} = +\frac{3}{2}$$

$\begin{bmatrix} \textbf{Solution 2:} \\ \textbf{Minus sign} \end{bmatrix}$
$$x = \frac{-b - \sqrt{b^2 - 4ac}}{2a}$$
$$= \frac{-(-5) - \sqrt{(-5)^2 - 4(2)(3)}}{2(2)}$$
$$= \frac{+5 - \sqrt{1}}{4} = +1$$

APPENDIX D

Exponents and Logarithms

Appendix A discusses powers of ten such as 10^3, which means ten times itself three times, or $10 \times 10 \times 10$. The three is referred to as an ***exponent.*** The use of exponents extends beyond powers of ten. In general, the term y^n means the factor y is multiplied by itself n times. For example, y^2, or y squared, is familiar and means $y \times y$. Similarly, y^5 means $y \times y \times y \times y \times y$.

The rules that govern algebraic manipulations of exponents are the same as those given in Appendix A (see Equations A-1, A-2, and A-3) for powers of ten:

$$\frac{1}{y^n} = y^{-n} \tag{D-1}$$

$$y^n y^m = y^{n+m} \qquad \text{(Exponents added)} \tag{D-2}$$

$$\frac{y^n}{y^m} = y^{n-m} \qquad \text{(Exponents subtracted)} \tag{D-3}$$

To the three rules above we add two more that are useful. One of these is

$$y^n z^n = (yz)^n \tag{D-4}$$

The reasoning behind this rule can be clarified by means of an example:

$$3^2 5^2 = (3 \times 3)(5 \times 5) = (3 \times 5)(3 \times 5) = (3 \times 5)^2$$

The other additional rule is

$$(y^n)^m = y^{nm} \qquad \text{(Exponents multiplied)} \tag{D-5}$$

To see why this rule applies, consider the following example:

$$(5^2)^3 = (5^2)(5^2)(5^2) = 5^{2+2+2} = 5^6 = 5^{2 \times 3}$$

Roots, such as a square root or a cube root, can be represented with fractional exponents. For instance,

$$\sqrt{y} = y^{1/2} \quad \text{and} \quad \sqrt[3]{y} = y^{1/3}$$

In general, the nth root of y is given by

$$\sqrt[n]{y} = y^{1/n} \tag{D-6}$$

The rationale for this way of representing roots can be explained using the fact that $(y^n)^m = y^{nm}$. For instance, the fifth root of y is the number that, when multiplied by itself five times, gives back y. As shown below, the term $y^{1/5}$ satisfies this definition:

$$(y^{1/5})(y^{1/5})(y^{1/5})(y^{1/5})(y^{1/5}) = (y^{1/5})^5 = y^{(1/5) \times 5} = y$$

Logarithms are closely related to exponents. To see the connection between the two, note that it is possible to express any number y as another number B raised to the exponent x. In other words,

$$y = B^x \tag{D-7}$$

The exponent x is called the **logarithm** of the number y. The number B is called the **base number.** One of two choices for the base number is usually used. If $B = 10$, the logarithm is known as the *common logarithm,* for which the notation "log" applies:

[Common logarithm] $$y = 10^x \quad \text{or} \quad x = \log y \tag{D-8}$$

If $B = e = 2.718 \ldots$, the logarithm is referred to as the *natural logarithm,* and the notation "ln" is used:

[Natural logarithm] $$y = e^x \quad \text{or} \quad x = \ln y \tag{D-9}$$

The two kinds of logarithms are related by

$$\ln y = 2.3026 \log y \tag{D-10}$$

Both kinds of logarithms are given on many calculators.

The logarithm of the product or quotient of two numbers A and C can be obtained from the logarithms of the individual numbers according to the rules below. These rules are illustrated here for natural logarithms, but they are the same for any kind of logarithm.

$$\ln AC = \ln A + \ln C \tag{D-11}$$

$$\ln \left(\frac{A}{C} \right) = \ln A - \ln C \tag{D-12}$$

Thus, the logarithm of the product of two numbers is the sum of the individual logarithms, and the logarithm of the quotient of two numbers is the difference between the individual logarithms. Another useful rule concerns the logarithm of a number A raised to an exponent n:

$$\ln A^n = n \ln A \tag{D-13}$$

Rules D-11, D-12, and D-13 can be derived from the definition of the logarithm and the rules governing exponents.

APPENDIX E

Geometry and Trigonometry

FIGURE E1

Listed below are some theorems of geometry and trigonometry that are useful in physics.

E1 GEOMETRY

ANGLES

Two angles are equal if

1. They are vertical angles (see Figure E1).
2. Their sides are parallel (see Figure E2).
3. Their sides are mutually perpendicular (see Figure E3).

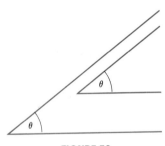

FIGURE E2

TRIANGLES

1. The *sum of the angles* of any triangle is 180° (see Figure E4).
2. A *right triangle* has one angle that is 90°.
3. An *isosceles triangle* has two sides that are equal.
4. An *equilateral triangle* has three sides that are equal. Each angle of an equilateral triangle is 60°.
5. Two triangles are *similar* if two of their angles are equal (see Figure E5). The corresponding sides of similar triangles are proportional to each other:

$$\frac{a_1}{a_2} = \frac{b_1}{b_2} = \frac{c_1}{c_2}$$

6. Two similar triangles are *congruent* if they can be placed on top of one another to make an exact fit.

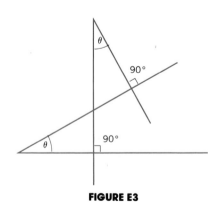

FIGURE E3

CIRCUMFERENCES, AREAS, AND VOLUMES OF SOME COMMON SHAPES

1. Triangle of base b and altitude h (see Figure E6): Area $= \frac{1}{2}bh$.
2. Circle of radius r: Circumference $= 2\pi r$, Area $= \pi r^2$.
3. Sphere of radius r: Surface area $= 4\pi r^2$, Volume $= \frac{4}{3}\pi r^3$.
4. Right circular cylinder of radius r and height h (see Figure E7):

   Surface area $= 2\pi r^2 + 2\pi rh$, Volume $= \pi r^2 h$.

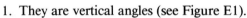

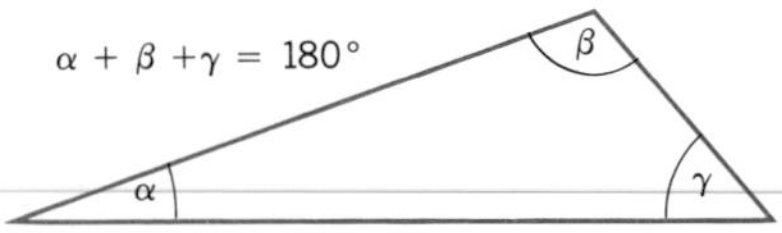

$\alpha + \beta + \gamma = 180°$

FIGURE E4

E2 TRIGONOMETRY

BASIC TRIGONOMETRIC FUNCTIONS

1. The sine, cosine, and tangent of an angle θ are as follows (see Figure E8):

$$\sin \theta = \frac{\text{Side opposite } \theta}{\text{Hypotenuse}} = \frac{h_o}{h}$$

$$\cos \theta = \frac{\text{Side adjacent to } \theta}{\text{Hypotenuse}} = \frac{h_a}{h}$$

$$\tan \theta = \frac{\text{Side opposite } \theta}{\text{Side adjacent to } \theta} = \frac{h_o}{h_a}$$

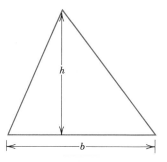

FIGURE E5

2. The secant, cosecant, and cotangent of an angle θ are defined as follows:

$$\sec \theta = \frac{1}{\cos \theta} \qquad \csc \theta = \frac{1}{\sin \theta} \qquad \cot \theta = \frac{1}{\tan \theta}$$

TRIANGLES AND TRIGONOMETRY

1. The **Pythagorean theorem** states that the square of the hypotenuse of a right triangle is equal to the sum of the squares of the other two sides (see Figure E8):

$$h^2 = h_o^2 + h_a^2$$

2. The **law of cosines** and the **law of sines** apply to any triangle, not just a right triangle, and they relate the angles and the lengths of the sides (see Figure E9):

[Law of cosines] $\qquad c^2 = a^2 + b^2 - 2ab \cos \gamma$

[Law of sines] $\qquad \dfrac{a}{\sin \alpha} = \dfrac{b}{\sin \beta} = \dfrac{c}{\sin \gamma}$

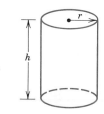

FIGURE E6

FIGURE E7

OTHER TRIGONOMETRIC IDENTITIES

1. $\sin \theta / \cos \theta = \tan \theta$.
2. $\sin^2 \theta + \cos^2 \theta = 1$.
3. $\sin (\alpha \pm \beta) = \sin \alpha \cos \beta \pm \cos \alpha \sin \beta$.

$$\text{If } \alpha = 90°, \sin (90° \pm \beta) = \cos \beta.$$

$$\text{If } \alpha = \beta, \sin 2\beta = 2 \sin \beta \cos \beta.$$

4. $\cos (\alpha \pm \beta) = \cos \alpha \cos \beta \mp \sin \alpha \sin \beta$.

$$\text{If } \alpha = 90°, \cos (90° \pm \beta) = \mp \sin \beta.$$

$$\text{If } \alpha = \beta, \cos 2\beta = \cos^2 \beta - \sin^2 \beta = 1 - 2 \sin^2 \beta.$$

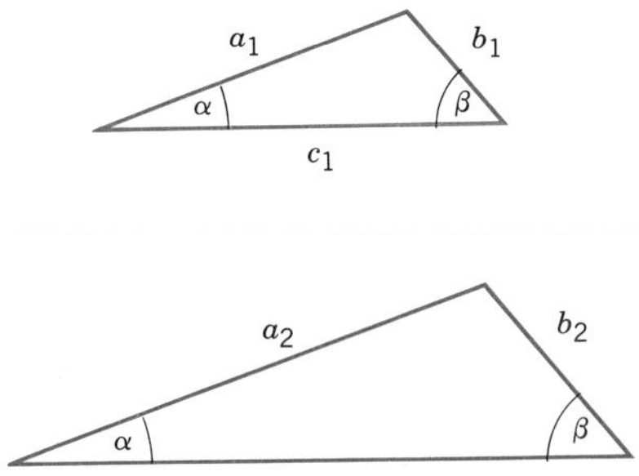

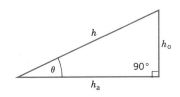

FIGURE E8

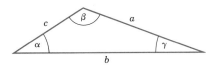

FIGURE E9

APPENDIX F

Trigonometric Table

Angle in Degrees	Angle in Radians	Sine	Cosine	Tangent	Angle in Degrees	Angle in Radians	Sine	Cosine	Tangent
0°	0.000	0.000	1.000	0.000					
1°	0.017	0.017	1.000	0.017	46°	0.803	0.719	0.695	1.036
2°	0.035	0.035	0.999	0.035	47°	0.820	0.731	0.682	1.072
3°	0.052	0.052	0.999	0.052	48°	0.838	0.743	0.669	1.111
4°	0.070	0.070	0.998	0.070	49°	0.855	0.755	0.656	1.150
5°	0.087	0.087	0.996	0.087	50°	0.873	0.766	0.643	1.192
6°	0.105	0.105	0.995	0.105	51°	0.890	0.777	0.629	1.235
7°	0.122	0.122	0.993	0.123	52°	0.908	0.788	0.616	1.280
8°	0.140	0.139	0.990	0.141	53°	0.925	0.799	0.602	1.327
9°	0.157	0.156	0.988	0.158	54°	0.942	0.809	0.588	1.376
10°	0.175	0.174	0.985	0.176	55°	0.960	0.819	0.574	1.428
11°	0.192	0.191	0.982	0.194	56°	0.977	0.829	0.559	1.483
12°	0.209	0.208	0.978	0.213	57°	0.995	0.839	0.545	1.540
13°	0.227	0.225	0.974	0.231	58°	1.012	0.848	0.530	1.600
14°	0.244	0.242	0.970	0.249	59°	1.030	0.857	0.515	1.664
15°	0.262	0.259	0.966	0.268	60°	1.047	0.866	0.500	1.732
16°	0.279	0.276	0.961	0.287	61°	1.065	0.875	0.485	1.804
17°	0.297	0.292	0.956	0.306	62°	1.082	0.883	0.469	1.881
18°	0.314	0.309	0.951	0.325	63°	1.100	0.891	0.454	1.963
19°	0.332	0.326	0.946	0.344	64°	1.117	0.899	0.438	2.050
20°	0.349	0.342	0.940	0.364	65°	1.134	0.906	0.423	2.145
21°	0.367	0.358	0.934	0.384	66°	1.152	0.914	0.407	2.246
22°	0.384	0.375	0.927	0.404	67°	1.169	0.921	0.391	2.356
23°	0.401	0.391	0.921	0.424	68°	1.187	0.927	0.375	2.475
24°	0.419	0.407	0.914	0.445	69°	1.204	0.934	0.358	2.605
25°	0.436	0.423	0.906	0.466	70°	1.222	0.940	0.342	2.747
26°	0.454	0.438	0.899	0.488	71°	1.239	0.946	0.326	2.904
27°	0.471	0.454	0.891	0.510	72°	1.257	0.951	0.309	3.078
28°	0.489	0.469	0.883	0.532	73°	1.274	0.956	0.292	3.271
29°	0.506	0.485	0.875	0.554	74°	1.292	0.961	0.276	3.487
30°	0.524	0.500	0.866	0.577	75°	1.309	0.966	0.259	3.732
31°	0.541	0.515	0.857	0.601	76°	1.326	0.970	0.242	4.011
32°	0.559	0.530	0.848	0.625	77°	1.344	0.974	0.225	4.331
33°	0.576	0.545	0.839	0.649	78°	1.361	0.978	0.208	4.705
34°	0.593	0.559	0.829	0.675	79°	1.379	0.982	0.191	5.145
35°	0.611	0.574	0.819	0.700	80°	1.396	0.985	0.174	5.671

APPENDIX F: Trigonometric Table—*Continued*

Angle in Degrees	Angle in Radians	Sine	Cosine	Tangent	Angle in Degrees	Angle in Radians	Sine	Cosine	Tangent
36°	0.628	0.588	0.809	0.727	81°	1.414	0.988	0.156	6.314
37°	0.646	0.602	0.799	0.754	82°	1.431	0.990	0.139	7.115
38°	0.663	0.616	0.788	0.781	83°	1.449	0.993	0.122	8.144
39°	0.681	0.629	0.777	0.810	84°	1.466	0.995	0.105	9.514
40°	0.698	0.643	0.766	0.839	85°	1.484	0.996	0.087	11.43
41°	0.716	0.656	0.755	0.869	86°	1.501	0.998	0.070	14.30
42°	0.733	0.669	0.743	0.900	87°	1.518	0.999	0.052	19.08
43°	0.750	0.682	0.731	0.933	88°	1.536	0.999	0.035	28.64
44°	0.768	0.695	0.719	0.966	89°	1.553	1.000	0.017	57.29
45°	0.785	0.707	0.707	1.000	90°	1.571	1.000	0.000	—

APPENDIX G

Selected Isotopes[a]

Atomic Number Z	Element	Symbol	Atomic mass Number A	Atomic Mass u	Percent Abundance, or Decay Mode if Radioactive	Half-life (If Radioactive)
0	(Neutron)	n	1	1.008 665	β^-	10.37 min
1	Hydrogen	H	1	1.007 825	99.985	
	Deuterium	D	2	2.014 102	0.015	
	Tritium	T	3	3.016 050	β^-	12.33 yr
2	Helium	He	3	3.016 030	0.000 138	
			4	4.002 603	≈ 100	
3	Lithium	Li	6	6.015 121	7.5	
			7	7.016 003	92.5	
4	Beryllium	Be	7	7.016 928	EC, γ	53.29 days
			9	9.012 182	100	
5	Boron	B	10	10.012 937	19.9	
			11	11.009 305	80.1	
6	Carbon	C	11	11.011 432	β^+, EC	20.39 min
			12	12.000 000	98.90	
			13	13.003 355	1.10	
			14	14.003 241	β^-	5730 yr
7	Nitrogen	N	13	13.005 738	β^+, EC	9.965 min
			14	14.003 074	99.634	
			15	15.000 108	0.366	
8	Oxygen	O	15	15.003 065	β^+, EC	122.2 s
			16	15.994 915	99.762	
			18	17.999 160	0.200	
9	Fluorine	F	18	18.000 937	EC, β^+	1.8295 h
			19	18.998 403	100	
10	Neon	Ne	20	19.992 435	90.51	
			22	21.991 383	9.22	
11	Sodium	Na	22	21.994 434	β^+, EC, γ	2.602 yr
			23	22.989 767	100	
			24	23.990 961	β^-, γ	14.659 h
12	Magnesium	Mg	24	23.985 042	78.99	
13	Aluminum	Al	27	26.981 539	100	
14	Silicon	Si	28	27.976 927	92.23	
			31	30.975 362	β^-, γ	2.622 h
15	Phosphorus	P	31	30.973 762	100	
			32	31.973 907	β^-	14.282 days
16	Sulfur	S	32	31.972 070	95.02	
			35	34.969 031	β^-	87.51 days

APPENDIX G: Selected Isotopes — *Continued*

Atomic Number Z	Element	Symbol	Atomic Mass Number A	Atomic Mass u	Percent Abundance, or Decay Mode if Radioactive	Half-life (If Radioactive)
17	Chlorine	Cl	35	34.968 852	75.77	
			37	36.965 903	24.23	
18	Argon	Ar	40	39.962 384	99.600	
19	Potassium	K	39	38.963 707	93.2581	
			40	39.963 999	β^-, EC, γ	1.277×10^9 yr
20	Calcium	Ca	40	39.962 591	96.941	
21	Scandium	Sc	45	44.955 910	100	
22	Titanium	Ti	48	47.947 947	73.8	
23	Vanadium	V	51	50.943 962	99.750	
24	Chromium	Cr	52	51.940 509	83.789	
25	Manganese	Mn	55	54.938 047	100	
26	Iron	Fe	56	55.934 939	91.72	
27	Cobalt	Co	59	58.933 198	100	
			60	59.933 819	β^-, γ	5.271 yr
28	Nickel	Ni	58	57.935 346	68.27	
			60	59.930 788	26.10	
29	Copper	Cu	63	62.939 598	69.17	
			65	64.927 793	30.83	
30	Zinc	Zn	64	63.929 145	48.6	
			66	65.926 034	27.9	
31	Gallium	Ga	69	68.925 580	60.1	
32	Germanium	Ge	72	71.922 079	27.4	
			74	73.921 177	36.5	
33	Arsenic	As	75	74.921 594	100	
34	Selenium	Se	80	79.916 520	49.7	
35	Bromine	Br	79	78.918 336	50.69	
36	Krypton	Kr	84	83.911 507	57.0	
			89	88.917 640	β^-, γ	3.16 min
			92	91.926 270	β^-, γ	1.840 s
37	Rubidium	Rb	85	84.911 794	72.165	
38	Strontium	Sr	86	85.909 267	9.86	
			88	87.905 619	82.58	
			90	89.907 738	β^-	28.5 yr
			94	93.915 367	β^-, γ	1.235 s
39	Yttrium	Y	89	88.905 849	100	
40	Zirconium	Zr	90	89.904 703	51.45	
41	Niobium	Nb	93	92.906 377	100	
42	Molybdenum	Mo	98	97.905 406	24.13	
43	Technecium	Tc	98	97.907 215	β^-, γ	4.2×10^6 yr
44	Ruthenium	Ru	102	101.904 348	31.6	
45	Rhodium	Rh	103	102.905 500	100	
46	Palladium	Pd	106	105.903 478	27.33	
47	Silver	Ag	107	106.905 092	51.839	
			109	108.904 757	48.161	
48	Cadmium	Cd	114	113.903 357	28.73	
49	Indium	In	115	114.903 880	95.7; β^-	4.41×10^{14} yr
50	Tin	Sn	120	119.902 200	32.59	

APPENDIX G: Selected Isotopes—*Continued*

Atomic Number Z	Element	Symbol	Atomic Mass Number A	Atomic Mass u	Percent Abundance, or Decay Mode if Radioactive	Half-life (If Radioactive)
51	Antimony	Sb	121	120.903 821	57.3	
52	Tellurium	Te	130	129.906 229	33.8; β^-	2.5×10^{21} yr
53	Iodine	I	127	126.904 473	100	
			131	130.906 114	β^-, γ	8.040 days
54	Xenon	Xe	132	131.904 144	26.9	
			136	135.907 214	8.9	
			140	139.921 620	β^-, γ	13.6 s
55	Cesium	Cs	133	132.905 429	100	
			134	133.906 696	β^-, EC, γ	2.062 yr
56	Barium	Ba	137	136.905 812	11.23	
			138	137.905 232	71.70	
			141	140.914 363	β^-, γ	18.27 min
57	Lanthanum	La	139	138.906 346	99.91	
58	Cerium	Ce	140	139.905 433	88.48	
59	Praseodymium	Pr	141	140.907 647	100	
60	Neodymium	Nd	142	141.907 719	27.13	
61	Promethium	Pm	145	144.912 743	EC, α, γ	17.7 yr
62	Samarium	Sm	152	151.919 729	26.7	
63	Europium	Eu	153	152.921 225	52.2	
64	Gadolinium	Gd	158	157.924 099	24.84	
65	Terbium	Tb	159	158.925 342	100	
66	Dysprosium	Dy	164	163.929 171	28.2	
67	Holmium	Ho	165	164.930 319	100	
68	Erbium	Er	166	165.930 290	33.6	
69	Thulium	Tm	169	168.934 212	100	
70	Ytterbium	Yb	174	173.938 859	31.8	
71	Lutecium	Lu	175	174.940 770	97.41	
72	Hafnium	Hf	180	179.946 545	35.100	
73	Tantalum	Ta	181	180.947 992	99.988	
74	Tungsten (wolfram)	W	184	183.950 928	30.67	
75	Rhenium	Re	187	186.955 744	62.60; β^-	4.6×10^{10} yr
76	Osmium	Os	191	190.960 920	β^-, γ	15.4 days
			192	191.961 467	41.0	
77	Iridium	Ir	191	190.960 584	37.3	
			193	192.962 917	62.7	
78	Platinum	Pt	195	194.964 766	33.8	
79	Gold	Au	197	196.966 543	100	
			198	197.968 217	β^-, γ	2.6935 days
80	Mercury	Hg	202	201.970 617	29.80	
81	Thallium	Tl	205	204.974 401	70.476	
			208	207.981 988	β^-, γ	3.053 min
82	Lead	Pb	206	205.974 440	24.1	
			207	206.975 872	22.1	
			208	207.976 627	52.4	
			210	209.984 163	α, β^-, γ	22.3 yr
			211	210.988 735	β^-, γ	36.1 min
			212	211.991 871	β^-, γ	10.64 h

APPENDIX G: Selected Isotopes — *Continued*

Atomic Number Z	Element	Symbol	Atomic Mass Number A	Atomic Mass u	Percent Abundance, or Decay Mode if Radioactive	Half-life (If Radioactive)
			214	213.999 798	β^-, γ	26.8 min
83	Bismuth	Bi	209	208.980 374	100	
			211	210.987 255	α, β^-, γ	2.14 min
			212	211.991 255	β^-, α, γ	1.0092 h
84	Polonium	Po	210	209.982 848	α, γ	138.376 days
			212	211.988 842	α, γ	45.1 s
			214	213.995 176	α, γ	163.69 μs
			216	216.001 889	α, γ	150 ms
85	Astatine	At	218	218.008 684	α, β^-	1.6 s
86	Radon	Rn	220	220.011 368	α, γ	55.6 s
			222	222.017 570	α, γ	3.825 days
87	Francium	Fr	223	223.019 733	α, β^-, γ	21.8 min
88	Radium	Ra	224	224.020 186	α, γ	3.66 days
			226	226.025 402	α, γ	1.6×10^3 yr
			228	228.031 064	β^-, γ	5.75 yr
89	Actinium	Ac	227	227.027 750	α, β^-, γ	21.77 yr
			228	228.031 015	β^-, γ	6.13 h
90	Thorium	Th	228	228.028 715	α, γ	1.913 yr
			231	231.036 298	β^-, γ	1.0633 days
			232	232.038 054	100; α, γ	1.405×10^{10} yr
			234	234.043 593	β^-, γ	24.10 days
91	Protactinium	Pa	231	231.035 880	α, γ	3.276×10^4 yr
			234	234.043 303	β^-, γ	6.70 h
			237	237.051 140	β^-, γ	8.7 min
92	Uranium	U	232	232.037 130	α, γ	68.9 yr
			233	233.039 628	α, γ	1.592×10^5 yr
			235	235.043 924	0.7200; α, γ	7.037×10^8 yr
			236	236.045 562	α, γ	2.342×10^7 yr
			238	238.050 784	99.2745; α, γ	4.468×10^9 yr
			239	239.054 289	β^-, γ	23.47 min
93	Neptunium	Np	239	239.052 933	β^-, γ	2.355 days
94	Plutonium	Pu	239	239.052 157	α, γ	2.411×10^4 yr
			242	242.058 737	α, γ	3.763×10^5 yr
95	Americium	Am	243	243.061 375	α, γ	7.380×10^3 yr
96	Curium	Cm	245	245.065 483	α, γ	8.5×10^3 yr
97	Berkelium	Bk	247	247.070 300	α, γ	1.38×10^3 yr
98	Californium	Cf	249	249.074 844	α, γ	350.6 yr
99	Einsteinium	Es	254	254.088 019	α, γ, β^-	275.7 days
100	Fermium	Fm	253	253.085 173	EC, α, γ	3.00 days
101	Mendelevium	Md	255	255.091 081	EC, α	27 min
102	Nobelium	No	255	255.093 260	EC, α	3.1 min
103	Lawrencium	Lr	257	257.099 480	α, EC	646 ms
104	Rutherfordium	Rf	261	261.108 690	α	1.08 min
105	Hahnium	Ha	262	262.113 760	α	34 s

[a] Data for atomic masses are taken from *Handbook of Chemistry and Physics,* 66th ed., CRC Press, Boca Raton, FL. The masses are those for the neutral atom, including the Z electrons. Data for percent abundance, decay mode, and half-life are taken from E. Browne and R. Firestone, *Table of Radioactive Isotopes,* V. Shirley, Ed., Wiley, New York, 1986. α = alpha particle emission, β^- = negative beta emission, β^+ = positron emission, γ = γ-ray emission, EC = electron capture.

APPENDIX H

Periodic Table of the Elements

Group I	Group II	\<Transition elements\>										Group III	Group IV	Group V	Group VI	Group VII	Group 0
H 1 1.00794 $1s^1$																	**He** 2 4.00260 $1s^2$
Li 3 6.941 $2s^1$	**Be** 4 9.01218 $2s^2$											**B** 5 10.81 $2p^1$	**C** 6 12.011 $2p^2$	**N** 7 14.0067 $2p^3$	**O** 8 15.9994 $2p^4$	**F** 9 18.9984 $2p^5$	**Ne** 10 20.179 $2p^6$
Na 11 22.9898 $3s^1$	**Mg** 12 24.305 $3s^2$											**Al** 13 26.9815 $3p^1$	**Si** 14 28.0855 $3p^2$	**P** 15 30.9738 $3p^3$	**S** 16 32.06 $3p^4$	**Cl** 17 35.453 $3p^5$	**Ar** 18 39.948 $3p^6$
K 19 39.0983 $4s^1$	**Ca** 20 40.08 $4s^2$	**Sc** 21 44.9559 $3d^14s^2$	**Ti** 22 47.88 $3d^24s^2$	**V** 23 50.9415 $3d^34s^2$	**Cr** 24 51.996 $3d^54s^1$	**Mn** 25 54.9380 $3d^54s^2$	**Fe** 26 55.847 $3d^64s^2$	**Co** 27 58.9332 $3d^74s^2$	**Ni** 28 58.69 $3d^84s^2$	**Cu** 29 63.546 $3d^{10}4s^1$	**Zn** 30 65.39 $3d^{10}4s^2$	**Ga** 31 69.72 $4p^1$	**Ge** 32 72.59 $4p^2$	**As** 33 74.9216 $4p^3$	**Se** 34 78.96 $4p^4$	**Br** 35 79.904 $4p^5$	**Kr** 36 83.80 $4p^6$
Rb 37 85.4678 $5s^1$	**Sr** 38 87.62 $5s^2$	**Y** 39 88.9059 $4d^15s^2$	**Zr** 40 91.224 $4d^25s^2$	**Nb** 41 92.9064 $4d^45s^1$	**Mo** 42 95.94 $4d^55s^1$	**Tc** 43 (98) $4d^55s^2$	**Ru** 44 101.07 $4d^75s^1$	**Rh** 45 102.906 $4d^85s^1$	**Pd** 46 106.42 $4d^{10}5s^0$	**Ag** 47 107.868 $4d^{10}5s^1$	**Cd** 48 112.41 $4d^{10}5s^2$	**In** 49 114.82 $5p^1$	**Sn** 50 118.71 $5p^2$	**Sb** 51 121.75 $5p^3$	**Te** 52 127.60 $5p^4$	**I** 53 126.905 $5p^5$	**Xe** 54 131.29 $5p^6$
Cs 55 132.905 $6s^1$	**Ba** 56 137.33 $6s^2$	57–71	**Hf** 72 178.49 $5d^26s^2$	**Ta** 73 180.948 $5d^36s^2$	**W** 74 183.85 $5d^46s^2$	**Re** 75 186.207 $5d^56s^2$	**Os** 76 190.2 $5d^66s^2$	**Ir** 77 192.22 $5d^76s^2$	**Pt** 78 195.08 $5d^96s^1$	**Au** 79 196.967 $5d^{10}6s^1$	**Hg** 80 200.59 $5d^{10}6s^2$	**Tl** 81 204.383 $6p^1$	**Pb** 82 207.2 $6p^2$	**Bi** 83 208.980 $6p^3$	**Po** 84 (209) $6p^4$	**At** 85 (210) $6p^5$	**Rn** 86 (222) $6p^6$
Fr 87 (223) $7s^1$	**Ra** 88 226.025 $7s^2$	89–103	**Rf** 104 (261) $6d^27s^2$	**Ha** 105 (262) $6d^37s^2$	106 (263)	107 (262)		109 (266)									

Lanthanide series (57–71)

La 57 139.906 $5d^16s^2$	**Ce** 58 140.12 $4f^16s^2$	**Pr** 59 140.908 $4f^36s^2$	**Nd** 60 144.24 $4f^46s^2$	**Pm** 61 (145) $4f^56s^2$	**Sm** 62 150.36 $4f^66s^2$	**Eu** 63 151.96 $4f^76s^2$	**Gd** 64 157.25 $5d^14f^76s^2$	**Tb** 65 158.925 $4f^96s^2$	**Dy** 66 162.50 $4f^{10}6s^2$	**Ho** 67 164.930 $4f^{11}6s^2$	**Er** 68 167.26 $4f^{12}6s^2$	**Tm** 69 168.934 $4f^{13}6s^2$	**Yb** 70 173.04 $4f^{14}6s^2$	**Lu** 71 174.967 $5d^14f^{14}6s^2$

Actinide series (89–103)

Ac 89 227.028 $6d^17s^2$	**Th** 90 232.038 $6d^27s^2$	**Pa** 91 231.036 $5f^26d^17s^2$	**U** 92 238.029 $5f^36d^17s^2$	**Np** 93 237.048 $5f^46d^17s^2$	**Pu** 94 (244) $5f^66d^07s^2$	**Am** 95 (243) $5f^76d^07s^2$	**Cm** 96 (247) $5f^76d^17s^2$	**Bk** 97 (247) $5f^96d^07s^2$	**Cf** 98 (251) $5f^{10}6d^07s^2$	**Es** 99 (252) $5f^{11}6d^07s^2$	**Fm** 100 (257) $5f^{12}6d^07s^2$	**Md** 101 (258) $5f^{13}6d^07s^2$	**No** 102 (259) $6d^07s^2$	**Lr** 103 (260) $7s^27p^1$

Key:
Symbol — **Cl** 17 — Atomic number
Atomic mass* — 35.453
$3p^5$ — Electron configuration

* Atomic mass values averaged over isotopes in percentages they occur on earth's surface. For many unstable elements, mass number of the most stable known isotope is given in parentheses.

Source. Handbook of Chemistry and Physics, 68th ed., CRC Press, Boca Raton, FL. Reprinted by permission.

CHAPTER 1

1. (a) 5700 s (b) 86 400 s   3. 27.4 km   5. 4048 m^2
7. 80.1 km, at 25.9° south of west   9. 0.707 m
11. 340 m   13. 35.3°
15. 7.80 km, at 35.0° north of west
17. (a) 5.75 km (b) 50.8°, west of south
19. (a) 67.9 units, at 45° south of west
(b) 67.9 units, at 45° north of west
21. (a) 1300 lb (b) Along the dashed line
23. (a) Vector **B** (b) Vector **B**   25. (a) 57° (b) 240 lb
27. (a) 41.2 m/s (b) 52.8 m/s
29. (a) $A_x = 650$ units (b) $A_{x'} = 570$ units
$\quad\quad A_y = 380$ units $\quad\quad A_{y'} = -480$ units
31. 4.80 km, at 24.0° north of east   33. 30.2 yd, $-10.2°$
35. (a) 192 units (b) 49.7 units   37. 6.88 km, $-26.9°$
39. 44.2 ft   41. 115 m^3
43. 3.00 m, at 42.8° above the $-x$ axis
45. 3.0×10^1 degrees, 51°, 99°
47. (a) 130 N, at 53° south of east
(b) 130 N, at 53° north of west

CHAPTER 2

1. 4.69 s   3. 7.1 m/s   5. 6.25 m/s, down
7. 34 km/h, due north   9. 9.78 ft/s^2
11. 11.3 ft/s, due west
13. 8.0 m/s, The motorcycle with the 2.0 m/s^2 acceleration was initially traveling faster.
15. (a) 2.0×10^1 m (b) 8.0 m/s, down the slope
17. 0.17 s
19. (Equation 2.7) $s - s_0 = \frac{1}{2}(v_0 + v)t$
(Equation 2.8) $s - s_0 = v_0 t + \frac{1}{2}at^2$
(Equation 2.9) $v^2 = v_0^2 + 2a(s - s_0)$
21. 17.7 m   23. (a) 15 s (b) 1100 m   25. 62 ft
27. (a) 15.4 m/s (b) 14.0 m/s (c) 62.5 m
29. (a) 153 m/s, upward (b) 1190 m (c) 15.6 s
31. (a) 40 m/s, upward (b) 80 m (c) 40 m/s, downward
33. 35.6 s   35. 2.0×10^1 m   37. 45 m   39. 8.16 m/s
41. A: 24 km/h B: 8 km/h C: -5 km/h   45. 2 m
47. (a) 4.0 s (b) 4.0 s   49. 33 m/s   51. 0.932 m/s
53. 19 m   55. (a) 26 s (b) 5.8 m/s, down
57. 1.4 m/s^2

CHAPTER 3

1. $x = 75.3$ km, $y = 143$ km   3. (a) 871 km/h
(b) 9.52° below the horizontal
5. (a) 4.42 s (b) 55.3 m   7. (a) 0.57 s (b) 630 m
9. 34 in.
11. (a) 7.82 s (b) 135 m/s (486 km/h) (c) 0.869 km
13. 24   15. 5.3 m/s
17. (a) 58 m (b) 9.4 m/s, 15° below the horizontal
19. (a) 5.5 s (b) 42 m/s (c) 38° below the horizontal
21. (a) 4500 ft (b) 510 ft/s (c) 66° below the horizontal
23. (a) 4.52 m/s, 59.4° above the horizontal (b) 1.27 s
25. (a) 55.0 m/s, 35.1° above the horizontal
(b) $x = 229$ m, $y = 33.7$ m
27. One building is four times higher than the other building.
29. 8.0×10^1 m   31. 5.7 m/s   33. 320 m
37. (a) 135 ft/s, east (b) 135 ft/s, west
39. 24.2 m/s, 21.1° south of east
41. (a) 11.4 m/s, due east (b) 8.30 m/s, 31.3° south of east
43. (a) 19.5° (b) 0.265 h   45. 1700 ft/s
47. (a) 77 ft/s (b) 53° above the horizontal
49. (a) 55°
(b) $t_1 = 18$ s (The projectile hits the plane on the way up.)
$t_2 = 47$ s (The projectile reaches its maximum height and then hits the plane on the way down.)
51. $y_1 = 1.92$ m (The arrow hits the target on the way up.)
$y_2 = 468$ m (The arrow reaches its maximum height and then hits the target on the way down.)
53. 10°   55. 14 ft/s

CHAPTER 4

1. 0.946 m/s^2 3. 3560 N   5. 2
7. (a) 72.1 N, at 56.3° above the $+x$ axis
(b) 92.7 N, at 27.2° above the $+x$ axis
(c) 100.0 N along the $+x$ axis
9. 11.7 lb at $\theta = 88.3°$
11. (a) 3.63×10^{-47} N
(b) The force on the proton is equal in magnitude, but opposite in direction, to the force on the electron.
13. 178   15. (a) 412 N (b) 68.0 N   17. 29 lb
19. $r_J = 11.0\ r_E$ 21. 2.83   23. Yes, $a = 3.72$ m/s^2
25. 1.57 m/s^2 27. 0.235
29. (a) 10.5 m/s^2 (b) 861 N (c) 1.07

31. 8.70×10^{-12} N   **33.** 2.64×10^6 m   **35.** 18.4 m
37. 0.665

CHAPTER 5

1. 58.8 N   **3.** 618 N   **5.** 0.444   **7.** 251 N
9. 286 N   **11.** 40.0 N, at 59.8° above the horizontal
13. 1730 N, due west   **15.** (a) 206 lb   (b) 185 lb
17. (a) 0.861 m/s²   (b) 12.9 m/s   (c) 32.3 N
19. (a) 36 200 N (Tension in the coupling between the engine and the first car)
(b) 9570 N (Tension in the coupling between the first and second cars)
21. 8.17 s   **23.** (a) 0.788 m/s²   (b) 255 N
25. 3.9 m/s²   **27.** (a) 13.7 N   (b) 1.37 m/s²
29. (a) 1.00×10^2 N   (b) 41.6 N
31. (a) 1.10×10^3 N   (b) 931 N   (c) 808 N   (d) 931 N
33. 0.20 m/s², up   **35.** 4290 N
37. (a) 10.0 lb   (b) 8.32 lb
39. (a) 3.56 m/s²   (b) 281 N
41. 56.0 lb (left rope), 71.3 lb (right rope)   **43.** 0.265 m
45. (a) $a = g \tan \theta$ (b) 1.73 m/s²   (c) 0°, No

CHAPTER 6

1. 18.8 in./s², toward the center of the take-up reel
3. One radius is four times greater than the other.
5. (a) $v = 464$ m/s, $a_c = 3.37 \times 10^{-2}$ m/s²
(b) $v = 328$ m/s, $a_c = 2.39 \times 10^{-2}$ m/s²
7. 3.22 m/s²   **9.** (a) 0.191 N
(b) No, the tension increases by a factor of four.
11. 0.75   **13.** 0.187   **15.** 23°
17. (a) $\tan \theta = v^2/rg$ (b) 25.2°   **19.** 4.20×10^4 m/s
21. 1.54×10^9 m   **23.** 7910 m/s
25. (a) 912 m   (b) 228 m   (c) 2.50 m/s²
27. (a) 1.70×10^3 N   (b) 1.65×10^3 N
29. 1310 N, directed toward the center of the circular arc
31. 14.3 m/s   **33.** (a) 3.00×10^4 m/s   (b) 2.02×10^{30} kg
35. 3.98 h (1.43×10^4 s)   **37.** 105 m   **39.** 32.3 m/s
41. (a) The centripetal force is provided by the normal force exerted by the wall on the rider.
(b) 1670 N   (c) 0.323

CHAPTER 7

1. (a) 1.7×10^5 J
(b) Positive, because the force and the displacement are in the same direction.
3. 1.88×10^7 J
5. (a) 1.80×10^3 J   (b) -1.20×10^3 J   (c) 6.0×10^2 J
(d) Because the normal force and the weight are perpendicular to the displacement.

7. 2.0×10^3 N   **9.** 2.33×10^3 ft · lb   **11.** 9.0×10^3 m/s
13. 0.094   **15.** 230 lb   **17.** (a) -32 ft · lb   (b) 32 ft · lb
19. (a) -1090 J
(b) -2.02 m, the skater is below the starting point.
21. 4130 N   **23.** 127 m

25.

Height	KE	PE	E
20.0 m	0	392 J	392 J
15.0 m	98 J	294 J	392 J
10.0 m	196 J	196 J	392 J
5.00 m	294 J	98 J	392 J
0 m	392 J	0	392 J

27. 4.8 m/s   **29.** 42 m   **31.** 6.32 m
33. 3.6×10^6 J   **35.** 73.5 s
37. (a) 990 W (1.3 hp)   (b) 240 W (0.33 hp)
39. 3.04×10^4 W (40.8 hp)   **41.** 6.4×10^5 J
43. 55.8 m/s   **45.** 3.0 m/s   **47.** 3.12 m/s   **49.** 0.327 m

CHAPTER 8

1. (a) $p = 3.00 \times 10^4$ kg · m/s, due north, KE $= 2.25 \times 10^5$ J
(b) The momentum increases by a factor of 3.
(c) The kinetic energy increases by a factor of 9.
3. (a) 1.0×10^8 kg · m/s, due north   (b) 6.8×10^4 m/s
(c) The car has the larger kinetic energy.
5. 1.3 kg · m/s, parallel to the velocity of the ball
(b) 220 N, parallel to the velocity of the ball
7. (a) 8.9 m/s   (b) 0.53 kg · m/s, down
(c) 0.48 kg · m/s, up   (d) 1.01 kg · m/s, up
(e) 2.0×10^1 N, up
9. (a) $m_A/m_B = 8$ (b) $v_A/v_B = \frac{1}{2}$
11. (a) 1.2 m/s
(b) 250 kg · m/s, opposite to the velocity of the swimmer
15. $v_B/v_A = 2.2$
17. (a) 5.25 m/s   (b) 223 J   (c) No   (d) 1.41 m
19. 182 m/s
21. (a) 3.17 m/s, parallel to the initial velocity of the person
(b) 0.0171
23. 0.56 m/s, to the right
25. 7.0 m/s, opposite its initial velocity (for the ball whose initial speed was 4.0 m/s); 4.0 m/s, opposite its initial velocity (for the ball whose initial speed was 7.0 m/s)
27. 12 000 N   **29.** 0.900 N, down   **31.** 313 N
33. (a) 1.86×10^4 kg · m/s, parallel to the initial velocity of the 1550-kg car (The momentum is the same both before and after the collision.)
(b) 1.12×10^5 J (The total kinetic energy is the same both before and after the collision.)
(c) 1.45 m/s (1550-kg car), 13.4 m/s (1220-kg car), (Both velocities are parallel to the initial velocity of the 1550-kg car.)
35. 0.330 m/s, opposite to the horizontal component of the velocity of the stone.

37. (a) 2.13 m/s, up   (b) 0.231 m

39. (a) 1.1 kg · m/s, parallel to the velocity of the bullet
(b) 3.0×10^{-3} s
(c) 370 N, parallel to the acceleration of the bullet

41. (a) 0.54 m/s, opposite to the velocity of the rock
(b) 1.1 m/s, opposite to the velocity of the rock   (c) No

CHAPTER 9

1. (a) 0.13 rev (0.79 rad)   (b) 0.50 rev (3.1 rad)
(c) 1.0 rev (6.3 rad)   (d) 1.3 rev (8.2 rad)

3. (a) 1.00 rev/min (0.105 rad/s), clockwise
(b) 1.67×10^{-2} rev/min (1.75×10^{-3} rad/s), clockwise
(c) 1.39×10^{-3} rev/min (1.45×10^{-4} rad/s), clockwise

5. 1.15 rad/s²   **7.** 1.50×10^3 rev/min

9. (a) 6 pieces   (b) 16.0°   **11.** 6.05 m

13. (a) 4.60×10^3 rad   (b) 2.00×10^2 rad/s²

15. (a) 1900 revolutions   (b) 110 s

17. 7.46 s   **19.** 7.4 s   **21.** 160 m/s (350 mi/h)

23. (a) 1.3 m/s   (b) 8.3 rev/s   **25.** 14.8 rad/s

27. (a) 31.4 rad/s   (b) 5.23 ft/s   (c) 164 ft/s²

29. 1.73 m   **31.** 0.213 s

33. (a) 45.5 rad/s   (b) 15.0 m/s   (c) 7.96 m/s

35. 0.300 m/s   **37.** 0.267 rad/s²

39. (a) 7.30×10^{-5} rad/s   (b) 1.99×10^{-7} rad/s

41. (a) 54.0 rad/s   (b) 486 rad

43. 267 rev/min   **45.** $L_1/L_2 = 1/\sqrt{3}$

47. 0.613 rad (Disk A), 0.088 rad (Disk B)

CHAPTER 10

1. (a) 84 N · m, counterclockwise   (b) 150 N · m, clockwise

3. 0.667 m

5. 0.100 m (Distance from right edge),
0.300 m (Distance from bottom edge)

7. (a) 2.70×10^2 lb   (b) 2.90×10^2 lb, down

9. $T = 56.4$ N, $F = 70.6$ N   **11.** −0.088 m

13. $F_N = 212$ N, The horizontal and vertical components of
the force exerted on her shoes are 212 N and 5.00×10^2 N,
respectively.

15. (a) 2310 N
(b) The horizontal and vertical components of the force that
the wall exerts on the beam are 1630 N and 1550 N, respectively.

17. 123 N   **19.** 1.42×10^{-3} kg · m²

21. (a) 0.131 kg · m²   (b) 3.63×10^{-4} kg · m²
(c) 0.149 kg · m²

23. 2.4 kg   **25.** 460 N   **27.** 0.36 N

29. 1.33 m, 1.67 m   **31.** 2.6×10^{29} J

33. (a) 3.8×10^5 J   (b) 6.2×10^3 J   (c) 3.9×10^5 J

35. $\sqrt{7/5}$ **37.** 4.15 ft/s   **39.** 1.08×10^7 m

41. (a) 0.14 rad/s
(b) An external torque must be applied in a direction that is
opposite in direction to the angular deceleration caused by the
baggage dropping onto the carousel.

43. 0.573 m

45. 765 lb (each front wheel), 510 lb (each rear wheel)

47. (a) 0.160 kg · m² (4.00-kg mass),
2.50 kg · m² (10.0-kg mass), 1.22 kg · m² (1.50-kg mass)
(b) 3.88 kg · m²
(c) The smallest mass does not necessarily contribute the
smallest amount to the moment of inertia.

49. (a) 2.52×10^4 N   (b) 2.35×10^4 N   **51.** 22.3 N

53. (a) 0.0800 rev/s, The disk rotates in a direction that is
opposite to the motion of the person.   (b) 2.99 s

55. 2.12 s

CHAPTER 11

1. 23 m   **3.** 1.6×10^{-3} rad (9.1×10^{-2} degrees)

5. (a) 1.6×10^8 N/m²   (b) 5.3×10^{-4} (c) 3.0×10^{11} N/m²

7. 1.6×10^5 N   **9.** 4 F

11. (a) 2.5×10^{-4} (b) 7.5×10^{-5} m   **13.** 2.0×10^{-3} m

15. 3.6×10^{-4} m (tungsten), 6.4×10^{-4} m (steel)

17. (a) 7.44 N   (b) 7.44 N

19. 53.4 lb   **21.** 0.371 m

23. (a) 0.407 m   (b) 397 N

25. 3.5×10^4 N/m   **27.** 0.069 m   **29.** 9.93×10^{-3} m

31. (a) 1.84 Hz   (b) 7.33×10^{-2} m

33. 7.18×10^{-2} m   **35.** 16 m/s (36 mi/h)

37. (a) 9.0×10^{-2} m   (b) 2.1 m/s

39. 0.138 m   **41.** 0.995 m   **43.** 0.996

45. $L = 7R/5$ **47.** 6.6×10^4 N   **49.** 2.6×10^{-2} m

51. 6.9×10^{-2} m

53. (a) $A = 1.41$ in., $f = 4.24$ Hz
(b) $A = 2.00$ in., $f = 4.24$ Hz

55. 0.240 m   **57.** 2.08 m/s   **59.** 0.144 m

CHAPTER 12

1. 12.3 cm   **3.** 3.08×10^{-6} m³   **5.** 5.68×10^{-6}

7. (a) 4.50×10^2 N   (b) 53.2 N

9. 9.84 lb   **11.** 103 bricks

13. (a) 6.0×10^4 Pa   (b) 8.7 lb/in.²   (c) 0.59 atm

15. 46.2 mm Hg

17. (a) 2.45×10^5 Pa   (b) 1.73×10^5 Pa

19. 0.741 m (mercury), 0.259 m (water)

21. $R_1 = 13.1$ m, $R_2 = 16.0$ m   **23.** 3.8×10^5 N

25. (a) 1.30×10^2 N   (b) 126 N   **27.** 2.2×10^3 kg

29. (a) 5.97 cm   (b) 7.12 cm   **31.** 2.04×10^{-3} m³

33. $r_1 = 5.28 \times 10^{-2}$ m, $r_2 = 6.20 \times 10^{-2}$ m   **35.** 4.89 m

37. 317 m²   **39.** (a) 762 mm   (b) 758 mm
41. 20 logs   **43.** 1.57 kg
45. (a) 1.07×10^7 N, down   (b) 1.07×10^7 N, up
(c) Net force = 0   (d) 1.41×10^5 N, down

CHAPTER 13

1. (a) Volume flow rate = 7.0×10^{-5} m³/s,
Mass flow rate = 7.4×10^{-2} kg/s   (b) 2.5×10^{-4} m/s
3. 8.10×10^6 gal/day
5. (a) 150 Pa
(b) The pressure inside the roof is greater than the pressure outside the roof, so there is a net outward force.
7. (a) 0.628 kg/s   (b) 22.2 m/s   (c) 3.47×10^5 Pa
9. (a) 7.00 m/s   (b) 1.21×10^5 Pa   **11.** 24 m/s
13. (a) The larger hole is near the top of the tank.   (b) 1.19
15. $v = \sqrt{2gy}$ (b) $y = 0$ (c) $P_A = P_0 - \rho g(y + h)$
17. (a) 0.5 m/s
(b) The ratio η/ρ is larger for blood than for water.
19. 2.25
21. (a) 2.48×10^5 Pa   (b) 1.01×10^5 Pa   (c) 0.343 m³/s
23. 2.9×10^3 Pa

CHAPTER 14

1. (a) -12.2 °C and 41 °C   (b) 261 K and 314 K
3. (a) 1.50×10^2 C°   (b) 270 F°   (c) 1.50×10^2 K
5. 18 °C   **7.** 1300 m   **9.** 0.080%   **11.** 110 C°
13. 41 °C   **15.** 41.5 N   **17.** 0.19 gal
19. 6.5×10^{-3} ft³
21. (a) 3.1×10^{-3} cm   (b) 8.9×10^{-2} cm³   (c) No
25. (a) The sphere at 25 °C weighs more.   (b) 18 N
27. 76.5 cm   **29.** $T_F = T_R - 459.67$ **31.** 5.8 m
33. 79 °C
35. (a) A temperature drop turns the light off   (b) 24 C°

CHAPTER 15

1. 36.2 °C   **3.** $226.00   **5.** 78.2 °C   **7.** 0.11 kg
9. 3100 W   **11.** 13.4 °C   **13.** 9.3 kg   **15.** 0.223
17. 3.50×10^2 m/s
19. (a) 145 BTU/lb   (b) 2.0×10^3 lb   (c) 1 ton
21. 0.237 kg   **23.** Vapor phase   **25.** 125 °C
27. 10 °C   **29.** 28%   **31.** 940 °C   **33.** 24%
35. 44 h   **37.** 1.2×10^5 J

CHAPTER 16

1. 17 °C and 25 °C   **3.** 12 J   **5.** 7.8 cm
7. 5.0×10^1 J/s
9. (a) 130 °C   (b) 830 J   (c) 237 °C

11. 283 °C   **13.** 0.74   **15.** 5800 K
17. (a) 1.19   (b) 1.32
19. (a) 70 W   (b) 60 Calories   **21.** 558 °C
23. Aluminum, copper, silver   **25.** 1.67
27. 4.60×10^2 °C

CHAPTER 17

1. 1.07×10^{-22} kg   **3.** 142 g
5. (a) 7.65×10^{-26} kg   (b) 2.11×10^{25} molecules
7. 0.550 kg   **9.** 3.2×10^4 Pa
11. (a) 8×10^{-18} mol/m³   (b) 2×10^{-16} Pa
13. (a) 3200 Pa   (b) 24 mm
15. 10.3 m   **17.** 0.795   **19.** 5.61×10^5 Pa
21. 5.1×10^{-16} kg
23. (a) 31 m²/s²   (b) 25 m²/s²   **25.** 399 J
27. (a) 120 N   (b) 120 N   (c) 4.0×10^5 Pa
29. 1.34×10^{-7} kg
31. (a) 5.00×10^{-13} kg/s   (b) 5.8×10^{-3} kg/m³
(c) 5.00×10^{-13} kg/s
33. (a) 5.8×10^{13} molecules   (b) 66 years
35. (a) 893.45 u   (b) 2680 g
37. (a) 2.3×10^6 Pa   (b) 7.3×10^4 Pa   (c) 2.4×10^6 Pa
39. 0.53 m   **41.** 3.1 kg   **43.** 307 K

CHAPTER 18

1. 32 miles
3. (a) 3.0×10^1 J, increase   (b) -130 J, decrease
(c) -1.0×10^2 J, decrease
5. 0.17 m   **7.** -2.0×10^{-3} m³, decrease
9. (a) 3.0×10^3 J   (b) The work is done by the system.
(c) The work is positive.   (d) -3.0×10^3 J
11. 3.0 J
13. -8.00×10^4 J, The flow of heat is out of the gas.
15. 0.59 J   **17.** 321 K   **19.** 18.0 : 1
21. $T_f = 327$ K, $V_f = 0.132$ m³   **23.** 0.075 kcal
25. 8.37×10^5 J   **27.** 15 K
29. (a) 60.0%   (b) 40.0%
(c) No heat energy is used to increase the internal energy of the gas, and 100% of the heat energy is used for doing work.
31. 434 K   **33.** 3.0×10^5 Pa
35. (a) $T_i V_i^{\gamma-1} = T_f V_f^{\gamma-1}$ (b) $P_i^{1-\gamma} T_i^{\gamma} = P_f^{1-\gamma} T_f^{\gamma}$
37. (a) -3100 J   (b) Work is done on the gas.
(c) The work is negative.   (d) -3100 J
(e) Heat flows out of the gas.

CHAPTER 19

1. (a) 4.35 kcal   (b) 0.757
3. 0.631   **7.** 0.34   **9.** 20

11. The greatest improvement is made by lowering the temperature of the cold reservoir.

13. (*a*) 0.36   (*b*) 1.3×10^{13} J

15. 0.424   **17.** 5.7 C°   **19.** 279 K   **21.** 1.4

23. 0.30¢   **25.** 279 s

27. (*a*) The entropy change of the environment must be greater than or equal to 25 J/K, depending on whether the process is irreversible or reversible.
(*b*) The disorder of the environment increases.

29. 293 K   **31.** (*a*) 2200 J   (*b*) 1.0×10^3 J

33. (*a*) 3680 J/K   (*b*) 18 200 J/K
(*c*) The vapor phase has more disorder.

35. 9.03   **37.** (*a*) 1260 K   (*b*) 4.17 kcal

39. $\frac{2}{9}$ (Engine A), $\frac{1}{3}$ (Engine B)   **41.** 0.123

CHAPTER 20

1. 0.083 Hz   **3.** 0.49 m

5. (*a*) 0.022 s   (*b*) 0.49 m
(*c*) The amplitude cannot be determined.

7. 78 cm   **9.** (*a*) 1.33 m/s   (*b*) 5.3 m

11. 0.211 s   **13.** 1.2 m/s²   **15.** 3.3×10^{-4} kg/m

17. (*a*) $+x$ direction   (*b*) -0.080 m

19. (*a*) $A = 0.45$ m, $f = 4.0$ Hz, $\lambda = 2.0$ m, $v = 8.0$ m/s
(*b*) $-x$ direction

21. $y = (2.00$ mm$) \sin (314t - 8.38)$

23. (*a*) 5.00 s   (*b*) 0.200 Hz   (*c*) 4.00 m/s

25. 0.25 m   **27.** 0.29 m/s

CHAPTER 21

1. (*a*) 20 m (at $f = 20$ Hz), 0.02 m (at $f = 20$ kHz)
(*b*) The statement is incorrect.

3. (*a*) 4.30×10^2 m/s   (*b*) 3.21×10^2 m/s

5. 2.8×10^{-4} s   **7.** 2.06

9. (*a*) steel, water, air   (*b*) 0.059 s, 0.339 s

11. Tungsten   **13.** 2.55 m

15. 43% (neon), 57% (argon)   **17.** 6.5 W

19. 1.0×10^{-6} W/m²   **21.** 63.1 W/m²

23. 2.0×10^5 **25.** 56.2 W

27. $r_1 = 3.8$ m and $r_2 = 4.8$ m   **29.** 786 Hz   **31.** 31 m/s

33. 0.26 m   **35.** (*a*) 3.9×10^{-3} m   (*b*) 1.9×10^{-3} m

37. (*a*) 3.9×10^{-3} W   (*b*) 8.5×10^{-4} W

39. -6.0 dB, The sound level decreases.

41. (*a*) 6.98×10^{-5} m   (*b*) 1.43×10^4 wavelengths

43. $f' = f \left(\dfrac{1 + \dfrac{v_c}{v}}{1 - \dfrac{v_c}{v}} \right)$

CHAPTER 22

1. (*b*) 2.5 s   **3.** (*b*) 2 s

5. (*a*) 44° (for $f = 2$ kHz), 13° (for $f = 6$ kHz)   (*b*) 0.10 m

7. 3.4°   **9.** 445 Hz   **11.** 3 Hz or 7 Hz

13. 1.4 m/s   **15.** 7.65×10^{-3} s   **17.** 0.16 m

19. 0.49   **21.** (*a*) 64.5 kg   (*b*) 40.3 kg

23. (*a*) 800, 1200, and 1600 Hz
(*b*) 800, 1200, and 1600 Hz   (*c*) 1200, 2000, and 2800 Hz

25. 1.96 m   **27.** 2.66×10^{-25} kg   **29.** 11.4 m

31. 1.2 m/s

33. (*a*) 8.6 m and 8.6×10^{-3} m   (*b*) 4.3 m and 4.3×10^{-3} m

35. 570 m/s   **37.** $n = 65$ **39.** 1.68×10^5 Pa

CHAPTER 23

1. 2.4×10^{13} **3.** 26 N   **5.** 3.09×10^3 C

7. 17.3 N, at 38.7° south of east

9. (*a*) 4.56×10^{-8} C   (*b*) 3.25×10^{-6} kg

11. 2.59×10^{12} **13.** 0.366 N

15. (*a*) 8.00 μC and 1.00 μC   (*b*) 9.82 μC and -0.82 μC

17. 18 N/C

19. (*a*) 0.45 N, due west   (*b*) 0.45 N, due east

21. 0.071 m²

23. 5.8×10^3 N/C, directed from the positive charge toward the negative charge.

25. $q_A = 0.716\, q$ and $q_B = 0.0894\, q$

27. 0.364   **29.** 3.25×10^{-8} C   **31.** 9

33. (*a*) 0.38 N, at 49° below the $-x$ axis
(*b*) 0.38 N, at 49° above the $+x$ axis

35. (*a*) The charges have like polarities.   (*b*) 1.7×10^{-16} C

37. $+3.3 \times 10^{-6}$ C

CHAPTER 24

1. (*a*) 2.00×10^{-14} J   (*b*) 2.00×10^{-14} J

3. 3.2×10^{-3} J   **5.** (*a*) 1500 V   (*b*) Point Y **7.** 1300 kg

9. (*a*) $d/3$ (*b*) d
(*c*) The positive potential due to the charge $+2q$ is always greater than the negative potential due to the charge $-q$.

11. 45 V, Point B is at the higher potential.

13. 0.18 m to the left of the negative charge, and 0.20 m to the right of the negative charge.

15. 3.42 cm   **17.** -0.747 J

19. (*a*) 0.187 m   (*b*) 3.67

21. (*a*) 4200 V/m   (*b*) 4200 V/m, from B to A

23. 9.0×10^3 V

25. (*a*) 2.54 m and 3.54 m
(*b*) 3.54 V (for $r = 2.54$ m), 2.54 V (for $r = 3.54$ m)

27. 2×10^{-8} F   **29.** 7.0×10^{13} **31.** 0.10 mm

33. (*a*) 41 J   (*b*) 8200 W   **35.** 3.98×10^{-15} C

37. (*a*) 2.9   (*b*) 8 V, decrease
39. No work is required.   **41.** 8.0×10^2 eV
43. (*a*) 159 V   (*b*) 125 V   (*c*) 135 V
45. 2.77×10^6 m/s

CHAPTER 25

1. 5.4×10^4 C   **3.** 22 A
5. (*b*) 880 A   (*c*) 350 A   (*d*) 7.9×10^5 C
7. 0.57 Ω   **9.** 1×10^{-3} (C°)$^{-1}$
11. (*a*) tungsten   (*b*) 37.8 °C   **13.** 4   **15.** 39.5 °C
17. $4.4   **19.** (*a*) 4.4 Ω   (*b*) 2.8 A   **21.** 96%
23. 1.77 A   **25.** (*a*) 9.0×10^2 W   (*b*) 1.8×10^3 W
27. 1.8 h
29. (*a*) 2.0×10^3 W   (*b*) 8.3 A   (*c*) 4.5×10^2 s
31. (*a*) π/4   (*b*) ½   **33.** 0.21 A   **35.** 50 m

CHAPTER 26

1. 4.0 V (8.0 Ω resistor), 8.00 V (16 Ω resistor)
3. (*a*) 1.6 A
(*b*) 14 V (9.0 Ω resistor), 8.0 V (5.0 Ω resistor),
1.6 V (1.0 Ω resistor)
(*c*) 23 W (9.0 Ω resistor), 13 W (5.0 Ω resistor),
2.6 W (1.0 Ω resistor)
5. 2   **7.** 5.3 Ω   **9.** 64
11. (*a*) 3.6 Ω   (*b*) 33 A, breaker will open   **13.** 6.76 Ω
15. (*a*) 3.00 A   (*b*) 1.50 A   (*c*) 56.3 W   **17.** 600 Ω
19. (*a*) 23.8 Ω   (*b*) 4.6 Ω   (*c*) 14.6 Ω
21. (*a*) 10.9 V   (*b*) 79.3 W
23. 0.449 Ω   **25.** 16 Ω   **27.** 25 V
29. (*a*) 256 Ω   (*b*) 304 Ω   **31.** (*a*) 2.0 A
33. (*a*) 0.73 A, from right to left   (*b*) 4.3 W
35. 1.7 A, from bottom to top   **37.** 9.23 μF   **39.** 25 V
43. (*a*) 10.0 V   (*b*) 3.50×10^{-4} J   **45.** 58 μC
47. (*a*) 0.15 s   (*b*) 0.99 V
49. 6.9   **51.** $V_{AB} = \dfrac{VR_1}{R_1 + R_2}$ **53.** 3430 Ω
55. (*a*) 2.60×10^{-2} A   (*b*) 3.8%
57. 0.054   **59.** 0.29   **61.** 0.11A

CHAPTER 27

1. 8.1×10^{-5} T
3. (*a*) 0.11 T, perpendicular to the **v**, **F** − plane and directed
away from the surface of the earth.
(*b*) 0.11 T, The direction is opposite to that in part (a).
5. (*a*) 4.1×10^{-12} N   (*b*) 4.6×10^{17}
7. (*a*) 4.3×10^2 m   (*b*) 7.8×10^5 m
9. 3.27×10^{-25} kg   **11.** 8.2×10^{-3} m   **13.** 120 eV

15. 1.50×10^7 C/kg   **17.** 0.32 m/s
19. (*a*) 5% larger   (*b*) 26% smaller
21. 8.1 N   **23.** 57.6°   **25.** 0.326 kg
27. (*a*) 11 A   (*b*) Away from the battery.
29. (*a*) 24 A · m²   (*b*) 4.8 N · m
(*c*) The normal to the coil is perpendicular to **B**.
31. (*a*) 2.3×10^{-2} N · m   (*b*) 2.9×10^{-2} N · m
33. 1.2×10^{-5} A · m²   **35.** 1.21×10^{-2} N, toward the wire
37. 1.3×10^{-2} T   **39.** 1.04×10^{-2} T
41. 0.50 m, to the left of wire A
43. 6.8 A, the direction of the current is opposite to the current
in the inner coil.
45. A: 4.3×10^{-5} T, B: 5.3×10^{-5} T, C: 6.0×10^{-5} T
47. (*a*) 4.46×10^{-4} N · m
(*b*) The normal to the loop should be perpendicular to **B**.
49. 1.5×10^{-8} s   **51.** $H = R/\pi$ **53.** 9.0×10^{-8} T
55. ½   **57.** 0.64 T, due north

CHAPTER 28

1. 0.47 T
3. (*a*) The driver's side   (*b*) 2.0 m
(*c*) The driver's side would still be positive.
5. (*a*) 3.3 m/s   (*b*) 4.6 N   **7.** (*a*) 5.6×10^{-3} Wb   (*b*) 0
9. (*a*) 7.3×10^{-4} Wb   (*b*) 0   (*c*) 4.7×10^{-3} Wb
11. 1.2 V   **13.** 0.15 T/s   **15.** 0.34 V   **17.** 0.459 T
19. (*a*) Left to right   (*b*) Left to right
21. (*a*) Clockwise   (*b*) No current exists.
(*c*) Counterclockwise   (*d*) No current exists.
23. The direction in each of the four positions is clockwise.
25. (*a*) $\mathscr{E}_0 = 180$ V, $T = 0.42$ s   (*b*) $\mathscr{E}_0 = 360$ V, $T = 0.21$ s
27. 36 V   **29.** 0.150 m
31. (*a*) 3.12 Ω   (*b*) 104 V   (*c*) 26.4 A
33. (*a*) 1.4 V   (*b*) 0.93 A   (*c*) 0.026 J
35. 9.27×10^{-7} Wb   **37.** 1.5×10^9 J
41. Step-down transformer, 1:12   **43.** 0.42 A
45. (*a*) 4.8×10^5 W   (*b*) 48 W
47. (*a*) 0.65 V   (*b*) 0.11 A   **49.** 170 V   **51.** 230 V
53. (*a*) 4.0 A   (*b*) 4.0×10^1 A   (*c*) 22 A
55. (*a*) 3.6×10^{-3} V   (*b*) 2.0×10^{-3} m²/s, Area must shrink.
57. 2.4×10^{-3} A

CHAPTER 29

1. 126 Hz
3. (*a*) 13.0 μF   (*b*) 17 V
(*c*) 0.094 A (2.0 μF capacitor), 0.19 A (4.0 μF capacitor),
0.33 A (7.0 μF capacitor)
5. 0.98 mm   **7.** (*a*) 0.13 A   (*b*) 0.32 A

9. 5.0×10^2 Hz   **11.** 3.0 W

13. (a) 19.0 V   (b) $-52.5°$

15. (a) 0.925 A   (b) 31.7°

17. (b) Crossover frequency = 500 Hz

19. 9.3 Ω   **21.** 0.651 W   **23.** 2.0×10^{-9} F

25. 0.81 W   **27.** 174 V

29. (a) 1.4×10^{-3} m²   (b) 5.7 MHz

31. $X_L/R = 4/\sqrt{3}$, $X_C/R = 1/\sqrt{3}$

33. (a) 2.00 μF   (b) 0.77 A
(c) 8.0×10^1 V (3.00 μF capacitor),
4.0×10^1 V (6.00 μF capacitor)

35. 0.44 A

37. (a) $Z = 211$ Ω, $\phi = -59.8°$ (Current leads voltage)
(b) $Z = 113$ Ω, $\phi = 21°$ (Voltage leads current)

39. 7.51×10^3 Hz, 3.12×10^3 Hz

CHAPTER 30

1. 8.3 min   **3.** 6.8×10^{-11} F   **7.** 11.118 m

9. 473 nm (Blue), 606 nm (Orange)

11. (a) 1.28 s   (b) 1.9×10^2 s

13. 24.1 rev/s   **15.** 1.6×10^3 mi   **17.** 8.44×10^2 J

19. (a) 1.3×10^4 N/C   (b) 6.2×10^{-9} J

21. 3.93×10^{26} W   **23.** (a) 0°   (b) 45°

25. (a) 611 W/m²   (b) 682 W/m²   **27.** 20

29. (a) 9.63×10^5 N/C   (b) 3.21×10^{-3} T

31. 8.37   **35.** 8.5×10^{-5} V

CHAPTER 31

1. 55°   **3.** 10°

9. The upright image is 3.0×10^1 cm behind the mirror and has a height of 2.0×10^1 cm.

11. The inverted image is 24 cm in front of the mirror and has a height of 4.4 cm.

13. (a) 24 cm

15. (a) -1.8 m   (b) Virtual   (c) 0.19 m

17. (a) Concave   (b) 2.3 cm   (c) 2.9   (d) Upright

19. (a) 6.0×10^1 cm

21. (a) 46.7 cm   (b) $d_0 = 27.5$ cm

27. (a) 62.1°   (b) 80.0°

CHAPTER 32

1. (a) 2.00×10^8 m/s   (b) 2.29×10^8 m/s

3. 5.89×10^{-7} m (air), 4.43×10^{-7} m (water), 2.44×10^{-7} m (diamond)

5. 1.40   **7.** 41.0°   **9.** 1.97×10^8 m/s

11. 2.05 cm   **13.** 2.1 cm   **15.** 3.23 cm   **19.** 1.54

21. (a) Point B (b) Point A **23.** 1.45

25. (a) 11.3 cm   (b) 3.8 cm   (c) Larger

27. 33.7°   **31.** 52.7° (red), 56.2° (violet)

33. (a) -24 cm   (b) 1.6

35. (a) 2.0×10^1 cm   (b) -6.0×10^1 cm

37. (a) -16 cm   (b) 0.57   (c) Virtual   (d) Upright
(e) Reduced

39. 13.7 m   **41.** 8.02 cm

43. (a) -4.00 cm   (b) $-\frac{1}{6}$ (c) Virtual   (d) Inverted
(e) Smaller

45. 60.0 cm

47. (a) -8.1 cm   (b) Virtual   (c) Upright   (d) Reduced

49. (a) 19.6 cm   (b) 0.87 cm

51. 0.011 m   **53.** (b) $d_i = -75.0$ cm, $m = 2.50$

55. 57.0°

57. (a) $1.15 \le n \le 2.00$ (b) $1.54 \le n \le 2.00$

CHAPTER 33

1. (a) 61.0 mm   (b) 220 mm

3. 16   **5.** 6.00 m   **7.** 138 mm   **9.** 87.0 cm

11. 38.0 cm   **13.** 26.9 cm

15. (a) 31.3 cm   (b) 2.43 m

17. (a) 5.1   (b) 4.1   **19.** 0.40 m

21. (a) 2.0×10^1 diopters   (b) 5.6 mm

23. 0.64 rad   **25.** 0.81 cm   **27.** $L' = L - f_e + f_o$

29. -2.0×10^2 **31.** 0.261 cm

33. (a) 19.0   (b) -194 (c) 7.6×10^{-2} mm
(d) 2.0×10^6 m

35. 3.0 cm   **37.** 13

39. (a) Converging   (b) Farsighted   (c) 96.3 cm

41. 3.3 mm

43. $M_o = \dfrac{-(L - f_e)}{f_o}$, $M_e = \dfrac{N}{f_e}$ **45.** 181

CHAPTER 34

1. (a) Constructive   (b) Destructive   (c) Destructive
(d) Constructive   (e) Constructive

3. 0.060 mm   **5.** 6   **7.** 304 nm   **9.** 612 nm

11. 198   **13.** 3.06×10^{-7} m   **15.** 480 nm

17. (a) 2.26 cm   (b) 3.58 cm

19. 485 nm   **21.** 0.013   **23.** 14 000 m

25. (a) 2.59 m   (b) 1.58 m

27. (a) 1.22 λ   (b) Shorter wavelength

29. 4.0×10^{-6} m   **31.** 4th order   **33.** 1.97 m

35. (a) 24°   (b) 33°   (c) 39°   **37.** 7.0×10^1 degrees

39. (a) 3.0×10^2 m   (b) No

41. (a) 1.0×10^{-7} m   (b) 2.1×10^{-7} m

43. (a) 7.9° (Violet), 13° (Red)
(b) 16° (Violet), 26° (Red)   (c) 24° (Violet), 41° (Red)
(d) 2nd and 3rd orders overlap.
45. (a) 2   (b) $m_A = 2$ and $m_B = 4$, $m_A = 3$ and $m_B = 6$

CHAPTER 35

1. 2.4×10^8 m/s   **3.** (a) 45.6 h   (b) 45.6 h
5. 5.57 s   **7.** 384 m
9. Length = 34 m, width = 15 m   **11.** 3.0 m $\times$ 1.3 m
13. 1940 kg   **15.** 3.4×10^5 **17.** 1.1 kg
19. (a) 3.8×10^{-15} J   (b) 1.047   **21.** $0.31c$
23. (a) c (b) $0.994c$ (c) $0.200c$ (d) $0.194c$
25. (a) 2.93×10^8 m/s   (b) 10.1×10^{-25} kg
27. 3.5×10^{-8} s   **29.** 2.6×10^8 m/s
31. (a) 6.70×10^5 J   (b) 7.45×10^{-12} kg

CHAPTER 36

1. 310 nm   **3.** 7.7×10^{29} **5.** 2.5×10^{21} **7.** 138
9. Aluminum (4.08 eV)   **11.** 1.4×10^5 m
13. (a) 7760 N/C   (b) 2.59×10^{-5} T
15. (a) 4.86×10^{-12} m   (b) 2.43×10^{-12} m
(c) 3.26×10^{-13} m
17. (a) 0.1819 nm   (b) 1.092×10^{-15} J
(c) 1.064×10^{-15} J   (d) 2.8×10^{-17} J
19. 7.79×10^{-13} J   **21.** 7.38×10^{-11} m
23. 8.5×10^{-12} m
25. 1.6×10^{-4} m (electron), 3.3×10^{-33} m (golf ball)
27. 37 J · s
29. x-ray (photon A), Microwave (photon B), Infrared (photon C)
31. (a) 1.97×10^{-24} kg · m/s   (b) 1.95×10^{-24} kg · m/s
33. (a) 1.10×10^3 m/s
(b) Speed of electron accelerated through a potential difference of 1 Volt is 5.93×10^5 m/s.
35. 42.8

CHAPTER 37

1. (a) 6.2×10^{-31} m³   (b) 4×10^{-45} m³
(c) 6×10^{-13} percent
3. 6.88×10^{-15} m   **5.** 434.1 nm
7. (a) 7457 nm   (b) 2279 nm   (c) Infrared
9. 1.23 eV (1.97×10^{-19} J)
11. (a) 16   (b) $\frac{1}{4}$ (c) 1
15. -1.11 eV
17. (a) $v_n = \dfrac{2\pi k e^2 Z}{hn}$ (b) 2.18×10^6 m/s   (c) 1.09×10^6 m/s
(d) Yes

19.

n	ℓ	m_ℓ	m_s
3	0	0	$\frac{1}{2}$
3	0	0	$-\frac{1}{2}$
3	1	1	$\frac{1}{2}$
3	1	1	$-\frac{1}{2}$
3	1	0	$\frac{1}{2}$
3	1	0	$-\frac{1}{2}$
3	1	-1	$\frac{1}{2}$
3	1	-1	$-\frac{1}{2}$
3	2	2	$\frac{1}{2}$
3	2	2	$-\frac{1}{2}$
3	2	1	$\frac{1}{2}$
3	2	1	$-\frac{1}{2}$
3	2	0	$\frac{1}{2}$
3	2	0	$-\frac{1}{2}$
3	2	-1	$\frac{1}{2}$
3	2	-1	$-\frac{1}{2}$
3	2	-2	$\frac{1}{2}$
3	2	-2	$-\frac{1}{2}$

21. $n = 5$, $\ell = 4$
23. (a) $L = h/2\pi$ (Bohr model), $L = 0$ (Quantum mechanics)
(b) $L = 2h/2\pi$ (Bohr model)

$\left. \begin{array}{l} \text{for } \ell = 0,\, L = 0 \\ \text{for } \ell = 1,\, L = \sqrt{2}h/2\pi \end{array} \right\}$ (Quantum mechanics)

25. (a) 32   (b) 50
27. 2.0×10^{-11} m   **29.** 7.230×10^{-11} m   **31.** 5.2×10^{17}
33. (a) The semiconductor laser   (b) 13.4
35. 10 700 V

37.

n	ℓ	m_ℓ	m_s
4	3	3	$\frac{1}{2}$
4	3	3	$-\frac{1}{2}$
4	3	2	$\frac{1}{2}$
4	3	2	$-\frac{1}{2}$
4	3	1	$\frac{1}{2}$
4	3	1	$-\frac{1}{2}$
4	3	0	$\frac{1}{2}$
4	3	0	$-\frac{1}{2}$
4	3	-1	$\frac{1}{2}$
4	3	-1	$-\frac{1}{2}$
4	3	-2	$\frac{1}{2}$
4	3	-2	$-\frac{1}{2}$
4	3	-3	$\frac{1}{2}$
4	3	-3	$-\frac{1}{2}$

39. 9.7×10^{47} **41.** 4   **43.** $n_i = 6$ and $n_f = 2$
45. 30.39 nm

CHAPTER 38

1. (a) 10 neutrons, 8 protons   (b) 18 neutrons, 17 protons
(c) 30 neutrons, 26 protons   (d) 70 neutrons, 50 protons

5. 1.2 N

7. (*a*) 2.3×10^{17} kg/m³   (*b*) 1.2×10^{10} kg   (*c*) 80

9. 127.6 MeV

11. (*a*) 10.55 MeV   (*b*) 7.55 MeV   (*c*) The neutron

13. (*a*) $^{14}_{6}C \longrightarrow ^{14}_{7}N + ^{0}_{-1}e$ (*b*) 0.156 MeV

15. 6.68×10^{-12} m   **17.** $^{212}_{84}Po$ **19.** $^{228}_{89}Ac$

21. 5.67×10^{3} m/s   **23.** 1.78×10^{-9} s⁻¹

25. 2.1×10^{-8} g   **27.** 0.837

29. (*a*) 1.1×10^{-6} g   (*b*) 3.2 g

31. 13 000 yr   **33.** 3.3×10^{9} yr

35. 6900 yr, Maximum error is 870 yr.

37. Platinum (Pt), Sulfur (S), Copper (Cu), Boron (B), Plutonium (Pu)

39. 19.9   **41.** $3^{14} = 4\ 782\ 969$ electrons

43. 1.59×10^{7} m/s

CHAPTER 39

1. 0.26 R   **3.** 30 rd   **5.** 100 mrem   **7.** $A = 30$

9. $\gamma + ^{17}_{8}O \longrightarrow ^{12}_{6}C + ^{4}_{2}He + ^{1}_{0}n$

11. Zinc, $Z = 30$, $A = 64$ **13.** Two neutrons

15. 232.7851 u   **17.** 3×10^{6} kg   **19.** 1.8×10^{4} kg/s

21. 1200 kg   **23.** 2.2 MeV

25. (*a*) 1.0×10^{22} (*b*) 8×10^{9} kg   **27.** 816 MeV

29. (*a*) 1.9×10^{-20} kg · m/s   (*b*) 3.5×10^{-14} m

31. 3.7×10^{-6} J   **33.** Oxygen, $Z = 8$, $A = 16$

35. 10^{9} W   **37.** 1.1×10^{-4} kg   **39.** 190 MeV

INDEX

PHOTO CREDITS

Chapter 30
Opener: Courtesy NASA. Figure 30.14: Courtesy Bausch & Lomb.

Chapter 31
Opener: George Holton/Photo Researchers. Figure 31.3*a*: Courtesy Professors Raymond Serway and Jerry Faughn. Figure 31.5*b*: Joel Gordon. Figure 31.17*b*: John Lei/Omni-Photo Communications, Inc. Figure 31.18*b*: Courtesy Campus Crafts, Inc.

Chapter 32
Opener: Courtesy Diamond Promotion Service. Figure 32.11: Ray Ellis/Photo Researchers. Page 695: Courtesy Dr. Hiromi Shinya, Beth Israel Hospital, NY.

Chapter 33
Page 715: Courtesy Pentax. Figures 33.13 and 33.14*a*: Courtesy Tasco Sales. Figure 33.14*b*: Courtesy Bausch & Lomb.

Chapter 34
Opener: G. Ronald Austing/National Audubon Society/Photo Researchers. Figure 34.7: From *Atlas of Optical Phenomena,* Michel Cagnet, Springer-Verlag, Berlin. Figures 34.13 and 34.14: Courtesy Bausch & Lomb. Figure 34.18: Courtesy Education Development Center. Figure 34.23 and 34.28*b*: From *Atlas of Optical Phenomena,* Michel Cagnet, Springer-Verlag, Berlin. Figure 34.37: Science Source/Photo Researchers.

Chapter 35
Opener: Ernst Haas/Globe Photos.

Chapter 36
Opener: Professor G.F. Leedale/Biophoto Associates/Photo Researchers. Figures 36.1: Courtesy Professor C. Jönsson. Figure 36.6*a*: From "Physics Review" 73, 527 (1948), by Wollan, Shull & Marney. Figure 36.6*b*: From *Fundamentals of College Physics* by W. Wallace McCormick. Figure 36.7: Courtesy Dr. Elisha Huggins.

Chapter 37
Opener: AP/Wide World Photos. Page 817: E. Celotti/Overseas/Phototake.

Chapter 38
Opener: Photo by Karl S. Luttrell, Science Diving & Environmental Company, assisted by Joe Stancanpanio, NGS. Page 837: M.E. Mosley/Anthro-Photo. Figure 38.13: Courtesy CERN.

Chapter 39
Opener: Courtesy Fermi National Accelerator Laboratory. Pages 856 and 857: Courtesy Department of Energy.

Color Photos

Chapter 14
Science Photo Library/Photo Researchers.

Chapter 37
Figure 37.3: Courtesy Bausch and Lomb.

SI UNITS

Quantity	Name of Unit	Symbol	Expression in Terms of Other SI Units	Quantity	Name of Unit	Symbol	Expression in Terms of Other SI Units
Length	meter	m	Base unit	Viscosity	—	—	Pa·s
Mass	kilogram	kg	Base unit	Electric charge	coulomb	C	A·s
Time	second	s	Base unit	Electric field	—	—	N/C
Electric current	ampere	A	Base unit	Electric potential	volt	V	J/C
Temperature	kelvin	K	Base unit	Resistance	ohm	Ω	V/A
Amount of substance	mole	mol	Base unit	Capacitance	farad	F	C/V
Velocity	—	—	m/s	Inductance	henry	H	V·s/A
Acceleration	—	—	m/s²	Magnetic field	tesla	T	N·s/C·m
Force	newton	N	kg·m/s²				
Work, energy	joule	J	N·m	Magnetic flux	weber	Wb	T·m²
Power	watt	W	J/s	Specific heat capacity	—	—	J/(kg·K) or J/(kg·C°)
Impulse, momentum	—	—	kg·m/s				
Plane angle	radian	rad	m/m	Thermal conductivity	—	—	J/(s·m·K) or J/(s·m·C°)
Angular velocity	—	—	rad/s	Entropy	—	—	J/K
Angular acceleration	—	—	rad/s²	Radioactive activity	becquerel	Bq	s⁻¹
Torque	—	—	N·m	Absorbed dose	gray	Gy	J/kg
Frequency	hertz	Hz	s⁻¹				
Density	—	—	kg/m³	Exposure	—	—	C/kg
Pressure, stress	pascal	Pa	N/m²				

THE GREEK ALPHABET

Alpha	A	α	Iota	I	ι	Rho	P	ρ	
Beta	B	β	Kappa	K	κ	Sigma	Σ	σ	
Gamma	Γ	γ	Lambda	Λ	λ	Tau	T	τ	
Delta	Δ	δ	Mu	M	μ	Upsilon	Y	υ	
Epsilon	E	ϵ	Nu	N	ν	Phi	Φ	ϕ	
Zeta	Z	ζ	Xi	Ξ	ξ	Chi	X	χ	
Eta	H	η	Omicron	O	o	Psi	Ψ	ψ	
Theta	Θ	θ	Pi	Π	π	Omega	Ω	ω	